Advances in Mathematical Methods for Electromagnetics

The ACES Series on Computational and Numerical Modelling in Electrical Engineering

Andrew F. Peterson, PhD – Series Editor

The volumes in this series will encompass the development and application of numerical techniques to electrical and electronic systems, including the modelling of electromagnetic phenomena over all frequency ranges and closely related techniques for acoustic and optical analysis. The scope includes the use of computation for engineering design and optimization, as well as the application of commercial modelling tools to practical problems. The series will include titles for senior undergraduate and postgraduate education, research monographs for reference, and practitioner guides and handbooks.

Titles in the Series

K. Warnick, **"Numerical Methods for Engineering,"** 2010
W. Yu, X. Yang and W. Li, **"VALU, AVX and GPU Acceleration Techniques for Parallel FDTD Methods,"** 2014.
A.Z. Elsherbeni, P. Nayeri and C.J. Reddy, **"Antenna Analysis and Design Using FEKO Electromagnetic Simulation Software,"** 2014.
A.Z. Elsherbeni and V. Demir, **"The Finite-Difference Time-Domain Method in Electromagnetics with MATLAB Simulations 2nd Edition,"** 2015.
M. Bakr, A.Z. Elsherbeni and V. Demir, **"Adjoint Sensitivity Analysis of High Frequency Structures with MATLAB,"** 2017.
O. Ergul, **"New Trends in Computational Electromagnetics,"** 2019.
D. Werner, **"Nanoantennas and Plasmonics: Modelling, design and fabrication,"** 2020.

Advances in Mathematical Methods for Electromagnetics

Edited by
Kazuya Kobayashi and Paul Denis Smith

The Institution of Engineering and Technology

Published by SciTech Publishing, an imprint of The Institution of Engineering and Technology, London, United Kingdom

The Institution of Engineering and Technology is registered as a Charity in England & Wales (no. 211014) and Scotland (no. SC038698).

First published 2020

The Institution of Engineering and Technology
Michael Faraday House Six
Hills Way, Stevenage Herts,
SG1 2AY, United Kingdom

www.theiet.org

British Library Cataloguing in Publication Data
A catalogue record for this product is available from the British Library

ISBN 978-1-78561-384-5 (hardback)
ISBN 978-1-78561-385-2 (PDF)

Typeset in India by MPS Limited

Contents

About the editors xxi

Introduction 1
Kazuya Kobayashi and Paul D. Smith

References 16

1 New insights in integral representation theory for the solution of complex canonical diffraction problems 17
J.M.L. Bernard

1.1 Representations of spectral function in frequency and time domain, for the scattering by a polygonal region 18
1.1.1 Basic elements in Sommerfeld–Maliuzhinets representation and properties 18
1.1.2 Spectral functions $f_{\pm}$ attached to the radiation of a single face and simple relation to f 20
1.1.3 Spectral function f from far field radiation of one face with arbitrary shape 22
1.1.4 Exact causal time domain representation of a field above a dispersive wedge-shaped region 23
1.2 Several orders asymptotic representation for scattering by a curved impedance wedge 25
1.2.1 Asymptotic representation in a region without creeping waves 25
1.2.2 Several orders asymptotic expressions in a region with creeping waves terms 27
1.2.3 Some validations concerning the expression of $f = \sum_{n\geq 0} f_n/k^n$ for arbitrary wedge angle 30
1.3 A novel expression of the field for arbitrary bounded sources above a passive or active impedance plane 32
1.3.1 Formulation of the problem 32
1.3.2 An expression of potentials ($\mathscr{E}_{inc}, \mathscr{H}_{inc}$) for bounded sources J and M 34
1.3.3 Expression of the potentials ($\mathscr{E}_s, \mathscr{H}_s$) for an impedance plane 34

1.4 Spectral representation of the field for 3D conical scatterers 36
1.4.1 Formulation 36
1.4.2 Expression of the potentials with Kontorovich–Lebedev integrals 38
1.4.3 Potentials and properties for an incident plane wave 41
References 44

2 Scattering of electromagnetic surface waves on imperfectly conducting canonical bodies 47
Mikhail A. Lyalinov and Ning Yan Zhu

2.1 Introduction and survey of some known results 47
2.1.1 Electromagnetic surface waves on impedance surfaces 49
2.1.2 Electromagnetic surface waves on a right circular conical surface 54
2.2 Excitation of an electromagnetic surface wave by a dipole located near a plane impedance surface 56
2.3 Scattering of a skew incident surface wave by the edge on an impedance wedge 59
2.3.1 Integral equations for the spectra 61
2.3.2 Far-field expansion 62
2.3.3 Reflection and refraction of an incident surface wave at the edge of an impedance wedge 63
2.3.4 Beyond the critical angle of edge diffraction 65
2.4 Conclusion 67
Appendix A 68
A.1 Brewster angles 69
Acknowledgement 69
References 69

3 Dielectric-wedge Fourier series 73
Svend Berntsen

3.1 Introduction 73
3.2 The diffraction problem 75
3.2.1 The Hilbert-space problem 75
3.2.2 The singular-field problem 76
3.3 Integral equations 77
3.4 The solution of the integral equation 79
3.5 The Bessel–Hankel Fourier series 81
3.6 Incident plane waves 87
3.7 Numerical results 88
3.8 Summary 92
Appendix A 92
Acknowledgment 93
References 94

4 Green's theorem, Green's functions and Huygens' principle in discrete electromagnetics **97**
John M. Arnold

4.1 Introduction 97
4.2 Green's theorem for adjacency matrices 99
4.2.1 Adjacency matrix 99
4.2.2 Weighted adjacency matrix 100
4.2.3 Matrix of adjacency 1 100
4.2.4 The vertex Laplacian matrix 100
4.2.5 Incidence matrices 101
4.3 Green's theorem on topological vector space 101
4.4 Difference forms and discrete exterior calculus 104
4.4.1 Simplicial decomposition 104
4.4.2 Dual forms 105
4.4.3 Hypercube decomposition 106
4.4.4 Contextual algebraic notation of forms 106
4.4.5 Manifolds, graphs and lattices 107
4.4.6 Essentials of cell decomposition 108
4.5 Higher order Green's theorem and Green's functions 108
4.5.1 Green's theorem for r-forms 108
4.5.2 Kirchhoff's theorem for r-forms 109
4.5.3 Green's function for r-forms 109
4.6 Dynamical systems on topological vector spaces: Maxwell's equations 109
4.6.1 Discrete Maxwell equations 110
4.6.2 Electromagnetic fields as differential forms 111
4.7 Time-domain Green's functions for dynamical systems 112
4.8 Discrete time 114
4.9 Discrete Green's theorem and Green's functions in computational field theory 116
4.9.1 Exterior–interior connection 116
4.9.2 Diakoptics 117
4.10 Conclusion 117
References 117

5 The concept of generalized functions and universal properties of the Green's functions associated with the wave equation in bounded piece-wise homogeneous domains **119**
M. Idemen

5.1 A short historical background 119
5.1.1 A remark 123
5.2 Some basic properties of the δ distribution 124
5.2.1 Distributions involving $\delta(R)$ and $\delta(R_1 - R_2)$ 125

5.3 Green's functions associated with the wave equation in bounded partially homogeneous domains 128
5.3.1 The outgoing Green's function 130
5.3.2 The ingoing Green's function 131
5.3.3 Some universal properties of the outgoing and ingoing Green's functions 132
5.4 Proofs of the theorems 133
5.4.1 Proof of Theorem 1 133
5.4.2 Proof of Theorem 2 133
5.4.3 Proof of Theorem 3 134
5.4.4 Proof of Theorem 4 134
5.4.5 Proof of Theorem 5 135
5.4.6 Proof of Theorem 6 136
5.5 Application: An inverse initial-value problem connected with the photoacoustic tomography in bounded non-homogeneous domains 138
5.5.1 Extension of the inverse initial value problem to the range $(-\infty) < t < \infty$ 139
5.5.2 Solution of the extended problem 140
5.5.3 Proof of (5.64) 142
References 143

6 Elliptic cylinder with a strongly elongated cross-section: high frequency techniques and function theoretic methods 145
Frédéric Molinet and Ivan Andronov

6.1 Introduction 145
6.2 Asymptotic currents on an elliptic cylinder with a truncated strongly elongated cross-section 147
6.2.1 Analysis of the interactions 147
6.2.2 Asymptotic field in the boundary layer due to a magnetic line current 148
6.2.3 Radiated field and total field 152
6.2.4 Spectral decomposition of the field in the boundary layer 154
6.2.5 Diffraction by the edge of a truncated elliptic cylinder 156
6.2.6 Asymptotic field in the boundary layer due to an incident plane wave 157
6.2.7 Asymptotic currents 160
6.3 Asymptotic currents on a cylinder with an ogival cross-section composed of two symmetric arcs of a strongly elongated ellipse 161
6.3.1 Presentation of the geometry and analysis of the problem 161
6.3.2 Asymptotic currents outside grazing incidence 162
6.3.3 Grazing incidence 163
6.4 Conclusion 166
References 168

7 High-frequency hybrid ray–mode techniques **169**
Hiroshi Shirai

7.1 Introduction 169
7.2 Ray–mode conversion technique 170
7.3 Modal excitation at the aperture 171
7.3.1 Formulation 171
7.3.2 Numerical results 174
7.4 Diffraction by a slit on a thick conducting screen 177
7.4.1 Background 177
7.4.2 Formulation 177
7.4.3 Diffraction by a thin slit 182
7.4.4 Diffraction by a thick and loaded slit 182
7.4.5 Diffraction by a trough 185
7.5 Conclusions 186
Acknowledgments 186
References 186

8 Scattering and diffraction of scalar and electromagnetic waves using spherical-multipole analysis and uniform complex-source beams **189**
Ludger Klinkenbusch and Hendrik Brüns

8.1 Introduction 189
8.2 Solution of Maxwell's equations in sphero-conal coordinates 191
8.2.1 Sphero-conal coordinates 191
8.2.2 Solution of the Helmholtz equation in sphero-conal coordinates 192
8.2.3 Vector spherical-multipole expansion of the electromagnetic field in the presence of a PEC semi-infinite elliptic cone 196
8.3 Complex-source beams 198
8.3.1 Converging and diverging CSB 198
8.3.2 Uniform CSB 200
8.4 Green's function of the semi-infinite elliptic cone for an incident uniform complex-source beam 202
8.4.1 Scalar Green's function 202
8.4.2 Dyadic Green's function 203
8.5 Numerical evaluation 204
8.5.1 Convergence analysis 204
8.5.2 Numerical results for an acoustically soft or hard semi-infinite elliptic cone 206
8.5.3 Numerical results for a perfectly conducting semi-infinite elliptic cone 208
8.6 Conclusions 212
References 213

9 Changes in the far-field pattern induced by rounding the corners of a scatterer: dependence upon curvature **215**
Audrey J. Markowskei and Paul D. Smith

9.1 Problem formulation 218
9.2 Numerical results and discussion 221
9.3 Analytic bounds for the far-field difference 222
9.3.1 Integral equations for the difference in surface quantities 222
9.3.2 Approximate integral equation for the difference Δ 223
9.3.3 The far-field difference 236
9.4 Conclusion 239
References 239

10 Radiation from a line source at the vertex of a right-angled dielectric wedge **241**
Anthony D. Rawlins

10.1 Introduction 241
10.2 Formulation of the boundary value problem 242
10.3 Singular integral equation for the double Laplace transform of the electric field 244
10.4 Approximate solution of the singular integral equation 246
10.5 Calculation of $\boldsymbol{E}^{(1)}(0,0)$ 248
10.6 Radiated far field 249
10.7 Conclusions 250
References 253

11 Wiener-Hopf analysis of the diffraction by a thin material strip **255**
Takashi Nagasaka and Kazuya Kobayashi

11.1 Introduction 255
11.2 The case of E polarization 256
11.2.1 Formulation of the problem 256
11.2.2 Factorization of the Kernel functions 259
11.2.3 Formal solution of the Wiener–Hopf equation 261
11.2.4 Asymptotic solution of a certain integral equation in the complex plane 263
11.2.5 High-frequency asymptotic solution 266
11.2.6 Scattered far field 268
11.3 The case of H polarization 269
11.4 Numerical results and discussion 272
11.5 Conclusions 276
Acknowledgment 276
References 276

12 The Wiener-Hopf Fredholm factorization technique to solve scattering problems in coupled planar and angular regions **279**
Vito G. Daniele and Guido Lombardi

12.1 Introduction 279
12.2 The WH equations of the problem 281
12.3 Reduction of the WH equations to FIEs 283
12.3.1 The Fredholm equation of the region (c) 284
12.3.2 The Fredholm equations of the region (b) 286
12.3.3 The Fredholm equation of the angular region (a) 288
12.4 Solution of the FIE 292
12.5 Analytical continuation of the numerical solution 294
12.6 A novel test case 298
12.7 Conclusion 301
Appendix A 301
References 301

13 On the analytical regularization method in scattering and diffraction **303**
Yury A. Tuchkin

13.1 Introduction 303
13.2 Instability in the numerical solution of infinite algebraic systems 304
13.3 The ARM: when is it necessary? 308
13.4 Potentials and their pseudodifferential representations 309
13.5 Solution of the key diffraction problems 313
13.5.1 Dirichlet BVP 313
13.5.2 Neumann BVP 315
13.6 Diffraction by a semi-transparent obstacle 316
13.6.1 The BVP description 316
13.6.2 Integral representation for $u^{s(+)}$ and $\partial_n u^{s(+)}$ 317
13.6.3 Reduction of the BVP to a system of integral equations 318
13.6.4 Reduction of the system of integral equations to an infinite system of linear algebraic equations 319
13.7 Diffraction of waves with complex frequencies and spectral theory of open cavities 320
13.7.1 Description of the BVP 320
13.7.2 Dirichlet BVP for complex-valued wave numbers 322
13.7.3 Qualitative features of the Dirichlet BVP 322
13.7.4 Numerical calculation of complex-valued eigen-wavenumbers and eigenmodes 322
13.8 ARM: considerations for implementation 323
13.9 ARM: various applications and conclusion 324
References 325

14 Resonance scattering of E-polarized plane waves by two-dimensional arbitrary open cavities: spectrum of complex eigenvalues **329**
Elena D. Vinogradova

14.1 Introduction 329
14.1.1 Preliminary remarks 329
14.1.2 Development of a systematic approach 330
14.2 Mathematical background 331
14.2.1 Schematic description of the MAR 331
14.2.2 Scheme for finding the complex eigenvalues 333
14.3 Computation of the complex eigenvalues for various open cavities 336
14.3.1 Circular cylinder with longitudinal slit 336
14.3.2 Elliptic cavity with moveable longitudinal slit 337
14.3.3 Open rectangular cavity with finite flanges 343
14.4 Resonance response of slotted cavities 347
14.4.1 Surface current calculations 348
14.4.2 Far-field calculations 350
14.5 Conclusion 355
References 357

15 Numerical solutions of integral equations for electromagnetics **359**
Roberto D. Graglia and Andrew F. Peterson

15.1 The EFIE and MFIE for perfectly conducting bodies 359
15.2 Some alternative formulations to remediate fictitious internal resonances 361
15.3 Integral equations for homogeneous dielectric bodies 361
15.4 Formulations that remediate fictitious internal resonances for dielectric targets 363
15.5 Single-source integral equations for dielectric bodies 364
15.6 Low-frequency breakdown of integral equations 365
15.7 Numerical solution of integral equations 365
15.8 Vector basis functions 366
15.9 Interpolatory and hierarchical vector basis functions 367
15.10 Singular vector basis functions 369
15.11 Summary 371
References 372

16 Electromagnetic modelling at arbitrarily low frequency via the quasi-Helmholtz projectors **381**
Adrien Merlini, Alexandre Dély, Kristof Cools and Francesco P. Andriulli

16.1 Introduction 381
16.2 Notation and background 383

16.2.1 Frequency domain 383
16.2.2 Time domain 388
16.3 The low-frequency breakdown in the FD 391
16.3.1 Illustration of the problem 391
16.3.2 Analysis of the low-frequency breakdown 392
16.3.3 Traditional LS decomposition 396
16.4 The large time step breakdown in the TD 398
16.5 DC instabilities 398
16.6 The qH projectors 401
16.7 An effective solution to the low-frequency breakdown for the EFIE 402
16.7.1 Leveraging the qH projectors 402
16.7.2 Implementation details 404
16.8 Solution to the large time step breakdown and the DC instability for the TD-EFIE 405
16.8.1 Preconditioning 406
16.8.2 Time discretization 407
16.8.3 Numerical results 409
16.9 Conclusions 411
References 411

17 Resistive and thin dielectric disk antennas with axially symmetric excitation analyzed using the method of analytical regularization 417
Nataliya Y. Bliznyuk and Alexander I. Nosich

17.1 Introduction 417
17.2 Formulation and GBC 419
17.3 Singular IEs and solution by MAR 421
17.3.1 Hyper-singular IE for a VED-excited resistive disk in free space 421
17.3.2 Eigenfunctions of the IE operator static limit for VED-excited PEC and resistive disks 422
17.3.3 Matrix equation and DIE for a VED-excited disk 423
17.3.4 Log-singular IE for a VMD-excited resistive disk in free space 424
17.4 Resistive disk MSA excited by VED 426
17.4.1 Dual IEs for a resistive disk MSA 426
17.4.2 Matrix equation for a resistive disk MSA 426
17.5 Thin disk DA excited by VED 428
17.5.1 Coupled set of DIEs for a thin disk DA 428
17.5.2 Matrix equation for a thin disk DA 428
17.6 Numerical results 429
17.6.1 Radiation characteristics of resistive MSA 430
17.6.2 Radiation characteristics of thin disk DA 432

17.7 Conclusions 435
References 435

18 Scattering and guiding problems of electromagnetic waves in inhomogeneous media by improved Fourier series expansion method 439
Tsuneki Yamasaki

18.1 Introduction 440
18.1.1 Formulation 440
18.1.2 Numerical results 445
18.1.3 Conclusions 449
18.2 Slanted layer and rhombic media with strips 449
18.2.1 Slanted layer 449
18.2.2 Rhombic media with strips 450
18.2.3 Conclusions 453
18.3 Elliptically layered, columnar, and rectangular media 454
18.3.1 Elliptically layered and columnar media 454
18.3.2 Rectangular media 455
18.3.3 Conclusions 457
18.4 Energy distribution of defect layers 458
18.4.1 Conclusions 459
18.5 Mixed positive and negative media 460
18.5.1 Conclusions 465
References 466

19 Methods and fast algorithms for the solution of volume singular integral equations 471
Alexander B. Samokhin

19.1 Introduction 471
19.2 Formulation of the problems 472
19.3 Spectrum of integral operator 475
19.3.1 Spectrum for low-frequency case 476
19.4 Stationary iteration methods 479
19.4.1 Generalized simple iteration method 479
19.4.2 Generalized Chebyshev iteration method 482
19.5 Nonstationary iteration methods 484
19.6 Discretization of integral equations 486
19.7 Fast algorithms 487
19.8 Numerical results 488
19.9 Conclusion 489
References 490

20 Herglotz functions and applications in electromagnetics **491**
Mitja Nedic, Casimir Ehrenborg, Yevhen Ivanenko, Andrei Ludvig-Osipov, Sven Nordebo, Annemarie Luger, Lars Jonsson, Daniel Sjöberg, and Mats Gustafsson

20.1 Introduction 491
20.2 Basics about Herglotz functions 492
20.3 Passive systems 499
20.4 Sum rules and physical bounds 503
20.5 Convex optimization and physical bounds 507
20.6 Conclusions 511
Acknowledgments 511
References 511

21 Scattering and guidance by layered cylindrically periodic arrays of circular cylinders **515**
Vakhtang Jandieri and Kiyotoshi Yasumoto

21.1 Introduction 515
21.2 Formulation of the problem 517
21.2.1 Field expressions 517
21.2.2 Calculation of the scattering amplitudes 518
21.2.3 Reflection and transmission matrices 520
21.2.4 Hertzian dipole source radiation in the layered cylindrical structure 522
21.2.5 Plane wave scattering by the layered cylindrical structure 523
21.2.6 Guidance in the layered cylindrical structure 525
21.3 Numerical results and discussions 526
21.3.1 Directivity of radiation of a dipole source coupled to the cylindrical EBG structure 526
21.3.2 Light scattering by the metal-coated dielectric nanocylinders with angular periodicity 534
21.3.3 Modal analysis of specific microstructured optical fibers 539
21.4 Conclusion 542
Acknowledgment 543
References 543

22 Analytical and numerical solution techniques for forward and inverse scattering problems in waveguides **547**
E.D. Derevyanchuk, Yu.V. Shestopalov and Yu.G. Smirnov

22.1 Introduction 547
22.2 Inverse problems 549
22.2.1 General statement for inverse problems 550
22.3 Inverse problems. Class I (Isotropic case) 553

22.3.1 Statement of the inverse problem for isotropic one-sectional diaphragm (Class I) 553
22.3.2 Explicit solution to the inverse problem 553
22.4 Inverse problems. Class AnI (Anisotropic case) 557
22.4.1 Statements of inverse problems for anisotropic one-sectional diaphragm (Class AnI) 557
22.4.2 Explicit formulas for the transmission coefficient 557
22.4.3 The existence and uniqueness of the solution to the inverse problem 559
22.5 Inverse problem for multi-sectional diaphragm. Class M 560
22.6 Numerical results 562
22.6.1 Example 1. Inverse problem for one-sectional anisotropic diaphragm 563
22.6.2 Example 2. Extraction of the complex permittivity of each section of three-sectional isotropic diaphragm 563
22.6.3 Example 3. Extraction of permittivity and permeability of one-sectional anisotropic diaphragm 564
22.6.4 Example 4. Extraction of permittivity tensor of two-sectional anisotropic diaphragm 565
22.6.5 Example 5. Inverse problem $P_{\varepsilon_1}^{\mathbb{C}}$ 566
22.7 Conclusion 566
Acknowledgments 567
References 567

23 Beam-based local diffraction tomography **571**
Ram Tuvi, Ehud Heyman and Timor Melamed

23.1 Introduction and overview 571
23.2 The UWB-PS-BS method 572
23.2.1 BS methods: an overview 572
23.2.2 The UWB-PS-BS method: a frequency domain formulation 574
23.2.3 The phase-space pulsed BS method: a TD formulation 578
23.3 UWB tomographic inverse scattering 582
23.3.1 Tomographic inverse scattering: frequency domain formulation 582
23.3.2 Time-domain diffraction tomography 587
23.4 Beam-based TD-DT 589
23.4.1 The beam-domain data 589
23.4.2 The beam-domain data-object relation within the Born approximation 590
23.4.3 Backpropagation and local reconstruction of $O(\mathbf{r})$ 590
23.4.4 Numerical examples 592
23.5 Conclusions 594
Acknowledgments 595
References 595

24 Modal expansions in dispersive material systems with application to quantum optics and topological photonics **599**
Mário G. Silveirinha

24.1 Introduction 599
24.2 Electrodynamics of dispersive media 600
24.3 Hermitian formulation in the time domain 602
24.4 Poynting theorem and stored energy 603
24.5 Canonical momentum 605
24.6 Modal expansions 607
24.7 Green's function 608
24.8 Positive and negative frequency components of the Green function 610
24.9 Application to topological photonics 611
24.10 Application to quantum optics 613
24.11 Summary 616
Acknowledgments 617
References 617

25 Multiple scattering by a collection of randomly located obstacles distributed in a dielectric slab **621**
Gerhard Kristensson and Niklas Wellander

25.1 Introduction 621
25.2 The geometry 622
25.3 Integral representation 623
25.4 Exploiting the integral representations 625
25.5 Expansions of surface fields 630
25.6 The transmitted and reflected fields 634
25.7 Statistical problem—ensemble average 635
25.8 Approximations 641
25.9 Conclusions 644
Appendix A 645
Appendix B 645
Appendix C 647
Appendix D 647
Appendix E 648
References 650

26 Electromagnetics of complex environments applied to geophysical and biological media **653**
Akira Ishimaru, Yasuo Kuga and Max Bright

26.1 Introduction 653
26.2 Stochastic wave theories 653
26.3 Time-reversal imaging 654

26.4 Imaging through random multiple scattering clutter 656
26.5 Geophysical remote sensing and imaging, and super resolution 656
26.6 Wigner distribution function and specific intensity 658
26.7 Biomedical electromagnetism and optics 660
26.8 Heat diffusion in tissues 660
26.9 Ultrasound in tissues and blood 661
26.10 Low coherence interferometry and optical coherence tomography (OCT) 663
26.11 Waves in metamaterials and electromagnetic and acoustic Brewster's angle 665
26.12 Coherence in multiple scattering 665
26.13 Porous media 667
26.14 Seismic coda 669
26.15 Conclusion 670
Acknowledgments 670
References 671

27 Innovative tools for SI units in solving various problems of electrodynamics **673**
Oleg A. Tretyakov, Oleksandr Butrym and Fatih Erden

27.1 Introduction 673
27.2 Novel format of Maxwell's equations in SI units: Energetic and mechanical field characteristics 674
27.2.1 Novel format of Maxwell's equations in SI units 674
27.2.2 Energetic characteristics of the electromagnetic field 676
27.2.3 Mechanical equivalents of the energetic field characteristics 677
27.3 Exact solutions for polarization of Lorentz media associated with a signal of finite duration 678
27.3.1 Rearrangement of the motion equation to its equivalent matrix format and solving a vector Cauchy problem 678
27.3.2 Exact explicit solutions for the amplitudes of the polarization vector 681
27.4 Upgrading the evolutionary approach to electrodynamics (*EAE*) 684
27.4.1 Comparison of two alternative approaches to the electromagnetic field theory 684
27.4.2 Separation of a self-adjoint operator from the vectorial Maxwell's equations 686
27.4.3 Normalization of the eigenvectors of operator $\mathfrak{R}$ 691
27.4.4 Configurational orthonormal modal basis in the space of solutions $\mathcal{L}_2$ 692
27.4.5 Projecting the field vectors and Maxwell's equations onto the modal basis 693

27.5 Present state of art and recent advances 697
27.6 Ongoing and future research 701
References 701

Index **709**

About the editors

Kazuya Kobayashi is a professor in the Department of Electrical, Electronic, and Communication Engineering at Chuo University, Japan. He has received a number of awards including The President's Award (2020) from URSI (International Union of Radio Science) and the M.A. Khizhnyak Award (2016) at the 16th International Conference on Mathematical Methods in Electromagnetic Theory. He is a Fellow of The Electromagnetics Academy and a Fellow of URSI. He has held various positions in the international radio science, electromagnetics, and optics communities including URSI Assistant Secretary-General for AP-RASC (since 2015); Chair of URSI Commission B (since 2017); Chair of the AP-RASC Standing Committee (since 2015); President of the Japan National Committee of URSI (2008-2018); Chair of the PIERS Young Scientists Award Committee, The Electromagnetics Academy (since 2018); Editor of *Radio Science* (since 2019); and Series Editor of *Springer Series in Optical Sciences* (since 2020). His research areas include developments of rigorous mathematical techniques as applied to electromagnetic wave problems; radar cross section; and scattering and diffraction.

Paul Denis Smith is a professor of mathematics at the Macquarie University, Australia. His awards include the best paper award at the 1987 International Symposium on Electromagnetic Compatibility. He served as associate editor of *J. IEEE Antennas and Propagation* from 2004 to 2011 and is currently an associate editor of *Radio Science* and a Board Member for *Proceedings of the Royal Society(A)*. He is president of the Australian URSI Committee and is a member of the Australian Academy of Science National Committee for Space and Radio Science. His research areas include analytical and semi-analytical techniques for wave scattering and diffraction.

Introduction

Kazuya Kobayashi[1] and Paul D. Smith[2]

Electromagnetic theory has a long history of development. Its modern formulation developed out of the investigations by scientists of the seventeenth, eighteenth and nineteenth centuries. Between 1861 and 1862, the mathematician and physicist James Clerk Maxwell formulated a version of the set of partial differential equations that are now associated with his name. The theory expounded in his treatise [1], provided a complete and unified framework for electrical, magnetic and optical phenomena. The present commonly employed form of Maxwell's equations derives from Heaviside's vector calculus formalism developed in the latter part of the nineteenth century. They also possess a relativistic formulation that links them to Einstein's development of the special theory of relativity at the start of the twentieth century. Einstein himself stated that 'the formulation of these equations is the most important event in physics since Newton's time, not only because of their wealth of content, but also because they form a pattern for a new type of law … Maxwell's equations are laws representing the structure of the field … All space is the scene of these laws and not, as for mechanical laws, only points in which matter or charges are present' [2].

In various forms, Maxwell's equations have been extensively investigated from both mathematical and physical points of view and their study has instigated the development of a number of sophisticated methods of analysis. They continue to play a central role in a variety of scientific and technological developments which demand the interdisciplinary interaction of mathematicians, physicists, engineers and other diverse specialists; these equations are still used in almost every application of electrical engineering, communications technology and optics.

Accepting that Maxwell's equations provide the basis for the theoretical understanding of electromagnetism, it is not surprising that a multiplicity of mathematical approaches has evolved to analyse the very diverse set of electro-magneto-optical phenomena of the field. The purpose of this book is to cover recent achievements in the area of advanced analytical and associated numerical methods as applied to problems arising in various branches of electromagnetics. The unifying theme is the application of advanced or novel mathematical techniques to produce analytical solutions (to canonical problems) or effective analytical–numerical methods for computational electromagnetics addressing more general problems.

[1]Department of Electrical, Electronic, and Communication Engineering, Chuo University, Tokyo, Japan
[2]Department of Mathematics and Statistics, Macquarie University, Sydney, Australia

The idea for this book came from several sessions on Mathematical Methods in Electromagnetics held in recent years at the General Assemblies of the International Union of Radio Science (URSI) and other URSI-sponsored meetings, including the Asia-Pacific Radio Science meetings, the Atlantic Radio Science meetings and the annual International Conference on Electromagnetics in Advanced Applications. We invited several of the prominent regular contributors to these sessions to contribute chapters elucidating some important methods of analysis and applications of contemporary relevance. As already indicated the unifying theme is the application of advanced or novel mathematical techniques to produce analytical solutions or effective analytical–numerical methods. A further aim is to encourage the continued investigation of important problems in electromagnetics with these and related procedures and to further the development of these tools. Accordingly, the book consists of a number of chapters authored by different experts, addressing a number of themes in electromagnetics. All manuscripts were prepared by the authors themselves and no attempt has been made to establish a uniformly common notation.

Perhaps the most fundamental aspect is that of electromagnetic theory. This concerns the development of fundamental principles providing insight into the structure of Maxwell's equations – in differential form, integral form or tensor form – and the field quantities, into the nature and type of the constitutive relations governing the interaction of fields and media, into the nature and type of boundary conditions characterising obstacles or interfaces and so on.

Scattering and diffraction of electromagnetic waves in various guises takes central stage. It is perhaps useful to characterise the manifold methods employed in its study in various ways. The size of the scattering obstacles in wavelengths is very important in determining what scattering mechanisms are dominant and will suggest the applicability of either a high-frequency, short-wavelength approach, or a low-frequency, long-wavelength Rayleigh scattering procedure, or an intermediate (resonance) regime where the scatterer is around one or a few wavelengths in size. If more than one scatterer is present, multiple scattering techniques are to be considered, and, if appropriate, multi-scale approaches may require a blend of low-/intermediate-/high-frequency techniques. Complex media (e.g. biological media) require their own techniques as do non-linear media; homogenisation techniques present one aspect of characterising bulk materials with small scale or microstructure and random media require the synthesis of probabilistic and statistical techniques with the underlying analysis of the electromagnetic scattering phenomena. Additional consideration must be given to the formulation when the scattering obstacles are embedded in a medium (or multilayered medium) that may usefully be regarded as of infinite extent and, in some contexts, the study of surface waves becomes important. Analysis of the guiding or propagation of waves employs diverse techniques to study classes of problems, ranging from waveguides to periodic structures, such as gratings of infinite extent. Although the design of antennas (and antenna arrays) is principally addressed in terms of scattering and diffraction, aspects of guiding wave energy and its directed propagation demand attention also.

Methods may be categorised as purely analytical, semi-analytical or numerical; most of the chapters in this volume describe purely analytical or semi-analytical techniques. Analytical techniques usually focus on canonical problems with the aim of the development of a closed-form solution or a rigorously correct solution for which either good approximations, or more usually, asymptotics (high or low frequency) can be developed, or alternatively, for which a minimal amount of inherently well-posed computation is needed to elucidate the features sought. A judicious choice of a good canonical scattering problem can be very illuminating of the dominant scattering features of a more general class of problems – for example scattering by sharp edges and corners is informed by the diffraction from a half plane or a wedge, respectively. Bodies that can be expressed in coordinate systems admitting full or partial separation of the partial differential equations embodied in Maxwell's equations possess exact analytical or rigorous asymptotic solutions [3]; the class of problems related to such canonical closed surfaces continues to expand and now embraces open surfaces and cavity structures [4].

A further division of methods differentiates between frequency-domain and time-domain analysis. The former examines the steady state response of the system when the excitation or illumination of the system is single-frequency or time-harmonic, whereas the latter includes transient phenomena before the attainment of either quiescence or a steady state. The Fourier transform and its inverse provides the means of determining the time harmonic response from the time domain response and vice-versa, and the choice of domain for analysis is usually determined by considerations of efficiency.

Canonical problems play an invaluable role in providing rigorously correct benchmark solutions for the purpose of validating general purpose computer codes based upon numerical methods. Focussing for the moment on time harmonic scattering problems in the low and intermediate frequency range, numerical methods usually discretise the partial differential equations by some means such as finite differences or finite element methods deployed on a volumetric mesh that is terminated, at some finite distance from the scatterer or scattering medium, by a boundary condition related to the radiation condition, with the aim of minimising artificial reflections on the terminating surface. In a number of circumstances, chiefly involving perfectly electrically conducting obstacles or homogeneous dielectric scatterers, it is advantageous to employ an alternative formulation that leads an integral equation for the surface current or an appropriate surface distribution. A suitable Green's function is deployed to convert the differential Maxwell equations to an equivalent integral equation; there are several such variants of the two most commonly deployed, the magnetic field integral equation (MFIE) and the electric field integral equation (EFIE). Similar treatments for one-dimensional objects such as thin wires exist, as does a volumetric integral equation for inhomogeneous scatterers. As well as eliminating the problem of (numerical) dispersion, there are two notable advantages over the differential approach: first, the Sommerfeld radiation condition is automatically enforced, thus ensuring that the solution is physically meaningful in supporting only outwardly travelling waves as the observation point recedes to infinity; and second, excepting the volumetric integral equation, all the unknowns to be determined are two-dimensional (2D) surface or one-dimensional contour line quantities, rather than the fully three-dimensional (3D) quantities encountered in finite difference or finite element methods – this represents a very significant reduction in the number of unknowns required in any discretisation, when compared to that in finite difference or finite element methods. By a suitable choice of a basis to represent the surface quantities plus a testing procedure such as the Petrov–Galerkin method or the method of moments [5], a set of linear algebraic equations for the desired coefficients of the basis functions is obtained. Although the basis may be finite or infinite (and in this case, capable of representing the solution exactly), practical numerical computations require truncation to obtain a finite linear system; the order of finite or truncated infinite bases bears directly on the final accuracy of the computed solution, or equivalently, on the convergence to the exact solution. All of these approaches to time harmonic scattering problems have direct counterparts in the time domain, with similar attendant advantages and limitations.

At high frequencies, methods in which the field behaves in a ray-like manner predominate: in free space, the rays travel in straight lines; rules for their reflection and diffraction from scattering surfaces distinguish the various methods. Physical optics (PO) is the simplest such method, but its limitations become apparent at angles well away from normal incidence or when edge or corner contribution are significant. The geometric theory of diffraction (GTD) developed by Keller [6] and others markedly improves the computed approximation by incorporating diffraction coefficients modifying the amplitude and phase of diffracted rays as they strike a corner or edge of the scattering obstacle. At high frequencies, it enables one to calculate the radar cross-section of a target, the dimensions of which are large compared to the wavelength; in this regime, the lower frequency numerical methods outlined earlier become infeasible due to the computer resources required for large matrix inversion or other procedures. Later developments to address some of its limitations (mainly the problem of caustics where GTD fails) include the uniform theory of diffraction and the physical theory of diffraction (PTD).

In contrast to the numerical methods outlined earlier, semi-analytic or analytical–numerical methods – as the name indicates – usually employ some form of analytic transformation of the underlying set of differential equations or integral equations so that the resultant system is much more well-conditioned for numerical solution. In particular the EFIE, which is a first-kind integral equation, has the generic difficulty of ill-posedness associated with all first-kind equations. Consequently, the solution to the discretised equation cannot be assumed to converge to the exact solution as the underlying mesh is refined; the failing becomes particularly marked for structures involving cavities and sharp edges, so that some form of regularisation is essential to obtain solutions of any reliability. Regularisation may take the form of Tikhonov regularisation applied to the discretised EFIE, but it is much more satisfactory to transform the underlying integral equation in an analytic fashion. A variant of this idea is to choose the basis for expressing the solution that in some way that regularises the original formulation, so that the resultant system is much more well conditioned for numerical solution. Alternatively, constraints that may be imposed on the solution (e.g. that it necessarily obeys edge conditions) imply a form of regularisation.

So-called function-theoretical methods also proceed by employing formulations that allow powerful analytical tools to be employed. Classic examples include the

Wiener–Hopf (WH) method, and its extensions from scalar to matrix versions, to solve such problems as diffraction from a pair or multiple set of semi-infinite parallel planes. Another example concerns the solution of problems of scattering and diffraction from the half plane, wedges and cones by the use of a contour integral representation in the complex plane, taking the form $\int_{C_1+C_2} G(\alpha)s(\alpha+\phi)d\phi$ where C_1 and C_2 are known as Sommerfeld contours, and the kernel $G(\alpha)$ is determined by the source. The representation is known as the Sommerfeld integral or Sommerfeld–Malyuzhinets integral in recognition of Malyuzhinets' development of an inversion in the spectral domain of complex angles that could be applied to many general diffraction problems. Third, we may mention the use of the Riemann–Hilbert problem that is well known in the theory of analytic functions to effect the inversion of a singular part of an integral operator arising in the solution of such problems as diffraction from an open circular cylindrical cavity; this is effectively an analytical regularisation in the sense described earlier.

Inverse scattering problems, where the scattering obstacle or medium is to be determined by the measurement of observation of the field scattered by known sources of illumination, are intimately linked to the corresponding direct scattering problem (where the obstacle is known and the scattered field is to be determined). A common methodology poses an integral equation linking the obstacle and scattered fields. In contrast to the direct problem, in which this integral equation is linear in its inputs and outputs, the inverse problem generates an inherently non-linear equation that is usually solved by postulating an initial approximation to the scattering object (specified by parameters relating to shape or medium) and which is subsequently improved in an iterative process. Without any form of regularisation, the equation is highly ill posed, so that small perturbations in the measured field may induce very large discrepancies in the computed solution, and some form of regularisation is essential to the development of usable algorithms. Two forms of regularisation are commonly employed: the scatterer or medium may be constrained to some specified format, and a (possibly non-linear) regularisation term may be added to constrain the solution to the problem when posed as a least squares problem (or alternatively another norm might be employed in the formulation). A number of such processes exist in the literature.

The first group of chapters deals with **canonical problems and function-theoretic methods**. The chapter by Bernard considers the problem of scattering by 2D and 3D canonical objects with imperfectly conducting surfaces, employing the Sommerfeld–Malyuzhinets integral. It presents this spectral method in a new perspective, by giving some novel exact general expressions and properties of the associated spectral function attached to the total field, a novel spectral expression of free space Green's function, the single face representation and its consequences. This is used to calculate an improved 2D diffraction coefficient for an impedance wedge with curved faces. The final part deals with the spectral representation of the field for 3D conical scatterers to obtain the representation of the fields diffracted by an imperfectly conducting cone.

A somewhat different extension of the Sommerfeld–Malyuzhinets approach is employed in the chapter by Lyalinov and Zhu to study the scattering of an electromagnetic plane wave which is skew incident at the edge of an impedance wedge. The reduction of the problem to a system of functional Malyuzhinets equations and their transformation to integral equations enables an efficient calculation of the far-field asymptotics from the corresponding Sommerfeld representation.

Electromagnetic wave propagation from one medium into a different wedge-shaped medium is recognised to be a difficult problem, and that no explicit closed-form solution of the wedge-diffraction problem with an arbitrary refractive index has been obtained so far. The chapter by Berntsen formulates an integral equation for the H-polarisation diffraction problem (where the magnetic field vector is parallel to the edge of the wedge). Care is taken to address the field singularity at the tip of the wedge and results are achieved by two closely related representations of the diffraction problem. The solution of the diffraction problem is based on a new Fourier series expansion of the field. The Fourier expansion applies to the near-field construction. An integral equation for the Fourier transform of the trace field is formulated and solved iteratively. The exact solution H of the diffraction problem is obtained as the limit of a convergent sequence of functions H_n.

Turning to the theme of **electromagnetic theory**, Arnold's chapter explores the formulation of Green's Theorem, Green's functions and Huygens' principle in discrete electromagnetics. It applied to a function φ over a discrete space, for example defined only at a discrete subset of points x_j of $\mathbf{R}^n$, with values $\varphi_j = \varphi(x_j)$, and differential operators are replaced by difference operators. Although this situation arises naturally as a consequence of the discretisation of partial differential equations such as the Poisson equation or wave equation into finite-difference equations for numerical solution by computational methods, the adopted approach here is a consideration of dynamical systems on a graph or cell complex, so that the framework in which discrete equivalents of Maxwell's equations and electromagnetic potentials appear is *a priori* discrete. The development of Green's functions is shown to have a very practical outcome in the appearance of Huygens' principle for the dynamical evolution of discrete fields.

The following chapter by Idemen examines universal properties of the Green functions associated with the wave equation. The discovery of the relation between the Green function and the delta function played a very important role in the development of the theory of boundary-value problems connected with partial differential equations and their engineering applications in the last century. However, the complexity of the explicit expressions of the Green functions has always been a severe handicap in various engineering applications especially when the domain is bounded and non-homogeneous. The chapter expounds some universal properties of the Green functions which overcome the difficulty in question. An illustrative example shows their use in some inverse problems: an inverse initial-value problem connected with photoacoustic tomography in bounded non-homogeneous domains is considered. One reason for the intense interest in acoustic tomography is the possibility of the detection of cancerous tissues because their contributions to the induced pressure wave are much stronger than that of normal tissues.

The next pair of chapters considers the development and application of **high-frequency techniques**. The method of asymptotic currents is an extension of PO and PTD that avoids the previously mentioned drawbacks of GTD. It takes into account the currents on the shadow side of the target by introducing creeping waves. Moreover, in the lit region, close to the shadow boundary, the PO currents are replaced by more accurate transition zone solutions. In their chapter, Molinet and Andronov apply function-theoretic methods to calculate the asymptotic currents on elliptic cylinders with a strongly elongated cross-section and extend to more general cylindrical configurations where the cross-section is a truncated ellipse or composed of arcs of ellipses. When an object is strongly elongated geometrically, the length of the Fock domain can be small for sufficiently high frequency and classical asymptotics apply. However, when the Fock domain becomes large, such asymptotics become inaccurate and eventually become inapplicable when it covers the whole object. Their approach gives the total field in the boundary layer along the elongated object (strongly elongated spheroids and elliptic cylinders with strongly elongated cross-sections) as well as the asymptotic currents on the surface. Solutions to the model problem can then be extended to treat more general configurations.

The chapter by Hiroshi Shirai considers hybrid ray–mode techniques. Rays and modes are alternative descriptions for the electromagnetic field, and they can be usefully regarded as Fourier transform pairs which have complementary convergence properties. A modal description is usually preferable for an internal structure, while a ray description is normally more appropriate for an exterior region. When a scatterer contains an open waveguide cavity structure, one wants to retain the advantages of fast converging approaches in their respective domains, namely rays for the exterior and modes for the interior regions, with a mechanism for conversion at the opening of the structure. This ray–mode conversion has its mathematical basis in the Poisson summation formula; this technique is found to be one of the most powerful tools for analysing high-frequency scattering problems. Modal excitation conversion at an aperture of a parallel-plane waveguide is formulated with this technique, and its validity is discussed. Some applications and examples, such as plane wave diffraction from a slit on a conducting screen, or from a trough on a ground, are given to show the validity of this technique.

The next three chapters present analytic treatments of various problems in **scattering and diffraction**. The chapter by Klinkenbusch and Brüns considers scattering and diffraction of scalar and electromagnetic waves using spherical-multipole analysis and uniform complex-source beams (CSBs). Spherical-multipole analysis belongs to the set of classical methods that analytically solves 3D scalar and electromagnetic problems for a certain class of canonical geometries. With sphero-conal coordinates – which can be understood as generalised spherical coordinates – that class of canonical geometries widens and includes the sphere, the right-circular cone, the elliptic cone, the wedge, the half plane and the plane angular sector. Usually in scattering and diffraction, an incident plane wave is chosen to deduce the corresponding characteristics of the object under investigation. One disadvantage is that all parts of the scattering object are illuminated, and consequently the scattered field contains all of the corresponding information. Moreover, a plane wave incident on semi-infinite structures usually leads to solutions with non-converging series where problem-specific integral transforms and/or special summation techniques are needed to find asymptotically valid limiting-value results of the scattered fields.

CSBs with a complex point-source produce a beam-like field which in the vicinity of the beam axis (paraxially) can be interpreted as one half of a Gaussian beam, a localised inhomogeneous plane wave. Such beams are perfectly suited to probe the scattering object by exclusively illuminating desired areas of the structure (such as tips or edges), to extract the fields scattered, and to derive the corresponding diffraction coefficients. In contrast to using a homogeneous plane wave as the incident field, the resulting spherical-multipole series converges so that diffraction and scattering characteristics of any desired part of the scattering object are obtained without applying series transformations as have been found necessary using homogeneous plane waves. A convergence analysis as well as calculation of the total near-fields and scattered far-fields for the acoustically soft, the acoustically hard and the electrically perfectly conducting semi-infinite elliptic cone illuminated by a uniform CSB is presented.

As already indicated, scattering by sharp edges and corners is informed by the canonical problem of diffraction from the half-plane and the wedge (of infinite extent). Although analytical and numerical methods have evolved to account for the field singularities and enable accurate modelling, these methods can be time consuming to implement and at times become very specialised. When numerical methods are employed, a common approach used when dealing with domains with corners is to round the corners, producing a smooth surface. This eliminates the field singularities introduced by the corners and allows for standard numerical quadratures to be used, but with no clear estimate of the error induced by the rounding.

A systematic treatment of the transition from sharp cornered objects to those with smooth boundaries is required, in particular as the radius of curvature of the rounded corner points tends to zero. In their chapter, Markowskei and Smith rigorously analyse the changes in the far-field pattern induced by rounding the corners of a 2D scatterer. An appropriate integral equation for the surface distribution on a sharp cornered obstacle and its rounded counterpart is given. The lemniscate (having a right-angled corner) and its rounded counterpart is used as a test case to establish analytic bounds for the maximum difference in the scattered far-field under plane wave illumination. An integral equation is obtained for the difference in the surface distributions on each obstacle; its approximate solution is shown to be $O((k\rho)^{2/3})$, as $k\rho \to 0$; here k denotes wavenumber and ρ the radius of curvature at the rounded corner. It then follows that the non-dimensionalised far-field difference is $O((k\rho)^{4/3})$, as $k\rho \to 0$. This is in accord with well-known results on the fields in the vicinity of the right-angled wedge (of infinite extent).

As already mentioned, the problem of the electromagnetic diffraction by a dielectric wedge is still an unsolved problem, in the sense that a closed-form solution as complex integrals, such as obtained, for example, for an imperfectly or perfectly conducting wedge, is still not known. The special case of a right-angled dielectric wedge was thought to be amenable to the WH technique, but the final solution unsatisfactorily ends up with the numerical solution of a Fredholm integral equation, or an analytic perturbation solution. In his chapter, Rawlins shows that the use of the WH approach is unnecessary, and a much simpler use of the double Laplace transform suffices.

The appropriate partial differential equation and boundary conditions can be converted to an equivalent singular integral equation (SIE) by means of Green's theorem; this equation is subsequently transformed to another more convenient integral equation by using the double Laplace transform in two complex variables. Although the exact solution of the resulting SIE is not yet possible, nevertheless a solution is constructed as a power series in the index of refraction. This series is convergent when the index of refraction is near unity. Using this solution, the far-field electric-field amplitude is derived to the first order of magnitude.

The **WH technique** is known as a rigorous, function-theoretic approach for electromagnetic wave problems related to canonical geometries and can be a very powerful tool to solve field problems in the presence of discontinuities. Two chapters provide contemporary exemplars. The chapter of Kobayashi and Nagasaka analyses the electromagnetic wave diffraction from a thin material strip with arbitrary permittivity and permeability, a problem that has importance from both the theoretical and engineering viewpoints. It is mathematically rigorous in the sense that the edge condition required for the uniqueness of the solution is explicitly incorporated into the analysis. Introducing the Fourier transform of the scattered field and applying approximate boundary conditions in the transform domain, the problem is formulated in terms of the WH equations, which are solved exactly via the factorisation and decomposition procedure that is central to the method. The final solutions are valid over a broad frequency range. Numerical examples are presented for various physical parameters, and the far-field scattering characteristics are discussed.

The chapter by Daniele and Lombardi introduces a generalisation of the WH technique that allows the study of geometries where coupled planar and angular region are present. Since exact solutions with closed-form factorisations are available only in few cases, most problems require an alternative approximate technique such as the Fredholm factorisation which reduces the solution of WH equations to a system of Fredholm integral equations of second kind which are amenable to very efficient numerical solution. The deduction is presented for a relatively simple yet novel problem. The numerical solution of Fredholm integral equations provides an analytical element of the spectra that, in general, is not sufficient to evaluate the different components of the diffracted field. To obtain the whole spectrum of the unknowns, analytical continuation and recursive equations obtained from the WH equations are presented. The chapter concludes with numerical simulations for the novel scattering problem.

The next group of chapters concerns the ideas of **analytical regularisation**. As already discussed, a widely used formulation of electromagnetic scattering from open scatterers is based on the EFIE. This first-kind equation has the generic difficulty of ill-posedness associated with all first-kind equations, with the consequence that the solution to the discretised equation does not converge to the true solution as the underlying mesh is refined. The practical difficulty manifests itself most severely with structures that exhibit a resonant or near-resonant response, such as cavities and other classes of structures. Analytical regularisation refers to those techniques that analytically transform the EFIE (and related first-kind equations) to well-posed operator equations, so that upon discretisation the linear algebraic systems of equations are well conditioned for standard numerical solution procedures.

The principal ideas and more advanced techniques of the analytical regularisation method (ARM) and the semi-inversion procedure (SIP) as they apply to the integral and integro-differential equations of scattering and diffraction problems in electromagnetics are considered in the chapter by Tuchkin. These methods draw upon some relatively sophisticated applications of functional analysis, including the theory of numerical methods, the theory of distributions (generalised functions), Sobolev spaces, Tikhonov regularisation and the propagation and amplification of round-off errors in numerical processes. Their principal advantage lies in the transformation of the underlying operator equations arising in diffraction theory – which typically are ill posed for the class of problem considered – to well-posed operator equations, with the attendant advantage of well-conditioned discretisations. The chapter presents a unified treatment of the ARM and SIP that draws upon some recent developments using the language of pseudo-differential operators, while clarifying its rigorous application to boundary value problems of electromagnetics.

The chapter by Vinogradova applies the ARM to a problem requiring guaranteed accuracy and convergence. The scattering of electromagnetic waves by open cavities usually causes a resonant response, arising when the frequency of excitation approaches one of the frequencies belonging to the spectrum of eigenvalues of such a cavity. Perfectly conducting 2D hollow cylinders of arbitrary shape with longitudinal slits are considered: in the context of propagating waves, these objects are models for waveguides, or in the context of standing waves, for resonators. The eigenvalues of the closed structure are real and represent cut-off frequencies, but the opening of apertures or slits causes the spectrum of eigenvalues to become complex valued, relating to radiation losses. Such open structures may be modelled by the EFIE, but its direct discretisation is unreliable for determining the imaginary part of the eigenvalues with any accuracy. The chapter presents a practical implementation of the ARM that converts the EFIE to an infinite second-kind system of linear algebraic equations. The system is solved numerically by the truncation method wherein the matrix and vectors are truncated to a finite size. As the system is progressively increased in size, the fast convergence of the solutions to the exact solution, along with the ability to guarantee any predetermined accuracy of calculations, makes this method an ideal instrument for the calculation of the spectrum of the complex eigenvalues.

The next group of three chapters addresses **effective numerical methods** for solving the various integral equations encountered in electromagnetics. Graglia and Peterson review the development of the EFIE and the MFIE for perfectly conducting targets and homogeneous dielectric targets. When applied to certain closed surfaces, the original equations exhibit uniqueness difficulties at frequencies where the target surface coincides with a resonant cavity. In addition, the original EFIE and MFIE also fail under certain circumstances for electrically small bodies. Alternate integral equations to address the impact of these remediate fictitious internal resonances are discussed. In addition, robust approaches to deal with the low-frequency breakdown of these integral equations are discussed. As already mentioned, the most common procedure for performing a discretisation is the method of moments, which is essentially the same as the boundary element method, the weighted residual method or the finite element method. The procedure involves approximating the quantities to be determined – the surface currents – by expansion in terms of a linearly independent basis of functions; the coefficients of these functions become the unknowns to be determined. After insertion of the basis expansion in the integral equation, a testing procedure produces a linear system of linear equations (with corresponding impedance matrix), the numerical solution of which yields the desired coefficients. The chapter surveys the types of bases and testing procedures employed in the field, reports the progress made in recent years associated with the use of higher order vector basis functions and describes recent improvements utilising singular basis functions.

The next chapter by Merlini, Dèly, Cools and Andriulli examines electromagnetic modelling with the EFIE. Despite its advantages, the EFIE suffers from issues limiting its applicability when employed naively. The condition number of the impedance matrix of the EFIE suffers from two types of ill conditioning: low-frequency breakdown which causes the conditioning of the matrix to grow as k^{-2} where k is the wavenumber, and high-refinement breakdown which causes the conditioning to grow as h^{-2} where h is the average size of the boundary elements. Both contributions cause iterative solvers to require more iterations before converging and to yield less accurate solutions or, in more extreme cases, prevent numerical resolution of the problem altogether. The difficulties of the EFIE are not confined to the frequency domain: the corresponding time-domain integral equation has problems that can be traced to similar causes.

A new family of techniques, based on so-called quasi-Helmholtz (qH) projectors, is introduced to cure the low-frequency breakdown without any of the limitations of the traditional solution. The projectors can be computed in near-linear complexity by leveraging multi-grid pre-conditioners, which means that their usage does not increase the complexity of the pre-conditioned formulation or significantly slows down existing solvers. High-refinement breakdown of the EFIE can be cured by Calderón-based techniques combined with the qH projectors to obtain a fully regularised formulation immune from both the high-refinement and low-frequency breakdowns.

The third chapter by Bliznyuk and Nosich synthesises ideas of the ARM with numerical tools in microstrip antenna modelling – in particular, resistive printed discs which are attractive, being more broadband than metallic ones, and low-loss, high-permittivity disc dielectric antennas which have further advantages. Such devices require efficient wide-band modelling so that effective numerical optimisation can be undertaken. The chapter presents an accurate numerical method and computational results in the modelling of axisymmetric electromagnetic wave scattering by such resistive and thin-dielectric circular disc-on-substrate antennas, simulated by using two-sided generalised boundary conditions and SIEs. The numerical analysis is based on the method of analytical regularisation exploiting a Galerkin method with a judicious choice of basis functions which convert the SIE into a Fredholm second-kind infinite matrix equation. This guarantees convergence and enables one to compute the solution with controlled accuracy, even near sharp resonances. Numerical results demonstrate the effect of the disc losses and transparency on the antenna bandwidth and radiation efficiency.

The next two chapters concern **inhomogeneous media**. Yamasaki considers scattering and guiding problems of electromagnetic waves in inhomogeneous media such as photonic crystals; these are of great interest in many areas of physics and engineering in nanoscale optical technology. Dielectric gratings are now widely used in integrated optics, such as in optical gratings, optical couplers, optical waveguide filters, and are used as frequency selective surfaces. However, the conventional method for tackling these problems is based on a Fourier series expansion method in inhomogeneous media and cannot be applied to profiles with discontinuities. This chapter introduces a new so-called improved Fourier series expansion method that can be applied to discontinuous profiles for scattering and guiding problems. The major components of the process are as follows: the inhomogeneous region is approximated by an assembly of stratified thin layers with step profile; the electromagnetic fields in each layer are expanded approximately by a finite Fourier series and finally all layers are combined with appropriate boundary conditions to get the fields in the inhomogeneous media. The convergence theory depending upon the modal truncation number and the Fourier truncation number is discussed, and the methods of numerical computation for the scattering and guiding of electromagnetic waves in inhomogeneous dielectric gratings with periodic surface relief are explained, for both TM- and TE-cases. The range of applicability to inhomogeneous media is much wider than that of conventional methods.

A number of scattering and guiding problems for several inhomogeneous media are considered, including as slanted layer and rhombic media with strips; elliptically layered, columnar and rectangular media; the energy distribution in defect layers and mixed positive and negative media. Cases in which the method is especially effective are discussed.

In the following chapter, Samokhin considers methods and fast algorithms for the solution of volume singular integral equations (VSIEs). These arise in finding the electromagnetic field excited by an external field in a medium which is a finite 3D domain characterised by a dielectric permittivity tensor function, outside of which the permittivity is constant; the permeability is constant everywhere. The corresponding problem when the same domain lies over a perfectly conducting plane may equally be considered. These problems can be reduced to VSIEs with respect to the electric field in the domain.

The spectrum of the operators of these VSIEs is studied, and methods and fast algorithms for the numerical solution of the integral equations are presented. Taking into account that the kernels of VSIEs depend only on the difference and sum of arguments when using a regular grid for the discretisation, the integral equations are converted to a system of linear algebraic equations with matrices possessing symmetry properties. The computer memory requirements are proportional to the matrix size. By using fast Fourier transform techniques, fast algorithms for the multiplication of a matrix by a vector may be constructed. This technique enables effective iterative methods for the solution of the linear system to be exploited. Some numerical results demonstrate its effectiveness.

The next chapter, by Nedic, Ehrenborg, Ivanenko, Ludvig-Osipov, Nordebo, Luger, Jonsson, Sjöberg and Gustafsson discusses **Herglotz functions** and their applications in electromagnetics. Their importance arises because they are the pertinent functions to model passive systems and thus appear in the modelling of electromagnetic phenomena in circuits, antennas, materials and scattering. Herglotz functions can be represented by an integral representation solely depending on scalar parameters and a positive measure. This representation is very powerful and is the starting point for the derivation of many results, and in particular, identities that relate weighted integrals of a Herglotz function with its asymptotic expansion. These identities are often referred to as sum rules and have many applications in electromagnetics.

In this chapter, the basic theory is reviewed and applied to determine sum rules and physical bounds for passive systems. Sum-rules and convex optimisation utilise the inherent constraints due to passivity, linearity and time-translational invariance of the electromagnetic applications to obtain fundamental physical bounds. Furthermore, these techniques not only give physical bounds but in several cases have been shown to be predictive as a tool in the design of antennas and other electromagnetic structures.

Periodic structures – dielectric or metallic – are a subject of continuing interest because of their wide use for practical devices in microwaves and optical waves. A periodic array of circular rods is typical of a discrete periodic structure. Various analytical or numerical techniques have been developed over the years to formulate the electromagnetic scattering and guidance by periodic arrays, mostly concerned with planar arrays.

The chapter by Jandieri and Yasumoto considers alternative configurations based upon a cylindrical array formed by circular rods periodically distributed on a circular ring; potential applications include the design of photonic crystal fibres, directive antennas or beam-switching antennas, plasmonic crystals and terahertz waveguides. It discusses a semi-analytical approach for electromagnetic radiation, scattering and guidance by cylindrical arrays composed of circular rods periodically distributed along a concentrically layered circular ring; the rods may be dielectrics, perfect conductors, air-holes, metals or magnetised ferrites.

The method uses the transition matrix (T-matrix) approach: it combines the T-matrix of a circular rod in isolation, the reflection and transmission matrices of a cylindrical array based on an expansion in cylindrical harmonics and the generalised reflection and transmission matrices for a cylindrically layered structure. The formulation is rigorous, and the only approximation involved in the solutions is the truncation of the cylindrical harmonic expansion and the translation matrices related to Graf's addition theorem, which is necessary to solve a finite set of linear equations by matrix inversion. The semi-analytical approach is general, yet efficient and takes into account all cylindrical space harmonics and their multiple interactions through the scattering process. The high-computational efficiency of the method is important for designing and optimising cylindrical periodic and bandgap structures. The formulation may be generalised to treat various configurations of layered cylindrical arrays with different types and locations of the excitation sources. A number of applications are considered.

The next two chapters deal with **inverse scattering**, where the scattering obstacle or medium is to be determined by the measurement of observation of the field scattered by known sources of illumination. The most general problems are usually ill-posed, and various constraints or regularisations need to be imposed to determine acceptable solutions. Algorithms are often based upon iterative schemes employing successive forward and inverse scattering steps.

In their chapter Derevyanchuk, Shestopalov and Smirnov describe inverse scattering problems in the more constrained context of waveguides. The determination of electromagnetic parameters of dielectric bodies that have complicated geometry or structure arises when, for example, nano composite or artificial materials and media are used as elements of various devices. Since these parameters cannot be directly measured (because of the composite character of the material and small size of samples), indirect methods must be employed. One approach is to measure the response of parallel-plane dielectric diaphragms and layered structures placed in a rectangular waveguide with perfectly conducting walls. This necessitates the development of appropriate mathematical models and numerical solution of the corresponding forward and inverse problems. For bodies of complicated shape, techniques are often sought in the resonance frequency range; this is the case when the permittivity of nanocomposite materials must be reconstructed. It is important to determine permittivity and conductivity not only of a composite as a solid body but also of its components, for example nanotubes.

This chapter analyses a set of benchmark problems related to forward and inverse scattering by parallel-plane dielectric diaphragms and layered structures placed in a rectangular waveguide with perfectly conducting walls, with the aim of reconstructing permittivity of the inclusions from the transmission coefficient. The unique solvability of these inverse problems is proved by establishing the domains where the transmission coefficient is a one-to-one function of permittivity and other problem parameters. The robustness and accuracy of the developed numerical solution is discussed.

A rather different perspective of inverse scattering is presented in the following chapter by Tuvi, Heyman and Melamed. It considers beam waves which are advantageous for local probing (for similar reasons to those given by Klinkenbusch and Brüns). Beam waves are highly collimated wave functions that propagate along ray trajectories, either in the time-harmonic context where the propagators are typically Gaussian beams, or in time-domain fields where the propagators are the so-called iso-diffracting pulsed beams. A beam-based local diffraction tomography inversion scheme, as an alternative to the conventional Green function or plane wave inversion schemes, is considered. The advantages of beam-based schemes include the ability to focus the imaging scheme on any given subdomain of interest in the image domain by retaining only those beam back propagators that pass through that domain. Thus, the algorithm not only reduces the overall complexity, but it also filters out irrelevant data and noise arriving from regions outside the domain of interest.

Modal expansions are ubiquitous in many branches of physics and engineering: they enable one to construct formal solutions of many linear partial differential equations of mathematical physics and to interpret and comprehend more easily the associated physical effects. Common applications include solutions of radiation problems in waveguides, cavities or photonic crystals, analytical expressions for Green's functions and others. The chapter by Silveirinha discusses modal expansions in dispersive material systems with application to quantum optics and topological photonics. The objective is to highlight that for lossless material platforms formed by arbitrary inhomogeneous bi-anisotropic and possibly non-reciprocal materials the natural modes of oscillation form a complete set of expansion functions.

It is shown that Maxwell's equations in dispersive systems can always be reduced to a generalised dynamical problem whose time evolution is described by a Hermitian operator. The effects of material dispersion are taken into account by introducing additional variables that may model the internal degrees of freedom of the material. Formal expansions of the electromagnetic field in terms of the normal modes are constructed, although the expansion coefficients are not unique; a modal expansion of the system Green function is obtained. Based on the equivalence between the dispersive Maxwell's equations and the generalised dynamical problem, definitions of energy and momentum in a dispersive system are discussed. In particular, it is shown that for time-harmonic solutions the energy and the total momentum can be written exclusively in terms of the electromagnetic degrees of freedom. An application of the proposed formalism is illustrated in the contexts of topological photonics and quantum optics. In particular, the theory enables the topological classification of a generic bi-anisotropic and nonreciprocal electromagnetic continuum. A simple and intuitive theory for the quantisation of the electromagnetic fields in dispersive material systems and for the calculation of the quantum field correlations is developed.

The next two chapters concern **waves in complex media**. Kristensson and Wellander discuss multiple scattering by a collection of randomly located obstacles distributed in a dielectric slab. It begins with a deterministic analysis of the multiple scattering of electromagnetic waves by a discrete collection of scatterers. The main tool to solve the problem is the integral representation which is exploited in the various homogeneous regions of the problem. The analysis solves the boundary value with an arbitrary number of general scatterers inside a slab with different material parameters. It enables the calculation of the transmitted and reflected coherent fields. The transmitted and reflected intensities are conveniently represented as the sum of two terms – the coherent and the incoherent contribution.

The stochastic description of the many-body problem is then considered. The complex scattering problem of randomly located obstacles in a slab with different material is solved by a systematic use of two main tools: the integral representation of the solution of the Maxwell equations, and the decomposition of the Green dyadic for the electric field in free space into spherical and planar vector waves. Finally, the tenuous media approximation and low-frequency approximation are both discussed.

Electromagnetic and related wave theories have a wide range of practical applications in complex environments, such as microwave remote sensing of the Earth, object detection and imaging in clutter, medical optics and ultrasound imaging, characterisation of metamaterials and composite and porous media, and communication through complex clutter environments. The chapter by Ishimaru, Kuga and Bright gives a review of wave theories applied to imaging in geophysical and biological media, including imaging through air turbulence and particulate matter, imaging near-ocean rough surfaces and communication and signal processing in clutter, coherence in multiple scattering and super resolution, time-reversal imaging, radiative transfer, waves in porous media, seismic CODA waves and the memory effect.

Previous chapters have focussed on methods in the frequency domain, where the electromagnetic fields are presumed to vary in a time-harmonic fashion. The final chapter by Tretyakov, Butrym and Erden takes a **time-domain** perspective. Quantities occurring in Maxwell's equations, the constitutive equations and other associated quantities are rescaled so that the various relationships appear in a dimensionally simpler format. This perspective is implemented in the evolutionary approach to electrodynamics (EAE) previously proposed by the authors to develop an adequate time-domain theory of transient processes in cavities and of propagation in waveguides of non-harmonic signals (including digital). As the new format of the equations is rather simpler than the standard one, the upgraded version of the EAE is more convenient for practical applications. Energetic and mechanical field characteristics are obtained as functions of time, and the concept of modes in the time harmonic setting for oscillations in cavities and waves in waveguides has its counterpart in time-dependent modal amplitudes. A number of applications are discussed.

In conclusion, we wish to thank our fellow contributors for their efforts to write their chapters. We would like to express our gratitude to the reviewers for their careful review of the manuscripts; their comments and criticism were instrumental in improving content and style. We would also like to acknowledge and thank Dr Elena D Vinogradova for her assistance at various stages in the progress of the book. Finally, we wish to express our sincere thanks to the publishers for assisting us with patience in the completion of the final text.

Kazuya Kobayashi and Paul Smith,
Editors
June 2020

References

[1] J.C. Maxwell, *A Treatise on Electricity and Magnetism*. Reprint of the 3rd edition from Dover Publications, New York, 1954.

[2] A. Einstein and L. Infeld, *The Evolution of Physics From Early Concepts to Relativity and Quanta*, Clarion edition, New York, 1938.

[3] J.J. Bowman, T.B.A. Senior and P.L.E. Uslenghi, *Electromagnetic and Acoustic Scattering by Simple Shapes*, Revised printing, Hemisphere Publishing Corporation, New York, 1987.

[4] S.S. Vinogradov, P.D. Smith and E.D. Vinogradova, *Canonical Problems in Scattering and Potential Theory*, Part I: *Canonical Problems in Potential Theory* and Part II: *Acoustic and Electromagnetic Diffraction by Canonical Structures*, Chapman and Hall/CRC, Boca Raton, Florida, 2001, 2002.

[5] R.F. Harrington, *Field Computations by Moment Methods*, Macmillan, New York, 1968.

[6] J.B. Keller, Geometric Theory of Diffraction, *J. Opt. Soc. Am.*, 1962, 52, 116–130.

Chapter 1

New insights in integral representation theory for the solution of complex canonical diffraction problems

J.M.L. Bernard[1,2]

The problem of scattering by two-dimensional (2D) and three-dimensional (3D) canonical objects with imperfectly conducting surfaces requires some particular efforts on the representation of scattered and incident fields, and we present here some remarkable aspects of them in complex situations and their applications.

We begin with the study of 2D problems. The Sommerfeld–Maliuzhinets integral and its inversion in the spectral domain of complex angles has opened a new way of investigation on diffraction by a wedge-shaped domain. In this frame, we present this spectral method in a new perspective, by giving some novel exact general expressions and properties of the associated spectral function attached to the total field: novel spectral expression of free-space Green function, single-face representation and its consequences, uniqueness, existence, reciprocity, spectral causal representation of field in time domain. We then analyze in a distinct section the diffraction by an impedance wedge with curved faces. The diffraction coefficient, when limited to its principal order in curvature, leads us to some miscalculation when the wedge approaches the discontinuity of curvature case, for example. Thus, we give here a general asymptotic representation taking into account a development at several orders for arbitrary wedge angle, which is uniformly valid from the discontinuity of curvature to the curved half-plane.

Concerning 3D problems, we begin with a section devoted to an efficient exact solution for the radiation of a point source above an impedance plane in electromagnetism. In classic approaches, we need to consider vertical and horizontal dipole cases separately, and we present here a method, consisting in filling a gap in representation of incident fields by potential theory, to finally permit an efficient global representation of the diffracted field for arbitrary primary sources in a direct manner. We conclude this chapter with the representation of fields for the diffraction by an imperfectly conducting cone. We present a general exact expression of fields with Debye potentials and Kontorovich–Lebedev (KL) integrals, and new general properties of

[1]CEA-DIF, Arpajon, France
[2]LRC MESO, CMLA, ENS Cachan, Cachan, France

spectral functions attached to them. We then insist on a novel compact representation of plane waves, which does not need any use of spherical harmonics series.

1.1 Representations of spectral function in frequency and time domain, for the scattering by a polygonal region

1.1.1 Basic elements in Sommerfeld–Maliuzhinets representation and properties

1.1.1.1 Basic integral representation

Let us consider the case of diffraction in free space of an incident plane wave

$$u^i(\rho, \varphi) = U^i(\omega)e^{ik\rho\cos(\varphi-\varphi_\circ)}, \tag{1.1}$$

by a scatterer enclosed in a wedge-shaped region (Figure 1.1), defined in cylindrical coordinates (ρ, φ, z) as the domain outside the free-space angular sector with origin O, $-\Phi_r \leq \varphi \leq \Phi_l$. An implicit harmonic dependence on time $e^{i\omega t}$ is understood and henceforth suppressed, and $k = \omega/c$ denotes the wave number of the exterior medium with $|\arg(ik)| < \pi/2$. Physically, $|\arg(ik)| < \pi/2$ means that there are some losses in free space, and $|\arg(ik)| = \pi/2$ is considered as a limit case. The function $U^i(\omega)$ is assumed $O(\omega^{-b})$ for $|\omega| \to \infty, b > 1$ so that the scattered field u is $O(\omega^{-v})$ with $v > 1$ as $|\mathrm{Re}\omega| \to \infty$ for $\mathrm{Im}\omega \leq 0$ and is analytic at any ω with $\mathrm{Im}\omega \leq 0$. The characteristics of the scatterer are supposed to be independent of z coordinate. We assume that the total field in the free-space region, $u = u_s + u^i$, satisfies the Helmholtz equation:

$$(\Delta + k^2)u(\rho, \varphi) = 0, \tag{1.2}$$

and that u is analytic with respect to ρ, φ and $\varphi_\circ$, except possibly at the origin, and that there exists a constant $s_\circ$ such that $\int_0^\infty |u(\rho, \varphi)e^{-s_\circ\rho}|d\rho < \infty$.

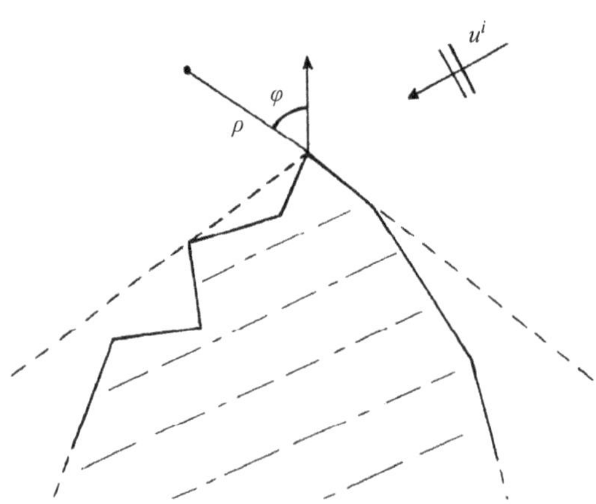

Figure 1.1 Geometry of a wedge-shaped region

The total field u for $-\Phi_r \leq \varphi \leq \Phi_l$ is then represented as a Sommerfeld–Maliuzhinets integral [1],

$$u(\rho,\varphi) = \frac{1}{2\pi i}\int_{\gamma} f(\alpha+\varphi)e^{ik\rho\cos\alpha}d\alpha, \tag{1.3}$$

which satisfies the Helmholtz equation. In this representation, f is an analytic function and the path γ consists of two branches: one, named γ_+, going from $(i\infty+\arg(ik) + (a_1 + (\pi/2)))$ to $(i\infty+\arg(ik) - (a_2 + (\pi/2)))$ with $0 < a_{1,2} < \pi$, as $\mathrm{Im}\alpha \geq d > 0$, above all the singularities of the integrand, and the other, named γ_-, obtained by inversion of γ_+ with respect to $\alpha = 0$. Stationary phase methods [2,3] can be applied to (1.3) to find the far field diffraction coefficient $F(\alpha,\varphi_\circ) = f(\pi+\alpha) - f(-\pi+\alpha)$ [4].

1.1.1.2 Basic properties of the total field and its spectral function *f*

Some elementary properties can be assumed to hold for the field:

- (a′) the only incoming plane wave, from the free-space sector with origin O, $-\Phi_r \leq \varphi \leq \Phi_l$, is the incident field;
- (b′) the limit of the field u as $\rho \to 0$ is finite and does not depend on φ, while the derivatives $\partial_\rho u$ and $\partial_\varphi u/\rho$ are locally summable with respect to ρ in the vicinity of the origin. This property applies for an origin taken at any point out of or upon the scatterer;
- (c′) the field, except possibly its geometrical optics part when $\mathrm{Im}(k) \neq 0$, does not grow at infinity. In addition, some bounds on the far field are assumed. We consider here that the field is $O(e^{ik\rho\cos(\varphi-\varphi_\circ)})$ for large ρ, $|\arg(ik)| < \pi/2$, which is a standard assumption in scattering theory.

These properties lead us to the following conditions on f [1,4,5]:

- (a) $(f(\alpha) - u^i(O)/(\alpha-\varphi_\circ))$ is regular at points with $\mathrm{Re}(\alpha)$ belonging to the free space angular sector with origin O, $-\Phi_r \leq \mathrm{Re}(\alpha) \leq \Phi_l$, which ensures (a′).
- (b) there exist some constants $g^\pm$, some analytic function h, and some Maliuzhinets contour γ such that $|f(\alpha+\varphi) \mp f(-\alpha+\varphi) - g^\pm| < |h(\alpha)|$ on and inside the loop formed by the upper branch γ_+ of γ, when $-\Phi_r \leq \varphi \leq \Phi_l$, the function h being summable on γ_+, regular on and within it. In this respect, we notice that $(f(i|\ln\rho|) - f(-i|\ln\rho|)) = -iu(0,\varphi) + O(\rho\partial u/\partial\rho)$, as $\rho \to 0$, with $\rho\partial u/\partial\rho = o((\ln\rho)^{-1})$ and $u(0,\varphi) = ig^+$. Since γ is odd, we can add a constant to f without changing u, which implies that we can define f with $f(i\infty) = -f(-i\infty)$. This ensures (b′).
- (c) $f(\alpha+\varphi)$ has no singularity, except possibly those associated with incident, reflected or transmitted plane waves not vanishing at infinity, in the zone defined by $\mathrm{Re}(ik\cos\alpha) > 0$ as $|\mathrm{Re}(\alpha)| < \pi$, $-\Phi_r \leq \varphi \leq \Phi_l$, $|\arg(ik)| < \pi/2$. Considering that the far field is $O(e^{ik\rho\cos(\varphi-\varphi_\circ)})$, $f(\alpha+\varphi)$ has no singularity in this region when $\mathrm{Re}(ik\cos(\varphi-\varphi_\circ)) < 0$, i.e., $\pi/2 < |\varphi-\varphi_\circ| < 3\pi/2$. This ensures (c′).

1.1.2 *Spectral functions $f_\pm$ attached to the radiation of a single face and simple relation to f*

1.1.2.1 Radiation of a single face of a wedge-shaped region

To simplify the notation without losing generality, we take $\Phi_r = \Phi_l = \Phi$ with $0 < \Phi < \pi$. From the properties (b′), (c′) on u, the scattered field in free space $u_s = u - u^i$ for $|\varphi| < \Phi$ can be written as the sum $u_+ + u_-$ of the radiations of equivalent surface currents carried by the faces $\varphi = +\Phi$ and $-\Phi$,

$$u_\pm(\rho,\varphi) = \frac{-i}{4} \lim_{\rho_0\to 0^+} \int_{\rho_0}^{\infty} \left(u(\rho',\varphi') \frac{\partial H_0^{(2)}(kR)}{\partial n} - \frac{\partial u(\rho',\varphi')}{\partial n} H_0^{(2)}(kR) \right) \Bigg|_{\varphi'=\pm\Phi} d\rho' \tag{1.4}$$

with $R = \sqrt{\rho^2 + \rho'^2 - 2\rho\rho' \cos(\varphi - \varphi')}$, $\partial(\cdot)/\partial n = \widehat{n}\nabla(\cdot) = \mp\partial(\cdot)/\rho'\partial\varphi'$, $\widehat{n}$ the outward normal to the face $\varphi' = \pm\Phi$, $|\varphi| < \Phi$, $|\arg(ik)| < \pi/2$. We show that it is possible to simply express the spectral function $f_\pm$ attached to $u_\pm$ with f as $0 < \Phi < \pi$. For this, we use an original representation of $H_0^{(2)}(kR)$.

1.1.2.2 Spectral function attached to the Sommerfeld–Maliuzhinets representation of $H_0^{(2)}(kR)$

As shown in [6], the spectral function for $H_0^{(2)}(kR)$, where $|\varphi'| > \pi/2$, $|\varphi| < |\varphi'|$, is given by

$$f_{H_0^{(2)}}(\alpha) = \frac{-1}{2\pi} \int_{\mathscr{S}} e^{ik\rho' \cos(\alpha' - \varphi')} \left(\tan\left(\frac{1}{2}(\alpha - \alpha')\right) - g_\circ(\alpha') \right) d\alpha', \tag{1.5}$$

for $\alpha \in]\mathscr{S} - \pi, \mathscr{S} + \pi[$ (i.e., the domain limited by $\mathscr{S} - \pi$ and $\mathscr{S} + \pi$), where $\mathscr{S}$ is the path from $-i\infty - \arg(ik)$ to $i\infty + \arg(ik)$ with $\mathrm{Im}(k \sin\alpha) = 0$. By analyticity, the path $\mathscr{S}$ can be deformed continuously, as long as the integrand remains bounded, without changing $f_{H_0^{(2)}}$, and this expression can be continued, for $|\varphi'| \leq \pi/2$ by deforming $\mathscr{S}$, or by $f_{H_0^{(2)}}(\pi + \alpha) - f_{H_0^{(2)}}(-\pi + \alpha) = -2ie^{ik\rho' \cos(\alpha - \varphi')}$, for α outside $]\mathscr{S} - \pi, \mathscr{S} + \pi[$. The terms $g_\circ$, normally unnecessary because γ is odd, is chosen as $g_\circ(\alpha) = -\tan(\alpha/2)$ in order to ameliorate the convergence of the integral.

1.1.2.3 Simple exact expression of single-face spectral function $f_\pm$

Therefore, we can write the spectral function corresponding to $H_0^{(2)}(kR_\varepsilon)$, with $R_\varepsilon = \sqrt{\rho^2 + \rho'^2 - 2\rho\rho' \cos(\varphi - \varepsilon\Phi)}$, $\varepsilon \equiv +$ or $-$, in a more general form,

$$f^{\varepsilon}_{H_0^{(2)}}(\alpha) = \frac{-1}{2\pi} \int_{\mathscr{S}_\varepsilon} e^{ik\rho' \cos(\alpha' - \varepsilon\Phi)} \left(\tan\left(\frac{1}{2}(\alpha - \alpha')\right) - g_\circ(\alpha') \right) d\alpha', \tag{1.6}$$

where $\mathscr{S}_\varepsilon = \mathscr{S} + a_\varepsilon$, with a_ε a constant satisfying $\Phi - 3\pi/2 < \varepsilon a_\varepsilon < \Phi - \pi/2$ for $0 < \Phi < \pi$.

The expression (1.6) is used for $H_0^{(2)}(kR_\varepsilon)$ and $\partial_n H_0^{(2)}(kR_\varepsilon)$ in (1.4), and we then obtain a Sommerfeld–Maliuzhinets representation of $u_\pm$ with f [4]–[6]:

$$u_\pm(\rho,\varphi) = \frac{1}{2\pi i}\int_\gamma f_\pm(\alpha+\varphi)e^{ik\rho\cos\alpha}d\alpha, \tag{1.7}$$

where

$$f_\varepsilon(\alpha) = \frac{1}{4\pi i}\int_{\mathscr{S}_\varepsilon} \varepsilon f(\varepsilon\pi+\alpha')\left(\tan\left(\frac{1}{2}(\alpha-\alpha')\right) - g_\circ(\alpha')\right)d\alpha', \tag{1.8}$$

for α between $\mathscr{S}_\varepsilon - \pi$ and $\mathscr{S}_\varepsilon + \pi$, provided $\pi/2 < \Phi - \varepsilon a_\varepsilon < 3\pi/2$, $\pi/2 < \Phi - \varepsilon\varphi_\circ < 3\pi/2$ and $g^- = f(i\infty)+f(-i\infty) = 0$, $\varepsilon \equiv +$ or $-$, $0 < \Phi < \pi$. Let us notice that the conditions (b) and (b′), which concern u and f, do not apply to u_ε and f_ε. However, we remark that $f_\varepsilon(\alpha+\varphi) - f_\varepsilon(-\alpha+\varphi)$ is bounded for large α on γ, even if $f_\varepsilon(\alpha)$ for $\text{Im}\alpha \to \infty$ (resp. u_ε for $\rho \to 0$) diverge when $f(\pm i\infty) \neq 0$ (resp. $u \neq 0$ at $\rho = 0$).

Since f is an analytic function, (1.8) can be analytically continued in the whole complex plane. We note, in particular, that, taking into account of the poles of $\tan(1/2(\alpha-\alpha'))$ which can be captured by $\mathscr{S}_\varepsilon$ as α varies, $f_\pm$ satisfies

$$f_\pm(\pi+\alpha) - f_\pm(-\pi+\alpha) = \pm f(\pm\pi+\alpha). \tag{1.9}$$

Concerning the dependence on $\varphi_\circ$ (or Φ), the expression (1.8) has been determined for $\pi/2 < \Phi - \varepsilon\varphi_\circ < 3\pi/2$, but we can consider $f_\varepsilon(\alpha)$ outside this domain by analytical continuation on $\varphi_\circ$ (or Φ), which corresponds to taking account of the contribution of any singularity that would go through $\mathscr{S}_\varepsilon$ as $\varphi_\circ$ (or Φ) goes into these regions.

1.1.2.4 Exact expression of f from diffraction coefficient F and consequences

We have $f(\alpha) = f_+(\alpha) + f_-(\alpha) + f_i(\alpha)$, with $f_i(\alpha) = (\text{Res} f|_{\alpha=\varphi_\circ}/2)\cot((\alpha-\varphi_\circ)/2)$ and $f_\pm$ verifying (1.8) and (1.9). Thus, we can write, when both conditions $\pi/2 < \Phi \mp \varphi_\circ < 3\pi/2$ (which implies $\Phi > \pi/2$) are satisfied,

$$f(\alpha) = \frac{1}{4\pi i}\int_{\mathscr{S}} F(\alpha',\varphi_\circ)\left(\tan\left(\frac{1}{2}(\alpha-\alpha')\right) - g_\circ(\alpha')\right)d\alpha' + f_i(\alpha), \tag{1.10}$$

for $\alpha \in]\mathscr{S}-\pi, \mathscr{S}+\pi[$. This expression can be extended, by analytical continuation on α, from $f(\pi+\alpha') - f(-\pi+\alpha') = F(\alpha',\varphi_\circ)$, and on $\varphi_\circ$ and Φ (for $\Phi \mp \varphi_\circ < \pi/2$ and $\Phi < \pi/2$ by example) considering suitable deformations of $\mathscr{S}$. As a matter of fact, this expression uniquely defines f from its far field diffraction coefficient, which implies uniqueness and existence. From general properties of F, a continuation on $\varphi_\circ$ adds real singularities (geometrical optics terms), and we can assume that $f(\alpha)$ has only singularities on the real axis for $\alpha \in]\mathscr{S}-\pi, \mathscr{S}+\pi[$, $|\varphi_\circ| < \Phi$. Besides, we remark that, if the system is reciprocal, i.e., $F(\alpha',\varphi_\circ) = F(\varphi_\circ,\alpha')$, and if equations on $\varphi_\circ$ variations of F are known, (1.10) implies equations on $\varphi_\circ$ variations of f [6].

1.1.3 *Spectral function f from far field radiation of one face with arbitrary shape*

1.1.3.1 Simple exact expression of the spectral function *f* derived from fields on a single planar face and properties

Stationary phase methods [2,3] can be applied to (1.7) to find the far field radiation of the face $\varphi = \pm\Phi$, also denoted $\varphi = \varepsilon\Phi$. From the regularity of f_ε, we can deform γ to stationary phase points $\alpha = +\pi$ and $-\pi$, when $\pi/2 < \Phi - \varepsilon\varphi_\circ < 3\pi/2$ and $\pi/2 < \Phi - \varepsilon\varphi < 3\pi/2$, and thus, out of the reflected and shadowed regions $2\Phi - \varepsilon(\varphi + \varphi_\circ) < \pi$ and $|\varphi - \varphi_\circ| > \pi$, we can write, for large $k\rho$ and using (1.9),

$$-\sqrt{2\pi k\rho}e^{i(k\rho+(\pi/4))}u_\pm(\rho,\varphi) \sim f_\pm(\pi+\varphi) - f_\pm(-\pi+\varphi) = \pm f(\pm\pi+\varphi), \tag{1.11}$$

which can be considered as an important generalization of the result obtained by Michaeli for Dirichlet boundary conditions in [7]. In other respects, we can use in (1.4) the formulas [8],

$$H_0^{(2)}(kR) \sim \sqrt{\frac{2}{\pi kR}}\, e^{-ikR+i\pi/4},\ \partial_n H_0^{(2)}(kR) \sim \sqrt{\frac{2}{\pi kR}}\, e^{-ikR-i\pi/4}\partial_n(kR), \tag{1.12}$$

with $R = \sqrt{\rho^2 + \rho'^2 - 2\rho\rho'\cos(\varphi - \varphi')}$, $\partial_n(kR) = \mp\partial(kR)/\rho'\partial\varphi' = (\pm k\rho\sin(\varphi - \varphi')/R)$ at $\varphi' = \pm\Phi$, and $\cos(\varphi \mp \Phi) < 0$. Taking into account that

$$R = \rho - \rho'\cos(\varphi \mp \Phi) + \frac{(\rho'\sin(\varphi \mp \Phi))^2}{R + \rho - \rho'\cos(\varphi \mp \Phi)}, \tag{1.13}$$

and considering the properties (b′) and (c′) on u, we then obtain another expression for the far field that, compared with (1.11), gives us

$$f(\pm\pi+\varphi) = \frac{1}{2}\int_0^\infty \left(iku(\rho',\pm\Phi)\sin(\varphi \mp \Phi) \pm \frac{\partial u}{\partial n}(\rho',\pm\Phi)\right) e^{ik\rho'\cos(\varphi\mp\Phi)}d\rho', \tag{1.14}$$

as $\pi/2 < \Phi \mp \varphi_\circ < 3\pi/2$ and $\pi/2 < \Phi \mp \varphi < 3\pi/2$, $|\arg(ik)| < \pi/2$, $0 < \Phi < \pi$.

By changing Φ for $\Phi_e^\pm$ with $0 < \Phi_e^\pm < \pi$ and letting $\pm\pi + \varphi = \alpha \pm \Phi_e^\pm$, we then notice that

$$f(\alpha \pm \Phi_e^\pm) = \frac{1}{2}\int_0^\infty \left(-iku(\rho',\pm\Phi_e^\pm)\sin\alpha \pm \frac{\partial u}{\partial n}(\rho',\pm\Phi_e^\pm)\right) e^{-ik\rho'\cos\alpha}d\rho', \tag{1.15}$$

when $\pi/2 < \Phi_e^\pm \mp \varphi_\circ < 3\pi/2$ and $\pi/2 < \mp\alpha + \pi < 3\pi/2$, which is also valid, by analytic continuation, as $\mathrm{Re}(ik(\cos\alpha - \cos(\Phi_e^\pm \mp \varphi_\circ))) > 0$, $|\mathrm{Re}\alpha| < \pi$ (note: on the straight semi-line $\varphi = \pm\Phi_e^\pm$, we have $\pm(\partial u(\rho',\varphi')/\partial n) = -(\partial u(\rho',\varphi')/\rho'\partial\varphi')$).

Thus, a knowledge of the field and of its normal derivative on one single face (a semi-line) gives a unique definition of the analytical spectral function f attached to the "total" radiated field (uniqueness and existence). By the way, we also notice that, in accordance with the property (b), we have from (1.15), $(f(ix \pm \Phi_e^{\pm}) - f(-ix \pm \Phi_e^{\pm})) = -iu(0, \pm\Phi_e^{\pm}) + O((\partial u/\partial \ln \rho_e)|_{|\ln \rho_e|=x})$ as $x \to \infty$.

1.1.3.2 Deformation and simple exact expression of the spectral function f from fields on a piecewise smooth single face

Using Green's theorem, we note that the contour of integration along $\varphi = \pm\Phi$ in (1.4) can be deformed into a new path $L^{\pm}_{0,\infty}$ from the origin to infinity without changing the field $u_{\pm}$ (except at points captured by the path during its deformation), provided that the integral remains bounded and no source passes through the path during the deformation. Thus, we can write (1.14) in a more general form, with integration along a piecewise smooth path $L^{\pm}_{0,\infty}$, following

$$f(\pm\pi + \varphi) = \frac{1}{2} \int_{L^{\pm}_{0,\infty}} \left(iku(\rho', \varphi') \sin(\varphi - \varphi'_t) \pm \frac{\partial u}{\partial n}(\rho', \varphi') \right) e^{ik\rho' \cos(\varphi - \varphi')} dl', \tag{1.16}$$

where dl' and φ'_t, both depending on (ρ', φ'), are respectively the element of length and the tangent angle along the piecewise smooth semi-line $L^{\pm}_{0,\infty}$.

If we divide the semi-infinite paths $L^{\pm}_{0,\infty}$ into $L^{\pm}_{0,\Delta^{\pm}}$ (i.e. $0 < l' < \Delta^{\pm}$) and $L^{\pm}_{\Delta,\infty}$ (i.e. $l' > \Delta^{\pm}$), we have

$$\begin{aligned} f(\pm\pi + \varphi) =& \frac{1}{2} \int_{L^{\pm}_{0,\Delta^{\pm}}} \left(iku(\rho', \varphi') \sin(\varphi - \varphi'_t) \pm \frac{\partial u}{\partial n}(\rho', \varphi') \right) e^{ik\rho' \cos(\varphi - \varphi')} dl'(\rho', \varphi') \\ &+ f_{L^{\pm}_{\Delta^{\pm},\infty}}(\pm\pi + \varphi), \end{aligned} \tag{1.17}$$

where $f_{L^{\pm}_{\Delta^{\pm},\infty}}(\alpha) = e^{-ik\rho_{\Delta^{\pm}} \cos(\alpha - \varphi_{\Delta^{\pm}})} f_e^{\pm}(\alpha)$, $f_e^{\pm}(\alpha)$ is the spectral function related to a shift of the origin at $l' = \Delta^{\pm}$. This implies, by analytic continuation,

$$\begin{aligned} f(\alpha) =& \frac{1}{2} \int_{L^{\pm}_{0,\Delta^{\pm}}} \left(-iku(\rho', \varphi') \sin(\alpha - \varphi'_t) \pm \frac{\partial u}{\partial n}(\rho', \varphi') \right) e^{-ik\rho' \cos(\alpha - \varphi')} dl'(\rho', \varphi') \\ &+ f_{L^{\pm}_{\Delta^{\pm},\infty}}(\alpha). \end{aligned} \tag{1.18}$$

Our expression can then lead us to a suitable solution for complex geometries and boundary conditions, in particular for the scattering by an impedance polygon [4].

1.1.4 Exact causal time domain representation of a field above a dispersive wedge-shaped region

According to the behavior of f (in particular for ω with large $\mathrm{Im}(\omega) = -\sigma \leq 0$), we can assume that the path γ can be deformed into a contour $\mathscr{D}$ that consists of two branches $\mathscr{D}_{\pm}$: $\mathscr{D}_{+}$ with $\mathrm{Im}\alpha > 0$, given by $](i\infty + \pi), \pi + i0^{+}] \cup [\pi + i0^{+},$

$-\pi + i0^+] \cup [-\pi + i0^+, i\infty - \pi[$, above all the singularities of the integrand as $|\mathrm{Re}\alpha| < \pi$, and $\mathscr{D}_-$, obtained by inversion of $\mathscr{D}_+$ with respect to $\alpha = 0$. Let us prove, from the single-face expression of the function f, that

$$\mathscr{F}_c(\alpha + \varphi, \tau) = \frac{1}{2\pi} \int_{-\infty - i\sigma}^{+\infty - i\sigma} f(\alpha + \varphi, \omega) e^{i\omega\tau} d\omega, \tag{1.19}$$

vanishes as $\tau < 0$ from field causality, so that $f(\alpha + \varphi)$ is regular and $O(\omega^{-\nu'})$ on $\mathscr{D}$ with $\nu' > 1$ as $\mathrm{Im}\omega \leq 0$. The time domain field $\overline{u}(t)$, deriving from the Fourier transform of u, will be then simplified and expressed in an efficient explicitly causal expression on a finite subset $\mathscr{H}$ of $\mathscr{D}$ [6,9].

1.1.4.1 Causality of $\mathscr{F}_c(\alpha + \varphi, \tau)$

Let us use in (1.19) the single-face expression (1.15) of $f(\alpha \pm \Phi_e^{\pm})$ for $\mathrm{Re}\alpha = 0$. Changing the order of the integration (permitted), we can write

$$\mathscr{F}_c(\alpha' \pm \Phi_e^{\pm}, \tau)$$

$$= \frac{1}{4\pi} \int_0^{\infty} d\rho' \int_{-\infty - i\sigma}^{+\infty - i\sigma} d\omega \left(-iku(\rho', \pm\Phi_e^{\pm}) \sin\alpha \pm \frac{\partial u}{\partial n}(\rho', \pm\Phi_e^{\pm}) \right) e^{i\omega t'}, \tag{1.20}$$

for $\mathrm{Re}\alpha' = 0$ with $t' = -\rho\cos\alpha'/c + \tau, \pi/2 < \Phi_e^{\pm} \mp \varphi_\circ < 3\pi/2, \Phi_e^{\pm} \leq \Phi$. The field u is nonsingular and $O(\omega^{-\nu})$, with $\nu > 1$, as $\mathrm{Im}\omega < 0$ and $\overline{u}(t)$ is causal with respect to the front of the incident plane wave, so that the contour of integration in ω can be closed at infinity for $\mathrm{Im}(\omega) < 0$ when $t' < -\rho/c$, which implies that $\mathscr{F}_c(\alpha' + \varphi, \tau) = 0$ for $\mathrm{Re}\alpha' = 0$ as $\tau < 0$. Considering the analyticity of f, $\mathscr{F}_c(\alpha' + \varphi, \tau)$ is assumed to be an analytic function of α' as τ is fixed, when $|\mathrm{Re}(\alpha' + \varphi)| \leq \pi + \Phi$ and $|\mathrm{Im}\alpha'| > 0^+$, and we then have

$$\mathscr{F}_c(\alpha' + \varphi, \tau) = 0,\ \tau < 0,\ |\mathrm{Re}(\alpha' + \varphi)| \leq \pi + \Phi. \tag{1.21}$$

1.1.4.2 Spectral causal expression of the field in time domain

From (1.3) and the causality of $\mathscr{F}_c$, the field in time domain is then [6,9]

$$\overline{u}(t) = \frac{1}{2\pi i} \int_{\mathscr{H}} \mathscr{F}_c(\alpha + \varphi, \tau_c(\alpha)) d\alpha, \tag{1.22}$$

where $\mathscr{H} = \mathscr{H}_+ \cup \mathscr{H}_-$ is a finite subset of $\mathscr{D}$ with $\tau_c(\alpha) = \rho\cos\alpha/c + t > 0$. Let $iv(\rho, t) = \left|{}^{i\ln(ct/\rho + \sqrt{(ct/\rho)^2 - 1}),\, ct \geq \rho}_{-\arccos(ct/\rho),\, -\rho < ct < \rho}\right.$, $\mathscr{H}_+$ is then given by $[\pi + iv + i0^+, -\pi - iv + i0^+]$ as $|ct| < \rho$, and $]\pi + iv(\rho, t), \pi + i0^+] \cup [\pi + i0^+, -\pi + i0^+] \cup [-\pi + i0^+, iv(\rho, t) - \pi[$ as $ct \geq \rho$, above all the real singularities of the integrand as $|\mathrm{Re}\alpha| < \pi$, and $\mathscr{H}_-$ is its inverse relatively to the origin. Considering the generality of (1.21), this expression can be used for a large class of a causal dispersive wedge-shaped region, in particular for a wedge with frequency-dependent surface impedances [6,9].

1.2 Several orders asymptotic representation for scattering by a curved impedance wedge

1.2.1 Asymptotic representation in a region without creeping waves

1.2.1.1 Maliuzhinets type representation and asymptotic boundary conditions on curved faces

We consider a wedge with tangent planes at the edge defined by $\varphi = \pm\Phi$, in cylindrical coordinates (ρ, φ), and curved faces of radius $a^{\pm}$ (Figure 1.2), with $\Phi > \pi/2$ (note: $\Phi = \pi/2$ is obtained by letting $\Phi \to \pi/2$). The electric and magnetic potentials E_z and H_z, indifferently denoted u, verify outside the wedge $(\Delta + k^2)u = 0$, with $k = \omega/c$ the wave number of the free space, $|\arg(ik)| < \pi/2$, and $e^{i\omega t}$ time dependence. As in the previous section, we consider a plane wave illumination, with φ_o the angle of incidence, and the total field u is represented as an integral of Sommerfeld–Maliuzhinets (1.3) with a contour γ composed of two infinite loops in the complex plane [10–13], for $|\varphi| < \Phi$, with f to be defined. After deformation of γ to the steepest descent path SDP, we can write

$$u(\rho,\varphi) = u_i + \sum u_r^{\pm} + \sum u_s^{\pm} + \frac{\mathrm{e}^{-ik\rho}}{2\pi i}\int_{SDP} f(\alpha + \varphi)\,\mathrm{e}^{ik\rho(\cos\alpha+1)}\,d\alpha, \qquad (1.23)$$

where the term u_i is the incident field in the illuminated zone and zero in the shadow zone; the terms $u_r^{\pm}$ correspond to geometrical optics field reflected by each faces $\varphi = \pm\Phi$ (note: some secular terms are present in the asymptotic development of this term at high order); the terms $u_s^{\pm}$ are terms of guided waves excited by the edge, which are the contribution of complex pôles $\alpha_s^{\pm}$ of $f(\alpha + \varphi)$; the last term, named u_d, is principally radiated cylindrically when $\rho \to \infty$. Approximating $f(\alpha + \varphi)$ on $SDP_{\pm}$ by its value at the stationary phase points $\alpha = \pm\pi$, we obtain

$$u_d \sim \frac{-e^{-i\pi/4-ik\rho}}{\sqrt{2\pi k\rho}}[f(\pi + \varphi) - f(-\pi + \varphi)], \qquad (1.24)$$

where $F(\varphi) = f(\pi + \varphi) - f(-\pi + \varphi)$ is the diffraction (or far field) coefficient.

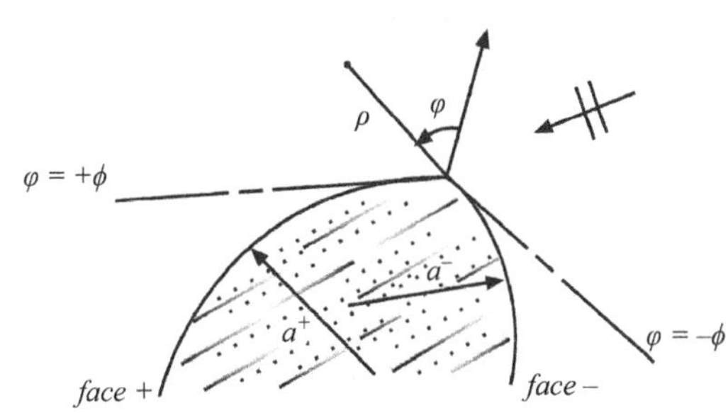

Figure 1.2 Geometry of a curved impedance wedge

We assume impedance boundary conditions on the surface of the wedge,

$$\left(\frac{\partial}{\partial n} - ik\sin\theta^{\pm}\right)u|_{\text{curved face}\pm} = 0, \tag{1.25}$$

with n the outward normal, and $\sin\theta^{\pm}$ some constants with $\mathrm{Re}\theta^{\pm} > 0$.

We approximate (1.25) by asymptotic conditions in entire powers of the curvature $1/a^{\pm}$ of each face, $\mathscr{B}^{\pm}u = 0$ on tangent plane at the edge $\varphi = \pm\Phi$, which gives [10,12,14,15]:

$$\left(\mp\frac{\partial}{\rho\partial\varphi} - ik\sin\theta^{\pm} - \frac{1}{2a^{\pm}}\left(\frac{\partial^2}{\partial\varphi^2} - \frac{\rho\partial}{\partial\rho} \pm ik\sin\theta^{\pm}\frac{\rho\partial}{\partial\varphi}\right) + O\left(\frac{\rho}{(a^{\pm})^2}\right)\right)u = 0. \tag{1.26}$$

From the Maliuzhinets inversion theorem applied to $\mathscr{B}^{\pm}u = 0$, we can write

$$B^{\pm}_{\alpha\mp\frac{\pi}{2}}f(\alpha\pm\Phi) - B^{\pm}_{-\alpha\mp(\pi/2)}f(-\alpha\pm\Phi) = 0, \tag{1.27}$$

where $B^{\pm}_{\alpha+2\pi} = B^{\pm}_{\alpha}$ (which implies $B^{\pm}_{\alpha\pm(\pi/2)}F(\alpha\pm\Phi) + B^{\pm}_{-\alpha\pm(\pi/2)}F(-\alpha\pm\Phi) = 0$), and

$$B^{\pm}_{\alpha\mp\frac{\pi}{2}} = \left(\sin\alpha \pm \sin\theta^{\pm} \pm \frac{D^{\pm}_{\alpha}(\,.\,)}{ika^{\pm}} + O\left(\frac{1}{(ka^{\pm})^2}\right)\right),$$

$$D^{\pm}_{\alpha}(.) = \frac{1}{2}\left[\frac{\partial^2(.)}{\partial\alpha^2} - \frac{\partial}{\partial\alpha}(\mathrm{cotg}\alpha(.)) \pm \sin\theta^{\pm}\frac{\partial}{\partial\alpha}\left(\frac{1}{\sin\alpha}\frac{\partial}{\partial\alpha}(.)\right)\right]. \tag{1.28}$$

We then let $f = \sum_{n\geq 0} f_n/k^n$ in the functional equation. Considering the terms of same powers in k, we then obtain a succession of functional equations of order m on the $f_{n\leq m}$, which we can solve, considering $f_{\circ}$ as the spectral function for a wedge with plane faces [13] (note: the constants $\sin\theta^{\pm}$ are generally different from the ones for plane faces).

1.2.1.2 The first term of f influenced by the curvatures

Let us detail the integral expression given initially in [10,11] for f_1, which is the first term influenced by the curvatures. For this, we let $f_{\circ}(\alpha) = \Psi(\alpha)\sigma(\alpha)/\Psi(\varphi_{\circ})$ where $\sigma(\alpha) = \mu\cos(\mu\varphi_{\circ})/(\sin(\mu\alpha) - \sin(\mu\varphi_{\circ}))$, $\mu = \pi/2\Phi$, $\varphi_{\circ}$ the angle of incidence, and $f_1(\alpha)/k = \Psi(\alpha)\chi(\alpha)/k$. Taking the initial expression of χ in [10], we can shift the integration path with arbitrary real quantities $\pm d^{\pm}$ along the real axis, for each term concerning the faces $\pm$, and, considering poles captured during the deformation, we obtain

$$\chi(\alpha') = \frac{i}{8\Phi\Psi(\varphi_{\circ})}\sum_{\pm}\frac{1}{ia^{\pm}}\left(\int_{-i\infty\pm d^{\pm}}^{i\infty\pm d^{\pm}} d\alpha\left[(-W^{\pm}\partial_{\alpha}\ln((-\sin\alpha\pm\sin\theta^{\pm})\Psi(-\alpha\pm\Phi))\right.\right.$$

$$\left.\left. - \partial_{\alpha}\sigma(\alpha\pm\Phi))\frac{1}{\sin\alpha}\right] \mp 2\pi i\,\mathrm{sgn}d^{\pm}\sum_{0<\pm\mathrm{sgn}d^{\pm}Re\alpha'_s<|d^{\pm}|}\mathrm{Residue}[\ldots]|_{\alpha=\alpha'_s}\right), \tag{1.29}$$

where $W^{\pm} = \mathrm{tg}((\pi/4\Phi)(\alpha' \pm \Phi - \alpha))$, Ψ is defined in [13] for wedge with plane faces. We note that this expression is the analytical continuation of the one given in [10] for $\mathrm{Re}(\alpha') \in]-3\Phi + d^+, \Phi + d^+[\cap] - \Phi - d^-, 3\Phi - d^-[$, which allows (also completed by (1.27)) a definition for any α'.

1.2.2 Several orders asymptotic expressions in a region with creeping waves terms

1.2.2.1 Plane wave illumination and observation at infinity

We note, from (1.27) and (1.29), that $f_1(\pm(\pi + \Phi))$ and then the expression (1.24) of u_d is singular for $\varphi = \pm\Phi$. In fact, near the tangents of the faces $\pm\Phi$, it is preferable to consider the radiation function F as the radiation of equivalent surface currents on a cylinder of radius $a^{\pm}$, which we characterize from the previous asymptotics of F, thereafter named $F_<$. Therefore, we write [10–12], when $|\pm\varphi - \Phi| < \pi/2$, $|\pm\varphi_\circ - \Phi| > \pi$,

$$F(\varphi) \equiv \frac{-e^{ika^{\pm}\sin(\pm\varphi-\Phi)}}{i\pi ka^{\pm}} \int_{\mathscr{E}} \frac{e^{-i\nu(\pm\varphi-\Phi)}\mathscr{M}^{\pm}(\pm\nu)}{H_\nu^{(2)\prime}(ka^{\pm}) + Q^{\pm}H_\nu^{(2)}(ka^{\pm})} d\nu, \tag{1.30}$$

with $Q^{\pm} = -i\sin\theta^{\pm}$, and $\mathscr{E}$ a path from $-\infty - i\varepsilon$ to $+\infty - i\varepsilon$ with $\mathrm{Im}\nu < 0$, while noting φ and φ_0 can be exchanged. Letting $\nu = ka^{\pm}\sin\alpha'$ with $|\mathrm{Re}\alpha'| \leq \pi/2$ and $|\arg(ik)| < \pi/2$, and α' going from $-i\infty - \arg(ik) + 0^+$ to $i\infty + \arg(ik) + 0^+$ on the integration path, an asymptotic expansion of $\mathscr{M}^{\pm}$ is given by

$$\mathscr{M}^{\pm}(\pm ka^{\pm}\sin\alpha') = \frac{i\pi ka^{\pm}}{2\cos\alpha'} \mathscr{O}^{\pm}_{\pm\alpha'}(F_<(\pm(\alpha' - \pi/2 + \Phi))), \tag{1.31}$$

where $\mathscr{O}^{\pm}_{\alpha} = (2/i\pi ka^{\pm})((1 + \sum_{p\geq 1}(\mathscr{P}^{\pm}_p(\cdot)/(a^{\pm})^{p+2}))B^{\pm}_{\alpha})$, and $\mathscr{P}^{\pm}_p$ are differential operators of $\sin\alpha$ [12]. The path $\mathscr{E}$ is such that the singularities of the integrand in (1.30), due to $\mathscr{M}^{\pm}$ and to zeros of $H_\nu^{(2)\prime}(ka^{\pm}) + Q^{\pm}\, H_\nu^{(2)}(ka^{\pm})$, are respectively above and below $\mathscr{E}$. We have, letting $\alpha = \alpha' - \pi/2$ in (1.31) and using (1.28),

$$\begin{aligned}\mathscr{M}^{\pm}(\pm ka^{\pm}\cos\alpha) &= \frac{1}{\sin\alpha}\left[\frac{-i\pi ka^{\pm}}{2}\mathscr{O}^{\pm}_{\pm(\alpha+\pi/2)}(\cdot)\right]F_<(\pm(\alpha+\Phi)) \\ = \frac{1}{\sin\alpha}&\left[(\sin\alpha - \sin\theta^{\pm})\cdot(\cdot) - \frac{D^{\pm}_{\pm\alpha+\pi}(\cdot)}{ika^{\pm}} + 0\left(\frac{1}{(ka^{\pm})^2}\right)\right] F_<(\pm(\alpha+\Phi)).\end{aligned} \tag{1.32}$$

1.2.2.2 Plane wave illumination and observation at finite distance

To consider an observation point (ρ,φ) at finite distance, we can then use that

$$u(\rho,\varphi) = -\frac{1}{2\pi i}\int_{\mathscr{S}_\circ} F(\alpha \pm (\Phi - \pi/2))e^{-ik\rho\cos(\alpha\pm(\Phi-\pi/2)-\varphi)}d\alpha, \tag{1.33}$$

is valid for $0 < \pm\varphi < \Phi$ when no source in the zone $\Phi - \pi < \pm\varphi < \Phi$, with correctly chosen $\mathscr{S}_\circ$ from $-\arg(ik) - i\infty$ to $\arg(ik) + i\infty$. We then use (1.30) in (1.33) with $\mathscr{L}_<^\pm \equiv (\mathscr{M}^\pm/i\pi ka^\pm)$, following

$$u(\rho,\varphi) = \frac{1}{2\pi i}\int_{\mathscr{E}} d\nu \frac{\mathscr{L}_<^\pm(\pm\nu)}{H_\nu^{(2)'}(ka^\pm) + Q^\pm H_\nu^{(2)}(ka^\pm)} \times \int_{\mathscr{S}_\circ} d\alpha e^{-ik\rho\cos(\alpha\pm(\Phi-\pi/2)-\varphi)-i\nu(\pm\alpha-\pi/2)-ika^\pm\cos\alpha}$$

$$= \int_{\mathscr{E}} d\nu \frac{\mathscr{L}_<^\pm(\pm\nu)}{H_\nu^{(2)'}(ka^\pm) + Q^\pm H_\nu^{(2)}(ka^\pm)} \times \frac{e^{\mp i\nu\Omega}}{2i}\left(\frac{e^{i\nu\pi/2}}{\pi}\int_{\mathscr{S}_\circ} e^{-ikR\cos(\alpha-\Omega)-i\nu(\pm(\alpha-\Omega))}d\alpha\right), \tag{1.34}$$

and we obtain

$$u(\rho,\varphi) = \frac{1}{2i}\int_{\mathscr{E}} \frac{\mathscr{L}_<^\pm(\pm\nu)H_\nu^{(2)}(kR)\, e^{\mp i\nu\Omega}}{H_\nu^{(2)'}(ka^\pm) + Q^\pm H_\nu^{(2)}(ka^\pm)} d\nu, \tag{1.35}$$

where relatively to the curvature center of each face, Ω is the angle from the edge to the point (ρ,φ), and $R = \sqrt{\rho^2 + (a^\pm)^2 + 2\rho a^\pm\cos(\Phi - \pi/2 \mp \varphi)}$ is the distance to (ρ,φ). The domain of validity of (1.35) is $|\pm\varphi - \Phi| < \pi/2$.

When we go beyond the frontier $\varphi = \pm\Phi$, there is no more stationary phase point for this integral, and considering the residues terms at zeros $\nu_p^\pm = ka^\pm\cos\alpha_p^\pm$ of $[H_\nu^{(2)'}(ka^\pm) + Q^\pm H_\nu^{(2)}(ka^\pm)]$ in the zone $\mathrm{Im}\nu < 0$, then closing the contour $\mathscr{E}$ at infinity, we obtain, when $\pm\varphi > \Phi$,

$$u(\rho,\varphi) = \sum_p \frac{-\pi\mathscr{L}_<^\pm(\pm\nu_p^\pm)H_{\nu_p^\pm}^{(2)}(kR)e^{\mp i\nu_p^\pm\Omega}}{\frac{\partial}{\partial\nu}(H_\nu^{(2)'}(ka^\pm) + Q^\pm H_\nu^{(2)}(ka^\pm))|_{\nu_p^\pm}}. \tag{1.36}$$

The expressions of F and u are then uniform in the region $\varphi \sim \pm\Phi$. It permits the continuous transformation of the field into creeping waves (a development in entire powers then changes continuously to one in fractional powers as φ varies), for any wedge angle. This result has been taken up and confirmed for first terms by Borovikov in his paper on impedance wedge with curved sides [14].

We remark that, contrary to expressions of Albertsen and Christiansen [16], the operators go beyond first order. We can then consider a second-order development of f without making diverge the expression (1.35), and we can study, for example, at low contrasts, the case of a curved plane with a discontinuity of material "and" of curvature.

Remark: Noticing that the diffraction coefficient $F_<$ for a cylinder $\mathscr{C}_{a^\pm,\infty}$ of large radius $a^\pm$ without discontinuity is null, we can analytically continue the solution for $\Phi - \pi < \pm\varphi_\circ < \Phi$, by adding the exact contribution of $\mathscr{C}_{a^\pm,\infty}$ in (1.36).

Remark: From (1.27), $\mathscr{O}^\pm_{\alpha\pm(\pi/2)}F_<(\alpha \pm \Phi) + \mathscr{O}^\pm_{-\alpha\pm(\pi/2)}F_<(-\alpha \pm \Phi) = 0$ [12], and $\mathscr{M}^\pm(\pm ka^\pm \cos\alpha)$ has no singularity at $\alpha = 0$.

Remark: For $\nu \approx ka^\pm$, $H_\nu^{(2)'}(ka^\pm) + Q^\pm H_\nu^{(2)}(ka^\pm) \sim (i/((m^\pm)^2\sqrt{\pi}))\ (w_2'(\tau) - q^\pm w_2(\tau))$, where $m^\pm = (ka^\pm/2)^{1/3}, q^\pm = -im^\pm \sin\theta^\pm$, $\nu = ka^\pm + m^\pm\tau$, $w_2 = \sqrt{\pi}(Bi - iAi)$.

1.2.2.3 Point source illumination and observation at finite distance

We can then derive an expression at finite distance for observation point (ρ', φ') and point source (ρ, φ) radiating $u^i = (i/2)H_\circ^{(2)}(k\sqrt{\rho^2 + (\rho')^2 + 2\rho\rho'\cos(\varphi - \varphi')})$. For this, we consider reciprocity, exchange the directions of observation and of incidence in $F_<$, let $\mathscr{L}^\pm(\pm ka^\pm \cos\alpha') = \mathscr{A}_{\alpha'}((1/2\pi i)\int_\gamma f(\alpha + \varphi, \pm(\alpha' + \Phi))e^{ik\rho\cos\alpha}d\alpha)$, with $\mathscr{A}_\alpha = (-\mathscr{O}^\pm_{\pm(\alpha+\pi/2)}/2\sin\alpha)$, and obtain, when $|\pm\varphi - \Phi| > \pi$ and $\varphi' \sim \pm\Phi$,

$$u(\rho', \varphi', \rho, \varphi) = \frac{1}{2i}\int_{\mathscr{E}} \frac{\mathscr{L}^\pm(\pm\nu')\, H_{\nu'}^{(2)}(kR')\, e^{\mp i\nu'\Omega'}}{H_{\nu'}^{(2)'}(ka^\pm) + Q^\pm H_{\nu'}^{(2)}(ka^\pm)} d\nu', \tag{1.37}$$

where $R' = \sqrt{\rho'^2 + (a^\pm)^2 + 2\rho' a^\pm \cos(\Phi - \pi/2 \mp \varphi')}$, and Ω' is the angle, relatively to the cylinder center, from the edge to the observation point (ρ', φ'), while noting reciprocity can apply.

When we go beyond the frontier $\varphi' = \pm\Phi$, there is no more stationary phase point for this integral, and considering the residues terms at zeros $\nu_p^\pm$ of $[H_\nu^{(2)'}(ka^\pm) + Q^\pm H_\nu^{(2)}(ka^\pm)]$ in the zone $\mathrm{Im}\nu < 0$, we obtain, when $\pm\varphi' > \Phi$,

$$u(\rho', \varphi', \rho, \varphi) = \sum_p \frac{-\pi\mathscr{L}^\pm(\pm\nu_p^\pm)H_{\nu_p^\pm}^{(2)}(kR')e^{\mp i\nu_p^\pm\Omega'}}{\frac{\partial}{\partial\nu}(H_\nu^{(2)'}(ka^\pm) + Q^\pm H_\nu^{(2)}(ka^\pm))|_{\nu_p^\pm}}, \tag{1.38}$$

where $\mathscr{L}^\pm(\pm ka^\pm \cos\alpha_p) = \mathscr{A}_{\alpha_p}((1/2\pi i)\int_\gamma f(\alpha + \varphi, \pm(\alpha_p + \Phi))e^{ik\rho\cos\alpha}d\alpha)$ can be developed as $\mathscr{A}_{\alpha_p}(u_{i,p} + \sum u^\pm_{r,p} + \sum u^\pm_{s,\varepsilon,p} + (\mathrm{e}^{-ik\rho}/2\pi i)\ \int_{SDP} f(\alpha + \varphi, \pm(\alpha_p + \Phi))$ $\mathrm{e}^{ik\rho(\cos\alpha+1)}\, d\alpha)$.

For $l\varphi > \Phi$ et $l'\varphi' > \Phi$ (l,l' are equal to + or −1), the problem is more complex and we need to express $\mathscr{L}^\pm(\pm\nu')$ as an integral of the previous form. We then obtain for the diffracted part of the field:

$$\begin{aligned} u_{da} = \pi^2 \sum_{p,q} \mathscr{A}^l_{\alpha^l_p}\mathscr{A}^{l'}_{\alpha^{l'}_q}(f(\pi + l(\alpha^l_p + \Phi), l'(\alpha^{l'}_q + \Phi)) \\ - f(-\pi + l(\alpha^l_p + \Phi), l'(\alpha^{l'}_q + \Phi)))\frac{H_{\nu^l_p}^{(2)}(kR)\, e^{-li\nu^l_p\Omega}}{\frac{\partial}{\partial\nu}(H_\nu^{(2)'}(ka^l) + Q^l H_\nu^{(2)}(ka^l))|_{\nu=\nu^l_p}} \\ \times \frac{H_{\nu^{l'}_q}^{(2)}(kR')e^{-l'i\nu^{l'}_q\Omega'}}{\frac{\partial}{\partial\nu}(H_\nu^{(2)'}(ka^{l'}) + Q^{l'} H_\nu^{(2)}(ka^{l'}))|_{\nu=\nu^{l'}_q}}. \end{aligned} \tag{1.39}$$

Let us notice that this expression has a form similar to the one defined by Thompson for the particular case of a discontinuity of materials on a circular cylinder.

Remark: If we denote $f_{a^\pm,\infty}$ the asymptotic spectral function for a cylinder $\mathscr{C}_{a^\pm,\infty}$ of radius $a^\pm$ without discontinuity, we can analytically continue (1.37) and (1.38) for $|\pm\varphi-\Phi|<\pi$ by letting $f(\alpha)-f_{a^\pm,\infty}(\alpha\mp(\Phi-(\pi/2)))$ in place of $f(\alpha)$ and add the exact contribution of $\mathscr{C}_{a^\pm,\infty}$ in (1.37) and (1.38).

1.2.3 Some validations concerning the expression of $f=\sum_{n\geq 0} f_n/k^n$ for arbitrary wedge angle

1.2.3.1 Perfectly conducting case: curved half plane to discontinuity of curvature

We now compare the results obtained from (1.29) with the expressions known for some particular wedge angles, with $\Phi\geq\pi/2$, in the limit case $\sin\theta^\pm\to\infty$ so that $\Psi=1$. Letting $\varepsilon=\mathrm{sgn}(\mathrm{Re}\alpha')$, we choose to take $d^+=(1+\varepsilon)\Phi$ and $d^-=(1-\varepsilon)\Phi$. Denoting $a^\delta=a^+$ or a^- when $\delta=+1$ or -1, we can write

$$\chi(\alpha')=\left(\frac{1}{8\Phi}\partial_{\alpha'}\partial_{\varphi_\circ}\int_{-i\infty}^{i\infty}d\alpha\left[\mathrm{tg}\left(\frac{\pi}{4\Phi}(\alpha'-\varepsilon\Phi-\alpha)\right)\frac{\frac{\pi}{2\Phi}\sin\frac{\pi}{2\Phi}\alpha}{\left(\cos\frac{\pi}{2\Phi}\alpha+\varepsilon\sin\frac{\pi}{2\Phi}\varphi_\circ\right)}\right.\right.$$
$$\left.\left.\times\left(\frac{1}{a^{-\varepsilon}\sin\alpha}+\frac{1}{a^{\varepsilon}\sin(\alpha+2\varepsilon\Phi)}\right)\right]\right)+\left(i\frac{\varepsilon}{a^\varepsilon}\frac{\pi}{4\Phi}\partial_{\alpha'}\partial_{\varphi_\circ}\left[\frac{1}{\sin(\varepsilon\Phi-\varphi_\circ)}\right.\right.$$
$$\left.\left.\times\,\mathrm{tg}\left(\frac{\pi}{4\Phi}(\alpha'+\varphi_\circ)\right)+\mathrm{tg}\left(\frac{\pi}{4\Phi}(\alpha'+\varepsilon(\Phi-\pi))\right)\frac{\frac{\varepsilon\pi}{2\Phi}\sin\frac{\pi}{2\Phi}\pi}{\left(\cos\frac{\pi}{2\Phi}\pi-\varepsilon\sin\frac{\pi}{2\Phi}\varphi_\circ\right)}\right]\right),\tag{1.40}$$

for $\mathrm{Re}(\alpha')\in\,]-3\Phi,3\Phi[$ with $\varepsilon=\mathrm{sgn}(\mathrm{Re}\alpha')$.

In the case of a curved half-plane, $\Phi=\pi$ and $\varepsilon a^\varepsilon=a$, and so we obtain

$$\chi(\alpha')=(0)+i\frac{1}{4a}\partial_{\alpha'}\partial_{\varphi_\circ}\left[\frac{1}{\sin(\varphi_\circ)}\mathrm{tg}\left(\frac{1}{4}(\alpha'+\varphi_\circ)\right)-\frac{1}{2}\mathrm{tg}\left(\frac{1}{4}\alpha'\right)\frac{1}{\sin(\frac{1}{2}\varphi_\circ)}\right].\tag{1.41}$$

This gives us the second-order term for far field function

$$\frac{(f_1(\pi+\varphi)-f_1(-\pi+\varphi))}{k}=\frac{i}{4ka}\partial_\varphi\partial_{\varphi_\circ}\left(\frac{\mathrm{tg}((\varphi+\varphi_\circ)/2)}{\cos(\varphi/2)\cos(\varphi_\circ/2)}\right),\tag{1.42}$$

which is conform to the result of Borovikov [17], first developed by V.B. Filippov.

For the discontinuity of curvature, $\Phi = \frac{\pi}{2}$ and then the integral term for χ is

$$\frac{1}{4\pi}\partial_{\alpha'}\partial_{\varphi_\circ}\int_{-i\infty}^{i\infty} d\alpha\left[\frac{\cos\alpha'}{(\cos\alpha+\varepsilon\sin\alpha')}\frac{1}{(\cos\alpha+\varepsilon\sin\varphi_\circ)}\left(\frac{1}{a^+}-\frac{1}{a^-}\right)\right]$$
$$= \frac{i}{2\pi}\partial_{\alpha'}\partial_{\varphi_\circ}\frac{\varepsilon\cos\alpha'}{(\sin\varphi_\circ-\sin\alpha')}\left(\frac{\pi/2-\varepsilon\alpha'}{\cos\alpha'}-\frac{\pi/2-\varepsilon\varphi_\circ}{\cos\varphi_\circ}\right)\left(\frac{1}{a^+}-\frac{1}{a^-}\right), \tag{1.43}$$

for $-\pi/2 < \varepsilon\mathrm{Re}\alpha' < 3\pi/2$. The expression for χ then gives us

$$\frac{(f_1(\pi+\varphi)-f_1(-\pi+\varphi))}{k} = -2i\frac{\cos\varphi\cos\varphi_\circ}{(\sin\varphi+\sin\varphi_\circ)^3}\left(\frac{1}{ka^+}-\frac{1}{ka^-}\right), \tag{1.44}$$

which is conform to the diffraction coefficient of Kaminetzky and Keller [15].

1.2.3.2 Discontinuity of curvature in an impedance surface

In this case, we have $\Phi = \pi/2$ et $\sin\theta^\pm = \sin\theta$, and $F(\varphi) = (f_1(\pi+\varphi) - f_1(-\pi+\varphi))/k$. Instead of directly using (1.29), we prefer here to rewrite the functional equation in f_1, letting $\alpha = \pm\pi/2+\varphi$,

$$f_1(\pm\pi+\varphi)\frac{\cos\varphi+\sin\theta}{-\cos\varphi+\sin\theta} - f_1(-\varphi)$$
$$= \frac{D^\pm_{\pm\pi/2+\varphi}(f_\circ(\pm\pi+\varphi)) - D^\pm_{\mp\pi/2-\varphi}(f_\circ(-\varphi))}{ia^\pm(\cos\varphi-\sin\theta)}. \tag{1.45}$$

From $\sin\theta^\pm = \sin\theta$ and $f_\circ(\pi+\varphi) = f_\circ(-\pi+\varphi)$, we have

$$D^+_{\pi/2+\varphi}(f_\circ(\pi+\varphi)) - D^+_{-\pi/2-\varphi}(f_\circ(-\varphi))$$
$$= D^-_{-\pi/2+\varphi}(f_\circ(\pi+\varphi)) - D^-_{\pi/2-\varphi}(f_\circ(-\varphi)) \tag{1.46}$$

and thus,

$$f_1(\pi+\varphi) - f_1(-\pi+\varphi) = i\left(\frac{1}{a^+}-\frac{1}{a^-}\right)$$
$$\times\frac{D^+_{\pi/2+\varphi}(f_\circ(\pi+\varphi)) - D^+_{-\pi/2-\varphi}(f_\circ(-\varphi))}{(\cos\varphi+\sin\theta)}. \tag{1.47}$$

Using $f_\circ(\alpha) = ((\cos\alpha+\sin\theta)/(\cos\varphi_\circ+\sin\theta))\ (\cos\varphi_\circ/(\sin\alpha-\sin\varphi_\circ))$, and expanding D, we obtain

$$f_1(\pi+\varphi) - f_1(-\pi+\varphi) = \left(\frac{1}{a^+}-\frac{1}{a^-}\right)$$
$$\times\frac{2i\cos\varphi\cos\varphi_\circ(1+\sin\varphi\sin\varphi_\circ-\sin^2\theta)}{(\cos\varphi_\circ+\sin\theta)(\cos\varphi+\sin\theta)(\sin\varphi+\sin\varphi_\circ)^3}, \tag{1.48}$$

which is perfectly in agreement with the result of Kaminetzky and Keller [15].

1.3 A novel expression of the field for arbitrary bounded sources above a passive or active impedance plane

1.3.1 Formulation of the problem

We consider the scattering by an imperfectly reflective plane when it is illuminated by the field radiated by a bounded source, composed of arbitrary electrical and magnetic currents J and M (see Figure 1.3). The plane is defined by $z = 0$ in Cartesian coordinates (x, y, z). A harmonic time dependence $e^{i\omega t}$, from now on assumed, is suppressed throughout. Each component of the scattered field is assumed to be regular in the domain $z > 0$, and $O(e^{-\gamma|OP|})$ at $P(x, y, z)$, $\gamma > 0$, as z or $\rho \to \infty$, when $|\arg(ik)| < \pi/2$.

Following Harrington [18, p.131] (see also Jones [2]), we can write the electric field E and the magnetic field H satisfying the Maxwell equations, with two scalar potentials $\mathscr{E}$ and $\mathscr{H}$, following

$$E = -ik\text{curl}(\mathscr{H}\widehat{z}) + (\text{grad}(\text{div}(.)) + k^2)(\mathscr{E}\widehat{z})$$

$$\sqrt{\frac{\mu_0}{\varepsilon_0}}H = ik\text{curl}(\mathscr{E}\widehat{z}) + (\text{grad}(\text{div}(.)) + k^2)(\mathscr{H}\widehat{z}), \tag{1.49}$$

where $(\Delta + k^2)\mathscr{E} = 0$ and $(\Delta + k^2)\mathscr{H} = 0$ outside the sources of radiation (i.e., outside J, M, and the scatterer), with $k = \omega\sqrt{\mu_0\varepsilon_0}$, the constants ε_0 and μ_0 being, respectively, the permittivity and the permeability of the medium above the plane, $|\arg(ik)| \leq \pi/2$. Following the theory of this representation, the constant vector $\widehat{z}$ can be chosen regardless of the sources, and $\mathscr{E}$ (or $\mathscr{H}$)$\equiv e^{\pm ikz}$ has no influence on (E, H). Thereafter, we denote $(\mathscr{E}_{inc}, \mathscr{H}_{inc})$ and $(\mathscr{E}_s, \mathscr{H}_s)$ the potentials corresponding to the incident field (incoming wave) and the scattered field (outgoing wave) and write (1.49) in the compact form $(E, \sqrt{(\mu_0/\varepsilon_0)}H) = \mathscr{L}(\widehat{z}\mathscr{E}, \widehat{z}\mathscr{H})$.

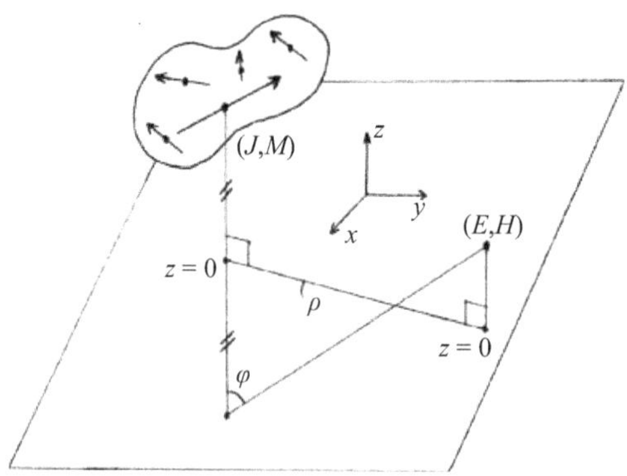

Figure 1.3 Geometry: sources (J, M) and observation point above the plane $z = 0$

In [19], we considered multimode boundary conditions on an isotropic plane,

$$\prod_{j=1}^{N}\Big(\frac{\partial}{\partial z}-ikg_j^e\Big)E_{z,tot}|_{z=0}=0,\quad \prod_{j=1}^{P}\Big(\frac{\partial}{\partial z}-ikg_j^h\Big)H_{z,tot}|_{z=0}=0, \tag{1.50}$$

which corresponds to the reflection coefficients of a plane wave for the principal polarizations *TM* (components of *E* in the plane of incidence) and *TE* (components of *H* in the plane of incidence),

$$R_{TM}(\beta)=\prod_{j=1}^{N}\frac{\cos\beta-g_j^e}{\cos\beta+g_j^e},\quad R_{TE}(\beta)=\prod_{j=1}^{P}\frac{\cos\beta-g_j^h}{\cos\beta+g_j^h}, \tag{1.51}$$

where β is the angle of incidence with the normal $\hat{z}$ [19], $g_j^{(e,h)}$ are complex constants, N and P are two positive numbers. This class of problem corresponds to the reflection by a multilayer [19,21,22], composed of isotropic media, or more generally, of uniaxial anisotropic media with the principal axis along z, backed with a perfectly reflective plane. From the symmetry at normal incidence, we notice that the condition $R_{TE}(0)=-R_{TM}(0)$ has also to be satisfied. This implies that $g_1^e=1/g_1^h$ for monomode conditions, when $N=P=1$.

We now restrict ourselves to this latter case, commonly named the impedance case, and take an arbitrary complex number $g_1^e=g^e$, corresponding to a passive ($\mathrm{Re}g^e>0$) or active ($\mathrm{Re}g^e<0$) plane. Using

$$E_z=\frac{\partial^2\mathscr{E}}{\partial z^2}+k^2\mathscr{E},\quad \sqrt{\frac{\mu_0}{\varepsilon_0}}H_z=\frac{\partial^2\mathscr{H}}{\partial z^2}+k^2\mathscr{H}, \tag{1.52}$$

we will search the potentials $\mathscr{E}_s$ and $\mathscr{H}_s$, satisfying the Helmholtz equation as $z>0$, regular and vanishing as $z\to\infty$ when $|\arg(ik)|<\pi/2$, which verify

$$\begin{aligned}\left(\frac{\partial}{\partial z}-ikg^e\right)\mathscr{E}_s(z)&=\left(\frac{\partial}{\partial z}+ikg^e\right)\mathscr{E}_{inc}(-z),\\ \left(\frac{\partial}{\partial z}-ikg^h\right)\mathscr{H}_s(z)&=\left(\frac{\partial}{\partial z}+ikg^h\right)\mathscr{H}_{inc}(-z)\end{aligned} \tag{1.53}$$

as $z\geq 0$, with $g^h=1/g^e$. Therefore, we need first a definition of $(\mathscr{E}_{inc},\mathscr{H}_{inc})$ as $z\leq 0$.

Remark: The boundary conditions on normal components can be simply deduced from boundary conditions on the tangential components after differentiation of them.

1.3.2 An expression of potentials ($\mathscr{E}_{inc}, \mathscr{H}_{inc}$) for bounded sources J and M

Let us consider the incident field (E, H) at r of coordinates (x, y, z), radiated by arbitrary electric and magnetic bounded sources J and M [2],

$$E = \text{curl}(G * M) + \frac{i}{\omega\varepsilon_0}(\text{grad}(\text{div}(.)) + k^2)(G * J)$$

$$\sqrt{\frac{\mu_0}{\varepsilon_0}}H = -\sqrt{\frac{\mu_0}{\varepsilon_0}}\text{curl}(G * J) + \frac{i}{k}(\text{grad}(\text{div}(.)) + k^2)(G * M) \tag{1.54}$$

where $G(r) = -(e^{-ik|r(x,y,z)|}/4\pi\,|r(x, y, z)|)$ with $|r| = \sqrt{x^2 + y^2 + z^2}$, and $*$ is the convolution product. The potentials ($\mathscr{E}_{inc}, \mathscr{H}_{inc}$) for this field, satisfying the Helmholtz equation outside the sources, and vanishing at infinity when $|\arg(ik)| < \pi/2$ as $\pm z \to \infty$, have a particularly compact expression, which we develop in [19] for arbitrary sources. It is given by

$$(\mathscr{E}_{inc}, \mathscr{H}_{inc}) = \frac{\widehat{z}}{8\pi k^2}(\sqrt{\frac{\mu_0}{\varepsilon_0}}(\text{grad}(\text{div}(J)) + k^2 J, ik\,\text{curl}(J))$$

$$+ (-ik\;\text{curl}(M), \text{grad}(\text{div}(M)) + k^2 M)) * \mathscr{W} = \frac{\widehat{z}}{8\pi k^2}\mathscr{L}\left(\sqrt{\frac{\mu_0}{\varepsilon_0}}J, M\right) * \mathscr{W}, \tag{1.55}$$

where we can take, to solve (1.53) as $z > 0$,

$$\mathscr{W}(r) = (e^{ik|z|}E_1(ik(|r|+|z|)) + e^{-ik|z|}(E_1(ik(|r|-|z|)) + 2\ln\rho)) \tag{1.56}$$

with $\rho = \sqrt{x^2 + y^2}$, E_1 being the exponential integral function [20], and the notation $(A, B) * C \equiv (A * C, B * C)$. The reader can verify by inspection our expression, considering that the following conditions are satisfied when $\pm z > 0$,

$$\left(\frac{\partial^2}{\partial z^2} + k^2\right)\frac{\mathscr{W}(r)}{8\pi ik} = G(r), (\Delta + k^2)\mathscr{W}(r) = 0 \tag{1.57}$$

and that all derivatives of $\mathscr{W}$ are regular in these domains.

Remark: It is worth noticing that we have $\Delta_{xy}\ln(\rho) = 0$ for $\rho \neq 0$, which implies that $\ln\rho$ in (1.56) has no influence on the field for $z \neq 0$, except by its singularity at $\rho = 0$. Another way to write the expression would be to suppress a vertical tube from the sources centered on the observation point, with a radius that we let tend to 0.

1.3.3 Expression of the potentials ($\mathscr{E}_s, \mathscr{H}_s$) for an impedance plane

Using our expression of ($\mathscr{E}_{inc}, \mathscr{H}_{inc}$) for the radiation of J and M, we can express the potentials $\mathscr{E}_s$ and $\mathscr{H}_s$ which satisfy the impedance boundary conditions (1.53), from the method developed in [19]. So, letting $N = P = 1$ in [19, prop. (5.2)], we obtain, as $z \geq 0$,

$$\begin{aligned}
&\mathscr{E}_s(x,y,z)\\
&= \mathscr{E}_{inc}(x,y,-z) + \Big(\Big(\frac{\widehat{z}}{\omega\varepsilon_0}\frac{\mathrm{grad}(\mathrm{div}(J)) + k^2 J}{8\pi k} + \frac{\widehat{z}}{k}\frac{(-ik\,\mathrm{curl}(M))}{8\pi k}\Big)\\
&\times \sum_{\varepsilon'=-1,1} \frac{a^e}{(g^e - \varepsilon')}(\mathscr{V}_{\varepsilon'} + \varepsilon' \mathscr{K}_{g^e})\Big)(x,y,-z)\\
&= \mathscr{E}_{inc}(x,y,-z) + \Big(\Big(\frac{\widehat{z}}{\omega\varepsilon_0}\frac{\mathrm{grad}(\mathrm{div}(J)) + k^2 J}{8\pi k} + \frac{\widehat{z}}{k}\frac{(-ik\,\mathrm{curl}(M))}{8\pi k}\Big)\\
&\times \sum_{\varepsilon'=-1,1} \Big(\Big(\frac{\varepsilon' + g^e}{\varepsilon' - g^e} - 1\Big)\mathscr{V}_{\varepsilon'} + \frac{\varepsilon' a^e \mathscr{K}_{g^e}}{(g^e - \varepsilon')}\Big)\Big)(x,y,-z) \qquad (1.58)
\end{aligned}$$

and

$$\begin{aligned}
&\mathscr{H}_s(x,y,z)\\
&= \mathscr{H}_{inc}(x,y,-z) + \Big(\Big(\frac{\widehat{z}}{\omega\varepsilon_0}\frac{(ik\,\mathrm{curl}(J))}{8\pi k} + \frac{\widehat{z}}{k}\frac{(\mathrm{grad}(\mathrm{div}(M)) + k^2 M)}{8\pi k}\Big)\\
&\times \sum_{\varepsilon'=-1,1} \frac{a^h}{(g^h - \varepsilon')}(\mathscr{V}_{\varepsilon'} + \varepsilon' \mathscr{K}_{g^h})\Big)(x,y,-z)\\
&= -\ \mathscr{H}_{inc}(x,y,-z) + \Big(\Big(\frac{\widehat{z}}{\omega\varepsilon_0}\frac{(ik\,\mathrm{curl}(J))}{8\pi k} + \frac{\widehat{z}}{k}\frac{(\mathrm{grad}(\mathrm{div}(M)) + k^2 M)}{8\pi k}\Big)\\
&\times \sum_{\varepsilon'=-1,1} \Big(\Big(\frac{\varepsilon' + g^h}{\varepsilon' - g^h} + 1\Big)\mathscr{V}_{\varepsilon'} + \frac{\varepsilon' a^h \mathscr{K}_{g^h}}{(g^h - \varepsilon')}\Big)\Big)(x,y,-z) \qquad (1.59)
\end{aligned}$$

where $g^e = 1/g^h$, $a^{e,h} = -2g^{e,h}$. In these expressions, the functions $\mathscr{V}_{\varepsilon'}$, $\mathscr{K}_g$ satisfy

$$\begin{aligned}
&\mathscr{V}_{\varepsilon'}(x,y,-z) = e^{\varepsilon' ikz}(E_1(ik(|r| + \varepsilon' z)) + (1 - \varepsilon')\ln\rho),\\
&\mathscr{K}_g(x,y,-z) = e^{ikgz}\,\mathscr{J}_g(\rho,-z) \qquad (1.60)
\end{aligned}$$

where $\rho = \sqrt{x^2 + y^2}$, $g = g^e$ or $g = g^h$, and $\mathscr{J}_g(\rho,-z)$ is given by the integral

$$\mathscr{J}_g(\rho,-z) = \frac{e^{-ikgz}}{2}\int_{\mathscr{D}} \frac{H_0^{(2)}(k\rho\sin\beta)e^{-ikz\cos\beta}}{\cos\beta + g}\sin\beta d\beta, \qquad (1.61)$$

with $\mathrm{Re}(ik\sin\beta) = 0$ on $\mathscr{D}$ from $-i\infty - \arg(ik)$ to $i\infty + \arg(ik)$, which is a Fourier–Bessel integral commonly encountered in scattering theory [21, p. 234], also called a Sommerfeld-type integral. Letting $g = \sin\theta_1$ with $|\mathrm{Re}(\theta_1)| \leq \pi/2$, we notice the presence of a cut with $\mathrm{Re}(ik\cos\theta_1) = 0$ in active case ($\mathrm{Re}g < 0$), which is due to poles of $(\cos\beta + g)^{-1}$ that can go through $\mathscr{D}$. Thus, as discussed in [19], any transformations that do not consider the presence of this branch cut can be wrong in active case.

To avoid any restriction, we have developed in [19,26,27], a novel expression for arbitrary $g = \sin\theta_1$ with $|\text{Re}(\theta_1)| \leq \pi/2$ as $|\arg(ik)| \leq \pi/2$,

$$\mathscr{I}_g(\rho, -z) = i \int_b^{i\infty} e^{-a\cos\alpha} d\alpha \tag{1.62}$$

where the parameters a and b, with $|Reb| \leq \pi$, $\text{Re}a > 0$, are defined following

$$\begin{aligned} e^{\mp ib} &= \frac{ikR}{a}(1 \pm \sin\theta_1)(1 \pm \cos\varphi),\ a = \varepsilon ikR \sin\varphi\cos\theta_1, \\ -ia\sin b &= ikR(\cos\varphi + \sin\theta_1), a\cos b = ikR(1 + \sin\theta_1\cos\varphi) \end{aligned} \tag{1.63}$$

and $z = R\cos\varphi$, $\rho = R\sin\varphi$, $R = \sqrt{\rho^2 + z^2}$, $\varepsilon = \text{sgn}(\text{Re}(ikR\sin\varphi\cos\theta_1))$ ($\text{Re}(a) = 0$ being a limit case), $0 < \varphi < \pi/2$. So defined, we have $\text{sgn}(\text{Re}b) = -\varepsilon\,\text{sgn}(\text{Im}(\sin\theta_1))$, and $\text{sgn}(\text{Im}a) = -\text{sgn}(\arg(ik))$ when $\varepsilon = -1$.

We notice that, as g varies in the complex plane, this expression has a correct cut as ε changes of sign for $\text{Re}g \leq 0$, is singular for $g = -1$, and is regular elsewhere (note: for $\text{Re}g > 0$, the change of sign of ε does not induce a cut as g varies).

Let us remark that this integral was also given in [23] for passive impedance case but it was with a definition of parameters that restricts its application (see details in Section 2 of [19]). It is intimately related to the incomplete cylindrical function in the Poisson form [24] and to the leaky aquifer function [25]. The reader can refer to [19,26,27] for its calculus in passive or active impedance cases.

Remark: The reader can verify by inspection that, when $z > 0$,

$$\frac{\partial \mathscr{I}_g(\rho, -z)}{\partial z} = \frac{e^{-ik(R+gz)}}{R},\ (\Delta + k^2)(e^{ikgz}\mathscr{I}_g(\rho, -z)) = 0. \tag{1.64}$$

Remark: Let us notice that the solution for an impedance plane in acoustics given by Ingard [28] presents, for some values of g with $\text{Re}(g) > 0$ (passive), nonphysical discontinuities as φ varies (due to errors in surface waves contributions reported by Wenzel [29]), which Thomasson [30] has corrected.

1.4 Spectral representation of the field for 3D conical scatterers

We describe here some new aspects of integral representations in problems with conical boundary conditions, encountered in electromagnetic scattering theory. We carefully analyze some new points in the general expression of Debye potentials in electromagnetism and give a novel efficient representation of incident plane waves for solving the problems of scattering with conical geometries.

1.4.1 Formulation

We consider the diffraction of an electromagnetic plane wave by an imperfectly conducting cone C, in the domain Ω of the 3D Euclidean space R^3 bounded by C.

A dependence on time $e^{i\eta t}$ is assumed and suppressed throughout. The Maxwell equations for the scattered wave field are given in Ω by

$$ik\vec{E} = \text{rot}(Z_0\vec{H}),\ ikZ_0\vec{H} = -\text{rot}\vec{E}, \tag{1.65}$$

where $k = \eta\sqrt{\varepsilon_0\mu_0}$ and Z_0 are, respectively, the wave number and the impedance of the free space, $\vec{E}$ and $\vec{H}$ are, respectively, the electric and the magnetic field, and k is complex with $|\arg(ik)| \leq \pi/2$. We introduce the representation of fields outside any sources with Debye potentials $(\mathscr{E}, \mathscr{H})$ [2],

$$\vec{E} = \text{rot rot}(\vec{r}\mathscr{E}) - ik\text{rot}(\vec{r}\mathscr{H}),\ Z_0\vec{H} = \text{rot rot}(\vec{r}\mathscr{H}) + ik\text{rot}(\vec{r}\mathscr{E}), \tag{1.66}$$

where rot rot $= \nabla\text{div} - \Delta$, and the potentials $\mathscr{E}, \mathscr{H}$ satisfy the Helmholtz equations. We then have the expressions

$$\begin{aligned}\vec{E} &= \vec{e}_r(\partial_r^2(r\mathscr{E}) + k^2 r\mathscr{E}) + r^{-1}\nabla_\omega\partial_r(r\mathscr{E}) - ik\nabla_\omega\mathscr{H}\wedge\vec{e}_r,\\ Z_0\vec{H} &= \vec{e}_r(\partial_r^2(r\mathscr{H}) + k^2 r\mathscr{H}) + r^{-1}\nabla_\omega\partial_r(r\mathscr{H}) + ik\nabla_\omega\mathscr{E}\wedge\vec{e}_r,\end{aligned} \tag{1.67}$$

where $\nabla_\omega = \vec{e}_\theta\partial_\theta + (\vec{e}_\varphi/\sin\theta)\partial_\varphi$, (r,θ,φ) are the spherical coordinates, and $(\vec{e}_r, \vec{e}_\theta, \vec{e}_\varphi)$ are the unit vectors basis of the spherical coordinate system, which gives us,

$$\begin{aligned}&E_r = k^2 r\mathscr{E} + \frac{\partial^2}{\partial r^2}(r\mathscr{E}),\ E_\theta = \frac{1}{r}\frac{\partial^2}{\partial r\partial\theta}(r\mathscr{E}) - \frac{ik}{\sin\theta}\frac{\partial\mathscr{H}}{\partial\varphi},\\ &E_\varphi = \frac{1}{r\sin\theta}\frac{\partial}{\partial r}(r\frac{\partial\mathscr{E}}{\partial\varphi}) + ik\frac{\partial\mathscr{H}}{\partial\theta},\end{aligned} \tag{1.68}$$

and

$$\begin{aligned}&Z_\circ H_r = k^2 r\mathscr{H} + \frac{\partial^2}{\partial r^2}(r\mathscr{H}),\ Z_\circ H_\theta = \frac{1}{r}\frac{\partial^2}{\partial r\partial\theta}(r\mathscr{H}) + \frac{ik}{\sin\theta}\frac{\partial\mathscr{E}}{\partial\varphi},\\ &Z_\circ H_\varphi = \frac{1}{r\sin\theta}\frac{\partial}{\partial r}\left(r\frac{\partial\mathscr{H}}{\partial\varphi}\right) - ik\frac{\partial\mathscr{E}}{\partial\theta}.\end{aligned} \tag{1.69}$$

Let the 2D unit sphere S^2 be centered at the vertex T of the convex cone C (Figure 1.4). We denote $\sigma = S^2 \cap C$ the boundary of the domain $\Sigma \subset S^2$ which is cut out on the unit sphere by the cone C. We assume that the curve σ is smooth and that Σ is geodesically convex and belongs to a hemisphere of S^2. The vector $\vec{r} = (r, \omega)$ refers to the observation point, $r = |\vec{r}|$ is the distance from T to this point, and ω is the point of the unit sphere defined by $\vec{\omega} = \vec{r}/r$, described by two angular coordinates (θ, φ) which have the traditional meaning as in the spherical coordinate system. In what follows, $\widehat{\theta}(\omega, \omega_0) = \text{dist}(\omega, \omega_0) = \arccos <\vec{\omega}, \vec{\omega}_0>_{R^3}$, is the geodesic distance between the points ω and ω_0 on the unit sphere S^2.

The conditions at infinity and at the vertex are those given in [33, sect.2]. To exhibit the properties of scattered fields for large kr, we rather choose to consider illumination by a point source at $\vec{r}_0$ that we let tend to ∞, kr being large but fixed. Denoting $\widehat{\theta}'(\omega, \omega_0) = \inf_{\omega'\in\sigma}(\widehat{\theta}(\omega, \omega') + \widehat{\theta}(\omega', \omega_0))$ the length of the geodesics (broken for the reflected wave), we note that the reflection and shadow regions, in the sense of geometrical optics, are defined by the inequality $\widehat{\theta}'(\omega, \omega_0) \leq \pi$, while $\widehat{\theta}'(\omega, \omega_0) > \pi$ is the zone where the behavior for large $|kr|$ and $|kr_0|$ is dictated by the spherically

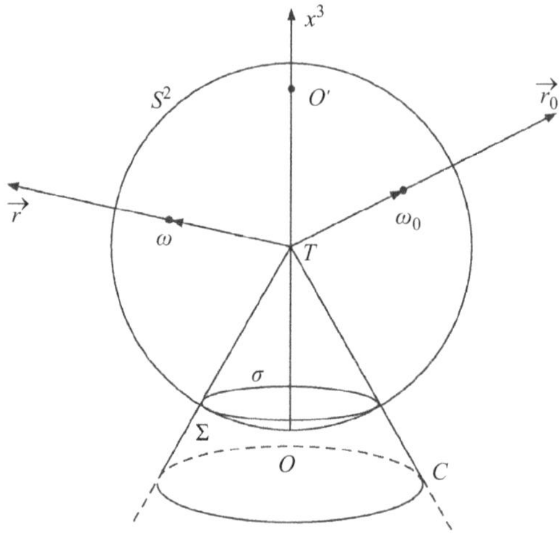

Figure 1.4 Geometry of a conical region

diffracted wave excited by the tip of the cone. We can then assume that the scattered field in Ω verifies for large kr as $r_0 \to \infty$, when $|\arg(ik)| < \pi/2$ [33],

$$ikr_0 \exp(ikr_0)(\vec{E}, Z_0\vec{H}) = O(\exp(ikr\cos(\pi/2 + \varepsilon(\omega, \omega_0)))), \tag{1.70}$$

where $\varepsilon(\omega, \omega_0) = \widehat{\theta}'(\omega, \omega_0) - \pi/2$ when $\widehat{\theta}'(\omega, \omega_0) \leq \pi$, with $\widehat{\theta}'(\omega, \omega_0) = \widehat{\theta}(\omega, \omega_*)$ in reflection zone and $\widehat{\theta}'(\omega, \omega_0) = \widehat{\theta}(\omega, \omega_0)$ in shadow zone, and $\varepsilon(\omega, \omega_0) = \pi/2$ as $\widehat{\theta}'(\omega, \omega_0) > \pi$.

1.4.2 Expression of the potentials with Kontorovich–Lebedev integrals

1.4.2.1 Integral and nonintegral terms

Considering the general conditions at the apex given in [33], the wave field is $O((kr)^m)$ with $m > -3/2$ as r tends to 0. Thus, assuming the Debye potentials and their angular derivatives are $O(r^l)$ in vicinity of the tip of cone, we obtain, from the expressions of fields with potentials, that $(l+1)r^{l-1} = O((kr)^m)$, $m > -3/2$. The latter condition is satisfied if $l > -1/2$ or $l = -1$. We then look for the representations of the scattered wave field potentials in the form,

$$(\mathscr{E}, \mathscr{H}) = \frac{2}{k}\sqrt{\frac{2}{\pi}}\left(\int_{-i\infty}^{i\infty} \frac{\nu \sin \pi\nu}{\nu^2 - 1/4} \frac{K_\nu(ikr)}{\sqrt{ikr}} g_{e,h}(\omega, \omega_0, \nu) d\nu \right.$$
$$\left. + i\pi \frac{K_{1/2}(ikr)}{\sqrt{ikr}} w_{e,h}(\omega, \omega_0)\right), \tag{1.71}$$

where K_ν is a modified Bessel function, and $K_{1/2}(ikr) = \sqrt{(\pi/2ikr)}\exp(-ikr)$. The integral terms are of KL type, and $g_{e,h}(\omega,\omega_0,\nu)$ are even analytical functions of ν that we assume definite at $\nu = \pm 1/2$. We can derive, from the theory of KL transform [33, sect.3.3 and app.A], that these expressions are valid as $|\arg(ik)| < \varepsilon(\omega,\omega_0)$ when $\widehat{\theta}'(\omega,\omega_0) \leq \pi$, and as $|\arg(ik)| \leq \varepsilon(\omega,\omega_0)$ when $\widehat{\theta}'(\omega,\omega_0) > \pi$, which is equivalent to $\widehat{\theta}'(\omega,\omega_0) > \pi/2 + |\arg(ik)|$.

We can write the Helmholtz equation for the potentials expressed with (1.71). Multiplying this equation by r^3, and assuming that angular derivatives of the integral term in (1.71) remain $O(r^{-1/2})$ as r tends to 0, we notice that $\Delta_\omega w_{e,h}(\omega,\omega_0) = O(r^{1/2})$ as $r \to 0$, where Δ_ω is the Laplacian on the unit sphere (also called the Laplace–Beltrami operator), and thus $\Delta_\omega w_{e,h}(\omega,\omega_0) = 0$. In consequence, the KL integral term and the nonintegral term in (1.71) satisfies the Helmholtz equation independently, and, we obtain that

$$\left(\Delta_\omega + \nu^2 - \frac{1}{4}\right) g_{e,h}(\omega,\omega_0,\nu) = 0,\ \Delta_\omega w_{e,h}(\omega,\omega_0) = 0. \tag{1.72}$$

1.4.2.2 The equality $w_{e,h}(\omega,\omega_0) = g_{e,h}(\omega,\omega_{0,1/2})$, and the compatibility conditions on $g_{e,h}(\omega,\omega_{0,1/2})$

Considering that the energy flux through a small sphere surrounding the vertex vanishes as its surface collapses [2], we can show [33, sect.2] that we have no pure term of the form $p(\omega,\omega_0)\exp(-ikr)/ikr$ which satisfies the Maxwell equations in the expression of the field. To express this condition on $g_{e,h}$, we begin to introduce the linear combinations $\vec{C}_\pm = \vec{E} \pm iZ_0\vec{H}$ and $\mathscr{M}_\pm = \mathscr{E} \pm i\mathscr{H}$ which satisfy

$$\vec{C}_\pm = \text{rot rot}(\vec{r}\mathscr{M}_\pm) \mp k\text{rot}(\vec{r}\mathscr{M}_\pm), \tag{1.73}$$

or, developed in the spherical coordinates,

$$\begin{aligned} C_{\pm,r} &= k^2 r\mathscr{M}_\pm + \frac{\partial^2}{\partial r^2}(r\mathscr{M}_\pm),\ C_{\pm,\theta} = \frac{1}{r}\frac{\partial^2}{\partial r\partial\theta}(r\mathscr{M}_\pm) \mp \frac{k}{\sin\theta}\frac{\partial \mathscr{M}_\pm}{\partial\varphi}, \\ C_{\pm,\varphi} &= \frac{1}{r\sin\theta}\frac{\partial}{\partial r}\left(r\frac{\partial\mathscr{M}_\pm}{\partial\varphi}\right) \pm k\frac{\partial\mathscr{M}_\pm}{\partial\theta}. \end{aligned} \tag{1.74}$$

Considering combinations of the terms in (1.74), we obtain

$$\begin{aligned} C_{\pm,\theta} \pm iC_{\pm,\varphi} &= \left(\frac{\partial}{\partial\theta} \pm i\frac{\partial}{\sin\theta\partial\varphi}\right)\left(\frac{\partial}{r\partial r}(r\mathscr{M}_\pm) + ik(\mathscr{M}_\pm)\right), \\ C_{\pm,\theta} \mp iC_{\pm,\varphi} &= \left(\frac{\partial}{\partial\theta} \mp i\frac{\partial}{\sin\theta\partial\varphi}\right)\left(\frac{\partial}{r\partial r}(r\mathscr{M}_\pm) - ik(\mathscr{M}_\pm)\right). \end{aligned} \tag{1.75}$$

Using (1.71), we can derive field terms with strict $\exp(-ikr)/ikr$ dependence in (1.75). Concerning integral terms, we exploit the relations $2\nu K_\nu(z)/z = K_{\nu+1}(z) - K_{\nu-1}(z)$ and $2K'_\nu(z) = -K_{\nu+1}(z) - K_{\nu-1}(z)$. We then deform the contours of integration so that we capture residue terms attached to simple poles of $K_{\nu\pm1}(z)/(\nu^2 - 1/4)$ at $\nu = \mp 1/2$. We add to these terms the contribution of nonintegral terms of (1.71) and isolate the complete field term with strict $\exp(-ikr)/ikr$ dependence in (1.75). Writing that it must vanish, we derive the compatibility conditions,

$$\frac{\partial g_e(\omega,\omega_0,1/2)}{\partial\theta}=\frac{\partial g_h(\omega,\omega_0,1/2)}{\sin\theta\,\partial\varphi},\ \frac{\partial g_h(\omega,\omega_0,1/2)}{\partial\theta}=-\frac{\partial g_e(\omega,\omega_0,1/2)}{\sin\theta\,\partial\varphi},\tag{1.76}$$

and

$$\begin{aligned}\frac{\partial(g_e(\omega,\omega_0,1/2)-w_e(\omega,\omega_0))}{\partial\theta}&=-\frac{\partial(g_h(\omega,\omega_0,1/2)-w_h(\omega,\omega_0))}{\sin\theta\,\partial\varphi},\\ \frac{\partial(g_h(\omega,\omega_0,1/2)-w_h(\omega,\omega_0))}{\partial\theta}&=\frac{\partial(g_e(\omega,\omega_0,1/2)-w_e(\omega,\omega_0))}{\sin\theta\,\partial\varphi}.\end{aligned}\tag{1.77}$$

All choices of $w_{e,h}$ satisfying this condition and Laplace–Beltrami equation give the same fields $\vec{E}$ and $\vec{H}$, and we can choose $w_{e,h}(\omega,\omega_0)=g_{e,h}(\omega,\omega_0,1/2)$. We then obtain for the potentials,

$$\begin{aligned}(\mathscr{E},\mathscr{H})=\frac{2}{k}\sqrt{\frac{2}{\pi}}\Bigg(&\int_{-i\infty}^{i\infty}\frac{\nu\sin\pi\nu}{\nu^2-1/4}\frac{K_\nu(ikr)}{\sqrt{ikr}}g_{e,h}(\omega,\omega_0,\nu)d\nu\\ &+i\pi\frac{K_{1/2}(ikr)}{\sqrt{ikr}}g_{e,h}\left(\omega,\omega_0,\frac{1}{2}\right)\Bigg),\end{aligned}\tag{1.78}$$

with (1.76) which ensures that $\vec{E}=(\vec{C}_+ + \vec{C}_-)/2$ and $Z_0\vec{H}=(\vec{C}_+ - \vec{C}_-)/2i$ do not contain any nonintegral terms with strict $\exp(-ikr)/r$ radial dependence.

It is useful to note that the equalities in (1.76) can be rewritten as

$$(\vec{e}_\varphi\mp i\vec{e}_\theta)\nabla_\omega\left(g_e\left(\omega,\omega_0,\frac{1}{2}\right)\pm ig_h\left(\omega,\omega_0,\frac{1}{2}\right)\right)=0,\tag{1.79}$$

or

$$\begin{aligned}&\left(\frac{\partial}{\partial\varphi}\mp i\frac{\sin\theta\,\partial}{\partial\theta}\right)\left(g_e\left(\omega,\omega_0,\frac{1}{2}\right)\pm ig_h\left(\omega,\omega_0,\frac{1}{2}\right)\right)\\ &=2\frac{\partial}{\partial z_\pm}\left(g_e\left(\omega,\omega_0,\frac{1}{2}\right)\pm ig_h\left(\omega,\omega_0,\frac{1}{2}\right)\right)=0,\end{aligned}\tag{1.80}$$

where $z_\pm=\varphi\pm i\ln(\tan(\theta/2)), z_+=z_-^*$. We can generalize in a simple manner the conditions (1.79) and (1.80) on $g_{e,h}(\omega,\omega_0,1/2)$ for a new orthogonal vectors basis $(\vec{e}_1,\vec{e}_2=\hat{r}\wedge\vec{e}_1)$ on the unit sphere, either by noticing that $(\vec{e}_\theta\pm i\vec{e}_\varphi)=((\vec{e}_1\pm i\vec{e}_2)/(\vec{e}_1\cdot(\vec{e}_\theta\mp i\vec{e}_\varphi)))$ or by using a conformal transformation in (1.80), and we have in particular,

$$(\vec{e}_2\mp i\vec{e}_1)\nabla_\omega(g_e(\omega,\omega_0,1/2)\pm ig_h(\omega,\omega_0,1/2))=0.\tag{1.81}$$

Because of the complexity of the expressions of the fields with potentials defined with (1.78), the properties of the spectral functions are more difficult to derive than in acoustics case [31,32]. For details, the reader can refer to [33,34]. We now insist on an efficient expression of the potentials attached to an incident electromagnetic plane wave.

Remark: Notice that $\Delta_\omega g_{e,h}(\omega, \omega_0, 1/2) = 0$ can be written, with the variables $z_\pm$, $\partial_{z_+}\partial_{z_-} g_{e,h}(\omega, \omega_0, 1/2) = 0$.

Remark: We easily deduce from (1.76) that the fields, respectively attached to the nonintegral terms of $\mathscr{E}$ and $\mathscr{H}$ in (1.71), are equal. Another remarkable consequence of (1.76) is the fact that changing $\exp(-ikr)$ into $\exp(ikr)$ in the nonintegral terms of the potentials lead their contributions to vanish. This implies that we can take $-2i\sin(kr)$ in place of $\exp(-ikr)$ in the expressions of nonintegral terms.

Remark: Functional equations on $g_{e,h}$ can be derived from (1.78) and boundary conditions on the cone, for some domain of complex k, then be used for arbitrary k from analyticity. Once $g_{e,h}$ determined, we can calculate (1.78), directly in the oasis region with $\widehat{\theta}'(\omega, \omega_0) > \pi$, or more generally after subtracting some approximation of $g_{e,h}$ as $\nu \to i\infty$ due to geometrical optics contributions (with physical optics approximation, for example) (see [34, p.63]).

1.4.3 Potentials and properties for an incident plane wave

1.4.3.1 An efficient expression

We here show how to obtain simple explicit expressions of potentials $(\mathscr{E}^i, \mathscr{H}^i)$ for an incident plane wave electromagnetic field [33], without usual series of spherical harmonics, and we detail some of their complex properties.

Let us consider a plane wave, described in spherical coordinates, following

$$\begin{aligned} \vec{E}^i &= (\widehat{\theta_0}\sin\beta + \widehat{\varphi_0}\cos\beta)e^{ikr\cos\widehat{\theta}(\omega,\omega_0)}, \\ Z_0\vec{H}^i &= (\widehat{\theta_0}\cos\beta - \widehat{\varphi_0}\sin\beta)e^{ikr\cos\widehat{\theta}(\omega,\omega_0)}, \end{aligned} \tag{1.82}$$

with $\omega_0 = (\theta_0, \varphi_0)$ the point on the unit sphere S^2 attached to the direction of incidence (i.e., the direction from which the wave comes), $(\widehat{\theta_0}, \widehat{\varphi_0})$ the spherical vectors associated with ω_0, and $\cos\widehat{\theta}(\omega, \omega_0) = \cos\theta\cos\theta_0 + \sin\theta\sin\theta_0\cos(\varphi - \varphi_0)$. Concerning the exponential term in (1.82), we notice [32] that

$$u^i := e^{ikr\cos\widehat{\theta}(\omega,\omega_0)} = \frac{4}{i\sqrt{2\pi}}\int_{-i\infty}^{+i\infty} \nu\sin(\pi\nu)u^i_\nu(\omega, \omega_0)\frac{K_\nu(ikr)}{\sqrt{ikr}}d\nu, \tag{1.83}$$

where $u^i_\nu(\omega, \omega_0) = -(P_{\nu-1/2}(-\cos\widehat{\theta}(\omega, \omega_0))/4\cos(\pi\nu))$, with $(\Delta_\omega + \nu^2 - 1/4)$ $g_{e,h}(\omega, \omega_0, \nu) = \delta(\omega - \omega_0)$ [32,35,36] and the integral in (1.83) converging provided $\widehat{\theta}(\omega, \omega_0) > \pi/2 + |\arg(ik)|$. We then derive the radial components of the plane wave field from (1.82), and express them in the form,

$$E^i_r = \mathscr{D}_e\Big(\frac{e^{ikr\cos\widehat{\theta}(\omega,\omega_0)}}{ikr}\Big), \quad Z_0H^i_r = \mathscr{D}_h\Big(\frac{e^{ikr\cos\widehat{\theta}(\omega,\omega_0)}}{ikr}\Big), \tag{1.84}$$

with $\mathscr{D}_e = \sin\beta(\partial/\partial\theta_0) + \cos\beta(\partial/\sin\theta_0\partial\varphi_0)$ and $\mathscr{D}_h = \cos\beta(\partial/\partial\theta_0) - \sin\beta(\partial/\sin\theta_0\partial\varphi_0)$. Denoting $C^i_{e,h} = (E^i_r, Z_0H^i_r)$ and $\mathscr{D}_{e,h} = (\mathscr{D}_e, \mathscr{D}_h)$, we can exploit the representation for the modified Bessel function $K_\nu(ikr)/ikr = (1/4i\nu\sin\pi\nu)\int_\gamma e^{ikr\cos\alpha}$

$\sin\alpha\cos(\nu\alpha)d\alpha$, in (1.83) and express (1.84) with Sommerfeld integrals. By means of an inverse Sommerfeld–Maliuzhinets transformation [1], we then obtain

$$\frac{-ik\sin\alpha}{2}\int_0^{\infty}\sqrt{ikr}C^i_{e,h}e^{-ikr\cos\alpha}dr$$
$$=\frac{2}{i\sqrt{2\pi}}\int_{-i\infty}^{+i\infty}\mathscr{D}_{e,h}(u^i_\nu(\omega,\omega_0))\pi\sin\alpha\cos\nu\alpha d\nu. \tag{1.85}$$

On the other hand, we consider the Debye potentials $(\mathscr{E}^i,\mathscr{H}^i)$, attached to a representation of the incident field with (1.68) and (1.69), in a form similar to (1.78),

$$(\mathscr{E}^i,\mathscr{H}^i)=\frac{4}{k\sqrt{2\pi}}\left(\int_{-i\infty}^{+i\infty}\frac{\nu\sin(\pi\nu)}{\nu^2-1/4}g^i_{e,h}(\omega,\omega_0,\nu)\frac{K_\nu(ikr)}{\sqrt{ikr}}d\nu\right.$$
$$\left.+i\pi\frac{K_{1/2}(ikr)}{\sqrt{ikr}}g^i_{e,h}\left(\omega,\omega_0,\frac{1}{2}\right)\right), \tag{1.86}$$

where the functions $g^i_{e,h}$ are even functions of ν. Exploiting the expressions (1.68)–(1.69) for E_r, H_r and substituting the expressions (1.86), we verify that

$$C^i_{e,h}=(E^i_r,Z_0H^i_r)=\left(k^2+\frac{\partial^2}{\partial r^2}\right)(r\mathscr{E}^i,r\mathscr{H}^i)$$
$$=\frac{4i}{\sqrt{2\pi}}\int_{-i\infty}^{i\infty}\nu\sin\pi\nu\, g^i_{e,h}(\omega,\omega_0,\nu)\frac{K_\nu(ikr)}{(ikr)^{3/2}}d\nu, \tag{1.87}$$

noticing that the contribution of $K_{1/2}(ikr)/\sqrt{ikr}$ term for the radial component is zero. We then apply an inverse Sommerfeld–Maliuzhinets transformation [1] to the expression of $\sqrt{ikr}C^i_{e,h}$ derived from (1.87), and obtain

$$\frac{-ik\sin\alpha}{2}\int_0^{\infty}\sqrt{ikr}C^i_{e,h}e^{-ikr\cos\alpha}dr$$
$$=\frac{2}{i\sqrt{2\pi}}\int_{-i\infty}^{+i\infty}(-g^i_{e,h}(\omega,\omega_0,\nu))\pi\sin\alpha\cos\nu\alpha d\nu. \tag{1.88}$$

Comparing the expressions (1.85) and (1.88), we have

$$\int_{-i\infty}^{+i\infty}(-g^i_{e,h}(\omega,\omega_0,\nu))\cos\nu\alpha d\nu=\int_{-i\infty}^{+i\infty}\mathscr{D}_{e,h}(u^i_\nu(\omega,\omega_0))\cos\nu\alpha d\nu. \tag{1.89}$$

To reduce (1.89), we have to consider the elementary properties of functions involved in it. They are derived from (1.83)–(1.84) and (1.87) according to the behavior of the field at the apex and at infinity: the even functions $\mathscr{D}_{e,h}(u^i_\nu(\omega,\omega_0))$ and $g^i_{e,h}(\omega,\omega_0,\nu)$ are regular in the strip $|\mathrm{Re}(\nu)| < 3/2$ and are $O(1/\cos(\nu\widehat{\theta}(\omega,\omega_0)))$ as Imν tends to infinity. These properties permit to use Fourier transformation for (1.89) and to obtain $g^i_{e,h}(\omega,\omega_0,\nu) = -\mathscr{D}_{e,h}(u^i_\nu(\omega,\omega_0))$, which can also be written as

$$g^i_e(\omega,\omega_0,\nu) = -\Big(\sin\beta\frac{\partial}{\partial\theta_0} + \cos\beta\frac{\partial}{\sin\theta_0\partial\varphi_0}\Big)(u^i_\nu(\omega,\omega_0)),$$
$$g^i_h(\omega,\omega_0,\nu) = -\Big(\cos\beta\frac{\partial}{\partial\theta_0} - \sin\beta\frac{\partial}{\sin\theta_0\partial\varphi_0}\Big)(u^i_\nu(\omega,\omega_0)). \tag{1.90}$$

It is worth noticing that the linear operator $\mathscr{D}_{e,h}$ applies only on the variables θ_0 and φ_0, which signifies a direct and easy work on u^i_ν for the integral representation of the incident field, and that the condition $\widehat{\theta}(\omega,\omega_0) > \pi/2 + |\arg(ik)|$ also applies for the convergence of integrals in the representation (1.86) of $(\mathscr{E}^i,\mathscr{H}^i)$.

1.4.3.2 Complex properties of $g^i_{e,h}$

Letting $\mathscr{D}^{\pm}_0 \equiv (\frac{\partial}{\partial\varphi_0} \pm i(\sin\theta_0\partial/\partial\theta_0))$, the functions $g^i_{e,h}(\omega,\omega_0,\nu)$ verify

$$g^i_e(\omega,\omega_0,\nu) \pm ig^i_h(\omega,\omega_0,\nu) = -\frac{e^{\mp i\beta}}{\sin\theta_0}\mathscr{D}^{\pm}_0(u^i_\nu(\omega,\omega_0)) \tag{1.91}$$
$$= -\frac{e^{\mp i\beta}}{\sin\theta_0}\mathscr{D}^{\pm}_0(\cos(\widehat{\theta}(\omega,\omega_0)))\frac{\nu^2-1/4}{\cos(\pi\nu)}\frac{(1-x^2)^{1/2}P^{-1}_{\nu-1/2}(x)}{(x^2-1)},$$
$$g^i_e(\omega,\omega_0,1/2) \pm ig^i_h(\omega,\omega_0,1/2) = \frac{e^{\mp i\beta}/\sin\theta_0}{\pi(1-\cos(\widehat{\theta}(\omega,\omega_0)))}\mathscr{D}^{\pm}_0(\cos\widehat{\theta}(\omega,\omega_0)),$$

with $x = -\cos(\widehat{\theta}(\omega,\omega_0))$, $0 \le \widehat{\theta}(\omega,\omega_0) \le \pi$. We can then consider the transform by the differential operator $\mathscr{D}^{\pm} \equiv ((\partial/\partial\varphi) \pm i(\sin\theta\partial/\partial\theta))$, following

$$\sin\theta_0 e^{\pm i\beta}\mathscr{D}^{\mp}(g^i_e(\omega,\omega_0,\nu) \pm ig^i_h(\omega,\omega_0,\nu)) = \mathscr{D}^{\mp}\mathscr{D}^{\pm}_0 u^i_\nu(\omega,\omega_0) \tag{1.92}$$
$$= \frac{1}{4\pi}\Bigg(\mathscr{D}^{\mp}\mathscr{D}^{\pm}_0(\cos(\widehat{\theta}(\omega,\omega_0)))\frac{\partial P_{\nu-1/2}(x)}{\partial x}$$
$$-\mathscr{D}^{\mp}(\cos(\widehat{\theta}(\omega,\omega_0)))\mathscr{D}^{\pm}_0(\cos(\widehat{\theta}(\omega,\omega_0)))\frac{\partial^2 P_{\nu-1/2}(x)}{\cos(\pi\nu)\partial x^2}\Bigg)$$
$$= \frac{\mathscr{D}^{\mp}\mathscr{D}^{\pm}_0(\cos(\widehat{\theta}(\omega,\omega_0)))}{4\pi}\Big(\frac{\partial P_{\nu-1/2}(x)}{\partial x} + (1-\cos(\widehat{\theta}(\omega,\omega_0)))\frac{\partial^2 P_{\nu-1/2}(x)}{\cos(\pi\nu)\partial x^2}\Big),$$

where we have used that

$$\mathscr{D}^{\pm}_0(\cos(\widehat{\theta}(\omega,\omega_0)))\mathscr{D}^{\mp}(\cos(\widehat{\theta}(\omega,\omega_0))) \tag{1.93}$$
$$= -(1-\cos(\widehat{\theta}(\omega,\omega_0)))\mathscr{D}^{\mp}\mathscr{D}^{\pm}_0(\cos(\widehat{\theta}(\omega,\omega_0))),$$

and recover the compatibility condition (1.80) at $\nu = 1/2$.

Remark: In the equations on $g_e^i(\omega,\omega_0,\nu)\pm ig_h^i(\omega,\omega_0,\nu)$, we have used that

$$\begin{aligned}\mathscr{D}_0^{\pm}(\cos(\widehat{\theta}(\omega,\omega_0))) &= \sin\theta_0(\sin\theta\sin(\varphi-\varphi_0)\\ &\pm i(\sin\theta\cos\theta_0\cos(\varphi-\varphi_0)-\cos\theta\sin\theta_0)),\\ \mathscr{D}^{\mp}(\cos(\widehat{\theta}(\omega,\omega_0))) &= -\sin\theta(\sin\theta_0\sin(\varphi-\varphi_0)\\ &\pm i(\sin\theta_0\cos\theta\cos(\varphi-\varphi_0)-\cos\theta_0\sin\theta)),\\ \mathscr{D}^{\mp}\mathscr{D}_0^{\pm}(\cos(\widehat{\theta}(\omega,\omega_0))) &= \sin\theta\sin\theta_0[\cos(\varphi-\varphi_0)(1+\cos\theta\cos\theta_0)\\ &+\sin\theta\sin\theta_0\mp i(\cos\theta_0+\cos\theta)\sin(\varphi-\varphi_0))],\end{aligned} \tag{1.94}$$

and that

$$\begin{aligned}&\left(\frac{1-x}{1+x}\right)^{\mu/2}P^{\mu}_{\nu-1/2}(x)=\left(\frac{z-1}{1+z}\right)^{\mu/2}P^{\mu}_{\nu-1/2}(z)|_{z=x\pm i0^+},\\ &\frac{1}{\cos(\pi\nu)}\frac{\partial P_{\nu-1/2}(z)}{\partial z}=\frac{\nu^2-1/4}{\cos(\pi\nu)}\frac{(z^2-1)^{1/2}P^{-1}_{\nu-1/2}(z)}{z^2-1},\\ &\frac{(1-z^2)}{\cos(\pi\nu)}\frac{\partial^2P_{\nu-1/2}(z)}{\partial z^2}=\frac{2z}{\cos(\pi\nu)}\frac{\partial P_{\nu-1/2}(z)}{\partial z}-\frac{\nu^2-1/4}{\cos(\pi\nu)}P_{\nu-1/2}(z)\\ &\quad=\frac{\nu^2-1/4}{\cos(\pi\nu)}\Big(\frac{2z(z^2-1)^{1/2}}{z^2-1}P^{-1}_{\nu-1/2}(z)-P_{\nu-1/2}(z)\Big),\\ &P_0^{-1}(z)=\frac{z-1}{(z^2-1)^{1/2}},\quad \lim_{\nu\to1/2}\frac{\partial P_{\nu-1/2}(z)}{\cos(\pi\nu)\partial z}=\frac{-1}{\pi(z+1)},\\ &\lim_{\nu\to1/2}\frac{\partial^2P_{\nu-1/2}(z)}{\cos(\pi\nu)\partial z^2}=\frac{1}{\pi(z^2-1)}\Big(\frac{2z}{z+1}-1\Big)=\frac{1}{\pi}\frac{1}{(z+1)^2}.\end{aligned} \tag{1.95}$$

with $z=x\pm i0^+$, $x=-\cos(\widehat{\theta}(\omega,\omega_0))$.

Remark: The function $u^i_\nu(\omega,\omega_0)$ satisfies the Helmholtz equation on the unit sphere with the Dirac source [33]–[36], and we can write that the spectral functions attached to the incident field satisfy

$$\Big(\Delta_\omega+\nu^2-\frac{1}{4}\Big)(g^i_{e,h})(\omega,\omega_0,\nu)=-\mathscr{D}_{e,h}\delta(\omega-\omega_0). \tag{1.96}$$

References

[1] G.D. Maliuzhinets, 'Inversion formula for the Sommerfeld integral', *Sov. Phys. Dokl.* 3, pp. 52–56, 1958.

[2] D.S. Jones, 'The theory of electromagnetism', Pergamon Press, London, 1964.

[3] L.B. Felsen and N. Marcuvitz, 'Radiation and scattering of waves', Prentice Hall, Englewood Cliffs, NJ, 1973.

[4] J.M.L. Bernard, 'A spectral approach for scattering by impedance polygons', *Q. J. Mech. Appl. Math.*, 59, 4, pp. 518–548, 2006.

[5] J.M.L. Bernard, 'Diffraction at skew incidence by an anisotropic impedance wedge in electromagnetism theory: a new class of canonical cases', *J. Phys. A: Math. Gen.*, 31, pp. 595–613, 1998.

[6] J.M.L. Bernard, 'Progresses on the diffraction by a wedge: transient solution for line source illumination, single face contribution to scattered field, and new consequence of reciprocity on the spectral function', *Rev. Tech. Thomson-CSF*, 25, 4, pp. 1209–1220, 1993.

[7] A. Michaeli, 'Contribution of a single face to the wedge diffracted field', *IEEE Trans. Antennas Propag.*, 33 (2), pp. 221–223, 1985.

[8] I.S. Gradshteyn and I.M. Ryzhik, 'Table of integrals, series and products', Academic Press, Boston, 1980.

[9] J.M.L. Bernard, 'On the time-domain scattering by a passive classical frequency dependent wedge-shaped region in a lossy dispersive medium', *Ann. Telecommun.*, 49, 11–12, pp. 673–683, 1994. (errata: exchange fig.3b for fig.4, read p.677 $\mathrm{Im}\alpha = 0$ instead of $\mathrm{Im}\alpha \neq 0$).

[10] J.M.L. Bernard, *Rev. Tech. Thomson-CSF*, 23, 2, pp. 321–330, 1991.

[11] J.M.L. Bernard, 'The diffraction by a curved impedance wedge: diffracted and creeping waves', Proceedings of URSI Conf. of Maastricht, 8–23 Août 2002.

[12] J.M.L. Bernard, PhD Thesis, Paris-Sud University, 1995.

[13] G.D. Maliuzhinets, *Sov. Phys. Dokl.*, 3, pp. 752–755, 1958.

[14] V.A. Borovikov, *J. Commun. Technol. Electron.*, 43, 12, pp. 1337–1346, 1998.

[15] L. Kaminetzky and J.B. Keller, *SIAM J. Appl. Math.*, 22, 1, pp. 109–134, 1972.

[16] N.C. Albertsen and P.L. Christiansen, 'Hybrid diffraction coefficients for first and second order discontinuities of two dimensional scatterers', *SIAM J. Appl. Math.*, 34, 2, pp. 398–414, 1978.

[17] V.A. Borovikov, *Akust. Zh.*, 25, 6, pp. 825–835, 1979.

[18] R.F. Harrington, 'Time-harmonic electromagnetic field', McGraw-Hill, New York, 1961.

[19] J.M.L. Bernard, 'On the expression of the field scattered by a multimode plane', *Q. J. Mech. Appl. Math.*, 63, 3, pp. 237–266, 2010 (erratum: p.255, read '...for monomode conditions ($N = P = 1$), that $g_1^{(e,h)} = 1/g_1^{(h,e)}$.').

[20] M. Abramowitz and I. Stegun, 'Handbook of mathematical functions', Dover, Inc., NY, 1972.

[21] L.M. Brekhovskikh, 'Waves in layered media', Academic Press, Inc., NY, 1980.

[22] J.M.L. Bernard, PHD thesis, Orsay University, 1995 (available at http://hal.archives-ouvertes.fr/tel-00221353/).

[23] I.S. Koh and J.G. Yook, 'Exact closed-form expression of a Sommerfeld integral for the impedance plane problem', *IEEE Trans. AP*, 54, 9, pp. 2568–2576, 2006.

[24] M.M. Agrest and M.S. Maksimov, 'Theory of incomplete cylindrical functions and their applications', Springer, New York, NY, 1971.

[25] F.E. Harris, 'Incomplete Bessel, generalized incomplete gamma, or leaky aquifer functions', *J. Comput. Appl. Math.*, 215, pp. 260–269, 2008.

[26] J.M.L. Bernard, 'On a novel expression of the field scattered by an arbitrary constant impedance plane', *Wave Motion*, 48, 7, pp. 635–646, 2011 (erratum: p.638, in second line of equ. 3.3, read $\sqrt{\pi} W_m$ instead of W_m).

[27] J.M.L. Bernard, 'Propagation over a constant impedance plane: arbitrary primary sources and impedance, analysis of cut in active case, exact series, and complete asymptotics', *IEEE Trans. AP*, 66, 12, pp. 6596–6605, 2018.

[28] U. Ingard, 'On the reflection of a spherical sound wave from an infinite plane', *J. Acoust. Soc. Am.*, 23, 3, pp. 329–335, 1951 (see corrections in [29]-[30]).

[29] A.R. Wenzel, 'Propagation of waves along an impedance boundary', *J. Acoust. Soc. Am.*, 55, 5, pp. 956–963, 1974.

[30] S.I. Thomasson, 'Reflection of waves from a point source by an impedance boundary', *J. Acoust. Soc. Am.*, 59, 4, pp. 780–785, 1976.

[31] D.S. Jones, 'Scattering by a cone', *Q. J. Mech. Appl. Math.*, 50, pp. 499–523, 1997.

[32] J.M.L. Bernard and M.A. Lyalinov, 'Diffraction of acoustic waves by an impedance cone of an arbitrary cross-section', *Wave Motion*, 33, pp. 155–181, 2001 (erratum: p.177 replace $O(1/\cos(\pi(\nu - b)))$ by $O(\nu^d \sin(\pi\nu)/\cos(\pi(\nu - b)))$).

[33] J.M.L. Bernard and M.A. Lyalinov, 'Electromagnetic scattering by a smooth convex impedance cone', *IMA J. Appl. Math.*, 69, 3, pp. 285–333, 2004 (erratum: p.322 multiply $\sin\zeta$ by $\mathrm{sgn}(n)$ in D.20).

[34] J.M.L. Bernard, 'Advanced theory of diffraction by a semi-infinite impedance cone', Alpha Sciences international Ltd, Oxford, 2014.

[35] V.P. Smyshlyaev, 'The high frequency diffraction of electromagnetic waves by cones of arbitrary cross section', *SIAM Appl. Math.*, 53, 3, pp. 670–688, 1993 (see also Preprint of LOMI, E-9-89, Leningrad, (1989)).

[36] V.M. Babich, V.P. Smyshlyaev, D.B. Dement'ev, and B.A. Samokish, 'Numerical calculations of the diffraction coefficients for an arbitrary shaped perfectly conducting cone', *IEEE Trans. AP*, 44, 5, pp. 740–747, 1996.

Chapter 2
Scattering of electromagnetic surface waves on imperfectly conducting canonical bodies

Mikhail A. Lyalinov[1] *and Ning Yan Zhu*[2]

An imperfectly conducting surface may support surface waves provided appropriate impedance boundary conditions (Leontovich conditions) are satisfied. Electromagnetic surface waves propagate along an impedance surface and interact with its singular points such as edges or conical vertices giving rise to the reflection and transmission of such surface waves as well as to those diffracted into the space surrounding the canonical body. In this work, we discuss a mathematical approach describing some physical processes dealing with the diffraction of surface waves by canonical singularities like wedges and cones. We develop a mathematically justified theory of such processes with the attention centred on diffraction of a skew incident surface wave at the edge of an impedance wedge. Questions of excitation of the electromagnetic surface waves by a Hertzian dipole are also addressed as well as the Geometrical Optics laws of reflection and transmission of a surface wave across the edge of an impedance wedge.

2.1 Introduction and survey of some known results

It is well known that in many situations imperfectly conducting surfaces may support propagation of electromagnetic surface waves; see for instance a recent tutorial [1]. From the theoretical point of view, these waves are some special (asymptotic or exact) solutions of Maxwell's equations satisfying appropriate boundary conditions and exponentially vanishing as the observation point goes away from the surface. On the other hand, different technical problems encountered, e.g. in plasmonics or in the theory of antenna design, dealing with the scattering of surface waves, require efficient study of the corresponding wave processes. This study becomes much more difficult when a surface wave interacts with different kinds of geometrical (edges, conical points) and (or) material (e.g. abrupt change of the surface impedance) irregularities

[1]Department of Mathematics and Mathematical Physics, Saint-Petersburg University, Saint Petersburg, Russia
[2]Institute of Radio Frequency Technology, University of Stuttgart, Stuttgart, Germany

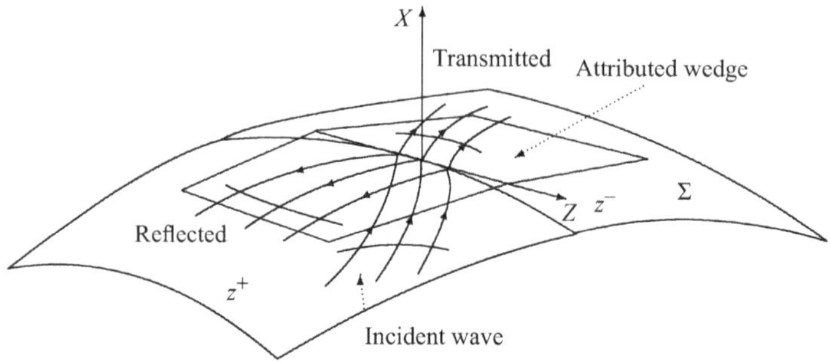

Figure 2.1 Surface wave on a curved wedge

on the supporting surface. Some appropriate short survey on this subject is given in this section. In this work, we also discuss some of these problems and offer an adequate and mathematically justified theory of the corresponding scattering phenomena.

Let us imagine that an electromagnetic surface wave propagates along a curved surface Σ (curved wedge, see Figure 2.1) on which the impedance-type boundary conditions are postulated. It is assumed that the surface may have an edge and, besides, the line of the edge is also the line of the jump of the surface impedance. The incident surface wave approaches the edge, reflects from the edge and transmits across it. In this process, also the edge-diffracted wave arises. Provided the radii of the curvature of the wedge's faces are much greater than the wavelength, we can make use of the localisation principle. The reflection and transmission coefficients as well as the diffraction coefficient are then obviously specified by the local characteristics of the scattering surface and by the incident angle of the surface wave. So one may consider an attributed wedge with planar faces and their common edge being tangential to the primary 'curved' wedge at the point of diffraction.

There are some recent results describing interaction of the incident surface waves with geometrical (edges, tips) or (and) material irregularities of the surfaces with the canonical shape (wedges, cones), see [2,3] [4, Chapter 4], [5,6]. It is also worth mentioning the works [7–12], where some other results and different applications in plasmonics are considered.

> In [3], a recent advance in applications of the Sommerfeld–Malyuzhinets technique to the problem of diffraction of a surface wave by an angular break of a thin material slab are discussed. The solution is represented by Sommerfeld integrals, which are then substituted into boundary conditions. The unknown spectral functions satisfy coupled Malyuzhinets functional equations. The latter are reduced to Fredholm integral equations of the second kind, which are solved numerically. The scattering diagram of the cylindrical wave arising from the edge of the structure is computed.
>
> Reference [2] deals with the scattering of an incident surface wave propagating to the vertex of a circular impedance cone. Diffraction of the incident

surface wave by the vertex gives rise to the reflected surface wave as well to the spherical wave from the vertex of the cone in the far-field approximation. The study is based on the Sommerfeld integrals, Fourier transform and on the incomplete separation of the spherical variables for the problem at hand. The analytic formulae for the reflected surface wave and for the diffraction coefficient of the spherical wave are obtained.

It is worth noting that some experimental results dealing with the diffraction of a surface wave by the conical tip are considered in the work [13]. Diffraction of a plane wave by a penetrable cone is discussed in [14]. The authors of the work [11] study surface waves on a conductive right-circular cone; however, their approach can hardly be justified from the mathematical point of view. Electromagnetic surface waves on the conical surface with Leontovich impedance boundary conditions are consequently derived and discussed in Chapter 6 of [4].

We study the process of scattering of the incident surface wave at the edge of a wedge. The corresponding recent extensions of the Sommerfeld–Malyuzhinets technique given in [4] (Chapter 2), [15,16] enable us to develop an analytical–numerical procedure of calculations of the reflection (transmission) coefficients as well as those for the edge diffraction coefficients in the case of an incident surface wave. In the short-wavelength approximation, we may expect that such computed coefficients will successfully serve in order to describe the scattering in the common case of the wedge with the curved faces. It is worth noting, however, that the problem of diffraction of an electromagnetic plane wave, which is skew incident at the edge of an impedance wedge, can also be treated by the use of the analytical–numerical technique based on the mentioned extensions of the Sommerfeld–Malyuzhinets approach. The explicit (i.e. in quadratures) solution has been obtained only in some particular or degenerated cases (see e.g. [17,18]).

We propose an efficient approach based on the reduction of the problem to a system of functional Malyuzhinets equations and their transformation to the integral ones completing the study by the calculation of the asymptotics of the far field from the corresponding Sommerfeld representation. Moreover, we exploit the plane wave expansion of the field from a Hertzian dipole located close to one of the impedance faces of the wedge. Such a dipole induces surface waves propagating to the edge of the wedge. We consequently develop the theory of the reflection and transmission of such a wave as well as of the space edge wave. The expressions for the reflection/transmission and diffraction coefficients are also addressed.

2.1.1 *Electromagnetic surface waves on impedance surfaces*

We describe some traditional approaches dealing with analytical or asymptotic constructions of the electromagnetic surface waves. In the next two sections, we make use of [19] (see also [20–23]) and describe the results in a form which is convenient for the rest of this chapter.

2.1.1.1 Electromagnetic surface waves supported by planar impedance surfaces

In this section, we present some known elementary results on construction of exact solutions of Maxwell's equations. These solutions are localised in some vicinity of the supporting surface Σ.

Consider solutions of time-harmonic Maxwell's equations (with omitted $e^{-i\omega t}$ time dependence)

$$\begin{aligned} \operatorname{curl} \mathbf{E} &= i\frac{\omega\mu}{c} Z_0 \mathbf{H}, \\ \operatorname{curl}(Z_0\mathbf{H}) &= -i\frac{\omega\varepsilon}{c}\mathbf{E} + Z_0\mathbf{J}_0, \end{aligned} \tag{2.1}$$

where $\mathbf{J}_0$ is a current ($\mathbf{J}_0 = 0$, μ, ε, the relative permeability and permittivity of the medium, are assumed to be constant in this section), c denotes the speed of light in free space and Z_0 is the intrinsic impedance of free space. An electromagnetic wave is

$$\begin{aligned} Z_0\mathscr{H}(q^1, q^2, n, t) &= Z_0\mathbf{H}(q^1, q^2, n)\, e^{-i\omega t} = Z_0\mathbf{H}_0\, e^{-i\omega t + i(k_1 q^1 + k_2 q^2 + i\gamma n)}, \\ \mathscr{E}(q^1, q^2, n, t) &= \mathbf{E}(q^1, q^2, n)\, e^{-i\omega t} = \mathbf{E}_0 e^{-i\omega t + i(k_1 q^1 + k_2 q^2 + i\gamma n)}, \\ \gamma > 0&, \quad \omega > 0, \end{aligned} \tag{2.2}$$

$q^1, q^2, q^3 = n$ are the Cartesian coordinates, $q^3 = n > 0$ is a half space Ω of R^3 (see Figure 2.1), where the solution (2.2) of the homogeneous equations (2.1) with $\mathbf{J}_0 = 0$ is sought. The solution (2.2) has the form of a plane wave with unknown parameters ω, k_1, k_2, γ in the phase and with constant amplitudes $\mathbf{E}_0, Z_0\mathbf{H}_0$. The wave field described by (2.2) exponentially vanishes as $n \to +\infty$. We look for the solution in the form of (2.2) which satisfies Leontovich impedance boundary conditions

$$\mathbf{E} - (\mathbf{n}, \mathbf{E})\mathbf{n} = \eta\, \mathbf{n} \times (Z_0\, \mathbf{H}) \tag{2.3}$$

on the surface Σ defined by the equation $q^3 = 0$, η is the (with respect to Z_0) normalised surface impedance.

The complex wave vector $\mathbf{K}$ is taken such that $\mathbf{K} = (k, 0, i\gamma)^T$ (T means transpose), i.e. the Cartesian coordinate system is oriented so that $k_1 = k$, $k_2 = 0$, k is the wave number. Making use of Maxwell's equations, we arrive at

$$\begin{aligned} \mathbf{K} \times \mathbf{E}_0 &= \frac{\omega\mu}{c} Z_0\, \mathbf{H}_0, \\ -\mathbf{K} \times Z_0\mathbf{H}_0 &= \frac{\omega\varepsilon}{c} \mathbf{E}_0. \end{aligned} \tag{2.4}$$

From the boundary condition (2.3), we find that

$$E_{01} = -\eta\, Z_0 H_{02}, \quad E_{02} = \eta\, Z_0 H_{01}.$$

The linear system of equations can be considered in C^6 as that for determination of the eigenvectors $\mathbf{E}_0 = (E_{01}, E_{02}, E_{03})^T$, $\mathbf{H}_0 = (H_{01}, H_{02}, H_{03})^T$ corresponding to the eigenvalue ω. We multiply the first equation in (2.4) by $\mathbf{E}_0$ and take into account that

$\mathbf{K} \times \mathbf{E}_0$ is orthogonal to $\mathbf{E}_0$ then $(\mathbf{E}_0, \mathbf{H}_0) = 0$. In view of the boundary conditions on Σ, we have $E_{01}H_{01} + E_{02}H_{02} = \eta^{-1}(E_{01}E_{02}/Z_0 - E_{02}E_{01}/Z_0) = 0$. As a result,

$$E_{03}\, H_{03} \;=\; 0.$$

We consider only two cases: (1) $E_{03} = 0,\ H_{03} \neq 0$ and (2) $E_{03} \neq 0,\ H_{03} = 0$ because $E_{03} = 0,\ H_{03} = 0$ lead to a trivial solution $\mathbf{E}_0 = \mathbf{H}_0 = \mathbf{0}$. In case 1, we write the equations in (2.4) in the coordinate form

$$\begin{cases} A\,H_{01} = A\,H_{02} = 0 \\ B\,H_{02} = -k\,H_{02} = 0 \\ B\,H_{01} = k\,H_{03} \\ \omega\dfrac{\mu}{c}\,H_{03} = k\,\eta\,H_{01} \end{cases}$$

with

$$A = \omega\frac{\mu}{c} + \mathrm{i}\gamma\eta\,, \quad B = \omega\frac{\varepsilon\eta}{c} + \mathrm{i}\gamma\,.$$

From the latter system, it is obvious that $H_{02} = 0$ and $A = 0$ so that $\omega = -(\mathrm{i}\gamma\eta c/\mu)$. Because ω and γ are positive, we conclude that

$$\eta = \mathrm{i}|\eta|.$$

We construct a non-trivial solution so that from the last two equations of the system, the expression for ω is determined in an explicit form by means of

$$\frac{c\eta}{\mu\omega}\,k^2 = B = \omega\frac{\varepsilon\eta}{c} + \mathrm{i}\gamma$$

and $\mathrm{i}\gamma = -(\mu\omega/c\eta)$. As a result, we arrive at

$$\omega \;=\; \frac{c|\eta|k}{\sqrt{\varepsilon\mu(|\eta|^2 + \mu/\varepsilon)}}\,. \tag{2.5}$$

It is worth saying that the phase and group velocities of such a wave coincide

$$\mathbf{v}_{\mathrm{ph}} = \left(\frac{c|\eta|}{\sqrt{\varepsilon\mu(|\eta|^2 + \mu/\varepsilon)}}, 0, 0\right)^{\mathrm{T}},$$

$$\mathbf{v}_{\mathrm{g}} = \left(\frac{\partial\omega}{\partial k}, 0, 0\right)^{\mathrm{T}} = \left(\frac{c|\eta|}{\sqrt{\varepsilon\mu(|\eta|^2 + \mu/\varepsilon)}}, 0, 0\right)^{\mathrm{T}}.$$

After elementary calculations, we find

$$\mathbf{E}_0 = C_*(0, \eta k, 0)^{\mathrm{T}}, \quad Z_0\,\mathbf{H}_0 = C_*\left(k, 0, \frac{\mathrm{i}\omega}{c}(\varepsilon|\eta| + \mu/|\eta|)\right)^{\mathrm{T}},$$

where C_* is an arbitrary complex constant. The constructed solution can be naturally called *magnetic* surface wave.

In a similar manner, we can construct an *electric* surface wave as $\eta = -\mathrm{i}|\eta|$. Thus, we have

$$\mathbf{v}_{\mathrm{ph}} = \mathbf{v}_{\mathrm{g}} = \left(\frac{c}{\sqrt{\varepsilon\mu(|\eta|^2\varepsilon/\mu + 1)}}, 0, 0 \right)^{\mathrm{T}},$$

$$\mathbf{E}_0 = C^* \left(k, 0, \frac{\mathrm{i}\omega}{c}(\varepsilon|\eta| + \mu/|\eta|) \right)^{\mathrm{T}}, \quad Z_0\,\mathbf{H}_0 = C^*(0, -k/\eta, 0)^{\mathrm{T}},$$

$$\gamma = \frac{\varepsilon\omega|\eta|}{c}, \quad \omega = \frac{ck}{\sqrt{\varepsilon\mu(|\eta|^2\varepsilon/\mu + 1)}}.$$

A propagating surface wave is localised near a surface with a purely imaginary surface impedance. The sign of the imaginary part of the surface impedance η specifies the type of the propagating surface wave, i.e. *electric* or *magnetic.* In some situations, e.g. for an axially anisotropic surface impedance (see [4], Chapter 2), the surface waves of both types can be excited simultaneously.

Remark 2.1. *It should be mentioned, however, that sometimes in applications the impedances with small positive real parts are also considered for some surfaces with absorption. In this case, the corresponding formal solution should be appropriately interpreted. Traditionally this interpretation also means that such a surface wave rapidly attenuates in the direction of propagation.*

In the following section, we study generalisation of the solutions obtained. We consider a curved supporting surface with inhomogeneous impedance and with varying electromagnetic constants of the medium above the surface then construct expressions for the surface waves in the short-wavelength approximation.

2.1.1.2 Electromagnetic surface waves on a curved surface with varying surface impedance in an inhomogeneous medium

Consider a regularly curved surface Σ in R^3 with its parametric equation $\mathbf{r}_s = \mathbf{r}_s(q^1, q^2)$, Figure 2.2. In a vicinity of Σ, we make use of the coordinates $q^1, q^2, q^3 = n$, where n is measured along the normal $\mathbf{n}$ to Σ pointing into the domain with the medium having the electromagnetic properties described by ε, μ which depend on the coordinates $q^1, q^2, q^3 = n$. The surface impedance η in the boundary conditions

$$E_1 = -\eta Z_0 H_2, \quad E_2 = \eta Z_0 H_1 \quad \text{on } \Sigma$$

may also depend on the coordinates q^1, q^2.

It is assumed that the wavelength l of a harmonic wave is much less than the characteristic radius $\mathscr{R}_0$ of curvature of Σ as well as of characteristic length of the variations of ε, μ and η so that $k\mathscr{R}_0 \gg 1$, $k = 2\pi/l \sim \omega/c$. In what follows, we imply that the coordinates are normalised by $\mathscr{R}_0$ and we use q^i in place of $q^i/\mathscr{R}_0$ with the notation k for the large parameter ($k \gg 1$) in place of $k\mathscr{R}_0$.

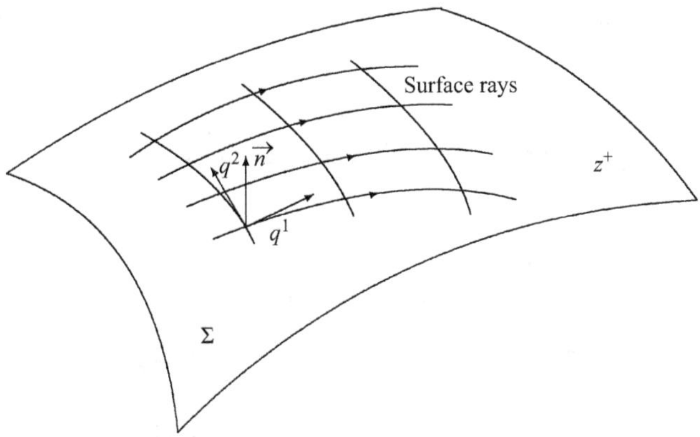

Figure 2.2 Surface wave on a curved surface

The asymptotic (ray) solution of Maxwell's equations (2.1) satisfying the boundary conditions and $(\mathbf{E}_0, Z_0\mathbf{H}_0) \to 0$ as $\nu = kn \to +\infty$ is found in the form of

$$\begin{aligned} Z_0\mathscr{H}(q^1, q^2, n, t) &= \mathrm{e}^{-\mathrm{i}\omega t + \mathrm{i}k\Theta(q^1,q^2)}\, Z_0\mathbf{H}_0(q^1, q^2, \nu)\left(1 + O\left(\frac{1}{k}\right)\right) \\ \mathscr{E}(q^1, q^2, n, t) &= \mathrm{e}^{-\mathrm{i}\omega t + \mathrm{i}k\Theta(q^1,q^2)}\, \mathbf{E}_0(q^1, q^2, \nu)\left(1 + O\left(\frac{1}{k}\right)\right), \end{aligned} \tag{2.6}$$

where $\nu = kn$ and

$$\begin{aligned} Z_0\mathbf{H}_0 &= \mathrm{e}^{-\nu\lambda(q^1,q^2)} A_0(q^1, q^2)\, \mathbf{h}_0, \\ \mathbf{E}_0 &= \mathrm{e}^{-\nu\lambda(q^1,q^2)} A_0(q^1, q^2)\, \mathbf{e}_0, \end{aligned} \tag{2.7}$$

with

$$\mathbf{h}_0 = \nabla\Theta + \mathbf{n}\frac{\mathrm{i}\omega}{c}\left(\varepsilon|\eta| + \frac{\mu}{|\eta|}\right), \quad \mathbf{e}_0 = \eta\,\mathbf{n} \times \nabla\Theta, \quad \lambda = \frac{\mu}{|\eta|}$$

for the magnetic surface wave,

$$\mathbf{e}_0 = \nabla\Theta + \mathbf{n}\frac{\mathrm{i}\omega}{c}\left(\varepsilon|\eta| + \frac{\mu}{|\eta|}\right), \quad \mathbf{h}_0 = -\eta^{-1}\mathbf{n} \times \nabla\Theta, \quad \lambda = |\eta|\varepsilon$$

for the electric surface wave.

In expressions (2.6), the surface eikonal solves the equation

$$c_s\,|\nabla\Theta| = c, \tag{2.8}$$

where $|\nabla\Theta| = \sqrt{g^{ij}\Theta_i\Theta_j}$, g_{ij} are the coefficients of the first quadratic form of Σ, g^{ij} is the inverse to the matrix g_{ij}. It is worth mentioning that in (2.8), we imply that

$$c_s = c_s^{\mathrm{m}}, \quad c_s^{\mathrm{m}} = \frac{c|\eta|}{\sqrt{\varepsilon\mu(|\eta|^2 + \mu/\varepsilon)}}$$

for magnetic wave and

$$c_s = c_s^{\mathrm{e}}, \quad c_s^{\mathrm{e}} = \frac{c}{\sqrt{\varepsilon\mu(|\eta|^2\varepsilon/\mu + 1)}}$$

for the electric surface wave. The eikonal equation (2.8) is solved by use of the classical method of characteristics for the Hamilton–Jacobi equation [22] which specifies the dependence of the ray coordinates a, s on Σ with those q^1, q^2; $q^1 = q^1(a,s)$, $q^2 = q^2(a,s)$ with non-degenerate Jacobian J (spreading).

The complex amplitude $A_0(q^1, q^2)$ is given by the expression

$$A_0(q^1, q^2) = |A_0(q^1, q^2)|\, \mathrm{e}^{\mathrm{i}V_0(q^1,q^2)}$$

with

$$|A_0(q^1, q^2)| = C(a)\sqrt{\frac{\lambda}{\mathrm{E}(\mathbf{e}_0, \mathbf{h}_0)J}}, \quad \mathrm{E}(\mathbf{e}_0, \mathbf{h}_0) = \frac{\varepsilon|\mathbf{e}_0|^2 + \mu|\mathbf{h}_0|^2}{8\pi}.$$

The *geometrical* phase $V_0(q^1, q^2)$ takes the form

$$\begin{aligned} V_0 = V_0|_{s=0} + \int_0^s \mathrm{d}s \Bigg(& \frac{1}{4\lambda^2} \frac{\frac{\partial\varepsilon}{\partial n}|_{n=0}|\mathbf{e}_0|^2 + \frac{\partial\mu}{\partial n}|_{n=0}|\mathbf{h}_0|^2}{\varepsilon_0|\mathbf{e}_0|^2 + \mu_0|\mathbf{h}_0|^2} \\ & + \frac{c}{2\lambda^2(\varepsilon_0|\mathbf{e}_0|^2 + \mu_0|\mathbf{h}_0|^2)}\Big(2M\mathrm{Re}(\nabla\Theta \times \mathbf{e}_0 \cdot \mathbf{h}_0) + 2\mathrm{Re}(B\nabla\Theta \times \mathbf{h}_0 \cdot \mathbf{e}_0) \\ & + \mathrm{Re}[\nabla\Theta \times B\mathbf{h}_0 \cdot \mathbf{e}_0 - \nabla\Theta \times B\mathbf{e}_0 \cdot \mathbf{h}_0]\Big)\Bigg), \end{aligned}$$

where M is the mean curvature, B is a tensor defined by $(B\mathbf{e})_j = g_{jk}b^{ki}e_i$, b^{ki} are the contravariant components of the second quadratic form of Σ. Remark that $C(a)$ and $V_0|_{s=0}$ are specified from some 'initial' data for the surface wave.

Remark 2.2. *In the latter integral for the geometrical phase, the summands in the integrand describe the influence of different geometrical and material characteristics of the surface and of the medium on the propagating surface wave. It is worth mentioning that the geometrical phase frequently arises in the asymptotic analysis of the wave phenomena and in the quantum theory.*

2.1.2 Electromagnetic surface waves on a right circular conical surface

Let us consider, in this section, the spherical coordinate system (r, θ, φ) connected with the axis X_3 (Figure 2.3) of the cone having the origin at the vertex of the cone. Let $\eta = \sin\zeta_u$ be the surface impedance with $\mathrm{Im}\,\zeta_u < 0$, $\eta^{-1} = \sin\zeta_v$. For the conical surface, we use the equation $\theta = \theta_1$. We notice that under the sufficient conditions

$$\theta - \theta_1 - \mathrm{Re}\,\zeta_u - \mathrm{gd}\,(\mathrm{Im}\,\zeta_u) > 0, \quad \mathrm{Im}\,\zeta_u < 0, \quad \mathrm{Re}\,\zeta_u = 0,$$

where $\mathrm{gd}(x)$ stands for the Gudermann function, from the integral representation of the wave field [4], Chapter 6 the expressions for electromagnetic surface wave can

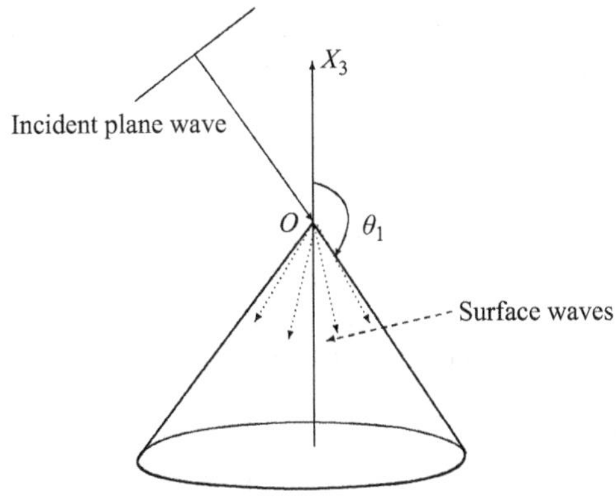

Figure 2.3 Surface wave on a right circular impedance cone

be derived. In these conditions, the electrical-type surface waves are excited provided the incident electromagnetic plane wave interacts with the conical surface [4].

After some calculations, in the leading approximation, we obtain

$$
\begin{aligned}
E_r^{\text{sw}}(kr,\theta,\varphi) &= \text{i}kT_u(\varphi,\theta)\,\sin^2(\theta_1-\theta+\zeta_u)\,\frac{\exp[\text{i}kr\cos(\theta_1-\theta+\zeta_u)]}{(-\text{i}kr)^{1/2+\lambda_u}} \\
&\quad\times\left[1+\text{O}\left(\frac{1}{kr}\right)\right], \\
E_\theta^{\text{sw}}(kr,\theta,\varphi) &= -\text{i}kT_u(\varphi,\theta)\sin(\theta_1-\theta+\zeta_u)\cos(\theta_1-\theta+\zeta_u) \\
&\quad\times\frac{\exp[\text{i}kr\cos(\theta_1-\theta+\zeta_u)]}{(-\text{i}kr)^{1/2+\lambda_u}}\left[1+\text{O}\left(\frac{1}{kr}\right)\right], \\
H_\varphi^{\text{sw}}(kr,\theta,\varphi) &= -\text{i}kT_u(\varphi,\theta)\sin(\theta_1-\theta+\zeta_u)\frac{\exp[\text{i}kr\cos(\theta_1-\theta+\zeta_u)]}{(-\text{i}kr)^{1/2+\lambda_u}} \\
&\quad\times\left[1+O\left(\frac{1}{kr}\right)\right],
\end{aligned}
\tag{2.9}
$$

with

$$
T_u(\varphi,\theta)=\frac{C_{0,u}(\varphi)}{\pi\text{i}}\,\frac{(\text{e}^{2\pi\text{i}\mu}-1)\,\Gamma(\mu)}{[-\sin(\theta_1-\theta+\zeta)]^{\mu}}\sqrt{\frac{\sin\theta_1}{\sin\theta}},
$$

with

$$
C_{0u}(\varphi)=\sum_{n=-\infty}^{\infty}\text{i}^n\,\text{e}^{-\text{i}n\varphi}C_{0u}(n),
$$

and $\lambda_u = -(1/2)\cot\theta_1 \tan\zeta_u$, $\mu = 1 + \lambda_u$. A procedure of derivation of the constants $C_{0u}(n)$ in the excitation coefficients of the surface waves is described in Chapter 6 of [4].

In accordance with the introduced terminology, it is natural to call the wave (2.9) the electrical-type surface wave, which is excited, in particular, provided $\mathrm{Im}\,\zeta_u < 0$, $\mathrm{Im}\,\zeta_v > 0$. Contrary to this case, provided $\mathrm{Im}\,\zeta_v < 0$, $\mathrm{Im}\,\zeta_u > 0$, the magnetic-type surface wave propagates along the impedance cone.

From (2.9), we obtain some simple physical conclusions. The surface wave propagates with the velocity $c/\cosh(|\zeta_u|)$, that is slower than the spherical wave from the vertex having the velocity c. On the conical surface, its amplitude decreases as $O(1/(kr)^{1/2+\lambda_u})$. The factor $(kr)^{-\lambda_u}$ written in the form $\exp[-\lambda_u \log(kr)]$ enables us to interpret it as being responsible for the geometrical phase $\Phi_g(kr) = \mathrm{i}\lambda_u \log(kr)$. For the surface waves on an impedance cone, the geometrical phase is specified by the mean curvature of the circular cone, which also follows from [19] and [21].

Calculation of the excitation coefficients of the surface wave along a right-circular conical surface requires solution of an integral equation and should be a subject of a special study. In this way, validation of the results might be performed by considering limiting cases when approximate analytic solution of the integral equation is available, e.g. in the case of very large impedances.

2.2 Excitation of an electromagnetic surface wave by a dipole located near a plane impedance surface

In this section, we briefly describe a way in order to find an expression for the surface wave and its excitation coefficient, provided that this wave is generated by the current $\mathbf{J}_0 = \mathbf{P}\,\delta(x - x_0)\delta(y - y_0)\delta(z)$ in (2.1) of a Hertzian dipole located near a plane surface $x = 0$ with impedance η^+, see Figure 2.4. The dipole is arbitrarily oriented, $\mathbf{P} = (P_x, P_y, P_z)^{\mathrm{T}}$. The cylindrical coordinates (r, φ, z) are introduced for $x \geq 0$, Figure 2.4, $x_0 = r_0 \cos\varphi_0$, $y_0 = r\sin\varphi_0$, $z_0 = 0$. 0 As discussed in [4], Chapter 3, [5], in order to asymptotically evaluate the integrals representing the solution, we deformed the contours, took into account crossed polar singularities and applied steepest descent (or other) techniques. However, provided, say, $\varphi_0 \sim \Phi$, some additional contributions to the far field come into being. In this case, the point source is close to the wedge's face $\varphi = \Phi$ then the factor $\exp(-\mathrm{i}kr_0 \cos[\Phi - \varphi_0 - \zeta^+(\beta)])$ in (3.53) of [4] is not small even, if $kr_0 \gg 1$, and the corresponding primary surface wave is excited by the dipole.

This integral representation is determined from the double integral (see (3.42) in [4]) representing the reflected wave from the impedance plane surface with the dipole being close to it. As is well known the surface wave generated by the Hertzian dipole over the surface with the impedance $\sin\zeta^+$ is due to the contribution of a pole of the reflection coefficient $\overline{\overline{R}}^{+}(\alpha, \beta)$ which is captured in the process of deformation of the contour into the steepest descent path (SDP). These polar singularities are due to zeros of denominators $D^{\pm}(-\alpha, \beta) = [\sin(-\alpha \pm \Phi) \pm \sin\theta^{\pm}(\beta)]$

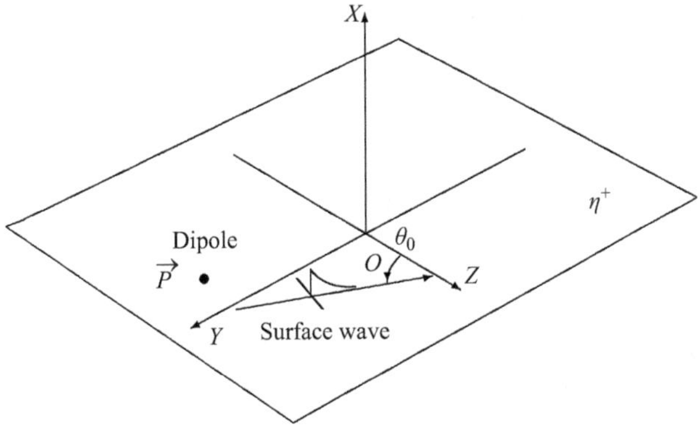

Figure 2.4 Surface wave from a dipole over an impedance plane

$[\sin(-\alpha \pm \Phi) \pm \sin \chi^{\pm}(\beta)]$ of the reflection coefficient $\overline{\overline{R}}^{+}(\alpha, \beta)$. We consider the polar singularity

$$\alpha_*(\beta) = \Phi + \zeta^+(\beta)$$

(with $\zeta^+ = \{\theta^+(\beta), \chi^+(\beta)\}$).

We assume that $\varphi_0 \sim \Phi$ and either $\zeta^+ = \theta^+$ with $\mathrm{Im}\theta^+ < 0$ or $\zeta^+ = \chi^+$ with $\mathrm{Im}\chi^+ < 0$ recalling that $\zeta^{\pm} = \{\theta^{\pm}, \chi^{\pm}\}$. Notice that by definition (see [4], Sect. 3.3.2)

$$\sin \zeta^+(\beta) = \frac{z^+}{\sin \beta},$$

where $z^{\pm} = \{\eta^{\pm}, (\eta^{\pm})^{-1}\}$. We assume $\eta^{\pm} = \mathrm{i}\,\mathrm{Im}(\eta^{\pm})$ purely imaginary, $|\,\mathrm{Im}(\eta^{\pm})| = |\eta^{\pm}|$ which implies that surface waves can actually propagate along the impedance surface provided that either $\mathrm{Im}(\eta^{\pm}) < 0$ or $\mathrm{Im}(\eta^{\pm})^{-1} < 0$.

The expression for the surface wave on an impedance surface excited by a dipole takes on the form (see also Sect. 3.5.1 in [4])

$$\begin{bmatrix} Z_0 H_z^{\mathrm{sw}} \\ E_z^{\mathrm{sw}} \end{bmatrix}_{z^+} = -\frac{k^3}{4\pi} \int\limits_{\Gamma(\pi/2)} \mathrm{d}\beta \ \mathrm{res}_{\alpha_*(\beta)} \overline{\overline{R}}^{+}(\alpha, \beta) \cdot \overline{U}_0(\alpha_*(\beta), \beta) \sin^2\!\beta \times \mathrm{e}^{\mathrm{i}k\{z \cos\beta + \sin\beta[r_0 \cos(\alpha_*(\beta) - \varphi_0) - r \,\mathrm{coscos}(\alpha_*(\beta) - 2\Phi + \varphi)]\}},$$

$\alpha_*(\beta) = \Phi + \zeta^+(\beta)$. Here the reflection coefficient $\overline{\overline{R}}^{+}(\alpha, \beta)$ is explicitly given by (3.40) from [4],

$$\overline{U}_0(\alpha, \beta) = [U_{10}(\alpha, \beta),\ U_{20}(\alpha, \beta)]^{\mathrm{T}},$$

$$U_{10}(\alpha, \beta) = \sin\alpha P_x - \cos\alpha P_y, \tag{2.10}$$

$$U_{20}(\alpha, \beta) = \cos\beta \left(\cos\alpha P_x + \sin\alpha P_y\right) + \sin\beta P_z.$$

After the change of the variable $\tau = \cos\beta$, we find

$$\begin{bmatrix} Z_0 H_z^{\mathrm{sw}} \\ E_z^{\mathrm{sw}} \end{bmatrix}_{z^+} = -\frac{k^3}{4\pi} \mathrm{e}^{\mathrm{i}k\{-r_0 \sin(\Phi-\varphi_0)z^+ - r\sin(\varphi-\Phi)z^+\}} \times \int\limits_{R'} \mathrm{d}\tau\, \overline{F}_+^p(\tau)\, \mathrm{e}^{\mathrm{i}k\rho_0\left\{z\tau/\rho_0+\sqrt{1+|z^+|^2-\tau^2}\right\}},$$

with

$$\rho_0 = r_0 \cos[\Phi - \varphi_0] - r\cos[\Phi - \varphi] > 0$$

and

$$\overline{F}_+^p(\tau)|_{\tau=\cos\beta} = -\sin\beta\ \mathrm{res}_{\alpha_*(\beta)} \overline{\overline{R}}^+(\alpha,\beta) \cdot \overline{U}_0(\alpha_*(\beta),\beta).$$

The contour R' goes along the real axis comprising the branch points $\pm\sqrt{1+|z^+|^2}$ from below (+) and above (−), the branch cuts are conducted from $\pm\sqrt{1+|z^+|^2}$ to $\pm\infty$ and $\sqrt{1+|z^+|^2-\tau^2} > 0$ as $\tau = 0$.

The stationary point of the latter integral, as $k\rho_0 \gg 1$,

$$\tau_p = \frac{z}{\rho_0} \frac{\sqrt{1+|z^+|^2}}{\sqrt{1+(z/\rho_0)^2}}$$

solves the equation

$$z = [r_0 \cos(\Phi - \varphi_0) - r\cos(\Phi - \varphi)] \frac{\tau_p}{\sqrt{1+|z^+|^2 - \tau_p^2}},$$

and the asymptotic expression for the leading term is written as

$$\begin{bmatrix} Z_0 H_z^{\mathrm{sw}} \\ E_z^{\mathrm{sw}} \end{bmatrix}_{z^+} = \frac{k^3 \mathrm{e}^{\mathrm{i}3\pi/4}}{4\pi^2} \sqrt{\frac{2\pi}{k\rho_0}} \frac{\rho_0^{3/4}(1+|z^+|^2)^{1/4}}{(z^2+\rho_0^2)^{3/4}} \overline{F}_+^p(\tau_p) \times \mathrm{e}^{\mathrm{i}k\sqrt{1+|z^+|^2}\sqrt{z^2+\rho_0^2}}\ \mathrm{e}^{\mathrm{i}k\{-r_0\sin(\Phi-\varphi_0)z^+ - r\sin(\Phi-\varphi)z^+\}} \times \left(1 + O\left(\frac{1}{k\rho_0}\right)\right) \tag{2.11}$$

remarking that $r_0 \sin(\Phi - \varphi_0) + r\sin(\Phi - \varphi) \geq 0$.

It is convenient to introduce the angle β_0 by the equality

$$\cos\beta_0 = \frac{z}{\sqrt{\rho_0^2 + z^2}} < 1.$$

We assume that $r_0 \cos(\Phi - \varphi_0) \gg r\cos(\Phi - \varphi)$ and also define the angle of 'incidence' θ_0 by

$$\cos\theta_0(\beta_0) = \sqrt{1+|z^+|^2}\cos\beta_0.$$

As a result, expression (2.11) can also be written in the form of

$$\begin{bmatrix} Z_0 H_z^{\mathrm{sw}} \\ E_z^{\mathrm{sw}} \end{bmatrix}_{z^+} = \overline{\mathscr{A}}(r_0, \varphi_0, \theta_0)\, \mathrm{e}^{\mathrm{i}k[z\cos\theta_0 - r\sin(\Phi-\varphi)z^+ - r\cos(\Phi-\varphi)\sqrt{\sin^2\theta_0 + |z^+|^2}]}$$

with the complex amplitude

$$\overline{\mathscr{A}}(r_0, \varphi_0, \theta_0) = \frac{k^3 \mathrm{e}^{\mathrm{i}3\pi/4}}{4\pi^2} \sqrt{\frac{2\pi}{k\rho_0}} \frac{\rho_0^{3/4}(1+|z^+|^2)^{1/4}}{(z^2+\rho_0^2)^{3/4}} \overline{F}_+^p(\cos\theta_0)$$
$$\times\, \mathrm{e}^{-\mathrm{i}k[r_0\sin(\Phi-\varphi_0)z^+ + r_0\cos(\Phi-\varphi_0)\sqrt{\sin^2\theta_0+|z^+|^2}]}.$$

It is worth remarking that in our assumptions, we have $\rho_0 \approx r_0\cos(\Phi-\varphi_0)$ and

$$\frac{\rho_0}{(z^2+\rho_0^2)^{1/2}} = \sin\beta_0.$$

As a result, the complex amplitude $\overline{\mathscr{A}}(r_0, \varphi_0, \theta_0)$ of the excited surface wave is specified by $r_0, \varphi_0, \theta_0, z^+$, i.e. by the position and orientation of the Hertzian dipole. It is also useful to introduce the complex angle

$$\varphi_0(\theta_0) = \Phi - \zeta^+(\theta_0),$$

where

$$\sin\zeta^+(\theta_0) = \frac{z^+}{\sin\theta_0}.$$

The surface wave excited by the dipole is then written as

$$\begin{bmatrix} Z_0 H_z^{\mathrm{sw}} \\ E_z^{\mathrm{sw}} \end{bmatrix}_{\zeta^+} = \overline{\mathscr{A}}\, \mathrm{e}^{\mathrm{i}k[z\cos\theta_0 - r\sin\theta_0\cos(\varphi_0(\theta_0)-\varphi)]} \tag{2.12}$$

with $\sin\theta_0\cos(\varphi_0(\theta_0)-\varphi) = \sin(\Phi-\varphi)z^+ + \cos(\Phi-\varphi)\sqrt{\sin^2\theta_0+|z^+|^2}$. The wave (2.12) can be interpreted as a skew incident at OZ plane wave with the angles $(\theta_0, \varphi_0(\theta_0))$ (actually φ_0 is complex) specifying the direction of incidence. In the next section, we make use of this simple observation and study the problem of scattering of a surface wave (2.12) by an impedance wedge (see also [15,16]).

2.3 Scattering of a skew incident surface wave by the edge on an impedance wedge

The impedance wedge under study is most conveniently described in a cylindrical coordinate system (r, φ, z), with its edge coinciding with the z-axis and its upper and lower faces being the half-planes $\varphi = \pm\Phi$ (Figure 2.5). The boundary conditions to be met by the electromagnetic field components turn out from (2.3)

$$E_z(r, \pm\Phi, z) = \pm Z_0\eta^\pm H_r(r, \pm\Phi, z), \quad E_r(r, \pm\Phi, z) = \mp Z_0\eta^\pm H_z(r, \pm\Phi, z).$$

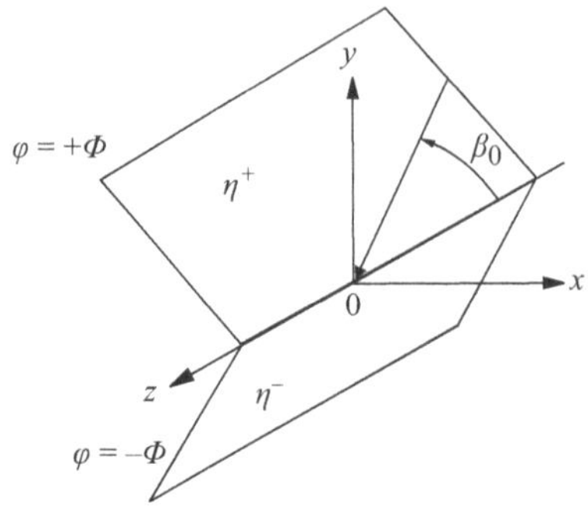

Figure 2.5 Diffraction of a surface wave at an impedance wedge

Now assume that the upper face of the wedge is purely inductive, i.e. $\eta^+ = -\mathrm{i}\left|\mathrm{Im}\,\eta^+\right|$. Let an E-mode electromagnetic surface wave of the type (2.12), but with a constant amplitude $\overline{U}_0$, move on the upper face of the wedge towards its edge under the angles (ϑ_0, φ_0)

$$\begin{bmatrix} Z_0 H_z^{\mathrm{inc}} \\ E_z^{\mathrm{inc}} \end{bmatrix} = \overline{U}_0\, \mathrm{e}^{\mathrm{i}[k'' z - k' r \cos(\varphi - \varphi_0)]}, \quad k' = k \sin\vartheta_0, \quad k'' = k\cos\vartheta_0,$$

with $\overline{U}_0 = [U_{10} \quad U_{20}]^{\mathrm{T}} = \left[-Z_0 H_0 \sin\beta_0 \quad -\eta^+ Z_0 H_0 \cos\beta_0\right]^{\mathrm{T}}$ and β_0 being the angle subtended by the wedge's edge and the direction of propagation. Taking this direction as the positive q^1-axis of the Cartesian coordinates (q^1, q^2, q^3) for the upper face of the wedge (see Section 2.1.1.1), the magnetic field of the incident surface wave in that coordinate system reads

$$Z_0 \mathbf{H}^{\mathrm{inc}} = (0, Z_0 H_0, 0)\mathrm{e}^{\mathrm{i}k\left(n^+ q^1 - q^3 \eta^+\right)}, \quad n^+ = \sqrt{1 - (\eta^+)^2}.$$

Being the ratio of the speed of light to that of the surface wave, n^+ is termed the refractive index of the surface wave supported by the upper face of the wedge. The respective electric field can be given as in Section 2.1.1.1.

The incident angles (ϑ_0, φ_0) depend upon β_0 and η^+ (via n^+)

$$\vartheta_0 = \arccos(n^+ \cos\beta_0), \quad \varphi_0 = \Phi - \vartheta^+, \quad \vartheta^{\pm} = \arcsin\left(\frac{\eta^{\pm}}{\sin\vartheta_0}\right). \tag{2.13}$$

In this section, ϑ_0 is assumed at first to be real, limiting β_0 to $|\cos\beta_0| < 1/n^+$. The last subsection tackles the case with $\cos\beta_0 > 1/n^+$.

Next we make use of an early work [15]. To this end, we consider the z-components of the total field

$$[Z_0 H_z(r, \varphi; z)\; E_z(r, \varphi; z)]^{\mathrm{T}} = \overline{U}(r, \varphi) \exp\left(\mathrm{i}k'' z\right). \tag{2.14}$$

$\overline{U}(r,\varphi) = [U_1(r,\varphi)\ U_2(r,\varphi)]^{\mathrm{T}}$ solves the two-dimensional Helmholtz equation in free space outside the wedge

$$\left[\frac{1}{r}\frac{\partial}{\partial r}\left(r\frac{\partial}{\partial r}\right) + \frac{1}{r^2}\frac{\partial^2}{\partial\varphi^2} + (k')^2\right]\overline{U}(r,\varphi) = 0$$

and satisfies the respective conditions on the faces of the wedge

$$\overline{\overline{I}}\frac{\mathrm{i}\partial\overline{U}(r,\varphi)}{kr\partial\varphi}\Bigg|_{\varphi=\pm\Phi} = \mp\sin^2\vartheta_0\overline{\overline{\mathscr{A}}}^{\pm}\overline{U}(r,\pm\Phi) + \cos\vartheta_0\overline{\overline{\mathscr{B}}}\frac{\mathrm{i}\partial\overline{U}(r,\pm\Phi)}{k\partial r}, \tag{2.15}$$

with

$$\overline{\overline{I}} = \begin{bmatrix} 1 & 0 \\ 0 & 1 \end{bmatrix}, \quad \overline{\overline{\mathscr{A}}}^{\pm} = \begin{bmatrix} \eta^{\pm} & 0 \\ 0 & 1/\eta^{\pm} \end{bmatrix}, \quad \overline{\overline{\mathscr{B}}} = \begin{bmatrix} 0 & -1 \\ 1 & 0 \end{bmatrix}.$$

In line with the Meixner edge condition, $\overline{U}(r,\varphi)$ remains finite as $r \to 0$

$$\overline{U}(r,\varphi) = \left[C_1 + \mathrm{O}(r^{\delta}) \quad C_2 + \mathrm{O}(r^{\delta})\right]^{\mathrm{T}}, \quad \delta > 0,$$

$C_{1,2}$ being constant. Furthermore, it is subject to the radiation conditions, given most concisely for the spectra of $\overline{U}(r,\varphi)$ in Section 2.3.1.

2.3.1 Integral equations for the spectra

As usual, $\overline{U}(r,\varphi)$ is expressed in terms of the Sommerfeld integrals:

$$\overline{U}(r,\varphi) = \frac{1}{2\pi\mathrm{i}}\int_{\gamma}\overline{f}(\alpha+\varphi)\,\mathrm{e}^{-\mathrm{i}k'r\cos\alpha}\mathrm{d}\alpha, \tag{2.16}$$

where γ denotes the Sommerfeld double-loop and $\overline{f}(\alpha) = [f_1(\alpha)\ f_2(\alpha)]^{\mathrm{T}}$ the spectra to be determined. The radiation condition demands that $\overline{f}(\alpha) - \overline{U}_0/\ (\alpha - \varphi_0)$ be regular in the strip $|\mathrm{Re}\,\alpha| \leq \Phi$, where φ_0 is defined in (2.13).

Inserting (2.16) into the boundary condition (2.15) and inverting the Sommerfeld integrals, we get a system of equations for the spectra. For example, the equation for $f_1(\alpha)$ reads

$$f_1(\alpha+2\Phi) - \frac{b_2^{+}(\alpha)}{b_2^{-}(\alpha)}f_1(\alpha-2\Phi) = q_1(\alpha)f_1(\alpha), \tag{2.17}$$

with the coefficients $b_2^{+}(\alpha), b_2^{-}(\alpha)$ and $q_1(\alpha)$ given in Appendix A.

On use of $f_1(\alpha) = F_0(\alpha)\mathscr{F}_1(\alpha)$, the previous functional equation can be simplified to

$$\mathscr{F}_1(\alpha+2\Phi) + \mathscr{F}_1(\alpha-2\Phi) = Q_1(\alpha)\mathscr{F}_1(-\alpha), \tag{2.18}$$

with $Q_1(\alpha) = q_1(\alpha)F_0(-\alpha)/F_0(\alpha+2\Phi)$ and the auxiliary function $F_0(\alpha)$ given in closed form in Appendix A.

By making use of the S-integrals and taking into account the edge and radiation conditions, an integral equivalent of (2.18) in the strip $|\mathrm{Re}\,\alpha| \leq 2\Phi$ for $\Phi > \pi/2$ reads

$$\mathscr{F}_1(\alpha) = \frac{\nu U_{10}/F_0(\varphi_0)}{\sin \nu(\alpha - \varphi_0)} + A_1^+ \mathrm{e}^{-\mathrm{i}\nu\alpha} + A_1^- \mathrm{e}^{\mathrm{i}\nu\alpha}$$
$$-\frac{\mathrm{i}}{8\Phi} \int_{-\mathrm{i}\infty}^{+\mathrm{i}\infty} \frac{Q_1(-t)\mathscr{F}_1(t)}{\cos \nu(\alpha + t)} \mathrm{d}t, \ \nu = \frac{\pi}{4\Phi}. \tag{2.19}$$

The constants $A_1^{\pm}$ are fixed by deleting non-physical poles

$$f_1\left(\pm\Phi - \frac{\pi}{2}\right) = b_1^{\pm}\left(\mp\Phi - \frac{\pi}{2}\right) f_1\left(\pm\Phi + \frac{\pi}{2}\right). \tag{2.20}$$

The coefficients $b_1^{\pm}(\alpha)$ are given in Appendix A.

Relation (2.19), together with (2.20), amounts to an integral equation for $\mathscr{F}_1(\alpha)$ on the imaginary axis of the complex α-plane. These values can be obtained by solving numerically the integral equation and then extrapolated into the strip $|\mathrm{Re}\,\alpha| \leq 2\Phi$ on use of (2.19). In a similar manner, the second spectrum $f_2(\alpha)$ can be deduced. Inserting them into the Sommerfeld integrals (2.16) leads to an exact solution, although not in an explicit form, to the problem under study.

2.3.2 Far-field expansion

Deforming the path of integration γ in (2.16), $\overline{U}(r, \varphi)$ can be rewritten as

$$\overline{U}(r, \varphi) = \overline{U}^{\mathrm{go}}(r, \varphi) + \overline{U}^{\mathrm{sw}}(r, \varphi) + \overline{U}^{\mathrm{d}}(r, \varphi). \tag{2.21}$$

The geometrical–optical part is related to the incident surface wave according to

$$\overline{U}^{\mathrm{go}}(r, \varphi) = \mathrm{H}(\varphi - \Phi + \pi - \mathrm{gd}(\mathrm{Im}\,\varphi_0))\overline{U}_0 \mathrm{e}^{-\mathrm{i}k'r\cos(\varphi - \varphi_0)},$$

where $\mathrm{H}(\,\cdot\,)$ stands for the Heaviside unit-step function.

Of particular importance are the surface waves

$$\overline{U}^{\mathrm{sw}}(r, \varphi) = \sum_{\ell=1}^{4} \mathrm{H}(A_\ell)\overline{R}_\ell \mathrm{e}^{-\mathrm{i}k'r\cos\alpha_\ell}, \tag{2.22}$$

where the poles and residues of $\overline{f}(\alpha + \varphi)$ related to surface waves are given in Table 2.1. And $\mathrm{gd}(x) = \arctan(\sinh x)$ stands for the Gudermann function.

For large $k'r$, the diffracted part $\overline{U}^{\mathrm{d}}(r, \varphi)$ is given by

$$\overline{U}^{\mathrm{d}}(r, \varphi) \sim \frac{\overline{Q}(\varphi)\,\mathrm{e}^{\mathrm{i}k'r}}{\sqrt{r}} \tag{2.23}$$

with the non-uniform diffraction coefficient (scattering diagram)

$$\overline{Q}(\varphi) = \left[\overline{f}(\varphi - \pi) - \overline{f}(\varphi + \pi)\right]\sqrt{\frac{\mathrm{i}}{(2\pi k')}}. \tag{2.24}$$

A uniform expression for the diffracted field can be given in a similar way as in [15].

Obviously, for the edge-diffracted rays the components of the wave vector are

$$K_z = k \cos \vartheta_0, \quad K_r = k \sin \vartheta_0,$$

implying that in the far field, these waves are located on a cone axis of which is the edge of the wedge and interior semivertex angle of which is ϑ_0. Therefore, the law of edge diffraction excited by an incident surface wave at the edge of an impedance wedge is given by the first of (2.13) rewritten as

$$\sin \kappa_{\rm d} = n^+ \sin \kappa^+, \quad \kappa_{\rm d} = \frac{\pi}{2} - \vartheta_0, \quad \kappa^+ = \frac{\pi}{2} - \beta_0. \tag{2.25}$$

The similarity of the previous relation to the Snell law of refraction for wave transmission through an interface between two different media is due to the edge-diffracted rays being excited by a slower incident surface wave. As a consequence, there exists a critical angle corresponding to $\kappa_{\rm d} = \pi/2$ (hence $\vartheta_0 = 0$)

$$\kappa_{\rm c}^+ = \arcsin(1/n^+)$$

beyond which, that is for $\kappa^+ > \kappa_{\rm c}^+$ the 'edge-diffracted' rays propagate along the z-axis, being localised in a neighbourhood of the edge; see Section 2.3.4.

As the surface waves given in (2.22) are excited by the incident surface wave, they can be regarded as the reflection and refraction of the latter at the edge of an impedance wedge.

2.3.3 *Reflection and refraction of an incident surface wave at the edge of an impedance wedge*

In close connection with the reflection and refraction of waves are their respective angles. The explicitly given poles α_ℓ (see Table 2.1) are very useful in this respect. On the upper face, the angle of incidence with respect to the normal to the edge is $\kappa^+ = \pi/2 - \beta_0$, and the corresponding components of the wave vector are

$$K_z = k \cos \vartheta_0, \quad K_r = -k \sin \vartheta_0 \sqrt{1 - \left(\frac{\eta^+}{\sin \vartheta_0}\right)^2}.$$

Table 2.1 Surface-wave poles and residues of $\overline{f}(\alpha + \varphi)$

ℓ	**Poles α_ℓ**	**Residues $\overline{R}_\ell$**	**Arguments A_ℓ**
1	$\pi + \Phi + \vartheta^+ - \varphi$	$\overline{\overline{R}}^+_{\vartheta^+} \cdot \overline{f}(\Phi - \pi - \vartheta^+)$	$-\Phi + \varphi - \operatorname{Re}\vartheta^+ - \operatorname{gd}(\operatorname{Im}\vartheta^+)$
2	$-\pi - \Phi - \vartheta^- - \varphi$	$\overline{\overline{R}}^+_{\vartheta^-} \cdot \overline{f}(-\Phi + \pi + \vartheta^-)$	$-\Phi - \varphi - \operatorname{Re}\vartheta^- - \operatorname{gd}(\operatorname{Im}\vartheta^-)$
3	$\pi + \Phi + \chi^+ - \varphi$	$\overline{\overline{R}}^+_{\chi^+} \cdot \overline{f}(\Phi - \pi - \chi^+)$	$-\Phi + \varphi - \operatorname{Re}\chi^+ - \operatorname{gd}(\operatorname{Im}\chi^+)$
4	$-\pi - \Phi - \chi^- - \varphi$	$\overline{\overline{R}}^+_{\chi^-} \cdot \overline{f}(-\Phi + \pi + \chi^-)$	$-\Phi - \varphi - \operatorname{Re}\chi^- - \operatorname{gd}(\operatorname{Im}\chi^-)$

For the excited surface wave on the upper face, we get from (2.22) that

$$K_z = k\cos\vartheta_0, \quad K_r = k\sin\vartheta_0\sqrt{1-\left(\frac{\eta^+}{\sin\vartheta_0}\right)^2},$$

implying that the angle of reflection κ_{r}^+ equals that of incidence κ^+, namely

$$\kappa_{\mathrm{r}}^+ = \kappa^+. \tag{2.26}$$

Similarly, for the surface wave excited on the lower face of the wedge, the respective components of the wave vector are

$$K_z = k\cos\vartheta_0, \quad K_r = k\sin\vartheta_0\sqrt{1-\left(\frac{z^-}{\sin\vartheta_0}\right)^2},$$

where $z^\pm = \{\eta^\pm, 1/\eta^\pm\}$ with $\mathrm{Im}\, z^\pm < 0$. Hence, the angle of refraction is defined according to

$$\sin\kappa^- = \frac{K_z}{\sqrt{K_z^2+K_r^2}} = \frac{\cos\vartheta_0}{n^-},$$

where $n^- = \sqrt{1-(z^-)^2}$ stands for the refractive index for the surface wave supported by the lower face of the wedge.

On use of (2.13), we arrive at a formula

$$n^+\sin\kappa^+ = n^-\sin\kappa^-, \tag{2.27}$$

which reminds us of the Snell law of refraction for wave transmission through a smooth interface between two different media. In spite of this similarity, we should not forget that in the present case with $\kappa^+ < \kappa_{\mathrm{c}}^+$, one more wave is born at the edge, namely the diffracted space-wave $\overline{U}^{\mathrm{d}}$ given in (2.23).

Another aspect of the study concerns the type of reflected and refracted surface waves. As can be expected, the reflected surface wave is of the same type as that of the incident surface wave; in this example also an E-type wave. The type of the refracted part is determined by the electrical property of the lower face of the wedge, as implied in the previous study of the law of refraction: if the lower face is also inductive with $\mathrm{Im}\,\eta^- < 0$, the refracted surface wave is of the same E-type; if the lower face is but capacitive ($\mathrm{Im}\,\eta^- > 0$), then the refracted part is of the other type, here the H-type. Obviously, such a conversion of the type of electromagnetic surface waves happens only under skew incidence, as shown by the explicit expressions of $\overline{\overline{R}}^{\pm}_{\vartheta^\pm,\chi^\pm}$

$$\overline{\overline{R}}^{\pm}_{\vartheta^\pm} = \pm\frac{2\tan\vartheta^\pm}{\sin\vartheta^\pm - \sin\chi^\pm}\begin{bmatrix} -(\csc^2\vartheta_0\sin\vartheta^\pm - \sin\chi^\pm) & \pm\cot\vartheta_0\csc\vartheta_0\cos\vartheta^\pm \\ \mp\cot\vartheta_0\csc\vartheta_0\cos\vartheta^\pm & -\cot^2\vartheta_0\sin\vartheta^\pm \end{bmatrix},$$

$$\overline{\overline{R}}^{\pm}_{\chi^\pm} = \pm\frac{2\tan\chi^\pm}{\sin\chi^\pm - \sin\vartheta^\pm}\begin{bmatrix} -\cot^2\vartheta_0\sin\chi^\pm & \pm\cot\vartheta_0\csc\vartheta_0\cos\chi^\pm \\ \mp\cot\vartheta_0\csc\vartheta_0\cos\chi^\pm & -(\csc^2\vartheta_0\sin\chi^\pm - \sin\vartheta^\pm) \end{bmatrix}.$$

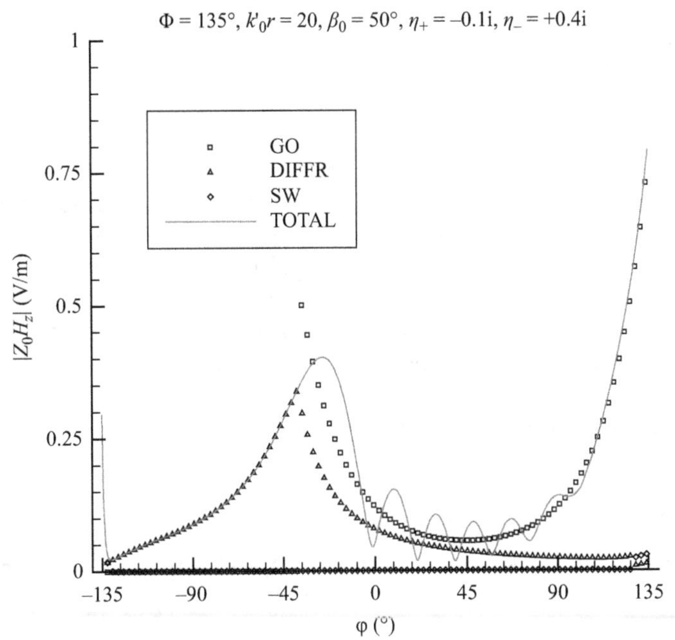

Figure 2.6 Diffraction of an E-type surface wave at an impedance wedge (cf. Figure 2.5) with an inductive upper face and a capacitive lower face

Now let us look at a numerical example shown in Figure 2.6. This figure displays $|Z_0H_z|$ of the different wave ingredients as well as the total field excited by an incident electromagnetic surface wave of the E-type by an impedance wedge with an inductive upper face and a capacitive lower face. Hence, the surface wave propagating away from the edge along the lower face of the wedge is of the H-type. It is noted that the reflection and refraction coefficients are calculated numerically, by solving at first the Fredholm integral equations of the second kind and then inserting the spectra $\overline{f}(\alpha)$ in the residues $\overline{R}_\ell$, as shown in Table 2.1.

2.3.4 Beyond the critical angle of edge diffraction

Now consider incidence of a surface wave under the condition $\kappa_c^+ < \kappa^+ < \pi/2$ with

$$\kappa_c^+ = \arcsin\left(\frac{1}{n^+}\right), \quad \kappa^+ = \frac{\pi}{2} - \beta_0.$$

In line with (2.13), ϑ_0 becomes purely imaginary with

$$\vartheta_0 = \mathrm{i}\,\mathrm{arccosh}(n^+ \sin\kappa^+).$$

Therefore, we get

$$\vartheta^+ = -\frac{\pi}{2} - \mathrm{i}\,\mathrm{arccosh}\left(\frac{|\eta^+|}{\sqrt{(n^+ \sin\kappa^+)^2 - 1}}\right),$$

$$\varphi_0 = \Phi - \vartheta^+.$$

The Sommerfeld integral (2.16) still expresses $\overline{U}(r,\varphi)$. But for the sake of convergence, the contour of integration runs now along $(\pi + \mathrm{i}\infty, \pi + \mathrm{i}\delta] \cup [\pi + \mathrm{i}\delta, -\pi + \mathrm{i}\delta] \cup [-\pi + \mathrm{i}\delta, -\pi + \mathrm{i}\infty)$ and its mirror image with respect to the origin of the complex α-plane. The positive constant δ is chosen in such a way that the two loops of integration contain no singularities of the spectra.

Even without determining the spectra, a formal asymptotic analysis of the Sommerfeld integral (2.16) affords useful insights into the related wave phenomena. Under this circumstance, the saddle points remain at $\alpha = \mp\pi$ with the SDPs given by

$$\mathrm{SDP}(-\pi):\ (-\pi - \mathrm{i}\infty, -\pi + \mathrm{i}\infty),$$

$$\mathrm{SDP}(\pi):\quad (\pi + \mathrm{i}\infty, \pi - \mathrm{i}\infty).$$

As a result, the formulae derived in Section 2.3.2 remain valid, except that the Gudermann functions appearing in the arguments A_ℓ in Table 2.1 are set to zero.

Remark 2.3. *In the case of a complex-valued k', the Gudermann function* gd(x) *is to be replaced by its generalisation* Gd(x, arg k') *given by (see, for example [6])*

$$\mathrm{Gd}(x,y) = \arctan\frac{\sinh x \cos y}{1 + \cosh x \sin y},$$

with the two special cases used in this section

$$\mathrm{Gd}(x, y=0) = \mathrm{gd}(x), \quad \mathrm{Gd}\left(x, y = \frac{\pi}{2}\right) = 0.$$

Hence, the SDPs for the Sommerfeld integral (2.16) in the general case are

$$\mathrm{SDP}_{\arg k'}(\mp\pi):\quad \mathrm{Re}\,\alpha = \mp\pi - \mathrm{Gd}(\mathrm{Im}\,\alpha, \arg k').$$

According to (2.23), the 'edge-diffracted' component $\overline{U}^{\mathrm{d}}(r,\varphi)$ decreases exponentially away from the edge with r, leading to

$$\left[Z_0 H_z^{\mathrm{d}}(r,\varphi;z)\ E_z^{\mathrm{d}}(r,\varphi;z)\right]^{\mathrm{T}} = \overline{U}^{\mathrm{d}}(r,\varphi)\mathrm{e}^{\mathrm{i}kz\cos\vartheta_0} \sim \overline{Q}(\varphi)\frac{\mathrm{e}^{k(\mathrm{i}z\cos\vartheta_0 - r\sinh|\vartheta_0|)}}{\sqrt{r}},$$

being a wave clung to and propagating along the edge. Furthermore, the larger the angle of incidence κ^+, the stronger the concentration of this wave ingredient to the edge.

On the lower face of the wedge, the Brewster angle ζ^- is given by

$$\zeta^- = \begin{cases} -\frac{\pi}{2} - \mathrm{i}\,\mathrm{arccosh}\,\frac{|z^-|}{\sqrt{(n^+\sin\kappa^+)^2-1}}, & \text{for } \kappa_{\mathrm{c}}^+ < \kappa^+ < \kappa_{\mathrm{t.r.}}^+, \\ -\arcsin\frac{|z^-|}{\sqrt{(n^+\sin\kappa^+)^2-1}}, & \text{for } \kappa_{\mathrm{t.r.}}^+ < \kappa^+ < \frac{\pi}{2}, \end{cases}$$

in the case of $|z^-| < |z^+|$ with

$$\kappa^+_{\text{t.r.}} = \arcsin \frac{n^-}{n^+}$$

and

$$\zeta^- = -\frac{\pi}{2} - \text{i}\,\text{arccosh}\,\frac{|z^-|}{\sqrt{(n^+ \sin\kappa^+)^2 - 1}}, \text{ for } \kappa^+_{\text{c}} < \kappa^+ < \frac{\pi}{2}$$

otherwise.

It is worth taking a close look at the case $|z^-| < |z^+|$. For $\kappa^+_{\text{c}} < \kappa^+ < \kappa^+_{\text{t.r.}}$, the law of the refraction for surface waves at an edge (2.27) holds good. If the angle of incidence κ^+ exceeds $\kappa^+_{\text{t.r.}}$, the 'transmitted' surface wave at the lower face of the wedge behaves in line with (2.22):

$$\text{e}^{k\{\text{i}z\cos\vartheta_0 - [r\cos(\Phi+\varphi)\sinh|\vartheta_0|\cos\zeta^- + r\sin(\Phi+\varphi)|z^-|]\}},$$

again clung to and moving along the edge! It implies that the incident surface wave is completely reflected at the edge, and the very angle $\kappa^+_{\text{t.r.}}$ is called the angle of total reflection.

The geometrical properties of an incident surface wave at the edge of an impedance wedge, like the laws of reflection (2.26) and refraction (2.27) and possible total reflection, have been known for scalar waves since 1965 [8]. The conversion of surface waves of the E-type to those of the H-type and vice versa at the edge of an impedance wedge, however, seems to be unique for vectorial waves such as electromagnetic waves studied in this chapter.

2.4 Conclusion

In this chapter, we made use of the concept of the Leontovich (impedance) boundary conditions and discussed excitation and propagation of surface waves as well as their interaction with some canonical singularities of the surface like edges or conical points. Although the validity of the impedance boundary conditions fails in a close neighbourhood of the singular points, nevertheless these conditions are widely applicable in practice and the corresponding applications give reliable and accurate results.

An adequate use of the mathematical methods enabled us to give a motivated description of the wave phenomena arising in the process of propagation and scattering of the surface waves supported by impedance surfaces. In this way, we could efficiently describe scattering of an incident surface wave at the edge, calculate amplitude and phases of the reflected, transmitted and diffracted waves. In a simple manner, the Geometrical Optics laws of reflection and transmission of the surface wave at the edge of the wedge are also deduced as well as some analysis of the depolarisation of the surface waves is given.

These results have been obtained for the angle of incidence of the surface wave which is less than the first critical angle defined earlier. However, the study of the reflection and transmission coefficient for the other angles of incidence of the surface wave will be given elsewhere.

Appendix A

To make this chapter self-sustained, several functions used in Section 2.3 are explicitly given as follows:

$$b_1^{\pm}(\alpha) = \frac{-2\cot^2\vartheta_0 \sin^2(\alpha \pm \Phi) - [\sin(\alpha \pm \Phi) \mp \sin\vartheta^{\pm}][\sin(\alpha \pm \Phi) \pm \sin\chi^{\pm}]}{[\sin(\alpha \pm \Phi) \pm \sin\vartheta^{\pm}][\sin(\alpha \pm \Phi) \pm \sin\chi^{\pm}]},$$

$$b_2^{\pm}(\alpha) = \frac{2\cot\vartheta_0 \csc\vartheta_0 \sin(\alpha \pm \Phi)\cos(\alpha \pm \Phi)}{[\sin(\alpha \pm \Phi) \pm \sin\vartheta^{\pm}][\sin(\alpha \pm \Phi) \pm \sin\chi^{\pm}]},$$

$$q_1(\alpha) = b_1^{+}(\alpha) - \frac{b_2^{+}(\alpha)}{b_2^{-}(\alpha)} b_1^{-}(\alpha).$$

$$F_0(\alpha) = \frac{\Psi_0(\alpha)}{\sin[\nu(\alpha - \Phi - \pi/2)]\sin[\nu(\alpha + \Phi + \pi/2)]}, \qquad \nu = \frac{\pi}{4\Phi},$$

$$\begin{aligned}\Psi_0(\alpha) &= \frac{\chi_\Phi(\alpha + \Phi - \pi)\chi_\Phi(\alpha + \Phi)\chi_\Phi(\alpha + \Phi + \pi/2)}{\chi_\Phi(\alpha - \Phi + \pi)\chi_\Phi(\alpha - \Phi)\chi_\Phi(\alpha - \Phi - \pi/2)} \\ &\quad\times \frac{\chi_\Phi(\alpha + \Phi - \pi/2)\chi_\Phi(\alpha - \Phi - \chi^- + \pi)\chi_\Phi(\alpha - \Phi + \chi^-)}{\chi_\Phi(\alpha - \Phi + \pi/2)\chi_\Phi(\alpha + \Phi + \chi^+ - \pi)\chi_\Phi(\alpha + \Phi - \chi^+)} \\ &\quad\times \frac{\chi_\Phi(\alpha - \Phi - \vartheta^- + \pi)\chi_\Phi(\alpha - \Phi + \vartheta^-)}{\chi_\Phi(\alpha + \Phi + \vartheta^+ - \pi)\chi_\Phi(\alpha + \Phi - \vartheta^+)},\end{aligned}$$

where $\chi_\Phi(\alpha)$ stands for a special function introduced by Bobrovnikov and is defined by the first-order functional difference equation [24]

$$\chi_\Phi(\alpha + 2\Phi) = \cos\left(\frac{\alpha}{2}\right)\chi_\Phi(\alpha - 2\Phi).$$

More on this special function, and especially its efficient computation, can be found for instance in [4].

A.1 Brewster angles

The second Brewster angle for the upper face χ^+ may be either complex valued or purely real, depending upon $|\eta^+|$ and κ^+. In the case of $|\eta^+| > 1$, we have

$$\chi^+ = \begin{cases} \dfrac{\pi}{2} + \mathrm{i}\,\mathrm{arccosh}\dfrac{1}{|\eta^+|\sqrt{(n^+ \sin\kappa^+)^2 - 1}}, & \text{for } \kappa_c^+ < \kappa^+ < \arcsin\frac{1}{|\eta^+|}, \\ \arcsin\dfrac{1}{|\eta^+|\sqrt{(n^+ \sin\kappa^+)^2 - 1}}, & \text{for } \arcsin\frac{1}{|\eta^+|} < \kappa^+ < \frac{\pi}{2}, \end{cases}$$

in the case of $|\eta^+| < 1$, there is

$$\chi^+ = \frac{\pi}{2} + \mathrm{i}\,\mathrm{arccosh}\frac{1}{|\eta^+|\sqrt{(n^+ \sin\kappa^+)^2 - 1}}$$

for $\kappa_c^+ < \kappa^+ < \pi/2$.

The second Brewster angle of the lower face $\tilde{\xi}^-$ takes the form

$$\tilde{\xi}^- = \begin{cases} \frac{\pi}{2} + \mathrm{i}\,\mathrm{arccosh}\,\frac{1}{|z^-|\sqrt{(n^+ \sin\kappa^+)^2 - 1}}, & \text{for } \kappa_c^+ < \kappa^+ < \arcsin\sqrt{\frac{1+1/|z^-|^2}{1+|z^+|^2}}, \\ \arcsin\,\frac{1}{|z^-|\sqrt{(n^+ \sin\kappa^+)^2 - 1}}, & \text{for } \arcsin\sqrt{\frac{1+1/|z^-|^2}{1+|z^+|^2}} < \kappa^+ < \frac{\pi}{2}, \end{cases}$$

in the case of $1/|z^-| < |z^+|$ and

$$\tilde{\xi}^- = \frac{\pi}{2} + \mathrm{i}\,\mathrm{arccosh}\,\frac{1}{|z^-|\sqrt{(n^+ \sin\kappa^+)^2 - 1}}, \quad \text{for } \kappa_c^+ < \kappa^+ < \frac{\pi}{2}$$

otherwise.

Acknowledgement

One of the authors (MAL) was supported in part by the grant of the Russian Science Foundation, RSCF 17-11-01126.

References

[1] Frezza F and Tedeschi N. Electromagnetic inhomogeneous waves at planar boundaries: tutorial. J Opt Soc Am A. 2015;32(8):1485–1501. Available from: http://josaa.osa.org/abstract.cfm?URI=josaa-32-8-1485.

[2] Lyalinov MA. Scattering of an acoustic axially symmetric surface wave propagating to the vertex of a right-circular impedance cone. Wave Motion. 2010;47(4):241–252.

[3] Grikurov VE and Lyalinov MA. Diffraction of the surface H-polarized wave by an angular break of a thin dielectric slab. J Math Sci. 2008;155(3):390–396.

[4] Lyalinov MA and Zhu NY. Scattering of Waves by Wedges and Cones with Impedance Boundary Conditions, in: Mario Boella Series on Electromagnetism in Information and Communication. 1st ed. Uslenghi P, editor. Edison, NJ: SciTech-IET; 2012.
[5] Lyalinov MA and Zhu NY. Electromagnetic scattering of a dipole-field by an impedance wedge, Part I: Far-field space waves. IEEE Trans Antennas Propag. 2013;61(1):329–337.
[6] Lyalinov MA and Zhu NY. Scattering of an electromagnetic surface wave from a Hertzian dipole by the edge of an impedance wedge. J Math Sci. 2017;226(6):768–778.
[7] Starovoytova RP, Bobrovnikov MS, and Kislitsina VN. Diffraction of a surface wave at the break of an impedance plane. Radio Eng Electron Phys. 1962;7:232–240.
[8] Bobrovnikov MS, Ponomareva V, Myshkin V, *et al.* Diffraction of a surface wave incident at an arbitrary angle on the edge of a plane. Sov Phys J. 1965;8(1): 117–121.
[9] Zon VB. Reflection, refraction, and transformation into photons of surface plasmons on a metal wedge. J Opt Soc Am B. 2007;24:1960–1967.
[10] Zon VB, Zon VA, Klyuev AN, *et al.* New method for measuring the IR surface impedance of metals. Opt Spectrosc. 2010;108:637–639.
[11] Zon VB and Zon VA. Terahertz surface plasmon polaritons on a conductive right circular cone: analytical description and experimental verification. Phys Rev A. 2007;84:013816.
[12] Kotelnikov IA, Gerasimov VV, and Knyazev BA. Diffraction of a surface wave on a conducting rectangular wedge. Phys Rev A. 2013;87(2):023828.
[13] Ropers C, Neacsu CC, Elsaesser T, Albrecht M, Raschke MB, and Lienau C. Grating-coupling of surface plasmons onto metallic tips: a nanoconfined light source. Nanoletters. 2007;7(9):2784–2788.
[14] Lyalinov MA. Acoustic scattering of a plane wave by a circular penetrable cone. Wave Motion. 2011;48(1):62–82.
[15] Lyalinov MA and Zhu NY. Diffraction of a skew incident plane electromagnetic wave by an impedance wedge. Wave Motion. 2006;44(1):21–43.
[16] Lyalinov MA and Zhu NY. Diffraction of a skew incident plane electromagnetic wave by a wedge with axially anisotropic impedance faces. Radio Sci. 2007;42(6):RS6S03.
[17] Maliuzhinets GD. Excitation, reflection and emission of surface waves from a wedge with given impedances. Sov Phys Dokl. 1958;3:753–755.
[18] Babich VM, Lyalinov MA, and Grikurov VE. Diffraction Theory: The Sommerfeld-Malyuzhinets Technique. 1st ed. Oxford, UK: Alpha Science; 2008.
[19] Babich VM and Kuznetsov AV. Propagation of surface electromagnetic waves similar to Rayleigh waves in the case of Leontovich boundary conditions. J Math Sci. 2006;138(2):5483–5490.
[20] Babich VM and Kirpichnikova NY. A new approach to the problem of the Rayleigh wave propagation along the boundary of a nonhomogeneous elastic body. Wave Motion. 2004;40:209–223.

[21] Grimshaw R. Propagation of surface waves at high frequencies. IMA J Appl Math. 1968;4(2):174–193.

[22] Babich VM, Buldyrev VS, and Molotkov IA. Space-time Ray Method: Linear and Non-linear Waves. 1st ed. Leningrad, SU: Leningrad Univ Press; 1985.

[23] Bilow HJ. Guided waves on a planar tensor impedance surface. IEEE Transactions on Antennas and Propagation. 2003;51(10): 2788–2792.

[24] Bobrovnikov MS. Diffraction of cylindrical waves by an ideally conducting wedge in an anisotropic plasma. Sov Phys J. 1968;11(5):8–13. Available from: http://dx.doi.org/10.1007/BF00816591.

Chapter 3

Dielectric-wedge Fourier series

Svend Berntsen[1,2]

3.1 Introduction

Electromagnetic wave propagation from one medium into a different wedge-shaped medium is a difficult problem. No explicit closed-form solution of the wedge-diffraction problem with an arbitrary refractive index has been obtained so far. For equal wave numbers, the exact solution was obtained in [1]. Several authors have used the Kontorovich–Lebedev transform for a formulation of the wedge-diffraction problem [2–6]. Diffraction by a wedge with impedance boundary conditions was investigated in [7–9]. A general way of proving existence and uniqueness is to find a representation of the diffraction problem by integral equations. An integral equation method in combination with geometrical optics for the wedge problem is described in [10,11]. A Fredholm second-kind integral equation was used in [12], and its generalization to the case of a skew incidence is described in [13]. A generalized Wiener–Hopf representation for a formulation of the diffraction problem by a Fredholm equation was used in [14]. Numerical solution of the equations can be found in these papers. The reader interested in a border introduction into the subject of diffraction by wedges is referred to the books [15–18].

The present chapter formulates an integral equation for the H-polarization diffraction problem (the magnetic field vector is parallel to the edge of the wedge). This equation has a unique solution in the space of functions with finite energy; see [13] and references therein. The behavior of an H-polarized field at the edge of a dielectric wedge has been investigated by many authors, e.g., [19–22]. It is suggested that the electric field is singular at the tip of the wedge. Namely, the electric field $\mathbf{E}$ behaves like the static field in polar coordinates:

$$\lim_{r\to 0} r^{1-s} \mid \mathbf{E}(r,\theta) \mid = c(\theta), \tag{3.1}$$

where $c(\theta)$ is proportional to a sinus function and s is the root in the interval $[0.5, 1]$ of the following equation:

$$(1 - \varepsilon_r)\sin s(2\nu - \pi) = (\varepsilon_r + 1)\sin s\pi, \quad \varepsilon_r = \frac{\varepsilon_1}{\varepsilon_0}, \tag{3.2}$$

[1] Department of Mathematical Sciences, Aalborg University, Aalborg, Denmark
[2] Prof. Berntsen passed away on 20.10.2015.

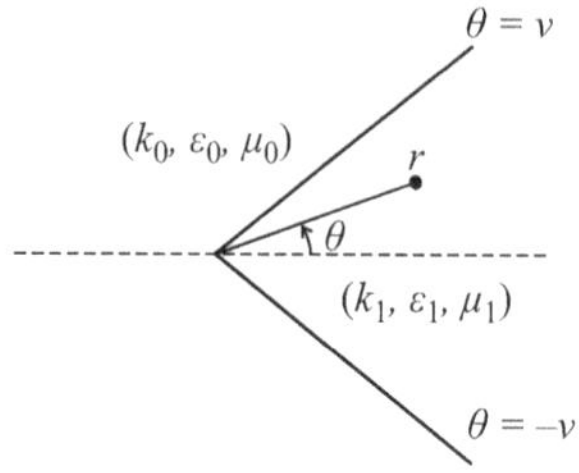

Figure 3.1 Wedge with the wedge angle $2\nu < \pi$

with 2ν being the wedge angle and ε_r the relative permittivity of the wedge (Figure 3.1). However, it was not proved that the dynamic and static fields have the same singularity at $r = 0$. The singular field has not been calculated. It is an important observation of the present chapter that the canonical assumptions used to formulate the diffraction problem by a Hilbert space integral equation are not sufficient for the determination of the field for $k_1 r \ll 1$. We will construct the solution of the integral equation, which satisfies an additional asymptotic boundary condition and for which the electric field behaves like (3.1) at $r = 0$. But the behavior of a field at a point is not a well-defined concept in a Hilbert space. And it is not a great surprise that the unique solution of the diffraction problem in the Hilbert space may be formulated by a model, where the E field is regular at $r = 0$.

In the present chapter, the scattered field is proved to be a superposition of the geometrical optics wave and an outgoing cylindrical wave. The exact solution H of the diffraction problem is the limit of a convergent sequence of function H_n. An exact formula for the error of the field $\|H - H_n\|$ is found. These results are achieved by two closely related representations of the diffraction problem. An integral equation for the Fourier transform of the trace field is formulated and solved iteratively. It is proved that the iterations converge. The physical interpretation of the iteration is multiple reflection of the incident wave at the boundaries of the wedge. The reflection and transmission coefficients are expressed as operators on the incident field. These coefficients may be calculated with any given error by explicit formulas. Alternatively the solution of the diffraction problem can be based on a new Fourier series expansion of the field. The Fourier expansion applies to the near-field construction. An exact error estimate for the far field is obtained for the integral equation method. The Fourier series method leads to an exact error estimate of the far field as well as the l^2 error of the field near the tip of the wedge.

The classical theory of diffraction of waves by circular cylinders, spheres, etc. employees an expansion of the scattered field on an orthogonal basis of functions. The basis functions used in these examples are eigenfunctions of differential operators obtained by separation of the variables. The Fourier method for the wedge problem has not to the best of author's knowledge been published. This is due to the fact that the eigenfunctions of the differential equations are Bessel or Hankel functions times trigonometric functions. The Bessel modes do not satisfy the radiation condition at infinity and Hankel functions times trigonometric functions are not bounded at $r = 0$. That is, these modes cannot be used as basis functions in a Fourier series. The "Bessel–Hankel modes" introduced in this chapter is an orthogonal basis of functions in the space of functions with finite energy. These functions are solutions of the Helmholtz equations. The far field behaves like the outgoing Hankel function $H_0^{(1)}$. The near field of the nth mode behaves like J_{s+2n}. We will prove that the Fourier coefficients of the "Bessel–Hankel" Fourier series are determined uniquely by the far field. The mapping of the far field onto the Fourier series is proved to be an isometric mapping of the space of functions with finite energy norm onto the space of Fourier series with the l^2 norm. This observation is used to construct a formula for the l^2 error of the near field.

3.2 The diffraction problem

A general theory of diffraction of a magnetically polarized time-harmonic wave by a dielectric wedge will be formulated. The incident wave is generated by given sources outside and/or inside a wedge. Alternatively the incident field can be a plane wave. The wedge angle is assumed to be less than π (Figure 3.1). In cylindrical coordinates, the wedge is the cylinder:

$$W = \{(r, \theta, x_3) \in R_+ \times [-\nu, \nu] \times R\}. \tag{3.3}$$

3.2.1 The Hilbert-space problem

Let $\mathbf{H}_{in} = (0, 0, H_{in}(r, \theta))$ be an incident magnetically polarized field that is independent of x_3. The incident field may be represented by the sum of two waves:

$$H_{in} = H_{in,sym} + H_{in,asym}, \tag{3.4}$$

where $H_{in,sym}$ is symmetric with respect to the symmetry plane of the wedge and $H_{in,asym}$ is antisymmetric. The scattered field generated by a symmetric/antisymmetric incident wave is symmetric/antisymmetric.

The diffraction problem will be solved for symmetric or antisymmetric fields. Let $H = H_{in} + H_{sc}$ be the sum of the third components of the incident and scattered fields. The scattered field satisfies the Helmholtz equation:

$$[\Delta + k^2(\mathbf{x})]H_{sc} = 0. \tag{3.5}$$

The wave numbers inside and outside the wedge are constants:

$$k^2(\mathbf{x}) = \begin{cases} k_1^2 = \mu_1 \varepsilon_1 \omega^2, & \mathbf{x} \in W \\ k_0^2 = \mu_0 \varepsilon_0 \omega^2, & \mathbf{x} \notin W \end{cases}, \tag{3.6}$$

where ω is the circular frequency, $\varepsilon_{0,1}$ are the permittivities and $\mu_{0,1}$ are the permeabilities inside and outside the wedge.

Let $\mathscr{E}$ be the Banach space with an energy norm which will be defined later (see (3.44)). Using polar coordinates, the boundary conditions at the upper boundary are as follows:

$$\|H(r, v+0_+) - H(r, v-0_+)\|_{\mathscr{E}} = 0, \tag{3.7}$$

$$\left\| \frac{\varepsilon_1}{r}\frac{\partial H}{\partial \theta}(r, v+0_+) - \frac{\varepsilon_0}{r}\frac{\partial H}{\partial \theta}(r, v-0_+) \right\|_{\mathscr{E}} = 0, \tag{3.8}$$

where 0_+ denotes an arbitrarily small positive number. That is, we are not assuming that the Dirichlet and Neumann data satisfy the boundary conditions point wise. Finally H_{sc} is an outgoing field at infinity*:

$$H_{sc}(r, \theta) = H_0^{(1)}(k_q r)\phi(\theta) + O(k_q r)^{-(3/2)}, \quad \theta \neq \pm v, \tag{3.9}$$

where $q = 0, 1$ and $H_0^{(1)}(k_q r)$ is the Hankel function.

The Hilbert-space diffraction problem is for a given incident field H_{in} to construct the scattered field H_{sc} which satisfies the following conditions: (a) the function H_{sc} is a solution of the Helmholtz equation (3.5) inside and outside the wedge; (b) the total field $H = H_{in} + H_{sc}$ satisfies the boundary and radiation conditions (3.7), (3.8) and (3.9). This problem is proved to have a unique solution.

3.2.2 The singular-field problem

An important point of this chapter is that although the Hilbert-space problem has a unique solution, the field singularity at the tip of the wedge is not uniquely determined by the solution of the Hilbert-space problem. However, we will show that there exists a representation of the unique solution of this problem by a function that has the same singularity at $r = 0$ as the static field problem. By definition, the singular-field problem is to construct the unique solution of the Hilbert-space problem, for which the asymptotic Dirichlet condition (3.10) and the asymptotic Neumann condition (3.11) hold the following:

$$H \in C^0([0, \infty[\times [-\pi, \pi]), \quad \left\| r^{-1}\frac{\partial H}{\partial \theta}(r, v \pm 0_+) \right\|_{\infty} = \infty, \tag{3.10}$$

$$\lim_{r \to 0} \left\{ \frac{\varepsilon_1}{r}\frac{\partial H}{\partial \theta}(r, v+0_+) - \frac{\varepsilon_0}{r}\frac{\partial H}{\partial \theta}(r, v-0_+) \right\} = 0. \tag{3.11}$$

We will prove that there exists a unique solution of the Hilbert-space problem, which satisfies the additional asymptotic boundary conditions (3.10) and (3.11).

*For plane-wave incidence with real-valued wave numbers, the scattered field H_{sc} contains a geometrical optics field H_{go} of several outgoing plane waves. In this case, it is the difference $H_{sc} - H_{go}$ that satisfies (3.9).

3.3 Integral equations

We now formulate an integral equation for the Fourier transform of the boundary field at the upper boundary $\theta = \nu$. The integral equation for this function is constructed such that (3.5), (3.7), (3.8) and (3.9) are satisfied in the space of functions with finite energy. The scattered outgoing waves inside and outside the wedge are represented by the following equations:

$$H_i(r,\theta) = \int_{-\infty}^{\infty} \hat{H}_i(p')\mathrm{e}^{ir\beta_1(p',\nu-\theta)}dp' \quad \text{for } \theta \in [-\nu,\nu], \tag{3.12}$$

$$H_e(r,\theta) = \int_{-\infty}^{\infty} \hat{H}_e(p')\mathrm{e}^{ir\beta_0(p',\theta-\nu)}dp' \quad \text{for } \theta \in [\nu,\pi], \tag{3.13}$$

$$\beta_q(p',\theta') = p'\cos\theta' + \alpha_q(p')\sin\theta', \tag{3.14}$$

$$\alpha_q(p) = \sqrt{k_q^2 - p^2}, \quad \sqrt{r\mathrm{e}^{i\chi}} = \sqrt{r}\exp\left\{\frac{i}{2}\chi\right\}, \quad \chi \in]-\pi,\pi]. \tag{3.15}$$

The function H_i is the down-going part of the scattered wave inside the wedge, and H_e is the up-going part of the scattered wave for $\theta \in [\nu,\pi]$. The total field inside and outside the wedge is given by the following functions:

$$H_{i,tot} = H_i + \Phi_i, \tag{3.16}$$

$$H_{e,tot} = H_e + \Phi_e. \tag{3.17}$$

The functions (Φ_i, Φ_e) are superpositions of the incident field H_{in} generated by sources and the scattered field generated at the lower boundary:

$$\Phi_i(r,\theta) = \pm H_i(r,-\theta) + H_{in}(r,\theta), \qquad \theta \in [-\nu,\nu], \tag{3.18}$$

$$\Phi_e(r,\theta) = \pm H_e(r,-\theta) + H_{in}(r,\theta), \qquad \theta \in [\nu,\pi]. \tag{3.19}$$

In the symmetric case, $+$ is used, the $-$ sign applies to antisymmetric waves.

The following notation will be used for the $\mathscr{F}_+$-Fourier transform:

$$\tilde{\Phi}(p) = \mathscr{F}_+\Phi(p) = \frac{1}{2\pi}\int_0^{\infty} \mathrm{e}^{-ipr}\,\Phi(r)dr. \tag{3.20}$$

Define the $\mathscr{F}_+$-Fourier transform of the trace field $\tilde{\boldsymbol{\Phi}} = (\tilde{\Phi}_+, \tilde{\Phi}_-, \tilde{\Phi}_{\nu+}, \tilde{\Phi}_{\nu-})$:

$$\tilde{\Phi}_-(p) = \mathscr{F}_+\Phi_i(p,\nu-0_+), \qquad \tilde{\Phi}_+(p) = \mathscr{F}_+\Phi_e(p,\nu+0_+), \tag{3.21}$$

$$\tilde{\Phi}_{\nu-}(p) = \mathscr{F}_+\left\{r^{-1}\frac{\partial\Phi_i}{\partial\theta}\right\}(p,\nu-0_+), \tag{3.22}$$

$$\tilde{\Phi}_{\nu+}(p) = \mathscr{F}_+\left\{r^{-1}\frac{\partial\Phi_e}{\partial\theta}\right\}(p,\nu+0_+). \tag{3.23}$$

The Dirichlet and Neumann boundary conditions (3.7) and (3.8) are satisfied if

$$\hat{H}_i(p) + \tilde{\Phi}_-(p) = \hat{H}_e(p) + \tilde{\Phi}_+(p), \tag{3.24}$$

$$\varepsilon_0\{-i\alpha_1(p)\hat{H}_i(p) + \tilde{\Phi}_{\nu-}(p)\} = \varepsilon_1\{i\alpha_0(p)\hat{H}_e(p) + \tilde{\Phi}_{\nu+}(p)\}. \tag{3.25}$$

The boundary conditions (3.24) and (3.25) hold if and only if the functions $\hat{H}_i$ and $\hat{H}_e$ are related to $\tilde{\mathbf{\Phi}}$ by equations similar to the Stratton formulas for diffraction of a wave at a plane:

$$\hat{\mathbf{H}} = (\hat{H}_i, \hat{H}_e) = \mathbf{K}\tilde{\mathbf{\Phi}} \tag{3.26}$$

with $\mathbf{K}\tilde{\mathbf{\Phi}} = (\mathbf{K}_i\tilde{\mathbf{\Phi}}, \mathbf{K}_e\tilde{\mathbf{\Phi}})$ defined by

$$\mathbf{K}_i\tilde{\mathbf{\Phi}} = \frac{1}{\varepsilon_1\alpha_0 + \varepsilon_0\alpha_1}\{\varepsilon_1\alpha_0(\tilde{\Phi}_+ - \tilde{\Phi}_-) + i\varepsilon_1\tilde{\Phi}_{\nu+} - i\varepsilon_0\tilde{\Phi}_{\nu-}\}, \tag{3.27}$$

$$\mathbf{K}_e\tilde{\mathbf{\Phi}} = \frac{1}{\varepsilon_1\alpha_0 + \varepsilon_0\alpha_1}\{\varepsilon_0\alpha_1(\tilde{\Phi}_- - \tilde{\Phi}_+) + i\varepsilon_1\tilde{\Phi}_{\nu+} - i\varepsilon_0\tilde{\Phi}_{\nu-}\}. \tag{3.28}$$

In order to formulate (3.26), (3.27) and (3.28) as integral equations, we will express $\tilde{\Phi}_i$ and $\tilde{\Phi}_e$ through the functions $\hat{H}_i$ and $\hat{H}_e$. The $\mathscr{F}_+$-Fourier transform of (3.12) and (3.13) is as follows:

$$\mathscr{F}_+H_i(p,\theta) = \mathbf{Q}_1\hat{H}_i(p,\theta) = \int_{-\infty}^{\infty} \hat{H}_i(p')\mathscr{F}_+1[p - \beta_1(p', \nu - \theta)]dp', \tag{3.29}$$

$$\begin{aligned}\mathscr{F}_+H_e(p,\theta) &= \mathbf{R}_0\hat{H}_e(p, \theta - \nu)\\ &= \int_{-\infty}^{\infty} \hat{H}_e(p')\mathscr{F}_+1[p - \beta_0(p', \theta - \nu)]dp',\end{aligned} \tag{3.30}$$

where $1(\cdots)$ denotes the unit-step function. Equation (3.30) holds for $\theta \in [\nu, \pi]$. For $\theta \in [-\pi, -\nu]$, we find that

$$\mathscr{F}_+H_e(p,\theta) = \mathbf{Q}_0\hat{H}_e(p,\theta) = \mathbf{R}_0[\mathbf{R}_0\hat{H}_e(p', \pi - \nu)](p, \theta + \pi). \tag{3.31}$$

Let $\tilde{\mathbf{H}}_{in}$ be the Dirichlet and Neumann data of the incident field at $\theta = \nu$:

$$\tilde{\mathbf{H}}_{in} = \mathscr{F}_+\left[H_{in}(r, \nu_+), H_{in}(r, \nu_-), r^{-1}\frac{\partial H_{in}}{\partial\theta}(r, \nu_+), r^{-1}\frac{\partial H_{in}}{\partial\theta}(r, \nu_-)\right] \tag{3.32}$$

with $\nu_\pm = \nu \pm 0_+$ for brevity. Then the Dirichlet and Neumann data of $\tilde{\mathbf{\Phi}}$ at $\theta = \nu$ is determined by (3.18), (3.19), (3.29), (3.30) and (3.31) as

$$\begin{aligned}[\tilde{\mathbf{\Phi}} - \tilde{\mathbf{H}}_{in}](p, \nu) &= \pm\mathbf{Q}\hat{\mathbf{H}}(p, -\nu)\\ &= \pm(\mathbf{Q}_0\hat{H}_e, \mathbf{Q}_1\hat{H}_i, \mathbf{Q}_{0,\nu}\hat{H}_e, \mathbf{Q}_{1,\nu}\hat{H}_i)(p, -\nu),\end{aligned} \tag{3.33}$$

where $+$ is used in the symmetric and $-$ in the antisymmetric case. The operators $(\mathbf{Q}_{0,\nu}, \mathbf{Q}_{1,\nu})$ are determined by the relations:

$$\mathbf{Q}_{1,\nu}\hat{H}_i(p, -\nu) = -i\mathbf{Q}_1\left[\frac{\partial\beta_1}{\partial\theta}(p', \nu - \theta)\hat{H}_i\right](p, -\nu), \tag{3.34}$$

$$\mathbf{Q}_{0,\nu}\hat{H}_e(p, -\nu) = i\mathbf{R}_0\left[\frac{\partial\beta_0}{\partial\theta}(p', \pi - \nu)\mathbf{R}_0\hat{H}_e\right](p, \pi - \nu). \tag{3.35}$$

Using $\hat{\mathbf{H}} = \mathbf{K}\tilde{\mathbf{\Phi}}$ in (3.33), we find the integral equation for the function $\tilde{\mathbf{\Phi}}$:

$$\tilde{\mathbf{\Phi}} = \tilde{\mathbf{H}}_{in} \pm \mathbf{QK}\tilde{\mathbf{\Phi}}, \tag{3.36}$$

where $+$ applies to symmetric and $-$ to antisymmetric fields. The corresponding Stratton formula (3.26) for the trace field is the integral equation:

$$\hat{\mathbf{H}} = \mathbf{K}\tilde{\mathbf{H}}_{in} \pm \mathbf{KQ}\hat{\mathbf{H}}. \tag{3.37}$$

These results prove that if (H_e, H_i) is a solution of the diffraction problem, then $\tilde{\mathbf{\Phi}}$ is a solution of the integral equation (3.36). Let conversely $\tilde{\mathbf{\Phi}}$ be a solution of the integral equation (3.36). We will show that this $\tilde{\mathbf{\Phi}}$ may be used to construct a solution of the diffraction problem. Define (H_i, H_e) by (3.12) and (3.13) with $(\hat{H}_i, \hat{H}_e) = \mathbf{K}\tilde{\mathbf{\Phi}}$. Then $(\hat{\mathbf{H}}, \tilde{\mathbf{\Phi}})$ satisfy the boundary conditions (3.24) and (3.25). Define the scattered field by the following equations:

$$\begin{aligned}\mathbf{H}_{sc}(r,\theta) &= [H_i(r,\theta) \pm H_i(r,-\theta), H_e(r,\theta) \pm H_e(r,-\theta)] \\ &= [H_i(r,\theta), H_e(r,\theta)] \pm \mathcal{F}^{-1}(Q_1\mathbf{K}_i\tilde{\mathbf{\Phi}}, Q_0\mathbf{K}_e\tilde{\mathbf{\Phi}})(r,\theta).\end{aligned} \tag{3.38}$$

This field is a solution of the Helmholtz equation. Using (3.36), the Fourier transform of this equation and the normal derivative of $\mathbf{H}_{sc}$ at $\theta = \nu$ are reduced to the boundary conditions (3.24) and (3.25), which proves that $\mathbf{H}_{sc}$ is a solution of the diffraction problem.

3.4 The solution of the integral equation

First we will specify the space of incident waves, for which the solution of the diffraction problem is found explicitly. That is, we will express the scattered field by a formula for any incident field in a given linear space.

The scattered field 1. Let $\mathbf{\Phi} = (\Phi_i, \Phi_e)$ be any given symmetric or antisymmetric solution of Helmholtz equation. Define the incident wave by $\tilde{\mathbf{H}}_{in} = \tilde{\mathbf{\Phi}} \mp \mathbf{QK}\tilde{\mathbf{\Phi}}$. Then H defined by (3.26), (3.12),(3.13), (3.16) and (3.17) is the explicit solution of the diffraction problem with this incident wave.

Proof: The function $\tilde{\mathbf{\Phi}}$ is a solution of the integral equation for the trace field (3.36). It was proved in Section 3.3 that the field $\mathbf{H}_{sc}$ defined by (3.38) is a solution of the diffraction problem with the incident field $\tilde{\mathbf{H}}_{in}$.

The canonical diffraction problem requires the construction of the scattered wave for a given incident field. The solution strategy is to solve the integral equation (3.36) iteratively. The nth-order iteration of this equation is given by

$$\tilde{\mathbf{\Phi}}^{(n)} = \sum_{q=0}^{n-1} (\pm\, \mathbf{QK})^q \tilde{\mathbf{H}}_{in}, \tag{3.39}$$

where $+$ is used for symmetric fields and $-$ applies in the antisymmetric case. Using this function, the exact solution of the diffraction problem with the incident field

$$\tilde{\mathbf{H}}_{in}^{(n)} = \tilde{\mathbf{H}}_{in} - (\pm\, \mathbf{QK})^n \tilde{\mathbf{H}}_{in} \tag{3.40}$$

is the scattered field given by the following equation:

$$(\hat{H}_i, \hat{H}_e)^{(n)} = \mathbf{K} \sum_{q=0}^{n-1} (\pm \mathbf{QK})^q \tilde{\mathbf{H}}_{in}. \tag{3.41}$$

Summarizing we arrive at the statement:

The scattered field 2. The exact explicit solution of the diffraction problem with the incident wave given by the function (3.40) is the functions $(H_i, H_e)^{(n)}$ determined by (3.12) and (3.13) with $(\hat{H}_i, \hat{H}_e)^{(n)}$ defined by (3.41).

We will now prove that the sequence (3.39) converges in the space of functions with finite energy norm. Define inner products and the energy norm of the Fourier transform of the trace field by the following expressions:

$$< \tilde{H}_i(p), \phi(p) >_i = \int_{-k_1}^{k_1} \alpha_1(p) \tilde{H}_i(p)[\phi(p)]^* dp, \tag{3.42}$$

$$< \tilde{H}_e(p), \psi(p) >_e = \int_{-k_0}^{k_0} \alpha_0(p) \tilde{H}_e(p)[\psi(p)]^* dp, \tag{3.43}$$

$$\|(\tilde{H}_i, \tilde{H}_e)\|^2_{\mathscr{E}} = < \tilde{H}_i(p), \tilde{H}_i(p) >_i + < \tilde{H}_e(p), \tilde{H}_e(p) >_e. \tag{3.44}$$

It is proved in the appendix that the norm of the operator **QK** is less than 1 in the space of functions with finite energy norm. With this assumption, we will construct the solution of the diffraction problem and an exact formula for the far-field error. The energy norm of an incident plane wave is not finite. But there exists a number N, for which the energy norm $\|(\mathbf{QK})^N \tilde{\mathbf{H}}_{in}\|_{\mathscr{E}}$ is finite for a plane incident wave. With this assumption on the incident wave, the strategy for the solution of the diffraction problem is as follows:

The scattered field 3. Let H_{in} be a given symmetric or antisymmetric incident field. Assume that **QK** is an operator with the norm less than 1 in the space of functions with finite energy norm. Assume that there exists a number N, for which the energy norm of $(\mathbf{QK})^N \tilde{\mathbf{H}}_{in}$ is finite. Then the unique solution of the wedge-diffraction problem is the convergent series (3.45) with (3.46) and (3.47). The exact formula for the far-field error defined as the energy norm of the error function is expressed by formula (3.48).

$$\tilde{\mathbf{\Phi}} = \sum_{q=0}^{\infty} (\pm \mathbf{QK})^q \tilde{\mathbf{H}}_{in}, \tag{3.45}$$

$$(\hat{H}_i, \hat{H}_e) = \lim_{n \to \infty} (\hat{H}_i, \hat{H}_e)^{(n)}, \tag{3.46}$$

$$(\tilde{H}_i, \tilde{H}_e)(p) = \left(\mathbf{Q}_1 \hat{H}_i(p, v), \mathbf{R}_0 \hat{H}_e(p, 0) \right). \tag{3.47}$$

The error of the approximation of the scattered field by $(\tilde{H}_i, \tilde{H}_e)^{(n)}$ is defined for any n larger than N by the energy norm (3.44). Energy conservation is expressed by the Pointing theorem:

$$\|(\tilde{H}_i, \tilde{H}_e) - (\tilde{H}_i, \tilde{H}_e)^{(n)}\|^2_{\mathscr{E}} = \|\tilde{H}^{(n)}_{in} - \tilde{H}_{in}\|^2_{\mathscr{E}} = \|(\mathbf{QK})^n \tilde{H}_{in}\|^2_{\mathscr{E}}. \tag{3.48}$$

An important feature of this equation is that the error of $(\tilde{H}_i, \tilde{H}_e)^{(n)}$ is expressed explicitly through the incident field.

3.5 The Bessel–Hankel Fourier series

By definition, the Bessel–Hankel Fourier method is an expansion of the solution of the diffraction problem in a series of orthogonal Bessel–Hankel modes. This series is proved to converge in the space of finite energy. We will prove that the Fourier coefficients are uniquely determined by the far field. That is, there exists a mapping of the far field onto the near field. This mapping is the Bessel–Hankel Fourier series. The far-field near-field mapping defined on the space of finite energy into the l^2 space of Fourier coefficients is proved to be isometric. It is emphasized that the mapping of the far field onto the L^∞ near-field space is not continuous. An important result of this section is that the error of the near field may be calculated explicitly.

Let $\phi_{n,q}$ be the Fourier transform of the Bessel functions $J_{2n+s}(k_q r)$ with $n = 0, 1, \ldots$ and of the complex conjugate of the Bessel functions for negative n:

$$\phi_{n,q}(p) = \mathscr{F}_+ J_{2n+s}(k_q r)(p - i0_+) = \frac{1}{2\pi} \frac{[\alpha_q(p - i0_+) - ip]^{2n+s}}{k_q^{2n+s}\alpha_q(p - i0_+)}, \tag{3.49}$$

$$\phi_{-n,q}(p) = \frac{\phi_{0,q}(p)}{k_q^{2n}}[\alpha_q(-p - i0_+) + ip]^{2n}, \qquad n = 1, 2, \ldots. \tag{3.50}$$

The constant s is a root of (3.2) in the interval $[0.5, 1]$.

The Bessel–Hankel modes $(F_{i,n}, F_{e,n})$ are defined as outgoing solutions of the Helmholtz equation, which are Bessel functions at $\theta = \nu$,

$$F_{i,n}(r, \theta) = \int_{-\infty}^{\infty} \phi_{n,1}(p') \mathrm{e}^{ir\beta_1(p', \nu-\theta)} dp' \;\text{ for }\; \theta \in [-\nu, \nu], \tag{3.51}$$

$$F_{e,n}(r, \theta) = \int_{-\infty}^{\infty} \phi_{n,0}(p') \mathrm{e}^{ir\beta_0(p', \theta-\nu)} dp' \;\text{ for }\; \theta \in [\nu, \pi], \tag{3.52}$$

$$F_{e,n}(r, \theta) = \int_{-\infty}^{\infty} \mathscr{F}_+[F_{e,n}(r, \pi)](p') \mathrm{e}^{ir\beta_0(p', \theta+\pi)} dp' \;\text{ for }\; \theta \in [-\pi, -\nu]. \tag{3.53}$$

Alternatively we may use the basis functions $\phi_{n,1}$ in the Bessel–Hankel modes outside the wedge. For angles in $[\nu, \pi]$, the modes are given by

$$F_{e1,n}(r,\theta) = \int_{-\infty}^{\infty} \phi_{n,1}(p')e^{ir\beta_0(p',\theta-\nu)}dp' \quad \text{for } \theta \in [\nu, \pi], \tag{3.54}$$

whereas for $\theta \in [-\pi, -\nu]$ these modes are obtained from (3.53) upon the substitution of $F_{e1,n}$ for $F_{e,n}$.

Substituting $p' = k_1 \sin u$ in (3.51) transforms this integral to a contour integral

$$F_{i,n}(r,\theta) = \int_c k_1 \cos u \, \phi_{n,1}(k_1 \sin u)e^{irk_1 \sin(u+\nu-\theta)}du, \tag{3.55}$$

where

$$\int_c = \int_{-\pi/2+i\infty}^{-\pi/2} + \int_{-\pi/2}^{\pi/2} + \int_{\pi/2}^{\pi/2-i\infty} . \tag{3.56}$$

The far field of $F_{i,n}$ is equal to the integral (3.55) over the interval $[-\pi/2, \pi/2]$. For real arguments, the integrand in (3.55) is proportional to the exponential function

$$k_1 \cos u \, \phi_{n,1}(k_1 \sin u) = \frac{1}{2\pi} \exp\{-i(s+2n)u\} \quad \text{for } u \in \left[-\frac{\pi}{2}, \frac{\pi}{2}\right], \tag{3.57}$$

and using this, the far field of F_i is expressed as

$$F_{i,n}(r,\theta) \approx \frac{1}{2\pi} \int_{-\pi/2}^{\pi/2} \exp\{-i(s+2n)u + irk_1 \sin(u+\nu-\theta)\}du. \tag{3.58}$$

The Hankel function is defined by the contour integral

$$H_m^{(1)}(z) = -\frac{1}{\pi} \int_c e^{iz\sin(u+\gamma)-im(u+\gamma)}du \tag{3.59}$$

with γ being any constant in the interval $]0, \pi[$ and m an arbitrary parameter. The far field of $H_m^{(1)}$ is the integral over the interval $[-\pi/2, \pi/2]$. Using $\gamma = \nu - \theta$ in (3.59), we obtain the far-field expansion of the Bessel–Hankel mode for $k_1 r \gg 1$ as

$$F_{i,n}(r,\theta) \approx -\frac{1}{2} \exp\{i(s+2n)(\nu-\theta)\}H_{s+2n}^{(1)}(k_1 r) \quad \text{for } \theta \in]-\nu, \nu[. \tag{3.60}$$

In the same way, the expression for the functions $F_{e,n}$ given by (3.52) and (3.53) is reduced to the following formula:

$$F_{e,n}(r,\theta) = \int_c k_0 \cos u \, \phi_{n,0}(k_0 \sin u)e^{irk_0 \sin(u+\theta-\nu)}du, \tag{3.61}$$

implying that the far field of the functions $F_{e,n}$ is proportional to the Hankel function $H_{s+2n}^{(1)}(k_0 r)$. The far-field expansions hold for any $|\theta| \neq \nu$, but not for $\theta = \pm\nu$.

The semi-static approximation of $F_{i,n}$ for $k_1 r \ll 1$ is defined as the solution of the Laplace equation, for which the static and dynamic fields coincide at $\theta = v$:

$$F_{i,ns}(r,\theta) = \int_{-\infty}^{\infty} \phi_{n,1}(p')[\mathrm{e}^{ir\beta_1(p',v-\theta)}]_{k_1=0}dp' = J_{s+2n}[k_1 r \mathrm{e}^{i(v-\theta)}]. \tag{3.62}$$

From the asymptotic behavior of the functions $\phi_{n,1}(p)$ for p large, it follows that the function

$$\psi(r,\theta) = \partial_r^{2n+s+1}[F_{i,ns}(r,\theta) - F_{i,n}(r,\theta)] \tag{3.63}$$

satisfies the relations

$$p^2 \mathscr{F}_+ \psi(p,\theta) \in L^\infty \Rightarrow \mathscr{F}_+ \psi \in L^1 \Rightarrow \psi \in L^\infty \tag{3.64}$$

for $n \geq 0$ and any fixed θ, thus proving that $F_{i,ns}$ and $F_{i,n}$ have the same singularity at $r = 0$:

$$F_{i,n}(r,\theta) = J_{s+2n}(k_1 r)\mathrm{e}^{i(s+2n)(v-\theta)} + O(k_1 r)^{2n+s+1}. \tag{3.65}$$

The asymptotic formula

$$F_{e,n}(r,\theta) = J_{s+2n}(k_0 r)\mathrm{e}^{i(s+2n)(\theta-v)} + O(k_0 r)^{2n+s+1} \tag{3.66}$$

for the modes defined outside the wedge and for the Bessel–Hankel modes with n negative is obtained in the same way.

Using (3.57), the functions $\phi_{n,1}$ are proved to be orthogonal basis functions:

$$< \phi_{n,1}(p), \phi_{m,1}(p) >_i = \frac{1}{4\pi}\delta_{n,m}. \tag{3.67}$$

The Fourier series of $\tilde{H}_i(p)$ will be expressed through the field calculated by the integral-equation method. Since the functions $\tilde{H}_i$ and $\hat{H}_i$ are related as $\tilde{H}_i(p) = \mathbf{Q}_1 \hat{H}_i(p, v - 0_+)$, we obtain the Fourier series

$$\tilde{H}_i(p) = \sum_{n=-\infty}^{\infty} c_n \phi_{n,1}(p), \tag{3.68}$$

$$H_i(r,\theta) = \sum_{n=-\infty}^{\infty} c_n F_{i,n}(r,\theta), \tag{3.69}$$

where

$$c_n = 4\pi < \mathbf{Q}_1 \hat{H}_i(p, v - 0_+), \phi_{n,1} >_i. \tag{3.70}$$

If the energy norm of the function H_i is finite, then the Fourier series converge in l^2 and the energy norm is related to the Fourier coefficients by the formula

$$\|\tilde{H}_i\|^2_{\mathscr{E}} =< \tilde{H}_i(p), \tilde{H}_i(p) >_i = \frac{1}{4\pi}\sum_{n=-\infty}^{\infty} |c_n|^2. \tag{3.71}$$

By the same arguments, the outgoing wave from the boundary $\theta = \nu$ can be expressed by the series

$$H_e(r,\theta) = \sum_{n=-\infty}^{\infty} d_n F_{e,n}(r,\theta) \tag{3.72}$$

with

$$d_n = 4\pi < \mathbf{R}_0 \hat{H}_e(p, \nu + 0_+), \phi_{n,0} >_e . \tag{3.73}$$

If the energy norm is finite, then the Fourier series converges in l^2 and the energy norm is related to the l^2 norm by

$$\|\tilde{H}_e\|_{\mathscr{E}}^2 = < \tilde{H}_e(p), \tilde{H}_e(p) >_e = \frac{1}{4\pi} \sum_{n=-\infty}^{\infty} | d_n |^2 . \tag{3.74}$$

Alternatively we may use the expansion on the modes $F_{e1,n}$:

$$H_e(r,\theta) = \sum_{n=-\infty}^{\infty} g_n F_{e1,n}(r,\theta), \tag{3.75}$$

where

$$g_n = 4\pi < \mathbf{R}_0 \hat{H}_e(p, \nu + 0_+), \phi_{n,1} >_i . \tag{3.76}$$

We summarize the results in the following statement:

The far-field near-field mapping 1. Let the energy norm of the functions $\tilde{H}_i$ and $\tilde{H}_e$ be finite. Then the far-field near-field mapping is defined as the mapping of $(\tilde{H}_i, \tilde{H}_e)$ onto the Fourier series (3.69) and (3.72) with Fourier coefficients (3.70) and (3.73). The Fourier series converge in l^2 norm, and the mapping $(\tilde{H}_i, \tilde{H}_e) \to (1/2\sqrt{\pi})(\mathbf{c}, \mathbf{d})$ is isometric.

An application of this result is as follows. The far field may be calculated from the approximate solution of the diffraction problem given by (3.45), (3.46) and (3.47). Let $(\mathbf{c}^{(n)}, \mathbf{d}^{(n)})$ denote the Fourier coefficients of the functions $(\tilde{H}_i, \tilde{H}_e)^{(n)}$ given by

$$c_{q,n} = 4\pi < \mathbf{Q}_1 \hat{H}_i^{(n)}(p, \nu - 0_+), \phi_{q,1} >_i , \tag{3.77}$$

$$d_{q,n} = 4\pi < \mathbf{R}_0 \hat{H}_e^{(n)}(p, \nu + 0_+), \phi_{q,0} >_e . \tag{3.78}$$

Let $(\mathbf{c}, \mathbf{d})$ denote the Fourier coefficients of the exact solution $(\tilde{H}_i, \tilde{H}_e)$ of the wedge-diffraction problem with an incident field with a finite energy norm. Define the l^2 error of the Fourier coefficients by

$$\|(\mathbf{c} - \mathbf{c}^{(n)}, \mathbf{d} - \mathbf{d}^{(n)})\|^2 = \sum_{q=-\infty}^{\infty} \{| c_{q,n} - c_q |^2 + | d_{q,n} - d_q |^2\}. \tag{3.79}$$

Then the error of the nth-order Fourier series is as follows:

$$\|(\mathbf{c} - \mathbf{c}^{(n)}, \mathbf{d} - \mathbf{d}^{(n)})\|^2 = \frac{1}{4\pi} \|(\mathbf{QK})^n \tilde{\mathbf{H}}_{in}\|_{\mathscr{E}}^2 , \tag{3.80}$$

$$\max\left(\mid c_q - c_q^{(n)} \mid^2, \mid d_q - d_q^{(n)} \mid^2\right) \leq \frac{1}{4\pi} \|(\mathbf{QK})^n \tilde{\mathbf{H}}_{in}\|_{\mathscr{E}}^2, \tag{3.81}$$

$$\lim_{n\to\infty} (c_q - c_q^{(n)}, d_q - d_q^{(n)}) = (0, 0). \tag{3.82}$$

The results can be summarized as follows:

The far-field near-field mapping 2. Let the energy norm of the incident field be finite. Let $(\hat{H}_i, \hat{H}_e)^{(n)}$ be the nth-order iterative solution. Then the Fourier coefficients $(c_q^{(n)}, d_q^{(n)})$ converge to the Fourier coefficients of the exact solution, and the Fourier series of the exact solution of the diffraction problem converges. The norm of the error of the Fourier series (3.69) and (3.72) satisfies (3.80) and (3.81).

The Fourier series (3.69) and (3.72) converge in the energy norm for any s in the interval $]0.5, 1]$ to the solution of the diffraction problem. Consequently the Fourier theory applies for any s in the interval $]0.5, 1]$. The unique solution of the diffraction problem may be represented by a Fourier series for any $s \in]0.5, 1]$. If s is not the root of (3.2), then the convergent Fourier series will not satisfy the asymptotic boundary conditions (3.10) and (3.11).

We will now show that if s is the root of (3.2), then the convergent Fourier series satisfies the asymptotic boundary conditions. The Fourier transform of the total field and the normal derivative of this function can be expanded on the orthogonal basis functions $\phi_{n,1}$. Let the incident field be generated by sources outside the wedge. Then the Fourier series of the total field in the limit from inside the wedge and from outside the wedge coincide,

$$H(r, v + 0_+) = H(r, v - 0_+) = \sum_{n=-\infty}^{\infty} f_n \mathscr{F}^{-1}[\phi_{n,1}(p)](r), \tag{3.83}$$

$$f_n = 4\pi < \mathscr{F}_+[H(r, v)], \phi_{n,1} >_i = c_n \pm 4\pi < \mathbf{Q}_1 \hat{H}_i(p, -v), \phi_{n,1} >_i \tag{3.84}$$

with $+$ for symmetric functions and $-$ else. The Bessel functions are continuous, and the Fourier series (3.83) defines a continuous function in the interval $[0, \infty[$.

For any N, the following finite sum has the following asymptotic expansion for $k_1 r \ll 1$:

$$H_N(r) = \sum_{n=-N}^{N} f_n \mathscr{F}^{-1}[\phi_{n,1}(p)](r) = f_0 J_s(k_1 r) + O(k_1 r)^{s+2}. \tag{3.85}$$

The corresponding asymptotic expansion of the antisymmetric field $H_N(r, \theta)$ inside the wedge is given by

$$\begin{aligned} H_N(r, \theta) &= \sum_{n=-N}^{N} f_n \mathscr{F}^{-1}[\phi_{n,1}(p)](r) \frac{\sin(s + 2n)\theta}{\sin(s + 2n)v} + O(k_1 r)^{s+2} \\ &= \frac{\sin s\theta}{\sin sv} f_0 J_s(k_1 r) + O(k_1 r)^{s+2}. \end{aligned} \tag{3.86}$$

The limit of this equation for $N \to \infty$ is proved to be the asymptotic equation of the total field inside the wedge for $k_1 r \ll 1$:

$$H(r,\theta) = \lim_{N\to\infty} H_N(r,\theta) = \frac{\sin s\theta}{\sin s\nu} f_0 J_s(k_1 r) + O(k_1 r)^{s+2}, \quad \theta \in [-\nu, \nu]. \quad (3.87)$$

A corresponding equation for symmetric functions holds with cos instead of sin.

The Fourier expansion (3.83) for $H(r, \nu + 0_+)$ is the same as the limit of (3.85), which can be alternatively represented by the asymptotic formula:

$$H(r, \nu + 0_+) = \frac{k_1^s}{k_0^s} f_0 J_s(k_0 r) + O(k_0 r)^{s+2}. \quad (3.88)$$

The asymptotic expansion of the antisymmetric total field outside the wedge for $k_0 r \ll 1$ is the function

$$H(r,\theta) = \frac{k_1^s \sin s(\pi - \theta)}{k_0^s \sin s(\pi - \nu)} f_0 J_s(k_0 r) + O(k_0 r)^{s+2}, \quad \theta \in [\nu, \pi]. \quad (3.89)$$

If the incident field is symmetric, then the corresponding formula for the asymptotic expansion is given by (3.89) with sin replaced with cos and the Fourier coefficient of the symmetric function instead of f_0.

We will now prove that the Fourier series satisfy the asymptotic boundary conditions (3.10) and (3.11). At $r = 0$, the normal derivative $r^{-1}\partial_\theta H(r,\theta)$ is proportional to r^{s-1}, and the sup norm of the electric field is not bounded. The following estimation shows that the Neumann data satisfy the asymptotic boundary condition (3.11):

$$\begin{aligned} &\frac{\varepsilon_1}{r}\frac{\partial H}{\partial \theta}(r, \nu + 0_+) - \frac{\varepsilon_0}{r}\frac{\partial H}{\partial \theta}(r, \nu - 0_+) \\ &\quad = -s f_0 \varepsilon_0 r^{-1} J_s(k_1 r) \frac{(\varepsilon_r + 1)\sin \pi s + (\varepsilon_r - 1)\sin(2s\nu - \pi s)}{2 \sin s\nu \sin s(\pi - \nu)} + O(k_1 r)^{s+1}. \end{aligned} \quad (3.90)$$

Indeed, if s is the root of (3.2), then the first term in this equation vanishes and we have proved the relation (3.11):

$$\frac{\varepsilon_1}{r}\frac{\partial H}{\partial \theta}(r, \nu + 0_+) - \frac{\varepsilon_0}{r}\frac{\partial H}{\partial \theta}(r, \nu - 0_+) = O(k_1 r)^{s+1}. \quad (3.91)$$

The results are summarized in the following statement:

The unique solution 1. Let the energy norm of the incident field be finite. Then the unique solution of the diffraction problem, which satisfies the boundary conditions (3.7), (3.8), (3.10) and (3.11), is given by the Fourier series (3.69) and (3.72). For an incident antisymmetric wave, the asymptotic field for $k_1 r \ll 1$ is given by the functions (3.87) and (3.89) with f_0 being the Fourier coefficient of $\mathscr{F}_+ H(r, \nu)$.

3.6 Incident plane waves

We will now investigate the diffraction problem for an incident plane wave: $H_{in}(\mathbf{x}) = \exp\{i\mathbf{q}\cdot\mathbf{x}\}$. Define the corresponding symmetric or antisymmetric incident field at the upper boundary by one of the following functions:

$$H_{in}(r,\theta) = e^{ir\beta_0(q_1,-\theta)} \pm e^{ir\beta_0(q_1,\theta)} \quad \text{for } q_1 \in\,]k_0\cos\nu, k_0[, \tag{3.92}$$

$$H_{in}(r,\theta) = e^{ir\beta_0(q_1,-\theta)} \pm \int_{-\infty}^{\infty} \mathscr{F}_+1(p'+q_1)e^{ir\beta_0(p',\pi-\theta)}dp' \quad \text{for } q_1 < k_0\cos\nu, \tag{3.93}$$

depending on whether the boundary is illuminated by the wave $\exp[ir\beta_0(q_1,\theta)]$ or not.

Let ν_{in} be defined by

$$\beta_0(q_1,-\nu) = k_1\cos\nu_{in}, \quad \nu_{in} \in\,]0,\pi]. \tag{3.94}$$

Then $\hat{H}_i$ has poles at the points $p_j = k_1\cos(\nu_{in}+2j\nu)$ for $j = 0,\ldots,N$ with N being an integer number defined by the conditions:

$$2N\nu \le \nu_{in} < 2(1+N)\nu, \quad N \ge 0. \tag{3.95}$$

The number N is equal to the number of times the geometrical optics field is reflected inside the wedge.

Let us define the modified geometrical optics (MGO) scattered field as the exact solution of the wedge-diffraction problem with the incident field

$$\tilde{H}_{in}^{MGO} = [\mathbf{I} - (\mathbf{QK})^{N+1}]\tilde{H}_{in}. \tag{3.96}$$

Numerical calculations show that the energy norm of $(\mathbf{QK})^{N+1}\tilde{H}_{in}$ is very small. That is, the Fourier transform of the scattered field at the wedge faces may be calculated with very small errors by formulas similar to reflection and transmission at a plane material boundary, and there exist a transmission coefficient $\mathbf{T}$ and a reflection coefficient $\mathbf{R}$, which relate the incident field to the Fourier transform of the scattered field at the wedge faces:

$$\hat{H}_i^{MGO} = \mathbf{T}\tilde{H}_{in}, \quad \mathbf{T} = \mathbf{K}_i\sum_{q=0}^{N}(\pm\mathbf{QK})^q, \tag{3.97}$$

$$\hat{H}_e^{MGO} = \mathbf{R}\tilde{H}_{in}, \quad \mathbf{R} = \mathbf{K}_e\sum_{q=0}^{N}(\pm\mathbf{QK})^q. \tag{3.98}$$

Using these formulas, the scattered field may be calculated by the formulas of the previous sections. The scattered field (H_i^{MGO}, H_e^{MGO}) is a superposition of the geometrical optics field and a cylindrical wave. It is easy to prove that the incident error field $(\mathbf{QK})^{N+1}\tilde{H}_{in}$ is a continuous function inside the wedge except for a $1/\alpha_1(p)$ singularity at $p = k_1$. Outside the wedge, this function is continuous except for a

$1/\alpha_0(p)$ singularity at $p = k_0$. Consequently this incident field has finite energy, and the sequence $(\hat{H}_i^{(n)}, \hat{H}_e^{(n)})$ converges to the exact solution of the diffraction problem.

The scattered field 4. Let H_{in} be one of the functions (3.92) or (3.93). Let N be the integer determined by (3.94) and (3.95). Then the solution of the wedge-diffraction problem is the convergent series (3.45). The transmission and reflection coefficients, which relate the incident field to $(\hat{H}_i^{MGO}, \hat{H}_e^{MGO})$, are given by (3.97) and (3.98). The exact far-field error is

$$\|H_{sc} - H_{sc}^{MGO}\|_{\mathscr{E}} \leq 2\|(\mathbf{QK})^{N+1}\tilde{H}_{in}\|_{\mathscr{E}}. \tag{3.99}$$

3.7 Numerical results

The numerical solution of the integral equation and the Fourier expansion of the field are investigated. We consider a wedge angle $v = 0.5$ and an antisymmetric incident wave. In order to construct the numerical solution, we will use the damped incident plane wave shown in Figure 3.2. This function is defined by $\mathscr{F}^{-1}\mathscr{F}_+H_{in}(p - i\delta)$ with H_{in} being an incident antisymmetric plane wave given by (3.93) with $q_1 = 0$.

The iterated solution $(\tilde{H}_i, \tilde{H}_e)^{(n)}$ was calculated for $n = 2, 3$ and 4. The numerical error of the iterated solutions for $k_1 = 2k_0$ and $q_1 = 0$ is summarized in Table 3.1.

The MGO field is the function $H^{(2)}$. The energy norm of the error of the MGO field is small for any v and any q_1. But higher order iterations must be calculated if v is very small (multiple reflections in the interior of the wedge) or if q_1 is close to $-k_0 \cos v$. Smaller error norms are obtained for larger values of q_1. The conclusion of

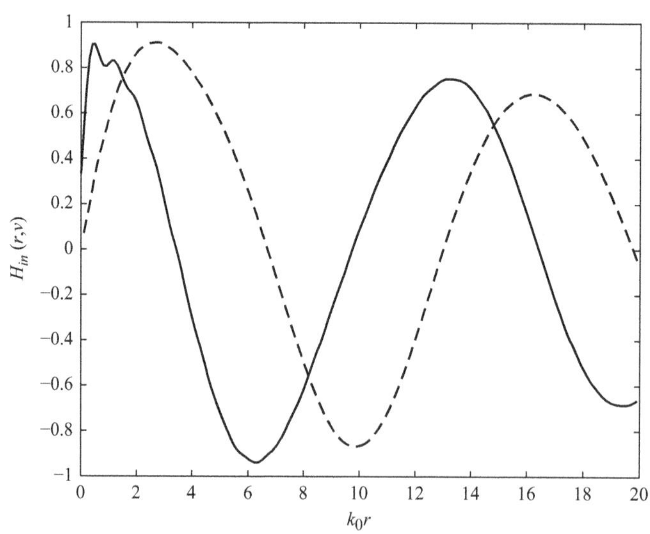

Figure 3.2 The real and imaginary parts of the damped incident plane wave on the upper boundary of the wedge as a function of k_0r

the numeric results (see the following figures) is that the error of the far field based on the geometrical optics field is very small. The l^2 error of the Fourier series is equal to 0.005 and much smaller for higher iterations. That is, from the point of view of applications the Fourier series is the exact field for all values of r.

An interesting result is that the error of the integral equation solutions $H(r, v)^{(n)}$ is very small for $k_0 r > 1$, already with $n = 2$, see Figure 3.3. The error of $H(r, v)^{(2)}$ for all values of r is plotted in Figure 3.4.

The numerical solution is furthermore justified by the fact that it correctly recovers the physical structure of the wave field. For example, the down-going part of the scattered wave inside the wedge is a superposition of the direct transmitted wave generated by the incident field and the scattered field from the lower boundary, see Figure 3.5. Outside the wedge, the scattered field is a superposition of the direct scattered wave and a cylindrical wave, see Figure 3.6. The field $H_{sc}(r, \theta)$ inside the wedge for $\theta \in [-v, v]$ and $k_1 r = 10$ is shown in Figure 3.7.

Table 3.1 Numerical error of iterated solutions for $v = 0.5$, $k_1 = 2k_0$ and $q_1 = 0$

n	2	3	4
$\|(\tilde{H}_i, \tilde{H}_e) - (\tilde{H}_i, \tilde{H}_e)^{(n)}\|^2_{\mathscr{E}}$	0.005	0.0002	0.00002

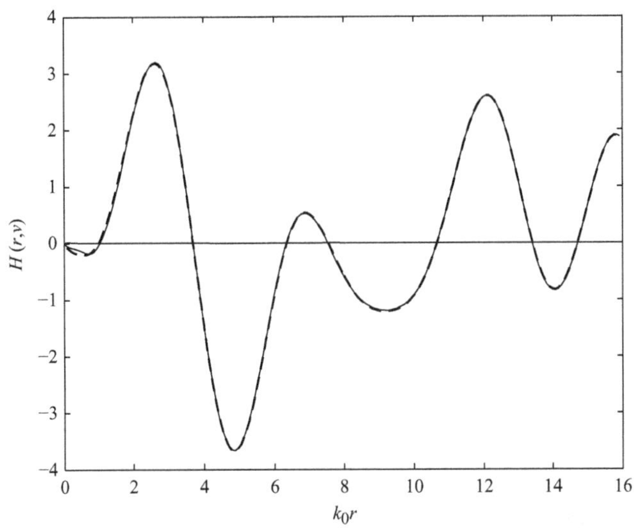

Figure 3.3 Total field on the face of the wedge as a function of $k_0 r$. The curves based on the second-order integral equation solution and on its Fourier series expansion almost coincide for $k_0 r > 1$.

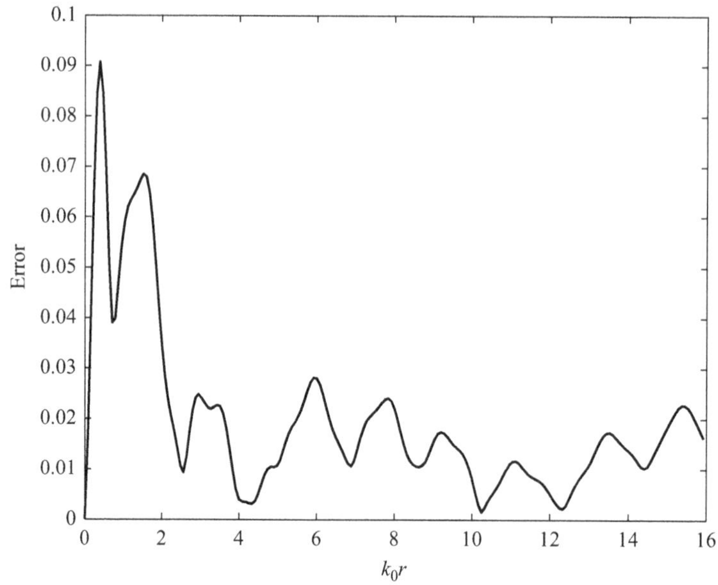

Figure 3.4 The error function $|H(r,v)-H(r,v)^{(2)}|$ *as a function of* k_0r

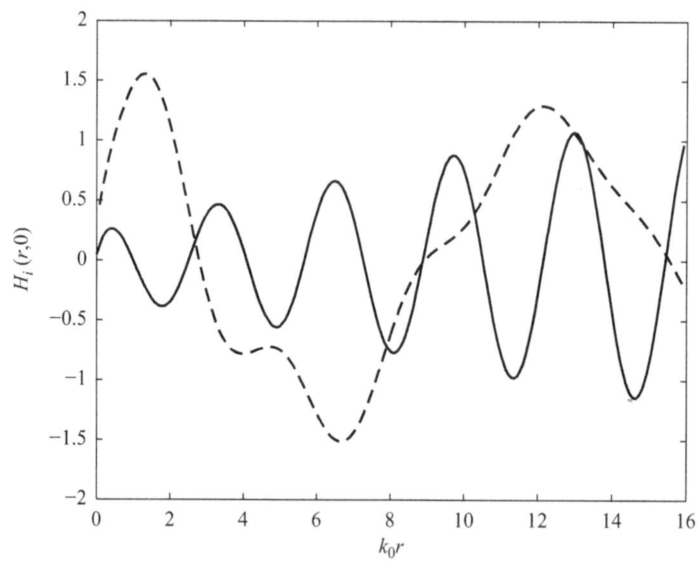

Figure 3.5 The scattered field $H_i(r,0)$ *generated at the upper boundary and calculated inside the wedge at* $\theta = 0$. *This wave is a superposition of the direct transmitted wave (dashed) and the two-times scattered wave (solid).*

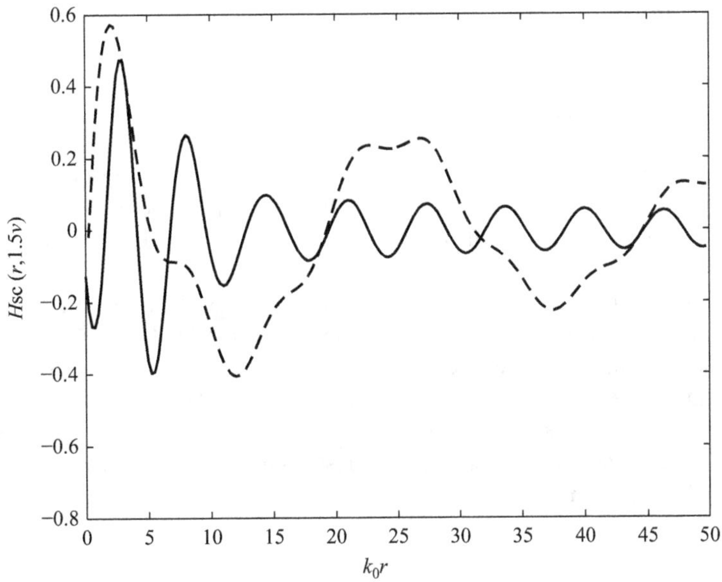

Figure 3.6 *The scattered field outside the wedge at $\theta = 1.5v$. The field is a superposition of the reflected field (dashed) and a cylindrical wave (solid). The latter is calculated as the difference between the exact field and the reflected wave.*

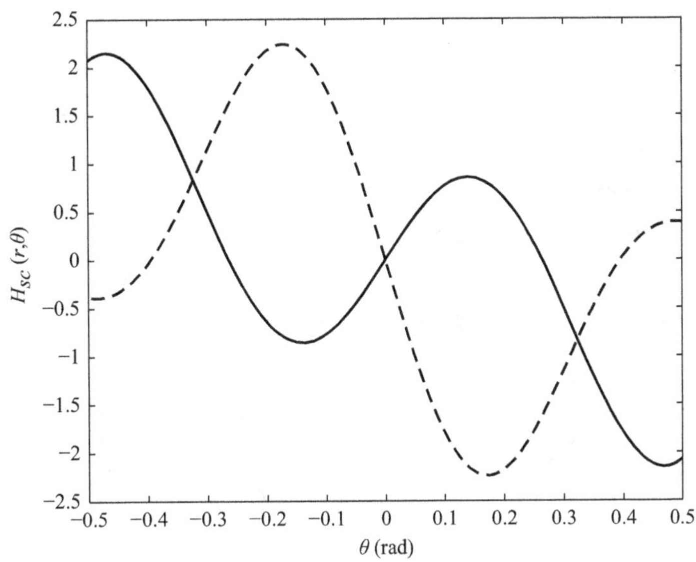

Figure 3.7 *The real and imaginary part of the scattered field inside the wedge for $k_1r = 10$ and $-v \leq \theta \leq v$*

3.8 Summary

Diffraction of a magnetically polarized incident field by a dielectric wedge has been considered. The diffraction problem has been reduced to an integral equation which has been solved by iterations in the form of a converging series. The error of the nth-order iteration has been explicitly calculated. Furthermore, the chapter has introduced a new Fourier series of orthogonal Bessel–Hankel modes. These modes behave as Bessel functions near the tip of the wedge and are outgoing Hankel functions for large values of kr. The scattered magnetic field represented by the Fourier series satisfies the radiation condition at infinity and is finite at the tip of the wedge. The Fourier series has been constructed by a mapping of the far field onto the Fourier series. The error in approximating the exact solution of the diffraction problem by n terms of the Fourier series has been explicitly evaluated. The first term in the Fourier series behaves as the Bessel function $J_s(k_q r)$ with r^s being the static field singularity at the tip of the wedge. The theory has been tested numerically, and examples of field plots have been presented.

Appendix A

The iterated solution of the integral equation converges if the norm of the operator **QK** is less than 1. The physical interpretation of this statement is that there exists no incident wave Φ, for which all the energy radiated from the upper boundary $\theta = v$ arrives at the lower boundary and no energy is radiated in directions $\theta \in [-v, 2\pi - v]$. Let $\tilde{\mathbf{\Phi}} = (\tilde{\Phi}_i, \tilde{\Phi}_e)$ denote an incident wave with the energy norm equal to 1:

$$\|\tilde{\mathbf{\Phi}}\|^2 = \|(\tilde{\Phi}_i, \tilde{\Phi}_e)\|^2 = <\tilde{\Phi}_i, \tilde{\Phi}_i>_i + <\tilde{\Phi}_e, \tilde{\Phi}_e>_e = 1. \tag{A.1}$$

The energy conservation law (Pointing theorem) relates this norm to the energy norm of the outgoing wave:

$$\|(\mathbf{K}_i\tilde{\Phi}_i, \mathbf{K}_e\tilde{\Phi}_e)_{\theta=v}\|^2 = \|(\tilde{\Phi}_i, \tilde{\Phi}_e)\|^2 = 1. \tag{A.2}$$

The far-field energy norm of the outgoing wave $H_i(r, \theta)$ for $\theta \in [-v, v]$ is given by

$$\|\mathbf{Q}_1\mathbf{K}_i\tilde{\mathbf{\Phi}}\|_F^2 = \int_{k_1 \cos 2v}^{k_1} \alpha_1(p) \mid \mathbf{K}_i\tilde{\mathbf{\Phi}}(p) \mid^2 dp. \tag{A.3}$$

The function $\mathbf{K}_i\tilde{\mathbf{\Phi}}(p)$ is analytic in the lower half of the complex p-plane. This function vanishes for all p if and only if the function vanishes in the interval $[k_1 \cos 2v, k_1]$. This proves that (A.3) defines a norm. The remaining energy radiated from the upper boundary of the wedge inside the wedge is equal to the energy flow into the lower boundary:

$$\|\mathbf{Q}_1\mathbf{K}_i\tilde{\mathbf{\Phi}}\|_{\theta=-v}^2 = \int_{-k_1}^{k_1 \cos 2v} \alpha_1(p) \mid \mathbf{K}_i\tilde{\mathbf{\Phi}}(p) \mid^2 dp. \tag{A.4}$$

Outside the scatterer, the norms of the outgoing field generated at the upper boundary will satisfy the following inequalities:

$$\|\mathbf{Q}_0\mathbf{K}_e\tilde{\mathbf{\Phi}}\|_F^2 \geq \int_{-k_0\cos v}^{k_0} \alpha_0(p) \mid \mathbf{K}_e\tilde{\mathbf{\Phi}}(p) \mid^2 dp \tag{A.5}$$

for the far-field energy flow outside the wedge and

$$\|\mathbf{Q}_0\mathbf{K}_e\tilde{\mathbf{\Phi}}\|_{\theta=-v}^2 \leq \int_{-k_0}^{-k_0\cos v} \alpha_0(p) \mid \mathbf{K}_e\tilde{\mathbf{\Phi}}(p) \mid^2 dp \tag{A.6}$$

for the energy flow of the field $\mathbf{Q}_0\mathbf{K}_e\tilde{\mathbf{\Phi}}$ at the lower boundary. The relation of the norms of the incident field and the scattered field is expressed by the energy conservation:

$$\|\tilde{\mathbf{\Phi}}\|^2 = \|\mathbf{Q}_0\mathbf{K}_e\tilde{\mathbf{\Phi}}\|_F^2 + \|\mathbf{Q}_1\mathbf{K}_i\tilde{\mathbf{\Phi}}\|_F^2 + \|\mathbf{Q}_0\mathbf{K}_e\tilde{\mathbf{\Phi}}\|_{\theta=-v}^2 + \|\mathbf{Q}_1\mathbf{K}_i\tilde{\mathbf{\Phi}}\|_{\theta=-v}^2. \tag{A.7}$$

We will now prove that the norm of the operator defined by (A.8) is less than 1,

$$\|\mathbf{QK}\| = \sup_{\tilde{\mathbf{\Phi}}} \frac{\|\mathbf{Q}_0\mathbf{K}_e\tilde{\mathbf{\Phi}}\|_{\theta=-v}^2 + \|\mathbf{Q}_1\mathbf{K}_i\tilde{\mathbf{\Phi}}\|_{\theta=-v}^2}{\|\tilde{\mathbf{\Phi}}\|} < 1. \tag{A.8}$$

Obviously the norm of $\mathbf{QK}$ is less than or equal to 1. Assume that $\|\mathbf{QK}\| = 1$. Then there exists a sequence of incident waves $\tilde{\mathbf{\Phi}}_n$ with norm 1, for which the following norm converges to 0:

$$\|\mathbf{Q}_0\mathbf{K}_e\tilde{\mathbf{\Phi}}_n\|_F^2 + \|\mathbf{Q}_1\mathbf{K}_i\tilde{\mathbf{\Phi}}_n\|_F^2 \to 0 \quad \text{for} \quad n \to \infty. \tag{A.9}$$

This equation together with (A.3) and (A.5) shows that $\mathbf{K}_i\tilde{\mathbf{\Phi}}_n(p)$ converge to 0 in the interval $p \in [k_1 \cos 2v, k_1]$ and $\mathbf{K}_e\tilde{\mathbf{\Phi}}_n(p)$ converge to 0 in the interval $p \in [-k_0 \cos v, k_0]$. The limiting functions are analytic and vanish in the specified intervals. Consequently they vanish for all p. Then the limit of the right-hand side of (A.7) with $\tilde{\mathbf{\Phi}} := \tilde{\mathbf{\Phi}}_n$ vanishes, which is not consistent with the assumption that the norm of any function $\tilde{\mathbf{\Phi}}_n$ is equal to 1. This conclusion proves that the norm of the operator $\mathbf{QK}$ is less than 1.

Acknowledgment

The family of Svend Berntsen would like to express their sincere appreciation to Dr. Andrey Osipov who has kindly helped to prepare this chapter, based on the last scientific work of our husband/father, who suddenly passed away in October 2015. Throughout many years, Dr. Andrey Osipov has been a dear and respected colleague of our husband/father, who always appreciated their conversations and collaboration. The work presented here was of particular importance for Svend Berntsen, and we know how grateful he would have been that this work is now available to the community. Without the generous help of Dr. Andrey Osipov, this would never have been possible.

References

[1] Knockaert L, Olyslager F, and de Zutter D. The diaphanous wedge. IEEE Trans Antennas Propag. 1997;45:1374–1381.
[2] Rawlins AD. Electromagnetic diffraction by wedge shaped obstacles [Doctoral thesis]. University of Surrey, Guildford, UK; 1972.
[3] Jones DS. The Kontorovich-Lebedev transform. J Inst Math Appl. 1980;26:133–141.
[4] Osipov AV. On the method of Kontorovich-Lebedev integrals in problems of wave diffraction in sectional media. Problems of Diffraction and Wave Propagation, St Petersburg State University, Saint Petersburg, Russia. 1993;25:173–219.
[5] Rawlins AD. Diffraction by, or diffusion into, a penetrable wedge. Proc R Soc A. 1999;455:2655–2686.
[6] Salem MA, Kamel AH, and Osipov AV. Electromagnetic fields in the presence of an infinite dielectric wedge. Proc R Soc A. 2006;462:2503–2522.
[7] Senior TBA. Diffraction by an imperfectly conducting wedge. Commun Pure Appl Math. 1959;12:337–372.
[8] Budaev BV and Bogy DB. Two-dimensional diffraction by a wedge with impedance boundary conditions. IEEE Trans Antennas Propag. 2005;53: 2073–2080.
[9] Osipov AV. A hybrid technique for the analysis of scattering by impedance wedges. In: Proceedings of URSI International Symposium on Electromagnetic Theory; 2004 May 23-27; Pisa, Italy; 2004. p. 1140–1142.
[10] Vasilev EN and Solodukhov VV. Diffraction of electromagnetic waves by a dielectric wedge. Radiophys Quantum Electron. 1976;17:1161–1169.
[11] Vasilev EN, Solodukhov VV, and Fedorenko AI. The integral equation method in the problem of electromagnetic waves diffraction by complex bodies. Electromagnetics. 1991;11:161–182.
[12] Berntsen S. Diffraction of an E-polarized wave by a dielectric wedge. SIAM J Appl Math. 1983;43:186–211.
[13] Bergljung C and Berntsen S. Diffraction by a dielectric wedge at skew incidence. Q J Mech Appl Math. 2001;54:549–583.
[14] Daniele V. The Wiener-Hopf formulation of the dielectric wedge problem: Part 1. Electromagnetics. 2010;30:625–643.
[15] Budaev BV. Diffraction by wedges. Longman Scientific & Technical, Harlow, Essex, England; 1995.
[16] Daniele V and Zich R. The Wiener-Hopf method in electromagnetics. SciTech Publishing, Edison, NJ; 2014.
[17] Lyalinov MA and Zhu NY. Scattering of waves by wedges and cones with impedance boundary conditions. SciTech Publishing, Edison, NJ; 2013.
[18] Osipov AV and Tretyakov SA. Modern electromagnetic scattering theory with applications. Wiley Chichester, West Sussex, United Kingdom; 2017.
[19] Bach Andersen J and Solodukhov VV. Field behavior near a dielectric wedge. IEEE Trans Antennas Propag. 1978;26:598–602.

[20] van Bladel J. Singular electromagnetic fields and sources. Oxford University Press, New York; 1991.
[21] Makarov GI and Osipov AV. Structure of Meixner's series. Radiophys Quantum Electron. 1986;29:544–549.
[22] Budaev BV and Bogy DB. On the electromagnetic field singularities near the vertex of a dielectric wedge. Radio Sci. 2007;42:RS6S08, 1–6.

Chapter 4

Green's theorem, Green's functions and Huygens' principle in discrete electromagnetics

John M. Arnold[1]

4.1 Introduction

Green's theorems are a cornerstone of field theories of differentiable functions over $\mathbb{R}^\nu$, being integral expressions of the form

$$\int_D (\phi\nabla^2\psi - \psi\nabla^2\phi)\mathrm{d}V = \int_{\partial D} (\phi\nabla\psi - \psi\nabla\phi)\cdot\mathrm{d}\mathbf{S} \tag{4.1}$$

for two functions $\psi(\mathbf{x}), \phi(\mathbf{x})$ on a bounded domain D of $\mathbb{R}^\nu$ with boundary ∂D, with $\mathrm{d}V$ the element of volume on D and $\mathrm{d}\mathbf{S}$ the element of surface area on ∂D oriented outwards from the closed volume D. The integral relation (4.1) is equivalent to the differential identity

$$\nabla\cdot(\phi\nabla\psi - \psi\nabla\phi) = \phi\nabla^2\psi - \psi\nabla^2\phi \tag{4.2}$$

along with Gauss's theorem to transform the volume integral of a divergence to a surface integral over the boundary of the domain. The theorem has many and varied applications, most notably in the reduction of differential equations over extended volumes of space to relations on surfaces bounding volumes. Suppose that ϕ and ψ are solutions of Poisson's equation

$$-\nabla^2\phi = f \tag{4.3}$$

$$-\nabla^2\psi = s \tag{4.4}$$

with sources f and s, and suppose further that the source s for ψ is localised at a single point $\mathbf{x}_0$ so that $\int_D \phi s\,\mathrm{d}V = \phi(\mathbf{x}_0)$ if $\mathbf{x}_0 \in D$ and is 0 otherwise. Then application of Green's theorem gives [1]

$$\phi(\mathbf{x}_0) = \int_D \psi f\,\mathrm{d}V - \int_{\partial D} (\phi\nabla\psi - \psi\nabla\phi)\cdot\mathrm{d}\mathbf{S}, \quad \mathbf{x}_0 \in D. \tag{4.5}$$

We seek the equivalent formulations of Green's theorem and Green's functions for the case that ϕ is a function over a discrete space, for example defined only at a

[1] School of Engineering, University of Glasgow, Glasgow, United Kingdom

discrete subset of $\mathbb{R}^\nu$, $\phi_j = \phi(\mathbf{x}_j)$, and the differential operators are replaced by difference operators. This situation arises naturally as a consequence of the discretisation of partial differential equations such as the Poisson equation or wave equation into finite-difference equations for numerical solution by computational methods. However, the adopted approach here is a consideration of dynamical systems on a graph or cell complex, so that the framework is *a priori* discrete; discrete equivalents of Maxwell's equations and electromagnetic potentials are shown using the formalism of discrete exterior calculus and discrete differential forms, as elaborated by Teixeira and Chew [2], Tonti [3], Bossavit [4,5] and Desbrun *et al.* [6]. Discrete Green's functions for stationary systems on a lattice have been proposed and studied by Chung and Yau [7], and earlier by Mugler [8], but the topic is quite rare in graph theory. Here we show how the development of Green's functions has a very practical outcome in the appearance of Huygens' principle for the dynamical evolution of discrete fields on the elements of a generally structured cell complex, with the appearance of Green's theorem as the means of attaching the elementary wavelets of Huygens' principle to the separator of an arbitrary partition of the discrete set on which the dynamical system is defined.

In the following it is shown, using the adjacency property on a graph as a template, that a discrete relation similar to (4.1) can be derived for a large class of matrices (symmetric and local in the adjacency topology), with ϕ_j being the components of a vector ϕ living in a vector space V. Suitable matrices represent operators on a Hilbert space with an inner product $\langle\ ,\rangle$. The Hilbert space contains vectors components of which are the real values of discrete functions ϕ_j, and the construction extends to any grade of r-forms on a simplicial or cell decomposition of $\mathbb{R}^\nu$. Hence, there exists a Green's theorem for any symmetric local operator on the vector space of r-forms, in the sense of evaluating the sum of an antisymmetric bilinear form over a set of r-elements D in terms of a sum over the elements of a cut-set S located at the separator between D and its complement $\bar{D}$.

The achievement of a general Green's theorem for discrete geometrical structures is nothing more than a simple identity of linear algebra. Given an operator F that is local symmetric, there exist field theories of the form

$$\mathsf{F}\phi = f \tag{4.6}$$

(e.g. discrete Poisson's equation) or dynamical systems of discrete fields such as

$$\mathsf{F}\phi + c^{-2}\partial_t^2\phi = f \tag{4.7}$$

(e.g. potential wave equations). The operators that invert these systems (F^{-1}, $(\mathsf{F} + c^{-2}\partial_t^2)^{-1}$, etc.) are constructed by auxiliary functions solving the same equation with local sources, and the elements of the matrix representation of the inverse operator are discrete Green functions. Using the discrete Green theorem with a discrete Green function for the auxiliary field, ψ leads to representations of the field ϕ in a domain D in terms of its boundary values on the cut-set S, similar to the Kirchhoff–Huygens (KH) representation of classical (continuum) field theory. Dynamical systems in discrete time explicitly exhibit Huygens' principle for the advancement of solutions by one time step when the values at earlier times are prescribed on a cut-set S. This last property can be used to construct a method for truncating an infinite domain into a finite computational domain.

A final observation is that the operators F appearing in these constructions need not be discretisations of spatial derivatives of an underlying continuum, although operators such as the Laplacian that are so derived do indeed qualify, being symmetric and local.

4.2 Green's theorem for adjacency matrices

4.2.1 Adjacency matrix

A prototype for discrete Green's theorem is provided by an elementary relation for the adjacency matrix a of a graph. Let the points x_j be the vertices of a graph in $\mathbb{R}^\nu$, and let the real values ϕ_j and ψ_j be prescribed at each vertex x_j. The adjacency matrix a has matrix elements $a_{jj'} = 1$ if vertices j and j' are connected by an edge, and $a_{jj'} = 0$ otherwise. Then it is a simple calculation to verify that

$$\sum_{j\in D}\sum_{j'\in M} a_{jj'}(\phi_j\psi_{j'} - \psi_j\phi_{j'}) = \sum_{j\sim j'\in S}(\phi_j\psi_{j'} - \psi_j\phi_{j'}) \tag{4.8}$$

where M is the complete set of vertices, D is a subset of M, $\bar{D}$ is the complement of D in M and S is the set of pairs (j,j') such that $j \in D, j' \in \bar{D}$ and j and j' are connected by an edge, and we use $j \sim j' \in S$ to denote the pairing. Clearly, the set of pairs of vertices S, or equivalently the edges of the graph that join the pairs, one in D and one in $\bar{D}$, are the 'surface' that separates the vertex domains D and $\bar{D}$.

Powers of the adjacency matrix a^q have elements equal to the number of paths of length q connecting the vertices j and j'

$$(a^q)_{jj'} = N_{jj'}(q) = \text{ number of paths over } q \text{ edges from } j \text{ to } j'. \tag{4.9}$$

It is convenient to introduce a measure of distance on the graph, namely the smallest number of edges in a path between two vertices (the Manhattan metric)

$$\text{dist}(j,j') = \inf_q\{q : N_{jj'}(q) \neq 0\}. \tag{4.10}$$

Then the adjacency power a^q can be assigned a range $r(q)$ which is the maximum distance over which the elements of the matrix are nonvanishing. Green's theorem for a^q is then

$$\sum_{j\in D}\sum_{j'\in M} (a^q)_{jj'}(\phi_j\psi_{j'} - \psi_j\phi_{j'}) = \sum_{j\sim j'\in S} N_{jj'}(q)(\phi_j\psi_{j'} - \psi_j\phi_{j'}) \tag{4.11}$$

4.2.2 Weighted adjacency matrix

The weighted adjacency matrix w replaces the unit values of the adjacency matrix a with weights $w_{jj'}$ and retains the 0 values of $a_{jj'}$ for nonadjacent vertices. Then the Green theorem (4.8) is modified to

$$\sum_{j\in D}\sum_{j'\in M} w_{jj'}(\phi_j\psi_{j'} - \psi_j\phi_{j'}) = \sum_{j\sim j'\in S} w_{jj'}(\phi_j\psi_{j'} - \psi_j\phi_{j'}) \tag{4.12}$$

Powers of the weighted adjacency matrix w^q have elements equal to the sum over all paths of length q of the product of path weights along each path

$$(w^q)_{jj'} = \sum_{\text{paths}(q)} w_{jj_1}w_{j_1j_2}\cdots w_{j_{q-1}j'} = W_{jj'}(q). \tag{4.13}$$

Green's theorem for powers of the weighted adjacency matrix is

$$\sum_{j\in D}\sum_{j'\in M} (w^q)_{jj'}(\phi_j\psi_{j'} - \psi_j\phi_{j'}) = \sum_{j\sim j'\in S} W_{jj'}(q)(\phi_j\psi_{j'} - \psi_j\phi_{j'}). \tag{4.14}$$

Now the set S is the set of all pairs of points $\{(j,j') : j \in D, j' \in \bar{D}\}$ that can be joined by a path consisting of q contiguous edges.

4.2.3 Matrix of adjacency 1

Let F be a symmetric matrix with adjacency range $q = 1$, which means that the elements of F vanish for $\text{dist}(j,j') > 1$. Then Green's theorem for F is

$$\sum_{j\in D}\sum_{j'\in M} F_{jj'}(\phi_j\psi_{j'} - \psi_j\phi_{j'}) = \sum_{j\sim j'\in S} F_{jj'}(\phi_j\psi_{j'} - \psi_j\phi_{j'}) \tag{4.15}$$

where S is the set of pairs (j,j'), $j \in D$, $j' \in \bar{D}$ such that $\text{dist}(j,j') = 1$. This set is parameterised by the nonvanishing elements of the adjacency matrix a:

$$j \sim j' \in S \text{ iff } a_{jj'} = 1, j \in D, j' \in \bar{D}.$$

It is obvious that the diagonal elements of F do not contribute because of the antisymmetry of the summand of the right-hand side.

4.2.4 The vertex Laplacian matrix

A matrix of particular importance is the discrete (graph) Laplacian L, in which the diagonal element of the row j is equal to the sum of the weights on the off-diagonal elements: $n_j = \sum_{j'} w_{jj'}$

$$L_{jj'} = n_j\delta_{jj'} - w_{jj'} \tag{4.16}$$

When all the non-zero weights are equal to 1, so that $w = a$, then n_j is an integer equal to the number of edges connected to vertex j, called its *degree* n_j. A *degree-regular* graph has identical $n_j = n$ at all vertices in the graph. The discrete Laplacian matrix with positive weights $w_{jj'}$ is symmetric positive semi-definite, with a zero eigenvalue whose eigenvector is $\phi_j = 1$.

4.2.5 Incidence matrices

The incidence matrix for a graph is a nonsquare matrix d whose entries d_{kj} are ± 1 when the vertex j is a boundary of edge k, and 0 otherwise

$$d_{kj} = \begin{cases} \pm 1, & \text{edge } k \text{ has boundary vertex } j \\ 0, & \text{otherwise.} \end{cases} \tag{4.17}$$

Each oriented edge of the graph has two vertices with opposite orientations, so each row of the incidence matrix has two nonvanishing entries, with opposite signs. Each column of the incidence matrix has n_j nonvanishing entries where n_j is the degree of the vertex j. The adjacency matrix a is constructed from the incidence matrix as

$$a_{jj'} = \begin{cases} -\sum_k d_{kj} d_{kj'}, & j \neq j' \\ 0, & j = j' \end{cases} \tag{4.18}$$

the weighted adjacency matrix w is

$$w_{jj'} = \begin{cases} -\sum_k d_{kj} *_k d_{kj'}, & j \neq j' \\ 0, & j = j' \end{cases} \tag{4.19}$$

and the Laplacian matrix L is

$$L_{jj'} = \begin{cases} \sum_k d_{kj} *_k d_{kj'}, & j \neq j' \\ \sum_k d_{kj} *_k d_{kj}, & j = j' \end{cases} \tag{4.20}$$

where $*_k = w_{jj'}$ is a weight function on edges, edge k being terminated by vertices j and j' and will reappear later as a Hodge factor in a cell complex. It follows from these expressions that

$$L = \Lambda - w, \quad L = d^T * d \tag{4.21}$$

where Λ is a diagonal matrix with diagonal entries equal to the degree of the vertex, superscript T represents the matrix transpose and $*$ is the diagonal matrix with entries $*_k \delta_{kk'}$.

We remark here that when the vertices x_j form a cubic lattice in $\mathbb{R}^\nu$ with edges along orthogonal Cartesian axes then the Laplacian matrix is identical to the finite-difference stencil of the continuum differential operator $-\nabla^2$ by Taylor expansion up to second-order

$$-\nabla^2 \phi(\mathbf{x}_j) \sim \text{vol}^{-1} \sum_{j'} L_{jj'} \phi(\mathbf{x}_{j'}) + O(h^2)$$

with edge length h, $\text{vol} = h^\nu$, $*_k = h^{\nu-2}$.

4.3 Green's theorem on topological vector space

We consider a real linear vector space V with basis $e_j : j = 1, 2, \ldots, N$ such that an element ϕ of V is $\phi = \sum_j \phi_j e_j$ with components ϕ_j. We suppose that the base space for V is endowed with a discrete topology such as a map to points on $\mathbb{R}^\nu$ $j \to x_j \in \mathbb{R}^\nu$ and a symmetric distance function $\text{dist}(j, j') > 0$, which makes the base space a discrete manifold M with vertices labelled by j and N the total number of vertices. We use the topological property of adjacency to define the distance function, not an embedding of the manifold in $\mathbb{R}^{\nu}$. Let there be two elements $\phi \in V$ and $\psi \in V$ with components ϕ_j and ψ_j, respectively, and define the Hilbert space inner product of these two vectors to be $\langle \phi, \psi \rangle = \sum_{j' \in M} \phi_j *_j \psi_j$, where $*_j$ is a positive definite weight. We shall later interpret $*$ as a Hodge operator and $*_j$ as the diagonal elements of its matrix representation, which act as volume elements for the inner product. Finally define linear operator $\mathsf{F} : V \to V$ which is self-adjoint with respect to the inner product $\langle \, , \, \rangle$. Such a linear operator has a matrix representation $*_j^{-1} F_{jj'}$ such that $\psi_j = \sum_{j' \in M} *_j^{-1} F_{jj'} \phi_{j'}$ represents the linear action $\psi = \mathsf{F}\phi$ of F on V. The matrix $F_{jj'}$ is symmetric for a symmetric operator F with a non-uniform weight function $*$ in the inner product. Observe here that the symmetric *operator* F is distinguished from its symmetric *matrix* representative F.

Now let the base manifold M be partitioned into two disjoint sets D and $\bar{D}$, corresponding to an orthogonal projection of the space V onto two subspaces. Given two vectors ψ and ϕ and the linear operator F form the antisymmetric bilinear form

$$\Psi_{jj'} = F_{jj'}(\phi_j \psi_{j'} - \psi_j \phi_{j'}). \tag{4.22}$$

Ψ is antisymmetric with respect to the interchange either of ϕ and ψ, or of the indices j and j'.

Then the following results are established:

$$\sum_{j \in D} \sum_{j' \in D} \Psi_{jj'} = 0 \tag{4.23}$$

$$\sum_{j \in D} \sum_{j' \in M} \Psi_{jj'} = \sum_{j \in D} \sum_{j' \in \bar{D}} \Psi_{jj'} \tag{4.24}$$

Now suppose that F is a *local* operator, by which we mean that $F_{jj'} = 0$ for $\operatorname{dist}(j,j') > q$, where q is a finite integer. Then the double sum on the right of (4.24) has nonvanishing contributions only from pairs of points in D and $\bar{D}$ that are sufficiently close in the discrete topology of the base manifold M, i.e. they lie close to the cut separating the two partitions of M. Denoting this set of pairs by S, we have finally Green's theorem

$$\sum_{j \in D} \sum_{j' \in M} \Psi_{jj'} = \sum_{j \sim j' \in S} \Psi_{jj'} \tag{4.25}$$

where $S = \{(j,j') : \operatorname{dist}(j,j') \leq q, j \in D, j' \in \bar{D}\}$. Each pair in S is included only once in the sum.

While the local operator F can be anything, a prototype for all discrete local operators is the discrete Laplacian, L. The unweighted Laplacian matrix is derived from an adjacency matrix, $L_{jj'} = n_j \delta_{jj'} - a_{jj'}$, having entries of -1 if the elements j and j' are adjacent, n_j when $j = j'$ where n_j is the number of vertices adjacent to vertex j, and equal to 0 otherwise. The weighted Laplacian has matrix elements $-w_{jj'}$ for adjacent elements, $\sum_{j'} w_{jj'}$ for the diagonal elements and 0 for all nonadjacent elements. The adjacency range q of the unweighted Laplacian is the largest distance between nonvanishing off-diagonal elements in its rows or columns, $q = \max\{\text{dist}(j,j')\}$, $j \neq j', w_{jj'} \neq 0$ and is equal to 1. When the operator F is the Laplacian, weighted or unweighted, then the set S consists of adjacent pairs of points of the manifold M for which one point is in D and the other is in $\bar{D}$. Evidently, powers of the Laplacian L^m and polynomials $\sum_{q=0}^{m} a_q \mathsf{L}^q$ are local also, with increasing adjacency range m.

Let the two vectors ϕ and ψ now be solutions of

$$\mathsf{F}\phi = f \Rightarrow \sum F_{jj'}\phi_{j'} = *_j f_j \tag{4.26}$$

$$\mathsf{F}\psi = s \Rightarrow \sum F_{jj'}\psi_{j'} = *_j s_j \tag{4.27}$$

where F is a local symmetric linear operator on V, and the inhomogeneous terms f and s are source vectors.

First, apply Green's theorem over the whole space M, so that

$$\sum_{j \in M} \sum_{j' \in M} \Psi_{jj'} = 0 \tag{4.28}$$

$$\Rightarrow \sum_{j \in M} (\phi_j *_j s_j - \psi_j *_j f_j) = 0 \tag{4.29}$$

Now require $s_j = 0$ on all points in M except for one, say $j = j_0$, on which $*_{j_0} s_{j_0} = 1$, then

$$\phi_{j_0} = \sum_{j \in M} \psi_j *_j f_j = \langle \psi, f \rangle. \tag{4.30}$$

Denoting the auxiliary function ψ as the Green function G,

$$\psi_j = G_{j_0 j} \tag{4.31}$$

then

$$\phi_{j_0} = \sum_{j \in M} G_{j_0 j} *_j f_j \quad \Rightarrow \phi = \mathsf{G}f \tag{4.32}$$

which is the standard inversion, or reproducing, formula for Poisson's equation from continuum field theory [1], adapted to the discrete spatial domain with an *exact* discrete Green's function $G_{j_0 j}$ corresponding to an arbitrary symmetric operator F. The reproducing formula represents the field throughout the region M as a superposition of contributions from elementary point sources. It is straightforward to demonstrate that

$$\sum_{j' \in M} F_{jj'} G_{j'j''} = \sum_{j' \in M} G_{jj'} F_{j'j''} = \delta_{jj''} \tag{4.33}$$

$$G_{jj'} = G_{j'j} \tag{4.34}$$

which are to say that the matrix G is symmetric and is the inverse of the matrix F. Green's operator $\mathsf{G} : V \to V$ has the matrix representative $G_{jj'} *_{j'}$, and as operators $\mathsf{FG} = \mathsf{GF} = I$.

Now suppose that the discrete supports of these sources are in opposite partitions of M,

$$s_j = 0, j \in \bar{D}, \quad f_j = 0, j \in D.$$

Then application of Green's theorem leads directly to

$$\sum_{j \in D} (\phi_j *_j s_j - \psi_j *_j f_j) = \sum_{j \sim j' \in S} \Psi_{jj'} \tag{4.35}$$

and since $f_j = 0$ for $j \in D$ the second term in the sum on the left vanishes, to give

$$\sum_{j \in D} \phi_j *_j s_j = \sum_{j \sim j' \in S} \Psi_{jj'}. \tag{4.36}$$

If we again require $s_j = 0$ on all points in D except for one, say $j = j_0$, on which $*_{j_0} s_{j_0} = 1$, then

$$\phi_{j_0} = \sum_{j \sim j' \in S} F_{jj'}(\phi_j \psi_{j'} - \psi_j \phi_{j'}), \quad j_0 \in D \tag{4.37}$$

This is the equivalent of Kirchhoff's formula in continuous field theory, in which the value of a field ϕ satisfying Poisson's equation in a source region $\bar{D}$ is determined in the complementary region D from its values on the boundary S of D using the Green function ψ.

4.4 Difference forms and discrete exterior calculus

The previous treatment involved a vector space V on a base space which formed a discrete manifold M, when it was envisaged that each 1-dimensional subspace of V was mapped over a discrete point of $\mathbb{R}^\nu$. Any vector space fitting this topological picture will be a candidate for Green's theorem for its local symmetric operators F. The situation can be generalised by considering simplicial or cell complexes that refine the decomposition of $\mathbb{R}^\nu$. Here we make an informal summary of the elements of discrete exterior calculus, using the notation and structural geometry for discrete forms described by Desbrun *et al.* [6]. Discrete difference forms and their associated exterior calculus are similar to the continuum case [9], except that the discretisation of space enables the introduction of difference operators whose coefficients are simple integers, often ± 1.

4.4.1 Simplicial decomposition

We consider in $\mathbb{R}^\nu$ a simplicial complex, in which vertices are joined in pairs by edges, edges surround faces, faces enclose volumes and so on. Each level or grade of this complex has dimension $0 \leq r \leq \nu$. At dimension r, each object is specified by $r + 1$ points determining a linear r-dimensional subspace of $\mathbb{R}^\nu$. Thus, a vertex ($r = 0$) is given by 1 point in $\mathbb{R}^\nu$, an edge ($r = 1$) is a linear path between 2 vertices, a face ($r = 2$) is a plane containing 3 points and so on. Given such a simplicial complex, functions can be defined on each grade by assigning a value in $\mathbb{R}$ to each r-dimensional element, giving a sequence of vector spaces V_r for $r = 0, 1, \ldots, \nu$ whose base spaces are the r-dimensional elements. A local topology can be assigned within each grade r using the adjacency concept, and a topology among the different grades can be assigned by the concept of incidence. Thus, two r-dimensional simplexes σ_j^r and $\sigma_{j'}^r$ are adjacent if they are both in the boundary of a single $r+1$ simplex σ_k^{r+1} and are nonadjacent otherwise. An r-simplex and an $(r+1)$-simplex are incident if the r-simplex is an element of the boundary of the $(r+1)$-simplex. Incidence matrices map r-simplexes to $(r+1)$-simplexes, with an entry ± 1 in the (k,j) element place if σ_j^r is incident with σ_k^{r+1}, and 0 entries for nonincident pairs; thus,

$$d_{kj} = \pm 1 \text{ or } 0. \tag{4.38}$$

The choice of sign is dependent on orientation. Each geometrical element has two orientations. A local orientation can be arbitrarily assigned to each element, and a separate cyclic orientation to each boundary of an element. Then the sign in d_{kj} is chosen positive if the local orientation matches the cyclic orientation, and negative if not.

There are two adjacency measures for higher order elements. Two edges can be adjacent either by being in the boundary of the same face, or by being attached to the same vertex. In general, two r-simplices are adjacent either by common incidence on an $r+1$ simplex or by common incidence on an $(r-1)$-simplex. For $r = 1$, these adjacency matrices are

$$a_{k\,k'} = \begin{cases} \sum_l d_{lk} d_{lk'}, & k \neq k' \\ 0, \quad k = k' \end{cases} \tag{4.39}$$

$$b_{k\,k'} = \begin{cases} \sum_j d_{jk} d_{jk'}, & k \neq k' \\ 0, \quad k = k' \end{cases} \tag{4.40}$$

where j, k, l are indices for vertices, edges and faces, respectively.

The incidence matrices are so far conceived of as matrix maps. However, they can also act as operators on the vector spaces V_r, by means of

$$\psi_k = \sum_j d_{kj} \phi_j \Rightarrow \psi = d\phi, \quad \phi \in V_r, \psi \in V_{r+1} \tag{4.41}$$

The operator d, which maps r-forms to $(r+1)$-forms, is the discrete de Rham (coboundary) operator. It is straightforward to verify that when the operator d is applied twice, the result is 0: $dd\phi = 0$.

4.4.2 Dual forms

Given a simplicial decomposition of $\mathbb{R}^\nu$, there exists a dual decomposition of the same space, obtained by associating r-elements of the primary system to $(\nu - r)$-elements of the dual system, so that r-forms in the primary system are $(\nu - r)$-forms in the dual system. This may be accomplished by the Hodge operator $*$ so that, if ϕ is an r-form, then $*\phi$ is its dual $(\nu - r)$-form. The components of the dual form $*\phi$ are related to those of the primary form ϕ by $\phi_j \rightarrow *_j\phi_j$, where $*_j$ are the diagonal entries of a diagonal matrix representation of the Hodge operator. For a reasonable embedding of the primary and dual meshes together as decompositions of $\mathbb{R}^\nu$, the Hodge matrices carry metric information in the form of ratios of sizes of the mapped elements. Since the vector space structure is conserved by this mapping, which is an isomorphism, then V_r and $*V_{\nu-r}$ are isomorphic vector spaces and the same topological adjacency measures apply in the dual spaces; thus, if two elements j and j' are adjacent in the primary complex V_r, then their Hodge images are adjacent in the dual $*V_{\nu-r}$. In the simplicial primary decomposition, the dual elements can be located at the circumcentres of their corresponding primary elements; thus, the dual 0-elements are at the centres of the primary ν-elements, dual 1-elements are located at the circumcentres of primary $(\nu - 1)$-elements and so on. Alternatively, once the dual 0-elements are located as vertices of the dual complex, they are linked by dual 1-elements (dual edges) that intersect the corresponding primary $(\nu - 1)$-elements. The dual complex is not in general simplicial and is a more general cell complex. The elements of the Hodge matrix $*_j$ are the volume elements for the inner product in V_r.

The simplicial decomposition of $\mathbb{R}^\nu$ is essentially the Delaunay–Voronoi tessellation method. The simplicial method simplifies in some cases of regularity of the distribution of vertices such as lattices and regular graphs.

4.4.3 Hypercube decomposition

In the hypercubic decomposition of $\mathbb{R}^\nu$, vertices are assigned to points of a simple cubic lattice in $\mathbb{R}^\nu$, and edges are assigned to the line segments joining the vertices along orthogonal Cartesian axes. Then the r-simplex is replaced by a linear subspace on 2^r vertices, rather than r vertices as for the simplicial decomposition. The dual hypercube complex can be constructed by assigning dual vertices to the circumcentres of primary ν-hypercubes, then constructing the dual complex exactly as for the primary complex. Because of the orthogonality of intersection of primary and dual elements, and the strong regularity of the lattice construction, the Hodge factors $*$ are more uniform in the hypercubic than in the simplicial decomposition. If the edge length is h, then the dimensional volume of each r-cell is h^r, and the Hodge factors are diagonal matrices $* = h^{\nu-2r}I$ for r-forms to dual $(\nu - r)$-forms; if the edge length is normalised to $h = 1$, then all the Hodge factors are also equal to 1, and effectively disappear from the calculations. More general hyperquad complexes can also be constructed with nonlattice vertices and nonorthogonal intersections.

The hypercube decomposition is the dual Yee cell decomposition for finite-difference electromagnetics (the Finite-Difference Time Domain (FDTD) method) [10–12].

4.4.4 Contextual algebraic notation of forms

The following notational conventions are adopted for dealing with forms of different grades. Operators are denoted by F and forms by ϕ, with no notational reference to the particular grade, so the notation is contextual. The linear action of operator F on r-form ϕ is the r-form $\mathsf{F}\phi$. When the components of a form ϕ have to be specified, these are indexed by an integer, ϕ_j. Thus, the action $\psi = \mathsf{F}\phi$ is expressed in components by $\psi_j = *_j^{-1} \sum_{j'} F_{jj'}\phi_{j'}$ where $*_j$ is the volume element for the inner product $\langle\, ,\rangle$ on the appropriate grade r, and $F_{jj'}$ are the matrix elements of the matrix F which is a matrix representative of the operator F. The index symbol j usually denotes the generic index over r-elements; however, when it is necessary to mix the form grades in a single expression, then separate index symbols $i, j, k, \ldots$, etc. may be used for each grade. Of course, the dimensions of the spaces V_r are different for different r. The operators d, $\tilde{d}$ and $*$ are contextual, $*_j$ represents the generic action of $*$ in mapping from primary r-forms to dual $(\nu - r)$-forms so the dual $(\nu - r)$-form $*\phi$ has components $*_j\phi_j$. We also allow $*$ to represent the converse Hodge mapping from dual r-forms to primary $(\nu - r)$-forms, and the composition $** = (-1)^{r(\nu - r)}$, which is 1 for ν odd. This method of notation carries the conventions of differential forms over to difference forms in a way that is compact in not requiring a proliferation of indices denoting the precise grade of form in generic expressions. It does however require some care in unpacking expressions involving sequences of operators, such as $d * d$, to realise that each operator in the sequence is a grade-changing operation and therefore has different algebraic and numerical expressions when applied to forms of various grades, primary and dual.

4.4.5 Manifolds, graphs and lattices

Generally a collection of discrete points $x_j \in \mathbb{R}^\nu$ is described as a manifold M. The total set of points in M is partitioned into two disjoint sets D and $\bar{D}$, and in the construction of Kirchhoff's formula (4.37) it is assumed that the observation point at which the field is required to be computed lives in the partition D and the sources f live in the partition $\bar{D}$. When the set M is finite, as the vertices in a finite graph, then the partitioning is unambiguous. In modelling electromagnetic fields using the graph as a discrete coordinate system, it may be necessary to consider that some points of the set M lie on boundaries over which special conditions on the fields need to be incorporated, and it may be desired to incorporate part or all of these boundary sets in the partitions. The inclusion of such boundary value conditions can readily be incorporated in the formalism that we describe, at the cost of more complex notation and analytical computation. Also, it may be that the set M is infinite, in describing a discretisation of all space $\mathbb{R}^\nu$. This case is assumed to be so here that M is a discretisation of the points of all $\mathbb{R}^\nu$, or of a finite subdomain of $\mathbb{R}^\nu$ without specific boundary conditions. In its simplest form, M is just free space.

In formulating the discretisation of space as a graph, edges of the graph are also introduced. It is shown subsequently that edges also support fields (1-forms). Edge fields can be treated in exactly the same formalism as vertex fields, and indeed there exist Green's theorem, Green's functions and Huygens' principle for edge fields. In these cases, the set M will be the set of all edges. In electromagnetics, the scalar potential ϕ is a vertex field, and the vector potential $\mathbf{A}$ is represented as an edge field.

Higher order discrete decompositions of $\mathbb{R}^\nu$ lead to a hierarchy of fields, through simplicial or hypercube decomposition. Strictly speaking the manifold for the decomposition is the cell complex K, but we can continue to refer to a subcomplex of any particular dimension as M.

An infinite division of space which has the vertex points arranged on a lattice has a high degree of regularity, which leads to the analytic computability of lattice Green's functions by combinatorics or Fourier methods.

4.4.6 Essentials of cell decomposition

The essential structures provided by any method of cell decomposition of $\mathbb{R}^\nu$ are a hierarchy of r-elements (vertices, edges, faces, etc.), with incidence matrices d and adjacency matrices a, b; a system of functions on those discrete r-elements (r-forms) forming a vector space V_r at each grade r; a Hilbert-space inner product $\langle\ ,\ \rangle$ on each V_r; volume elements $*$ for the inner product at each grade; dual cells and dual forms; a discrete de Rham operator d which maps r-forms to $(r+1)$-forms, taking the form of a pure finite-difference operator with integer coefficients obtained directly from the incidence structure of the decomposition. The simplicial and hypercube decompositions are simply algebraic machinery for generating these objects.

We assume henceforth that there exists a hierarchy of vector spaces $V_r, r = 1, \ldots, \nu$ each with an inner product $\langle\ ,\ \rangle$ and associated Hodge weight $*$, which also acts as a dualising operator $* : V_r \to *V_{\nu-r}$.

4.5 Higher order Green's theorem and Green's functions

4.5.1 Green's theorem for r-forms

The results of Section 4.3 can now be extended to each of the vector spaces V_r in turn. The vectors in V_r are called r-forms, the set D in the Green theorem is a set of r-elements and the boundary set $S(q)$ is the set of pairs of r-elements, one in D and one in $\bar{D}$, separated by a distance no greater than q.

Let V_r be the linear vector space of r-forms on a cell complex K, equipped with inner product $\langle , \rangle$, and let ϕ and ψ be two r-forms in V_r. Let F be a local linear operator on V_r, which is self-adjoint with respect to the inner product $\langle , \rangle$, having a symmetric matrix representation $F_{jj'}$. Define the antisymmetric bilinear form

$$\Psi_{jj'} = F_{jj'}(\phi_j\psi_{j'} - \psi_j\phi_{j'}), \tag{4.42}$$

where j, j' are generic indices for r-elements. Then

$$\sum_{j\in D}\sum_{j'\in M} \Psi_{jj'} = \sum_{j\sim j'\in S} \Psi_{jj'}, \tag{4.43}$$

where $S = \{(j,j') : \mathrm{dist}(j,j') \le a, j \in D, j' \in \bar{D}\}$, each pair in S being counted once in the sum.

4.5.2 *Kirchhoff's theorem for r-forms*

With the definitions and conditions of Section 4.5.1, suppose also that

$$\mathsf{F}\phi = f \Rightarrow \sum F_{jj'}\phi_{j'} = *_j f_j \tag{4.44}$$

$$\mathsf{F}\psi = s \Rightarrow \sum F_{jj'}\psi_{j'} = *_j s_j \tag{4.45}$$

with r-form sources f and s which vanish in D and $\bar{D}$, respectively. Then

$$\sum_{j \in D} \phi_j *_j s_j = \sum_{j \sim j' \in S} \Psi_{jj'}. \tag{4.46}$$

If, further, the source s is a delta function $*_j s_j = \delta_{jj_0}$, then ψ is a Green's function, and

$$\phi_{j0} = \sum_{j \sim j' \in S} \Psi_{jj'} \tag{4.47}$$

$$= \sum_{j \sim j' \in S} F_{jj'}(\phi_j \psi_{j'} - \psi_j \phi_{j'}). \tag{4.48}$$

4.5.3 *Green's function for r-forms*

The special case of ψ being generated by a point source is given a singular recognition as a Green's function. An extra notational index is added to indicate the position of the source, so we have

$$\psi_j \rightarrow G_{j_0 j}. \tag{4.49}$$

In terms of the Green function, the Kirchhoff theorem becomes

$$\phi_{j0} = \sum_{j \sim j' \in S} F_{jj'}(\phi_j G_{j_0 j'} - G_{j_0 j} \phi_{j'}), \quad j_0 \in D. \tag{4.50}$$

The fundamental property of the Green function $G_{j_0 j}$ is that it is the matrix representation of the operator inverse to F,

$$\mathsf{FG} = \mathsf{GF} = I \tag{4.51}$$

$$\sum_{j'} F_{jj'} G_{j'j''} = \delta_{jj''_j}. \tag{4.52}$$

4.6 Dynamical systems on topological vector spaces: Maxwell's equations

A very great variety of physical field theories are representable as dynamical systems on a topological vector space. Dynamical systems are of the basic form

$$\mathsf{F}\phi + \alpha_p \partial_t^p \phi = f \tag{4.53}$$

where the integer p is the order of the dynamical system, ϕ is a field over geometrical elements (vertices, edges, r-forms, etc.), f is a source r-form matching the form grade of the field ϕ, F is the time-independent system operator and α_p are constant coefficients arising from the particular physical context. Basic types are classified by order p:

$$\mathsf{F}\phi = f: \quad \text{static or Poisson;} \tag{4.54}$$

$$\mathsf{F}\phi + \alpha_1 \partial_t \phi = f: \quad \text{diffusion or heat;} \tag{4.55}$$

$$\mathsf{F}\phi + \alpha_2 \partial_t^2 \phi = f: \quad \text{wave.} \tag{4.56}$$

We describe here the particular example of Maxwell's equations of electrodynamics, in order to indicate the ways in which the Green and Kirchhoff theorems apply. The Maxwell equations are the minimal linear dynamical system involving primary and dual 1-form fields, coupling between the fields through primary and dual 2-form fluxes, or alternatively requiring both 0- and 1-form potentials ϕ and A for their full description, each satisfying a wave equation of type (4.56).

4.6.1 Discrete Maxwell equations

The discrete Maxwell equations for electromagnetic fields on a discrete spatial cell complex, in continuous time t, are

$$\partial_t B = -dE \tag{4.57}$$

$$\partial_t D = dH - J \tag{4.58}$$

$$dB = 0 \tag{4.59}$$

$$dD = \rho \tag{4.60}$$

$$dJ + \partial_t \rho = 0 \tag{4.61}$$

along with the free-space Hodge constitutive relations $D = \varepsilon * E, B = \mu * H$. Here E is a 1-form (electric intensity), B is a 2-form (magnetic flux), H is a dual 1-form (magnetic intensity), D is a dual 2-form (electric flux), J is a dual 2-form (current), ρ is a dual 3-form (charge). ε and μ are the permittivity and permeability, respectively, of vacuum. The electromagnetic energy is

$$W_{\mathrm{em}} = \tfrac{1}{2}\varepsilon\langle E, E\rangle + \tfrac{1}{2}\mu\langle H, H\rangle \tag{4.62}$$

and in the absence of sources, $J = 0$, W_{em} is conserved: $\partial_t W_{\mathrm{em}} = 0$. When $J \neq 0$, then $\partial_t W_{\mathrm{em}} = -\langle E, *J\rangle$.

Because $dB = 0$, there exists a 1-form potential A such that $B = dA$ (discrete Poincare theorem), and $E = -\partial_t A - d\phi$ where ϕ is a 0-form discrete scalar potential. Then the potentials satisfy discrete wave equations

$$-*d*d\phi + c^{-2}\partial_t^2 \phi = \varepsilon^{-1} * \rho \tag{4.63}$$

$$*d*dA - d*d*A + c^{-2}\partial_t^2 A = \mu * J \tag{4.64}$$

with a Lorentz gauge condition

$$*d*A + c^{-2}\partial_t\phi = 0 \tag{4.65}$$

and $c = (\mu\varepsilon)^{-1/2}$ is the wave speed. The discrete-space wave systems (4.63) and (4.64) decouple the form grades of the two potentials, so that (4.63) is for the 0-form ϕ and (4.64) is for the 1-form A. The discrete Laplacian operators forming the system operators F for these equations can be identified from (4.63) and (4.64) as $\mathsf{F} = \mathsf{L} = (-1)^{\nu r+1}(*d*d - d*d*)$ with $\nu = 3$ and $r = 0, 1$; when acting on 0-forms the second term vanishes, and when acting on ν-forms the first term vanishes.

In addition to the Maxwell equations for the fields, the sources ρ and J may themselves be subject to a dynamical system driven by the fields, typically in the form of oscillator equations such as

$$\partial_t^2 P + \omega^2 P = \gamma * E \quad \text{(Lorentz atomic oscillator)} \tag{4.66}$$

with $J = \partial_t P$, $\rho = -dP$ for dual 2-form polarisation P.

4.6.2 Electromagnetic fields as differential forms

The question of the discretisation of Maxwell's differential equations for the purposes of computation has a long history, which begins for our purposes with the Yee dual grid scheme [10], in which the spatial derivatives of the Maxwell differential equations are discretised as finite difference approximations with fields defined by samples located on r-elements of a cubic lattice and its associated dual. The Yee scheme was reformulated by Chew [11], who introduced discrete differencing operators that are similar to the de Rham d operators defined earlier. Tonti [3] reformulated the integral Maxwell equations and interpreted the resulting systems as discretisations of differential forms. Teixeira and Chew [2] considered carefully the geometrical and topological characterisation of Maxwell's equations discretised in the form of discrete differential forms.

It is convenient to represent the discrete fields by quantities that would be the physical continuum fields integrated over the appropriate geometrical elements. The 0-form potential ϕ is the electromagnetic scalar potential sampled at vertices, the 1-form electric intensity E is the physical electric field $\mathbf{E}$ integrated over the line element of the edge $E \sim \int_{\text{edge}} \mathbf{E}\cdot d\mathbf{x}$, 2-form B is the physical magnetic flux density integrated over a face $B \sim \int_{\text{face}} \mathbf{B}\cdot d\mathbf{S}$ and so on. The integral form of Faraday's law for the continuum fields is

$$\int_{\partial S} \mathbf{E}\cdot d\mathbf{x} = -\partial_t \int_S \mathbf{B}\cdot d\mathbf{S}$$

where ∂S indicates the circuital boundary of an element of surface S. This is true for any surface S arbitrarily drawn in R^3. If we choose S to be an elemental polygonal cell face f bounded by edge elements $\partial f = e_1 + e_2 + \cdots + e_K$, then

$$\sum_{k=1}^{K} \int_{e_k} \mathbf{E} \cdot \mathrm{d}\mathbf{x} = -\partial_t \int_f \mathbf{B} \cdot \mathrm{d}\mathbf{S} \Rightarrow \sum_{k=1}^{K} E_k = -\partial_t B_l$$

when the integrals are replaced by the form values, k indexes edges in the boundary and l indexes the face. The d-operator on 1-forms carries out the sum over edge values on the left-hand side, so one gets exactly

$$dE = -\partial_t B \quad \Rightarrow \sum_k d_{lk} E_k = -\partial_t B_l$$

for the global forms when the local expressions are applied at each face l. The matrix d_{lk} is exactly the incidence matrix for edges to faces, having value ± 1 when the edge k is incident on the face l, with sign according to orientation, and 0 otherwise.

The question of the construction and geometrical meaning of the discrete Hodge operators $*$ has been debated by a number of authors in the context of electromagnetic field discretisation. We are here adopting the purely geometrical convention that the Hodge operator $*$ carries out the metrical scalings necessary to map from r-forms to $\nu - r$-dual forms with consistency in the inner products that appear in the formalism. Thus, for instance, in the continuum theory for $\nu = 3$, the inner product $\int_M \psi \phi \mathrm{d}^3\mathbf{x}$ of two potentials $\psi(\mathbf{x})$ and $\phi(\mathbf{x})$ would be represented in the discrete theory by $\sum_{j \in M} \psi_j \phi_j *_j$, with the potentials sampled at the vertices indexed by j, and $*_j$ representing the discrete element of volume enclosing the vertex, identifiable as the volume of the dual 3-cell, centre of which is the vertex x_j. One then defines the dual 3-form $\tilde{\phi}_j = *_j \phi_j$ at each cell j, and the global dual 3-form $\tilde{\phi} = *\phi$. A similar computation can be carried out for any grade r of r-form. For 1-forms, let ψ and ϕ be global 1-forms with values on edge k of ψ_k and ϕ_k. Then the sum $\sum_k \psi_k \; *_k \phi_k$ defines the 1-form inner product $\langle \psi, \phi \rangle$ with an appropriate distribution $*_k$ for the edges over which the sum is taken.

4.7 Time-domain Green's functions for dynamical systems

The potential wave equations (4.63) and (4.64) are of the form

$$\mathsf{F}\phi + c^{-2}\partial_t^2 \phi = f \tag{4.67}$$

where ϕ is an r-form and F is a symmetric local operator on V_r, the vector space of r-forms with an inner product $\langle , \rangle$ and a local adjacency operator. The concept of Green's theorems and Green's functions can be directly extended to the time-domain for these systems.

For the second-order system (4.56), we intend to integrate this system as an initial-value problem in time $t > 0$, so let us suppose that initial conditions are prescribed so that

$$\phi(0) = 0, \quad \partial_t\phi(0) = 0, \quad \phi(t) = 0 \text{ for } t < 0, \quad f(t) = 0 \text{ for } t < 0. \tag{4.68}$$

Let u be the r-form whose component values vanish on all r-elements of the complex except for one, $j = j_0$, where it has component value $*_{j0}^{-1}$,

$$u_j = *_{j0}^{-1}\delta_{jj0} = \begin{cases} *_{j0}^{-1}, & j = j_0 \\ 0, & j \neq j_0. \end{cases} \tag{4.69}$$

Then $\langle \phi, u \rangle = \phi_{j0}$, so that u behaves similarly to a distributional test-function, projecting the r-form ϕ onto its value on the element j_0. Then consider the auxiliary function ψ which satisfies the system

$$\mathsf{F}\psi + c^{-2}\partial_t^2\psi = s \tag{4.70}$$

with s a delta-function source r-form which is impulsive in time at $t = t_0$, so that

$$s(t) = u\delta(t - t_0) \tag{4.71}$$

$$s_j(t) = *_{j0}^{-1}\delta_{jj0}\delta(t - t_0) = \begin{cases} *_{j0}^{-1}\delta(t - t_0), & j = j_0 \\ 0, & j \neq j_0 \end{cases} \tag{4.72}$$

and initial conditions

$$\psi(t_0) = 0, \quad \partial_t\psi(t_0) = -c^2 u, \quad \psi(t) = 0 \text{ for } t \geq t_0. \tag{4.73}$$

(Note the time-reversal of the domain of definition of the auxiliary function.) The formal solution of the auxiliary system (4.70) is

$$\psi(t) = c\mathsf{F}^{-1/2}\sin(c(t_0 - t)\mathsf{F}^{1/2})u, \quad t \leq t_0 \tag{4.74}$$

$$= 0, \quad t > t_0 \tag{4.75}$$

where the operator function can be explicitly computed by matrix algebra for a time invariant positive semi-definite operator F, for example as

$$\mathsf{F}^{-1/2}\sin(ct\mathsf{F}^{1/2}) = \sum_{m=0}^{\infty}(-1)^m\frac{(ct)^{2m+1}}{(2m+1)!}\mathsf{F}^m, \quad t \geq 0, \tag{4.76}$$

(so square-root operators need not be defined). The solution of the original dynamical system (4.56) on the unbounded vector space V_r is then

$$\phi(t) = c\int_0^t \mathsf{F}^{-1/2}\sin(c(t - t')\mathsf{F}^{1/2})f(t')\mathrm{d}t'. \tag{4.77}$$

Let the matrix representation of the operator F be

$$(\mathsf{F}\phi)_j = \sum_{j'} *_j^{-1} F_{jj'}\phi_{j'} \tag{4.78}$$

$$\Rightarrow \langle \phi', \mathsf{F}\phi \rangle = \sum_j \sum_{j'} \phi'_j F_{jj'}\phi_{j'} \quad \forall \phi, \phi' \in V_r. \tag{4.79}$$

Form the antisymmetric bilinear form

$$\Psi_{jj'} = F_{jj'}(\phi_j \psi_{j'} - \psi_j \phi_{j'}). \tag{4.80}$$

Apply Kirchhoff's theorem (4.46) and integrate with respect to time, to obtain

$$\phi_{j_0}(t_0) = \sum_{j\sim j' \in S} F_{jj'} \int_0^{t_0} (\phi_j(t)\psi_{j'}(t) - \psi_j(t)\phi_{j'}(t))\mathrm{d}t \tag{4.81}$$

As before, the notation should indicate that the auxiliary function $\psi(t)$ depends on the position j_0 of the delta-source s, and also on the time t_0. Accordingly, let $\psi_j(t) \rightarrow G_{j_0 j}(t_0 - t)$. Then we have the time-domain Kirchhoff theorem for r-forms

$$\phi_{j_0}(t_0) = \sum_{j\sim j' \in S} F_{jj'} \int_0^{t_0} (\phi_j(t)G_{j_0 j'}(t_0 - t) - G_{j_0 j}(t_0 - t)\phi_{j'}(t))\mathrm{d}t \tag{4.82}$$

where S is the cut-set of pairs of elements, one being $j \in D$ and one being $j' \in \bar{D}$, separated by an adjacency distance $\mathrm{dist}(j, j') \leq q$, where q is determined by the operator F. The kernel operator $\mathsf{G}(t)$ is explicitly given by

$$\mathsf{G}(t) = \begin{cases} c\sin(ct\mathsf{F}^{1/2})\mathsf{F}^{-1/2}, & t \geq 0 \\ 0, \quad t \leq 0 \end{cases} \tag{4.83}$$

and it is easily verified directly that

$$(\mathsf{F} + c^{-2}\partial_t^2)\mathsf{G}(t) = \mathsf{I}\delta(t). \tag{4.84}$$

4.8 Discrete time

In practice, computational realisations of field theories are advanced in discrete time rather than the continuous time assumed in the previous section. In order to do this, the time derivative in the dynamical equations is replaced by a time-difference operator that acts on discrete-time samples of the field ϕ. In second-order systems, the simplest such replacement is obtained by sampling the fields at discrete times $m\tau, m \in Z$ and replacing the second-order time derivative by

$$\partial_t^2 \phi \rightarrow \tau^{-2}(\phi(m+1) - 2\phi(m) + \phi(m-1)),$$

with time interval τ between samples. Then the dynamical systems are expressed in recurrent time-update form

$$\phi(m+1) = 2\phi(m) - \phi(m-1) - c^2\tau^2(\mathsf{F}\phi(m) - f(m)), \quad m \geq 0 \tag{4.85}$$

with some initial conditions, for example $\phi(m) = 0$ for $m \leq 0$, and $f(m) = 0$ for $m < 0$ to represent the switching on of the source at discrete time $m = 0$. The auxiliary field ψ is a solution of the time-reversed discrete system with impulsive point source excitation

$$\psi(m-1) = 2\psi(m) - \psi(m+1) - c^2\tau^2(\mathsf{F}\psi(m) - s(m)), \quad m \leq m_0, \tag{4.86}$$

and initial conditions $\psi(m) = 0$ for $m \geq m_0$, $\psi(m_0 - 1) = c^2\tau^2 u$, where u is an r-form delta source located at the element $j = j_0$ as in (4.69)

$$s_j(m) = u_j\delta_{mm_0} = \begin{cases} *_{j_0}^{-1}\delta_{mm_0}, & j = j_0 \\ 0, & j \neq j_0. \end{cases} \tag{4.87}$$

$$u_j = *_{j_0}^{-1}\delta_{jj_0} = \begin{cases} *_{j_0}^{-1}, & j = j_0 \\ 0, & j \neq j_0. \end{cases} \tag{4.88}$$

The Kirchhoff theorem for discrete time then has the form

$$\phi_{j_0}(m_0) = \sum_{j\sim j' \in S} F_{jj'} \sum_{m=0}^{m_0} (\phi_j(m)G_{j_0j'}(m_0 - m) - G_{j_0j}(m_0 - m)\phi_{j'}(m)) \tag{4.89}$$

with the Green function

$$G_{j_0j}(m_0 - m) = \psi_j(m). \tag{4.90}$$

Solving (4.86) for the auxiliary r-form ψ, from which the discrete time-domain Green function is constructed, ψ is

$$\psi(m) = c^2\tau^2 \frac{\sin((m_0 - m)\Theta)}{\sin\Theta} u, \quad m < m_0 \tag{4.91}$$

$$= 0, \quad m \geq m_0, \tag{4.92}$$

where Θ is the operator

$$\cos\Theta = 1 - \tfrac{1}{2}c^2\tau^2\mathsf{F}. \tag{4.93}$$

Using the identity

$$\frac{\sin m\Theta}{\sin\Theta} = U_{m-1}(\cos\Theta), \quad m \geq 1 \tag{4.94}$$

where $U_m(x)$ is the Tchebychev polynomial of the second kind [13], we see that the operator in (4.91) is essentially a polynomial in F of finite order equal to $m_0 - m - 1$, and ψ can be written as

$$\psi(m) = c^2\tau^2 U_{m_0-m-1}\left(1 - \tfrac{1}{2}c^2\tau^2\mathsf{F}\right)u, \quad m_0 - m \geq 1. \tag{4.95}$$

This in turn means that when F is local, with an adjacency range of 1, the values of $\psi_j(m)$ vanish on all points of the discrete manifold M outside an adjacency range of size $m_0 - m$ around the source point j_0. This property makes the Kirchhoff theorem into an expression of Huygens' principle in the (discrete) time-domain.

In forming the auxiliary function $\psi(m)$, an initial condition $\psi(m_0) = 0$ was used (after (4.86)). This condition means that $G_{jj'}(0) = 0$ for any pair (j, j'), which deletes the term $m = m_0$ from the sum in the Discrete KH representation (4.89), modifying it to

$$\phi_{j_0}(m_0) = \sum_S F_{jj'} \sum_{m=0}^{m_0-1} (\phi_j(m) G_{j_0 j'}(m_0 - m) - G_{j_0 j}(m_0 - m)\phi_{j'}(m)). \tag{4.96}$$

This modification makes the KH theorem an *explicit updating* scheme: the field value $\phi_{j_0}(m_0)$ at a position j_0 in D and time m_0 is determined by a superposition of contributions from fields in the cut-set S at times *earlier* than m_0.

4.9 Discrete Green's theorem and Green's functions in computational field theory

4.9.1 Exterior–interior connection

An important practical application of Green's theorems in computational field theory is concerned with the estimation of certain components of the field outside a finite region on which the computations are defined. Suppose we have a dynamical system of the kind described in the previous section, for example the r-form potentials ϕ and A for Maxwell's equations, to be executed by the time-recursive method of (4.85) (the FDTD method), from initial conditions at some discrete time labelled as $m = 0$. In this method, in order to obtain the fields in a region $\bar{D}$ at time step $m + 1$, it is necessary to know the values of the field at time step m not only throughout the same region $\bar{D}$, but also some values from outside $\bar{D}$ are needed, from D, due to the adjacency range of the spatial operators [14,15]. This is essentially what the Kirchhoff formulae achieve: the prediction of values of a field ϕ in region D from its values on the cut-set between D and $\bar{D}$, given a previously known auxiliary field ψ in D, which is itself a solution of the same discrete dynamical system but with a specific source distribution consisting of a single point source in region D. Thus, the time-recursive update is carried out in two steps: generate the fields in region $\bar{D}$ at time $m + 1$ from the fields at time m using (4.85); predict the fields in D at time $m + 1$ using the KH formula (4.96); then repeat the recurrence for the next time step $m + 1$.

It is clear that this two-step recurrence will only be beneficial if the Green functions $G_{jj'}(m)$ can be computed rapidly at the relatively small number of pairs of points $j \sim j' \in S$ where they are actually required by the KH theorem, rather than by the brute-force method of computing the whole operator G as point sources radiating into a large open mesh by the time-recursive method. Therefore, explicit rapidly computable representations are required for the point-to-point discrete time Green functions. The formula (4.95) derived here is a basis for such explicit representations and has the merit of being exact (in principle) for any suitable primary-dual cell decomposition, including arbitrary irregular Delaunay–Voronoi decompositions, with only finite amounts of arithmetic required for exact evaluation.

4.9.2 *Diakoptics*

A development of the diakoptic principle can be applied to multicomponent systems. Suppose the region $\bar{D}$ consists of two (or more) disconnected components $\bar{D} = \cup_i \bar{D}_i$, each forming an inclusion in the surrounding space D. Each inclusion $\bar{D}_i$ contains its own dynamical system, each of whose outermost vertices is paired with a vertex in the exterior medium in the cut-set S_i. The discrete Green functions described here can also be used to propagate the fields through the 'white space' between the objects $\bar{D}_i$, *via* the KH theorems.

4.10 Conclusion

Green's theorem and Green's functions are applied to general dynamical systems on discrete structures in discrete time. The discrete structures may arise from spatial discretisations of continuum fields described by PDEs but are not restricted to this source and may be purely graph-theoretical in origin. The case of Maxwell's equations, spatially discretised by finite-differences on a simplicial cell complex, is introduced as an example leading to coupled discrete potentials that each satisfies a second-order discrete dynamical system, one on vertices and the other on edges. The main result is a generic expression (4.95) for the second-order discrete-time Green operator G as a polynomial in the system operator F whose order m is equal to the time step. The discrete-time matrix elements $G_{jj'}(m)$ in this expression are exactly computable in a finite number of arithmetic steps when the operator F has finite adjacency measure. The typical candidate for F is the discrete Laplacian L, which has adjacency 1. A fully discrete form of Huygens' principle is obtained for this system, which predicts the field in an exterior region D from its values at earlier times in the cut-set S consisting of pairs of elements, one in D and one in $\bar{D}$, using explicit time-stepping. The representation of Huygens' principle in this form is a superposition of expanding wavelets each radiated from an element of the cutset S. While the general setting for the description of the dynamical system is a cell complex with Hodge dual, deletion of the metric Hodge data from the Laplacian weights leaves behind a topological dynamical system, that is essentially exploring the connectivity of a graph in discrete time steps.

References

[1] Roach GF. *Green's Functions*. Cambridge: Cambridge University Press; 1970.

[2] Teixeira FL and Chew WC. 'Lattice electromagnetic theory from a topological viewpoint'. *Journal of Mathematical Physics*. 1999;**40**:169–187.

[3] Tonti E. 'On the geometrical structure of electromagnetism'. In: *Gravitation, Electromagnetism and Geometrical Structures*. Bologna: Pitagora Editrice; 1995. p. 281–308.

[4] Bossavit A. *Computational Electromagnetism*. Boston: Academic Press; 1998.

[5] Bossavit A. 'Generalised finite differences in computational electromagnetics'. *Progress in Electromagnetics Research*. 2001;**32**:45–64.

[6] Desbrun M, Kanso E, and Tong Y. 'Discrete differential forms for computational modelling'. In: Bobenko AI, Schröder P, Sullivan JM, Ziegler GM, editors. *Discrete Differential Geometry*. No. **38** in Oberwolfach Seminars. Berlin: Birkhauser Verlag; 2008. p. 287–323.

[7] Chung FRK and Yau ST. 'Discrete Green's functions'. *Journal of Combinatorial Theory (A)*. 2000;**91**:191–214.

[8] Mugler DH. 'Green's functions for the finite difference heat, Laplace and wave equations'. In: Butzer PL, Stens RL and Nagy-Sz. B, editors. *Anniversary Volume on Approximation Theory and Functional Analysis*. Oberwolfach Seminars. Basel: Birkhauser; 1984. p. 543–554.

[9] Frankel T. *The Geometry of Physics: An Introduction*. Cambridge: Cambridge University Press; 1997.

[10] Yee K. 'Numerical solution of initial boundary value problems involving Maxwell's equations in isotropic media'. *IEEE Transactions on Antennas and Propagation*. 1966;**14**:302–307.

[11] Chew WC. 'Electromagnetic theory on a lattice'. *Journal of Applied Physics*. 1994;**75**:4843–4850.

[12] Mansfield EL and Hydon PE. 'Difference forms'. *Foundations of Computational Mathematics*. 2008;**8**:427–467.

[13] Abramowitz M and Stegun I. *Handbook and Tables of Mathematical Functions*. London: Dover; 2000.

[14] Vazquez J and Parini CG. 'Discrete Green's function formulation of FDTD method for electromagnetic modelling'. *Electronics Letters*. 1999;**35**: 554–555.

[15] Holtzman R and Kastner R. 'On the time-domain Green's function at the FDTD grid boundary'. *IEEE Transactions on Antennas and Propagation*. 2001;**49**:1079–1093.

Chapter 5

The concept of generalized functions and universal properties of the Green's functions associated with the wave equation in bounded piece-wise homogeneous domains

Mithat Idemen[1]

5.1 A short historical background

It goes without saying that the mathematics which flourished in the eighteenth century was the main pillar of the glorious scientific achievements gained in the next two centuries. In this struggle, the basic role was played by the concepts of continuous functions and their derivatives. Almost all physical laws discovered in that period were stated and generalized in terms of differential equations. Among them we can mention, for example, fluid mechanics, gas dynamics, thermodynamics, electromagnetic theory, quantum mechanics, general theory of relativity, etc. Those pure scientific achievements also excited countless grandiose engineering applications that created the actual civilization. It is interesting to observe that in that course sometimes the mathematics drag the natural scientific ideas and engineering applications while sometimes the engineering applications, conversely, showed the directions in which mathematics have to expand. Among them we can mention, for example, the birth of the variational calculus, complex analysis, Fourier analysis, Wiener–Hopf techniques, functional analysis, etc.

In those days mathematics, physics and engineering had been running happily together. But in the last two decades of the nineteenth century, one observed that the classical concepts of function and its derivative were not capable to formulate and analyse some new problems of the theoretical physics and engineering. This lack of capability should be observed, of course, by many researchers at that time. But we know only G. R. Kirchhoff, a famous German physicist, as the brave pioneer who enforced the boundary of mathematics to overcome the difficulty. Indeed, both in his lectures and a book published in 1876 [1], he defined some functions that have unusual properties. These kinds of functions also took place in his study submitted to the Berlin Royal Academy of Sciences in 1882 [2]. It was defined as follows: 'F is a function such that for all positive and negative finite values of its argument it is equal

[1]Electronics Engineering Faculty, Technical University of Istanbul, Istanbul, Turkey

to zero while its integral from a finite negative limit to a finite positive limit is equal to 1'. By using this function, he derived some interesting results. But mathematicians of those days had not taken seriously the existence of such a function as well as the legitimacy of mathematical operations applied on it. In the second volume of his book, published posthumously in 1891 under the editorship of Kurt Hensel, we see some words added by Hensel* to legitimate the use of the previously mentioned function F [3]. We read in page 24: 'To conceive that such a function indeed exists, consider, for example:

$$F = \left(\frac{\mu}{\sqrt{\pi}}\right) e^{-\mu^2 x^2}, \tag{5.1}$$

where μ is a very large positive number'. It is obvious that the integral of F with respect to x on the interval $(-\infty, \infty)$ (not on a finite interval!) is equal to 1 for all values of μ. But the condition $F = 0$ for all positive and negative values of x requires $\mu \to \infty$, which causes a confusion for the integral of the limit function.

Nearly three decades later, a similar attempt seemed in England. In 1925 Paul Dirac, a young physicist, introduced a function $\delta(x)$ with the same properties similar to the function F of Kirchhoff [4]. Dirac used his function extensively and obtained some results that have been welcomed enthusiastically by many physicists. Although it does not exist as a function from classical point of view, $\delta(x)$ was also extensively used by engineers and became known with his name: Dirac delta function. Hard disputes between mathematicians, physicists and engineers which lasted more than two decades recalled the birth of the complex analysis one century ago.

I think many mathematicians, as the pioneers of the complex calculus in eighteenth century, should attempt to establish a rigorous mathematical basis for $\delta(x)$. We do not know them but Laurent Schwartz from France. In 1950, Schwartz tried to introduce $\delta(x)$ (and similar new entities called 'generalized functions') through the linear functionals defined on a special test function space D [5]. D involves functions of bounded support having derivatives of all orders, namely $D \equiv C_0^\infty$. Schwartz named these new entities as 'distributions'.† The space of distributions defined on D is denoted by D'. Since in the Hilbert space a linear functional is always identical to the inner product, for a distribution f defined on $\Re = (-\infty, \infty)$ one also writes formally

$$\langle f, \varphi \rangle = \int f(x)\varphi(x)dx, \quad \varphi(x) \in D. \tag{5.2}$$

Here and in what follows, all the integrals $\int$ written without limits mean integrals on $\Re$ (or $\Re^3$). From (5.2), one concludes that every locally integrable function belongs to D' which is also called 'the space of generalized functions'. These kinds of generalized functions are said to be the 'regular distributions'.

*The author acknowledges Prof. Dr. Ning Yan Zhu from Stutgart University for his kind private communication about the early use of the delta-type functions by Kirchhoff as well as the contribution of Dr. Hensel.

†In this chapter, the terms 'distribution' and 'generalized function' will be considered interchangeable.

The already known concepts such as the sum, multiplication by a constant or an ordinary function, change of variables, derivatives, etc. are all extended to D' by considering (5.2) with regular distributions. For example, for the derivative of a known function f one writes

$$\langle f',\varphi\rangle = \int_a^b f'(x)\varphi(x)dx = -\int_a^b f(x)\varphi'(x)dx = -\langle f,\varphi'\rangle, \varphi \in D. \tag{5.3}$$

Here (a, b) stands for the support of the test function $\varphi \in D$, which yields $\varphi(a) = \varphi(b) = 0$. The relation $\langle f',\varphi\rangle = -\langle f,\varphi'\rangle$ is assumed to be the definition of the derivative 'in the sense of distribution'. It is worthwhile to remark that the knowledge of f as a distribution suffices to the existence and knowledge of its derivatives of all orders, namely:

$$\langle f^{(n)},\varphi\rangle = (-1)^n\langle f,\varphi^{(n)}\rangle. \tag{5.4}$$

This property is not valid, in general, for the derivative in the classical sense. To grasp the importance of it, consider for example the Heaviside unit step function $H(x) \in L_0^1(\Re)$ which has not a derivative in the classical sense on $\Re$. But in the sense of distribution, it has a well-defined derivative:

$$\langle H',\varphi\rangle = -\int_0^b H(x)\varphi'(x)dx = \varphi(0), \quad \forall\varphi \in D, b > 0. \tag{5.5}$$

Notice that in the theory of distributions established by Schwartz, the Dirac function $\delta(x)$ is introduced by

$$\langle \delta,\varphi\rangle = \varphi(0), \quad \forall\varphi \in D. \tag{5.6}$$

A comparison of (5.6) with (5.5) shows that $H'(x) = \delta(x)$ and $H^{(n)}(x) = \delta^{(n-1)}(x)$ for all $n \in N$.

From the results obtained earlier, one concludes that the concepts of generalized functions and derivatives in the sense of distribution have many properties which differ from those pertinent to classical functions. Therefore, the question that follows is of crucial importance:

Are the differential equations of the classical physics valid in the sense of distribution?

It is worthwhile to remark here that an affirmative response to this question does not only mean that the derivatives have to be computed as distributions but also the field components themselves involve singular parts which cannot be expressed in terms of classical functions.

By adopting the validity of the Maxwell equations in the sense of distribution, in the studies [6–13] the author has derived many results connected with the discontinuities of the electromagnetic field. Among them we can mention, for example, the 'universal boundary conditions' for the electromagnetic field, 'compatibility conditions' of the singular field components, 'critical frequencies' for the validity of the impedance type conditions, etc. At present, these concepts constitute basis for the analysis and synthesis of the so-called 'metasurfaces' [14–17].

Another important effect of the concept of generalized functions was observed in the clarification and extension of the so-called Green's function. The first idea leading to the concept of Green's function has been rooted in a book by George Green published in England in 1828 [18]. In that book, he introduced a function now identified as what Riemann later coined the 'Green's function'. Perhaps due to his little formal education, the method was not well-known during his lifetime in the mathematical community. But after two decades, it was rediscovered and popularized by William Thomson (Lord Kelvin).[‡] The main aim of the present chapter is to show the crucial role of the concept of generalized functions played in the study of Green's functions.

In books published before the birth of the distribution concept in 1950, the Green's function had been introduced as follows [19]: Let $u \in C^2(A,B)$ satisfies the non-homogeneous differential equation

$$a(x)u'' + b(x)u' + c(x)u = f(x), \quad A < x < B \tag{5.7a}$$

under the homogeneous boundary conditions

$$u(A) = 0, \;\; u(B) = 0. \tag{5.7b}$$

Here $a(x)$,$b(x)$, $c(x)$ and $f(x)$ are given functions while A and B stand for fixed constants. The Green's function associated with this boundary-value problem is the function $G(x,\xi)$ that satisfies the following relations for all fixed $\xi \in (A,B)$:

$$a(x)G_{xx} + b(x)G_x + c(x)G = 0, \;\; x \neq \xi, \;\; x,\xi \in (A,B), \tag{5.8a}$$

$$G(A,\xi) = 0, \;\; G(B,\xi) = 0, \tag{5.8b}$$

$$G(\xi + 0,\xi) - G(\xi - 0,\xi) = 0, \tag{5.8c}$$

$$G_x(\xi + 0,\xi) - G_x(\xi - 0,\xi) = -1/a(\xi). \tag{5.8d}$$

If one has an explicit expression of $G(x,\xi)$, then one can easily check that the function

$$u(x) = \int_A^B G(x,\xi)f(\xi)d\xi \tag{5.9}$$

satisfies the non-homogeneous equation (5.7a) under the boundary conditions (5.7b). This is a great success. But the philosophy behind this method is not easily comprehensible, which makes it impossible to guess how to generalize it to the case of partial differential equations.

[‡]Nearly a century later, A. Einstein declared that Green had been 20 years ahead of his time.

Fortunately, nearly one century and a half later, after the introduction of the concept of generalized function $\delta(x)$, one had observed that the previously mentioned Green's function can also be defined as follows:

$$a(x)G_{xx} + b(x)G_x + c(x)G = -\delta(x - \xi), \quad x, \xi \in (A, B), \tag{5.10a}$$

$$G(A, \xi) = 0, \quad G(B, \xi) = 0. \tag{5.10b}$$

Now the derivatives in (5.10a) are in the sense of distributions, which causes the inclusion of the conditions (5.8c) and (5.8d) into (5.10a). Equation (5.10a) is obviously appropriate to generalize to the case of partial differential equations stated in domains of higher dimensions. It will be done in Section 5.3.

The discovery of the relation between the Green and delta functions played very important role in the development of the theory of boundary-value problems connected with partial differential equations and their engineering applications in the last century. However, the complexity of the explicit expressions of the Green's functions has always been a severe handicap in various engineering applications especially when the domain is bounded and non-homogeneous. In what follows, we will give some universal properties of the Green's functions, which overcome the difficulty in question. An illustrative example shows their use in some inverse problems.

5.1.1 A remark

Earlier we have introduced the generalized functions as continuous linear functionals. But it is not the unique approach suitable to this purpose. The generalized functions can also be introduced through the sequence of functionals [20, 21]. To conclude this historical introduction, we present briefly the essential points of this approach which we will have occasion to use later on.

Let $\{f_n\}$ be any sequence of locally integrable functions on $\Re$. If the limit of $\int f_n(x)\varphi(x)dx$ *exists for all* $\varphi \in D$, then the sequence *is said to converge to the distribution* f defined by

$$<f, \varphi> = \lim_{n\to\infty} \int f_n(x)\varphi(x)dx, \quad \forall \varphi \in D. \tag{5.11a}$$

To indicate this fact, one writes

$$n \to \infty \Rightarrow f_n(x) \to f(x) \in D'. \tag{5.11b}$$

As a simple example reconsider the Kirchhoff–Hensel function F given in (5.1). One can easily check that for all $\varphi(x) \in D$ one has

$$\lim_{n\to\infty} \int_{-\infty}^{\infty} \frac{n}{\sqrt{\pi}} e^{-n^2x^2} \varphi(x)dx = \varphi(0) = \langle \delta(x), \varphi(x) \rangle,$$

which yields

$$\frac{n}{\sqrt{\pi}} e^{-n^2x^2} \to \delta(x). \tag{5.12}$$

This example shows the close connection between the ideas of Kirchhoff–Hensel and Dirac.

5.2 Some basic properties of the δ distribution

For a locally integrable function $f(x)$and any real constant a, from (5.2) one gets

$$\begin{aligned}\langle f(x-a),\varphi(x)\rangle &= \int f(x-a)\varphi(x)dx = \int f(x)\varphi(x+a)dx \\ &= \langle f(x),\varphi(x+a)\rangle. \end{aligned} \tag{5.13}$$

This simple result is generalized to define the rule for the change of variables in D'. For example, one has

$$H'(x-a)=\delta(x-a), \quad \langle\delta(x-a),\varphi(x)\rangle = <\delta(x),\varphi(x+a) = \varphi(a). \tag{5.14}$$

Similarly, the first two derivatives (in the sense of distribution) of a function $f(x)$ that has bounded derivatives in $(-\infty, a]$ and $[a, \infty)$ are

$$f'(x)=\{f'(x)\}+[\![f]\!]\delta(x-a) \tag{5.15a}$$

and

$$f''(x)=\{f''(x)\}+[\![f']\!]\delta(x-a)+[\![f]\!]\delta'(x-a). \tag{5.15b}$$

Here the functions in the curved brackets such as $\{f'\}$ denote the classical derivatives that are not defined at $x=a$ while the quantities in double square brackets such as $[[f]]$ stand for the jump discontinuities at $x=a$, namely:

$$[\![f]\!]=f(a+0)-f(a-0). \tag{5.15c}$$

The terms in (5.15a) and (5.15b), written in curved brackets, are called the ‘regular parts’ of the derivatives while the remaining terms are their ‘singular parts’.

From (5.15a), one concludes that the assumption of the validity of a physical law in the sense of distribution also yields the assumption of the existence of singular terms in the field components.[§]

Another property to be derived from (5.2) is

$$\langle f(ax),\varphi(x)\rangle = \int f(ax)\varphi(x)dx = \int f(x)\varphi(x/a)dx/|a| = \left\langle \frac{1}{|a|}f(x),\varphi(x/a)\right\rangle. \tag{5.16a}$$

Thus one has

$$\delta(ax)=\frac{1}{|a|}\delta(x), \quad \delta(-x)=\delta(x). \tag{5.16b}$$

From the latter, one concludes that $\delta(x)$ is an even function of its argument.

[§]For the case of Maxwell equations, see [13].

The extension of $\delta(x)$, defined for $x \in \Re$, to $x \in \Re^3$ can be made in a straightforward manner. For a test function $\varphi(x_1, x_2, x_3) \equiv \varphi(x)$, one writes[‖]

$$\langle \delta(x), \varphi(x) \rangle = \varphi(0) \equiv \varphi(0,0,0) = \langle\, \delta(x_1), \langle \delta\,(x_2), \langle \delta\,(x_3), \varphi(x_1, x_2, x_3\,)\rangle\rangle\rangle \tag{5.17a}$$

From this one concludes that

$$\delta(x) = \delta(x_1)\delta(x_2)\delta(x_3) \tag{5.17b}$$

The point $O(0,0,0)$ is called the 'support' of the distribution $\delta(x)$.

Practical applications enforce sometimes the use of spherical coordinates which requires the computations of integrals of the form

$$\int_0^\infty \delta(R)\psi(R)dR,$$

which differs from (5.2) because the support of $\delta(R)$ is now located at the lower end of the integration domain. Therefore, one has a difficulty in the interpretation of this type of integrals. To overcome the difficulty, we consider the continuation of the function $\psi(R)$, defined for $0 \leq R < \infty$, into $(-\infty) < R < 0$ as an even function. Thus the extended function, say $\tilde{\psi}(R)$, becomes continuous at $R = 0$ and permits us to write

$$\psi(0) = \tilde{\psi}(0) = \left\langle \delta(R), \tilde{\psi}(R) \right\rangle = \int \delta(R)\tilde{\psi}(R)dR = 2\int_0^\infty \delta(R)\psi(R)dR, \tag{5.18a}$$

which yields

$$\int_0^\infty \delta(R)\psi(R)dR = \frac{1}{2}\psi(0). \tag{5.18b}$$

In writing the last term in (5.18a), we took into account the evenness of $\delta(x)$ stated by (5.16b).

5.2.1 Distributions involving $\delta(R)$ and $\delta(R_1 - R_2)$

Let x, y and η be points in $\Re^3$ while R_1 and R_2 denote the distances of x and y to η, namely: $R_1 = |y - \eta|$ and $R_2 = |x - \eta|$. As we will see later on, the distributions $\delta(R_1)/R_1^2$, $\delta'(R_1)/R_1$ and $\delta'(R_1 - R_2)/R_1R_2$ play important roles in the study of some inverse problems. Now we give some properties of them in two lemmas.

Lemma 1. Let $\eta \in \Re^3$ be a fixed point while $|x - \eta|$ stands for the distance of x to η. Then one has

$$\frac{\delta(|x-\eta|)}{|x-\eta|^2} = 2\pi\delta(x-\eta), \qquad \frac{\delta'(|x-\eta|)}{|x-\eta|} = -2\pi\delta(x-\eta). \tag{5.19}$$

[‖]The Cartesian components of any $x \in \Re^3$ will be denoted by (x_1, x_2, x_3).

Lemma 2 [22]. Let D be a convex domain bounded by a closed regular surface S while $x \in D$ and $y \in D$ are two arbitrarily chosen fixed points. Then one has

$$\int_D \frac{\delta'(R_2 - R_1)}{R_1 R_2} dV_\eta = 0, \tag{5.20}$$

where $R_1 = |y - \eta|$, $R_2 = |x - \eta|$ and dV_η stands for the volume element at the point $\eta \in D$.

Proof of Lemma 1

Let us define a polar coordinate system (R, θ, ϕ) with origin at the point η and consider the distribution $\delta(R)/R^2$. If $\varphi(x) \equiv \varphi(x_1, x_2, x_3)$ denotes any test function, then we write

$$\left\langle \frac{\delta(R)}{R^2}, \varphi(x) \right\rangle = \iiint \frac{\delta(R)}{R^2} \varphi(\eta_1 + R \sin\theta \cos\phi, \ldots) R^2 \sin\theta dR d\theta d\phi \tag{5.21a}$$

$$= \int_0^{2\pi}\int_0^{\pi}\int_0^{\infty} \delta(R)\varphi(\eta_1 + R\sin\theta\cos\phi, \ldots)\sin\theta dR d\theta d\phi \tag{5.21b}$$

$$= 2\pi\varphi(\eta) = \langle 2\pi\delta(x - \eta), \varphi(x)\rangle, \tag{5.21c}$$

(5.18b) being taken into account. From the latter, one gets the first equality in (5.19).

To prove the second equality in (5.19), we write successively

$$\left\langle \frac{\delta'(R)}{R}, \varphi(x) \right\rangle = \iiint \frac{\delta'(R)}{R} \varphi(\eta_1 + R \sin\theta \cos\phi, \ldots) R^2 \sin\theta dR d\theta d\phi$$

$$= \int_0^{2\pi}\int_0^{\pi}\int_0^{\infty} \delta'(R)\{R\varphi(\eta_1 + R\sin\theta\cos\phi, \ldots)\}\sin\theta dR d\theta d\phi$$

$$= -\int_0^{2\pi}\int_0^{\pi}\int_0^{\infty} \delta(R)\frac{\partial}{\partial R}\{R\varphi(\eta_1 + R\sin\theta\cos\phi, \ldots)\}\sin\theta dR d\theta d\phi$$

$$= -2\pi\varphi(\eta) = \langle -2\pi\delta(x - \eta), \varphi(x)\rangle \tag{5.22}$$

This proves the second identity in (5.19).

Proof of Lemma 2 [22]

First of all, we have to prove that the function multiplied by δ' in the volume integral in (5.20) is integrable (then in every bounded interval it can be approximated by test functions[¶]). To this end consider first the apparent singularity for $\eta \to x$ and define a polar coordinate system (R_2, θ_2, ϕ_2) with the origin at the point x, which yields $dV_\eta = R_2^2 \sin\theta_2 dR_2 d\theta_2 d\phi_2$ (see Figure 5.1). If we insert this expression of the volume element into (5.20), then we see that the function multiplied by δ' is $O(R_2)$

[¶] D is dense in the space of locally integrable functions. For detail, see [13], p. 13.

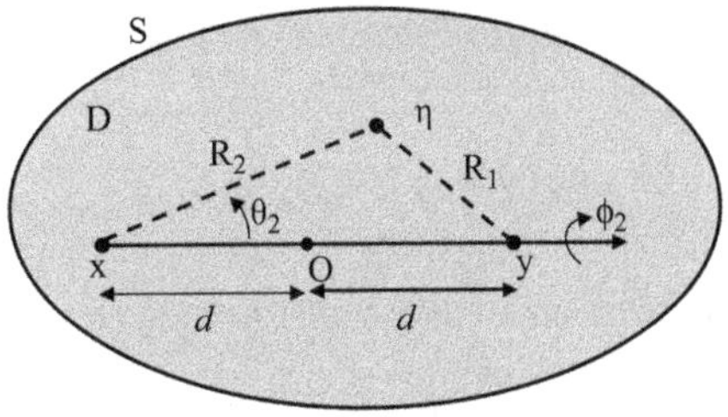

Figure 5.1 Parameters pertinent to the system (R_2, θ_2, ϕ_2)

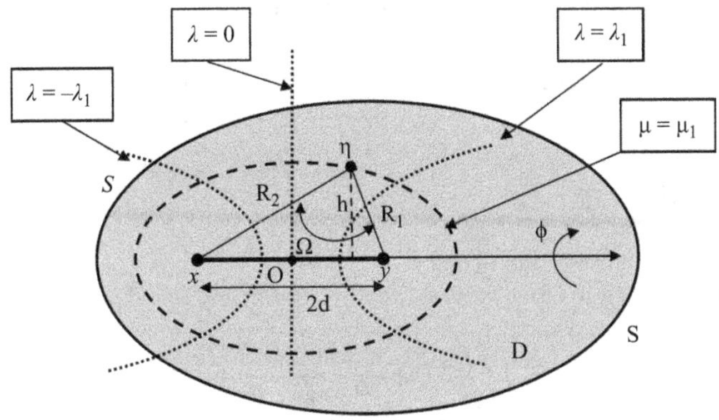

Figure 5.2 Coordinate surfaces in the system (λ, μ, ϕ)

when $R_2 \to 0$. Similar result is also valid for $R_1 \to 0$. This means that the apparent singularity of the integrand for $R_{1,2} \to 0$ is integrable.

To evaluate the integral in (5.20), let us consider the coordinate system (λ, μ, ϕ) defined as follows:

$$(R_2 - R_1) = \lambda \in [-2d, 2d], \tag{5.23a}$$

$$(R_2 + R_1) = \mu \in [2d, \infty), \tag{5.23b}$$

$$\phi \in [0, 2\pi) = \text{polar angle about } xy. \tag{5.23c}$$

This is obviously an orthogonal system whose coordinate surfaces are rotational hyperboloids ($\lambda =$ constant), rotational ellipsoids ($\mu =$ constant) and half-planes ($\phi =$ constant) (see Figure 5.2).

The volume element dV_η in this system of coordinates can easily be computed as follows:

$$dV_\eta = \frac{1}{|\mathrm{grad}\lambda|}\frac{1}{|\mathrm{grad}\mu|}hd\lambda d\mu d\phi \equiv Jd\lambda d\mu d\phi. \tag{5.24a}$$

Here J stands for the Jacobian of the transformation $(x, y, z) \to (\lambda, \mu, \phi)$. By a straightforward computation, one gets

$$|\mathrm{grad}\lambda||\mathrm{grad}\mu| = \left|\frac{x-\eta}{|x-\eta|} - \frac{y-\eta}{|y-\eta|}\right|\left|\frac{x-\eta}{|x-\eta|} + \frac{y-\eta}{|y-\eta|}\right| = 2\sin\Omega, \tag{5.24b}$$

where Ω is the angle between $(x-\eta)$ and $(y-\eta)$ (see Figure 5.2). The symmetry existing in Figure 5.2 shows that both the height h and $\sin\Omega$ are even functions of λ. This means that the Jacobian $J \equiv J(\lambda, \mu)$ taking place in (5.24a) is even with respect to λ.

As to the product R_1R_2 taking place in (5.20), from the definitions of λ and μ it is equal to

$$R_1R_2 = \frac{1}{4}(\mu^2 - \lambda^2), \tag{5.25}$$

which is also an even function of λ.

Now let us return to (5.20) and write it in terms of (λ, μ, ϕ). Thus one gets

$$\begin{aligned}\int_D \frac{\delta'(R_2 - R_1)}{R_1R_2}dV_\eta &= \iiint_D \delta'(\lambda)\frac{2h(\lambda,\mu)}{(\mu^2-\lambda^2)\sin\Omega}d\lambda d\mu d\phi \\ &= -2\iint\left[\frac{\partial}{\partial\lambda}\frac{h(\lambda,\mu)}{(\mu^2-\lambda^2)\sin\Omega}\right]_{\lambda=0} d\mu d\phi \\ &= -2\iint\left[\frac{\partial}{\partial\lambda}\left(\frac{1}{8d}\right)\right]_{\lambda=0} d\mu d\phi = 0,\end{aligned} \tag{5.26}$$

which proves the lemma.

5.3 Green's functions associated with the wave equation in bounded partially homogeneous domains

Let $V \subset \Re^3$ denote a bounded region composed of n subdomains having different (constant) constitutive parameters, say ε_j and μ_j $(j = 1, \ldots, n)$ as shown in Figure 5.3. The constitutive parameters of the surrounding space, say V_0, will be denoted by ε_0 and μ_0. Let a function $p(x, t)$, defined for $(x, t) \in \Re^3 \times \Re$, satisfy the wave equation

$$\mathrm{div}\left(\frac{1}{\varepsilon}\mathrm{grad}p\right) - \mu\frac{\partial^2 p}{\partial t^2} = -p_0(x, t), \quad x \in \Re^3, t \in \Re \tag{5.27a}$$

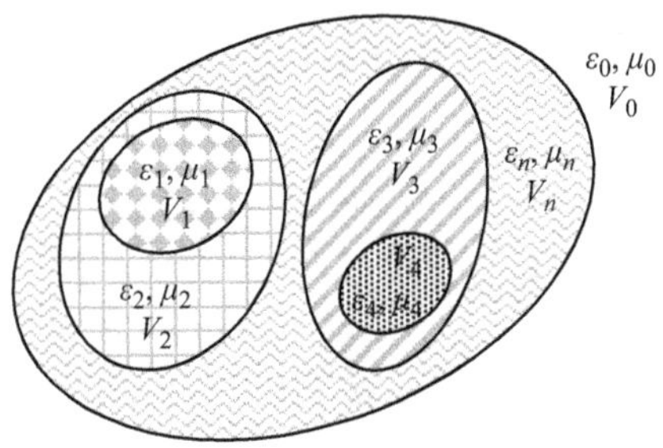

Figure 5.3 A piece-wise homogeneous bounded domain

in the sense of distribution. Here $p_0(x, t)$ stands for the density of the source which will be assumed located *inside* V while $\varepsilon = \varepsilon(x)$ and $\mu = \mu(x)$ denote the previously mentioned constitutive parameters:

$$\varepsilon = \varepsilon(x) = \begin{cases} \varepsilon_j, & x \in V_j, \quad j = 1, ..., n \\ \varepsilon_0, & x \in V_0 \end{cases} \tag{5.27b}$$

$$\mu = \mu(x) = \begin{cases} \mu_j, & x \in V_j, \quad j = 1, ..., n \\ \mu_0, & x \in V_0. \end{cases} \tag{5.27c}$$

To determine $p(x, t)$ uniquely, one has to add to (5.27a) the so-called radiation conditions too. These will be clarified later (see (5.31a) and (5.31b)).

If $p_0(x, t)$ does not contain singular sources distributed on the interfaces between the subregions, which will be assumed in this work, the validity of (5.27a) 'in the sense of distribution' enforces the following boundary conditions on the intersections of the subregions:

$$[\![p]\!] = 0, \quad \left[\!\!\left[\frac{1}{\varepsilon(x)}\frac{\partial p}{\partial n}\right]\!\!\right] = 0. \tag{5.28}$$

Here $\partial p/\partial n$ stands for the derivative in the normal direction on the interface in connection. Notice that the relations in (5.28) consist of the extension of (5.15c) to the functions defined in $\Re^3$ and assume the boundedness of grad$p(x, t)$ [6, 13]. The assumption made about the validity of (5.28) is evidently an extra assumption put on the physical phenomenon described by (5.27a). When the source density $p_0(x, t)$ involves also distributions concentrated at a certain time, say $t = t_0$, relations similar to (5.28) are also written for $p(x, t)$ and its time derivatives at $t = t_0$ (see (5.42)).

In what follows, we will assume that $p(x, t)$ is Fourier transformable with respect to t. The transform of each quantity is denoted by a hat on that quantity. For example, one has

$$\widehat{p}(x, \omega) = \frac{1}{2\pi}\int_{-\infty}^{\infty} p(x, t)e^{i\omega t}dt, \quad \omega \in \Re \tag{5.29a}$$

and, inversely

$$\frac{1}{2}[p(x,t+0)+p(x,t-0)] = \int_{-\infty}^{\infty} \widehat{p}(x,\omega)e^{-i\omega t}d\omega, \quad t \in \Re. \tag{5.29b}$$

Thus the transform of (5.27a) is as follows:

$$\operatorname{div}\left(\frac{1}{\varepsilon}\operatorname{grad}\hat{p}\right) + \omega^2\mu\hat{p} = -\hat{p}_0(x,\omega). \tag{5.30}$$

The radiation conditions satisfied by $\widehat{p}(x,\omega)$ are

$$\widehat{p}(x,\omega) = O(1/|x|) \quad as \quad |x| \to \infty \tag{5.31a}$$

and

$$\partial\widehat{p}/\partial|x| - i(\omega/c_0)\widehat{p} = O(1/|x|^2) \quad as \quad |x| \to \infty, \tag{5.31b}$$

where we put

$$c_0 = 1/\sqrt{\varepsilon_0\mu_0}. \tag{5.31c}$$

In connection with the reduced wave equation (5.30) (or the wave equation (5.27a)), we will define two kinds of Green functions: the outgoing and ingoing ones. The first one permits us to find an explicit expression of the solution to the problem stated by (5.30)–(5.31c) (the direct problem) while the other plays an important role in inverse problems. We will consider them in detail later.

5.3.1 The outgoing Green's function

The outgoing Green's function is the function $G_1(x,\eta,\omega)$ determined uniquely through the following relations:

$$\operatorname{div}_x\left(\frac{1}{\varepsilon(x)}\operatorname{grad}_x G_1\right) + \omega^2\mu(x)G_1 = -\delta(x-\eta) \tag{5.32a}$$

$$G_1(x,\eta,\omega) = O(1/|x|) \quad \text{as} \quad |x| \to \infty \tag{5.32b}$$

$$\partial G_1/\partial|x| - i(\omega/c_0)G_1 = O(1/|x|^2) \quad \text{as} \quad |x| \to \infty. \tag{5.32c}$$

Here $\eta \in \Re^3$ stands for any point while $\delta(x-\eta)$ is the Dirac distribution concentrated at $x=\eta$.

The sub-index x in div_x and grad_x, which appear in (5.32a), means that the derivatives are to be computed with respect to the components of x. Notice that due to the radiation condition (5.32c), the wave associated with G_1 propagates in the direction $\eta \to x$. This Green's function permits one to transport the data known at the source point η to the observation point x and write

$$\hat{p}(x,\omega) = \int_{\Re^3} G_1(\eta,x,\ \omega)\hat{p}_0(\eta,\omega)dv_\eta = \int_V G_1(\eta,x,\omega)\hat{p}_0(\eta,\omega)dv_\eta. \tag{5.33}$$

To prove (5.33), multiply first (5.30) with G_1 and (5.32a) with $\widehat{p}$, then subtract the resulting expressions side by side to get

$$\mathrm{div}_x\left[\frac{1}{\varepsilon(x)}\left(G_1 \mathrm{grad}_x \hat{p} - \hat{p}\mathrm{grad}_x G_1\right)\right] = -\hat{p}_0(x,\omega)G_1(x,\eta,\omega) + \hat{p}(x,\omega)\delta(x-\eta).$$

If we integrate both sides in a sphere of radius R and transform the first side to the surface integral on $|x| = R$, then we write

$$\int\limits_{|\mathbf{x}|=R} \frac{1}{\varepsilon(\mathbf{x})}(G_1 \mathrm{grad}_x \widehat{p} - \widehat{p}\mathrm{grad}_x G_1) \cdot dS_x = \widehat{p}(\eta,\omega) - \int\limits_{|x|\le R} G_1(x,\eta,\omega)\widehat{p}_0(x,\omega)dv_x.$$

Now let us make $R \to \infty$. Due to the radiation conditions satisfied by G_1 and $\widehat{p}$, the surface integral tends to zero and yields (5.33) with the substitution $\eta \leftrightarrow x$.

5.3.2 The ingoing Green's function

The ingoing Green's function $G_2(x,\eta,\omega)$ satisfies

$$\mathrm{div}_x\left(\frac{1}{\varepsilon(x)}\mathrm{grad}_x G_2\right) + \omega^2\mu(x)G_2 = -\delta(x-\eta) \tag{5.34a}$$

in the sense of distribution under the radiation conditions

$$G_2(x,\eta,\omega) = O(1/|x|) \quad \text{as} \quad |x| \to \infty \tag{5.34b}$$

and

$$\partial G_2/\partial|x| + i(\omega/c_0)G_2 = O(1/|x|^2) \quad \text{as} \quad |x| \to \infty. \tag{5.34c}$$

Notice that (5.34a) and (5.34b) are quite identical to (5.32a) and (5.32b) while (5.34c) differs from (5.32c) only with the sign of (i). Due to this difference, the wave associated with $G_2(x,\eta,\omega)$ propagates in the direction $x \to \eta$ and serves to transport the data measured at x to the source point η. Therefore, in inverse source problems one has to use both of the Green's functions together. In what follows, we give some *universal* properties of these functions, which play important roles in inverse problems connected with the wave equation in bounded partially homogeneous regions. Here the universality means that the properties in question do not depend on the geometrical shapes and constitutive parameters of the subdomains. Thus the results will also be valid in the case of an arbitrary non-homogeneous domain whenever the functions $\varepsilon(x)$, $\varepsilon'(x)$ and $\mu(x)$ change continuously and $\varepsilon(x)$ differs from zero.

5.3.3 Some universal properties of the outgoing and ingoing Green's functions

Observe first that the Green's functions defined earlier differ from the *classical* ones by the factor $\varepsilon(x)$ (see [23]). The previously mentioned *universal* properties are stated in the following theorems (cf. [27]).

Theorem 1. For all $x, \eta \in \Re^3$ the Green's function $G_1(x, \eta, \omega)$ is symmetric with respect to x and η:

$$G_1(x,\eta,\omega) \equiv G_1(\eta,x,\omega). \tag{5.35}$$

Theorem 2. For $\omega \in \Re$, the Green's functions $G_1(x, \eta, \omega)$ and $G_2(x, \eta, \omega)$ are interrelated with the following functional relations:

$$G_2(x,\eta,\omega) \equiv G_1(x,\eta,-\omega), \tag{5.36a}$$

$$G_2(x,\eta,\omega) \equiv G_1^*(x,\eta,\omega), \tag{5.36b}$$

$$G_2(x,\eta,\omega) \equiv G_2(\eta,x,\omega). \tag{5.36c}$$

Here (*) means the complex conjugate. The relations in (5.36a) and (5.36c) are also valid for complex ω.

Theorem 3. For all $x, \eta \in \Re^3$ one has

$$n \to \infty \Rightarrow \int_{-\Omega_n}^{\Omega_n} G_1(x,\eta,\omega)\omega d\omega \to \frac{i\pi}{\mu(\eta)}\delta(x-\eta) \tag{5.37a}$$

$$\to \frac{i}{2\mu(\eta)}\frac{\delta(|x-\eta|)}{|x-\eta|^2} \tag{5.37b}$$

$$\to -\frac{i}{2\mu(\eta)}\frac{\delta'(|x-\eta|)}{|x-\eta|}. \tag{5.37c}$$

Here $\{\Omega_n\}$ stands for any sequence such that $\Omega_n \to \infty$ as $n \to \infty$.

Theorem 4. For all $x, \eta \in \Re^3$ one has

$$n \to \infty \Rightarrow \int_{-\Omega_n}^{\Omega_n} [G_2(x,\eta,\omega) - G_1(x,\eta,\omega)]\omega d\omega \to -\frac{2i\pi}{\mu(\eta)}\delta(x-\eta) \tag{5.38a}$$

$$\to -\frac{i}{\mu(\eta)}\frac{\delta(|x-\eta|)}{|x-\eta|^2} \tag{5.38b}$$

$$\to \frac{i}{\mu(\eta)}\frac{\delta'(|x-\eta|)}{|x-\eta|}. \tag{5.38c}$$

Here $\{\Omega_n\}$ stands for any sequence such that $\Omega_n \to \infty$ as $n \to \infty$.

Theorem 5. For all $x \in V$, $\eta \in V_0$ and $\alpha \in [-\pi/2, \pi/2]$ one has

$$n \to \infty \Rightarrow G_1(x, \eta, \Omega_n \sin\alpha) \to \frac{\varepsilon_0}{4\pi|x-\eta|} e^{i\Omega_n \sin\alpha |x-\eta|/c_0} \tag{5.39a}$$

$$n \to \infty \Rightarrow G_2(x, \eta, \Omega_n \sin\alpha) \to \frac{\varepsilon_0}{4\pi|x-\eta|} e^{-i\Omega_n \sin\alpha |x-\eta|/c_0}. \tag{5.39b}$$

Here $\{\Omega_n\}$ stands for any sequence such that $\Omega_n \to \infty$ as $n \to \infty$.

Theorem 6. For all $x, y \in V$ and $\eta \in V_0$ one has

$$n \to \infty \Rightarrow \int_{-\Omega_n}^{\Omega_n} G_1(x, \eta, \omega) G_2(y, \eta, \omega) \omega d\omega \to -\frac{i}{8\pi} \frac{\varepsilon_0}{\mu_0} \frac{\delta'(R_2 - R_1)}{R_1 R_2}. \tag{5.40a}$$

Here $\{\Omega_n\}$ stands for any sequence such that $\Omega_n \to \infty$ as $n \to \infty$ and

$$R_1 = |y - \eta|, \quad R_2 = |x - \eta|. \tag{5.40b}$$

5.4 Proofs of the theorems

Now we will give the proofs of the previously mentioned theorems in turn.

5.4.1 *Proof of Theorem 1*

Now consider the function $\widehat{p}_1(x, \omega)$ defined as follows:

$$\widehat{p}_1(x, \omega) = \int_{\Re^3} G_1(x, \eta, \omega) \widehat{p}_0(\eta, \omega) dV\eta.$$

One can easily check that $\widehat{p}_1$ satisfies (5.30) as well as the radiation conditions (5.31a) and (5.31b). Therefore, it is equal to $\widehat{p}(x, \omega)$. If we compare this new expression of $\widehat{p}(x, \omega)$ with (5.33), then we also write

$$\int_{\Re^3} [G_1(x, \eta, \omega) - G_1(\eta, x, \omega)] \widehat{p}_0(\eta, \omega) dV\eta \equiv 0.$$

Since $\widehat{p}_0(\eta, \omega)$ is arbitrary, from the latter one gets (5.35).

5.4.2 *Proof of Theorem 2*

Assume $\omega \in \Re$ and make the change $\omega \to (-\omega)$ in the differential equation and the radiation conditions satisfied by G_1. One obviously obtains the equations satisfied by G_2. From this one concludes (5.36a).

If under the assumption $\omega \in \Re$ one considers the complex conjugates of the equations satisfied by G_1, one sees that the resulting equations are nothing but those satisfied by G_2. This proves (5.36b).

As to (5.36c), it is a direct consequence of (5.36a) and Theorem 1. Since the proofs of (5.36a) and (5.36c) are independent from the nature of the parameter ω, the corresponding relations are also valid for complex ω.

5.4.3 Proof of Theorem 3

Consider (5.27a) for the case when $p_0(x, t) = p_0(x)T_0(t)$ with $T_0(t) \equiv 0$ for $t < 0$. In this case, one obviously has $p(x, t) \equiv 0$ for $t < 0$. Thus from (5.29a) and (5.29b) written for $t = 0$ one gets

$$\frac{1}{2}\frac{\partial p}{\partial t}(x, +0) = -i \lim_{n\to\infty} \int_V \int_{-\Omega_n}^{\Omega_n} \omega G_1(x, \eta, \omega) p_0(\eta) \hat{T}_0(\omega) d\omega dv_\eta. \tag{5.41}$$

Assume first that $T_0(t) = \delta(t)$ which yields $\hat{T}_0(\omega) = 1/(2\pi)$. In this case from (5.27a) one concludes

$$\frac{\partial p}{\partial t}(x, +0) = \frac{p_0(x)}{\mu(x)}. \tag{5.42}$$

Comparison of (5.42) with (5.41) gives (5.37a). Equations (5.37b) and (5.37c) are direct results of (5.37a) and the relations given in Lemma 1.

Remark. All the results obtained earlier can easily be checked in the simplest case of homogeneous space for which one has

$$G_1(x, \eta, \omega) = \frac{\varepsilon_0}{4\pi R} e^{i\omega R/c_0}, \quad G_2(x, \eta, \omega) = \frac{\varepsilon_0}{4\pi R} e^{-i\omega R/c_0}. \tag{5.43}$$

To this end, one has to observe that in $\Re^3$ one has

$$R = |x - \eta| \Rightarrow \frac{\delta(R)}{R} = 0, \quad \frac{\varepsilon_0}{2iR}\delta'\left(\frac{R}{c_0}\right) = i\frac{\pi}{\mu_0}\delta(x - \eta).$$

5.4.4 Proof of Theorem 4

By considering (5.36a), we get directly

$$\begin{aligned}\int_{-\Omega_n}^{\Omega_n} G_2(x, \eta, \omega)\omega d\omega &= \int_{-\Omega_n}^{\Omega_n} G_1(x, \eta, -\omega)\omega d\omega \\ &= -\int_{-\Omega_n}^{\Omega_n} G_1(x, \eta, \omega)\omega d\omega\end{aligned}$$

Thus from (5.37a), (5.37b) and (5.37c) one gets directly (5.38a), (5.38b) and (5.38c).

5.4.5 Proof of Theorem 5

The function $G_1(x, \eta, \omega)$ is symmetrical with respect to x and η (see (5.35)) and, as a function of $\eta \in V_0$, satisfies the Helmholtz equation

$$\Delta_\eta G_1 + k_0^2 G_1 = 0, \quad \eta \in V_0, \quad x \in V \tag{5.44a}$$

under the radiation conditions

$$G_1 = O(1/|\eta|), \quad \frac{\partial G_1}{\partial |\eta|} - ik_0 G_1 = O(1/|\eta|^2) \quad \text{as} \quad |\eta| \to \infty. \tag{5.44b}$$

Here we put $k_0 = \omega\sqrt{\varepsilon_0 \mu_0}$. Now let us define a polar coordinate system (R, θ, ϕ) whose origin is at the point x and $R = |\eta - x|$. In this system, (5.44a) and (5.44b) become

$$\frac{\partial^2 G_1}{\partial R^2} + \frac{2}{R}\frac{\partial G_1}{\partial R} + \frac{1}{R^2 \sin\theta}\frac{\partial}{\partial \theta}\left(\sin\theta \frac{\partial G_1}{\partial \theta}\right) + \frac{1}{R^2 \sin^2\theta}\frac{\partial^2 G_1}{\partial \phi^2} + \frac{\omega^2}{c_0^2} G_1 = 0 \tag{5.45a}$$

and

$$R \to \infty \Rightarrow G_1 = O(1/R), \quad \frac{\partial G_1}{\partial R} - i\frac{\omega}{c_0} G_1 = O(1/R^2). \tag{5.45b}$$

Now for each Ω_n let us make the substitutions

$$\omega = \Omega_n \sin\alpha, \alpha \in [-\pi/2, \pi/2] \Rightarrow \frac{1}{\Omega_n} k_0 = \frac{\sin\alpha}{c_0} \equiv \tilde{k}_0 \tag{5.46a}$$

and

$$\Omega_n R = r \Rightarrow \frac{\partial G_1}{\partial R} = \Omega_n \frac{\partial G_1}{\partial r}, \frac{\partial^2 G_1}{\partial R^2} = \Omega_n^2 \frac{\partial^2 G_1}{\partial r^2}, \tag{5.46b}$$

$$\frac{1}{\Omega_n} G_1(x, \eta, \omega) = G_{1n}(r, \theta, \phi, \alpha), \tag{5.46c}$$

which reduce (5.45a) and (5.45b) to

$$\frac{\partial^2 G_{1n}}{\partial r^2} + \frac{2}{r}\frac{\partial G_{1n}}{\partial r} + \frac{1}{r^2 \sin\theta}\frac{\partial}{\partial \theta}\left(\sin\theta \frac{\partial G_{1n}}{\partial \theta}\right) + \frac{1}{r^2 \sin^2\theta}\frac{\partial^2 G_{1n}}{\partial \phi^2} + \tilde{k}_0^2 G_{1n} = 0. \tag{5.47a}$$

and

$$r \to \infty \Rightarrow G_{1n} = O(1/r), \quad \frac{\partial G_{1n}}{\partial r} - i\tilde{k}_0 G_{1n} = O(1/r^2). \tag{5.47b}$$

Notice that (5.47a) is satisfied outside the region V, and G_{1n} may depend on Ω_n (and α) also through the boundary conditions on ∂V.

It is obvious that (5.47a) and (5.47b) define the *radiating* (= outgoing) solution to the Helmholtz equation. When $\Omega_n \to \infty$, one has $r = \Omega_n R \to \infty$, which reduces the expression of G_{1n} to the so-called *far-field pattern* of $\tilde{G}_{1n}$ [24]:

$$n \to \infty \Rightarrow \Omega_n \to \infty \Rightarrow r \to \infty \Rightarrow G_{1n} = \frac{1}{r} e^{i\tilde{k}_0 r}\left\{A_\infty + O\left(\frac{1}{r}\right)\right\}. \tag{5.48}$$

Here A_∞ stands for the *scattering amplitude*. It does not depend on r. By using (5.48), we write successively

$$n \to \infty \Rightarrow \int_{-\Omega_n}^{\Omega_n} G_1(x,\eta,\omega)\omega d\omega = \int_{-\pi/2}^{\pi/2} [\Omega_n G_{1n}(r,\theta,\phi,\alpha)]\Omega_n^2 \sin\alpha\cos\alpha d\alpha$$

$$\to \int_{-\pi/2}^{\pi/2} \left[\Omega_n A_\infty \frac{1}{r} e^{i\tilde{k}_0 r}\right]\Omega_n^2 \sin\alpha\cos\alpha d\alpha$$

$$= \frac{1}{R}\int_{-\Omega_n}^{\Omega_n} A_\infty e^{i\omega R/c_0}\omega d\omega$$

$$= \frac{c_0^2}{R}\int_{-\Omega_n/c_0}^{\Omega_n/c_0} A_\infty e^{i\omega R}\omega d\omega. \tag{5.49a}$$

If we compare (5.49a) with (5.37c) written as follows

$$\int_{-\infty}^{\infty} G_1(x,\eta,\omega)\omega d\omega = -\frac{i}{2\mu_0}\frac{\delta'(R)}{R} = \frac{1}{4\pi\mu_0}\frac{1}{R}\int_{-\infty}^{\infty} e^{i\omega R}\omega d\omega \tag{5.49b}$$

we conclude that

$$A_\infty \equiv \frac{\varepsilon_0}{4\pi} \tag{5.50}$$

and the explicit expression of (5.48) is

$$n \to \infty \Rightarrow G_{1n}(x,\eta,\Omega_n\sin\alpha) \to \frac{\varepsilon_0}{4\pi}\frac{1}{r}e^{ir\sin\alpha/c_0} \tag{5.51a}$$

which yields

$$n \to \infty \Rightarrow G_1(x,\eta,\Omega_n\sin\alpha) \to \frac{\varepsilon_0}{4\pi}\frac{1}{R}e^{i\Omega_n\sin\alpha R/c_0}. \tag{5.51b}$$

The latter proves (5.39a). The proof of (5.39b) is a direct result of (5.39a) and (5.36a).

It is important to remark here that (5.51b) is not the far-field expression of G_1 because R is finite. It states merely that when computing the integral appearing on the left-hand side of (5.49a), for $n \to \infty$, G_1 can be replaced by (5.51b).

5.4.6 *Proof of Theorem 6*

Let us now define two polar coordinates (R_1, θ_1, ϕ_1) and (R_2, θ_2, ϕ_2) whose origins are located at the points x and y, respectively. If we make the substitutions

$$R_1 = |x-\eta|, R_2 = |y-\eta|, r_1 = \Omega_n R_1, r_2 = \Omega_n R_2, \omega = \Omega_n\sin\alpha \tag{5.52a}$$

and

$$\frac{1}{\Omega_n}G_1(x,\eta,\omega) = G_{1n}(r_1,\theta_1,\phi_1,\alpha), \frac{1}{\Omega_n}G_2(y,\eta,\omega) = G_{2n}(r_2,\theta_2,\phi_2,\alpha), \tag{5.52b}$$

then for $n \to \infty$ we write (see (5.51b) and (5.36a))

$$n \to \infty \Rightarrow \Omega_n G_{1n} \to \frac{\varepsilon_0}{4\pi}\frac{1}{R_1}e^{i\omega R_1/c_0} \tag{5.53a}$$

and

$$n \to \infty \Rightarrow \Omega_n G_{2n} \to \frac{\varepsilon_0}{4\pi}\frac{1}{R_2}e^{-i\omega R_2/c_0}. \tag{5.53b}$$

Now let us define the *distribution* defined through the sequence

$$W_n = \int_{-\Omega_n}^{\Omega_n} G_1(x,\eta,\omega))G_2(y,\eta,\omega)\omega d\omega, \quad n = 1,2,\ldots$$

By using (5.53a) and (5.53b) and making $n \to \infty$ we write successively

$$W_n = \int_{-\pi/2}^{\pi/2} [\Omega_n G_{1n}(r_1,\theta_1,\phi_1,\alpha)][\Omega_n G_{2n}(r_2,\theta_2,\phi_2,\alpha)]\Omega_n^2 \sin\alpha\cos\alpha d\alpha \tag{5.53c}$$

$$\to \int_{-\pi/2}^{\pi/2} \left[\frac{\varepsilon_0}{4\pi R_1}e^{i\omega R_1/c_0}\right]\left[\frac{\varepsilon_0}{4\pi R_2}e^{-i\omega R_2/c_0}\right]\Omega_n^2 \sin\alpha\cos\alpha\, d\alpha \tag{5.53d}$$

$$\to \left(\frac{\varepsilon_0}{4\pi}\right)^2 \frac{1}{R_1 R_2}\int_{-\Omega_n}^{\Omega_n} e^{i\omega(R_1-R_2)/c_0}\omega d\omega \tag{5.53e}$$

$$\to \left(\frac{\varepsilon_0}{4\pi}\right)^2 \frac{1}{R_1 R_2}(-2\pi i)\delta'\left(\frac{R_1-R_2}{c_0}\right) \tag{5.53f}$$

$$\to -\frac{i}{8\pi}\frac{\varepsilon_0}{\mu_0}\frac{\delta'(R_1-R_2)}{R_1 R_2}, \tag{5.53g}$$

$\delta'((R_1-R_2)/c_0) = c_0^2\delta'(R_1-R_2)$ being taken into account. This proves (5.40a).

Remark. Equations (5.53g) as well as (5.51a) and (5.51b) obtained earlier can easily be checked in the simplest case of homogeneous space for which G_1 and G_2 are given by (5.43).

5.5 Application: An inverse initial-value problem connected with the photoacoustic tomography in bounded non-homogeneous domains

Let $V \subset \Re^3$ denote a bounded non-homogeneous biological domain with constitutive parameters $\varepsilon(x)$ and $\mu(x)$. The parameters of the surrounding space are constants ε_0 and μ_0. In what follows, we will assume *first* that $\varepsilon(x) \in C^1(\Re^3)$, $\mu(x) \in C(\Re^3)$ and $\varepsilon(x) \neq 0$. The results to be obtained under this assumption will automatically be generalized to the case when V is filled with a partially homogeneous material as shown in Figure 5.3 (see the remark later). Let the function $p(x,t)$ defined for $x \in \Re^3$ and $t > 0$ satisfy the wave equation

$$\left\{\operatorname{div}\left(\frac{1}{\varepsilon}\operatorname{grad} p\right)\right\} - \left\{\mu\frac{\partial^2 p}{\partial t^2}\right\} = 0, x \in \Re^3, t > 0 \tag{5.54a}$$

under the initial conditions

$$p(x,+0) = p_0(x), \quad \frac{\partial}{\partial t}p(x,+0) = v_0(x), \quad x \in \Re^3. \tag{5.54b}$$

The so-called photoacoustic (and thermo-acoustic) tomography aims the determination of the initial values $p(x,+0)$ and $v_0(x)$ in V by using the values of $p(x,t)$ measured on a closed surface S involving V inside (see Figure 5.4). This means that the tomography problem in question consists of an inverse initial value problem. In acoustic tomography, $p(x,t)$ stands for the pressure density excited in $\Re^3$ by an electromagnetic (or thermal) impulse which is concentrated at time $t = 0$ and illuminates the region V. The energy absorbed by the biological tissue filling the volume V creates then the pressure wave $p(x,t)$ which satisfies (5.54a) for $t > 0$. The initial value $p(x,+0)$ gives an idea about the susceptibility of the tissue at the point $x \in V$ and thus permits one to reveal the configuration of the region V. One of the reasons for the intensive interest in the acoustic tomography is the hope of the detection of cancerous tissues because their contributions to the induced pressure wave are much stronger than that of the normal tissues.

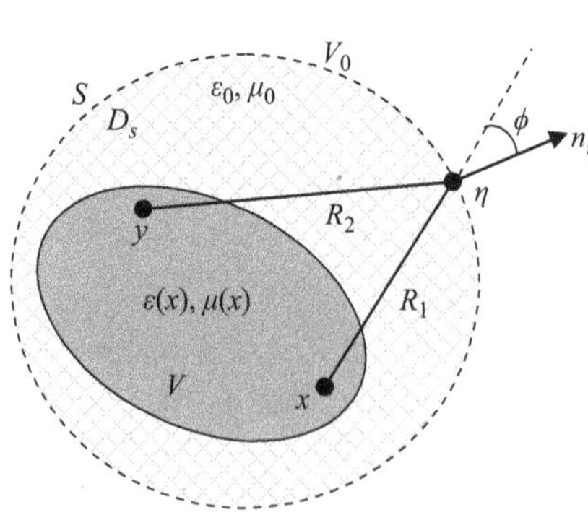

Figure 5.4 Various parameters connected with the tomography problem

5.5.1 *Extension of the inverse initial value problem to the range $(-\infty) < t < \infty$*

Now consider the following function defined for all $t \in (-\infty, \infty)$:

$$p(x,t) = \begin{cases} p(x,t), & t > 0 \\ 0, & t < 0. \end{cases} \tag{5.55a}$$

This function satisfies (5.54a) for both $t \in (0, \infty)$ and $t \in (-\infty, 0)$ while it has a jump discontinuity at $t = 0$ given by

$$[[p]] = p_0(x), \quad [[\partial p/\partial t]] = v_0(x), \quad x \in \Re^3. \tag{5.55b}$$

On the other hand, according to (5.15b) and the assumptions made on $\varepsilon(x)$ and $\mu(x)$ one has

$$\left\{\mu\frac{\partial^2 p}{\partial t^2}\right\} = \mu\frac{\partial^2 p}{\partial t^2} - \mu p_0(x)\delta'(t) - \mu v_0(x)\delta(t), x \in \Re^3, t \in (-\infty, \infty) \tag{5.55c}$$

and

$$\left\{\operatorname{div}\left(\frac{1}{\varepsilon}\operatorname{grad} p\right)\right\} = \operatorname{div}\left(\frac{1}{\varepsilon}\operatorname{grad} p\right), \quad x \in \Re^3, t \in (-\infty, \infty). \tag{5.55d}$$

The derivatives taking place on the right-hand sides of (5.55c) and (5.55d) are now in the sense of distributions. Thus the basic equation (5.54a), written for $t > 0$, yields

$$\operatorname{div}\left(\frac{1}{\varepsilon}\operatorname{grad} p\right) - \mu\frac{\partial^2 p}{\partial t^2} = -\mu p_0(x)\delta'(t) - \mu v_0(x)\delta(t), \quad x \in \Re^3, t \in \Re. \tag{5.56a}$$

Notice that the latter also involves the initial conditions (5.54b). It is worthwhile to remark that in classical formulation (5.54a) and (5.54b) the functions $p_0(x)$ and $v_0(x)$ take place as initial conditions while in distributional formulation (5.56a) they appear as source density.

Now let us write the second side of (5.56a) as

$$-\mu p_0(x)\delta'(t) - \mu v_0(x)\delta(t) = -\frac{\partial}{\partial t}[\mu p_0(x)\delta(t) + \mu v_0(x)H(t)], \tag{5.56b}$$

where $H(t)$ stands for the Heaviside unit step function. The terms inside the square bracket is interpreted as the density of the exciting source [25, 26]. The existence of $H(t)$ means that the source is not an impulse, which contradicts the basic assumption. Hence one has $v_0(x) \equiv 0$ which reduces (5.56a) finally to

$$\operatorname{div}\left(\frac{1}{\varepsilon}\operatorname{grad} p\right) - \mu\frac{\partial^2 p}{\partial t^2} = -\mu p_0(x)\delta'(t), \quad x \in \Re^3, t \in \Re. \tag{5.57}$$

Remark 1. If the boundary conditions (5.28) are assumed to be valid, (5.55d) also becomes valid for the configuration shown in Figure 5.3. In this case various sub-regions show, for example, the bone, skin, muscle, fat, cancerous tissue, epileptic tumours, etc.

Remark 2. When $\varepsilon(x)$ and $\mu(x)$ are known, (5.54b) shows that the knowledge of $p_0(x)$ suffices to the knowledge of $p(x, +0)$. On the other hand, (5.57) involves the latter in its second side as a source term. For this reason, the tomography problem (stated in (5.54a) and (5.54b) as an inverse initial value problem) is sometimes considered to be an inverse source problem (stated by (5.57)). For the historical development, see the review papers [25, 26].

5.5.2 Solution of the extended problem

Let $G_1(x, \eta, \omega)$ and $G_2(x, \eta, \omega)$ be the Green's functions associated with the configuration shown in Figure 5.3 (see also Figure 5.4). The pressure $p(\eta, t)$ measured at a point $\eta \in S \subset V_0$ is connected to the source density $p_0(x)$ through the outgoing Green's function $G_1(x, \eta, \omega)$ as given in formula (5.33), where $\hat{p}_0(x, \omega) = -i(\omega/2\pi)\mu(x)p_0(x)$. Now, by using the ingoing Green's function $G_2(x, \eta, \omega)$ we transport the data known at the point $\eta \in S$ to the point $x \in V$ through the following function $q(x)$:

$$q(x) = \int_{-\infty}^{\infty}\int_{S} \frac{1}{\varepsilon(\eta)}\hat{p}(\eta, \omega)\frac{\partial}{\partial n_\eta}G_2(x, \eta, \omega)dS_\eta d\omega. \tag{5.58}$$

Here n_η stands for the unit outward normal vector to the surface S at the point η and $\partial/\partial n_\eta$ in (5.58) means the derivative in this direction. We will show that

$$q(x) = -\frac{1}{2}p_0(x) = -\frac{1}{2}p(x, +0), x \in V. \tag{5.59}$$

To prove (5.59), let us replace $\hat{p}(\eta,\omega)$in (5.58) by (5.33) and write

$$\begin{aligned} q(x) &= -\frac{1}{2\pi}\lim_{\Omega\to\infty}\int_{-\Omega}^{\Omega}\left\{\int_{S}\frac{i}{\varepsilon(\eta)}\left[\int_{V}\mu(y)p_0(y)G_1(y, \eta, \omega)dv_y\right]\frac{\partial G_2}{\partial n_\eta}(x, \eta, \omega)dS_\eta\right\}\omega d\omega \\ &= \frac{1}{2\pi}\lim_{\Omega\to\infty}\int_{V} p_0(y)P(y, x, \Omega)dv_y, \end{aligned} \tag{5.60}$$

where we put

$$P(y, x, \Omega) = -i\mu(y)\int_{-\Omega}^{\Omega}\left\{\int_{S}\frac{1}{\varepsilon(\eta)}G_1(y, \eta, \omega)\frac{\partial G_2}{\partial n_\eta}(x, \eta, \omega)dS_\eta\right\}\omega d\omega. \tag{5.61a}$$

The latter can also be written as follows (in what follows D_S shows the domain bounded by S):

$$
\begin{aligned}
P(y,x,\Omega) &= -i\mu(y)\int_{-\Omega}^{\Omega}\left\{\int_S G_1(y,\eta,\omega)\frac{1}{\varepsilon(\eta)}\mathrm{grad}_\eta G_2(x,\eta,\omega)\cdot \mathbf{dS}_\eta\right\}\omega d\omega \\
&= -i\mu(y)\int_{-\Omega}^{\Omega}\left\{\int_{D_S}\mathrm{div}_\eta[G_1(y,\eta,\omega)\frac{1}{\varepsilon(\eta)}\mathrm{grad}_\eta G_2(x,\eta,\omega)]dv_\eta\right\}\omega d\omega \\
&= -i\mu(y)\int_{-\Omega}^{\Omega}\left\{\int_{D_S}\left[G_1(y,\eta,\omega)\mathrm{div}_\eta\left\{\frac{1}{\varepsilon(\eta)}\mathrm{grad}_\eta G_2(x,\eta,\omega)\right\}\right.\right. \\
&\quad \left.\left. +\,\mathrm{grad}_\eta G_1(y,\eta,\omega)\cdot\frac{1}{\varepsilon(\eta)}\mathrm{grad}_\eta G_2(x,\eta,\omega)\right]dv_\eta\right\}\omega d\omega. \qquad (5.61b)
\end{aligned}
$$

Now reconsider the equations satisfied by G_1 and G_2 (see (5.32a) and (5.34a)) to write

$$
\begin{aligned}
G_1(y,\eta,\omega)\mathrm{div}_\eta\left[\frac{1}{\varepsilon(\eta)}\mathrm{grad}_\eta G_2(\eta,x,\omega)\right] &= -\omega^2\mu(\eta)G_2(\eta,x,\omega)G_1(y,\eta,\omega) \\
&\quad -\delta(\eta-x)G_1(y,\eta,\omega) \qquad (5.62a)
\end{aligned}
$$

$$
\begin{aligned}
G_2(\eta,x,\omega)\mathrm{div}_\eta\left[\frac{1}{\varepsilon(\eta)}\mathrm{grad}_\eta G_1(\eta,y,\omega)\right] &= -\omega^2\mu(\eta)G_1(\eta,y,\omega)G_2(\eta\,,x,\omega) \\
&\quad -\delta(\eta-y)G_2(\eta,x,\omega). \qquad (5.62b)
\end{aligned}
$$

Subtracting these equations side by side, we get

$$
\begin{aligned}
G_1(y,\eta,\omega)\mathrm{div}_\eta\left[\frac{1}{\varepsilon(\eta)}\mathrm{grad}_\eta G_2(\eta,x,\omega)\right] &= G_2(\eta,x,\omega)\mathrm{div}_\eta\left[\frac{1}{\varepsilon(\eta)}\mathrm{grad}_\eta G_1(\eta,y,\omega)\right] \\
&\quad +\delta(\eta-y)G_2(\eta,x,\omega)-\delta(\eta-x)G_1(y,\eta,\omega). \qquad (5.62c)
\end{aligned}
$$

Now let us add the left-hand side again to both sides to obtain a more symmetrical expression

$$
\begin{aligned}
2G_1(y,\eta,\omega)\mathrm{div}_\eta\left[\frac{1}{\varepsilon(\eta)}\mathrm{grad}_\eta G_2(\eta,x,\omega)\right] &= G_2(\eta,x,\omega)\mathrm{div}_\eta\left[\frac{1}{\varepsilon(\eta)}\mathrm{grad}_\eta G_1(\eta,y,\omega)\right] \\
&\quad + G_1(y,\eta,\omega)\mathrm{div}_\eta\left[\frac{1}{\varepsilon(\eta)}\mathrm{grad}_\eta G_2(\eta,x,\omega)\right] \\
&\quad +\delta(\eta-y)G_2(\eta,x,\omega) \\
&\quad -\delta(\eta-x)G_1(y,\eta,\omega). \qquad (5.62d)
\end{aligned}
$$

By using the latter in (5.61b), one obtains

$$P(y,x,\omega)=P_1(y,x,\omega)+P_2(y,x,\omega) \qquad (5.63a)$$

with

$$P_1(y,x,\Omega) = -\frac{i}{2}\mu(y)\int\limits_{-\Omega}^{\Omega} \{G_2(y,x,\omega) - G_1(y,x,\omega)\}\omega d\omega \tag{5.63b}$$

and

$$\begin{aligned} P_2(y,x,\Omega) = -\frac{i}{2}\mu(y)\int\limits_{-\Omega}^{\Omega}\Bigg[\int\limits_{D_S}\Bigg\{ & G_2(\eta,x,\omega)\mathrm{div}_\eta\left[\frac{1}{\varepsilon(\eta)}\mathrm{grad}_\eta G_1(\eta,y,\omega)\right] \\ & + G_1(y,\eta,\omega)\mathrm{div}_\eta\left[\frac{1}{\varepsilon(\eta)}\mathrm{grad}_\eta G_2(x,\eta,\omega)\right] \\ & + \frac{2}{\varepsilon(\eta)}\mathrm{grad}_\eta G_1(y,\eta,\omega)\cdot\mathrm{grad}_\eta G_2(x,\eta,\omega)\Bigg\} dv_\eta\Bigg]\omega d\omega. \end{aligned} \tag{5.63c}$$

In what follows we will show that if D_S is *convex* (see Section 5.3), then

$$\lim_{\Omega\to\infty} P_2(y,x,\Omega) = 0, \tag{5.64}$$

which means that the contribution of P_2 to $q(x)$ is naught. As to the contribution of P_1, it is obvious from (5.60), (5.63b) and (5.38a). Indeed, by (5.38a) one has

$$\lim_{\Omega\to\infty} P_1(y,x,\Omega) = -\pi\delta(x-y), \tag{5.65a}$$

which reduces (5.60) to

$$q(x) = -\frac{1}{2}p_0(x) = -\frac{1}{2}p(x,+0). \tag{5.65b}$$

This is the main formula of the present work. It can be used in photoacoustic and thermo-acoustic tomography problems connected with non-homogeneous media [27].

Remark. When the space is homogeneous, the original inverse initial value problem stated by (5.54a) and (5.54b) becomes meaningless while the inverse source problem stated by (5.57) is still meaningful. In this case, (5.58) gives the solution to this problem (cf. [22]).

5.5.3 *Proof of (5.64)*

One can easily check that the integrand in (5.63c) is equal to

$$\mathrm{div}_\eta\left(\frac{1}{\varepsilon(\eta)}\mathrm{grad}_\eta[G_1(y,\eta,\omega)G_2(x,\eta,\omega)]\right), \quad x,y\in V, \quad \eta\in D_S. \tag{5.66}$$

Thus, by using the Gauss–Ostrogradski theorem one can also write

$$P_2(y,x,\Omega) = -\frac{i}{2}\mu(y)\int\limits_{-\Omega}^{\Omega}\int\limits_{S}\frac{1}{\varepsilon_0}\mathrm{grad}_\eta[G_1(y,\eta,\omega)G_2(x,\eta,\omega)]\cdot dS_\eta\omega d\omega \tag{5.67a}$$

or

$$\lim_{\Omega\to\infty} P_2(y,x,\Omega) = -\frac{i}{2\varepsilon_0}\mu(y)\int_S \operatorname{grad}_\eta\left\{\int_{-\infty}^{\infty} G_1(y,\eta,\omega)G_2(x,\eta,\omega)\omega d\omega\right\}\cdot dS_\eta$$

$$= -\frac{i}{2\varepsilon_0}\mu(y)\int_S \operatorname{grad}_\eta\left\{-\frac{i}{8\pi}\frac{\varepsilon_0}{\mu_0}\frac{\delta'(R_2-R_1)}{R_1R_2}\right\}\cdot dS_\eta$$

$$= -\frac{\mu(y)}{16\pi\mu_0}\int_{D_S} \operatorname{div}_\eta\operatorname{grad}_\eta\left[\frac{\delta'(R_2-R_1)}{R_1R_2}\right]dv_\eta, \tag{5.67b}$$

(5.40a) being taken into account. Now it is important to observe that the function in the bracket is a function of the form

$$\frac{\delta'(R_2-R_1)}{R_1R_2} = f(x-\eta, y-\eta)$$

and

$$\operatorname{div}_\eta\operatorname{grad}_\eta f(x-\eta,y-\eta) \equiv (\operatorname{div}_x + \operatorname{div}_y)(\operatorname{grad}_x + \operatorname{grad}_y)f.$$

This shows that (5.67b) can also be written as follows:

$$\lim_{\Omega\to\infty} P_2(y,x,\Omega) = -\frac{\mu(y)}{16\pi\mu_0}(\operatorname{div}_x + \operatorname{div}_y)(\operatorname{grad}_x + \operatorname{grad}_y)\int_{D_S}\frac{\delta'(R_2-R_1)}{R_1R_2}dv_\eta. \tag{5.67c}$$

By Lemma 2 in Section 5.2.1, the integral taking place in (5.67c) is equal to naught whenever the domain D_S is convex. This proves (5.64).

References

[1] Kirchhoff, G. R. *Vorlesungen über mathematische Physik, Band-1Mechanik.* Leipzig: Teubner; 1876, p. 315.

[2] Kirchhoff, G. R. *Zur Theory der Lichtstrahlen.* Sitzungsberichte der königlich preussichen Akademie der Wissenschaften zu Berlin; 1882, vol. 30, pp. 641–69.

[3] Kirchhoff, G. R. *Vorlesungen über mathematische Optik Band-2* (herausgegeben von Dr. Kurt Hensel). Leipzig: Teubner; 1891, p. 24.

[4] Dirac, P. A. M. *The Principles of Quantum Mechanics, Fourth Edition (Revised).* Oxford: Clarendon Press; 1967, p. 58.

[5] Schwartz, L. *Théorie des distributions.* Paris: Hermann; tome 1 1950, tome 2 1951.

[6] Idemen, M. The Maxwell's equations in the sense of distributions, *IEEE Trans. Antennas Propagat.* 1973, AP-21, 736–8.

[7] Idemen, M. Necessary and sufficient conditions for a surface to be an impedance boundary. *AEÜ.* 1981, 35(2), 84–6.

[8] Idemen, M. and Serbest, H. Boundary conditions of the electromagnetic field. *Electron. Lett.* 1987, 23(13), 704–5.
[9] Idemen, M. Straightforward derivation of the boundary conditions on a sheet simulating an anisotropic thin layer, *Electron. Lett.* 1988, 24, 663–5.
[10] Idemen, M. Universal boundary relations of the electromagnetic field. *J. Phys. Soc. Jpn.* 1990, 59(1), 71–80.
[11] Büyükaksoy, A. and Idemen, M. Generalized boundary conditions for a material sheet with both sides coated by a dielectric layer. *Electron. Lett. 1990, 26(23)*, 1967–9.
[12] Idemen, M. Universal boundary conditions and Cauchy data for the electromagnetic field. In Lakhtakia, A. (ed.), *Essays on the Formal Aspect of Electromagnetic Theory*. Singapore: World Scientific; 1993, pp. 657–98.
[13] Idemen, M. Discontinuities in the Electromagnetic Field. *New Jersey: John-Wiley; 2011.*
[14] Achouri, K., Salem, M. A., and Caloz, C. General metasurface synthesis based on susceptibility tensors. *IEEE Trans. Antennas Propagate.* 2015, 63(7), 2977–91.
[15] Hillion, P. Reflection from a flat dielectric film with negative refractive index. *Braz. J. Phys.* 2007, 37(4), 1–8.
[16] Albooyeh, M., Kwon, D.-H., Capolino, F., and Tretyakov, S. A. Equivalent realizations of reciprocal metasurfaces. *Phys. Rev. B.* 2017, 9, 115435.
[17] Pfeifer, C. and Grnic, A. Emulating nonreciprocity with spatially dispersive metasurfaces excited at oblique incidence. *Phys. Rev. Lett.* 2016, 117, 077401.
[18] Green, G. *Essay on the Application of Mathematical Analysis to the Theory of Electricity and Magnetism.* Nottingham; 1828.
[19] Ince, E. L. *Ordinary Differential Equations*. New York: Dover Publ.; 1956, p. 254.
[20] Richtmyer, R. D. *Principles of Advanced Mathematical Physics*, Vol. 1. New York: Springer-Verlag; 1978, p. 29.
[21] Gel'fand, I. M. and Shilov, G. E. *Generalized Functions*, Vol. 2. New York: Academic Press; 1968, p. 99.
[22] Idemen, M. and Alkumru, A. On an inverse source problem connected with photo-acoustic and thermo-acoustic tomography. *Wave Motion* 2012, 49, 595–604.
[23] Jones, D. S. *The Theory of Electromagnetism*. Oxford: Pergamon Press; 1964, p. 60.
[24] Colton, D. and Kress, R. *Inverse Acoustic and Electromagnetic Scattering Theory*. Berlin: Springer Verlag; 1992.
[25] Xia, J., Yao, J., and Wang, L. H. W. Photoacoustic tomography: Principles and advances (invited). *PIERS.* 2014, 147, 1–22.
[26] Beard, P. Biomedical photoacoustic imaging (review). *Interface Focus.* 2011, 1, 602–31.
[27] Idemen, M. Some universal properties of the Green's functions associated with the wave equation in bounded partially-homogeneous domains and their use in acoustic tomography. *Appl. Math.* 2017, 8, 483–99.

Chapter 6

Elliptic cylinder with a strongly elongated cross-section: high-frequency techniques and function theoretic methods

Frédéric Molinet[1] and Ivan Andronov[2]

6.1 Introduction

In this chapter, we apply the function theoretic methods developed by Andronov [1] to calculate the asymptotic currents on elliptic cylinders with a strongly elongated cross-section to more general cylindrical configurations where the cross-section is a truncated ellipse or composed of arcs of ellipses. Our objective is to establish the formulas for the asymptotic currents taking into account the diffraction by the edges.

The method of asymptotic currents is an extension of physical optics (PO) and the physical theory of diffraction (PTD). At high frequencies, it allows to calculate the radar cross-section (RCS) of a target, the dimensions of which are large compared with the wavelength by avoiding the drawbacks of the geometrical theory of diffraction (GTD), namely the problem of caustics where GTD fails. It also avoids the time-consuming procedure of matrix inversion in numerical methods.

Compared with PO and PTD, it takes into account the currents on the shadow side of the target by introducing the creeping waves. Moreover, in the lit region, close to the shadow boundary, the PO currents are replaced by more accurate transition zone solutions.

The method of asymptotic currents has been widely used to solve diffraction and radiation problems and excellent results have been obtained in many applications where the transition zone also known as the Fock domain is a small domain. In that case, the radius of curvature along the geodesic followed by the wave in the Fock domain can be decomposed in a Taylor series $\rho(s) = \rho_o + s\rho_1 + \dots$ with respect to the curvilinear abscissa$s$. Moreover in the Maxwell equations transformed into a parabolic equation by separation of the dominant phase term $\exp(iks)$ from the total field, ρ can be replaced by ρ_o. However, in the case of a geometrical strongly elongated object like a spheroid with a small transverse radius of curvature, the transition zone in the Fock domain for paraxial incidence may extend to almost the whole surface. This fact can be justified by the relation between the angular extension α of the transition

[1] Retired, Theoretical studies and mathematical simulations, Société, Mothesim, France
[2] Computational Physics, University of St. Petersburg, Russian Federation

region corresponding to the Fock domain and the local radius of curvature ρ_o of the surface in the plane of incidence: $\alpha = m^{-1}$, where $m = \left(\frac{k\rho_o}{2}\right)^{1/3}$. The length ℓ of the Fock domain is therefore given by:

$$\ell = \alpha\rho_o = \rho_o^{2/3}\left(\frac{k}{2}\right)^{-1/3} \tag{6.1}$$

Equation (6.1) shows that the Fock domain depends on the geometry of the scatterer at the light-shadow boundary and on the frequency of the incident wave. For a fixed frequency it augments with the radius of curvature of the surface, in the plane of incidence, at the light-shadow boundary and can therefore extend to the whole surface for a geometrical strongly elongated object. But it also decreases when the frequency augments. Hence, even when the object is geometrically strongly elongated, the length of the Fock domain can be small for a sufficiently high frequency and the classical asymptotics applies. However, when the Fock domain becomes large classical asymptotics becomes inaccurate, and when it covers the whole object it is inapplicable. For a spheroid with half length b and maximum radius a, we have $\rho_o = b^2/a$ and (6.1) for $\ell = hb$ gives:

$$\chi = kb\frac{a^2}{b^2} = \frac{2}{h^3} \tag{6.2}$$

The parameter χ takes values between 2 and 0.25 when ℓ varies from b to $2b$. This parameter has been introduced by Andronov [1] where a strongly prolate body has been defined by $\chi = 0(1)$. When χ becomes large, the Fock domain becomes small and the classical asymptotics applies.

In the study by Andronov and colleagues [1–3], we have considered some typical problems that we call etalon or model problems. They concern plane wave diffraction by strongly elongated spheroids and by elliptic cylinders with strongly elongated cross-sections. For solving these problems, we expressed Maxwell's equations in spheroidal or elliptic coordinates, respectively, and introduced the condition $\chi = 0(1)$ in the parabolic equation method. By variable separation, we obtained in the leading order a solution expressed in terms of an integral of Whittaker functions instead of the well-known Airy functions of the classical asymptotics.

Our approach gives the total field in the boundary layer along the elongated object as well as the asymptotic currents on the surface. An extension of our modal problems, which is important in practical applications, consists to treat the diffraction by a strongly elongated spheroid truncated by a plane perpendicular to its axis (Figure 6.1(a)) or by an elliptic cylinder with a truncated strongly elongated cross-section (Figure 6.1(b)) or more generally by a body that is composed of an assembly of elements of strongly elongated elliptic cylinders delimited by sharp edges (Figure 6.1(c) and (d)).

The objective of the present chapter is to demonstrate how our model problem solutions can be extended to treat these more general configurations. We limit our investigation to two-dimensional bodies and derive explicit formulas for the

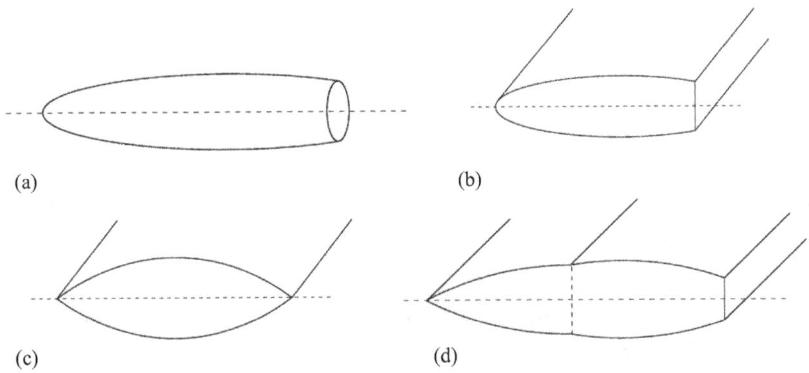

Figure 6.1 Some typical problems: (a) truncated spheroid, (b) truncated elliptic cylinder, (c) ogival cylinder with a cross-section composed of two arcs of an ellipse and (d) cylinder with a more complicated cross-section composed of arcs of ellipses

asymptotic currents on an elliptic cylinder with a truncated elliptic cross-section (Figure 6.1(b)) and on an ogival cylinder with an ogival cross-section (Figure 6.1(c)) composed of two symmetric arcs of an ellipse.

After the introduction, which corresponds to Section 6.1, the chapter comprises three other sections.

In Section 6.2, we derive the formulas for the asymptotic currents on an elliptic cylinder with a truncated cross-section, corresponding to Figure 6.1(b).

In Section 6.3, we treat the diffraction by a cylinder with an ogival cross-section, composed of two symmetric arcs of a strongly elongated ellipse (Figure 6.1(c)). Particular attention is devoted to the derivation of an asymptotic solution valid at grazing incidence on one of the edges.

In Section 6.4, we conclude with a presentation of some possible extensions of the method to a larger class of geometries and give some information concerning the calculation of the Whittaker functions.

6.2 Asymptotic currents on an elliptic cylinder with a truncated strongly elongated cross-section

6.2.1 Analysis of the interactions

We consider an elliptic cylinder truncated by a plane perpendicular to its longitudinal symmetric plane (Figure 6.1(b)), illuminated by an incident plane wave propagating along or close to the longitudinal axis of its cross-section with its magnetic field parallel to the generatrixes. Let M be an arbitrary point on the elliptic cross-section. The electromagnetic field at M can be decomposed in different contributions comprising the field due to the direct wave, the single- and double-edged diffracted waves and

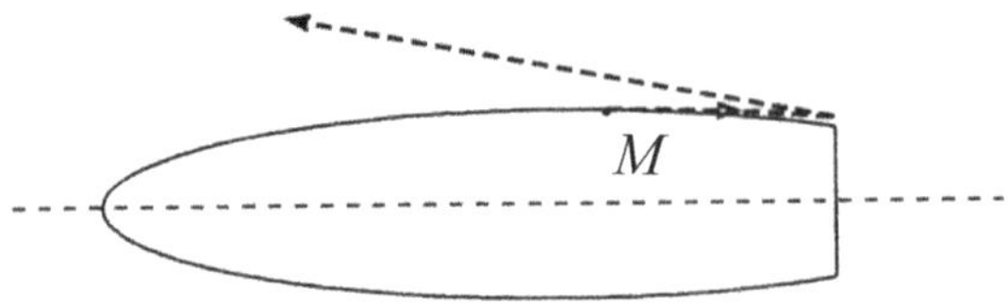

Figure 6.2 Reciprocal problem: field radiated by a line source and diffracted by the edge

higher order contributions due to multiple forward and backward waves. We consider only the direct and single-edged diffracted waves and neglect the higher order contributions.

The formulas giving the field at M due to the direct wave have been derived by Andronov [1]. To calculate the field at M due to the diffraction of the incident wave by the edge, we solve the reciprocal problem of a line source at M radiating a field that is diffracted by the edge and observed close to the axial direction at large distance from the cylinder (Figure 6.2). The solution of this problem together with that of plane wave diffraction is needed in all types of bodies composed of one or more strongly elongated two-dimensional elliptic surfaces delimited by sharp edges.

The model problem concerning the diffraction of an incident plane wave by an elliptic cylinder with a strongly elongated cross-section has been solved by Andronov [1]. In Section 6.2.6, we give a short presentation of the results and apply the method presented in Section 6.2.4 for decomposing the field into a spectrum of plane waves permitting to solve the diffraction by the truncation.

The method for solving the field radiated by a line current close to or on the surface is developed in Section 6.2. It is similar to that developed by Andronov [1], but the representation of the incident field in terms of integrals of Whittaker functions is new as well as the spectral decomposition of the total field on the surface.

6.2.2 Asymptotic field in the boundary layer due to a magnetic line current

The problem is two-dimensional and can be solved in the cross-section of the elliptic cylinder.

6.2.2.1 The incident field

The magnetic field radiated by a magnetic line source parallel to the generatrixes is given by (the time dependence is $e^{-i\varpi t}$):

$$\vec{H} = \frac{i}{\varpi \mu_o} \nabla \times \nabla \times \vec{F} = \frac{i\varpi \varepsilon_o}{4} H_o^{(1)}(k|\vec{\rho} - \vec{\rho}'|)\hat{z}M \tag{6.3}$$

where $\vec{F}$ is the magnetic vector potential, $H_o^{(1)}$ the Hankel function of the first kind and order zero, M the intensity of the magnetic line current, $\vec{z}$ the unit vector in the direction of the generatrixes and $\vec{\rho}'$, $\vec{\rho}$ the position vectors $\vec{\rho}' = \overrightarrow{ON'}$ and $\vec{\rho} = \overrightarrow{ON}$ of

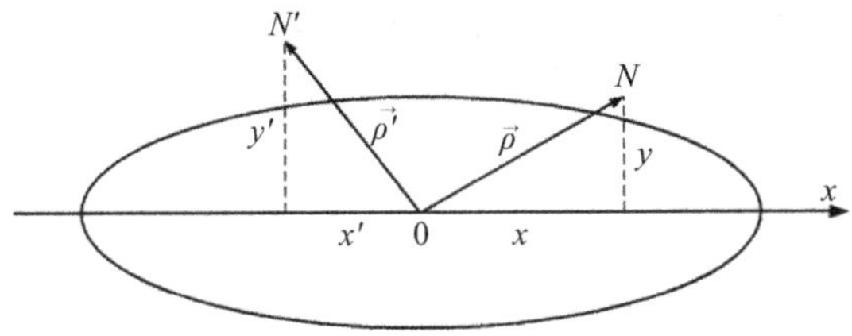

Figure 6.3 *Coordinates defining the position of the source N′ and the observation point N*

the source and the observation point with respect to the centre of the elliptic cross-section. The axis *ox* is in the direction of the major axis of the ellipse and *oy* is in the direction of the minor axis (Figure 6.3).

We suppose that we are at high frequencies and that $k|\vec{\rho}-\vec{\rho}'|\gg 1$, so that the Hankel function can be approximated by its Debye approximation:

$$H_o^{(1)}(k|\vec{\rho}-\vec{\rho}'|)\simeq\sqrt{\frac{2}{\pi k|\vec{\rho}-\vec{\rho}'|}}\exp\left(ik|\vec{\rho}-\vec{\rho}'|\right)e^{-\frac{i\pi}{4}} \tag{6.4}$$

Let (x',y') and (x,y) be the Cartesian coordinates of N' and N in the cross-section of the elliptic cylinder. The elliptic coordinates of $N$and N' are given by:

$$y=p\sqrt{\xi^2-1}\sqrt{1-\eta^2},y'=p\sqrt{\xi'^2-1}\sqrt{1-\eta'^2},x=p\xi\eta,x'=p\xi'\eta' \tag{6.5}$$

where the focal distance is determined by $p=\sqrt{b^2-a^2}$, a and b being the minor and major semi-axis of the ellipse.

The curves $\xi=Cte$ are cofocal ellipses with the elliptic cross-section of the cylinder, the axis *ox* corresponding to $\xi=1$ and the curves $\eta=Cte$ are cofocal hyperboles with the same focus as the elliptic cross-section, the axis *oy* corresponding to $\eta=0$. We introduce the condition describing a strongly elongated cross-section:

$$\chi=kb\frac{a^2}{b^2}\simeq 1 \tag{6.6}$$

and a stretched coordinate τ defined by:

$$\xi^2-1\simeq\frac{\tau\chi}{kp} \tag{6.7}$$

$\tau=0$ on the axis *ox* and according to (6.6) we have $\tau\simeq 1$ on the contour of the elliptic cross-section. Moreover, the x coordinate of the observation point N in the boundary layer is approximated by:

$$x\simeq p\eta+\frac{\tau\chi}{2k}\eta \tag{6.8}$$

By neglecting the terms of order $\chi/2k$ in the boundary layer, the magnetic field radiated by the magnetic line current, which is the incident field in our problem, is given by:

$$H_z^i = C_1 \frac{e^{ikp\left[(\eta-\eta')+\frac{a^2}{2b^2}(\tau\eta-\tau'\eta')\right]}}{\sqrt{\eta-\eta'}} \times \exp\left[\frac{i\chi}{2}\frac{\tau(1-\eta^2)+\tau'(1-\eta'^2)-2\sqrt{\tau\tau'(1-\eta^2)(1-\eta'^2)}}{\eta-\eta'}\right] \tag{6.9}$$

with:

$$C_1 = \frac{i\varpi\varepsilon_o}{4}\sqrt{\frac{2}{\pi kb}}Me^{-\frac{i\pi}{4}} \tag{6.10}$$

We separate the incident field in even and odd parts with respect to y and y':

$$H_z^i = u_e^i + u_o^i \tag{6.11}$$

where:

$$\begin{bmatrix} u_e^i \\ u_o^i \end{bmatrix} = C_1 \frac{e^{ikp\left[(\eta-\eta')+\frac{a^2}{2b^2}(\tau\eta-\tau'\eta')\right]}}{\sqrt{\eta-\eta'}} \exp\left[\frac{i\chi}{2}\frac{\tau(1-\eta^2)+\tau'(1-\eta'^2)}{\eta-\eta'}\right] \times \begin{bmatrix} \cos \\ -isin \end{bmatrix} \frac{\chi\sqrt{\tau\tau'(1-\eta^2)(1-\eta'^2)}}{\eta-\eta'} \tag{6.12}$$

6.2.2.2 Representation of the incident field in terms of Whittaker functions

The general solution of the Helmholtz equation:

$$\Delta u + k^2 u = 0 \tag{6.13}$$

in elliptic coordinates, under the conditions (6.6) and (6.7), has been established by Andronov [1] following the parabolic equation method. By extracting the factor that describes the rapid oscillations of the field:

$$u = e^{ikp(\eta-\eta'')}U(\eta,\tau) \tag{6.14}$$

and substituting (6.14) in the Helmholtz equation expressed in elliptic coordinates:

$$\sqrt{\xi^2-1}\frac{\vartheta}{\vartheta\xi}\left(\sqrt{\xi^2-1}\frac{\vartheta U}{\vartheta\xi}\right) + \sqrt{1-\eta^2}\frac{\vartheta}{\vartheta\eta}\left(\sqrt{1-\eta^2}\frac{\vartheta U}{\vartheta\eta}\right) + (k\rho)^2(\xi^2-\eta^2)U = 0 \tag{6.15}$$

we get, at the leading order, the parabolic equation:

$$4\tau\frac{\vartheta^2 U}{\vartheta\tau^2} + \frac{2\vartheta U}{\vartheta\tau} + 2i\chi(1-\eta^2)\frac{\vartheta U}{\vartheta\eta} - i\chi\eta U + \chi^2\tau U = 0 \tag{6.16}$$

By separation of variables, we get the elementary solutions:

$$U = (x\tau)^{-\frac{1}{4}}(1-\eta^2)^{-\frac{1}{4}}\left(\frac{1-\eta}{1+\eta}\right)^{it} F((-i\chi\tau)) \tag{6.17}$$

where t is the parameter of variable separation and where F satisfies the Whittaker equation:

$$F''(z) + \left(-\frac{1}{4} + \frac{it}{z} + \frac{3}{16z^2}\right)F(z) = 0 \tag{6.18}$$

The expression of the incident field gives some rules for the choice of the solutions of (6.18). According to (6.12), the even part of the incident field tends to a constant when τ tends to zero whereas the odd part behaves like $\tau^{1/2}$. This implies that $F(-i\chi\tau)$ in (6.17) behaves like $z^{1/4}$ for the even part of U and like $z^{3/4}$ for the odd part, when z tends to zero. The solution of (6.18) which satisfies these conditions are the regular Whittaker functions $M_{it,-\frac{1}{4}}(z)$ and $M_{it,\frac{1}{4}}(z)$ [4]. According to (6.17), we can represent the solutions of (6.16) in the form of an integral of the parameter of variable separation t and write the incident field in the form:

$$\begin{bmatrix} u_e^i \\ u_o^i \end{bmatrix} = \frac{e^{ikp(\eta-\eta')}}{(\chi\tau)^{1/4}(1-\eta^2)^{1/4}} \int_{-\infty}^{+\infty} \left(\frac{1-\eta}{1+\eta}\right)^{it} \begin{bmatrix} A_e(t) \\ A_o(t) \end{bmatrix} M_{it,\mp\frac{1}{4}}(-i\chi\tau)dt \tag{6.19}$$

where we have chosen the path of integration along the real t axis and where $A_e(t)$ and $A_o(t)$ are functions of the integration variable t which can be determined by identifying (6.19) with (6.12) for $\tau = 0$.

By replacing $M_{it,\pm\frac{1}{4}}(z)$ in (6.19) by their behaviour when z tends to zero [4]:

$$\left.\begin{array}{l} M_{it,-\frac{1}{4}}(z) \sim z^{1/4} \\ M_{it,\frac{1}{4}}(z) \sim z^{3/4} \end{array}\right| z \to 0 \tag{6.20}$$

and by identifying (6.19) with (6.12), we obtain the integral equations:

$$\int_{-\infty}^{+\infty} \left(\frac{1-\eta}{1+\eta}\right)^{it} A_e(t)dt = C_1 e^{\frac{i\pi}{8}} \frac{(1-\eta^2)^{1/4}}{\sqrt{(\eta-\eta')}} e^{\frac{i\chi}{2}\tau'\frac{(1-\eta'^2)}{\eta-\eta'}} e^{-\frac{i\chi\eta'\tau'}{2}} \tag{6.21}$$

$$\int_{-\infty}^{+\infty} \left(\frac{1-\eta}{1+\eta}\right)^{it} A_o(t)dt = C_1 e^{-\frac{i\pi}{8}} \frac{(1-\eta^2)^{3/4}}{(\eta-\eta')^{3/2}} \chi^{1/2}$$

$$\times\sqrt{\tau'(1-\eta'^2)} e^{\frac{i\chi}{2}\tau'\frac{(1-\eta'^2)}{\eta-\eta'}} e^{-\frac{i\chi\eta'\tau'}{2}} \tag{6.22}$$

where C_1 is given by (6.10). To solve these equations, we proceed as in the study [1] by changing the variable η to $s = (1+\eta)/(1-\eta)$. Then (6.21) and (6.22) take the form of an inverse Mellin transform:

$$\int_{-\infty}^{+\infty} s^{-it} A(t)\, dt = F(s) \tag{6.23}$$

and the solution is given by direct Mellin transform:

$$A(t) = \frac{1}{2\pi}\int_{o}^{\infty} F(s)\, s^{it-1} ds \tag{6.24}$$

Applying this transformation to (6.21) and (6.22) and introducing $\theta = (1+\eta')(1-\eta')^{-1}$ gives:

$$A_e(t) = \frac{C_1 e^{\frac{i\pi}{8}} i e^{-\frac{i\chi\eta'\tau'}{2}}}{\pi\sqrt{2}(1+\eta')^{1/2}} \int_{o}^{\infty} \left(\frac{\sqrt{s}}{1+\theta s}\right)^{1/2} s^{it-1} \exp\left(-\frac{i\chi}{2}\tau'(1-\eta')\frac{s+1}{s\theta+1}\right) ds \tag{6.25}$$

$$A_o(t) = -\frac{2C_1 e^{-\frac{i\pi}{8}} i e^{-\frac{i\chi\eta'\tau'}{2}}}{\pi\sqrt{2}} \chi^{1/2} \frac{\sqrt{\tau'(1-\eta'^2)}}{(1+\eta')^{3/2}} \int_{o}^{\infty} \left(\frac{\sqrt{s}}{1+\theta s}\right)^{3/2} s^{it-1}$$
$$\times \exp\left(-\frac{i\chi}{2}\tau'(1-\eta')\frac{s+1}{s\theta+1}\right) ds \tag{6.26}$$

The second member of (6.25) and (6.26) can be further simplified by changing the variable s to:

$$s = \frac{1}{\theta}\left(\frac{1-u}{1+u}\right) \quad \text{or} \quad u = \frac{2}{s\theta+1} - 1 \tag{6.27}$$

and using the integral representation of the Whittaker functions $M_{\lambda,\mu}(z)$ given by formula 9.221 in the study [4], we obtain after some elementary calculations:

$$A_e(t) = C_2\left(\frac{\eta'+1}{\eta'-1}\right)^{it} \Gamma\left(\frac{1}{4}+it\right)\Gamma\left(\frac{1}{4}-it\right) M_{it,-\frac{1}{4}}(-i\chi\tau') \tag{6.28}$$

$$A_o(t) = 4iC_2\left(\frac{\eta'+1}{\eta'-1}\right)^{it} \Gamma\left(\frac{3}{4}+it\right)\Gamma\left(\frac{3}{4}-it\right) M_{it,\frac{1}{4}}(-i\chi\tau') \tag{6.29}$$

where $\Gamma(u)$ is the gamma function and where:

$$C_2 = \frac{iC_1}{\pi\sqrt{2\pi}} \frac{(\chi\tau')^{-1/4}}{(1-\eta'^2)^{1/4}} \tag{6.30}$$

6.2.3 Radiated field and total field

The diffracted or radiated field must satisfy the radiation condition. A solution of (6.18) which verifies this condition is the irregular Whittaker function $W_{it,-\frac{1}{4}}(z) \cong W_{it,\frac{1}{4}}(z)$, the behavior of which, when $|z| \to \infty$, is given by formula 9.227 in the study [4]:

$$W_{it,\frac{1}{4}}(z) \sim z^{it} e^{-\frac{z}{2}}, \quad |z| \to \infty, \quad \arg(z) < \pi \tag{6.31}$$

According to (6.17), the diffracted field can therefore be written:

$$u^d = u_e^d + u_o^d \tag{6.32}$$

with:

$$\begin{bmatrix} u_e^d \\ u_o^d \end{bmatrix} = \frac{e^{ik\rho(\eta-\eta')}}{(\chi\tau)^{1/4}(1-\eta^2)^{1/4}} \int_{-\infty}^{+\infty} \left(\frac{1-\eta}{1+\eta}\right)^{it} \begin{bmatrix} B_e(t) \\ B_o(t) \end{bmatrix} W_{it,\frac{1}{4}}(-i\chi\tau)dt \tag{6.33}$$

The total field in the boundary layer is given by the sum of the incident field and the radiated field:

$$\begin{aligned} u &= u_e^i + u_e^d + u_o^i + u_o^d \\ &= \frac{e^{ik\rho(\eta-\eta')}}{(\chi\tau)^{1/4}(1-\eta^2)^{1/4}} \int_{-\infty}^{+\infty} \left(\frac{1-\eta}{1+\eta}\right)^{it} \Big\{A_e(t)\, M_{it,-\frac{1}{4}}(-i\chi\tau) \\ &\quad + B_e(t) W_{it,\frac{1}{4}}(-i\chi\tau) + A_o(t) M_{it,\frac{1}{4}}(-i\chi\tau) + B_o(t) W_{it,\frac{1}{4}}(z)\Big\}\, dt \end{aligned} \tag{6.34}$$

The coefficients $B_e(t)$ and $B_o(t)$ are determined by the boundary conditions that are verified separately by the even and odd parts of the total field. For a perfectly conducting surface, they are given by:

$$\frac{\vartheta}{\vartheta\tau}\left(u_e^i + u_e^d\right)|_{\tau=1} = 0, \quad \frac{\vartheta}{\vartheta\tau}\left(u_o^i + u_o^d\right)|_{\tau=1} = 0 \tag{6.35}$$

Hence:

$$\begin{aligned} \frac{B_e(t)}{A_e(t)} &= -\frac{4i\chi \dot{M}_{it,-\frac{1}{4}}(-i\chi) + M_{it,-\frac{1}{4}}(-i\chi)}{4i\chi \dot{W}_{it,\frac{1}{4}}(-i\chi) + W_{it,\frac{1}{4}}(-i\chi)} \\ \frac{B_o(t)}{A_o(t)} &= -\frac{4i\chi \dot{M}_{it,\frac{1}{4}}(-i\chi) + M_{it,\frac{1}{4}}(-i\chi)}{4i\chi \dot{W}_{it,\frac{1}{4}}(-i\chi) + W_{it,\frac{1}{4}}(-i\chi)} \end{aligned} \tag{6.36}$$

where the dot designates the derivations of $M_{it,\pm\frac{1}{4}},(z)$ and $W_{it,\frac{1}{4}}(z)$ with respect to the argument z.

Inserting (6.36) into (6.34) gives the total field which we write:

$$u = \frac{e^{ikp(\eta-\eta')}}{(1-\eta^2)^{1/4}} \int_{-\infty}^{+\infty} \left(\frac{1-\eta}{1+\eta}\right)^{it} [A_e(t) I_e(\tau,t) + A_o(t) I_o(\tau,t)]dt \tag{6.37}$$

$$\begin{aligned} I_e(\tau,t) &= (\chi\tau)^{-\frac{1}{4}} \\ &\quad \times \left[M_{it,-\frac{1}{4}}(-i\chi\tau) - \frac{4i\chi \dot{M}_{it,-\frac{1}{4}}(-i\chi) + M_{it,-\frac{1}{4}}(-i\chi)}{4i\chi \dot{W}_{it,\frac{1}{4}}(-i\chi) + W_{it,\frac{1}{4}}(-i\chi)} W_{it,\frac{1}{4}}(-i\chi\tau)\right] \end{aligned} \tag{6.38}$$

$$I_o(\tau,t) = (\chi\tau)^{-\frac{1}{4}} \left[M_{it,\frac{1}{4}}(-i\chi\tau) - \frac{4i\chi \dot{M}_{it,\frac{1}{4}}(-i\chi) + M_{it,\frac{1}{4}}(-i\chi)}{4i\chi \dot{W}_{it,\frac{1}{4}}(-i\chi) + W_{it,\frac{1}{4}}(-i\chi)} W_{it,\frac{1}{4}}(-i\chi\tau)\right] \tag{6.39}$$

6.2.4 Spectral decomposition of the field in the boundary layer

For solving the problem of diffraction of the total field in the boundary layer, given by (6.37), by an edge, we apply Michaeli's procedure [5, 6] consisting in decomposing the total field u in a spectrum of local plane waves and applying the spectral theory of diffraction (STD) [7].

By expending $I_e(\tau, t)$ and $I_o(\tau, t)$ close to the surface ($\tau = 1$) in Taylor series with respect to $\tau - 1$and by using the expression of the Wronskian:

$$M_{it,\mu}(z)\,\dot{W}_{it,\mu}(z) - \dot{M}_{it,\mu}(z)\,W_{it;\mu}(z) = -\frac{\Gamma(1+2\mu)}{\Gamma\left(\frac{1}{2}+\mu-it\right)} \tag{6.40}$$

we obtain for the first three terms of the series:

$$I_e(\tau,t) = -\chi^{-\frac{1}{4}}\frac{4i\chi}{4i\chi\dot{W}+W}\frac{\sqrt{\pi}}{\Gamma\left(\frac{1}{4}-it\right)} \times\left[1-\frac{(\tau-1)^2}{8}(\chi^2+4t\chi)+O((\tau-1)^3)\right] \tag{6.41}$$

$$I_o(\tau,t) = -\chi^{-\frac{1}{4}}\frac{4i\chi}{4i\chi\dot{W}+W}\frac{\sqrt{\pi}}{2\Gamma\left(\frac{3}{4}-it\right)} \times\left[1-\frac{(\tau-1)^2}{8}(\chi^2+4t\chi)+O((\tau-1)^3)\right] \tag{6.42}$$

In the brackets on the right-hand side of (6.41) and (6.42), we recognize the Taylor series expansion of $cos[(\tau-1)\,\gamma]$, where:

$$\gamma = \frac{1}{2}\sqrt{\chi^2+4t\chi} \tag{6.43}$$

Substituting (6.41) and (6.42) into (6.37) and replacing the cosine function by exponentials, we obtain a spectral decomposition of the field in the boundary layer given by:

$$u = -C_2\frac{\sqrt{\pi}}{2}\frac{e^{ikp(\eta-\eta'')}}{\chi^{1/4}\left(1-\eta^2\right)^{1/4}}$$

$$\int_{-\infty}^{+\infty}\left(\frac{\eta'-1}{\eta'+1}\right)^{-it}\left(\frac{1-\eta}{1+\eta}\right)^{it}\frac{4i\chi S(it,-i\chi\tau')}{4i\chi\dot{W}_{it,\frac{1}{4}}(-i\chi)+W_{it,\frac{1}{4}}(-i\chi)} \times\left(e^{i\gamma(t)(\tau-1)}+e^{-i\gamma(t)(\tau-1)}\right)dt \tag{6.44}$$

where C_2 is given by (6.30) and where:

$$S(it,-i\chi\tau') = \Gamma\left(\frac{1}{4}+it\right)M_{it,-\frac{1}{4}}(-i\chi\tau') + 2i\Gamma\left(\frac{3}{4}+it\right)M_{it,\frac{1}{4}}(-i\chi\tau') \tag{6.45}$$

Now, we consider a point $M_o(\eta_o, 1)$ on the surface and analyse the spectral decomposition (6.44) in the vicinity of that point. Later on, this point will be the position of a singularity of the surface like an edge or a curvature discontinuity. Let $M(\eta, \tau)$ be a point in the boundary layer close to M_o. At that point, we can write:

$$\eta = \eta_O + \delta\eta, \left(\frac{1-\eta}{1+\eta}\right)^{it} = \left(\frac{1-\eta_o}{1+\eta_O}\right)^{it} exp\left(-\frac{2it}{1-\eta_o^2}\delta\eta\right) \tag{6.46}$$

By using the metric coefficients of the elliptic cylinder:

$$d\ell = p\left(\frac{\xi^2-\eta^2}{1-\eta^2}\right)^{\frac{1}{2}} d\eta, \quad dn = p\left(\frac{\xi^2-\eta^2}{\xi^2-1}\right)^{\frac{1}{2}} d\xi \tag{6.47}$$

and the relation (6.7) between ξ and the stretched coordinate τ, the phase terms in the integral of (6.44) can be written:

$$\begin{aligned} &\exp\left(i\left[\left(kp - \frac{2t}{1-\eta_O^2}\right)\delta\eta \pm \gamma(t)(\tau-1)\right]\right) \\ &\quad = \exp\left(ik\left[1 - \frac{\beta^2\gamma^2 m^2}{2k^4a^2}\right]\delta\ell \pm \frac{m\beta\gamma}{k^2a}\delta n\right) \end{aligned} \tag{6.48}$$

where:

$$\beta = \frac{k}{m}\frac{1}{\sqrt{1-\eta_o^2}} \tag{6.49}$$

Let:

$$\theta = \frac{m\beta\gamma}{k^2a} \tag{6.50}$$

Then the phase terms in (6.48) take the form:

$$\phi^{\pm}(t) = \exp\left[ik\left(1-\frac{\theta^2}{2}\right)\delta\ell \pm \theta\delta n\right] \tag{6.51}$$

$\theta = \theta(t)$ is a function of t and of the elongation parameter χ defined in (6.2):

$$\theta(t) = \frac{m\beta\gamma(t)}{k^2a} = 2^{2/3}\frac{1}{m}\frac{1}{\sqrt{1-\eta_o^2}}\chi^{1/3}\sqrt{\frac{1}{4}+\frac{t}{\chi}} \tag{6.52}$$

Since m is large and since the integral (6.44) which may also be written:

$$\int_{-\infty}^{+\infty}\left(\frac{1+\eta'}{1-\eta'}\right)^{it}\left(\frac{1-\eta_o}{1+\eta_o}\right)^{it} e^{\pi t}\frac{4i\chi S(it,-i\chi\tau')}{4i\chi\dot{W}_{it,\frac{1}{4}}(-i\chi)+W_{it,\frac{1}{4}}(-i\chi)}[\phi^+(t)+\phi^-(t)]dt \tag{6.53}$$

converges rapidly when $|t| \to \infty$, $\theta(t)$ is small and $(\phi^{\pm}(t))$ can be replaced by:

$$\phi^{\pm}(t) = \exp[ik(\delta\ell\cos\theta \pm \delta n\sin\theta)] \tag{6.54}$$

Since $\delta\ell$ is the curvilinear arc along the cross-section of the elliptic cylinder and δn the height above the surface of the cylinder, the spectral decomposition of the field (6.44) can be written in terms of plane waves.

In the vicinity of the point M_o, the total field can therefore be written as a spectrum of incident and reflected plane waves:

$$u(M) = -\frac{C_2}{2}\sqrt{\pi}\frac{e^{ikp(\eta-\eta')}}{\chi^{1/4}(1-\eta^2)^{1/4}}\int_{-\infty}^{+\infty}\left(\frac{1-\eta_o}{1+\eta_o}\right)^{it}\left(\frac{\eta'+1}{\eta'-1}\right)^{it}$$
$$\times\frac{4i\chi S(it,-i\chi\tau')}{4i\chi\dot{W}_{it,\frac{1}{4}}(-i\chi)+W_{it,\frac{1}{4}}(-i\chi)}$$
$$(\exp[ik\,(\delta\ell\cos\theta+\delta n\sin\theta)]+\exp[ik\,(\delta\ell\cos\theta-\delta n\sin\theta)])\,dt \tag{6.55}$$

6.2.5 Diffraction by the edge of a truncated elliptic cylinder

We consider now the field in the boundary layer of a strongly elongated elliptic cylinder, due to a magnetic line source parallel to the generatrixes and located on the cylinder ($\tau' = 1$) with the objective to formulate its diffraction by the edge of a truncation of the cylinder by a plane perpendicular to the longitudinal axis of its cross-section (Figure 6.4). Let M_o be the diffraction point on the cross-section of the truncated elliptic cylinder and P the observation point at large distance from M_o, in the plane of the cross-section. The diffracted magnetic field at P is obtained by applying the STD to (6.55), which gives:

$$H_z^d(P) = \frac{C_2(\tau'=1)}{2}\sqrt{\pi}\frac{e^{ikp(\eta_o-\eta')}}{\chi^{1/4}(1-\eta_o^2)^{1/4}}\int_{-\infty}^{+\infty}\left(\frac{1-\eta_o}{1+\eta_o}\right)^{it}\left(\frac{\eta'+1}{\eta'-1}\right)^{it}$$
$$\times\frac{4i\chi S(it,-i\chi)}{4i\chi\dot{W}_{it,\frac{1}{4}}(-i\chi)+W_{it,\frac{1}{4}}(-i\chi)}\times D_h(\phi,\theta)\frac{e^{ikM_oP}}{\sqrt{M_oP}}dt \tag{6.56}$$

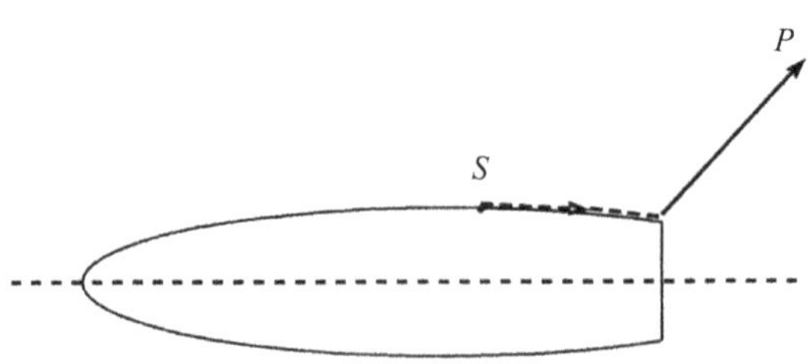

Figure 6.4 STD by the edge of a truncated elliptic cylinder

where $C_2(\tau' = 1)$ is given by (6.30) and where D_h is Keller's diffraction coefficient for a hard boundary condition:

$$D_h(\phi,\theta) = \frac{e^{\frac{i\pi}{4}} sin\frac{\pi}{n}}{n\sqrt{2\pi k}} \left(\frac{1}{\cos(\frac{\pi}{n}) - \cos[(\frac{\phi-\theta}{n})]} + \frac{1}{\cos(\frac{\pi}{n}) - \cos[(\frac{\phi+\theta}{n})]} \right) \tag{6.57}$$

where $n\pi$ is the exterior angle of the wedge formed by the plane tangent to the surface at M_o and the plane of the truncation. Since this coefficient depends only on trigonometric functions, it varies slowly with $\theta(t)$ as compared with the integrant of (6.56), for ϕ not too close to π, which happens at grazing observation. Outside this transition region which can be defined by $\pi - \phi \leq O(m^{-1})$, we can therefore replace Keller's diffraction coefficient by the first term of its Taylor series expansion with respect to θ:

$$D_h(\phi,\theta) \simeq D_h(\phi, O) = \frac{2e^{\frac{i\pi}{4}} \sin\frac{\pi}{n}}{n\sqrt{2\pi k}} \frac{1}{\cos\frac{\pi}{n} - \cos\frac{\phi}{n}} \tag{6.58}$$

and write (6.56) in the form:

$$H_z^d(P) = \frac{ikM}{Z_o} e^{ikp(\eta_o-\eta')} D_h(\phi, O) N(\eta_o, \eta') \frac{e^{ikM_oP}}{\sqrt{M_oP}} \tag{6.59}$$

where $N(\eta_o, \eta'))$ is given by:

$$N(\eta_o, \eta') = \frac{-e^{\frac{i\pi}{4}}}{8\pi\sqrt{\pi kb}} \frac{\chi^{-\frac{1}{2}} e^{i\chi\frac{\eta'}{2}}}{(1-\eta'^2)^{\frac{1}{4}}(1-\eta_o^2)^{\frac{1}{4}}} \int_{-\infty}^{+\infty} \left(\frac{1-\eta_o}{1+\eta_o}\right)^{it} \left(\frac{\eta'+1}{\eta'-1}\right)^{it}$$
$$\times \frac{4i\chi S(it, -i\chi)}{4i\chi \dot{W}_{it,\frac{1}{4}}(-i\chi) + W_{it,\frac{1}{4}}(-i\chi)} dt \tag{6.60}$$

In accordance to the classical asymptotics, we call the function $N(\eta_o, \eta')$ a Nicholson function for elliptic cylinders with a strongly elongated cross-section.

6.2.6 *Asymptotic field in the boundary layer due to an incident plane wave*

To solve this problem, only the incident field has to be modified in the parabolic equation method followed in Section 6.2.2.2. Equation (6.19) is still valid, but the expressions of the incident field are different, leading to new expressions for $A_e(t)$ and $A_o(t)$

6.2.6.1 The incident field

Consider a plane wave propagating in a direction normal to the generatrixes of an elliptic cylinder and making a small angle φ with the main axis of its elliptic cross-section (Figure 6.5). In the Cartesian coordinates introduced in Section 6.2.2.1, where

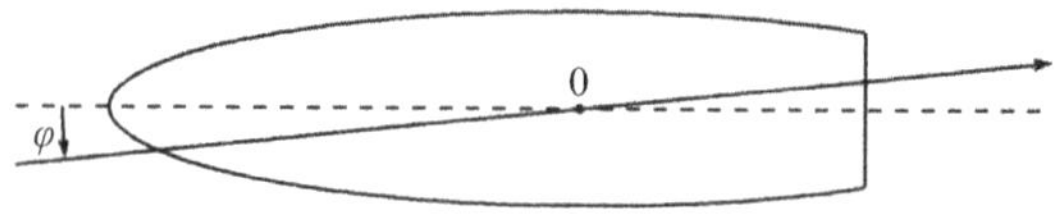

Figure 6.5 Incident plane wave on a truncated elliptic cylinder

o is the center of the ellipse and the ox axis is in the direction of the major axis, we obtain for the even and odd parts with respect to y, the following expressions for the field of an incident plane wave of amplitude unity:

$$u_e^i = e^{ikx\cos\varphi}\cos(ky\sin\varphi), \quad u_o^i = e^{ikx\cos\varphi} i\sin(ky\sin\varphi) \tag{6.61}$$

Replacing x and y by elliptic coordinates and using (6.6) and (6.7) in (6.5) we obtain:

$$\begin{aligned} u_e^i &= e^{ikp\eta} e^{\frac{i}{2}(\tau\chi - kp\varphi^2)\eta} \cos\left[\sqrt{\chi\tau(1-\eta^2)}\,(kp)^{\frac{1}{2}}\,\varphi\right] \\ u_o^i &= e^{ikp\eta} e^{\frac{i}{2}(\tau\chi - kp\varphi^2)\eta} i\sin\left[\sqrt{\chi\tau(1-\eta^2)}\,(kp)^{\frac{1}{2}}\,\varphi\right] \end{aligned} \tag{6.62}$$

where we have neglected the term $\exp(-i\frac{\tau\chi}{4}\eta\phi^2)$. We see that ϕ appears only in the form $\alpha = \phi\sqrt{kp}$ and that $u_o^i = 0$ for $\phi = 0$, but takes non-negligible values for small angles ϕ such that $\alpha \simeq 1$.

6.2.6.2 Representation of the incident field in terms of Whittaker functions

By extracting the phase factor that describes the rapid oscillations of the field in the boundary layer:

$$u = e^{ikp\eta} U(\eta, \tau) \tag{6.63}$$

we obtain for the solution of the Helmholtz equation, in the leading order, the parabolic equation (6.16) and a representation of the incident field in terms of Whittaker functions identical to (6.19) with the phase term replaced by $\exp(ikp\eta)$. The unknown functions $A_e(t)$ and $A_o(t)$ of the integration variable t verify:

$$\begin{aligned} \int_{-\infty}^{+\infty} A_e(t)\left(\frac{1-\eta}{1+\eta}\right)^{it} dt &= e^{\frac{i\pi}{8}}(1-\eta^2)^{1/4} e^{-\frac{i}{2}\eta\alpha^2} \\ \int_{-\infty}^{+\infty} A_o(t)\left(\frac{1-\eta}{1+\eta}\right)^{it} dt &= e^{+\frac{3i\pi}{8}}(1-\eta^2)^{3/4} e^{-\frac{i}{2}\eta\alpha^2}\alpha \end{aligned} \tag{6.64}$$

Proceeding as in Section 6.2.2.2 and changing the variable η to $s = (1+\eta)(1-\eta)$, the integral equations (6.64) take the form of inverse Mellin transforms (6.23) and their solution is given by the direct Mellin transform (6.24). After some elementary calculations, using the integral representation of the Whittaker functions $M_{it,\pm\mu}(z)$ given by formula 9.221 in [4], we obtain:

$$A_e(t) = \frac{1}{\sqrt{2}\pi^{3/2}} \frac{M_{it,-\frac{1}{4}}(i\alpha^2)}{\sqrt{\alpha}} \Gamma\left(\frac{1}{4} + it\right) \Gamma\left(\frac{1}{4} - it\right) \tag{6.65}$$

$$A_o(t) = \frac{2i\sqrt{2}}{\pi^{3/2}} \frac{M_{it,\frac{1}{4}}(i\alpha^2)}{\sqrt{\alpha}} \Gamma\left(\frac{3}{4} + it\right) \Gamma\left(\frac{3}{4} - it\right) \tag{6.66}$$

We see that when the angle of incidence $\varphi = 0$ which corresponds to an incident wave propagating along the main axis of the elliptic cross-section, we have $A_o(t) = 0$ and:

$$A_e(t) = e^{\frac{i\pi}{8}} \frac{1}{\sqrt{2}\pi^{3/2}} \Gamma\left(\frac{1}{4} + it\right) \Gamma\left(\frac{1}{4} - it\right) \tag{6.67}$$

6.2.6.3 Total field in the boundary layer and spectral decomposition

Apart the phase term $\exp[ikp(\eta - \eta')]$ which has to be replaced by $\exp(ikp\eta)$, the diffracted field and the total field for $y > 0$ are, respectively, given by (6.33) and (6.34). Moreover, by applying the boundary conditions on the surface which is supposed to be perfectly conducting, the magnetic field verifies (6.35) and the total magnetic field is given by:

$$u = \frac{e^{ikp\eta}}{(1-\eta^2)^{\frac{1}{4}}} \int_{-\infty}^{+\infty} \left(\frac{1-\eta}{1+\eta}\right)^{it} [A_e(t) I_e(\tau, t) + A_o(t) I_o(\tau, t)] dt \tag{6.68}$$

where $I_e(\tau, t)$ and $I_o(\tau, t)$ are given by (6.38) and (6.39), respectively, and since they are unchanged compared with the line source problem, their spectral decomposition is also unchanged and given by (6.41) and (6.42), respectively. Finally, the total field in the boundary layer is given by:

$$u = \frac{Ce^{ikp\eta}}{(1-\eta^2)^{\frac{1}{4}}} \int_{-\infty}^{+\infty} \left(\frac{1-\eta}{1+\eta}\right)^{it} Q(it, \alpha) \left[\left(e^{i\gamma(t)(\tau-1)} + e^{-i\gamma(t)(\tau-1)}\right)\right] dt \tag{6.69}$$

where

$$C = -\frac{i\sqrt{2}\chi^{3/4}}{\pi} \tag{6.70}$$

and:

$$Q(it, \alpha) = \frac{1}{\sqrt{\alpha}} \frac{M_{it,-\frac{1}{4}}(i\alpha^2)\Gamma\left(\frac{1}{4} + it\right) + 2iM_{it,\frac{1}{4}}(i\alpha^2)\Gamma\left(\frac{3}{4} + it\right)}{4i\chi \dot{W}_{it,\frac{1}{4}}(-i\chi) + W_{it,\frac{1}{4}}(-i\chi)} \tag{6.71}$$

In the vicinity of an observation point $M_o(\eta_o, 1)$ on the surface, we can proceed as in Section 6.2.4 (see (6.46)–(6.52)) and write:

$$\begin{aligned} u(M_o) = {} & \frac{Ce^{ikp\eta}}{(1-\eta_o^2)^{1/4}} \int_{-\infty}^{+\infty} \left(\frac{1-\eta_o}{1+\eta_o}\right)^{it} Q(it, \alpha)(\exp[ik\,(\delta\ell \cos\theta + \delta n \sin\theta)] \\ & + \exp[ik(\delta\ell \cos\theta - \delta n \sin\theta)]) dt \end{aligned} \tag{6.72}$$

where $\theta = \theta(t)$ is given by (6.52).

6.2.6.4 Diffraction of a plane wave with a small incident angle by the edge of a truncated elliptic cylinder

Let M_o be the diffraction point on the cross-section of the truncated elliptic cylinder and P the observation point at large distance from M_o. We suppose that the incident magnetic field is parallel to the generatrixes of the cylinder. Then, by applying the STD [7] to (6.72), we obtain:

$$H_z^d(P) = \frac{Ce^{ikp\eta_o}}{(1-\eta_o^2)^{1/4}} \left(\int_{-\infty}^{+\infty} \left(\frac{1-\eta_o}{1+\eta_o} \right)^{it} Q(it,\alpha)\, D_h(\phi,\theta)\, dt \right) \frac{e^{ikM_oP}}{\sqrt{M_oP}} \tag{6.73}$$

where $D_h(\phi,\theta)$ is the edge diffraction coefficient given by (6.57). Proceeding as in Section 6.2.5 and replacing $D_h(\phi,\theta)$ by (6.58), we obtain:

$$H_z^d(P) = e^{ikp\eta_o} D_h(\phi,0)\, G(\eta_o) \frac{e^{ikM_oP}}{\sqrt{M_oP}} \tag{6.74}$$

where $G(\eta)$ is a new Fock function given by:

$$\begin{aligned} G(\eta) &= \frac{C}{\sqrt{\alpha}(1-\eta^2)^{\frac{1}{4}}} \int_{-\infty}^{+\infty} \left(\frac{1-\eta}{1+\eta} \right)^{it} \\ &\times \frac{M_{it,-\frac{1}{4}}(i\alpha^2)\Gamma\left(\frac{1}{4}+it\right) + 2iM_{it,\frac{1}{4}}(i\alpha^2)\Gamma\left(\frac{3}{4}+it\right)}{4i\chi\, \dot{W}_{it,\frac{1}{4}}(-i\chi) + W_{it,\frac{1}{4}}(-i\chi)} dt \end{aligned} \tag{6.75}$$

where C is given by (6.70).

6.2.7 Asymptotic currents

The new Nicholson and Fock functions established in Sections 6.2.5 and 6.2.6 enable us to establish the formulas giving the asymptotic currents on the surface of the elliptic cylinder. As mentioned in Section 6.2.1, to calculate the field at a point M on the surface due to the diffraction of an incident wave by the edge, we consider the reciprocal problem of a line current at M the field of which is diffracted on the edge and observed in the opposite direction at an observation point P far in wavelengths from the elliptic cylinder (Figure 6.2). The solution of this problem is obtained by combining the solution of the problem of Figure 6.4 with the solution of the reciprocal problem of Figure 6.5. According to (6.74) and (6.59), we obtain directly:

$$H_z^d(P) = \frac{ikM}{z_0} e^{ikp(\eta_o-\eta')} N(\eta_o,\eta') D_h(o,o) e^{ikp\eta_o} G(-\eta_o) \frac{e^{ikOP}}{\sqrt{OP}} \tag{6.76}$$

Applying again the reciprocity theorem to (6.76), we obtain at the observation point (η):

$$H_z^d(M) = H_z^i(O)\, e^{ikp(\eta_o-\eta)} N(\eta,\eta_o)\, G(\eta_o)\, e^{ikp\eta_o} D_h(o,o)\, S^{-1} \tag{6.77}$$

where S is the source factor for an isotropic line source [6], $S = (8\pi k)^{-1/2} e^{\frac{i\pi}{4}}$ and where $H_Z(O)$ is the amplitude of the incident plane wave at the origin O, which in Section 6.2.3 has been taken equal to unity.

The asymptotic current at M due to the edge diffracted wave is therefore given by:

$$\vec{J} = \hat{n} \times \vec{H}_z^d = H_z^d \hat{t} \tag{6.78}$$

where $\hat{n}$ and $\hat{t}$ are the unit vectors along the normal and the tangent to the elliptic cross-section at M, respectively.

To the current induced at M by the edge diffracted wave, we have to add the current due to the direct wave reaching M. According to (6.69), we obtain for $\tau = 1$, the field at an observation point M on the surface:

$$H_z(M) = H_z^i(O)\, e^{ikp\eta} G(\eta) \tag{6.79}$$

where O is the centre of the elliptic cross-section and $G(\eta)$ is the new Fock function given by (6.75).

The asymptotic current at M due to the direct wave is therefore given by:

$$\vec{J}(M) = \hat{n} \times \vec{H}(M) = H_z(M)\, \hat{t} \tag{6.80}$$

And the total asymptotic current $\vec{J}_t$ at M is given by:

$$\vec{J}_t(M) = \hat{n} \times [\vec{H}^d(M) + \vec{H}(M)] = [H_z^d(M) + H_Z(M)]\hat{t} \tag{6.81}$$

6.3 Asymptotic currents on a cylinder with an ogival cross-section composed of two symmetric arcs of a strongly elongated ellipse

6.3.1 Presentation of the geometry and analysis of the problem

The cross-section of the ogival cylinder has two axes of symmetry and is described by the intersection of two elliptic curves (Figure 6.1(c)). The cross-section is supposed to be strongly elongated, which means that $\chi = kba^2/b^2 \simeq O(1)$, where a and b are the half axes of the primary ellipse. The incident field is a plane electromagnetic wave propagating at a small angle with the major axis. The asymptotic currents on the surface of the ogival cylinder are composed of two main contributions corresponding to the direct field and the edge diffracted field, respectively. The current due to the direct wave is given by the solution described in Section 6.2.6. We have shown that this solution can be written in the same form as on a non-elongated elliptic cylinder by introducing new Fock functions that depend on Whittaker functions instead of Airy functions. The currents due to the edge diffracted waves are derived from the solution of the reciprocal problem, namely the diffraction by the edge of the wave emitted by a line current parallel to the generatrixes. The solution to this problem is described in Section 6.2.2. The resulting formulas are given in the form of modified Nicholson functions expressed in terms of Whittaker functions. However, a new problem arises for the determination of the asymptotic currents due to the edge diffracted field at grazing incidence on the surface at one of the edges. In classical GTD analysis, this problem has been solved by Michaeli [8] who combined the STD with Ufimtsev's theory [9]. The latter consists in decomposing the field close to the edge, associated with each component of the spectrum of plane waves, into a uniform part corresponding to the current induced by this component on the face of the wedge as if the edge was absent, and the fringe part was radiated by correction currents. The latter is calculated by replacing Keller's diffraction coefficients by those of Ufimtsev, in the STD procedure developed in Sections 6.2.4 and 6.2.5. The contribution of the uniform part to the diffracted field is obtained by replacing the diffraction problem by a radiation problem using the surface currents established in Section 6.2.7 and (6.78), (6.79) which correspond to the PO contribution.

In this chapter, we apply a similar procedure to an ogival cylinder with a strongly elongated cross-section and derive new formulas for the induced asymptotic currents.

Section 6.3 is organized as follows:

In Section 6.3.2, we give the formulas for the asymptotic currents on the surface of the ogival cylinder in the case where the direction of the incident wave is outside the critical situation of near-grazing incidence.

In Section 6.3.3, we treat the problem of grazing incidence starting with the fringe current contribution to the field radiated by a line source parallel to the generatrixes of the ogival cylinder and diffracted by the edge followed by the PO contribution to the diffracted field. Then we apply the reciprocity theorem to the different components of the solution of the radiation problem and establish the expression for the asymptotic surface currents.

6.3.2 Asymptotic currents outside grazing incidence

As mentioned in Section 6.3.1, the asymptotic currents on the surface of the ogival cylinder are composed of the current due to the direct wave and the current due to the edge diffracted wave. The formulation of the latter is different for an illuminated edge and a shadowed edge.

6.3.2.1 Illuminated edge

The formula giving the asymptotic current due to the diffraction of an incident plane wave propagating in a direction normal to the generatrixes of the ogival cylinder, by the illuminated edge, is deduced by reciprocity from the formula giving the field radiated by a magnetic line current parallel to the generatrixes and diffracted by the illuminated edge into a space wave. The solution to this problem has been established in Section 6.2 and the magnetic field at an observation point P far from the edge is given by (6.59). By reciprocity, the field at a point M on the surface of the ogival cylinder due to the diffraction of the incident plane wave by the illuminated edge is given by:

$$H_z^d(M) = H_z^i(M_o)\, e^{ikp(\eta-\eta_o)} D_h(0,\phi)\, N(\eta,\eta_o)\, S^{-1} \tag{6.82}$$

where S is the source factor (see Section 6.2.7), η and η_o are the elliptic coordinates of M and of the diffraction point Q on the edge, respectively, $D_h(0,\phi)$ is

Keller's diffraction coefficient for the hard boundary condition and $N(\eta, \eta_o)$ is the new Nicholson function given by (see (6.60)):

$$N(\eta, \eta_o) = \frac{-e^{i\frac{\pi}{4}}}{8\pi\sqrt{\pi kb}} \frac{\chi^{-\frac{1}{2}} e^{i\chi\frac{\eta}{2}}}{(1-\eta_o^2)^{1/4}(1-\eta^2)^{1/4}} \int_{-\infty}^{+\infty} \left(\frac{1-\eta}{1+\eta}\right)^{it} \left(\frac{\eta_o+1}{\eta_o-1}\right)^{it} \times \frac{4i\chi S(it, -i\chi)}{4i\chi \dot{W}_{it,\frac{1}{4}}(-i\chi) + W_{it,\frac{1}{4}}(-i\chi)} dt \quad (6.83)$$

where $S(it, -i\chi)$ is given by (6.45) with $\tau' = 1$.

The asymptotic current corresponding to (6.82) is given by:

$$\vec{J} = (\hat{n} \times \hat{z}) H_z^d(M) \quad (6.84)$$

In addition to the edge diffracted wave, we have also the current due to the direct wave which is given by the same formula as (6.80) in which $H_z(M)$ is given by (6.79) with the Fock function $G(\eta)$ given by (6.75). It is important to mention that the phase reference in (6.79) is not at the centre of the ogive, but at the centre of the elliptic curve on which the observation point M is located (see Figure 6.6).

6.3.2.2 Edge excited currents on shadowed edge

When the edge is in the shadow region of the incident field, its contribution to the currents at a point M on the surface is due to a backward propagating wave generated by the diffraction of the incident wave by the shadowed edge. The solution to this problem is identical to that derived in Section 6.4 and is given by formula (6.77) which we write:

$$H_z^d(M) = H_Z(O)\, e^{ikp(\eta_1-\eta)} N(\eta, \eta_1)\, G(\eta_1)\, e^{ikp\eta_1} D_h(o, o)\, S^{-1} \quad (6.85)$$

where η_1 and η are the elliptic coordinates along the surface of the diffraction point on the shadowed edge and of the observation point M, respectively.

6.3.3 Grazing incidence

We consider now the case of grazing incidence and more generally the case where the transition regions of the direct and reflected rays at the edge overlap. As mentioned in the treatment of the problem in Section 6.1 to solve this problem, we follow Michaeli's

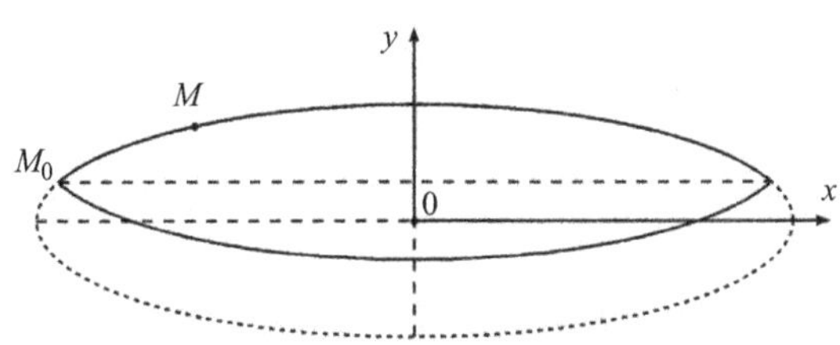

Figure 6.6 Primary elliptic cross-section associated with the upper face of the ogival cross-section and phase reference O

procedure [8] that consists to decompose the problem into two parts: the fringe current contribution and the PO contribution.

6.3.3.1 Fringe current contribution to the diffracted field

The fringe current contribution of the reciprocal problem is given by (6.59) in which Keller's diffraction coefficient has to be replaced by Ufimtsev's diffraction coefficient, which for the hard boundary condition is given by:

$$D_h^u(\phi,\theta) = D_h(\phi,\theta) - D_h^{PO}(\phi,\theta) \tag{6.86}$$

where:

$$D_h^{PO}(\phi,\theta) = -\frac{2\sin\phi}{\cos\theta + \cos\phi} \tag{6.87}$$

$D_h^u(\phi,\theta)$ is not singular at grazing incidence $\theta = 0$, $\phi = \pi$, whereas $D^{PO}(\phi,\theta)$ and $D_h(\phi,\theta)$ are both singular.

Applying the reciprocity theorem, we obtain for the fringe current contribution at the observation point $M(\eta, 1)$ on the surface:

$$H_z^F(M) = H_Z(M_o)\, e^{ikp(\eta-\eta_o)} D_h^u(\phi,\theta)\, S^{-1} N(\eta,\eta_o) \tag{6.88}$$

where S is the source factor (see Section 6.2.7) and the Nicholson function $N(\eta,\eta_o)$ is given by (6.60).

6.3.3.2 Radiation of the PO current

The magnetic field radiated by the PO current induced by a magnetic line source is given by:

$$U_h(P) = \frac{i}{4}\int_o^{\infty} \frac{\vartheta}{\vartheta n} H_o^{(1)}(k|\vec{r}-\vec{\rho}|)u_h(M)d\ell \tag{6.89}$$

where $u_h(M)$ is the field that would exist at a point M of the ogival cylinder if the edge was absent and where P is the observation point away from the surface, with:

$$\overrightarrow{M_oP} = \vec{r}, \quad \overrightarrow{M_oM} = \vec{\rho}, \quad \ell = \mathrm{arc}M_oM \tag{6.90}$$

Using the Debye approximation of the Hankel function for $|\vec{r}-\vec{\rho}|$ large:

$$H_o^{(1)}(k|\vec{r}-\vec{\rho}|) = \sqrt{\frac{2}{\pi kr}}e^{-\frac{i\pi}{4}} e^{ikr} e^{-ik\hat{r}\cdot\vec{\rho}} \tag{6.91}$$

and applying the relation $\frac{\vartheta}{\vartheta n} = \hat{n}\cdot\frac{\vartheta}{\vartheta\vec{\rho}}$, we obtain:

$$U_h(P) = \sqrt{\frac{k}{8\pi r}}e^{-\frac{i\pi}{4}} e^{ikr}\int_o^{\infty} (\hat{r}\cdot\hat{n})e^{-ik(\hat{r}\cdot\vec{\rho})}u_h(M)d\ell \tag{6.92}$$

We proceed as Michaeli [8] and write the integral (6.89) in the form:

$$\int_o^{\infty} (\cdot)\, d\ell = \int_o^{-\infty} (\cdot)\, d\ell + \int_{-\infty}^{+\infty} (\cdot)\, d\ell \tag{6.93}$$

where the integration is carried out on the face of the ogive containing the observation point M of the asymptotic surface current.

The second integral on the right-hand side of (6.93) corresponds to the field radiated by a regular surface. By reciprocity, this problem corresponds to the current induced by an incident plane wave on a regular surface for which an asymptotic solution has been established in Section 6.2.6.

For the calculation of the first integral on the right-hand side of (6.93), we use a circular cylindrical approximation of the surface close to M_o with $\ell < 0$:

$$\hat{r}\cdot\hat{n} = \sin\left(\psi - \frac{\ell}{q}\right), \quad \hat{r}\cdot\vec{p} = q\left[\sin\left(\psi - \frac{\ell}{q}\right) - \sin\psi\right] \tag{6.94}$$

where q is the radius of the circular cylinder and ψ is the angle belonging to the interval $(o, \frac{\pi}{2})$ between $\overrightarrow{M_oP}$ and the tangent to the upper surface of the illuminated edge. It is important to note that grazing incidence on an ogival cylinder can appear only for four different directions of the incident wave, which can be deduced from one of them by symmetry with respect to the symmetry axes of the ogival cylinder. Near grazing incidence, ψ is small.

We replace $u_h(M)$ by its spectral decomposition in the vicinity of M_o, given by (6.55) with $\vartheta n = 0, \delta\ell = -\ell$, and according to (6.48), we have:

$$\exp(-ik\ell\cos\theta) \simeq \exp\left[-ik\ell\left(1 - \frac{\theta^2}{2}\right)\right] = \exp\left[-ik\ell\left(1 - \frac{m^2\beta^2\gamma^2}{2k^4a^2}\right)\right] \tag{6.95}$$

Interchanging the order of integration in (6.92) and replacing ℓ by the new integration variable $= \psi - \ell/q$, we obtain after expending $\sin u$ and $\sin\psi$ in Taylor series to the third power in the phase terms and to the first power in the amplitude terms:

$$U_h(P) = \frac{ikM}{Z_o} e^{ikp(\eta_o - \eta')} P(\eta_o, \eta', \psi) S \frac{e^{ikr}}{\sqrt{r}} \tag{6.96}$$

where S is the source factor (see Section 6.2.7) and:

$$P(\eta_o, \eta', \psi) = \frac{C_3 e^{-ikq\frac{\psi^3}{6}}}{\chi^{1/2}(1-\eta'^2)^{1/4}(1-\eta_o^2)^{1/4}}$$

$$\int\limits_{-\infty}^{+\infty} \left(\frac{1-\eta_o}{1+\eta_o}\right)^{it} \left(\frac{\eta'+1}{\eta'-1}\right)^{it} \frac{4i\chi S(it, -i\chi)}{4i\chi \dot{W}_{it,\frac{1}{4}}(-i\chi) + W_{it,\frac{1}{4}}(-i\chi)} e^{ikq\frac{\theta^2}{2}\psi} I(t, \psi)dt \tag{6.97}$$

and:

$$I(t, \psi) = \int\limits_{\psi}^{\infty} \exp\left[ikq\left(\frac{u^3}{6} - \frac{\theta^2}{2}u\right)\right] u\, du \tag{6.98}$$

$$C_3 = \frac{kqe^{-\frac{i\pi}{4}}}{4\pi\sqrt{\pi kb}} e^{\frac{i\chi\eta'}{2}} \tag{6.99}$$

In (6.98), θ is a function of t given by (6.52). The function $P(\eta_o, \eta, \psi)$ is a new transition function.

We obtain the result for the reciprocal problem by replacing ikM by $Z_o H_Z(M_o)$ and by suppressing the factor $\frac{e^{ikr}}{\sqrt{r}}$ and the source factor S in (6.96).

6.3.3.3 Asymptotic current for grazing incidence

To get the asymptotic current for grazing incidence, we have to add three contributions: the fringe current contribution given by (6.88), the regularized surface contribution given by (6.78) and the PO current contribution of the fictive cylindrical prolongation of the surface beyond the edge given by the reciprocal form of (6.96). If t'' is the unit vector of the tangent to the cross-section of the surface at the observation point M, the asymptotic current is given by:

$$\vec{J} = H_z(M)\hat{t} \tag{6.100}$$

where:

$$\begin{aligned} H_z(M) &= H_z(M_o)\left\{e^{ikp(\eta-\eta_o)} N(\eta,\eta_o)\left[D_h(\psi,o) - D_h^{OP}(\psi,o)\right]S^{-1}\right. \\ &\quad \left. + e^{ikp(\eta-\eta_o)} P(\eta,\eta_o,\psi)\right\} + H_z(o)\, e^{ikp\eta} G(\eta) \end{aligned} \tag{6.101}$$

where η_o and η correspond to the value of the elliptic coordinate η at the edge M_o and at the observation point M, respectively The origin O of the phase of the regularized surface contribution is the centre of the ellipse to which belongs the upper face of the ogival cylinder.

6.4 Conclusion

In this chapter, we have combined the function theoretic methods on elliptic cylinders with a strongly elongated cross-section with high-frequency techniques. New solutions have been obtained for an elliptic cylinder with a truncated cross-section and for an ogival cylinder, the cross-section of which is composed of two symmetric arcs of an ellipse. The technique is based on a spectral decomposition of the field in the boundary layer of the surface obtained by the function theoretic method, followed by the application of the STD.

The formulas obtained for the asymptotic currents are also applied to cylinders with more general cross-sections, composed of arcs of ellipses, an example of which is shown on Figure 6.1(d). Only the formulas for the $H_{//}$ component have been presented. Those for the $E_{//}$ component can be deduced in a similar way by replacing the magnetic line current by an electric line current and by applying the Dirichlet boundary condition to the total field.

At the time, when this chapter has been written, no numerical results for a cylinder with a truncated strongly elongated cross-section were available. In principle, the computation of the new Fock and Nicholson functions present no special difficulty, and the Whittaker functions appearing under the integral sign can be expressed in terms of Coulomb functions [10] for which a computer program is available in the literature [11]. Moreover, numerical control of the new Fock function has already been performed since it appears in the expression of the currents on a non-truncated elliptic cylinder for which numerical results have been published by Andronov [1] and shown in Figures 6.7 and 6.8 for $H_{//}$ (left) and $E_{//}$ (right) polarizations. For $k = 100$, the discrepancies are due to an insufficient mesh refinement in the numerical method.

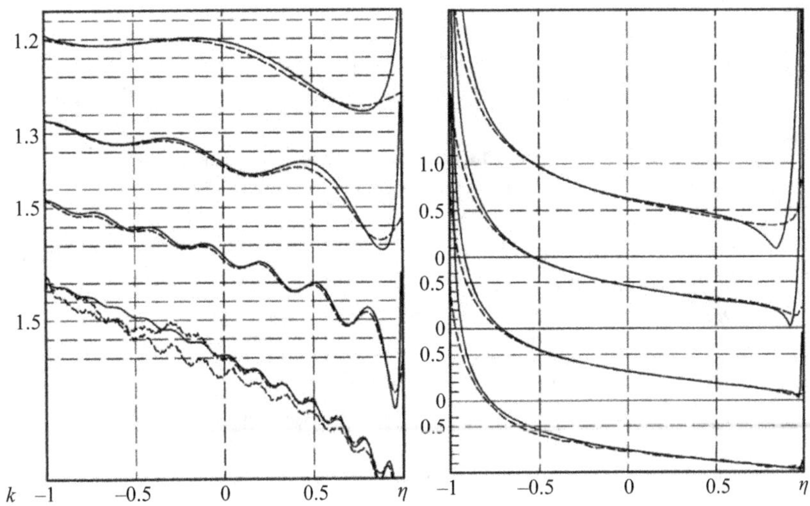

Figure 6.7 $H_{//}$ *(left) and* $E_{//}$ *(right) currents on the elliptic cylinder with* $a = 0.05$ *and* $b = 0.2$ *for* $k = 10, 20, 50$ *and 100 (downwards). Solid (–) asymptotics, dashed (–) numerical*

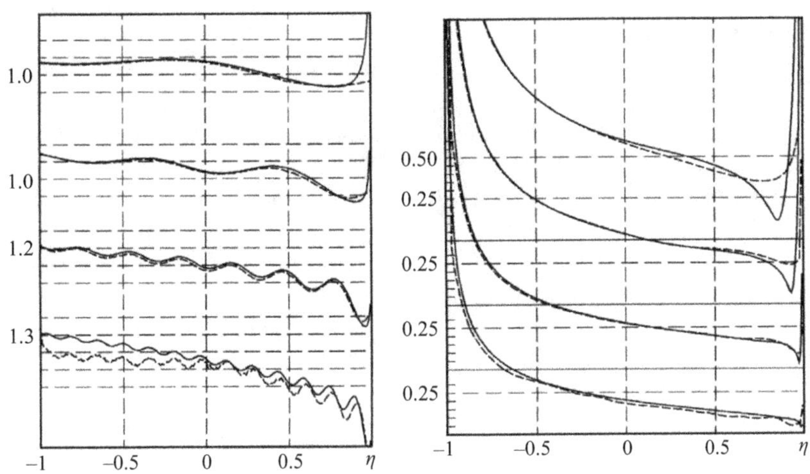

Figure 6.8 $H_{//}$ *(left) and* $E_{//}$ *(right) currents on the elliptic cylinder with* $a = 0.02$ *and* $b = 0.2$ *for* $k = 10, 20, 50$ *and 100 (downwards). Solid (–) asymptotics, dashed (–) numerical*

Nevertheless, numerical results for the asymptotic currents on a truncated elliptic cylinder and on an ogival cylinder as a function of the frequency and the elongation parameter and comparison with results obtained by a numerical method are desirable.

References

[1] Andronov, I. V. 'High-frequency diffraction by an elliptic cylinder: the near field', *Journal of Electromagnetic Waves and Applications*, 2014, **28**, 2318–2326.

[2] Andronov, I. V. 'High-frequency acoustic scattering from prolate spheroids with high aspect ratio', *Journal of the Acoustic Society of America,* 2013, **134**(6), 4307–4316.

[3] Andronov, I. V., Bouche, D., and Durufle, M. 'High-frequency diffraction of a plane electromagnetic wave by an elongated spheroid', *IEEE Transactions on Antennas and Propagation*, 2012, **60**, 5286–5295.

[4] Gradshteyn, I. S. and Ryzhik, I. M. *Table of Integrals, Series and Products*, Elsevier, Academic Press, New York, 2007.

[5] Michaeli, A. 'Transition functions for high-frequency diffraction by a curved, perfectly conducting wedge, Part II: a partially uniform solution for a general wedge angle', *IEEE Transactions on Antennas and Propagation*, 1989, **37**, 1080–1085.

[6] Molinet, F., Andronov, I. V., and Bouche, D. 'Asymptotic and Hybrid Methods in Electromagnetics', *IEE Electromagnetic Waves Series*, United Kingdom, 2005, **51**, Chap. 3.

[7] Rhamat-Samii, Y. and Mittra, R. 'A spectral domain interpretation of high-frequency diffraction', *IEEE Transactions on Antennas and Propagation*, 1977, **AP-25**, 676–687.

[8] Michaeli, A. 'Transition functions for high-frequency diffraction by a curved, perfectly conducting wedge, Part III: extension to overlapping transition regions', *IEEE Transactions on Antennas and Propagation*, 1989, **37**, 1086–1092.

[9] Ufimtsev, P. Ya. *Fundamentals of the Physical Theory of Diffraction*, John Wiley & Sons, USA, 2007.

[10] Abramowitz, M. and Stegun, I. *Handbook of Mathematical Functions*, Dover Publications, Inc., New York, 1972.

[11] Thomson, I. J. and Barnett, A. R. 'COULCC: A continued-fraction algorithm for Coulomb functions of complex order with complex arguments', *Computer Physics Communication*, 1985, **36**, 363–372.

Chapter 7
High-frequency hybrid ray–mode techniques

Hiroshi Shirai[1]

Rays and modes are two alternative descriptions for electromagnetic field, and they are regarded as Fourier transform pairs that have complementary convergence properties. Accordingly, a modal description would be preferable for an internal structure, while a ray description would be appropriate for an exterior region. When a scatterer contains an open waveguide cavity structure, one wants to retain the fast converging description, namely, rays for the exterior and modes for the interior regions, respectively, and the conversion should be considered at the opening of the structure. This ray–mode conversion between them may be found from a mathematical formula called the Poisson summation formula, and this technique is found to be one of the powerful tools for analyzing high-frequency scattering problems. Some application examples, such as plane wave diffraction from a slit on a conducting screen and a trough on a ground, are given to show the validity of this technique.

7.1 Introduction

A ray description can be an efficient way to represent scattered and diffracted fields by objects whose dimensions are relatively large compared with the wavelength. Since a progressing ray property can be determined by a local feature of the surrounding environment, the ray tracing is rather easy and the total scattered field can be written as a summation of rays that experience multiple scattering process, such as reflection, transmission, diffraction, and diffusion. The situation, however, may be different, when a scatterer contains an open resonance structure [1]. One may trace all the rays bouncing from the internal walls of the structure, but it is tedious to trace all the edge-diffracted multiply reflected rays. Therefore, the solution obtained by summing these bouncing rays would be inaccurate and it would be numerically inefficient to predict the field, particularly in the diffraction region.

A modal description, if available, would be preferable for the internal guiding structures, while a ray description would be preferable for the exterior region [1,2]. When the internal field can be expressed in terms of the corresponding waveguide modes, formally infinite modal summation may be truncated by the propagating modes, or only selected *significant* modes may be summed to express the field [3–5].

[1]Department of Electrical, Electronic, and Communication Engineering, Chuo University, Tokyo, Japan

It would also be much easier to treat some discontinuities inside the waveguide by the scattering matrix approach. Accordingly, to calculate the diffracted field efficiently, each description should be retained in suitable regions. Then, ray–mode conversion between the earlier two alternative descriptions must be considered at the opening. Since rays and modes are regarded as Fourier transformation pairs, one can utilize the Poisson summation formula to establish an alternative ray or mode description or a hybrid form of both [1]. Thus, one can construct the solution while maintaining the advantages of both descriptions.

An idea of the ray–mode conversion was successfully applied to analyze a waveguide modal reflection at the open-end by Yee *et al.* [6]. Then a full conversion equivalence between rays and modes was established for a duct propagation problem by Felsen and Ishihara [7], and for a propagation in a parallel-plane waveguide by Felsen and Kamel [8]. While the ray–mode conversion is mainly used in the frequency domain, it can also be applied in the time domain where rays are related to wave fronts, and modes to sinusoidal oscillations [9,10]. Review for early works on the ray–mode conversion can be found by Felsen [1].

In the following discussion, the Poisson summation formula is briefly discussed, then a modal excitation at an aperture of a parallel-plane waveguide is formulated by ray–mode conversion technique, and the validity is discussed. Then the derived formula is applied to find a diffracted field by a slit and a trough structures.

The time-harmonic factor $e^{-i\omega t}$ is assumed and suppressed throughout the text.

7.2 Ray–mode conversion technique

The ray–mode conversion method may be found from a mathematical formula called the Poisson summation formula that establishes the equivalence between a series of f_n and those of its Fourier transformed constituents as [1]

$$\sum_{n=-\infty}^{\infty} f_n = \sum_{m=-\infty}^{\infty} F_m, \tag{7.1}$$

where

$$F_m = \mathscr{F}[f_n] = \frac{1}{2\pi}\int_{-\infty}^{\infty} f(\tau = 2\pi n)e^{im\tau}\,d\tau, \tag{7.2}$$

and a discrete integer index n is replaced by a real number τ by $\tau = 2\pi n$. It is known that the distributions of function f_n and its Fourier spectral function F_m exhibit rather complementary characteristics, as one might remember of a relation between an impulsive signal and its broad spectrum. Accordingly, if a series $\sum_{n=-\infty}^{\infty} f_n$ converges slowly, then the corresponding transformed series $\sum_{m=-\infty}^{\infty} F_m$ converges fast. Rays and modes are found to be exactly this Fourier transform pairs relation [1]. In general, ray formulation is rather easy for the exterior scattering problem, while it may be difficult for analyzing the interior scattering such as waveguide and cavity structures, since one has to consider multiple reflected and diffracted rays due to the interior boundary of the structure. This ray–mode conversion technique was first applied to obtain the modal reflection coefficients at an open end of a parallel-plane waveguide [6] then successfully extended for analyzing scattering problems from open cavity/waveguide structures [3–8,11–16].

From the previous investigation [8], it is interesting to observe a fact that using the Poisson summation formula, one can recover a summation of the *exact* intrinsic waveguide modes, even one starts from a summation of rays that are derived from high-frequency asymptotic assumption.

More efficient hybrid representation may be derived from a partial Poisson summation formula [1]:

$$\sum_{n=n_1}^{n_2} f_n = \frac{1}{2} f_{n_1} + \frac{1}{2} f_{n_2} + \sum_{m=-\infty}^{\infty} F_m(n_1, n_2), \tag{7.3}$$

$$F_m(n_1, n_2) = \frac{1}{2\pi} \int_{\tau_1=2\pi n_1}^{\tau_2=2\pi n_2} f(\tau = 2\pi n) e^{im\tau} d\tau. \tag{7.4}$$

7.3 Modal excitation at the aperture

7.3.1 Formulation

Let us first discuss about waveguide modal excitation at an aperture of an open-ended parallel-plane waveguide, as shown in Figure 7.1. An E-polarized plane wave expressed as

$$u^i (= E_y^i) = e^{-ik(x \cos\theta_0 + z \sin\theta_0)} \tag{7.5}$$

illuminates an open-ended aperture of a parallel-plane waveguide with an incident angle θ_0. The width of the aperture is a, and k is the free-space wavenumber.

When an incident plane wave impinges on the edges of the upper aperture ($x = \pm a/2$, $z = 0$), edge-diffracted rays are generated. These edge-diffracted rays yield dominant fields in the upper half space, and modal fields in the waveguide. According

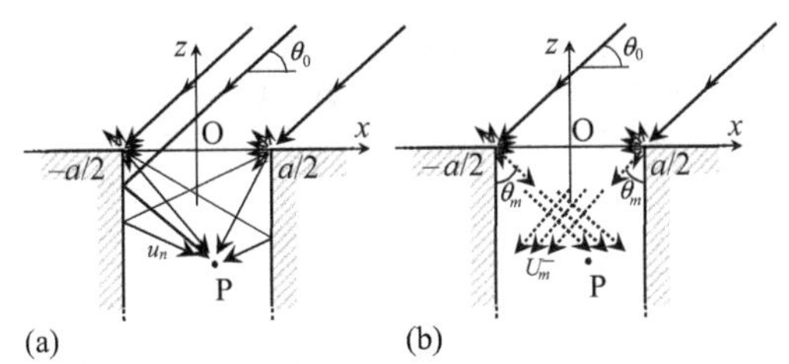

Figure 7.1 A plane wave coupling into an open-ended parallel-plane waveguide: (a) ray description and (b) waveguide modal description

to the geometrical theory of diffraction (GTD) [17], edge-diffracted ray u excited at $x = a/2$ can be written for the far field at the observation point $(\rho,\ \theta)$ as

$$u = C(k\rho)D_{-1}\left(\theta,\theta_0;\tfrac{3}{2}\pi\right)e^{-ika(\cos\theta+\cos\theta_0)/2}, \tag{7.6}$$

where $C(\chi)$ is the following asymptotic far-field expression for the 2D free-space Green function:

$$C(\chi) = (8\pi\chi)^{-1/2}e^{i(\chi+\pi/4)}, \tag{7.7}$$

and $D_\tau(\phi,\phi_0;\phi_w)$ is Keller's edge diffraction coefficient for a perfectly electric conducting wedge with wedge angle $(2\pi - \phi_w)$, with incident angle ϕ_0 and observation angle ϕ,

$$D_\tau(\phi,\phi_0;\phi_w) = \tfrac{2\pi}{\phi_w}\sin\tfrac{\pi^2}{\phi_w}\left\{\left(\cos\tfrac{\pi^2}{\phi_w} - \cos\tfrac{\phi-\phi_0}{\phi_w}\right)^{-1} + \tau\left(\cos\tfrac{\pi^2}{\phi_w} - \cos\tfrac{\phi+\phi_0}{\phi_w}\right)^{-1}\right\}, \tag{7.8}$$

where $\tau = +1\,(-1)$ for $H(E)$ polarization, respectively.

The field in the waveguide may be written as a summation of multiple bouncing reflected rays that are excited by edge diffraction at the aperture, as in Figure 7.1(a). Since there exists an infinite number of multiple bouncing rays between the waveguide walls, and these rays propagate without reflection loss at the walls, this infinite ray series exhibits very slow convergence property. On the other hand, a field description by an alternative waveguide modal summation in Figure 7.1(b), which is formally an infinite summation as well, can be truncated to a finite summation, since the higher order waveguide modes become eventually evanescent ones which rapidly decay as they propagate in the waveguide. In order to obtain the excitation coefficients of the waveguide modes, one may apply the Poisson summation formula in (7.3) where f_n represents edge-diffracted ray u_n with n-time internal reflections. Then the evaluation of the Fourier integral in (7.4) gives us an alternative waveguide modal field U_m. Detailed derivation may be found in [8].

The same results can intuitively be derived from an assumption that edge-diffracted wave excited at the aperture edge may be treated to be radiated by an anisotropic source at the edge location in the waveguide. A two-dimensional Green's function $G(x,z;x_0,z_0)$ for the parallel-plane waveguide with an isotropic line source at (x_0,z_0) satisfies the wave equation:

$$\left(\tfrac{\partial^2}{\partial x^2} + \tfrac{\partial^2}{\partial z^2} + k^2\right)G(x,z;x_0,z_0) = -\delta(x-x_0)\delta(z-z_0), \tag{7.9}$$

with an appropriate Dirichlet boundary condition at the waveguide walls:

$$G(x,z;x_0,z_0) = 0 \quad \text{at} \quad x = \pm\tfrac{a}{2}. \tag{7.10}$$

The solution may be written in terms of parallel-plane waveguide modes $U_m^{\pm}$ as for $z < z_0$ [18],

$$G(x, z; x_0, z_0) = \sum_{m=1}^{\infty} \tilde{A}_m(x_0, z_0) U_m^-(x, z), \tag{7.11}$$

$$U_m^{\pm}(x, z) = \sin \tfrac{m\pi}{a} \left(x + \tfrac{a}{2}\right) e^{\pm i\zeta_m z}, \tag{7.12}$$

$$\tilde{A}_m(x_0, z_0) = \tfrac{i}{\zeta_m a} \sin \tfrac{m\pi}{a} \left(x_0 + \tfrac{a}{2}\right) e^{i\zeta_m z_0} = \tilde{A}_m^+(x_0, z_0) + \tilde{A}_m^-(x_0, z_0), \tag{7.13}$$

$$\tilde{A}_m^{\pm}(x_0, z_0) = \tfrac{\pm 1}{2\zeta_m a} \exp\left\{\pm \tfrac{im\pi}{a} \left(x_0 + \tfrac{a}{2}\right) + i\zeta_m z_0\right\}, \tag{7.14}$$

where $\tilde{A}_m^{\pm}(x_0, z_0)$ is an excitation coefficient of waveguide mode U_m^- toward $\pm x$-direction, and

$$\zeta_m = \sqrt{k^2 - \left(\frac{m\pi}{a}\right)^2} = k \cos\theta_m \tag{7.15}$$

is the modal propagation constant along the z-direction with the modal propagation angle θ_m:

$$\theta_m = \sin^{-1}\left(\tfrac{m\pi}{ka}\right). \tag{7.16}$$

The previous results are obtained for an isotropic impulsive source but may be extended for anisotropic equivalent sources that radiate diffracted fields at the aperture edges ($x = \pm a/2$, $z = 0$). Then, the internal field can be written as in (7.11) but with a different modal excitation coefficient A_m:

$$\begin{aligned} A_m &= \tilde{A}_m^-\left(x_0 = \tfrac{a}{2}, z_0 = 0\right) D_{-1}\left(\tfrac{3}{2}\pi - \theta_m, \theta_0; \tfrac{3}{2}\pi\right) e^{-ika(\cos\theta_0)/2} \\ &\quad + \tilde{A}_m^+\left(x_0 = -\tfrac{a}{2}, z_0 = 0\right) D_{-1}\left(\theta_m, \theta_0 + \tfrac{\pi}{2}; \tfrac{3}{2}\pi\right) e^{ika(\cos\theta_0)/2} \\ &= \frac{1}{2\zeta_m a} \left\{(-1)^m D_{-1}\left(\tfrac{3}{2}\pi - \theta_m, \theta_0, \tfrac{3}{2}\pi\right) e^{-ika(\cos\theta_0)/2} \right. \\ &\quad \left. - D_{-1}\left(\theta_m, \theta_0 + \tfrac{\pi}{2}, \tfrac{3}{2}\pi\right) e^{ika(\cos\theta_0)/2}\right\}. \end{aligned} \tag{7.17}$$

This coefficient A_m is valid for propagating modes with $m < ka/\pi$. It is noted that each diffraction coefficient in (7.17) diverges when $\theta_0 = \pi/2 \pm \theta_m$, but the sum of two terms becomes a finite value. Accordingly, this excitation coefficient A_m in (7.17) is valid for all incident angles. At $m = ka/\pi$, the mode U_m becomes cutoff and one obtains $\zeta_m = 0$. Then the coefficient A_m in (7.17) diverges. Accordingly, our solution fails to predict the correct field when the aperture width ka becomes close to a multiple of π. On the other hand, for $m > ka/\pi$, the propagation constant ζ_m in (7.15) along the z-direction becomes purely imaginary. Then the mode U_m^- becomes evanescent and decays exponentially as it propagates along the z-direction. Accordingly, the modal summation in (7.11) may be truncated at the final propagating mode U_M^-.

It is, however, sometimes necessary to include the evanescent modes. In order to find the evanescent modal excitation coefficient by the ray–mode conversion method, one needs to define the complex modal propagation angle $\hat{\theta}_m$ via analytic continuation into the complex angular domain. This is possible by defining the complex modal propagation angle $\hat{\theta}_m$ as

$$\hat{\theta}_m = \frac{\pi}{2} - i\cosh^{-1}\left(\tfrac{m\pi}{ka}\right). \tag{7.18}$$

It has already been found that the thus-derived excitation coefficient A_m in (7.17) is also valid for evanescent modes except near the cutoff frequency [19].

Multiple edge diffractions between the aperture edges occur, and this effect *cannot* be ignored when the slit aperture becomes narrower. The effect of these multiple edge-diffracted rays can also be formulated by the GTD [16], and the following terms may be added to (7.17):

$$\begin{aligned}\bar{A}_m = {}& \frac{1}{2\zeta_m a}\left\{(-1)^m D_{-1}\left(\tfrac{\pi}{2}, \theta_0 + \tfrac{\pi}{2}; \tfrac{3}{2}\pi\right) D_{-1}\left(\tfrac{3}{2}\pi - \theta_m, \pi; \tfrac{3}{2}\pi\right) e^{ika(\cos\theta_0)/2}\right.\\ & \left. - D_{-1}\left(\pi, \theta_0; \tfrac{3}{2}\pi\right) D_{-1}\left(\theta_m, \tfrac{\pi}{2}; \tfrac{3}{2}\pi\right) e^{-ika(\cos\theta_0)/2}\right\}\\ & \times \sum_{s=1}^{\infty}\left(-\tfrac{1}{2}\right)^{2(s-1)} C((2s-1)ka)\\ & + \frac{1}{2\zeta_m a}\left\{(-1)^m D_{-1}\left(\pi, \theta_0; \tfrac{3}{2}\pi\right) D_{-1}\left(\tfrac{3}{2}\pi - \theta_m, \pi; \tfrac{3}{2}\pi\right) e^{-ika(\cos\theta_0)/2}\right.\\ & \left. - D_{-1}\left(\tfrac{\pi}{2}, \theta_0 + \tfrac{\pi}{2}; \tfrac{3}{2}\pi\right) D_{-1}\left(\theta_m, \tfrac{\pi}{2}; \tfrac{3}{2}\pi\right) e^{ika(\cos\theta_0)/2}\right\}\\ & \times \sum_{t=1}^{\infty}\left(-\tfrac{1}{2}\right)^{2t-1} C(2tka).\end{aligned} \tag{7.19}$$

7.3.2 Numerical results

In order to show the validity of the formulation, some numerical results are shown here. Figure 7.2 shows the frequency change of the excitation coefficient A_m of the parallel-plane waveguide mode U_m^- in (7.12) for $\theta = 50.0°$. Results by (7.17) are plotted as sGTD, and those with multiple edge-diffracted rays in (7.19) are plotted as mGTD. As a reference, the figure also includes results obtained from the Kobayashi Potential (KP) method [20], which is an analytical eigenfunction expansion method utilizing Weber–Schafheitlin discontinuous integrals. One observes that the results (sGTD) agree pretty well with the reference values even below the cut-off frequency, although multiple diffraction effect is not included. If one adds multiple edge interaction terms in (7.19), our solutions (mGTD) coincide with the reference solutions except near the cut-off frequencies.

Figure 7.3 shows angular change of excitation coefficients. We have shown here two aperture widths, $ka = 7.00$ in Figure 7.3(a) and (b), and $ka = 30.0$ in Figure 7.3(c), respectively. One observes that the excitation has a peak near the

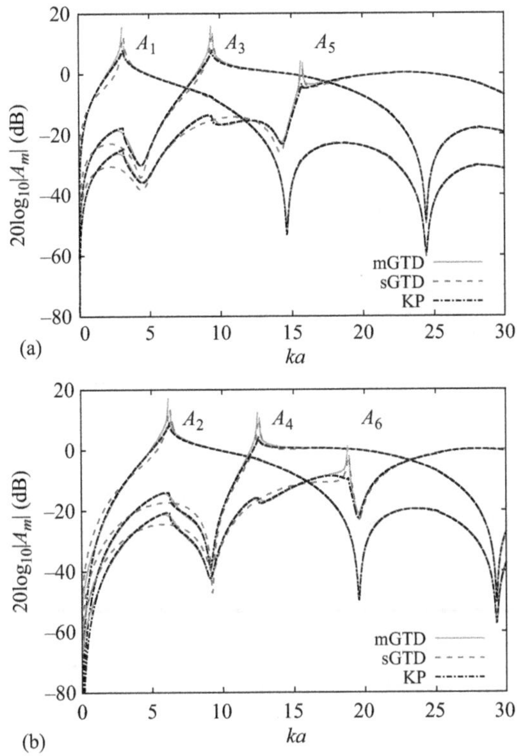

Figure 7.2 Frequency change of excitation coefficient A_m for the parallel-plane waveguide modes U_m^-. $\theta_0 = 50.0°$. (a) Odd modes ($m = 1,3,5$) and (b) even modes ($m = 2,4,6$)

modal propagation angle θ_m in (7.16) of each waveguide modes. Namely, if a plane wave is incident from angle θ_0 into an open-ended waveguide, then all modes are excited, but those whose modal propagation angle θ_m is near $\pi/2 \pm \theta_0$ will be excited strongly (see Figure 7.1(b)). This phenomena is closely related with the incident and reflection shadow boundaries of the edge diffraction. Accordingly, one may be able to predict the internal waveguide field by these modes. These modes are sometimes called *significant* modes [2–5]. Since our solution is based on the high-frequency asymptotic approximation, the result in Figure 7.3(c) for the wide aperture case matches better than the narrow case in Figure 7.3(a) and (b), as one expects. One may also notice that the excitation coefficient by sGTD does not predict well as the incident angle θ_0 approaches zero. Excitation coefficients A_m ($m = 3, 4, 5$) in Figure 7.3(b) are those for evanescent modes and are derived by extending a real modal propagation angle θ_m into a complex modal propagation angle $\hat{\theta}_m$. It is remarkable to observe that the GTD diffraction coefficient is still valid for such complex angles.

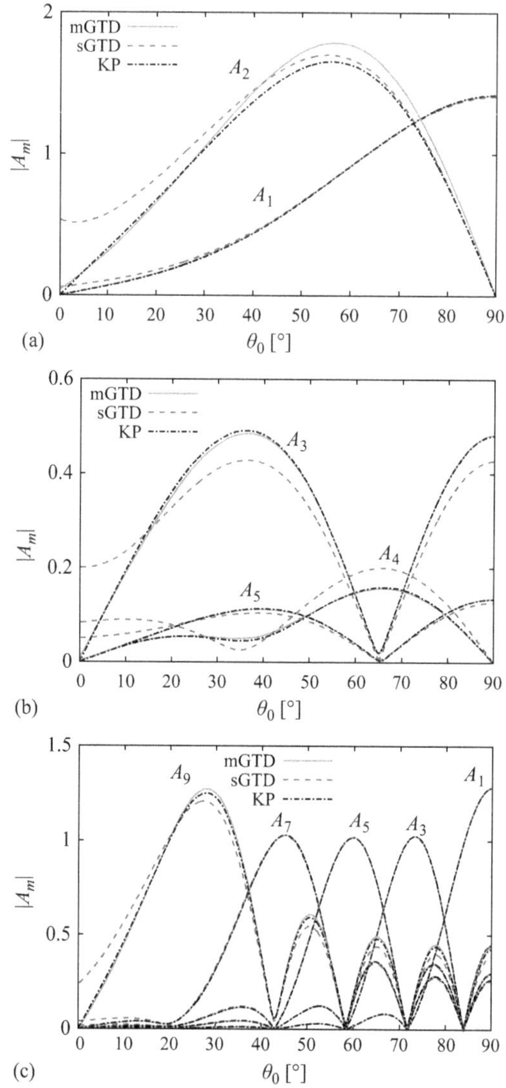

Figure 7.3 Angular change of excitation coefficients A_m for the parallel-plane waveguide modes U_m^-: (a) $ka = 7.00$, propagating modes $m = 1,2$, (b) $ka = 7.00$, evanescent modes $m = 3,4,5$, (c) $ka = 30.0$, $m = 1,3,5,7,9$

While this formulation results seem to be sufficient for the later application, better ray optical solution may be found by applying an elaborate uniform asymptotic theory [21] or uniform geometrical theory of diffraction [22].

7.4 Diffraction by a slit on a thick conducting screen

7.4.1 Background

Let us now apply the hybrid ray–mode technique to a plane wave diffraction by a slit on a thick conducting screen. This problem is a canonical problem to estimate a field through an aperture of a thick screen and may be directly connected with outdoor–indoor wireless communication through building walls and windows. As the frequency increases, electromagnetic waves decay more rapidly as they pass through concrete walls. Therefore, the windows on building walls can be considered to be primary gates for such transmitting waves, and a reliable means is required for estimating the reflection/transmission property through a window that is sufficiently large compared with the wavelength.

Diffraction by a slit is a classical problem in electromagnetic diffraction analysis that has been studied by many authors [23–25]. For a slit on an infinitely thin conducting screen, an eigenfunction expansion solution in terms of Mathieu functions [23] or the use of the KP method utilizing Weber–Schafheitlin discontinuous integrals [24] may be possible. These results can be useful references for small apertures but may exhibit a series convergence problem for large aperture cases.

The diffraction by a wide slit on an infinitely thin conducting screen [25] is easy to formulate, since one only needs to consider two edges for the diffraction, and the GTD may be a powerful tool for such analysis. On the other hand, the diffraction by a thick slit is rather difficult to solve, although it may be analyzed by the KP method [27,28], the Wiener–Hopf [29] and generalized matrix techniques [30,31], integral equation approaches [32,33], and Fourier transform techniques [34,35]. These previously published studies have mainly dealt with relatively narrow apertures, and few detailed numerical results for scattering patterns have been given for wide apertures.

Since the slit region in the thick conducting screen can be considered as an open-ended parallel-plane waveguide cavity, one may be able to apply the ray–mode conversion technique to formulate the field in the waveguide by its eigenmodes, and the scattering field can be calculated from the multiply reflected modal reradiation fields.

In the following discussion, we first formulate the diffracted field using the edge-diffracted rays. Then the ray–mode conversion method is utilized to obtain the modal excitation and reflection coefficients at the slit apertures. Thus-obtained coefficients are combined through a matrix formulation to synthesize the modal reradiation from the aperture. Thus, derived formulation can be easily extended to an infinitely thin screen case, and a loaded slit case, as well as a plane wave diffraction by a trough on a ground.

7.4.2 Formulation

Let us now consider that an E-polarized electromagnetic plane wave in (7.5) illuminates a slit on a thick conducting screen, as depicted in Figure 7.4(a). Here, the

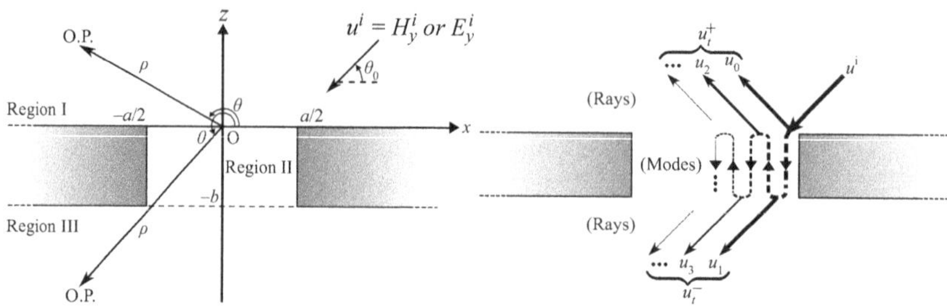

Figure 7.4 A plane wave diffraction by a slit on a thick conducting screen: (a) geometry of the problem and (b) successive excitation of diffracted rays and waveguide modes

width and the thickness of the slit are assumed to be a and b, respectively. For convenience, the entire region is divided into three regions: (I) semi-infinite upper half space ($z > 0$), (II) the slit region ($-b < z < 0$), and (III) the lower half space ($z < -b$).

The slit structure may be considered as an open-ended parallel-plane waveguide cavity excited from the outer Region I. One observes the reflected plane wave:

$$u^r = -e^{-ik(x\cos\theta_0 - z\sin\theta_0)} \tag{7.20}$$

in Region I and edge-diffracted rays excited at the aperture edges. While the reflected plane wave contributes dominantly in Region I, our interest here is to find the diffraction effect from the slit geometry. Accordingly, we shall omit the contribution by the reflected wave for later discussion. Let us now treat each diffraction component separately.

7.4.2.1 Edge-diffracted rays

When an incident plane wave in (7.5) impinges on the edges of the upper aperture ($x = \pm a/2$, $z = 0$), edge-diffracted rays are generated. These primary edge-diffracted rays u_0 give a dominant field in Region I and a modal excitation in Region II. According to the GTD [17], edge-diffracted rays u_0 can be written for the far field at the observation point (ρ, θ) in Region I as

$$u_0 = u_0^+ + u_0^-, \tag{7.21}$$

$$u_0^+ = C(k\rho)D_{-1}\left(\theta, \theta_0; \tfrac{3}{2}\pi\right) e^{-ika(\cos\theta + \cos\theta_0)/2}, \tag{7.22}$$

$$u_0^- = C(k\rho)D_{-1}\left(\theta + \tfrac{\pi}{2}, \theta_0 + \tfrac{\pi}{2}; \tfrac{3}{2}\pi\right) e^{ika(\cos\theta + \cos\theta_0)/2}. \tag{7.23}$$

An individual edge-diffracted ray, u_0^+ or u_0^-, diverges at the reflection shadow boundary direction $\theta = \pi - \theta_0$. However, their combination $u_0 (= u_0^+ + u_0^-)$ becomes a finite value owing to the cancellation of each diverging feature.

The effect of multiple edge-diffracted rays can also be formulated by the GTD [16], and the following terms:

$$
\begin{aligned}
&C(k\rho)\big\{D_{-1}\left(\pi, \theta_0; \tfrac{3}{2}\pi\right) D_{-1}\left(\tfrac{\pi}{2}+\theta, \tfrac{\pi}{2}; \tfrac{3}{2}\pi\right) e^{ika(\cos\theta-\cos\theta_0)/2} \\
&\quad + D_{-1}\left(\tfrac{\pi}{2}, \tfrac{\pi}{2}+\theta_0; \tfrac{3}{2}\pi\right) D_{-1}\left(\theta, \pi; \tfrac{3}{2}\pi\right) e^{-ika(\cos\theta-\cos\theta_0)/2}\big\} \\
&\quad \times \sum_{s=1}^{\infty} \left(-\tfrac{1}{2}\right)^{2(s-1)} C((2s-1)ka) \\
&\quad + C(k\rho)\big\{D_{-1}\left(\pi, \theta_0; \tfrac{3}{2}\pi\right) D_{-1}\left(\theta, \pi; \tfrac{3}{2}\pi\right) e^{-ika(\cos\theta+\cos\theta_0)/2} \\
&\quad + D_{-1}\left(\tfrac{\pi}{2}, \tfrac{\pi}{2}+\theta_0; \tfrac{3}{2}\pi\right) D_{-1}\left(\tfrac{\pi}{2}+\theta, \tfrac{\pi}{2}; \tfrac{3}{2}\pi\right) e^{ika(\cos\theta+\cos\theta_0)/2}\big\} \\
&\quad \times \sum_{t=1}^{\infty} \left(-\tfrac{1}{2}\right)^{2t-1} C(2tka)
\end{aligned}
\tag{7.24}
$$

may be added to (7.21).

7.4.2.2 Modal excitation

As depicted in Figure 7.4(b), a part of the primary edge-diffracted waves also propagates into the slit Region II and reradiates after several internal reflections and multiple diffractions. Since there exists an infinite number of multiple bouncing rays between the waveguide walls, it would be convenient to use the modal description for the field inside the slit, as discussed in Section 7.3. According to the ray–mode conversion [13,15], the parallel-plane waveguide modes U_m^- propagating along negative z-direction are excited by the primary edge-diffracted waves. The initial ray–mode conversion is performed at the upper end of the aperture ($x = \pm a/2$, $z = 0$) by extending the slit depth b to infinity. Then the scenario is exactly the same as the modal excitation in Section 7.3, and the modal summation $\dot{u}$ is given by

$$
\dot{u} = \sum_{m=1}^{\infty} A_m U_m^-, \tag{7.25}
$$

where A_m denotes the modal excitation coefficient for mode U_m^- and is exactly the same as in (7.17) by primary edge-diffracted wave, plus (7.19), if the multiple edge interaction is considered.

7.4.2.3 Modal reradiation and reflection coupling

A mode U_m^- propagates toward the lower open end $z = -b$, at which modal radiation and reflection occur, as shown in Figure 7.4(b). The modal radiation field u_1 in Region III can be formed by collecting all modal radiation fields generated by modes U_m^- at the lower aperture. At the same time, modal reflection occurs to generate waveguide modes U_p^+ propagating along the positive z-direction with new excitation coefficients. Note that modal coupling exists between different waveguide modes. Such modal reradiation and coupling can also be obtained by the high-frequency asymptotic method [15]. The complex propagation angle $\hat{\theta}_m$ in (7.18) enables us to add the contribution from the non-propagating evanescent modes, which may play an important role in the case of a narrow or a thin slit.

7.4.2.4 Total diffracted field

The successive process of modal radiation and reflection/coupling continues to generate modal radiation fields u_{2n} in Region I ($z > 0$) and u_{2n+1} in Region III ($z < -b$). This modal radiation continues until all the energy of the bouncing waveguide modes is dissipated. While the field description of the modal reradiation may be complicated as the modal coupling occurs at every reflection process, it can be written in a compact matrix form. The total of the radiation fields u_t in Region I can be given as

$$\begin{aligned} u_t^+ &= u_0 + \sum_{n=1}^{\infty} u_{2n} = u_0 + C(k\rho)[R^+] \sum_{n=0}^{\infty} [B]^{2n+1}[A] \\ &= u_0 + C(k\rho)[R^+] \left[[I] - [B]^2 \right]^{-1} [B][A], \end{aligned} \tag{7.26}$$

and in Region III as

$$u_t^- = \sum_{n=0}^{\infty} u_{2n+1} = C(k\rho)[R^-] \sum_{n=0}^{\infty} [B]^{2n}[A] = C(k\rho)[R^-] \left[[I] - [B]^2 \right]^{-1} [A]. \tag{7.27}$$

In the previous equations, $[R^{\pm}]$ denotes the modal radiation row vector at the upper (+) and lower (−) aperture edges, and $[A]$ is the modal excitation column vector due to the primary edge diffraction. Also, matrix $[B]$ is the modal coupling matrix at each open end and $[I]$ is a unit matrix. The components of the previous matrices can be found in [16]. For these components, the effect of multiple edge diffraction between the upper and lower aperture edges can be included. Thus, the sum of the multiply reflected modes at the top and bottom open ends has a closed-form solution, as in (7.26) and (7.27), and one does not need to perform an actual summation, and the formulation should be valid, essentially regardless of the thickness b.

Figure 7.5 shows the far field diffraction pattern for different aperture widths. Here, we have shown two aperture widths, $ka = 7.00$ in Figure 7.5(a) and $ka = 30.0$ in Figure 7.5(b), respectively. A common radiation factor $C(k\rho)$ in the formulation is omitted here. For Figure 7.5, there are only two propagation modes, while nine modes for Figure 7.5(b). Results with evanescent modal contribution and multiple diffracted rays are plotted as emGTD, and those without the evanescent modes are plotted as mGTD. Results by the primary edge diffraction effect are also plotted as sGTD. As a reference, the figure also includes results obtained from the KP method [28]. One observes that our present results agree pretty well with the reference value, even multiple diffraction effect is not included. As one can expect, the result in Figure 7.5(b) for the wide slit aperture case matches better than the narrow case in Figure 7.5(a).

It should be mentioned that each modal coupling element can be derived directly from the edge diffraction without solving the matrix equations to match the boundary

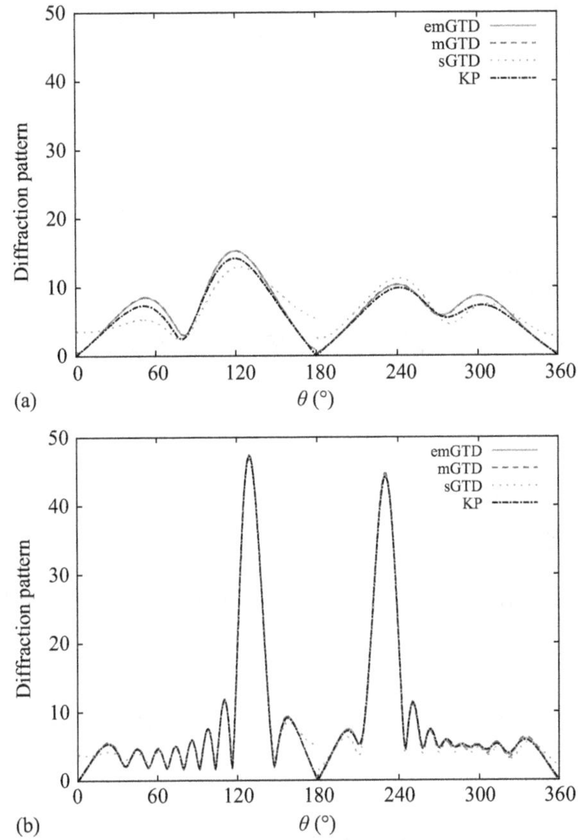

Figure 7.5 Comparison of far-field diffraction patterns (E polarization case). $\theta_0 = 50.0°$, $kb = 2.00$. —: Present GTD results with multiple edge-diffracted waves and evanescent modes (emGTD); – – –: GTD results without the evanescent modes (mGTD); · · · · · · ·: GTD results only by the primary edge diffraction effect (sGTD); – · – · – · –: KP method [28]. (a) $ka = 7.00$ and (b) $ka = 30.0$

condition. Accordingly, the calculation is very fast even for the case of a large aperture. One may also reduce the computational time by not using all the modes but selecting only the significant modes [2–4].

A similar formulation is possible for H polarized plane wave incidence case [15], and the corresponding results are shown in Figure 7.6. A main feature of the diffraction pattern is the same as for E polarization. Better agreement can be found, at the screen boundary directions: $\theta = 0°,\ 180°,\ 360°$, if one includes the multiple edge diffraction and the evanescent modal effects.

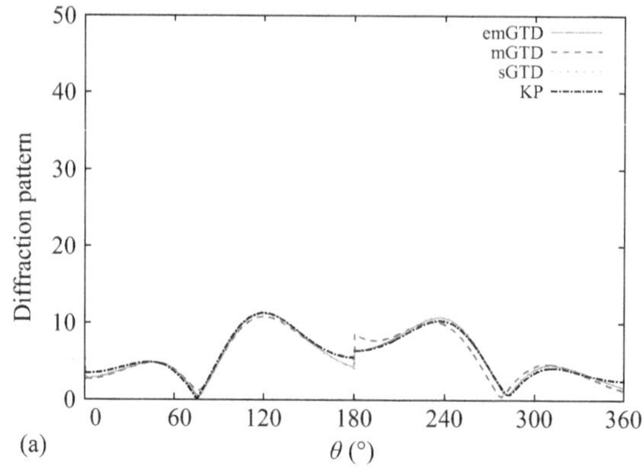

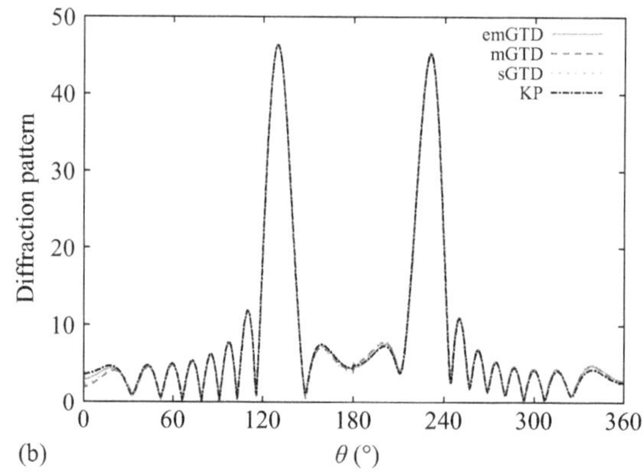

Figure 7.6 Comparison of far-field diffraction patterns (H polarization case). Parameters are the same as in Figure 7.5. (a) ka = 7.00 (b) ka = 30.0

7.4.3 Diffraction by a thin slit

While the present formulation is based on a thick slit geometry with the slit aperture composed of four right-angle conducting wedges, it may be interesting to take the limit of $b \to 0$ to investigate the case of an infinitely thin slit. For this case, the diffraction pattern is known to be symmetric with respect to the screen [25], and the GTD formulation is rather easy, since there are only two edges to consider and no modal excitation is involved.

Figure 7.7 shows comparison between our limiting case (emGTD, mGTD), the result for a slit on an infinitely thin screen (thinGTD) [25], and the result by the KP method [27]. While our formulation gives us a different edge condition, our limiting results are in good agreement with those for an infinitely thin slit. An evanescent modal field is expected to play an important role for this limiting case.

7.4.4 Diffraction by a thick and loaded slit

Formulation can be extended to a case for which a discontinuity exists in the waveguide. Figure 7.8 shows a case for which a material load of the relative permittivity

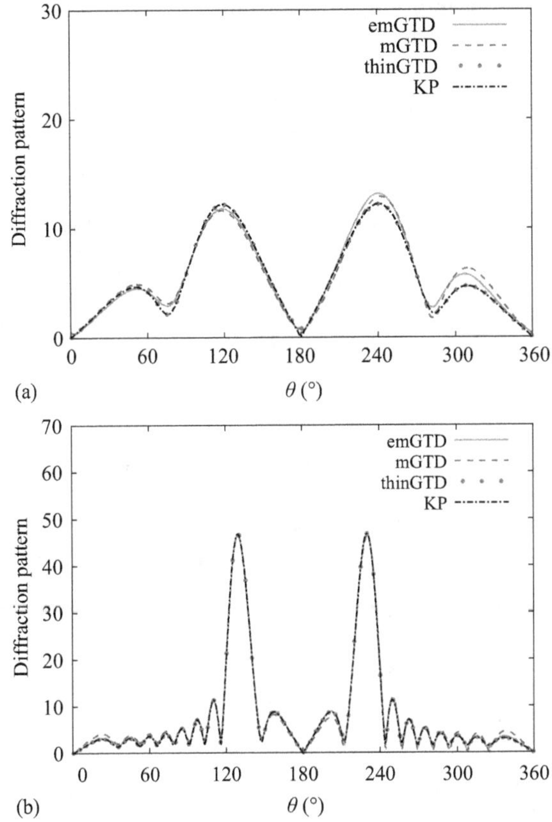

Figure 7.7 *Far-field diffraction patterns from a slit in an infinitely thin conducting screen (E polarization case). $kb \to 0$, $\theta_0 = 50.0°$. (a) $ka = 7.00$ and (b) $ka = 30.0$*

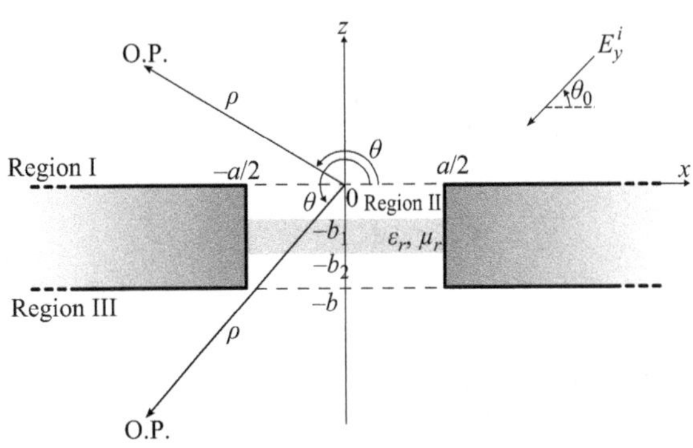

Figure 7.8 *A plane wave diffraction from a loaded slit in a thick conducting screen*

ε_r and the relative permeability μ_r is inserted in a slit on a thick conducting screen. For this case, there are modal reflection and transmission, but no modal coupling at the material interface at $z = -b_1, \ -b_2$. Accordingly, one has to modify the matrix formulation in (7.26) and (7.27) with additional reflection and transmission terms.

Figure 7.9 shows far-field diffraction pattern from a dielectric-loaded slit. This configuration is made for simulating a diffraction effect by a window glass in a

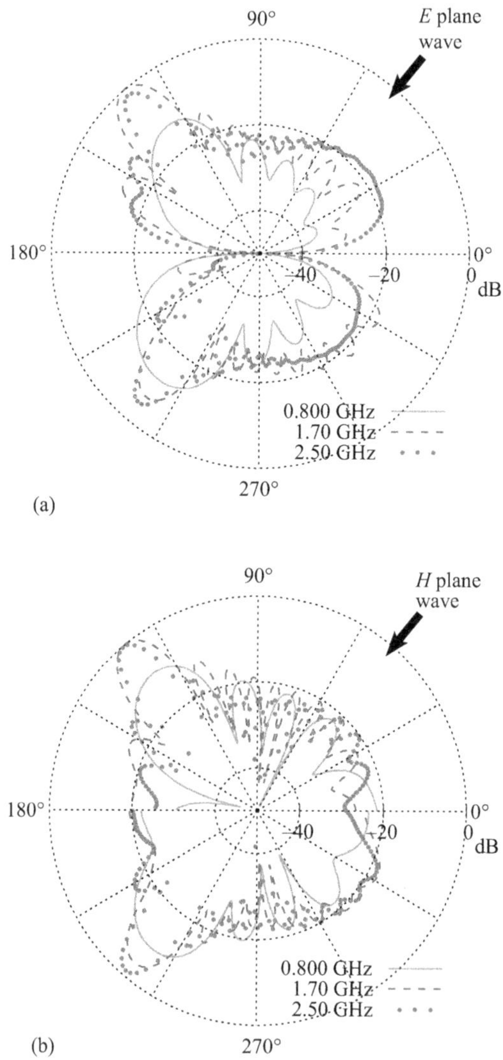

Figure 7.9 Far-field diffraction patterns from a dielectric-loaded slit. $a = 1{,}000$ mm, $b = 150$ mm, $b_2 - b_1 = 5.00$ mm, $\theta_0 = 50.0°$, $\varepsilon_r = 7.50$, $\mu_r = 1.00$. ——: $f = 0.800$ GHz; - - - -: $f = 1.70$ GHz; • • •: $f = 2.50$ GHz. (a) E polarization case and (b) H polarization case

conducting building wall. Here window width a is 1,000 mm, wall thickness b is 150 mm, glass thickness $b_2 - b_1$ is 5.00 mm, glass material constant (ε_r, μ_r) is (7.50, 1.00), and incident angle θ_0 is 50.0°, respectively. Three frequencies, 0.800, 1.70, and 2.50 GHz, are chosen as communication channels for cellular mobile phones. As the frequency increases, diffraction beams become sharp, since the window aperture becomes relatively wide. While calculation is made for E polarization in Figure 7.9(a), and for H polarization in Figure 7.9(b), both results have almost the same main diffraction lobes, and a difference arises at the slit boundary directions owing to the difference in the boundary condition.

7.4.5 *Diffraction by a trough*

A plane wave diffraction by a trough on a ground can similarly be formulated by setting a conducting wall in the waveguide region in Figure 7.4. For this instance, we have a diffraction field in the upper half space $z > 0$ only. While a direct formulation for the trough is possible [13], it may be interesting to obtain a comparable result from the case in which a very lossy material is loaded inside a thick slit as in Figure 7.10. Figure 7.11 shows the diffraction patterns calculated by the formulation for a trough on

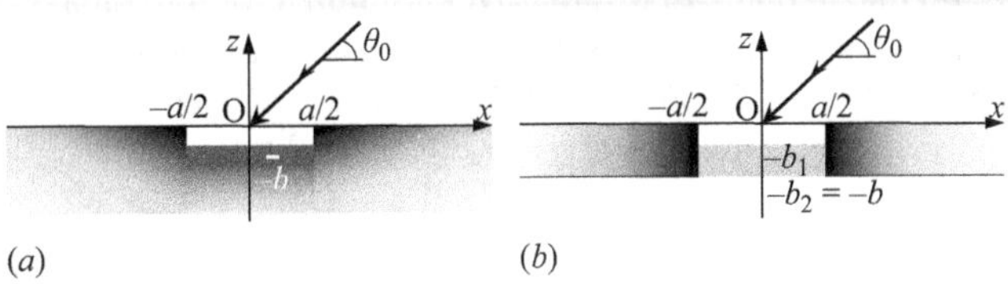

Figure 7.10 A plane wave diffraction by a trough: (a) direct trough formulation and (b) a thick slit with a conducting wall in the slit

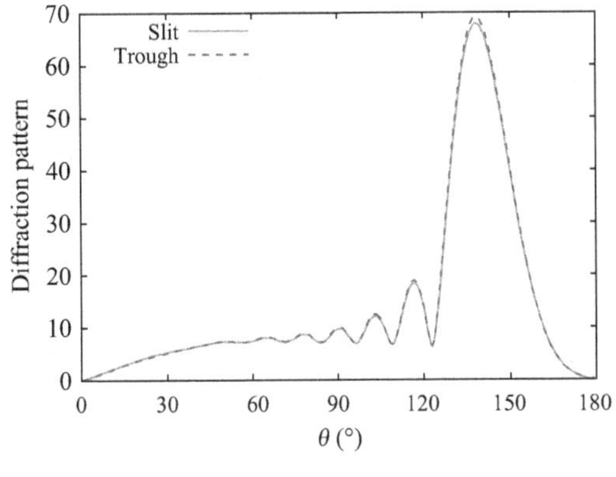

Figure 7.11 Comparison of far-field diffraction patterns from a trough (width a, depth $\bar{b}$) and a lossy loaded slit (width a, depth b, lossy load $\varepsilon_r = 15.0 + i9{,}500$, $\mu_r = 1.00$). $ka = 30.0$, $k\bar{b} = kb_1 = 2.00$, $kb_2 = kb = 10.0$, $\theta_0 = 40.0°$

a ground, and a thick slit partially filled with a lossy ($\varepsilon_r = 15.0 + i9,500$, $\mu_r = 1.00$) material. Two results compare well for all observation angles.

7.5 Conclusions

In this chapter, the ray–mode conversion technique that can be a useful tool for analyzing the high-frequency scattering by open waveguide/cavity structures is discussed. As an example, a modal excitation at an aperture of an open-ended parallel-plane waveguide has been formulated and the result has been applied to a plane wave diffraction by a slit in a thick conducting screen. By including the contribution from the multiple edge diffraction effect and evanescent modes, our results match well with the reference result. By extending the ray–mode conversion method, one may be able to solve more complicated high-frequency scattering problems.

Acknowledgments

The author would like to thank to Mr. Masayuki Shimizu for providing the numerical calculations for the figures. A part of this work has been supported by a Scientific Research Grant-In-Aide (15K06083, 2015) from Japan Society for Promotion of Science, and 2017 Chuo University Overseas Research Program.

References

[1] Felsen L.B. 'Progressing and oscillatory waves for hybrid synthesis of source excited propagation and diffraction'. *IEEE Trans. Antennas Propag.* 1984, vol. 32(8), pp. 775–96.

[2] Pathak P.H. and Burkholder R.J. 'Modal, ray, and beam techniques for analyzing the EM scattering by open-ended waveguide cavities'. *IEEE Trans. Antennas Propag.* 1989, vol. 37(5), pp. 635–47.

[3] Shirai H. and Felsen L.B. 'Rays, modes and beams for plane wave coupling into a wide open-ended parallel plane waveguide'. *Wave Motion.* 1987, vol. 9(4), pp. 301–17.

[4] Altintas A., Pathak P.H., and Liang M.C. 'A selective modal scheme for the analysis of EM coupling into or radiation from large open-ended waveguides'. *IEEE Trans. Antennas Propag.* 1988, vol. 36(1), pp. 84–96.

[5] Pathak P.H. and Altintas A. 'An efficient high-frequency analysis of modal reflection and transmission coefficients for a class of waveguide discontinuities'. *Radio Sci.* 1988, vol. 23(6), pp. 1107–19.

[6] Yee H.Y., Felsen L.B., and Keller J.B. 'Ray theory of reflection from the open end of a waveguide'. *SIAM J. Appl. Math.* 1968, vol. 16, pp. 268–300.

[7] Felsen L.B. and Ishihara T. 'Hybrid ray-mode formulation of ducted propagation'. *J. Acoust. Soc. Am.* 1979, vol. 65(3), pp. 595–607.

[8] Felsen L.B. and Kamel A. 'Hybrid ray-mode formulation of parallel plate waveguide Green's functions'. *IEEE Trans. Antennas Propag.* 1981, vol. 29(7), pp. 637–49.

[9] Heyman E. and Felsen L.B. 'Creeping waves and resonances in transient scattering by smooth convex objects'. *IEEE Trans. Antennas Propag.* 1983, vol. 31(3), pp. 426–37.

[10] Shirai H. and Felsen L.B. 'Wavefront and resonance analysis of scattering by a perfectly conducting flat strip'. *IEEE Trans. Antennas Propag.* 1986, vol. 34(10), pp. 1196–207.

[11] Shirai H. and Felsen L.B. 'Rays and modes for plane wave coupling into a large open-ended circular waveguide'. *Wave Motion.* 1987, vol. 9(7), pp. 461–82.

[12] Shirai H. and Hirayama K. 'Ray mode coupling analysis of plane wave scattering by a trough'. *IEICE Trans. Commun.* 1993, vol. E76-B(12), pp. 1558–63.

[13] Shirai H. 'Ray mode coupling analysis of EM wave scattering by a partially filled trough'. *J. Electromagn. Waves Appl.* 1994, vol. 8, pp. 1443–64.

[14] Shirai H., Watanabe K., Hasegawa N., and Sekiguchi H. 'Plane wave scattering with multiple diffraction by finite parallel plate waveguide cavities'. *IEICE Trans. Electron.* 1999, vol. J82-C-I(11), pp. 625–32 (in Japanese).

[15] Shirai H. and Sato R. 'High frequency ray-mode coupling analysis of plane wave diffraction by a wide and thick slit on a conducting screen'. *IEICE Trans. Electron.* 2012, vol. E95-C(1), pp. 10–5.

[16] Shirai H., Shimizu M., and Sato R. 'Hybrid ray-mode analysis of E-polarized plane wave diffraction by a thick slit'. *IEEE Trans. Antennas Propag.* 2016, vol. 64(11), pp. 4828–35.

[17] Keller J.B. 'Geometrical theory of diffraction'. *J. Opt. Soc. Am.* 1962, vol. 52(2), pp. 116–30.

[18] Felsen L.B. and Marcuvitz N. *Radiation and Scattering of Waves*. Englewood Cliffs, NJ: Prentice-Hall; 1973. p. 924

[19] Shirai H., Matsuda Y., and Sato R. 'GTD analysis for evanescent modal excitation'. *IEICE Trans. Electron.* 1997, vol. E80-C(1), pp. 190–2.

[20] Hongo K. and Ogawa Y. 'Receiving characteristics of a flanged parallel-plate waveguide'. *IEEE Trans. Antennas Propag.* 1977, vol. 25(3), p. 424.

[21] Boersma J. 'Ray optical analysis of reflection in an open-ended parallel plane waveguide. I: TM case'. *SIAM J. Appl. Math.* 1975, vol. 29, pp. 164–95.

[22] Kouyoumjian R.G. and Pathak P.H. 'A uniform geometrical theory of diffraction for an edge in a perfectly conducting surface'. *Proc. IEEE.* 1974, vol. 62(11), pp. 1448–61.

[23] Morse P.M. and Rubenstein P.J. 'The diffraction of waves by slits'. *Phys. Rev.* 1938, vol. 54(11), pp. 895–8.

[24] Nomura Y. and Katsura S. 'Diffraction of electromagnetic waves by ribbon and slit. I'. *J. Phys. Soc. Jpn.* 1957, vol. 12, pp. 190–200.

[25] Karp S.N. and Russek A. 'Diffraction by a wide slit'. *J. Appl. Phys.* 1956, vol. 27(8), pp. 886–94.

[26] Bowman J.J., Senior T.B.A., and Uslenghi P.L.E (eds.). *Electromagnetic and Acoustic Scattering by Simple Shapes*. New York, NY: Hemisphere Publishing; 1969. p. 728.

[27] Hongo K. 'Diffraction of electromagnetic plane wave by infinite slit perforated in a conducting screen with finite thickness'. *Trans. IECE.* 1971, vol. 54-B(7), pp. 419–25 (in Japanese).

[28] Hongo K. and Ishii G. 'Diffraction of an electromagnetic plane wave by a thick slit'. *IEEE Trans. Antennas Propag.* 1978, vol. 26(3), pp. 494–9.

[29] Aoki K. and Tanaka Y. 'Diffraction of a electromagnetic wave by a slit in thick screen'. *Tech. Rep. Kyushu Univ.* 1983, vol. 56(6), pp. 811–6 (in Japanese).

[30] Kashyap S.C. and Hamid M.A.K. 'Diffraction characteristics of a slit in a thick conducting screen'. *IEEE Trans. Antennas Propag.* 1971, vol. 19(4), pp. 499–507.

[31] Kashyap S.C., Hamid M.A.K., and Mostowy N.J. 'Diffraction pattern of a slit in a thick conducting screen'. *J. Appl. Phys.* 1971, vol. 42, pp. 894–5.

[32] Morita N. 'Diffraction of electromagnetic wave by a two-dimensional aperture with arbitrary cross section shape'. *Trans. IECE.* 1971, vol. 54-B(5), pp. 218–22 (in Japanese).

[33] Neerhoff F.L. and Mur G. 'Diffraction of a plane electromagnetic wave by a slit in a thick screen placed between two different media'. *Appl. Sci. Res.* 1973, vol. 28, pp. 73–88.

[34] Kang S.H., Eom H.J., and Park T.J. 'TM scattering from a slit in a thick conducting screen: Revisited'. *IEEE Trans. Microwave Theory Tech.* 1993, vol. 41(5), pp. 895–9.

[35] Park T.J., Kang S.H., and Eom H.J. 'TE scattering from a slit in a thick conducting screen: Revisited'. *IEEE Trans. Antennas Propag.* 1994, vol. 42(1), pp. 112–4.

Chapter 8

Scattering and diffraction of scalar and electromagnetic waves using spherical-multipole analysis and uniform complex-source beams

Ludger Klinkenbusch[1] *and Hendrik Brüns*[1]

The chapter deals with a spherical-multipole expansion of scalar or electromagnetic fields in the presence of a semi-infinite elliptic cone and a uniform complex-source beam (CSB) as the incident field. The analysis is performed in sphero-conal coordinates that can be understood as generalized spherical coordinates. The corresponding coordinate surfaces include the elliptic cone, the plane angular sector, and the wedge. As the uniform CSB paraxially approximates a Gaussian beam, its waist represents a localized inhomogeneous plane wave. Hence, it is possible to exclusively illuminate a desired area of the structure—for instance the tip of the cone—by a localized plane wave. Differently from using a homogeneous plane wave as the incident field in that case the resulting spherical-multipole series converges even if the scattered far field is evaluated. Therefore, the proposed technique should allow to extract diffraction and scattering characteristics of any desired part of the scattering objects without applying series transformations as it has been found to be necessary in the case of a homogeneous plane wave.

8.1 Introduction

Spherical-multipole analysis belongs to the classical methods to analytically solve three-dimensional scalar and electromagnetic problems for a certain class of canonical geometries. With sphero-conal coordinates which can be understood as generalized spherical coordinates that class of canonical geometries widens and includes the sphere, the right-circular cone, the elliptic cone, the wedge, the half plane, and the plane angular sector. Usually in scattering and diffraction, an incident plane wave has been chosen to deduce the corresponding characteristics of the object under investigation. This is a realistic approach as the scattering objects are often

[1]Institute of Electrical and Information Engineering, Kiel University, Kiel, Germany

in the far field of the original sources and equally is a systematic approach as any incident field can be composed of a suitably chosen set of plane waves. Moreover, the plane wave scattering is the basis for incorporating scattered field results into asymptotically valid high-frequency methods like the geometrical theory of diffraction (GTD) or the uniform theory of diffraction (UTD) [1]. One disadvantage of an incident plane wave is that it illuminates all parts of the scattering object, and that consequently the scattered field contains all of the corresponding information. Moreover, a plane wave incident on semi-infinite structures usually leads to solutions with non-converging series where problem-specific integral transforms and/or special summation techniques are needed to find asymptotically valid limiting-value results of the scattered fields [2]. It should be noted that another approach introduced by Smyshlyaev was presented in [3]. It deals with a method to extend the classical Sommerfeld solution on the scattering of a plane wave by a half plane to semi-infinite conical structures and results in an integral equation that has to be solved numerically.

Recently, CSBs have been applied as incident fields for such problems. CSBs are obtained by simply replacing a real-valued coordinate indicating the location of the source by a complex value. For a point source, this results in a beam-like field that in the vicinity of the beam axis (*paraxially*) can be interpreted as one half of a Gaussian beam. While the real part of the complex source coordinates describes the location of the beam's waist, the imaginary part corresponds to the beam's focal (Rayleigh) length. However, unlike a Gaussian beam a CSB is (except for the waist) an analytic function and an exact solution of Maxwell's equations. A uniform CSB can be interpreted as a full Gaussian beam, where the field is analytic even in the waist [4]. Moreover, in that waist the CSB has a plane wave front and can be used as a localized inhomogeneous plane wave. It is hence perfectly suited to probe a certain part of the scattering object and thus an excellent candidate to extract fields scattered by interesting parts like tips and edges and to derive corresponding diffraction coefficients. One idea how this could be accomplished is by applying techniques similar to those ones that have been applied to the case of an inhomogeneous plane wave [5].

The chapter is organized as follows. First, the solution of Maxwell's equations in sphero-conal coordinates is summarized. Then the uniform CSB is introduced as a combination of a—with respect to the waist—purely outgoing (*diverging*) and a purely incoming (*converging*) CSB. The uniform CSB is incorporated as a source into the spherical-multipole solutions for an elliptic cone. Since the field scattered by the cone still has to satisfy the radiation condition (i.e., only outgoing parts in the far field), special emphasis will be laid on the question how to correctly choose the complex source coordinate to achieve a causal solution. Finally the numerical evaluation includes a convergence analysis as well as total near fields and scattered far fields for the acoustically soft and hard and for the electrically perfectly conducting semi-infinite elliptic cone illuminated by a uniform CSB.

8.2 Solution of Maxwell's equations in sphero-conal coordinates

8.2.1 Sphero-conal coordinates

Sphero-conal (or conical) coordinates r, ϑ, φ belong to those 11 three-dimensional systems where the Helmholtz equation is fully separable [6,7] . They can be related to Cartesian coordinates by

$$x = r \sin\vartheta \cos\varphi \tag{8.1}$$

$$y = r\sqrt{1 - k^2 \cos^2\vartheta} \sin\varphi \tag{8.2}$$

$$z = r \cos\vartheta \sqrt{1 - k'^2 \sin^2\varphi}. \tag{8.3}$$

The ranges of variability are given by $0 \leq r,\ 0 \leq \vartheta \leq \pi,\ 0 \leq \varphi \leq 2\pi$. k and k' are positive real-valued parameters satisfying $k^2 + k'^2 = 1$. Figure 8.1 depicts three typical coordinate surfaces. While $r = r_0$ describes a spherical surface with radius r_0, $\vartheta = \vartheta_0$ represents an elliptic cone symmetrically located around the positive (for $0 < \vartheta < \pi/2$) or negative (for $\pi/2 < \vartheta < \pi$) z-axis. The half-opening angles of that elliptic cone are given by $\vartheta_x = \vartheta_0$ in the xz-plane and $\vartheta_y = \arccos(k \cos\vartheta_0)$ in the yz-plane. For $\vartheta = 0$ $(\vartheta = \pi)$, the elliptic cone degenerates into a plane angular sector with the half-opening angle $\vartheta_y = \arccos(k)$ in the yz-plane symmetrically located around the positive (negative) z-axis. In case that the sector-like surfaces at $\vartheta = 0$ and/or at $\vartheta = \pi$ belong to the free space, both sides of that sectors describe the same point in space, i.e., $(r, 0, \varphi) = (r, 0, \pi - \varphi)$. Consequently, for any continuously differentiable function $u(r, \vartheta, \varphi) = R(r)\Theta(\vartheta)\Phi(\varphi)$ it holds

$$\left.\frac{\partial\Theta(\vartheta)}{\partial\vartheta}\right|_{\vartheta=0} = 0 \quad \text{if } \Phi(\varphi) = \Phi(\pi - \varphi) \tag{8.4}$$

$$\Theta(\vartheta = 0) = 0 \quad \text{if } \Phi(\varphi) = -\Phi(\pi - \varphi). \tag{8.5}$$

The conditions (8.4) and (8.5) are referred to as *geometrical boundary conditions.*

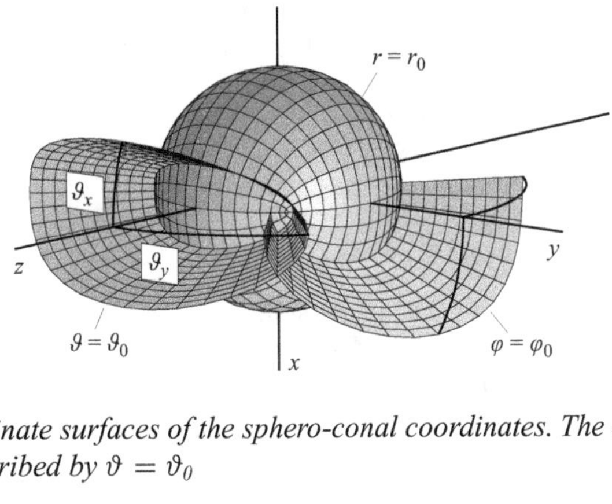

Figure 8.1 Coordinate surfaces of the sphero-conal coordinates. The elliptic cone is described by $\vartheta = \vartheta_0$

In the case of $k = 1$, the sphero-conal coordinates turn into ordinary spherical coordinates that will be denoted by $r, \bar{\vartheta}, \bar{\varphi}$ in the following. The normalized metric scaling coefficients of the sphero-conal coordinates are found as

$$s_\vartheta = \frac{1}{r}\left|\frac{\partial \vec{r}}{\partial \vartheta}\right| = \sqrt{\frac{k^2 \sin^2 \vartheta + k'^2 \cos^2 \varphi}{1 - k^2 \cos^2 \vartheta}} \tag{8.6}$$

$$s_\varphi = \frac{1}{r}\left|\frac{\partial \vec{r}}{\partial \varphi}\right| = \sqrt{\frac{k^2 \sin^2 \vartheta + k'^2 \cos^2 \varphi}{1 - k'^2 \sin^2 \varphi}}, \tag{8.7}$$

while the nabla operator reads

$$\vec{\nabla} \equiv \hat{r}\frac{\partial}{\partial r} + \hat{\vartheta}\frac{1}{r s_\vartheta}\frac{\partial}{\partial \vartheta} + \hat{\varphi}\frac{1}{r s_\varphi}\frac{\partial}{\partial \varphi} \tag{8.8}$$

with $\hat{r}, \hat{\vartheta}, \hat{\varphi}$ being the corresponding unit vectors.

8.2.2 Solution of the Helmholtz equation in sphero-conal coordinates

In sphero-conal coordinates, the Helmholtz equation

$$\Delta\Psi(r, \vartheta, \varphi) + \kappa^2\Psi(r, \vartheta, \varphi) = 0 \tag{8.9}$$

with wave number $\kappa = 2\pi/\Lambda$ and wave length Λ reads

$$\frac{1}{r^2}\frac{\partial}{\partial r}\left(r^2\frac{\partial\Psi(r, \vartheta, \varphi)}{\partial r}\right) + \frac{1}{r^2}(\vec{r} \times \vec{\nabla})^2\Psi(r, \vartheta, \varphi) + \kappa^2\Psi(r, \vartheta, \varphi) = 0. \tag{8.10}$$

A first separation ansatz

$$\Psi_\nu(r, \vartheta, \varphi) = z_\nu(\kappa r) Y_\nu(\vartheta, \varphi) \tag{8.11}$$

with a separation constant $\nu(\nu + 1)$ leads to the differential equation of the spherical Bessel functions of order ν

$$\frac{1}{(\kappa r)^2}\frac{d}{d(\kappa r)}\left[(\kappa r)^2\frac{dz_\nu(\kappa r)}{d(\kappa r)}\right] + \left(1 - \frac{\nu(\nu + 1)}{(\kappa r)^2}\right) z_\nu(\kappa r) = 0 \tag{8.12}$$

which are related to (ordinary) Bessel functions by

$$z_\nu(\kappa r) = \sqrt{\frac{\pi}{2\kappa r}} Z_{\nu+1/2}(\kappa r), \tag{8.13}$$

and to the eigenvalue equation of the *Lamé* products

$$-\left(\vec{r} \times \vec{\nabla}\right)^2 Y_\nu(\vartheta, \varphi) = \nu(\nu + 1) Y_\nu(\vartheta, \varphi). \tag{8.14}$$

We are particularly interested in the solutions of the Helmholtz equation where $Y_\nu(\vartheta, \varphi)$ is regular in $0 \le \vartheta \le \vartheta_0 \le \pi; 0 \le \varphi \le 2\pi$, 2π-periodic in φ, which satisfies one of the following boundary conditions

$$Y_\sigma(\vartheta, \varphi)|_{\vartheta=\vartheta_0} = 0 \quad \text{(Dirichlet condition)}, \tag{8.15}$$

$$\left.\frac{\partial Y_\tau(\vartheta, \varphi)}{\partial \vartheta}\right|_{\vartheta=\vartheta_0} = 0 \quad \text{(Neumann condition)}. \tag{8.16}$$

In these cases, the operator $(\vec{r} \times \vec{\nabla})^2$ is symmetric [8,9] according to

$$\langle(\vec{r} \times \vec{\nabla})^2 f, g\rangle = \langle f, (\vec{r} \times \vec{\nabla})^2 g\rangle \tag{8.17}$$

with $\{f, g\} \in \{Y_\sigma\}$ or $\{f, g\} \in \{Y_\tau\}$ and with the standard definition of the (Hilbert space) scalar product

$$\langle f, g\rangle = \int_0^{\vartheta_0}\int_0^{2\pi} f(\vartheta, \varphi) g^*(\vartheta, \varphi) s_\vartheta s_\varphi d\vartheta d\varphi. \tag{8.18}$$

The asterisk denotes the complex conjugation. From the symmetry, it follows that all eigenvalues are countable, real, and positive according to

$$\sigma_i(\sigma_i + 1) > 0 \to \sigma_i > 0 \quad (i = 1, 2, 3, \ldots) \tag{8.19}$$

$$\tau_i(\tau_i + 1) \geq 0 \to \tau_i \geq 0 \quad (i = 1, 2, 3, \ldots). \tag{8.20}$$

Moreover, the accordingly normalized eigenfunctions form an orthonormal and complete set on the spherical surface

$$\langle Y_{\sigma_p}, Y_{\sigma_q}\rangle = \langle Y_{\tau_p}, Y_{\tau_q}\rangle = \delta_{pq} \tag{8.21}$$

with the Kronecker symbol $\delta_{pq} = 1$ if $p = q$ and $\delta_{pq} = 0$ if $p \neq q$. Any function from the corresponding Hilbert space can be expanded by means of a convergent series

$$f(\vartheta, \varphi) = \sum_{p=1}^{\infty} a_p Y_{\nu_p}(\vartheta, \varphi) \text{ with } a_p = \langle f, Y_{\nu_p}\rangle \quad (\nu_p = \sigma_p \text{ or } \nu_p = \tau_p). \tag{8.22}$$

A second separation ansatz

$$Y_\nu(\vartheta, \varphi) = \Theta_\nu(\vartheta)\Phi_\nu(\varphi) \tag{8.23}$$

with a second separation constant $\tilde{\lambda}$ leads to the ordinary differential equations of the periodic Lamé functions

$$\sqrt{1 - k'^2 \sin^2\varphi}\frac{d}{d\varphi}\left(\sqrt{1 - k'^2 \sin^2\varphi}\frac{d\Phi_\nu}{d\varphi}\right) + \left[\nu(\nu + 1)k'^2 \cos^2\varphi + \tilde{\lambda}\right]\Phi_\nu = 0 \tag{8.24}$$

and of the nonperiodic Lamé functions

$$\sqrt{1 - k^2 \cos^2\vartheta}\frac{d}{d\vartheta}\left(\sqrt{1 - k^2 \cos^2\vartheta}\frac{d\Theta_\nu}{d\vartheta}\right) + \left[\nu(\nu + 1)k^2 \sin^2\vartheta - \tilde{\lambda}\right]\Theta_\nu = 0. \tag{8.25}$$

With the substitution

$$\tilde{\lambda} = \lambda - k'^2\nu(\nu + 1) \tag{8.26}$$

we finally derive

$$\sqrt{1-k'^2\sin^2\varphi}\frac{d}{d\varphi}\left(\sqrt{1-k'^2\sin^2\varphi}\frac{d\Phi_\nu}{d\varphi}\right)+\left[\lambda-\nu(\nu+1)k'^2\sin^2\varphi\right]\Phi_\nu=0 \tag{8.27}$$

and

$$\sqrt{1-k^2\cos^2\vartheta}\,\frac{d}{d\vartheta}\left(\sqrt{1-k^2\cos^2\vartheta}\,\frac{d\Theta_\nu}{d\vartheta}\right)+\left[\nu(\nu+1)\left(1-k^2\cos^2\vartheta\right)-\lambda\right]\Theta_\nu=0. \tag{8.28}$$

For the solution of this two-parametric eigenvalue problem with two coupled Lamé differential equations, the periodic Lamé functions are expressed by four types of Fourier expansions. It will be useful to have for each of these types two different kinds of expansions [10]

$$\Phi_\nu^{(1)}(\varphi)=\begin{cases}\sum\limits_{i=0}^{\infty}A_{2i}\cos 2i\varphi\\ \sqrt{1-k'^2\sin^2\varphi}\sum\limits_{i=0}^{\infty}C_{2i}\cos 2i\varphi\end{cases} \tag{8.29}$$

$$\Phi_\nu^{(2)}(\varphi)=\begin{cases}\sum\limits_{i=0}^{\infty}A_{2i+1}\cos(2i+1)\varphi\\ \sqrt{1-k'^2\sin^2\varphi}\sum\limits_{i=0}^{\infty}C_{2i+1}\cos(2i+1)\varphi\end{cases} \tag{8.30}$$

$$\Phi_\nu^{(3)}(\varphi)=\begin{cases}\sum\limits_{i=0}^{\infty}B_{2i+2}\sin(2i+2)\varphi\\ \sqrt{1-k'^2\sin^2\varphi}\sum\limits_{i=0}^{\infty}D_{2i+2}\sin(2i+2)\varphi\end{cases} \tag{8.31}$$

$$\Phi_\nu^{(4)}(\varphi)=\begin{cases}\sum\limits_{i=0}^{\infty}B_{2i+1}\sin(2i+1)\varphi\\ \sqrt{1-k'^2\sin^2\varphi}\sum\limits_{i=0}^{\infty}D_{2i+1}\sin(2i+1)\varphi.\end{cases} \tag{8.32}$$

Note that these four function types can also be interpreted as four regular Sturm–Liouville problems with consequences regarding the number of zeros on the interval $(0,\pi/2)$, orthogonality, completeness, etc. Inserting these eight expansions into (8.27) leads to a three-term-recurrence relation for each set of coefficients. Each of the recurrence relations can be written as a matrix equation representing an algebraic eigenvalue problem with $\tilde{\lambda}$ being the eigenvalues, while the matrix elements depend on k' and ν. Numerically solving the matrix equations for a fixed k' leads for every ν to a set of eigenvalues λ_m $(m=0,1,2,\ldots)$. For the cases considered here and described by the conditions (8.15) and (8.16), it holds [9]

$$0\le\lambda\le\nu(\nu+1). \tag{8.33}$$

Hence, a finite number of λ_m is obtained for each ν. Finally, for each value of k', the relations between ν and λ_m can be represented as *eigenvalue curves* shown in Figure 8.2 as an example. Note that the two types of expansions for each function type lead to the same eigenvalues, but generally to different eigenfunctions. For the nonperiodic Lamé functions, we have to consider the geometrical boundary conditions (8.4) and (8.5) as well as the fact that at $\vartheta = \vartheta_0$ Dirichlet or Neumann conditions have to be satisfied. Consequently the following four expansions of the nonperiodic Lamé functions are used [10]

$$\Theta_\nu^{(1)}(\vartheta) = \sum_{i=0}^{\infty} A_{2i} T(2i) P_\nu^{2i}(\cos\vartheta) \tag{8.34}$$

$$\Theta_\nu^{(2)}(\vartheta) = \sum_{i=0}^{\infty} A_{2i+1} T(2i+1) P_\nu^{2i+1}(\cos\vartheta) \tag{8.35}$$

$$\Theta_\nu^{(3)}(\vartheta) = \frac{\sqrt{1-k^2\cos^2\vartheta}}{\sin\vartheta} \sum_{i=0}^{\infty} B_{2i+2}(2i+2) T(2i+2) P_\nu^{2i+2}(\cos\vartheta) \tag{8.36}$$

$$\Theta_\nu^{(4)}(\vartheta) = \frac{\sqrt{1-k^2\cos^2\vartheta}}{\sin\vartheta} \sum_{i=0}^{\infty} B_{2i+1}(2i+1) T(2i+1) P_\nu^{2i+1}(\cos\vartheta). \tag{8.37}$$

The algebraic factors $T(i)$ are obtained recursively according to

$$\begin{aligned} T(i) &= -(\nu-1)(\nu+i+1)T(i+2) \\ T(0) &= T(1) = 1. \end{aligned} \tag{8.38}$$

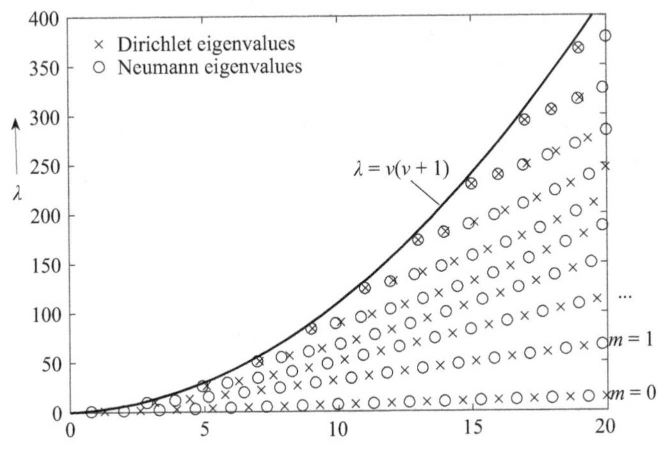

Figure 8.2 Dirichlet and Neumann eigenvalues on the eigenvalue curves of the periodic Lamé functions. $k'^2 = 0.5$; *function type 2*

In (8.34)–(8.37), the $P_\nu^m(\cos\vartheta)$ denote associated Legendre functions of the first kind to real degree ν and integer order m. Inserting these four function types into the differential equation (8.28) yields the same three-term-recurrence relations and consequently the same matrix equations and expansion coefficients as obtained for the corresponding four types of expansions of the periodic Lamé functions. Moreover, any valid combination of ν and λ of the nonperiodic Lamé differential equation lies on the same eigenvalue curve (Figure 8.2) as for the periodic Lamé function. A numerically performed search on each of the eigenvalue curves finally yields the unique eigenvalue pairs (σ,λ) and (τ,λ) according to the Dirichlet condition (8.15) and to the Neumann condition (8.16).

In summary, we can write any solution of the Helmholtz equation in the presence of a semi-infinite elliptic cone as

$$\Psi(\vec{r}) = \sum_{p=1}^{\infty} c_{\nu_p} z_{\nu_p}(\kappa r) Y_{\nu_p}(\vartheta,\varphi) \tag{8.39}$$

with expansion coefficients c_{ν_p} where $\nu_p = \sigma_p(\tau_p)$ for an acoustically soft (hard) cone surface according to (8.15) and (8.16). Note that the notation Y_ν for the Lamé products generally includes all four function types and that only products of periodic and nonperiodic Lamé functions $\Theta_\nu^{(l)}(\vartheta)\Phi_\nu^{(l)}(\varphi)$ of the same function type l are valid solutions of the boundary-value problems considered in this chapter.

8.2.3 Vector spherical-multipole expansion of the electromagnetic field in the presence of a PEC semi-infinite elliptic cone

At a time factor $\exp(j\omega t)$, the Maxwell curl-equations for the phasors in a homogeneous, isotropic, and linear medium read

$$\vec{\nabla} \times \vec{H}(\vec{r}) = \vec{J}(\vec{r}) + j\omega\varepsilon\vec{E}(\vec{r}) \tag{8.40}$$

$$\vec{\nabla} \times \vec{E}(\vec{r}) = -j\omega\mu\vec{H}(\vec{r}), \tag{8.41}$$

where $\vec{H}$ and $\vec{E}$ represent the magnetic and electric field intensities, ε and μ are the complex-valued permittivity and permeability, respectively, while $\vec{J}$ is the impressed electric current density. Immediately the *curl–curl* equation for the electric field intensity follows as

$$\vec{\nabla} \times \vec{\nabla} \times \vec{E}(\vec{r}) - \kappa^2\vec{E}(\vec{r}) = -j\omega\mu\vec{J}(\vec{r}) \tag{8.42}$$

with the wave number $\kappa = \omega\sqrt{\varepsilon\mu}$. Outside of the domain containing the impressed currents, the electric field is solenoidal, and the fundamental system of the corresponding homogeneous curl–curl equation can be constructed by [11]

$$\vec{M}_\nu(\vec{r}) = \left(\vec{r} \times \vec{\nabla}\right)\Psi_\nu(\vec{r}) \tag{8.43}$$

$$\vec{N}_\nu(\vec{r}) = \left[\vec{\nabla} \times \left(\vec{r} \times \vec{\nabla}\right)\right]\Psi_\nu(\vec{r}) \tag{8.44}$$

where Ψ_ν is a suitable elementary solution of the homogeneous Helmholtz equation. In sphero-conal coordinates, the functions defined in (8.43) and (8.44) are referred to as the vector spherical-multipole functions. They are calculated by

$$\vec{M}(\vec{r}) = z_\nu(\kappa r) \underbrace{\left[-\frac{1}{s_\varphi} \frac{\partial Y_\nu(\vartheta,\varphi)}{\partial \varphi} \hat{\vartheta} + \frac{1}{s_\vartheta} \frac{\partial Y_\nu(\vartheta,\varphi)}{\partial \vartheta} \hat{\varphi} \right]}_{=\vec{m}_\nu(\vartheta,\varphi)} \tag{8.45}$$

$$\begin{aligned} \vec{N}(\vec{r}) = & -\frac{z_\nu(\kappa r)}{\kappa r} \nu(\nu+1) Y_\nu(\kappa r) \hat{r} \\ & -\frac{1}{\kappa r} \frac{d}{dr} [r z_\nu(\kappa r)] \underbrace{\left[\frac{1}{s_\vartheta} \frac{\partial Y_\nu(\vartheta,\varphi)}{\partial \vartheta} \hat{\vartheta} + \frac{1}{s_\varphi} \frac{\partial Y_\nu(\vartheta,\varphi)}{\partial \varphi} \hat{\varphi} \right]}_{=\vec{n}_\nu(\vartheta,\varphi)} . \end{aligned} \tag{8.46}$$

The *angular vector functions* $\vec{m}_\nu(\vartheta,\varphi)$ and $\vec{n}_\nu(\vartheta,\varphi)$ defined in (8.45) and (8.46) form a complete and orthogonal set according to [9]

$$\begin{aligned} \langle \vec{m}_{\nu_p}, \vec{m}_{\nu_q} \rangle &= \langle \vec{n}_{\nu_p}, \vec{n}_{\nu_q} \rangle = \nu_p(\nu_p+1)\delta_{pq} \ ; \ \langle \vec{n}_{\nu_p}, \vec{m}_{\nu_q} \rangle = 0 \\ \nu_p &= \sigma_p \text{ or } \nu_p = \tau_p \quad \forall p, q \in \{1,2,3,\ldots\}. \end{aligned} \tag{8.47}$$

Finally, outside of the impressed-current domains the electromagnetic field in the presence of a perfectly electrically conducting (PEC) semi-infinite elliptic cone described by $\vartheta = \vartheta_0$ can be represented by the vector spherical-multipole expansion

$$\vec{E}(\vec{r}) = \sum_{p=1}^{\infty} a_p \vec{N}_{\sigma_p}(\vec{r}) + \frac{Z}{j} \sum_{p=1}^{\infty} b_p \vec{M}_{\tau_p}(\vec{r}) \tag{8.48}$$

$$\vec{H}(\vec{r}) = \frac{j}{Z} \sum_{p=1}^{\infty} a_p \vec{M}_{\sigma_p}(\vec{r}) + \sum_{p=1}^{\infty} b_p \vec{N}_{\tau_p}(\vec{r}), \tag{8.49}$$

where $Z = \sqrt{\mu/\varepsilon}$ is the intrinsic impedance of the medium, while a_p and b_p are referred to as the electric multipole amplitudes (SI unit V/m) and the magnetic multipole amplitudes (SI unit A/m), respectively. The perfect symmetry between electric and magnetic quantities in the vector spherical-multipole expansion is also expressed by its invariance to the following extended Fitzgerald-transform [9]

$$\begin{aligned} \vec{E} &\rightarrow \vec{H} \\ \vec{H} &\rightarrow -\vec{E} \\ a_p &\rightarrow b_p \\ b_p &\rightarrow -a_p \\ \varepsilon &\leftrightarrow \mu \\ \sigma &\leftrightarrow \tau \end{aligned} \tag{8.50}$$

which turns the expansions (8.48) and (8.49) into themselves.

8.3 Complex-source beams

As shown in [2,12,13], the attempt to derive converging far-field expressions for a plane wave incident on a semi-infinite cone using the spherical-multipole expansion fails. Suitable summation techniques like the Cesàro or the Shanks transform have been employed to obtain approximately valid numerical results. Since we are mainly interested in the interaction of the incident plane wave with special parts of the elliptic cone like its tip, we will use a CSB as the incident field [14–16]. As will be shown, a uniform CSB can approximately represent a Gaussian beam that is not illuminating the entire semi-infinite elliptic cone and consequently is leading to convergent spherical-multipole far-field expressions.

8.3.1 Converging and diverging CSB

We start from the closed form of the scalar Green function to the Helmholtz equation in free space

$$G(\vec{r},\vec{r}\,') = \frac{1}{4\pi}\frac{\exp[-j\kappa\,|\vec{r}-\vec{r}\,'|]}{|\vec{r}-\vec{r}\,'|} = \frac{1}{4\pi}\frac{\exp\left[-j\kappa\sqrt{(\vec{r}-\vec{r}\,')^2}\right]}{\sqrt{(\vec{r}-\vec{r}\,')^2}} \tag{8.51}$$

and recall that for real-valued $\vec{r}$ and $\vec{r}\,'$ the positive value (main value) of the root has to be chosen to comply with the radiation condition. Now suppose that the source coordinate is complex-valued according to

$$\vec{r}\,' = \vec{r}_r - j\vec{r}_i \quad (\vec{r}_r,\vec{r}_i) \in \mathbb{R}^3 \ ; \ r_i > 0. \tag{8.52}$$

As usually done by numerical algorithms, we assume that the main value of the complex root in (8.51) is chosen, i.e., that for any complex-valued expression A we have $\Re\sqrt{A} \geq 0$.

As an example, we set the source coordinate to $\vec{r}\,' = (z_r - jz_i)\hat{z}$ with $z_i > 0$, let $\rho^2 = x^2 + y^2$, and derive with a first-order Taylor approximation for $\rho \ll \sqrt{\tilde{z}^2 + z_i^2}$ for the main value of the root

$$\sqrt{(\vec{r}-\vec{r}\,')^2} = \sqrt{\rho^2 + (\tilde{z}+jz_i)^2} \approx \begin{cases} (\tilde{z}+jz_i) + \frac{1}{2}\frac{\rho^2}{(\tilde{z}+jz_i)} & \text{if } \tilde{z} > 0 \\ -(\tilde{z}+jz_i) - \frac{1}{2}\frac{\rho^2}{(\tilde{z}+jz_i)} & \text{if } \tilde{z} < 0 \end{cases} \tag{8.53}$$

where $\tilde{z} = z - z_r$. The branch line (where the root is zero) is given by a circle at ($\tilde{z} = 0$; $\rho = z_i$); hence, we define a branch disk in between that circles at $z = z_r$; $\rho \leq z_i$. We insert (8.53) into (8.51) and derive for the Green function in a *paraxial approximation*

$$G(\vec{r},\vec{r}\,') \approx \begin{cases} \frac{1}{4\pi(\tilde{z}+jz_i)}\exp\left(-j\kappa(\tilde{z}+jz_i) - \frac{j\kappa\rho^2}{2(\tilde{z}+jz_i)}\right) & \text{if } \tilde{z} > 0 \\ -\frac{1}{4\pi(\tilde{z}+jz_i)}\exp\left(+j\kappa(\tilde{z}+jz_i) + \frac{j\kappa\rho^2}{2(\tilde{z}+jz_i)}\right) & \text{if } \tilde{z} < 0. \end{cases} \tag{8.54}$$

Note that this case distinction also ensures causality, i.e., that we only have outgoing fields with respect to the branch disc. After some algebraic transformation, we are led to the representation

$$G(\vec{r},\vec{r}\,') \approx \frac{1}{4\pi}\frac{\exp(\kappa z_i)}{jz_i}\frac{1}{\sqrt{1+\frac{\tilde{z}^2}{z_i^2}}}\exp\left(j\arctan\left(\frac{\tilde{z}}{z_i}\right)\right)$$
$$\exp(-j\kappa\tilde{z})\exp\left(-\frac{j\kappa\rho^2}{2\left(\tilde{z}+\frac{z_i^2}{\tilde{z}}\right)}\right)\exp\left(-\frac{\rho^2}{2\frac{z_i}{\kappa}\left(1+\frac{\tilde{z}^2}{z_i^2}\right)}\right) \quad (\tilde{z}>0). \tag{8.55}$$

A comparison to the standard form of a Gaussian beam

$$A(\tilde{z},\rho) = A_0\frac{w_0}{w(\tilde{z})}\exp(-j\kappa\tilde{z})\exp\left(-\frac{j\kappa\rho^2}{2R(\tilde{z})}\right)\exp\left(-\frac{\rho^2}{[w(\tilde{z})]^2}\right) \tag{8.56}$$

reveals that for $\tilde{z} > 0$, the paraxial approximation of a CSB can be interpreted as the field of a Gaussian beam outgoing in $+\hat{z}$-direction from its waist at $\tilde{z} = 0$ ($z = z_r$) with the following parameters:

$$\begin{aligned}
\text{Amplitude:}\ & A_0 = \frac{1}{4\pi}\frac{\exp(\kappa z_i)}{jz_i}\\
\text{Beam radius:}\ & w(\tilde{z}) = w_0\sqrt{\left(1+\frac{\tilde{z}^2}{z_i^2}\right)}\\
\text{Beam radius at the waist:}\ & w_0 = \sqrt{2\frac{z_i}{\kappa}}\\
\text{Rayleigh length and direction of propagation:}\ & z_i\hat{z}\\
\text{Radius of curvature:}\ & R(\tilde{z}) = \tilde{z}+\frac{z_i^2}{\tilde{z}}\\
\text{Gouy phase:}\ & \arctan\left(\frac{\tilde{z}}{z_i}\right).
\end{aligned} \tag{8.57}$$

While for $\tilde{z} > 0$ the amplitude increases exponentially $\sim\exp(\kappa z_i)$, it decreases everywhere for $\tilde{z} < 0$ according to $\exp(-\kappa z_i)$ and can practically be neglected for $\kappa z_i \gg 1$, i.e., for any focused beam. Therefore, the similarity to a Gaussian beam is valid only for its outwardly traveling half (for $\tilde{z} > 0$) and nearby the axis. This kind of a CSB outgoing from the waist is also referred to as a *diverging* CSB [17]. It is worth noting that the field due to the complex source coordinate in (8.51) is except for the branch disk analytic and an exact solution of the Helmholtz equation.

By just changing the sign in the exponent of (8.51), the field should become inwardly traveling with respect to the source. Accordingly, for a complex source coordinate (8.52) the paraxial approximation of the resulting Green function reads

$$G(\vec{r},\vec{r}\,') \approx \begin{cases} \frac{1}{4\pi(\tilde{z}+jz_i)} \exp\left(j\kappa(\tilde{z}+jz_i) + \frac{j\kappa\rho^2}{2(\tilde{z}+jz_i)}\right) & \text{if } \tilde{z} > 0 \\ -\frac{1}{4\pi(\tilde{z}+jz_i)} \exp\left(-j\kappa(\tilde{z}+jz_i) - \frac{j\kappa\rho^2}{2(\tilde{z}+jz_i)}\right) & \text{if } \tilde{z} < 0. \end{cases} \tag{8.58}$$

Obviously, for $\tilde{z} < 0$ (8.58) can be interpreted as a half of a Gaussian beam inwardly traveling toward the branch disc. Moreover, for $\tilde{z} < 0$ the amplitude increases exponentially $\sim\exp(\kappa z_i)$, while it decreases for $\tilde{z} > 0$ according to $\exp(-\kappa z_i)$; therefore, this part can practically be neglected for any focused beam. This kind of a CSB incoming to the waist is also referred to as a *converging* CSB [17].

8.3.2 Uniform CSB

We now consider a focused beam ($\kappa r_i \gg 1$) and subtract from the Green function (8.51) for an outwardly traveling wave the Green function for an inwardly traveling wave

$$G(\vec{r},\vec{r}\,') = \frac{1}{4\pi}\left\{\frac{\exp\left[-j\kappa\sqrt{(\vec{r}-\vec{r}\,')^2}\right]}{\sqrt{(\vec{r}-\vec{r}\,')^2}} - \frac{\exp\left[+j\kappa\sqrt{(\vec{r}-\vec{r}\,')^2}\right]}{\sqrt{(\vec{r}-\vec{r}\,')^2}}\right\}. \tag{8.59}$$

The paraxial approximation of the resulting Green function reads uniformly

$$G(\vec{r},\vec{r}\,') \approx \frac{1}{4\pi(\tilde{z}+jz_i)} \exp\left(-j\kappa(\tilde{z}+jz_i) - \frac{j\kappa\rho^2}{2(\tilde{z}+jz_i)}\right) \tag{8.60}$$

where we have neglected the exponentially decreasing field parts. Obviously, this paraxial approximation of this field can be interpreted as a full Gaussian beam. It is analytic everywhere even at the waist and of course an exact solution of the Helmholtz equation [18,19]. This combination of a diverging and converging CSBs will be referred to as a *uniform* CSB [4].

Finally, we conclude that (8.59) represents a uniform CSB with a waist centered at $\vec{r}_r$, propagating in the direction $\hat{r}_i$, and with a Rayleigh length r_i. For the special case that the uniform CSB with a Rayleigh length r_i, with a waist centered at $\vec{r}_r$ and which is propagating toward the origin, we have in sphero-conal coordinates

$$\vec{r}_r = (r_r, \vartheta_r, \varphi_r), \ \vec{r}_i = (r_i, \pi - \vartheta_r, \pi + \varphi_r). \tag{8.61}$$

In Cartesian coordinates, it follows

$$\begin{aligned}
\vec{r}' &= \vec{r}_r - j\vec{r}_i \\
&= \begin{pmatrix} r_r \sin\vartheta_r \cos\varphi_r \\ r_r\sqrt{1-k^2\cos^2(\vartheta_r)}\sin\varphi_r \\ r_r\cos\vartheta_r\sqrt{1-k'^2\sin^2\varphi_r} \end{pmatrix} - j \begin{pmatrix} r_i\sin(\pi-\vartheta_r)\cos(\pi+\varphi_r) \\ r_i\sqrt{1-k^2\cos^2(\pi-\vartheta_r)}\sin(\pi+\varphi_r) \\ r_i\cos(\pi-\vartheta_r)\sqrt{1-k'^2\sin^2(\pi+\varphi_r)} \end{pmatrix} \\
&= \begin{pmatrix} (r_r+jr_i)\sin\vartheta_r\cos\varphi_r \\ (r_r+jr_i)\sqrt{1-k^2\cos^2\vartheta_r}\sin\varphi_r \\ (r_r+jr_i)\cos\vartheta_r\sqrt{1-k'^2\sin^2\varphi_r} \end{pmatrix}.
\end{aligned} \tag{8.62}$$

Consequently, by choosing just for the radial source coordinate a complex value according to $r' = r_r + jr_i$ and leaving the angular coordinates ϑ' and φ' real, the uniform CSB is traveling toward the center of the coordinate system, see Figure 8.3. For $r_i < 0$, the result is complex conjugate; thus, the uniform CSB propagates in the reverse direction.

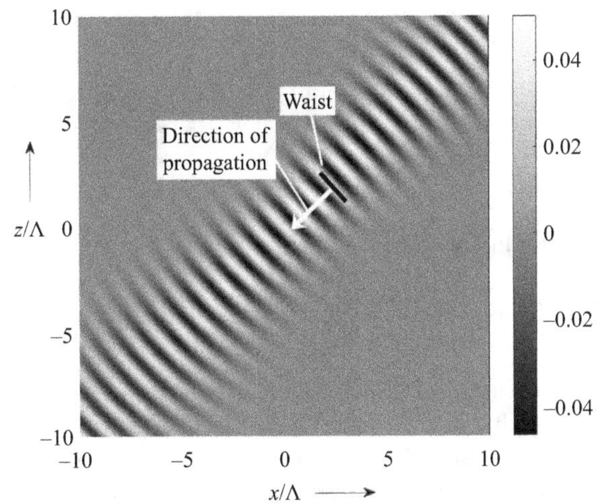

Figure 8.3 Uniform complex-source beam traveling toward the center of coordinate system. The spherical coordinates of the source are $(r', \bar{\vartheta}', \bar{\varphi}') = ((3+j10)\Lambda, 45°, 0)$

8.4 Green's function of the semi-infinite elliptic cone for an incident uniform complex-source beam

8.4.1 Scalar Green's function

The scalar Green function of the inhomogeneous Helmholtz equation $\Delta\Psi_{\substack{s\\h}}(\vec{r}) + \kappa^2\Psi_{\substack{s\\h}}(\vec{r}) = q(\vec{r})$ in a lossless medium bounded by an acoustically soft or hard semi-infinite elliptic cone at $\vartheta = \vartheta_0$ is defined as a solution of the differential equation

$$\Delta G_{\substack{s\\h}}(\vec{r},\vec{r}\,') + \kappa^2 G_{\substack{s\\h}}(\vec{r},\vec{r}\,') = \delta(\vec{r} - \vec{r}\,') \tag{8.63}$$

subject to the boundary conditions

$$G_s(\vec{r},\vec{r}\,')\big|_{\vartheta=\vartheta_0} = 0 \quad \text{(soft cone; Dirichlet condition)} \tag{8.64}$$

$$\left.\frac{\partial G_h(\vec{r},\vec{r}\,')}{\partial\vartheta}\right|_{\vartheta=\vartheta_0} = 0 \quad \text{(hard cone, Neumann condition),} \tag{8.65}$$

respectively. Due to energy constraints of the corresponding total field, the Green function has to be square-integrable in any finite partial volume but particularly around the tip of the cone. However, for a uniform CSB the Green function does *not* satisfy the radiation condition. Instead we consider that for an incident uniform CSB case, there is no energy loss from any bounded domain. We choose a spherical domain centered at the origin and note that the only functions that guarantee such a condition of energy conservation are the spherical Bessel functions $j_\nu(\kappa r)$. Hence, the scalar Green function reads (for a detailed derivation see [20])

$$G_{\substack{s\\h}}(\vec{r},\vec{r}\,') = 2j\kappa \sum_i j_{\substack{\sigma_i\\\tau_i}}(\kappa r) j_{\substack{\sigma_i\\\tau_i}}(\kappa r') Y_{\substack{\sigma_i\\\tau_i}}(\vartheta,\varphi) Y^*_{\substack{\sigma_i\\\tau_i}}(\vartheta',\varphi'). \tag{8.66}$$

The scalar field can then be represented by the multipole expansion

$$\Psi_{\substack{s\\h}}(\vec{r}) = \sum_{i=1}^{\infty} a_{i,\substack{s\\h}} j_{\substack{\sigma_i\\\tau_i}}(\kappa r) Y_{\substack{\sigma_i\\\tau_i}}(\vartheta,\varphi) \tag{8.67}$$

with the multipole amplitudes

$$a_{i,\substack{s\\h}} = 2j\kappa \int_V q(\vec{r}\,') j_{\substack{\sigma_i\\\tau_i}}(\kappa r') Y^*_{\substack{\sigma_i\\\tau_i}}(\vartheta',\varphi') dv'. \tag{8.68}$$

Note that in the case of a uniform CSB, $q(\vec{r}\,')$ in (8.68) is not a source distribution in the common meaning. It rather could be interpreted as a distribution of waists of uniform CSBs. Moreover, the corresponding condition of no energy loss from any bounded domain does not guarantee a causal solution. As has been first demonstrated in [4] for comparing the scattering of converging and diverging CSBs by a circular cone and later investigated in [21] for the wedge problem, the complex source coordinates have to be carefully chosen. Consider, for example, a uniform CSB of Rayleigh length r_i with the waist centered at $(r_r, \vartheta_r, \varphi_r)$ illuminating directly the tip of a semi-infinite elliptic cone. As already has been shown in (8.62), the complex source coordinate is then found to be $(r_r + jr_i, \vartheta_r, \varphi_r)$. In that case, the tip is illuminated by the diverging (outgoing from the waist) part of the uniform CSB. For a focused beam that part is built up by spherical Hankel functions of the second kind; hence, it satisfies the radiation condition. The field scattered by the cone then also satisfies the radiation condition, and the total field is causal.

To treat the case that the uniform CSB is traveling from the tip toward the waist, we could have the idea to simply use the complex source coordinates $(r_r - jr_i,\ \vartheta_r,\ \varphi_r)$ with $r_i > 0$. However, because of $j_\nu(z^*) = j_\nu^*(z)$ we then just obtain the conjugate complex result, i.e., the *time-reversal* (noncausal) solution of the case of the uniform CSB is traveling from the waist toward the tip. To obtain a causal solution for that case as well we instead include negative values of r_r exploiting the fact that the sphero-conal coordinates $(r_r,\ \vartheta_r,\ \varphi_r)$ and $(-r_r,\ \pi - \vartheta_r,\ \pi + \varphi_r)$ describe the same point (x, y, z), see (8.1). Using the representation with a negative r_r and a derivation similar to (8.62), the resulting causal solution for a uniform CSB traveling from the origin toward the waist centered at r_r, ϑ_r, φ_r can be represented by the complex source coordinates $(-r_r + jr_i,\ \pi - \vartheta_r,\ \pi + \varphi_r)$ with $r_i > 0$.

8.4.2 Dyadic Green's function

The dyadic Green function to the inhomogeneous curl–curl equation $\vec{\nabla} \times \vec{\nabla} \times \vec{E}(\vec{r}) - \kappa^2 \vec{E}(\vec{r}) = -j\omega\mu \vec{J}(\vec{r})$ in a lossless medium bounded by a PEC semi-infinite elliptic cone at $\vartheta = \vartheta_0$ is defined as a solution of the differential equation

$$\vec{\nabla} \times \vec{\nabla} \times \tilde{\Gamma}(\vec{r}, \vec{r}\,') - \kappa^2 \tilde{\Gamma}(\vec{r}, \vec{r}\,') = -\tilde{I}\delta(\vec{r} - \vec{r}\,') \tag{8.69}$$

with the boundary condition

$$\hat{\vartheta} \times \tilde{\Gamma}(\vec{r}, \vec{r}\,')\big|_{\vartheta = \vartheta_0} = 0, \tag{8.70}$$

respectively. Again, for an illuminating uniform CSB defined by a complex coordinate $\vec{r}\,'$, the dyadic Green function does not satisfy the radiation condition but a condition of energy conservation in any bounded domain. Following otherwise a derivation similar to that one used in [2], the dyadic Green function reads

$$\tilde{\Gamma}(\vec{r}, \vec{r}\,') = j\kappa \left\{ \sum_{p=1}^{\infty} \frac{\vec{N}_{\sigma_p}(\vec{r}) \vec{N}_{\sigma_p}(\vec{r}\,')}{\sigma_p(\sigma_p + 1)} + \sum_{p=1}^{\infty} \frac{\vec{M}_{\tau_p}(\vec{r}) \vec{M}_{\tau_p}(\vec{r}\,')}{\tau_p(\tau_p + 1)} \right\}. \tag{8.71}$$

Note that for a uniform CSB, there is no additional source term as derived in [2] for a real-valued source coordinate as the field is solenoidal even at the waist. The values of σ_p and τ_p in (8.71) are defined as in (8.15) and (8.16), respectively, while the vector spherical-multipole functions in the dyadic products in (8.71) are built up using spherical Bessel functions of the first kind only. Now the electric field can be written as

$$\vec{E}(\vec{r}) = j\kappa Z \iiint\limits_V \tilde{\Gamma}(\vec{r}, \vec{r}\,') \cdot \vec{J}(\vec{r}\,') dv'. \tag{8.72}$$

Again, for a uniform CSB the source distribution $\vec{J}(\vec{r}\,)$ is not a usual current but a distribution of waists of uniform CSBs. For instance, for a Hertzian dipole characterized by $\vec{c}_e$ located at the complex location $\vec{r}_d = \vec{r}_r - j\vec{r}_i$, the corresponding current is given by $\vec{J}(\vec{r}) = \vec{c}_e\delta(\vec{r} - \vec{r}_d)$. For the electric field, we immediately obtain with the filter property of the δ distribution $\vec{E}(\vec{r}) = j\kappa\tilde{\Gamma}(\vec{r}, \vec{r}_d) \cdot \vec{c}_e$. Inserting of (8.71) into (8.72) yields the spherical-multipole expansion

$$\vec{E}(\vec{r}) = \sum_{p=1}^{\infty} a_p \vec{N}_{\sigma_p}(\vec{r}) + \frac{Z}{j} \sum_{p=1}^{\infty} b_p \vec{M}_{\tau_p}(\vec{r}) \tag{8.73}$$

$$\vec{H}(\vec{r}) = \frac{j}{Z} \sum_{p=1}^{\infty} a_p \vec{M}_{\sigma_p}(\vec{r}) + \sum_{p=1}^{\infty} b_p \vec{N}_{\tau_p}(\vec{r}) \tag{8.74}$$

with the multipole amplitudes

$$a_p = \frac{j\kappa}{\sigma_p(\sigma_p + 1)} \iiint_V \vec{N}_{\sigma_p}(\vec{r}\,') \cdot \vec{J}(\vec{r}\,')dv' \tag{8.75}$$

$$b_p = \frac{-\kappa}{Z\tau_p(\tau_p + 1)} \iiint_V \vec{M}_{\tau_p}(\vec{r}\,') \cdot \vec{J}(\vec{r}\,')dv'. \tag{8.76}$$

Similar to the scalar case, for an incident uniform CSB the complex source coordinate has to be carefully chosen to achieve a causal solution.

8.5 Numerical evaluation

8.5.1 Convergence analysis

First we investigate the convergence properties of the resulting multipole series for evaluating the far field. For this purpose, we start from the case that a plane angular sector as shown in Figure 8.4 is illuminated by a uniform CSB. The scattered far field, however, can only consist of outgoing waves and has to satisfy the radiation

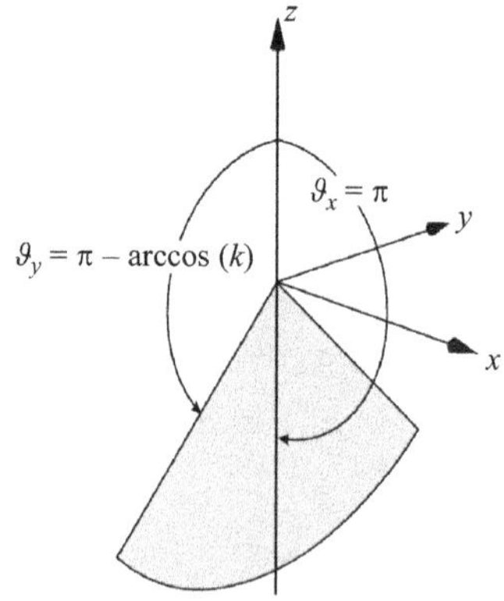

Figure 8.4 Geometry of the plane angular sector $\vartheta = \pi$

condition which is mathematically ensured by using spherical Hankel functions of the second kind, $h_\nu^{(2)}(\kappa r)$ for the scattered fields. To this end, we first subtract from the total field the incident field, i.e., we subtract that uniform CSB that would be present without the sector. Second, we exploit the relation $j_\nu(\kappa r) = (1/2)\left[h_n^{(1)}(\kappa r) + h_n^{(2)}(\kappa r)\right]$ to separate the scattered far field that satisfies the radiation condition [22]. Practically, this can be easily achieved if the vector spherical-multipole functions in (8.73) are built up using spherical Hankel functions of the second kind. With their corresponding approximation for large values of the argument, we end up with the far-field spherical-multipole expansion

$$\vec{E}_\infty(\vec{r}) = \frac{\exp(-j\kappa r)}{\kappa r}\left[-\sum_{p=1}^{\infty} j^{\sigma_p} a_p \vec{n}_{\sigma_p}(\vartheta,\varphi) + \frac{Z}{j}\sum_{p=1}^{\infty} j^{\sigma_p+1} b_p \vec{m}_{\tau_p}(\vartheta,\varphi)\right] \quad (8.77)$$

with the multipole amplitudes given by (8.75) and (8.76). The scattered far field is then obtained by evaluating the term inside the brackets, i.e., without the preceding spherical-wave term in (8.77) and subtracting the corresponding far-field expansion for an undisturbed CSB. The results in Figure 8.5 demonstrate that the scattered electromagnetic far field for the sector can be achieved by an (absolutely) convergent spherical-multipole expansion. Moreover, we clearly observe the development of the

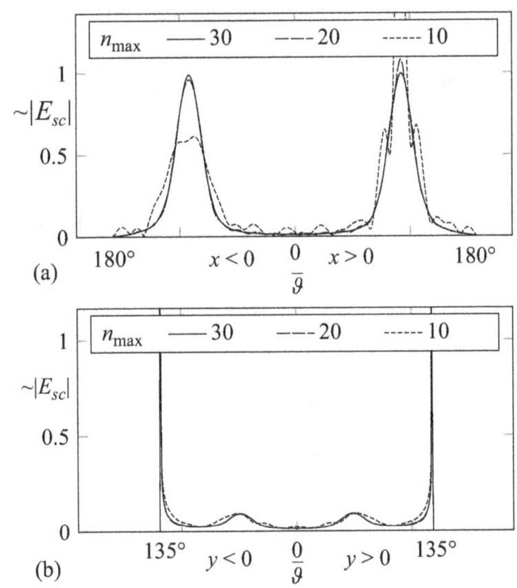

Figure 8.5 Normalized amount of the scattered far field in the xz-plane (a) and in the yz-plane (b) of a PEC quarter (see Figure 8.4) illuminated by a uniform CSB for different maximum numbers of the eigenfunctions. The complex-valued spherical coordinates of the source point of the y-polarized Hertzian dipole are given by $r' = (0.001 + 10)\Lambda, \bar{\vartheta}' = 75°, \bar{\varphi} = 0°$

singularity at the edges of the waist, while the field converges at all other observation points. Since the tangential field vanishes on the PEC sector and due to the symmetry of the incident CSB field there is no normal field component on the sector so the amount of the scattered field is zero there (see Figure 8.5(b)). Similarly, excellent convergence properties of the far-field expansions in the case of an incident uniform CSB are also observed for other geometries of the semi-infinite elliptic cone.

8.5.2 Numerical results for an acoustically soft or hard semi-infinite elliptic cone

For the following examples, we consider a semi-infinite elliptic cone as sketched in Figure 8.6. For all numerical results, we consider a cone with complementary (outer) half-opening angles, $\vartheta_x = 135°, \vartheta_y = 120°$. The center of the waist of the incident uniform CSB is located immediately at the cone's tip in the xz-plane at $r_r = 0.001\,\Lambda, \bar{\vartheta}_r = 75°, \bar{\varphi}_r = 0°$. The uniform CSB is traveling toward the tip of the elliptic cone with a Rayleigh length of $r_i = 10\Lambda$. Figures 8.7–8.10 show the diffracted (total) fields in the xz-plane (Figures 8.7 and 8.8) and yz-plane (Figures 8.9 and 8.10) of a focused uniform CSB incident on the tip of an acoustically hard or soft elliptic cone. While the incident beam is clearly identified in the xz-planes (where the waist of the uniform CSB is located), the yz-planes are dominated by the field scattered by the illuminated area, including the field scattered by the tip.

Figures 8.11–8.13 represent the corresponding scattered far fields in the xz-, yz-, and xz-planes. All far fields are obtained following the previously described procedure (see Section 8.5.1). The scattered far fields are normalized with respect to the maximum value obtained for the soft cone in the xz-plane (Figure 8.11).

For both cases, the soft and the hard elliptic cones, we observe a maximum of the scattered far field in the transition region; however, the maxima are not at the same direction. A second maximum at an identical angle for both cases is observed at the angle of reflection.

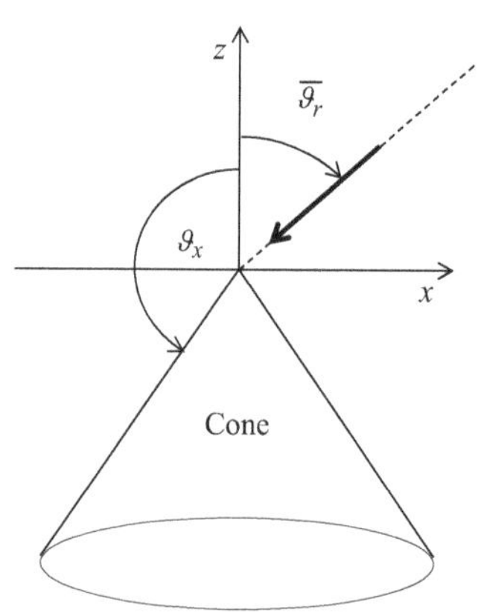

Figure 8.6 Geometry of the elliptic cone illuminated by an incident uniform CSB

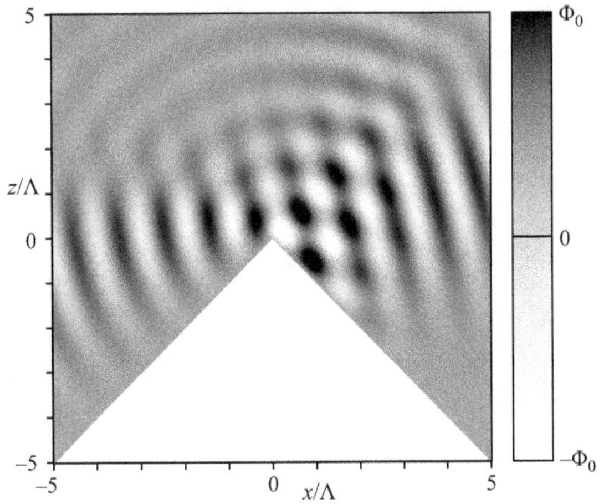

Figure 8.7 *Snapshot of a uniform CSB diffracted by an acoustically soft elliptic cone in the xy-plane (geometry see Figure 8.6). The cone's complementary half-opening angles are $\vartheta_x = 135°, \vartheta_y = 120°$. The waist of the incident uniform CSB is located in the xz-plane at $r_r = 0.001\Lambda, \bar{\vartheta}_r = 75°, \bar{\varphi}_r = 0°$. The uniform CSB is traveling directly toward the tip of the elliptic cone with a Rayleigh length of $r_i = 10\Lambda$*

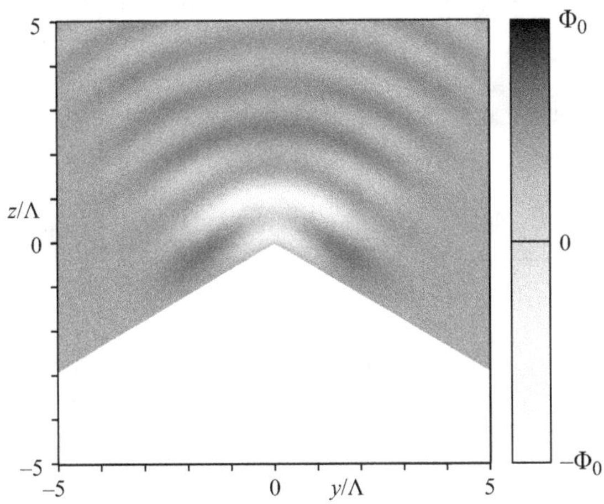

Figure 8.8 *Snapshot of a uniform CSB diffracted by an acoustically soft elliptic cone in the yz-plane. (See the caption of Figure 8.7 for more details.)*

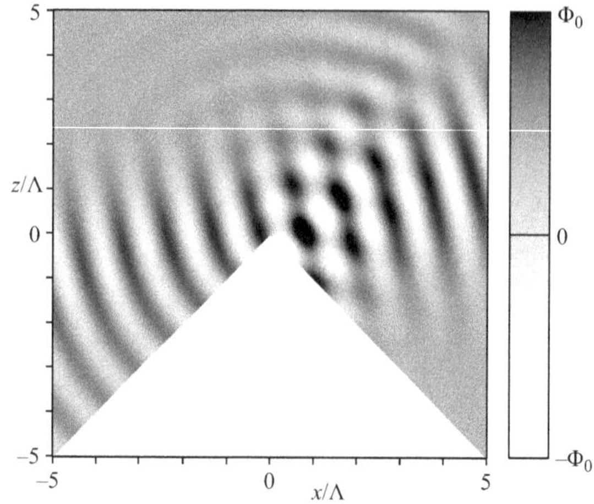

Figure 8.9 *Snapshot of a uniform CSB diffracted by an acoustically hard elliptic cone in the xz-plane. (See the caption of Figure 8.7 for more details.)*

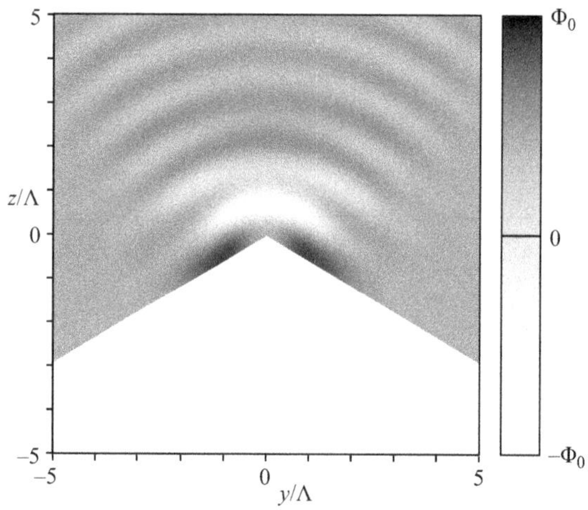

Figure 8.10 *Snapshot of a uniform CSB diffracted by an acoustically hard elliptic cone in the yz-plane. (See the caption of Figure 8.7 for more details.)*

8.5.3 Numerical results for a perfectly conducting semi-infinite elliptic cone

For the electromagnetic case, we consider that the source of the vector uniform CSB is given by a Hertzian dipole located at the complex location with the spherical coordinates $((0.001 + j10)\Lambda,\ 75^\circ,\ 0^\circ)$ polarized either in the $\bar{\varphi}$- ($= y$-) or in

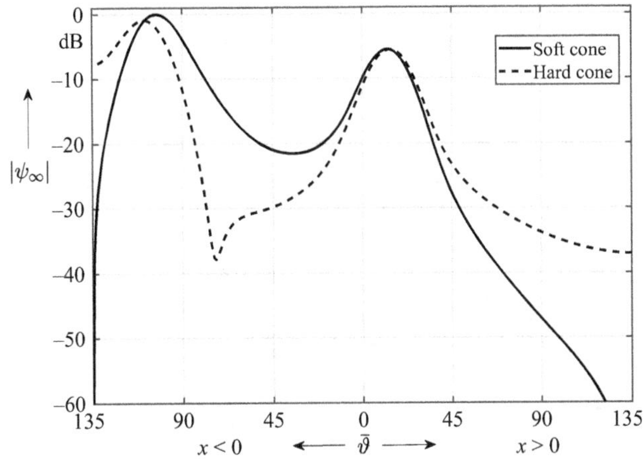

Figure 8.11 Normalized scattered far field of a uniform CSB illuminating an acoustically soft or hard elliptic cone in the xz-plane. (See the caption of Figure 8.7 for more details.)

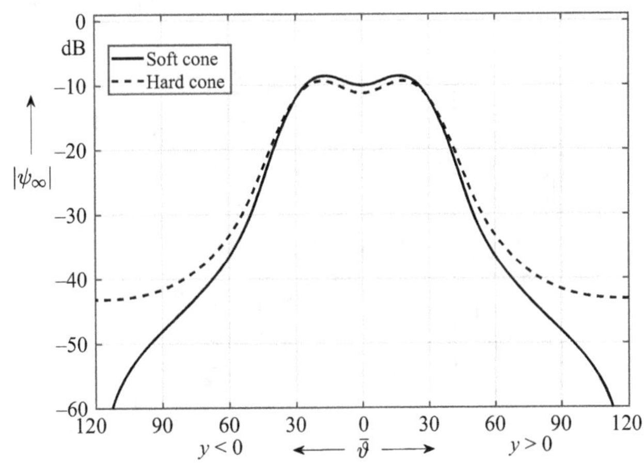

Figure 8.12 Normalized scattered far field of a uniform CSB illuminating an acoustically soft or hard elliptic cone in the yz-plane. (See the caption of Figure 8.7 for more details.)

the $\bar{\vartheta}$-direction. Note that the polarization of the Hertzian dipole leads to an equivalent polarization of the uniform CSB. The other data are chosen identically to the acoustic case, see the caption of Figure 8.7.

Figures 8.14 and 8.15 show the E_y component in the *xz*- and *yz*-planes if the Hertzian dipole is *y* polarized.

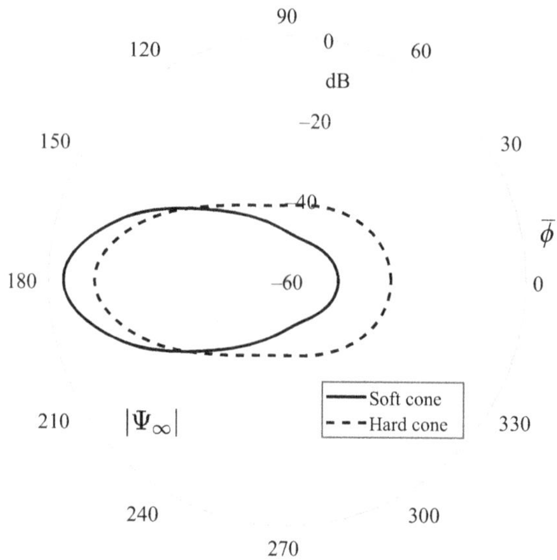

Figure 8.13 *Normalized scattered far field of a uniform CSB illuminating an acoustically soft or hard elliptic cone in the xy-plane. (See the caption of Figure 8.7 for more details.)*

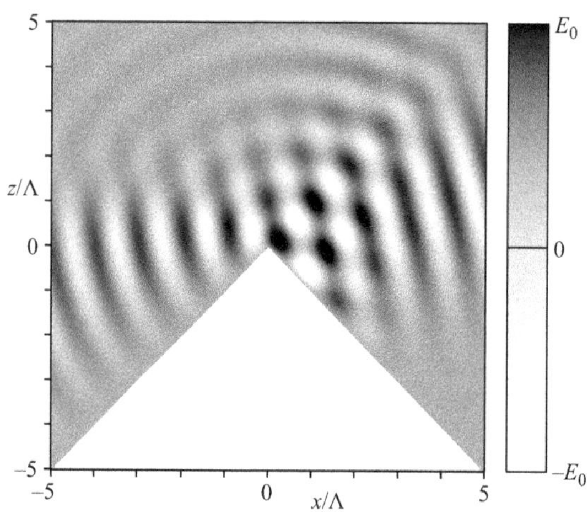

Figure 8.14 *Snapshot of the E_y-component in the xz-plane of a uniform CSB polarized in the y-direction ($=\hat{\varphi}$ direction) diffracted by a PEC elliptic cone. (See the caption of Figure 8.7 for more details.)*

Finally, Figures 8.16–8.18 represent the scattered electric far fields in the *xz*-, *yz*-, and *xy*-planes, each for two orthogonal polarizations of the incident uniform CSB. All scattered electric far fields are normalized with respect to the maximum value obtained for the result in the *xz*-plane obtained for a $\hat{\varphi}$-polarization of the

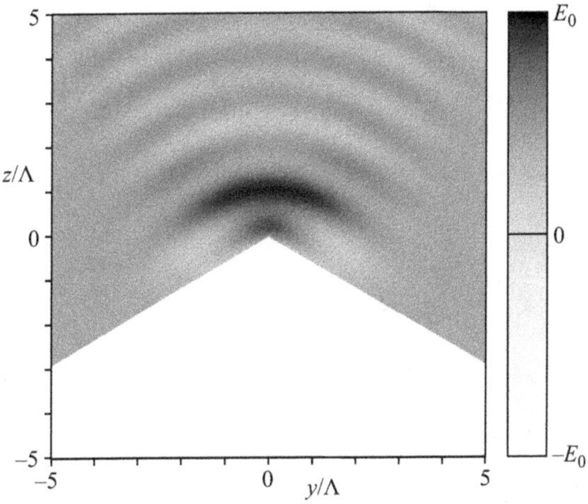

Figure 8.15 Snapshot of the E_y-component in the yz-plane of a uniform CSB polarized in the y-direction diffracted by a PEC elliptic cone. (See the caption of Figure 8.7 for more details.)

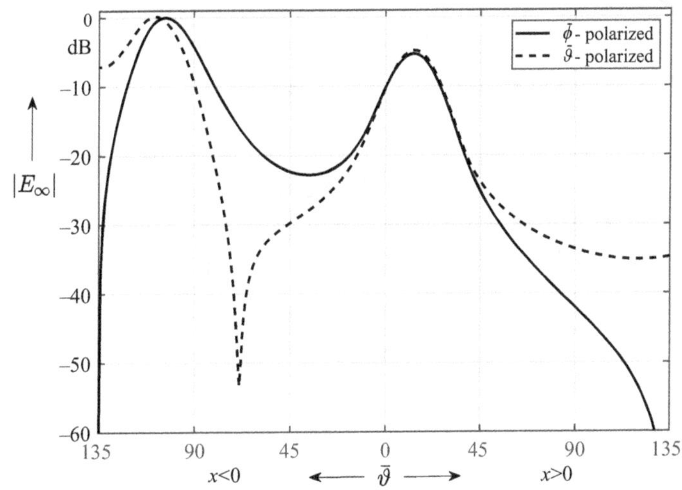

Figure 8.16 Normalized scattered far field in the xz-plane of a uniform CSB illuminating a PEC elliptic cone for two different polarizations of the incident CSB. (See the caption of Figure 8.7 for more details.)

Hertzian dipole (Figure 8.16). In the three orthogonal planes, we observe that the scattered electric far field for a $\bar{\varphi}$-polarized Hertzian dipole is similar to the scattered acoustic far field in the case of a soft cone, while the scattered electric far field for a $\bar{\vartheta}$-polarized Hertzian dipole has similarity to the scattered acoustic far field in the case of a hard cone.

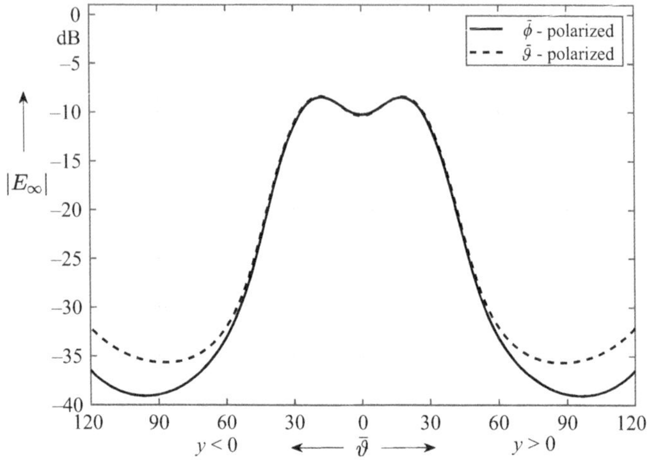

Figure 8.17 Normalized scattered far field in the yz-plane of a uniform CSB illuminating a PEC elliptic cone for two different polarizations of the incident CSB. (See the caption of Figure 8.7 for more details.)

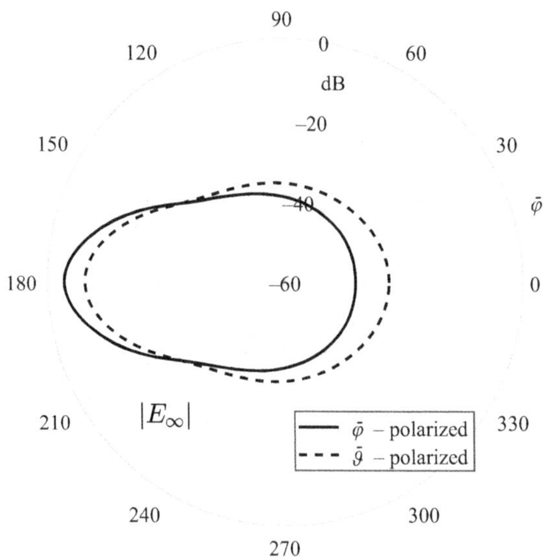

Figure 8.18 Normalized scattered far field in the xy-plane of a uniform CSB illuminating a PEC elliptic cone for two different polarizations of the incident CSB. (See the caption of Figure 8.7 for more details.)

8.6 Conclusions

In this chapter, we successfully investigated the combination of the spherical-multipole expansion of scalar or electromagnetic fields in the presence of a semi-infinite elliptic cone with the uniform CSB as the incident field. Special emphasis has to be laid on the correct choice of the complex source coordinates in order to obtain a causal solution. It has been shown that the multipole expansion with a uniform CSB as the incident field is absolutely convergent and yields stable and correct results even for obtaining the scattered far fields. The outcomes can be used as reference results for both, numerical codes and—in particular—for asymptotic methods like the GTD/UTD. Finally, the developed methods and results will be the basis for further investigations aiming at a separation of the tip-diffracted parts from the exact scattered far fields found in this chapter.

References

[1] D. Bouche, F. Molinet, and R. Mittra, *Asymptotic Methods in Electromagnetics*. Berlin-Heidelberg-New York: Springer, 1997.

[2] S. Blume and L. Klinkenbusch, "Spherical-multipole analysis in electromagnetics," in *Frontiers in Electromagnetics* (D. Werner and R. Mittra, eds.), pp. 553–608, Piscataway (NJ): IEEE Press, 2000.

[3] V. Babich, V. Smyshlyaev, D. Dement'ev, and B. Samokish, "Numerical calculation of the diffraction coefficients for an arbitrary shaped perfectly conducting cone," *IEEE Transactions on Antennas and Propagation*, vol. 44, no. 5, pp. 740–747, 1996.

[4] L. Klinkenbusch and H. Brüns, "Diffraction of a uniform complex-source beam by a circular cone," in *Proc. 2014 International Conference on Electromagnetics in Advanced Application, Aruba, August 3–8*, pp. 470–472, 2014.

[5] R. G. Kouyoumjian, T. Celandroni, G. Manara, and P. Nepa, "Inhomogeneous electromagnetic plane wave diffraction by a perfectly electric conducting wedge at oblique incidence," *Radio Science*, vol. 42, no. 06, pp. 1–10, 2007.

[6] L. Eisenhart, "Stäckel systems in conformal Euclidian space," *Annals of Mathematics*, vol. 36, pp. 57–70, 1935.

[7] P. Moon and D. Spencer, *Field Theory Handbook (2nd ed.)*. Berlin-Heidelberg-New York: Springer Verlag, 1971. pp. 136–143.

[8] G. Hellwig, *Differential Operators in Mathematical Physics (in German, original title: Differentialoperatoren der mathematischen Physik)*. Berlin-Göttingen-Heidelberg: Springer Verlag, 1964.

[9] L. Klinkenbusch, *Spherical-multipole analysis of the scattering and diffraction by an elliptically contoured spherical shell (in German)*. Dissertation, Ruhr-Universität Bochum, Bochum, Germany, 1991.

[10] J. Jansen, *Simple-periodic and non-periodic Lamé functions and their application in the theory of conical waveguides*. Dissertation, Eindhoven University of Technology, Eindhoven, The Netherlands, 1976.

[11] J. Stratton, *Electromagnetic Theory*. New York: McGraw-Hill, 1941.

[12] S. Blume and U. Uschkerat, "The radar cross section of the semi-infinite elliptic cone: numerical evaluation," *Wave Motion*, vol. 22, no. 3, pp. 311–324, 1995.

[13] L. Klinkenbusch, "Electromagnetic scattering by semi-infinite circular and elliptic cones," *Radio Science*, vol. 42, no. RS6S10, 2007.

[14] L. Felsen, "Evanescent waves," *Journal of the Optical Society of America*, vol. 66, pp. 751–760, 1976.

[15] S. Shin and L. Felsen, "Gaussian beam modes by multipoles with complex source points," *Journal of the Optical Society of America*, vol. 67, pp. 699–700, 1977.

[16] S. Vinogradov, P. Smith, and E. Vinogradova, *Canonical Problems in Scattering and Potential Theory, Part II*. Boca Raton (FL): Chapman & Hall/CRC Press, 2002. Chapter 4.5.

[17] M. Katsav, E. Heyman, and L. Klinkenbusch, "Diverging and converging beam diffraction by a wedge. Part II: Plane wave spectral solutions and complex ray solutions," in *Proceedings of the Progress in Electromagnetics Research Symposium (PIERS), Prague, July 6–9*, pp. 1204–1208, 2015.

[18] C. Scheppard and S. Saghafi, "Beam modes beyond the paraxial approximation: A scalar treatment," *Physical Review A*, vol. 57, no. 4, pp. 2971–2979, 1998.

[19] A. Tagirdzhanov and A. Kiselev, "Complexified spherical waves and their sources: A review," *Optics and Spectroscopy*, vol. 119, no. 2, pp. 257–267, 2015.

[20] L. Klinkenbusch and H. Brüns, "Combined complex-source beam and spherical-multipole analysis for the electromagnetic probing of conical structures," *Comptes Rendus Physique*, vol. 17, pp. 960–965, 2016.

[21] M. Katsav and E. Heyman, "Converging and diverging beam diffraction by a wedge: a complex-source formulation and alternative solutions," *IEEE Transactions on Antennas and Propagation*, vol. 64, no. 7, pp. 3080–3093, 2016.

[22] H. Brüns and L. Klinkenbusch, "Acoustic scattering of a complex-source beam by the edge of a plane angular sector," *Advances in Radio Science*, vol. 12, pp. 179–186, 2014.

Chapter 9

Changes in the far-field pattern induced by rounding the corners of a scatterer: dependence upon curvature

Audrey J. Markowskei[1] *and Paul D. Smith*[1]

Diffraction of acoustic or electromagnetic waves by canonical shapes and structures of more general and arbitrary shape is of enduring interest. The choice of an appropriate canonical structure to model the dominant features of a scattering scenario can be very illuminating. Scattering by sharp edges and corners is informed by, for example, the diffraction from the half-plane and the wedge (of infinite extent). The nature of the singularities in the field and its derivatives is described in [1–3]. There has been much work to develop analytical and numerical methods to account for these singularities which enable accurate modelling; however, these methods can be time consuming to implement and at times become very specialised. When numerical methods are employed, a common approach used when dealing with domains with corners is to round the corners, producing a smooth surface. This eliminates the singularities introduced by the corners and allows for standard numerical quadratures to be used, though leaving the researcher with no clear estimate of the error or difference induced by the rounding.

The earliest study of scattering from a sharp object could be said to begin with Sommerfeld's solution to the half-plane problem [4,5]. It was the first recognition that the dependencies of the scattered field differed to that of a smooth body. The next major development was the study of the wedge [6–8] where the dependence of the scattered field on the angle of the wedge was made explicit. These canonical problems admitted analytical solutions that demonstrated explicitly the dependence of field quantities on the corner angle.

There is a vast literature on numerical approaches to scattering calculations that may be classified in terms of the diameter $2a$ of the scatterer in wavelengths. At long wavelengths, the scattered field may be regarded as a perturbation of a corresponding static problem and is expanded in a series of powers of ka (k – wavenumber), while at short wavelengths ($ka \gg 1$), ray tracing techniques are often deployed.

In the intermediate or resonance regime, integral equation approaches provide the usual basis for numerical methods. Over the last couple of decades, the advantages

[1] Department of Mathematics and Statistics, Macquarie University, Sydney, Australia

of integral equation formulations – well-posed second-kind equations, unknowns to be found only on the scattering surface (rather than the surrounding space) and automatic incorporation of the radiation condition obviating the need for terminating the volumetric grid characteristic of differential equation methods – have become well established for the reasons explained in [9, Ch 3].

Although there is an extensive literature on scattering and diffraction from sharp-cornered objects as well as those with smooth boundaries, there does not seem to be a systematic treatment of the transition from one to the other, in particular, as the radius of curvature of the rounded corner points tends to zero.

In earlier work [10–12], we described a numerical method that is suitable for examining the scattering of acoustic or appropriately polarised electromagnetic plane waves by cylindrical structures possessing some points of small or zero curvature (i.e., having a sharp corner) and compared the convergence of solutions for different discretisations of the surfaces of the scatterers and then assessed the impact on near- and far-field scattering, as a function of the radius of curvature in the vicinity of the rounded corner point. We implemented the Nyström method expounded by [9] for the soft boundary condition for a scatterer with a single corner to obtain numerical solutions of this integral equation. We then developed other (similar) Nyström methods for the hard and impedance boundary conditions and adapted these methods for scatterers with two and four corners to obtain numerical solutions of the respective integral equations.

These numerical methods were used to compare the convergence of solutions for different discretisations of the surfaces of the scatterers, and then to assess the impact on near- and far-field scattering, as a function of the radius of curvature in the vicinity of the rounded corner point. We numerically demonstrated that the field scattered by the rounded structure converges, in both the L^2 and L^∞ norm, to that scattered by the corresponding sharp-cornered object as the radius of curvature in the vicinity of the corner tends to zero. (L^2 and L^∞ denote the function spaces of square integrable functions and bounded functions, respectively.) A variety of single-cornered convex and concave structures with a range of interior angles was examined in [13].

In [14], we numerically examined the quantity

$$\sqrt{k}\,\left\|u_0^\infty - u_\rho^\infty\right\|_\infty, \tag{9.1}$$

being the maximum difference (using the L^∞ norm) in the far-field quantity, where u_0^∞ is the far-field resulting from scattering of a time-harmonic plane wave travelling in direction θ_0 with wavenumber k, from a closed obstacle with a corner, ∂D_0, and u_ρ^∞ is the far-field from the structure resulting from rounding the corner; ρ is the radius of curvature of the rounded corner. (The $\sqrt{k}$ factor correctly non-dimensionalises the far-field quantities.) It was demonstrated that the maximum differences between the far-fields occur in the back-scatter region, and that the magnitude of these differences is dependent on the radius of curvature, ρ, used for the rounding, the wavenumber, k, and the angle of the incident plane wave, θ_0. Of particular interest was the behaviour of the quantity (9.1) as $k\rho$ approaches 0. The study examined E-polarised, H-polarised and impedance-loaded structures, with one, two or four right-angled corners. In all the cases studied, as $k\rho \to 0$,

$$\sqrt{k}\,\left\|u_0^\infty - u_\rho^\infty\right\|_\infty \le C\,(\theta_0)\,(k\rho)^{4/3}, \tag{9.2}$$

and in the E-polarised case

$$\sqrt{k}\,\left\|u_0^\infty - u_\rho^\infty\right\|_\infty \approx C\,(\theta_0)\,(k\rho)^{4/3}, \tag{9.3}$$

for some constant C dependent on the direction of the plane wave, θ_0.

The rounding used in all papers [10]-[15] replaces cornered scatterers (both curvilinear and polygonal) by a smooth object that is extremely close to the original except in the vicinity of the corners. The rounded curve is nearly a hyperbola with sides of the corner being its asymptotes, or similar. Recently, Epstein and O'Neil [16] have developed a different method for rounding the corner of polygons. They have considered the Dirichlet and Neumann problem, and although their approach differs, the numerical results are consistent with ours. The reported relative errors are of a similar order to the earlier results [11,12,15] and they also demonstrate numerically that the difference in the scattered potentials converges as the radius of curvature decreases and that the rate of convergence is dependent on the radius of curvature. For the two examples used in their paper – a structure with interior right-angles and one with interior angles of $\pi/3$ – the rates of convergence correspond to ours. In the thesis [17], the precise dependence of the rate of convergence on the interior angle at the corner, as well as on the radius of curvature, incident wave number and boundary conditions, has been established and numerically quantified.

Another work specifically addressing corner rounding is that of Engineer *et al.* [18], examining diffraction of high-frequency sound waves from two-dimensional (2D) curved slender bodies with Neumann boundary condition as the radius of curvature changes. For this high-frequency study, they used ray theory and identified additional creeping wave propagation features associated with high-frequency problems. A more sophisticated approach to the scattering from soft cylindrical structures with sharp corners is given by [19]. It employs the so-called recursively compressed inverse preconditioning method, and as the authors note in their survey of the 2D scattering literature, it alone addresses the problem of accurate near-field evaluation in scatterers with corners.

In this chapter, we rigorously examine the differences in the far-field patterns (9.1) for an E-polarised scatterer with a single corner. The scattering of a plane wave by such a scatterer is formulated in Section 9.1, and an appropriate integral equation for the surface distribution on the obstacle is given. As a motivation for the analysis to follow, a brief discussion of numerical results is given in Section 9.2. In Section 9.3, the lemniscate (having a right-angled corner) and its rounded counterpart is used as a test case to establish analytic bounds for the maximum difference in the far field. An integral equation is obtained for the difference in the surface distributions on each obstacle; its approximate solution is shown to be $O((k\rho)^{2/3})$, as $k\rho \to 0$ (Theorem 9.1). It then follows that the non-dimensionalised far-field difference $\sqrt{k}\,\left\|u_0^\infty - u_\rho^\infty\right\|_\infty$ is $O((k\rho)^{4/3})$, as $k\rho \to 0$ (Theorem 9.2).

9.1 Problem formulation

Consider an infinitely long cylinder with uniform cross-section, with axis parallel to the z-axis. The cylinder is illuminated by an incident plane wave propagating with direction parallel to the x–y plane. The cross-section D lying in the x–y plane has a closed boundary ∂D parametrised by

$$\mathbf{x}(t) = (x_1(t), x_2(t)), \qquad t \in [0, 2\pi]. \tag{9.4}$$

In this work, we consider scatterers with one acute-angled corner and the family of scatterers that result after the corner has been rounded (see Figure 9.1). The single-cornered scatterer (lemniscate) has parametric representation

$$\mathbf{x} = \mathbf{x}(t) = a\,(2\sin(t/2), -\tan(\alpha/2)\sin t), \quad t \in [0, 2\pi], \tag{9.5}$$

where a is a parameter, henceforth set equal to 1 length unit, and $\alpha \in [0, \pi/2]$ is the interior angle of the (acute) corner which occurs at $t = 0$. In the analysis described in the next section, we will specifically examine the right-angled structure corresponding to the value $\alpha = \pi/2$. The families of curves in which the corners have been rounded are parametrised by the quantity ε $(0 \le \varepsilon \le 1)$. The rounded lemniscate has representation

$$\mathbf{x} = \mathbf{x}(t) = a\left(2\sqrt{\varepsilon^2 + (1 - \varepsilon^2)\sin^2(t/2)}, -\tan(\alpha/2)\sin t\right), \quad t \in [0, 2\pi]. \tag{9.6}$$

The radius of curvature ρ at points $\mathbf{x}(t)$ is

$$\rho(t) = \left|\frac{\left(x_1'(t)^2 + x_2'(t)^2\right)^{3/2}}{x_1'(t)x_2''(t) - x_2'(t)x_1''(t)}\right| \tag{9.7}$$

so that at the rounded corner point $\mathbf{x}(0)$, the radius of curvature is

$$\rho(0) = 2\varepsilon\tan^2\frac{\alpha}{2} + O(\varepsilon^3), \qquad \text{as } \varepsilon \to 0. \tag{9.8}$$

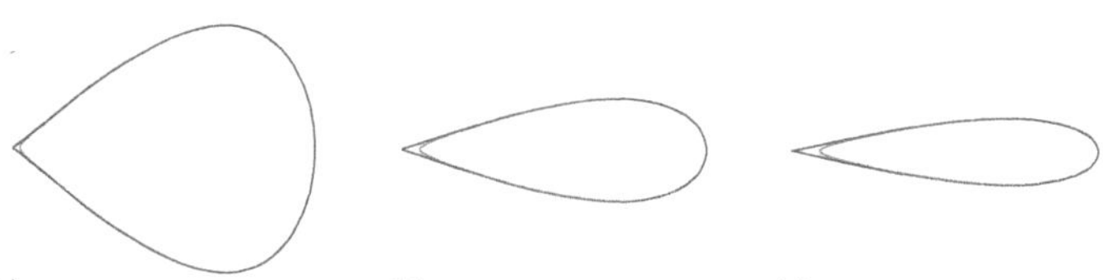

(a) (b) (c)

Figure 9.1 Some lemniscates, (a) α = π/2, (b) α = π/4, (c) α = π/6, considered in this work shown in blue; rounded lemniscate with radius of curvature ρ ≈ 0.05 shown in red

The incident field illuminating the scatterer induces a scattered field. We assume that the incident and scattered fields are time harmonic with a temporal factor $e^{-i\omega t}$. The spatial component $u^{\text{inc}}(x,y)$ of the incident wave travelling in the direction of the unit vector $\mathbf{d} = (\cos\theta_0, \sin\theta_0)$ takes the form

$$u^{\text{inc}}(x,y) = e^{ik\mathbf{x}\cdot\mathbf{d}} \tag{9.9}$$

and satisfies the Helmholtz equation

$$\Delta u(x,y) + k^2 u(x,y) = 0, \qquad (x,y) \in \mathbb{R}^2, \tag{9.10}$$

where $k = \omega/c$ is the wavenumber and c is the speed of waves in the medium or of light in free space.

The spatial component $u^{\text{sc}}(x,y)$ of the scattered field satisfies the Helmholtz equation (9.10) at all points (x,y) exterior to the body; it obeys the 2D form of the Sommerfeld radiation condition

$$\lim_{|\mathbf{x}|\to\infty} \sqrt{|\mathbf{x}|}\left(\frac{\partial u^{\text{sc}}(\mathbf{x})}{\partial \mathbf{x}} - iku^{\text{sc}}(\mathbf{x})\right) = 0, \qquad \mathbf{x} \in \mathbb{R}^2 \backslash D, \tag{9.11}$$

as well as the finiteness of energy condition in the vicinity of the corners.

The nature of the scatterer imposes certain conditions on the total field $u^{\text{tot}} = u^{\text{inc}} + u^{\text{sc}}$ at the boundary of the scatterer ∂D. This work considers the E-polarised (Dirichlet, sound-soft) boundary condition:

$$u^{\text{tot}}(\mathbf{x}) = 0, \qquad \mathbf{x} \in \partial D. \tag{9.12}$$

We define two operators associated with the single- and double-layer potentials of a continuous density $\phi(\mathbf{y})$ defined on the boundary ∂D,

$$(\mathscr{S}\phi)(\mathbf{x}) = 2\int_{\partial D} G(\mathbf{x},\mathbf{y})\phi(\mathbf{y})\,\mathrm{d}s(\mathbf{y}), \quad (\mathscr{K}\phi)(\mathbf{x}) = 2\int_{\partial D} \frac{\partial G(\mathbf{x},\mathbf{y})}{\partial \mathbf{n}(\mathbf{y})}\phi(\mathbf{y})\,\mathrm{d}s(\mathbf{y}), \tag{9.13}$$

where G is the 2D free-space Green function

$$G(\mathbf{x},\mathbf{y}) = \frac{i}{4}H_0^{(1)}(k\,|\mathbf{x}-\mathbf{y}|\,), \tag{9.14}$$

and $H_0^{(1)}$ denotes the Hankel function of first kind and order zero.

The solution to the exterior Dirichlet problem is based on representing the scattered field as a combination of the single $((1/2)\mathscr{S}\phi(\mathbf{x}))$ and double-layer $((1/2)\mathscr{K}\phi(\mathbf{x}))$ potentials via

$$u^{\text{sc}}(\mathbf{x}) = \int_{\partial D}\left\{\frac{\partial G(\mathbf{x},\mathbf{y})}{\partial \mathbf{n}(\mathbf{y})} - i\eta G(\mathbf{x},\mathbf{y})\right\}\phi(\mathbf{y})\,\mathrm{d}s(\mathbf{y}), \qquad x \in \mathbb{R}^2\backslash\bar{D}, \tag{9.15}$$

where η is a coupling parameter, provided the continuous density $\phi(\mathbf{x})$ is a solution to the following integral equation on ∂D:

$$\phi + \mathscr{K}\phi - i\eta\mathscr{S}\phi = -2g, \tag{9.16}$$

where $g = u^{\text{inc}}$. Uniqueness and solubility conditions are detailed in [20,21]. To ensure uniqueness, the parameter η is chosen to be positive number (usually equal to k).

The numerical solution to (9.16) is obtained by inserting the parametrisation (9.4) into (9.16) to obtain an integral equation over $[0, 2\pi]$. This can then be discretised subsequently by the use of a quadrature rule or a Galerkin method or similar to produce a linear set of equations for the coefficients that represent the unknown surface quantity ϕ. In the work [15] and all related papers, we used the Nyström method outlined in [9] as the numerical method used to approximate the solution to the integral equation (9.15). In [12], we demonstrated that when the scatterer possesses rounded corners of small radii of curvature, it is essential to use an appropriate quadrature scheme – a *graded mesh* – in order to obtain numerical results efficiently. Use of an equispaced mesh is at best grossly inefficient and at worst produces non-convergence.

Once the solution to (9.16) has been obtained for a scatterer with surface ∂D obeying the Dirichlet boundary condition, the far-field pattern is calculated as [9]

$$u^{\infty}(\hat{\mathbf{x}}) = \frac{e^{-i\frac{\pi}{4}}}{\sqrt{8\pi k}} \int_{\partial D} \left\{ k\, \mathbf{n}(\mathbf{y}) \cdot \hat{\mathbf{x}} + \eta \right\} e^{-ik\hat{\mathbf{x}}\cdot\mathbf{y}}\, \phi(\mathbf{y})\, \mathrm{d}s(\mathbf{y}), \qquad |\hat{\mathbf{x}}| = 1. \tag{9.17}$$

Converting to a line integral by inserting the relevant surface parametrisation $\mathbf{y} = \mathbf{y}(t) = (y_1(t), y_2(t))$ for $t \in [0, 2\pi]$ produces

$$u^{\infty}(\hat{\mathbf{x}}) = \frac{e^{-i\frac{\pi}{4}}}{\sqrt{8\pi k}} \int_{0}^{2\pi} \left\{ k\, \mathbf{n}(\mathbf{y}(t)) \cdot \hat{\mathbf{x}} + \eta \right\} e^{-ik\hat{\mathbf{x}}\cdot\mathbf{y}(t)}\, \psi(t)\, \mathrm{d}t, \tag{9.18}$$

where $\psi(t) = \psi(\mathbf{y}(t)) = \phi(\mathbf{y}(t))\, |\mathbf{y}'(t)|$.

Let u_0^{∞} denote the far-field of the scatterer ∂D_0 with a single corner and surface parameterisation $\mathbf{x}_0 = \mathbf{x}_0(t)$. Let u_ρ^{∞} denote the far-field of the corresponding scatterer ∂D_0 after rounding the corner with radius of curvature ρ (see (9.6)), possessing surface parameterisation $\mathbf{x}_\rho = \mathbf{x}_\rho(t)$. Then the maximum difference in the far-field pattern between unrounded and rounded scatterers is (with the choice $\eta = k$)

$$\begin{aligned} \sqrt{k}\, \left\| u_0^{\infty}(\hat{\mathbf{x}}) - u_\rho^{\infty}(\hat{\mathbf{x}}) \right\|_\infty = \sup_{\hat{\mathbf{x}} \in [0,2\pi]} \Bigg| \frac{e^{-i\frac{\pi}{4}}}{\sqrt{8\pi}} \int_0^{2\pi} \Big\{ & k\left\{ \mathbf{n}(\mathbf{y}_0(t)) \cdot \hat{\mathbf{x}} + 1 \right\} e^{-ik\hat{\mathbf{x}}\cdot\mathbf{y}_0(t)}\, \psi_0(t) \\ & - k\left\{ \mathbf{n}(\mathbf{y}_\rho(t)) \cdot \hat{\mathbf{x}} + 1 \right\} e^{-ik\hat{\mathbf{x}}\cdot\mathbf{y}_\rho(t)}\, \psi_\rho(t) \Big\} \mathrm{d}t \Bigg|. \end{aligned} \tag{9.19}$$

Note that the $\sqrt{k}$ factor correctly non-dimensionalises the far-field quantities.

9.2 Numerical results and discussion

In [14], we investigated the rate of convergence of far-field solutions as the radius of curvature ρ at the rounded corner approaches zero, by examining the non-dimensional quantity (9.19). Using a least squares fit to the logarithms of the data, for $k\rho \leq 0.25$, it was found that for scatterers with interior right-angled corners with Dirichlet boundary conditions,

$$\sqrt{k}\left\|u_0^\infty - u_\rho^\infty\right\|_\infty \approx C\left(\theta_0\right)\left(k\rho\right)^{4/3}, \tag{9.20}$$

as $k\rho \to 0$, for some constant C dependent on the angle of the incident wave θ_0.

An example of the far-field data differences computed for the lemniscate studied in this work, illuminated in the direction $\theta_0 = \pi/4$, is shown in Figure 9.2. In each example, the data was collated cover a range of wave numbers k and radii of curvature ρ.

Subsequently this work was extended [17] to examine lemniscates with interior corner angles α, where $\pi/36 < \alpha \leq \pi/2$. Again, a least squares fit to the logarithms of the data, for $k\rho \ll 1$, shows that, as $k\rho \to 0$,

$$\sqrt{k}\left\|u_0^\infty - u_\rho^\infty\right\|_\infty \approx C\left(\theta_0, \alpha\right)\left(k\rho\right)^m, \tag{9.21}$$

where the constant C is dependent on the angle of the incident wave, θ_0 and the interior angle α, and the exponent m is dependent on the interior angle α. Moreover, the exponent m equals (to an excellent approximation) $2/\nu$, where $\nu = (2\pi - \alpha)/\pi$.

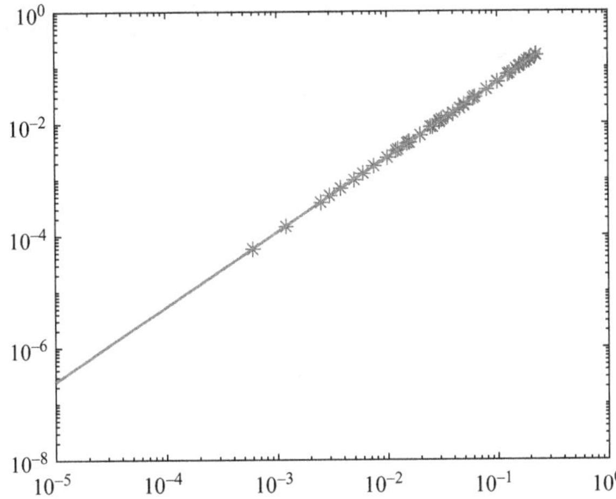

Figure 9.2 Logarithmic plot: $x = k\rho$, $y = \sqrt{k}\left\|u_0^\infty - u_\rho^\infty\right\|_\infty$. The data points used are represented by the blue asterisks, the least squares line of fit is shown in red.

The quantity ν is a key parameter occurring in the analysis of the infinite wedge with interior angle α (see [1,3]).

9.3 Analytic bounds for the far-field difference

In the previous section, an integral equation formulation was used for the numerical studies of the scattering of plane waves by an E-polarised obstacle (Dirichlet boundary condition); its solution provides a continuous surface density from which all physical quantities such as far-field pattern can be calculated. This surface density is different from the physical surface quantities described earlier and has no simple physical interpretation.

In this section, we analyse the underlying integral equations obtained by introducing suitable surface parametrisations for rounded and unrounded scatterers; we deduce an approximate integral equation for the difference in the surface density at corresponding points, in terms of the difference in the illuminating incident field on each scatterer and of the surface quantity on the sharp-cornered object. The lemniscate scatterer with right-angled corner, introduced in Section 9.1 and shown in Figure 9.1(a), is taken as a test case. The approximate solution of the integral equation will be shown to be $O((k\rho)^{2/3})$, from which it is deduced that $\sqrt{k}\,\left\|u_0^\infty - u_\rho^\infty\right\|_\infty = O((k\rho)^{4/3})$, as $k\rho \to 0$.

9.3.1 Integral equations for the difference in surface quantities

Thus, we consider the integral equation (9.15). For simplicity of presentation, we will suppose that $\eta = 0$; then the integral equation has a unique solution at all except countably many wave numbers k.

The lemniscate, a PEC cylinder with uniform cross-section and axis parallel to the z-axis, is illuminated by a time-harmonic incident plane wave $u^{\mathrm{inc}}(\mathbf{x})$ propagating with direction parallel to the x–y plane. The cross-section D lying in the x–y plane has a closed boundary ∂D parametrised by

$$\mathbf{x}_0(t) = (x_{0,1}(t), x_{0,2}(t)) = (2\sin(|t|/2), -\sin t), \tag{9.22}$$

and the corresponding rounded object depending upon a (small) positive parameter ε is parametrised by

$$\mathbf{x}_\varepsilon(t) = (x_{\varepsilon,1}(t), x_{\varepsilon,2}(t)) = \left(2\sqrt{\varepsilon^2 + (1-\varepsilon^2)\sin^2(t/2)}, -\sin t\right), \tag{9.23}$$

for $t \in [-\pi, \pi]$. The radius of curvature ρ at the rounded corner point (corresponding to $t = 0$) is very close to 2ε, for small ε.

The outward pointing unit normal at the point $\mathbf{x}_0(t)$ on the lemniscate is

$$\mathbf{n}(\mathbf{x}_0(t)) = \frac{\left(x'_{0,2}(t), -x'_{0,1}(t)\right)}{\sqrt{\left(x'_{0,1}(t)\right)^2 + \left(x'_{0,2}(t)\right)^2}}; \tag{9.24}$$

the outward pointing unit normal at the point $\mathbf{x}_\varepsilon(t)$ on the rounded object is similarly defined.

Inserting these parametrisations in (9.13) determines double-layer potentials $\mathscr{K}_0$ and $\mathscr{K}_\varepsilon$. The integral equations governing the surface quantities $\varphi(\mathbf{x}_0)$ and $\varphi(\mathbf{x}_\varepsilon)$ are

$$\varphi(\mathbf{x}_0) + \mathscr{K}_0\varphi(\mathbf{x}_0) = -2u^{\text{inc}}(\mathbf{x}_0), \tag{9.25}$$

and

$$\varphi(\mathbf{x}_\varepsilon) + \mathscr{K}_\varepsilon\varphi(\mathbf{x}_\varepsilon) = -2u^{\text{inc}}(\mathbf{x}_\varepsilon). \tag{9.26}$$

More concretely, if we set $\varphi_0(t) = \varphi_0(\mathbf{x}_0(t))$, $\varphi_\varepsilon(t) = \varphi_\varepsilon(\mathbf{x}_\varepsilon(t))$, $g_0(t) = 2u^{\text{inc}}(\mathbf{x}_0(t))$ and $g_\varepsilon(t) = 2u^{\text{inc}}(\mathbf{x}_\varepsilon(t))$, then

$$\varphi_\varepsilon(t) - \int_{-\pi}^{\pi} H_\varepsilon(t,\tau)\varphi_\varepsilon(\tau)\,\mathrm{d}\tau = g_\varepsilon(t), \tag{9.27}$$

and

$$\varphi_0(t) - \int_{-\pi}^{\pi} H_0(t,\tau)\varphi_0(\tau)\,\mathrm{d}\tau = g_0(t), \tag{9.28}$$

for $t \in [-\pi,\pi]$, where (see [9])

$$H_0(t,\tau) = \frac{1}{2}ik\,\frac{H_1^{(1)}(k|\mathbf{x}_0(t)-\mathbf{x}_0(\tau)|)}{|\mathbf{x}_0(t)-\mathbf{x}_0(\tau)|}\,\mathbf{n}\,(\mathbf{x}_0(\tau))\cdot(\mathbf{x}_0(t)-\mathbf{x}_0(\tau))\;|\mathbf{x}_0'(\tau)|, \tag{9.29}$$

and $H_\varepsilon(t,\tau)$ is similarly defined.

Thus, setting $\Delta(t) = \varphi_\varepsilon(t) - \varphi_0(t)$, we obtain

$$\Delta(t) - \int_{-\pi}^{\pi} H_\varepsilon(t,\tau)\Delta(\tau)\,\mathrm{d}\tau = \int_{-\pi}^{\pi} (H_\varepsilon(t,\tau) - H_0(t,\tau))\,\varphi_0(\tau)\,\mathrm{d}\tau + g_\varepsilon(t) - g_0(t). \tag{9.30}$$

Thus, if we regard $\varphi_0(\tau)$ as known, then this is an integral equation to be solved for the unknown difference $\Delta(t)$.

9.3.2 Approximate integral equation for the difference Δ

1 Now let I be a symmetrical subinterval of $[-\pi,\pi]$ containing 0, and set $J = [-\pi,\pi]\setminus I$. The intervals I and J will be fully specified later (in fact

$I = [-\varepsilon^{2/3}, \varepsilon^{2/3}]$), but the requirements are as follows:

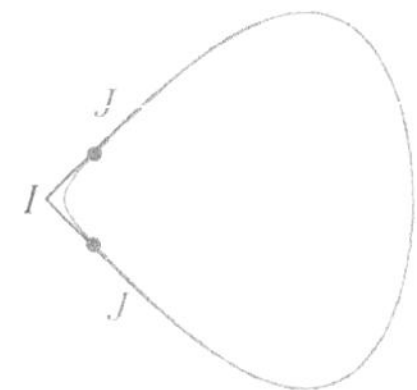

1. on the set J, the difference in the parametrisations (9.22) and (9.23), and in their derivatives, is negligibly small. In fact, a straightforward calculation shows that

$$\|\mathbf{x}_0 - \mathbf{x}_\varepsilon\|_{\infty,J} = \max_{t\in J} |\mathbf{x}_0(t) - \mathbf{x}_\varepsilon(t)| \leq 4\varepsilon^{4/3}, \tag{9.31}$$

and

$$\|\mathbf{x}_0' - \mathbf{x}_\varepsilon'\|_{\infty,J} = \max_{t\in J} |\mathbf{x}_0'(t) - \mathbf{x}_\varepsilon'(t)| \leq 2\varepsilon^{2/3}; \tag{9.32}$$

2. on the set J, the maximum difference $\|\varphi_\varepsilon - \varphi_0\|_{\infty,J}$, in the surface quantities on either object, is negligibly small. We make this analytic assumption which is numerically justified in Appendix C.2 of [17];
3. the interval I is small, that is, $k|I|$ is small, so that we may use small argument approximations for the Hankel function $H_1^{(1)}$. This is equivalent to approximating the Green function $G(\mathbf{x}, \mathbf{y})$ at points $\mathbf{x}, \mathbf{y}$, parametrised by I, by the corresponding values of static Green's function $G_0(\mathbf{x}, \mathbf{y})$.

Thus, we make the approximation $\Delta(t) = 0$ for $t \in J$. Inserting this in (9.30) produces, when $t \in I$,

$$\begin{aligned}\Delta(t) - \int_I H_\varepsilon(t,\tau)\Delta(\tau)\,\mathrm{d}\tau &= \int_I (H_\varepsilon(t,\tau) - H_0(t,\tau))\varphi_0(\tau)\,\mathrm{d}\tau \\ &\quad + \int_J (H_\varepsilon(t,\tau) - H_0(t,\tau))\varphi_0(\tau)\,\mathrm{d}\tau \\ &\quad + g_\varepsilon(t) - g_0(t).\end{aligned} \tag{9.33}$$

Define the operator L by

$$L\varphi(t) = \int_I H_\varepsilon(t,\tau)\varphi(\tau)\,\mathrm{d}\tau, \tag{9.34}$$

for $t \in I$, and for each $\varphi \in C[-\pi, \pi]$. Then denoting the right-hand side of (9.33) by $h(t)$, we obtain

$$(I - L)\Delta(t) = h(t). \tag{9.35}$$

We will proceed to obtain an estimate for $\Delta(t)$ by obtaining bounds on the operator L and on the function $h(t)$. First we establish

Lemma 9.1. *The operator L has the property* $||L||_\infty < 1$.

Proof. Without loss of generality, we fix $t \in I$ for some $t > 0$. Consider

$$
\begin{aligned}
|L\varphi(t)| &= \left| \int_I H_\varepsilon(t,\tau)\,\varphi(\tau)\,\mathrm{d}\tau \right| \le \int_I |H_\varepsilon(t,\tau)\,\varphi(\tau)|\,\mathrm{d}\tau \\
&\le \int_I |H_\varepsilon(t,\tau)|\,\mathrm{d}\tau\,\|\varphi\|_\infty, \qquad \text{where } \|\varphi\|_\infty = \sup_{t\in I}|\varphi(t)|, \\
&\le \left(\int_{I,\tau>0} |H_\varepsilon(t,\tau)|\,\mathrm{d}\tau + \int_{I,\tau<0} |H_\varepsilon(t,\tau)|\,\mathrm{d}\tau \right) \|\varphi\|_\infty,
\end{aligned} \tag{9.36}
$$

where the first integral is taken over the interval $\{\tau \in I \mid \tau > 0\}$ and the second integral similarly. We will evaluate each of the integral contributions separately.

Contribution when $\tau > 0$:

We use the small argument approximation for $H_1^{(1)}$ [22], thus,

$$
\begin{aligned}
&\int_{I,\tau>0} |H_\varepsilon(t,\tau)|\,\mathrm{d}\tau \\
&= \int_0^{\varepsilon^{2/3}} \left| \frac{ik}{2} \frac{H_1^{(1)}(k|\mathbf{x}_\varepsilon(t) - \mathbf{x}_\varepsilon(\tau)|)}{|\mathbf{x}_\varepsilon(t) - \mathbf{x}_\varepsilon(\tau)|} \mathbf{n}(\mathbf{x}_\varepsilon(\tau)) \cdot (\mathbf{x}_\varepsilon(t) - \mathbf{x}_\varepsilon(\tau))\,|\mathbf{x}_\varepsilon'(\tau)| \right| \mathrm{d}\tau \\
&= \int_0^{\varepsilon^{2/3}} \left| \frac{ik}{2} \left(\frac{-2i}{\pi} \right) \frac{\mathbf{n}(\mathbf{x}_\varepsilon(\tau)) \cdot (\mathbf{x}_\varepsilon(t) - \mathbf{x}_\varepsilon(\tau))}{k|\mathbf{x}_\varepsilon(t) - \mathbf{x}_\varepsilon(\tau)|^2} |\mathbf{x}_\varepsilon'(\tau)| \right| \mathrm{d}\tau \\
&\le \frac{1}{\pi} \int_0^{\varepsilon^{2/3}} |\mathbf{x}_\varepsilon'(\tau)| \frac{|\,\mathbf{n}(\mathbf{x}_\varepsilon(\tau)) \cdot (\mathbf{x}_\varepsilon(t) - \mathbf{x}_\varepsilon(\tau))\,|}{|\mathbf{x}_\varepsilon(t) - \mathbf{x}_\varepsilon(\tau)|^2}\,\mathrm{d}\tau.
\end{aligned} \tag{9.37}
$$

It may be verified [17] that $|\mathbf{x}_\varepsilon'(\tau)| \le \sqrt{2}$; also $\mathbf{n}(\mathbf{x}_\varepsilon(\tau)) \cdot \mathbf{x}_\varepsilon'(\tau) = 0$ for all τ. Thus, the integral (9.37) is bounded by

$$
\frac{\sqrt{2}}{\pi} \int_0^{\varepsilon^{2/3}} \frac{|\,\mathbf{n}(\mathbf{x}_\varepsilon(\tau)) \cdot (\mathbf{x}_\varepsilon(t) - \mathbf{x}_\varepsilon(\tau) - (t-\tau)\mathbf{x}_\varepsilon'(\tau))\,|}{|\mathbf{x}_\varepsilon(t) - \mathbf{x}_\varepsilon(\tau)|^2}\,\mathrm{d}\tau. \tag{9.38}
$$

Applying the mean value theorem to the denominator, we obtain

$$
|\mathbf{x}_\varepsilon(t) - \mathbf{x}_\varepsilon(\tau)| = |t-\tau|\,|\mathbf{x}_\varepsilon'(\tau^*)|, \tag{9.39}
$$

for some $\tau^* \in (t, \tau)$. Taking the standard central finite difference choice $\tau^* = (1/2)(t+\tau)$ and approximating the numerator with the second-order Taylor series expansion with the same choice of τ^*,

$$
\mathbf{x}_\varepsilon(t) = \mathbf{x}_\varepsilon(\tau) + (t-\tau)\mathbf{x}_\varepsilon'(\tau) + \frac{1}{2}(t-\tau)^2\mathbf{x}_\varepsilon''(\tau^*), \tag{9.40}
$$

then expression (9.38) is approximated by

$$\begin{aligned}
&\frac{1}{\sqrt{2}\pi}\int_{0}^{\varepsilon^{2/3}}\frac{|\mathbf{n}(\mathbf{x}_\varepsilon(\tau))\cdot((t-\tau)^2\,\mathbf{x}_\varepsilon''(\tau^*)\,|}{|(t-\tau)|^2|\mathbf{x}_\varepsilon'(\tau^*)|^2}\,\mathrm{d}\tau \\
&= \frac{1}{\sqrt{2}\pi}\int_{0}^{\varepsilon^{2/3}}\frac{|(t-\tau)|^2\,|\mathbf{n}(\mathbf{x}_\varepsilon(\tau))\cdot\mathbf{x}_\varepsilon''\left(\frac{t+\tau}{2}\right)|}{|(t-\tau)|^2|\mathbf{x}_\varepsilon'\left(\frac{t+\tau}{2}\right)|^2}\,\mathrm{d}\tau \\
&= \frac{1}{\sqrt{2}\pi}\int_{0}^{\varepsilon^{2/3}}\frac{|\mathbf{n}(\mathbf{x}_\varepsilon(\tau))\cdot\mathbf{x}_\varepsilon''\left(\frac{t+\tau}{2}\right)|}{|\mathbf{x}_\varepsilon'\left(\frac{t+\tau}{2}\right)|^2}\,\mathrm{d}\tau. \qquad (9.41)
\end{aligned}$$

Note that $\mathbf{x}_\varepsilon(\tau)=(x_{\varepsilon,1}(\tau),-\sin\tau)\sim(x_{\varepsilon,1}(\tau),-\tau)$ when τ is small. As such,

$$\begin{aligned}
\mathbf{x}_\varepsilon'(\tau) &\sim (x_{\varepsilon,1}'(\tau),-1)\\
|\mathbf{x}_\varepsilon'(\tau)| &\sim (1+(x_{\varepsilon,1}'(\tau))^2)^{1/2}\\
\mathbf{x}_\varepsilon''(\tau) &\sim (x_{\varepsilon,1}''(\tau),0). \qquad (9.42)
\end{aligned}$$

Applying these results, the integral (9.41) is less than

$$\begin{aligned}
&\frac{1}{\sqrt{2}\pi}\int_{0}^{\varepsilon^{2/3}}\frac{|-x_{\varepsilon,1}''\left(\frac{t+\tau}{2}\right)|}{|1+\left(x_{\varepsilon,1}'\left(\frac{t+\tau}{2}\right)\right)^2|}\,\mathrm{d}\tau = \frac{1}{\sqrt{2}\pi}\int_{0}^{\varepsilon^{2/3}}\frac{x_{\varepsilon,1}''\left(\frac{t+\tau}{2}\right)}{1+\left(x_{\varepsilon,1}'\left(\frac{t+\tau}{2}\right)\right)^2}\,\mathrm{d}\tau \\
&= \frac{1}{\sqrt{2}\pi}\left[2\arctan\left(x_{\varepsilon,1}'\left(\tfrac{t+\tau}{2}\right)\right)\right]_0^{\varepsilon^{2/3}} \le \frac{\sqrt{2}}{\pi}\arctan\left(x_{\varepsilon,1}'\left(\frac{t+\varepsilon^{2/3}}{2}\right)\right)\\
&\le \frac{\sqrt{2}}{\pi}\arctan(1), \qquad (9.43)
\end{aligned}$$

since $|x_{\varepsilon,1}'(t^*)|<1$ for all t^*. Thus,

$$\int_{I,\tau>0}|H_\varepsilon(t,\tau)|\,\mathrm{d}\tau \le \frac{1}{2\sqrt{2}}. \qquad (9.44)$$

Contribution when $\tau<0$:

In this case, the argument applied earlier may be employed to show

$$\int_{I,\tau<0}|H_\varepsilon(t,\tau)|\,\mathrm{d}\tau \le \frac{1}{2\sqrt{2}}. \qquad (9.45)$$

However, it might be considered that the Taylor series substitution is not suitable for this case because with $t>0$ and $\tau<0$ in (9.40), the term $\mathbf{x}_\varepsilon''(\tau^*)$ may be very large; indeed $\mathbf{x}_\varepsilon''(0)\sim 1/\varepsilon$. Instead, we approximate $H_\varepsilon(t,\tau)$ using $H_0(t,\tau)$. Thus, $\mathbf{x}_0(t)=(t,t)$, $\mathbf{x}_0(\tau)=(\tau,-\tau)$, $\mathbf{n}(\mathbf{x}_0(\tau))\cdot\mathbf{x}_0(\tau)=0$ for all τ, and $\mathbf{n}(\mathbf{x}_0(\tau))\cdot\mathbf{x}_0(t)=-|\mathbf{x}_0(t)|$.

Using the small argument approximation for $H_1^{(1)}$ [22],

$$\int\limits_{I,\tau<0} |H_\varepsilon(t,\tau)|\,\mathrm{d}\tau \approx \int\limits_{I,\tau<0} |H_0(t,\tau)|\,\mathrm{d}\tau$$

$$= \int\limits_{-\varepsilon^{2/3}}^{0} \left| \frac{ik}{2} \frac{H_1^{(1)}(k|\mathbf{x}_0(t)-\mathbf{x}_0(\tau)|)}{|\mathbf{x}_0(t)-\mathbf{x}_0(\tau)|} \mathbf{n}(\mathbf{x}_0(\tau))\cdot(\mathbf{x}_0(t)-\mathbf{x}_0(\tau))\,|\mathbf{x}_0'(\tau)| \right| \mathrm{d}\tau$$

$$\simeq \int\limits_{-\varepsilon^{2/3}}^{0} \left| \frac{ik}{2}\left(\frac{-2i}{\pi}\right) \frac{\mathbf{n}(\mathbf{x}_0(\tau))\cdot(\mathbf{x}_0(t)-\mathbf{x}_0(\tau))}{k|\mathbf{x}_0(t)-\mathbf{x}_0(\tau)|^2}\,|\mathbf{x}_0'(\tau)| \right| \mathrm{d}\tau$$

$$= \frac{\sqrt{2}}{\pi} \int\limits_{-\varepsilon^{2/3}}^{0} \frac{|\mathbf{x}_0(t)|}{|\mathbf{x}_0(t)-\mathbf{x}_0(\tau)|^2}\,\mathrm{d}\tau$$

$$= \frac{1}{\pi}\int\limits_{0}^{\varepsilon^{2/3}} \frac{t}{t^2+\tau^2}\,\mathrm{d}\tau = \frac{1}{\pi}\arctan\frac{\varepsilon^{2/3}}{t} \le \frac{1}{2}. \tag{9.46}$$

Thus,

$$|L\varphi(t)| \le \left(\int\limits_{I,\tau>0} |H_\varepsilon(t,\tau)|\,\mathrm{d}\tau + \int\limits_{I,\tau<0} |H_\varepsilon(t,\tau)|\,\mathrm{d}\tau \right) \|\varphi\|_\infty$$

$$\le \left(\frac{1}{2\sqrt{2}} + \frac{1}{2}\right) \|\varphi\|_\infty < \|\varphi\|_\infty. \tag{9.47}$$

We have thus established that $\|L\|_\infty < 1$. □

Applying Lemma 9.1, $\Delta(t)$ may be recovered from the convergent Neumann series

$$\Delta(t) = (I-L)^{-1}h(t) = \sum_{n=0}^{\infty} L^n h(t). \tag{9.48}$$

Moreover,

$$|\Delta(t)| \le \|(I-L)^{-1}\|_\infty\,\|h\|_\infty, \tag{9.49}$$

for all t; also

$$\|(I-L)^{-1}\|_\infty \le (1-\|L\|_\infty)^{-1}. \tag{9.50}$$

We now consider estimates for $h(t)$. First an estimate for the function $\varphi_0(t)$ for $t \in I$ is needed.

Lemma 9.2.

$$\varphi_0(r) = A + B(kr)^{2/3} + O(kr),$$

where A and B are suitable constants, as $kr \to 0$.

Proof. We approximate this by the surface current induced by the same illuminating field on the infinite wedge whose edges are tangent to the corner point (corresponding to parameter value $t = 0$). We use the results of [1] concerning infinite wedges of internal angle 2Ω (and specialise later to the case $2\Omega = \pi/2$).

Employ cylindrical polars (r, θ) with centre at the tip of the wedge of internal angle 2Ω, sides of which are defined by $\theta = \pm\Omega$. Suppose the wedge is illuminated by a unit strength field incident from direction θ_0, so that

$$u^{\mathrm{inc}} = \exp(-ikr\cos(\theta - \theta_0)). \tag{9.51}$$

Set $\nu = (2\pi - 2\Omega)/\pi$.

Then in the vicinity of the tip [1], the total field is

$$u_0^{\mathrm{tot}}(r,\theta) = \frac{4}{\Gamma(1/\nu)}\left(\frac{1}{2}kr\right)^{1/\nu} e^{-i\pi/\nu}\sin\left(\frac{\theta-\Omega}{\nu}\right)\sin\left(\frac{\theta_0-\Omega}{\nu}\right) + O((kr)^{\min(2/\nu,1)}), \tag{9.52}$$

as $kr \to 0$. Thus, the scattered field has the form

$$u_0^{\mathrm{sc}}(r,\theta) = u_0^{\mathrm{tot}}(r,\theta) - u_0^{\mathrm{inc}}(r,\theta) = C_\nu\left(\frac{1}{2}kr\right)^{1/\nu} - 1 + O((kr)^{\min(2/\nu,1)}), \tag{9.53}$$

as $kr \to 0$, for a suitable constant C_ν.

The scattered field at an off-body point $\mathbf{x}$ is related to the corresponding surface distribution φ_0^w on the wedge ∂D by

$$u_0^{\mathrm{sc}}(\mathbf{x}) = \int_{\partial D} \frac{\partial G(\mathbf{x},\mathbf{y})}{\partial \mathbf{n}(\mathbf{y})}\varphi_0^w(\mathbf{y})\,\mathrm{d}s(\mathbf{y}). \tag{9.54}$$

We use the parameterisation $\mathbf{x}_0(\tau) = (|\tau|\cos\Omega, \tau\sin\Omega)$ for the wedge so that the unit outward normal is $\mathbf{n}(\mathbf{x}_0(\tau)) = (-\sin\Omega, (\mathrm{sgn}\tau)\cos\Omega,)$.

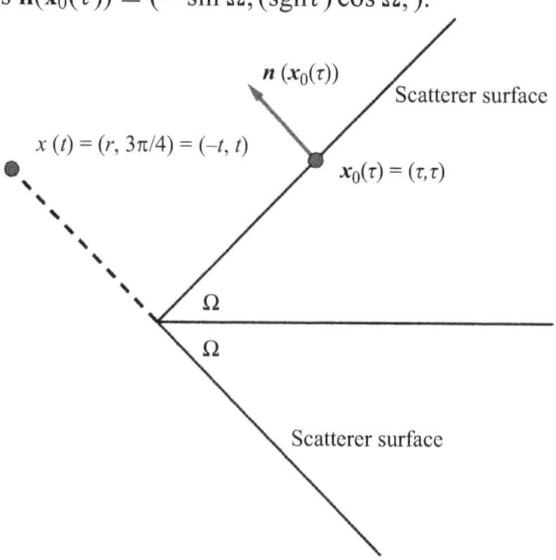

Notice that $\mathbf{n}(\mathbf{x}_0(\tau)) \cdot \mathbf{x}_0(\tau) = 0$, for all τ, and for an off-body point with polar coordinates $(r, 3\pi/4)$, $\mathbf{n}(\mathbf{x}_0(\tau)) \cdot \mathbf{x}(t) = 0$ when $\tau < 0$. Thus, for the right-angled wedge, the scattered field representation at that off-body point may be written as

$$\frac{ik}{2} \int_{\partial D} \frac{H_1^{(1)}(k|\mathbf{x}_0(t) - \mathbf{x}_0(\tau)|)}{|\mathbf{x}_0(t) - \mathbf{x}_0(\tau)|} \, \mathbf{n}\,(\mathbf{x}_0(\tau)) \cdot (\mathbf{x}_0(t) - \mathbf{x}_0(\tau)) \, |\mathbf{x}_0'(\tau)| \, \varphi_0(\tau) \, \mathrm{d}\tau$$

$$= \frac{ik}{2} \int_0^\infty \frac{H_1^{(1)}(k|\mathbf{x}_0(t) - \mathbf{x}_0(\tau)|)}{|\mathbf{x}_0(t) - \mathbf{x}_0(\tau)|} \, \mathbf{n}\,(\mathbf{x}_0(\tau)) \cdot (\mathbf{x}_0(t) - \mathbf{x}_0(\tau)) \, |\mathbf{x}_0'(\tau)| \, \varphi_0(\tau) \, \mathrm{d}\tau$$

$$= \frac{ikr}{\sqrt{2}} \int_0^\infty \frac{H_1^{(1)}(k\sqrt{2}(r^2 + \tau^2)^{1/2})}{(r^2 + \tau^2)^{1/2}} \, \varphi_0(\tau) \, \mathrm{d}\tau, \tag{9.55}$$

and setting $\tau = ru$ yields,

$$\frac{ikr}{\sqrt{2}} \int_0^\infty \frac{H_1^{(1)}(\sqrt{2}kr(u^2 + 1)^{1/2})}{(u^2 + 1)^{1/2}} \, \varphi_0(ru) \, \mathrm{d}u. \tag{9.56}$$

This must equal (9.53), upon setting $2\Omega = \pi/2$. We note that this integral is convergent, since, using the limiting values of the Hankel function [22],

$$\frac{H_1^{(1)}(\sqrt{2}kr(u^2 + 1)^{1/2})}{(u^2 + 1)^{1/2}} \sim \sqrt{\frac{2}{\pi}} \, e^{i3\pi/4} \, \frac{e^{-i\sqrt{2}kr\,(u^2+1)^{1/2}}}{(\sqrt{2}kr)^{1/2}(u^2 + 1)^{3/4}} = O\left(u^{-3/2}\right) \tag{9.57}$$

as $u \to \infty$.

We split the contribution to the integral as the sum over into the two intervals $[0, 1/(kr)]$ and $[1/(kr), \infty)$. When $\varphi_0(r)$ has the form $(kr)^\alpha$ for some α (restricted to $0 \le \alpha < 1$), we can show that the integral takes the values $K_\alpha (kr)^\alpha + O(kr)$ as $kr \to 0$, where

$$K_\alpha = \frac{1}{\pi} \int_0^\infty \frac{u^\alpha}{u^2 + 1} \, \mathrm{d}u. \tag{9.58}$$

The details are as follows:

$$\frac{ikr}{\sqrt{2}} \int_0^{1/kr} \frac{H_1^{(1)}(\sqrt{2}kr(u^2 + 1)^{1/2})}{(u^2 + 1)^{1/2}} \, \varphi_0(ru) \, \mathrm{d}u$$

$$= \frac{ikr}{\sqrt{2}} \int_0^{1/kr} \frac{H_1^{(1)}(\sqrt{2}kr(u^2 + 1)^{1/2})}{(u^2 + 1)^{1/2}} \, (kru)^\alpha \, \mathrm{d}u, \tag{9.59}$$

and using the small argument approximation for $H_1^{(1)}$ [22],

$$\sim \frac{ikr}{\sqrt{2}} \int_0^{1/kr} \frac{-2i}{\pi} \frac{1}{\sqrt{2}kr(u^2+1)^{1/2}} \frac{(kru)^\alpha}{(u^2+1)^{1/2}} \mathrm{d}u$$

$$= \frac{(kr)^\alpha}{\pi} \int_0^{1/kr} \frac{u^\alpha}{u^2+1} \mathrm{d}u$$

$$= K_\alpha (kr)^\alpha (1 + \mathrm{o}(1)), \tag{9.60}$$

as $kr \to 0$. Evaluating the second contribution on $[1/(kr), \infty)$,

$$\left| \frac{ikr}{\sqrt{2}} \int_{1/kr}^{\infty} \frac{H_1^{(1)}(\sqrt{2}kr(u^2+1)^{1/2})}{(u^2+1)^{1/2}} \varphi_0(ru) \right| \mathrm{d}u$$

$$\leq \frac{kr}{\sqrt{2}} \int_{1/kr}^{\infty} \frac{|H_1^{(1)}(\sqrt{2}kr(u^2+1)^{1/2})|}{|(u^2+1)^{1/2}|} |\varphi_0(ru)| \, \mathrm{d}u, \tag{9.61}$$

and using the large argument approximation for $H_1^{(1)}$ [22],

$$\sim \frac{kr}{\sqrt{2}} \int_{1/kr}^{\infty} \sqrt{\frac{2}{\pi}} \frac{1}{\sqrt{\sqrt{2}kr(u^2+1)^{1/2}}} \frac{|\varphi_0(ru)|}{(u^2+1)^{1/2}} \mathrm{d}u$$

$$= \sqrt{\frac{kr}{\pi\sqrt{2}}} \int_{1/kr}^{\infty} \frac{1}{(u^2+1)^{3/4}} \mathrm{d}u \, \|\varphi_0\|_\infty$$

$$\leq \sqrt{kr}\, \|\varphi_0\|_\infty \int_{1/kr}^{\infty} \frac{1}{(u^2+1)^{3/4}} \mathrm{d}u, \tag{9.62}$$

since $\varphi_0(u)$ is bounded by $\|\varphi_0\|_\infty$ [1]. Thus, the integral is bounded by

$$\sqrt{kr}\, \|\varphi_0\|_\infty \int_{1/kr}^{\infty} \frac{1}{u^{3/2}} \mathrm{d}u = 2kr\, \|\varphi_0\|_\infty, \tag{9.63}$$

which is $O(kr)$ as $kr \longrightarrow 0$. Combining these two results, as $kr \to 0$, the integral takes the values $K_\alpha (kr)^\alpha + O(kr)$. For the right-angled lemniscate, $\alpha = 1/\nu = 2/3$. In other words, we may take the surface distribution to be

$$\varphi_0(r) = A + B(kr)^{2/3} + O(kr), \tag{9.64}$$

where A and B are suitable constants, as $kr \to 0$. □

We are now in a position to evaluate the function $h(t)$ in (9.35) by inserting (9.64) in the first term on the right of (9.33) and approximating the other two.

Let $h_1(t)$, $h_2(t)$ denote the first and second terms on the right-hand side of (9.33). Thus,

$$h_1(t) = \int_I (H_\varepsilon(t,\tau) - H_0(t,\tau))\varphi_0(\tau)\,\mathrm{d}\tau, \tag{9.65}$$

and

$$h_2(t) = \int_J (H_\varepsilon(t,\tau) - H_0(t,\tau))\varphi_0(\tau)\,\mathrm{d}\tau. \tag{9.66}$$

The following lemmas establish that both $h_1(t)$ and $h_2(t)$ are $O\left((k\varepsilon)^{2/3}\right)$ as $k\varepsilon \to 0$, for $t \in I$. Lemmas 9.3, 9.4, 9.5 establish the result for $h_1(t)$ by inserting (9.64) into (9.65). Lemma 9.6 establishes the result for $h_2(t)$.

Lemma 9.3. *Suppose $t \in I$ and $\tau \in J$. Then*

$$H_\varepsilon(t,\tau) - H_0(t,\tau) = O\left(\varepsilon^{2/3}\right), \qquad as\ \varepsilon \to 0.$$

Proof. Without loss of generality, we assume $t > 0$.

$$\begin{aligned} H_\varepsilon(t,\tau) &= \frac{ik}{2} \frac{H_1^{(1)}(k|\mathbf{x}_\varepsilon(t) - \mathbf{x}_\varepsilon(\tau)|)}{|\mathbf{x}_\varepsilon(t) - \mathbf{x}_\varepsilon(\tau)|} (x'_{\varepsilon,2}(\tau), -x'_{\varepsilon,1}(\tau)) \cdot (\mathbf{x}_\varepsilon(t) - \mathbf{x}_\varepsilon(\tau)) \\ &= \frac{ik}{2} H_1^{(1)}(k|\mathbf{x}_\varepsilon(t) - \mathbf{x}_\varepsilon(\tau)|)\,|\mathbf{x}_\varepsilon(t) - \mathbf{x}_\varepsilon(\tau)| \\ &\quad \times \frac{(x'_{\varepsilon,2}(\tau), -x'_{\varepsilon,1}(\tau)) \cdot (\mathbf{x}_\varepsilon(t) - \mathbf{x}_\varepsilon(\tau))}{|\mathbf{x}_\varepsilon(t) - \mathbf{x}_\varepsilon(\tau)|^2}. \end{aligned} \tag{9.67}$$

The function $kz\,H^{(1)}(kz)$ is analytic and is bounded on the sets of points

$$\{z = |\mathbf{x}_\varepsilon(t) - \mathbf{x}_\varepsilon(\tau)| : \quad t, \tau \in [-\pi, \pi]\}, \tag{9.68}$$

and

$$\{z = |\mathbf{x}_0(t) - \mathbf{x}_0(\tau)| : \quad t, \tau \in [-\pi, \pi]\}; \tag{9.69}$$

also as $\varepsilon \to 0$,

$$kH_1^{(1)}(k|\mathbf{x}_\varepsilon(t) - \mathbf{x}_\varepsilon(\tau)|)\,|\mathbf{x}_\varepsilon(t) - \mathbf{x}_\varepsilon(\tau)| \to kH_1^{(1)}(k|\mathbf{x}_0(t) - \mathbf{x}_0(\tau)|)\,|\mathbf{x}_0(t) - \mathbf{x}_0(\tau)|. \tag{9.70}$$

If $\tau \in J$ and $\varepsilon^{2/3} < \tau < 2\varepsilon^{2/3}$, we use the Taylor series approximation

$$\frac{1}{2}\left(x'_{\varepsilon,2}(\tau), -x'_{\varepsilon,1}(\tau)\right) \cdot \frac{\mathbf{x}''_\varepsilon\left(\frac{t+\tau}{2}\right)}{\left|\mathbf{x}'_\varepsilon\left(\frac{t+\tau}{2}\right)\right|^2}, \tag{9.71}$$

together with the estimates

$$\mathbf{x}'_\varepsilon(\tau) \sim \left(\frac{\tau}{\sqrt{4\varepsilon^2+\tau^2}}, -1\right) \sim \mathbf{x}'_0(\tau) + O\left(\frac{\varepsilon^2}{\tau^2}\right) \sim \mathbf{x}'_0(\tau) + O\left(\varepsilon^{2/3}\right)$$

$$\mathbf{x}''_\varepsilon(\tau) \sim \left(\frac{4\varepsilon^2}{\left(4\varepsilon^2+\tau^2\right)^{3/2}}, 0\right) \sim \mathbf{x}''_0(\tau) + O\left(\varepsilon^{2/3}\right). \tag{9.72}$$

Noting that $|\mathbf{x}'_\varepsilon(\tau)|$ is bounded below by a constant independent of ε and τ, we deduce that the quantity (9.71) differs from

$$\frac{1}{2}\left(x'_{0,2}(\tau), -x'_{0,1}(\tau)\right) \cdot \frac{\mathbf{x}''_0\left(\frac{t+\tau}{2}\right)}{\left|\mathbf{x}'_0\left(\frac{t+\tau}{2}\right)\right|^2}, \tag{9.73}$$

by an amount that is $O\left(\varepsilon^{2/3}\right)$, as $\varepsilon \to 0$.

On the other hand, if $\tau \in J$ and $\tau > 2\varepsilon^{2/3}$, the quantity $\mathbf{x}_\varepsilon(\tau)$ is much bigger than $\mathbf{x}_\varepsilon(t)$ when $t \in I$. It follows that by expanding each of the terms involving $\mathbf{x}_\varepsilon(t)$ or $\mathbf{x}'_\varepsilon(t)$ in

$$\frac{(x'_{\varepsilon,2}(\tau), -x'_{\varepsilon,1}(\tau)) \cdot (\mathbf{x}_\varepsilon(t) - \mathbf{x}_\varepsilon(\tau))}{|\mathbf{x}_\varepsilon(t) - \mathbf{x}_\varepsilon(\tau)|^2}, \tag{9.74}$$

that this quantity differs from

$$\frac{(x'_{0,2}(\tau), -x'_{0,1}(\tau)) \cdot (\mathbf{x}_0(t) - \mathbf{x}_0(\tau))}{|\mathbf{x}_0(t) - \mathbf{x}_0(\tau)|^2}, \tag{9.75}$$

by an amount of $O\left(\varepsilon^{2/3}\right)$ as $\varepsilon \to 0$.

Turning to the case where $\tau \in J$ and $\tau < 0$, a similar argument again shows that this difference is $O\left(\varepsilon^{2/3}\right)$ as $\varepsilon \to 0$. The result of the lemma now follows.

Corollary 9.1. *The result of the previous lemma holds when the relevant Green function is replaced by the static Green function, that is, if*

$$H^0_\varepsilon(t,\tau) = \frac{(x'_{\varepsilon,2}(\tau), -x'_{\varepsilon,1}(\tau)) \cdot (\boldsymbol{x}_\varepsilon(t) - \boldsymbol{x}_\varepsilon(\tau))}{|\boldsymbol{x}_\varepsilon(t) - \boldsymbol{x}_\varepsilon(\tau)|^2},$$

and $H^0_0(t,\tau)$ is similarly defined, then

$$H^0_\varepsilon(t,\tau) - H^0_0(t,\tau) = O\left(\varepsilon^{2/3}\right),$$

as $\varepsilon \to 0$ for $t \in I$, $\tau \in J$.

□

Lemma 9.4. *Let A be a constant. Then for $t \in I$,*

$$\int_I (H_\varepsilon(t,\tau) - H_0(t,\tau))A\, d\tau = O\left(\varepsilon^{2/3}\right), \qquad \text{as } \varepsilon \to 0.$$

Proof. The integral may be approximated by

$$\int_I \left(H_\varepsilon^0(t,\tau) - H_0^0(t,\tau)\right) A \, d\tau$$

$$= \int_{-\pi}^{\pi} \left(H_\varepsilon^0(t,\tau) - H_0^0(t,\tau)\right) A \, d\tau - \int_J \left(H_\varepsilon^0(t,\tau) - H_0^0(t,\tau)\right) A \, d\tau. \tag{9.76}$$

The result of the previous lemma and corollary show that the contribution of the integral over J is $O\left(\varepsilon^{2/3}\right)$ as $\varepsilon \to 0$.

On the other hand, a fundamental property of the static Green function is [21],

$$2\int_{\partial D} \frac{\partial G_0(\mathbf{x},\mathbf{y})}{\partial \mathbf{n}(\mathbf{y})} \, ds(\mathbf{y}) = -1, \qquad \mathbf{x} \in \partial D, \tag{9.77}$$

where D is any domain. Thus,

$$\int_{-\pi}^{\pi} H_\varepsilon^0(t,\tau) \, d\tau = \int_{-\pi}^{\pi} H_0^0(t,\tau) \, d\tau = -\frac{1}{2}, \tag{9.78}$$

so that

$$\int_{-\pi}^{\pi} \left(H_\varepsilon^0(t,\tau) - H_0^0(t,\tau)\right) A \, d\tau = 0. \tag{9.79}$$

The result is now proven. □

Lemma 9.5. *Let $t \in I$. Then*

$$\int_I (H_\varepsilon(t,\tau) - H_0(t,\tau))(k\tau)^{2/3} \, d\tau = O\left((k\varepsilon)^{2/3}\right), \qquad \text{as } k\varepsilon \to 0.$$

Proof. We may suppose $t > 0$ and consider the contribution to the integral when $\tau \in I$ and $\tau > 0$. First, the contribution from $H_0(t,\tau)$ may be neglected because using the Taylor approximation,

$$\begin{aligned} H_0(t,\tau) &= \frac{\frac{1}{2}\left(x'_{0,2}(\tau), -x'_{0,1}(\tau)\right) \cdot \mathbf{x}''_0\left(\frac{t+\tau}{2}\right)}{\left|\mathbf{x}'_0\left(\frac{t+\tau}{2}\right)\right|^2} \\ &= O\left((k\varepsilon)^{2/3}\right), \qquad \text{as } k\varepsilon \to 0. \end{aligned} \tag{9.80}$$

The contribution from $H_\varepsilon(t,\tau)$ may be evaluated as follows:

$$\begin{aligned} \int_{I,\tau>0} H_\varepsilon(t,\tau)(k\tau)^{2/3} \, d\tau = \int_0^{\varepsilon^{2/3}} \frac{ik}{2} \frac{H_1^{(1)}(k|\mathbf{x}_\varepsilon(t) - \mathbf{x}_\varepsilon(\tau)|)}{|\mathbf{x}_\varepsilon(t) - \mathbf{x}_\varepsilon(\tau)|} \, \mathbf{n}(\mathbf{x}_\varepsilon(\tau)) \cdot (\mathbf{x}_\varepsilon(t) - \mathbf{x}_\varepsilon(\tau)) \\ \times |\mathbf{x}'_\varepsilon(\tau)| \, (k\tau)^{2/3} \, d\tau. \end{aligned} \tag{9.81}$$

Using the small argument approximation for $H_1^{(1)}$ [22], and following the steps (9.37)–(9.41) from Lemma 9.1, the integral (9.81) is less than

$$\frac{1}{\sqrt{2}\pi}\int_0^{\varepsilon^{2/3}} \frac{x''_{\varepsilon,1}\left(\frac{t+\tau}{2}\right)}{1+\left(x'_{\varepsilon,1}\left(\frac{t+\tau}{2}\right)\right)^2}(k\tau)^{2/3}\,\mathrm{d}\tau. \tag{9.82}$$

We note that for small t,

$$\begin{aligned}
\mathbf{x}_\varepsilon(t) &\sim \left(\sqrt{4\varepsilon^2+t^2}, -t\right),\\
\mathbf{x}'_\varepsilon(t) &\sim \left(\frac{t}{\sqrt{4\varepsilon^2+t^2}}, -1\right),\\
\mathbf{x}''_\varepsilon(t) &\sim \left(\frac{4\varepsilon^2}{(4\varepsilon^2+t^2)^{3/2}}, 0\right).
\end{aligned} \tag{9.83}$$

Inserting these into (9.82) produces

$$\frac{1}{\sqrt{2}\pi}\int_0^{\varepsilon^{2/3}} \frac{4\varepsilon^2\,(k\tau)^{2/3}}{\sqrt{4\varepsilon^2+\left(\frac{t+\tau}{2}\right)^2}\left(4\varepsilon^2+2\left(\frac{t+\tau}{2}\right)^2\right)}\,\mathrm{d}\tau. \tag{9.84}$$

Applying the substitutions $t = t_1\varepsilon$ and $\tau = \tau_1\varepsilon$, (9.84) is equal to

$$\begin{aligned}
&\frac{(k\varepsilon)^{2/3}}{\sqrt{2}\pi}\int_0^{\varepsilon^{-1/3}} \frac{\tau_1^{2/3}}{\sqrt{4+\left(\frac{t_1+\tau_1}{2}\right)^2}\left(4+2\left(\frac{t_1+\tau_1}{2}\right)^2\right)}\,\mathrm{d}\tau_1\\
&\leq \frac{(k\varepsilon)^{2/3}}{\sqrt{2}\pi}\int_0^{\varepsilon^{-1/3}} \frac{4\tau_1^{2/3}}{\sqrt{16+\tau_1^2}\left(16+2\tau_1^2\right)}\,\mathrm{d}\tau_1, \qquad \text{since } t>0,\\
&\leq \frac{\sqrt{2}(k\varepsilon)^{2/3}}{\pi}\int_0^{\varepsilon^{-1/3}} \frac{\tau_1^{2/3}}{(16+\tau_1^2)^{3/2}}\,\mathrm{d}\tau_1\\
&\leq \frac{\sqrt{2}(k\varepsilon)^{2/3}}{\pi}\int_0^{\infty} \frac{\tau_1^{2/3}}{(16+\tau_1^2)^{3/2}}\,\mathrm{d}\tau_1 = C(k\varepsilon)^{2/3},
\end{aligned} \tag{9.85}$$

for some constant C.

We turn to the contribution when $\tau < 0$. Since $t > 0$ and $\tau < 0$, $|\mathbf{x}_\varepsilon(t) - \mathbf{x}_\varepsilon(\tau)|$ and $|\mathbf{x}_0(t) - \mathbf{x}_0(\tau)|$ never vanish, and for τ large enough we may approximate $H_\varepsilon(t,\tau)$ by $H_0(t,\tau)$; equivalently the dominant contribution to the integral comes from an interval of τ-values near the origin. We split the interval of integration $[-\varepsilon^{2/3}, 0] = [-\varepsilon^{2/3}, -10\varepsilon] \cup [-10\varepsilon, 0]$ and neglect the contribution from the first interval and evaluate

$$\int_{-10\varepsilon}^{0} \left(H_\varepsilon(t,\tau) - H_0(t,\tau)\right)(k\tau)^{2/3}\,\mathrm{d}\tau$$

$$= \int_{-10\varepsilon}^{0} H_\varepsilon(t,\tau)(k\tau)^{2/3}\,\mathrm{d}\tau - \int_{-10\varepsilon}^{0} H_0(t,\tau)(k\tau)^{2/3}\,\mathrm{d}\tau. \tag{9.86}$$

The first term is approximated by its Taylor expansion

$$\frac{1}{\pi}\int_{-10\varepsilon}^{0} \frac{x''_{\varepsilon,1}\left(\frac{t+\tau}{2}\right)}{1 + x'_{\varepsilon,1}\left(\frac{t+\tau}{2}\right)^2}(k\tau)^{2/3}\,\mathrm{d}\tau. \tag{9.87}$$

Now

$$\left|x''_{\varepsilon,1}\left(\frac{t+\tau}{2}\right)\right| \le \frac{1}{\varepsilon}, \tag{9.88}$$

so the absolute value of the integral is bounded by

$$\frac{10\varepsilon}{\pi}\frac{1}{\varepsilon}(10k\varepsilon)^{2/3} = O\left((k\varepsilon)^{2/3}\right), \qquad \text{as } k\varepsilon \to 0. \tag{9.89}$$

The second term is approximated by

$$\frac{1}{\pi}\int_{-10\varepsilon}^{0} \frac{t}{t^2+\tau^2}(k\tau)^{2/3}\,\mathrm{d}\tau \le \frac{(k\varepsilon)^{2/3}}{\pi}\left[\arctan\left(\frac{\tau}{t}\right)\right]_{-10\varepsilon}^{0} \le \frac{1}{2}(k\varepsilon)^{2/3}. \tag{9.90}$$

This conclude the proof of the lemma. □

Lemmas 9.3–9.5 establish the desired bound for $h_1(t)$:

$$h_1(t) = O\left((k\varepsilon)^{2/3}\right), \qquad \text{as } k\varepsilon \to 0. \tag{9.91}$$

Turning to $h_2(t)$, we have

Lemma 9.6.

$$h_2(t) = O\left((k\varepsilon)^{2/3}\right), \qquad \textit{as } k\varepsilon \to 0, \textit{ for } t \in I.$$

Proof.

$$|h_2(t)| \le \|\varphi_0\|_\infty \int_I |H_\varepsilon(t,\tau) - H_0(t,\tau)|\,\mathrm{d}\tau, \tag{9.92}$$

where we recall that $\varphi_0(t)$ is the surface distribution on the sharp-cornered lemniscate. Also we have established

$$H_\varepsilon(t,\tau) - H_0(t,\tau) = O\left((k\varepsilon)^{2/3}\right), \qquad \text{as } k\varepsilon \to 0. \tag{9.93}$$

The result immediately follows. □

Finally, it is easily seen that $g_\varepsilon(t) - g_0(t) = O(k\varepsilon)$, as $k\varepsilon \to 0$, and since $h(t) = h_1(t) + h_2(t) + g_\varepsilon(t) - g_0(t)$, we conclude

Lemma 9.7.

$$h(t) = O\left((k\varepsilon)^{2/3}\right), \qquad \textit{as } k\varepsilon \to 0, \textit{ for } t \in I.$$

We can now summarise the main result of this section which directly follows from the insertion of Lemma 9.7 into (9.49).

Theorem 9.1. *The difference* $\Delta(t) = \varphi_0(t) - \varphi_\varepsilon(t)$ *between the surface distribution on the lemniscate* (9.22) *and its rounded counterpart* (9.23) *satisfies*

$$\Delta(t) = O\left((k\varepsilon)^{2/3}\right), \qquad \textit{for } t \in I, \textit{ as } k\varepsilon \to 0.$$

9.3.3 The far-field difference

The far-field patterns u_0^∞ and u_ε^∞ of the lemniscate and its rounded version, respectively, may be expressed in terms of the corresponding surface quantities φ_0 and φ_ε induced by the illuminating field [9]. Recalling that $\hat{\mathbf{x}} = \hat{\mathbf{x}}(\hat{\theta}) = (\cos\hat{\theta}, \sin\hat{\theta})$ is the unit vector with $\hat{\theta}$ being the angle of observation in the far field,

$$\begin{aligned} u_0^\infty(\hat{\mathbf{x}}) &= \frac{e^{-i\pi/4}}{\sqrt{8\pi k}} \int_{-\pi}^{\pi} k\, \mathbf{n}(\mathbf{x}_0(t)) \,.\, \hat{\mathbf{x}}\, e^{-ik\hat{\mathbf{x}}.\mathbf{x}_0(t)}\, \varphi_0(t)\, |\mathbf{x}_0'(t)|\, \mathrm{d}t \\ &= \frac{\sqrt{k}\, e^{-i\pi/4}}{\sqrt{8\pi}} \int_{-\pi}^{\pi} m(\hat{\mathbf{x}}; \mathbf{x}_0)\, \varphi_0(t)\, \mathrm{d}t \end{aligned} \tag{9.94}$$

where $m(\hat{\mathbf{x}}; \mathbf{x}) = \mathbf{n}(\mathbf{x}(t)) \cdot \hat{\mathbf{x}}\, e^{-ik\hat{\mathbf{x}}\cdot\mathbf{x}(t)}\, |\mathbf{x}'(t)|$. Note that $|m(\hat{\mathbf{x}}; \mathbf{x})| \le ||\mathbf{x}'||_\infty$. The far field for the rounded object $u_\varepsilon^\infty(\hat{\mathbf{x}})$ is similarly defined. Thus,

$$\begin{aligned} \sqrt{k}|u_\varepsilon^\infty(\hat{\mathbf{x}}) - u_0^\infty(\hat{\mathbf{x}})| &\le \frac{k}{\sqrt{8\pi}} \left| \int_{-\pi}^{\pi} \{m(\hat{\mathbf{x}}; \mathbf{x}_\varepsilon(t))\, \varphi_\varepsilon(t) - m(\hat{\mathbf{x}}; \mathbf{x}_0(t))\, \varphi_0(t)\}\, \mathrm{d}t \right| \\ &\le \frac{k}{\sqrt{8\pi}} \left| \int_{-\pi}^{\pi} m(\hat{\mathbf{x}}; \mathbf{x}_\varepsilon)\, (\varphi_\varepsilon(t) - \varphi_0(t))\, \mathrm{d}t \right| \\ &\quad + \frac{k}{\sqrt{8\pi}} \left| \int_{-\pi}^{\pi} (m(\hat{\mathbf{x}}; \mathbf{x}_\varepsilon) - m(\hat{\mathbf{x}}; \mathbf{x}_0))\, \varphi_0(t)\, \mathrm{d}t \right| \end{aligned} \tag{9.95}$$

We consider the final two terms in (9.95). Recalling that $I = [-\varepsilon^{2/3}, \varepsilon^{2/3}]$, and that $\Delta(t)$ is assumed to vanish for t outside I,

$$\frac{k}{\sqrt{8\pi}}\left|\int_{-\pi}^{\pi} m(\hat{\mathbf{x}};\mathbf{x}_\varepsilon)\,(\varphi_\varepsilon(t)-\varphi_0(t))\,\mathrm{d}t\right| \leq \frac{k}{\sqrt{8\pi}}\|\mathbf{x}_\varepsilon'\|_\infty \int_{-\varepsilon^{2/3}}^{\varepsilon^{2/3}} |\Delta(t)|\,\mathrm{d}t. \tag{9.96}$$

Theorem 9.1 shows that $\Delta(t) = O((k\varepsilon)^{2/3})$ as $k\varepsilon \to 0$ and so the contribution (9.96) is thus $O((k\varepsilon)^{4/3})$ as $k\varepsilon \to 0$. It remains to show that the other contribution to (9.95) is of smaller order as $k\varepsilon \to 0$.

Lemma 9.8. *The quantity*

$$\frac{k}{\sqrt{8\pi}}\left|\int_{-\pi}^{\pi} (m(\hat{x};x_\varepsilon) - m(\hat{x};x_0))\,\varphi_0(t)\,dt\right|$$

is $o((k\varepsilon)^{4/3})$ as $k\varepsilon \to 0$.

Proof. The quantity is bounded by

$$\frac{k}{\sqrt{8\pi}}\left|\int_{-\pi}^{\pi} (\mathbf{v}(\mathbf{x}_\varepsilon(t)) - \mathbf{v}(\mathbf{x}_0(t)))\cdot\hat{\mathbf{x}}\,e^{-ik\hat{\mathbf{x}}\cdot\mathbf{x}_\varepsilon(t)}\,\varphi_0(t)\,\mathrm{d}t\right|$$
$$+\frac{k}{\sqrt{8\pi}}\left|\int_{-\pi}^{\pi} \mathbf{v}(\mathbf{x}_0(t))\cdot\hat{\mathbf{x}}\left(e^{-ik\hat{\mathbf{x}}\cdot\mathbf{x}_\varepsilon(t)} - e^{-ik\hat{\mathbf{x}}\cdot\mathbf{x}_0(t)}\right)\varphi_0(t)\,\mathrm{d}t\right|, \tag{9.97}$$

where $\mathbf{v}(\mathbf{x}(t)) = (x_2'(t), -x_1'(t))$. For convenience, let $\widetilde{\mathbf{v}}(\mathbf{x}(t))$ denote $(x_2(t), -x_1(t))$.

The first contribution to the inequality (9.97) is, after an integration by parts,

$$\frac{k}{\sqrt{8\pi}}\left[(\widetilde{\mathbf{v}}(\mathbf{x}_\varepsilon(t)) - \widetilde{\mathbf{v}}(\mathbf{x}_0(t)))\cdot\hat{\mathbf{x}}e^{-ik\hat{\mathbf{x}}\cdot\mathbf{x}_\varepsilon(t)}\varphi_0(t)\right]_{-\pi}^{\pi}$$
$$-\frac{k}{\sqrt{8\pi}}\int_{-\pi}^{\pi} (\widetilde{\mathbf{v}}(\mathbf{x}_\varepsilon(t)) - \widetilde{\mathbf{v}}(\mathbf{x}_0(t)))\cdot\hat{\mathbf{x}}\,\frac{d}{dt}\left(e^{-ik\hat{\mathbf{x}}\cdot\mathbf{x}_\varepsilon(t)}\varphi_0(t)\right)\mathrm{d}t. \tag{9.98}$$

The first term is zero, and we estimate the second term by splitting it as a sum over three subintervals $[-\pi,-\delta]\cup[-\delta,\delta]\cup[\delta,\pi]$, where the positive quantity δ will be chosen shortly, satisfying $\delta \to 0$ as $k\varepsilon \to 0$. On the first and third subintervals, the derivative term is bounded by the maximum value M of $2|\varphi_0'(t)| + 2k|\varphi_0(t)|$ on these subintervals; this is k times a term that is $O((k\delta)^{-(1/3)})$ as $\varepsilon \to 0$. Using the estimate

$$|\widetilde{\mathbf{v}}(\mathbf{x}_\varepsilon(t)) - \widetilde{\mathbf{v}}(\mathbf{x}_0(t))| = |\mathbf{x}_\varepsilon(t) - \mathbf{x}_0(t)| \leq \frac{\varepsilon^2}{\sin(t/2)}, \tag{9.99}$$

the contribution to the integral over the first and third intervals is bounded by

$$\frac{2Mk}{\sqrt{8\pi}}\int_{\delta}^{\pi} \frac{\varepsilon^2}{\sin(t/2)}\,\mathrm{d}t = \frac{Mk\varepsilon^2}{\sqrt{8\pi}}\left[-\log\cot\frac{t}{4}\right]_{\delta}^{\pi} = \frac{Mk\varepsilon^2}{\sqrt{8\pi}}\log\cot\frac{\delta}{4}. \tag{9.100}$$

On the other hand, the contribution over the interval $[-\delta, \delta]$ may be obtained by summing the contribution over $[-\delta, 0] \cup [0, \delta]$; the treatment of the first subinterval is similar to the second which we now consider

$$\frac{k}{\sqrt{8\pi}} \int_0^\delta (\widetilde{\mathbf{v}}(\mathbf{x}_\varepsilon(t)) - \widetilde{\mathbf{v}}(\mathbf{x}_0(t))) \cdot \hat{\mathbf{x}} \frac{d}{dt} \left(e^{-ik\hat{\mathbf{x}}\cdot\mathbf{x}_\varepsilon(t)} \varphi_0(t)\right) \mathrm{d}t$$

$$\leq \frac{2k\varepsilon}{\sqrt{8\pi}} \int_0^\delta \left(|\varphi_0'(t)| + |\varphi_0(t)|\right) \mathrm{d}t \leq \frac{2k\varepsilon}{\sqrt{8\pi}} \sqrt{\delta} \left(||\varphi_0'||_2 + ||\varphi_0||_2\right) \quad (9.101)$$

by an application of the Cauchy–Schwarz inequality. Choosing $\delta = \varepsilon^{3/2}$ makes both the bounds for (9.100) and (9.101) to be $o((k\varepsilon)^{4/3})$ as $k\varepsilon \to 0$.

We now establish a similar result for the second contribution to the inequality (9.97), which is bounded by

$$\frac{2k}{\sqrt{8\pi}} ||\varphi||_\infty \int_{-\pi}^{\pi} | e^{-ik\hat{\mathbf{x}}\cdot\mathbf{x}_\varepsilon(t)} - e^{-ik\hat{\mathbf{x}}\cdot\mathbf{x}_0(t))} | \,\mathrm{d}t. \quad (9.102)$$

Noting that the integrand is $2|\sin\left\{\frac{k}{2}\hat{\mathbf{x}}\cdot(\mathbf{x}_\varepsilon(t) - \mathbf{x}_0(t))\right\}|$, the integral is bounded by

$$\frac{2k^2}{\sqrt{8\pi}} ||\varphi||_\infty \int_{-\pi}^{\pi} | \mathbf{x}_\varepsilon(t) - \mathbf{x}_0(t)) | \,\mathrm{d}t. \quad (9.103)$$

We split the integral over the intervals $I_1 = [-\varepsilon, \varepsilon]$ and $I_2 = [-\pi, \pi] \setminus I$. Now

$$\int_{I_1} | \mathbf{x}_\varepsilon(t) - \mathbf{x}_0(t) | \,\mathrm{d}t \leq \int_{-\varepsilon}^{\varepsilon} 2\varepsilon \,\mathrm{d}t = (2\varepsilon)^2, \quad (9.104)$$

and using the estimate (9.99)

$$\int_{I_2} | \mathbf{x}_\varepsilon(t) - \mathbf{x}_0(t) | \,\mathrm{d}t \leq 2\varepsilon^2 \int_{\varepsilon}^{\pi} \frac{1}{\sin(t/2)} \,\mathrm{d}t = -2\varepsilon^2 \log\cot\frac{\varepsilon}{4}$$

$$= -2\varepsilon^2 \log\varepsilon + O\left(\varepsilon^2\right), \quad (9.105)$$

as $\varepsilon \to 0$. The result stated now follows. □

Combining the result of Lemma 9.8 and of inequality (9.96), we deduce

Theorem 9.2. *The maximum difference in the non-dimensionalised far-field patterns for the right-angled lemniscate and its rounded counterpart is $O((k\varepsilon)^{4/3})$ as $k\varepsilon \to 0$:*

$$\sqrt{k}\|u_\varepsilon^\infty - u_0^\infty\|_\infty = \sqrt{k} \max_{\hat{x}\in[-\pi,\pi]} |u_\varepsilon^\infty(\hat{x}) - u_0^\infty(\hat{x})| = O\left((k\varepsilon)^{4/3}\right). \quad (9.106)$$

9.4 Conclusion

Our previous numerical studies demonstrated that the maximum difference in the non-dimensionalised far-field patterns of a lemniscate with Dirichlet boundary condition (perfectly electrically conducting) and its rounded counterpart is, to a good approximation, equal to $C(\theta_0)(k\rho)^{4/3}$, for some constant $C(\theta_0)$ dependent on the direction of the incident plane wave θ_0; here ρ denotes the radius of curvature in the vicinity of the rounded corner point.

In this chapter, a theoretical basis for these numerical results was derived. An approximate integral equation for the difference in the surface quantities on the lemniscate and its rounded counterpart was obtained, and it was shown that the difference is $O((k\varepsilon)^{2/3})$ as $k\varepsilon \to 0$. As a consequence, the maximum difference in the non-dimensionalised far-field patterns is $O((k\varepsilon)^{4/3})$ as $k\varepsilon \to 0$, in accord with the computed results.

These results are readily capable of extension to structures with sharp corners of interior angles other than 90°. A similar approach to the Neumann case will be discussed in future publications.

References

[1] Bowman JJ, Senior TBA, and Uslenghi PLE. Electromagnetic and acoustic scattering by simple shapes. Revised ed. New York: Hemisphere Publishing Corp.; 1987.

[2] Van Bladel J. Singular electromagnetic fields and sources. Oxford, UK: Oxford University Press; 1991.

[3] Babich VM, Lyalinov MA, and Grikurov VE. Diffraction theory: the Sommerfeld-Malyuzhinets technique. Oxford, UK: Alpha Science; 2008.

[4] Sommerfeld A. Mathematische theorie der diffraction. Math Ann. 1896; 47(2–3):317–314.

[5] Sommerfeld A. Mathematical theory of diffraction. In: Mathematical theory of diffraction. Progress in Mathematics, vol. 35. Birkhäuser, Boston MA.

[6] Macdonald H. Electric waves. Cambridge, UK: Cambridge University Press; 2013.

[7] Wiegrefe A. Über einige mehrwertige Lösungen der Wellengleichung $\Delta u + k^2 u = 0$. Ann Phys. 1912;344(12):449–484.

[8] Carslaw H. Diffraction of waves by a wedge of any angle. Proc Lond Math Soc. 1920;2(1):291–306.

[9] Colton DL and Kress R. Inverse acoustic and electromagnetic scattering theory. 3rd ed. Applied mathematical sciences. New York: Springer; 2013.

[10] Smith PD, Markowskei A, and Rawlins AD. Two-dimensional diffraction by impedance loaded structures with corners. In: Electromagnetics in Advanced Applications (ICEAA), 2014 International Conference on. Aruba; 2014. p. 644–647.

[11] Smith PD and Markowskei AJ. The diffractive effect of rounding the corners of scattering structures. In: Electromagnetics in Advanced Applications (ICEAA), 2015 International Conference on. Turin, Italy; 2015. p. 1592–1595.

[12] Smith PD and Markowskei AJ. What effect does rounding the corners have on diffraction from structures with corners? In: Bashir S, editor. Advanced electromagnetic waves. Croatia: InTech; 2015. p. 1–28.

[13] Markowskei AJ and Smith PD. Scattering from structures with acute, obtuse and reflex corners. In: Electromagnetics in Advanced Applications (ICEAA), 2016 International Conference on; Cairns, Australia: IEEE Xplore; 2016.

[14] Markowskei AJ and Smith PD. Measuring the effect of rounding the corners of scattering structures. Radio Sci. 2017;2017RS006276. Available from: http://dx.doi.org/10.1002/2017RS006276.

[15] Smith PD and Markowskei AJ. The effect of rounding vertices on the diffraction from polygons and other scatterers. In: 2016 URSI International Symposium on Electromagnetic Theory (EMTS). Espoo, Finland; 2016. p. 236–239.

[16] Epstein CL and O'Neil M. Smoothed corners and scattered waves. SIAM J Sci Comput. 2016;38(5):A2665–A2698.

[17] Markowskei AJ. Two-dimensional diffraction by structures with corners. Sydney, Australia: Macquarie University; 2019.

[18] Engineer JC, King JR, and Tew RH. Diffraction by slender bodies. Eur J Appl Math. 1998;9(2):129–158.

[19] Helsing J and Karlsson A. An accurate boundary value problem solver applied to scattering from cylinders with corners. IEEE Trans Antennas Propag. 2013;61(7):3693–3700.

[20] Colton DL and Kress R. Integral equation methods in scattering theory. 1st ed. Pure and applied mathematics. New York: Wiley; 1983.

[21] Kress R. Linear integral equations. vol. 82. 3rd ed. New York: Springer; 2013.

[22] Olver FW, Lozier DW, Boisvert RF, *et al.* NIST handbook of mathematical functions. Cambridge, UK: Cambridge University Press; 2010.

Chapter 10
Radiation from a line source at the vertex of a right-angled dielectric wedge

Anthony D. Rawlins[1]

10.1 Introduction

The problem of the electromagnetic diffraction by a dielectric wedge is still an unsolved problem. Unsolved in the sense that a closed-form solution as complex integrals as given, for example, for an imperfectly or perfectly conducting wedge is still not known. The special case of a right-angled dielectric wedge was thought to offer a possible solution because of the simple orthogonal nature of the bounding surfaces and the possibility of applying the Wiener–Hopf technique. However, success in the sense outlined earlier has been limited. Either the final solution ends up with the numerical solution of a Fredholm integral equation, or an analytic perturbation solution. The application of the Wiener–Hopf method for the right-angled wedge heretofore has involved the theory of functions of two complex variables and the resultant Wiener–Hopf factorization in the quarter planes of two complex variables. These approaches derive a basic two-dimensional integral equation that is solved numerically, or an analytic perturbation solution obtained. The object of the present work is to show that the use of the Wiener–Hopf approach is unnecessary. A much simpler use of the double Laplace transform suffices. A full history, and an extensive number of references up to 1999, to various attempts is given in Rawlins [1], see also Salem *et al.* [2]. As an application, a perturbation solution of the relevant integral equation is obtained for the situation of a line source situated at the vertex of the right-angled wedge. This physical problem is interesting because it is of possible use in the numerical algorithms used by Groth *et al.* [3] when dealing with the meteorological problems of scattering by ice crystals. An interpretation of their algorithm is that at the corners of their scattering bodies, cylindrical waves are generated by the incident wave. These cylindrical corner waves are far enough away from the singular corner to produce plane waves that propagate along the flat surfaces of the scatterer.

In Section 10.2, we shall formulate the mathematical boundary value problem that describes the physical problem of the radiation of an E-polarized line source located at the edge of a dielectric rectangular wedge. This mathematical formulation

[1]Department of Mathematics, Brunel University, Uxbridge, UK

is in terms of a partial differential equation and boundary conditions derived from Maxwell's equations. The solution of this partial differential equation and the boundary conditions can be converted to an equivalent singular integral equation by means of Green's theorem. In Section 10.3, we shall convert this integral equation to another more convenient transformed integral equation by using the double Laplace transform in two complex variables. This approach is simpler than the quadruple Wiener–Hopf approach used by Kraut and Lehman [4]. The exact solution of the resulting singular integral equation is not possible yet. Instead, a solution of the singular integral equation is derived and constructed as a power series in the index of refraction. This series has been shown by Kraut and Lehman [4] to converge when the index of refraction is near unity. Using this solution, the far-field electric-field amplitude is derived to the first order of magnitude. Finally, the work ends with conclusions that discuss certain aspects of the work and further work and generalizations.

10.2 Formulation of the boundary value problem

We consider a piecewise-homogeneous, isotropic, conducting, dielectric medium of infinite extent referred to rectangular Cartesian coordinates (x_1, x_2, x_3). The right-angled wedge occupying the region $x_1 > 0, x_2 > 0, |x_3| \leq \infty$ is characterized by a constant electrical conductivity σ_d, a constant magnetic permittivity μ_d, and a constant electric permittivity ε_d. The medium external to this right-angled wedge is similarly characterized by different constant physical parameters $\sigma_v, \mu_v, \varepsilon_v$. It will be further realistically assumed that both regions have the magnetic permeability of free space μ_0, that is, $\mu_v = \mu_d = \mu_0$, and that the relative dielectric constant n of the right-angled wedge is unity or greater so that $n = \varepsilon_d/\varepsilon_v \geq 1$. The mathematical description of the physical situation follows from Maxwell's equations relating the electric- and magnetic-field intensities $\mathbf{E}(\mathbf{r}, t)$ and $\mathbf{H}(\mathbf{r}, t)$ with the electric displacement $\mathbf{D}(\mathbf{r}, t)$ and magnetic field $\mathbf{B}(\mathbf{r}, t)$ at the point $\mathbf{r}$ and time t. Maxwell's equations in the MKS system are

$$\begin{aligned} &\nabla \times \mathbf{E} + \frac{\partial \mathbf{B}}{\partial t} = 0, \qquad \nabla \cdot \mathbf{B} = 0, \\ &\nabla \times \mathbf{H} - \frac{\partial \mathbf{D}}{\partial t} = \mathbf{J}, \qquad \nabla \cdot \mathbf{D} = \rho, \end{aligned} \tag{10.1}$$

with a continuity equation relating the charge density ρ and the current density $\mathbf{J}$ given by

$$\nabla \cdot \mathbf{J} + \frac{\partial \rho}{\partial t} = 0 \tag{10.2}$$

We define piecewise-constant electrical permittivity and conductivity parameters by

$$\begin{aligned} \varepsilon(\mathbf{r}) &= \varepsilon_d, \quad x_1 \geq 0 \cap x_2 \geq 0 \cap |x_3| \leq \infty, \\ &= \varepsilon_v, \quad x_1 \leq 0 \cup x_2 \leq 0 \cap |x_3| \leq \infty, \end{aligned} \tag{10.3}$$

$$\sigma(\mathbf{r}) = \sigma_d, \quad x_1 \geq 0 \cap x_2 \geq 0 \cap |x_3| \leq \infty,$$
$$= \sigma_v, \quad x_1 \leq 0 \cup x_2 \leq 0 \cap |x_3| \leq \infty. \tag{10.4}$$

In terms of (10.4), the constitutive equations of the medium are

$$\mathbf{D}(\mathbf{r},t) = \varepsilon(\mathbf{r})\mathbf{E}(\mathbf{r},t), \tag{10.5}$$

$$\mathbf{J}(\mathbf{r},t) = \sigma(\mathbf{r})\mathbf{E}(\mathbf{r},t), \tag{10.6}$$

$$\mathbf{B}(\mathbf{r},t) = \mu_0\mathbf{H}(\mathbf{r},t). \tag{10.7}$$

For a harmonic time dependence of the form

$$\mathbf{E}(\mathbf{r},t) = \mathbf{E}(\mathbf{r})\exp(-\imath\omega t),$$

and

$$\mathbf{H}(\mathbf{r},t) = \mathbf{H}(\mathbf{r})\exp(-\imath\omega t),$$

Maxwell's equations give

$$\nabla \times \mathbf{H} + \imath\omega\hat{\varepsilon}(\mathbf{r})\mathbf{E} = 0, \tag{10.8}$$

$$\nabla \times \mathbf{E} - \imath\omega\mu_0\mathbf{H} = 0, \tag{10.9}$$

where

$$\hat{\varepsilon}(\mathbf{r}) = \varepsilon(\mathbf{r}) + \imath\sigma(\mathbf{r})\omega^{-1}. \tag{10.10}$$

Taking the curl of (10.9) and by using (10.8) gives

$$(\nabla^2 + \omega^2\mu_0\hat{\varepsilon}_v)\mathbf{E} = \nabla\nabla\cdot\mathbf{E} - \omega^2\mu_0\hat{\varepsilon}_v\left[\hat{\varepsilon}/\hat{\varepsilon}_v - 1\right]\mathbf{E}, \tag{10.11}$$

where

$$\hat{\varepsilon}_v = \varepsilon_v + \imath\sigma_v\omega^{-1}. \tag{10.12}$$

With the aid of the divergence of (10.8), we can express $\nabla\nabla\cdot\mathbf{E}$ as

$$\nabla\nabla\cdot\mathbf{E} = -\nabla\nabla\cdot\left[\left(\frac{\hat{\varepsilon}(\mathbf{r})}{\hat{\varepsilon}_v}\right) - 1\right]\mathbf{E}, \tag{10.13}$$

and by using this result in (10.11) gives

$$(\nabla^2 + \omega^2\mu_0\hat{\varepsilon}_v)\mathbf{E} = -(\nabla\nabla + \omega^2\mu_0\hat{\varepsilon}_v)\left[\left(\frac{\hat{\varepsilon}(\mathbf{r})}{\hat{\varepsilon}_v}\right) - 1\right]\mathbf{E}. \tag{10.14}$$

We shall restrict our consideration to electric fields that are polarized parallel to the vertex of the wedge, that is, along the x_3 axis, and which do not depend on the x_3 coordinate. For such fields given by

$$\mathbf{E}(x_1,x_2) = [0, 0, E(x_1,x_2)], \tag{10.15}$$

(10.14) reduces to the scalar two-dimensional form

$$(\nabla^2 + \omega^2\mu_0\hat{\varepsilon}_v)E(x_1,x_2) = (-\,\omega^2\mu_0\hat{\varepsilon}_v)\left[\left(\frac{\hat{\varepsilon}(\mathbf{r})}{\hat{\varepsilon}_v}\right) - 1\right]E(x_1,x_2). \tag{10.16}$$

It is convenient to define the complex wave vector $k(\mathbf{r}) = \omega^2\mu_0\hat{\varepsilon}(\mathbf{r})$, and wave numbers k_d, and k_v by

$$\begin{aligned} k^2(\mathbf{r}) &= k_d^2, \quad x_1 \geq 0 \cap x_2 \geq 0 \cap |x_3| \leq \infty, \\ &= k_v^2, \quad x_1 \leq 0 \cup x_2 \leq 0 \cap |x_3| \leq \infty, \end{aligned} \tag{10.17}$$

in terms of which (10.16) becomes

$$\left(\nabla^2 + k_v^2\right) E\left(x_1, x_2\right) = -\left[\left(k^2\left(\mathbf{r}\right)\right) - k_v^2\right] E\left(x_1, x_2\right). \tag{10.18}$$

By using Green's theorem, (10.18) can be converted to an integral equation for $E(x_1, x_2)$:

$$\begin{aligned} E(x_1, x_2) &= E^{(0)}(x_1, x_2) + \frac{\imath}{4}\left(k_d^2 - k_v^2\right) \\ &\times \int_0^\infty \int_0^\infty H_0^{(1)}\left(k_v\left[(x_1 - \xi_1)^2 + (x_2 - \xi_2)^2\right]^{1/2}\right) E(\xi_1, \xi_2) d\xi_1 d\xi_2, \end{aligned} \tag{10.19}$$

where $E^{(0)}(x_1, x_2)$ represents the electric field incident on the wedge $x_1 \geq 0 \cap x_2 \geq 0 \cap |x_3| \leq \infty$ that we will assume is a line source and therefore satisfies the following equation

$$\left(\nabla^2 + k_v^2\right) E^{(0)}(x_1, x_2) = -\delta\left(x_1 - x_1^0\right) \delta\left(x_2 - x_2^0\right), \tag{10.20}$$

so that $E^{(0)}(x_1, x_2) = \frac{\imath}{4} H_0^{(1)}(k_v\sqrt{(x_1 - x_1^0)^2 + (x_2 - x_2^0)^2})$, where without loss of generality we assume $x_1^0 \leq 0 \cap |x_2^0| \leq \infty$. The kernel in (10.19) is a Hankel function of the first kind of order zero, where k_v and k_d are chosen to have positive imaginary parts

$$Im(k_v) > 0, Im(k_d) > 0. \tag{10.21}$$

In order to solve the integral equation (10.19), we convert it into a singular integral equation for the Fourier transform of $E(x_1, x_2)$.

10.3 Singular integral equation for the double Laplace transform of the electric field

We now take the double Laplace transform of the integral equation (10.19) to produce an integral equation that is more amenable to approximate evaluation. Before this is done, we make use of the following representation for the Hankel function kernel:

$$\begin{aligned} &\frac{\imath}{4} H_0^{(1)}(k_v[(x_1 - \xi_1)^2 + (x_2 - \xi_2)^2]^{1/2}) \\ &= \frac{1}{(2\pi)^2} \int_{-\infty}^{\infty} \int_{-\infty}^{\infty} \frac{\exp\left[\imath z_1(\xi_1 - x_1) + \imath z_2(\xi_2 - x_2)\right]}{z_1^2 + z_2^2 - k_v^2} dz_1 dz_2, \end{aligned} \tag{10.22}$$

valid for $Im[k_v] > 0$, see Morse and Feshbach [5] (vol 1, page 817). Substituting this expression into (10.19), and interchanging the order of integration, this is allowed because the resulting expressions are uniformly valid before and after the interchange. We have

$$E(x_1, x_2) \\ = E^{(0)}(x_1, x_2) + \frac{(k_d^2 - k_v^2)}{(2\pi)^2} \int_{-\infty}^{\infty}\int_{-\infty}^{\infty} \exp\left[-\imath z_1 x_1 - \imath z_2 x_2\right] E_{++}(z_1, z_2) dz_1 dz_2, \quad (10.23)$$

where $E_{++}(z_1, z_2)$ is the double Laplace transform of $E(\xi_1, \xi_2)$ and is given by

$$E_{++}(z_1, z_2) = \int_0^{\infty}\int_0^{\infty} \exp\left[\imath z_1 \xi_1 + \imath z_2 \xi_2\right] E(\xi_1, \xi_2) d\xi_1 d\xi_2, \quad (10.24)$$

with $Im[z_1] \geq 0, Im[z_2] \geq 0$. The function $E_{++}(z_1, z_2)$ is a regular analytic function of the two complex variables in the complex domain $Im[z_1] \geq 0, Im[z_2] \geq 0$. Now we take the double Laplace transform of (10.23) by operating across the equation by $\int_0^{\infty}\int_0^{\infty} \exp\left[\imath k_1 x_1 + \imath k_2 x_2\right] \cdots dx_1 dx_2$ giving the double Laplace transformed integral equation

$$E_{++}(k_1, k_2) \\ = E_{++}^{(0)}(k_1, k_2) + \frac{(k_d^2 - k_v^2)}{(2\pi\imath)^2} \int_{-\infty}^{\infty}\int_{-\infty}^{\infty} \frac{E_{++}(z_1, z_2) dz_1 dz_2}{\left(z_1^2 + z_2^2 - k_v^2\right)(z_1 - k_1)(z_2 - k_2)}, \quad (10.25)$$

where $Im[k_1] \geq 0$ and $Im[k_2] \geq 0$. This integral equation is identical to that derived by Kraut and Lehman [4] by a more complicated Wiener–Hopf method involving four unknown functions of two complex variables. The transformed integral equation (10.25) involves a much simpler algebraic kernel than the untransformed integral equation (10.19) that involves Bessel functions. Thus, the resulting integrals that will occur in a perturbation solution will presumably be easier to evaluate.

The integral equation (10.25) has been analyzed by Kraut and Lehman [4] in Section 4 of their paper, and they show that provided the L_2 norm of $E_{++}^{(0)}$ is bounded, that is

$$||E_{++}^{(0)}||_2 = \int_{-\infty}^{\infty}\int_{-\infty}^{\infty} E_{++}^{(0)}|(k_1, k_2)|^2 dk_1 dk_2 < \infty, \quad (10.26)$$

and that the complex wave numbers satisfy*

$$|k_d^2 - k_v^2| < 2|Im(k_v)Re(k_v)| \quad (10.27)$$

*In terms of the complex refractive index n (10.26) becomes $|n^2 - 1| < 2|Im(k_v/|k_v|)||Re(k_v/|k_v|)|$ and the maximum value of the right-hand side of this inequality is taken when $k_v = |k_v| \exp \imath\pi/4$ and the iterative solution of (10.26) will converge in this case for wedges, relative refractive dielectric constants of which satisfy $0 \leq |n^2 - 1| < 1$ or $1 \leq |n| < \sqrt{2}$.

then the solution of the integral equation is the limit of a sequence of successive approximations converging in L_2 norm to E_{++}. That is, the successive approximations to the solution

$$E_{++}^{(0)}(k_1,k_2), E_{++}^{(1)}(k_1,k_2), E_{++}^{(2)}(k_1,k_2), \ldots, E_{++}^{(m)}(k_1,k_2) \tag{10.28}$$

take the form

$$E_{++}^{(m+1)}(k_1,k_2) = E_{++}^{(0)}(k_1,k_2) + \frac{\left(k_d^2-k_v^2\right)}{(2\pi\imath)^2}\int_{-\infty}^{\infty}\int_{-\infty}^{\infty}\frac{E_{++}^{(m)}(z_1,z_2)dz_1dz_2}{\left(z_1^2+z_2^2-k_v^2\right)(z_1-k_1)(z_2-k_2)}, \tag{10.29}$$

where $m = 0, 1, 2, \ldots$, $Im(k_1) = 0, Im(k_2) = 0$ and $(k_d^2 - k_v^2)$ satisfies (10.27). The solution of (10.25) for real k_1 and k_2 is the limit of the sequence (10.29):

$$E_{++}(k_1,k_2) = \lim_{n\to\infty} E_{++}^{(m)}(k_1,k_2). \tag{10.30}$$

10.4 Approximate solution of the singular integral equation

We shall now derive an approximate solution of the integral equation for the situation where the line source is at the vertex of the right-angled wedge or at infinity, and the refractive index is such that $1 < |n| < \sqrt{2}$. From the reciprocity theorem Clemmow [6], the field is the same in both these situations since inherent in Maxwell's equation the solution for a line source is unaffected if the location of the source is interchanged with that of the point of observation. We may therefore consider a line source at infinity and calculate the field at the origin. This will correspond to a plane wave of a certain magnitude incident on the dielectric right-angled wedge. The electric field at the vertex of the wedge is given by

$$E(0,0) = \lim_{n\to\infty} E^{(m)}(0,0) \tag{10.31}$$

where

$$E^{(m)}(0,0) = \lim_{x_1\to0^+}\lim_{x_2\to0^+}\frac{1}{(2\pi)^2}\int_{-\infty}^{\infty}\int_{-\infty}^{\infty}E_{++}^{(m)}(k_1,k_2)\exp\left(-\imath k_1x_1-\imath k_2x_2\right)dk_1dk_2. \tag{10.32}$$

To carry out this limiting process, it is convenient to use the initial value theorem for the double Laplace transform, that is

$$\lim_{x_1\to0^+}\lim_{x_2\to0^+}\frac{1}{(2\pi\imath)^2}\int_{-\infty}^{\infty}\int_{-\infty}^{\infty}E_{++}^{(m)}(k_1,k_2)\exp(-\imath k_1x_1-\imath k_2x_2)dk_1dk_2$$

$$= \lim_{Im(k_1)\to\infty}\lim_{Im(k_2)\to\infty}\frac{1}{(2\pi\imath)^2}\int_{-\infty}^{\infty}\int_{-\infty}^{\infty}\frac{k_1k_2E_{++}^{(m)}(z_1,z_2)dz_1dz_2}{(z_1-k_1)(z_2-k_2)}. \tag{10.33}$$

It follows from (10.32) and (10.33) that

$$E_{++}^{(m)}(0,0) = -\lim_{Im(k_1)\to\infty}\lim_{Im(k_2)\to\infty} k_1k_2E_{++}^{(m)}(k_1,k_2) \tag{10.34}$$

for the *m*th-order approximation to the electric field at the vertex of the wedge. To first order in $k_d - k_v$, by using (10.29), (10.25), and (10.30) we get

$$E_{++}^{(1)}(k_1,k_2) = E_{++}^{(0)}(k_1,k_2) + \frac{\left(k_d^2-k_v^2\right)}{(2\pi\imath)^2}\int_{-\infty}^{\infty}\int_{-\infty}^{\infty}\frac{E_{++}^{(0)}(z_1,z_2)dz_1dz_2}{\left(z_1^2+z_2^2-k_v^2\right)(z_1-k_1)(z_2-k_2)}, \tag{10.35}$$

and, from (10.34),

$$E^{(1)}(0,0) = E_0(0,0) + \frac{\left(k_v^2-k_d^2\right)}{(2\pi\imath)^2}\int_{-\infty}^{\infty}\int_{-\infty}^{\infty}\frac{E_{++}^{(0)}(z_1,z_2)dz_1dz_2}{\left(z_1^2+z_2^2-k_v^2\right)}. \tag{10.36}$$

When the line source tends to infinity with $x_1^0 < 0, x_2^0 < 0$, the asymptotic form of the Hankel function line source (10.20) for large argument $r_0 = \sqrt{(x_1^0)^2+(x_2^0)^2} \to \infty$ gives

$$E^{(0)}(x_1,x_2) = E_0 \exp\imath(a_1x_1 + a_2x_2) \tag{10.37}$$

where $a_1 = k_v\cos\theta_0, a_2 = k_v\sin\theta_0, 0 < \theta_0 < \pi/2$; and $E_0 = (1/2\sqrt{2\pi r_0})\exp\imath(k_vr_0 + \pi/4)$. The double Laplace transform of (10.37) gives

$$E_{++}^{(0)}(z_1,z_2) = -\frac{E_0}{(z_1+a_1)(z_2+a_2)}. \tag{10.38}$$

Hence, (10.36) becomes

$$E^{(1)}(0,0) = E_0\left(1 + \frac{\left(k_d^2-k_v^2\right)}{(2\pi\imath)^2}\int_{-\infty}^{\infty}\int_{-\infty}^{\infty}\frac{dz_1dz_2}{\left(z_1^2+z_2^2-k_v^2\right)(z_1+a_1)(z_2+a_2)}\right), \tag{10.39}$$

where $Im(k_v + a_1) > 0, Im(k_v + a_2) > 0$ must be taken for the integrals to exist for all $0 < \theta_0 < \pi/2$.

In an analogous manner, it can be shown that the second-order field at the edge is given by

$$E^{(2)}(0,0) = E^{(1)}(0,0) + \frac{(k_d^2-k_v^2)^2}{(2\pi\imath)^4}E_0\int_{-\infty}^{\infty}\int_{-\infty}^{\infty}\frac{dz_1dz_2}{\left(z_1^2+z_2^2-k_v^2\right)} \times\int_{-\infty}^{\infty}\int_{-\infty}^{\infty}\frac{d\zeta_1d\zeta_2}{\left(\zeta_1^2+\zeta_2^2-k_v^2\right)(\zeta_1+a_1)(\zeta_2+a_2)(\zeta_1-z_1)(\zeta_2-z_2)}, \tag{10.40}$$

where $Im(k_v) > 0, Im(\zeta_1 + a_1) > 0, Im(\zeta_2 + a_2) > 0$ must be taken for the integrals to exist.

10.5 Calculation of $E^{(1)}$(0,0)

In order to calculate $E^{(1)}(0,0)$, we must evaluate the double integral occurring in (10.39) for $Im(k_v) > 0$, that is

$$I(a_1, a_2) = \int_{-\infty}^{\infty}\int_{-\infty}^{\infty} \frac{dz_1 dz_2}{\left(z_1^2 + z_2^2 - k_v^2\right)(z_1 + a_1)(z_2 + a_2)}. \tag{10.41}$$

Let $z_1 = r\cos\theta, z_2 = r\sin\theta \quad r > 0, 0 < \theta < 2\pi$ then

$$I(a_1, a_2) = \int_0^{\infty}\int_0^{2\pi} \frac{rdrd\theta}{\left(r^2 - k_v^2\right)(r\cos\theta + a_1)(r\sin\theta + a_2)}; \tag{10.42}$$

then using the easily proved result

$$\int_0^{2\pi} \frac{d\theta}{(r\cos\theta + a_1)(r\sin\theta + a_2)} = 2\pi\imath\left(\frac{a_2}{\left(r^2 - k_v^2\right)\sqrt{r^2 - a_1^2}} + \frac{a_1}{\left(r^2 - k_v^2\right)\sqrt{r^2 - a_2^2}}\right), \tag{10.43}$$

we have

$$I(a_1, a_2) = 2\pi\imath a_1 \int_0^{\infty} \frac{rdr}{\left(r^2 - k_v^2\right)^2\sqrt{r^2 - a_2^2}} + 2\pi\imath a_2 \int_0^{\infty} \frac{rdr}{\left(r^2 - k_v^2\right)^2\sqrt{r^2 - a_1^2}}. \tag{10.44}$$

The only singularities† of these integrals in the complex r-plane occur in the first and third quadrant. Thus, we can use analytic continuation to put these integrals into a more convenient form by rotating clockwise the path of integration to run from $(0, -\imath\infty)$. Then making an obvious change of variable we obtain:

$$I(a_1, a_2) = 2\pi a_1 \int_0^{\infty} \frac{r\,dr}{\left(r^2 + k_v^2\right)^2\sqrt{r^2 + a_2^2}} + 2\pi a_2 \int_0^{\infty} \frac{r\,dr}{\left(r^2 + k_v^2\right)^2\sqrt{r^2 + a_1^2}}. \tag{10.45}$$

†In the work that follows, we shall be working in the cut plane for which the square root function $\kappa(z) = \sqrt{k_v^2 - z^2}$ will be defined such that $\kappa(z) = k_v$ for $z = 0$, and $Im(\kappa(z)) > 0$.

These integrals can now be evaluated by analytic continuation with Mathematica for real k_v giving

$$
\begin{aligned}
I(a_1, a_2) &= 2\pi a_1 \left(\frac{-\sqrt{k_v^2 a_2^2 - a_2^4} + k_v^2 \arccos\left(\frac{\sqrt{a_2^2}}{k_v}\right)}{2k_v^2\left(k_v^2 - a_2^2\right)^{3/2}} \right) \\
&\quad + 2\pi a_2 \left(\frac{-\sqrt{k_v^2 a_1^2 - a_1^4} + k_v^2 \arccos\left(\frac{\sqrt{a_1^2}}{k_v}\right)}{2k_v^2\left(k_v^2 - a_1^2\right)^{3/2}} \right), \\
&= -\frac{|\sin\theta_0|}{k_v^2 \cos\theta_0} + \frac{\arccos|\sin\theta_0|}{k_v^2 \cos\theta_0 |\cos\theta_0|} - \frac{|\cos\theta_0|}{k_v^2 \sin\theta_0} + \frac{\arccos|\cos\theta_0|}{k_v^2 \sin\theta_0 |\sin\theta_0|}.
\end{aligned}
\tag{10.46}
$$

Hence, substituting this result (10.46) into expression (10.39) we have

$$
\begin{aligned}
E^{(1)}(0,0) &= E_0 \left(1 - \frac{(n^2-1)}{(4\pi)} \left(\sin\theta_0 \left[\frac{(\pi/2 - |\arcsin(\cos\theta_0)|)}{|\sin^3\theta_0|} - \frac{|\cos\theta_0|}{\sin^2\theta_0} \right] \right.\right. \\
&\quad \left.\left. + \cos\theta_0 \left[\frac{(\pi/2 - |\arcsin(\sin\theta_0)|)}{|\cos^3\theta_0|} - \frac{|\sin\theta_0|}{\cos^2\theta_0} \right] \right) \right).
\end{aligned}
\tag{10.47}
$$

10.6 Radiated far field

Since the incident plane wave $e^{\imath k_v(x\cos\theta_0 + y\sin\theta_0)} = e^{\imath k_v r\cos(\theta-\theta_0)}$, where $0 < \theta_0 < \pi/2$, it produces the field $E(0,0) = E_z(\theta_0)$ at the origin $r = 0$. For a line source at the edge of the wedge, we have the integral representation

$$
H_0^{(1)}(k_v r) = \frac{1}{\pi} \int\limits_{S(\theta)} e^{\imath k_v r cos(\theta-\theta_0)} d\theta_0,
\tag{10.48}
$$

where $S(\theta)$ is the path of the steepest descent, Clemmow [6]. Thus, the field radiated by a line source at the origin is given by integrating over the incident angle of the plane wave solution by $(1/\pi)\int_{S(\theta)} e^{\imath k_v r cos(\theta-\theta_0)} d\theta_0$, giving

$$
E_z(r,\theta) = \frac{1}{\pi} \int\limits_{S(\theta)} E_z(\theta_0) e^{\imath k_v r cos(\theta-\theta_0)} d\theta_0.
\tag{10.49}
$$

This result also follows directly from an application of the reciprocity theorem, Clemmow [6]. An application of the saddle point method then gives

$$
E_z(r,\theta) \sim E_z(\theta_0) H_0^{(1)}(k_v r) \sim \sqrt{\frac{2}{\pi k_v r}} E_z(\theta) e^{(k_v r - (\imath\pi/4))},
\tag{10.50}
$$

provided the function $E_z(\theta_0)$ in the integrand has no pole singularities at $\theta_0 = \theta$. Nonetheless, there is a mathematical inconsistency in that the exponential wave-number cannot match on the boundaries of the dielectric wedge and inside the dielectric wedge. However, the solution is approximate in terms of powers of $(k_d^2 - k_v^2)$, and therefore for mathematical consistency we should expand k_v in terms of $(k_d^2 - k_v^2)$. To this end, we note that

$$\begin{aligned} k_v &= \sqrt{\frac{k_v^2 + k_d^2}{2} + \frac{k_v^2 - k_d^2}{2}} = \sqrt{\frac{k_v^2 + k_d^2}{2}}\left(1 + \frac{k_v^2 - k_d^2}{k_d^2 + k_v^2}\right)^{1/2} \\ &= k + \frac{1}{4k}\left(k_v^2 - k_d^2\right) - \frac{1}{32k^3}\left(k_v^2 - k_d^2\right)^2 + O\left(\left(k_d^2 - k_v^2\right)^3\right) \end{aligned} \tag{10.51}$$

where $k = \sqrt{(k_v^2 + k_d^2)/2}$.

Then expression (10.50) becomes

$$E_z(r, \theta) \sim \sqrt{\frac{2}{\pi kr}} E_z(\theta) e^{(kr - (\imath\pi/4))}, \tag{10.52}$$

so that the first-order field everywhere is given by

$$E_z(r, \theta) \sim \sqrt{\frac{2}{\pi kr}} E^{(1)}(0, 0) e^{(kr - (\imath\pi/4))} \tag{10.53}$$

with θ_0 replaced by θ in expression (10.47). However, there is a restriction on r and k given by $|(1 - 2kr)(k_v^2 - k_d^2)| \ll |4(k_v^2 + k_d^2)|$. This aspect has already been discussed in Rawlins [1].

Some graphical plots for the refractive index taking the values $n = 1, \sqrt{2}, 1.5, 2.5$ are given in Figures 10.1–10.4. The first two values fall within the range of convergence of the approximation. However, this range of convergence although sufficient is not necessarily the full range of convergence. It is possible that the approximation is valid for a much larger range of the absolute values of the refractive index.

10.7 Conclusions

We have shown that the use of the double Wiener–Hopf technique involving the use of two complex variables and the complicated concomitant problems of factorization in the two complex variables is not necessary for the problem of the diffraction by a right-angled dielectric wedge. The simpler approach of using directly the double complex Laplace or Fourier transform is sufficient to derive the basic transformed double integral equation that can be approximately solved as a Neumann series solution.

As an application, the approximate solution for a line source at the edge of the wedge is derived and graphical plots show the symmetric nature of the far field radiation pattern. We have not been able to give precise estimates of the accuracy of the graphs in Figures 10.2–10.4. However, within the refractive index range $0 < n < \sqrt{2}$ we expect there to be good agreement with the physical field behavior. The

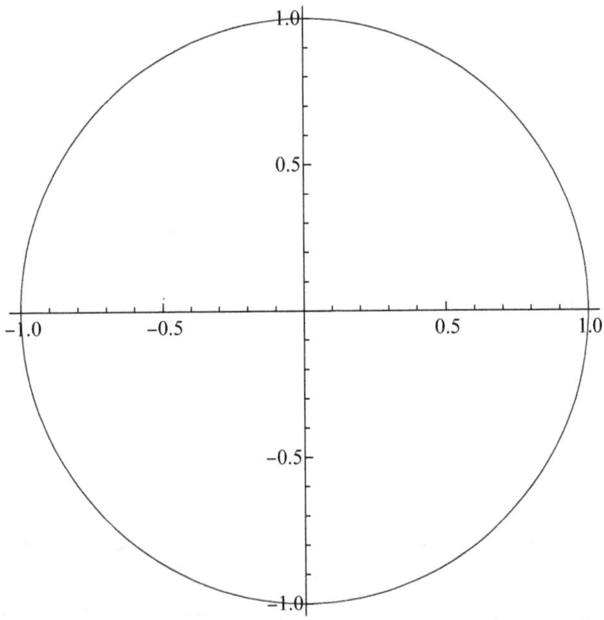

Figure 10.1 *Refractive index* $n = 1$

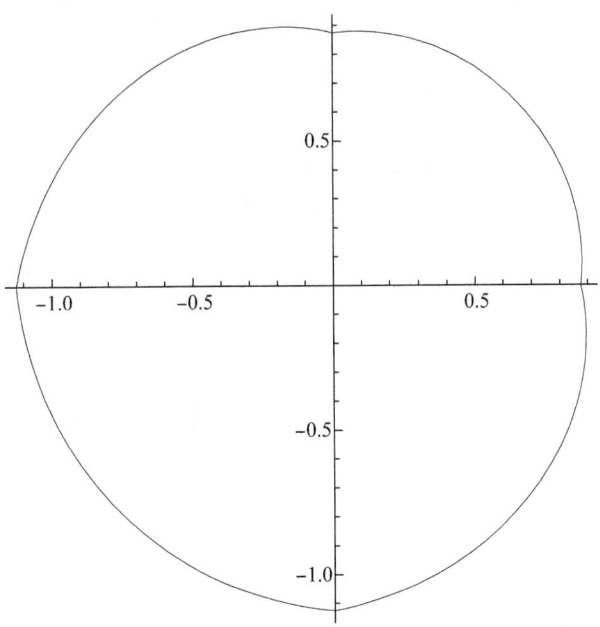

Figure 10.2 *Refractive index* $n = \sqrt{2}$

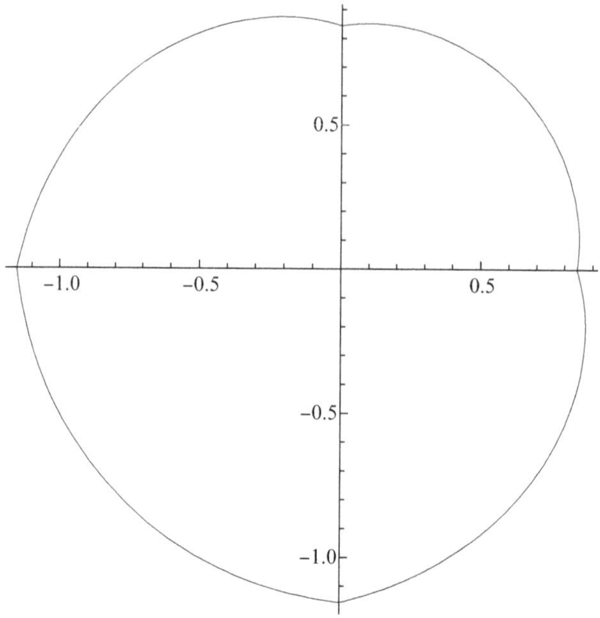

Figure 10.3 Refractive index $n = 1.5$

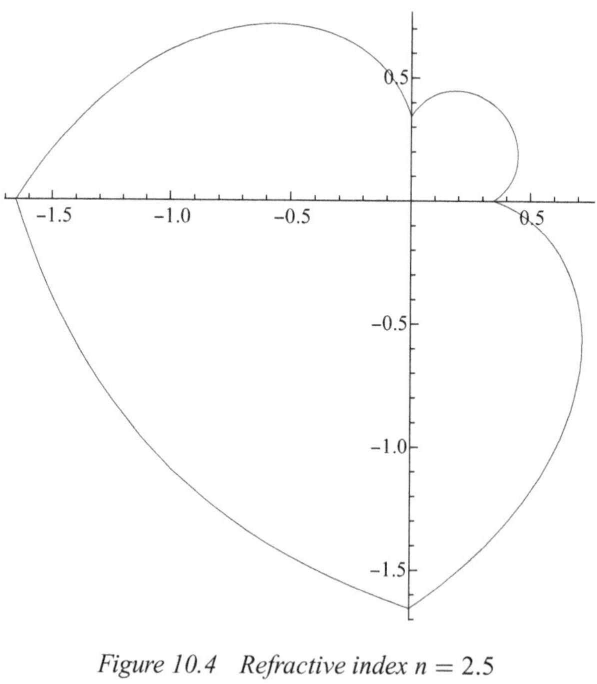

Figure 10.4 Refractive index $n = 2.5$

graphical results show the radiation pattern symmetry one would expect about the diagonal line $x_1 = x_2$. If we let the magnitude of refractive index n increase beyond $n = \sqrt{2}$, the field inside the dielectric wedge gets smaller, and most of the field is radiated outside the wedge. The structure is behaving like an impedance wedge as one would expect. However, as the refractive index increases there is an increase in the tangential discontinuity in the slope of these figures on the coordinate axes x_1, x_2. This is reasonable on the faces of the dielectric wedge $x_1 > 0, x_2 = 0$, and $x_1 = 0, x_2 > 0$, but not so for $x_1 < 0, x_2 = 0$. A possible explanation of this anomaly is that the asymptotics are breaking down because of singularities in the integrand approaching the saddle point. The nature of these singularities is difficult to specify a priori. We could go further, within the refractive index range $0 < n < \sqrt{2}$, and evaluate the second-order perturbation effect. This would involve the evaluation of a quadruple complex integral. It is the author's contention that this is possible, if somewhat laborious to fulfill. Finally, it might be possible to extend this method to deal with a point source at the vertex of a three-dimensional corner by considering oblique incidence of a plane wave and the triple Laplace transformation of the analogous triple variable integral equation to (10.25).

References

[1] Rawlins, A. D. Diffraction by, or Diffusion Into, a Penetrable Wedge. *Proceedings of the Royal Society of London.* A455, 2655–2686, 1999.

[2] Salem, M. A., Kamel, A. H. and Osipov, A. V. Electromagnetic Fields in the Presence of an Infinite Dielectric Wedge. *Proceedings of the Royal Society of London.* A462, 2503–2522, 2006.

[3] Groth, S. P., Hewett, D. P. and Langdon, S. Hybrid Numerical-Asymptotic Approximation for High-Frequency Scattering by Penetrable Convex Polygons. *IMA Journal of Applied Mathematics.* 80(2), 324–353, 2015.

[4] Kraut, E. A. and Lehman, G. W. Diffraction of Electromagnetic Waves by a Right-Angle Dielectric Wedge. *Journal of Mathematical Physics.* 10(8), 1340–1348, 1969.

[5] Morse, P. M. and Feshbach, H. Methods of Theoretical Physics. McGraw-Hill, New York, 1953.

[6] Clemmow, P. C. The Plane Wave Spectrum Representation of Electromagnetic Fields. New Jersey: IEEE Press. The IEEE/OUP Series on Electromagnetic Wave Theory, 1996.

Chapter 11

Wiener–Hopf analysis of the diffraction by a thin material strip

Takashi Nagasaka[1] and Kazuya Kobayashi[1]

The analysis of wave scattering and diffraction problems involving canonical objects is one of the important subjects in electromagnetic theory and radar cross section (RCS) studies. Various analytical and numerical methods have been developed so far and the scattering problems have been investigated for many kinds of two- and three-dimensional structures. Among a number of analysis methods, the Wiener–Hopf technique is known as a rigorous, function-theoretic approach for electromagnetic wave problems related to canonical geometries. In this chapter, we shall consider a thin material strip that is important from both the theoretical and engineering viewpoints, and analyze the electromagnetic wave diffraction by means of the Wiener–Hopf technique. It is shown that our final solutions are valid over a broad frequency range. Numerical examples are presented for various physical parameters, and the far field scattering characteristics are discussed in detail. Some comparisons with other existing methods are also given.

11.1 Introduction

The aim of this work is to analyze the diffraction by two-dimensional obstacles having various physical parameters applying the Wiener–Hopf technique [1–4]. The Wiener–Hopf technique is one of the powerful, rigorous approaches for analyzing wave scattering and diffraction problems related to canonical obstacles, which is mathematically rigorous in the sense that the edge condition required for the uniqueness of the solution is explicitly incorporated into the analysis. In this chapter, we shall consider a material strip with arbitrary permittivity and permeability, and analyze the plane wave diffraction using the Wiener–Hopf technique.

It is worthwhile to remark that in the open literature there are many works devoted to strip problems. For example, Volakis [5] analyzed the H-polarized plane wave diffraction by a thin material strip using the dual integral equation approach [6] and

[1]Department of Electrical, Electronic, and Communication Engineering, Chuo University, Bunkyo-ku, Tokyo, Japan

the extended spectral ray method [7] together with approximate boundary conditions [8]. In his study [5], Volakis first solved rigorously the diffraction problem involving a single material half-plane, and subsequently obtained a high-frequency solution to the original strip problem by superposing the singly diffracted fields from the two independent half-planes and the doubly/triply diffracted fields from the edges of the two half-planes. Therefore his analysis is not rigorous from the viewpoint of boundary value problems, and may not be applicable unless the strip width is relatively large compared with the wavelength. The same problem was reconsidered more recently by Shapoval *et al.* [9, 10] by using the generalized boundary conditions and the singular integral equation.

In this chapter, we shall consider Volakis's problem [5] and analyze the plane wave diffraction for H and E polarizations by a thin material strip using Wiener–Hopf technique. Assuming that the thickness is small compared with the wavelength, the original problem is replaced by a strip of zero thickness satisfying the approximate boundary conditions [8]. Introducing the Fourier transform of the scattered field and applying approximate boundary conditions in the transform domain, the problem is formulated in terms of the Wiener–Hopf equations, which are solved exactly via the factorization and decomposition procedure. However, the solution is formal in the sense that branch-cut integrals with unknown integrands are involved. By employing a rigorous asymptotic method [11] together with the special function introduced in our previous papers [12, 13], we shall derive a high-frequency solution of the Wiener–Hopf equations, which is valid for the strip width greater than about the incident wavelength.

The scattered field in the real space is evaluated asymptotically by taking the Fourier inverse of the solution in the transform domain and applying the saddle point method of integration. Our final solution is uniformly valid in arbitrary incidence and observation angles. Numerical examples of the RCS are presented for various physical parameters and far field scattering characteristics of the strip are discussed. Some comparisons with other methods are also provided.

The time factor is assumed to be $e^{-i\omega t}$ and suppressed throughout this chapter.

11.2 The case of E polarization

11.2.1 Formulation of the problem

We consider the diffraction of an E-polarized plane wave by a thin material strip as shown in Figure 11.1, where the relative permittivity and permeability of the strip are denoted by ε_r and μ_r, respectively. Let the total electric field $\phi^t(x,z)[\equiv E_y^t(x,z)]$ be

$$\phi^t(x,z) = \phi^i(x,z) + \phi(x,z), \tag{11.1}$$

where $\phi^i(x,z)$ is the incident field given by

$$\phi^i(x,z) = e^{-ik(x\sin\theta_0 + z\cos\theta_0)}, \quad 0 < \theta_0 < \pi/2 \tag{11.2}$$

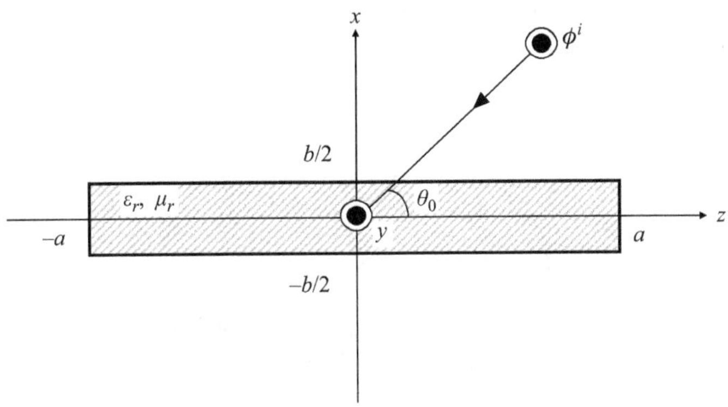

Figure 11.1 Geometry of the problem

with $k[=\omega(\varepsilon_0\mu_0)^{1/2}]$ being the free-space wavenumber. The term $\phi(x, z)$ is the unknown scattered field and satisfies the following two-dimensional Helmholtz equation outside the material strip:

$$(\partial^2/\partial x^2 + \partial^2/\partial z^2 + k^2)\phi(x, z) = 0. \tag{11.3}$$

Nonzero components of the scattered electromagnetic fields outside the material strip are derived from the following relation:

$$(E_y, H_x, H_z) = \left(\phi, \frac{i}{\omega\mu_0}\frac{\partial\phi}{\partial z}, \frac{1}{i\omega\mu_0}\frac{\partial\phi}{\partial x}\right). \tag{11.4}$$

If the strip thickness b is small compared with the wavelength, the material strip is approximately replaced by a strip of zero thickness satisfying the second-order impedance boundary conditions [8]. On the strip surface, the total electromagnetic fields satisfy the approximate boundary conditions as given by

$$H_z^t(+0, z) + H_z^t(-0, z) = -2R_m[E_y^t(+0, z) - E_y^t(-0, z)], \tag{11.5}$$

$$\left[\frac{1}{R_e} + \frac{1}{\tilde{R}_m}\left(1 + \frac{1}{k^2}\frac{\partial^2}{\partial x^2}\right)\right][E_y^t(+0, z) + E_y^t(-0, z)]$$
$$= -2[H_z^t(+0, z) - H_z^t(-0, z)], \tag{11.6}$$

where

$$R_e = \frac{iZ_0}{kb(\varepsilon_r - 1)}, \quad R_m = \frac{iY_0}{kb(\mu_r - 1)}, \quad \tilde{R}_m = \frac{iZ_0\mu_r}{kb(\mu_r - 1)} \tag{11.7}$$

with $Z_0[\equiv(\mu_0/\varepsilon_0)^{1/2}]$ and $Y_0[\equiv(\varepsilon_0/\mu_0)^{1/2}]$ being the intrinsic impedance and admittance of free space, respectively.

For convenience of analysis, we assume the medium to be slightly lossy as in

$$k = k_1 + ik_2, \quad 0 < k_2 \ll k_1. \tag{11.8}$$

We can recover the case for real k by letting $k_2 \to +0$ at the end of analysis. It follows from the radiation condition that

$$\phi(x,z) = O\left(e^{-k_2|z|\cos\theta_0}\right), \quad |z| \to \infty. \tag{11.9}$$

We define the Fourier transform $\Phi(x,\alpha)$ of the scattered field $\phi(x,z)$ with respect to z as

$$\Phi(x,\alpha) = (2\pi)^{-1/2}\int_{-\infty}^{\infty} \phi(x,z)e^{i\alpha z}dz, \tag{11.10}$$

where $\alpha(\equiv\text{Re}\alpha + i\text{Im}\alpha) = \sigma + i\tau$. In view of (11.9), it is found that $\Phi(x,\alpha)$ is regular in the strip $|\tau| < k_2\cos\theta_0$ of the complex α-plane. Introducing the Fourier integrals as

$$\Phi_\pm(x,\alpha) = \pm(2\pi)^{-1/2}\int_{\pm a}^{\pm\infty} \phi(x,z)e^{i\alpha(z\mp a)}dz, \tag{11.11}$$

$$\Phi_1(x,\alpha) = (2\pi)^{-1/2}\int_{-a}^{a} \phi(x,z)e^{i\alpha z}dz, \tag{11.12}$$

we can express $\Phi(x,\alpha)$ as

$$\Phi(x,\alpha) = e^{-i\alpha a}\Phi_-(x,\alpha) + \Phi_1(x,\alpha) + e^{i\alpha a}\Phi_+(x,\alpha). \tag{11.13}$$

In (11.13), $\Phi_+(x,\alpha)$ and $\Phi_-(x,\alpha)$ are regular in the half-planes $\tau > -k_2\cos\theta_0$ and $\tau < k_2\cos\theta_0$, respectively, whereas $\Phi_1(x,\alpha)$ is an entire function.

Taking the Fourier transform of the two-dimensional Helmholtz equation (11.3) and making use of (11.9), we find that

$$(d^2/dx^2 - \gamma^2)\Phi(x,\alpha) = 0 \tag{11.14}$$

for any α in the strip $|\tau| < k_2\cos\theta_0$, where $\gamma = (\alpha^2 - k^2)^{1/2}$. Equation (11.14) is called the *transformed wave equation*. Since γ is a double-valued function of α, we choose a proper branch of γ such that $\text{Re}\gamma > 0$ in the α-plane with branch cuts. The solution of (11.14) can be written as

$$\begin{aligned} \Phi(x,\alpha) &= A(\alpha)e^{-\gamma x}, \quad x > 0, \\ &= B(\alpha)e^{\gamma x}, \quad x < 0 \end{aligned} \tag{11.15}$$

by imposing the condition that $\Phi(x,\alpha)$ be bounded as $|x| \to \infty$, where

$$A(\alpha) = \frac{1}{2}\left[J_m(\alpha) - \frac{ikZ_0}{\gamma}J_e(\alpha)\right], \tag{11.16}$$

$$B(\alpha) = -\frac{1}{2}\left[J_m(\alpha) + \frac{ikZ_0}{\gamma}J_e(\alpha)\right] \tag{11.17}$$

with

$$J_e(\alpha) = -\frac{i}{kZ_0}\left[\frac{d\Phi_1(+0,\alpha)}{dx} - \frac{d\Phi_1(-0,\alpha)}{dx}\right], \tag{11.18}$$

$$J_m(\alpha) = \Phi_1(+0,\alpha) - \Phi_1(-0,\alpha). \tag{11.19}$$

Taking into account the boundary conditions on the strip, we derive, after some manipulations, that

$$M(\alpha)J_e(\alpha) = e^{-i\alpha a}U_-(\alpha) + e^{i\alpha a}U_{(+)}(\alpha), \tag{11.20}$$

$$-K(\alpha)J_m(\alpha) = 2[e^{-i\alpha a}V_-(\alpha) + e^{i\alpha a}V_{(+)}(\alpha)], \tag{11.21}$$

where

$$M(\alpha) = 1 - \frac{ikZ_0}{2\gamma}\left[\frac{1}{R_e} + \frac{1}{\tilde{R}_m}\left(1 + \frac{\gamma^2}{k^2}\right)\right], \tag{11.22}$$

$$K(\alpha) = \gamma - 2ikZ_0R_m, \tag{11.23}$$

$$\left.\begin{matrix} U_{(+)}(\alpha) \\ U_-(\alpha) \end{matrix}\right\} = \tilde{\Phi}_\pm(\alpha) \mp \frac{A_{1,2}}{\alpha - k\cos\theta_0}, \tag{11.24}$$

$$\left.\begin{matrix} V_{(+)}(\alpha) \\ V_-(\alpha) \end{matrix}\right\} = \Phi'_\pm(\alpha) \mp \frac{B_{1,2}}{\alpha - k\cos\theta_0} \tag{11.25}$$

with

$$\tilde{\Phi}_\pm(\alpha) = \left(\frac{1}{R_e} + \frac{1}{\tilde{R}_m}\right)\Phi_\pm(0,\alpha) + \frac{1}{\tilde{R}_m k^2}\frac{d^2\Phi_\pm(0,\alpha)}{dx^2}, \tag{11.26}$$

$$\Phi'_\pm(\alpha) = \frac{d\Phi_\pm(0,\alpha)}{dx}, \tag{11.27}$$

$$A_{1,2} = \frac{1}{(2\pi)^{1/2}i}\left(\frac{1}{R_e} + \frac{\cos^2\theta_0}{\tilde{R}_m}\right)e^{\mp ika\cos\theta_0}, \tag{11.28}$$

$$B_{1,2} = -\frac{k\sin\theta_0}{(2\pi)^{1/2}}e^{\mp ika\cos\theta_0}. \tag{11.29}$$

Equations (11.20) and (11.21) are the Wiener–Hopf equations satisfied by unknown spectral functions, where $J_e(\alpha)$ and $J_m(\alpha)$ are the Fourier transforms of the unknown electric and magnetic surface currents on the strip, respectively. In the above notation, the subscripts "+" and "−" imply that the functions are regular in the upper ($\tau > -k_2\cos\theta_0$) and the lower ($\tau < k_2\cos\theta_0$) half-planes, respectively, whereas the subscript "(+)" implies that the functions are regular in $\tau > -k_2\cos\theta_0$ except for a simple pole at $\alpha = k\cos\theta_0$.

11.2.2 Factorization of the Kernel functions

In this subsection, we shall factorize the kernel functions $M(\alpha)$ and $K(\alpha)$ defined by (11.22) and (11.23) with the aid of Noble's approach [2]. The factorization is to split $M(\alpha)$ and $K(\alpha)$ into the multiplication form as in

$$M(\alpha) = M_+(\alpha)M_-(\alpha) = M_+(\alpha)M_+(-\alpha), \tag{11.30}$$

$$K(\alpha) = K_+(\alpha)K_-(\alpha) = K_+(\alpha)K_+(-\alpha). \tag{11.31}$$

To factorize $M(\alpha)$ and $K(\alpha)$, let us introduce the auxiliary functions $N_n(\alpha)$ as

$$N_n(\alpha) = 1 + \frac{i}{k\delta_n}\gamma, \quad n = 1, 2, 3, \tag{11.32}$$

where

$$\delta_{1,2} = -\frac{\tilde{R}_m}{Z_0}\left\{1 \pm \left[1 + \frac{Z_0^2}{\tilde{R}_m}\left(\frac{1}{R_e} + \frac{1}{\tilde{R}_m}\right)\right]^{1/2}\right\}, \qquad \delta_3 = 2Z_0R_m. \tag{11.33}$$

Using the notations (11.32) and (11.33), it follows that

$$M(\alpha) = \frac{kZ_0}{2i}\left(\frac{1}{R_e} + \frac{1}{\tilde{R}_m}\right)\frac{N_1(\alpha)N_2(\alpha)}{\gamma}, \tag{11.34}$$

$$K(\alpha) = -2ikZ_0R_mN_3(\alpha). \tag{11.35}$$

Assuming that $N_n(\alpha)$ in (11.32) can be factorized as

$$N_n(\alpha) = N_{n+}(\alpha)N_{n-}(\alpha) = N_{n+}(\alpha)N_{n+}(-\alpha), \tag{11.36}$$

the split functions $N_{n\pm}(\alpha)$ are expressed as follows:

$$N_{n\pm}(\alpha) = N_n^{1/2}(0)\exp\left\{\int_0^\alpha \frac{d}{d\beta}[\ln N_{n\pm}(\beta)]\,d\beta\right\}. \tag{11.37}$$

We can show that $N_n(\alpha)$ can be written as

$$\frac{d}{d\alpha}[\ln N_n(\alpha)] = \frac{1}{2}\left(\frac{1}{\alpha - id_n} + \frac{1}{\alpha + id_n}\right) + \frac{ik\delta_n}{2}\left(\frac{1}{\alpha - id_n} + \frac{1}{\alpha + id_n}\right)L(\alpha), \tag{11.38}$$

where

$$L(\alpha) = \gamma^{-1} = (\alpha^2 - k^2)^{-1/2}, \tag{11.39}$$

$$d_n = k(\delta_n^2 - 1)^{1/2}. \tag{11.40}$$

Equation (11.39) can be decomposed as

$$L(\alpha) = L_+(\alpha) + L_-(\alpha) = L_+(\alpha) + L_+(-\alpha), \tag{11.41}$$

where

$$L_\pm(\alpha) = \frac{1}{\pi\gamma}\arccos(\pm\alpha/k). \tag{11.42}$$

We find from (11.38) and (11.41) that

$$\frac{d}{d\beta}[\ln N_{n\pm}(\alpha)] = -\frac{ik\delta_n}{2}\left[\frac{L_\pm(id_n)}{\alpha - id_n} + \frac{L_\mp(id_n)}{\alpha + id_n}\right] + ik\delta_n\frac{\alpha L_\pm(\alpha)}{\alpha^2 + d_n^2}. \tag{11.43}$$

Substituting (11.43) into (11.37), $N_n(\alpha)$ for $n = 1, 2, 3$ are factorized as

$$\begin{aligned} N_{n\pm}(\alpha) &= (1+\delta_n^{-1})^{1/2} \exp\left\{\frac{1}{4}\ln\left[1+\frac{\alpha^2}{k^2(\delta_n^2-1)}\right]\right. \\ &\quad \pm \frac{i}{2\pi}\ln\left[\delta_n+(\delta_n^2-1)^{1/2}\right]\ln\left[\frac{ik(\delta_n^2-1)^{1/2}+\alpha}{ik(\delta_n^2-1)^{1/2}-\alpha}\right] \\ &\quad \left. -\frac{\delta_n}{\pi}\int_{\pi/2}^{\arccos(\pm\alpha/k)}\frac{t\cos t}{\sin^2 t-\delta_n^2}dt\right\}. \end{aligned} \tag{11.44}$$

From (11.44), we find that the split functions $M_\pm(\alpha)$ and $K_\pm(\alpha)$ are expressed as

$$M_\pm(\alpha) = \left[\frac{kZ_0}{2}\left(\frac{1}{R_e}+\frac{1}{\tilde{R}_m}\right)\right]^{1/2}\frac{N_{1\pm}(\alpha)N_{2\pm}(\alpha)}{(k\pm\alpha)^{1/2}}, \tag{11.45}$$

$$K_\pm(\alpha) = (2kZ_0R_m)^{1/2}e^{-i\pi/4}N_{3\pm}(\alpha), \tag{11.46}$$

where $M_\pm(\alpha)$ and $K_\pm(\alpha)$ are regular and nonzero in the half-plane $\tau \gtrless \mp k_2\cos\theta_0$, and show an algebraic behavior at infinity.

11.2.3 Formal solution of the Wiener–Hopf equation

We multiply both sides of (11.20) by $e^{\pm i\alpha a}/M_\mp(\alpha)$ and apply the decomposition procedure with the aid of the edge condition. This leads to

$$\begin{aligned} &e^{-i\alpha a}M_-(\alpha)J_e(\alpha) + \frac{A_1}{M_+(k\cos\theta_0)(\alpha-k\cos\theta_0)} + \frac{1}{2\pi i}\int_{C_2}\frac{U_-(\beta)e^{-2i\beta a}}{M_+(\beta)(\beta-\alpha)}d\beta \\ &= \frac{U_{(+)}(\alpha)}{M_+(\alpha)} + \frac{A_1}{M_+(k\cos\theta_0)(\alpha-k\cos\theta_0)} + \frac{1}{2\pi i}\int_{C_1}\frac{U_-(\beta)e^{-2i\beta a}}{M_+(\beta)(\beta-\alpha)}d\beta, \end{aligned} \tag{11.47}$$

$$\begin{aligned} &\frac{U_-(\alpha)}{M_-(\alpha)} - \frac{1}{2\pi i}\int_{C_2}\frac{e^{2i\beta a}U_{(+)}(\beta)}{M_-(\beta)(\beta-\alpha)}d\beta \\ &= e^{i\alpha a}M_+(\alpha)J_e(\alpha) - \frac{1}{2\pi i}\int_{C_1}\frac{e^{2i\beta a}U_{(+)}(\beta)}{M_-(\beta)(\beta-\alpha)}d\beta, \end{aligned} \tag{11.48}$$

where C_1 and C_2 are infinite integration paths running parallel to the real axis, as shown in Figure 11.2. We see that the left-hand sides of (11.47) and (11.48) are regular in lower ($\tau < k_2\cos\theta_0$) half-plane and the right-hand sides of (11.47) and (11.48) are regular in upper ($\tau > -k_2\cos\theta_0$) half-plane, and both sides of (11.47) and (11.48) have a common strip of regularity $|\tau| < k_2\cos\theta_0$. Hence, the argument of analytic continuation shows that there exists an entire function. Taking into account

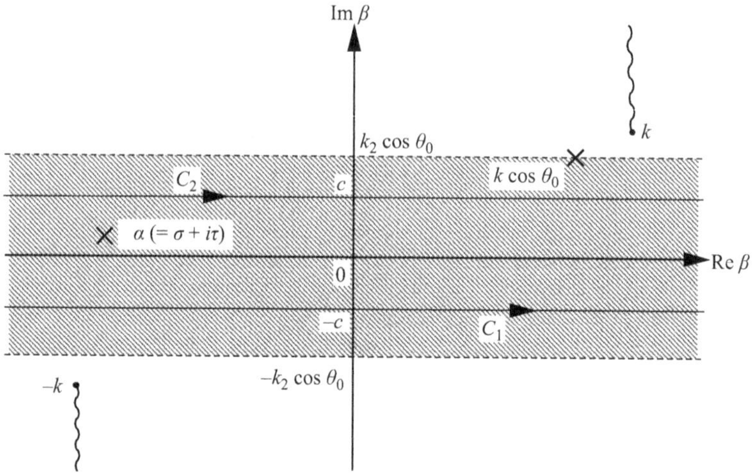

Figure 11.2 Integral paths C_1 and C_2 for decomposition ($|\tau| < c < k_2 \cos\theta_0$)

Liouville's theorem, it follows that both sides of (11.47) and (11.48) are equal to zero. Thus we obtain that

$$\frac{U_{(+)}(\alpha)}{M_+(\alpha)} + \frac{A_1}{M_+(k\cos\theta_0)(\alpha - k\cos\theta_0)} + \frac{1}{2\pi i}\int_{C_1} \frac{e^{-2i\beta a}U_-(\beta)}{M_+(\beta)(\beta-\alpha)}d\beta = 0, \quad (11.49)$$

$$\frac{U_-(\alpha)}{M_-(\alpha)} - \frac{1}{2\pi i}\int_{C_2} \frac{e^{2i\beta a}U_{(+)}(\beta)}{M_-(\beta)(\beta-\alpha)}d\beta = 0. \quad (11.50)$$

Equations (11.49) and (11.50) form a set of two coupled integral equations for $U_{(+)}(\alpha)$ and $U_-(\alpha)$. However, they may decoupled in the following manner. We make a change of variable $\beta \to -\beta$ in (11.49) and set $\alpha \to -\alpha$ in (11.50). Then taking the sum and difference of the resultant equations, we obtain that

$$\frac{U^{s,d}_{(+)}(\alpha)}{M_+(\alpha)} + \frac{A_1}{M_+(k\cos\theta_0)(\alpha - k\cos\theta_0)} \mp \frac{1}{2\pi i}\int_{C_2} \frac{e^{2i\beta a}U^{s,d}_{(+)}(\beta)}{M_-(\beta)(\beta+\alpha)}d\beta = 0, \quad (11.51)$$

where

$$U^{s,d}_{(+)}(\alpha) = U_{(+)}(\alpha) \pm U_-(-\alpha). \quad (11.52)$$

The term $M_-(\beta)$ associated with the integral in (11.51) has a branch point at $\beta = k$, as is seen from (11.45). We now choose the branch cut emanating from $\beta = k$ as a straight line parallel to the imaginary axis, extending to infinity in the upper half-plane as shown in Figure 11.2. Evaluating the integral in (11.51) by deforming the contour into the upper half-plane yields,

$$U_{(+)}^{s,d}(\alpha) = M_+(\alpha)\Bigg[-\frac{A_1}{M_+(k\cos\theta_0)(\alpha - k\cos\theta_0)} \mp \frac{A_2}{M_-(k\cos\theta_0)(\alpha + k\cos\theta_0)} \pm u_{s,d}(\alpha)\Bigg], \tag{11.53}$$

where

$$u_{s,d}(\alpha) = \frac{1}{\pi i}\int_k^{k+i\infty} \frac{e^{2i\beta a}(\beta - k)^{1/2}}{\beta + \alpha} U_{(+)}^{s,d}(\beta) F_+(\beta) d\beta \tag{11.54}$$

with

$$F_+(\beta) = \frac{ikZ_0}{2}\frac{[1/R_e + \beta^2/(\tilde{R}_m k^2)](\beta + k)^{1/2}}{\beta^2 - k^2 + k^2 Z_0^2[1/R_e + \beta^2/(\tilde{R}_m k^2)]^2/4} M_+(\beta). \tag{11.55}$$

In (11.54), the contour is a straight path on the right-hand side of the branch cut.

We also multiply both sides of (11.21) by $e^{\pm i\alpha a}/K_\mp(\alpha)$ and carry out the decomposition procedure similar to that leading to (11.53). Omitting the details, we obtain that

$$V_{(+)}^{s,d}(\alpha) = K_+(\alpha)\Bigg[-\frac{B_1}{K_+(k\cos\theta_0)(\alpha - k\cos\theta_0)} \mp \frac{B_2}{K_-(k\cos\theta_0)(\alpha + k\cos\theta_0)} \pm v_{s,d}(\alpha)\Bigg], \tag{11.56}$$

where

$$V_{(+)}^{s,d}(\alpha) = V_{(+)}(\alpha) \pm V_-(-\alpha), \tag{11.57}$$

$$v_{s,d}(\alpha) = \frac{1}{\pi i}\int_k^{k+i\infty} \frac{e^{2i\beta a}(\beta - k)^{1/2}}{\beta + \alpha} V_{(+)}^{s,d}(\beta) T_+(\beta) d\beta \tag{11.58}$$

with

$$T_+(\beta) = \frac{(\beta + k)^{1/2} K_+(\beta)}{\beta^2 - k^2 + 4k^2 Z_0^2 R_m^2}. \tag{11.59}$$

Equations (11.53) and (11.56) provide the exact solutions to the Wiener–Hopf equations (11.20) and (11.21), respectively, but they are formal in the sense that the branch-cut integrals with unknown integrands $u_{s,d}(\alpha)$ and $v_{s,d}(\alpha)$ are involved.

11.2.4 Asymptotic solution of a certain integral equation in the complex plane

In this subsection, we shall consider a certain integral equation in the complex plane that often arises in the Wiener–Hopf analysis of canonical scattering problems and discuss a method of solution in detail.

The Wiener–Hopf analysis often leads to the following exact solution in the complex plane:

$$f(\alpha) = h(\alpha)\left[g(\alpha) + \frac{C}{\pi i}\int_k^{k+i\infty} e^{i\beta l}\frac{(\beta-k)^\nu G(\beta)f(\beta)}{\beta+\alpha}d\beta\right]. \tag{11.60}$$

In (11.60), $f(\alpha)$ is the unknown function to be determined, and all the other quantities are known constants or functions.

Let $f(\beta)$ be a function of a complex variable β satisfying the following conditions:

(i) *$f(\beta)$ is an analytic function of β regular in $|\beta - k| < \varepsilon < \infty$, where $k = k_1 + ik_2$ with $k_1 > 0$, $k_2 > 0$, and $\varepsilon \not\approx 0$.*

(ii) *$f(\beta)$ satisfies $O[(\beta-k)^\delta]$ for any β such that $|\beta - k| \geq R$ with $\varepsilon < R < \infty$, where δ is some real constant.*

(iii) *$f(\beta)$ is a continuous function of β on any bounded part of the semi-infinite straight path from k to $k + i\infty$ in the β-plane.*

Let α be a complex variable such that $|\alpha + k| > 0$ and $-\pi/2 < \arg(\alpha + k) < 3\pi/2$, and introduce

$$F_m^\nu(l,\alpha) = \frac{1}{\pi i}\int_k^{k+i\infty} e^{i\beta l}\frac{(\beta-k)^\nu G(\beta)f(\beta)}{(\beta+\alpha)^m}d\beta \tag{11.61}$$

for $l > 0$, and $Re\nu > -1$, positive integer m, where $\arg(\beta - k) = \pi/2$. In (11.61), $G(\beta)$ is regular in the neighborhood of $\beta = k$, and shows an algebraic behavior as $\beta \to \infty$ in the upper half-plane.

We define the region in the α-plane as follows:

$$D = \{\alpha : |\alpha + k| > 0,\ -\pi/2 < \arg(\alpha + k) < 3\pi/2\}. \tag{11.62}$$

Then it is shown that the function $F_m^\nu(l,\alpha)$ defined by (11.61) is uniformly convergent in any bounded closed region in D and hence regular in D.

We can prove the following theorem on the asymptotic expansion for $F_m^\nu(l,\alpha)$ large l.

Theorem. *The function $F_m^\nu(l,\alpha)$ has an asymptotic expansion*

$$F_m^\nu(l,\alpha) \sim \frac{e^{ikl}}{\pi}\sum_{n=0}^{\infty} f_n \frac{i^{\nu-m+n}}{l^{\nu-m+n+1}}\Gamma_m^g[\nu+n+1, -i(\alpha+k)l] \tag{11.63}$$

as $l \to \infty$, where

$$f_n = \frac{1}{n!}\frac{d^n f(\beta)}{d\beta^n}\bigg|_{\beta=k}. \tag{11.64}$$

In (11.63), *$\Gamma_m^g(\cdot,\cdot)$ is the special function defined by*

$$\Gamma_m^g(u,w) = \int_0^\infty \frac{t^{u-1}e^{-t}}{(t+w)^m}G(k+it/l)dt \tag{11.65}$$

for $Reu > 0$, $|w| > 0$, $|\arg w| < \pi$, and positive integer m.

Using the function $F_m^\nu(l,\alpha)$, (11.60) can be written as

$$f(\alpha) = h(\alpha)[g(\alpha) + CF_1^\nu(l,\alpha)], \tag{11.66}$$

where it is assumed that $f(\alpha)$ satisfies the conditions (i)–(iii), and $g(\alpha)$ and $h(\alpha)$ are regular in region Δ contained in D. Then we can apply theorem to derive an asymptotic expansion of $F_1^\nu(l,\alpha)$ for large l. Thus we obtain from (11.66) that

$$f(\alpha) \sim h(\alpha)\left[g(\alpha) + C\sum_{n=0}^{\infty} f_n \xi_{0n}^g(\nu,l,\alpha)\right] \tag{11.67}$$

as $l \to \infty$, where f_n is defined by (11.64), and

$$\xi_{0n}^g(\nu,l,\alpha) = \frac{e^{ikl}}{\pi}\frac{i^{n+\nu-1}}{l^{n+\nu}}\Gamma_1^g[\nu+n+1, -i(\alpha+k)l]. \tag{11.68}$$

Equation (11.67) is the asymptotic solution of the integral equation (11.66) for large l, where an infinite number of unknowns f_n with $n = 0, 1, 2, \ldots$ are contained. Hence, it is required to derive matrix equations for these unknowns.

Setting $\alpha = k$ in (11.67), and using the notation (11.64), we find that

$$f_0 \sim h(k)\left[g(k) + C\sum_{n=0}^{\infty} f_n \xi_{0n}^g(\nu,l,k)\right]. \tag{11.69}$$

We now differentiate both sides of (11.66) m times ($m = 1, 2, 3, \ldots$) with respect α by taking into account the regularity of $F_m^\nu(l,\alpha)$ in D, and setting $\alpha = k$ in the resultant equation, we obtain that

$$f_m \sim \sum_{p=0}^{m} \frac{h^{(m-p)}(k)}{p!(m-p)!}\left[g^{(p)}(k) + C\sum_{n=0}^{\infty} f_n \xi_{pn}^g(\nu,l,k)\right] \tag{11.70}$$

for $m = 1, 2, 3, \ldots$, where

$$h^{(m-p)}(k) = \left.\frac{d^{m-p}h(\alpha)}{d\alpha^{m-p}}\right|_{\alpha=k}, \tag{11.71}$$

$$g^{(p)}(k) = \left.\frac{d^p g(\alpha)}{d\alpha^p}\right|_{\alpha=k}, \tag{11.72}$$

$$\xi_{pn}^g(\nu,l,k) = \frac{e^{ikl}}{\pi}(-1)^p p!\frac{i^{n-p+\nu-1}}{l^{n-p+\nu}}\Gamma_{p+1}^g(\nu+n+1, -2ikl). \tag{11.73}$$

Hence it follows from (11.69) and (11.70) that

$$f_m - C\sum_{n=0}^{\infty} A_{mn} f_n \sim B_m, \quad m = 0, 1, 2, \ldots \tag{11.74}$$

for large l, where

$$A_{mn} = \sum_{p=0}^{m} \frac{h^{(m-p)}(k)\xi_{pn}^g(\nu,l,k)}{p!(m-p)!}, \quad B_m = \sum_{p=0}^{m} \frac{h^{(m-p)}(k)g^{(p)}(k)}{p!(m-p)!}. \tag{11.75}$$

Equation (11.74) provides the desired matrix equation for determining the unknowns f_n with $n = 0, 1, 2, \ldots$ in (11.67), and it is valid for large l.

11.2.5 High-frequency asymptotic solution

In this section, we shall apply the method established in the previous section to solve (11.53) and (11.56) asymptotically. To eliminate the singularities of $U^{s,d}_{(+)}(\alpha)$ in (11.53) at $\alpha = k\cos\theta_0$, we introduce the auxiliary functions $\tilde{\Phi}^{s,d}_{+}(\alpha)$ as

$$\begin{aligned}\tilde{\Phi}^{s,d}_{+}(\alpha) &= \tilde{\Phi}_{+}(\alpha) \pm \tilde{\Phi}_{-}(-\alpha)\\ &= U^{s,d}_{(+)}(\alpha) + \frac{A_1}{\alpha - k\cos\theta_0} \pm \frac{A_2}{\alpha + k\cos\theta_0},\end{aligned} \tag{11.76}$$

where $\tilde{\Phi}^{s,d}_{+}(\alpha)$ is regular in the upper half-plane $\tau > -k_2\cos\theta_0$. Then (11.53) can be written as

$$\tilde{\Phi}^{s,d}_{+}(\alpha) = M_{+}(\alpha)[\chi_{us,ud}(\alpha) + C_{s,d}F^{u}_{s,d}(\alpha)], \tag{11.77}$$

where

$$\chi_{us,ud}(\alpha) = A_1[P_1(\alpha) \pm \eta_{f2}(\alpha)] + A_2[\eta_{f1}(\alpha) \pm P_2(\alpha)], \tag{11.78}$$

$$C_{s,d} = \pm 1, \tag{11.79}$$

$$F^{u}_{s,d}(\alpha) = \frac{1}{\pi i}\int_{k}^{k+i\infty} \frac{e^{2i\beta a}(\beta - k)^{1/2}}{\beta + \alpha}\tilde{\Phi}^{s,d}_{+}(\beta)F_{+}(\beta)d\beta \tag{11.80}$$

with

$$\eta_{f1,f2}(\alpha) = \frac{\xi^{f}_{00}(\alpha) - \xi^{f}_{00}(\pm k\cos\theta_0)}{\alpha \mp k\cos\theta_0}, \tag{11.81}$$

$$P_{1,2}(\alpha) = \frac{1}{\alpha \mp k\cos\theta_0}\left[\frac{1}{M_{+}(\alpha)} - \frac{1}{M_{\pm}(k\cos\theta_0)}\right], \tag{11.82}$$

$$\xi^{f}_{00}(\alpha) = \frac{e^{2ika}}{\pi}\frac{i^{-1/2}}{(2a)^{1/2}}\Gamma^{f}_{1}[3/2, -2i(\alpha + k)a]. \tag{11.83}$$

In (11.83), $\Gamma^{f}_{m}(\cdot,\cdot)$ is the special function defined by

$$\Gamma^{f}_{m}(u, w) = \int_{0}^{\infty} \frac{t^{u-1}e^{-t}}{(t + w)^{m}}F_{+}[k + it/(2a)]dt \tag{11.84}$$

for $\mathrm{Re}\,u > 0$, $|w| > 0$, $|\arg w| < \pi$, and positive integer m, which accounts for multiple diffraction effects.

Applying theorem in Section 11.4, we can obtain a high-frequency asymptotic expansion of (11.77) with the result that

$$\tilde{\Phi}^{s,d}_{+}(\alpha) \sim M_{+}(\alpha)\left[\chi_{us,ud}(\alpha) + C_{s,d}\sum_{n=0}^{N} f^{us,ud}_{n}\xi^{f}_{0n}(\alpha)\right] \tag{11.85}$$

for $ka \to \infty$, where N denotes the truncation number of the infinite asymptotic series, and

$$f_n^{us,ud} = \frac{1}{n!} \frac{d^n \tilde{\Phi}_+^{s,d}(\alpha)}{d\alpha^n}\bigg|_{\alpha=k}, \tag{11.86}$$

$$\xi_{0n}^f(\alpha) = \frac{e^{2ika}}{\pi} \frac{i^{n-1/2}}{(2a)^{n+1/2}} \Gamma_1^f[3/2+n, -2i(\alpha+k)a]. \tag{11.87}$$

Taking into account (11.74) and carrying out some manipulations, we can show the unknowns $f_n^{us,ud}$ in (11.85) are determined by solving the matrix equation

$$f_m^{us,ud} - C_{s,d} \sum_{n=0}^{N} A_{mn}^u f_n^{us,ud} \sim B_m^{us,ud} \tag{11.88}$$

for $m = 0, 1, 2, \ldots, N$, where

$$A_{mn}^u = \sum_{p=0}^{m} \frac{M_+^{(m-p)}(k)\xi_{pn}^f(k)}{p!(m-p)!}, \quad B_m^{us,ud} = \sum_{p=0}^{m} \frac{M_+^{(m-p)}(k)\chi_{us,ud}^{(p)}(k)}{p!(m-p)!} \tag{11.89}$$

with

$$M_+^{(m-p)}(k) = \frac{d^{m-p}M_+(\alpha)}{d\alpha^{m-p}}\bigg|_{\alpha=k}, \tag{11.90}$$

$$\xi_{pn}^f(k) = \frac{e^{2ika}}{\pi}(-1)^p p! \frac{i^{n-p-1/2}}{(2a)^{n-p+1/2}} \Gamma_{p+1}^f(3/2+n, -4ika), \tag{11.91}$$

$$\chi_{us,ud}^{(p)}(k) = \frac{d^p \chi_{us,ud}(\alpha)}{d\alpha^p}\bigg|_{\alpha=k}. \tag{11.92}$$

Making use of the above results and carrying out further manipulations, we finally arrive at an explicit asymptotic solution to the Wiener–Hopf equation (11.20) with the result that

$$\left.\begin{matrix} U_{(+)}(\alpha) \\ U_-(\alpha) \end{matrix}\right\} \sim M_\pm(\alpha)\left[\mp \frac{A_{1,2}}{M_\pm(k\cos\theta_0)(\alpha - k\cos\theta_0)} + A_{2,1}\eta_{f1,f2}(\pm\alpha) + \frac{1}{2}\sum_{n=0}^{N}(f_n^{us} \mp f_n^{ud})\xi_{0n}^f(\pm\alpha)\right] \tag{11.93}$$

as $ka \to \infty$.

A similar procedure may also be applied to (11.56) for a high-frequency solution. Omitting the details, we can obtain

$$\left.\begin{matrix} V_{(+)}(\alpha) \\ V_-(\alpha) \end{matrix}\right\} \sim K_\pm(\alpha)\left[\mp \frac{B_{1,2}}{K_\pm(k\cos\theta_0)(\alpha - k\cos\theta_0)} + B_{2,1}\eta_{t1,t2}(\pm\alpha) + \frac{1}{2}\sum_{n=0}^{N}(f_n^{vs} \mp f_n^{vd})\xi_{0n}^t(\pm\alpha)\right] \tag{11.94}$$

for $ka \to \infty$, where

$$\eta_{t1,t2}(\alpha) = \frac{\xi^t_{00}(\alpha) - \xi^t_{00}(\pm k\cos\theta_0)}{\alpha \mp k\cos\theta_0}, \tag{11.95}$$

$$f_n^{vs,vd} = \frac{1}{n!}\frac{d^n \Phi'^{s,d}_+(\alpha)}{d\alpha^n}\bigg|_{\alpha=k}, \tag{11.96}$$

$$\xi^t_{0n}(\alpha) = \frac{e^{2ika}}{\pi}\frac{i^{n-1/2}}{(2a)^{n+1/2}}\Gamma^t_1[3/2+n, -2i(\alpha+k)a] \tag{11.97}$$

with

$$\Phi'^{s,d}_+(\alpha) = V^{s,d}_{(+)}(\alpha) + \frac{B_1}{\alpha - k\cos\theta_0} \pm \frac{B_2}{\alpha + k\cos\theta_0}, \tag{11.98}$$

$$\Gamma^t_m(u,w) = \int_0^\infty \frac{t^{u-1}e^{-t}}{(t+w)^m}T_+[k+it/(2a)]dt. \tag{11.99}$$

Equations (11.93) and (11.94) provide complete, high-frequency asymptotic solutions to the Wiener–Hopf equations. It is to be noted that the above results rigorously take into account the multiple diffraction between the edges of the strip.

11.2.6 Scattered far field

Taking into account (11.15)–(11.21), the scattered field in the Fourier transform domain is expressed as

$$\Phi(x,\alpha) = \Phi_{a,b}(\alpha)e^{\mp\gamma x}, \quad x \gtrless 0, \tag{11.100}$$

where

$$\begin{aligned}\Phi_{a,b}(\alpha) &= \frac{1}{2}\left[-\frac{ikZ_0}{\gamma}J_e(\alpha) \pm J_m(\alpha)\right] \\ &= -ikZ_0\frac{e^{-i\alpha a}U_-(\alpha)+e^{i\alpha a}U_{(+)}(\alpha)}{2\gamma M(\alpha)} \mp \frac{e^{-i\alpha a}V_-(\alpha)+e^{i\alpha a}V_{(+)}(\alpha)}{K(\alpha)}.\end{aligned} \tag{11.101}$$

The scattered field $\phi(x,z)$ in the real space is obtained by taking the inverse Fourier transform of (11.100) according to the formula

$$\phi(x,z) = (2\pi)^{-1/2}\int_{-\infty+ic}^{\infty+ic}\Phi_{a,b}(\alpha)e^{-\gamma|x|-i\alpha z}d\alpha, \tag{11.102}$$

where c is a constant such that $|c| < k_2\cos\theta_0$. Since $J_e(\alpha)$ and $J_m(\alpha)$ in (11.101) are entire functions, the singularities of the integrand of (11.102) are only branch point at $\alpha = \pm k$. Introducing the cylindrical coordinate (ρ, θ) centered at the origin as

$$x = \rho\sin\theta, \quad z = \rho\cos\theta, \quad -\pi < \theta < \pi \tag{11.103}$$

and applying the saddle point method of integration, we derive a far field asymptotic expression with the result that

$$\phi(\rho,\theta) \sim \pm\Phi_{a,b}(-k\cos\theta)\,k\sin\theta\frac{e^{i(k\rho-\pi/4)}}{(k\rho)^{1/2}}, \quad x \gtrless 0 \tag{11.104}$$

as $k\rho \to \infty$. Equation (11.104) is uniformly valid for arbitrary incidence and observation angles.

11.3 The case of H polarization

In this section, we shall consider the same diffraction problem for the H-polarized plane wave incidence. Since the method is similar to that followed in the E-polarized case, only the main results will be summarized.

If we define the total magnetic field $\phi^t(x,z)[\equiv H_y^t(x,z)]$ by (11.1), the scattered field $\phi(x,z)$ satisfies the two-dimensional Helmholtz equation as given by (11.3). Once we find the solution of (11.3) subject to appropriate boundary conditions, nonzero components of the scattered electromagnetic fields outside the material strip are derived from

$$(H_y, E_x, E_z) = \left(\phi,\ \frac{1}{i\omega\varepsilon_0}\frac{\partial\phi}{\partial z},\ \frac{i}{\omega\varepsilon_0}\frac{\partial\phi}{\partial x}\right). \tag{11.105}$$

As in previous section, we shall introduce a slight loss in the medium so that k has a small positive imaginary part shown in (11.8). The solution for real k is obtained by letting $k_2 \to +0$ at the end of analysis.

Assuming that the strip thickness is small compared with the wavelength, the material strip is approximately replaced by a strip of zero thickness satisfying the second-order impedance boundary conditions as given by

$$E_z^t(+0,z) + E_z^t(-0,z) = 2R_e[H_y^t(+0,z) - H_y^t(-0,z)], \tag{11.106}$$

$$\begin{aligned}\left[\frac{1}{R_m} + \frac{1}{\tilde{R}_e}\left(1 + \frac{1}{k^2}\frac{\partial^2}{\partial x^2}\right)\right]&[H_y^t(+0,z) + H_y^t(-0,z)]\\ &= 2[E_z^t(+0,z) - E_z^t(-0,z)],\end{aligned} \tag{11.107}$$

where

$$R_e = \frac{iZ_0}{kb(\varepsilon_r - 1)}, \quad \tilde{R}_e = \frac{iY_0\varepsilon_r}{kb(\varepsilon_r - 1)}, \quad R_m = \frac{iY_0}{kb(\mu_r - 1)}. \tag{11.108}$$

Taking into account the boundary conditions on the strip as in Section 11.2.1, we obtain that

$$-M(\alpha)J_m(\alpha) = e^{-i\alpha a}U_-(\alpha) + e^{i\alpha a}U_{(+)}(\alpha), \tag{11.109}$$

$$-K(\alpha)J_e(\alpha) = 2[e^{-i\alpha a}V_-(\alpha) + e^{i\alpha a}V_{(+)}(\alpha)], \tag{11.110}$$

where

$$M(\alpha) = M_+(\alpha)M_-(\alpha) = 1 - \frac{ikY_0}{2\gamma}\left[\frac{1}{R_m} + \frac{1}{\tilde{R}_e}\left(1 + \frac{\gamma^2}{k^2}\right)\right], \tag{11.111}$$

$$K(\alpha) = K_+(\alpha)K_-(\alpha) = \gamma - 2ikY_0R_e, \tag{11.112}$$

$$\left.\begin{matrix} U_{(+)}(\alpha) \\ U_-(\alpha) \end{matrix}\right\} = \tilde{\Phi}_\pm(\alpha) \mp \frac{A_{1,2}}{\alpha - k\cos\theta_0}, \tag{11.113}$$

$$\left.\begin{matrix} V_{(+)}(\alpha) \\ V_-(\alpha) \end{matrix}\right\} = \Phi'_\pm(\alpha) \mp \frac{B_{1,2}}{\alpha - k\cos\theta_0}, \tag{11.114}$$

$$J_m(\alpha) = \frac{i}{kY_0}\left[\frac{d\Phi_1(+0,\alpha)}{dx} - \frac{d\Phi_1(-0,\alpha)}{dx}\right], \tag{11.115}$$

$$J_e(\alpha) = \Phi_1(+0,\alpha) - \Phi_1(-0,\alpha) \tag{11.116}$$

with

$$\tilde{\Phi}_\pm(\alpha) = \left(\frac{1}{R_m} + \frac{1}{\tilde{R}_e}\right)\Phi_\pm(0,\alpha) + \frac{1}{\tilde{R}_e k^2}\frac{d^2\Phi_\pm(0,\alpha)}{dx^2}, \tag{11.117}$$

$$\Phi'_\pm(\alpha) = \frac{d\Phi_\pm(0,\alpha)}{dx}, \tag{11.118}$$

$$A_{1,2} = \frac{1}{(2\pi)^{1/2}i}\left(\frac{1}{R_m} + \frac{\cos^2\theta_0}{\tilde{R}_e}\right)e^{\mp ika\cos\theta_0}, \tag{11.119}$$

$$B_{1,2} = -\frac{k\sin\theta_0}{(2\pi)^{1/2}}e^{\mp ika\cos\theta_0}, \tag{11.120}$$

$$M_\pm(\alpha) = \left[\frac{kY_0}{2}\left(\frac{1}{R_m} + \frac{1}{\tilde{R}_e}\right)\right]^{1/2}\frac{N_{1\pm}(\alpha)N_{2\pm}(\alpha)}{(k\pm\alpha)^{1/2}}, \tag{11.121}$$

$$K_\pm(\alpha) = (2kY_0R_e)^{1/2}e^{-i\pi/4}N_{3\pm}(\alpha). \tag{11.122}$$

In (11.121) and (11.122), $N_{n\pm}(\alpha)$ for $n = 1, 2, 3$ can be written as

$$\begin{aligned} N_{n\pm}(\alpha) = (1+\delta_n^{-1})^{1/2}\exp\Bigg\{&\frac{1}{4}\ln\left[1 + \frac{\alpha^2}{k^2(\delta_n^2 - 1)}\right] \\ &\pm\frac{i}{2\pi}\ln\left[\delta_n + (\delta_n^2-1)^{1/2}\right]\ln\left[\frac{ik(\delta_n^2-1)^{1/2}+\alpha}{ik(\delta_n^2-1)^{1/2}-\alpha}\right] \\ &-\frac{\delta_n}{\pi}\int_{\pi/2}^{\arccos(\pm\alpha/k)}\frac{t\cos t}{\sin^2 t - \delta_n^2}dt\Bigg\}, \end{aligned} \tag{11.123}$$

where

$$\delta_{1,2} = -\frac{\tilde{R}_e}{Y_0}\left\{1 \pm \left[1 + \frac{Y_0^2}{\tilde{R}_e}\left(\frac{1}{R_m} + \frac{1}{\tilde{R}_e}\right)\right]^{1/2}\right\}, \quad \delta_3 = 2Y_0R_e. \tag{11.124}$$

Equations (11.109) and (11.110) are the Wiener–Hopf equations satisfied by unknown spectral functions.

Applying the asymptotic method established in Section 11.2, we can derive a high-frequency representation of (11.113) for large ka with the result that

$$\left.\begin{array}{l} U_{(+)}(\alpha) \\ U_{-}(\alpha) \end{array}\right\} \sim M_{\pm}(\alpha)\left[\mp\frac{A_{1,2}}{M_{\pm}(k\cos\theta_0)(\alpha - k\cos\theta_0)}\right.$$

$$\left. + A_{2,1}\eta_{f1,f2}(\pm\alpha) + \frac{1}{2}\sum_{n=0}^{N}(f_n^{us} \mp f_n^{ud})\xi_{0n}^{f}(\pm\alpha)\right], \tag{11.125}$$

where N is truncation number, and

$$\eta_{f1,f2}(\alpha) = \frac{\xi_{00}^{f}(\alpha) - \xi_{00}^{f}(\pm k\cos\theta_0)}{\alpha \mp k\cos\theta_0}, \tag{11.126}$$

$$f_n^{us,ud} = \frac{1}{n!}\frac{d^n\tilde{\Phi}_{+}^{s,d}(\alpha)}{d\alpha^n}\bigg|_{\alpha=k}, \tag{11.127}$$

$$\xi_{0n}^{f}(\alpha) = \frac{e^{2ika}}{\pi}\frac{i^{n-1/2}}{(2a)^{n+1/2}}\Gamma_1^{f}[3/2+n, -2i(\alpha+k)a] \tag{11.128}$$

with

$$\tilde{\Phi}_{+}^{s,d}(\alpha) = U_{(+)}^{s,d}(\alpha) + \frac{A_1}{\alpha - k\cos\theta_0} \pm \frac{A_2}{\alpha + k\cos\theta_0}, \tag{11.129}$$

$$\Gamma_m^{f}(u,w) = \int_0^{\infty}\frac{t^{u-1}e^{-t}}{(t+w)^m}F_{+}[k + it/(2a)]dt, \tag{11.130}$$

$$F_{+}(\beta) = \frac{ikY_0}{2}\frac{[1/R_m + \beta^2/(\tilde{R}_e k^2)](\beta+k)^{1/2}}{\beta^2 - k^2 + k^2Y_0^2[1/R_m + \beta^2/(\tilde{R}_e k^2)]^2/4}M_{+}(\beta). \tag{11.131}$$

A similar procedure may also be applied to (11.114) for a high-frequency solution. Omitting the details, we can obtain

$$\left.\begin{array}{l} V_{(+)}(\alpha) \\ V_{-}(\alpha) \end{array}\right\} \sim K_{\pm}(\alpha)\left[\mp\frac{B_{1,2}}{K_{\pm}(k\cos\theta_0)(\alpha - k\cos\theta_0)}\right.$$

$$\left. + B_{2,1}\eta_{t1,t2}(\pm\alpha) + \frac{1}{2}\sum_{n=0}^{N}(f_n^{vs} \mp f_n^{vd})\xi_{0n}^{t}(\pm\alpha)\right] \tag{11.132}$$

for $ka \to \infty$, where N is truncation number, and

$$\eta_{t1,t2}(\alpha) = \frac{\xi_{00}^{t}(\alpha) - \xi_{00}^{t}(\pm k\cos\theta_0)}{\alpha \mp k\cos\theta_0}, \tag{11.133}$$

$$f_n^{vs,vd} = \frac{1}{n!}\frac{d^n\Phi_{+}^{\prime s,d}(\alpha)}{d\alpha^n}\bigg|_{\alpha=k}, \tag{11.134}$$

$$\xi_{0n}^{t}(\alpha) = \frac{e^{2ika}}{\pi}\frac{i^{n-1/2}}{(2a)^{n+1/2}}\Gamma_1^{t}[3/2+n, -2i(\alpha+k)a] \tag{11.135}$$

with

$$\Phi_{+}^{\prime s,d}(\alpha) = V_{(+)}^{s,d}(\alpha) + \frac{B_1}{\alpha - k\cos\theta_0} \pm \frac{B_2}{\alpha + k\cos\theta_0}, \tag{11.136}$$

$$\Gamma_m^t(u,w) = \int_0^\infty \frac{t^{u-1}e^{-t}}{(t+w)^m} T_+[k + it/(2a)]dt, \tag{11.137}$$

$$T_+(\beta) = \frac{(\beta+k)^{1/2}K_+(\beta)}{\beta^2 - k^2 + 4k^2Y_0^2R_e^2}. \tag{11.138}$$

Taking into account the earlier discussion in Section 11.2.4, the unknown constants $f_n^{us,ud}$ in (11.125) and $f_n^{vs,vd}$ in (11.132) are determined by solving the matrix equation.

Introducing the cylindrical coordinate $x = \rho\sin\theta, z = \rho\cos\theta$ for $-\pi < \theta < \pi$ and applying the saddle point method, we derive a far field asymptotic expression with the result that

$$\phi(\rho,\theta) \sim \pm\Phi_{a,b}(-k\cos\theta)k\sin\theta\frac{e^{i(k\rho-\pi/4)}}{(k\rho)^{1/2}}, \quad x \gtrless 0 \tag{11.139}$$

as $k\rho \to \infty$, where

$$\Phi_{a,b}(\alpha) = -ikY_0\frac{e^{-i\alpha a}U_-(\alpha) + e^{i\alpha a}U_{(+)}(\alpha)}{2\gamma M(\alpha)} \mp \frac{e^{-i\alpha a}V_-(\alpha) + e^{i\alpha a}V_{(+)}(\alpha)}{K(\alpha)}. \tag{11.140}$$

Equation (11.139) is uniformly valid for arbitrary incidence and observation angles.

11.4 Numerical results and discussion

In this section, we shall present numerical results on the RCS for E polarization and discuss far field scattering characteristics of the strip in detail. The normalized RCS per unit length is defined by

$$\sigma/\lambda = \lim_{\rho\to\infty}\left(k\rho\left|\phi/\phi^i\right|^2\right) \tag{11.141}$$

with λ being the free-space wavelength. All the results are plotted in decibels (dB) by computing $10\log_{10}(\sigma/\lambda)$.

In computing (11.141), we have used the high-frequency asymptotic expressions as given by (11.93), (11.94), (11.125), and (11.132) for $U_{(+)}(\alpha), U_-(\alpha), V_{(+)}(\alpha)$, and $V_-(\alpha)$, where the truncation number N for the asymptotic series is contained. Let $\sigma^{(N)}$ and $\sigma^{(N+1)}$ be the RCS with the truncation numbers being N and $N+1$, respectively. In numerical computation, we have employed the convergence criteria $\left|\sigma^{(N+1)} - \sigma^{(N)}\right| < 10^{-3}$ to determine the desired truncation number N. By careful numerical investigation, we have verified that the choice of $N = 3$ satisfies the aforementioned convergence criteria and hence provides sufficiently accurate solutions.

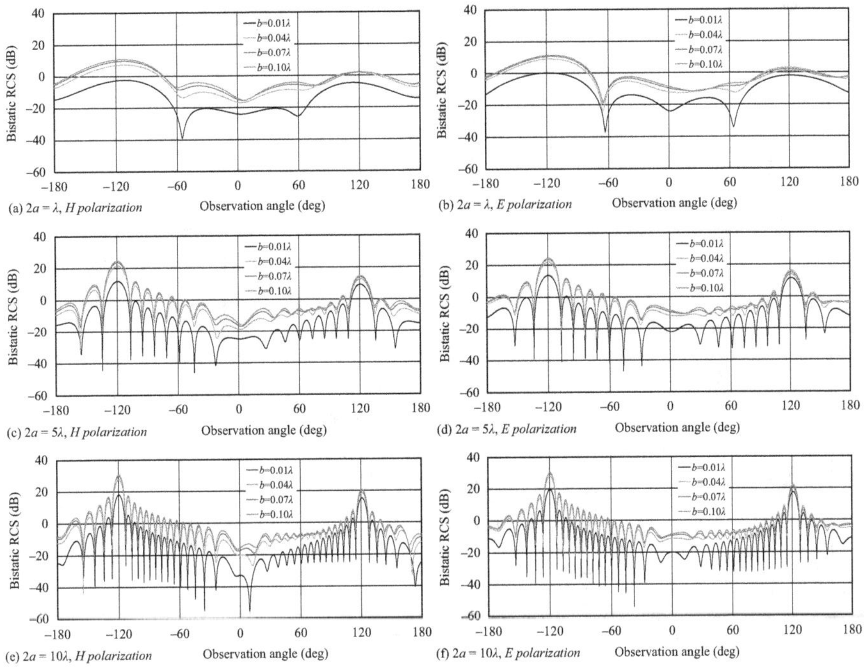

Figure 11.3 Bistatic RCS $\sigma^{(N)}/\lambda$ versus observation angle θ for $\theta_0 = 60°$, $\varepsilon_r = 12.0 + i0$, $\mu_r = 1.4 + i4.5$, $N = 3$

Figure 11.3 shows the bistatic RCS as a function of observation angle θ, where the strip width is $2a = \lambda, 5\lambda$, and 10λ, the strip thickness is $b = 0.01\lambda, 0.04\lambda, 0.07\lambda$, and 0.10λ, and the incidence angle θ_0 is fixed as 60°. As an example of existing lossy materials, we have chosen the ferrite with $\varepsilon_r = 12.0 + i0$ and $\mu_r = 1.4 + i4.5$ in numerical computation [14]. It is seen from the figure that the RCS shows noticeable peaks along the reflected and incident shadow boundaries at $\theta = 120°$ and $\theta = -120°$, respectively. We observe by comparing the results for $2a = \lambda$, 5λ, and 10λ that the RCS shows sharp oscillation for larger strip width $2a$. We also notice that, for fixed $2a$, the RCS level becomes larger with an increase of the strip thickness b except in the neighborhood of the reflected shadow boundary at $\theta = 120°$. Scattering characteristics near $\theta = 120°$ need some considerations. We see that, for $b = 0.04\lambda, 0.07\lambda$, and 0.10λ, the RCS level around $\theta = 120°$ becomes lower with an increase of the strip thickness. This is because ferrite is an electromagnetic wave absorber, and hence the RCS reduction around the specular reflection direction becomes noticeable for larger strip thickness b.

Figure 11.4 shows the monostatic RCS as a function of incidence angle θ_0, where the strip width is $2a = \lambda, 5\lambda$, and 10λ, the strip thickness is $b = 0.01\lambda, 0.04\lambda, 0.07\lambda$, and 0.10λ, and the same material parameters as in Figure 11.3 have been chosen for computation. The truncation number is $N = 3$. It is obvious that the peaks at $\theta_0 = 90°$ in the figure correspond to the specular reflection from the strip. Comparing the

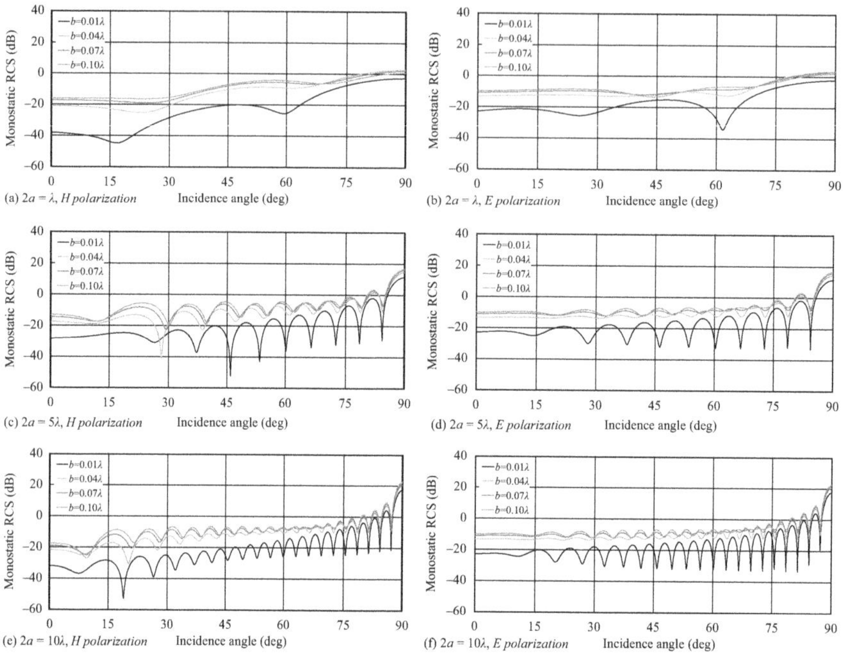

Figure 11.4 Monostatic RCS $\sigma^{(N)}/\lambda$ versus incidence angle θ_0 for $\varepsilon_r = 12.0 + i0$, $\mu_r = 1.4 + i4.5$, $N = 3$

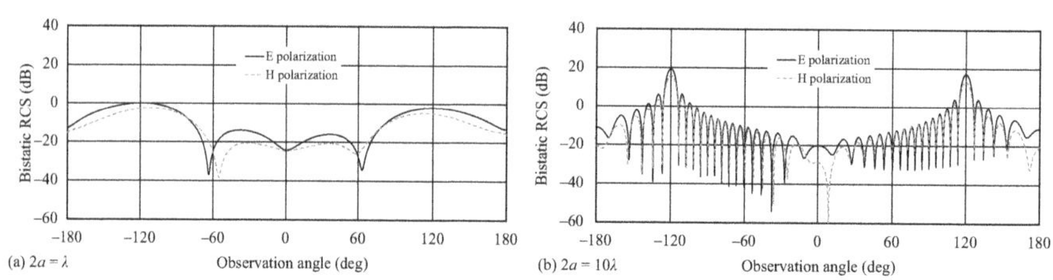

Figure 11.5 Bistatic RCS $\sigma^{(N)}/\lambda$ versus observation θ for E polarization, $b = 0.01\lambda$, $\theta_0 = 60°$, $\varepsilon_r = 12.0 + i0$, $\mu_r = 1.4 + i4.5$, $N = 3$, and its comparison with H polarization

RCS characteristics for $2a = \lambda$, 5λ, and 10λ, we observe that the RCS exhibits sharp oscillation for larger strip width $2a$. We also notice that for fixed $2a$, the RCS level becomes larger with an increase of the strip thickness b except in the neighborhood of $\theta_0 = 90°$, whereas the RCS becomes lower with an increase of the strip thickness for $b = 0.04\lambda, 0.07\lambda$, and 0.10λ, around $\theta_0 = 90°$. This is due to the electromagnetic energy absorption for the ferrite material with larger strip thickness b, as has been investigated earlier for the bistatic RCS.

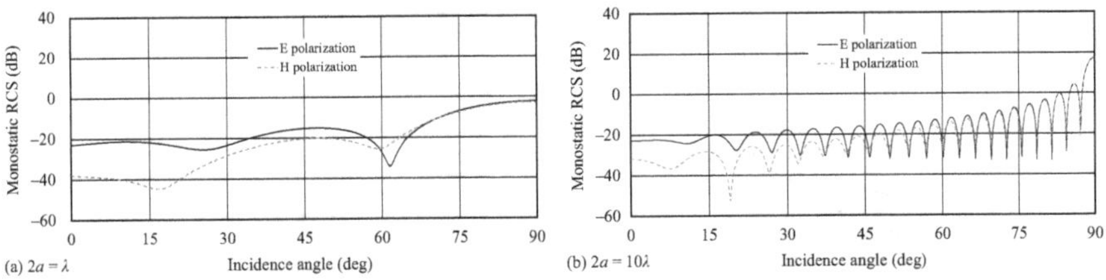

Figure 11.6 Monostatic RCS $\sigma^{(N)}/\lambda$ versus incidence angle θ_0 for E polarization, $b = 0.01\lambda$, $\varepsilon_r = 12.0 + i0$, $\mu_r = 1.4 + i4.5$, $N = 3$, and its comparison with H polarization

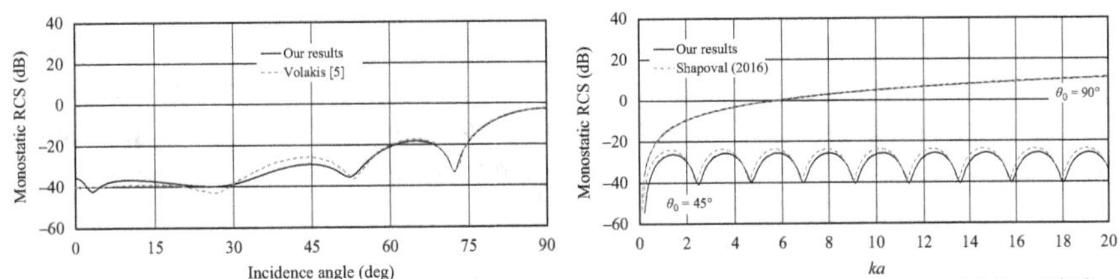

(a) *Comparison with Volakis [5]*, $2a = 1.7\lambda$, $b = 0.01\lambda$, $\varepsilon_r = 7.4 + i1.11$, $\mu_r = 1.4 + i0.672$, $N = 3$.

(b) *Comparison with Shapoval [private communication, 2016]*, $\theta_0 = 45°, 90°$, $b = 0.025\lambda$, $\varepsilon_r = 4 + i0.4$, $\mu_r = 1$, $N = 3$.

Figure 11.7 Monostatic RCS $\sigma^{(N)}/\lambda$ versus incidence angle θ_0 and frequency parameter ka for H polarization

We shall now make some comparison between two different polarizations. Figures 11.5 and 11.6 show the bistatic as a function of observation angle θ and monostatic RCS as a function of incidence angle θ_0, respectively, where the strip dimension is $2a = \lambda, 10\lambda, b = 0.01\lambda, \varepsilon_r = 12.0 + i0, \mu_r = 1.4 + i4.5$, and in the bistatic RCS, the incidence angle θ_0 has been 60°. Comparing the RCS characteristics for E polarization with those for H polarization, we find that the RCS level for H polarization is lower than that for E polarization over the whole range of the incidence angle θ_0, as shown in Figures 11.5 and 11.6.

Figures 11.7 and 11.8 show the monostatic RCS versus incidence angle θ_0 and the frequency parameter *ka* for both E and H polarizations, respectively. In the figures, the results obtained by Volakis [5] and Shapoval [private communication, September 2016] are also plotted. It is observed from Figures 11.7 and 11.8 that our results agree well with Shapoval's results. It is also seen from Figure 11.7(a) that our results agree reasonably well with Volakis's results over $40° < \theta_0 < 90°$, but there are some discrepancies for $0° < \theta_0 < 40°$. These discrepancies are perhaps due to the fact that Volakis's solution is derived on the basis of the solutions for the two independent half-planes and becomes less accurate at narrow strip width as in $2a = 1.7\lambda$. It is therefore inferred that Volakis's solution is not applicable at low frequencies and for near-gazing incidence ($\theta_0 \approx 0°$).

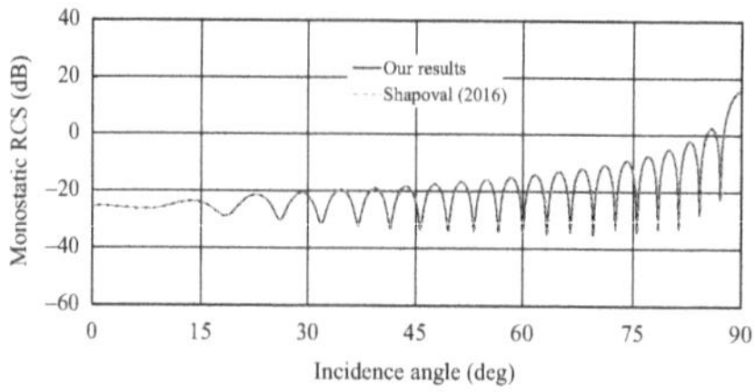

(a) *Comparison with Shapoval [private communication, 2016]*, $2a = 10\lambda$, $b = 0.01\lambda$, $\varepsilon_r = 3.4 + i10$, $\mu_r = 1$, $N = 3$.

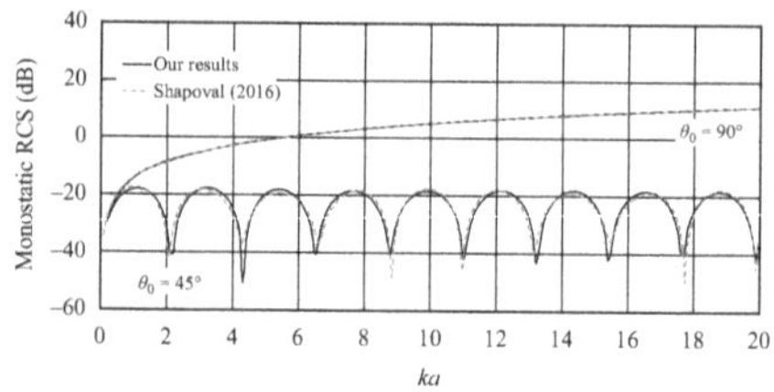

(b) *Comparison with Shapoval [private communication, 2016]*, $\theta_0 = 45°, 90°$, $b = 0.025\lambda$, $\varepsilon_r = 4 + i0.4$, $\mu_r = 1$, $N = 3$.

Figure 11.8 Monostatic RCS $\sigma^{(N)}/\lambda$ versus incidence angle θ_0 and frequency parameter ka for E polarization

We shall now make general remarks on the Wiener–Hopf solution. In carrying out the Wiener–Hopf analysis, we have employed a high-frequency asymptotic method under the condition that the strip width is large compared with wavelength. Hence, our final solution becomes highly accurate with an increase of the strip width.

11.5 Conclusions

In this chapter, we have solved the plane wave diffraction by a thin material strip using the Wiener–Hopf technique and approximate boundary conditions. Employing a rigorous asymptotics, a high-frequency solution for large strip width has been obtained. Illustrative numerical examples on the RCS are presented, and the far field scattering characteristics of the strip have been discussed in detail. Some comparisons with the other existing method have also been provided.

Acknowledgment

The authors would like to thank Dr. Olga V. Shapoval of the Institute of Radio-Physics and Electronics of the National Academy of Sciences of Ukraine (IRE NASU) for providing various numerical results based on generalized boundary conditions and the singular integral equation.

References

[1] Wiener, N. and Hopf, E., "Über eine Klasse singulärer Integralgleichungen," *S. B. Preuss. Akad. Wiss.*, 1931, pp. 696–706.

[2] Noble, B., *Methods Based on the Wiener–Hopf Technique for the Solution of Partial Differential Equations*, London: Pergamon, 1958.

[3] Mittra, R. and Lee, S.-W., *Analytical Techniques in the Theory of Guided Waves*, New York: Macmillan, 1971.

[4] Daniele, V. G. and Zich, R. S., *The Wiener–Hopf Method in Electromagnetics*, Edison, New Jersey: SciTech Publishing, 2014.

[5] Volakis, J. L., “High-frequency scattering by a thin material half plane and strip,” *Radio Science*, 23, 3, 1988, pp. 450–462.

[6] Clemmow, P. C., “A method for the exact solution of a class of two-dimensional diffraction problems,” *Proceedings of the Royal Society of London Series A*, 205, 1081, February 1951, pp. 286–308.

[7] Herman, M. I. and Volakis, J. L., “High-frequency scattering by a resistive strip and extensions to conductive and impedance strips,” *Radio Science*, 22, 3,1987, pp. 335–349.

[8] Senior, T. B. A. and Volakis, J. L., *Approximate Boundary Conditions in Electromagnetics*, London: IEE, 1995.

[9] Shapoval, O. V., Sauleau, R., and Nosich, A. I., “Scattering and absorption of waves by flat material strips analyzed using generalized boundary conditions and Nystrom-type algorithm,” *IEEE Transactions on Antennas and Propagation*, 59, 9, September 2011, pp. 3339–3346.

[10] Sukharevsky, I. O., Shapoval, O. V., Altintas, A., and Nosich, A. I., “Validity and limitations of the median-line integral equation technique in the scattering by material strips of sub-wavelength thickness,” *IEEE Transactions on Antennas and Propagation*, 62, 7, July 2014, pp. 3623–3631.

[11] Kobayashi, K., “Solutions of wave scattering problems for a class of the modified Wiener–Hopf geometries,” *IEEJ Transactions on Fundamentals and Materials*, 133, 5, 2013, pp. 233–241.

[12] Nagasaka, T. and Kobayashi, K., “Wiener–Hopf analysis of the plane wave diffraction by a thin material strip,” *IEICE Transactions on Electronics, E100-C*, 1, January 2017, pp. 11–19.

[13] Nagasaka, T. and Kobayashi, K., “Wiener–Hopf analysis of the plane wave diffraction by a thin material strip: the case of E polarization,” *IEICE Transactions on Electronics*, E100-C, 1, January 2018, pp. 12–19.

[14] Komoni, T., Yasuda, F., and Saito, K., “Development of the demountable damped cavity,” *Proceedings of 15th International Conference on RF Superconductivity (SRF 2011)*, Chicago, IL, USA, July 2011, pp. 172–176.

Chapter 12

The Wiener–Hopf Fredholm factorization technique to solve scattering problems in coupled planar and angular regions

Vito G. Daniele[1] and Guido Lombardi[1]

The Wiener–Hopf (WH) technique is a very powerful tool in the spectral domain to solve field problems in the presence of discontinuities. This chapter introduces a generalization of this technique that allows to study geometries where coupled planar and angular region are present. Since exact solutions with closed form factorizations are available only in few cases, most of the problems require an alternative approximate technique as the Fredholm factorization. The Fredholm factorization reduces the solution of WH equations to a system of Fredholm integral equations (FIE) of second kind amenable of very efficient numerical solution. The deduction of the FIE is presented in this chapter for a relatively simple novel problem. The numerical solution of FIEs provides an analytical element of the spectra, which in general is not sufficient to evaluate the different components of the diffracted field. To obtain the whole spectrum of the unknowns, analytical continuations and recursive equations deduced by the WH equations are presented. The work ends with a short description of the numerical simulations for the novel scattering problem.

12.1 Introduction

The aim of this work is to present an extension of the WH technique (WHT) to electromagnetic problems constituted of coupled angular and planar regions. This possibility arises from two recent advances:

- The generalization of the WHT for the solution of wedge problems (GWHT) [1, 2].
- The reduction of the factorization problem to the solution of FIE of second kind (Fredholm factorization) [3–5].

To illustrate the technique reported in this chapter, we consider the electromagnetic scattering by a novel canonical structure constituted of a PEC wedge

[1] DET, Politecnico di Torino, Torino, Italy

with a staggered PEC half-plane (Figure 12.1). The application of the Wiener-Hopf Factorization Technique (WHFT) is based on five steps:

1. Deduction of the WH equations of the problem,
2. Reduction of the WH equations to FIE,
3. Solution of the FIE,
4. Analytical continuation of the numerical solution of the FIE,
5. Evaluation of the physical field components if present: reflected and refracted plane waves, diffracted fields, surface waves, lateral waves, leaky waves, mode excitations, near field and so on.

In practice, steps (2)–(4) substitute the fundamental step of the classical WH technique: the factorization of the kernel of the system of WH equations. For reason of space, some deductions are only outlined in this text and complete details are considered in the study by Daniele [12]. For the same reason, the step (5) is not discussed here because it is extensively reported in several papers and books [6–17]. This procedure has been reframed for several problems by resorting to network modelling [12, 15–17]; however, for reason of space the network framework is not presented here.

To illustrate the general technique, with reference to the problem of Figure 12.1, we assume that time variations $\exp(j\omega t)$ are omitted, and that two sets of polar coordinates (ρ_i, φ_i, z) will be used together with two sets of Cartesian coordinates (x_i, y_i, z) with $i = 1, 2$ that refers to the two origins. (Alternatively, sometimes we consider: $(\rho_1, \varphi_1, z) = (\rho, \varphi, z), (x_1, y_1, z) = (x, y, z)$ in some definition for reuse).

For the sake of simplicity, the two PEC structures are immersed in the free space with impedance $Z_o = 1/Y_o$ and propagation constant k. The structure is illuminated by an incident E-polarized plane wave that propagates inward from the angular region (a) with direction φ_o:

$$E_z^i = E_o e^{jk\,\rho\cos(\varphi-\varphi_o)} = E_o e^{jk\cos\varphi_o x + jk\sin\varphi_o y} \tag{12.1}$$

$$H_\rho^i = Y_o \sin(\varphi - \varphi_o) E_o e^{jk\rho\cos(\varphi-\varphi_o)} \tag{12.2}$$

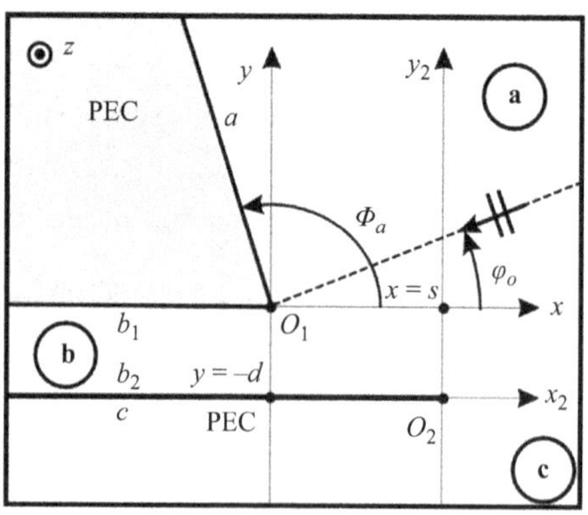

Figure 12.1 The geometry considered in this chapter to illustrate the WHFT

Since all the quantities are invariant with respect to $z(\partial/\partial z = 0)$, the total electric field is polarized in the z-direction: $\mathbf{E} = \hat{z}\, E_z(x,y)$. The generalization to skew incident plane wave sources doubles the number of the unknowns but does not introduce new conceptual difficulties.

The text is divided into six sections including the Introduction and an Appendix A. Sections 12.2–12.5 report the first four steps of the procedure. Section 12.6 applies the proposed procedure to the problem of Figure 12.1 by considering particular geometrical parameters. Finally Appendix A shows an useful extension of the Jordan's lemma.

12.2 The WH equations of the problem

With reference to Figure 12.1, we subdivide the entire problem in planar and angular regions bounded by half-plane interfaces. These subdomains are the angular region (a), and the planar layers (b) and (c). In the all subdomains the electromagnetic fields are represented by the spectra. They are the unilateral Laplace transforms in the region (a) and the bilateral Laplace transforms in the regions (b) and (c).

Region (a): $0 \le \varphi \le \Phi_a$

$$\begin{aligned} V_{1+}(\eta,\varphi) &= \int_0^{\infty} E_z(\rho,\varphi)e^{j\eta\rho}d\rho \\ I_{1+}(\eta,\varphi) &= \int_0^{\infty} H_\rho(\rho,\varphi)e^{j\eta\rho}d\rho \end{aligned} \tag{12.3}$$

Region (b): $-d \le y \le 0$

$$\begin{aligned} v(\eta,y) &= \int_{-\infty}^{\infty} E_z(x,y)e^{j\eta x}dx \\ i(\eta,y) &= \int_{-\infty}^{\infty} H_x(x,y)e^{j\eta x}dx \end{aligned} \tag{12.4}$$

Region (c): $-\infty \le y \le -d$

$$\begin{aligned} v_2(\eta,y) &= \int_{-\infty}^{\infty} E_z(x,y)e^{j\eta y}dy \\ i_2(\eta,y) &= \int_{-\infty}^{\infty} E_z(x,y)e^{j\eta y}dy \end{aligned} \tag{12.5}$$

To avoid singularities on the real axis η, quasi-free loss medium is assumed with vanishing negative imaginary parts of the propagation constant. For this reason, remote sources such as plane waves present unbounded values from the direction of incidence.

In the Laplace representations, the subscript + indicates plus functions, i.e. functions having regularity half-planes that are upper half-planes in the η-plane. Conversely the subscript–indicates minus functions, i.e. functions having regularity half-planes that are lower half-planes in the η-plane. In the presence of plane waves, singularities located in the half-plane $\mathrm{Im}[\eta] > 0$ ($\mathrm{Im}[\eta] < 0$) are present in some plus (minus) spectra and we label them not standard (n.s.) contributions due to these singularities. A very important property is that the n.s. contributions are known a priori by geometrical optical (GO) analysis without the necessity of solving the equations of the problem [6–17].

Let us focus the attention on the spectral WH unknowns (12.6–12.11) which are the Laplace transforms located along the interfaces of the subdomains:

$$V_{1+}(\eta) = V_{1+}(\eta, 0), \qquad I_{1+}(\eta) = I_{1+}(\eta, 0) \tag{12.6}$$

$$I_{a+}(\eta) = I_{1+}(\eta, \Phi_a), \qquad I_{1-}(\eta) = \int_{-\infty}^{0} H_x(x, 0)e^{j\eta x}dx \tag{12.7}$$

$$V_{2+}(\eta)\, e^{j\eta s} = \int_{s}^{\infty} E_z(x, -d)e^{j\eta x}dx \tag{12.8}$$

$$I_{2+}(\eta)e^{j\eta s} = \int_{s}^{\infty} H_x(x, -d)e^{j\eta x}dx \tag{12.9}$$

$$I_{2-}(\eta)e^{j\eta s} = \int_{-\infty}^{s} H_x(x, -d_+)e^{j\eta x}dx \tag{12.10}$$

$$I_{c-}(\eta)e^{j\eta s} = \int_{-\infty}^{s} H_x(x, -d_-)e^{j\eta x}dx \tag{12.11}$$

The above spectra generally are multivalued functions of η. From here on unless otherwise indicated, the principal branches of the multivalue functions are those evaluated by the computer code Mathematica (Wolfram Research, Inc., Champaign, Illinois, USA) where the principal branch of

$$\xi(\eta) = \sqrt{k^2 - \eta^2} \tag{12.12}$$

is assumed with $\mathrm{Re}[\xi(\eta)] \geq 0$ for each value of η.

While in the planar layers the WH equations are classical since they involve plus and minus functions defined in the same complex plane η [5], for angular regions we have generalized WH equations (GWHEs) that relate the spectral unknowns defined into different complex planes.

In particular, in region (a), the generalization of the WH technique to angular regions [1, 2] allows to write a WH equation as follows:

$$Y_\infty(\eta)V_{1+}(\eta) - I_{1+}(\eta) = -I_{a+}(-m_a(\eta)) \tag{12.13}$$

where

$$\begin{aligned} Y_\infty(\eta) &= \sqrt{k^2 - \eta^2}/(k\, Z_o), \\ m_a(\eta) &= -\eta \cos \Phi_a + \sqrt{k^2 - \eta^2} \sin \Phi_a \end{aligned} \tag{12.14}$$

Note that the GWHE (12.13) can be reduced to a classical WH equation using a suitable mapping (12.15) [1, 2]

$$\eta = \eta(\alpha) = -k \cos\left[\frac{\Phi_a}{\pi} \arccos\left(-\frac{\alpha}{k}\right)\right] \tag{12.15}$$

that yields in the α-plane:

$$\bar{Y}_{ao}(\alpha)\bar{V}_{1+}(\alpha) - \bar{I}_{1+}(\alpha) = -\bar{I}_{a+}(-\alpha) \tag{12.16}$$

where

$$Y_{\infty}(\eta) = \bar{Y}_{ao}(\alpha), \quad V_{1+}(\eta) = \bar{V}_{1+}(\alpha), \quad I_{1+}(\eta) = \bar{I}_{1+}(\alpha), \quad I_{a+}(-m_a) = \bar{I}_{a+}(-\alpha) \tag{12.17}$$

The functional equations of the region (b) are derived from circuital representation of the slab $-d < y < 0$ [5]:

$$i(\eta, 0) = -Y_{11}(\eta)v(\eta, 0) - Y_{12}(\eta)v(\eta, -d) \tag{12.18}$$

$$i(\eta, -d) = Y_{21}(\eta)v(\eta, 0) + Y_{22}(\eta)v(\eta, -d) \tag{12.19}$$

where

$$Y_{11}(\eta) = Y_{22}(\eta) = -jY_{\infty}(\eta)\cot[\xi(\eta)d] \tag{12.20}$$

$$Y_{12}(\eta) = Y_{21}(\eta) = jY_{\infty}(\eta)/\sin[\xi(\eta)d] \tag{12.21}$$

Taking into account, the definitions (12.6)–(12.11), (12.18)–(12.19) can be rewritten as:

$$-I_{1+}(\eta) - I_{1-}(\eta) = Y_{11}(\eta)V_{1+}(\eta) + Y_{12}(\eta)V_{2+}(\eta)\exp(j\eta s) \tag{12.22}$$

$$\exp(j\eta s)(I_{2+}(\eta) + I_{2-}(\eta)) = Y_{21}(\eta)V_{1+}(\eta) + Y_{22}(\eta)V_{2+}(\eta)\exp(j\eta s) \tag{12.23}$$

In region (c), the functional equations of the half space $-\infty \leq y \leq -d$ are derived again from a circuital representation described in the study by Daniele and Zich [5, 12]:

$$i_2(\eta, -d) = -Y_{\infty}(\eta)v_2(\eta, -d) \tag{12.24}$$

Taking into account the definitions (12.6)–(12.11), it yields

$$I_{c-}(\eta)e^{j\eta s} + I_{2+}(\eta)e^{j\eta s} = -Y_{\infty}(\eta)V_{2+}(\eta)e^{j\eta s} \tag{12.25}$$

12.3 Reduction of the WH equations to FIEs

The reduction of the WH equations to FIEs constitutes the central problem of the WHFT. While the classical factorization separates the minus unknowns from the plus unknowns by working on all the equations of the WH system at the same time, the Fredholm factorization simply eliminates time after time the minus unknown present in any single equation obtained in step (1) (Section 12.2). This elimination is accomplished through a Cauchy decomposition and of course it is neither so beautiful nor so powerful as the very ingenious factorization ideated by Wiener and Hopf in 1931. However, it is always possible. Furthermore, it is derived independently of the geometrical form of the region located out of the considered subdomain. One of the main advantages is that the Fredholm equations can be obtained once and for all, by studying time after time one single subregion. These characteristics of the Fredholm factorization notably simplify the deduction of the FIE of the whole problem. In particular, network representations [5, 12, 15–17] turn out to be very useful since they model the whole problem as an electrical network having as components the multi-ports representing the single subdomains; however, for reason of space the network framework is not presented here.

We anticipate that in the final Fredholm equations the not standard parts of the spectra constitute the known second members. For this purpose, we resort to (12.26) and (12.27) that are valid for real values of η

$$\begin{aligned} \frac{1}{2\pi j}\int_{\gamma_{1\eta}} \frac{F_+(\eta')}{\eta'-\eta}d\eta' &= F_+(\eta) - F_+^{n.s.}(\eta), \\ \frac{1}{2\pi j}\int_{\gamma_{1\eta}} \frac{F_-(\eta')}{\eta'-\eta}d\eta' &= F_-^{n.s.}(\eta) \end{aligned} \tag{12.26}$$

$$\begin{aligned} \frac{1}{2\pi j}\int_{\gamma_{2\eta}} \frac{F_-(\eta')}{\eta'-\eta}d\eta' &= -F_-(\eta) + F_-^{n.s.}(\eta) \\ \frac{1}{2\pi j}\int_{\gamma_{2\eta}} \frac{F_+(\eta')}{\eta'-\eta}d\eta' &= -F_+^{n.s.}(\eta) \end{aligned} \tag{12.27}$$

where $F_+^{n.s.}(\eta)(F_-^{n.s.}(\eta))$ is the not standard part of $F_+(\eta)(F_-(\eta))$ and $\gamma_{1\eta}$ and $\gamma_{2\eta}$ are, respectively, the 'smile' and the 'frown' integration line in the η-plane [3–5]. The Fredholm factorization yields integrals that couples the plus functions and whose kernel depends on the integration variable η' and the observation point η. From here until the final FIE equations (12.68), the observation point η will be assumed always real. The procedure to reduce the WH equations to a FIE is slightly different according to the considered subdomain. To face increasing difficulties, we reverse the order of the three subdomains c–b–a.

12.3.1 The Fredholm equation of the region (c)

Equation (12.25) can be rewritten

$$-I_{c-}(\eta) = +I_{2+}(\eta) + Y_\infty(\eta)V_{2+}(\eta) \tag{12.28}$$

The elimination of the minus function $I_{c-}(\eta)$ is readily obtained by using the second equation of (12.26) on (12.28). We get:

$$\frac{1}{2\pi j}\int_{\gamma_{1\eta}} \frac{I_{2+}(\eta') + Y_\infty(\eta')V_{2+}(\eta')}{\eta'-\eta}d\eta' = \frac{1}{2\pi j}\int_{\gamma_{1\eta}} \frac{-I_{c-}(\eta')}{\eta'-\eta}d\eta' = 0 \tag{12.29}$$

The second member is vanishing because no GO contribution is present on the face c with the assumption $\varphi_o > 0$. Taking into account the first of (12.26), we get:

$$I_{2+}(\eta) - I_{2+}^{n.s.}(\eta) + \frac{1}{2\pi j}\int_{\gamma_{1\eta}} \frac{Y_\infty(\eta')V_{2+}(\eta')}{\eta'-\eta}d\eta' = 0 \tag{12.30}$$

While the 'smile' decomposition eliminates the minus function $I_{c-}(\eta)$, (12.30) is not yet the final form of the integral representation since its kernel is singular at

$\eta' = \eta$. To get Fredholm equations of second kind, we must modify the integral. From (12.27), for this purpose, we consider

$$\frac{Y_\infty(\eta)}{2\pi j}\int_{\gamma_{2\eta}} \frac{V_{2+}(\eta')}{\eta'-\eta}d\eta' = -Y_\infty(\eta)V_{2+}^{n.s.}(\eta) \tag{12.31}$$

where $\gamma_{2\eta}$ is the 'frown' integration line [3–5]. Subtracting (12.31) from (12.30)

$$I_{2+}(\eta) = -\frac{1}{2\pi j}\int_{\gamma_{1\eta}} \frac{Y_\infty(\eta')V_{2+}(\eta')}{\eta'-\eta}d\eta' + \frac{1}{2\pi j}\int_{\gamma_{2\eta}} \frac{Y_\infty(\eta)V_{2+}(\eta')}{\eta'-\eta}d\eta' - I_{lc}(\eta) \tag{12.32}$$

where

$$I_{lc}(\eta) = -(I_{2+}^{n.s.}(\eta) + Y_\infty(\eta)V_{2+}^{n.s.}(\eta)) \tag{12.33}$$

Now taking into account that:

$$\begin{aligned}
&\frac{1}{2\pi j}\int_{\gamma_{1\eta}} \frac{Y_\infty(\eta')V_{2+}(\eta')}{\eta'-\eta}d\eta' - \frac{1}{2\pi j}\int_{\gamma_{2\eta}} \frac{Y_\infty(\eta)V_{2+}(\eta')}{\eta'-\eta}d\eta' \\
&= \frac{1}{2\pi j}P.V.\int_{-\infty}^{\infty} \frac{Y_\infty(\eta')V_{2+}(\eta')}{\eta'-\eta}d\eta' + \frac{1}{2}Y_\infty(\eta)V_{2+}(\eta) \\
&\quad -\left[\frac{1}{2\pi j}P.V.\int_{-\infty}^{\infty} \frac{Y_\infty(\eta)V_{2+}(\eta')}{\eta'-\eta}d\eta' - \frac{1}{2}Y_\infty(\eta)V_{2+}(\eta)\right] \\
&= \frac{1}{2\pi j}\int_{-\infty}^{\infty} \frac{[Y_\infty(\eta') - Y_\infty(\eta)]V_{2+}(\eta')}{\eta'-\eta}d\eta' + Y_\infty(\eta)V_{2+}\eta)
\end{aligned}$$

(see also (12.39) and (12.40)) we get:

$$I_{2+}(\eta) = -Y_\infty(\eta)V_{2+}(\eta) - \mathscr{Y}_\infty[V_{2+}(\eta')] - I_{lc}(\eta) \tag{12.34}$$

where $\mathscr{Y}_\infty[\ldots] = \frac{1}{2\pi j}\int_{-\infty}^{\infty} y_c(\eta,\eta')[\ldots]d\eta'$ is an integral operator with kernel $y_c(\eta,\eta') = \frac{Y_\infty(\eta')-Y_\infty(\eta)}{\eta'-\eta}$ not singular along the real axis.

We recall that while Y_∞ is evaluated independently on the region out the layer (c), the term $I_{lc}(\eta)$ is known and depends on the n.s. GO contributions of the whole problem [5, 12, 15–17]. Equation (12.34) is the FIE of the subdomain (c). The evaluation of the n.s. parts of the GO contribution depends on the incident plane wave and the plane wave reflected by the face a. Taking into account that the reflected wave has the direction $\varphi_{ra} = 2\Phi_a - \varphi_o$, we deal with a n.s. incident wave and a n.s. reflected wave provided that $\varphi_o < \pi/2$ and $\varphi_{ra} < \pi/2$, respectively. Hence, we get:

$$V_{2+}^{n.s.}(\eta) = j\frac{E_o e^{j\psi_o}}{\eta-\eta_o}u\left(\frac{\pi}{2}-\varphi_o\right) - j\frac{E_o e^{j\psi_a}}{\eta-\eta_{ra}}u\left(\frac{\pi}{2}-2\Phi_a+\varphi_o\right) \tag{12.35a}$$

$$I_{2+}^{n.s}(\eta) = -j\frac{Y_o E_o \sin\varphi_o e^{j\psi_o}}{\eta-\eta_o}u\left(\frac{\pi}{2}-\varphi_o\right) + j\frac{Y_o E_o \sin\varphi_{ra} e^{j\psi_a}}{\eta-\eta_{ra}}u\left(\frac{\pi}{2}-2\Phi_a+\varphi_o\right) \tag{12.35b}$$

where $u(\delta)$ is the step function and

$$\eta_o = -k\cos\varphi_o, \quad \eta_{ra} = -k\cos\varphi, \quad \psi_o = ks\cos\varphi_o - kd\sin\varphi_o,$$
$$\psi_a = ks\cos\varphi_{ra} - kd\sin\varphi_{ra}$$

12.3.2 The Fredholm equations of the region (b)

The WH equations of the region (b) are (12.22) and (12.23) which present the two minus functions $I_{1-}(\eta)$ and $I_{2-}(\eta)$. Again these minus functions can be eliminated by using the Cauchy decomposition. However in these equations the presence of $\exp(j\eta s)$ requires a discussion on the possibility to close the lines γ_{1n} and γ_{2n} with half circles with radius $|\eta'| \to \infty$. In particular, Appendix A shows an extension of Jordan's lemma that shows this possibility. Since again no GO contribution is present on the face b_1, applying the 'smile' integration to (12.22) we get:

$$\frac{1}{2\pi j}\int_{\gamma_{1\eta}} \frac{I_{1+}(\eta') + Y_{11}(\eta')V_{1+}(\eta') + Y_{12}(\eta')V_{2+}(\eta')e^{j\eta' s}}{\eta' - \eta}d\eta'$$
$$= \frac{1}{2\pi j}\int_{\gamma_{1\eta}} \frac{I_{1-}(\eta')}{\eta' - \eta}d\eta' = 0 \tag{12.36}$$

Taking into account the frown real axis γ_{2n} and the not standard part of $V_{1+}(\eta)$, $I_{1+}(\eta)$, $V_{2+}(\eta)$ as given in (12.26) and (12.27), it yields

$$\frac{1}{2\pi j}\int_{\gamma_{2\eta}} \frac{I_{1+}(\eta') + Y_{11}(\eta)V_{1+}(\eta') + Y_{12}(\eta)V_{2+}(\eta')e^{j\eta s}}{\eta' - \eta}d\eta' = -I_{l1}(\eta) \tag{12.37}$$

where

$$I_{l1}(\eta) = I_{1+}^{n.s.}(\eta) + Y_{11}(\eta)V_{1+}^{n.s.}(\eta) + Y_{12}(\eta)V_{2+}^{n.s.}(\eta)e^{j\eta s} \tag{12.38}$$

Taking into account that

$$\frac{1}{2\pi j}\int_{\gamma_{1\eta}} f_1(\eta')d\eta' = \frac{1}{2}\mathrm{Res}[f_1(\eta')]_{\eta'=\eta} + \frac{1}{2\pi j}P.V.\int_{-\infty}^{\infty} f_1(\eta')d\eta' \tag{12.39}$$

$$\frac{1}{2\pi j}\int_{\gamma_{2\eta}} f_2(\eta')d\eta' = -\frac{1}{2}\mathrm{Res}[f_2(\eta')]_{\eta'=\eta} + \frac{1}{2\pi j}P.V.\int_{-\infty}^{\infty} f_2(\eta')d\eta' \tag{12.40}$$

and subtracting (12.38) from (12.37) it yields

$$I_{1+}(\eta) + Y_{11}(\eta)V_{1+}(\eta) + Y_{12}(\eta)V_{2+}(\eta)e^{j\eta s} + \frac{1}{2\pi j}\int_{-\infty}^{\infty} y_{11}(\eta,\eta')V_{1+}(\eta')d\eta'$$
$$+ \frac{1}{2\pi j}\int_{-\infty}^{\infty} y_{12}(\eta,\eta')V_{2+}(\eta')d\eta' = I_{l1}(\eta) \tag{12.41}$$

with regular kernels:

$$y_{11}(\eta,\eta') = \frac{(Y_{11}(\eta') - Y_{11}(\eta))}{\eta' - \eta}$$

$$y_{12}(\eta,\eta') = \frac{Y_{12}(\eta')e^{j\eta' s} - Y_{12}(\eta)e^{j\eta s}}{\eta' - \eta}$$

Equation (12.41) is the FIE relevant the WH equation (12.22).

The same procedure applies to (12.23) rewritten in the form:

$$I_{2-}(\eta) = -I_{2+}(\eta) + Y_{21}(\eta)V_{1+}(\eta)\exp(-j\eta s) + Y_{22}(\eta)V_{2+}(\eta) \tag{12.42}$$

Taking into account that the eventual GO contribution on the face b_2 has a finite support, the n.s. terms are absent in $I_{2-}(\eta)$ and the 'smile' integration yields:

$$\frac{1}{2\pi j}\int_{\gamma_{1\eta}} \frac{-I_{2+}(\eta') + Y_{21}(\eta')V_{1+}(\eta')e^{-j\eta' s} + Y_{22}(\eta')V_{2+}(\eta')}{\eta' - \eta}d\eta'$$
$$= \frac{1}{2\pi j}\int_{\gamma_{1\eta}} \frac{I_{2-}(\eta')}{\eta' - \eta}d\eta' = 0 \tag{12.43}$$

Taking into account the effect of the presence of not standard functions on the 'frown' integration:

$$\frac{1}{2\pi j}\int_{\gamma_{2\eta}} \frac{-I_{2+}(\eta') + Y_{21}(\eta)V_{1+}(\eta')e^{-j\eta s} + Y_{22}(\eta)V_{2+}(\eta')}{\eta' - \eta}d\eta' = -I_{l2}(\eta) \tag{12.44}$$

with

$$I_{l2}(\eta) = -I_{2+}^{n.s.}(\eta) + Y_{21}(\eta)V_{1+}^{n.s.}(\eta)e^{-j\eta s} + Y_{22}(\eta)V_{2+}^{n.s.}(\eta) \tag{12.45}$$

Finally taking into account (12.39) and (12.40), subtracting (12.44) from (12.43), it yields the FIE relevant to the WH equation (12.23):

$$I_{2+}(\eta) = Y_{21}(\eta)V_{1+}(\eta)e^{-j\eta s} + Y_{22}(\eta)V_{2+}(\eta) + \frac{1}{2\pi j}\int_{-\infty}^{\infty} y_{21}(\eta,\eta')V_{1+}(\eta')d\eta'$$
$$+\frac{1}{2\pi j}\int_{-\infty}^{\infty} y_{22}(\eta,\eta')V_{2+}(\eta')d\eta' - I_{l2}(\eta) \tag{12.46}$$

with regular kernels:

$$y_{22}(\eta,\eta') = \frac{Y_{22}(\eta') - Y_{22}(\eta)}{\eta' - \eta}$$

$$y_{212}(\eta,\eta') = \frac{Y_{21}(\eta')e^{-j\eta' s} - Y_{21}(\eta)e^{-j\eta s}}{\eta' - \eta}$$

In the evaluation of (12.38) and (12.45), the known quantities $V_{2+}^{n.s.}(\eta)$ and $I_{2+}^{n.s.}(\eta)$ are reported in (12.35). Similarly the known quantities $V_{1+}^{n.s.}(\eta)$ and $I_{1+}^{n.s.}(\eta)$ are:

$$V_{1+}^{n.s.}(\eta) = j\frac{E_o}{\eta - \eta_o}u\left(\frac{\pi}{2} - \varphi_o\right) - j\frac{E_o}{\eta - \eta_{ra}}u\left(\frac{\pi}{2} - 2\Phi_a + \varphi_o\right) \tag{12.47a}$$

$$I_{1+}^{n.s.}(\eta) = -j\frac{Y_o E_o \sin\varphi_o}{\eta-\eta_o}u\left(\frac{\pi}{2}-\varphi_o\right) + j\frac{Y_o E_o \sin\varphi_{ra}}{\eta-\eta_{ra}}u\left(\frac{\pi}{2}-2\Phi_a+\varphi_o\right) \tag{12.47b}$$

12.3.3 The Fredholm equation of the angular region (a)

In this subsection, we consider the deduction of the FIE relevant to the angular region (a). The WH equations of this region are reported in (12.13) or (12.16). The form (12.16) is a classical WH equation that allows the elimination of the minus function $I_{a+}(-m_a) = \bar{I}_{a+}(-\alpha)$ by using the smile integration $\gamma_{1\alpha}$ on the α-plane, thus from the second of (12.26) we get:

$$\begin{aligned}\frac{1}{2\pi j}\int_{\gamma_{1\alpha}}\frac{\bar{Y}_{ao}(\alpha')\bar{V}_{1+}(\alpha')-\bar{I}_{1+}(\alpha')}{\alpha'-\alpha}d\alpha' &= -\frac{1}{2\pi j}\int_{\gamma_{1\alpha}}\frac{\bar{I}_{a+}(-\alpha')}{\alpha'-\alpha}d\alpha' \\ &= -\bar{I}_{a+}^{n.s.}(-\alpha)\end{aligned} \tag{12.48}$$

Since the inverse of the mapping (12.15) is

$$\alpha = \alpha(\eta) = -k\cos\left[\frac{\pi}{\Phi_a}\arccos\left(-\frac{\eta}{k}\right)\right] \tag{12.49}$$

in the α-plane the pole $\eta_o = -k\cos\varphi_o$ of the plane wave is located at

$$\alpha_o = \alpha(\eta_o) = -k\cos\frac{\pi}{\Phi_a}\varphi_o \tag{12.50}$$

This pole is n.s. for the minus function $\bar{I}_{a+}(-\alpha)$ when $\varphi_o > \Phi_a/2$ thus

$$-\bar{I}_{a+}^{n.s.}(-\alpha) = -\frac{R_\alpha}{\alpha-\alpha_0}u\left(\varphi_0-\frac{\Phi_a}{2}\right) \tag{12.51}$$

where R_α is the residue of $\bar{I}_{a+}(-\alpha)$ at $\alpha = \alpha_o$:

$$R_\alpha = -\frac{2jE_o}{Z_o}\frac{\pi}{\Phi_e}\sin\left(\frac{\pi}{\Phi_e}\varphi_o\right) \tag{12.52}$$

Considering the terms $\frac{1}{2\pi j}\int_{\gamma_{1\alpha}}\frac{\bar{I}_{1+}(\alpha')}{\alpha'-\alpha}d\alpha'$ of the first member of (12.48), taking into account the first of (12.26) and next returning in the η-plane we get

$$\begin{aligned}\frac{1}{2\pi j}\int_{\gamma_{1\alpha}}\frac{\bar{I}_{1+}(\alpha')}{\alpha'-\alpha}d\alpha' &= \bar{I}_{1+}(\alpha) - \bar{I}_{1+}^{n.s.}(\alpha) \\ &= I_{1+}(\eta) - \frac{R_{\alpha o}}{\alpha(\eta)-\alpha(\eta_o)}u\left(-\varphi_o+\frac{\pi}{2}\right)\end{aligned} \tag{12.53}$$

where the residue of $\bar{I}_{1+}(\alpha)$ at $\alpha(\eta_o)$ is given by:

$$R_{\alpha o} = -\frac{j}{Z_o}\frac{\pi}{\Phi_a}\sin\left(\frac{\pi}{\Phi_a}\varphi_o\right) \tag{12.54}$$

The most difficult step is to estimate $\frac{1}{2\pi j}\int_{\gamma_{1\alpha}} \frac{\bar{Y}_{ao}(\alpha')\bar{V}_{1+}(\alpha')}{\alpha'-\alpha} d\alpha'$ as a smile integral in the η'-plane. The mapping (12.49) yields:

$$\frac{1}{2\pi j}\int_{\gamma_{1\alpha}} \frac{\bar{Y}_{\alpha o}(\alpha')\bar{V}_{1+}(\alpha')}{\alpha'-\alpha} d\alpha' = \frac{1}{2\pi j}\int_{\bar{\gamma}_{1\alpha}} \frac{\bar{Y}_{\alpha o}(\alpha(\eta'))\bar{V}_{1+}(\alpha(\eta'))}{\alpha(\eta')-\alpha(\eta)} \frac{d\alpha}{d\eta'} d\eta'$$
$$= \frac{1}{2\pi j}\int_{\bar{\gamma}_{1\alpha}} \frac{Y_{\infty}(\eta')V_{1+}(\eta')}{\alpha(\eta')-\alpha(\eta)} \frac{d\alpha}{d\eta'} d\eta' \tag{12.55}$$

where $\bar{\gamma}_{1\alpha}$ is the image on the η'-plane of the 'smile' line $\gamma_{1\alpha}$. The problem to face is the application of the residue theorem when we deform the line $\bar{\gamma}_{1\alpha}$ on the 'smile' line $\gamma_{1\eta}$ in the η' plane.

Figure 12.2 represents $\bar{\gamma}_{1\alpha}$ (except for the indentation at the observation point α) for different values of Φ_a. The curve $\bar{\gamma}_{1\alpha}$ depends on Φ_a but it remains always located in the upper half-plane $\text{Im}[\eta'] > 0$. In Figure 12.2, the upper half-plane $\text{Im}[\alpha] \geq 0$ is mapped into the convex upper left region $\bar{U}_\alpha$ bounded by $\bar{\gamma}_{1\alpha}$. We note that the contour indentation is also present in η' and because of the introduction of $\alpha(\eta)$ in (12.55), the observation point $\eta = \eta(\alpha)$ must be located on the centre of the notch of $\bar{\gamma}_{1\alpha}$. Next, we warp $\bar{\gamma}_{1\alpha}$ and at the same time we move η to deform the line $\bar{\gamma}_{1\alpha}$ to the smile line $\gamma_{1\eta}$ which is basically the real axis of the η'-plane except for the indentation at the observation point η (Figure 12.2).

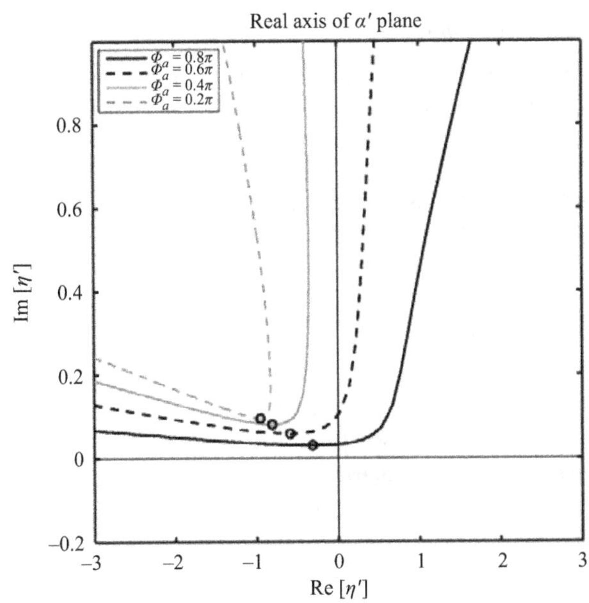

Figure 12.2 Real axis of α'-plane that constitutes part of $\bar{\gamma}_{1\alpha}$ in η'-plane except for the indentation at the observation point α for $k = 1 - 0.1j$ and for $\Phi_a = 0.8\pi$, 0.6π, 0.4π, and 0.2π. Circles represent the origins of α'-plane.

Once completed the warping, we need to consider the singularities of the integrand of (12.55) in region Ω that is the region between $\bar{\gamma}_{1\alpha}$ and $\gamma_{1\eta}$. The integrand depends on the functions $Y_\infty(\eta')$, $\alpha(\eta')$, $d\alpha(\eta')/d\eta'$, $V_{1+}(\eta')$ and present basically the following singularities: the branch point $\eta' = -k$, the GO source poles and the singularities $\alpha(\eta') = \alpha(\eta)$.

Theoretical and numerical considerations show that the branch point $\eta' = -k$ is located in $\bar{U}_\alpha$, therefore the warping does not yield any further contribution.

The GO source singularities due to $V_{1+}(\eta')$ are $\eta' = \eta_o$ and $\eta' = \eta_{ra}$ and they are located in Ω (and then captured) only if $\Phi_a/2 < \varphi_o, \varphi_{ra} < \pi/2$. Note that the residue theorem provides a contribution that together with $\bar{I}_{1+}^{n.s.}(\alpha(\eta))$ produces the known term $I_{ca}(\eta)$ in the final equation (12.67).

Finally, we need to consider the kernel singularities $\alpha(\eta') = \alpha(\eta)$. By varying the observation point η, they constitute singularity lines that require special attention. To locate these singularities in a simple way, it is convenient to introduce the angular complex planes w and w' through the mappings

$$\eta = -k\cos w, \quad \eta' = -k\cos w' \tag{12.56}$$

It yields:

$$\alpha(\eta) = -k\cos\frac{\pi}{\Phi_a}w, \quad \alpha(\eta') = -k\cos\frac{\pi}{\Phi_a}w' \tag{12.57}$$

The singularities are thus obtained in the angular complex plane:

$$w' = \pm w + 2n\Phi_a, \quad n \in \mathbb{N}_0 \tag{12.58}$$

We recall that we assume the principal branch of $\xi(\eta) = \sqrt{k^2 - \eta^2}$ as done in Wolfram Mathematica (i.e. with $\mathrm{Re}[\xi(\eta)] \geq 0$ for each value of η), and therefore in w'-plane the image of the principal sheet is for $-\pi < \mathrm{Re}[w'] < 0$.

According to this assumption, when Φ_a is obtuse, the only admissible singular values of w' (12.58) located in the principal sheet are $w' = -w(\eta) - 2\Phi_a$ where $w(\eta) = -\arccos(-\eta/k)$, $\mathrm{Im}[\eta] = 0$.

In the η'-plane, we obtain the singularity line:

$$\eta'(\eta) = -k\cos[-w(\eta) - 2\Phi_a] = \eta\cos(2\Phi_a) - \sqrt{k^2 - \eta^2}\sin 2\Phi_a, \mathrm{Im}[\eta] = 0 \tag{12.59}$$

Taking into account that for real η, the principal branch yields $\mathrm{Im}[\sqrt{k^2 - \eta^2}] < 0$, it follows that for $\Phi_a > \pi/2$, $\mathrm{Im}[\eta'(\eta)] = \mathrm{Im}[-\sqrt{k^2 - \eta^2}\sin 2\Phi_a] < 0$. It means that for obtuse angle there are no singularities lines of the kernel in Ω (Figure 12.3).

This situation changes completely when Φ_a is acute. In fact, values of $n < n_{at} = \lfloor \pi/2\Phi_a \rfloor$ make $\sin 2n\Phi_a > 0$, i.e. $\mathrm{Im}[\eta'(\eta)] > 0$, and thus we have kernel singularities at:

$$\eta'(\eta) = -k\cos[-w(\eta) - 2n\Phi_a] = p_n^{\Phi_a}[\eta] = \eta\cos 2n\Phi_a - \sqrt{k^2 - \eta^2}\sin 2n\Phi_a \tag{12.60}$$

The contribution of the singularity lines $p_n^{\Phi_a}[\eta]$ is captured when located in Ω. Since both $\bar{\gamma}_{1\alpha}$ and $p_n^{\Phi_a}[\eta]$ are below the branch point $\eta = -k$, we observe that the

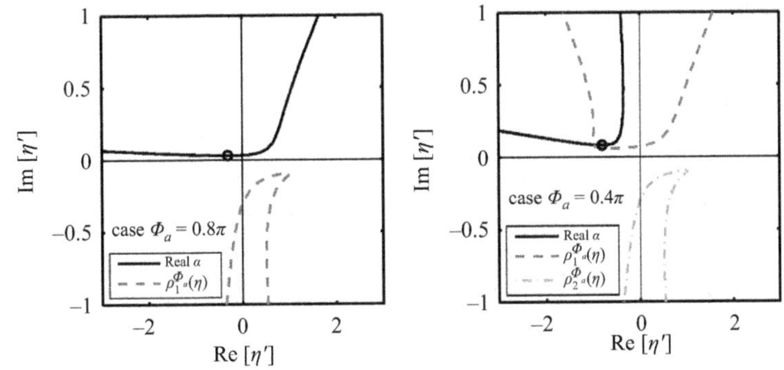

Figure 12.3 Real axis of α'-plane that constitutes part of $\bar{\gamma}_{1\alpha}$ in η'-plane together with the singularity lines (12.60) for $k = 1 - 0.1j$ and for $\Phi_a = 0.8\pi$, 0.4π. Circles represent the origins of α'-plane.

line $\bar{\gamma}_{1\alpha}$ can be deformed to enclose the lines $\eta' = p_n^{\Phi}[\eta], n = 1, 2, \ldots, n_{at}$, located in the principal sheet and having $\text{Im}[\eta'] > 0$. In particular, for a given real observation point η, a singularity $\eta' = p_n^{\Phi}[\eta]$ is captured with residue:

$$\frac{Y_\infty(p_n^{\Phi_a}[\eta])\, V_{1+}(p_n^{\Phi_a}[\eta])}{\partial_{\eta'}\alpha(\eta')\big|_{\eta'=p_n^{\Phi_a}[\eta]}}\,\partial_{\eta'}\alpha(\eta')\big|_{\eta'=p_n^{\Phi_a}[\eta]} = Y_\infty(p_n^{\Phi_a}[\eta])\, V_{1+}(p_n^{\Phi_a}[\eta])$$

$$= -q_n^{\Phi_a}[\eta] V_{1+}(p_n^{\Phi_a}[\eta]) \qquad (12.61)$$

where

$$q_n^{\Phi_a}[\eta] = \frac{1}{k\, Z_o}\left(\eta \sin 2n\Phi_a + \sqrt{k^2 - \eta^2}\cos 2n\Phi_a\right) \qquad (12.62)$$

Applying the residue theorem in the region Ω yields:

$$\int_{\bar{\gamma}_{1\alpha}} \frac{Y_\infty(\eta')\, V_{1+}(\eta')}{\alpha(\eta') - \alpha(\eta)}\frac{d\alpha}{d\eta'}d\eta'$$

$$= \frac{1}{2\pi j}\int_{\gamma_{1\eta}} \frac{Y_\infty(\eta')}{\alpha(\eta') - \alpha(\eta)}\frac{d\alpha}{d\eta'} V_{1+}(\eta') d\eta'$$

$$+ \frac{R_{\alpha o}}{\alpha(\eta) - \alpha(\eta_o)}\left(u\left(\frac{\Phi_a}{2} - \varphi_o\right) - u\left(\frac{\pi}{2} - \varphi_o\right)\right)$$

$$+ \sum_{n=1}^{n_{at}} q_n^{\Phi_a}(\eta) V_{1+}(p_n^{\Phi_a}(\eta)) u\left(\frac{\pi}{2} - n\Phi_a\right) \qquad (12.63)$$

Although (12.63) contains unknown quantities $V_{1+}(p_n^{\Phi_a}(\eta))$, since $\text{Im}[p_n^{\Phi}[\eta]] > 0$, we apply the Cauchy formula:

$$V_{1+}(p_n^{\Phi}(\eta)) = \frac{1}{2\pi j}\int_{-\infty}^{\infty} \frac{V_{1+}(\eta')}{\eta' - p_n^{\Phi}(\eta)} d\eta' + V_{1+}^{n.s.}(\mathrm{p}_n^{\Phi_a}(\eta)) \qquad (12.64)$$

and it yields

$$\frac{1}{2\pi j}\int_{\bar{\gamma}_{1\alpha}} \frac{Y_\infty(\eta')\,V_{1+}(\eta')}{\alpha(\eta')-\alpha(\eta)}\frac{d\alpha}{d\eta'}d\eta'$$
$$= \frac{1}{2\pi j}\int_{\gamma_{1\eta}}\left[\frac{Y_\infty(\eta')}{\alpha(\eta')-\alpha(\eta)}\frac{d\alpha}{d\eta'}+\sum_{n=1}^{n_{at}}\frac{q_n^{\Phi_a}(\eta)}{\eta'-\mathrm{p}_n^{\Phi_a}(\eta)}u\left(\frac{\pi}{2}-n\Phi_a\right)\right]V_{1+}(\eta')d\eta'$$
$$+\frac{R_{\alpha o}}{\alpha(\eta)-\alpha(\eta_o)}\left(u\left(\frac{\Phi_a}{2}-\varphi_o\right)-u\left(\frac{\pi}{2}-\varphi_o\right)\right)$$
$$+\sum_{n=1}^{n_{at}} q_n^{\Phi_a}(\eta)V_{1+}^{n.s.}(\mathrm{p}_n^{\Phi_a}(\eta))u\left(\frac{\pi}{2}-n\Phi_a\right) \tag{12.65}$$

The next step is to regularize the kernel in (12.65). Again, we make use of the 'frown' integration of $V_{1+}(\eta')$ (12.26) that yields:

$$\frac{1}{2\pi j}\int_{\gamma_{2\eta}}\frac{Y_\infty(\eta)\,V_{1+}(\eta')}{\eta'-\eta}d\eta' = -Y_\infty(\eta)V_{1+}^{n.s.}(\eta') \tag{12.66}$$

Subtracting (12.66) from (12.65), after algebraic manipulation it yields the final form of the Fredholm equation of the angular region (a):

$$I_{1+}(\eta) = Y_\infty(\eta)V_{1+}(\eta) + \mathscr{Y}_a[V_{1+}(\eta')] - I_{ca}(\eta), \eta\in\mathbb{R} \tag{12.67}$$

where $\mathscr{Y}_a[\ldots] \doteq \frac{1}{2\pi j}\int_{-\infty}^{\infty} y_a(\eta,\eta'][\ldots]d\eta$ with

$$y_a(\eta,\eta'] = \frac{Y_o(\eta')}{\alpha(\eta')-\alpha(\eta)}\frac{d\alpha}{d\eta'} - \frac{Y_o(\eta)}{\eta'-\eta} + \sum_{n=1}^{n_{at}}\frac{q_n^{\Phi_a}(\eta)}{\eta'-\mathrm{p}_n^{\Phi_a}(\eta)}u\left(\frac{\pi}{2}-n\Phi_a\right)$$

$$I_{ca}(\eta) = \left(\begin{array}{l} -\dfrac{R_\alpha}{\alpha(\eta)-\alpha(\eta_o)} + \left(\dfrac{R_{\alpha o}}{\alpha(\eta)-\alpha(\eta_o)} + \dfrac{jY_o(\eta)}{\eta-\eta_o}\right)u\left(-\varphi_o+\dfrac{\pi}{2}\right) + \\ -\displaystyle\sum_{n=1}^{n_{at}} q_n^{\Phi_a}(\eta)V_{1+}^{n.s.}(\mathrm{p}_n^{\Phi_a}(\eta))u\left(\frac{\pi}{2}-n\Phi_a\right)\end{array}\right)E_o$$

To verify the correctness of (12.67), these authors have introduced two completely different alternative deductions working in the w-plane (see [12, 15]).

12.4 Solution of the FIE

In Section 12.3, we have obtained four integral representations (12.34), (12.41), (12.46), (12.67) that contain only the plus unknowns $\mathbf{V}_+(\eta) = |V_{1+}(\eta)\ V_{2+}(\eta)|^t$ and $\mathbf{I}_+(\eta) = |I_{1+}(\eta)\ I_{2+}(\eta)|^t$ defined on the interfaces of the subregions. In fact, the spectra on the PEC faces a, b_1, b_2, c (Figure 12.1) do not occur since the (generalized) minus functions were eliminated by the Fredholm factorization.

By substitution, we eliminate the two unknowns $\mathbf{I}_+(\eta)$ and after algebraic manipulations we get the FIE in $\mathbf{V}_+(\eta)$:

$$\mathbf{V}_+(\eta) + \int_{-\infty}^{+\infty} \mathbf{M}(\eta, \eta')\mathbf{V}_+(\eta')d\eta' = \mathbf{N}(\eta), \quad \eta \in \mathbb{R} \tag{12.68}$$

where $\mathbf{M}(\eta, \eta')$ is the regular kernel:

$$\mathbf{M}(\eta, \eta') = \frac{1}{2\pi j} \begin{vmatrix} Z^e(\eta) & Z_m^e(\eta)e^{j\eta s} \\ Z_m^e(\eta)e^{-j\eta s} & Z^e(\eta) \end{vmatrix} \begin{vmatrix} y_a(\eta, \eta') + y_{11}(\eta, \eta') & y_{12}(\eta, \eta') \\ y_{21}(\eta, \eta') & y_c(\eta, \eta') + y_{11}(\eta, \eta') \end{vmatrix} \tag{12.69}$$

$$\mathbf{N}(\eta) = \begin{vmatrix} Z^e(\eta) & Z_m^e(\eta)e^{j\eta s} \\ Z_m^e(\eta)e^{-j\eta s} & Z^e(\eta) \end{vmatrix} \begin{vmatrix} I_{ca}(\eta) + I_{l1}(\eta) \\ I_{l2}(\eta) + I_{lc}(\eta) \end{vmatrix} \tag{12.70}$$

$$Z_m^e(\eta) = \frac{k\, Z_o}{2\xi(\eta)}, \quad Z^e(\eta) = \frac{k\, Z_o e^{-j\xi(\eta)d}}{2\xi(\eta)} \tag{12.71}$$

A detailed discussion on (12.68) allows to ascertain that in the absence of subregions filled by materials different from the free space, the singularity of $\mathbf{V}_+(\eta)$ is the branch point $\eta = k$ and the poles arisen from the GO contribution.

While the classical factorization is available only in relatively very few cases and requires some mathematical skill, the kernels of the obtained FIE involve simple functions and their approximate solutions can be obtained using well-known simple numerical methods of quadrature [6,19].

Taking into account the Meixner condition near the edges, as $\eta \to \infty$, the asymptotic behaviour of the unknowns $V_{i+}(\eta)(i = 1, 2)$ is: $V_{r+}(\eta) = O(\eta^{-(1+c_r)}), c_r > 0$. The fast convergence of the unknowns allows to estimate the integral in the FIEs on a limited spectrum band. To make a rigorous mathematical discussion and suggest other properties and/or solution techniques of (12.68), we can introduce the powerful apparatus of the functional analysis. In particular, it is possible to show that in the generalized Hilbert space $L_2(\mathbb{R}, \mu(\eta))$, where $\mu(\eta)$ is a suitable weight with $\mu(\eta) = O(1/\eta^{1/2}), \eta \to \pm\infty$, the kernel $\mathbf{M}(\eta, \eta')$ is a compact operator [15, 18].

We note that when singularities are near the integration line, to obtain fast convergence of (12.68), we need to warp the integration line on a suitable path $v(u)$that keeps the singularities far away. In this regard, we observe that in our problem the singularities of the kernel and the source term are located in the second and fourth quadrants, therefore we select the contour warp based on the line B_θ: $v(u) = u \exp[j\theta], -\infty < u < \infty, 0 < \theta < \pi/2$ that does not capture singularities of the integrand [3].

Unfortunately for the presence of factors $\exp[\pm j\eta' s]$, the closure at infinity between the real axis and B_θ with arcs located to infinity is possible only if $0 < \sin\theta < d/s$ (see Appendix A). Further studies to solve efficiently the FIE (12.63) appear necessary for vanishing values of d and $s \neq 0$.

Warping the real axis of η' on B_θ to avoid the capture of singularities of the kernel, we must change η so that it remains located on the integration line. We get the FIE:

$$\mathbf{V}_+(\eta) = -\int_{B_\theta} \mathbf{M}(\eta,\eta')\mathbf{V}_+(\eta')d\eta' + \mathbf{N}(\eta), \eta \in \mathrm{B}_\theta \tag{12.72}$$

To estimate (12.72), we apply simple sample and hold quadrature in a limited interval where A and h are truncation and step parameters, respectively:

$$\mathbf{V}_+(v(u)) + h\sum_{i=-A/h}^{A/h} \mathbf{M}(v(u), v(hi))\mathbf{V}_+(v(hi))v'(hi) = \mathbf{N}(v(u)) \tag{12.73}$$

The samples $\mathbf{V}_+(v(h\,i))$ allow to reconstruct an approximate version of $\mathbf{V}_+(\eta)$ through

$$\mathbf{V}_+(\eta) = -h\sum_{i=-A/h}^{A/h} \mathbf{M}(\eta, \eta(hi))\mathbf{V}_+(\eta(hi))\eta'(hi)) + \mathbf{N}(\eta) \tag{12.74}$$

This representation is valid on $\eta \in \mathrm{B}_\theta$ with $\eta' \in \mathrm{B}_\theta$ and by analytical continuation also for $\eta \notin \mathrm{B}_\theta$ until the singularity lines of the kernel $\mathbf{M}(\eta,\eta')$ are encountered.

12.5 Analytical continuation of the numerical solution

A very important drawback of the Fredholm factorization is that (12.74) is sometimes insufficient to accomplish the evaluation of fields (step 5). In fact (12.74) is an approximate analytical element of $\mathbf{V}_+(\eta)$, consequently an analytical continuation that overcomes the problems is needed.

Let us focus the attention on the starting spectral region where (12.74) is valid. Each of the i-component of the vector integral in (12.72) presents scalar terms of the form $\psi_{ij}(\eta) = \int_{B_\theta} M_{ij}(\eta,\eta')V_{j+}(\eta')d\eta'$ that are sectional analytic function in η-plane [20]. Moving the observation point η from the integration line $\eta \in \mathrm{B}_\theta$ provides an analytical continuation of $\psi_{ij}(\eta)$ till a singularity line of the kernel $M_{ij}(\eta,\eta')$ is crossed [20]. This singularity line depends on the integration line. For instance assuming $\eta' \in \mathrm{B}_\theta$ with $\theta = \pi/4$, Figure 12.4 reports the singularity lines of $M_{11}(\eta,\eta')$ that are the same of $y_a(\eta,\eta')$ (12.67) in the two cases $\Phi_a = 0.8\pi, \pi/5.5$ (note that for small value of Φ_a, we obtain more singularity lines in the principal sheet of η). Independently on the location of the kernel singularity lines, it is possible to extend the representation of $\mathbf{V}_+(\eta)$ in the half-plane U located upper B_θ (Figure 12.4) by applying (instead of (12.74)) the Cauchy integration formula (12.26) to $\mathbf{V}_+(\eta) - \mathbf{V}_+^{n.s.}(\eta)$ which is regular in this half-plane. In fact being regular in this half-plane:

$$\begin{aligned}\mathbf{V}_+(\eta) &= \frac{1}{2\pi j}\int_{B_\theta}\frac{V_+(\eta')}{\eta'-\eta}d\eta' + \mathbf{V}_+^{n.s.}(\eta)\\ &\approx \frac{h}{2\pi j}\sum_{i=-A/h}^{A/h}\frac{\mathbf{V}_+(v(hi))v'(hi)}{\eta - hi} + \mathbf{V}_+^{n.s.}(\eta), \eta \in U\end{aligned} \tag{12.75}$$

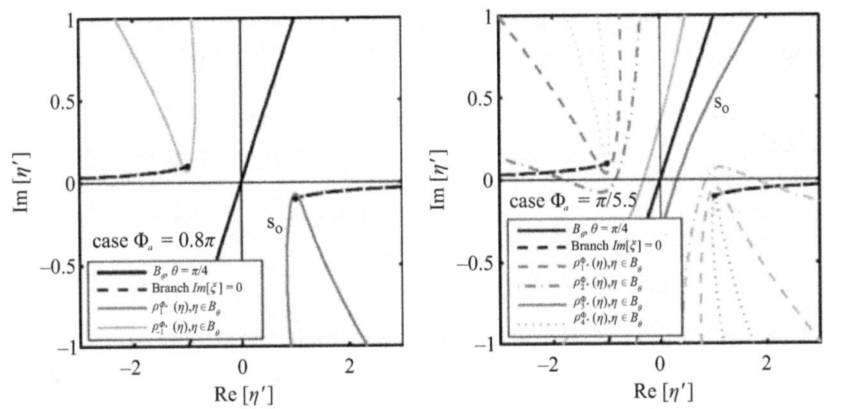

Figure 12.4 Singularities lines of the kernel $y_a(\eta, \eta')$ with $\eta' \in B_\theta$ on an angular region with aperture $\Phi_a = 0.8\pi$ (left), $\Phi_a = \pi/5.5$ (right) and for $k = 1 - 0.1j$. Note that some of the singularity lines cross the branches line and therefore they move from the principal sheet to the secondary sheet.

The same considerations apply to $\mathbf{I}_+(\eta)$. In particular, the samples $\mathbf{I}_+(v(hi))$ on B_θ can be obtained by using the Norton representations (12.41) and (12.46), or alternatively (12.34) and (12.67). Note that the use of (12.75) generates a singularity line at $\eta = \eta'(\eta' \in B_\theta)$, however the evaluation of the spectrum at $\eta = \eta'$ is obtained directly from (12.72).

Every kernel $M_{ij}(\eta, \eta')$ is related to a subdomain region and defines a starting spectral region. The intersection of the starting regions of the subdomains defines the cumulative starting spectral region $S_{\eta o}$ of the entire problem that is the region where the spectrum is known through the samples $\mathbf{V}_+(v(hi))$. With reference to the problem of Figure 12.1, $S_{\eta o}$ is the starting region of the subdomain a. For instance if s_o is the nearest singularity line of $y_a(\eta, \eta')$ to B_θ (Figure 12.4), the starting region is the region $S_{\eta o} = U \cup u$ where u is the region of $\bar{U}$ between B_θ and s_o. We recall that $\mathbf{V}_+(\eta)$ is expressed by (12.75) in region U and (12.72) in region u.

To obtain the missing spectral region of the principal sheet of η-plane, i.e. $U \cup \bar{U} - S_{\eta o}$, we can resort to the theory of sectional analytic functions [20] where the discontinuity of $\psi_{ij}(\eta)$ due to discontinuity line is evaluated. This method has been accomplished directly in w-plane in the study by Daniele *et al.* [15].

An alternative and probably more effective method is to resort to recursive equations in the w-plane, and this is the strategy applied in this work.

To get simple analytical continuation, it is convenient to define the spectra in the w-plane and therefore we introduce the following notation:

$$\hat{F}_+(w) = F_+(-k\cos w), \quad \hat{F}_d(w) = \sin w \hat{F}_+(w) \tag{12.76}$$

A fundamental property of the plus functions in the w-plane is that they are even function of w, i.e. $\hat{F}_+(w) = \hat{F}_+(-w), \hat{F}_d(-w) = -\hat{F}_d(w)$ [1]. This property allows to immediately double the strips where the spectrum is known. Another advantage is the possibility to immediately eliminate the minus functions in the WH equation (see [6–17]).

Starting from the GWHE of the angular region (a) (12.13), after some mathematical manipulation in w-plane, we obtain the following equation that contains only the plus spectra with the elimination of $I_{a+}(\eta)$:

$$\hat{I}_{1+}(w) = -Y_o\hat{V}_{1d}(w) + \hat{I}_{1+}(w - 2\Phi_a) - Y_o\hat{V}_{1d}(w - 2\Phi_a) \tag{12.77}$$

Similar equations can be obtained for the regions (b) and (c). Algebraic manipulations similar to those reported by the authors in previous works yield the general recursive equations for the problem of Figure 12.1:

$$\begin{aligned}\hat{V}_{1d}(w) = \; & \frac{1}{2}(-Z_o\hat{I}_{1+}(w+2\pi) + Z_o\hat{I}_{1+}(w+2\Phi_a) + \hat{V}_{1d}(w+2\pi) - \hat{V}_{1d}(w+2\Phi_a) \\ & - e^{-jks\cos w + jkd\sin w}(-Z_o\hat{I}_{2+}(w+2\pi) + Z_o\hat{I}_{2+}(w+2\pi) + \hat{V}_{2d}(w+2\pi) \\ & + \hat{V}_{2d}(w+\pi)))\end{aligned} \tag{12.78}$$

$$\begin{aligned}\hat{I}_{1+}(w) = \; & \frac{1}{2Z_o}(\hat{I}_{1+}(w+2\pi) + Z_o\hat{I}_{1+}(w+2\Phi_a) - \hat{V}_{1d}(w+2\pi) - \hat{V}_{1d}(w+2\Phi_a) \\ & - e^{-jks\cos w + jkd\sin w}(-Z_o\hat{I}_{2+}(w+2\pi) + Z_o\hat{I}_{2+}(w+2\pi) + \hat{V}_{2d}(w+2\pi) \\ & + \hat{V}_{2d}(w+2\pi)))\end{aligned} \tag{12.79}$$

$$\begin{aligned}\hat{V}_{2d}(w) = \; & \frac{1}{2}(Z_o\hat{I}_{2+}(w+2\pi) - Z_o\hat{I}_{2+}(w+2\pi) + \hat{V}_{2d}(w+2\pi) - \hat{V}_{2d}(w+2\pi) \\ & - e^{-jks\cos w + jkd\sin w}(Z_o\hat{I}_{1+}(w+2\pi) - Z_o\hat{I}_{1+}(w+2\Phi_a) + \hat{V}_{1d}(w+2\pi) \\ & + \hat{V}_{1d}(w+2\Phi_a)))\end{aligned} \tag{12.80}$$

$$\begin{aligned}\hat{I}_{2+}(w) = \; & \frac{1}{2Z_o}(Z_o\hat{I}_{2+}(w+2\pi) + Z_o\hat{I}_{2+}(w+2\pi) + \hat{V}_{2d}(w+2\pi) + \hat{V}_{2d}(w+2\pi) \\ & - e^{-jks\cos w + jkd\sin w}(Z_o\hat{I}_{1+}(w+2\pi) - Z_o\hat{I}_{1+}(w+2\Phi_a) + \hat{V}_{1d}(w+2\pi) \\ & + \hat{V}_{1d}(w+2\Phi_a)))\end{aligned} \tag{12.81}$$

Based on their definition and through numerical tests, we note that the recursive equations like (12.78)–(12.81) allow the evaluation of the spectra for arbitrary values of w provided the spectrum is known in the fundamental strip S_f:$-\pi < \text{Re}[w] < \pi$. Let us label S_o the image of $S_{\eta o}$ in w-plane.

To obtain $S_o \supseteq S_f$, in several works we have used different formulations of the WHF technique, using a new complex plane $\bar{w} = \pi w/\Phi$ [6–8] for each of the angular regions of the problem. Resorting to these variables, we prevent to have singularity lines that yield $S_o \subset S_f$. This definition constituted an important tool that has allowed to solve, for example, the dielectric wedge problem [7], the PEC problem in the presence of an infinite dielectric half-space [11,13] and concave impenetrable wedge [6, 21].

However when the angular regions are coupled with finite layer regions, there is one main reason that discourage the introduction of complex planes like $\bar{w}$: due to the presence of finite layers, infinite modal poles arise.

An alternative and probably more effective method is based on the study of properties in w-plane that allows to a validity spectral region S such that $S \supseteq S_f$.

With respect to η-plane, in the w-plane the singularity lines of the kernel assume simple forms (each line is parallel to each other). To locate them, we indicate with w_r their trace on the real axis $\text{Im}[w] = 0$. For instance, the line $\text{B}_\theta^{\text{W}}$ is identified by $w_i = -\pi/2$ when $\theta = \pi/4$. From here on, we also use the notation $w_m \subset w \subset w_n (m < n)$ to identify the (undulate) strip in the w-plane delimited by the lines w_m and w_n (see, e.g. Figure 13 of [6]). In particular, we note that all the singularity lines are parallel to the image $\text{B}_\theta^{\text{W}}: w' = -\arccos[-v\exp(j\theta)], v \in R$ of the integration line $\text{B}_\theta: \eta' = kv\exp(j\theta), v \in R$, since for an angular region of aperture Φ, they are defined by ($w' \subset \text{B}_\theta^{\text{W}}$), the integer $n_t = n_t(\Phi) = \lfloor \pi/(2\Phi) \rfloor$ with $w = -w' + 2n\,\Phi, n = -1, -2, \ldots -n_t, \pm(n_t + 1), \pm(n_t + 2), \ldots$ and $w = w' + 2n\Phi, n = \pm 1, \pm 2, \ldots$ We label with $-w_o: -\pi < -w_o < -\pi/2$ the image of the singularity line nearest B_θ (Figure 12.4), the starting strip is defined by $-w_o \subset w \subset 0$ and taking into account the symmetry by $S_o: -w_o \subset w \subset w_o$. It is possible to show that $w_o = w_o(\Phi)$ is given by the following equation and plotted in Figure 12.5:

$$w_o(\Phi) = -\frac{\pi}{2} + 2\Phi + 2\Phi\, n_t(\Phi) \tag{12.82}$$

We observe that the layers (b) and (c) do not introduce kernel singular lines in the principal sheet of η. It means that (12.34), and (12.41), (12.46) rewritten in the w-plane are valid in the fundamental strip S_f:

$$\hat{I}_{2+}(w) = Y_o V_{2d}(w) + \Delta_c(w), w \in S_f \tag{12.83}$$

$$\begin{vmatrix} -\hat{I}_{1+}(w) \\ \hat{I}_{2+}(w) \end{vmatrix} = -\frac{\hat{\mathbf{Y}}(w)}{\sin w} \begin{vmatrix} \hat{V}_{1d}(w) \\ \hat{V}_{2d}(w) \end{vmatrix} + \begin{vmatrix} \Delta_1(w) \\ \Delta_2(w) \end{vmatrix}, w \in S_f \tag{12.84}$$

where the three quantities $\Delta_c(w), \Delta_1(w), \Delta_2(w)$ are integral terms known in S_f. Conversely, the kernel $y_a(-k\cos w, \eta')$ related to region (a) presents singularity lines in $w \in S_f$ that yields the starting strip $S_o: -w_o \subset w \subset w_o$ (Figure 12.5). By using (12.77), we extend the validity strip to $-w_1 \subset w \subset w_1$ and we observe that while the functions $\hat{I}_{1+}(w), \hat{V}_{1d}(w)$ are known in S_o, the term

$$\begin{aligned} \Delta_a(w) &= \hat{I}_{1+}(w - 2\Phi_a) - Y_o \hat{V}_{1d}(w - 2\Phi_a)\Delta_a(w) \\ &= \hat{I}_{1+}(w - 2\Phi_a) - Y_o \hat{V}_{1d}(w - 2\Phi_a) \end{aligned} \tag{12.85}$$

is known in $w_{\Delta l} \subset w \subset w_{\Delta r}$ where $w_{\Delta l} = -w_o + 2\Phi$, $w_{\Delta r} = w_o + 2\Phi$ are functions of Φ as reported in Figure 12.5.

Taking into account (12.77) with (12.85), (12.83), (12.84), we can eliminate $\hat{I}_{1+}(w), \hat{I}_{2+}(w)$:

$$\hat{V}_{1d}(w) = \hat{Z}^e(w)\Delta_a(w) + \hat{Z}^e_m(w)e^{-j\cos wks}\Delta_c(w) + N_{w1}(w) \tag{12.86}$$

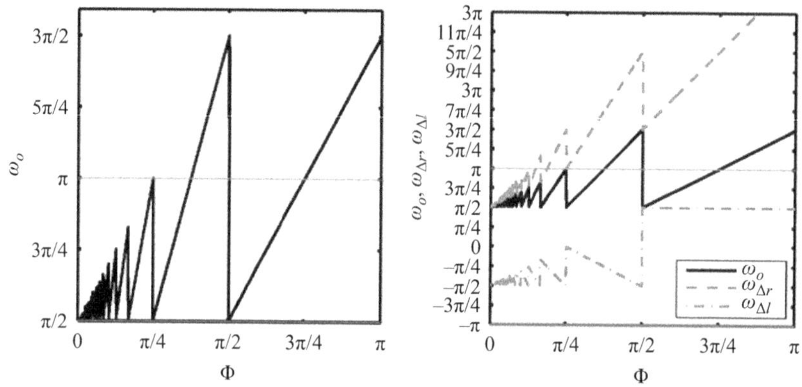

Figure 12.5 Left: Values of w_o of region the starting region S_o:$-w_o \subset w \subset w_o$ as a function of Φ. Right: Plots of $w_{\Delta r} = w_o + 2\Phi$, $w_{\Delta l} = -w_o + 2\Phi$ that delimits the validity of $\Delta_a(w)$:$w_{\Delta l} \subset w \subset w_{\Delta r}$ as a function of Φ.

$$\hat{V}_{2d}(w) = \hat{Z}^e(w)\Delta_c(w) + \hat{Z}^e_m(w)e^{j\cos wks}\Delta_a(w) + N_{w2}(w) \tag{12.87}$$

where $\mathbf{N}_w(w) = \left| N_{w1} \; N_{w2} \right|^t = \left(\frac{\hat{\mathbf{Y}}(w)}{\sin w} + Y_o\mathbf{1}\right)^{-1}\begin{vmatrix}\Delta_1(w)\\ \Delta_2(w)\end{vmatrix}$ is known for every value of w of the fundamental strip $w \subset S_f$.

We note that in the coupled equations (12.86) and (12.87) while $\Delta_c(w)$ is known in S_f, $\Delta_a(w)$ is known only in the strip $-w_o + 2\Phi \subset w \subset w_o + 2\Phi$. Finally, $\hat{V}_{1d}(w)$ and $\hat{V}_{2d}(w)$ can be valued with (12.74) and/or (12.75) in the starting strip S_o and with (12.86), (12.87) in the strip $w_o \subset w \subset w_o + 2\Phi \doteq w_1$. By applying symmetry, the validity strip can be enlarged to $-w_1 \subseteq w \subset w_1$. Recursive iterations further enlarge the validity strip up to the necessary S_f. The earlier method can be generalized in the presence of two or more angular regions, although this generalization can be cumbersome. Finally, we note that the starting spectrum usually assume small values in the region $S_f - S_o$ and therefore as first approximation we can omit the procedure presented in this section and directly apply the general recursive equations reported in (12.78)–(12.81).

12.6 A novel test case

To demonstrate the efficiency and the methodology of the proposed method, we present an original test case: the diffraction by a PEC wedge in the presence of a staggered PEC half-plane shown in Figure 12.1. The discretize version of the equations of this works allows to obtain an approximate solution in terms of plus spectral unknowns. With reference to Figure 12.1, several numerical simulation have been worked out with different values of the parameters Φ_a, φ_o, d, s. For space reason, we will not discuss here the details of these simulations in terms of convergence. In this study, we present effective results for a particular set of parameters in terms of excitation of modal fields in the planar waveguide ($x_1 < 0, -d < y_1 < 0$), the diffraction coefficients and the computation of total field in regions (a) and (c). The complete deduction formulas to evaluate the physical quantities reported earlier are also reported in various works (see, for instance, [6, 7, 11–15]). In particular, the excitation coefficients of the H_i modes in the planar waveguide ($x_1 < 0, -d < y_1 < 0$) with propagation constants $\eta_i = \sqrt{k^2 - (i\pi/d)^2}, i \in \mathbb{N}_0$ can be obtained from (12.22) [12]:

$$c[-\eta_i] = \mathrm{Res}[Z_o I_{1-}(\eta), -\eta_i] = -j\left(\frac{i\pi}{kd}\right)^2 \frac{k^2}{kd\eta_i}[V_{1+}(-\eta_i) + (-1)^{i-1}V_{2+}(-\eta_i)e^{-j\eta_i s}] \tag{12.88}$$

The radiation in the angular regions can be obtained by the saddle point method:

$$E_z^d(\rho, \varphi) = E_o\sqrt{\frac{1}{2\pi k\rho}}e^{-j(k\rho+\pi/4)}D(\varphi, \varphi_o), \quad k\rho \ll 1 \tag{12.89}$$

where $D(\varphi, \varphi_o)$ is the diffraction coefficient. For instance in region (a) we have [6, 8] $D(\varphi, \varphi_o) = -k\hat{V}_{\varphi d}(-\pi)/(jE_o)$ with

$$\hat{V}_{\varphi d}(w) = [Z_o(\hat{I}_{1+}(w - \varphi) - \hat{I}_{1+}(w + \varphi)) + \hat{V}_{1d}(w - \varphi) + \hat{V}_{1d}(w + \varphi)]/2 \tag{12.90}$$

The following numerical results are related to the parameters $E_o = 1V/m, \Phi_a = 0.8\pi, \varphi_o = 0.55\pi, kd = 5.5\pi, ks = 10, k = 1 - j10^{-7}$. Table 12.1 reports the excitation coefficient of the first 6 H_n modes (only 5 are propagating), Figure 12.6 the intensity of the first 30 excitation coefficients, Figure 12.7 the diffraction coefficient in dB for the regions (a) and (c), finally Figure 12.8 reports the total far field in regions (a) and (c).

Table 12.1 Excitation coefficients for $E_o = 1V/m, \Phi_a = 0.8\pi, \varphi_o = 0.55\pi, kd = 5.5\pi, ks = 10, k = 1 - j10^{-7}$

Mode	η_i	$Z_o\, c[-\eta_i]$
H_1	0.983332	$0.0000807492 + j0.0000587773$
H_2	0.931541	$-0.00328192 + j0.000137413$
H_3	0.83814	$-0.0145758 + j0.015822$
H_4	0.686349	$-0.0145215 + j0.022461$
H_5	0.416598	$-0.158276 - j0.000886479$
H_6	$-j0.435985$	$0.40502 - j0.0918935$

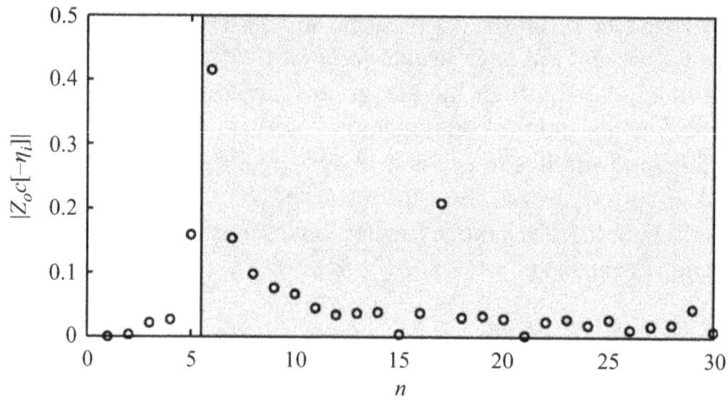

Figure 12.6 Intensity of the first 30 excitation coefficients (12.88). Grey background for not propagating modes.

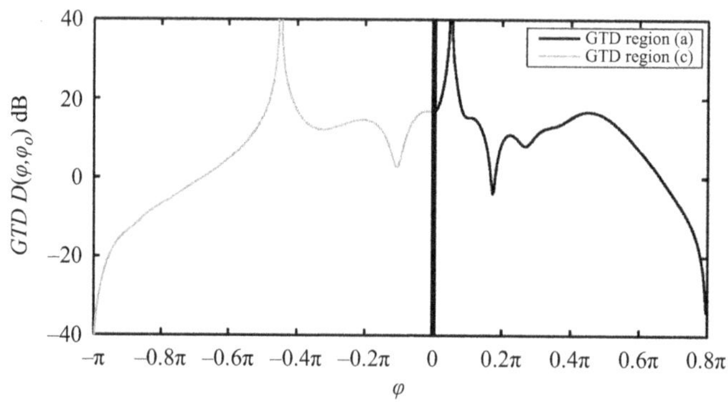

Figure 12.7 GTD diffraction coefficients for regions (a) and (c). Note that the two coefficients are computed with respect to the two reference systems reported in Figure 12.1.

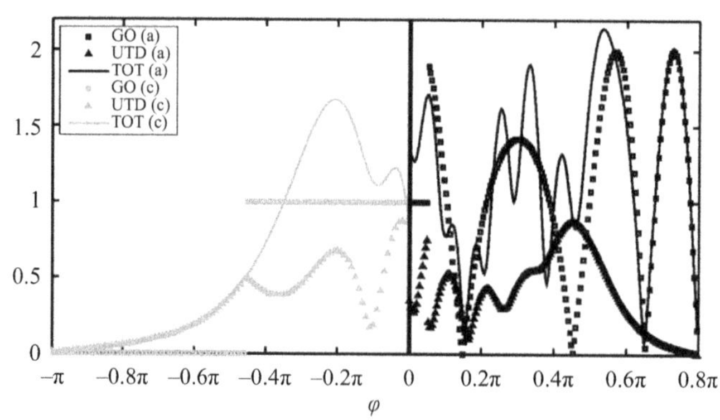

Figure 12.8 GO, UTD and total far field at $k\rho_1 = 10$, $k\rho_2 = 10$ for regions (a) and (c), respectively, taking into consideration the two reference systems as reported in Figure 12.1.

12.7 Conclusion

The GWHT is a powerful technique that has been successfully used in important wedge problems [1–17, 21]. In this study, we introduce a general methodology that provides the steps to obtain the FIE of a general scattering problem constituted of PEC planar and angular regions immersed in free space. Extension to the cases where general impenetrable as well as penetrable wedges or not free space angular and/or planar regions is possible without the introduction of new concepts or ideas. Although the methodology is similar, in more general cases the relevant FIEs contains more complicated kernels with different branch points and/or not simple elementary functions. For instance, Malyuzhinets function is involved when some face of the angular regions is defined by a surface impedance. Future studies will present the extension of this method to these more general cases.

Appendix A

While the spectrum $V_{2+}(\eta')$ has been defined such that it is vanishing at infinity in whole the complex plane, the function $V_{2+}(\eta)e^{j\eta s}$ diverges in the half-plane $\mathrm{Im}(\eta) < 0$ because it is the Laplace transform of a retarded function. Since in the (12.36) we close the smile line with a lower half circle Γ with radius $|\eta'| \to \infty$, we need to demonstrate if the integral $\int_\Gamma \frac{Y_{12}(\eta')V_{2+}(\eta')e^{j\eta' s}}{\eta'-\eta}d\eta'$ vanishes as $|\eta'| \to \infty$.

To ascertain it directly, we must take into account that $|Y_{12}(\eta)| \to \sqrt{|\eta|}e^{-|\eta|d}$ as $|\eta| \to \infty$. Therefore, $|\int_\Gamma \frac{Y_{12}(\eta')V_{2+}(\eta')e^{j\eta' s}}{\eta'-\eta}d\eta'| \to \frac{\sqrt{|\eta'|}}{|\eta'|}e^{-|\eta|d}\int_{-\pi}^{0} e^{+|\eta|\sin\vartheta s}d\vartheta$ $|\int_\Gamma \frac{Y_{12}(\eta')V_{2+}(\eta')e^{j\eta' s}}{\eta'-\eta}d\eta'| \to \frac{\sqrt{|\eta'|}}{|\eta'|}e^{-|\eta|d}\int_{-\pi}^{0} e^{+|\eta|\sin\vartheta s}d\vartheta$ with $\eta = |\eta|e^{j\vartheta}$ on Γ and it yields $\int_{-\pi}^{0} e^{+|\eta|\sin\vartheta s}d\vartheta = \pi(I_o(|\eta|s) - L_o(|\eta|s))$, where $I_o(x), L_o(x)$ are the modified Bessel and Struve function of order 0 with $I_o(x), L_o(x) I_o(x) - L_o(x) = O(x^{-1})$. It means that although $Y_{12}(\eta')V_{2+}(\eta')e^{j\eta' s}$ diverges for directions θ such that $d/s > \tan\theta$, $|\eta| \to \infty$ the integral is convergent, i.e. $\int_\Gamma \frac{Y_{12}(\eta')V_{2+}(\eta')e^{j\eta' s}}{\eta'-\eta}d\eta' \to 0$.

References

[1] Daniele V.G., 'The Wiener–Hopf technique for impenetrable wedges having arbitrary aperture angle'. *SIAM Journal of Applied Mathematics*. 2003; 63: 1442–1460.

[2] Daniele V.G. and Lombardi G., 'The Wiener–Hopf technique for impenetrable wedge problems'. *Proc. Days Diffraction Int. Conf.*, Saint Petersburg, Russia, June 2005, pp. 50–61.

[3] Daniele V.G., 'The Fredholm factorization', Report DET-2003-1, available at http://personal.delen.polito.it/vito.daniele/

[4] Daniele V.G. and Lombardi G., 'Fredholm factorization of Wiener–Hopf scalar and matrix kernels'. *Radio Science*. 2007; 42: RS6S01. doi:10.1029/2007RS003673

[5] Daniele V.G. and Zich R.S., *The Wiener Hopf Method in Electromagnetics*, Scitech Publishing, Edison, NJ, USA, 2014.

[6] Daniele V.G. and Lombardi G., 'Wiener–Hopf solution for impenetrable wedges at skew incidence'. *IEEE Transactions on Antennas and Propagation*. 2006; 54: 2472–2485.

[7] Daniele V.G., 'The Wiener–Hopf formulation of the penetrable wedge problem. Part III: The skew incidence case'. *Electromagnetics*. 2011; 31(8): 550–570.

[8] Daniele V.G., 'The Wiener–Hopf formulation of the dielectric wedge problem. Part I'. *Electromagnetics*. 2010; 30(8): 625–643.

[9] Daniele V.G., 'The Wiener–Hopf formulation of the dielectric wedge problem. Part II'. *Electromagnetics*. 2011; 31(1): 1–17.

[10] Daniele V.G., 'The Wiener–Hopf formulation of the penetrable wedge problem. Part III: The skew incidence case'. *Electromagnetics*. 2011; 31(8): 550–570.

[11] Daniele V.G., 'Electromagnetic fields for PEC wedge over stratified media. Part I'. *Electromagnetics*. 2013; 33: 179–200.

[12] Daniele V.G., 'Diffraction by two wedges', Report DET-2014-1, available at http://personal.delen.polito.it/vito.daniele/

[13] Daniele V.G. and Lombardi G., 'Arbitrarily oriented perfectly conducting wedge over a dielectric half-space: diffraction and total far field'. *IEEE Transactions on Antennas Propagation*. 2016; 64(4): 1416–1433.

[14] Daniele V.G., 'Diffraction by two wedges (Appendices)', Report DET-December 2016, available at http://personal.delen.polito.it/vito.daniele/

[15] Daniele V.G., Lombardi G., and Zich R.S., 'Network representations of angular regions for electromagnetic scattering'. *PLOS ONE*. 2017; 12(8): e0182763, 1–53. https://doi.org/10.1371/journal.pone.0182763

[16] Daniele V.G., Lombardi G., and Zich R.S., 'The electromagnetic field for a PEC wedge over a finite dielectric slab: Part I – formulation and validation'. *Radio Science*. 2017; 52(12): 1472–1491. doi: 10.1002/2017RS006355

[17] Daniele V.G., Lombardi G., and Zich R.S., 'The electromagnetic field for a PEC wedge over a finite dielectric slab: Part II – diffraction, modal field, surface waves and leaky waves'. *Radio Science*. 2017; 52(12): 1492–1509. doi: 10.1002/2017RS006388.

[18] Budaev B. *Diffraction by Wedges*. London, UK: Longman Scientific; 1995.

[19] Kantorovich L.V. and Krylov V.I. *Approximate Methods of Higher Analysis*. Groningen, the Netherlands: P. Noordhoff, 1964.

[20] Gakhov F.D., *Boundary Value Problems*. New York: Dover Publications, Inc., 1966.

[21] Lombardi G., 'Skew incidence on concave wedge with anisotropic surface impedance'. *IEEE Antennas and Wireless Propagation Letters*. 2012; 11: 1141–1145.

Chapter 13

On the analytical regularization method in scattering and diffraction

Yury A. Tuchkin[1]

13.1 Introduction

This chapter describes the principal ideas and more advanced techniques of the analytical regularization method (ARM) and the semi-inversion procedure (SIP) particularly as they apply to the integral and integro-differential equations that arise in scattering and diffraction problems in electromagnetics.[1]

The application of the SIP in diffraction problems was placed on a firm foundation nearly 50 years ago (see [1–4]) and was followed by the development of the ARM some 30 years later (see [5–13]). However, in the intervening period and later, a number of publications have appeared in which these methods do not seem to have been properly applied (including mismatching of the SIP and ARM), possibly because these methods draw on some relatively sophisticated application of functional analysis including topics from the theory of numerical methods, the theory of distributions (generalized functions), Sobolev spaces, Tikhonov regularization and the propagation and amplification of round-off errors in numerical processes (and even details of the rounding algorithm employed by arithmetic processors on various platforms).

The principal advantage of the SIP and ARM lies in the transformation of the underlying operator equations arising in diffraction theory – which typically are ill-posed for the class of problem considered – to well-posed operator equations, so that on discretization the linear algebraic systems of equations are well-conditioned, whereas a direct discretization of the original operator equation normally results in an ill-conditioned system that is unstable, and hence unreliable for numerical purposes. The difficulties associated with ill-posed diffraction problems are analysed, for example, in the study by Hsiao and Kleinman [14].

The purpose of this chapter is to present a clear and unified treatment of the ARM and SIP approaches that draws on some recent developments using the language of pseudo-differential operators, while clarifying its rigorous application to boundary value problems (BVP) of electromagnetics.

[1]Department of Electronics Engineering, Gebze Technical University, Gebze, Turkey

Section 13.2 discusses the origin of the numerical instability that arises in the solution of an infinite linear algebraic system by the process of truncation to systems of finite rank. The next section explains the principal ideas of the ARM and the SIP, their points of commonality and the advantage of the ARM over the SIP. Section 13.4 describes the rather general and relatively new description of the BVP that gives a solid theoretical background for the regularization of the associated operator. A new treatment of the corresponding operators as special pseudo-differential operators brings out the underlying generality and simplifies the construction of the necessary regularizations. The following two sections present the application of the ARM to the standard Dirichlet and Neumann BVPs, and to diffraction by semi-transparent obstacles (i.e. those with an interface boundary condition). Section 13.7 describes the use of the ARM in diffraction problems employing complex-valued wave numbers and frequencies, and the associated spectral theory of relevance to cavities with a small aperture. Section 13.8 presents some considerations for the practical implementation of the ARM, while Section 13.9 concludes with a brief survey of the most important applications of the ARM.

13.2 Instability in the numerical solution of infinite algebraic systems

Let B be an integral or differential–integral operator that appears in diffraction theory. In many publications, it is assumed (sometimes implicitly) that there is an operator B given and defined on some space W_B, that is,

$$B : W_B \to W_B \tag{13.1}$$

and this operator arises from the requirement to solve a functional equation of the kind

$$Bz = f, z, f \in W_B \tag{13.2}$$

with an unknown vector or function z, and a given right-hand side f. In many publications, W_B is not defined, or if it is, the choice of W_B as the space of square-summable sequences l_2 or the space of square-integrable functions L_2 has bad consequences [14]. Typically, the space W_B is infinite-dimensional. Since we cannot solve numerically an infinite-dimensional equation, finite-dimensional approximations must be used. The corresponding 'algebraization' of (13.2) can be effected by various direct methods: moment methods, finite difference methods, finite element methods and related methods. This leads to a set of algebraic equations

$$A_N x_N = b_N, x_N, \qquad b_N \in E_N, \qquad N = 1, 2, 3, \ldots, \tag{13.3}$$

where E_N is the standard real or complex-valued Euclidean space of dimension N. The matrix A_N and (13.3) directly or indirectly are truncations to a finite range of some infinite-dimensional matrix operator A and operator equation

$$Ax = b. \tag{13.4}$$

The solutions x_N are supposed to be approximations of x, which are desired to be more and more accurate, as $N \to \infty$. Unfortunately, such an optimistic expectation is not always correct. It may happen that numerical solutions $\tilde{x}_N$ of (13.3) are far from, or even have nothing common with the exact solutions of both (13.3) and (13.4). The main reason for this problem is well-understood to be the presence of round-off errors, their propagation and amplification in the computational process for matrices that are ill-conditioned. Whatever method of numerical solution is chosen (Gaussian elimination, LU decomposition or various iterative methods), there is no qualitative difference from the point of view of accuracy and stability of the solution. A fundamental investigation of this problem has been made in the study by Wilkinson [15]. In particular, it is shown that the number of correct binary figures m_e in the numerical solution $\tilde{x}_N$ typically equals

$$m_e(N) = m_c - \log_2 \nu (A_N), \tag{13.5}$$

where m_c is the binary length of the computer mantissa and $\nu (A_N)$ is the condition number of A_N given by

$$\nu (A_N) = \|A_N\| \cdot \left\|A_N^{-1}\right\|, \tag{13.6}$$

where the norm of the matrices corresponds to the Euclidean metric of real or complex-valued E^N. Thus, the bigger $\nu (A_N)$, the less accurate is the approximation $\tilde{x}_N$ of x_N. The situation in which $\|\tilde{x}_N - x_N\| > \|x_N\|$ is known as 'numerical catastrophe', and it occurs when $m_c < \log_2 \nu (A_N)$. Exceptions to this rule may apply for matrices of special shape such as diagonal and triangular [15]. Estimate (13.5) is derived under the assumption that the relative errors of the matrix and vector of the right-hand side are smaller than machine epsilon ε_∞. When this is not the case, estimate (13.5) is often too optimistic.

Consider a sequence of matrices A_N, as $N \to \infty$. For the sake of simplicity, we consider the most important situation when the matrices A_N are the finite-dimensional truncations of an infinite-dimensional matrix A and arising from solving an equation

$$Ax = b. \tag{13.7}$$

Instead of the original, one can try to solve numerically the system $A_N x_N = b_N$, hoping that the solutions $\tilde{x}_N$ are good enough approximations to x, at least for sufficiently big N. Three fundamental questions arise:

Q1. Does $\tilde{x}_N$ converge to x?
Q2. If the answer to Q1 is positive, can $\tilde{x}_N$ approximate x with accuracy near machine epsilon?
Q3. Does the approximation of x by $\tilde{x}_N$ become better and better, as $N \to \infty$?

Unfortunately, the answers to all these questions are negative in general, and, in particular, for matrices A arising from very many problems of diffraction theory.

Let us consider a special kind of matrix A in the space l_2:

$$A = I + H, \tag{13.8}$$

where I is the identity and H is a matrix-operator compact in l_2.

Such a matrix A obeys the Fredholm alternative, and if it is not singular, the equation

$$(I + H)x = b \tag{13.9}$$

has a unique solution. Let H_N be a sequence of finite rank matrix operators satisfying $\|H - H_N\| \to 0$. Then the solutions x_N to the truncated system obey $\|x - x_N\| \to 0$ as $N \to \infty$, and Q1 now has a positive answer. In the space l_2, the limit

$$\nu_\infty = \nu_\infty(A) = \lim_{N\to\infty} \nu(A_N) \tag{13.10}$$

exists and is finite. Consequently, all of $\nu(A_N)$ are uniformly bounded for large enough N. Thus, formula (13.5) where $\nu(A_N)$ is replaced by $\nu_\infty(A)$ remains valid for all big enough N For most practical problems in diffraction theory, the value of ν_∞ does not exceed a hundred in non-resonant cases (of course, it can be arbitrarily large near a resonance frequency where A becomes nearly singular).

Equation (13.9) is known as one of the *second kind*. When it is not representable in the form (13.9), the general equation (13.4) is classified as one of the *first kind*. Qualitative plots of the correlation of numerical solution error and condition numbers of equations of the first and second kinds in the space l_2 are shown in Figure 13.1.

The graphs (a) and (b) correspond to an equation of the first kind when the condition number tends to infinity. The error $\|x - \tilde{x}_N\|$ typically decays until some value at $N = N_0$, due to better approximation of A, but the growth of the condition number leads to an unbounded increase of the error for $N > N_0$ and when $N \to \infty$. The graphs (c) and (d) correspond to an equation of the second kind, having a uniformly bounded condition number for $N \to \infty$. In according with formulae (13.5) and (13.10), the condition number shown in graph (c) stabilizes at the value ν_∞.

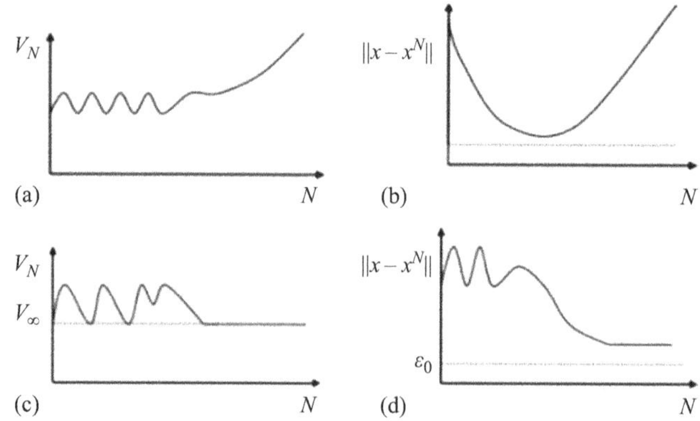

Figure 13.1 Condition number and numerical solution error of truncated systems as functions of the truncated matrix dimension N: (a) and (b) for equations of the first kind and (c) and (d) for equations of the second kind

It is necessary to emphasize the importance of (13.9) being of the second kind in the space l_2, where computer arithmetic provides numerical stability and convergence of $\tilde{x}_N$ to x within the small deviations mentioned earlier. For any other space, we may have the same qualitative characteristics of condition numbers and exact solutions x_N, but the computer 'treats' such equations as one in the space l_2, although the system is not precisely of the second kind in the space l_2. In such a case, the numerical solutions $\tilde{x}_N$ become unstable and divergent. If the space is 'close' to l_2, this undesirable effect may be mitigated and the numerical catastrophe may only appear for an extremely small predetermined error ε and a rather large $N = N(\varepsilon)$. Typically, it is hard to understand in this situation whether or not the solution is the result of the numerical catastrophe.

What to do if we need to solve an equation of the first kind? The most general tool is Tikhonov regularization [16]. This tool is powerful, but sometimes too roughly approximative, it needs non-trivial skill in its usage, and rather often, it does not give all the necessary qualitative features of the solutions. The alternative discussed herein is based on the idea of the transform of (13.4) of the first kind to an equivalent equation of the second kind (13.9). We understand any two equations to be equivalent if there is one to one correspondence between solutions of the equations.

We use the terminology of Fredholm operators. The standard definition of the Fredholm operator can be found in almost every book on functional analysis. Nevertheless, a simpler and more useful description is given by the Nikolskii–Atkinson criterion. Let H_1 and H_2 be Hilbert or Banach spaces. The operator $A : H_1 \to H_2$ is Fredholm if and only if it allows a representation $A = S + K$ where $K : H_1 \to H_2$ is a compact operator and both the operator $S : H_1 \to H_2$ and its inverse $S^{-1} : H_2 \to H_1$ are bounded. It follows from this criterion that the operators $S^{-1}K : H_1 \to H_1$ and $KS^{-1} : H_2 \to H_2$ are compact. Consequently, the operator S^{-1} is both a left-sided and right-sided regularizer for the operator A.

It is noteworthy that operators of integral or integral–differential equations arising in diffraction theory are typically of Fredholm type for a properly chosen pair H_1 and H_2, where $H_1 \neq H_2$. By contrast, the other typical situation occurs when $H_1 = H_2$ and the operator A is compact with unbounded inverse, or A is unbounded and its inverse is compact. A transformation to Fredholm type is not possible in the same space when (13.4) is not of Fredholm type. If such a transform from (13.4) to (13.9) were to be constructed in the same functional space, where (13.4) is not an equation of Fredholm type, then there are only two possibilities: the resulting equation is not equivalent to the initial one, or (13.9) is not of the second kind (i.e. operator H is not compact in the space).

A naive approach to the regularization described earlier by means of an operator such as S^{-1} is as follows. Let $D : H_2 \to H_1$ be any operator applied from the left to (13.4), producing the equation $(I + K_D)z = Db$, where $K_D = DA - I$. It might be argued that if D effectively inverts the main singularity of the kernel of the integral operator A then the operator K_D is compact, and so the transformed equation is stable for a satisfactory numerical procedure. However, the transformed operator is not of the second kind unless K_D is compact and, as mentioned earlier, this is in principle possible when A is of Fredholm type (but may not necessarily occur if the particular choice of D is based on inversion of the main singularity). Finally, it is important to note that K_D must be compact rather than merely bounded.

13.3 The ARM: when is it necessary?

In spite of the reservations in the previous section, the idea of the regularizing transform is useful and fruitful after some reformulation and understanding of its mathematical nature. First, one must reformulate (13.4) as one given on a pair of functional space H_1 and H_2, so that $b \in H_2$ and $A : H_1 \to H_2$. Second, the operator A must be the subject of special additive and multiplicative splitting requirements, namely,

$$A = A_0 + A_1 \tag{13.11}$$

where A_1 is subordinate to the operator $A_0 : H_1 \to H_2$. The spaces H_1 and H_2 should form a set of correctness of A_0 in the Tikhonov sense (see [16]). Third, the operator A_0 must have the representation $A_0 = L^{-1}R^{-1} : H_1 \to H_2$ where the operators

$$\begin{array}{l} R^{-1} : H_1 \to H_0, \;\; R : H_0 \to H_1 \\ L^{-1} : H_0 \to H_2, \;\; L : H_2 \to H_0 \end{array} \tag{13.12}$$

are known in closed analytical form. Here H_0 is some 'intermediate' space, and the operator $LA_1R : H_0 \to H_0$ is compact in H_0.

Due to the arguments given earlier about the advantage of posing equation (13.9) in the space l_2 and because of the reasons explained later, the best choice of the functional space H_0 is l_2. If the construction is implemented mathematically, it gives a direct and simple way of transforming (13.4) to (13.9): Introduce the new unknown $y = Rx$ and apply the operator L to (13.4) from the left to obtain.

$$(I + LA_1R)\,y = Lb, \quad (y, Lb \in l_2) \tag{13.13}$$

where the operator $K = LA_1R$ is compact in l_2.

This construction is known as the ARM, (13.13) is the regularized equation (13.4) and we refer to the pair (L, R) as a sandwich-type regularizer (because the sandwich-like construction of operator LAR possesses the desired property). Thus, the purpose of ARM is the transformation of (13.4) of the first kind to the equivalent one (13.9) of the second kind, thus providing a well-posed reformulation of (13.4).

The principal historical precursor of the ARM is the SIP [1–4]. It is a special form of the ARM when $R = I$ is the identity operator and L becomes a left-sided regularizer (or vice versa when $L = I$ and R is a right-sided regularizer). The graphical form of operator A from the ARM and the SIP points of view is illustrated in the commutative diagrams of Figure 13.2.

One can see that although the space H_0 may be freely chosen in the ARM, it should in fact be chosen as l_2. However, the SIP does not have this flexibility in the choice of the space H_0 not such a parameter and necessarily the space H_1 cannot be equal to l_2 or L_2.

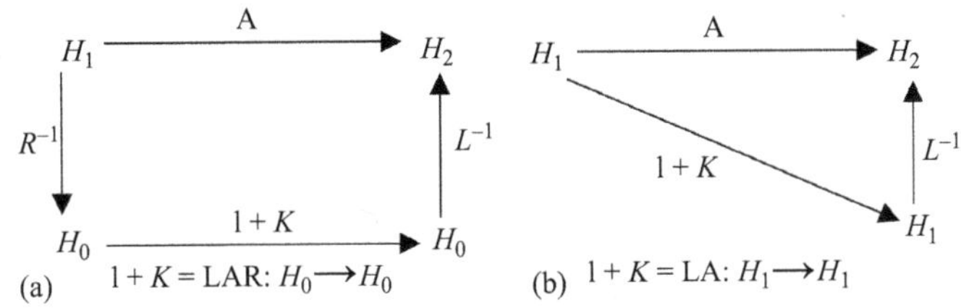

Figure 13.2 Graphical scheme of (a) ARM: $I+K=LAR: H_0 \rightarrow H_0$ and (b) SIP: $I+K=LA: H_I \rightarrow H_I$

13.4 Potentials and their pseudodifferential representations

Let us start with some necessary definitions. The Hilbert space $l_2(\lambda)$ of infinite sequences $\{c_n\}$. is

$$l_2(\lambda) = \left\{ \{c_n\}_{n=-\infty}^{\infty} : \sum_{n=-\infty}^{\infty} |c_n|^2 \tau_n^{2\lambda} < \infty \right\} \text{ where } \tau_n = \max\left(1, |n|^{1/2}\right) \quad (13.14)$$

with an obviously defined scalar product and norm. As well, we use the notation H^λ for the well-known Sobolev spaces of functions with Fourier coefficients belonging to $l_2(\lambda)$. For an arbitrary simple (and smooth) contour S, we understand the Sobolev spaces on the contour by means of its parametrization proportional to the contour arc-length.

Now, let a non-self-crossing contour S of class $C^{2,\alpha}$ be situated inside a two-dimensional open domain Ω. Here $C^{2,\alpha}$ denotes the class of twice differentiable functions with the second derivative of the Holder class (see [17]). For single- and double-layer potentials in Ω, we use the notations

$$P[v](q) \stackrel{def}{=} \int_S v(p)\, \varepsilon_2(q,p)\, dl_p, \quad q \in \Omega \text{ S} \quad (13.15)$$

$$Q[u](q) \stackrel{def}{=} \int_S u(p) \frac{\partial \varepsilon_2(q,p)}{\partial n_p} dl_p, \quad q \in \Omega S, \quad (13.16)$$

respectively, where $\varepsilon_2(q,p)$ is a Green's function of the domain Ω (which satisfies some boundary conditions on the boundary $\partial\Omega$). In particular, the possibility that $\Omega = R^2$ and $\varepsilon_2(q,p)$ satisfies the Sommerfeld radiation conditions or more general conditions, is included. The functions $u = u(p)$ and $v = v(p)$ satisfy the conditions $u \in C^{1,\alpha}(S), v \in C^{0,\alpha}(S)$ [17].

In a small enough open vicinity of S, any point $q \in V$ has unique representation $q = q_0 + hn_q$, where $q_0 \in S$, n_q is the unit outward normal to S at the point q_0 and $h \neq 0$. Thus, one can define in V the normal and tangential derivatives of the potentials P and Q that we denote with prefixes ∂_n and ∂_t, respectively. Let

$$[\partial_n Pv](q) = (\partial/\partial n_q)[Pv](q); \tag{13.17}$$

$$[\partial_t Pv](q) = (\partial/\partial t_q)[Pv](q); \tag{13.18}$$

$$[\partial_n Qu](q) = (\partial/\partial n_q)[Qu](q); \tag{13.19}$$

$$[\partial_t Qu](q) = (\partial/\partial t_q)[Qu](q), \tag{13.20}$$

where t_q is the unit tangential vector at the point $q_0 \in S$. For any function $f(q)$, we use the following notations for the limiting and direct values

$$f^{(\pm)}(q) = \lim_{h\to\pm 0} f(q \pm h n_q), \quad q \in S, \tag{13.21}$$

$$\bar{f}(q) = f(q), \quad q \in S \tag{13.22}$$

respectively, under the assumption that $f^{(\pm)}$ and $\bar{f}$ exist.

Relations between the limiting and direct values of the potentials (13.15)–(13.20) are the same as for the classic potentials [18, 19] but, of course, for $v(p)$ and $u(p)$ belonging to corresponding functional spaces:

$$\left[P^{(+)}v\right](q) = \left[P^{(-)}v\right](q) = \left[\bar{P}v\right](q), \quad q \in S \tag{13.23}$$

$$\left[\partial_n P^{(\pm)}v\right](q) = \left[\overline{\partial_n P}v\right](q) \pm \frac{1}{2}v(q), \quad q \in S \tag{13.24}$$

$$\left[\partial_t P^{(\pm)}v\right](q) = \left[\overline{\partial_t P}v\right](q), \quad q \in S \tag{13.25}$$

$$\left[Q^{(\pm)}u\right](q) = \left[\bar{Q}u\right](q) \mp \frac{1}{2}u(q), \quad q \in S \tag{13.26}$$

$$\left[\partial_n Q^{(+)}u\right](q) = \left[\partial_n Q^{(-)}u\right](q), \quad q \in S \tag{13.27}$$

$$\left[\partial_t Q^{(\pm)}u\right](q) = \left[\overline{\partial_t Q}\right](q) \pm \frac{1}{2}\frac{\partial}{\partial s}u(q), \quad q \in S \tag{13.28}$$

where $\partial/\partial s$ means the derivative with respect to arc-length on the contour, taken in the same direction as t_q. Formulae (13.23)–(13.28) are valid at least when $v \in C^{0,\alpha}(S)$ and $u \in C^{1,\alpha}(S)$ [17].

Let us assume now that the contour S is parametrized by a vector function $\eta(\theta) = (x(\theta), y(\theta))$, that $\eta(\theta) \in C^{2,\alpha}[-\pi, \pi]$ and in addition,

$$(\partial/\partial\theta)^k \eta(-\pi + 0) = (\partial/\partial\theta)^k \eta(\pi - 0), \kappa = 0, 1, 2.... \tag{13.29}$$

The parametrization must provide a one-to-one correspondence between points on the contour S and values $\theta \in (-\pi, \pi]$. A sufficient inequality to ensure this is

$$0 < l(\theta) < \infty, \text{ where } l(\theta) \overset{def}{=} \left\{[x'(\theta)]^2 + [y'(\theta)]^2\right\}^{1/2} \tag{13.30}$$

Let us introduce some notation to be employed. For any two functions $\alpha(\theta)$ and $\beta(\theta)$ of an arbitrary nature, we denote their product and composition, when it exists, by

$$\langle \alpha \cdot \beta \rangle (\theta) = \alpha(\theta)\, \beta(\theta), \tag{13.31}$$

$$(\alpha \circ \beta)(\theta) = \alpha(\beta(\theta)). \tag{13.32}$$

The notations F and F^{-1} are used the direct and inverse discrete Fourier transforms on $[-\pi, \pi]$, respectively. Finally, the matrices T and J are defined as the diagonal matrices

$$T = \mathrm{diag}\{\tau_n\}_{n=-\infty}^{\infty}, \tau_n = \max\left(1, |n|^{1/2}\right), \tag{13.33}$$

$$J = \mathrm{diag}\{\sigma_n\}_{n=-\infty}^{\infty}, \sigma_n = \mathrm{sign} n, n \neq 0, \sigma_0 = 1. \tag{13.34}$$

Adopting a pseudo-differential form of the integral and integro-differential operator in (13.23)–(13.28) we may conveniently express them as products of Fourier transforms (and their inverse) and some matrix operations. One can prove (see later) the existence of six matrices $M^P, M^Q, M^{\partial_n P}, M^{\partial_n Q}, M^{\partial_t P}, M^{\partial_t Q}$ such that the following identities hold:

$$\left[\overline{P}v\right] \circ \eta = -\frac{1}{2}\left[F^{-1}T^{-1}\left(I - 2M^P\right)T^{-1}F\left(l \cdot \langle v \circ \eta \rangle\right)\right], \tag{13.35}$$

$$\left[\overline{Q}u\right] \circ \eta = \left[F^{-1}T^{-1}M^Q TF\left(u \circ \eta\right)\right], \tag{13.36}$$

$$\left[\overline{\partial_n P}v\right] \circ \eta = \left\langle l^{-1} \cdot \left[F^{-1}TM^{\partial_n P}T^{-1}F\langle l \cdot (v \circ \eta)\rangle\right]\right\rangle, \tag{13.37}$$

$$\left[\partial_n Q^{(\pm)}u\right] \circ \eta = \frac{1}{2}\left\langle l^{-1} \cdot \left[F^{-1}T\left(I + 2M^{\partial_n Q}\right)TF\left(u \circ \eta\right)\right]\right\rangle, \tag{13.38}$$

$$\left[\partial_t Pv\right] \circ \eta = -\frac{1}{2}\left\langle l^{-1} \cdot \left[F^{-1}T\left(iJ - 2M^{\partial_t P}\right)T^{-1}F\langle l \cdot (v \circ \eta)\rangle\right]\right\rangle, \tag{13.39}$$

$$\left[\overline{\partial_t Q}u\right] \circ \eta = \left\langle l^{-1} \cdot \left[F^{-1}TM^{\partial_t Q}TF\langle u \circ \eta\rangle\right]\right\rangle. \tag{13.40}$$

The argument ϑ of the parametrization $\eta = \eta(\vartheta)$ is omitted in the identities (13.35)–(13.40). Moreover if $M = \{m_{sn}\}_{s,n=-\infty}^{\infty}$ denotes any of matrices $M^P, M^Q, M^{\partial_n P}, M^{\partial_n Q}, M^{\partial_t P}, M^{\partial_t Q}$, then

$$\sum_{s=-\infty}^{\infty}\sum_{n=-\infty}^{\infty}(1 + |s|)(1 + |n|)\,|m_{sn}|^2 < \infty. \tag{13.41}$$

The inequality (13.41) means that the coefficients m_{sn} are decaying when $s, n \to \infty$ even faster than that of a Hilbert–Schmidt matrix. In particular, such a matrix M defines a compact matrix operator in l_2.

To understand the representations (13.35)–(13.40), let us consider the simplest case of the potential $\bar{P}v$. Inserting the parametrization $\eta = \eta(\vartheta)$ into (13.15), it can be shown [12] that

$$[Pv \circ \eta](\theta) = \int_{-\pi}^{\pi} \varepsilon_2\left(\eta(\theta), \eta(\tau)\right) l(\tau)\, v\left(\eta(\tau)\right) d\tau, \theta \in [-\pi, \pi]. \tag{13.42}$$

The Green's function $\varepsilon_2(\eta(\theta),\eta(\tau))$ of two-dimensional free space admits the representation

$$\varepsilon_2(\eta(\theta),\eta(\tau)) = L(\theta-\tau) + p(\theta,\tau) \quad , \tag{13.43}$$

where

$$L(t) = \frac{1}{2\pi}\left\{-\frac{1}{2} + \ln\left|2\sin\frac{t}{2}\right|\right\} \tag{13.44}$$

$$= -\frac{1}{2}\sum_{n=-\infty}^{\infty}\frac{e^{in\theta}}{\tau_n^2}, t \in [-\pi,\pi]. \tag{13.45}$$

and the function $p(\theta,\tau)$ has continuous first partial derivatives and $\left(\partial^2/\partial\theta\,\partial\tau\right)p(\theta,\tau) \in L_2$; in particular, if p_{sn} denotes the Fourier coefficients of $p(\theta,\tau)$,

$$\sum_{s=-\infty}^{\infty}\sum_{n=-\infty}^{\infty}\tau_n^4\tau_s^4|p_{sn}|^2 < \infty \tag{13.46}$$

where the definition of τ_n is given in (13.33). The identity (13.35) and inequality (13.41) for the matrix $M = M^P$ follow immediately [12] from (13.42)–(13.46).

Turning to the remaining representations, the identity having the most complicated proof is (13.38). First of all [12],

$$\left(\partial_n Q^{(\pm)}u \circ \eta\right)(\theta) = A(\theta) + B(\theta), \tag{13.47}$$

with

$$A(\theta) = \frac{1}{l(\theta)}\frac{d^2}{d\theta^2}\int_{-\pi}^{\pi}u(\eta(\tau))\ln\left|2\sin\frac{\theta-\tau}{2}\right|d\tau + \frac{1}{l(\theta)}\int_{-\pi}^{\pi}u(\eta(\tau))d\tau, \tag{13.48}$$

and

$$B(\theta) = \frac{1}{l(\theta)}\int_{-\pi}^{\pi}u(\eta(\tau))W(\theta,\tau)\,d\tau, \tag{13.49}$$

where the only singularities occur in the kernel $W(\theta,\tau)$ and proportional to $\ln\left|2\sin\frac{\theta-\tau}{2}\right|$. Calculation of the limits $\partial_n Q^{(\pm)}u$ in (13.47)–(13.49) is relatively simple at a point on a locally flat part of the contour, when the contour is parametrized by its arc-length, but the proof becomes rather non-trivial in the general case [12, 20]. The operator $d^2/d\theta^2$ cannot be moved inside the first integral in (13.48) because the result is a divergent integral. Nevertheless, under the assumption $u \in C^{1,\alpha}(S)$, the expression for $A(\theta)$ is correct when differential and integral operators are applied in the written order. Due to the properties of $W(\theta,\tau)$ mentioned earlier, its Fourier coefficients w_{sn} satisfy the inequality

$$\sum_{s=-\infty}^{\infty}\sum_{n=-\infty}^{\infty}|w_{sn}|^2 < \infty. \tag{13.50}$$

Taking together (13.44) and (13.47)–(13.50), one arrives at the representation (13.38).

The other formulae, namely (13.36), (13.37) and (13.39), (13.40) can be proved in a very similar way as for (13.35) and (13.38).

13.5 Solution of the key diffraction problems

In this section, we consider two dimensional Dirichlet and Neumann diffraction BVP which are posed in a way similar to that studied by Colton and Kress [17].

13.5.1 Dirichlet BVP

Let S be a closed non-self-crossing contour in R^2. Suppose that S lies in $C^{2,\alpha}$, and that the contour is parametrized by the function $\eta(\theta) = (x(\theta), y(\theta))$, $\theta \in [-\pi, \pi]$ Let $V^{(-)}$ be an open bounded domain with boundary $\partial V^{(-)} = S$ and let $V^{(+)}$ be complementary to the domain $\overline{V}^{(-)} = V^{(-)} \cup S$ in R^2. Thus

$$\partial V^{(-)} = S; \quad V^{(+)} = R^2 \backslash \overline{V}^{(-)}. \tag{13.51}$$

The unknown function $u^s(q), q \in R^2 \backslash S$ (which has the physical sense of a scattered field) belongs to the intersection

$$C^2(R^2 \backslash S) \cap C^{1,\alpha}(\bar{V}^{(+)}) \cap C^{1,\alpha}(\overline{V}^{(-)}); \tag{13.52}$$

in particular, $u^s(q)$ and all its derivatives of the first order are continuous in $\overline{V}^{(+)}$and $\overline{V}^{(-)}$, but the limiting values $u^{s(+)}(q)$ and $u^{s(-)}(q)$ as well as $\partial_n u^{s(+)}(q)$ and $\partial_n u^{s(-)}(q)$ for $q \in S$ are not necessarily equal, where

$$\partial_n u^{s(\pm)}(q) = \left(\frac{\partial u^s}{\partial n_q}\right)^{(\pm)}. \tag{13.53}$$

The function $u^s(q)$ must satisfy the homogeneous Helmholtz equation in $V^{(\pm)}$, i.e.

$$(\Delta + k^2) u^s(q) = 0, \quad q \in R^2 \backslash S \tag{13.54}$$

Also, $u^s(q)$ must satisfy the well known Sommerfeld radiation condition

$$\lim_{|q| \to \infty} |q|^{1/2} \left(\frac{\partial u^s(q)}{\partial |q|} - iku^s(q)\right) = 0 \tag{13.55}$$

as well as the Dirichlet boundary condition

$$u^{s(+)}(q) = u^{s(-)}(q) = -u^i(q), \quad q \in S \tag{13.56}$$

where $u^i(q)$ is a known function (having the sense of incident field values on S).

Utilization of Green's formulae technique gives the identity [12, 17–19]

$$\left[\overline{P}(\delta\, \partial_n u)\right](q) = -u^{(i)}(q), \quad q \in S \tag{13.57}$$

where $\delta\, \partial_n u(q) = \partial_n u^{(+)} - \partial_n u^{(-)}, q \in S$; for external excitation it can be taken that the total field in the inner domain is zero, i.e. $\partial_n u^{s(-)} + u^i(q) = 0 \Rightarrow \partial_n(u^{s(-)} + u^i)(q) = 0$ (for non-resonant frequencies).

Taking identity (13.57) as the starting point, one can consider the integral equation for the unknown function $v(q) \in C^{0,\alpha}(S)$

$$\left[\overline{P}v\right](q) = -u^i(q), \quad q \in S \tag{13.58}$$

with the expectation that the function

$$U(q) = [Pv](q), \quad q \in R^2 \backslash S \tag{13.59}$$

provides a solution of the Dirichlet BVP (this assumption is not always fulfilled). It follows from (13.43) and (13.44) that the kernel of $\overline{P}$ is square integrable and, consequently, (13.58) is one of the first kind in space $L_2(S)$. As mentioned earlier, such an equation posed in L_2 has rather 'pathological' features. In particular, it does not obey the Fredholm alternative and may have many solutions depending on the choice of $u^i(q) \in L_2(S)$ Also, it is easy to construct a sequence of right-hand sides $u^i_m(q), m = 1, 2, 3, \ldots$ such that $\left\|u^i_m\right\|_{L_2} \to 0$, when $m \to \infty$, but the sequence v_m of the corresponding solutions is unbounded: $\|v_m\|_{L_2} \to \infty$ as $m \to \infty$.

To apply the ARM to (13.58), it is first necessary to specify a set of correctness (H_1, H_2) (see Section 13.3). The choice of Sobolev spaces $H_1 = H^{-1/2}(S)$ and $H_2 = H^{1/2}(S)$ is correct from a purely mathematical point of view, but for a suitable numerical implementation we choose the linear (incomplete) spaces

$$H_1 = H^{-1/2}(S) \cap C^{0,\alpha}(S), H_2 = H^{1/2}(S) \cap C^{1,\alpha}(S). \tag{13.60}$$

This choice is applicable under the assumption $S \in C^{2,\alpha}$. It must be noted that (13.58) remains valid for an open contour S, but it cannot be posed in L_2 simply because its solution cannot belong to L_2, but must have singularities that are not square-integrable.

Taking the composition of both sides of (13.58) with $\eta(\theta)$ and taking into account the representation (13.35), one arrives at the equation

$$-\frac{1}{2}\left[F^{-1}T^{-1}\left(I - 2M^P\right)T^{-1}\left(l \cdot \langle u \circ \eta\rangle\right)\right] = -\left(u^i \circ \eta\right) \tag{13.61}$$

where the dependence on $\theta \in [-\pi, \pi]$ is omitted. It is evident that (13.61) has the form of (13.11) with

$$L^{-1} = -\frac{1}{2}F^{-1}T^{-1} \tag{13.62}$$

$$R^{-1}v = T^{-1}F\left(l \cdot \langle v \circ \eta\rangle\right). \tag{13.63}$$

Acting on (13.61) from the left by the operator $L = -2TF$ and introducing the new unknown vector column z_D and righthand side f_D,

$$z_D = T^{-1}F\langle l \cdot (v \circ \eta)\rangle \tag{13.64}$$

$$f_D = -2TF\left(u^i \circ \eta\right) \tag{13.65}$$

one arrives at the equation

$$\left(I - 2M^P\right)z_D = f_D; z_D, f_D \in l_2. \tag{13.66}$$

Thus (13.58) is reduced to the equivalent infinite algebraic system (13.66) in l_2 with the operator $H = -2M^P$ being compact in l_2 (see (13.41)).

Any solution z_D of (13.66) defines $v = v(q)$ in an evident way. It is proved in [12] (including the more complicated case of an open contour S) that (13.59) gives a solution of the Dirichlet BVP when v is a solution of (13.58), and that there is the one-to-one correspondence between solutions of the Dirichlet BVP and solutions of (13.58) and (13.66). Also, if the parameters of the BVP provide the only trivial solution (i.e. $v = 0$ identically) for the homogeneous BVP (when $u^i(q) = 0$ identically) then the same is true for (13.66), namely that if f_D is identically zero, the unique solution is the zero function $z_D = 0$. Equation (13.66), as an equation of the second kind in l_2 satisfies the Fredholm alternative. Consequently, from the uniqueness of the solution of the homogeneous equation (13.66) follows the existence and uniqueness of the solution $z_D \in l_2$ for any $f_D \in l_2$. This equivalence establishes the existence (and uniqueness) of the solution to the Dirichlet BVP.

As one can see, (13.66) follows from the Dirichlet BVP in quite an elementary way, if representation (13.35) was constructed already.

13.5.2 Neumann BVP

The posing of the diffraction problems with Neumann boundary conditions seems similar to that of the Dirichlet problem since condition (13.56) must be changed to the Neumann boundary condition:

$$\partial_n u^{s(+)}(q) = \partial_n u^{s(-)}(q) = -\partial_n u^i(q), \quad q \in S \tag{13.67}$$

This seemingly small change results in a substantial qualitative difference in the solution properties and the way of its construction via the ARM. It is shown in [12, 20] that every solution $u^s(q)$ of the Neumann BVP can be represented, if it exists, as

$$u^s(q) = -[Q(\delta u^s)](q), \quad q \in R^2 S \tag{13.68}$$

$$\delta u^s(q) = u^{s(+)}(q) - u^{s(-)}(q), \quad q \in S \tag{13.69}$$

Consequently (see [12, 17] and (13.27), (13.52)) the following identity holds:

$$\left[\partial_n Q^{(\pm)} \delta u^s\right](q) = \partial_n u^i(q), \quad q \in S \tag{13.70}$$

In the same manner as for identity (13.57) one may introduce the unknown function $u(q)$ satisfying the equation

$$\left[\partial_n Q^{(+)} u\right](q) = \partial_n u^i(q), \quad q \in S \tag{13.71}$$

with the expectation that the function

$$U(q) = -[Qu](q), \quad q \in R^2 S \tag{13.72}$$

gives a solution of the Neumann BVP.

The qualitative properties of (13.71) are quite opposite to those of (13.57). If one poses (13.71) in L_2, the inverse operator to $\partial_n Q^{(\pm)}$ becomes bounded and even compact, but $\partial_n Q^{(\pm)}$ itself is an unbounded operator in L_2. Similar to that discussed for (13.57), a numerical instability described by formulae (13.5) and (13.6) emerges, not because of the factor $\|A_N\|$ in (13.6), but due to the factor $\left\|A_N^{-1}\right\|$ which tends to infinity when $N \to \infty$. Thus, we need to choose again a set of correctness to provide additive and multiplicative splitting of $\partial_n Q^{(\pm)}$, and so on.

Representation (13.38) gives an excellent hint for the purpose. Indeed, one can choose the opposite of that employed in (13.60):

$$H_1 = H^{1/2}(S) \cap C^{1,\alpha}(S); H_2 = H^{-1/2}(S) \cap C^{0,\alpha}(S) \tag{13.73}$$

Formulae (13.38) reduces (13.71) to

$$F^{-1}T\left(I + 2M^{\partial_n Q}\right)TF(u \circ \eta) = 2\langle l \cdot (u \circ \eta)\rangle \tag{13.74}$$

where the dependence $\theta \in [-\pi, \pi]$ is omitted. The regularization of (13.74) is trivial now. The operators L^{-1} and R^{-1} evidently are

$$L^{-1} = F^{-1}T; R^{-1}u = TF(u \circ \eta) \tag{13.75}$$

Applying operator $L = T^{-1}F$ from the left to (13.74) and introducing vector columns of the unknown z_N and the given g, where

$$z_N = TF(u \circ \eta) \tag{13.76}$$

$$g = 2T^{-1}F\langle l \cdot (u \circ \eta)\rangle \tag{13.77}$$

one arrives at the equation

$$\left[I + 2M^{\partial_n Q}\right]z_N = g; z_N, g \in l_2 \tag{13.78}$$

The connection between the Neumann BVP, the existence and uniqueness of its solution, and (13.71), (13.72) and (13.78) are exactly the same as those discussed earlier for the Dirichlet BVP and (13.58), (13.59) and (13.66) (see [12]).

13.6 Diffraction by a semi-transparent obstacle

In this section, we consider the ARM algorithm for the rather important problem of wave diffraction by an inhomogeneity nested in the medium with a constant material parameter, when the inhomogeneity has also constant, but different material parameters. This means that there is a jump of wavelength on the common boundary of nesting and nested media. A similar but more complicated problem for piecewise analytical material parameters has been solved in the study by Shestopalov *et al.* [12]. Herein we have made a slight modification and simplification of this solution on the basis of the technique presented in [21].

13.6.1 The BVP description

Let the open domains $V^{(\pm)}$ be the same as introduced for the Dirichlet and Neumann BVP (see Figure 13.3) except that the domains $V^{(+)}$ and $V^{(-)}$ have different material parameters, resulting in different but constant wave numbers $k^{(+)}$ and $k^{(-)}$, respectively for the same oscillation frequency. It is required to find the field

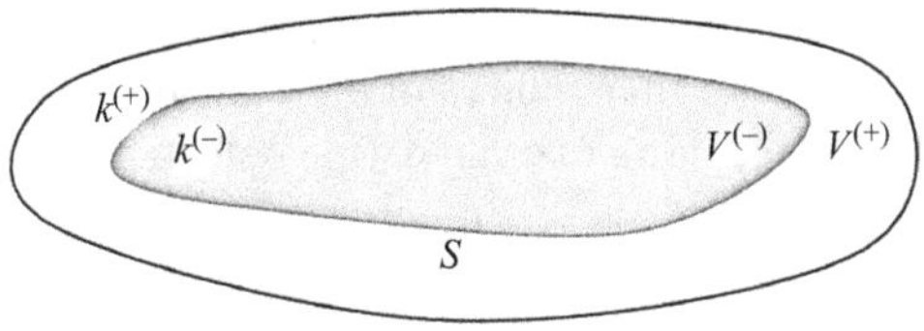

Figure 13.3 Schematic view of a semi-transparent obstacle

$u^s(q), q \in V^{(+)} \cup V^{(-)}$ which belongs to the class (13.52) but satisfies two Helmholtz equations with different wave numbers:

$$\left(\Delta + k^2\right) u^s(q) = 0, q \in V^{(-)} \cup V^{(+)} \tag{13.79}$$

where

$$k = k(q) = \begin{cases} k^{(+)}, q \in V^{(+)}, \\ k^{(-)}, q \in V^{(-)}. \end{cases} \tag{13.80}$$

In addition, the function $u^s(q)$ obeys the semi-transparent boundary conditions (also known as the interface boundary condition)

$$\alpha^{(+)}\left[u^{s(+)}(q) + f(q)\right] = \alpha^{(-)} u^{s(-)}(q), \quad q \in S \tag{13.81}$$

$$\beta^{(+)}\left[\partial_n u^{s(+)}(q) + g(q)\right] = \beta^{(-)} \partial_n u^{s(-)}(q), \quad q \in S \tag{13.82}$$

where $\alpha^{(\pm)}$, $\beta^{(\pm)}$ are given non-zero constants and $f(q)$, $g(q)$ are known functions, having the physical sense of functions related to the outer incident field $u^i(q)$, with

$$f(q) = u^i(q); g(q) = \partial_n u^i(q), \quad q \in S. \tag{13.83}$$

As well, we assume that $u^s(q)$ satisfies some boundary conditions on the outer domain $V^{(+)}$ of the form

$$\Gamma u^s(q) = 0, q \in V^{(+)} S \tag{13.84}$$

for a suitable operator Γ. In particular, for the unbounded domain $V^{(+)}$ the operator Γ can be chosen so that (13.84) is the Sommerfeld radiation condition the Sveshnikov–Reichardt partial radiation condition [22, 23] or similar.

13.6.2 *Integral representation for $u^{s(+)}$ and $\partial_n u^{s(+)}$*

Let open domains $V_{ext}^{(-)}$ and $V_{ext}^{(+)}$ be chosen so that

$$\overline{V^{(-)}} \subset V_{ext}^{(-)}; \overline{V^{(+)}} \subset V_{ext}^{(+)} \cup S \tag{13.85}$$

and furthermore suppose that the fundamental solutions $\varepsilon_{2(-)}$ and $\varepsilon_{2(+)}$ of the Helmholtz equations (13.79), (13.80) for $k = k^{(-)}$ and $k = k^{(+)}$ are known in the domains $V_{ext}^{(\pm)}$, respectively. Additionally, the fundamental solutions $\varepsilon_{2(\pm)}$ are required to satisfy the boundary conditions (13.81), (13.82) on the outer boundary $\partial V^{(+)} S$. In particular, the fundamental solutions may be the Green's functions of free space with $k = k^{(-)}$ and $k = k^{(+)}$. Much better choices of $\varepsilon_{2(\pm)}$ are also possible [17].

Similar to (13.15) and (13.16), one can define potentials

$$\left[P_{(\pm)}v\right](q) = \int_S \varepsilon_{2(\pm)}(q,p)\, v(p)\, dl_p, \quad q \in V^{(\pm)} \tag{13.86}$$

$$\left[Q_{(\pm)}u\right](q) = \int_S \frac{\partial \varepsilon_{2(\pm)}(q,p)}{\partial n_p} u(p)\, dl_p, \quad q \in V^{(\pm)} \tag{13.87}$$

where n_p is the unit outward normal to S at points $p \in S$ Here and below we use the suffices $(\pm)$ for values and functions associated with domains $V^{(\pm)}$, respectively. Using a standard Green's formula technique, one obtains the integral representations for u^s:

$$u^s(q) = \left[Q_{(-)}u^{s(-)}\right](q) - \left[P_{(-)}\partial_n u^{s(-)}\right](q), \quad q \in V^{(-)} \tag{13.88}$$

$$u^s(q) = -\left[Q_{(+)}u^{s(+)}\right](q) + \left[P_{(+)}\partial_n u^{s(+)}\right](q) =, \quad q \in V^{(+)} \tag{13.89}$$

Passing to the corresponding limits by using (13.23), (13.24) and (13.26), (13.27), one arrives at the following integral representations for $u^{s(\pm)}(q)$, $q \in S$:

$$u^{s(+)} = -2\overline{Q}_{(+)}u^{s(+)} + 2P_{(+)}\partial_n u^{s(+)}; \tag{13.90}$$

$$\partial_n u^{s(+)} = -2\partial_n Q^{(+)}_{(+)}u^{s(+)} + 2\overline{\partial_n P_{(+)}}\partial_n u^{s(+)}, \tag{13.91}$$

$$u^{s(-)} = 2\overline{Q_{(-)}}u^{s(-)} - 2\overline{P_{(-)}}\partial_n u^{s(-)}, \tag{13.92}$$

$$\partial_n u^{s(-)} = 2\partial_n Q^{(-)}_{(-)}u^{s(-)} - 2\overline{\partial_n P_{(-)}}\partial_n u^{s(-)}, \tag{13.93}$$

where the dependence on $q \in S$ is omitted for the sake of simplicity.

13.6.3 Reduction of the BVP to a system of integral equations

Now we introduce the notations

$$u = u^{s(-)}; v = \partial_n u^{s(-)}. \tag{13.94}$$

From boundary conditions (13.81), (13.82) it follows that

$$u^{s(+)} = \left(\alpha^{(-)}/\alpha^{(+)}\right)u - f, \tag{13.95}$$

$$\partial_n u^{s(+)} = \left(\beta^{(-)}/\beta^{(+)}\right)v - g. \tag{13.96}$$

Substitution of (13.94)–(13.96) in (13.90)–(13.93) gives the possibility to eliminate $u^{s(+)}(q)$ and $\partial_n u^{s(+)}$ from the right-hand sides of (13.90)–(13.93). Substitution of these integral representations in (13.81), (13.82) gives after elementary manipulations the following system of integral identities:

$$\begin{aligned} &2\alpha^{(-)}\beta^{(+)}\left[\overline{Q}_{(+)} + \overline{Q}_{(-)}\right]u - 2\left[\alpha^{(-)}\beta^{(+)}\overline{P}_{(-)} + \alpha^{(+)}\beta^{(-)}\overline{P_{(+)}}\right]v \\ &= \alpha^{(+)}\beta^{(+)}\left\{\left(I + 2\overline{Q}_{(+)}\right)f - 2\overline{P}_{(+)}g\right\}, \end{aligned} \tag{13.97}$$

$$2\left[\alpha^{(-)}\beta^{(+)}\partial_n Q^{(+)}_{(+)} + \alpha^{(+)}\beta^{(-)}\partial_n Q^{(-)}_{(-)}\right]u - 2\alpha^{(+)}\beta^{(-)}\left[\overline{\partial_n P}_{(+)} + \overline{\partial_n P}_{(-)}\right]v = \alpha^{(+)}\beta^{(+)}\left\{\left(I - 2\overline{\partial_n P}_{(+)}\right)g + 2\partial_n Q^{(+)}_{(+)} f\right\}. \tag{13.98}$$

The identities (13.97), (13.98) constitute a system of integral equations for the unknown functions $u = u(q)$ and $v = v(q)$. Once the equations are solved, the functions $u(q)$ and $v(q)$ can be used for calculations of $u^{s(\pm)}$ and $\partial_n u^{s(\pm)}$ (see (13.94)–(13.96)). Substitution of the functions $u^{s(\pm)}$ and $\partial_n u^{s(\pm)}$calculated in this way into integral representations (13.88), (13.89) gives the hope that such representations may bring a solution of BVP under consideration. These expectations are nearly the same as have considered for the Dirichlet and Neumann BVPs.

13.6.4 Reduction of the system of integral equations to an infinite system of linear algebraic equations

Our next step is based on the presentations (13.35)–(13.38) and the observation that system (13.97), (13.98) can be considered as a 2×2 block matrix system of integral equations and the operator of the system has the form (13.11), (13.12) with block-diagonal operator L, blocks of which coincide with (13.62) and (13.75) respectively. Operator R^{-1} is also 2×2 block matrix, elements of which are given by the operator R^{-1} in (13.63) and (13.75).

Substitution of (13.35)–(13.38) in (13.97), (13.98) results in the following 2×2 block-matrix equation, with each block as an infinite algebraic operator:

$$\lambda x + Kx = Wh, \tag{13.99}$$

$$\lambda = \alpha^{(+)}\beta^{(-)} + \alpha^{(-)}\beta^{(+)}, \tag{13.100}$$

where

$$x = \begin{pmatrix} x_1 \\ x_2 \end{pmatrix}, \quad \begin{aligned} x_1 &= T^{-1}F\langle l \cdot (v \circ \eta)\rangle, \\ x_2 &= TF(u \circ \eta); \end{aligned} \tag{13.101}$$

$$h = \begin{pmatrix} h_1 \\ h_2 \end{pmatrix}. \quad \begin{aligned} h_1 &= T^{-1}F\langle l \cdot (g \circ \eta)\rangle, \\ h_2 &= TF(f \circ \eta); \end{aligned} \tag{13.102}$$

and the block-matrices $K = \begin{bmatrix} K_{11} & K_{12} \\ K_{21} & K_{22} \end{bmatrix}$ and $W = \begin{bmatrix} W_{11} & W_{12} \\ W_{21} & W_{22} \end{bmatrix}$ comprise infinite algebraic operators defined as follows:

$$K_{11} = -2\left[\alpha^{(-)}\beta^{(+)}M^P_{(+)} + \alpha^{(+)}\beta^{(-)}M^P_{(-)}\right] \tag{13.103}$$

$$K_{12} = 2\alpha^{(-)}\beta^{(+)}\left[M^Q_{(+)} + M^Q_{(-)}\right] \tag{13.104}$$

$$K_{21} = -2\alpha^{(+)}\beta^{(-)}\left[M^{\partial_n P}_{(+)} + M^{\partial_n P}_{(-)}\right] \tag{13.105}$$

$$K_{22} = 2\left[\alpha^{(-)}\beta^{(+)}M^{\partial_n Q}_{(+)} + \alpha^{(+)}\beta^{(-)}M^{\partial_n Q}_{(-)}\right] \tag{13.106}$$

$$W_{11} = \alpha^{(+)}\beta^{(+)}\left(I + 2M^Q_{(+)}\right) \tag{13.107}$$

$$W_{12} = \alpha^{(+)}\beta^{(+)}\left(I - 2M^{P}_{(+)}\right) \tag{13.108}$$

$$W_{21} = \alpha^{(+)}\beta^{(+)}\left(I - 2M^{\partial_n P}_{(+)}\right) \tag{13.109}$$

$$W_{22} = \alpha^{(+)}\beta^{(+)}\left(I + 2M^{\partial_n Q}_{(+)}\right) \tag{13.110}$$

All the infinite matrices M in (13.103)–(13.110) are of the kind (13.41). Consequently, the same is true for the operators $K_{ij}, i,j = 1,2$ and for the operator K.

For the majority of physical applications (we know of no exceptions), the coefficients $\alpha^{(\pm)}$and $\beta^{(\pm)}$ obey

$$\lambda \neq 0 \tag{13.111}$$

Accepting that (13.111) holds, one can divide (13.99) by λ and obtain a system of the kind

$$(I + H)x = b \tag{13.112}$$

where $x, h \in l_2 \oplus l_2$ and the operator H is compact in $l_2 \oplus l_2$. Thus, the diffraction BVP for a semi-transparent obstacle is reduced to an algebraic system of the second kind (13.112) that can be solved efficiently by various methods of truncation.

The generalization of the solution for a few obstacles, including nested ones, seems to be evident.

13.7 Diffraction of waves with complex frequencies and spectral theory of open cavities

13.7.1 Description of the BVP

A rigorous treatment of the propagation and diffraction of waves having complex frequencies is interesting and useful from both mathematical and physical points of view. In particular, it helps to understand various resonant phenomena for open screens and cavities.

As is well known, any closed cavity has eigenfrequencies and corresponding eigenmodes. The eigenfrequencies can be complex-valued if energy losses are present in the cavity or on its surface. However, the presence of a small hole or slot connecting the cavity to free space leads immediately (from the point of view of the classic L_2 spectral theory) to the disappearance of discrete eigen-spectral points, and to the appearance of a continuous spectrum on the real non-negative axis of frequency values.

Being mathematically rigorous and perfect, this theory draws a picture that dramatically contradicts physical intuition. It seems strange that radiation of energy from the cavity to the external region does not lead to the same effect as the losses of any other nature inside the cavity. Physical intuition and natural experiments demonstrate nothing other than that the eigenfrequencies and eigenmodes are slightly perturbed when the losses are small. Thus, the necessity of a change of the mathematical model of the diffraction problem and corresponding spectral theory became evident to researchers in the field. A few attempts of such a change were made; but although mathematically correct, they also resulted in pathological physical consequences. In particular, various generalizations of Sommerfeld radiation conditions failed too.

Perhaps the first step in the proper direction has been made in [22], where partial radiation condition (for waveguides) is suggested; publication [23] is also essential. In the studies [11, 12, 24], it was understood that any model of open cavities and their resonances must be a part of a more general theory of diffraction of waves with complex-valued frequencies or wavenumbers.

Thus, we consider in this section almost the same BVPs as investigated earlier, but now, when the outer domain $V^{(+)}$ is a vicinity of infinity, we use the Sveshnikov–Reihardt partial radiation condition. This states that there exists a vicinity Ω_∞ of infinity such that the scattered field $u^s = u^s(q)$ allows the representation

$$u^s(q) = \sum_{n=-\infty}^{\infty} u_n^s H_n^{(1)}(kr_q)\, e^{in\phi_q}, q = (r_q, \phi_q) \in \Omega_\infty \tag{13.113}$$

in an axial coordinate system (r, ϕ), where u_n^s denote coefficients and $H_n^{(1)}(z)$ is the Hankel function of the first kind and order n.

It is important to define the domain under consideration that contains complex-valued k. The most natural domain of such a kind is the Riemann surface S_R of analytical continuation of $H_0^{(1)}(z)$, which is the same as that for the natural logarithm $\ln z$ with a cut along the real negative semi-axis. For the majority of physical applications, most important is the 'physical' sheet of S_R, i.e. the usual set C of complex numbers, but with the negative semi-axis excluded. For the 'physical' sheet D^o and its open upper and low half-planes $D^{(+)}$ and $D^{(-)}$, we use the following notations:

$$D^o = C[0,\infty), D^{(+)} = \{z \in D^o : Imz > 0\}, D^{(-)} = \{z \in D^o : Imz < 0\}. \tag{13.114}$$

For the sake of simplicity, we restrict attention to wavenumbers $k \in D^o$ only. The general case for $k \in S_R$ can be analysed in a similar way [11, 24].

The crucial difference of the condition (13.113) from the more usual Sommerfeld radiation condition (13.55) is that the latter restricts the asymptotic behaviour of $u^s(q)$ for $q \to \infty$, but (13.113) gives a point-wise restriction on $u^s(q)$ for every $q \in \Omega_\infty$. One can easily check that (13.113) dictates exponential decay of $u^s(q)$ at infinity for $Imk > 0$, but exponential growth for $Imk < 0$ and $q \to \infty$.

Due to such exponential growth, a fundamental theorem in [24], claiming that condition (13.113) is necessary and sufficient for the correctness of the well-known Green's formulae technique in an unbounded domain for every $k \in S_R$, is not quite evident. However all integral representations involving $\varepsilon_2(q,p)$ considered earlier (see (13.15), (13.16)) are valid, including identities (13.57), (13.70) and (13.88), (13.89) for $k \in S_R$.

It can also be shown [24] that conditions (13.113) and (13.55) are equivalent for $Imk \geq 0$. Summarizing, one can conclude that (13.113) is a 'proper' generalization of (13.55) for $k \in S_R$ in the sense that it includes (13.55) and also makes the Green's formulae technique valid for all $k \in S_R$.

13.7.2 Dirichlet BVP for complex-valued wave numbers

To be concrete, we consider in this section only the Dirichlet BVP (13.52)–(13.54), (13.113). The other BVP considered earlier allow almost an identical analysis leading qualitatively to the same results as those formulated later for the Dirichlet BVP. As mentioned earlier, the identity (13.57) is valid for $k \in S_R$, the representations for corresponding potentials as pseudo-differential operators for $k \in S_R$ are also valid, and (13.66) also follows from (13.57).

The wavenumber $k \in S_R$ is a parameter of the matrix operator M^P in (13.66). Taking this into account, we introduce and analyse the matrix functions

$$H(k) = M^P\,(k)\,; A(k) = I + H\,(k)\,, k \in D^o. \tag{13.115}$$

13.7.3 Qualitative features of the Dirichlet BVP

Five important statements about the Dirichlet BVP can be found in Poyedinchuk *et al.* [24]:

- For any fixed $k \in D^o$ there is one-to-one correspondence between solutions of the Dirichlet BVP and solutions of (13.66). In particular, this is true for solutions of both homogeneous problems, when $u^i\,(q) \equiv 0, q \in S$ and $f_D = 0$. Relations (13.64), (13.65) and their inversions demonstrate the correspondence.
- The set $\sigma \subset D^o$ of eigen-wavenumbers of the Dirichlet BVP (when homogeneous BVP has non-trivial solutions) is a countable discrete set in $\overline{D^{(-)}} = D^{(-)} \cup (0, \infty)$, i.e. $\sigma = \{k_1, k_2, k_3, ..., k_j, ...\} \in \overline{D^{(-)}}$ and the only possible accumulation point of σ is infinity.
- The standard Green's function $G\,(q,p;k)$ of the BVP (the resolvent) is analytical for $k \in D^{(+)}$, it has an analytical continuation in $\overline{D^{(-)}}$ as a meromorphic function, and the set of its poles coincides with σ. The residue of $G\,(q,p;k)$ at any simple pole $k = k_j$ has the form $C\left(k_j\right) u_j\,(q)\, u_j\,(p)$ where $C\left(k_j\right)$ is a non-zero constant and $u_j\,(q)$ is the eigenmode of the Dirichlet BVP corresponding to $k = k_j$.
- The operator function $A\,(k)$ occurring in (13.115) is an analytical matrix-operator function in l_2 and it is invertible in $D^{(+)}$. The matrix-operator function $A^{-1}\,(k)$ is a meromorphic operator function in l_2 for $k \in \overline{D^{(-)}}$.
- The characteristic values of $A\,(k)$ coincide with the eigen-wavenumbers of the Dirichlet BVP.

13.7.4 Numerical calculation of complex-valued eigen-wavenumbers and eigenmodes

From the relation (13.115) between the Dirichlet BVP and the matrix-operator function $A(k)$, the eigen-wavenumbers and corresponding eigenmodes can be evaluated from the characteristic values and corresponding characteristic (eigen-like) functions of $A(k)$. The analogy to finite-dimensional operator functions suggests that the eigen-wavenumbers $\{k_j\}$ can be calculated as roots of the one-dimensional equation involving the determinant of the matrix operator $A(k)$,

$$\det A(k) = 0, k \in D^o, \tag{13.116}$$

where the operator $A(k) : l_2 \to l_2$ and its determinant are formed from the compact operator $H(k) : l_2 \to l_2$ using the limit

$$\det A(k) = \lim_{N\to\infty} \det (I + H_N (k)), \tag{13.117}$$

where $H_N(k)$ is a sequence of operators such that

$$\lim_{N\to\infty} H_N (k) = H(k). \tag{13.118}$$

Unfortunately, this idea is not valid in general. Indeed, it is easy to find an example of a bounded, and even a Hilbert–Schmidt operator $H(k)$ for which the sequence $\det (I + H_N (k))$ is divergent no matter what sequence $\{H_N (k)\}_{N=1}^{\infty}$ is chosen. It means that (13.116) does not provide a sensible approach in general. This situation typically arises when the SIP is used for diffraction problems. Some workaround of this difficulty with compact $H = H(k)$ can be found in Poyedinchuk *et al.* [24], although this study demonstrates numerical instability arise when truncations of $H(k)$ are of large algebraic dimension.

Thus, stronger limitations on H and H_N are necessary. The definition (13.117) of the determinant of an infinite matrix is not the only possibility, but it is the simplest and the most suitable one from the numerical point of view. It turns out that this definition is correct if H is of trace (i.e. nuclear) class and if limit in (13.118) exists in the trace class norm [25].

In contrast to the SIP, the ARM gives a solid basis for the approach (13.116)–(13.118). Namely, all *the* matrices M in (13.35)–(13.40) are of trace class. The proof of this is similar to that made in [12] for the inequalities (13.41). Standard finite-dimensional truncations H_N of Hconverge to H in trace class norm and the limit (13.117) exists and is finite. Consequently, (13.115) can be used for the purpose of computing the characteristic values of $A(k)$ from the roots of $\det A(k)$.

It is essential that $\det A\,(k)$(as well as $\det (I + H_N (k))$ for any$N \geq 1$) is an analytical function of the complex variable $k \in D^o$. Very efficient numerical methods are available for finding roots of an analytical function (see, e.g. [26]). It is essential that such roots do not lie in $D^{(+)}$. When a root $k = k_j$ is found, the corresponding eigenvector (i.e. a characteristic vector of $A(k_j)$) can also be found. Among the best tools for this purpose is the inverse iteration method [15]. The corresponding eigenmode can now be calculated with help of the inversion (13.64) and (13.59).

13.8 ARM: considerations for implementation

We have presented in the preceding sections a rather general description of the ARM and the principal steps of its implementation based on the equivalent transform of the BVP operators to appropriate pseudo-differential operators in a form which is suitable for the further implementation. This description is too general for the immediate solution of any non-trivial scattering or diffraction problem to hand, and a significant amount of auxiliary analytical work should be done prior to numerical implementation.

The first task is a detailed analysis of the singular structure of the relevant integral equation. Knowledge of the singular structure guides the proper choice of the correct functional spaces H_1 and H_2 so that the operator A be of Fredholm type (see (13.11) and the related text). This is necessary for the proper selection and posing of the most appropriate canonical integral equation corresponding to the BVP under consideration. Such analytical work may be rather non-trivial but is an essential for the final implementation of stable and robust algorithms.

One of the critical parameters is the degree of smoothness of the obstacle surface. If the surface and its parametrization are not smooth enough (e.g. they are not of class $C^{2,\alpha}$ (see [17])), smoother or even infinitely smooth approximations should be constructed at first.

A third point concerns the construction of an efficient and accurate algorithm for Fourier Transform of singular functions with respect to various (exponential, polynomial, etc.) bases. Typically due consideration must be given to the restriction imposed by limited smoothness of the kernel under the transform.

Although these are the main analytical issues to address in any practical implementation, other minor analytical problems may require attention, depending on the precise nature of the diffraction problem to hand.

13.9 ARM: various applications and conclusion

The ARM was developed as a more powerful successor of the SIP by Tuchkin and Shestopalov [5–7]; it was named and formulated as a methodology by Poyedinchuk and colleagues [8, 11, 12]. During the last 30 years, a number of approaches have been based on the ARM and a huge amount of problems and related publications appeared.

In this chapter, we have focused consideration on closed contours S only. However, the ARM made its first appearance [5, 6, 8] in the context of open screens. Diffraction problems involving open screens have been tackled using dual series equations techniques, involving operators of fractional differentiation and integration (essentially the Abel transform) [27–31]. Another powerful technique is based on the construction of representations for potentials and their derivatives, very similar to (13.35)–(13.40), but using Fourier transforms involving various orthogonal polynomials. The ARM based on Fourier–Chebyshev transforms was suggested at first by Dikmen and colleagues [32, 33] for axially symmetrical obstacles excited by a scalar or electromagnetic wave.

Periodic structures may be treated by the ARM. For example, periodical gratings and periodically corrugated surfaces have been studied by Tuchkin and colleagues [13, 34–36], and an approach employing the ARM for wave diffraction by infinite flat gratings of metal strips placed on a layer of chiral media is constructed in the study by Panin *et al.* [37]. In another direction, wave diffraction by an arbitrary-shaped screen or body of revolution has been simulated utilizing the ARM constructed by Tuchkin and colleagues [38–41].

Finally, among the most interesting applications is the BVP with boundary conditions of the third kind

$$\alpha(q)\,\partial_n u^{s(+)} + \beta(q)\,\partial_n u^{s(+)} = f(q), \quad q \in S, \tag{13.119}$$

where $\alpha(q)$, $\beta(q)$ and $f(q)$ are known functions in certain functional spaces. Evidently, when $\alpha(q) \equiv 0$ or $\beta(q) \equiv 0$ one arrives at the Dirichlet or Neumann problems, respectively. If $\alpha(q)$ is not much smaller than $\beta(q)$ on S, then the BVP with (13.119) can be treated as the Neumann BVP, but with a regular perturbation proportional to $\beta(q)/\alpha(q)$. The appropriate ARM is constructed in [42]. It is similar to that for the Neumann problem but combined with the technique applied earlier (see Section 13.6.3) for the semi-transparent obstacle.

Unfortunately, if the condition $|\alpha(q)/\beta(q)| << 1$ occurs at even one point $q \in S$, this approach becomes impractical because the dimension of the truncated algebraic system must necessarily be bigger than $|\beta(q)/\alpha(q)|$. Such a situation appears, for example, in diffraction problems involving a cylinder of high but finite conductivity, illuminated by an E-polarized wave; the value $|\beta(q)/\alpha(q)| \cong 10^8$ is quite typical. The approach suggested by Tuchkin [42] treats (13.119) for small $\alpha(q)/\beta(q)$ as à singularly perturbed Dirichlet problem and involves the well-studied technique of solving singularly perturbed ordinary differential equations. The ARM for the result is constructed in a way similar to that for the Dirichlet BVP.

In conclusion, this chapter has presented a unified treatment of the ARM and its precursor, the SIP. The methodology has as its main purpose the transformation of the underlying ill-posed operator equations, arising in diffraction theory, to a well-posed formulation, that of a second-kind Fredholm operator equation. The methodology has a long history of development, and the various approaches that have appeared in the literature can be reconciled by using the formalism of pseudo-differential operators.

References

[1] Agranovich Z.S., Marchenko V.A., and Shestopalov V.P. 'Diffraction of electromagnetic waves on flat metallic gratings'. *Zhurnal Tekhnicheskoy Phisiki.* 1962;**4**:381–394 (in Russian).

[2] Shestopalov V.P. *Method of the Riemann–Hilbert Problem in Theory of Diffraction and Propagation of Electromagnetic Waves.* Kharkov: Publishing House of Kharkov University; 1971, p. 400 (in Russian).

[3] Shestopalov V.P., Litvinenko L.N., Masalov S.A., and Sologub V.G. *Wave Diffraction on Gratings.* Kiev: Naukova dumka; 1973, p. 278 (in Russian).

[4] Masalov S.A. and Sologub V.G. 'Exact solution of the problem of the diffraction of an electromagnetic wave at a strip structure'. *USSR Computational Mathematics and Mathematical Physics.* 1970;**10**(3):169–192.

[5] Tuchkin Y.A. 'Wave scattering by unclosed cylindrical screen of arbitrary profile with Dirichlet boundary condition'. *Doklady AN SSSR.* 1985;**285**(6): 1370–1373 (in Russian).

[6] Tuchkin Y.A. 'Wave scattering by unclosed cylindrical screen of arbitrary profile with Neuman boundary condition'. *Doklady AN SSSR*. 1987;**293**(2): 343–345 (in Russian).

[7] Tuchkin Y.A. and Shestopalov V.P. 'Wave scattering by finite system of cylindrical screens of arbitrary profile with Dirichlet boundary condition'. *Doklady AN SSSR*. 1985;**285**(5):1107–1109 (in Russian).

[8] Poyedinchuk A.Y., Tuchkin Y.A., and Shestopalov V.P. 'Diffraction by curve strip'. *Transactions on IEE Japan*. 1993;**113-A**(3):139–146.

[9] Tuchkin Y.A. 'A new method in wave diffraction theory by thins screens'. *Electromagnetics*. 1993; **13**(3): 319–338.

[10] Poyedinchuk A.Y., Tuchkin Y.A., and Shestopalov V.P. 'Wave diffraction in a locally heterogeneous medium'. *Computational Mathematics and Mathematical Physics*. 1996;**36**(7):935–946.

[11] Poyedinchuk A.Y., Tuchkin Y.A., and Shestopalov V.P. 'New numerical–analytical methods in diffraction theory'. *Mathematical and Computer Modeling*. 2000;**32**:1029–1046 (in Russian).

[12] Shestopalov V.P., Tuchkin Y.A., Poyedinchuk A.Y., and Sirenko Y.K. *New Methods for Solving Direct and Inverse Problems of Diffraction Theory, Part 1: Analytical Regularization of Boundary Value Problems of Electromagnetics*. Kharkov: Osnova Publishing House; 1997 (in Russian).

[13] Tuchkin Y.A. and Poyedinchuk A.Y. 'Analytic regularization methods', in Sirenko Y.K. and Ström S. (eds.), *Modern Theory of Gratings*. New York: Springer; 2010.

[14] Hsiao, G.C. and Kleinman, R.E. 'Mathematical foundations for error estimation in numerical solutions of integral equations in electromagnetics'. *IEEE Trans. Antennas and Propagation*, 1997;**47**(3):316–328.

[15] Wilkinson J.H. *Algebraic Eigenvalue Problem*. Oxford: Clarendon Press; 1965 (1988 reprinted edition).

[16] Tikhonov A.N. and Arsenin V.Y. *Methods for Solving Ill-Posed Problems*. Moscow: Nauka; 1986 (in Russian); Tikhonov. A.N. and Arsenin. V.Y. *Solutions of Ill-Posed Problems*. Washington D.C: V.H. Winston & Sons; 1977 (English translation).

[17] Colton D. and Kress R. *Integral Equation Methods in Scattering Theory*. New York: Wiley-Interscience; 1983.

[18] Gunter N.M. *Potential Theory and Its Applications to Basic Problems of Mathematical Physics*. New York: Frederick Ungar Publishing Company; 1967.

[19] Tikhonov A.N. and Samarskii A.A. *Equations of Mathematical Physics*. Oxford: Pergamon Press; 1963, p. 766.

[20] Tuchkin Y.A. 'Derivation of one integral–differential equation of theory of diffraction', *Doklady AN Ukrainskoy SSR*. 1987;series **A**(4):21–25 (in Russian).

[21] Tuchkin Y.A. 'Regularization of one class of integral equation in diffraction theory'. *Doklady AN Ukrainskoy SSR*. 1997;series **A**(10):47–51 (in Russian).

[22] Sveshnikov A.G. 'Radiation principle'. *Doklady AN SSSR*. 1950;**73**(5), 917–920 (in Russian).

[23] Reichardt H. 'Ausstrahlungsbedingungen fur die Wellengleihung'. *Abhandlungen aus dem Mathematischen Seminar der Universität Hamburg*. 1960;**24**; 41–53.

[24] Poyedinchuk A.Y., Tuchkin Y.A., and Shestopalov V.P. 'Regularization of spectral problems of wave scattering by unclosed screens'. *Doklady AN SSSR*. 1987;**295**(6):1358–1362 (in Russian).

[25] Simon M. and Reed B. *Methods of Modern Mathematical Physics, Functional Analysis*, Vol. 1. Cambridge, MA: Academic Press; 1981, p. 400.

[26] Traub J.F., *Iterative Methods for the Solution of Equations*, Englewood Cliffs, NJ: Prentice-Hall; 1964.

[27] Vinogradov S.S., Tuchkin Yu.A., and Shestopalov V.P. 'The effective solving of dual series equation involving associative Legendre functions'. *Doklady AN SSSR*, 1978;**242**(1):80–83 (in Russian).

[28] Vinogradov S.S., Tuchkin Yu.A., and Shestopalov V.P. 'Summator equations with kernels on the form of Jacoby polynomials'. *Soviet Physics Doklady*, 1980;**25**(7):531–532.

[29] Svischev Y.V., Tuchkin Y.A., and Shestopalov V.P. 'On a new class of dual series equations involving Jacoby polynomials'. *Doklady AN SSSR*. 1991;**317**(6):1369–1373 (in Russian).

[30] Vinogradov S.S., Tuchkin Y.A., and Shestopalov V.P. 'On Abel's transform in diffraction problem by thin strip'. *Doklady AN SSSR*. 1982;**267**(2):330–334 (in Russian).

[31] Vinogradov S.S., Smith P.D., and Vinogradova E.D. *Canonical Problems in Scattering and Potential Theory, Part 1: Canonical Structures in Potential Theory*. London: Chapman & Hall/CRC; 2001, p. 374.

[32] Dikmen F., Karacuha E., and Tuchkin Y. A. 'Scalar wave diffraction by a perfectly soft infinitely thin circular ring'. *Turkish Journal of Electrical Engineering and Computer Science*. 2001;**9**(2):199–219.

[33] Dikmen F. and Tuchkin Y.A., 'Analytical regularization method for electromagnetic wave diffraction by axially symmetrical thin annular strips'. *Turkish Journal of Electrical Engineering and Computational Science.* 2009;**17**: 107–124.

[34] Tuchkin Y.A. 'Reflecting grating with arbitrary smooth profile', in Shestopalov V.P. and Sirenko Y.K. (eds.), *Dynamic Theory of Gratings*. Kiev: Naukova Dumka; 1989.

[35] Krutin Y.I., Tuchkin Y.A., and Shestopalov V.P. 'Regularization of boundary value problem of wave diffraction by a grating which formed by rods of an arbitrary profile with Dirichlet boundary condition'. *Zhurnal Vychislitel'noy Matematiki i Matematicheskoy Fiziki*. 1991;**30**(6):864–875 (in Russian).

[36] Krutin Y.I., Tuchkin Y.A., and Shestopalov V.P. 'Diffraction of the E-polarized wave by periodic smooth wavy surfaces'. *Radiotekhnica i Electronica*. 1992;**37**(2):202–210 (in Russian).

[37] Panin S.B., Turetken B., Poyedinchuk A.Y., and Tuchkin Y.A. 'Diffraction from a grating on a chiral medium: Application of analytical regularization method'. *Progress In Electromagnetics Research B (PIER B)*. 2014;**59**:19–29.

[38] Tuchkin Y.A. 'Analytical regularization method for wave diffraction by a bowl-shaped screen of revolution', in Smith P.D. and Cloude S.D. (eds.), *Ultra-Wideband, Short-Pulse Electromagnetics 5*. New York: Kluwer Academic/Plenum Publishers, 2002, pp. 153–158.

[39] Panin S.B., Smith P.D., Vinogradova E.D., Tuchkin Y.A., and Vinogradov S.S. 'Regularization of the Dirichlet problem for Laplace's equation: surfaces of revolution'. *Electromagnetics*. 2009;**29**(1):53–76.

[40] Panin S.B., Smith P.D., Vinogradova E.D., Tuchkin Y.A., and Vinogradov S.S. 'Diffraction from arbitrary shaped bodies of revolution: analytical regularization'. *Journal of Engineering Mathematics*. 2009; **65**(2):125–141.

[41] Panin S.B., Smith P.D., Vinogradova E.D., Tuchkin Y.A., and Vinogradov S.S. 'Analytical regularization of the Dirichlet diffraction problem for screen of revolution'. *International Electronic Journal of Pure and Applied Mathematics*. 2011;**3**(4):289–308.

[42] Tuchkin Y.A. 'Electromagnetic wave scattering by smooth imperfectly conductive cylindrical obstacle', in Smith P.D. and Cloude S.D. (eds.), *Ultra-Wideband, Short-Pulse Electromagnetics 5*. New York: Kluwer Academic/Plenum Publishers, 2002, pp. 137–142.

Chapter 14

Resonance scattering of E-polarized plane waves by two-dimensional arbitrary open cavities: spectrum of complex eigenvalues

Elena D. Vinogradova[1]

14.1 Introduction

14.1.1 Preliminary remarks

The scattering of acoustic and electromagnetic waves by open cavities usually causes a resonant response, arising when the frequency of excitation approaches one of the frequencies belonging to the spectrum of eigenvalues of such a cavity. In radio-engineering practice, such frequencies are usually called the 'resonance' frequencies. In our investigation we will focus on investigation of resonant objects widely used in practice, representing two-dimensional (2D) hollow cylinders of arbitrary shape with longitudinal slits. In the context of propagating waves, these objects are treated as waveguides; in the context of standing waves, as resonators. The eigenvalues of the closed hollow circular, rectangular, and elliptical cavities are well-studied [1,2]. In the transmission regime, one calls them the 'cut-off frequencies'. In the absence of ohmic losses in the cavity walls, the cut-off frequencies are real-valued. When the surface of the cavity is disturbed by some openings, for example, by transverse or longitudinal slits, the spectrum of eigenvalues becomes complex-valued. This phenomenon (the appearance of an imaginary part in the eigenvalues) is explained by dissipation of the energy through the opening to the external space. The radiation of electromagnetic energy to external space causes the losses of energy inside the open cavity, which are called the 'radiation losses'. Furthermore, throughout the text, we will term such eigenvalues, containing an imaginary part, as *complex eigenvalues*.

The calculation of eigenvalues for arbitrary closed 2D cavities can be realized by a variety of methods. An incomplete list includes the frequency-domain finite-difference method [3], the method of fundamental solutions [4], the boundary collocation method with meshless concept of Chen *et al.* [5], integral equation technique [6], etc. The merits and limitations of each of them are well discussed in the literature. A detailed review of these approaches seems less relevant here, since

[1]Department of Mathematics and Statistics, Macquarie University, Sydney, Australia

our investigation is based on a quite different method, which is called the method of analytical regularization (MAR) and which is equally applicable to the spectral investigation of both closed and open cavities. To the best of our knowledge, a systematic investigation of the problem of calculating the spectrum of the complex eigenvalues for open arbitrary 2D cavities is absent in the literature, though some particular problems are discussed in some publications (for instance, see [7,8]). At the same time, the problem of the resonance scattering by slotted cylinders, mostly of canonical shape, has been investigated by many authors. For example, scattering from a cavity-backed slit in a ground plane in both E- and H-polarized cases are studied in Jeng and Tzeng [9,10], where the problems are analyzed by Harrington and Mautz's generalized network formulation. Radar cross-section (RCS) studies of a partially open rectangular box in the resonant region have been carried out by Wang *et al.* [11] using the method of moments (MoM) solution to the electric field integral equation (EFIE). Scattering and coupling properties of a slotted elliptic cylinder are treated in Hussein *et al.* [12] using an integral equation for the aperture field. The solution of the integral equation is carried out by expressing the aperture field in terms of a Fourier series expansion with unknown coefficients. Then Galerkin's technique is introduced to solve for the unknown aperture field coefficients. The previously mentioned papers represent only the small part of publications on this subject, which continues to be a focus of current investigations on this topic.

14.1.2 Development of a systematic approach

A widely used formulation of electromagnetic scattering from open scatterers is based on the EFIE. This first kind equation has the generic difficulty of ill-posedness associated with all first kind equations, with the consequence that the solution to the discretized equation does not converge to the true solution as the underlying mesh is refined.

An effective mathematical approach (see [13, 15, 16]), free from this difficulty, is entirely based on the rigorous so-called MAR. In this method, the EFIE is considered as the starting point for its transformation by a regularization technique. By use of this technique, the problem of wave diffraction by open cavities is reduced to the solving of an infinite second-kind system (or systems) of linear algebraic equations of the form $(I + H)X = B$ for the set of Fourier coefficients $X = \{x_n\}_{n=0}^{\infty}$, to be determined as a result of expansion of the two-sided surface current density in Fourier series. Here I is the identity operator, H is a completely continuous operator in the functional space of square summable sequences l_2, and $B = \{b_n\}_{n=0}^{\infty}$ is a known right-hand side of the equation, belonging to the same class l_2. The system is solved numerically by the truncation method wherein the matrix and vectors are truncated to a finite size. As the size is progressively increased, the fast convergence of the solutions to the truncated systems, along with the ability to guarantee any predetermined accuracy of calculations, makes this method an ideal instrument for calculation of the spectrum of the complex eigenvalues. The dispersion equation arises as a result of solution of the homogeneous equation, when $B \equiv 0$. To find the non-trivial solutions of the homogeneous equation it is necessary to require the fulfilment of the condition $\det(A) = 0$, where $A = \{A_{nm}\}_{n,m=0}^{N-1}$ is the truncated matrix, of order the truncation number N. The basics of the MAR are described by Vinogradov and others [13, 14], while its implementation for solving wave scattering problems for open arbitrary-shaped cylinders, for both E and H polarizations, is described by Tuchkin and others [15–20].

In this chapter, the algorithm based on the rigorous MAR is used to examine the spectrum of the complex eigenvalues of a 2D open perfectly conducting cavity of arbitrary shape. The TM-polarized eigenmodes are studied. In fact, even for open cavities of the classical shape (circular, elliptic and rectangular cylinder), comprehensive studies on this issue are largely absent in the literature. The partial study [20], related to the circular slotted cylinder, does not cover all aspects arising in the computation of the complex eigenvalues.

This chapter is divided into two parts. The first part deals with computation of the spectrum of complex eigenvalues. We start from the slotted circular cylinder, then study open elliptic and rectangular cylinders with a slit, which may be moved along their perimeters. Next we analyze the spectrum of some selected open cavities, being rectangular cavities with attached flanges. In the second part of this chapter, we use the data of the resonance spectrum for studies of the resonance response of these cavities when excited by an obliquely incident E-polarized plane wave. The study includes the computation of the surface current density, the far-field and the frequency dependence of the RCS.

14.2 Mathematical background

14.2.1 Schematic description of the MAR

We give later, in brief, the formal scheme of the MAR's implementation for solving the problem related to scattering of an E-polarized plane wave by an open perfectly electrically conducting (PEC) cylindrical cavity of arbitrary shape.

The contour of the cylindrical cross-section L of the closed structure is parameterized by the arc length function $L(\theta) = \{x(\theta), y(\theta)\}$ dependent on the parameter θ, where $\theta \in [-\pi, \pi]$. The approach requires the curve L to be smooth such that $L(\theta)$ is continuous function with continuous first derivative and the second derivative must exist. A cavity boundary L_0 is defined by the parameter subinterval $\theta \in [-\theta_0, \theta_0]$, while the aperture $L_1$is created by the removal from L of the segment $L_0 : L_1 = \{x(\theta), y(\theta)\}$, where $\theta \in [-\pi, -\theta_0] \cup [\theta_0, \pi]$.

A detailed account of the solution and its deduction is contained in the recent publication [19]. This problem is treated as a mixed boundary value problem for the Helmholtz equation. Using the classical potential theory [21], we seek the solution in the form of a single-layer potential formed from the unknown function representing the jump of the surface current density at the cavity surface. After imposing the boundary condition, we arrive at the well-known EFIE. In most cases, this equation is taken as the basis for numerical solution. The ill-posedness of this first kind Fredholm equation forces us to seek various ways to implement numerical algorithms which may produce adequate results. However, the reliability of these algorithms is doubtful since none of them overcomes the generic difficulties caused by ill-posedness of the initial equation. At this point, only further transform of the EFIE to a well-conditioned second-kind Fredholm equation provides the possibility to construct a robust numerical algorithm. The MAR is designed to provide a mathematically rigorous transform to a well-conditioned Fredholm equation of the second kind. The main steps in implementation of the MAR are as follows:

- The 'standard' EFIE in operator form is represented as $AZ = B$, where A is the integral operator associated with the EFIE and its Green's function kernel, Z (a function representing the jump of the surface current density at the cavity surface) is the solution to be found and B is the known right-hand side of the equation, given by an incident field.
- Split the integral kernel (Green's function) into the singular and smooth parts. Concretely, this is achieved by extraction of logarithmic singularity in the 2D Green's function.
- Represent all functions under the integral by their exponential Fourier series and dealing with the kernel singularity analytically, derive dual series equations for the Fourier coefficients of the unknown function U defined on two subintervals of the parameter $\theta \in [-\pi, \pi]$.
- Use the Abel's integral transform to reduce analytically the determination of the solution of the dual series equations to that of solving a coupled set of infinite systems of linear algebraic equations of the second kind:

$$\begin{aligned} &(I + H_{11})X + H_{12}Y = B_1 \\ &H_{21}X + (I + H_{22})\,Y = B_2 \end{aligned} \tag{14.1}$$

  where $X = \{X_n\}_{n=0}^{\infty}$ and $Y = \{Y_n\}_{n=1}^{\infty}$ are infinite sequences of unknown coefficients, I is the identity operator, H_{ij} $(i,j = 1,2)$ are completely continuous operators in the functional space of the square summable sequences l_2, $B_1 \equiv \{B_{1n}\}_{n=0}^{\infty}$ and $B_2 \equiv \{B_{2n}\}_{n=1}^{\infty}$ are the Fourier coefficients of the known functions generated by a plane wave exciting the open cavity. The set of Fourier coefficients of the unknown normalized surface current density is a simple linear combination of X and Y.

The Fredholm nature of the operators H_{ij} in (14.3) ensures the fulfilment of the *Fredholm alternative*. The consequence of that is the possibility to use the truncation method for numerical solution of (14.3) with guaranteed convergence. The specific features of the MAR allow us not only to achieve convergence in itself but also to provide the fast convergence of the truncation method. One more attractive feature of the MAR is high effectiveness of the matrix filling. In fact, the computation of the matrix elements is reduced to the usage of the 2D and 1D fast Fourier transform (FFT) and of recurrence formulas for incomplete scalar products of the normalized Jacobi polynomials. The time of the matrix filling varies from fractions of second to a few seconds (in a case of the large matrices) on a standard PC in the examples as follows.

14.2.2 Scheme for finding the complex eigenvalues

A consequence of this powerful MAR-based method is that there is no need to devise the additional instruments for finding the complex eigenvalues: one may proceed straightforwardly, using the well-known basics of the linear algebra associated with matrix equations.

Our procedure is as follows. We consider the homogeneous truncated matrix equations (14.3) jointly, setting $B_1 = B_2 = 0$. It is well-known that non-trivial solutions of this equation may exist when its determinant is equal to zero, i.e. $\det(A) = 0$, where $A = \{A_{nm}\}_{n,m=0}^{N-1}$ is the truncated joint matrix of size $N \times N$ and N is the truncation number. The natural spectral parameter is frequency f. However, in theoretical studies it is more convenient to use the relative wave number, ka $(k = 2\pi/\lambda)$, where a is a characteristic size of the cavity. We will use the term 'frequency dependence' when calculating some quantity as a function of ka $(k = 2\pi f/c$, $c = 3 \times 10^8$m/sec is light velocity). Leaving aside the arguments supporting our assertion, we simply note that spectral dependence of the condition number $\mathrm{cond}A = \mathrm{cond}A\,(ka)$ on ka is highly informative for primary studies of the resonance spectrum. Inside certain frequency intervals, the function $\mathrm{cond}A\,(ka)$ exhibits sharp peaks when values ka approach the 'quasi-singular' points of the matrix determinant $\det(A)$. We use the term 'quasi-singular' to stress that because the argument of the function $\mathrm{cond}A\,(ka)$ takes only real values ka it is impossible to fulfil the condition $\det(A) = 0$ precisely since the roots of this equation lie in the complex plane. However, for the eigenmodes with relatively high quality factors (Qfactors) the imaginary parts of the roots (say, γ'') of equation $\det(A) = 0$ are significantly less than their real parts, γ' $(\gamma'' \ll \gamma')$. This circumstance explains the emergence of the sharp peaks in the graph of the spectral dependence $\mathrm{cond}A\,(ka)$. In fact, the 'resonance' values ka, as determined from the dependence $\mathrm{cond}A\,(ka)$, deviate only slightly from the accurate values of γ' $(|ka - \gamma'| \ll 1)$. Thus maxima of $\mathrm{cond}A\,(ka)$ clearly reveal almost precisely the location of the eigenvalues and the number of eigenmodes contained in any prescribed interval of ka. This preliminary information is especially important for arbitrary cavities, for which beforehand we have no accurate information about the spectrum of eigenvalues.

Thus, finding complex eigenvalues starts with the computation of $\mathrm{cond}A\,(ka)$. Then we restrict the calculated sequence of the resonance values ka to a certain number of eigenmodes for further investigation. These resonance values $(ka \sim \gamma')$ are treated as starting values, which we use in MATLAB code for finding the complex roots of equation $\det(A_N) = 0$. The desired accuracy of calculation of any complex eigenvalue $\gamma = \gamma' - i\gamma''$ is achieved by successive increase of the truncation number N to that value which would provide the accuracy that has been prescribed in advance. In most cases, we calculate eigenvalues with an accuracy of 5–6 significant decimal digits.

Let us illustrate the proposed scheme for finding the complex eigenvalues with an example of a deep duct-like cavity, shown in Figure 14.1(a). The bounding contour L_0 of this cavity is defined parametrically by

$$x = \cos\theta, y = \tan^{-1}\left(\frac{3}{2}\cos\theta\right) + q\sin\theta, \; -\theta_0 \le \theta \le \theta_0,$$

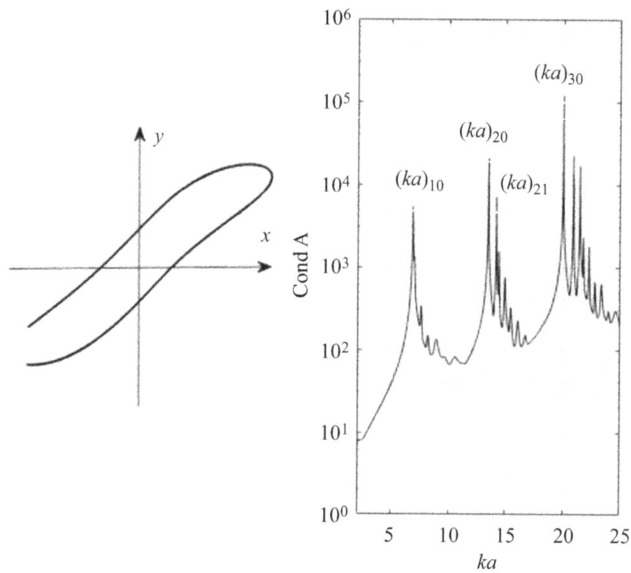

Figure 14.1 *(a) Duct-like cavity and (b) condition number condA versus wave number ka for the duct-like cavity: $\theta_0 = 150°, q = 0.4$*

with parameters $\theta_0 = 150°, q = 0.4$. It is worth emphasizing that before calculation of the function cond$A$ (ka), the distribution of the spectrum of eigenvalues along the axis ka is completely unknown. The calculated dependence cond$A$ (ka) removes this uncertainty and provides the first stage of the scheme when we fix the resonance values $(ka)_{n,m}$. For illustration purposes, we select three main peaks (resonances): $(ka)_{1,0} = 6.888$; $(ka)_{2,0} = 13.478$; $(ka)_{3,0} = 19.950$, as shown in Figure 14.1(b). Furthermore, we also fix one value of a supplementary resonance adjacent to $(ka)_{2,0}$, namely, $(ka)_{2,1} = 14.160$. These four values play the role of the first approximations or starting values for finding the complex eigenvalues $\gamma_{n,m} = \gamma''$, which are the complex roots of the function $\det(A_N) = 0$, where A_N is the truncated matrix (N being truncation number). Implementing this in MATLAB code, we find the roots $\left(\gamma_{n,m} = \gamma''\right)$ at increasing values N and stop the computational routine after stabilization of the pre-assigned number of significant digits. The results are shown in Table 14.1.

It can be seen from Table 14.1 that stabilization at six significant digits for the values of γ'' and γ'' occurs at truncation number $N = 2048$. This relatively large truncation value is characteristic for all shapes of deep cavities, since complex eigenvalues for deep cavities in absolute values are much larger than their counterparts for moderate or shallow cavities. It should be noticed that for most cavities of moderate depth, obtaining four significant decimal digits requires much smaller truncation numbers $N = 128$ or $N = 256$ (see Section 14.3). On current models of a standard PC, the computational time for finding the single complex eigenvalue $\gamma_{n,m}$ is not significant: for $N = 128$, it does not exceed 3 s, and for $N = 256$ it is less than 12 s.

Table 14.1 Selected complex eigenvalues $\gamma_{n,m}$ of the duct-like cavity with parameters $\theta_0 = 150^\circ, q = 0.4$ computed at increasing truncation number N with starting first approximations $(ka)_{n,m}$

N	$(ka)_{1,0} = 6.888$ $\gamma_{1,0}$	$(ka)_{2,0} = 13.478$ $\gamma_{2,0}$	$(ka)_{2,1} = 14.160$ $\gamma_{2,1}$	$(ka)_{3,0} = 19.950$ $\gamma_{3,0}$
128	6.887915 − i0.019127	13.477134 − i0.001768	14.159515 − i0.008852	19.950542 − i0.000235
256	6.889167 − i0.019200	13.484003 − i0.001689	14.171968 − i0.008917	19.968834 − i0.002183
512	6.889322 − i0.019210	13.484848 − i0.001681	14.173491 − i0.008930	19.971418 − i0.002193
1,024	6.889342 − i0.019211	13.484953 − i0.001680	14.173680 − i0.008932	19.971739 − i0.002195
2,048	6.889344 − i0.019211	13.484966 − i0.001680	14.173704 − i0.008932	19.971779 − i0.002195

14.3 Computation of the complex eigenvalues for various open cavities

In this section, we examine various open cavities. The same scheme for calculation of the complex eigenvalues is explained in detail in Section 14.2, so we restrict ourselves to short comments in the tables, where necessary.

14.3.1 Circular cylinder with longitudinal slit

The bounding contour L_0 of the circular cavity is defined parametrically by

$$x = a\cos\theta, y = a\sin\theta, \; -\theta_0 \le \theta \le \theta_0$$

It is known that eigenmodes of a closed hollow circular cylinder of radius a are double-degenerate so that one eigenvalue $\gamma_{nm}^{(0)}$ simultaneously describes two eigenmodes:

$$\begin{aligned} TM_{cnm} &\sim J_n\left(\gamma_{nm}^{(0)}\tfrac{\rho}{a}\right)\cos n\phi, n = 0,1,2,\ldots\ldots; \\ TM_{snm} &\sim J_n\left(\gamma_{nm}^{(0)}\tfrac{\rho}{a}\right)\sin n\phi, n = 1,2,\ldots, \end{aligned} \tag{14.2}$$

where $\gamma_{nm}^{(0)}$ is the m-th root of the Bessel function of order n, i.e. $J_n\left(\gamma_{nm}^{(0)}\right) = 0$. The integer m stands for the number of electromagnetic field variations along the radial coordinate $\rho \in (0,a)$ while the integer n stands for the number of variations along the angular variable $\phi \in (-\pi,\pi)$. For the closed circular hollow cylinder, there is no need to calculate the function cond$A\,(ka)$ as the sequence of eigenvalues has precisely readily accessed the roots $\gamma_{nm}^{(0)}$; however, the calculation of this function is very helpful when analyzing the spectrum of the complex eigenvalues for circular cylinder with a wide longitudinal slit. The geometry of the circular cylindrical cavity of radius a with longitudinal slit of aperture $w = 2a\sin(\pi - \theta_0) = 2a\sin(\theta_0)$ is shown in Figure 14.2 It is convenient to characterize width of the slit of the circular cavity by the angular semi-width $\theta_1 = \pi - \theta_0$ so the dimensionless width (defined as a fraction of the diameter of circle) is $w_{2a} = 2a\sin\theta_1/(2a) = \sin\theta_1$. In all numerical calculations we fix the radius a to be of a geometrical unit length, $a = 1$.

The appearance of the longitudinal slit immediately leads to splitting the eigenvalues γ_{nm} with non-zero index n $(n = 1,2,\ldots)$. The consequence of the splitting is emergence of two complex eigenvalues (γ_{cnm} and γ_{snm}) and two independent complex TM-modes: TM_{cnm} and TM_{snm}:

$$TM_{cnm} \sim J_n\left(\gamma_{cnm}\frac{\rho}{a}\right)\cos n\phi, \quad n,m = 1,2,\ldots, 0 \le \rho \le a,$$

$$TM_{snm} \sim J_n\left(\gamma_{snm}\frac{\rho}{a}\right)\sin n\phi, \quad n,m = 1,2,\ldots, 0 \le \rho \le a.$$

The results of calculation of the first ten complex eigenvalues of the open circular cavity are shown in Table 14.2 for four values of slit width $\theta_1 = 10°, 20°, 30°$and, $40°$. It can be seen from the data in Table 14.2 that the sequences of the complex eigenvalues γ_{cnm} and γ_{snm} are quite different from each other in their properties. The first and most significant distinction is that the value $\mathrm{Im}\,(\gamma_{snm})$ is very much smaller than

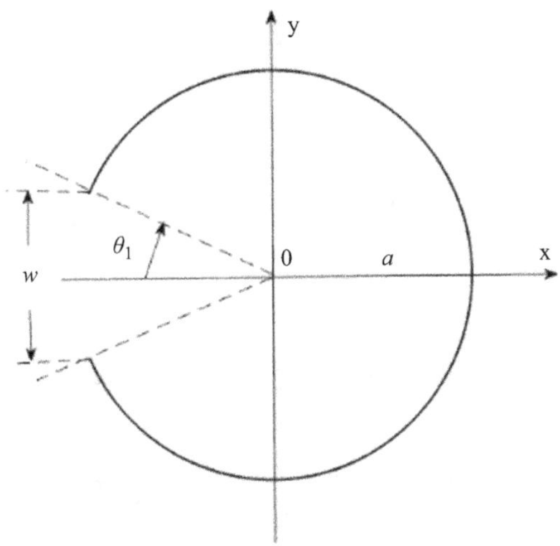

Figure 14.2 Slotted circular cavity of radius a with slit width $w = 2a\sin\theta_1$

$\mathrm{Im}(\gamma_{cnm})$. The reason for that lies in the specific distribution of the electromagnetic field describing the eigenmodes TM_{cnm} and TM_{snm}.

The maximum intensity of the electromagnetic field for TM_{cnm}-modes occurs at the wave crests, which are always near the slit, while the maximum intensity for all TM_{snm}-modes are shifted by 90° so that only the wave nodes face the slit. As a result, the presence of the slit distorts the TM_{cnm}-modes of the closed cylinder much more than the TM_{snm}-modes.

Further insight into the behaviour of the functions $\gamma_{cnm}(\theta_1)$ and $\gamma_{snm}(\theta_1)$ may be obtained from the graphs represented in Figure 14.3. The graphs present the real and negative imaginary parts of the seven complex eigenvalues as the angular slit semi-width θ_1 is increased from 5° to 60°. We restrict ourselves to one remark. In practice, the high values of the corresponding Qfactor, $Q_{snm}(\theta_1)$, which are about $10^7 - 10^8$ when $\theta_0 \sim 5°$, are not attainable since the *ohmic losses* significantly exceed the *radiation losses*. With the increase of slit width (or θ_1) the radiation losses at some value θ_1 become comparable with ohmic losses, and then dominate over ohmic losses, after passing the threshold value of a slit width. For this reason, in Table 14.2, we collected the computed values of the first ten complex eigenvalues for slit sizes not less than the minimum size, defined by the angle $\theta_1 = 10°$.

14.3.2 Elliptic cavity with moveable longitudinal slit

From geometrical and electromagnetic points of view, the slotted elliptic hollow cylinder is a much more complicated structure than the circular cylinder. If the slotted circular cylinder is characterized by slit width only, the description of a slotted elliptic cylinder (see Figure 14.4) requires additional parameters of eccentricity,

Table 14.2 Complex eigenvalues $\gamma_{c(s)nm}$ of open circular cavity of unit radius $a = 1$ with the slit of dimensionless width $w_{2a} = 2a \sin\theta_1 / (2a) = \sin\theta_1$

θ_1	0^0	10^0	20^0	30^0	40^0
γ_{c01}	2.404825	$2.394756 - i0.000526$	$2.359010 - i0.010238$	$2.301132 - i0.052355$	$2.245766 - i0.137520$
γ_{c11}	3.831705	$3.796472 - i0.004493$	$3.682232 - i0.083239$	$3.592331 - i0.291395$	$3.572761 - i0.556399$
γ_{s11}	3.831705	$3.831590 - i3.16 \times 10^{-7}$	$3.829642 - i0.00008219$	$3.819451 - i0.002146$	$3.790571 - i0.019571$
γ_{c21}	5.135622	$5.082494 - i0.013250$	$4.970887 - i0.208549$	$4.995897 - i0.568681$	$5.070815 - i0.998196$
γ_{s21}	5.135622	$5.134993 - i4.59 \times 10^{-6}$	$5.123660 - i0.0012667$	$5.068796 - i0.031244$	$4.994522 - i0.176267$
γ_{c02}	5.520078	$5.493768 - i0.006757$	$5.442347 - i0.057514$	$5.397607 - i0.085696$	$5.359233 - i0.081505$
γ_{c31}	6.380161	$6.313727 - i0.027105$	$6.291599 - i0.254718$	$6.508566 - i0.264041$	$6.456089 - i0.208372$
γ_{s31}	6.380161	$6.378371 - i0.00002772$	$6.344925 - i0.008056$	$6.244690 - i0.137402$	$6.251692 - i0.442474$
γ_{c12}	7.015586	$6.940606 - i0.035757$	$6.834610 - i0.274934$	$6.648444 - i0.834928$	$6.694294 - i1.498421$
γ_{s12}	7.015586	$7.015352 - i5.091 \times 10^{-6}$	$7.010037 - i0.001938$	$7.003250 - i0.028877$	$7.007679 - i0.060480$

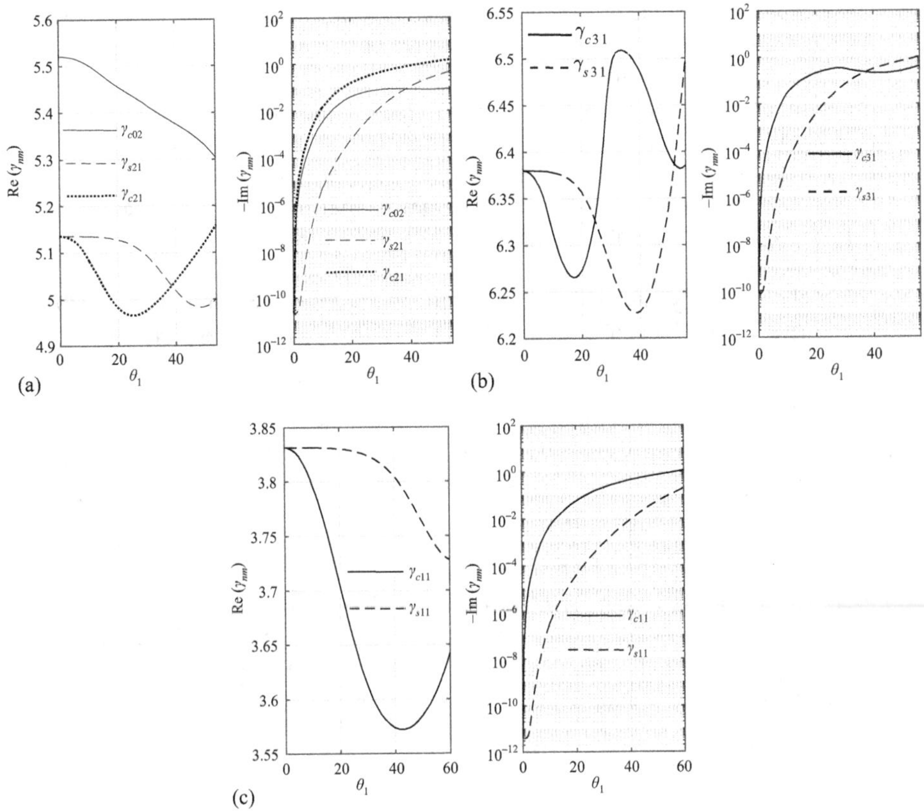

Figure 14.3 *Real and negative imaginary parts* $Re(\gamma_{c(s)nm})$, $-Im\left(\gamma_{c(s)nm}\right)$ *of complex eigenvalues* $\gamma_{c(s)nm}$ *versus semi-angular slit width* θ_1 *for a circular cavity: (a)* $\gamma_{c21}, \gamma_{s21}, \gamma_{c02}$*; (b)* $\gamma_{c31}, \gamma_{s31}$ *and (c)* $\gamma_{c11}, \gamma_{s11}$

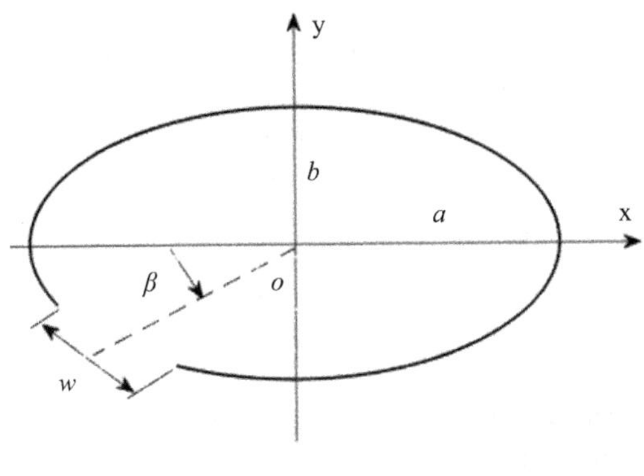

Figure 14.4 *Elliptic cavity with variably placed longitudinal slit of width w*

$e = \sqrt{1-(b/a)^2}$ (b is the minor semi-axis and a is the major semi-axis of ellipse), and position of a slit along the bounding contour. It is convenient to characterize a slit position by an angular parameter β, as shown in Figure 14.4.

The slit of dimensionless width $w_{2b} = w/(2b)$ (defined as a fraction of the minor axis of ellipse) is located along the ellipse circumference and the position of the slit middle is defined by the angle β measured from the negative x-axis ($\theta = \pi$) in an anticlockwise direction. Since the ellipse possesses fourfold symmetry, the slit shift may be considered in one of four quadrants of the $x-y$ plane. We choose the third quadrant, so that $0 \le \beta \le \pi/2$.

The elliptic cavity is specified by the following parameterization dependent on the parameters a and b(major and minor semi-axes of the ellipse, respectively):

$$x(\theta) = a\cos(\beta+\theta), y(\theta) = b\sin(\beta+\theta), \quad a = 1, b \le a, \theta \in [-\pi, \pi].$$

The screen L_0 is defined by the parameter subinterval $\theta \in [-\theta_0, \theta_0]$, while the slit $L_1$is created by the removal from L of the segment L_0: $L_1 = \{x(\theta), y(\theta)\}, \theta \in [-\pi, -\theta_0] \cup [\theta_0, \pi]$. With this parameterization the slit width w is given by $w = 2\sin\theta_0\sqrt{\sin^2\beta + b^2\cos^2\beta}$ so that $w = 2b\sin\theta_0$ if $\beta = 0°$. We find the semi-angular size of the screen (required in the solution algorithm) with slit w at the position β to be $\theta_0 = \pi - \sin^{-1}\left(w/\left(2\sqrt{\sin^2\beta + b^2\cos^2\beta}\right)\right)$.

The replacement of the circular cylinder by the elliptic cavity immediately leads to the disappearance of the degenerate eigenmodes that are characteristic of the circular cavity. Schematically, it can be denoted as

$$\gamma_{nm}^{(0)} \to \begin{cases} \gamma_{cnm}^{(0)} \\ \gamma_{snm}^{(0)} \end{cases}, \gamma_{cnm}^{(0)} \ne \gamma_{snm}^{(0)}$$

The multiparameter nature of the discussed problem forces us to restrict examination of a very few combinations of the parameters (e, w_{2b}, β). In the study by Zhang and Shen [22] the cut-off frequencies of the closed elliptical waveguide were calculated for eccentricities $e = 0.1, 0.5, 0.9$. The first five complex eigenvalues for the open elliptical cavity with eccentricity $e = 0.1$ and $e = 0.5$ are presented in Tables 14.3 and 14.4, respectively.

Several waveguides used in practice possess eccentricity e varying around 0.8, for example, $e = 0.825$ (EW20, HELIAX® Standard Elliptical Waveguide, 2.3–2.4 GHz, unjacketed); $e = 0.787$ (EWP180-180, HELIAX® Premium Elliptical Waveguide, 17.7–19.7 GHz, unjacketed); $e = 0.777$ (EW220, HELIAX Standard Elliptical Waveguide, 21.2–23.6 GHz, unjacketed) and $e = 0.8067$ (FLEXWELL® standard elliptical waveguide).

Fixing the value of eccentricity as $e = 0.8$ ($b/a = 0.6$), we calculate the first five eigenvalues of the closed elliptical cavity and open cavity with slit width $w_{2b} = 0.25$ and $w_{2b} = 0.5$ (see Table 14.5).

It can be seen from Tables 14.3–14.5 that eigenvalues with indices c ('cosine') are exposed to maximum influence or displacement when $\beta = 0°$, and minimum influence when $\beta = 90°$. For eigenvalues with indices s ('sine'), the situation is diametrically opposite: the maximum influence occurs at $\beta = 90°$ and minimum at

Table 14.3 *Complex eigenvalues $\gamma_{c(s)nm}$ of open elliptic cavity with eccentricity $e = 0.1$ $(b/a \approx 0.995)$, slit width relative to the minor axis of the ellipse $w_{2b} = 0,\ 0.2,\ 0.5$, slit location $\beta = 0°,\ 90°$*

	Closed cavity	$\beta = 0^0$		$\beta = 90^0$	
	$w_{2b} = 0$	$w_{2b} = 0.2$	$w_{2b} = 0.5$	$w_{2b} = 0.2$	$w_{2b} = 0.5$
γ_{c01}	2.410889	$2.397352 - i0.000950$	$2.308215 - i0.051791$	$2.397105 - i0.000903$	$2.306667 - i0.052116$
γ_{c11}	3.836538	$3.788565 - i0.008291$	$3.598788 - i0.288672$	$3.836331 - i0.000000$ (97)	$3.824311 - i0.002119$
γ_{s11}	3.846190	$3.845988 - i0.000000$ (97)	$3.834147 - i0.002124$	$3.797449 - i0.008438$	$3.604835 - i0.291496$
γ_{c21}	5.148339	$5.073794 - i0.025713$	$4.999959 - i0.566463$	$5.147430 - i0.000014$	$5.081766 - i0.030803$
γ_{s21}	5.148569	$5.147455 - i0.000014$	$5.082743 - i0.030611$	$5.077177 - i0.024798$	$5.016112 - i0.567051$

Table 14.4 Complex eigenvalues $\gamma_{c(s)nm}$ of open elliptic cavity with eccentricity $e = 0.5$ $(b/a \approx 0.866)$, slit width relative to the minor axis of the ellipse $w_{2b} = 0,\ 0.2,\ 0.5$, slit location $\beta = 0°,\ 90°$

	Closed cavity	$\beta = 0^0$		$\beta = 90^0$	
	$w_{2b} = 0$	$w_{2b} = 0.2$	$w_{2b} = 0.5$	$w_{2b} = 0.2$	$w_{2b} = 0.5$
γ_{c01}	2.596779	$2.587259 - i0.000623$	$2.519826 - i0.033799$	$2.580698 - i0.000921$	$2.477939 - i0.046687$
γ_{c11}	3.986407	$3.951560 - i0.005117$	$3.791382 - i0.220869$	$3.986177 - i0.000000$ (65)	$3.974748 - i0.001226$
γ_{s11}	4.282215	$4.282106 - i0.000000$ (52)	$4.274973 - i0.001260$	$4.227058 - i0.008499$	$3.986304 - i0.298467$
γ_{c21}	5.392343	$5.315885 - i0.020910$	$5.168234 - i0.493942$	$5.541569 - i0.000010$	$5.477148 - i0.021275$
γ_{s21}	5.542780	$5.542148 - i0.000009$	$5.500553 - i0.017153$	$5.365756 - i0.007487$	$5.402368 - i0.122855$

$\beta = 0°$. The explanation is related to the different locations of the *wave crests* and *wave nodes* for 'c-modes' and 's-modes', which are shifted relative to each other by 90°, so that maximum distortion of the wave crests for 'c-modes' occurs at the vertical location of a slit ($\beta = 0°$), but that for 's-modes' occurs at the horizontal location ($\beta = 90°$).

14.3.3 Open rectangular cavity with finite flanges

The deep rectangular cavity embedded in a ground plane has been frequently discussed in the literature. It may be regarded as a terminated rectangular waveguide with infinite flanges. We will study the more realistic structure incorporating finite flanges. Since the MAR requires the smoothness of the bounding contours, we parameterize the rectangular cavity rounding off the corners, as shown in Figure 14.5. The clearly

Table 14.5 Complex eigenvalues $\gamma_{c(s)nm}$ of open elliptic cavity with eccentricity $e = 0.8$ $(b/a = 0.6)$, slit width relative to the minor axis of the ellipse $w_{2b} = 0,\ 0.2,\ 0.5$, slit location $\beta = 0°,\ 45°,\ 90°$

	Closed cavity $w_{2b} = 0$	**$\beta = 0^0$ $w_{2b} = 0.25$**	**$\beta = 45^0$ $w_{2b} = 0.25$**	**$\beta = 90^0$ $w_{2b} = 0.5$**
γ_{c01}	3.293277	3.288247 – i0.000492	3.278019 – i0.001076	3.135481 – i0.039600
γ_{c11}	4.577284	4.555823 – i0.003792	4.536363 – i0.005496	4.566875 – i0.000350
γ_{c11}	5.825630	5.825596 – i0.0000004419	5.803504 – i0.007190	
γ_{c21}	5.944017	5.891059 – i0.015591	5.902001 – i0.007694	5.930531 – i0.001808
γ_{c21}	6.990664	6.990418 – i0.000005464	6.921508 – i0.024322	6.929108 – i0.010091

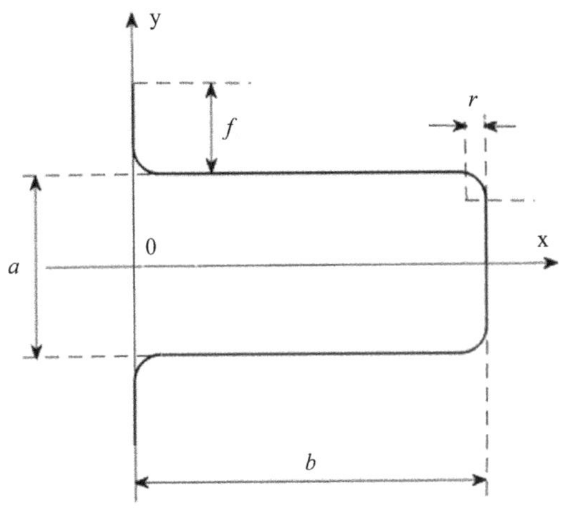

Figure 14.5 Flanged rectangular cavity with rounded off corners and flanges of length f

visible chamfering is only for illustrative purpose. In computations, the rounding off radius r is negligibly small with dimensions of the cavity (a, b); size of the flanges f may take any value.

It is well known that eigenvalues $\gamma_{nm}^{(0)}$ of the TM-modes in the closed rectangular cavity of dimensions a and b are given by the formula

$$\gamma_{nm}^{(0)} \equiv (ka)_{nm} = \pi\sqrt{m^2 + q^2n^2}, q = a/b, n, m \geq 1, \tag{14.3}$$

where m is the number of the field variations along the length a, and n that along the length b.

The computed frequency dependence cond$A$ (ka), presented in Figure 14.6 for three values of aspect ratio $(b/a = 2, 5, 20)$ for a rectangular cavity without flanges $(f \equiv 0)$, definitely reveals the coincidence of the resonant values cond$A$ (ka) with the results calculated by formula (14.3).

This coincidence becomes more accurate with growth of the cavity depth (see Table 14.6). The data, presented in Figure 14.6 and Table 14.6, force us to the conclusion that the main mechanism of the resonance spectrum formation is due to the field oscillations along the -coordinate between the horizontal walls of the deep rectangular cavity, with the dominance of the modes TM_{1m} $(m = 1, 2, \ldots)$. In fact, the deep cavity, in an electromagnetic sense, behaves similarly to the Fabry–Perot resonator. That is why the base of a cavity or its opening has no significant impact on the spectrum of complex eigenvalues. Flanges insignificantly affect cavities only of moderate depth, not exceeding the aspect ratio $b/a = 2$ (see Tables 14.7 and 14.8). The complex eigenvalues for cavities without flanges $(f \equiv 0)$ with aspect ratio $b/a = 2, 5, 20$ are presented in Table 14.7.

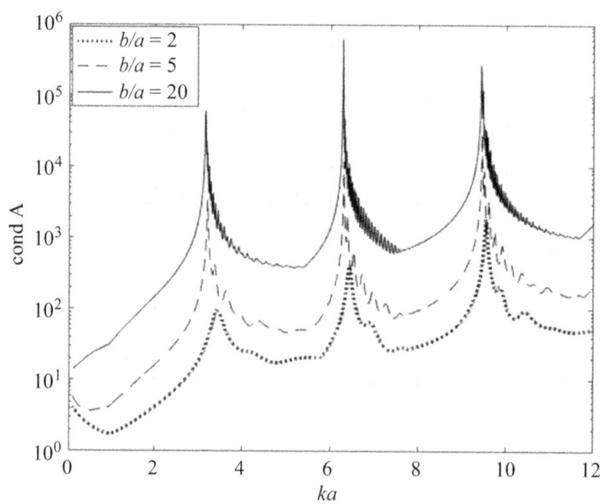

Figure 14.6 Condition number condA versus wave number ka for the open rectangular cavity without flanges: $b/a = 2, 5, 20$

Table 14.6 Resonance values cond (A) for the open rectangular cavity without flanges with aspect ratio $b/a = 5,\ 20$ and the real eigenvalues γ_{nm} calculated by formula (14.3)

	$b/a = 5$		$b/a=20$	
	Formula (14.3)	**cond (A) (resonances)**	**Formula (14.3)**	**cond (A) (resonances)**
γ_{11}	3.203	3.198	3.145	3.145
γ_{21}	3.383	3.363	3.157	3.157
γ_{31}	3.663	3.627	3.176	3.176
γ_{41}	4.023	3.949	3.203	3.203
γ_{51}	4.442	4.379	3.238	3.237
γ_{12}	6.314	6.313	6.285	6.284
γ_{22}	6.407	6.401	6.291	6.290
γ_{32}	6.559	6.545	6.314	6.314
γ_{42}	6.767	6.741	6.332	6.331
γ_{52}	7.024	6.987	6.353	6.352
γ_{13}	9.445	9.444	9.426	9.424
γ_{23}	9.508	9.504	9.430	9.428
γ_{33}	9.611	9.603	9.436	9.434
γ_{43}	9.754	9.740	9.445	9.443
γ_{53}	9.934	9.915	9.457	9.455

Table 14.7 Complex eigenvalues γ_{nm} of the open rectangular cavity with aspect ratio $b/a = 2,\ 5,\ 20$ without flanges ($f \equiv 0$)

	$b/a = 2$	$b/a = 5$	$b/a = 20$
γ_{11}	$3.4213 - i0.10155$	$3.1978 - i0.00878$	$3.1454 - i0.000155$
γ_{21}		$3.3615 - i0.03336$	$3.1569 - i0.000618$
γ_{31}		$3.6190 - i0.06937$	$3.1760 - i0.001381$
γ_{41}			$3.2025 - i0.002434$
γ_{51}			$3.2363 - i0.003758$
γ_{12}	$6.4433 - i0.03720$	$6.3124 - i0.00284$	$6.2851 - i0.000048$
γ_{22}	$6.9051 - i0.14044$	$6.3995 - i0.01121$	$6.3005 - i0.000431$
γ_{32}		$6.5422 - i0.02471$	$6.3140 - i0.000765$
γ_{42}		$6.7375 - i0.04280$	$6.3313 - i0.001193$
γ_{52}		$6.9808 - i0.06470$	$6.3524 - i0.001712$
γ_{13}	$9.5368 - i0.02008$	$9.4445 - i0.00147$	$9.4260 - i0.000024$
γ_{23}	$9.8660 - i0.07842$	$9.5037 - i0.00587$	$9.4299 - i0.000098$
γ_{33}		$9.6018 - i0.01312$	$9.4363 - i0.000220$
γ_{43}		$9.7374 - i0.02299$	$9.4454 - i0.000391$
γ_{53}		$9.9087 - i0.03536$	$9.4570 - i0.000610$

Table 14.8 Complex eigenvalues γ_{nm} of the open rectangular cavity with aspect ratio $b/a = 2$ with attached flanges of length f

f	γ_{11}	γ_{12}	γ_{22}	γ_{13}	γ_{23}
0	$3.4213 - i0.10155$	$6.4433 - i0.03720$	$6.9051 - i0.14044$	$9.5368 - i0.02008$	$9.8660 - i0.07842$
0.25	$3.4179 - i0.10588$	$6.4430 - i0.03882$	$6.9045 - i0.14613$	$9.5367 - i0.02088$	$9.8659 - i0.08144$
1	$3.4186 - i0.10529$	$6.4428 - i0.03872$	$6.9039 - i0.14611$	$9.5367 - i0.02094$	$9.8659 - i0.08168$
2	$3.4187 - i0.10530$	$6.4429 - i0.03873$	$6.9039 - i0.14611$	$9.5367 - i0.02093$	$9.8659 - i0.08167$
5	$3.4188 - i0.10529$	$6.4429 - i0.03873$	$6.9040 - i0.14607$	$9.5367 - i0.02093$	$9.8659 - i0.08166$

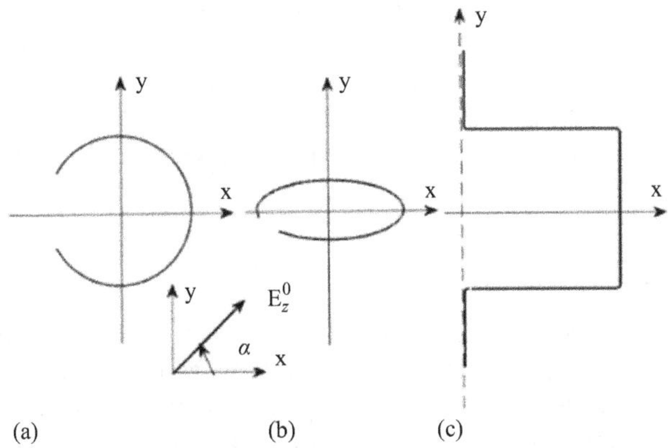

Figure 14.7 Plane wave excitation of slotted cavities: (a) circular cylinder; (b) elliptic cylinder and (c) flanged rectangular cavity

The results contained in Table 14.8 indicate that spectrum of the complex eigenvalues practically does not change when the relative size of flanges, f/a, exceeds the value 1. The important consequence of these calculations is that the numerous results in the literature for the rectangular cavity embedded in a ground plane adequately describes the electromagnetic field inside the deep cavity with finite flanges with no less than.

14.4 Resonance response of slotted cavities

In this section, we examine the resonance response of the slotted cavities studied in previous sections, when excited by an E-polarized plane wave, as shown in Figure 14.7. The plane wave strikes each of the slotted cavities at the incidence angle α, measured from the positive x-axis in anticlockwise direction. The geometrical parameters of the slotted cavities are as described in Section 14.3.

The investigation includes the calculation of the *current density jump* J over the cavity surface, the frequency dependence of the *monostatic* RCS, $\sigma_B(ka)$ and the normalized far-field $A(\theta)$. The representation of these characteristics in the form of a Fourier series is described by Vinogradova [19]. Before describing the calculations, we wish to stress the main principles, which make the calculations more informative.

First, the general oscillation theory for high*Q*factor resonance systems states that resonance peaks of the different physical characteristics (in our case, $J, \sigma_B(ka), A(\theta)$) occur at frequencies (or relative wave numbers ka), close to, but shifted from the real parts $\text{Re}(\gamma_{nm})$ of the complex eigenvalues γ_{nm}. The shift Δ is inversely proportional to the unloaded*Q*factor of a resonance system. This fact mostly concerns the computation in the far-field zone, whereas in near-field zone, setting $ka = \text{Re}(\gamma_{nm})$ guarantees the appearance of the resonance oscillations.

The second point to be mentioned concerns the process of excitation of the slotted cavities. In general, the most effective excitation is 'internal', when a compact source (such as a long electric dipole) is placed directly at the wave crest of one of eigenmodes we wish to excite. In the case of the plane wave excitation ('external' excitation), its effectiveness is dependent on the chosen incidence angle α.

14.4.1 Surface current calculations

In this subsection, we calculate the distribution of the surface current density at the fixed values $ka = \mathrm{Re}\,(\gamma_{nm})$, expecting to find the best possible way to excite the cavity by variation of the incidence angle α. Extensive calculations show that in general some 'optimal' incidence angle α does not exist. Instead of a single 'optimal' value, we have a certain interval of angles, inside which any value α is practically equivalent in the sense of producing 'optimal' excitation of a resonance cavity.

The first example is related with excitation of the four TM-modes in a slotted circular cavity (with $\theta_1 = 10°$, Table 14.2): (a) TM_{c01}, (b) TM_{c11}, (c) TM_{s11} and (d) TM_{c21}. The corresponding results for logarithmic surface current density $|J\,(\theta)|$ are presented in Figure 14.8(a)–(d). The graphics are chosen to demonstrate the 'dynamics' of the formation of the resonance distribution $|J\,(\theta)|$, when incidence angle α approaches the interval of the most effective excitation in steps of $\Delta\alpha = 10°$.

It should be noticed that the sharp peaks of the function $|J\,(\theta)|$ at the edges $(\theta = \pm 170°)$ are due to the so-called 'edge effect'. It can be seen that modes with a non-zero number of oscillations along the angular coordinate θ $(n = 1, 2, \ldots)$ are optimally excited in the range $50° < \alpha \leq 60°$. Figure 14.9 shows the best choice of the incidence angle α for exciting the modes TM_{s11} $(\alpha = 60°)$ and TM_{s21} $(\alpha = 55°)$ in the slotted elliptical cavity $(e = 0.8, w_{2b} = 0.25)$.

It is also instructive to trace the formation of the resonance distribution $|J\,(\theta)|$ in the narrow frequency range $\mathrm{Re}\,(\gamma_{nm}) - \delta \leq ka \leq \mathrm{Re}\,(\gamma_{nm}) + \delta$ for small δ An example is shown in Figure 14.10, where excitation of the TM_{11}-.mode in the deep $(b/a = 20)$ open rectangular cavity is illustrated by the resonance behaviour of the surface current density $|J\,(\theta)|$ at $ka = \mathrm{Re}\,(\gamma_{11}) = 3.1454$ (see Table 14.7). The graphs use the values $\alpha = 0°, \delta = 0.0016$. The function $|J\,(\theta)|$ has been calculated in the interval $3.1438 \leq ka \leq 3.1470$ with steps of $\Delta ka = 0.0001$. Figure 14.10 (left) contains 17 plots in the interval $3.1438 \leq ka \leq 3.1454$ while Figure 14.10 (right) contains the same number of plots, calculated in the interval $3.1454 \leq ka \leq 3.1470$.

For deep rectangular cavities, excitation of resonance modes TM_{nm} is most effective at different values of the incidence angle α, depending on the type of oscillations. For example, oscillations $TM_{n,1}$ are effectively excited at $\alpha = 0°$, $TM_{n,2}$ at $\alpha = 45°$ and $TM_{n,3}$ at $\alpha = 50°$. As an illustration of this assertion, consider the surface current density induced in a rectangular cavity with aspect ratio $b/a = 20$ with lowest index $m = 1$ for $n = 1, 2, 3, 4, 5$. The results of calculations are shown in Figure 14.11 for five distributions $|J\,(\theta)|$, when the relative wave number ka successively takes the values identified in Table 14.7: $\mathrm{Re}\,(\gamma_{11}) = 3.1454$, $\mathrm{Re}\,(\gamma_{21}) = 3.1569$, $\mathrm{Re}\,(\gamma_{31}) = 3.1760$, $\mathrm{Re}\,(\gamma_{41}) = 3.2025$ and $\mathrm{Re}\,(\gamma_{51}) = 3.2363$. In contrast to the oscillations of the type

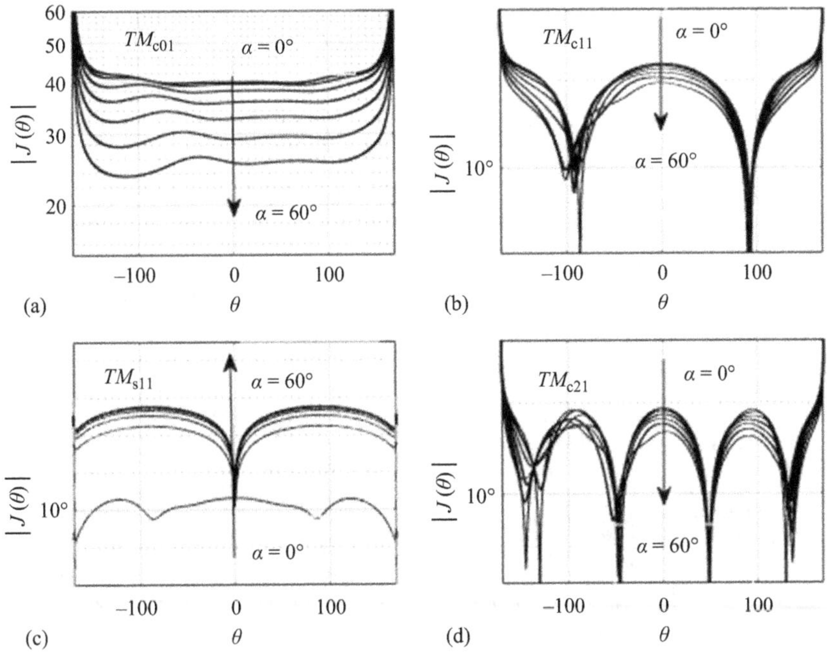

Figure 14.8 Effectiveness of TM-modes excitation upon incidence angle α in a slotted circular cavity ($\theta_1 = 10°$): logarithmic surface current density $|J(\theta)|$ for $\alpha = 0°, 10°, 20°, 30°, 40°, 50°$ and, $60°$

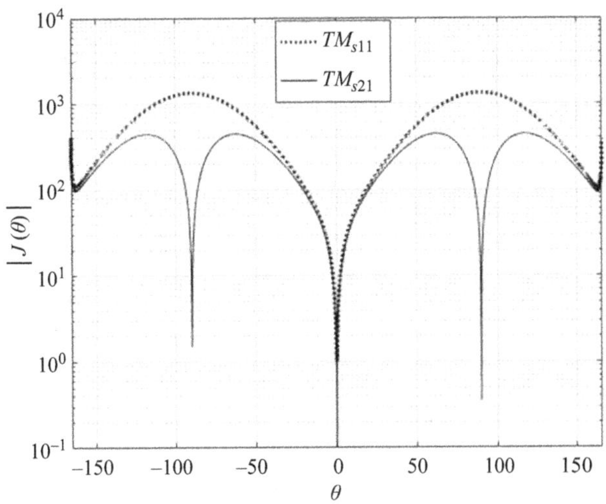

Figure 14.9 Logarithmic surface current density $|J(\theta)|$ for the slotted elliptical cavity ($e = 0.8, w_{2b} = 0.25, \beta = 0°$) at $ka = Re(\gamma_{s11})$ for $\alpha = 60°$ and $ka = Re(\gamma_{s21})$ for $\alpha = 55°$

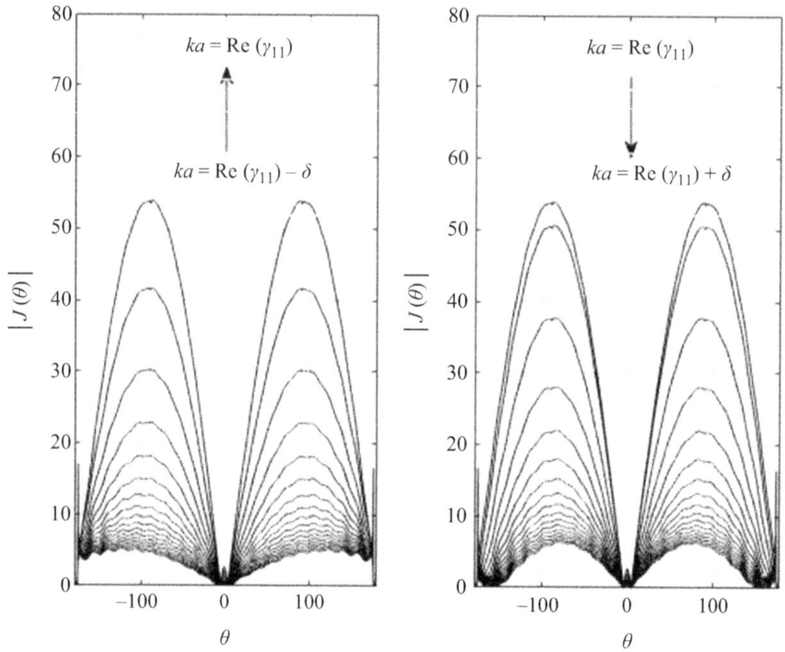

Figure 14.10 Surface current density $|J(\theta)|$ for an open deep rectangular cavity ($b/a = 20$) excited at $\alpha = 0°$ in vicinity of resonance value $ka = Re(\gamma_{11}) = 3.1454$: $Re(\gamma_{11}) - \delta \leq ka \leq Re(\gamma_{11})$ (left) and $Re(\gamma_{11}) \leq ka \leq Re(\gamma_{11}) + \delta$ (right); $\delta = 0.0016$, $\Delta ka = 0.0001$

$TM_{n,2p+1}$ ($p = 0, 1, 2, \ldots$), the eigenmodes $TM_{n,2p}$ ($p = 1, 2, \ldots$) are not excited at all, when $\alpha = 0°$. They are effectively excited in the interval $45° \leq \alpha \leq 60°$.

14.4.2 Far-field calculations

Among all aspects of resonance scattering, we now consider the most delicate one, which cannot be properly analyzed by the MoM and other 'brute force' methods discussed in Section 14.1.1; more accurate techniques are required for its analysis. We have in mind the so-called 'resonance doublet' or 'double extremum' (remittent maximum and minimum) [13], which arises in the narrow frequency band near those frequencies corresponding to real parts of the complex eigenvalues $\text{Re}(\gamma_{nm})$. It manifests itself when calculating the frequency dependence of the monostatic RCS $\sigma_B(ka)$. The method we use here is capable of resolving such doublets, i.e. of accurately obtaining the alternating maxima and minima in the extremely narrow intervals of ka values, of form $(\text{Re}(\gamma_{nm}) - \delta_1, \text{Re}(\gamma_{nm}) + \delta_2)$, where $\delta_1, \delta_2 \ll \text{Re}(\gamma_{nm})$ In most cases, the maximum $\sigma_B(ka)$ occurs at $ka = \text{Re}(\gamma_{nm}) - \delta_1$ and the minimum at $ka = \text{Re}(\gamma_{nm}) + \delta_2$.

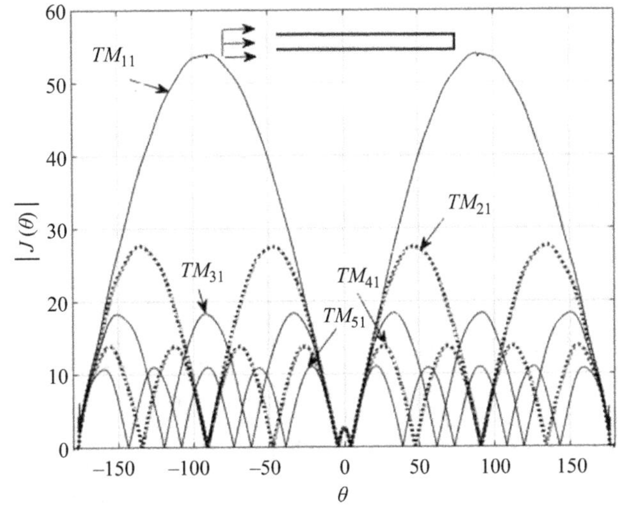

Figure 14.11 Resonance distribution of the surface current density $|J(\theta)|$ for a rectangular cavity ($b/a = 20$) in a case of excitation at $\alpha = 0°$ of $TM_{n,1}$-modes: $n = 1, 2, 3, 4, 5$

As an example of this phenomenon, the RCS $\sigma_B(ka)$ for the slotted ($\theta_1 = 10°$) circular cavity excited by an E-polarized plane wave ($\alpha = 0°$) is shown in Figure 14.12. The behaviour of $\sigma_B(ka)$ at the incidence angles $\alpha = 45°$ and $\alpha = 90°$ is shown in Figure 14.13(a) and (b), respectively. The efficiency and accuracy of the MAR allow the fine resolution of the RCS response to be obtained. Careful inspection of Figures 14.12 and 14.13 shows that the resonance doublets occur at the same wavenumber *ka*, or very nearly so; the accurate determination of the RCS in the vicinity is a unique advantage of the MAR approach.

Such 'resonance doublets' arise under excitation of eigenmodes with high Q factor values ($Q_{nm} = -2\mathrm{Re}(\gamma_{nm})/\mathrm{Im}(\gamma_{nm})$). Similar behaviour is demonstrated in the monostatic RCS, $\sigma_B(ka)$, for the slotted elliptical cavity ($e = 0.8, \beta = 0°$) at two incident angles: $\alpha = 0°$ (Figure 14.14(a)) and $\alpha = 60°$ (Figure 14.14(b)). Figure 14.15 contains the RCS, $\sigma_B(ka)$, for the same elliptical cavity ($e = 0.8$) and the same slit width ($w_{2b} = 0.25$), but with aperture shifted along the elliptic contour by 90° in an anticlockwise direction ($\beta = 90°$). The calculations are performed for two values of incident angle: $\alpha = 0°$ (see Figure 14.15(a)) and $\alpha = 90°$ (see Figure 14.15(b)). As observed for the circular cavity, the location of the resonance doublets is the same, or very nearly so, whatever the angle of incidence.

We conclude this set of calculations by presenting in Figure 14.16 the graph of $\sigma_B(ka)$ for an extremely deep ($b/a = 20$) rectangular cavity, irradiated by a plane wave at an incidence angle of $\alpha = 50°$.

All the graphs presented in Figures 14.12–14.16 feature the phenomenon of the resonance doublet, which we now consider in more detail, by magnifying the graphs at those parts of the dependence $\sigma_B(ka)$ where this phenomenon occurs, i.e.

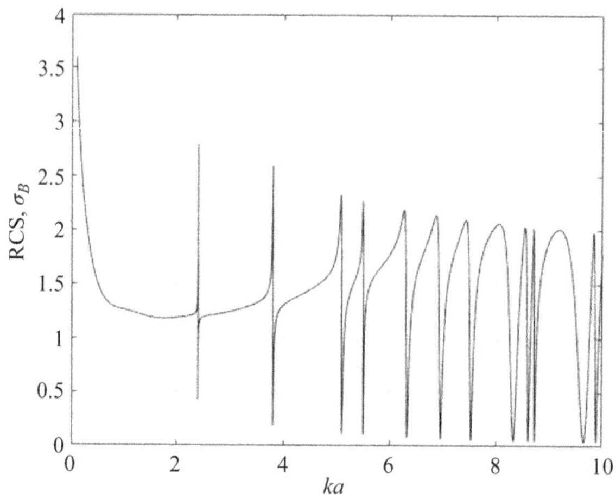

Figure 14.12 *Frequency dependence $\sigma_B(ka)$ for a slotted ($\theta_1 = 10°$) circular cavity excited at the incidence angle $\alpha = 0°$*

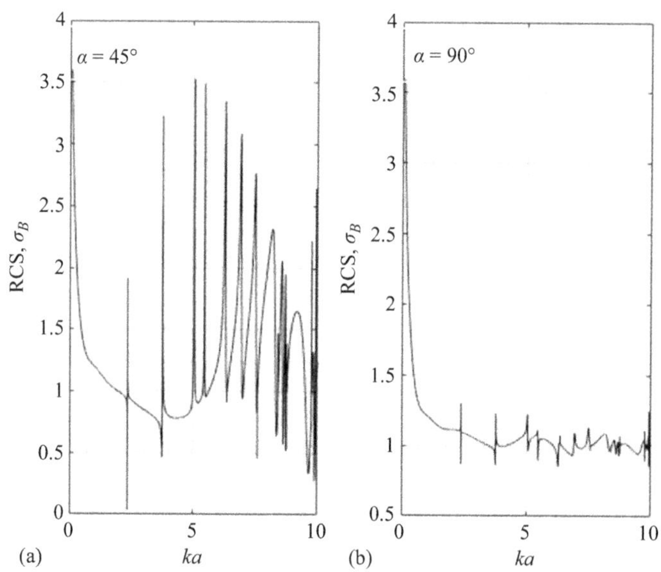

Figure 14.13 *Frequency dependence $\sigma_B(ka)$ for a slotted ($\theta_1 = 10°$) circular cavity excited at the incidence angles: (a) $\alpha = 45°$ and (b) $\alpha = 90°$*

for wavenumbers $ka \in (\mathrm{Re}(\gamma_{nm}) - \delta_1, \mathrm{Re}(\gamma_{nm}) + \delta_2)$, for suitable $\delta_1, \delta_2 \ll \mathrm{Re}(\gamma_{nm})$ To keep consistency, when comparing the *resonance doublet* for differently shaped cavities, we consider those values *ka* which lie in the neighbourhood of the real part of the lowest complex eigenvalue: γ_{11} for an open rectangular cavity and γ_{c01} for

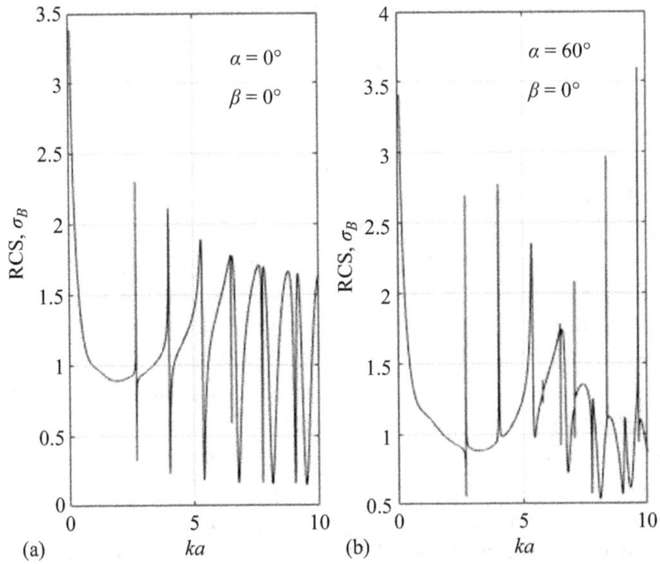

Figure 14.14 Frequency dependence σ_B (ka) for a slotted elliptical cavity ($e = 0.8, w_{2b} = 0.25$), the slit position $\beta = 0°$, excited at the incidence angles: (a) $\alpha = 0°$ and (b) $\alpha = 60°$

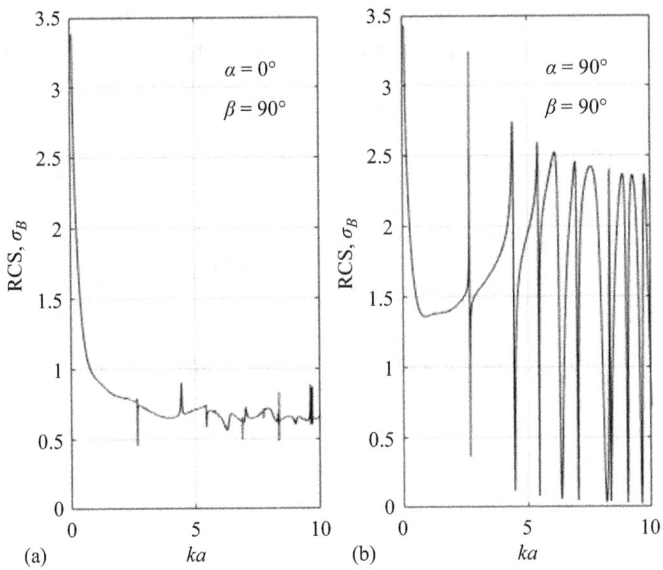

Figure 14.15 Frequency dependence σ_B (ka) for a slotted elliptical cavity ($e = 0.8, w_{2b} = 0.25$), the slit position $\beta = 90°$, excited at the incidence angles: (a) $\alpha = 0°$ and (b) $\alpha = 90°$

open circular and elliptical cavities, attributed to the eigenmodes TM_{11} and TM_{c01} respectively. For this purpose, we consider the part of the RCS $\sigma_B(ka)$, as plotted in Figure 14.13(a), but occupying the interval $2.392 \le ka \le 2.398$, which includes the value $ka = \mathrm{Re}(\gamma_{11}) = 2.3947$ (see Figure 14.17(a)).

The same is done for two cases of the slotted elliptical cavity (Figure 14.17(b) and (c)) and the deep ($b/a = 20$) rectangular cavity (Figure 14.17(d)). It should be noted that in the case of elliptical cavity with parameters $e = 0.8, w_{2b} = 0.25, \beta = 90°$ (Figure 14.17(c)), the value $\gamma_{c01} = 3.257804 - i0.002170$ has been calculated for this purpose, whereas the other values of γ_{c01} shown in Figure 14.17(a) and (b) were extracted from Tables 14.2 and 14.5, while the value of γ_{11} shown in Figure 14.17(d) was obtained from Table 14.7. The physical reasons for the appearance of the *resonance doublet* are well known and are explained in the general theory of oscillations. For resonant acoustic and electromagnetic oscillations, this effect has been exhaustively studied in the particular case of spherical cavity with a circular aperture by Vinogradov *et al.* [13]. It is of interest to know how the far-field pattern $A(\varphi) = A(ka, \varphi)$ behaves in the vicinity of those values ka which provide the sharp peaks of $\sigma_B(ka)$ in an extremely narrow frequency band.

The behaviour of the normalized far field at the three particular values of ka shown in Figure 14.17(a)–(c) is displayed in Figure 14.18 (slotted circular cavity) and Figure 14.19 (slotted elliptical cavities). For all cases, the backscattering is largest at the values of ka corresponding to the resonance maxima of σ_B. It can be seen from the graphs in Figure 14.19 that in spite of the small wave size ($a \cong 0.52\lambda$) we may observe well-collimated beams in the direction defined by the backscattering angle $\varphi = \pi + \alpha$ at $ka = 3.2880$ (Figure 14.19(a)) and $ka = 3.2555$. (Figure 14.19(b)). Small or almost insignificant shifts from these wavenumbers completely change the character of

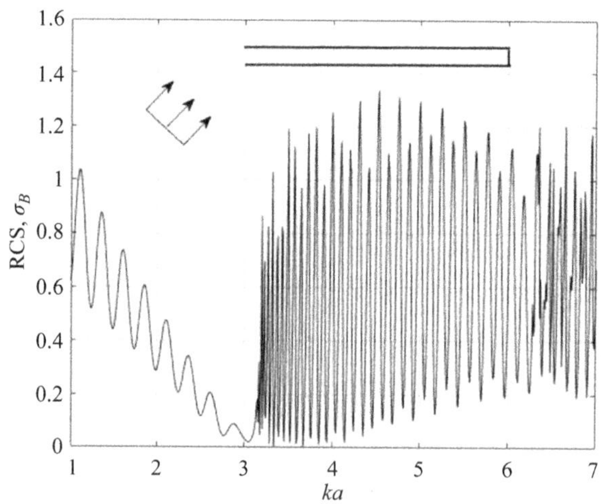

Figure 14.16 Frequency dependence $\sigma_B(ka)$ for a rectangular cavity ($b/a = 20$) excited at $\alpha = 50°$

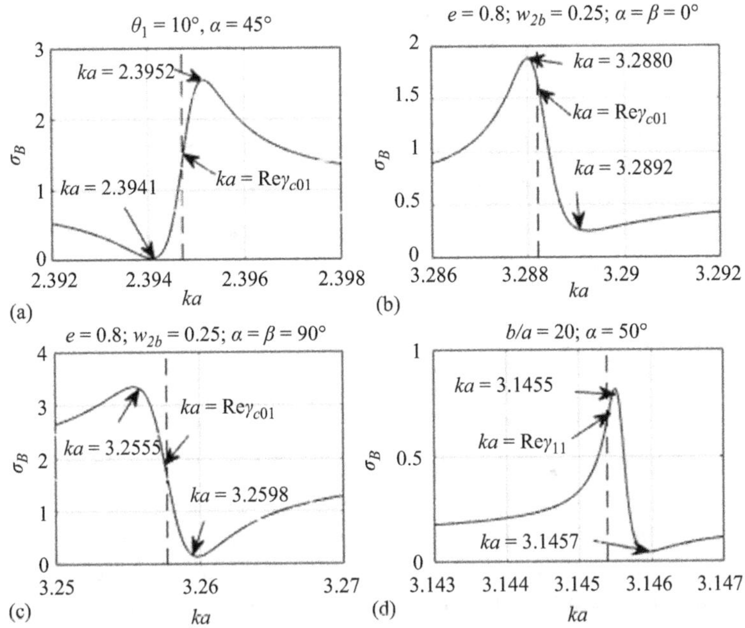

Figure 14.17 Magnified parts of the plots σ_B (ka) in the neighbourhood of the first resonance doublet for: (a) circular cavity, $\theta_1 = 10°$, $\alpha = 45°$; (b) elliptical cavity, $e = 0.8, w_{2b} = 0.25$, $\beta = 0°$, $\alpha = 0°$; (c) elliptical cavity, $e = 0.8, w_{2b} = 0.25$, $\beta = 90°$, $\alpha = 90°$ and (d) rectangular cavity, $b/a = 20, \alpha = 50°$

backscattering, particularly at those wavenumbers corresponding to minima of σ_B, namely at $ka = 3.2892$ (Figure 14.19(a)) and $ka = 3.2898$ (Figure 14.19(b)): at these wavenumbers, the RCS has a local minimum at the backscattering angle that is very much less than the corresponding maximum observed at $ka = 3.2880$ or $ka = 3.2555$.

14.5 Conclusion

A reliable scheme is developed for the accurate calculation of the complex eigenvalues of 2D open arbitrary cavities. It is based on the rigorous MAR, previously successfully implemented for the mathematically rigorous solution of 3D and 2D diffraction problems for hollow spherical cavities with circular apertures and 2D cavities with longitudinal slits. At the core of this method for the study of spectral problems lie some remarkable properties that make computation of the complex eigenvalues quite a routine and reliable procedure. First of all, the method possesses proven fast convergence, allowing us to control the computational accuracy. In addition, the matrix filling is based on computation of the matrix elements which relies on the FFT and recursive formulas, so that matrix fill time does not dominate total computation time.

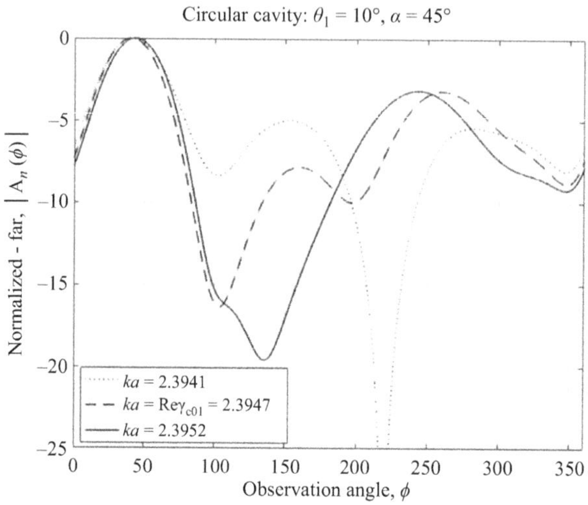

Figure 14.18 Normalized far-field patterns for slotted circular cavity ($\theta_1 = 10°$, $\alpha = 45°$) at the first resonance doublet of $\sigma_B\,(ka)$: ka = 2.3941 (minimum), ka= 2.3952 (maximum) near ka = $Re\gamma_{c01}$ = 2.3947

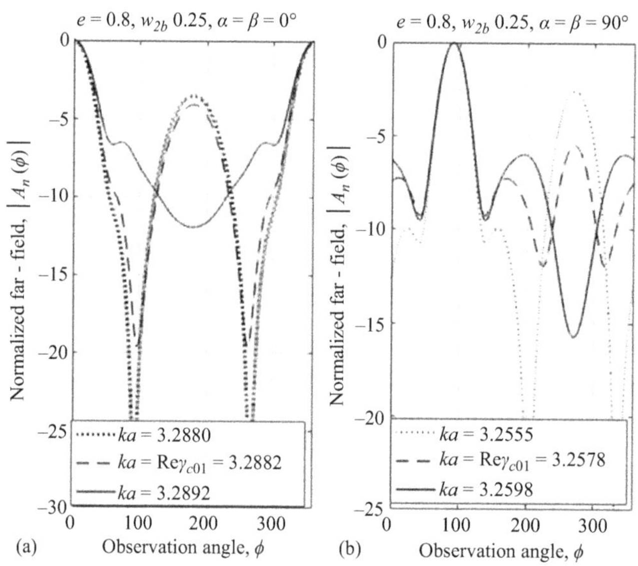

Figure 14.19 Normalized far-field patterns for slotted elliptical cavity ($e = 0.8, w_{2b} = 0.25$) at the first resonance doublet of $\sigma_B\,(ka)$: (a) $ka = 3.2880$ (maximum), $ka = 3.2892$ (minimum) near $ka = Re\gamma_{c01} = 3.2882, \beta = 0°, \alpha = 0°$ and (b) $ka = 3.2555$ (maximum), $ka = 3.2598$ (minimum), near $ka = Re\gamma_{c01} = 3.2578; \beta = 90°, \alpha = 90°$

Furthermore, the method has none of the restrictions that are often imposed on the shape of the cavities and slot widths in other studies in the literature. In this chapter, we examined the eigenmodes of the *TM*-type, computing the complex eigenvalues for a variety of cavities that are representative of the class of problems of interest. Future publications will address the analogous problem for eigenmodes of TE-type. As it stands, the approach described in this chapter may be implemented for the calculation of the eigenvalues for cavities with a number of inclusions inside the open cavities, for coupled resonators and for finite gratings because all of the corresponding wave scattering problems have previously been successfully solved by the MAR (see, e.g. [18,23]).

References

[1] Collin R.E. *Field Theory of Guided Waves*, 2nd ed. New York: Wiley-IEEE Press; 1990.

[2] Marcuvitz N. *Waveguide Handbook*. London: P. Peregrinus on behalf of the Institution of Electrical Engineers; 1986.

[3] Fanti A., Mazzarella G., Montisci G., and Casula G.A. 'Computation of the modes of elliptic waveguides with a curvilinear 2D frequency-domain finite-difference approach'. *Progress In Electromagnetics Research M*. 2012;**26**: 69–84.

[4] Karageorghis A. 'The method of fundamental solutions for the calculation of the eigenvalues of the Helmholtz equation'. *Applied Mathematical Letters*. 2001;**14**:837–842.

[5] Chen J.T., Chang M.H., Chen K.H., and Lin S.R. 'The boundary collocation method with meshless concept for acoustic eigen analysis of two-dimensional cavities using radial basis function'. *Journal of Sound and Vibration*. 2002;**257**(4):667–711.

[6] Tai G.R.C. and Shaw R.P. 'Helmholtz-equation eigenvalues and eigenmodes for arbitrary domains'. *The Journal of the Acoustical Society of America*. 1974;**56**(3):796–804.

[7] González L.M., Cobo P., Theofilis V., and Valero E. 'Acoustic resonances in 2D open cavities'. *Acta Acustica United With Acustica*. 2013;**99**: 572–581.

[8] Chernokozhin E.V. and Shestopalov Y.V. 'Resonant and non-resonant diffraction by open image-type slotted structures'. *IEEE Transactions on Antennas and Propagation*. 2001;**49**(5):793–800.

[9] Jeng S.K. and Tzeng S.T. 'Scattering from a cavity-backed slit in a ground plane-TM case'. *IEEE Transactions on Antennas and Propagation*. 1991;**39**(5):661–663.

[10] Jeng S.K. 'Scattering from a cavity-backed slit in a ground plane -TE case'. *IEEE Transactions on Antennas and Propagation*. 1990;**38**(10):1523–1529.

[11] Wang T.M., Cuevas A., and Ling H. 'RCS of a partially open rectangular box in the resonant region'. *IEEE Transactions on Antennas and Propagation*. 1990;**38**(9):1498–1504.

[12] Hussein M., Sebak A., and Hamid M. 'Scattering and coupling properties of a slotted elliptic cylinder'. *IEEE Transactions on Electromagnetic Compatibility*. 1994;**36**(1):76–81.

[13] Vinogradov S.S., Smith P.D., and Vinogradova E.D. *Canonical Problems in Scattering and Potential Theory. Part 2: Acoustic and Electromagnetic Diffraction by Canonical Structures*. Boca Raton, FL/New York/Washington, DC: Chapman & Hall/CRC; 2002.

[14] Smith P.D. 'Recent advances in regularization techniques for scattering and diffraction'. *Radio Science*. 2007;**42**(6). Available from https://doi.org/10.1029/2007RS003703

[15] Tuchkin Ya.A., 'Wave scattering by open cylindrical screens of arbitrary profile with Dirichlet boundary conditions'. *Soviet Physics Doklady* (Engl. Transl.). 1985;**30**(12):1027–1029.

[16] Tuchkin Ya.A. 'A new method in wave diffraction theory by thin screens'. *Electromagnetics*. 1993,**13**(3):319–338.

[17] Vinogradov S.S, Vinogradova E.D., Wilson C., Sharp I., and Tuchkin Yu.A. 'Scattering of an E-polarized plane wave by two-dimensional air-foils'. *Electromagnetics*. 2009;**29**(3):268–282. Available from https://doi.org/10.1080/02726340902718526

[18] Safonova G. and Vinogradova E.D., 'Accurate backscattering studies for double-layered array of infinitely long cylinders'. *Microwave and Optical Technology Letters*. 2013;**55**:2127–2131.

[19] Vinogradova E., 'Electromagnetic plane wave scattering by arbitrary two-dimensional cavities: Rigorous approach'. *Wave Motion*. 2017;**70**:47–64. Available from https://doi.org/10.1016/j.wavemoti.2016.06.009

[20] Vinogradova E., 'Reflectivity of two-dimensional airfoils illuminated by H-polarized plane wave: Rigorous approach'. *Electromagnetics*. 2017;**37**(6): 337–397. Available from https://doi.org/10.1080/02726343.2017.1359964

[21] Colton D. and Kress R. *Integral Equation Methods in Scattering Theory*. (SIAM, Reprint of: New York: Wiley, 1983); 2013.

[22] Zhang S. and Shen Y. 'Eigenmode sequence for an elliptical waveguide with arbitrary ellipticity'. *IEEE Transactions on Microwave Theory and Techniques*. 1995;**43**(1):227–230.

[23] Safonova G., Vinogradova E.D., and Tuchkin Yu.A. 'Accurate impedance calculations for transmission lines with adjustable inner conductor'. *Microwave and Optical Technology Letters*. 2014;**56**(9):2066–2070. Available from https://doi.org/10.1002/mop.28497

Chapter 15

Numerical solutions of integral equations for electromagnetics

Roberto D. Graglia[1] and Andrew F. Peterson[2]

Integral equations have proven their popularity for the electromagnetic analysis of radiation and scattering problems. The workhorse equations are the *electric-field integral equation* (*EFIE*) and the *magnetic-field integral equation* (*MFIE*) [1,2]. The development of these equations will be reviewed in the following for perfectly conducting targets and homogeneous dielectric targets. When applied to certain closed surfaces, the original equations exhibit uniqueness difficulties at frequencies where the target surface coincides with a resonant cavity [3]. In addition, the original EFIE and MFIE also fail under certain circumstances for electrically small bodies. Alternate integral equations were proposed to remedy those situations, and these will also be summarized in the following section. In addition, we describe the numerical solution of these equations and report the progress made in recent years associated with the use of hierarchical vector basis functions, and the recent use of singular basis functions.

15.1 The EFIE and MFIE for perfectly conducting bodies

Consider a perfect electric conducting (PEC) body in an infinite homogeneous environment of permittivity ε and permeability μ.* The body is illuminated by a sinusoidal steady-state source of electromagnetic radiation, having radian frequency ω, and all the time-varying quantities are represented by phasors with suppressed time dependence $\exp(j\omega t)$. In the absence of the body, the source would produce an electric field $\bar{E}^{\mathrm{inc}}(x,y,z)$ and a magnetic field $\bar{H}^{\mathrm{inc}}(x,y,z)$ throughout the surrounding space. (These are denoted the *incident* fields.) In the presence of the conducting body, the fields are perturbed from these to the *total* fields $\bar{E}^{\mathrm{tot}}(x,y,z)$ and $\bar{H}^{\mathrm{tot}}(x,y,z)$. The perturbation can be accounted for by the presence of equivalent induced currents on the body.

The induced surface current density can be expressed as $\bar{J}(s,t)$, where s and t are parametric variables on the surface. (Henceforth, all references to the variable t are intended to denote position, not time.) The current density, if treated as a source

[1] Dipartimento di Elettronica e Telecomunicazioni, Politecnico di Torino, Torino, Italy
[2] School of Electrical and Computer Engineering, Georgia Institute of Technology, Atlanta, GA, USA
*This material was adapted from [4].

function that exists in the absence of the conducting body, produces the *scattered* fields $\bar{E}^s(x,y,z)$ and $\bar{H}^s(x,y,z)$. The various fields are related by

$$\bar{E}^{\text{inc}} + \bar{E}^s = \bar{E}^{\text{tot}} \tag{15.1}$$

$$\bar{H}^{\text{inc}} + \bar{H}^s = \bar{H}^{\text{tot}} \tag{15.2}$$

Since the scattered fields are produced in infinite homogeneous space by $\bar{J}$, they are readily determined using any of the standard source-field relations such as

$$\bar{E}^s = \frac{\nabla(\nabla \bullet \bar{A}) + k^2\bar{A}}{j\omega\varepsilon} \tag{15.3}$$

$$\bar{H}^s = \nabla \times \bar{A} \tag{15.4}$$

where the wave number of the medium is given by $k = \omega\sqrt{\mu\varepsilon}$, and the magnetic vector potential function is

$$\bar{A}(x,y,z) = \int\int \bar{J}(s',t')\frac{e^{-jkR}}{4\pi R}ds'dt' \tag{15.5}$$

In (15.5), R is the distance from a point (s', t') on the surface to the point where the field is evaluated. In the following, primed coordinates will be used to describe the "source" of the electromagnetic field (the current density), while unprimed coordinates denote the "observer" location where that field is evaluated.

The total fields in (15.1) and (15.2) must satisfy the tangential-field electromagnetic boundary conditions on the surface of the perfect electric body:

$$\hat{n} \times \bar{E}^{\text{tot}}\Big|_{\text{surface}} = 0 \tag{15.6}$$

$$\hat{n} \times \bar{H}^{\text{tot}}\Big|_{\text{surface}} = \bar{J} \tag{15.7}$$

where $\hat{n}$ is the outward normal unit vector, and in (15.7), the surface is approached from the exterior. By combining (15.1), (15.3), and (15.6), we obtain

$$\hat{n} \times \bar{E}^{\text{inc}}\Big|_{\text{surface}} = -\hat{n} \times \frac{\nabla(\nabla \bullet \bar{A}) + k^2\bar{A}}{j\omega\varepsilon}\Bigg|_{\text{surface}} \tag{15.8}$$

This equation is one form of the *EFIE*. (While it is in fact an integro-differential equation, the term "integral equation" is commonly used for simplicity.) The equation can be solved in principle for the surface current density $\bar{J}$ appearing within the magnetic vector potential $\bar{A}$.

By combining (15.2), (15.4), and (15.7), we obtain the *MFIE*:

$$\hat{n} \times \bar{H}^{\text{inc}}\Big|_{\text{surface}} = \bar{J} - \hat{n} \times (\nabla \times \bar{A})\Big|_{\text{surface}} \tag{15.9}$$

Equation (15.9) is enforced in the limiting case from the exterior of the surface of the body. This equation may also be solved for $\bar{J}$. Once $\bar{J}$ is determined by a solution of (15.8) or (15.9), the fields and other observable quantities associated with the electromagnetic scattering problem can be determined by direct calculation.

The MFIE in (15.9) is only applicable to closed conducting bodies where (15.7) is a valid boundary condition, but the EFIE can be used for thin shells as well as solid structures where (15.6) holds. This restriction on the MFIE is due to the fact that (15.7) has been specialized to the situation where the magnetic field is zero on the inner side of the surface. The form of the MFIE in (15.9) is not valid if the surface is open and the field is nonzero on both sides of the surface. In addition, the MFIE may fail for multiply-connected surfaces if the overall structure is electrically small [5]. For these reasons, the MFIE is not as widely used as the EFIE.

15.2 Some alternative formulations to remediate fictitious internal resonances

The tangential-field EFIE and MFIE described above can fail for closed bodies whose surfaces coincide with resonant cavities [3]. In that situation, neither (15.6) nor (15.7) alone is sufficient to guarantee that the surface current density is the proper exterior solution. However, the two equations can be combined together to form the *combined-field integral equation* [6]. In this manner, the two boundary conditions are both invoked, and are sufficient to produce the proper solution at the resonant frequencies. An alternative equation is the *combined source integral equation* [7,8], where the equivalent electric surface current density is replaced by a combination of electric and magnetic surface currents. The combination of the two sources yields the correct representation of the exterior fields but prevents the cavity resonances from satisfying the resulting equation.

Other formulations have been proposed to remedy fictitious internal resonances for closed bodies. The *augmented electric and magnetic field equations* incorporate boundary conditions on the normal components of the fields on the target surface in addition to those imposed on the tangential fields [9]. The resulting system is an overdetermined set of equations but eliminates the interior solutions. Another approach has been to impose the conventional EFIE or MFIE boundary condition on a set of dual surfaces, one located entirely inside the target (ideally a quarter wavelength inside the actual surface) where the total fields must vanish [10–12]. These *dual surface integral equations* also eliminate fictitious resonances since the resonant fields cannot simultaneously satisfy boundary conditions on the two surfaces [13].

15.3 Integral equations for homogeneous dielectric bodies

Consider a homogeneous, penetrable body illuminated by an incident electromagnetic field.† Region 1 (the exterior) is free space (permeability $\mu_1 = \mu_0$ and permittivity $\varepsilon_1 = \varepsilon_0$) and Region 2 (the target interior) is characterized by a constant permeability $\mu_2 = \mu_d$ and constant permittivity $\varepsilon_2 = \varepsilon_d$. $\bar{E}_1$ and $\bar{H}_1$ denote the fields in Region 1, and $\bar{E}_2$ and $\bar{H}_2$ denote the fields throughout Region 2. Using the surface equivalence

†This material was adapted from [13].

principle, we define equivalent sources on the target surface that replicate the original fields in both regions.

For a homogeneous dielectric body, equivalent sources on the target's surface may be defined as

$$\bar{J}_1 = \hat{n} \times \bar{H}_1 \tag{15.10}$$

$$\bar{M}_1 = \bar{E}_1 \times \hat{n} \tag{15.11}$$

where $\hat{n}$ is the outward normal vector at points on the surface. These sources, radiating in conjunction with the original source, replicate the original exterior fields throughout Region 1. Null fields are produced in Region 2, allowing us to replace Medium 2 with free space without changing the fields in Region 1. Thus, the exterior part of the original problem only involves sources radiating in free space.

To describe the interior problem, a second equivalence relationship must be constructed. Sources $\bar{J}_2$ and $\bar{M}_2$ are defined according to

$$\bar{J}_2 = (-\hat{n}) \times \bar{H}_2 \tag{15.12}$$

$$\bar{M}_2 = \bar{E}_2 \times (-\hat{n}) \tag{15.13}$$

where $\hat{n}$ is still the normal vector pointing into Region 1 (out of the scatterer). Radiating *in the absence* of the original source, these equivalent sources replicate the original fields throughout Region 2 and produce null fields throughout the entirety of Region 1. Since the Region 1 fields vanish, we are free to insert material having $\mu = \mu_2 = \mu_d$ and $\varepsilon = \varepsilon_2 = \varepsilon_d$ throughout Region 1 in order to convert the problem to one involving infinite, homogeneous space. Thus, the equivalent interior problem involves sources $\bar{J}_2$ and $\bar{M}_2$ radiating in homogeneous space characterized by permittivity ε_2 and permeability μ_2.

The continuity of the tangential fields at the dielectric interface dictates that $\bar{J}_1 = -\bar{J}_2$ and $\bar{M}_1 = -\bar{M}_2$. Therefore, it suffices to work with $\bar{J}_1$ and $\bar{M}_1$ as the primary unknowns to be determined. Since we have two equivalent problems and two unknown sources, we must employ a system of two coupled equations. General source-field relationships similar to (15.3) and (15.4) are

$$\bar{E}^s = \frac{\nabla(\nabla \bullet \bar{A}_i) + k_i^2 \bar{A}_i}{j\omega\varepsilon_i} - \nabla \times \bar{F}_i \tag{15.14}$$

$$\bar{H}^s = \nabla \times \bar{A}_i + \frac{\nabla(\nabla \bullet \bar{F}_i) + k_i^2 \bar{F}_i}{j\omega\mu_i} \tag{15.15}$$

where the vector potential functions are defined

$$\bar{A}_i(x, y, z) = \int\int \bar{J}_1(s', t') \frac{e^{-jk_i R}}{4\pi R} ds' dt' \tag{15.16}$$

$$\bar{F}_i(x, y, z) = \int\int \bar{M}_1(s', t') \frac{e^{-jk_i R}}{4\pi R} ds' dt' \tag{15.17}$$

In these expressions, the subscript i denotes the medium, k_i denotes the corresponding wave number in that medium, and R is the distance from a point (s', t') on the surface to the point (x, y, z) where the field is evaluated.

Equations (15.14) and (15.15) can be combined with conditions (15.10)–(15.13) to produce the coupled *EFIEs*

$$\hat{n} \times \bar{E}^{\text{inc}}(\bar{r}) = -\bar{M}_1 - \hat{n} \times \left\{ \frac{\nabla(\nabla \bullet \bar{A}_1) + k_1^2 \bar{A}_1}{j\omega\varepsilon_1} - \nabla \times \bar{F}_1 \right\}_{S^+} \tag{15.18}$$

$$0 = \bar{M}_1 - \hat{n} \times \left\{ \frac{\nabla(\nabla \bullet \bar{A}_2) + k_2^2 \bar{A}_2}{j\omega\varepsilon_2} - \nabla \times \bar{F}_2 \right\}_{S^-} \tag{15.19}$$

where the subscripts S^+ and S^- indicate that (15.18) is evaluated in the limiting sense from the exterior of the surface, while (15.19) is evaluated in the limiting sense from the interior.

Similarly, we could obtain coupled *MFIEs* as

$$\hat{n} \times \bar{H}^{\text{inc}}(\bar{r}) = \bar{J}_1 - \hat{n} \times \left\{ \nabla \times \bar{A}_1 + \frac{\nabla(\nabla \bullet \bar{F}_1) + k_1^2 \bar{F}_1}{j\omega\mu_1} \right\}_{S^+} \tag{15.20}$$

$$0 = -\bar{J}_1 - \hat{n} \times \left\{ \nabla \times \bar{A}_2 + \frac{\nabla(\nabla \bullet \bar{F}_2) + k_2^2 \bar{F}_2}{j\omega\mu_2} \right\}_{S^-} \tag{15.21}$$

Either of these formulations could be used to represent homogeneous scatterers. However, because these equations may fail when the target surface coincides with the boundary of a resonant cavity, alternative approaches are usually employed.

15.4 Formulations that remediate fictitious internal resonances for dielectric targets

In common with the EFIE and MFIE for PEC scatterers, the coupled surface integral equations do not always guarantee a unique solution if the surface in question coincides with the boundary of a resonant cavity. However, alternative equations can be obtained that ensure a unique solution by constructing combinations of the equations in (15.18)–(15.21). One possibility is to combine the two exterior equations into one and the two interior equations into one to form a coupled *combined field integral equation* [14,15]. Despite the popularity of the combined field approach for perfectly conducting targets, this has not been a common way to treat dielectric targets within the electromagnetics community.

A second way to ensure unique solutions is to equate the magnetic current density in (15.18) with that in (15.19) to obtain one equation, and equate the electric current density in (15.20) with that in (15.21) to obtain another. This leads to the two coupled equations of the form

$$\hat{n} \times \bar{E}^{\text{inc}}(\bar{r}) = -\hat{n} \times \left\{ \frac{\nabla\nabla \bullet \bar{A}_1 + k_1^2 \bar{A}_1}{j\omega\varepsilon_1} - \nabla \times \bar{F}_1 \right\}_{S^+} - \hat{n} \times \left\{ \frac{\nabla\nabla \bullet \bar{A}_2 + k_2^2 \bar{A}_2}{j\omega\varepsilon_2} - \nabla \times \bar{F}_2 \right\}_{S^-} \tag{15.22}$$

$$\hat{n} \times \bar{H}^{\text{inc}}(\bar{r}) = -\hat{n} \times \left\{ \nabla \times \bar{A}_1 + \frac{\nabla\nabla \bullet \bar{F}_1 + k^2 \bar{F}_1}{j\omega\mu_1} \right\}_{S^+} - \hat{n} \times \left\{ \nabla \times \bar{A}_2 + \frac{\nabla\nabla \bullet \bar{F}_2 + k^2 \bar{F}_2}{j\omega\mu_2} \right\}_{S^-} \tag{15.23}$$

These equations, known as the *PMCHWT system of equations* after the authors of [2, 16,17], have unique solutions even at internally resonant frequencies. This formulation has been widely used throughout the electromagnetics community.

A third approach was proposed in [18], by combining (15.18) and (15.19) and combining (15.20) and (15.21) in such a way that their strongest integrand singularities (arising within the terms $\nabla\nabla \bullet \bar{A}_i$ and $\nabla\nabla \bullet \bar{F}_i$) are canceled. This approach, which also produces equations that eliminate the interior resonance problem, leads to the *Müller system of equations*

$$\hat{n} \times \bar{E}^{\text{inc}}(\bar{r}) = -\left(1 + \frac{\varepsilon_2}{\varepsilon_1}\right) \bar{M}_1 - \hat{n} \times \left\{ \frac{\nabla\nabla \bullet \bar{A}_1 + k_1^2 \bar{A}_1}{j\omega\varepsilon_1} - \nabla \times \bar{F}_1 \right\}_{S^+} + \hat{n} \times \left\{ \frac{\nabla\nabla \bullet \bar{A}_2 + k_2^2 \bar{A}_2}{j\omega\varepsilon_1} - \frac{\varepsilon_2}{\varepsilon_1} \nabla \times \bar{F}_2 \right\}_{S^-} \tag{15.24}$$

$$\hat{n} \times \bar{H}^{\text{inc}}(\bar{r}) = \left(1 + \frac{\mu_d}{\mu_0}\right) \bar{J}_1 - \hat{n} \times \left\{ \nabla \times \bar{A}_1 + \frac{\nabla\nabla \bullet \bar{F}_1 + k^2 \bar{F}_1}{j\omega\mu_1} \right\}_{S^+} + \hat{n} \times \left\{ \frac{\mu_2}{\mu_1} \nabla \times \bar{A}_2 + \frac{\nabla\nabla \bullet \bar{F}_2 + k^2 \bar{F}_2}{j\omega\mu_1} \right\}_{S^-} \tag{15.25}$$

When used for numerical solutions, the Müller formulation is apparently more stable than the PMCHWT formulation for low-contrast targets, although this may depend on the details of the specific discretization [19].

15.5 Single-source integral equations for dielectric bodies

The preceding formulations involve coupled systems of equations. In fact, it is possible to formulate integral equations for dielectric bodies in terms of a single equation and single unknown quantity, instead of the two ($\bar{J}_1$ and $\bar{M}_1$) used above. Formulations were proposed in [20–22], where [22] extended the work to the development of a *single-source combined field integral equation* to eliminate interior resonances. The cost of eliminating the second unknown and equation is that one integral operator is nested within another in the single-source formulations. An alternative approach that combines surface and volume integral equations is proposed in [23].

15.6 Low-frequency breakdown of integral equations

There is an additional difficulty with the preceding integral equations, namely that most of them become unstable at very low frequencies or when they are applied to electrically small targets [19,24–29]. The electric-field integral operator in (15.3) has two terms, one proportional to frequency that involves a current density source and the other proportional to the reciprocal of frequency, involving a charge density source. Integral equations that contain the EFIE operator become unstable at very low frequencies when these terms decouple. A common approach for remediating this difficulty is to employ a quasi-Helmholtz decomposition, which explicitly separates and rescales these two operators in the system of equations. Such a decomposition can be obtained in terms of so-called "loop-star" basis functions, "loop-tree" representations, or "loop-flower" bases [25–27].

Unfortunately, the Helmholtz decomposition does not usually produce a system of equations that is well-conditioned enough to facilitate a solution by iterative algorithms. In addition, a similar instability occurs in these same operators at very dense discretization levels, and is not remediated by a Helmholtz decomposition. Treating these difficulties led to the development of Calderon preconditioned integral equations [28–30] and several alternative formulations, including the use of an augmented integral equation that explicitly separated the current and charge source terms [31], and integral equations based on potential functions [32,33]. Today, robust formulations for electromagnetic scattering are based on these alternative integral equations. Please see Chapter 16 by Merlini *et al.* for more information on some of these approaches.

15.7 Numerical solution of integral equations

To solve any of the preceding integral equations numerically, the infinite dimensional space of continuous functions containing the solution to the integral equation must be projected to a finite dimensional space (in other words, the integral equation must be converted into a matrix equation). This process is known as discretization. In the electromagnetics community, the most common procedure for performing a discretization is known by the name *method of moments* (MoM), which is essentially the same as the *boundary element method*, the *weighted residual method*, or the *finite element method (FEM)*. Details of the approach may be found in textbooks such as [13].

The MoM procedure involves approximating the quantity to be determined, one or both of the surface currents in the preceding equations, by an expansion in linearly independent basis functions. The N coefficients of these functions become the unknowns to be determined. To obtain N linearly independent equations, the original equation is multiplied with suitable testing functions, and integrated over the surface of the body, to obtain an N by N linear system that can be organized into a matrix equation. The numerical solution of the matrix equation yields the unknown coefficients.

The computation of the entries of the system matrix is complicated by the singularities in the Green's functions of the integral operators. Space precludes a discussion of techniques for evaluating these expressions, and we refer the reader to several publications on this topic [4,13,34–49].

In general, the basis and testing functions used to represent the surface currents may be spread over the entire surface, or defined with a limited domain of support. Because they allow flexibility in the geometry, cells with a limited domain such as triangles or quadrilaterals are applicable to a broad class of problems and have been widely studied. Curved cells may be employed [4] to improve the geometric modeling accuracy, and are expected to be mandatory when higher-order basis functions are employed [34].

15.8 Vector basis functions

The quantities to be determined in the integral equations are vector in nature, so it is natural to employ vector representations within numerical solution procedures. Since their development in the late 1970s, vector basis functions are commonly defined on simple two-dimensional patches (elements) such as interconnected triangles or quadrilaterals. The most useful vector basis functions are the so-called divergence or curl-conforming (CC) bases, which have continuous normal or tangential components, respectively, across adjacent elements. In practice, divergence-conforming (DC) functions are appropriate for the EFIE operator while both DC and CC basis functions may be appropriate for the MFIE operator. In contrast to scalar bases, conforming vector bases not only provide appropriate normal or tangential continuity at element (material) interfaces, but also provide convenient means for imposing boundary conditions on unknown fields or currents along irregular interfaces.

Conforming vector functions are described in a fundamental work by Nedelec [50], published in 1980. The earliest use of (low-order) DC functions for numerical solutions of electromagnetic field problems can be traced to Raviart and Thomas [51], Rao *et al.* [52] (in two dimensions), and Schaubert *et al.* [53] (in three dimensions). Similarly, in the early 1980s, Bossavit and Verite [54], Hano [55], and Barton and Cendes [56] were some of the first users of CC functions in numerical solutions.

The lowest-order DC basis functions for triangular and tetrahedral cells are commonly referred to as the RWG and SWG functions, respectively. (The acronyms are formed by the initials of the surname of the authors of [52,53].) Other vector basis functions are described, for example, in [34], as is the use of quadrilateral cells, brick or hexahedral cells, and triangular prism cells. The popularity of RWG functions is due to the ease with which triangular meshes can be generated for arbitrary surfaces and the fact that the geometrical error obtained with patch dimensions not exceeding one tenth of the smallest wavelength under consideration is lower than the error due to the use of the RWG basis functions. Thus, the error is generally well controlled throughout an h-refinement process [57,58].

The second distinctive property of basis functions of the Nedelec type, in addition to their CC or DC nature, is that they require a reduced number of degrees of freedom for a fixed order of approximation. Nedelec apparently realized that a number of degrees of freedom could be eliminated for efficiency without a deleterious effect on accuracy. For example, the EFIE operator has separate terms involving the current density and its divergence, the charge density. Hence, given a basis for current density that is complete to polynomial order $p + 1$, the accuracy of the overall solution is limited by the terms involving the divergence of the basis, which can only be of complete polynomial order p. Hence, those terms of order $p + 1$, whose divergence vanishes, neither contribute to modeling the charge density nor to the overall accuracy of the solution. These unnecessary degrees of freedom can be discarded in forming DC bases, resulting in bases of mixed degree.

For some applications, it is necessary to partition a numerical field representation into solenoidal and irrotational parts. Such a partitioning may provide a more physical representation in situations where the field is supposed to be solenoidal or irrotational; it may also be necessary to isolate the null-space of the resulting matrix operator in various situations, or stabilize the matrix operator at very low frequencies when other scale factors produce instabilities. Unfortunately, an exact Helmholtz decomposition cannot be obtained in terms of either CC or DC representations alone: an irrotational field requires a CC expansion, while a solenoidal field requires a DC expansion. However, one may obtain a quasi-Helmholtz decomposition with either CC or DC bases, in terms of loop and star vector basis functions [25–27].

15.9 Interpolatory and hierarchical vector basis functions

There has been a consistent increase in the number of scientific publications devoted to the development and use of higher-order vector functions for the numerical solution of integral (and differential) equations for electromagnetic problems since the mid-1990s.[‡] This suggests that higher-order vector bases are beneficial for their better convergence properties. Higher-order vector functions compatible with the Nedelec representations are of interest since they eliminate some wasted degrees of freedom and, hence, are relatively efficient in a numerical solution. Because Nedelec's prescription can be applied in many different ways, numerous approaches to the construction of higher-order bases have been proposed in the recent past. The progress in the area since Nedelec's original work and up to 1995 is documented in the bibliography of [59]. A recent survey may be found in [60].

High-order bases can usually be categorized as *interpolatory* or *hierarchical*. Interpolatory bases interpolate the value of a field quantity at a number of interpolation

[‡]This material was adapted from [34,61].

points such that only one basis function is nonzero at a given interpolation point. This property allows a direct physical interpretation of the expansion coefficients, but does not facilitate mixing different orders throughout the computational domain. Conversely, an expansion of order p is said to be hierarchical if it includes, as a subset, all the basis functions forming the $(p-1)$th-order base. This property prevents one from associating the coefficient of a hierarchical basis function with the solution at a given point in the domain. However, hierarchical bases allow for p-refinement, that is for a selective expansion using different orders in different regions of the computational domain [57,58]. For example, the lowest-order hierarchical basis can be employed in sub-domains where the approximated quantity is expected to vary slowly, while higher-order bases can be employed in sub-domains where a rapid variation of the quantity is anticipated. In the numerical solution of integral equations, a selective choice of the order to be used on each sub-domain has the potential to reduce memory and CPU time as well as improve accuracy.

Fully interpolatory higher-order vector basis functions of the Nedelec type were defined for the most common element shapes in 1997 [61]. These interpolatory functions, known in the literature as GWP functions (the acronym is formed by the initials of the surname of the authors of [61]), have been widely used. They are easily obtained as the product of zeroth-order Nedelec representations and interpolatory scalar polynomials with specially arranged arrays of interpolation points. The completeness properties of these vector functions are given in [34] together with several tables useful for their construction, and examples of the results obtained in electromagnetic applications. In this regard, it should be stressed that higher-order functions almost always require the use of geometric models which are also of higher order (in other words, curved) to control the geometry error (i.e. the discrepancy between the model surface and the physical surface) as the representation order increases.

Hierarchical DC and CC vector bases for the most common two and three-dimensional cells are described in detail in [34]. In common with the interpolatory functions, these bases are constructed by using orthogonal scalar polynomials to multiply the zeroth-order vector basis functions in order to form redundant vector sets. The dependent elements of these sets are then discarded. By using orthogonal scalar polynomials to form the hierarchical bases, one avoids the need to orthogonalize the elements of the vector base afterward to ensure their linear independence. However, one of the drawbacks of hierarchical bases is that they tend to yield system matrices with a higher condition number (CN) than interpolatory bases of the same order.

The elements of a hierarchical vector base of order p are linear combinations of the interpolatory basis functions of the same order. Therefore, results obtained with a homogeneous order of either type of basis function will be the same. To illustrate the improved accuracy as the polynomial order increases, Figure 15.1 shows the two-norm error in the bistatic scattering cross section (SCS)

$$\sigma(\theta,\phi) = \lim_{r\to\infty} 4\pi r^2 \frac{\left|\bar{E}^S(\theta,\phi)\right|^2}{\left|\bar{E}^{inc}\right|^2} \tag{15.26}$$

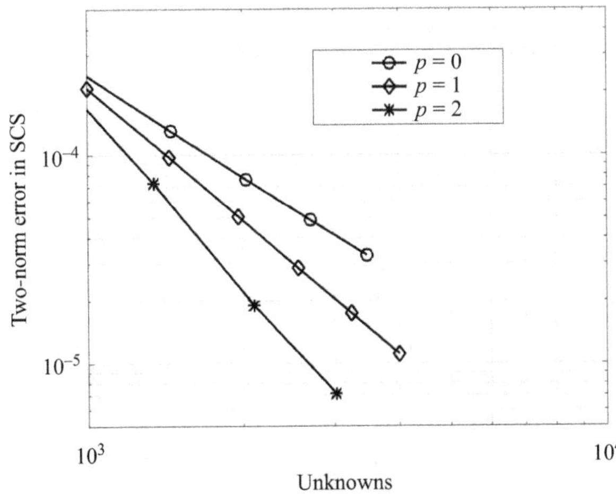

Figure 15.1 *Error in the bistatic SCS for a sphere of radius 1.0 λ, produced by the EFIE. Results are reported for p = 0, 1, and 2 DC basis and test functions, in conjunction with cubic-curved-patch models*

associated with a perfectly conducting sphere of radius 1.0 λ (where λ is the free-space wavelength). Numerical results were obtained by solving the EFIE in the frequency domain with DC hierarchical basis functions of orders $p = 0$, 1, and 2, and cubic curved patches. The SCS error is determined by comparing the numerical solutions at 5 degree increments in θ and ϕ in the far field with the exact solution from the Mie series. Other results of this type for sphere and toroid targets may be found in [62].

Higher-order DC vector functions also exhibit excellent performance for solving time domain (TD) problems [63–65]. Figure 15.2 shows the error in the bistatic RCS with respect to Mie series solution for a dielectric sphere 1 m in radius, at 70 MHz, from [65].

The literature reports a number of specific numerical solutions that use the hierarchical vector bases introduced in [34,66–69]. These include [70–80]. In addition, recent work is exploring the use of adaptive refinement for these problems [81,82]. However, higher-order polynomial representations fail in the presence of singularities in the current or charge densities at corners and edges of an object. Current research is exploring the treatment of edges and corners by combining hierarchical polynomial vector bases with singular basis functions as discussed in the following section.

15.10 Singular vector basis functions

The literature on the numerical treatment of field singularities in various disciplines, including electromagnetics, has been extensive since the 1970s. Functions

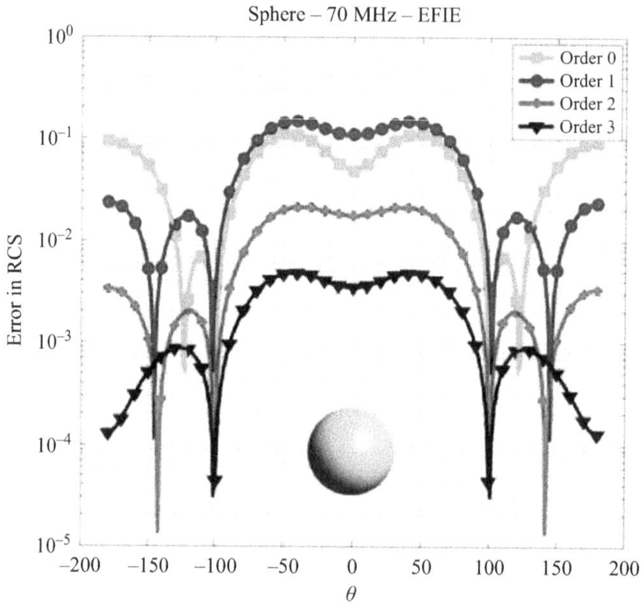

Figure 15.2 Error in the bistatic RCS with respect to Mie series solution at frequency 70 MHz upon solving TD single-source EFIE for a 1-m radius sphere with $\varepsilon_2 = 2\varepsilon_0$ illuminated by a Gaussian plane-wave pulse. The current density is modeled with DC basis functions of orders p = 0,1,2,3. ©2013 IEEE. Reprinted with permission from [65]

that model electromagnetic field singularities were proposed in [83–92].[§] Most of these approaches employ a single singular basis order that is used to replace some of the polynomial functions in the representation of the current or charge density when a singularity is present. This is known as a *substitutive* expansion. In recent years, high-order and even hierarchical singular representations have been proposed (see [34, Chapt. 7] and [93–96]). These representations are usually of the *additive* type since they are obtained by augmenting the regular polynomial vector bases with additional independent degrees of freedom to model the singular field behavior [97–99]. Once again, the (added) singular vector functions are constructed by using orthogonal scalar polynomials.

For high-order treatment, basis functions of the additive kind provide improved accuracy and additional flexibility since one can model appropriate field behavior even if the expected singularity is not excited by the source. Most of all, whenever the polynomial basis subset is hierarchical, additive bases allow p-refinement up to the polynomial order p for which the polynomial functions remain linearly independent

[§]The term "singular" is used to denote nonanalytic behavior, even if the resulting function is bounded.

from the singular ones. The CC functions developed for triangular and quadrilateral cells, and for tetrahedral, brick, and prism cells were proven to be very effective in the FEM context (see [100–107]).

Higher-order singular DC functions for the MoM solution of integral equations were proposed in [97,98] in conjunction with high-order interpolatory polynomial subsets. Recently, high-order hierarchical polynomial subsets were introduced [108]. One of the challenges to overcome is the integration of the singular basis functions for integrals that also contain a singular Green's kernel. Ad-hoc quadrature techniques to overcome this problem are discussed in [49].

The authors of the present chapter favor the use of singular additive basis functions because they facilitate p-refinement [57,58]. As an example, Figure 15.3 shows the magnitude of the $\hat{x}$ (co-polar) component of the current induced on a $1\lambda \times 1\lambda$ square plate by a normally incident plane wave with $H_y = 1$ A/λ. The result was obtained by solving the EFIE with Galerkin testing using a 5×5 quadrilateral cell mesh [108]. This current component is singular along the plate edges. It becomes unbounded at $y = \pm\lambda/2$ with a behavior that is inversely proportional to the square root of the distance from the edge. On the other two edges, the current tends to zero in proportion to the square root of the distance from the edge. The special singular basis functions used in the edge cells provide this behavior. Figure 15.3 at top shows the result obtained with a $p = 2$ order base plus singular basis functions in the 16 border cells (a total of 456 degrees of freedom (DoFs) are used). The results at bottom show the current $\hat{x}$-component along the $x = 0$ (at left) and $y = 0$ (at right) cuts obtained with bases of different order in the various cells. In particular, the $p = 0$ base is always used for the nine inner cells, while the basis order is varied in the 16 outer cells. The "background" results obtained with the purely polynomial $p = 0$ base are shown by dashed lines (40 DoFs). The dotted-line results are obtained after increasing the order in the outer cells to $p = 1$ (120 DoFs). (For the edges shared with an internal cell, we introduce only the $p = 0$ edge-based functions and discard the $p = 1$ ones so not to spoil the zeroth-order representation used on the inner cells.) Finally, the solid-line results (156 DoFs) are obtained by adding the singular set to the external cells (an extra 36 singular basis functions). Note that for the excitation considered it would be sufficient to add singular basis functions only on the two edges where the current is unbounded, reducing the number from 36 to 18 singular functions. In doing so, the results for the induced current remain essentially the same, but the result requires only 138 DoFs.

Possible future developments in the use of singular vector functions are traced in [99]. These include the development of singular basis functions for modeling corner singularities which cannot be properly modeled by the functions developed so far [99,109].

15.11 Summary

This chapter reviewed the development of the common surface integral equations of electromagnetics, and discussed some alternatives that address nonuniqueness

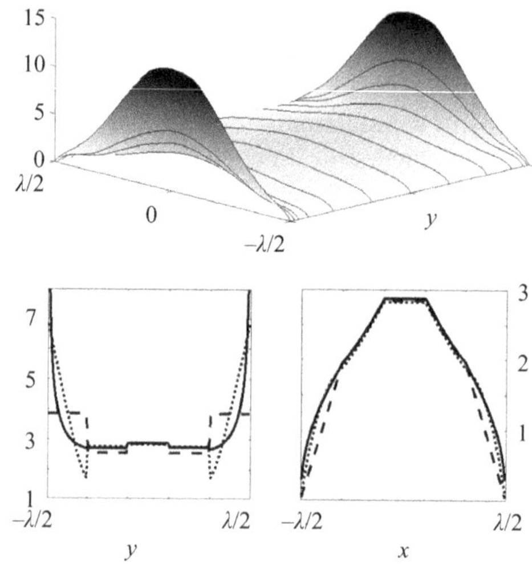

Figure 15.3 Magnitude of the x (co-polar) components of the current induced on a $1\lambda \times 1\lambda$ square plate by a normally incident plane wave with $H_y =$ 1 A/m obtained by solving the EFIE with Galerkin testing with a 5×5 quadrilateral cell mesh. The figure at the top shows the result obtained with 456 DoFs by using a $p=2$ order base plus singular basis functions added only to the 16 cells on the border that are introduced to model the possible unbounded behavior of the current component tangent to the edges. The results at bottom show the current x-component along the $x = 0$ (at left) and $y = 0$ (at right) cut obtained with p-refinement (that is with bases of different order in the various cells) for an increasing number of DoF: 40 DoFs for dashed-line results, and 120 DoFs for dotted-line results; by augmenting the polynomial base used for the 120 DoF results with 36 singular basis functions (for a total of 156 DoFs), we obtain the solid-line results. ©2018 IEEE. Reprinted with permission from [108].

issues and other difficulties. Numerical solution techniques are often based on vector representations, and hierarchical vector basis functions and singular basis functions that may facilitate the use of adaptive refinement schemes were also considered.

References

[1] A.W. Maue, "On the formulation of a general scattering problem by means of an integral equation," *Zeitschrift für Physik*, vol. 126, pp. 610–618, 1949.

[2] A.J. Poggio and E.K. Miller, "Integral equation solutions of three-dimensional scattering problems," in *Computer Techniques for Electromagnetics*, R. Mittra, ed. New York: Hemisphere, 1987 (Reprint of the original 1973 edition).

[3] A.F. Peterson, "The interior resonance problem associated with surface integral equations of electromagnetics: Numerical consequences and a survey of remedies," *Electromagnetics*, vol. 10, pp. 293–312, 1990.

[4] A.F. Peterson, *Mapped Vector Basis Functions for Electromagnetic Integral Equations*. San Rafael: Morgan & Claypool Synthesis Lectures, 2006.

[5] A.E. Ofluoglu, T. Ciftci, and O. Ergul, "Magnetic-field integral equation," *IEEE Antennas and Propagation Magazine,* vol. 57, no. 4, pp. 134–142, 2015.

[6] J.R. Mautz and R.F. Harrington, "H-field, E-field, and combined-field solutions for conducting bodies of revolution," *Arch. Elektron. Ubertragungstech.*, vol. 32, pp. 157–163, 1978.

[7] J.C. Bolomey and W. Tabbara, "Numerical aspects on coupling between complementary boundary value problems," *IEEE Trans. Antennas Propagat.*, vol. AP-21, pp. 356–363, 1973.

[8] J.R. Mautz and R.F. Harrington, "A combined-source formulation for radiation and scattering from a perfectly conducting body," *IEEE Trans. Antennas Propagat.*, vol. AP-27, pp. 445–454, 1979.

[9] A.D. Yaghjian, "Augmented electric- and magnetic-field equations," *Radio Science*, vol. 16, pp. 987–1001, 1981.

[10] M.B. Woodworth and A.D. Yaghjian, "Derivation, application, and conjugate gradient solution of dual-surface integral equations for three-dimensional, multi-wavelength perfect conductors," in *Application of Conjugate Gradient Method to Electromagnetics and Signal Analysis*, T.K. Sarkar, ed. New York: Elsevier, 1991.

[11] M.B. Woodworth and A.D. Yaghjian, "Multiwavelength three-dimensional scattering with dual-surface integral equations," *J. Optical Society America A*, vol. 11, pp. 1399–1413, 1994.

[12] R.A. Shore and A.D. Yaghjian, "Dual-surface integral equations in electromagnetic scattering," *IEEE Trans. Antennas Propagat.*, vol. 53, no. 5, pp. 1706–1709, 2005.

[13] A.F. Peterson, S.L. Ray, and R. Mittra, *Computational Methods for Electromagnetics.* New York: IEEE Press, 1998.

[14] J.R. Mautz and R.F. Harrington, "Electromagnetic scattering from a homogeneous material body of revolution," *Arch. Elektron. Ubertragungstech.*, vol. 33, pp. 71–78, 1979.

[15] B.M. Kolundzija and A.R. Djordjevic, *Electromagnetic Modeling of Composite Metallic and Dielectric Structures*. Boston: Artech House, 2002.

[16] Y. Chang and R.F. Harrington, "A surface formulation for characteristic modes of material bodies," *IEEE Trans. Antennas Propagat.*, vol. 25, pp. 789–795, 1977.

[17] T.K. Wu and L.L. Tsai, "Scattering from arbitrarily shaped lossy dielectric bodies of revolution," *Radio Science*, vol. 12, pp. 709–718, 1977.

[18] C. Müller, *Foundations of the Mathematical Theory of Electromagnetic Waves*. Berlin: Springer, 1969.

[19] P. Yla-Oijala and M. Taskinen, "Well-conditioned Muller formulation for electromagnetic scattering by dielectric objects," *IEEE Trans. Antennas Propagat.*, vol. 53, pp. 3316–3323, 2005.

[20] E. Marx, "Single integral equation for wave scattering," *J. Math. Physics*, vol. 23, pp. 1057–1065, 1982.

[21] A.W. Glisson, "An integral equation for electromagnetic scattering from homogeneous dielectric bodies," *IEEE Trans. Antennas Propagat.*, vol. 32, pp. 173–175, 1984.

[22] M.S. Yeung, "Single integral equation for electromagnetic scattering by three-dimensional homogeneous dielectric objects," *IEEE Trans. Antennas Propagat.*, vol. 47, pp. 1615–1622, 1999.

[23] A. Menshov and V. Okhmatovski, "New single source surface integral equations for scattering on penetrable cylinders and current flow modeling in 2-D conductors," *IEEE Trans. Microwave Theory Tech.*, vol. 61, pp. 341–349, 2013.

[24] D.R. Wilton, "Topological considerations in surface patch and volume cell modeling of electromagnetic scatterers," *Proceedings of the URSI Symposium on Electromagnetic Theory*, Santiago de Compostela (Spain), pp. 65–68, Aug. 23–26, 1983.

[25] G. Vecchi, "Loop star decomposition of basis functions in the discretization of the EFIE," *IEEE Trans. Antennas Propagat.*, vol. 47, pp. 339–346, Feb. 1999.

[26] F.P. Andriulli, "Loop-star and loop-tree decompositions: Analysis and efficient algorithms," *IEEE Trans. Antennas Propagat.*, vol. 60, pp. 2347–2356, 2012.

[27] G. Xiao, "Applying loop-flower basis functions to analyze electromagnetic scattering problems of PEC scatterers," *Int. J. Antennas Propag.*, vol. 2014, Article 905935, 2014.

[28] H. Contopanagos, B. Dembart, M. Epton, *et al.*, "Well-conditioned boundary integral equations for three-dimensional electromagnetic scattering," *IEEE Trans. Antennas Propagat.*, vol. 50, pp. 1824–1830, 2002.

[29] F.P. Andriulli, K. Cools, H. Bagci, *et al.*, "A multiplicative Calderon preconditioner for the electric field integral equation," *IEEE Trans. Antennas Propagat.*, vol. 56, pp. 2398–2412, 2008.

[30] M.B. Stephanson and J.-F. Lee, "Preconditioned electric field integral equation using Calderon identities and dual loop/star basis functions," *IEEE Trans. Antennas Propagat.*, vol. 57, pp. 1274–1279, 2009.

[31] Z.G. Qian and W.C. Chew, "Fast full-wave surface integral equation solver for multiscale structure modeling," *IEEE Trans. Antennas Propagat.*, vol. 57, pp. 3594–3601, 2009.

[32] W.C. Chew, "Vector potential electromagnetics with generalized gauge for inhomogeneous media: Formulation," *Progress in Electromagnetic Research*, vol. 149, pp. 69–84, 2014.

[33] U.M. Gur and O. Ergul, "Accuracy of sources and near-zone fields when using potential integral equations at low frequencies," *IEEE Antennas and Wireless Propagation Letters*, vol. 16, pp. 2783–2786, 2017.

[34] R.D. Graglia and A.F. Peterson, *Higher-order Techniques in Computational Electromagnetics,* 392 pages. Edison, NJ: SciTech Publishing/IET, 2016.

[35] D. Wilton, S. Rao, A. Glisson, D. Schaubert, O. Al-Bundak, and C. Butler "Potential integrals for uniform and linear source distributions on polygonal and polyhedral domains," *IEEE Trans. Antennas Propagat.*, vol. 32 , no. 3, pp. 276–281, 1984.

[36] R.D. Graglia, P.L.E. Uslenghi, and R.S. Zich, "Moment method with isoparametric elements for three-dimensional anisotropic scatterers," *Proceedings of the IEEE*, vol. 77, no. 5, pp. 750–760, 1989.

[37] R.D. Graglia, "On the numerical integration of the linear shape functions times the 3-D Green's function or its gradient on plane triangle," *IEEE Trans. Antennas Propag.*, vol. 41, no. 10, pp. 1448–1455, 1993.

[38] L. Rossi and P.J. Cullen, "On the fully numerical evaluation of the linear-shape function times the 3D Green's function on a plane triangle," *IEEE Trans. Microwave Theory Tech.*, vol. 47, no. 4, pp. 398–402, 1999.

[39] A. Herschlein, J. von Hagen, and W. Wiesbeck, "Methods for the evaluation of regular, weakly singular and strongly singular surface reaction integrals arising in method of moments," *Applied Computational Electromagnetics Society J.*, vol. 17, pp. 63–73, 2002.

[40] D.J. Taylor, "Accurate and efficient numerical integration of weakly singular integrals in Galerkin EFIE solutions," *IEEE Trans. Antennas Propagat.*, vol. 51, no. 7, pp. 1630–1637, 2003.

[41] M.A. Khayat and D.R. Wilton, "Numerical evaluation of singular and near-singular potential integrals," *IEEE Trans. Antennas Propagat.*, vol. 53, no. 10, pp. 3180–3190, 2005.

[42] R.D. Graglia and G. Lombardi, "Machine precision evaluation of singular and nearly singular potential integrals by use of Gauss quadrature formulas for rational functions," *IEEE Trans. Antennas Propagat.*, vol. 56, no. 4, pp. 981–998, 2008.

[43] G. Lombardi, "Design of quadrature rules for Müntz and Müntz-logarithmic polynomials using monomial transformation," *Int. J. Numer. Methods Eng.*, vol. 80, no. 13, pp. 1687–1717, 2009.

[44] M.S. Tong and W.C. Chew, "A novel approach for evaluating hypersingular and strongly singular surface integrals in electromagnetics," *IEEE Trans. Antennas Propagat.*, vol. 58, no. 11, pp. 3593–3601, 2010.

[45] A.G. Polimeridis, and J.R. Mosig, "Complete semi-analytical treatment of weakly singular integrals on planar triangles via the direct evaluation method," *International J. Numerical Methods in Engineering,* vol. 83, no. 12, pp. 1625–1650, 2010.

[46] F. Vipiana, D.R. Wilton, and W.A. Johnson, "Advanced numerical schemes for the accurate evaluation of 4-D reaction integrals in the method of moments," *IEEE Trans. Antennas Propag.*, vol. 61, no. 11, pp. 5559–5566, 2013.

[47] D.R. Wilton, F. Vipiana, and W.A. Johnson, "Evaluation of 4-D reaction integrals in the method of moments: Coplanar element case," *IEEE Trans. Antennas Propag.*, vol. 65, no. 5, pp. 2479–2493, 2017.

[48] D. Tihon and C. Craeye, "All-analytical evaluation of the singular integrals involved in the method of moments," *IEEE Trans. Antennas Propag.*, Vol. 66, No. 4, pp. 1925–1936, 2018.

[49] R.D. Graglia, A.F. Peterson, and P. Petrini, "Computation of EFIE matrix entries with singular basis functions," *IEEE Trans. Antennas Propagat.*, Vol. 66, No. 11, pp. 6217–6224, 2018.

[50] J.C. Nedelec, "Mixed finite elements in R3," *Numer. Mathem.*, vol. 35, pp. 315–341, 1980.

[51] P.A. Raviart and J.M. Thomas, "A mixed finite element method for 2nd order elliptic problems," in *Mathematical Aspects of Finite Element Methods,* A. Dold and B. Eckmann, eds. New York: Springer-Verlag, 1977.

[52] S.M. Rao, D.R. Wilton, and A.W. Glisson, "Electromagnetic scattering by surfaces of arbitrary shape," *IEEE Trans. Antennas Propagat.*, vol. AP-30, pp. 409–418, 1982.

[53] D.H. Schaubert, D.R. Wilton, and A.W. Glisson, "A tetrahedral modeling method for electromagnetic scattering by arbitrarily shaped inhomogeneous dielectric bodies," *IEEE Trans. Antennas Propagat.*, vol. AP-32, pp. 77–85, 1984.

[54] A. Bossavit and J.-C. Verite, "A mixed FEM-BIEM method to solve 3-D eddy current problems," *IEEE Trans. Magn.*, vol. MAG-18, pp. 431–435, 1982.

[55] M. Hano, "Finite-element analysis of dielectric-loaded waveguides," *IEEE Trans. Microwave Theory Tech.*, vol. MTT-32, pp. 1275–1279, 1984.

[56] M.L. Barton and Z.J. Cendes, "New vector elements for three-dimensional magnetic field computation," *J. Appl. Phys.*, vol. 61, no. 8, pp. 3919–3921, 1987.

[57] L. Demkowicz, *Computing with hp-Adaptive Finite Elements, vol. 1.* Boca Raton: Chapman & Hall/CRC, 2007.

[58] L. Demkowicz, *Computing with hp-Adaptive Finite Elements, vol. 2.* Boca Raton: Chapman & Hall/CRC, 2008.

[59] D. Sun, J. Manges, X. Yuan, and Z. Cendes, "Spurious modes in finite element methods," *IEEE Antennas Propagat. Mag.*, vol. 37, no. 5, pp. 12–24, 1995.

[60] A.F. Peterson and R.D. Graglia, "A review of hierarchical vector basis functions for triangular and quadrilateral domains," *Forum for Electromagnetic Research Methods and Application Technologies (FERMAT)*, vol. 18, 24 pages, November–December 2016.

[61] R.D. Graglia, D.R. Wilton and A.F. Peterson, "Higher order interpolatory vector bases for computational electromagnetics," special issue on "Advanced Numerical Techniques in Electromagnetics," *IEEE Trans. Antennas Propagat.*, vol. 45, no. 3, pp. 329–342, 1997.

[62] A.F. Peterson, "Cell curvature and far field superconvergence in numerical solutions of electromagnetic integral equations," *International Journal of*

Antennas and Propagation, vol. 2016, Article ID 9808637, 7 pages, 2016, doi:10.1155/2016/9808637.

[63] R.A. Wildman, G. Pisharody, D.S. Weile, S. Balasubramaniam, and E. Michielssen, "An accurate scheme for the solution of the time-domain integral equations of electromagnetics using higher order vector bases and bandlimited extrapolation," *IEEE Trans. Antennas Propagat.*, vol. 52, no. 11, pp. 2973–2984, 2014.

[64] G. Pisharody and D.S. Weile, "Electromagnetic scattering from homogeneous dielectric bodies using time-domain integral equations," *IEEE Trans. Antennas Propagat.*, vol. 54, no. 2, pp. 687–697, 2006.

[65] F. Valdes, F.P. Andriulli, H. Bagci, and E. Michielssen, "Time-domain single-source integral equations for analyzing scattering from homogeneous penetrable objects," *IEEE Trans. Antennas Propagat.*, vol. 61, no. 3, pp. 1239–1254, 2013.

[66] R.D. Graglia, A.F. Peterson, and F.P. Andriulli, "Curl-conforming hierarchical vector bases for triangles and tetrahedra," *IEEE Trans. Antennas Propagat.*, vol. 59, no. 3, pp. 950–959, 2011.

[67] R.D. Graglia and A.F. Peterson, "Hierarchical curl-conforming Nédélec elements for quadrilateral and brick cells," *IEEE Trans. Antennas Propagat.*, vol. 59, no. 8, pp. 2766–2773, 2011.

[68] R.D. Graglia and A.F. Peterson, "Hierarchical curl-conforming Nédélec elements for triangular-prism cells," *IEEE Trans. Antennas Propagat.*, vol. 60, no. 7, pp. 3314–3324, 2012.

[69] R.D. Graglia and A.F. Peterson, "Hierarchical divergence-conforming Nédélec elements for volumetric cells," *IEEE Trans. Antennas Propagat.*, vol. 60, no. 11, pp. 5215–5227, 2012.

[70] L.P. Zha, Y.Q. Hu, and T. Su, "Efficient surface integral equation using hierarchical vector bases for complex EM scattering problems," *IEEE Trans. Antennas Propagat.*, vol. 60, pp. 952–957, 2012.

[71] L. Zha, R. Chen, and T. Su, "Fast EM scattering analysis for the hard targets in a layered medium by using the hierarchical vector basis functions," *ACES J.*, vol. 30, pp. 1154–1160, 2015.

[72] P. Yang, Z. Nie, S. He, and T. Qing, "The application of higher-order basis functions with phase description for electromagnetic simulation of arbitrary structures," *Electromagnetics*, vol. 33, pp. 561–574, 2013.

[73] P. Yang, Z. Nie, S. He, and T. Qing, "Efficient surface integral equation using travelling-standing-wave vector basis function for complex EM scattering problems," *Applied Mechanics and Materials*, Vols. 325–326, pp. 1498–1501, 2013.

[74] D.J. Ludick, J. Van Tonder, and U. Jakobus, "Combining domain decomposition solution techniques with higher order hierarchical basis functions," *ICEAA Conference*, pp. 70–73, 2013.

[75] L. Weggler, J.J. Van Tonder, and U. Jakobus, "Electromagnetic scattering by isoparametric elements of higher order," *J. Computational Methods in Applied Mathematics*, vol. 14, no. 3, pp. 411–418, 2014.

[76] Q.M. Cai, Y.-W. Zhao, Y.-T. Zheng, M.-M. Jia, Z. Zhao, and Z.-P. Nie, "Volume integral equation with higher order hierarchical vector basis functions for analysis of dielectric electromagnetic scattering," *IEEE Trans. Antennas Propagat.*, vol. 63, pp. 4964–4975, 2015.

[77] Q.-M. Cai, Y.-W. Zhao, W.-F. Huang, *et al.*, "Volume surface integral equation method based on higher order hierarchical vector basis functions for EM scattering and radiation from composite metallic and dielectric structure," *IEEE Trans. Antennas Propagat.*, vol. 64, no. 12, pp. 5359–5372, 2016.

[78] W. Luo, Z. Nie, and Y.P. Chen, "Efficient higher-order analysis of electromagnetic scattering by objects above, below, or straddling a half-space," *IEEE Antennas Wireless Propag. Lett.*, vol. 15, pp. 332–335, 2016.

[79] L.-W. Guo, J. Hu, W. Luo, L.-N. Song, and Z. Nie, "Efficient higher-order analysis of electromagnetic scattering of objects in half-space by domain decomposition method with a hybrid solver," *Progress in Electromagnetics Research C*, vol. 68, pp. 201–209, 2016.

[80] G.S. Cheng, Z.H. Fan, D.Z. Ding, and R.S. Chen, "An efficient high order plane wave time domain algorithm for transient electromagnetic scattering analysis," *Engineering Analysis with Boundary Elements,* vol. 85, pp. 13–19, 2017.

[81] S.K. Kim and A.F. Peterson, "Evaluation of local error estimators for the RWG based EFIE," *IEEE Trans. Antennas Propagat.*, vol. 66, pp. 819–826, 2018.

[82] S.K. Kim and A.F. Peterson, "Adaptive *h*-refinement for the RWG based EFIE," *IEEE J. Multiscale and Multiphysics Comp. Tech.*, vol. 3, pp. 58–65, 2018.

[83] D.R. Wilton and S. Govind, "Incorporation of edge conditions in moment method solutions," *IEEE Trans. Antennas Propagat.*, vol. 25, no. 6, pp. 845–850, 1977.

[84] J. Richmond, "On the edge mode in the theory of TM scattering by a strip or strip grating," *IEEE Trans. Antennas Propagat.*, vol. 28, no. 6, pp. 883–887, 1980.

[85] Z. Pantic and R. Mittra, "Quasi-TEM analysis of microwave transmission lines by the finite–element method," *IEEE Trans. Microwave Theory Tech.*, vol. 34, no. 11, pp. 1096–1103, 1986.

[86] J.P. Webb, "Finite element analysis of dispersion in waveguides with sharp metal edges," *IEEE Trans. Microwave Theory and Tech.,* vol. 36, no. 12, pp. 1819–1824, 1988.

[87] Z. Pantic-Tanner, C.H. Chan, and R. Mittra, "The treatment of edge singularities in the full wave finite-element solution of waveguiding problems," *Abstracts of the 1988 URSI Radio Science Meeting*, Syracuse, NY, pages 336, 1988.

[88] T. Andersson, "Moment-method calculations on apertures using basis singular functions," *IEEE Trans. Antennas Propagat.,* vol. 41, no. 12, pp. 1709–1716, 1993.

[89] J.M. Gil and J. Zapata, "Efficient singular element for finite element analysis of quasi-TEM transmission lines and waveguides with sharp metal

edges," *IEEE Trans. Microwave Theory and Tech.,* vol. 42, no. 1, pp. 92–98, 1994.

[90] J.M. Gil and J.P. Webb, "A new edge element for the modeling of field singularities in transmission lines and waveguides," *IEEE Trans. Microwave Theory and Tech.,* vol. 45, no. 12, Part 1, pp. 2125–2130, 1997.

[91] Z. Pantic-Tanner, J.S. Savage, D.R. Tanner, and A.F. Peterson, "Two-dimensional singular vector elements for finite element analysis," *IEEE Trans. Microwave Theory Tech.*, vol. 46, pp. 178–184, 1998.

[92] W.J. Brown and D.R. Wilton, "Singular basis functions and curvilinear triangles in the solution of the electric field integral equation," *IEEE Trans. Antennas Propagat.,* vol. 47, no. 2, pp. 347–353, 1999.

[93] R.D. Graglia and G. Lombardi, "Singular higher-order complete vector bases for finite methods," *IEEE Trans. Antennas Propagat.,* vol. 52, pp. 1672–1685, 2004.

[94] R.D. Graglia, A.F. Peterson, and L. Matekovits, "Singular, hierarchical scalar basis functions for triangular cells," *IEEE Trans. Antennas Propagat.*, vol. 61, no. 7, pp. 3674–3692, 2013.

[95] R.D. Graglia, A.F. Peterson, L. Matekovits, and P. Petrini, "Hierarchical additive basis functions for the finite-element treatment of corner singularities," special issue on "Finite Elements for Microwave Engineering," *Electromagnetics*, vol. 34, pp. 171–198, 2014.

[96] R.D. Graglia, A.F. Peterson, L. Matekovits, and P. Petrini, "Singular hierarchical curl-conforming vector bases for triangular cells," *IEEE Trans. Antennas Propagat.*, vol. 62, no. 7, pp. 3632–3644, 2014.

[97] R.D. Graglia and G. Lombardi, "Singular higher order divergence-conforming bases of additive kind and moments method applications to 3D sharp-wedge structures," *IEEE Trans. Antennas Propagat.*, vol. 56, no. 12, pp. 3768–3788, 2008.

[98] G. Lombardi, and R.D. Graglia, "Modeling junctions in sharp edge conducting structures with higher order method of moments." *IEEE Trans. Antennas Propagat.,* vol. 62, pp. 5723–5731, 2014.

[99] A.F. Peterson, and R.D. Graglia, "Basis functions for vertex, edge, and corner singularities: A review," invited paper, *IEEE Journal on Multiscale and Multiphysics Computational Techniques*, vol. 1, pp. 161–175, 2016.

[100] R.D. Graglia, A.F. Peterson, L. Matekovits, and P. Petrini, "The performance of additive singular basis functions for triangles," *Proceedings of the 15th International Conference on Electromagnetics in Advanced Applications (ICEAA13)*, Sept. 9–13, 2013, Torino, Italy, pp 1557–1560, DOI: 10.1109/ICEAA.2013.6632514

[101] R.D. Graglia, A.F. Peterson, L. Matekovits, and P. Petrini "Modeling corner singularities in triangular cells with hierarchical bases." In: *The 29th International Review of Progress in Applied Computational Electromagnetics,* Monterey, USA, pp. 7–11, Mar. 24–28, 2013.

[102] R.D. Graglia, P. Petrini, A.F. Peterson, and L. Matekovits, "Full-wave analysis of inhomogeneous waveguiding structures containing corners with singular

hierarchical curl-conforming vector bases" *IEEE Antennas and Propagation Letters.,* vol. 13, pp. 1701–1704, 2014. DOI: 0.1109/LAWP.2014.2351573

[103] R.D. Graglia, A.F. Peterson, L. Matekovits, and P. Petrini, "Hierarchical and singular bases for finite methods," presented at 2015 *IEEE International Conference on Computational Electromagnetics (ICCEM),* Hong Kong, pp. 26–29, Feb. 2–5, 2015. DOI: 10.1109/COMPEM.2015.7052543

[104] R.D. Graglia, A.F. Peterson, P. Petrini, and L. Matekovits "Hierarchical functions for multiscale problems," invited paper at *2016 IEEE International Conference on Computational Electromagnetics (ICCEM)* in Guangzou, China, Feb. 23–25, 2016.

[105] R.D. Graglia, A.F. Peterson, P. Petrini, and L. Matekovits "Hierarchical functions for multiscale problems," presented at *The International Workshop on Finite Elements for Microwave Engineering*, Florence, May 16–18, 2016.

[106] R.D. Graglia, A.F. Peterson, and P. Petrini, "Hierarchical basis functions for the numerical modeling of edge-field singularities," Invited paper, *2017 IEEE International Conference on Computational Electromagnetics (ICCEM)*, Kumamoto, Japan, March 8–10, 2017.

[107] R.D. Graglia, A.F. Peterson, and P. Petrini "Efficient use of singular hierarchical vector functions for the FEM solution of multiscale problems," *2017 International Applied Computational Electromagnetics Society (ACES) Symposium*, Firenze, Italy, March 26–30, 2017.

[108] R.D. Graglia, A.F. Peterson, and P. Petrini, "Hierarchical divergence conforming bases for edge singularities in quadrilateral cells," *IEEE Trans. Antennas Propagat.*, vol. 66, no. 11, pp. 6191–6201, 2018.

[109] R.D. Graglia, A.F. Peterson, and P. Petrini, "Treatment of plate scattering using singular edge and corner bases," *2018 IEEE International Conference on Computational Electromagnetics*, Chengdu, China, March 26–28, 2018.

Chapter 16
Electromagnetic modelling at arbitrarily low frequency via the quasi-Helmholtz projectors

Adrien Merlini[1], Alexandre Dély[1], Kristof Cools[2], and Francesco P. Andriulli[1]

16.1 Introduction

The finite element method (FEM) and the boundary element method (BEM) are two of the most widely used techniques for the numerical evaluation of electromagnetic fields scattered by objects with arbitrary shape and various properties. While particularly well-adapted to the simulation of highly inhomogeneous media, the FEM requires the computation and inversion of large sparse matrices. On the other hand, the BEM has several advantages over FEM in case of piecewise homogeneous media: by construction its solutions satisfy the radiation conditions, it yields smaller interaction matrices because it only requires discretization of the boundary surfaces and, finally, these matrices can be applied in near linear time to arbitrary vectors by leveraging fast solvers [1–3]. Another family of solutions for electromagnetic scattering problems are the finite difference approaches that discretize the partial derivatives on a grid using finite difference approximations [4]. A particularly popular instance of these techniques is the finite difference time-domain (TD) method [5] which is widely used to simulate transient scattering scenarios. However, in general, these methods suffer from numerical dispersion [6].

The electric-field integral equation (EFIE), a formulation originally developed for problems of scattering by perfectly electrically conducting (PEC) surfaces, has been extended to a wide variety of problems, including penetrable bodies and surface impedance [7–11] either in the frequency-domain (FD) EFIE (FD-EFIE) or in the TD-EFIE. This versatility makes any advancement in the computation or inversion of the underlying electric-field integral operator (EFIO) particularly impacting and explains the continued focus of the community on this topic.

Despite its advantages, the FD-EFIE suffers from issues limiting its applicability when employed naively. The condition number of the impedance matrix (the ratio of its maximum and minimum singular values) of the FD-EFIE suffers from two types

[1]Department of Electronics and Telecommunications, Politecnico di Torino, Turin, Italy
[2]Department of Applied Mathematics, Delft University of Technology, Delft, The Netherlands

of ill-conditioning: (i) the low-frequency breakdown which causes the conditioning of the matrix to grow as k^{-2}, where k is the wave number [12], and (ii) the high-refinement breakdown which causes the conditioning to grow as $(h/\sqrt{LC})^{-2}$, where h is the average size of the boundary elements, C is the lowest singular value of the hypersingular operator and L^{-1} is the largest singular value of the singular operator [12]. While it is difficult to accurately estimate $\sqrt{LC}$, the diameter of the structure is a satisfying approximation for practical considerations. Both breakdowns cause iterative solvers to require more iterations before converging and to yield less accurate solutions or, in more extreme cases, prevent numerical resolution of the problem altogether.

Similarly, the TD-EFIE suffers from the same breakdowns, even though the low-frequency breakdown manifests itself as the so-called large time step breakdown which is an ill-conditioning of order $\mathcal{O}(\Delta t^2)$ [13,14], where Δt is the time step of the marching-on-in-time (MOT) algorithm [10,15]. On top of these limitations, the TD-EFIE also exhibits its own specific set of problems, such as the support of non-physical, source-less or divergence-free solutions [16,17], often called DC-instabilities.

While the EFIO-based formulations are not intrinsically inaccurate, when it is computed in limited precision a loss of significant digits in part of the solution vector will severely degrade the attainable accuracy of the formulations at low frequencies [18–22]. This loss of significant digits is caused by the different frequency scaling of the solenoidal and non-solenoidal parts of the solution: the smaller term will fall below the numerical cut-off of the computation buffer imposed by the larger one, making it impossible to recover after the fact. This loss is especially problematic when computing the far field scattered by the solution current densities, since both contributions have the same importance for certain types of excitation [23]. While the ill-conditioning of the impedance matrices and the loss of significant digits in the solution vector are distinct phenomena, they are often both referred to as the 'EFIE low-frequency breakdown' without distinction.

The traditional way of addressing the low-frequency breakdown, both in the TD and FD, has been to expose the solenoidal and non-solenoidal parts of the BEM matrices, and re-scale them independently with the adequate powers of k or Δt. In the TD, this re-scaling is coupled with a mixed discretization scheme to address the DC-instabilities [24–26]. While this Loop–Star (LS)[27–31] decomposition effectively addresses the conditioning degradation as well as the loss of significant digits, it suffers from two major drawbacks: (i) it requires the identification of the global loops of the simulated structure [27] which is a quadratic in complexity operation and (ii) it further degrades the high-refinement conditioning of the impedance matrices [31], because the LS matrices behave as a graph Laplacian [12,32]. Both issues severely limit the applicability of the technique to large and/or structurally complex problems. Other formulations that are immune to the low-frequency breakdown, while not requiring the detection of global loops, have been proposed [33].

A new family of techniques, based on quasi-Helmholtz (qH) projectors, has recently been introduced to cure the low-frequency breakdown without any of the limitations of the traditional solution [23]. In particular, the projectors do not require the search for global loops because only the non-solenoidal projector is directly computed and the solenoidal projector is derived as its complement. Furthermore, this technique does not further degrade the high-refinement breakdown because projectors have a flat spectrum and do not introduce further h dependencies to the equation. The projectors can be computed in near-linear complexity by leveraging multigrid preconditioners [23,34,35], which means that their usage does not increase the complexity of the preconditioned formulation or significantly slow down existing solvers.

While outside the scope of this chapter, the high-refinement breakdown of the EFIE can be cured by Calderón-based techniques [36–39] which can be combined with the qH projectors to obtain fully regularized EFIO-based formulations [23,40], immune from both the high-refinement and low-frequency breakdowns.

In this chapter, we start by setting notation and recall the FD-EFIE and TD-EFIE as well as their respective discretization strategies in Section 16.2. In Section 16.3, the low-frequency breakdown of the FD-EFIE is illustrated and analysed and the traditional LS technique is introduced and its limitations discussed. We focus our attention on the traditional LS decomposition for the sake of simplicity and for didactic purposes; however, a similar analysis could be applied to several other more recent strategies for tackling low-frequency problems. The large time step breakdown of the TD-EFIE is introduced in Section 16.4, while the DC instabilities are introduced in Section 16.5. The qH projectors and efficient ways to compute them are presented in Section 16.6. In Section 16.7 and 16.8, the qH-based low-frequency stable FD-EFIE and TD-EFIE are, respectively, introduced. Along with the developments required to derive the qH-based formulations, we present implementation details that will help the reader to swiftly and efficiently integrate the projectors into existing solvers without major modifications.

16.2 Notation and background

To make this chapter self-contained, we will recall the FD-EFIE and TD-EFIE as well as the related operators and discretization schemes. The notations used throughout this work will be set along these developments.

16.2.1 Frequency domain

In this section, the theory and notations for the TD-EFIE will be introduced. The time dependency $e^{-i\omega t}$, where ω is the simulated angular frequency, is left out for clarity but is assumed throughout our developments.

16.2.1.1 Definition of the EFIE

Let Γ be the boundary surface of a PEC body residing in a space characterized by the electric permittivity ε and magnetic permeability μ. Assuming that Γ is orientable, we denote $\hat{\boldsymbol{n}}_r$ the outgoing unitary vector normal to the surface in $\boldsymbol{r}$. If a complex incident electric field – or phasor – $\boldsymbol{e}^i(\boldsymbol{r})$ is introduced, it will induce a surface current density $\boldsymbol{j}(\boldsymbol{r})$ on Γ, which will be solution of the EFIE

$$\eta(\mathscr{T}\boldsymbol{j})(\boldsymbol{r}) = -\hat{\boldsymbol{n}}_r \times \boldsymbol{e}^i(\boldsymbol{r}) \tag{16.1}$$

with the EFIO $\mathscr{T}$ defined as a linear combination of the vector potential $\mathscr{T}_s$ and the scalar potential $\mathscr{T}_h$

$$(\mathscr{T}\boldsymbol{j})(\boldsymbol{r}) = \mathrm{i}k\, \mathscr{T}_s\boldsymbol{j}(\boldsymbol{r}) + \frac{1}{\mathrm{i}k}\, \mathscr{T}_h\boldsymbol{j}(\boldsymbol{r}) \tag{16.2}$$

$$(\mathscr{T}_s\boldsymbol{j})(\boldsymbol{r}) = \hat{\boldsymbol{n}}_r \times \int_\Gamma \frac{\mathrm{e}^{\mathrm{i}k\|\boldsymbol{r}-\boldsymbol{r}'\|}}{4\pi\|\boldsymbol{r}-\boldsymbol{r}'\|}\boldsymbol{j}(\boldsymbol{r}')\,\mathrm{d}\boldsymbol{r}' \tag{16.3}$$

$$(\mathscr{T}_h\boldsymbol{j})(\boldsymbol{r}) = -\hat{\boldsymbol{n}}_r \times \nabla \int_\Gamma \frac{\mathrm{e}^{\mathrm{i}k\|\boldsymbol{r}-\boldsymbol{r}'\|}}{4\pi\|\boldsymbol{r}-\boldsymbol{r}'\|}\nabla'\cdot\boldsymbol{j}(\boldsymbol{r}')\,\mathrm{d}\boldsymbol{r}' \tag{16.4}$$

where $k = \omega\sqrt{\varepsilon\mu}$ is the wave number and $\eta = \sqrt{\mu/\varepsilon}$ is the characteristic impedance of the medium.

16.2.1.2 Definition of the RWG elements

To numerically solve the EFIE, Γ is discretized with planar triangular mesh elements of average edge length h. We will denote the number of internal edges as N_e, the number of faces as N_f and the number of vertices as N_v.

The current density $\boldsymbol{j}$ is discretized with interpolatory basis functions defined on the triangular mesh elements. The Rao–Wilton–Glisson (RWG) [41] basis functions are one of the most common choices for discretizing this type of problem because they are div-conforming, meaning that their divergence is well defined, and because of their relatively simple implementation. They are defined as follows: for each pair of cells c_n^+ and c_n^- sharing an internal edge $\boldsymbol{e}_n$, we define the RWG basis function $\boldsymbol{f}_n$ as

$$\boldsymbol{f}_n(\boldsymbol{r}) = \begin{cases} \dfrac{\boldsymbol{r}-\boldsymbol{r}_n^+}{2A_n^+} & \text{for } \boldsymbol{r}\in c_n^+ \\ \dfrac{\boldsymbol{r}_n^- - \boldsymbol{r}}{2A_n^-} & \text{for } \boldsymbol{r}\in c_n^- \\ \boldsymbol{0} & \text{otherwise} \end{cases} \tag{16.5}$$

where $A_n^\pm$ denotes the area of cell $c_n^\pm$, $\boldsymbol{r}_n^+$ and $\boldsymbol{r}_n^-$ are, respectively, the vertices of c_n^+ and c_n^- that do not belong to $\boldsymbol{e}_n$ while $\boldsymbol{v}_n^+$ and $\boldsymbol{v}_n^-$ are the vertices that do (Figure 16.1). A consequence of this construction is that the basis functions are only defined on inner edges – edges that are not on the boundary of Γ.

16.2.1.3 Definition of the LS matrices

In this section, we introduce a decomposition of the RWG space into its solenoidal, harmonic and non-solenoidal subspaces to properly study the low-frequency properties of the FD-EFIE and TD-EFIE.

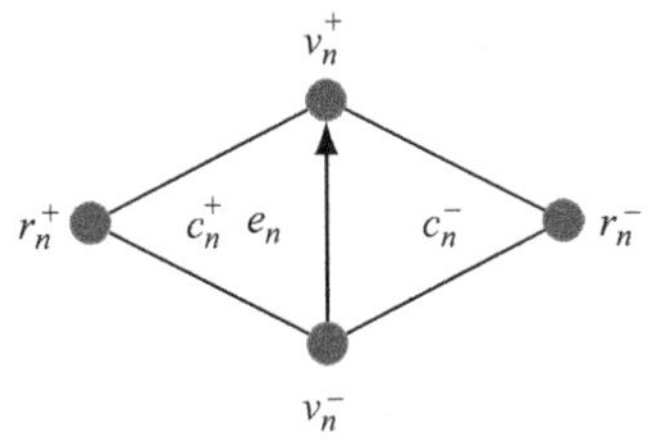

Figure 16.1 Illustration of the notations used to define an RWG basis function: the edge $\boldsymbol{e}_n$ between vertices $\boldsymbol{v}_n^+$ and $\boldsymbol{v}_n^-$ corresponds to any inner edge of the underlying mesh, and the cells c_n^+ and c_n^- are the two triangles attached to this edge and completed by the vertices $\boldsymbol{r}_n^+$ and $\boldsymbol{r}_n^-$, respectively

The solenoidal subspace Λ only includes divergence free functions ($\forall \boldsymbol{l} \in \Lambda$, $\nabla \cdot \boldsymbol{l} = 0$) which, in case of the current density, have been demonstrated [42] to form closed paths – or loops. This is why we define Λ as the subspace generated by the local loop basis functions $\{\boldsymbol{\lambda}_j\}, j \in [1, N_v]$ built on every vertex of the mesh connected to N_j edges by linearly combining the N_j RWG functions defined on these edges. Said otherwise, using the same notation as in Figure 16.1, for every vertex $\boldsymbol{v}_j$ of degree N_j we combine the N_j RWG functions $\{\boldsymbol{f}_i\}_i$ for which $\boldsymbol{v}_i^+ = \boldsymbol{v}_j$ or $\boldsymbol{v}_i^- = \boldsymbol{v}_j$

$$\boldsymbol{\lambda}_j(\boldsymbol{r}) = \sum_{i=1}^{N_e} \left[\tilde{\boldsymbol{\Lambda}}\right]_{ij} \boldsymbol{f}_i(\boldsymbol{r}), \quad j \in [1, N_v] \tag{16.6}$$

where the N_j non-zero $\{[\tilde{\boldsymbol{\Lambda}}]_{ij}\}$ are chosen such that $\nabla \cdot \boldsymbol{\lambda}_j = 0$. From this definition, one can build a mapping from the original RWG space to the solenoidal subspace which, in matrix form $\tilde{\boldsymbol{\Lambda}} \in \mathbb{R}^{N_e \times N_v}$, is expressed as

$$\left[\tilde{\boldsymbol{\Lambda}}\right]_{ij} = \begin{cases} 1 & \text{if vertex } j \text{ is } \boldsymbol{v}_i^+ \\ -1 & \text{if vertex } j \text{ is } \boldsymbol{v}_i^- \\ 0 & \text{otherwise} \end{cases} \tag{16.7}$$

The non-solenoidal – or star – subspace Σ can be constructed in different ways, with different sets of RWG basis functions. Our particular choice of star basis functions corresponds to the dual definition with respect to the one of the loop basis functions. Let Σ^S be one such construction generated by the set of functions $\{\boldsymbol{\sigma}_j^S\}, j \in [1, N_f]$, where each function $\boldsymbol{\sigma}_j^S$ is defined on cell c_j as a linear combination of the N_j RWG functions defined on this cell (basis functions for which $c^+ = c_j$ or $c^- = c_j$)

$$\boldsymbol{\sigma}_j^S(\boldsymbol{r}) = \sum_{i=1}^{N_e} \left[\tilde{\boldsymbol{\Sigma}}^S\right]_{ij} \boldsymbol{f}_i(\boldsymbol{r}), \quad j \in [1, N_f] \tag{16.8}$$

and, by construction, $N_j \in [1, 3]$ (on structures without junctions). The N_j non-zero coefficients $[\tilde{\boldsymbol{\Sigma}}^S]_{ij}$ are chosen to be ± 1, where the sign is chosen so that all currents are flowing out of the cell c_j. Similarly to Λ, from this definition a mapping matrix $\tilde{\boldsymbol{\Sigma}}^S \in \mathbb{R}^{N_e \times N_f}$ can be constructed as

$$\left[\tilde{\boldsymbol{\Sigma}}^S\right]_{ij} = \begin{cases} 1 & \text{if the cell } j \text{ is } c_i^+ \\ -1 & \text{if the cell } j \text{ is } c_i^- \\ 0 & \text{otherwise} \end{cases} \tag{16.9}$$

Finally, we define the harmonic subspace H to be generated by the divergence-free functions defined on the global loops of Γ. Consequently, H is non-empty if and only if Γ is not simply connected, meaning that it contains holes or handles. For instance, in case of a sphere H is empty while in case of a torus it contains two functions corresponding to the toroidal and poloidal directions illustrated in Figure 16.2. The determination of these functions is computationally more expensive than the ones for functions in Λ and Σ, because the detection of global loops in the structure is a superlinear operation in the general case. While no analytical expression can be obtained for arbitrary geometries, the mapping matrix $\tilde{\boldsymbol{H}}$ should contain the RWG coefficients of the functions spanning the space H as its columns. Even though the subspaces Λ, H and Σ^S are a decomposition of the original RWG space (i.e. $\Lambda \oplus H \oplus \Sigma$ is equal to the original RWG space), before obtaining an LS transformation matrix the linear dependency within Loop and within Star functions needs to be addressed: according to Euler–Poincaré formula [43], the relation between the number of inner edges N_e, inner vertices N_v, faces N_f, apertures N_a and handles N_h of each connected component of Γ is

$$N_f - N_e + N_v = 2 - N_a - 2N_h \tag{16.10}$$

Given the definitions of the matrices $\tilde{\boldsymbol{\Lambda}}$ and $\tilde{\boldsymbol{\Sigma}}^S$, N_e is also the number of RWG basis functions, N_f the number of columns of $\tilde{\boldsymbol{\Sigma}}^S$ and N_v the number of columns of $\tilde{\boldsymbol{\Lambda}}$. It follows that the Euler–Poincaré formula imposes, for each connected component of Γ, that (i) one star basis function must be removed, regardless of the geometry and that (ii) one loop basis function must be removed if the component is closed. For example, on a plate ($N_a = 1$, $N_h = 0$, $N_e = N_f - 1 + N_v$), one column of $\tilde{\boldsymbol{\Sigma}}^S$ needs

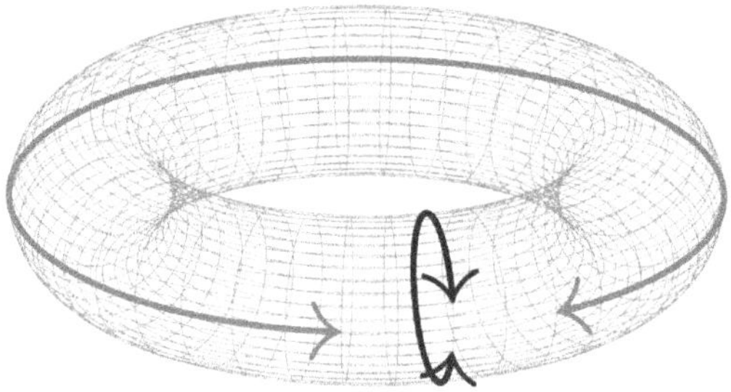

Figure 16.2 Representation of the global loops in a toroidal structure. The poloidal direction is indicated by the vertical (green) arrow and the toroidal direction by the horizontal (red) one

to be removed, no columns are removed from $\tilde{\boldsymbol{\Lambda}}$, and $\tilde{\boldsymbol{H}} = \boldsymbol{0}$. On a sphere ($N_a = 0$, $N_h = 0$, $N_e = N_f - 1 + N_v - 1$), instead, one column is to be removed from both $\tilde{\boldsymbol{\Sigma}}^S$ and $\tilde{\boldsymbol{\Lambda}}$.

Once the adequate number of basis function have been removed from the different LS matrices, the LS transformation matrix $\boldsymbol{B}$ is defined as

$$\boldsymbol{B} = \begin{bmatrix} \boldsymbol{\Lambda} & \boldsymbol{H} & \boldsymbol{\Sigma} \end{bmatrix} \tag{16.11}$$

where the matrices $\boldsymbol{\Lambda}$ and $\boldsymbol{\Sigma}$, respectively, correspond to the matrices $\tilde{\boldsymbol{\Lambda}}$ and $\tilde{\boldsymbol{\Sigma}}^S$ with the adequate number of columns removed. The matrix $\boldsymbol{H}$ is identical to $\tilde{\boldsymbol{H}}$ and is only introduced for clarity of the notation.

Finally, it should be noted that, since $\boldsymbol{\Sigma}^{\mathrm{T}}$ is a discretization of the divergence operator, $\boldsymbol{\Sigma}^{\mathrm{T}}\boldsymbol{\Lambda} = \boldsymbol{0}$ and $\boldsymbol{\Sigma}^{\mathrm{T}}\boldsymbol{H} = \boldsymbol{0}$. In other words, the non-solenoidal subspace is orthogonal to the solenoidal and harmonic subspaces.

16.2.1.4 Discretization strategy for the EFIE

Using the RWG basis functions introduced in Section 16.2.1.2, the EFIE can be discretized to proceed with the numerical resolution of the problem following a Galerkin scheme. First, the current density $\boldsymbol{j}(\boldsymbol{r})$ is approximated as a linear combination of the RWG basis functions $\{\boldsymbol{f}_m\}$, $m \in [1, N_e]$ defined on the meshed geometry

$$\boldsymbol{j}(\boldsymbol{r}) \approx \sum_{m=1}^{N_e} [\boldsymbol{j}]_m \boldsymbol{f}_m(\boldsymbol{r}) \tag{16.12}$$

Then, (16.12) is replaced into (16.3) and (16.4), which yields the expressions for the discretized potentials

$$(\mathcal{T}_s \boldsymbol{j})(\boldsymbol{r}) \approx \sum_{m=1}^{N_e} [\boldsymbol{j}]_m \hat{\boldsymbol{n}}_r \times \int_\Gamma \frac{\mathrm{e}^{\mathrm{i}k\|\boldsymbol{r}-\boldsymbol{r}'\|}}{4\pi\|\boldsymbol{r}-\boldsymbol{r}'\|} \boldsymbol{f}_m(\boldsymbol{r}')\,\mathrm{d}\boldsymbol{r}' \tag{16.13}$$

$$(\mathcal{T}_h \boldsymbol{j})(\boldsymbol{r}) \approx -\sum_{m=1}^{N_e} [\boldsymbol{j}]_m \hat{\boldsymbol{n}}_r \times \nabla \int_\Gamma \frac{\mathrm{e}^{\mathrm{i}k\|\boldsymbol{r}-\boldsymbol{r}'\|}}{4\pi\|\boldsymbol{r}-\boldsymbol{r}'\|} \nabla' \cdot \boldsymbol{f}_m(\boldsymbol{r}')\,\mathrm{d}\boldsymbol{r}' \tag{16.14}$$

These equations are then tested with the $\{\hat{\boldsymbol{n}}_r \times \boldsymbol{f}_m\}$, $m \in [1, N_e]$ basis functions, giving rise to the matrix system

$$\boldsymbol{Z}\boldsymbol{j} = \left(\mathrm{i}k\,\boldsymbol{Z}_s + \frac{1}{\mathrm{i}k}\boldsymbol{Z}_h\right)\boldsymbol{j} = \boldsymbol{v} \tag{16.15}$$

where the matrix elements are

$$[\boldsymbol{Z}_s]_{ij} = \langle \hat{\boldsymbol{n}}_r \times \boldsymbol{f}_i, \eta(\mathcal{T}_s \boldsymbol{f}_j)\rangle \tag{16.16}$$

$$[\boldsymbol{Z}_h]_{ij} = \langle \hat{\boldsymbol{n}}_r \times \boldsymbol{f}_i, \eta(\mathcal{T}_h \boldsymbol{f}_j)\rangle \tag{16.17}$$

$$[\boldsymbol{v}]_i = \langle \hat{\boldsymbol{n}}_r \times \boldsymbol{f}_i, -\hat{\boldsymbol{n}}_r \times \boldsymbol{e}^i\rangle \tag{16.18}$$

An approximation of $\boldsymbol{j}$ can then be obtained by solving this system with standard numerical techniques and replacing the coefficients $\{[\boldsymbol{j}]_i\}$ into (16.12).

Numerical quadrature rules adapted to triangular domains can be used to evaluate the different integrals required for building the system matrices. These rules approximate the integral by a weighted sum of the values of its kernel in specific points of the integration domain

$$\int f(\boldsymbol{r})\,\mathrm{d}S(\boldsymbol{r}) \approx S\sum_{n=1}^{N} w_n f(\boldsymbol{r}_n) \tag{16.19}$$

where N is the degree of the quadrature, S is the area of the triangle and $\{w_n, \boldsymbol{r}_n\}$, $n \in [1, N]$ can be computed as explained in [44]. While the quadrature rules are efficient for smooth function, in the neighbourhood of singularities numerical extraction techniques coupled with analytic formulas should be used for accurate evaluation of the integrals [45]. Further details on the implementation of these techniques can be found in the aforementioned references or in dedicated works [46].

16.2.2 Time domain

16.2.2.1 Definition of the TD-EFIE equation

Assuming the same geometry and notations as for the FD-EFIE, an analogous TD formulation can be derived. Because of the similarities between the formulations and for notation simplicity, we will use the same symbols for the FD and TD operators. At $t > 0$, an incident electric field $\boldsymbol{e}^i(\boldsymbol{r}, t)$ induces on Γ a surface electric current density $\boldsymbol{j}(\boldsymbol{r}, t)$ that is solution of the TD-EFIE

$$\eta(\mathcal{T}\boldsymbol{j})(\boldsymbol{r}, t) = -\hat{\boldsymbol{n}}_r \times \boldsymbol{e}^i(\boldsymbol{r}, t)\ , \quad \forall \boldsymbol{r} \in \Gamma\ , \quad \forall t > 0 \tag{16.20}$$

where $\mathcal{T}$ is a linear combination of the vector potential $\mathcal{T}_s$ and scalar potential $\mathcal{T}_h$

$$(\mathcal{T}\boldsymbol{j})(\boldsymbol{r}, t) = \mathcal{T}_s\boldsymbol{j}(\boldsymbol{r}, t) + \mathcal{T}_h\boldsymbol{j}(\boldsymbol{r}, t) \tag{16.21}$$

$$(\mathcal{T}_s\boldsymbol{j})(\boldsymbol{r}, t) = -\frac{1}{c}\hat{\boldsymbol{n}}_r \times \int_\Gamma \frac{\partial_t \boldsymbol{j}\left(\boldsymbol{r}', t - \frac{\|\boldsymbol{r}-\boldsymbol{r}'\|}{c}\right)}{4\pi\,\|\boldsymbol{r} - \boldsymbol{r}'\|}\,\mathrm{d}\boldsymbol{r}' \tag{16.22}$$

$$(\mathcal{T}_h\boldsymbol{j})(\boldsymbol{r}, t) = c\hat{\boldsymbol{n}}_r \times p.v. \int_\Gamma \nabla \frac{\partial_t^{-1}\nabla' \cdot \boldsymbol{j}\left(\boldsymbol{r}', t - \frac{\|\boldsymbol{r}-\boldsymbol{r}'\|}{c}\right)}{4\pi\,\|\boldsymbol{r} - \boldsymbol{r}'\|}\,\mathrm{d}\boldsymbol{r}' \tag{16.23}$$

where $c = 1/\sqrt{\epsilon\mu}$ is the speed of light in the medium, ∂_t denotes a temporal derivation and ∂_t^{-1} a temporal integration, defined as $\partial_t^{-1} f(\tau) = \int_{-\infty}^{\tau} f(u)\,\mathrm{d}u$.

As expected, (16.20) is very similar to its FD counterpart (16.1). From this comparison, we can notice that time derivatives ∂_t are analogous to multiplications by a factor $-ikc$, time integrations ∂_t^{-1} to multiplications by $(-ikc)^{-1}$, and time delays $\boldsymbol{j}(\boldsymbol{r}', t - \|\boldsymbol{r} - \boldsymbol{r}'\|/c)$ to phase shifts $\exp(ik\|\boldsymbol{r} - \boldsymbol{r}'\|)\boldsymbol{j}(\boldsymbol{r}')$.

Straightforward derivations show that $\boldsymbol{j}(\boldsymbol{r}, t)$ is only determined by (16.20) up to a constant solenoidal part (see Section 16.5 on DC instability). Hence we enforce causality, meaning that all fields are assumed to vanish in the neighbourhood of Γ when $t < 0$, to ensure that the solution is uniquely determined.

16.2.2.2 Spatial discretization of the TD-EFIE

The TD-EFIE can be spatially discretized using the same scheme as for the FD-EFIE in Section 16.2.1.4. Following our previous developments $\boldsymbol{j}(\boldsymbol{r}, t)$ is expanded as a linear combination of the RWG basis functions $\{\boldsymbol{f}_m\}$, $m \in [1, N_e]$

$$\boldsymbol{j}(\boldsymbol{r}, t) \approx \sum_{m=1}^{N_e} [\boldsymbol{\mathsf{j}}(t)]_m \boldsymbol{f}_m(\boldsymbol{r}) \tag{16.24}$$

with the notable difference that the RWG coefficients $[\boldsymbol{\mathsf{j}}(t)]_m$ have a time dependency. A Galerkin scheme is then applied to the TD-EFIE (16.20) with the rotated RWG basis functions $\{\hat{\boldsymbol{n}}_r \times \boldsymbol{f}_m\}$, $m \in [1, N_e]$ as testing, yielding the matrix system

$$\boldsymbol{Z}\boldsymbol{\mathsf{j}}(t) = \boldsymbol{\mathsf{e}}(t) \tag{16.25}$$

where $\forall t > 0$

$$[\boldsymbol{Z}_s \boldsymbol{\mathsf{j}}(t)]_m = \int_\Gamma (\hat{\boldsymbol{n}}_r \times \boldsymbol{f}_m(\boldsymbol{r})) \cdot (\eta(\mathcal{T}_s \boldsymbol{j})(\boldsymbol{r}, t))\,\mathrm{d}\boldsymbol{r} \tag{16.26}$$

$$[\boldsymbol{Z}_h \boldsymbol{\mathsf{j}}(t)]_m = \int_\Gamma (\hat{\boldsymbol{n}}_r \times \boldsymbol{f}_m(\boldsymbol{r})) \cdot (\eta(\mathcal{T}_h \boldsymbol{j})(\boldsymbol{r}, t))\,\mathrm{d}\boldsymbol{r} \tag{16.27}$$

$$[\boldsymbol{Z}\,\boldsymbol{\mathsf{j}}(t)]_m = \int_\Gamma (\hat{\boldsymbol{n}}_r \times \boldsymbol{f}_m(\boldsymbol{r})) \cdot (\eta(\mathcal{T} \boldsymbol{j})(\boldsymbol{r}, t))\,\mathrm{d}\boldsymbol{r} \tag{16.28}$$

$$[\boldsymbol{\mathsf{e}}(t)]_m = \int_\Gamma (\hat{\boldsymbol{n}}_r \times \boldsymbol{f}_m(\boldsymbol{r})) \cdot (-\hat{\boldsymbol{n}}_r \times \boldsymbol{e}^i(\boldsymbol{r}, t))\,\mathrm{d}\boldsymbol{r} \tag{16.29}$$

16.2.2.3 Temporal discretization of the TD-EFIE

Since the equation is also time dependent, an additional discretization of the temporal component is required for numerical resolution. The time axis is divided into N_t time steps of duration Δt, so that a solution for $\boldsymbol{j}(\boldsymbol{r}, t)$ is computed for $0 \le t \le N_t \Delta t$. At each time step $i \in [1, N_t]$, the pulse functions $p(t - i\Delta t)$ defined as

$$p(t) = \begin{cases} 1 & \text{if } -\Delta t \le t \le 0 \\ 0 & \text{otherwise} \end{cases} \tag{16.30}$$

and illustrated in Figure 16.3(a) are used to discretize the time dependency of the RWG coefficients in both the source and testing terms of (16.25) as

$$[\boldsymbol{\mathsf{j}}(t)]_m \approx \sum_{i=1}^{N_t} [\boldsymbol{\mathsf{j}}_i]_m p(t - i\Delta t) \tag{16.31}$$

where the vector $\boldsymbol{\mathsf{j}}_i \in \mathbb{R}^{N_e}$ contains the coefficients for the time step i. Thus, the current density can be approximated by combining (16.24) and (16.31)

$$\boldsymbol{j}(\boldsymbol{r}, t) \approx \sum_{m=1}^{N_e} \sum_{i=1}^{N_t} [\boldsymbol{\mathsf{j}}_i]_m \boldsymbol{f}_m(\boldsymbol{r}) p(t - i\Delta t) \tag{16.32}$$

The final linear systems are then obtained by applying a temporal Galerkin scheme, in which the pulse functions $p(t - j\Delta t)$ are used to test (16.25) at each time step $j \in [1, N_t]$, yielding

$$\sum_{i=0}^{j} \boldsymbol{Z}_i \boldsymbol{j}_{j-i} = \mathbf{e}_j \tag{16.33}$$

where

$$\begin{aligned}[\boldsymbol{Z}_i]_{mn} &= \int_{\mathbb{R}} p(t - i\Delta t) \int_{\Gamma} (\hat{\boldsymbol{n}}_r \times \boldsymbol{f}_m(\boldsymbol{r})) \cdot (\eta \mathscr{T}\{\boldsymbol{f}_n p\}(\boldsymbol{r}, t))\, \mathrm{d}\boldsymbol{r}\, \mathrm{d}t \\ &= \Delta t \int_{\Gamma} (\hat{\boldsymbol{n}}_r \times \boldsymbol{f}_m(\boldsymbol{r})) \cdot (\eta \mathscr{T}\{\boldsymbol{f}_n h\}(\boldsymbol{r}, i\Delta t))\, \mathrm{d}\boldsymbol{r}\end{aligned} \tag{16.34}$$

with the hat basis functions h, illustrated in Figure 16.3(b), defined as

$$h(t) = \begin{cases} 1 + \dfrac{t}{\Delta t} & \text{if } -\Delta t \le t \le 0 \\ 1 - \dfrac{t}{\Delta t} & \text{if } 0 \le t \le \Delta t \\ 0 & \text{otherwise} \end{cases} \tag{16.35}$$

and finally

$$\mathbf{e}_j = \int_{\mathbb{R}} p(t - j\Delta t)\mathbf{e}(t)\, \mathrm{d}t \tag{16.36}$$

Assuming that $\forall i \le 0, \boldsymbol{j}_i = \mathbf{0}$, if the current $\boldsymbol{j}_i$ is known for each time step $i \in [1, j-1]$, $\boldsymbol{j}_j$ can be computed by rewriting the system (16.33) into

$$\boldsymbol{Z}_0 \boldsymbol{j}_j = \mathbf{e}_j - \sum_{i=1}^{j} \boldsymbol{Z}_i \boldsymbol{j}_{j-i} \tag{16.37}$$

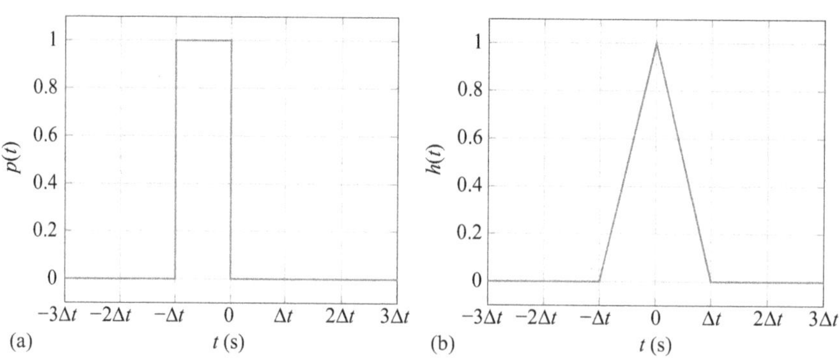

Figure 16.3 Illustrations of the basis functions used for the standard time discretization of the TD-EFIE. (a) Pulse basis function p. (b) Hat basis function h

which can be solved by the MOT method whose name arises from the fact that it computes each $\boldsymbol{j}_j$ step by step.

Because $\mathcal{T}_h$ involves a temporal integration, the number of non-zero $\boldsymbol{Z}_i$ is infinite, even though the Green's function has a compact support. A standard technique used to address this issue is to apply a time derivative on both sides of the TD-EFIE (16.20) and discretize the source terms of (16.31) with hats h instead of pulses p basis functions. This causes the number of non-zero interaction matrices in the convolution to be bounded and roughly equal to $a/(c\Delta t)$, where a is the diameter of Γ. However, this approach exacerbates the DC breakdown, as explained in Section 16.5. Denoting the number of non-zero matrices by Q, (16.37) can be rewritten

$$\boldsymbol{Z}_0\boldsymbol{j}_j = \mathbf{e}_j - \sum_{i=1}^{Q} \boldsymbol{Z}_i\boldsymbol{j}_{j-i} \tag{16.38}$$

The right-hand side of (16.38) contains a convolution of the sequence of interaction matrices with the sequence of current vectors. With N_t time steps and N_e unknowns, the total computational cost of this convolution is $\mathcal{O}(QN_tN_e^2)$ when performed naively. Fortunately, fast algorithms have been introduced to reduce this cost [47–49].

16.3 The low-frequency breakdown in the FD

One of the principal limitations of the FD-EFIE is its inaccuracy at low frequencies when computed in limited precision. In this section, we will illustrate the issue with numerical examples, identify and analyse the underlying problems, and finally explain why the historical solution is unsatisfactory. A more adequate solution based on qH projectors will be introduced in Section 16.7.

16.3.1 Illustration of the problem

To illustrate the low-frequency inaccuracies plaguing the FD-EFIE, we have simulated the radar cross section (RCS) of a PEC sphere of unit radius, excited by plane waves, at low and high frequencies. This choice of geometry allows us to compare our results against the Mie series which is valid at arbitrary frequencies. The RCS is defined as

$$\sigma(\boldsymbol{r}) = \lim_{r\to\infty} 4\pi r^2 \frac{\|e^s(\boldsymbol{r})\|^2}{\|e^i(\boldsymbol{r})\|^2} \tag{16.39}$$

While Figure 16.4(b) shows a perfect match between the RCS obtained from the FD-EFIE and the theoretical solution at high frequencies, Figure 16.4(a) demonstrates that at low frequencies the formulation becomes unreliable. We have used the extremely low frequency 1×10^{-40} Hz to show that the solutions presented later on behave well until arbitrarily low frequencies. However the problem shows severe ill-conditioning starting from the point where the characteristic dimension of the structure is small compared to the wavelength. In case of a sphere of 1 m radius, a 1 MHz simulation already shows signs of low-frequency breakdown as indicated by the very high condition number obtained at this frequency in Figure 16.4(c). In practice, these results

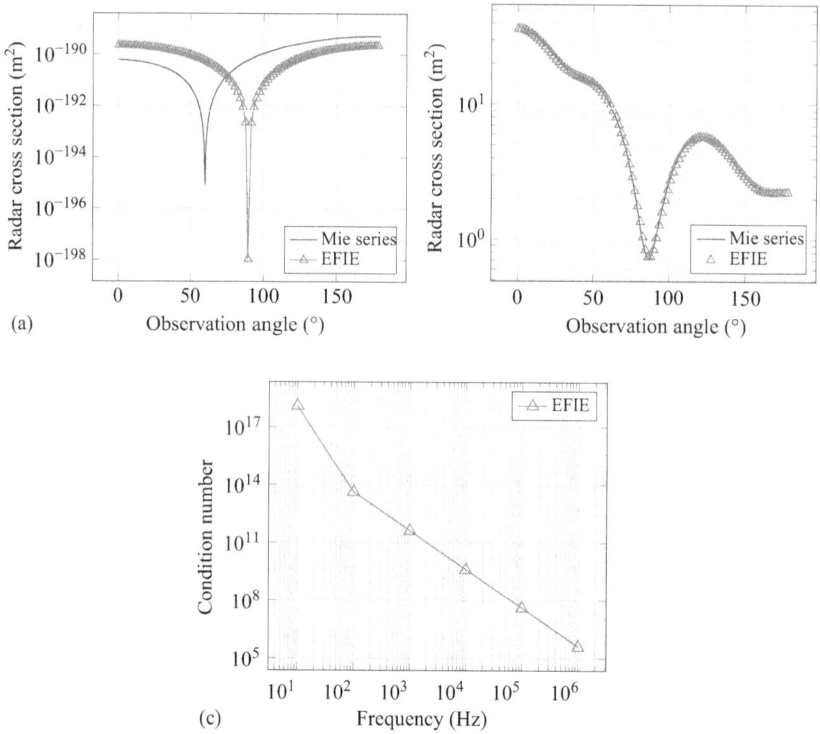

Figure 16.4 RCS of a PEC sphere of radius 1 m, excited by a plane wave oscillating at 1×10^{-40} Hz (a) and at one wavelength per diameter, i.e. 1.5×10^{8} Hz (b). (c) illustrates the evolution of the condition number of the corresponding EFIE matrices at different frequencies; very low-frequency points have been removed due to numerical limitations in the computation of high condition numbers

confirm that the EFIE cannot be used in the low-frequency regime without further treatment.

In addition to yielding the wrong solution, the conditioning of the EFIE matrix will grow quadratically with the inverse of the frequency, as illustrated in Figure 16.4(c). This ill-conditioning will cause, at first, iterative solvers to require more iterations, hence making simulations slower, and, in more pathological cases, not to converge or to converge to the wrong solution.

16.3.2 Analysis of the low-frequency breakdown

The low-frequency breakdown actually has two root causes: (i) an ill scaling of the matrix blocks causing the matrix to be severely ill-conditioned and (ii) a loss of significant digits in the solution vector when computed in finite precision. These two issues can be exposed by studying the frequency scaling of the solenoidal and non-solenoidal parts of the EFIE impedance matrix. Hence we will first apply the LS decomposition to (16.15)

$$\underbrace{\boldsymbol{B}^{\mathrm{T}}(\mathrm{i}k\boldsymbol{Z}_s + (\mathrm{i}k)^{-1}\boldsymbol{Z}_h)\boldsymbol{B}}_{\boldsymbol{Z}^{\Lambda H \Sigma}}\boldsymbol{y} = \boldsymbol{B}^{\mathrm{T}}\boldsymbol{v} \tag{16.40}$$

where $\boldsymbol{B}\boldsymbol{y} = \boldsymbol{j}$. The left-hand side of this equation is, in block representation,

$$\boldsymbol{Z}^{\Lambda H \Sigma} = \mathrm{i}k \begin{bmatrix} \boldsymbol{\Lambda}^{\mathrm{T}}\boldsymbol{Z}_s\boldsymbol{\Lambda} & \boldsymbol{\Lambda}^{\mathrm{T}}\boldsymbol{Z}_s\boldsymbol{H} & \boldsymbol{\Lambda}^{\mathrm{T}}\boldsymbol{Z}_s\boldsymbol{\Sigma} \\ \boldsymbol{H}^{\mathrm{T}}\boldsymbol{Z}_s\boldsymbol{\Lambda} & \boldsymbol{H}^{\mathrm{T}}\boldsymbol{Z}_s\boldsymbol{H} & \boldsymbol{H}^{\mathrm{T}}\boldsymbol{Z}_s\boldsymbol{\Sigma} \\ \boldsymbol{\Sigma}^{\mathrm{T}}\boldsymbol{Z}_s\boldsymbol{\Lambda} & \boldsymbol{\Sigma}^{\mathrm{T}}\boldsymbol{Z}_s\boldsymbol{H} & \boldsymbol{\Sigma}^{\mathrm{T}}(\boldsymbol{Z}_s - k^{-2}\boldsymbol{Z}_h)\boldsymbol{\Sigma} \end{bmatrix} \tag{16.41}$$

where we have used the following properties:

$$\boldsymbol{Z}_h\boldsymbol{\Lambda} = \boldsymbol{0}, \quad \boldsymbol{Z}_h\boldsymbol{H} = \boldsymbol{0}, \quad \boldsymbol{\Lambda}^{\mathrm{T}}\boldsymbol{Z}_h = \boldsymbol{0}, \quad \boldsymbol{H}^{\mathrm{T}}\boldsymbol{Z}_h = \boldsymbol{0} \tag{16.42}$$

immediately coming from the fact that $\forall \boldsymbol{l} \in \Lambda \oplus H,\ \nabla \cdot \boldsymbol{l} = 0$. Additionally, at low frequency, the bottom right block is dominated by the scalar potential contribution and can hence be approximated as

$$\boldsymbol{\Sigma}^{\mathrm{T}}(\boldsymbol{Z}_s - k^{-2}\boldsymbol{Z}_h)\boldsymbol{\Sigma} \underset{k\to 0}{\approx} -k^{-2}\boldsymbol{\Sigma}^{\mathrm{T}}\boldsymbol{Z}_h\boldsymbol{\Sigma} \tag{16.43}$$

This simplification illustrates the ill-scaling of the matrix terms, with all terms proportional to k except for a single one proportional to $1/k$. We will show that this $1/k$ contribution will have dramatic consequences on the numerical accuracy of the solution vector and on the conditioning of the overall matrix.

16.3.2.1 Numerical instability

As a means to study the loss of significant digits occurring in the solution vector, we will perform a scaling analysis of the solution of (16.40):

$$\boldsymbol{B}^{-1}\boldsymbol{j} = \left(\boldsymbol{Z}^{\Lambda H \Sigma}\right)^{-1}\boldsymbol{B}^{\mathrm{T}}\boldsymbol{v} \tag{16.44}$$

First, using the Schur complement [50] we can derive the frequency scaling of the blocks of the inverse of $\boldsymbol{Z}^{\Lambda H \Sigma}$

$$\left(\boldsymbol{Z}^{\Lambda H \Sigma}\right)^{-1} \underset{k\to 0}{=} \begin{bmatrix} \mathcal{O}(1/k) & \mathcal{O}(1/k) & \mathcal{O}(k) \\ \mathcal{O}(1/k) & \mathcal{O}(1/k) & \mathcal{O}(k) \\ \mathcal{O}(k) & \mathcal{O}(k) & \mathcal{O}(k) \end{bmatrix} \tag{16.45}$$

From this expression, we can deduce the scalings of the solution vector given the scalings of the right-hand side $\boldsymbol{B}^{\mathrm{T}}\boldsymbol{v}$. Table 16.1 summarizes the Right-Hand Side (RHS) frequency scalings for some common types of excitation; for the sake of brevity, we will not demonstrate these results here, but more details can be found in [20]. The plane wave excitation corresponds to an RHS of the form $\boldsymbol{r} \mapsto \exp(\mathrm{i}k\hat{\boldsymbol{k}} \cdot \boldsymbol{r})\boldsymbol{u}$, where $\hat{\boldsymbol{k}}$ is the direction of propagation and $\boldsymbol{u}$ is the polarization. The capacitive voltage gap excitation corresponds to a geometry in which the delta gap does not excite any global loops; a parallel plate capacitor with a delta gap on the connecting strip is a good example of such a configuration. Its dual excitation, the inductive voltage gap, does present a global loop excitation and could be obtained by adding a second connecting strip to the plate capacitor. In case of a plane wave excitation, we obtain

$$\boldsymbol{B}^{-1}\boldsymbol{j} \underset{k\to 0}{=} \begin{bmatrix}\mathcal{O}(1) & \mathcal{O}(1) & \mathcal{O}(k)\end{bmatrix}^{\mathrm{T}} \tag{16.46}$$

from which we deduce the low-frequency current scaling $\boldsymbol{j} = \boldsymbol{\Lambda}\cdot\mathcal{O}(1) + \boldsymbol{H}\cdot\mathcal{O}(1) + \boldsymbol{\Sigma}\cdot\mathcal{O}(k)$. Given its $\mathcal{O}(k)$ scaling, the star contribution will lose several significant digits because of the $\mathcal{O}(1)$ scaling of the loop contribution it is stored along with, when computed in finite precision. This non-reversible loss of part of the solution current is lamentable *per se* (since the information on the current-associated charge, $\boldsymbol{\Sigma}^{\mathrm{T}}\boldsymbol{j} = \boldsymbol{\Sigma}^{\mathrm{T}}\boldsymbol{\Sigma}\cdot\mathcal{O}(k)$, is lost at low frequency), but also has catastrophic consequences in the calculation of the associated far field. In far-field computations, the star currents have the same strength as the other two contributions, causing numerical noise to be amplified by the scattering process. This can illustrated by analysing the far field scattering operator

$$\mathcal{F}(\boldsymbol{j}) = (\hat{\boldsymbol{\theta}}\hat{\boldsymbol{\theta}} + \hat{\boldsymbol{\phi}}\hat{\boldsymbol{\phi}})\cdot \mathrm{i}k\mathcal{T}_s(\boldsymbol{j}) \tag{16.47}$$

where $\hat{\boldsymbol{\theta}}$ and $\hat{\boldsymbol{\phi}}$ are the polar and azimuthal unit vectors of the spherical coordinate system associated with Γ. Noting that the exponential can be written $\mathrm{e}^x = \mathrm{e}^x - 1 + 1$ and that, in the far field regime, $\|\boldsymbol{r}-\boldsymbol{r}'\| \approx r - \hat{\boldsymbol{r}}\cdot\boldsymbol{r}'$ and $1/\|\boldsymbol{r}-\boldsymbol{r}'\| \approx 1/r$, (16.3) can be rewritten into

$$\mathcal{T}_s(\boldsymbol{j}) = \hat{\boldsymbol{n}}_r \times \frac{\mathrm{e}^{\mathrm{i}kr}}{4\pi r}\int_\Gamma (\mathrm{e}^{-\mathrm{i}k\hat{\boldsymbol{r}}\cdot\boldsymbol{r}'} - 1)\boldsymbol{j}(\boldsymbol{r}')\,\mathrm{d}\boldsymbol{r}' + \hat{\boldsymbol{n}}_r \times \frac{\mathrm{e}^{\mathrm{i}kr}}{4\pi r}\int_\Gamma \boldsymbol{j}(\boldsymbol{r}')\,\mathrm{d}\boldsymbol{r}' \tag{16.48}$$

While nothing can be said about the scaling of the radiation of the star currents, in case of solenoidal currents $\boldsymbol{l} \in \Lambda \oplus H$, $\int_l \boldsymbol{l}(\boldsymbol{r}')\,\mathrm{d}\boldsymbol{r}' = \boldsymbol{0}$, according to Green's theorem. The $\mathrm{e}^x - 1$ term can then be further analysed in the low-frequency regime by expanding it into its Taylor series

$$\mathcal{T}_s(\boldsymbol{l}) \underset{k\to 0}{=} -\hat{\boldsymbol{n}}_r \times \frac{1}{4\pi r}\int_l (\mathrm{i}k\hat{\boldsymbol{r}}\cdot\boldsymbol{r}' + \mathcal{O}(k^2))\boldsymbol{l}(\boldsymbol{r}')\,\mathrm{d}\boldsymbol{r}',\ \forall \boldsymbol{l} \in \Lambda \oplus H \tag{16.49}$$

which demonstrates an $\mathcal{O}(k)$ scaling of the loop contribution in the far field. The discrete far field $\boldsymbol{F}$ operator will hence amplify the solenoidal contributions by a factor of k and leave the scaling of the non-solenoidal contributions unchanged.

The frequency behaviour of $\boldsymbol{F}$, combined with (16.46), allows us to determine that the far field $\boldsymbol{f}$ scattered by the plane wave will scale as

$$k^{-1}\cdot\boldsymbol{f} = \begin{bmatrix}\mathcal{O}(k) & \mathcal{O}(k) & \mathcal{O}(1)\end{bmatrix}\cdot\boldsymbol{B}^{-1}\boldsymbol{j} \underset{k\to 0}{=} \mathcal{O}(k) + \mathcal{O}(k) + \mathcal{O}(k) \tag{16.50}$$

Since the loop and star components contribute equally to the far field, even though the star contribution has been lost during the computation of the current density vector,

Table 16.1 Frequency scalings of different kinds of RHS excitations

Type of excitation	$\boldsymbol{\Lambda}$	$\boldsymbol{H}$	$\boldsymbol{\Sigma}$
Plane wave	$\mathcal{O}(k)$	$\mathcal{O}(k)$	$\mathcal{O}(1)$
Capacitive voltage gap	0	0	$\mathcal{O}(1)$
Inductive voltage gap	$\mathcal{O}(1)$	$\mathcal{O}(1)$	$\mathcal{O}(1)$

numerical noise will be amplified and the computed field will be incorrect. The voltage gap excitations are less pathological: the capacitive voltage gap has no loss of significant digits in the current and has a physical one in the scattered field

$$\boldsymbol{j} \underset{k\to 0}{=} \boldsymbol{\Lambda}\cdot\mathcal{O}(k)+\boldsymbol{H}\cdot\mathcal{O}(k)+\boldsymbol{\Sigma}\cdot\mathcal{O}(k) \tag{16.51}$$

$$k^{-1}\cdot\boldsymbol{f} \underset{k\to 0}{=} \mathcal{O}(k^2)+\mathcal{O}(k^2)+\mathcal{O}(k) \tag{16.52}$$

and the inductive voltage gap does have a loss of significant digits in the solution but the contribution of the inaccurate term is further reduced in the scattered field

$$\boldsymbol{j} \underset{k\to 0}{=} \boldsymbol{\Lambda}\cdot\mathcal{O}(1/k)+\boldsymbol{H}\cdot\mathcal{O}(1/k)+\boldsymbol{\Sigma}\cdot\mathcal{O}(k) \tag{16.53}$$

$$k^{-1}\cdot\boldsymbol{f} \underset{k\to 0}{=} \mathcal{O}(1)+\mathcal{O}(1)+\mathcal{O}(k) \tag{16.54}$$

meaning that in neither case is numerical noise amplified.

Our analysis demonstrates that the current density $\boldsymbol{j}$ may not be stored accurately in finite precision, since a critical part of the solution may be lost because of numerical effects, for certain types of excitation.

16.3.2.2 Low-frequency ill conditioning

The other cause of the low-frequency breakdown, namely the low-frequency ill-conditioning of impedance matrix, can be brought to light by left and right multiplying $\boldsymbol{Z}^{\Lambda H\Sigma}$ by the block-diagonal matrices $\boldsymbol{L} = \mathrm{diag}(1/ika, 1/ika, 1)$ and $\boldsymbol{R} = \mathrm{diag}(1, 1, ika)$, respectively, where a is a length scaling ensuring the correct dimensional homogeneity and the scale invariance of the scheme. The blocks of the decomposed matrix $\boldsymbol{Z}_{LR} = \boldsymbol{L}\boldsymbol{Z}^{\Lambda H\Sigma}\boldsymbol{R}$ are

$$\boldsymbol{Z}_{LR} = \begin{bmatrix} a^{-1}\boldsymbol{\Lambda}^{\mathrm{T}}\boldsymbol{Z}_s\boldsymbol{\Lambda} & a^{-1}\boldsymbol{\Lambda}^{\mathrm{T}}\boldsymbol{Z}_s\boldsymbol{H} & \mathrm{i}k\boldsymbol{\Lambda}^{\mathrm{T}}\boldsymbol{Z}_s\boldsymbol{\Sigma} \\ a^{-1}\boldsymbol{H}^{\mathrm{T}}\boldsymbol{Z}_s\boldsymbol{\Lambda} & a^{-1}\boldsymbol{H}^{\mathrm{T}}\boldsymbol{Z}_s\boldsymbol{H} & \mathrm{i}k\boldsymbol{H}^{\mathrm{T}}\boldsymbol{Z}_s\boldsymbol{\Sigma} \\ \mathrm{i}k\boldsymbol{\Sigma}^{\mathrm{T}}\boldsymbol{Z}_s\boldsymbol{\Lambda} & \mathrm{i}k\boldsymbol{\Sigma}^{\mathrm{T}}\boldsymbol{Z}_s\boldsymbol{H} & a\boldsymbol{\Sigma}^{\mathrm{T}}(\boldsymbol{Z}_h - k^2\boldsymbol{Z}_s)\boldsymbol{\Sigma} \end{bmatrix} \tag{16.55}$$

which becomes when $k \to 0$

$$\boldsymbol{Z}^0_{LR} = \lim_{k\to 0}\boldsymbol{Z}_{LR} = \begin{bmatrix} a^{-1}\boldsymbol{\Lambda}^{\mathrm{T}}\boldsymbol{Z}_s\boldsymbol{\Lambda} & a^{-1}\boldsymbol{\Lambda}^{\mathrm{T}}\boldsymbol{Z}_s\boldsymbol{H} & \boldsymbol{0} \\ a^{-1}\boldsymbol{H}^{\mathrm{T}}\boldsymbol{Z}_s\boldsymbol{\Lambda} & a^{-1}\boldsymbol{H}^{\mathrm{T}}\boldsymbol{Z}_s\boldsymbol{H} & \boldsymbol{0} \\ \boldsymbol{0} & \boldsymbol{0} & a\boldsymbol{\Sigma}^{\mathrm{T}}\boldsymbol{Z}_h\boldsymbol{\Sigma} \end{bmatrix} \tag{16.56}$$

Since $\boldsymbol{Z}^0_{LR}$ is frequency independent, $\boldsymbol{Z}_{LR}$ does not suffer from a low-frequency condition number breakdown. Additionally, because $\boldsymbol{Z}^{\Lambda H\Sigma} = \boldsymbol{L}^{-1}\boldsymbol{Z}_{LR}\boldsymbol{R}^{-1}$ and by properties of the conditioning of a matrix product

$$\operatorname{cond}\left(\boldsymbol{Z}^{\Lambda H\Sigma}\right) \leq \operatorname{cond}\left(\boldsymbol{Z}_{LR}\right) \operatorname{cond}\left(\boldsymbol{L}^{-1}\right) \operatorname{cond}\left(\boldsymbol{R}^{-1}\right) \tag{16.57}$$

The lower bound of the condition number can be determined by considering that $\boldsymbol{Z}_{LR} = \boldsymbol{D}\, \mathbf{Z}^{\Lambda H\Sigma} \boldsymbol{D}$ with $\boldsymbol{D} = \operatorname{diag}(1/\sqrt{\mathrm{i}ka}, 1/\sqrt{\mathrm{i}ka}, \sqrt{\mathrm{i}ka})$ and that $\operatorname{cond}(\boldsymbol{A}^{\mathrm{T}}\boldsymbol{A}) = \operatorname{cond}(\boldsymbol{A}\boldsymbol{A}^{\mathrm{T}})$ for any square matrix $\boldsymbol{A}$, which allows us to write, after a few passages, that

$$\operatorname{cond}\left(\boldsymbol{D}\left(\boldsymbol{Z}^{\Lambda H\Sigma}\right)^{-1/2} \boldsymbol{Z}^{\Lambda H\Sigma} \left(\boldsymbol{Z}^{\Lambda H\Sigma}\right)^{-1/2} \boldsymbol{D}\right) \leq \operatorname{cond}\left(\boldsymbol{Z}^{\Lambda H\Sigma}\right) \operatorname{cond}\left(\boldsymbol{Z}_{LR}\right) \tag{16.58}$$

This result, combined with (16.57), demonstrates in the low-frequency approximation that

$$\frac{1}{\gamma k^2} \leq \operatorname{cond}\left(\boldsymbol{Z}^{\Lambda H\Sigma}\right) \leq \frac{\gamma}{k^2} \tag{16.59}$$

where $\gamma = \operatorname{cond}(\boldsymbol{Z}^0_{LR})$. This is equivalent to saying that $\operatorname{cond}(\boldsymbol{Z}^{\Lambda H\Sigma}) = \mathcal{O}(k^{-2})$ when $k \to 0$ and that the standard EFIE indeed has a severe condition number breakdown at low frequency, in accordance with the observations from Figure 16.4(c).

16.3.3 Traditional LS decomposition

The issues plaguing the EFIE at low frequencies, identified in the previous section, have traditionally been addressed using the LS decomposition technique. In this section we will introduce the technique and detail its shortcomings. Several other more recent and sophisticated techniques have been proposed. However, we restrict here the scope our treatments to the traditional LS decomposition for the sake of simplicity and for didactic purposes, it should be noted, nonetheless, that a similar analysis to the one presented in this section could be applied to several other more recent strategies for tackling low-frequency problems.

The development of the LS decomposition is actually very similar to these of Section 16.3.2. In short, the technique relies on the $\boldsymbol{B}$ transform to expose the different scalings in the EFIE matrix and regularize them with a left and right multiplication by $\boldsymbol{L}$ and $\boldsymbol{R}$, as shown in (16.40) and (16.55). This gives rise to the LS-EFIE

$$\boldsymbol{L}\boldsymbol{B}^{\mathrm{T}}\left(\mathrm{i}k\boldsymbol{Z}_s + \frac{1}{\mathrm{i}k}\boldsymbol{Z}_h\right)\boldsymbol{B}\boldsymbol{R}\boldsymbol{y}_{LS} = \boldsymbol{L}\boldsymbol{B}^{\mathrm{T}}\boldsymbol{v} \tag{16.60}$$

where $\boldsymbol{y}_{LS}$ is such that $\boldsymbol{j} = \boldsymbol{B}\boldsymbol{R}\boldsymbol{y}_{LS}$.

The reasoning detailed in Section 16.3.2.2 can be used verbatim to demonstrate that this technique effectively addresses the low-frequency conditioning breakdown. However, the curative effect of the LS decomposition on the loss of significant digits of the non-solenoidal part of the current still has to be demonstrated. Once again we will leverage on previous developments by adapting them to the new LS equation. In particular, (16.44) becomes

$$\boldsymbol{R}^{-1}\boldsymbol{B}^{-1}\boldsymbol{j} = \left(\boldsymbol{R}^{-1}\boldsymbol{Z}^{\Lambda H \Sigma^{-1}}\boldsymbol{L}^{-1}\right)\boldsymbol{L}\boldsymbol{B}^{\mathrm{T}}\boldsymbol{v} \tag{16.61}$$

from which we can derive the solution scalings. For instance, in case of a plane wave excitation,

$$\boldsymbol{y}_{LS} = \boldsymbol{R}^{-1}\boldsymbol{B}^{-1}\boldsymbol{j} \underset{k\to 0}{=} \left[\mathcal{O}(1) \quad \mathcal{O}(1) \quad \mathcal{O}(1)\right]^{\mathrm{T}} \tag{16.62}$$

Since all the contributions of the LS solution $\boldsymbol{y}_{LS}$ have the same scaling, it can safely be stored in finite precision without any loss of information.

The curative effects of the LS decomposition have been illustrated in Figure 16.5(a) by performing the same simulation as in Figure 16.4(a), with the LS-EFIE. It is evident that the results yielded by the LS-EFIE are in accordance with the Mie series. Additionally we can see in Figure 16.5(b) that the low-frequency condition number breakdown has also been addressed. While the LS decomposition completely addresses the low-frequency breakdown of the EFIE, it introduces problems of its own: (i) as we have seen in Section 16.2.1.3, it requires the search for global loops, which makes the technique challenging to use on multiply connected structures, and (ii) it introduces a severe high-refinement breakdown that prevents it from being used with very detailed meshes. The issue (ii) has been illustrated in Figure 16.6(a) and (b) by computing the condition number of $\boldsymbol{\Lambda}$, $\boldsymbol{\Sigma}$ and of the LS-EFIE matrix on a spherical structure refined with an increasing number of elements. It is clear that, while the standard EFIE does suffer from a high-refinement ill-conditioning, the LS-EFIE conditioning has been further degraded by the multiplication by $\boldsymbol{\Lambda}$ and $\boldsymbol{\Sigma}$. The underlying cause is that $\boldsymbol{\Lambda}^{\mathrm{T}}\boldsymbol{\Lambda}$ and $\boldsymbol{\Sigma}^{\mathrm{T}}\boldsymbol{\Sigma}$ are graph-Laplacian operators and their spectra behave as operators of derivative order 2 [32].

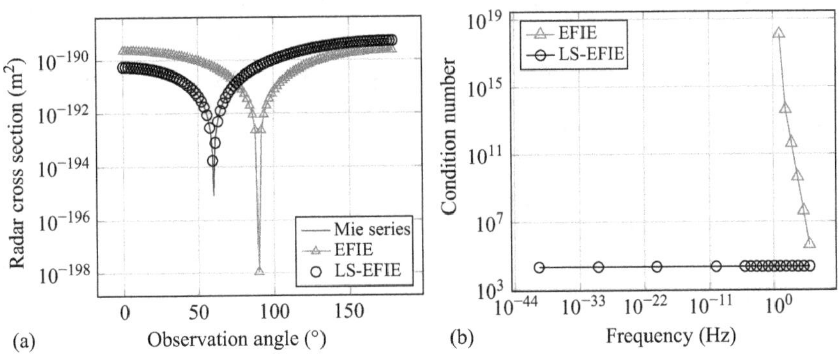

Figure 16.5 *Low-frequency behaviour of the LS-EFIE compared against the standard EFIE. (a) illustrates the RCS of a PEC sphere of radius 1 m, excited by a plane wave oscillating at 1×10^{-40} Hz, while (b) illustrates the evolution of conditioning of the impedance matrices of both formulations*

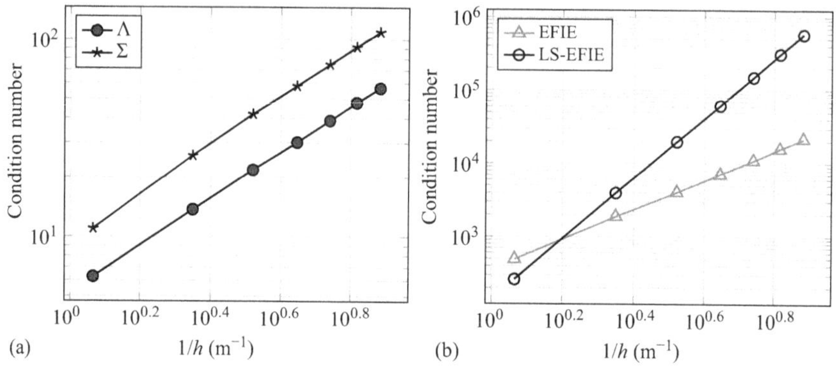

Figure 16.6 Illustration of the refinement behaviour of the conditioning of the LS matrices (a) and of the LS-EFIE compared against its standard counterpart (b)

16.4 The large time step breakdown in the TD

The large time step breakdown in TD is analogous to the low-frequency breakdown in FD. It corresponds to an ill-conditioning of the matrix $\boldsymbol{Z}_0$ when the time step increases. As the problem in FD has been extensively analysed in Section 16.3.2.2, similar derivations can be done for the large time step breakdown in TD. For a sake of brevity, only the main results are presented in this section.

The behaviours of the two parts of the EFIE operator $\mathcal{T} = \mathcal{T}_s + \mathcal{T}_h$ are studied when $\Delta t \to \infty$. The time derivative in the singular part $\mathcal{T}_s$ scales as $\mathcal{O}(1/\Delta t)$, while the time integration in the hypersingular part $\mathcal{T}_h$ scales as $\mathcal{O}(\Delta t)$. The $\mathcal{O}(\Delta t)$ scaling of $\mathcal{T}_h$ is dominant on the non-solenoidal subspace. Then, $\mathcal{T}_h$ being exclusively in the non-solenoidal subspace, the solenoidal subspace has the scaling of $\mathcal{T}_s$, i.e. $\mathcal{O}(1/\Delta t)$. Therefore, the ratio between the scaling of the two subspaces is Δt^2. After the discretization, it follows that the condition number also increases as $\mathcal{O}(\Delta t^2)$. Otherwise said, when $\Delta t \to \infty$, $\operatorname{cond}(\boldsymbol{Z}_0) = \mathcal{O}(\Delta t^2)$.

The reference simulation used to illustrate the issues caused by the low-frequency breakdown of the TD-EFIE will be that of a sphere discretized as in Figure 16.7. Figure 16.8 displays the graph of the condition number of the non-stabilized TD-EFIE system $\operatorname{cond}(\boldsymbol{Z}_0)$ as a function of the time step Δt. As the time step increases, the quadratic growth is clear from this graph.

16.5 DC instabilities

As mentioned in 16.2.2.1, if causality is not imposed, the solution of the TD-EFIE is only determined up to constant solenoidal currents – also called static loops. These currents $\boldsymbol{j}_{cs}(\boldsymbol{r})$ are both constant in time and divergence free – $\partial_t \boldsymbol{j}_{cs}(\boldsymbol{r}) = \boldsymbol{0}$

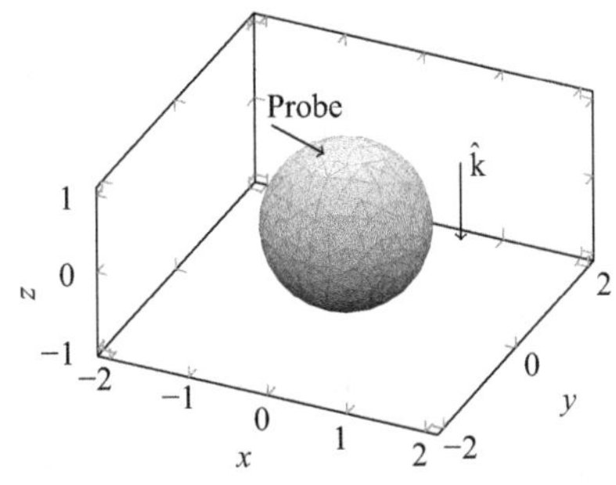

Figure 16.7 Reference unit sphere mesh used for the TD-EFIE simulations, discretized with 828 edges. The travelling direction of the impinging wave is illustrated as vector $\hat{\boldsymbol{k}}$ *and the probe location is (−0.186, −0.173, 0.957) m*

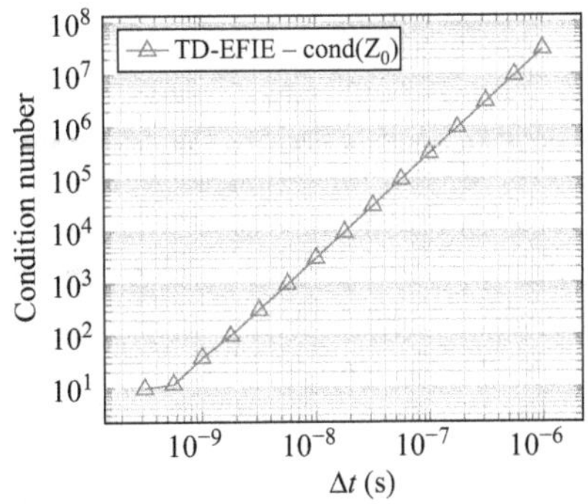

Figure 16.8 Conditioning of $\boldsymbol{Z}_0$ *as a function of the time step* Δt

and $\nabla \cdot \boldsymbol{j}_{cs}(\boldsymbol{r}) = 0$ – and, as a consequence, they are in the null-space of $\mathscr{T}$. It can be trivially demonstrated that, if $\boldsymbol{j}(\boldsymbol{r}, t)$ is solution of the TD-EFIE, then $\boldsymbol{j}(\boldsymbol{r}, t) + \boldsymbol{j}_{cs}(\boldsymbol{r})$ is also a solution. To eliminate this null-space, the fields are assumed to vanish in the neighbourhood of Γ when $t < 0$ so that the only possible constant current is $\mathbf{0}$. This effectively enforces uniqueness of the solution in the continuous formulation. However, due to numerical errors and the limited precision in the computation of the discretized system, static currents creep into the solution and grow exponentially. This effect is known as the DC instability.

A solution perturbed by a DC instability is illustrated in Figure 16.9, where the sphere displayed in Figure 16.7 is excited by a modulated Gaussian plane wave

$$\boldsymbol{e}^i(\boldsymbol{r}, t) = \hat{\boldsymbol{p}} e_0 \exp\left(-\frac{\left(t - t_0 - \frac{\hat{\boldsymbol{k}} \cdot \boldsymbol{r}}{c}\right)^2}{2\sigma^2} \right) \cos\left(2\pi f_0 \left(t - t_0 - \frac{\hat{\boldsymbol{k}} \cdot \boldsymbol{r}}{c}\right)\right) \tag{16.63}$$

where $\sigma = 6/(2\pi f_{bw})$, the polarization $\hat{\boldsymbol{p}} = \hat{\boldsymbol{x}}$, the wave vector $\hat{\boldsymbol{k}} = -\hat{\boldsymbol{z}}$, the central frequency $f_0 = 10$ MHz, the frequency bandwidth is $f_{bw} = 20$ MHz, the delay is $t_0 = 400$ ns and the peak amplitude is $e_0 = 1\,\mathrm{Vm}^{-1}$. The excitation field is vanishing when $t \to \infty$; however, the solution current grows slowly but exponentially when it would also vanish if the solution was correct.

This instability can be illustrated by studying the behaviour of the solution of the homogeneous ($\forall j$, $\mathbf{e}_j = \mathbf{0}$) system, by computing the polynomial eigenvalues associated with the sequence of matrices $\{\boldsymbol{Z}_i\}_{i\in[0,Q]}$ defined in (16.38). The polynomial eigenvalues are the eigenvalues of the companion matrix

$$\boldsymbol{Z}_C = \begin{bmatrix} -\boldsymbol{Z}_0^{-1}\boldsymbol{Z}_1 & -\boldsymbol{Z}_0^{-1}\boldsymbol{Z}_2 & \dots & -\boldsymbol{Z}_0^{-1}\boldsymbol{Z}_{Q-1} & -\boldsymbol{Z}_0^{-1}\boldsymbol{Z}_Q \\ \mathbf{I} & & & & \\ & \mathbf{I} & & & \\ & & \ddots & & \\ & & & \mathbf{I} & \end{bmatrix} \tag{16.64}$$

which satisfies the equation $\boldsymbol{J}_i = \boldsymbol{Z}_C \boldsymbol{J}_{i-1}$, where

$$\boldsymbol{J}_i = \left[\boldsymbol{j}_i \; \boldsymbol{j}_{i-1} \; \cdots \; \boldsymbol{j}_{i-Q+1}\right]^{\mathrm{T}} \tag{16.65}$$

When all the eigenvalues λ_n lie within the unit circle ($\forall n$, $\|\lambda_n\| < 1$), the system is stable. On the contrary, if an eigenvalue is outside ($\exists n$, $\|\lambda_n\| > 1$), the system is unstable and the solution grows exponentially. The eigenvalues associated with the static solenoidal currents form a cluster at the point (1, 0), as illustrated in Figure 16.10, on the boundary of the circle, which is not problematic in theory, however, in practice, because of numerical errors in the discretized system, some of them will spread outside the unit circle and cause the system to become unstable.

This problem is amplified when a time derivative is applied on both sides of the TD-EFIE, which is the traditional technique used to limit the number of non-zeros $\boldsymbol{Z}_i$, as explained in Section 16.2.2.3. This scheme causes the dimension of the space

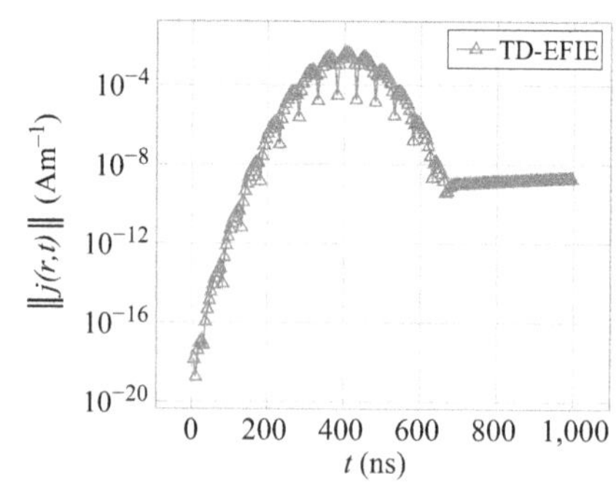

Figure 16.9 Norm of the surface current density as a function of time $N_t = 1000$, $\Delta t = 1$ ns and $\mathbf{r} = (-0.186, -0.173, 0.957)$ m

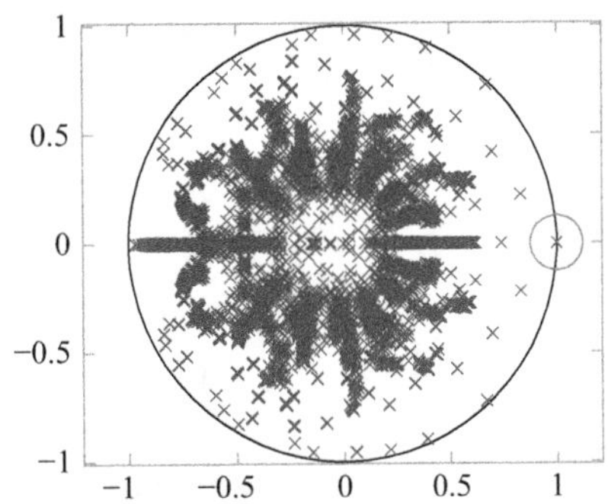

Figure 16.10 Polynomial eigenvalues associated with the differentiated TD-EFIE system showing a cluster at $(1, 0)$

of static solutions to double, since the linear in time solenoidal currents will also be solution of the homogeneous equation.

16.6 The qH projectors

As explained in Section 16.3.3, the LS decomposition has several limitations that include a deterioration of the condition number of the system, and the need for identification of the global loops, which is an expensive operation in general.

The qH projectors offer a more satisfactory alternative. Instead of leveraging on the Loop and Star transformation matrices directly, $\boldsymbol{\Sigma}$ is used to build a projector to the non-solenoidal space $\boldsymbol{P}_{\Sigma}$, while the solenoidal projector is deduced as its complement, and not explicitly computed. The projectors $\boldsymbol{P}_{\Sigma}$ and $\boldsymbol{P}_{\Lambda H} \in \mathbb{R}^{N_e \times N_e}$ are

$$\boldsymbol{P}_{\Sigma} = \boldsymbol{\Sigma}\left(\boldsymbol{\Sigma}^{\mathrm{T}}\boldsymbol{\Sigma}\right)^{+}\boldsymbol{\Sigma}^{\mathrm{T}} \tag{16.66}$$

$$\boldsymbol{P}_{\Lambda H} = \mathbf{I} - \boldsymbol{P}_{\Sigma} \tag{16.67}$$

where $^{+}$ denotes the Moore–Penrose pseudo-inverse. Since only $\boldsymbol{\Sigma}$ is required for computing both projectors, the detection of the global loops is not required. Interestingly, in case of simply connected geometries, H is empty and we have the relationship

$$\boldsymbol{P}_{\Lambda H} = \boldsymbol{P}_{\Lambda} = \boldsymbol{\Lambda}\left(\boldsymbol{\Lambda}^{\mathrm{T}}\boldsymbol{\Lambda}\right)^{+}\boldsymbol{\Lambda}^{\mathrm{T}} \tag{16.68}$$

To fully exploit the projectors a few of their properties should be detailed: by construction the projectors are complementary, which means that any vector of RWG coefficients $\boldsymbol{j}$ can be decomposed in its solenoidal and non-solenoidal part

$$\boldsymbol{j} = \boldsymbol{P}_{\Lambda H}\boldsymbol{j} + \boldsymbol{P}_{\Sigma}\boldsymbol{j} \tag{16.69}$$

In addition they are symmetric thus $\boldsymbol{P}_{\Sigma}^{\mathrm{T}} = \boldsymbol{P}_{\Sigma}$ and $\boldsymbol{P}_{\Lambda H}^{\mathrm{T}} = \boldsymbol{P}_{\Lambda H}$. Finally, while relying on the pseudo-inverse is convenient for our theoretical developments, it is too computationally expensive in practice and would outweigh the complexity gains from not having to detect global loops. The projectors can be evaluated in linear time by first regularizing ($\boldsymbol{\Sigma}^{\mathrm{T}}\boldsymbol{\Sigma}$) with $\mathbf{1}_{\boldsymbol{\Sigma}}$ (the all-one vector of dimension equaling the number of columns of $\boldsymbol{\Sigma}$); it can be easily verified that

$$\begin{aligned}\left(\boldsymbol{\Sigma}^{\mathrm{T}}\boldsymbol{\Sigma}+\frac{1}{N_c}\mathbf{1}_{\boldsymbol{\Sigma}}\mathbf{1}_{\boldsymbol{\Sigma}}^{\mathrm{T}}\right)^{+} &= \left(\boldsymbol{\Sigma}^{\mathrm{T}}\boldsymbol{\Sigma}+\frac{1}{N_c}\mathbf{1}_{\boldsymbol{\Sigma}}\mathbf{1}_{\boldsymbol{\Sigma}}^{\mathrm{T}}\right)^{-1} \\ &= \left(\boldsymbol{\Sigma}^{\mathrm{T}}\boldsymbol{\Sigma}\right)^{+}+\frac{1}{N_c}\mathbf{1}_{\boldsymbol{\Sigma}}\mathbf{1}_{\boldsymbol{\Sigma}}^{\mathrm{T}}\end{aligned} \tag{16.70}$$

and, given that $\mathbf{1}_{\boldsymbol{\Sigma}}$ is in the null-space of $\boldsymbol{\Sigma}$, we can prove that

$$\boldsymbol{\Sigma}\left(\boldsymbol{\Sigma}^{\mathrm{T}}\boldsymbol{\Sigma}\right)^{+}\boldsymbol{\Sigma}^{\mathrm{T}} = \boldsymbol{\Sigma}\left(\boldsymbol{\Sigma}^{\mathrm{T}}\boldsymbol{\Sigma}+\frac{1}{N_c}\mathbf{1}_{\boldsymbol{\Sigma}}\mathbf{1}_{\boldsymbol{\Sigma}}^{\mathrm{T}}\right)^{-1}\boldsymbol{\Sigma}^{\mathrm{T}} \tag{16.71}$$

Because the matrix ($\boldsymbol{\Sigma}^{\mathrm{T}}\boldsymbol{\Sigma}+1/N_c\mathbf{1}_{\boldsymbol{\Sigma}}\mathbf{1}_{\boldsymbol{\Sigma}}^{\mathrm{T}}$) is spectrally equivalent to a graph Laplacian [32], its inverse can be computed in linear time thanks to multigrid preconditioning [34,35].

16.7 An effective solution to the low-frequency breakdown for the EFIE

We have demonstrated in Section 16.3.3 that the traditional LS decomposition suffers from severe drawbacks, specifically: (i) the need for a prohibitively expensive global loop search and (ii) the further degradation of the high-refinement conditioning of the FD-EFIE. In this section, we will introduce a cure for the low-frequency breakdown of the FD-EFIE relying on the qH projectors that does not suffer from these limitations.

16.7.1 Leveraging the qH projectors

The LS technique effectively addresses the EFIE low-frequency breakdown because it independently re-scales the solenoidal and non-solenoidal subspaces, by leveraging the $\boldsymbol{\Lambda}$, $\boldsymbol{H}$ and $\boldsymbol{\Sigma}$ matrices, even though they are ill-conditioned. The same separation can be achieved using qH projectors that have, by definition, a flat spectrum. In place of the transformation matrix $\boldsymbol{B}$, we define the two projection matrices

$$\boldsymbol{M} = \tfrac{1}{\mathrm{i}ka}\boldsymbol{P}_{\Lambda H}+\boldsymbol{P}_{\boldsymbol{\Sigma}} \tag{16.72}$$

$$\boldsymbol{N} = \boldsymbol{P}_{\Lambda H}+\mathrm{i}ka\boldsymbol{P}_{\boldsymbol{\Sigma}} \tag{16.73}$$

that can be used to derive the qH-projected EFIE (qH-EFIE)

$$\boldsymbol{M}^{\mathrm{T}}\left(\mathrm{i}k\boldsymbol{Z}_s+\frac{1}{\mathrm{i}k}\boldsymbol{Z}_h\right)\boldsymbol{N}\boldsymbol{y}_{qH} = \boldsymbol{M}^{\mathrm{T}}\boldsymbol{v} \tag{16.74}$$

where $\boldsymbol{j} = \boldsymbol{N}\mathbf{y}_{qH}$. The left-hand side of the equation can then be expanded into

$$\begin{aligned}\boldsymbol{M}^{\mathrm{T}}\left(\mathrm{i}k\boldsymbol{Z}_s + \frac{1}{\mathrm{i}k}\boldsymbol{Z}_h\right)\boldsymbol{N} &= \left(a^{-1}\boldsymbol{P}_{\boldsymbol{\Lambda}H}\boldsymbol{Z}_s\boldsymbol{P}_{\boldsymbol{\Lambda}H} + a\boldsymbol{P}_{\boldsymbol{\Sigma}}\boldsymbol{Z}_h\boldsymbol{P}_{\boldsymbol{\Sigma}}\right)\\ &\quad + \left(\boldsymbol{P}_{\boldsymbol{\Lambda}H}\boldsymbol{Z}_s\boldsymbol{P}_{\boldsymbol{\Sigma}} + \boldsymbol{P}_{\boldsymbol{\Sigma}}\boldsymbol{Z}_s\boldsymbol{P}_{\boldsymbol{\Lambda}H}\right)\mathrm{i}k\\ &\quad + \left(\boldsymbol{P}_{\boldsymbol{\Sigma}}\boldsymbol{Z}_s\boldsymbol{P}_{\boldsymbol{\Sigma}}\right)a(\mathrm{i}k)^2\\ &= \left(a^{-1}\boldsymbol{P}_{\boldsymbol{\Lambda}H}\boldsymbol{Z}_s\boldsymbol{P}_{\boldsymbol{\Lambda}H} + a\boldsymbol{Z}_h\right) + \mathcal{O}(k)\end{aligned} \tag{16.75}$$

where we have used the properties

$$\boldsymbol{P}_{\boldsymbol{\Lambda}H}\boldsymbol{Z}_h = \boldsymbol{Z}_h\boldsymbol{P}_{\boldsymbol{\Lambda}H} = \boldsymbol{0} \tag{16.76}$$

for the same reason that $\boldsymbol{\Lambda}\boldsymbol{Z}_h = \boldsymbol{0}$ in Section 16.3.3, and $\boldsymbol{P}_{\boldsymbol{\Sigma}}\boldsymbol{Z}_h = \boldsymbol{Z}_h\boldsymbol{P}_{\boldsymbol{\Sigma}} = \boldsymbol{Z}_h$, which can be demonstrated by using (16.76) in $(\boldsymbol{P}_{\boldsymbol{\Sigma}} + \boldsymbol{P}_{\boldsymbol{\Lambda}H})\boldsymbol{Z}_h = \boldsymbol{Z}_h$. This expansion immediately allows us to notice that the qH-EFIE matrix has a stable conditioning until arbitrarily low frequencies, since

$$\lim_{k\to 0}\operatorname{cond}\left(\boldsymbol{M}^{\mathrm{T}}\boldsymbol{Z}\,\mathbf{N}\right) = \operatorname{cond}\left(a^{-1}\boldsymbol{P}_{\boldsymbol{\Lambda}H}\boldsymbol{Z}_s\boldsymbol{P}_{\boldsymbol{\Lambda}H} + a\boldsymbol{Z}_h\right) \tag{16.77}$$

is frequency independent. This property can be verified in Figure 16.11(b).

The numerical stability of the formulation, i.e. the lack of loss of significant digits in the non-solenoidal solution current, can be verified by considering, in case of a plane wave, that the physical solution behaves as $\boldsymbol{j} = \boldsymbol{\Lambda}\cdot\mathcal{O}(1) + \boldsymbol{H}\cdot\mathcal{O}(1) + \boldsymbol{\Sigma}\cdot\mathcal{O}(k)$ when $k \to 0$. Given that solenoidal and non-solenoidal parts of the physical solution $\boldsymbol{j}$ can be recovered with the help of the projectors as

$$\boldsymbol{j}^{\Lambda H} = \boldsymbol{P}_{\boldsymbol{\Lambda}H}\boldsymbol{N}\mathbf{y}_{qH} = \boldsymbol{P}_{\boldsymbol{\Lambda}H}\mathbf{y}_{qH} \tag{16.78}$$

$$\boldsymbol{j}^{\Sigma} = \boldsymbol{P}_{\boldsymbol{\Sigma}}\boldsymbol{N}\mathbf{y}_{qH} = \mathrm{i}ka\boldsymbol{P}_{\boldsymbol{\Sigma}}\mathbf{y}_{qH} \tag{16.79}$$

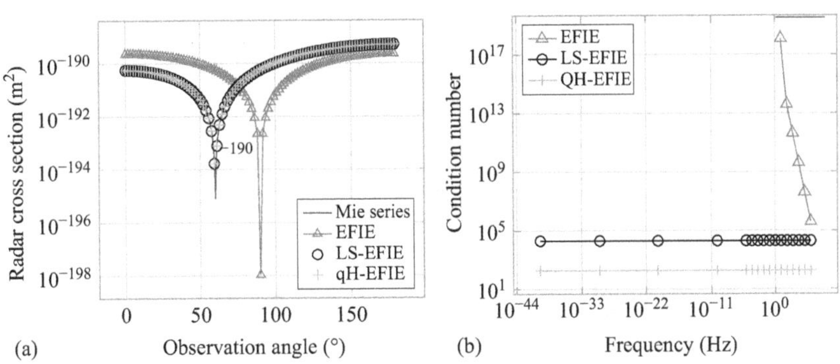

Figure 16.11 Low-frequency behaviour of the qH-EFIE compared against the standard and LS-EFIE. (a) illustrates the RCS of a PEC sphere of radius 1 m, excited by a plane wave oscillating at 1×10^{-40} Hz, while (b) illustrates the evolution of the conditioning of the impedance matrices of both formulations

and given that the projection matrices are frequency independent, the non-solenoidal (resp. solenoidal) part of the scaled solution scales as $\mathcal{O}(1)$ (resp. $\mathcal{O}(1)$). Consequently, $\boldsymbol{y}_{qH}$ is free from any loss of significant digits and can safely be stored in finite precision.

The same numerical test as in Section 16.3.3 confirms that the qH projectors deliver accurate results for the far field, as illustrated in Figure 16.11(a).

Our developments so far have demonstrated that the qH projectors can fix the EFIE low-frequency breakdown, and unlike traditional LS decompositions, (i) we did not need to detect global loops in Γ and (ii) since $\boldsymbol{P}_{\Lambda H}$ and $\boldsymbol{P}_{\Sigma}$ are projectors $\operatorname{cond}(\boldsymbol{M})$ and $\operatorname{cond}(\boldsymbol{N})$ are independent of h, meaning that our technique does not introduce any additional high-refinement ill-conditioning, as illustrated in Figure 16.12.

16.7.2 Implementation details

High accuracy numerical results require special care to be given to a few implementation details that will be presented in this section.

Due to the limited precision of computers, some of the mathematical relations seen throughout our developments should be explicitly enforced. First, the computation of the qH-EFIE impedance matrix $\boldsymbol{M}^{\mathrm{T}}\boldsymbol{Z}\,\mathbf{N}$ should be handled carefully. Since the terms in (16.76) will not exactly equal $\mathbf{0}$ when computed directly, they will introduce severe numerical errors at low frequencies. This effect can be illustrated with a numerical example : let us assume that, when computed directly,

$$\|\boldsymbol{P}_{\Lambda H}\boldsymbol{Z}_h\boldsymbol{P}_{\Lambda H}\| = \varepsilon_{MACH} \tag{16.80}$$

with a value of ε_{MACH} typically around 1×10^{-16} then, once scaled adequately in frequency, this term becomes

$$\|\frac{1}{\mathrm{i}k}\boldsymbol{P}_{\Lambda H}\boldsymbol{Z}_h\boldsymbol{P}_{\Lambda H}\| = \frac{\varepsilon_{MACH}}{k} \tag{16.81}$$

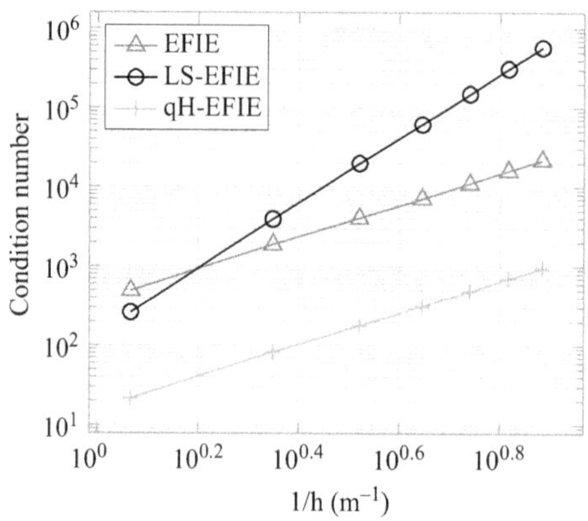

Figure 16.12 Comparison of the high-refinement conditioning of the qH-EFIE, LS-EFIE and associated operators and projectors against the standard EFIE

When going to extremely low frequency, the norm of this term will become considerably far from 0. For instance at 1×10^{-40} Hz its norm is in the order of 1×10^{31}. To address this issue, the impedance matrix should be expanded to explicitly cancel out the problematic terms, as was done in (16.75).

Another point to be considered is the computation of the RHS of (16.40). For clarity we will separate the solenoidal and non-solenoidal parts of the RHS

$$\mathbf{v}^{\Sigma} = \mathbf{P}_{\Sigma}\mathbf{v} \tag{16.82}$$

$$\mathbf{v}^{\Lambda H} = \frac{1}{ika}\mathbf{P}_{\Lambda H}\mathbf{v} \tag{16.83}$$

When $\mathbf{v}$ is the result of a plane wave excitation of the form $\hat{\mathbf{r}} \mapsto \mathbf{u}\mathrm{e}^{ik\cdot r}$, the term $\mathbf{v}^{\Lambda H}$ should be computed from the extracted exponential $\mathrm{e}^{ik\cdot r} - 1$ to avoid any loss of significant digits. The reason why this special treatment is required is fundamentally the same as the reason why the different parts of the far field scale differently. The expression of the plane wave induced RHS is

$$[\mathbf{v}]_i = -\int_{f_i} \mathbf{f}_i(\mathbf{r}') \cdot \mathbf{u}\mathrm{e}^{ik\cdot r'}\,\mathrm{d}\mathbf{r}' \tag{16.84}$$

This expression can be decomposed into two integrals with the kernels $\mathrm{e}^x - 1$ and 1. When the testing is performed with a solenoidal basis function, the integral with the constant kernel will be null as per Green's theorem, which means that the RHS will scale as $\mathcal{O}(k)$. However, at low frequencies and in finite precision, when the un-extracted kernel is used, this contribution will be numerically hidden by the 0th order element in the Taylor series expansion of the exponential. Implementation of this part of the RHS then requires the usage of a specialized subroutine performing accurate evaluations of $x \mapsto \mathrm{e}^x - 1$ for small arguments. Several numerical libraries offer their own versions (see the function *expm1* in MATLAB, NumPy, etc). The expression of $\mathbf{v}^{\Lambda H}$ becomes $(ik)^{-1}\mathbf{P}_{\Lambda H}\mathbf{v}^{\mathrm{ext}}$, where

$$[\mathbf{v}^{\mathrm{ext}}]_i = -\int_{f_i} \mathbf{f}_i(\mathbf{r}') \cdot \mathbf{u}\,\mathrm{expm1}(i\mathbf{k}\cdot\mathbf{r}')\,\mathrm{d}\mathbf{r}' \tag{16.85}$$

Finally, as we have seen in Section 16.6, linear-time computation of the projectors require the usage of multigrid preconditioning [34,35]. While this technique is outside the scope of this chapter, several implementations are readily available in well tested numerical packages such as *pyamg*[51] or AGMG[52].

16.8 Solution to the large time step breakdown and the DC instability for the TD-EFIE

In this section, the DC instability and the large time step breakdown of the TD-EFIE are addressed using qH projector-based solutions. An appropriate temporal differentiation and integration on the Helmholtz-decomposed system prevents the static loops from being a solution of the equation and an adequate temporal discretization is used in order to comply with the application of the preconditioner.

16.8.1 Preconditioning

Since the DC instability and the large time step breakdown are related to the time discretization, the starting point of our analysis is the semi-discretized TD-EFIE equation (16.25). Following the same strategy as in the FD, the EFIE operator $\mathcal{T}$ is decomposed into its singular part $\mathcal{T}_s$ and its hypersingular part $\mathcal{T}_h$. After spatial discretization, the semi-discretized TD-EFIE reads

$$\boldsymbol{Z}\boldsymbol{j}(t) = (\boldsymbol{Z}_s + \boldsymbol{Z}_h)\boldsymbol{j}(t) = \mathbf{e}(t) \tag{16.86}$$

which, once the qH projectors have been applied, gives in block form

$$(\boldsymbol{P}_{\Lambda H} + \boldsymbol{P}_{\Sigma})(\boldsymbol{Z}_s + \boldsymbol{Z}_h)(\boldsymbol{P}_{\Lambda H} + \boldsymbol{P}_{\Sigma}) = \begin{bmatrix}\boldsymbol{P}_{\Lambda H} & \boldsymbol{P}_{\Sigma}\end{bmatrix}\begin{bmatrix}\boldsymbol{Z}_s & \boldsymbol{Z}_s \\ \boldsymbol{Z}_s & \boldsymbol{Z}_s + \boldsymbol{Z}_h\end{bmatrix}\begin{bmatrix}\boldsymbol{P}_{\Lambda H} \\ \boldsymbol{P}_{\Sigma}\end{bmatrix} \tag{16.87}$$

where the TD counterpart of (16.76) has been used.

As explained in the Section 16.5, the static null-space of the equation is created by the time derivative in the singular part $\mathcal{T}_s$ of the EFIE operator, hence by introducing a time integration on the loop part of the left preconditioner it can be cancelled. Similarly, the problematic integration in the hypersingular operator can be cancelled by introducing a time derivative on the star part of the right preconditioner, this will keep the number of non-zero interaction matrices bounded. Finally, the ill-scaling between $\mathcal{T}_s$ and $\mathcal{T}_h$ can be suppressed by multiplying the $\mathcal{T}_s$ by c and the $\mathcal{T}_h$ by c^{-1}. This scheme is consistent with the preconditioner developed in the FD; the frequency factors $-ika$ and $(-ika)^{-1}$ correspond to $ac^{-1}\partial_t$ and $a^{-1}c\partial_t^{-1}$ in TD, respectively. This approach has been presented in [26] using a Loop-Tree decomposition and can be extended to leverage on qH projectors to benefit from their advantages. The left and right projectors are defined as

$$\boldsymbol{M}' = \left(\frac{c}{a}\partial_t^{-1}\boldsymbol{P}_{\Lambda H} + \boldsymbol{P}_{\Sigma}\right) \tag{16.88}$$

$$\boldsymbol{N}' = \left(\boldsymbol{P}_{\Lambda H} + \frac{a}{c}\partial_t\boldsymbol{P}_{\Sigma}\right) \tag{16.89}$$

which, when applied to the TD-EFIE, yield the qH-projected TD-EFIE

$$\boldsymbol{M}'^{\mathrm{T}}\boldsymbol{Z}\boldsymbol{N}'\mathbf{y}(t) = \mathbf{v}(t) \tag{16.90}$$

where

$$\boldsymbol{j}(t) = \boldsymbol{N}'\mathbf{y}(t) \tag{16.91}$$

$$\mathbf{v}(t) = \boldsymbol{M}'^{\mathrm{T}}\mathbf{e}(t) \tag{16.92}$$

The interaction matrix of the preconditioned equation can be expanded into

$$\boldsymbol{Z}' = \boldsymbol{M}'^{\mathrm{T}}\boldsymbol{Z}\boldsymbol{N}' = \begin{bmatrix}\boldsymbol{P}_{\Lambda H} & \boldsymbol{P}_{\Sigma}\end{bmatrix}\begin{bmatrix}\frac{c}{a}\partial_t^{-1}\boldsymbol{Z}_s & \boldsymbol{Z}_s \\ \boldsymbol{Z}_s & \frac{a}{c}\partial_t\boldsymbol{Z}_s + \frac{a}{c}\partial_t\boldsymbol{Z}_h\end{bmatrix}\begin{bmatrix}\boldsymbol{P}_{\Lambda H} \\ \boldsymbol{P}_{\Sigma}\end{bmatrix} \tag{16.93}$$

For decreasing frequency, or increasing time step, the norm of the $\boldsymbol{Z}_s$ contribution in the off-diagonal blocks and the $ac^{-1}\partial_t\boldsymbol{Z}_s$ contribution in the bottom right will decay to 0. The remaining contributions $a^{-1}c\partial_t^{-1}\boldsymbol{Z}_s$ and $ac^{-1}\partial_t\boldsymbol{Z}_h$ will dictate the low-frequency behaviour of the preconditioned equation. This low-frequency approximation of the impedance matrix can be used to verify the stability of its conditioning in the problematic domain.

While the overall scheme addresses the ill-conditioning and DC-instabilities of the TD-EFIE, because of the introduction of additional time derivatives or integrals, the time discretization should be adapted accordingly.

16.8.2 Time discretization

To maintain the pulses $p(t)$ discretization of the solution $\boldsymbol{j}(t)$, the discretization of the scaled solution $\mathbf{y}(t)$ should be adapted accordingly. Because the non-solenoidal part of the scaled solution is exposed to an additional time derivative with regards to the solenoidal part, a mixed discretization should be adopted. Since the time derivative of h can be expressed as a sum of pulses p

$$\partial_t h(t) = \frac{1}{\Delta t}(p(t) - p(t - \Delta t)) \tag{16.94}$$

it is an adequate choice of source basis function for the non-solenoidal part of the equation. An analogous reasoning is used to determine the adequate choice of testing basis functions. Noting that

$$\mathbf{Z}\boldsymbol{j}(t) = \left(\frac{a}{c}\partial_t \boldsymbol{P}_{\Lambda H} + \boldsymbol{P}_{\Sigma}\right)\mathbf{Z}'\mathbf{y}(t) \tag{16.95}$$

is tested with pulses $p(t)$, and the presence of the time derivative, the non-solenoidal part of the equation is still tested with the pulses $p(t)$, however the solenoidal part has to be tested with a Dirac delta distribution $\delta(t)$ (illustrated in Figure 16.13(a)). This

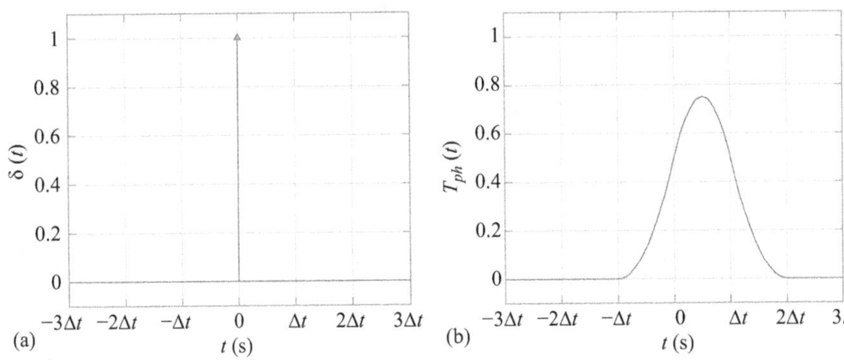

Figure 16.13 Illustrations of the basis functions used for the new time discretization of the qH TD-EFIE. The triangle in (a) denotes an infinite value at 0 and its ordinate corresponds to its value when the function is integrated. (a) Dirac δ distribution. (b) Piecewise polynomial basis function T_{ph}

is justified by the fact that the time derivative of p can be expressed as a sum of δ in the sense of distributions

$$\partial_t p(t) = \delta(t) - \delta(t + \Delta t) \tag{16.96}$$

In addition, since the testing of each Helmholtz component needs to have the same scaling the pulses should be scaled by a $1/\Delta t$ factor because

$$\int_{\mathbb{R}} \delta(t)\,\mathrm{d}t = 1 \tag{16.97}$$

while

$$\int_{\mathbb{R}} p(t)\,\mathrm{d}t = \Delta t \tag{16.98}$$

With this time discretization, the MOT of the qH-preconditioned TD-EFIE reads

$$\boldsymbol{Z}'_0 \mathbf{y}_j = \mathbf{v}_j - \sum_{i=1}^{j-1} \boldsymbol{Z}'_i \mathbf{y}_{j-i} \tag{16.99}$$

where $\boldsymbol{Z}'_x = \boldsymbol{M}'^{\mathrm{T}} \boldsymbol{Z}_x \boldsymbol{N}'$ and, according to (16.92) and our choice of temporal testing basis functions, the excitation vector $\mathbf{v}_j$ is

$$\mathbf{v}_j = \int_{\mathbb{R}} \left(\delta(t - j\Delta t) \boldsymbol{P}_{\Lambda H} + \frac{1}{\Delta t} p(t - j\Delta t) \boldsymbol{P}_{\Sigma} \right) \mathbf{v}(t)\,\mathrm{d}t \tag{16.100}$$

$$= \int_{\mathbb{R}} \left(\delta(t - j\Delta t) a^{-1} c \partial_t^{-1} \boldsymbol{P}_{\Lambda H} + \frac{1}{\Delta t} p(t - j\Delta t) \boldsymbol{P}_{\Sigma} \right) \mathbf{e}(t)\,\mathrm{d}t \tag{16.101}$$

Similarly (16.93) can be computed as

$$\boldsymbol{Z}'_i = \begin{bmatrix} \boldsymbol{P}_{\Lambda H} & \boldsymbol{P}_{\Sigma} \end{bmatrix} \begin{bmatrix} \boldsymbol{Z}_i'^{LL} & \boldsymbol{Z}_i'^{LS} \\ \boldsymbol{Z}_i'^{SL} & \boldsymbol{Z}_i'^{SS} \end{bmatrix} \begin{bmatrix} \boldsymbol{P}_{\Lambda H} \\ \boldsymbol{P}_{\Sigma} \end{bmatrix} \tag{16.102}$$

with

$$\begin{aligned} [\boldsymbol{Z}_i'^{LL}]_{mn} &= \int_{\mathbb{R}} \delta(t - i\Delta t) a^{-1} c \partial_t^{-1} \int_{\Gamma} (\hat{\boldsymbol{n}}_r \times \boldsymbol{f}_m(\boldsymbol{r})) \cdot (\eta \mathcal{T}_s\{\boldsymbol{f}_n p\}(\boldsymbol{r}, t))\,\mathrm{d}\boldsymbol{r}\,\mathrm{d}t \\ &= \eta \int_{\Gamma} (\hat{\boldsymbol{n}}_r \times \boldsymbol{f}_m(\boldsymbol{r})) \cdot \left(a^{-1} c \partial_t^{-1} \mathcal{T}_s\{\boldsymbol{f}_n p\}(\boldsymbol{r}, i\Delta t) \right)\,\mathrm{d}\boldsymbol{r} \end{aligned} \tag{16.103}$$

$$\begin{aligned} [\boldsymbol{Z}_i'^{LS}]_{mn} &= \int_{\mathbb{R}} \delta(t - i\Delta t) \int_{\Gamma} (\hat{\boldsymbol{n}}_r \times \boldsymbol{f}_m(\boldsymbol{r})) \cdot (\eta \mathcal{T}_s\{\boldsymbol{f}_n h\}(\boldsymbol{r}, t))\,\mathrm{d}\boldsymbol{r}\,\mathrm{d}t \\ &= \eta \int_{\Gamma} (\hat{\boldsymbol{n}}_r \times \boldsymbol{f}_m(\boldsymbol{r})) \cdot (\mathcal{T}_s\{\boldsymbol{f}_n h\}(\boldsymbol{r}, i\Delta t))\,\mathrm{d}\boldsymbol{r} \end{aligned} \tag{16.104}$$

$$
\begin{aligned}
[\boldsymbol{Z}_i'^{SL}]_{mn} &= \int_{\mathbb{R}} \frac{1}{\Delta t} p(t - i\Delta t) \int_{\Gamma} (\hat{\boldsymbol{n}}_r \times \boldsymbol{f}_m(\boldsymbol{r})) \cdot (\eta \mathscr{T}_s\{\boldsymbol{f}_n p\}(\boldsymbol{r}, t))\, \mathrm{d}\boldsymbol{r}\, \mathrm{d}t \\
&= \eta \int_{\Gamma} (\hat{\boldsymbol{n}}_r \times \boldsymbol{f}_m(\boldsymbol{r})) \cdot (\mathscr{T}_s\{\boldsymbol{f}_n h\}(\boldsymbol{r}, i\Delta t))\, \mathrm{d}\boldsymbol{r} \qquad (16.105)
\end{aligned}
$$

$$
\begin{aligned}
[\boldsymbol{Z}_i'^{SS}]_{mn} &= \int_{\mathbb{R}} \frac{1}{\Delta t} p(t - i\Delta t) \frac{a}{c} \partial_t \int_{\Gamma} (\hat{\boldsymbol{n}}_r \times \boldsymbol{f}_m(\boldsymbol{r})) \cdot (\eta \mathscr{T}\{\boldsymbol{f}_n h\}(\boldsymbol{r}, t))\, \mathrm{d}\boldsymbol{r}\, \mathrm{d}t \\
&= \eta \int_{\Gamma} (\hat{\boldsymbol{n}}_r \times \boldsymbol{f}_m(\boldsymbol{r})) \cdot (\frac{a}{c} \partial_t \mathscr{T}\{\boldsymbol{f}_n T_{ph}\}(\boldsymbol{r}, i\Delta t))\, \mathrm{d}\boldsymbol{r} \qquad (16.106)
\end{aligned}
$$

where T_{ph} is the piecewise polynomial basis function illustrated in Figure 16.13(b) and defined as

$$
T_{ph}(t) = \frac{1}{\Delta t} \int_{\mathbb{R}} p(\tau) h(t + \tau)\, \mathrm{d}\tau \qquad (16.107)
$$

$$
= \begin{cases}
\dfrac{1}{2} + \dfrac{t}{\Delta t} + \dfrac{1}{2}\left(\dfrac{t}{\Delta t}\right)^2 & \text{if } -\Delta t \le t \le 0 \\
\dfrac{1}{2} + \dfrac{t}{\Delta t} - \left(\dfrac{t}{\Delta t}\right)^2 & \text{if } 0 \le t \le \Delta t \\
2 - \dfrac{2t}{\Delta t} + \dfrac{1}{2}\left(\dfrac{t}{\Delta t}\right)^2 & \text{if } \Delta t \le t \le 2\Delta t \\
0 & \text{otherwise}
\end{cases} \qquad (16.108)
$$

Once the scaled solution $\mathbf{y}_j$ of the MOT (16.99) has been computed, the current coefficients are retrieved for the solenoidal and non-solenoidal parts, according to (16.91) and (16.94), as

$$
\mathbf{j}_j^{\Lambda H} = \boldsymbol{P}_{\Lambda H} \mathbf{y}_j \qquad (16.109)
$$

$$
\mathbf{j}_j^{\Sigma} = \frac{a}{c\Delta t} \boldsymbol{P}_{\Sigma} \left(\mathbf{y}_j - \mathbf{y}_{j-1}\right) \qquad (16.110)
$$

where $\mathbf{j}_j = \mathbf{j}_j^{\Lambda H} + \mathbf{j}_j^{\Sigma}$. Finally, the surface current density $\boldsymbol{j}(\boldsymbol{r}, t)$ can be recovered from $\{\mathbf{j}_j\}, j \in [1, N_t]$, using (16.32).

16.8.3 Numerical results

To prove the efficiency of our scheme, we have performed the same simulations as in Section 16.4 and 16.5 with the low-frequency stable formulation.

The qH-stabilized TD-EFIE requires the computation of the primitive $\boldsymbol{E}^i(\boldsymbol{r}, t) = \partial_t^{-1} \boldsymbol{e}^i(\boldsymbol{r}, t)$ that is $\mathbf{0}$ for $t \to -\infty$. The primitive that corresponds to the modulated Gaussian plane wave (16.63) is

$$
\boldsymbol{E}^i(\boldsymbol{r}, t) = \boldsymbol{\alpha} \Re \left(\mathrm{erfc}\left(-\frac{\beta(\boldsymbol{r}, t)}{\sqrt{2}\sigma}\right)\right) \qquad (16.111)
$$

where $\Re(\mathrm{erfc}(\cdot))$ denotes the real part of the complementary error function and

$$\boldsymbol{\alpha} = \hat{\boldsymbol{p}} e_0 \sqrt{\frac{\pi}{2}} \sigma \mathrm{e}^{-2\pi^2 f_0^2 \sigma^2} \tag{16.112}$$

$$\beta(\boldsymbol{r}, t) = t - t_0 - \frac{\hat{\boldsymbol{k}} \cdot \boldsymbol{r}}{c} + 2\mathrm{i}\pi f_0 \sigma^2 \tag{16.113}$$

First the stability of the conditioning of the qH TD-EFIE has been verified, and as shown in Figure 16.14(a), numerical results confirm that it remains constant when the MOT time step increases. Similarly Figure 16.14(b) shows that the problematic DC component of the standard formulation has been removed, which can also be explained by the disappearance of the eigenvalue cluster at $(1, 0)$ in Figure 16.14(c).

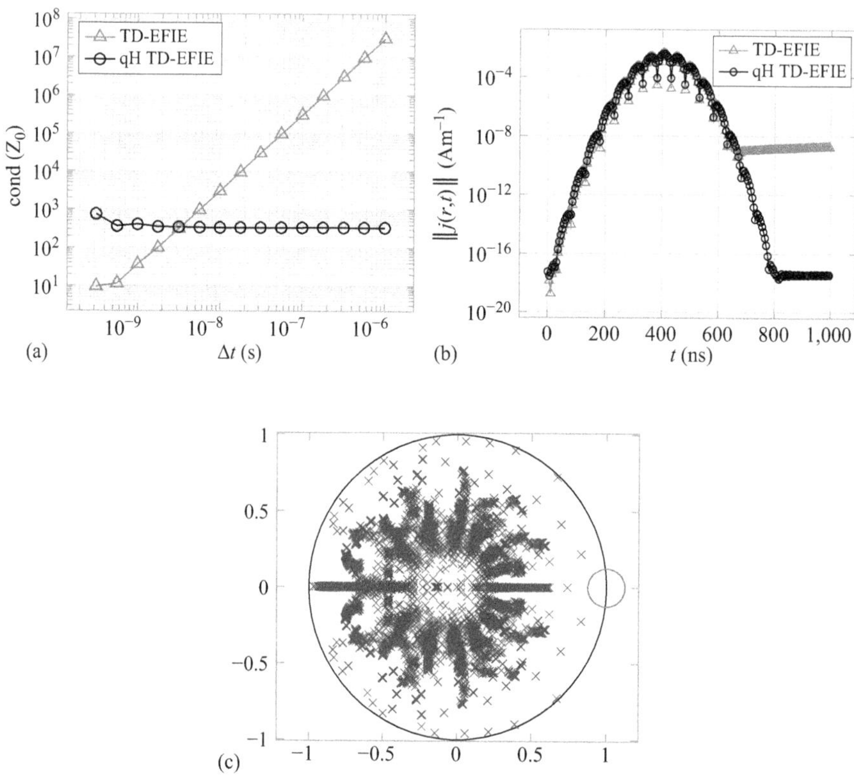

Figure 16.14 Illustration of the curative effects of the qH projectors and mixed-time discretization on the TD-EFIE: (a) stabilized condition number of $\boldsymbol{Z}_0$ as a function of the time step Δt, (b) absence of DC instabilities in the solution surface current density ($N_t = 1000$, $\Delta t = 1$ ns and $\boldsymbol{r} = (-0.186, -0.173, 0.957)$ m) and (c) absence of numerically unstable eigenvalues clustering at (1, 0)

These numerical results demonstrate a perfect agreement between the expected curative effects of the new formulation and the actual performance of its implementation.

16.9 Conclusions

Throughout this chapter, we have identified the sources of the different problems plaguing the EFIE at low frequencies in both the frequency and the TD, as well as their traditional cures. Despite their apparent effectiveness, these techniques have been shown to have a limited applicability because they introduce their own set of problems which include the high computational burden of the LS decomposition and its effect on the high-refinement conditioning of the FD-EFIE and the numerical instabilities introduced by the treatment of the TD-EFIE.

Techniques leveraging qH projectors, immune from the aforementioned side-effects, have been introduced to address the different aspects of the low-frequency breakdown of the FD formulation and of the large time step breakdown of its TD counterpart. In case of the FD, using projectors allows the same re-scaling of the solenoidal and non-solenoidal parts of the RWG space as traditional LS, but it has the added benefits of not requiring identification of the global loops of the structure as well as not introducing any further high-refinement ill-conditioning. In the TD case, the projectors are still used to separate the loop and star parts of the discretized space, but this separation is used to apply the correct derivative and integrative terms to the different parts of the operators. Coupled with an adequate mixed time-discretization scheme, this technique fully addresses the low-frequency limitations of the TD-EFIE. Along with presenting these purely theoretical concepts, we have provided implementation related hints, allowing the techniques presented in this chapter to be reliably and readily implemented into existing solvers. Finally, while we have addressed their low-frequency breakdown, both EFIE formulations still suffer from a high-refinement breakdown. While in standard low-frequency scenarios, a curing of low-frequency issues may suffice, for more pathological cases techniques addressing both breakdowns may be required. Strategies based on qH projectors and Calderón identities have recently been introduced for the frequency and TD formulations [23,40] and should be used in this case.

References

[1] Coifman R, Rokhlin V and Wandzura S. The fast multipole method for the wave equation: a pedestrian prescription. Antennas and Propagation Magazine, IEEE. 1993;35(3):7–12.

[2] Lu CC and Chew WC. A multilevel algorithm for solving a boundary integral equation of wave scattering. Microwave and Optical Technology Letters. 1994;7(10):466–470.

[3] Michielssen E and Boag A. A multilevel matrix decomposition algorithm for analyzing scattering from large structures. Antennas and Propagation, IEEE Transactions On. 1996;44(8):1086–1093.

[4] Jin JM. Theory and Computation of Electromagnetic Fields. 2nd ed. Piscataway, NJ: IEEE Press; 2015. OCLC: 931929424.

[5] Yee KS. Numerical solution of initial boundary value problems involving Maxwell's equations in isotropic media. IEEE Trans Antennas Propag. 1966;14(3):302–307.

[6] Juntunen JS and Tsiboukis TD. Reduction of numerical dispersion in FDTD method through artificial anisotropy. IEEE Transactions on Microwave Theory and Techniques. 2000;48(4):582–588.

[7] Schaubert DH, Wilton DR and Glisson AW. A tetrahedral modeling method for electromagnetic scattering by arbitrarily shaped inhomogeneous dielectric bodies. Antennas and Propagation, IEEE Transactions On. 1984;32(1):77–85.

[8] Poggio AJ and Miller EK. Integral Equation Solutions of Three-dimensional Scattering Problems. In Computer Techniques for Electro-magnetics, R. Mittra, Ed. Oxford, UK: Pergamon, 1973.

[9] Glisson AW. Electromagnetic scattering by arbitrarily shaped surfaces with impedance boundary conditions. Radio Science. 1992;27(6):935–943.

[10] Rao SM and Wilton DR. Transient scattering by conducting surfaces of arbitrary shape. Antennas and Propagation, IEEE Transactions On. 1991;39(1): 56–61.

[11] Shanker B, Lu M, Yuan J and Michielssen E. Time domain integral equation analysis of scattering from composite bodies via exact evaluation of radiation fields. IEEE Transactions on Antennas and Propagation. 2009;57(5):1506–1520.

[12] Andriulli FP, Tabacco A, and Vecchi G. Solving the EFIE at low frequencies with a conditioning that grows only logarithmically with the number of unknowns. IEEE Transactions on Antennas and Propagation. 2010;58(5): 1614–1624.

[13] Andriulli FP, Bagci H, Vipiana F, Vecci G and Michielssen E. Analysis and regularization of the TD-EFIE low-frequency breakdown. IEEE Transactions on Antennas and Propagation. 2009;57(7):2034–2046.

[14] Chen NW, Aygün K, and Michielssen E. Integral-equation-based analysis of transient scattering and radiation from conducting bodies at very low frequencies. IEE Proceedings - Microwaves, Antennas and Propagation. 2001;148(6): 381–387.

[15] Bennett CL, Weeks WL. A technique for computing approximate electromagnetic impulse response of conducting bodies. Purdue University. 1968.

[16] Andriulli FP, Cools K, Olyslager F and Michielssen E. Time domain Calderon identities and their application to the integral equation analysis of scattering by PEC objects Part II: stability. IEEE Transactions on Antennas and Propagation. 2009;57(8):2365–2375.

[17] Shi Y, Bağcı H and Lu M. On the static loop modes in the marching-on-in-time solution of the time-domain electric field integral equation. IEEE Antennas and Wireless Propagation Letters. 2014;13:317–320.

[18] Zhang Y, Cui TJ, Chew WC and Zhao JS. Magnetic field integral equation at very low frequencies. IEEE Transactions on Antennas and Propagation. 2003;51(8):1864–1871.
[19] Chew WC, Tong MS, and Hu B. Integral equation methods for electromagnetic and elastic waves. Synthesis Lectures on Computational Electromagnetics. 2008;3(1):1–241.
[20] Qian ZG and Chew WC. Enhanced A-EFIE with perturbation method. IEEE Transactions on Antennas and Propagation. 2010;58(10):3256–3264.
[21] Bogaert I, Cools K, Andriulli FP and De Zutter D. Low frequency scaling of the mixed MFIE for scatterers with a non-simply connected surface. In: Electromagnetics in Advanced Applications (ICEAA), 2011 International Conference On. IEEE; 2011. pp. 951–954.
[22] Bogaert I, Cools K, Andriulli FP, Peeters J and De Zutter D. Low frequency stability of the mixed discretization of the MFIE. In: Antennas and Propagation (EUCAP), Proceedings of the 5th European Conference On. IEEE; 2011. pp. 2463–2465.
[23] Andriulli FP, Cools K, Bogaert I and Michielssen E. On a well-conditioned electric field integral operator for multiply connected geometries. IEEE Transactions on Antennas and Propagation. 2013;61(4):2077–2087.
[24] Geranmayeh A, Ackermann W and Weiland T. Temporal discretization choices for stable boundary element methods in electromagnetic scattering problems. Applied Numerical Mathematics. 2009;59(11):2751–2773.
[25] Beghein Y, Cools K, Bagci H and De Zutter D. A space-time mixed Galerkin marching-on-in-time scheme for the time-domain combined field integral equation. IEEE Transactions on Antennas and Propagation. 2013;61(3):1228–1238.
[26] Pisharody G and Weile DS. Robust solution of time-domain integral equations using loop-tree decomposition and bandlimited extrapolation. IEEE Transactions on Antennas and Propagation. 2005;53(6):2089–2098.
[27] Wilton D and Glisson A. On improving the electric field integral equation at low frequencies. Proc URSI Radio Sci Meet Dig. 1981;24.
[28] Vecchi G. Loop-star decomposition of basis functions in the discretization of the EFIE. IEEE Transactions on Antennas and Propagation. 1999;47(2): 339–346.
[29] Zhao JS and Chew WC. Integral equation solution of Maxwell's equations from zero frequency to microwave frequencies. Antennas and Propagation, IEEE Transactions On. 2000;48(10):1635–1645.
[30] Lee JF, Lee R, and Burkholder RJ. Loop star basis functions and a robust preconditioner for EFIE scattering problems. IEEE Transactions on Antennas and Propagation. 2003;51(8):1855–1863.
[31] Eibert TF. Iterative-solver convergence for loop-star and loop-tree decompositions in method-of-moments solutions of the electric-field integral equation. Antennas and Propagation Magazine, IEEE. 2004;46(3):80–85.
[32] Andriulli FP. Loop-star and loop-tree decompositions: analysis and efficient algorithms. IEEE Transactions on Antennas and Propagation. 2012;60(5):2347–2356.

[33] Qian ZG and Chew WC. An augmented electric field integral equation for high-speed interconnect analysis. Microwave and Optical Technology Letters. 2008;50(10):2658–2662.

[34] Livne O and Brandt A. Lean algebraic multigrid (LAMG): fast graph Laplacian linear solver. SIAM Journal on Scientific Computing. 2012;34(4): B499–B522.

[35] Napov A and Notay Y. An algebraic multigrid method with guaranteed convergence rate. SIAM Journal on Scientific Computing. 2012;34(2): A1079–A1109.

[36] Andriulli FP, Cools K, Bagci H, *et al.* A multiplicative Calderon preconditioner for the electric field integral equation. IEEE Transactions on Antennas and Propagation. 2008;56(8):2398–2412.

[37] Beghein Y, Cools K, Andriulli FP, De Zutter D. and Michielssen E. A Calderon multiplicative preconditioner for the PMCHWT equation for scattering by chiral objects. IEEE Transactions on Antennas and Propagation. 2012;60(9):4239–4248.

[38] Adams RJ. Physical and analytical properties of a stabilized electric field integral equation. IEEE Transactions on Antennas and Propagation. 2004;52(2):362–372.

[39] Stephanson MB and Lee JF. Preconditioned electric field integral equation using Calderon identities and dual loop/star basis functions. IEEE Transactions on Antennas and Propagation. 2009;57(4):1274–1279.

[40] Beghein Y, Cools K and Andriulli FP. A DC stable and large-time step well-balanced TD-EFIE based on quasi-Helmholtz projectors. IEEE Transactions on Antennas and Propagation. 2015;63(7):3087–3097.

[41] Rao SM, Wilton D and Glisson AW. Electromagnetic scattering by surfaces of arbitrary shape. IEEE Transactions on Antennas and Propagation. 1982;30(3):409–418.

[42] Wu WL, Glisson AW and Kajfez D. A study of two numerical solution procedures for the electric field integral equation at low frequency. Applied Computational Electromagnetics Society Journal. 1995;10(3):69–80.

[43] Wilton DR. Topological consideration in surface patch and volume cell modeling of electromagnetic scatterers. In: Proc. URSI Int. Symp. Electromagn. Theory; 1983. pp. 65–68.

[44] Dunavant DA. High degree efficient symmetrical Gaussian quadrature rules for the triangle. International journal for numerical methods in engineering. 1985;21(6):1129–1148.

[45] Wilton DR, Rao SM, Glisson AW, Schaubert DH, Al-Bundak OM and Butler CM. Potential integrals for uniform and linear source distributions on polygonal and polyhedral domains. IEEE Transactions on Antennas and Propagation. 1984;30(3):276–281.

[46] Gibson WC. The Method of Moments in Electromagnetics New York: Chapman and Hall/CRC Press, Boca Raton, FL; 2014.

[47] Ergin AA, Shanker B and Michielssen E. The plane-wave time-domain algorithm for the fast analysis of transient wave phenomena. IEEE Antennas and Propagation Magazine. 1999;41(4):39–52.

[48] Yilmaz AE, Weile DS, Jin HM and Michielssen E. A hierarchical FFT algorithm (HIL-FFT) for the fast analysis of transient electromagnetic scattering phenomena. IEEE Transactions on Antennas and Propagation. 2002;50(7):971–982.
[49] Shanker B, Ergin AA, Lu M, *et al.* Fast analysis of transient electromagnetic scattering phenomena using the multilevel plane wave time domain algorithm. IEEE Transactions on Antennas and Propagation. 2003;51(3):628–641.
[50] Henderson HV and Searle SR. On deriving the inverse of a sum of matrices. SIAM Review. 1981;23(1):53–60.
[51] Olson LN and Schroder JB. PyAMG: Algebraic Multigrid Solvers in Python v4.0; 2018. Release 4.0. Available from: https://github.com/pyamg/pyamg.
[52] Notay Y. AGMG Software and Documentation;. Accessed: 2018-01-10. Available from: http://homepages.ulb.ac.be/~ynotay/AGMG.

Chapter 17

Resistive and thin dielectric disk antennas with axially symmetric excitation analyzed using the method of analytical regularization

Nataliya Y. Bliznyuk[1] *and Alexander I. Nosich*[2]

We present accurate numerical method and computational results in the modeling of the axisymmetric electromagnetic wave scattering by resistive and thin dielectric circular disk-on-substrate antennas, simulated by using the two-sided generalized boundary conditions (GBC) and singular integral equations (SIE). The numerical analysis is based on the method of analytical regularization (MAR) exploiting a Galerkin method (GM) with judicious basis functions, which convert SIE into a Fredholm second-kind infinite matrix equation. This guarantees convergence and enables one to compute the solution with controlled accuracy even near sharp resonances if they are present. Numerical results demonstrate the effect of the disk losses and transparency on the antenna bandwidth and radiation efficiency.

17.1 Introduction

Flat metallic disks are frequently met in microwave communications and radar as elements of phased array antenna. Such elements are known as microstrip antennas (MSA) [1]. If excited axisymmetrically, single circular disks are able to produce conical far-field patterns [2] used in the design of horizontally omnidirectional cellular and automobile antennas [3]. Here, resistive printed disks are attractive as more broadband MSAs than metallic ones [4]. Moreover, recently new technologies have made available low loss and high permittivity disk dielectric antennas (DAs) [5], which have further advantages.

Numerical tools of MSA and DA design consist of direct numerical and numerical analytical methods such as FDTD and method of moments (MM). FDTD is used to study many electromagnetic problems including inverse and time-domain ones. However, its application to antennas entails severe difficulties in how to limit the infinite space domain and how to mesh it properly. This leads to the violation of the

[1]Advanced Systems Group, Space Systems/Loral, Palo Alto, CA, USA
[2]Institute of Radio-Physics and Electronics NASU, Kharkiv, Ukraine

radiation condition and also to very large-sized matrices to be inverted. In contrast, MM is based on a numerical approximation of IEs derived for the equivalent currents. Although less flexible than FDTD, it is more economic. This is because only the boundaries need to be discretized, while the radiation condition is automatically satisfied due to the proper choice of the IE kernels. In the 1980s, numerous remarkable MM analyses of perfectly electric conducting (PEC) circular MSA were published [6–11]. Due to the disk symmetry, they worked mainly with the Hankel-transformed IEs and used transforms of the so-called "cavity modes" as basis functions.

However, electromagnetic problems are mostly reduced to the first-kind and/or strongly singular IEs, and thus the convergence of common MM algorithms, which end up with the dense first-kind matrices, generally cannot be guaranteed [12]. Especially significant losses of accuracy in MM and FDTD may happen near the sharp resonances [13]. As printed antennas are essentially resonant devices, this disadvantage is quite serious. Additionally, both MM and FDTD are time- and memory-consuming even for obtaining basic results. If the frequency is increased, the required computer resources rapidly grow up. Therefore MSA modeling in wide band and its numerical optimization is painful, especially if a supercomputer is not available.

Although currently there is no versatile theoretical technique able to compute arbitrary MSA/DA with arbitrary accuracy, it is possible to achieve this goal in the case of circular disks, both PEC and imperfect. Such a study can be valuable for a better insight in the fundamental wave phenomena in MSA and DA in general. For example, the effect of the patch partial transparency and losses on the antenna characteristics has not yet been well quantified. Provided that the dielectric disk thickness is a fraction of the wavelength, one may derive open-surface IEs either by using GBC and Green's theorems [14, 15] or via reduction from the volume IEs [16, 17].

Rapidly convergent IE algorithms can be built by using the concept of analytical regularization, which implies a transformation of the original singular IE (or a set of them) to the Fredholm second-kind matrix equation, with or without the intermediate stage of a Fredholm second-kind IE having a smooth kernel (see [18]). Theory of the Fredholm second-kind equations guarantees existence and uniqueness of exact solutions and convergence of corresponding numerical approximations. The first application of MAR to a PEC circular disk was done by Lugovoy and Sologub [19], where the dual integral equations (DIE) in the Hankel transform (HT) domain were considered instead of the singular IEs. They were transformed to a Fredholm second-kind IE and then numerically solved by the collocation algorithm (pulse basis—delta testing, in MM terminology). Alternatively, in [20] it was proposed to solve the same DIEs by using MAR in the form of specialized GM. By using the latter technique, radiation characteristics of an axisymmetrically excited PEC circular disk MSA were studied by Bliznyuk *et al.* [21]. Although in disguise, the same ideas can be found behind the later on developed weighted Chebyshev GM [22], Neumann series GM [23], and Helmholtz decomposition GM [24, 25].

For the sake of completeness, it should also be mentioned that rapidly convergent numerical solutions to electromagnetics IEs with smooth and singular kernels can be obtained using the Nystrom-type algorithms [26–28]. Such algorithms do not use partial inversion and, instead, rely on the theorems of approximation of integrals with quadratures. Their numerical performance is generally at the same level as those based on MAR although, unlike Fredholm second-kind IE, they cannot be used for obtaining analytic solutions in the form of asymptotic series in quasi-static limit.

In this chapter, the static-eigenfunction-GM version of MAR is generalized for the analysis of axisymmetrically excited circular disk resistive and thin DAs on infinite dielectric grounded substrates. Section 17.2 contains mathematical formulations. In Section 17.3, we solve singular IEs of two auxiliary problems about a resistive disk in free space excited by vertical electric and magnetic dipoles (VED and VMD) and show connections between singular IE and DIEs. Sections 17.4 and 17.5 present MAR-based analyses of the resistive MSA and thin DA, respectively. In Section 17.6, numerical results are presented followed by the conclusions.

17.2 Formulation and GBC

Figure 17.1 shows the geometry of a disk MSA or DA. A disk of the thickness b and radius a can be resistive with finite electron conductivity σ or dielectric with dielectric permittivity $\varepsilon_d\varepsilon_0$, $\varepsilon_d = \varepsilon_d' + i\varepsilon_d''$. It is placed on an infinite dielectric substrate of the thickness h and real-valued dielectric permittivity $\varepsilon_0\varepsilon$ backed with a PEC ground plane. This structure is excited by the given electric and magnetic currents, $\boldsymbol{J}^e$ and $\boldsymbol{J}^m$, which are assumed to have the exp $(i\omega t)$ time dependence, where ω is the frequency (this factor will be suppressed throughout this chapter). Free space and substrate material wavenumbers are $k_0 = \omega\sqrt{\varepsilon_0\mu_0} = 2\pi/\lambda_0$ and $k_\varepsilon = k_0\sqrt{\varepsilon_d} = 2\pi/\lambda_\varepsilon$, respectively.

The problem consists in determination of the total electromagnetic field $\{\vec{E}, \vec{H}\}$, which satisfies *inhomogeneous Maxwell's equations* with a piecewise-constant permittivity ε, *continuity conditions* for the tangential components at the substrate

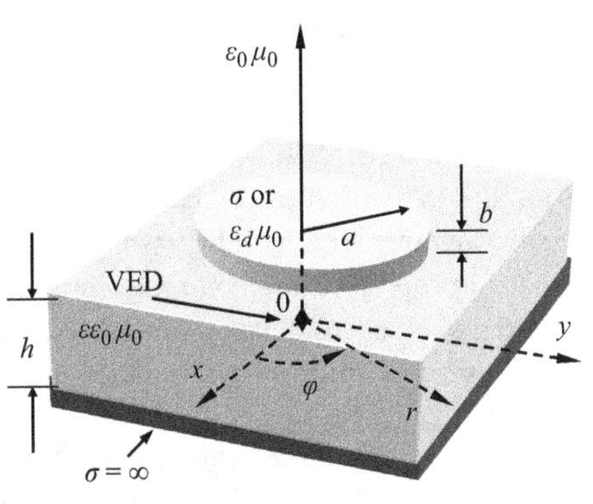

Figure 17.1 Geometry of circular disk MSA/DA axisymmetrically excited by VED

interface, *PEC boundary condition* at the ground plane, and *modified radiation condition* at $r \to \infty$ that takes into account both the spherical wave and the surface waves of the substrate [29]. Besides, the field must satisfy certain boundary conditions on the disk. As known, if a penetrable sheet is thin, $b \ll \lambda_0$, one can exclude the internal field from consideration by assuming the thickness to be zero, however imposing certain effective two-sided GBC [14, 15]. In the resistive disk case, we shall use GBC as

$$(1/2)\left[\boldsymbol{E}_t^{\varepsilon} + \boldsymbol{E}_t^0\right] = Z_0 R \boldsymbol{n} \times \left[\boldsymbol{H}_t^{\varepsilon} - \boldsymbol{H}_t^0\right], \tag{17.1}$$

$$\boldsymbol{E}_t^{\varepsilon} = \boldsymbol{E}_t^0, \tag{17.2}$$

where the indices 0 and ε refer to the free half-space and dielectric substrate, respectively, n is the unit-normal vector along the z-axis, subscript «t» means "tangential," $Z_0 = (\mu_0/\varepsilon_0)^{1/2}$ is the free space impedance, and R is the normalized electrical resistivity [14]. For instance, $R = (Z_0 b \sigma)^{-1}$ for a thinner than skin-depth metal sheet, where σ is the electron conductance. For a sheet of graphene, R is complex and both its real and imaginary parts can be taken from the Kubo formulas.

For the simulation of a thin dielectric disk, we shall use GBC in the following form [15]:

$$\left[\boldsymbol{E}_t^{\varepsilon} + \boldsymbol{E}_t^0\right] = 2 Z_0 R_d \boldsymbol{n} \times \left[\boldsymbol{H}_t^{\varepsilon} - \boldsymbol{H}_t^0\right], \tag{17.3}$$

$$\left[\boldsymbol{H}_t^{\varepsilon} + \boldsymbol{H}_t^0\right] = -2 Z_0^{-1} S_d \boldsymbol{n} \times \left[\boldsymbol{E}_t^{\varepsilon} - \boldsymbol{E}_t^0\right], \tag{17.4}$$

where R_d and S_d are the normalized electrical and magnetic resistivities, respectively. In the case of a single-layered sheet having $|\varepsilon_d| \gg 1$ and $b \ll \lambda_0$, they can be expressed as

$$R_d = \frac{i}{2} \varepsilon_d^{-1/2} \cot\left(\varepsilon_d^{1/2} \frac{k_0 b}{2}\right), \quad S_d = \frac{i}{2} \varepsilon_d^{1/2} \cot\left(\varepsilon_d^{1/2} \frac{k_0 b}{2}\right). \tag{17.5}$$

Although these GBC are derived for an infinite flat thin sheet, we shall apply them for finite disks. As remarked in the study by Bouchitte and Petit [14], using GBC is less ambiguous than using together PEC conditions and zero thickness of a screen. Besides, we shall include an appropriate *edge condition* into the problem formulation that is a request that the field power is finite in any bounded domain including domain containing the disk rim. This set of conditions guarantees the uniqueness of the disk antenna problem solution.

Note that in the resistive disk case, the edge condition entails that the normal component of the current vanishes as $(a - r)^{1/2}$ at the rim and the parallel component is finite [30].

17.3 Singular IEs and solution by MAR

17.3.1 Hyper-singular IE for a VED-excited resistive disk in free space

Consider now an auxiliary scattering problem of a resistive disk, which is placed in free space at the plane $z = 0$ and axisymmetrically excited by a VED. Decompose the total electromagnetic field $\{\boldsymbol{E}, \boldsymbol{H}\}$ into a sum of the incident field $\{\boldsymbol{E}^{in}, \boldsymbol{H}^{in}\}$ generated by the given current in the host medium without the disk, and the scattered field $\{\boldsymbol{E}^{sc}, \boldsymbol{H}^{sc}\}$ excited by the currents induced on the disk. In the axisymmetric problems, the fields can be conveniently derived via either H_φ of E_φ component. In the case considered, this enables one to look for the scattered field as a generalized double-layered potential:

$$H_\varphi(r,z) = \int_0^a j_r(r') \frac{\partial}{\partial z'} G^0(r,r',z,z') r' dr' \Big|_{z'=0}, \tag{17.6}$$

where

$$G^0(r,r',z,z') = \int_0^{2\pi} \frac{\exp(-ik_0 R_{pq})}{R_{pq}} \cos\psi d\psi, \tag{17.7}$$

$$R_{pq} = [(z-z')^2 + r^2 + r'^2 - 2rr'\cos\psi]^{1/2}, \tag{17.8}$$

$p = (r,z)$ and $q = (r',z')$ are the observation and source points, respectively, and $j_r(r)$ is the surface current to be found. Then GBC (17.1) and (17.2) lead to the following IE:

$$ik_0 R j_r(r) + \int_0^a j_r(r') \frac{\partial^2}{\partial z \partial z'} G^0(r,r',z,z') r' dr' = ik_0 Z_0^{-1} E_r^{in}(r), \tag{17.9}$$

where $z = z' = 0$ and E_r^{in} is a known function of excitation.

If $r \to r'$, the function $K(r,r') = \frac{\partial^2 G^0(r,r',z,z')}{\partial z \partial z'}\Big|_{z,z'=0}$ asymptotically behaves as

$$K(r,r') = K_{st}(r,r') + 2k_0^2[r^{-1}\ln\xi + O(1)], \tag{17.10}$$

where $\xi = r/(r-r')$, and the first, i.e., the static, term can be reduced to the elliptic functions showing that

$$K_{st}(r,r') = \frac{\partial}{\partial z}\frac{\partial}{\partial z'}\int_0^{2\pi} \frac{\cos\psi}{R_{pq}} d\psi \underset{r\to r'}{\sim} \frac{2}{r^3}\left(\xi^2 - \frac{3}{8}\ln\xi\right), \tag{17.11}$$

Thus, it is seen that IE (17.9) is hyper-singular, with the dominant singularity provided by the static part of kernel, and should be understood in Hadamard principal part sense. Hence, although it is a second-kind IE for any $R \neq 0$, this is not a Fredholm equation. As a result, convergence of its numerical solution through usual MM discretization schemes (e.g., pulse basis—delta testing) is not guaranteed, because they end up with ill-conditioned matrix equations. Therefore, we shall develop a MAR, which converts “bad” IE (17.9) into a “good” Fredholm second-kind matrix equation.

17.3.2 Eigenfunctions of the IE operator static limit for VED-excited PEC and resistive disks

In the wave scattering by PEC and imperfect thin screens, MAR is based on the analytical inversion of the static parts of IEs (see [14]). It is possible to combine regularization and discretization of the strongly singular IE like (17.9) in a single operation. To do this, the GM basis functions should be orthogonal eigenfunctions of the most singular part of the IE operator. Consider the functions defined as

$$\Phi_n(s) = \begin{cases} \varsigma_n s \mathrm{P}_n^{(1,1/2)}\left(1-2s^2\right), & s \le 1 \\ 0, & s > 1 \end{cases}, \tag{17.12}$$

where we have introduced dimensionless coordinate $s = r/a$, $n = 0, 1, \ldots$, $\varsigma_n = n!(2n+5/2)^{1/2}\Gamma^{-1}(n+3/2)$, and $\mathrm{P}_n^{(1,1/2)}(.)$ is a conventional Jacobi polynomial. In the considered case of axisymmetrically VED-excited circular disk, these functions are the eigenfunctions of static limit of singular integral operator of (17.9), with the weight $w(s) = (1-s^2)^{1/2}$. Namely,

$$\int_0^1 w(s')\Phi_n(s')K_{st}(s,s')s'ds' = \mu_n\Phi_n(s), \tag{17.13}$$

where $\mu_n = 2\Gamma(n+3/2)\Gamma(n+5/2)(n!(n+1)!)^{-1}$. The orthogonality of the functions (17.12) is given by

$$\int_0^1 w(s)\,\Phi_n(s)\,\Phi_p(s)\,sds = \mu_n^{-1}\delta_{np}, \tag{17.14}$$

where δ_{np} is the Kronecker delta. Furthermore, define direct and inverse HTs of the first order as

$$f(\kappa) = \kappa\int_0^\infty g(s)\,J_1(\kappa s)\,sds, \tag{17.15}$$

$$g(s) = \int_0^\infty f(\kappa)\,J_1(\kappa s)\,d\kappa, \tag{17.16}$$

respectively. The HT images of the weighted functions (17.12) can then be derived in the closed form as

$$\int_0^\infty w(s)\,\Phi_n(s)\,J_1(\kappa s)\,sds = \frac{\phi_n(\kappa)}{\kappa}, \tag{17.17}$$

$$\phi_n(\kappa) = \eta_n\kappa^{-1/2}J_{2n+5/2}(\kappa), \quad 0 \le \kappa < \infty, \tag{17.18}$$

$\eta_n = (4n+5)^{1/2}$, and $J_{2n+5/2}(\kappa)$ is the Bessel function. The orthogonality relation of the HTs is given by

$$\int_0^\infty \phi_n(\kappa)\,\phi_p(\kappa)\,d\kappa = \delta_{np}. \tag{17.19}$$

The functions (17.18) were first used as an HT-GM basis by Khizhnyak [20] because of the following properties: they are complete and orthogonal on the interval $0 < \kappa < \infty$, and their originals satisfy the edge condition at $s \to 1$. Now we shall see that their role is even more important as they invert the static (most singular) part of associated DIE. As can be seen from (9), zero resistivity of the disk does not change the singular part of IE operator. Thus, the discussed basis functions can be used both in PEC and resistive disk cases.

17.3.3 Matrix equation and DIE for a VED-excited disk

To discretize and partially invert IE (9), we expand the disk current in terms of the weighted functions (17.12) as

$$j_r(s) = w(s) \sum_{n=0}^{\infty} x_n \Phi_n(s), \tag{17.20}$$

where x_n are unknown coefficients. Taking advantage of (17.10) and (17.13), we test IE with each of the function $w(s)\Phi_p(s)$ and use (17.14) to come to the following matrix equation:

$$x_p + \sum_{n=0}^{\infty} x_n \tilde{A}_{np} = \tilde{B}_p, \quad p = 0, 1, 2 \ldots \infty, \tag{17.21}$$

$$A_{np} = \int_0^1 \int_0^1 ww' \Phi_p \Phi_n' \left[K\left(s, s'\right) - K_{st}\left(s, s'\right) + ik_0 aR\delta\left(s - s'\right) \right] ss' ds ds', \tag{17.22}$$

$$\tilde{B}_p = ik_0 Z_0^{-1} \int_0^1 E_r^{in}(s)\, w(s)\, \Phi_p(s)\, s ds, \tag{17.23}$$

where primed quantities depend on the primed argument. Large index behavior of the elements (17.22) is established as $O[(pn)^{1/2}(n^4 + p^4)^{-1}]$. Hence the corresponding operator is compact (i.e., it has a bounded l_2 norm: $||\tilde{A}||_{l_2}^2 = \sum_{p,n} |\tilde{A}_{pn}|^2 < \infty$). As the column vector (17.23) belongs to the same space l_2, matrix equation (17.21) is a Fredholm second-kind one.

Now we will show the connection between the hyper-singular IE and the DIEs, which are frequently used in disk scattering problems [7, 8, 10, 20–24]. By using the known representation of the axisymmetric free space Green's function,

$$G^0(s, s') = (4\pi)^{-1} \int_0^{\infty} \gamma_0^{-1}(\kappa)\, e^{\gamma_0 |z - z'|/a} J_1(\kappa s) J_1(\kappa s') \kappa d\kappa, \tag{17.24}$$

where $\gamma_0(\kappa) = (\kappa^2 - k_0^2 a^2)^{1/2}$, we obtain that

$$K(s, s') = \int_0^{\infty} \gamma_0(\kappa) J_1(\kappa s) J_1(\kappa s') \kappa d\kappa, \tag{17.25}$$

On continuing $j_r(s)$ by zero value out of the disk region, the substitution of (17.20) and (17.17), and (17.25) into (17.9) and integration with respect to r' over the whole domain yields DIE as

$$\begin{cases} \int_0^\infty X(\kappa)\left[\gamma_0(\kappa) + ik_0 aR\right] J_1(\kappa s)\, d\kappa = ik_0 Z_0^{-1} E_r^{in}(s), \ s \le 1 \\ \int_0^\infty X(\kappa) J_1(\kappa s)\, d\kappa = 0, \ s > 1 \end{cases} \quad . \tag{17.26}$$

Here $X(\kappa)$ is the HT of the "extended" $j_r(r)$, therefore

$$X(\kappa) = \sum_{n=0}^{\infty} x_n \phi_n(\kappa) \tag{17.27}$$

and we apply the operator $\int_0^\infty w(s)\,\Phi_p(s)\, s ds$ to both sides of (17.26). The second equation in (17.26), due to the properties of basis functions, is satisfied automatically [31]. As for the first equation, its *static limit* is solved explicitly. Indeed, a study of the function $\gamma_0(\kappa)$ behavior if $\kappa \to \infty$ shows that

$$\gamma_0(\kappa) \underset{\kappa\to\infty}{=} \kappa - 2^{-1}k_0^2 a^2 \kappa^{-1} + O(k_0^4 a^4 \kappa^{-3}). \tag{17.28}$$

Then, introducing the function $\Omega(\kappa) = \gamma_0(\kappa) - \kappa$, and taking advantage of (17.17) and (17.19), we arrive at the second-kind matrix equation (17.21), where now

$$\tilde{A}_{pn} = \int_0^\infty \left[\Omega(\kappa) + ik_0 aR\right] \phi_p(\kappa)\, \phi_n(\kappa)\, \kappa^{-1} d\kappa, \tag{17.29}$$

The equivalency between (17.22) and (17.29) follows from identity

$$K(s,s') - K_{st}(s,s') = \int_0^\infty \kappa\Omega(\kappa) J_1(\kappa s) J_1(\kappa s')\, d\kappa, \tag{17.30}$$

which enables obtaining one from another without HT. The same property holds for the right-hand parts.

Thus either way of building (17.21) requests analytical inversion of the static part of the corresponding IE. Here, to link MAR technique with the methods of Losada and others [22–25], we note that the Jacobi polynomial $R_n^{(1,1/2)}(1-2s^2)$ reduces to the Chebyshev one, $U_{2n+1}(s)$, and the series like (17.27) in terms of functions (17.18) are called the Neumann series.

17.3.4 Log-singular IE for a VMD-excited resistive disk in free space

Consider now another auxiliary problem of a resistive disk in free space axisymmetrically excited by a VMD. Then the scattered field is sought as a single-layered potential:

$$E_\varphi(r,z) = \int_0^a j_\varphi(r') G^0(r,r',z,0) r' dr', \tag{17.31}$$

and GBC (17.3) and (17.4) lead to the following IE for the unknown disk current, $j_\varphi(r)$:

$$ik_0^{-1}Rj_\varphi(r) + \int_0^a j_\varphi(r')G^0(r,r',0,0)r'dr' = ik_0^{-1}Z_0^{-1}E_\varphi^{in}(r), \tag{17.32}$$

One can verify that the kernel of (17.32) has a logarithmic singularity if $r \to r'$. Thus, for any $R \neq 0$ this is a Fredholm second-kind IE, which does not need an analytical regularization and can be discretized with any set of basis functions that are complete on $r \leq a$. Here, one can take the functions that satisfy the edge condition and whose HTs (17.15) and (17.16) can be found analytically.

The basis functions that satisfy these requirements in terms of dimensionless coordinate s are

$$\Psi_n(s) = \begin{cases} \chi_n s P_n^{(1,0)}\left(1-2s^2\right), & s \leq 1 \\ 0, & s > 1 \end{cases}, \tag{17.33}$$

where $n = 0, 1, \ldots, \chi_n = 2(n+1)^{1/2}$, $P_n^{(1,0)}(.)$ is the Jacobi polynomial. The corresponding HT image is

$$\psi_n(\kappa) = \chi_n J_{2n+2}(\kappa), \quad 0 \leq \kappa < \infty. \tag{17.34}$$

Hence, we expand the current and its HT image as

$$j_\varphi(s) = \sum_{n=0}^{\infty} y_n \Psi_n(s), \quad Y(\kappa) = \sum_{n=0}^{\infty} y_n \psi_n(\kappa), \tag{17.35}$$

respectively, and obtain DIEs for the VMD-excited disk as

$$\begin{cases} \displaystyle\int_0^\infty Y(\kappa)\left[\gamma_0^{-1}(\kappa) + i(k_0a)^{-1}R\right]J_1(\kappa s)\,d\kappa = i(k_0a)^{-1}Z_0^{-1}E_r^{in}(s), \; s \leq 1 \\ \displaystyle\int_0^\infty Y(\kappa)J_1(\kappa s)\,d\kappa = 0, \; s > 1 \end{cases}. \tag{17.36}$$

On discretizing (17.32) or (17.36) with the aid of (17.34) in the GM scheme, we obtain an infinite matrix equation as

$$y_p + \sum_{n=0}^{\infty} y_n \widehat{A}_{np} = \widehat{B}_p, \quad p = 0, 1, 2 \ldots \infty. \tag{17.37}$$

Similarly to the VED case, here the matrix elements can be written in two equivalent ways:

$$\widehat{A}_{np} = -ik_0aR^{-1}\int_0^1\int_0^1 \Psi_p(s)\,\Psi_n(s')G^0(s,s')\,ss'dsds' \tag{17.38}$$

$$\widehat{A}_{np} = -ik_0aR^{-1}\int_0^\infty \gamma_0^{-1}(\kappa)\,\kappa^{-1}\psi_p(\kappa)\,\psi_n(\kappa)\,d\kappa, \tag{17.39}$$

and the right-hand part column vector is determined by the excitation. Large index behavior of the matrix elements is established as $O[(pn)^{1/2}(n^2+p^2)^{-1}]$ that leads to the conclusion of compactness of corresponding operator.

17.4 Resistive disk MSA excited by VED

17.4.1 Dual IEs for a resistive disk MSA

Consider now a resistive disk MSA shown in Figure 17.1, i.e., excited by a VED located on the ground plane below the disk center. The current of such a feed is assumed to be

$$J_z^e(r',\varphi',z') = I^e l\delta(r')\delta(\varphi')r'^{-1}\delta(z'), \tag{17.40}$$

where $I^e l$ is the dipole moment and $\delta(.)$ is the Dirac delta.

Due to the earlier discussed equivalence, we are not obliged to derive a singular IE of antenna and thus build a complicated Green's function. Instead, tangential components of the electric and magnetic fields can be sought as inverse HT integrals similar to (17.2)–(17.5) by Bliznyuk *et al.* [21]. By enforcing GBC (17.1) and (17.2) on the disk, continuity conditions on the substrate interface, and the PEC boundary condition on the ground plane, we derive DIEs for the radial current on the resistive disk MSA:

$$\begin{cases} \displaystyle\int_0^\infty X(\kappa)\,\beta_{\mathrm{R}}(\kappa,k_0,\varepsilon,h,a,R)\,J_1(\kappa s)\,d\kappa = \Pi(s), \quad s\le 1 \\ \displaystyle\int_0^\infty X(\kappa)\,J_1(\kappa s)\,d\kappa = 0, s>1 \end{cases}, \tag{17.41}$$

where $X(\kappa)$ is unknown HT amplitude of the current. The spectral weight function and excitation term are, respectively,

$$\beta_R(\kappa) = \gamma_0\gamma_\varepsilon \sinh(\gamma_\varepsilon h/a)D^{-1}(\kappa) + ik_0aR; \tag{17.42}$$

$$\Pi(s) = v_0\int_0^\infty \gamma_0(\kappa)\,D^{-1}(\kappa)\,J_1(\kappa s)\,\kappa^2 d\kappa, \tag{17.43}$$

$$D(\kappa) = \gamma_\varepsilon \sinh(\gamma_\varepsilon h/a) + \varepsilon\gamma_0\cosh(\gamma_\varepsilon h/a), \tag{17.44}$$

where $v_0 = I^e l(2\pi)^{-1}$ and $\gamma_\varepsilon = (\kappa^2 - k_0^2a^2\varepsilon)^{1/2}$. Note that if $R=0$, DIE (17.41) reduces to the PEC case ones of [7, 21].

17.4.2 Matrix equation for a resistive disk MSA

To transform DIEs (17.41) into an infinite matrix equation of the Fredholm second kind, we must inspect large κ behavior of the function $\beta_R(\kappa)$. Expression (17.42) shows:

$$\beta_R(\kappa) = \zeta_0\kappa\left(1-2e^{-2\kappa\frac{h}{a}}\right) + ik_0aR + \zeta_1\frac{k_0^2a^2}{\kappa} + O\left(\frac{k_0^4a^4}{\kappa^3},\ \kappa e^{-4\kappa\frac{h}{a}}\right), \tag{17.45}$$

where $\zeta_0 = (1+\varepsilon)^{-1}$ and $\zeta_1 = 2^{-1}\zeta_0(1+\varepsilon^2)$. Here it is seen that only the term corresponding to the *static limit of the single disk case* grows up as $O(\kappa)$, that is a singular behavior in HT sense. Furthermore, we introduce the function

$$\Omega_R(\kappa) = \zeta_0^{-1}\kappa^{-1}[\gamma_0\gamma_\varepsilon \sinh(\gamma_\varepsilon h/a)D^{-1}(\kappa) - \kappa] \tag{17.46}$$

and observe that $\beta_R(\kappa) = \zeta_0\kappa[1 + \Omega_R(\kappa) + ik_0aR]$. Finally, we expand the function $X(\kappa)$ similar to (17.27) and retrace all the operations of the previous section to obtain the following equations:

$$x_p + \sum_{n=0}^{\infty} x_n A_{pn} = B_p, \quad p = 0, 1, 2, \ldots \tag{17.47}$$

$$A_{pn} = \int_0^\infty [\Omega(\kappa) + ik_0aR\kappa^{-1}]\phi_n(\kappa)\phi_p(\kappa)\,d\kappa, \tag{17.48}$$

$$B_p = v_0\zeta_0^{-1}\int_0^\infty D^{-1}(\kappa)\gamma_0(\kappa)\phi_p(\kappa)\,\kappa d\kappa. \tag{17.49}$$

One can easily show that the matrix operator with elements (17.48) is compact, and the column vector (17.49) belongs to the space l_2. Therefore, (17.47) is a Fredholm second-kind equation, which can be truncated and efficiently solved numerically. Figure 17.2(a) demonstrates the behavior of the normalized truncation error, by the norm in l_2,

$$\delta(N) = \left(\sum_{n=0}^{N}\left|x_n^N - x_n^{N-1}\right|^2\right)^{1/2}\left(\sum_{n=0}^{N}\left|x_n^N\right|^2\right)^{-1/2}, \tag{17.50}$$

as a function of the matrix truncation number N. Provided that all intermediate operations have been done with superior accuracy, $\delta(N)$ starts rapid fall if $N > k_0\varepsilon^{1/2}a$ and

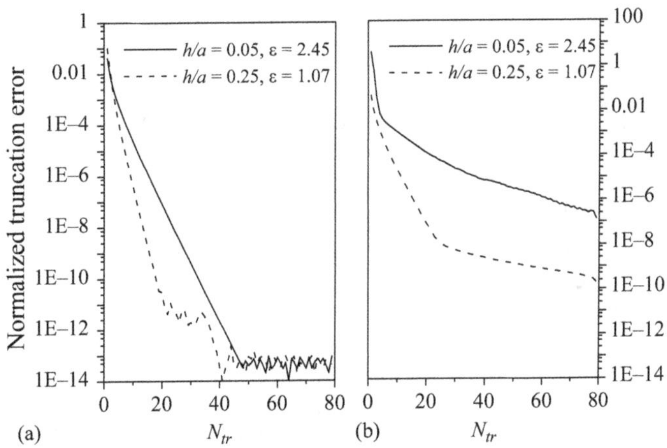

Figure 17.2 Truncation error at $a/\lambda_0 = 0.5$ for (a) resistive disk MSA, $R = 0.05$ and (b) thin disk DA. $\varepsilon_d = 152$, $\tan\delta = 0.001$, $b/a = 0.001$

reaches machine precision with small N (the smaller a/h and R the better). This is something impossible for conventional MM algorithms, which hit a "numerical wall" after a few correct digits.

Additionally, there is a way to reduce the time of filling in the matrix (17.47). First, a recurrence relation [20, 21] allows computing only $3N$ matrix elements (48) instead of N^2. Second, by integrating explicitly a few first inverse power κ terms of $\Omega_R(\kappa)$ with the aid of the Weber–Shafheitlin formulas [32], we end up with numerical integration of the functions that decay as κ^{-8} if $\kappa \to \infty$ (see [21] for details).

17.5 Thin disk DA excited by VED

17.5.1 Coupled set of DIEs for a thin disk DA

Consider now the case of a thin disk DA. Using GBC (17.3) and (17.4) and following the same scheme, we obtain a coupled set of DIEs:

$$\begin{cases} \int_0^\infty \left[X(\kappa)\beta_R^d(\kappa) + Y(\kappa)\sigma(\kappa)\right] J_1(\kappa s)\,d\kappa = \Pi_x(s), s \le 1 \\ \int_0^\infty X(\kappa) J_1(\kappa s)\,d\kappa = 0, s > 1 \end{cases} \tag{17.51}$$

$$\begin{cases} \int_0^\infty \left[Y(\kappa)\beta_S^d(\kappa) + X(\kappa)\sigma(\kappa)\right] J_1(\kappa s)\,d\kappa = \Pi_y(s), s \le 1 \\ \int_0^\infty Y(\kappa) J_1(\kappa s)\,d\kappa = 0, s > 1 \end{cases} \tag{17.52}$$

where the function Π_x is the same as in (17.43), and

$$\beta_R^d(\kappa) = \gamma_0\gamma_\varepsilon \sinh(\gamma_\varepsilon h/a) D^{-1}(\kappa) - ik_0 a R_d, \tag{17.53}$$

$$\beta_S^d(\kappa) = \Omega_S(\kappa) + iS_d(k_0 a)^{-1}, \tag{17.54}$$

$$\Omega_S(\kappa) = \varepsilon \cosh(\gamma_\varepsilon h/a) D^{-1}(\kappa), \tag{17.55}$$

$$\sigma(\kappa) = 2^{-1}[\gamma_\varepsilon \sinh(\gamma_\varepsilon h/a) - \gamma_0\varepsilon \cosh(\gamma_\varepsilon h/a)] D^{-1}(\kappa), \tag{17.56}$$

$$\Pi_y(s) = -v_0 \int_0^\infty D^{-1}(\kappa) J_1(\kappa s)\kappa^2 d\kappa. \tag{17.57}$$

Unlike $\beta_R^d(\kappa)$, the expansion of $\beta_S^d(\kappa)$ at $\kappa \to \infty$ starts from a constant that shows non-singular nature of DIEs (17.52). One can also notice that dielectric disk DIEs reduce to resistive disk ones if $S_d^{-1} = 0$ that entails $Y(\kappa) = 0$.

17.5.2 Matrix equation for a thin disk DA

To reduce (17.51) and (17.52) to the Fredholm second-kind matrix equation by means of MAR combined with GM, we should find appropriate sets of basis functions. Analysis shows that (17.51) has the same features as DIEs for the VED-excited resistive disk, and (17.52) is similar to the VMD case. Then it is clear that $X(\kappa)$and $Y(\kappa)$ should be expanded in the Neumann series of functions (17.18) and (17.34), respectively. This leads to the following block-type (2 × 2) matrix equation:

$$\begin{cases} x_p + \sum_{n=0}^{\infty} x_n A_{np} + \sum_{n=0}^{\infty} y_n F_{np}^x = B_p^x, & p = 0, 1, 2, \ldots, \infty \\ y_p + \sum_{n=0}^{\infty} x_n F_{np}^y + \sum_{n=0}^{\infty} y_n C_{np} = B_p^y, & p = 0, 1, 2, \ldots, \infty \end{cases} \tag{17.58}$$

where x_n and y_n are unknowns, A_{pn} and B_p^x are determined by (17.48) where R is replaced with R_d and (17.49), respectively,

$$F_{np}^x = (\varepsilon + 1) \int_0^{\infty} \sigma(\kappa)\, \phi_p(\kappa)\, \psi_n(\kappa)\, \kappa^{-1} d\kappa, \tag{17.59}$$

$$F_{np}^y = \tau (1 + \varepsilon)^{-1} F_{np}^y, \quad \tau = ik_0 a S_d^{-1} \tag{17.60}$$

$$C_{np} = -\tau \int_0^{\infty} \Omega_S(\kappa)\, \psi_p(\kappa)\, \psi_n(\kappa)\, \kappa^{-1} d\kappa, \tag{17.61}$$

$$B_p^y = v_0 \tau \int_0^{\infty} D^{-1}(\kappa)\, \psi_p(\kappa)\, \kappa d\kappa. \tag{17.62}$$

The matrices of each block of (17.58) have bounded l_2 norms, and right-hand parts belong to l_2. Hence, each block can be truncated to the order N and then (17.58) can be solved in the same way as for a resistive disk. Dashed curves in Figure 17.2(b) show truncation error, in the sense of the norm of the space $l_2^2 = l_2 \otimes l_2$, as a function of N. As seen, for DA the error starts a fall down if $N > k_0 \varepsilon^{1/2} a$ with the rate determined by h/a and the largest of the values R_d and S_d^{-1}.

17.6 Numerical results

The formulas for computing the radiation characteristics of resistive MSA and thin disk DA are similar to those for PEC MSA [21]. The only difference is the presence of losses in imperfect disk. Therefore, the power absorbed in a thin DA is

$$\frac{P_{abs}}{P_0} = -\frac{6\pi^2}{k_0^2 a^2} \left[ReR_d \int_0^{\infty} |X(\kappa)|^2 \frac{d\kappa}{\kappa} + \frac{ReS_d}{k_0^2 a^2} \int_0^{\infty} |Y(\kappa)|^2 \frac{d\kappa}{\kappa} \right], \tag{17.63}$$

where $P_0 = 2\pi v_0^2 Z_0 k_0^2 / 3$ is the power radiated by a VED into free half space. For a resistive disk MSA, the second term in the brackets is absent. Throughout the computations, we kept the integration and matrix truncation errors to be lower than 10^{-8} and 10^{-5}, respectively, and controlled the power balance (i.e., Poynting theorem) to be satisfied at the 10^{-8} level.

17.6.1 Radiation characteristics of resistive MSA

The plots in Figures 17.3–17.6 show electromagnetic performance of a resistive MSA on a thin substrate for different disk resistivities. The frequency dependences of the normalized space and surface-wave powers are depicted in Figures 17.3 and 17.4.

Note that in the studied frequency band only the principal surface wave of the substrate can propagate. As we can see, behavior of the far-field characteristics is generally similar to the PEC MSA. So far as resistivity is small, sharp maxima are observed at resonant frequencies. Additionally, the surface-wave power has sharp

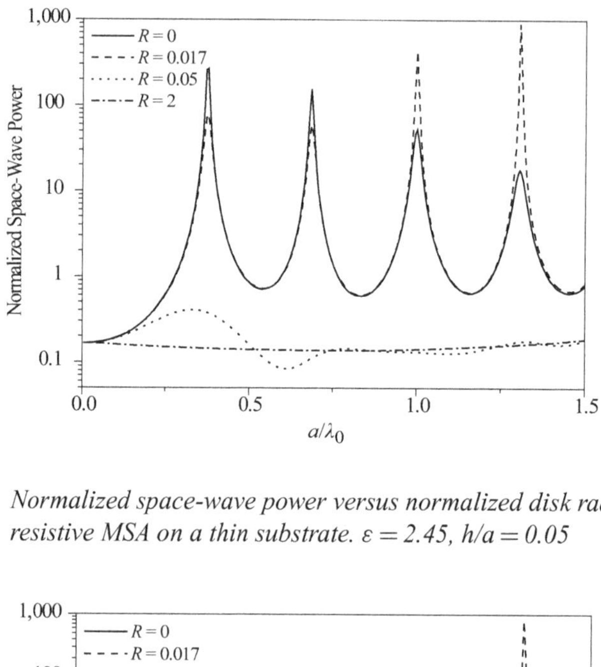

Figure 17.3 Normalized space-wave power versus normalized disk radius for a resistive MSA on a thin substrate. ε = 2.45, h/a = 0.05

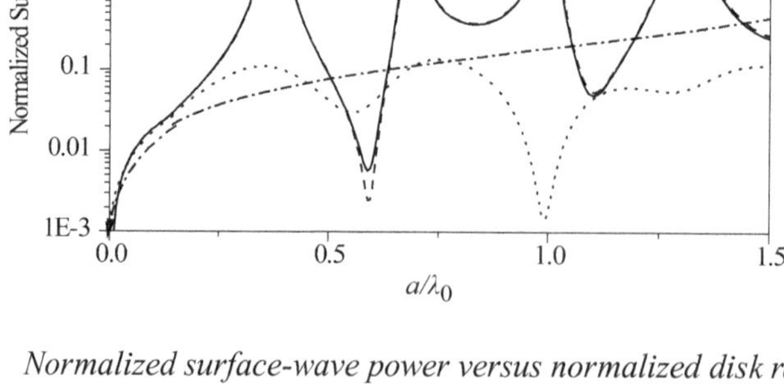

Figure 17.4 Normalized surface-wave power versus normalized disk radius for a resistive MSA on a thin substrate. ε = 2.45, h/a = 0.05

minima if the so-called surface-wave suppression condition is satisfied [33]. If the disk material is a poor conductor, the power curves get smoother.

A new feature is the absorption in resistive MSA that has maxima in the resonances (Figure 17.5). Due to normalization applied the curves go down if the frequency is increased, although the absorption itself goes up. In the low frequency limit, P_{abs} falls down less rapidly than the normalization value in (17.63) (i.e., P_0) and therefore the curves tend to infinity.

Eventually, Figure 17.6 demonstrates a comparison of radiation efficiencies of the PEC and resistive MSA built on a thin–dense substrate. The main effect of $R \neq 0$

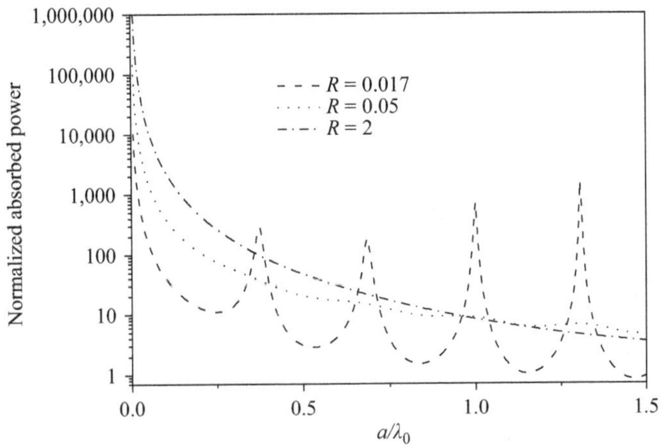

Figure 17.5 Normalized absorbed power versus normalized disk radius for a resistive MSA on the thin substrate. $\varepsilon = 2.45$*,* $h/a = 0.05$

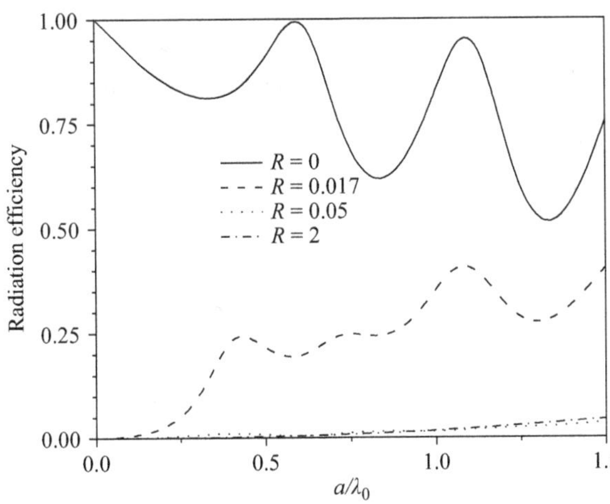

Figure 17.6 Radiation efficiency versus normalized disk radius for a resistive MA built on a thin substrate. $\varepsilon = 2.45$*,* $h/a = 0.05$

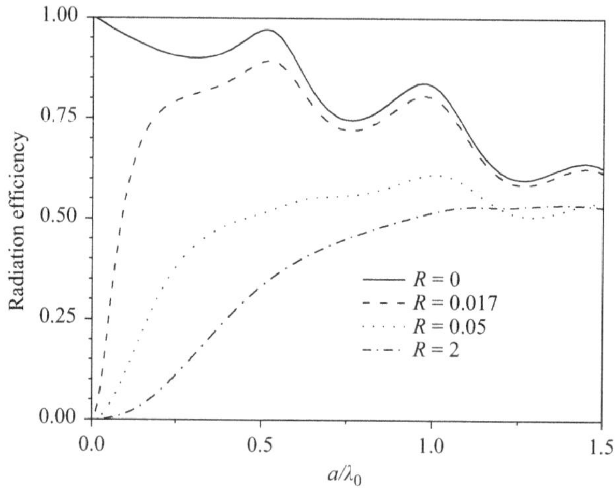

Figure 17.7 Radiation efficiency versus normalized disk radius for a resistive MSA on a thick foam substrate. $\varepsilon = 1.07$, $h/a = 0.25$

Table 17.1 1-st Resonance of Resistive MSA: Bandwidth and Space-Wave Power

MA	Resistivity	Bandwidth, %	Normalized Power
$\varepsilon = 1.07$	0	20	8.677
$h/a = 0.25$	0.017	19	10.386
$\varepsilon = 2.45$	0	4	266.89
$h/a = 0.05$	0.017	5.3	77.93

is that the presence of even small losses causes the radiation efficiency depression at $a/\lambda_0 \rightarrow 0$. Such a behavior is in contrast to the PEC MSA models, where efficiency of radiation tends to 100% at zero frequency. Figure 17.7 illustrates the same quantity for a resistive MSA on a thicker low permittivity substrate, which is attractive in applications due to reduced surface-wave losses.

Note that here the radiation efficiency is generally greater than in the case of a thin high permittivity substrate, although the power radiated in the first resonance is considerably lower. This fact is illustrated in Table 17.1. As it is seen, replacing PEC disk with a thin partially transparent one made from imperfect metal can result in either increasing the space-wave power or broadening the bandwidth.

17.6.2 Radiation characteristics of thin disk DA

Figures 17.8–17.11 demonstrate the performance of thin disk DAs made of very high ε material and built on a thin–dense substrate and a thick foam-like one, respectively. Comparison of Figures 17.8 and 17.3–17.5 shows that the structure of the resonances

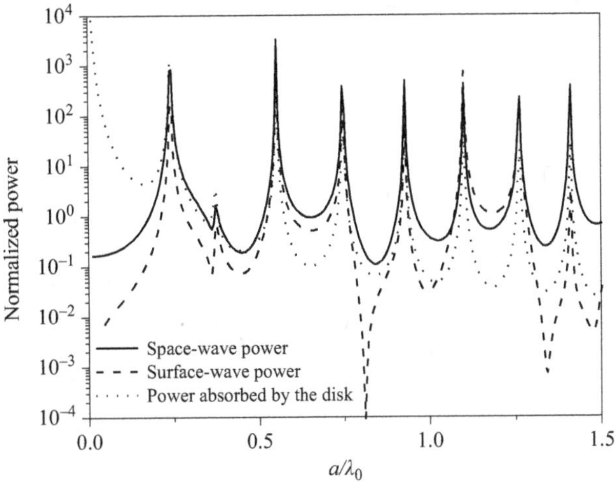

Figure 17.8 Power characteristics versus normalized disk radius for DA on a thin substrate. $\varepsilon = 2.45$, $h/a = 0.05$, $\varepsilon_d = 152$, $\tan\delta = 0.001$, $b/a = 0.01$

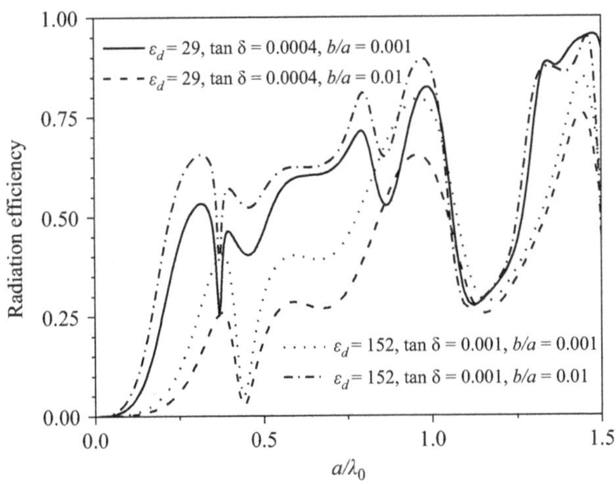

Figure 17.9 Radiation efficiency versus normalized disk radius for DA on a thin substrate. $\varepsilon = 2.45$, $h/a = 0.05$

changes considerably, with new resonances brought by a very dense dielectric disk. In resonances, the radiated power is higher than for a metal MSA, apparently due to the partial transparency. What is remarkable, the radiation efficiency does not gradually fall down with increasing frequency but displays bands of high values (Figure 17.9). In the case of a foam-like substrate material, very high permittivity

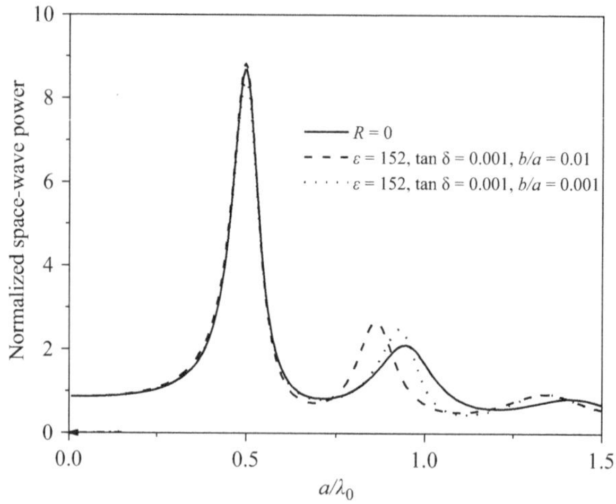

Figure 17.10 Power characteristics versus normalized disk radius for DA on a thick foam substrate. $\varepsilon = 1.07$, $h/a = 0.25$

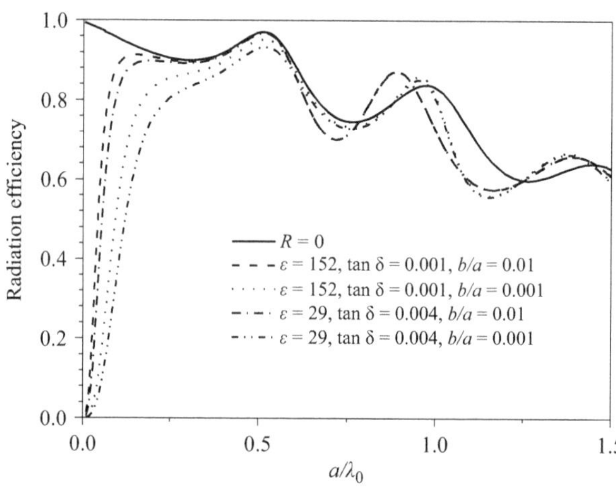

Figure 17.11 Radiation efficiency versus normalized disk radius for DA on a thick foam substrate. $\varepsilon = 1.07$, $h/a = 0.25$

disk DA behaves very similar to the resistive MSA—compare Figures 17.10 and 17.11 with Figures 17.3–17.5. Similarly to resistive MSA, a DA on a low permittivity substrate has generally higher values of radiation efficiency. What is remarkable, this improvement can be achieved together with a rise in radiated power which is not possible with metallic MSA. Table 17.2 presents characteristics of thin disk DA in the first resonance.

Table 17.2 1-st Resonance of DA: Bandwidth and Space-Wave Power

DA, $\varepsilon_d = 152$	Disk	Bandwidth, %	Normalized Power
$\varepsilon = 1.07$	$b/a = 0.001$	20	8.42
$h/a = 0.25$	$b/a = 0.01$	19.7	8.83
$\varepsilon = 2.45$	$b/a = 0.001$	4.2	98.29
$h/a = 0.05$	$b/a = 0.01$	4.2	816.65

17.7 Conclusions

We have demonstrated that MAR technique based on the analytical inversion of the static part of the isolated disk problem can be successfully applied to the imperfect disk antenna analysis. This implies the use of the generalized two-sided boundary conditions, which lead to the hyper-singular or logarithmic singular IEs or, equivalently, to the dual IEs in the HT domain. Regularization of SIEs is done through the Galerkin method diagonalization with the static case eigenfunctions or by the inverse HT, depending on the IE properties. This procedure can also be considered as analytical preconditioning and makes the numerical analysis very economic and accurate. We have quantified interesting features of imperfect antennas: broadening of the bandwidth and improvement of radiation efficiency, with the latter value falling to zero in the low frequency limit.

References

[1] Pozar DM and Shaubert DH. (Eds.), *Microstrip Antennas*. New York: IEEE Press, 1995.
[2] Huang J, Circularly polarized conical patterns from circular microstrip antennas. *IEEE Trans. Antennas Propagat.* 1984;32(9):991–994.
[3] Economou L and Langley RJ. Circular microstrip patch antennas on glass for vehicle applications. *IEE Proc. Microw. Antennas Propagat.* 1998;145(5): 418–420.
[4] Huang CY. Microstrip antenna with lossy patch. *Microwave Opt. Technology Lett.* 1998;15(2):228–230.
[5] Kishk AA, Zanoubi MR, and Kajfez D. A numerical study of a dielectric disk antenna above grounded dielectric surface. *IEEE Trans. Antennas Propagat.* 1993;41(6):813–821.
[6] Itoh T and Mittra R. Analysis of a microstrip disk resonator. *Arch. Elektron. Ubertrat. Techn.* 1973;27(11):456–458.
[7] Chew WC and Kong JA. Resonances of the axial-symmetric modes in microstrip disk resonators, *J. Math. Phys.* 1980;21(6):582–591.
[8] Chew WC and Kong JA. Analysis of a circular microstrip antenna with a thick dielectric substrate, *IEEE Trans. Antennas Propagat.* 1981;29(1):68–76.

[9] Yano S and Ishimaru A. A theoretical study of the input impedance of a circular microstrip disk antenna. *IEEE Trans. Antennas Propagat.* 1981;29(1):77–83.

[10] Itoh T and Araki K. Hankel transform domain analysis of open circular microstrip radiation structures. *IEEE Trans. Antennas Propagat.* 1981;29(1):84–89.

[11] Mosig J, Hall RL, and Gardiol F. Numerical analysis of microstrip patch antennas. Chap. 8 in James JR, Hall PS (Eds.), *Handbook of Microstrip Antennas*. London: Peter Peregrinus Publ., 1989.

[12] Dudley DG. Error minimization and convergence in numerical methods. *Electromagnetics.* 1985;5(2–3):89–97.

[13] Hower GL, Olsen RG, Earls JD, and Schneider JB. Inaccuracies in numerical calculation of scattering near natural frequencies of penetrable objects. *IEEE Trans. Antennas Propagat.* 1993;41(7):982–986.

[14] Bouchitte G and Petit R. On the concepts of a perfectly conducting material and of a perfectly conducting and infinitely thin screen. *Radio Science.* 1989;24(1):13–26.

[15] Bleszynski E, Bleszynski M, and Jaroszewicz T. Surface integral equations for electromagnetic scattering from impenetrable and penetrable sheets. *IEEE Antennas Propagat. Mag.* 1993;35(6):14–25.

[16] Koh IS and Sarabandi K. A new approximate solution for scattering by thin disks of arbitrary shape and size. *IEEE Trans. Antennas. Propagat.* 2005;53(6):1920–1926.

[17] Chiang IT and Chew WC. Thin dielectric sheet simulation by surface integral equation using modified RWG and pulse bases. *IEEE Trans. Antennas. Propagat.* 2006;54(7):1927–1934.

[18] Nosich AI. Method of analytical regularization in the wave-scattering and eingenvalue problems: foundations and review of solutions. *IEEE Antennas Propagat Mag.* 1999;41(3):34–49.

[19] Dunai BA, Lugovoi AV, and Sologub VG. Scattering of electromagnetic waves by a disk at the boundary of a dielectric half-space. *Sov. Phys. Tech. Phys.* 1975;20:1396–1398.

[20] Khizhnyak AN. Plane wave diffraction by a circular disk. *Sov. Phys. Acoustics.* 1989;35:539–541.

[21] Bliznyuk NY, Nosich AI, and Khizhnyak AN. Accurate computation of a circular-disk antenna axisymmetrically excited by an electric dipole. *Microwave Opt. Technology Lett.* 2000;25(3):211–216.

[22] Losada V, Boix RR, and Horno M. Resonant modes of circular microstrip patches on multilayered substrates. *IEEE Trans. Microwave Theory Techniq.* 1999;47(4):488–498.

[23] Di Murro F, Lucido M, Panariello G, and Schettino F. Guaranteed-convergence method of analysis of the scattering by an arbitrarily oriented zero-thickness PEC disk buried in a lossy half-space. *IEEE Trans. Antennas Propagat.* 2015;63(8):3610–3620.

[24] Lucido M, Schettino F, and Panariello G. Scattering by a zero-thickness PEC disk: a new analytically regularizing procedure based on Helmholtz decomposition and Galerkin method. *Radio Sci.* 2017;52(1):2–14.

[25] Lucido M, Schettino F, Migliore MD, Pinchera D, Di Murro F, and Panariello G. Electromagnetic scattering by a zero-thickness PEC annular ring: a new highly efficient MoM solution. *J. Electromagn. Waves Applic.* 2017;31(4):405–416.

[26] Bulygin VS, Gandel YV, and Nosich AI. Nystrom-type method in three-dimensional electromagnetic diffraction by a finite PEC rotationally symmetric surface. *IEEE Trans. Antennas Propagat.* 2012;60(11):4710–4718.

[27] Bulygin VS, Gandel YV, Benson TM, and Nosich AI. Full-wave analysis and optimization of a TARA-like shield-assisted paraboloidal reflector antenna using a Nystrom-type method. *IEEE Trans. Antennas Propagat.* 2013;61(10):4981–4989.

[28] Tsalamengas JL. Quadrature rules for weakly singular, strongly singular, and hypersingular integrals in boundary integral equation methods. *J. Computat. Phys.* 2015;303(12):498–513.

[29] Nosich AI. Radiation conditions, limiting absorption principle, and general relations in open waveguide scattering. *J. Electromagn. Waves Applicat.* 1994;8(3):329–353.

[30] Braver IM, Fridberg PS, Garb KL, and Yakover IM. The behavior of the electromagnetic field near the edge of a resistive half-plane. *IEEE Trans. Antennas Propagat.* 1988;36(12):1760–1768.

[31] Gradstein IS and Ryzhik IM. *Tables of Integrals, Series and Products*. London: Academic Press, 1994.

[32] Abramovitz M and Stegan IA. *Handbook of Mathematical Functions*. New York: Dover Publ., 1979.

[33] Jackson D, Williams JT, Bhattacharyya AK, Smith RL, Buchheit SJ, and Long SA, Microstrip patch designs that do not excite surface waves. *IEEE Trans. Antennas Propagat.* 1993;41(10):1026–1037.

Chapter 18

Scattering and guiding problems of electromagnetic waves in inhomogeneous media by improved Fourier series expansion method

Tsuneki Yamasaki[1]

Scattering and guiding problems of electromagnetic waves in inhomogeneous media such as photonic crystals are very interesting in many areas of physics and engineering in nanoscale optical technology. However, the conventional method that is based on the Fourier series expansion method in inhomogeneous media cannot be applied to profiles with discontinuity. Our new method that is called IFEM (improved Fourier series expansion method) can be applied to profiles with discontinuity for scattering and guiding problems. The major process of IFEM is as follows:

1. The inhomogeneous region is approximated by an assembly of stratified thin layers with step profile.
2. The electromagnetic fields in each layer are expanded approximately by a finite Fourier series.
3. All layers are combined with appropriate boundary conditions to get the electromagnetic fields in the inhomogeneous media.

The range of applicability to inhomogeneous media is much wider than that of conventional methods.

In this chapter, first we introduce the IFEM, the convergence theory of the modal truncation number N and Fourier truncation number N_f and explain the methods of numerical computation in detail for the scattering and guiding of electromagnetic waves in inhomogeneous dielectric gratings with periodic surface relief, using the IFEM and multilayer method (MLM) for both TM and TE cases. Next we discuss scattering and guiding problems for several inhomogeneous media as follows: (1) slanted layer and rhombic media with strips; (2) elliptically layered, columnar and rectangular media; (3) energy distribution of defect layers; and (4) mixed positive and negative media.

[1]College of Science and Technology, Nihon University, Tokyo, Japan

18.1 Introduction

Dielectric gratings are now widely used in integrated optics [1,2], such as in optical gratings, optical couplers, optical waveguide filters, and are used as frequency-selective surfaces. They are generally divided into two types [3]: (1) surface-relief type and (2) modulated refractive index type. For the surface-relief type, various analytical and numerical methods that are applicable to arbitrarily shaped gratings for the scattering and guiding problem have been proposed [4–16], because the thickness of the grating affects the accuracy of analysis rather than that of modulated refractive index type.

For the modulated refractive index type, the method based on the space-harmonics expansion method [17] cannot be applied to the grating, the profiles of which have discontinuities. We pointed out this issue and gave an improved method that is based on the Fourier series expansion method [18]. Our IFEM can also be applied to surface-relief type [19] by approximating it as an assembly of stratified thin modulated refractive index layers; the order of the matrix size for the simultaneous equation system depends on the modal truncation number, but it does not depend on the number of layers. Therefore, our method is also especially effective in the case of the planar slanted grating for scattering [20] and guiding problems [21], because the electromagnetic fields in each layer can easily be obtained by shifting the solution in the first layer.

Recently, the refractive index can easily be controlled to make periodic structures by the developments in manufacturing technology of optical devices. Accordingly, the numerical methods that are applicable to inhomogeneous dielectric gratings have been proposed [22–24].

In this chapter, first we introduce the formulation of IFEM for the scattering [25] and guiding [26] of electromagnetic waves in inhomogeneous dielectric gratings with periodic surface relief, the interior distribution of permittivity of which is of modulated type in the direction of depth, and discuss the convergence theory of the modal truncation number N and Fourier truncation number N_f for TM waves. Next we explain scattering and guiding problems for several inhomogeneous media as follows: (1) slanted layer and rhombic media with strips; (2) elliptically layered, columnar and rectangular media; (3) energy distribution of defect layers; and (4) mixed positive and negative media.

18.1.1 Formulation

We consider the inhomogeneous media with periodic surface-relief grating as shown in Figure 18.1(a). The grating is uniform in the y-direction, and the permittivity $\varepsilon(x, z)$ is an arbitrary periodic function of z with period p in the surface $s(z)$. Figure 18.1(a) shows the configuration, the surface profile of which is triangular and parameterized by the position (w) along the z-direction, and the interior distribution of permittivity of which is sinusoidally modulated in the direction of grating depth (d). The permeability is assumed to be μ_0. The time dependence is $\exp(-i\omega t)$ and is suppressed throughout. In the formulation, the TM (the magnetic field has only the y-component) is discussed.

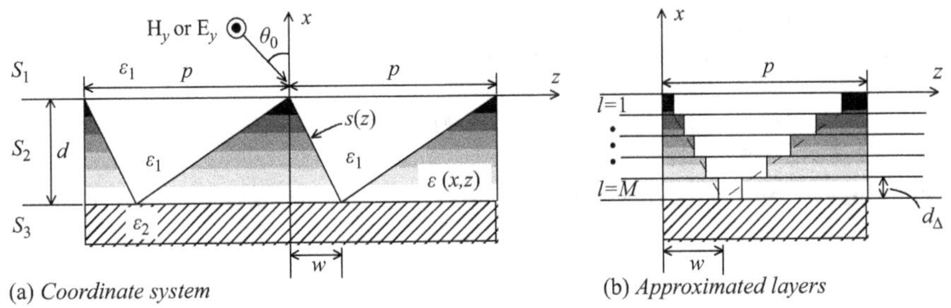

Figure 18.1 Structure of inhomogeneous dielectric grating with periodic surface relief: (a) coordinate system and (b) approximated inhomogeneous layers

For the TE (the electric field has only the y-component) cases, only numerical results are presented.

18.1.1.1 Scattering problem

When the plane wave is assumed to be incident from $x > 0$ at the angle θ_0, the scattering fields in the regions $S_1(x \geq 0)$ and $S_3(x \leq -d)$ are expressed [26,27] as

$$\underline{S_1(x \geq 0)\ :\ } H_y^{(1)} = e^{ik_1(z\sin\theta_0 - x\cos\theta_0)} + e^{ik_1 z\sin\theta_0} \sum_{n=-N}^{N} r_n^{(1)} e^{i\left\{k_n^{(1)}x + 2\pi nz/p\right\}}, \tag{18.1}$$

$$\underline{S_3(x \leq -d)\ :\ } H_y^{(3)} = e^{ik_1 z\sin\theta_0} \sum_{n=-N}^{N} t_n^{(3)} e^{-i\left\{k_n^{(3)}(x+d) - 2\pi nz/p\right\}}, \tag{18.2}$$

$$E_z^{(j)} = (-i\omega\varepsilon_j)^{-1}\left(\frac{\partial H_y^{(j)}}{\partial x}\right), \tag{18.3}$$

$$k_n^{(j)} := \sqrt{k_0^2\varepsilon_j/\varepsilon_0 - (k_1\sin\theta_0 + 2\pi n/p)^2}; \qquad j = 1, 3, \quad k_k{:=}\omega\sqrt{\varepsilon_k\mu_0},$$

$$k_0 := \frac{2\pi}{\lambda}, \quad k = 1, 2,$$

where λ is the wavelength in free space, and $r_n^{(1)}$ and $t_n^{(3)}$ are unknown coefficients to be determined from boundary conditions.

In the inhomogeneous grating region S_2, the layer with surface-relief profile $(-d < x < 0)$ is divided into thin layers $[d_\Delta(= d/M)]$, as shown in Figure 18.1(b). The dielectric distribution of thin layer is approximated by a step index profile [19] $(\varepsilon^{(l)}(z){:}l = 1 \sim M)$. In each layer $(l = 1 \sim M)$, the magnetic field can be written as

$H_y^{(l,2)} = H^{(l)}(z) \cdot e^{ih^{(l)}x}$, where $h^{(l)}$ is the propagation constant in the x-direction, and $H^{(l)}(z)$ must satisfy the following wave equation:

$$\frac{d^2H^{(l)}(z)}{dz^2} - \frac{1}{\varepsilon^{(l)}(z)} \frac{d\varepsilon^{(l)}(z)}{dz} \frac{dH^{(l)}(z)}{dz} + \left[k_0^2\varepsilon^{(l)}/\varepsilon_0 - \left\{h^{(l)}\right\}^2\right] H^{(l)}(z) = 0. \quad (18.4)$$

Taking into account Floquet's theorem, $H^{(l)}(z)$ can be approximated by the finite Fourier series as

$$H^{(l)}(z) = e^{ik_1 z \sin\theta_0} \sum_{n=-N}^{N} u_n^{(l)} e^{i2\pi nz/p}. \quad (18.5)$$

If $\varepsilon^{(l)}(z)$ is decomposed as $\varepsilon^{(l)}(z) := f^{(l)}(z)/g^{(l)}(z)$, for example, $\varepsilon^{(l)}(z) := \varepsilon_0 \operatorname{sech}(1 + z/p)$, the analysis is rather easier than that of [17], so that substituting (18.5) into (18.4) and multiplying both sides by $g^{(l)}(z) f^{(l)}(z) \cdot e^{-i2\pi nz/p}$, and rearranging after integrating with respect to z in the interval $0 \le z < p$, we get the following eigenvalue equation in regard to $h^{(l)}$:

$$\Lambda_1 \mathbf{U}^{(l)} = \left\{h^{(l)}\right\}^2 \Lambda_2 \mathbf{U}^{(l)}, \quad \Lambda_1 := \left[\eta_{m,n}^{(l)}\right], \quad \Lambda_2 := \left[\zeta_{m,n}^{(l)}\right]; \quad l = 1 \sim M, (18.6)$$

where

$$\mathbf{U}^{(l)} := \left[u_{-N}^{(l)}, \ldots, u_0^{(l)}, \ldots u_N^{(l)}\right]^T, \quad T: \textit{transpose},$$

$$\zeta_{n,m}^{(l)} := k_0^2 \xi_{n,m}^{(l)} - \gamma_n \left\{\gamma_n \eta_{n,m}^{(l)} + 2\pi\,(n-m)\,\eta_{n,m}^{(l)}/p - \varphi_{n,m}^{(l)}\right\},$$

$$\eta_{n,m}^{(l)} := \frac{1}{p} \int_0^p \left\{f^{(l)}(z) g^{(l)}(z)\right\} e^{i2\pi(n-m)z/p} dz,$$

$$\varphi_{n,m}^{(l)} := \frac{2i}{p} \int_0^p \left\{f^{(l)}(z) \frac{dg^{(l)}(z)}{dz}\right\} e^{i2\pi(n-m)z/p} dz,$$

$$\xi_{n,m}^{(l)} := \frac{1}{p} \int_0^p \left\{f^{(l)}(z)\right\}^2 e^{i2\pi(n-m)z/p} dz, \quad \gamma_n := (k_1 \sin\theta_0 + 2\pi n/p),$$

$$m, n = (-N, \ldots, 0, \ldots N).$$

However, since $f^{(l)}(z)$ *or* $g^{(l)}(z)$ contain discontinuities such as the step function, $(n-m)\,\eta_{n,m}^{(l)}$ does not converge, because $\eta_{n,m}^{(l)}$ is $O\,(1/|n-m|\,;\; n \neq m)$ as $|n-m| \to \infty$ [$\left|\eta_{n,m}^{(l)}\right|$ is less than $K/|n-m|$, where K is independent of $|n-m|$], so that the solution of (18.6) also does not converge to the correct value. Therefore, the method cannot directly be applied to the grating with a step distribution [18].

To solve this difficulty, the function containing the discontinuity is approximated by a Fourier series of N_f terms [18]:

$$f^{(l)}(z) \textit{ or } g^{(l)}(z) = \sum_{n=-N_f}^{N_f} \tau_n^{(l)} e^{i2\pi nz/p}. \quad (18.7)$$

We have experienced that $N = 1.5N_f$ is sufficient when N_f is related to the modal truncation number N [18,19].

The electromagnetic fields in each inhomogeneous layer are expressed as

$\underline{S_2(-d < x < 0):}$

$$
\begin{aligned}
H_y^{(l,2)} &= \sum_{\nu=1}^{2N+1} \left[A_\nu^{(l)} e^{ih_\nu^{(l)}\{x+(l-1)d_\Delta\}} + B_\nu^{(l)} e^{ih_\nu^{(l)}(x+ld_\Delta)} \right] \cdot f_\nu^{(l)}(z); \qquad (18.8)\\
d_\Delta &= d/M, \quad l = 1 \sim M,\\
E_z^{(l,2)} &= \left\{ -i\omega\varepsilon^{(l)}(z) \right\}^{-1} \frac{\partial H_y^{(l,2)}}{\partial x}, \quad f_\nu^{(l)}(z) := e^{ik_1 z \sin\theta_0} \sum_{n=-N}^{N} u_{\nu,n}^{(l)} e^{i2\pi nz/p},
\end{aligned}
$$

where $A_\nu^{(l)}, B_\nu^{(l)}$ are unknown coefficients to be determined by boundary conditions.

In the inhomogeneous layer $(-d < x < 0)$, we can obtain the matrix relations between $A^{(1)}, B^{(1)}$ and $A^{(M)}, B^{(M)}$. From the boundary condition at $x = -l \cdot d_\Delta$ $(l = 1 \sim M-1)$, we get

$$
\begin{aligned}
\begin{pmatrix} \mathbf{A}^{(1)} \\ \mathbf{B}^{(1)} \end{pmatrix} &= \begin{pmatrix} \mathbf{S}_1^{(1)} & \mathbf{S}_2^{(1)} \\ \mathbf{S}_3^{(1)} & \mathbf{S}_4^{(1)} \end{pmatrix} \begin{pmatrix} \mathbf{S}_1^{(2)} & \mathbf{S}_2^{(2)} \\ \mathbf{S}_3^{(2)} & \mathbf{S}_4^{(2)} \end{pmatrix} \cdots \begin{pmatrix} \mathbf{S}_1^{(M)} & \mathbf{S}_2^{(M)} \\ \mathbf{S}_3^{(M)} & \mathbf{S}_4^{(M)} \end{pmatrix} \begin{pmatrix} \mathbf{A}^{(M)} \\ \mathbf{B}^{(M)} \end{pmatrix}\\
&= \begin{pmatrix} \mathbf{S}_1 & \mathbf{S}_2 \\ \mathbf{S}_3 & \mathbf{S}_4 \end{pmatrix} \begin{pmatrix} \mathbf{A}^{(M)} \\ \mathbf{B}^{(M)} \end{pmatrix}, \qquad (18.9)
\end{aligned}
$$

where

$$
\begin{aligned}
\mathbf{S}_k^{(l)} &:= \left[{}^{(l)}s_{n,\nu}^{(k)} \right]; \quad k = 1 \sim 4, \; l = 1 \sim M,\\
{}^{(l)}s_{n,\nu}^{(1)} &:= \frac{1}{2}\left[v_{n,\nu}^{(l)} + \theta_{n,\nu}^{(l)} h_{n+N+1}^{(l+1)} / h_\nu^{(l)} \right] e^{-ih_\nu^{(l)} d_\Delta},\\
{}^{(l)}s_{n,\nu}^{(2)} &:= \frac{1}{2}\left[v_{n,\nu}^{(l)} - \theta_{n,\nu}^{(l)} h_{n+N+1}^{(l+1)} / h_\nu^{(l)} \right] e^{i\left(h_{n+N+1}^{(l+1)} - h_\nu^{(l)}\right) d_\Delta},\\
{}^{(l)}s_{n,\nu}^{(3)} &:= \frac{1}{2}\left[v_{n,\nu}^{(l)} - \theta_{n,\nu}^{(l)} h_{n+N+1}^{(l+1)} / h_\nu^{(l)} \right],\\
{}^{(l)}s_{n,\nu}^{(4)} &:= \frac{1}{2}\left[v_{n,\nu}^{(l)} + \theta_{n,\nu}^{(l)} h_{n+N+1}^{(l+1)} / h_\nu^{(l)} \right] e^{ih_{n+N+1}^{(1)} d_\Delta},\\
\mathbf{V} &:= \left[\mathbf{v}_{n,\nu}^{(l)} \right] = \left[\mathbf{U}^{(l)} \right]^{-1} \cdot \left[\mathbf{U}^{(l+1)} \right], \; \mathbf{U}^{(l)} := \left[\mathbf{u}_{n,\nu}^{(l)} \right],\\
\Theta &:= \left[\theta_{\nu,n}^{(l)} \right] = \left[\Psi^{(l)} \right]^{-1} \cdot \left[\Phi^{(l+1)} \right], \; \Psi^{(l)} := \left[\psi_{\nu,n}^{(l)} \right], \; \Phi^{(l)} := \left[\phi_{\nu,n}^{(l)} \right],\\
\psi_{\nu,n}^{(l)} &:= \sum_{m=-N}^{N} u_{\nu,m}^{(l)} \chi_{m,n}^{(l,l+1)}, \; \phi_{\nu,n}^{(l)} := \sum_{m=-N}^{N} u_{\nu,m}^{(l+1)} \chi_{m,n}^{(l+1,l)},\\
\chi_{m,n}^{(s,q)} &:= \frac{1}{p} \int_0^p \left\{ f^{(s)}(z) g^{(q)}(z) \right\} e^{i2\pi(m-n)z/p} dz.
\end{aligned}
$$

From final boundary conditions at $x = 0$ and $x = -d$, we get the following homogeneous matrix equation in regard to $A^{(M)}$ by matrix algebra [25]:

$$\mathbf{W} \cdot \mathbf{A}^{(M)} = \mathbf{F}; \quad \mathbf{W} := [\mathbf{Q}_1\mathbf{S}_1 + \mathbf{Q}_2\mathbf{S}_3 - (\mathbf{Q}_1\mathbf{S}_2 + \mathbf{Q}_2\mathbf{S}_4)\mathbf{Q}_4^{-1}\mathbf{Q}_3], \tag{18.10}$$

where

$$\mathbf{A}^{(M)} := \left[A_1^{(M)}, A_2^{(M)}, \ldots, A_{2N+1}^{(M)}\right]^T,$$

$$\mathbf{F} := \left[2k_0^{(1)}\eta_{0,-N}^{(1)}, \ldots, 2k_0^{(1)}\eta_{0,0}^{(1)}, \ldots, 2k_0^{(1)}\eta_{0,N}^{(1)}\right]^T,$$

$$\mathbf{Q}_k := \left[q_{n,\nu}^{(k)}\right]; \quad k = 1 \sim 4, \quad n = (-N, \ldots, 0, \ldots, N), \quad \nu = 1 \sim (2N+1),$$

$$q_{n,\nu}^{(1)} := \left[\rho_{\nu,n}^{(1)} + \sigma_{\nu,n}^{(1)}\right], \quad q_{n,\nu}^{(2)} := -\left[\rho_{\nu,n}^{(1)} + \sigma_{\nu,n}^{(1)}\right]e^{ih_\nu^{(1)}d_\Delta},$$

$$q_{n,\nu}^{(3)} := -\left[\rho_{\nu,n}^{(M)} - \sigma_{\nu,n}^{(M)}\right]e^{ih_\nu^{(M)}d_2},$$

$$q_{n,\nu}^{(4)} := -\left[\rho_{\nu,n}^{(M)} + \sigma_{\nu,n}^{(M)}\right], \quad \rho_{\nu,n}^{(k)} := h_\nu^{(k)} \sum_{m=-N}^{N} u_{\nu,m}^{(k)}\alpha_{\nu,m}^{(k)}; \quad k = 1, M,$$

$$\sigma_{\nu,n}^{(1)} := \sum_{m=-N}^{N} \varepsilon_0 k_m^{(1)} u_{\nu,m}^{(1)} \beta_{m,n}^{(1)}/\varepsilon_1, \quad \sigma_{\nu,n}^{(M)} := \sum_{m=-N}^{N} \varepsilon_0 k_m^{(3)} u_{\nu,m}^{(M)} \beta_{m,n}^{(M)}/\varepsilon_3,$$

$$\alpha_{m,n}^{(k)} := \frac{1}{p}\int_0^p \left\{f^{(k)}(z)\right\}e^{i2\pi(m-n)z/p}dz,$$

$$\beta_{m,n}^{(k)} := \frac{1}{p}\int_0^p \left\{g^{(k)}(z)\right\}e^{i2\pi(m-n)z/p}dz, \quad k = 1, M.$$

The mode power transmission coefficients $|T_n^{(TM)}|^2$ are given by

$$|T_n^{(TM)}|^2 := \varepsilon_1 \mathrm{Re}\left\{k_n^{(3)}\right\}|t_n^{(3)}|^2 \Big/ \left(\varepsilon_3 k_0^{(1)}\right), \tag{18.11}$$

where superscript (TM) indicates TM wave case.

18.1.1.2 Guiding problem

We consider the case of incidence from the left (along $+z$) as shown in Figure 18.1(a). The electromagnetic fields in the regions S_1, S_2, and S_3 are expressed [19] as

$$\underline{S_1\ (x \geq 0)}\ :\ H_y^{(1)} = e^{i\gamma z}\sum_{n=-N}^{N} r_n^{(1)} e^{i\left\{k_n^{(1)}x + 2\pi nz/p\right\}}; \quad k_1 := \omega\sqrt{\varepsilon_1\mu_0}, \tag{18.12}$$

$$\underline{S_2\ (-d < x < 0)}\ :\ H_y^{(l,2)} = \sum_{\nu=1}^{2N+1}\left[A_\nu^{(l)} e^{ih_\nu^{(l)}\{x+(l-1)d_\Delta\}} + B_\nu^{(l)} e^{ih_\nu^{(l)}(x+ld_\Delta)}\right] \cdot f_\nu^{(l)}(z), \tag{18.13}$$

$$E_z^{(l,2)} = \left\{-i\omega\varepsilon^{(l)}(z)\right\}^{-1}\left(\frac{\partial H_y^{(1,2)}}{\partial x}\right), \quad f_\nu^{(l)}(z) := e^{i\gamma z}\sum_{n=-N}^{N} u_{\nu,n}^{(l)} e^{i2\pi nz/p},$$

$$l = 1 \sim M, \quad d_\Delta = \frac{d}{M},$$

$$\underline{S_3\ (x \leq -d)\ :}\ H_y^{(3)} = e^{i\gamma z}\sum_{n=-N}^{N} t_n^{(3)} e^{-i\left\{k_n^{(3)}(x+d)-2\pi nz/p\right\}}, \tag{18.14}$$

$$E_z^{(j)} = \left(-i\omega\varepsilon_j\right)^{-1}\left(\frac{\partial H_y^{(j)}}{\partial x}\right), \quad k_n^{(j)} := \sqrt{k_0^2\varepsilon_k/\varepsilon_0 - (\gamma + 2\pi n/p)^2};$$

$$j = 1(k = 1),\ 3(k = 2),$$

where $\gamma\ (:= \beta + i\alpha, \alpha \geq 0)$ is the unknown propagation constant in the z-direction and $k_n^{(j)}\ (j = 1,3)$ are propagation constants in the x-direction, and $r_n^{(1)}$, $t_n^{(3)}$, $A_\nu^{(l)}$, and $B_\nu^{(l)}$ are unknown coefficients to be determined from the boundary conditions at $x = 0, x = -l \cdot d_\Delta\ (l = 1 \sim M - 1)$ and $x = -d$.

The signs of $k_n^{(1)}$, $k_n^{(3)}$ are chosen as in the guiding problem, so that the real and imaginary parts of $k_n^{(1)}$, $k_n^{(3)}$ are both positive [Re $\left(k_n^{(j)}\right) \geq 0$, Im $\left(k_n^{(j)}\right) \geq 0\ (j = 1,3)$], except in the case $n < 0$ and $(\beta + 2\pi n/p) > 0$ for which the imaginary parts of $k_n^{(1)}$, $k_n^{(3)}$ are negative [Im $\left(k_n^{(j)}\right) < 0\ \ (j = 1,3)$] [28].

From the boundary conditions, we get the following homogeneous matrix equation with respect to $A^{(M)}$ [19]:

$$\mathbf{W} \cdot \mathbf{A}^{(M)} = \mathbf{0}, \quad \mathbf{W} := \left[\mathbf{Q}_1\mathbf{S}_1 + \mathbf{Q}_2\mathbf{S}_3 - (\mathbf{Q}_1\mathbf{S}_2 + \mathbf{Q}_2\mathbf{S}_4)\,\mathbf{Q}_4^{-1}\mathbf{Q}_3\right], \tag{18.15}$$

where the elements of matrix $\mathbf{W}$ are obtained in (18.10) by replacing the term $k_1 \sin\theta_0$ with γ.

For (18.15), the order of matrix $\mathbf{W}$ depends on the modal truncation number but does not depend on the number of layers. Therefore, our method is effective for the guiding problem.

For nontrivial solution to exist in (18.15), the following characteristic equation holds

$$\det(\mathbf{W}) = 0; \quad \det : \textit{determinant}. \tag{18.16}$$

The propagation constant $\gamma\ (:= \beta + i\alpha, \alpha \geq 0)$ should be determined to satisfy (18.6) replacing the term $k_1 \sin\theta_0$ with γ and (18.15) simultaneously. Numerical computation to find γ is carried out by using Muller's method [29], iterations of which are continued until it converges to at least seven significant digits for fixed N and M [19].

18.1.2 Numerical results

We consider the following profile of the grating as follows:

1. Surface profile $(0 < z \leq p)$; triangular profile

$$h(z) := \begin{cases} -z/w & ;0 \leq x < w \\ -(p-z)/(p-w) & ;w \leq x < p \end{cases}. \tag{18.17}$$

2. Distribution of permittivity $\varepsilon(x, z)$; modulated profile

$$\varepsilon(x,z) := \begin{cases} \varepsilon_1 & ; s(z) \le x < 0 \\ \varepsilon_A\left[1 + \delta\cos\left\{\dfrac{\pi}{d}(x)\right\}\right] & ; -d \le x < s(z) \end{cases} \tag{18.18}$$

For the previous case, we put $g^{(l)}(z) = 1$ in (18.7), and the values of parameters are chosen as $\varepsilon_1 = \varepsilon_2 = \varepsilon_0$, $\varepsilon_A/\varepsilon_0 = 2$, $\delta = 0.5$, $d/p = 2/3$.

The validity of our method has been checked for a larger grating depth by increasing the number of modal expansion terms and the number of multilayers. For the sinusoidal profile case, the phase constants obtained by various authors agree quite well with our results, while the agreement in the attenuation constants is bad, especially for TM case [19].

First, we investigate the scattering problem for the influences of the incident angle θ_0 and normalized frequency p/λ on the $[n]$th-mode transmission coefficients for both TM and TE waves. The validity of our method has been carefully checked by varying the number N of modal expansion terms and the number M of multilayers. For the TM wave, the results are computed with $N = 9(N_f = 6)$ and $M = 20$. For the TE wave, the results are computed with $N = 10$ and $M = 20$. In this case, the relative errors are less than about 0.1% and the energy error is less than about 10^{-3} for both TM and TE waves [26].

18.1.2.1 Characteristic of incident angle

Figure 18.2(a) and (b) shows $|T_n^{(TM)}|^2$ and $|T_n^{(TE)}|^2$ for various values of incident angle θ_0 at $w/p = 0.5$ and $p/\lambda = 0.6$. From Figure 18.2(a) and (b), the effect of changing the distribution of permittivity from low to high ($\delta = 0.5$) or high to low ($\delta = -0.5$) is seen clearly at Wood's anomaly [$\theta_0 = 41.8°$; it appears first mode] and at the minimum of coupling resonance ($|T_n^{(TM)\ or\ (TE)}|^2 \cong 0$). We note that the θ_0 dependence of Wood's anomaly is more significant for the TE wave than that of the TM wave. Figure 18.3(a) and (b) shows $|T_n^{(TM)}|^2$ and $|T_n^{(TE)}|^2$ versus θ_0 for various values of incident angle θ_0 and the normalized position w/p at $p/\lambda = 0.6$ for the case of the [0]th mode.

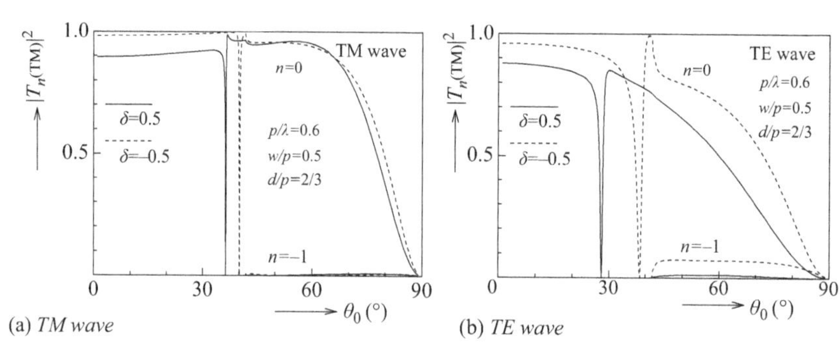

Figure 18.2 Modal power transmission coefficients $|T_n^{(TM)}|^2$ and $|T_n^{(TE)}|^2$ versus incident angle θ_0 for the case $\delta = 0.5$ and $\delta = -0.5$: (a) TM wave and (b) TE wave

From Figure 18.3(a) and (b), we note that the minimum point of the curve for coupling resonance for the TM wave moves toward larger θ_0 as w/p decreases. On the other hand, for TE wave the minimum point of the curve for coupling resonance moves toward smaller θ_0 as w/p decreases.

18.1.2.2 Characteristic of normalized frequency

Figure 18.4(a) and (b) shows $|T_0^{(TM)}|^2$ and $|T_0^{(TE)}|^2$ for (p/λ) and w/p at $\theta_0 = 45°$ with the same parameters as in Figure 18.3.

Comparing the TM wave with the TE one, we note that (1) the characteristic tendencies for the effect of normalized position w/p are approximately same, but for $0.9 < p/\lambda < 1.0$, the effect of the normalized position is more significant for TE waves; (2) the minimum point of the curves for coupling resonance for TM wave moves toward larger p/λ as w/p decreases. On the other hand, for TE wave the

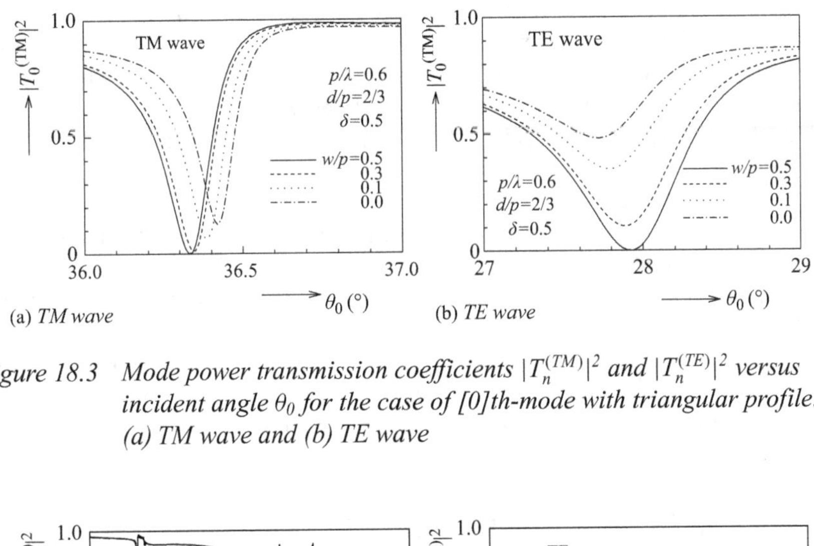

Figure 18.3 Mode power transmission coefficients $|T_n^{(TM)}|^2$ and $|T_n^{(TE)}|^2$ versus incident angle θ_0 for the case of [0]th-mode with triangular profile: (a) TM wave and (b) TE wave

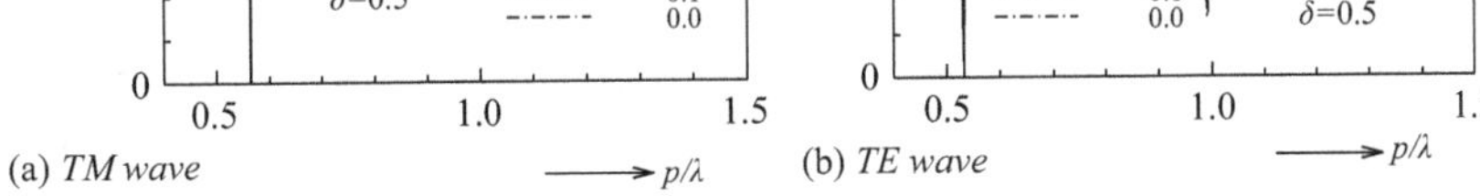

Figure 18.4 Mode power transmission coefficients $|T_n^{(TM)}|^2$ and $|T_n^{(TE)}|^2$ versus a normalized frequency for the case of (p/λ) [0]th-mode with triangular profile: (a) TM wave and (b) TE wave

minimum point of the curve for coupling resonance moves toward smaller p/λ as w/p decreases.

Second, we investigate the guiding problem for the differences of the propagation constants between the homogeneous ($\delta = 0$) and inhomogeneous medium for both the TM and TE modes. Propagating modes considered here are the fundamental TM_0 and TE_0 modes. We can compute the propagation constants (γ) with the relative error less than 0.1% both for TE ($N = 10$, $M = 20$) and TM ($N = 9 = 1.5N_f$ [18], $M = 30$) modes [19].

18.1.2.3 Propagation characteristics

We consider the lowest guided TM_0 and TE_0 modes. Figure 18.5(a) and (b) shows the Brillouin diagrams as a function of δ in the case of a symmetric triangular surface profile ($w/p = 0.5$) for TM mode. Figure 18.5(a) shows the normalized phase constants $\beta p/(2\pi)$, while Figure 18.5(b) shows the normalized attenuation constants $\alpha p/(2\pi)$. From Figure 18.5, we note that (1) the values $\beta p/(2\pi)$ for ($\delta = 0.5$) are different from ($\delta = 0$), because the equivalent permittivity $\bar{\varepsilon}(x, z)/\varepsilon_0$ in the surface region is smaller than that of ($\delta = 0.5$), so that $\bar{\varepsilon}(x, z)/\varepsilon_0$ becomes $[0.5 + \varepsilon_A(0.5 + \delta/\pi^2)] < 2.0$; (2) the maximum of the values $\alpha p/(2\pi)$ for ($\delta = 0$) at the Bragg region is larger than that of ($\delta = 0$); therefore, the effect of the inhomogeneous medium is more significant at the higher order Bragg region; and (3) in the first-order Bragg region, the maximum of the values $\alpha p/(2\pi)$ for $\delta = 0.5$ is smaller than that of homogeneous medium ($\delta = 0$).

On the other hand, Figure 18.6 (a) and (b) shows the Brillouin diagrams for the TE mode with the same conditions as Figure 18.5. Comparing the figures for TM mode (Figure 18.5), we can see the following features: (1) the characteristic tendencies for $\beta\, p/2\pi$ and $\alpha p/(2\pi)$ are approximately the same, but the maximum of the values $\alpha p/(2\pi)$ at the first-order Bragg region is larger than that of the homogeneous medium ($\delta = 0$); this is attributed to the effect of polarization; (2) the effects of the inhomogeneous medium of the values $\alpha p/(2\pi)$ are more significant in the Bragg region both for TE and TM modes.

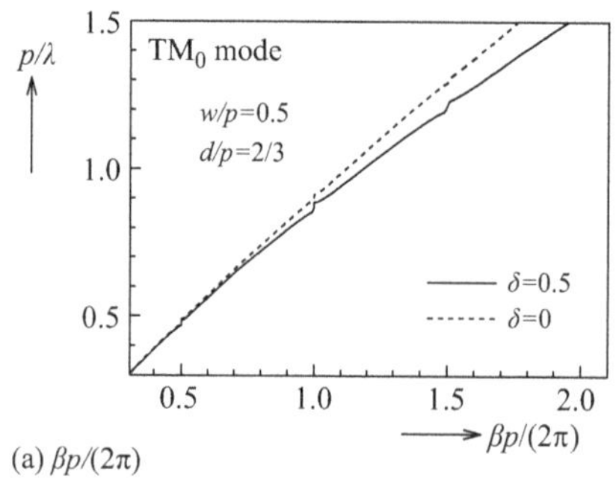

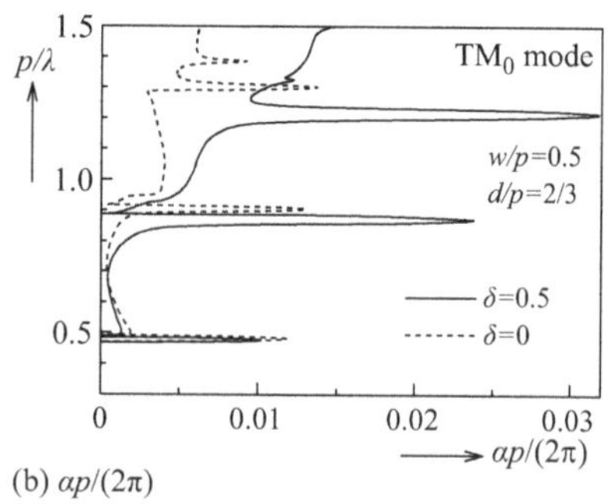

Figure 18.5 Brillouin diagrams as a function of δ in the case of a symmetric triangular surface profile for TM_0 mode: (a) normalized phase constants $\beta p/(2\pi)$ and (b) normalized attenuation constants $\alpha p/(2\pi)$

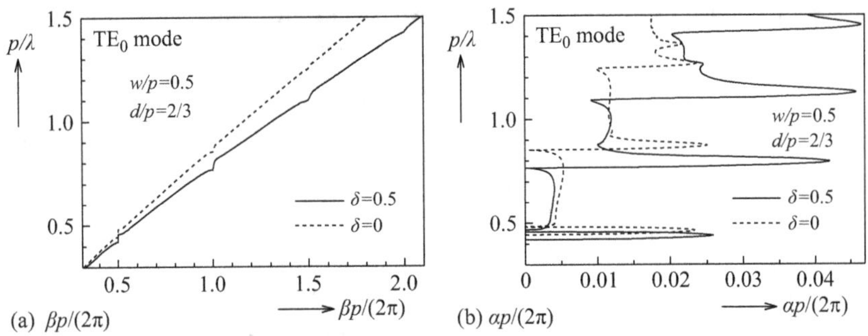

Figure 18.6 Brillouin diagrams as a function of δ in the case of a symmetric triangular surface profile for TE_0 mode: (a) normalized phase constants $\beta p/(2\pi)$ and (b) normalized attenuation constants $\alpha p/(2\pi)$

18.1.3 Conclusions

In this section, we have proposed a new method for scattering and guiding problems of inhomogeneous dielectric gratings of arbitrary profiles and media with periodic surface. The main points of our IFEM are as follows:

1. The order of the matrix size for the characteristic equation depends on the modal truncation number, but it does not depend on the number of layers used in the approximation.
2. For the stratified approximation profile, it is effective to approximate with a step function.
3. The validity of our method has been checked by increasing the number of modal expansion terms and the number of multilayers.
4. There are remarkable differences in propagation characteristics between the surface-relief grating and the modulated dielectric grating in the Bragg region for TM mode and TE mode.

In the following sections, we explain the advantage of the IFEM for several inhomogeneous media in scattering and guiding problems as follows: Section 18.2—slanted layered and rhombic media with strips; Section 18.3—elliptically layered, columnar, and rectangular media; Section 18.4—energy distribution in defect layers; and Section 18.5—mixed positive and negative media.

18.2 Slanted layer and rhombic media with strips

18.2.1 Slanted layer

IFEM can be applied to a wide range of periodic structures and is especially effective in the case of planar slanted gratings [20,21], because the electromagnetic fields in each layer can be obtained more easily than with the previous method [30,31] by shifting the solution in the first layer, so that, from (18.6), since $\varepsilon(x, z)$ in each region is shifted according to the slant angle ϕ, we obtain the following relation for the eigenvalues $h^{(l)}(l = 1 \sim M)$ and eigenvectors $u_n^{(l)}(l = 1 \sim M)$.

$$\begin{aligned} h^{(l)} &= h^{(1)},\ u_n^{(l)} = u_n^{(1)} e^{-i2\pi n\delta(l)/p}; \quad \delta(l) = W_l - W_1, \\ W_l &:= (l - 0.5)\, d_l \tan\phi/M. \end{aligned} \tag{18.19}$$

The results [20,21] of our method are good agreement with those of Moharam and Gaylord [31].

18.2.2 Rhombic media with strips

Next, we have proposed a new method for the scattering problems of electromagnetic waves by a rhombic grating with strips [32] for matrix formulation of the Point Matching Method (PMM [33,34]).

We consider the following profiles of the inhomogeneous rhombic grating with strips in Figure 18.7(a), and the approximated inhomogeneous media in Figure 18.7(b),

$$\begin{aligned} &\frac{\varepsilon_2(x,z)}{\varepsilon_0} \triangleq \begin{cases} 1.0{:}(0 \le \phi \le \alpha \text{ and } \beta) \\ 1.5{:}(\alpha \text{ and } \beta < \phi \le \pi/2) \end{cases} \\ &\varepsilon_1 = \varepsilon_3 = \varepsilon_o, b/p = 0.5 \text{ and } D/p = 0.3. \end{aligned} \tag{18.20}$$

We obtain the matrix form for the combination of metallic region and the dielectric region using the boundary condition $Z_j = (j-1)p/[(2N+1)];\ j = 1 - (2N+1)$ at the matching points on 0, and on D for the TE wave.

The boundary conditions using the PMM [33,34] are as follows:

$$\begin{aligned} Z_j \in C_1;\ & \left[E_z^{(1)} = 0,\ E_z^{(2,1)} = 0\right]_{x=0},\quad Z_j \in \overline{C_1};\ \left[H_y^{(1)} = H_y^{(2,1)}\right]_{x=0}, \\ & \left[E_z^{(1)} = E_z^{(2,1)}\right]_{x=0}, \end{aligned} \tag{18.21}$$

$$\begin{aligned} Z_j \in C_2;\ & \left[E_z^{(3)} = 0\right]_{x=D},\ \left[E_z^{(2,M)} = 0\right]_{x=D},\quad Z_j \in \overline{C_2};\ \left[H_y^{(3)} = H_y^{(2,M)}\right]_{x=D}, \\ & \left[E_z^{(3)} = E_z^{(2,M)}\right]_{x=D}. \end{aligned} \tag{18.22}$$

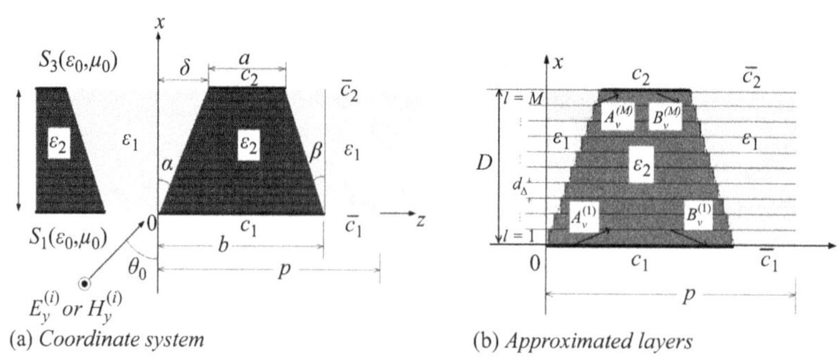

Figure 18.7 Structure of inhomogeneous dielectric gratings loaded with strips. (a) coordinate system and (b) approximated inhomogeneous layers.

If the boundary condition of (18.21), and (18.22) in electric fields, is satisfied in all matching points, by using the orthogonality properties of $\{e^{-i2\pi mz/p}\}$ we get the following equation in regard to matrices $\mathbf{A}^{(1)}, \mathbf{B}^{(1)}$ and $\mathbf{A}^{(\mathrm{M})}$, $\mathbf{B}^{(\mathrm{M})}$,

$$\mathbf{Q}_1\mathbf{A}^{(1)} + \mathbf{Q}_2\mathbf{B}^{(1)} = \mathbf{F}, \quad \mathbf{Q}_3\mathbf{A}^{(M)} + \mathbf{Q}_4\mathbf{B}^{(M)} = \mathbf{0}, \tag{18.23}$$

where

$$\mathbf{F} \triangleq \left[0(Z_k \in C_1), 2k_0^{(1)}(Z_k \in \overline{C_1})\right]^T,$$

$$\mathbf{A}^{(k)} := \left[A_1^{(k)}, A_2^{(k)}, \ldots, A_{2N+1}^{(k)}\right]^T, k = 1, M,$$

$$\mathbf{B}^{(k)} := \left[B_1^{(k)}, B_2^{(k)}, \ldots, B_{2N+1}^{(k)}\right]^T, k = 1, M$$

$$\begin{aligned}
&\mathbf{Q}_1 := \mathbf{X}^C\mathbf{U}^{(1)} + \mathbf{K}^{(1)}\mathbf{X}^{\overline{C}}\mathbf{U}^{(1)} + \mathbf{X}^{\overline{C}}\mathbf{U}^{(l)}\mathbf{H}^{(1)},\\
&\mathbf{Q}_2 := (\mathbf{X}^C\mathbf{U}^{(1)} + \mathbf{K}^{(1)}\mathbf{X}^{\overline{C}}\mathbf{U}^{(1)} - \mathbf{X}^{\overline{C}}\mathbf{U}^{(l)}\mathbf{H}^{(1)})\mathbf{D},\\
&\mathbf{Q}_3 := \mathbf{X}^C\mathbf{U}^{(M)}\mathbf{H}^{(M)} + \mathbf{K}^{(3)}\mathbf{X}^{\overline{C}}\mathbf{U}^{(M)} - \mathbf{X}^{\overline{C}}\mathbf{U}^{(M)}\mathbf{H}^{(M)},\\
&\mathbf{Q}_4 := -\mathbf{X}^C\mathbf{U}^{(M)}\mathbf{H}^{(M)}\mathbf{D} + \mathbf{K}^{(3)}\mathbf{X}^{\overline{C}}\mathbf{U}^{(M)} + \mathbf{X}^{\overline{C}}\mathbf{U}^{(M)}\mathbf{H}^{(M)},
\end{aligned}$$

$$n = (-N, \ldots, 0, \ldots, N), \quad \nu = 1 \sim (2N+1)$$

$$\mathbf{X}^C \triangleq \begin{bmatrix} \vdots & \vdots & \vdots & \vdots & \vdots \\ e^{-iNZ_j} & \cdots & e^{in(=0)Z_j} & \cdots & e^{iNZ_j} \\ \vdots & \vdots & \vdots & \vdots & \vdots \\ 0 & \cdots & 0 & \cdots & 0 \end{bmatrix} \begin{matrix} \left.\begin{matrix} \\ \\ \end{matrix}\right\} Z_j \in C_1 \\ \cdot\mathbf{U}^{(l)} \\ \} Z_j \in \overline{C_1} \end{matrix}$$

$$\mathbf{X}^{\overline{C}} \triangleq \begin{bmatrix} \vdots & \vdots & \vdots & \vdots & \vdots \\ 0 & & 0 & & 0 \\ \vdots & \vdots & \vdots & \vdots & \vdots \\ e^{-iNZ_j} & & e^{i0Z_j} & & e^{iNZ_j} \end{bmatrix} \begin{matrix} \left.\begin{matrix} \\ \\ \end{matrix}\right\} Z_j \in C_1 \\ \cdot\mathbf{U}^{(l)} \\ \} Z_j \in \overline{C_1} \end{matrix}$$

$$\mathbf{U}^{(l)} := \left[{}^{(l)}u_n^{(\nu)}\right], \quad \mathbf{D}^{(l)} := \left[e^{ih_\nu^{(l)}d_\Delta} \cdot \delta_{\nu,\nu}\right], \quad \mathbf{H}^{(l)} := \left[h_\nu^{(l)} \cdot \delta_{\nu,\nu}\right],$$

$$\mathbf{K}^{(l)} := \left[k_n^{(l)} \cdot \delta_{\nu,\nu}\right], \quad l = 1, \ M, \quad \nu = 1 \sim (2N_f + 1).$$

By using the matrix relationship of (18.23) and (18.9), we get the following homogeneous matrix equation in regard to $A_\nu^{(M)}$ $(\nu = 1 \sim (2N+1))$ [32]:

$$\mathbf{W} \cdot \mathbf{A}^{(M)} = \mathbf{F}, \tag{18.24}$$

where $\mathbf{W}{:} = \left[\mathbf{Q}_1\mathbf{S}_1 + \mathbf{Q}_2\mathbf{S}_3 - (\mathbf{Q}_1\mathbf{S}_2 + \mathbf{Q}_2\mathbf{S}_4)\,\mathbf{Q}_4^{-1}\mathbf{Q}_3\right]$.

The mode power transmission coefficients ρ_t are given by

$$\rho_t^{(TE)} := \sum_{n=-N}^{N} \mathrm{Re}\left[k_n^{(3)}\right]|b_n|^2/k_0^{(1)}, \quad \rho_r^{(TE)} := \sum_{n=-N}^{N} \mathrm{Re}\left[k_n^{(1)}\right]|a_n|^2/k_0^{(1)}. \tag{18.25}$$

The values of parameters chosen are $\varepsilon_1 = \varepsilon_3 = \varepsilon_0$, $(a = b)/p = 0.5$, $\varepsilon_2/\varepsilon_0 = 1.5$, and $D/P = 0.3$, $\varepsilon_1/\varepsilon_0 = 1.0$, $\theta_0 = \alpha = \beta = 18.43°$. The relative errors are less than about 0.1% and the energy error $\varepsilon_e^{(TE)}\left[:= 1 - (\rho_t^{(TE)} + \rho_r^{(TE)})\right]$ is less than about 10^{-3} for TE and TM waves when we computed with $N = 15$ at $a/b = 1$ and $p/\lambda = 1.5$ [32].

Figure 18.8 shows values of ρ_t for various values of normalized frequency for (p/λ) for $\varepsilon_1/\varepsilon_0 = 1.0$ and $\varepsilon_2/\varepsilon_0 = 1.5$ at ①$\alpha(=\beta) = 0°$, ②$\alpha = (-\beta) = 18.43°$, and ③$\alpha = (\beta) = 18.43°$ with $\theta_0 = 0°$. Figure 18.8(a) shows the TE wave and Figure 18.8(b) the TM wave.

From Figure 18.8, we can see the following features:

1. Characteristic tendencies of coupling resonance cases of ①, ②, and ③ are approximately same for the TE and TM waves at $1.4 < p/\lambda < 1.5$.
2. For the TE wave, the peak of $\rho_t^{(\mathrm{TE})}$ in $p/\lambda < 1.0$ is suppressed and the resonant peaks of frequency decrease in the case of ③.
3. For the TM wave in $\rho/\lambda < 1.0$, the resonant peaks of frequency increase ③.

Figure 18.9 shows values of ρ_t for various values of normalized frequency (p/λ) at $\theta_0 = \alpha = 18.43$ for the same parameters as in Figure 18.8. We can see the following features:

1. The characteristic tendencies for the effect of equivalent permittivity are approximately the same.

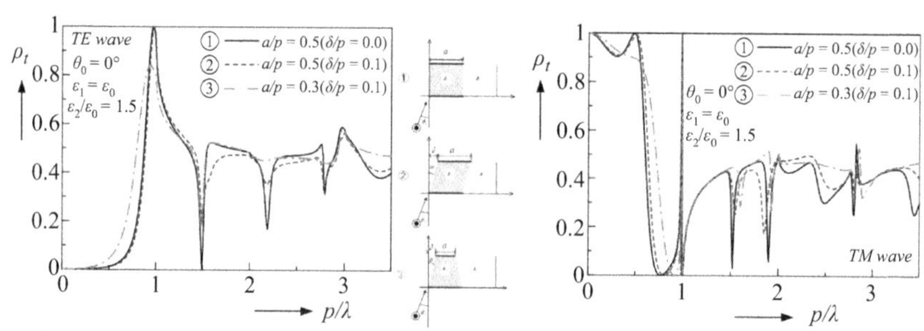

(a) *TE wave* (b) *TM wave*

Figure 18.8 Power transmission coefficients ρ_t versus normalized frequency p/λ at $\theta_0 = 0°$: (a) TE wave and (b) TM wave

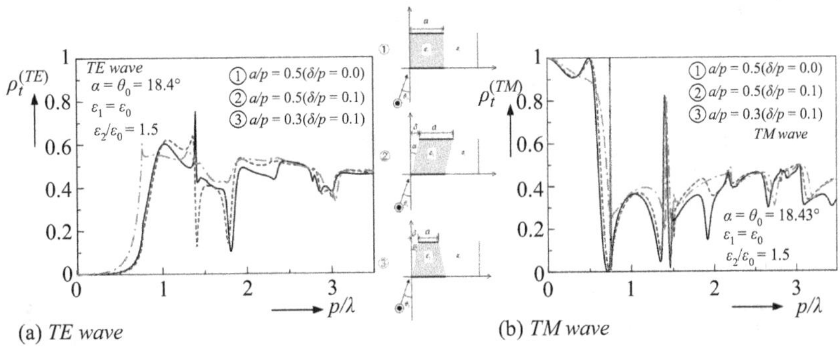

Figure 18.9 Power transmission coefficients ρ_t versus normalized frequency p/λ at $\theta_0 = 18.43°$: (a) TE wave and (b) TM wave

2. For the TE wave, when the incidence angle θ_0 is equal to $\alpha = 18.43$, and the value of p/λ is around 1.4, resonant peaks appears, and if p/λ is such that $0.7 < \lambda/p < 1.4$ in the case of the trapezoid type ③, it is interesting that becomes about constant.
3. For the TM wave, it does not change into resonant peaks around $p/\lambda = 0.75$ and $p/\lambda = 1.4$ in case ③. But its peaks move toward smaller values of p/λ.

18.2.3 Conclusions

In this section, we have proposed a new method for the scattering of electromagnetic waves by a slanted layer and by rhombic media loaded with perfectly conducting strips, using the combination of IFFM and the Point Matching Method. The main points of the results are as follows:

1. In the slanted layer, the thin modulated one in each layer is a simple shift of the first one. Therefore, the eigenvalues and eigenvectors can easily be computed using the solution in the first layer, and the matrix formulation effectively leads to accurate and fast computations. We also show the influences of slant angle on the diffraction efficiencies at the Wood's anomaly and coupling resonance frequency.
2. In the rhombic media with strips, we have proposed a new formulation for the scattering of electromagnetic waves by inhomogeneous dielectric gratings loaded with perfectly conducting strips, for TE and TM cases. We also show the significant effects in the normalized frequency band $0.7 < \lambda/p < 1.4$ at $\theta_0 = \alpha = (\beta) = 18.43°$ for the TE wave. This method also can be applied to inhomogeneous dielectric gratings having arbitrarily periodic structures combination of dielectric and metallic materials.

18.3 Elliptically layered, columnar, and rectangular media

18.3.1 Elliptically layered and columnar media

We consider the following structures of inhomogeneous elliptically layered [35] and columnar [36] gratings shown in Figure 18.10(a).

1. Grating shape; elliptic cylinder

$$\left[\frac{(x+d/2)}{(d/2)}\right]^2+\left(\frac{z}{a}\right)^2=1. \tag{18.26}$$

2. Distribution of permittivity $\varepsilon(x,z)$; elliptically layered media

$$\varepsilon(x,z):=\begin{cases}\varepsilon_2[1-b\{(2(x+d/2)/d)^2+(z/a)^2\}];\\ \qquad b=1-\varepsilon_1\varepsilon_2 : \text{inside of elliptic cylinder}\\ \varepsilon_1 : \text{outside of elliptic cylinder}\end{cases} \tag{18.27}$$

Generally, the permittivity profile $\varepsilon(x,z)$ of (18.27) is not separable with respect to the x and z coordinates. Therefore, it is difficult to obtain the analytical solution, such as in terms of Mathieu functions, in the case of (18.27).

However, in the IFEM, the inhomogeneous region is divided into thin layers $[d_\Delta(=d/M)]$, as shown in Figure 18.10(b), with respect to only the z components.

First we show again the relationship between the matrices $\mathbf{A}^{(1)}$, $\mathbf{B}^{(1)}$ and $\mathbf{A}^{(M)}$, $\mathbf{B}^{(M)}$ for inhomogeneous layer in the following previous relation:

$$\begin{aligned}\begin{pmatrix}\mathbf{A}^{(1)}\\ \mathbf{B}^{(1)}\end{pmatrix}&=\begin{pmatrix}\mathbf{S}_1^{(1)} & \mathbf{S}_2^{(1)}\\ \mathbf{S}_3^{(1)} & \mathbf{S}_4^{(1)}\end{pmatrix}\begin{pmatrix}\mathbf{S}_1^{(2)} & \mathbf{S}_2^{(2)}\\ \mathbf{S}_3^{(2)} & \mathbf{S}_4^{(2)}\end{pmatrix}\cdots\begin{pmatrix}\mathbf{S}_1^{(M)} & \mathbf{S}_2^{(M)}\\ \mathbf{S}_3^{(M)} & \mathbf{S}_4^{(M)}\end{pmatrix}\begin{pmatrix}\mathbf{A}^{(M)}\\ \mathbf{B}^{(M)}\end{pmatrix}\\ &=\begin{pmatrix}\mathbf{S}_1 & \mathbf{S}_2\\ \mathbf{S}_3 & \mathbf{S}_4\end{pmatrix}\begin{pmatrix}\mathbf{A}^{(M)}\\ \mathbf{B}^{(M)}\end{pmatrix}.\end{aligned} \tag{18.28}$$

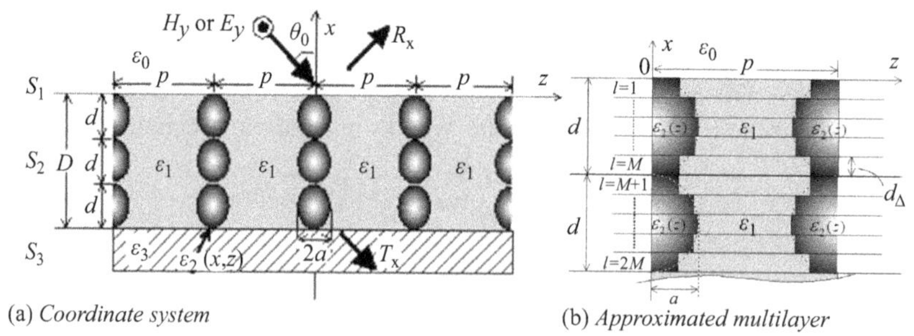

Figure 18.10 Structure of the multilayered dielectric grating with elliptically layered media. (a) coordinate system, (b) approximated multilayer

For the case of n multilayered dielectric gratings, we obtain the relationship between $A_v^{(1)}, B_v^{(1)}(d_\Delta < x < 0)$ and $A_v^{(M_T)}, B_v^{(M_T)}$, $[M_T = nM, \ -nd < x < -(nd - d_\Delta)]$,

$$\begin{pmatrix} \mathbf{A}^{(1)} \\ \mathbf{B}^{(1)} \end{pmatrix} = \begin{pmatrix} \mathbf{S}_1 & \mathbf{S}_2 \\ \mathbf{S}_3 & \mathbf{S}_4 \end{pmatrix} \left[\begin{pmatrix} \mathbf{S}_1^{(M+1)} & \mathbf{S}_2^{(M+1)} \\ \mathbf{S}_3^{(M+1)} & \mathbf{S}_4^{(M+1)} \end{pmatrix} \begin{pmatrix} \mathbf{S}_1 & \mathbf{S}_2 \\ \mathbf{S}_3 & \mathbf{S}_4 \end{pmatrix} \right]^n \begin{pmatrix} \mathbf{A}^{(M_T)} \\ \mathbf{B}^{(M_T)} \end{pmatrix}$$
$$= \begin{pmatrix} \mathbf{S'}_1 & \mathbf{S'}_2 \\ \mathbf{S'}_3 & \mathbf{S'}_4 \end{pmatrix} \begin{pmatrix} \mathbf{A}^{(M_T)} \\ \mathbf{B}^{(M_T)} \end{pmatrix}. \quad (18.29)$$

For the TM wave, to check the validity of our method, the results are computed with $N = 9\,(N_f = 6)$ and $M = 30$. For the TE wave, the results are computed with $N = 10$ and $M = 20$. In this case, the relative errors are less than about 0.1% and the energy error is less than about 10^{-3} for both TM and TE waves [36]. In the case of a circular cylinder, our results are in good agreement with those of Jandieri and Yasumoto [37].

Figure 18.11(a) and (b) shows values of $|T_0^{(TM)}|^2$ and $|T_0^{(TE)}|^2$ for various values of incident angle θ_0 with $2a/d = 0.8$, 1.0, and 1.2, $p/\lambda = 1.0$ and; the circular cylinder occurs at $2a/d = 1.0$. In Figure 18.11(a) and (b), the effect of the grating shape is seen clearly at the minimum of coupling resonance $|T_n^{(TM),(TE)}|^2$, and it moves toward smaller values of θ_0 as $2a/d$ increases. On the other hand, for the TE case, $|T_0^{(TM)}|^2$ has a symmetric shape $\theta_0 = 30°$ around because of the Bragg angle. Therefore, we note that the θ_0 dependence at coupling resonance is more significant for the TE case than for the TM case.

Figure 18.12(a) and (b) shows values of $|T_0^{(TM)}|^2$, and $|T_0^{(TE)}|^2$ for various values of normalized frequency ($p/lambda$) at $\theta_0 = 30°$ with the same parameters as in Figure 18.11. Comparing the TM case with the TE case, we note that the characteristic tendencies for the effect of grating shape are approximately the same when $p\lambda < 0.7$, but if $p\lambda > 0.7$, the effect of the grating shape is more significant for the TE case.

18.3.2 Rectangular media

In Figure 18.13, we consider the rectangular cylinders [38,39] in the middle layer sandwiched between two multilayers, yielding the basic characteristic of resonance peak for switching or frequency-selective devices for both TM and TE cases, under

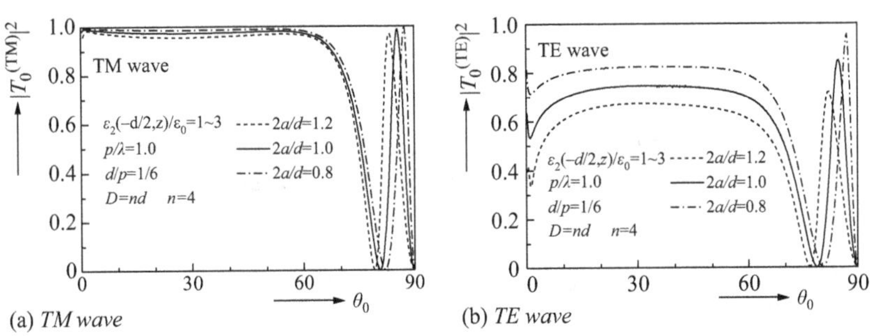

Figure 18.11 $|T_0^{(TM)}|^2$ *and* $|T_0^{(TM)}|^2$ *versus* θ_0*: (a) TM wave and (b) TE Wave*

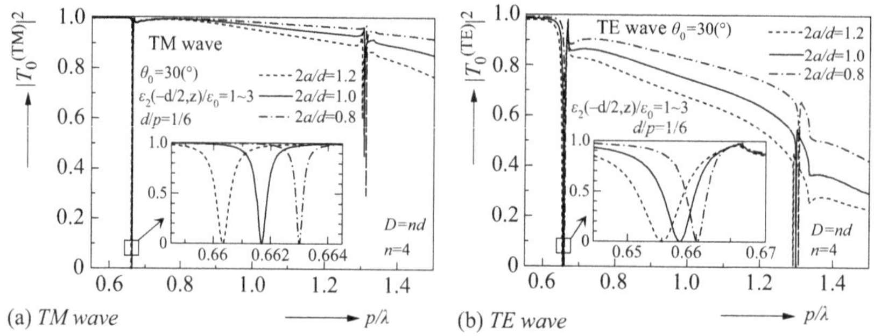

Figure 18.12 $|T_0^{(TM)}|^2$ *and* $|T_0^{(TM)}|^2$ *versus* p/λ*: (a) TM wave and (b) TE Wave*

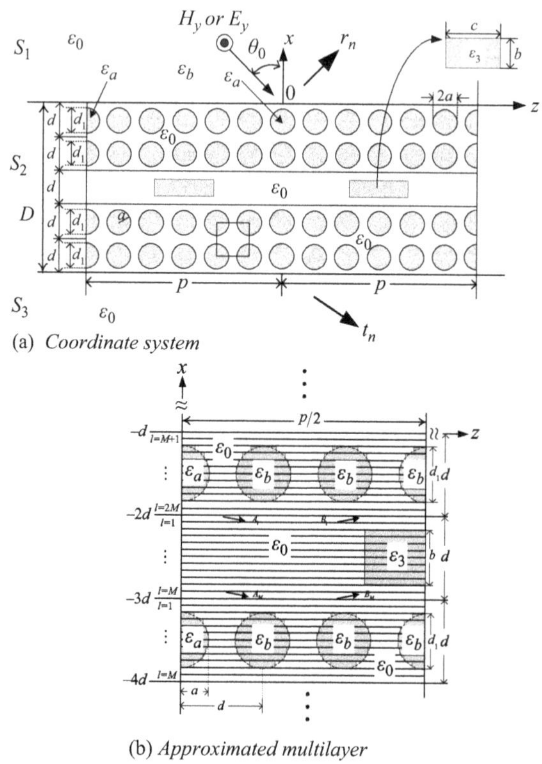

Figure 18.13 *Structure of dielectric gratings with rectangular cylinders sandwiched between two multilayers: (a) coordinate system and (b) approximated multilayer*

the conditions $\varepsilon_b/\varepsilon_0 = 3$, $D = 4d$, $b/d = 1/2$, $c/p = 1/2$, $2a/d = 1$, $d_1/d = 1$, $d/p = 1/6$, $D/p = 5/6$.

Figure 18.14(a) and (b) show the values of $|T_0^{(TM)}|^2$ and $|T_0^{(TE)}|^2$ for various values of normalized frequency (p/λ) at $\theta_0 = 30°$. From Figure 18.14(a) and (b), we can see

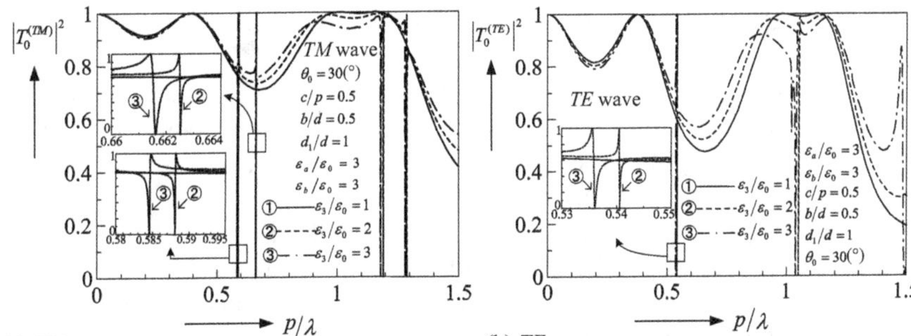

(a) *TM wave* (b) *TE wave*

Figure 18.14 Mode power transmission coefficients $|T_0^{(TM)}|^2$ and $|T_0^{(TE)}|^2$ versus a normalized frequency p/λ for the case of $\varepsilon_3/\varepsilon_0$[0]th-mode: (a) TM wave and (b) TE wave

the following features:

1. The two resonance peaks appear for both low- and high-frequency regions for the TM and TE waves; the effect in the switching region $0.5 < p/\lambda < 0.7$ of the resonance peak point does not occur for $\varepsilon_3/\varepsilon_0 = 1$.
2. The graphs of $|T_0^{(TM)}|^2$ and $|T_0^{(TE)}|^2$ have a flat maximum in the region $0.9 < p/\lambda < 1.2$ at $\varepsilon_3/\varepsilon_0 = 1$ for both TM and TE waves.

As a result, it is possible to control the characteristics by loading with rectangular cylinders in the middle layer because the resonance peaks appear.

18.3.3 Conclusions

In this section, we have analyzed a new technique for the scattering of electromagnetic waves by elliptically layered, columnar and rectangular media using a combination of the MLM and IFFM. The main points of the results are as follows:

1. In the elliptically layered and columnar media, our method is effective because the inhomogeneous region is divided into thin layers with respect to only the z components. The influences on grating shape are more significant for the inhomogeneous case than that of the homogeneous one.
2. In the rectangular media, we have obtained the basic characteristics of the power transmission coefficients and reflection coefficients in switching or frequency-selective devices for both TM and TE waves for the connection with the propagation constants in the free mode. As a result, it is possible to control the characteristics by loading the middle layer with rectangular cylinders because the resonance peaks appear. This result applies to the stop band region control of a photonic crystal waveguide discussed in the next session.

18.4 Energy distribution of defect layers

As applications of our structure, we consider optical devices that are applicable to electromagnetic wave propagation along a waveguide with periodical defects. For example, we can apply the type of another structure such as a band-pass resonant filter, channel drop filter, or electromagnetic wave filter.

In previous papers [40–42], we have analyzed dielectric waveguides composed of a dielectric circular cylinder array with an inhomogeneous dielectric structure such as triangular cylindrical or rhombic dielectric structures in the middle layer and investigated the influence of complex propagation constants at the first stop band region to examine the energy distribution analysis for the TE mode at an excited normalized frequency.

We proposed a new structure [43] that combines dielectric cylinders with an air-hole cylindrical array and investigated the influence of dielectric structures in the middle layer by IFEM. It was shown that the energy can be concentrated in the defect area by using dielectric circular cylinders in the middle layer compared with both hollow dielectric cylinders and rhombic dielectric cylinders with equivalent cross section.

In order to examine the energy distribution using the propagation constants in the guided region, the Poynting vector is defined by the following equations [43]:

$$\mathbf{S} := \mathbf{a}_x S_x^{(TE)} + \mathbf{a}_z S_z^{(TE)}, \tag{18.30}$$

$$S_x^{(TE)} := \frac{Re[E_y^{(2,l)}(x,z) \times \{H_z^{(2,l)}(x,z)\}^*]}{2},$$

$$S_z^{(TE)} := -\frac{Re[E_y^{(2,l)}(x,z) \times \{H_x^{(2,l)}(x,z)\}^*]}{2}, \tag{18.31}$$

where $\mathbf{a}_x$, $\mathbf{a}_z$ are the unit vector in the x- and z-directions, respectively. In the electromagnetic fields of the middle layer, we evaluate the unknown coefficients by solving the simultaneous equation system for $A_{\nu\neq 1}^{(l=l_m)}/A_{\nu=1}^{(l=l_m)}$ from (18.15), by setting with $A_{\nu=1}^{(l=l_m)} = 1$ at the center ($l_m := (M+1)/2$) of the middle layer. Then the coefficients $A_\nu^{(l\neq l_m)}$ and $B_\nu^{(l\neq l_m)}$ are obtained by using boundary conditions at $x = -ld_\Delta$. The energy distribution is given by

$$P^{(TE)} := \sqrt{\{S_x^{(TE)}\}^2 + \{S_z^{(TE)}\}^2}. \tag{18.32}$$

We consider the lowest guided TE_0 mode and the symmetric periodic dielectric waveguide structure which is formed by an air cylinder array in the background medium of dielectric constant ε_a, a dielectric circular cylinder array with permittivity ε_a for the second layer, and a hollow dielectric cylinder of dielectric constant ε_3^m for the middle layer. The values of parameters chosen are $D/p = 5/6$, $\varepsilon_a/\varepsilon_0 = 3$, $d_1/d = d_2/d = 0.5$, and $b/d = 1$ in Figure 18.15.

Figure 18.16(a) and (b) shows the normalized frequency p/λ versus the normalized phase constants ($\beta p/(2\pi)$) and normalized attenuation constants ($\alpha p/(2\pi)$) at the first stop band region. The results of type A are shown as a blue dashed line, that of

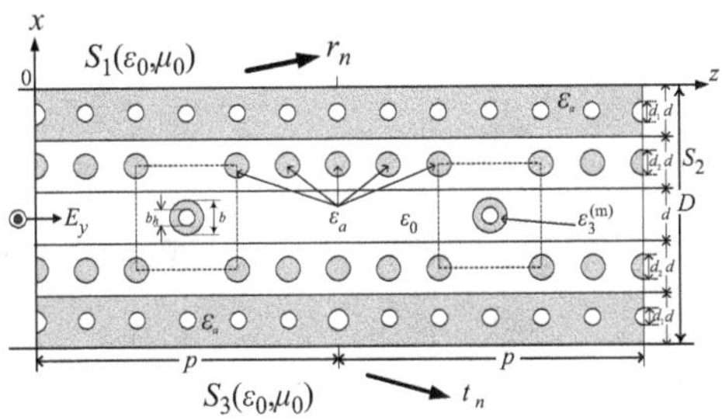

Figure 18.15 Structure and coordinate system of periodically dielectric waveguide formed by arbitrary shape of dielectric constants

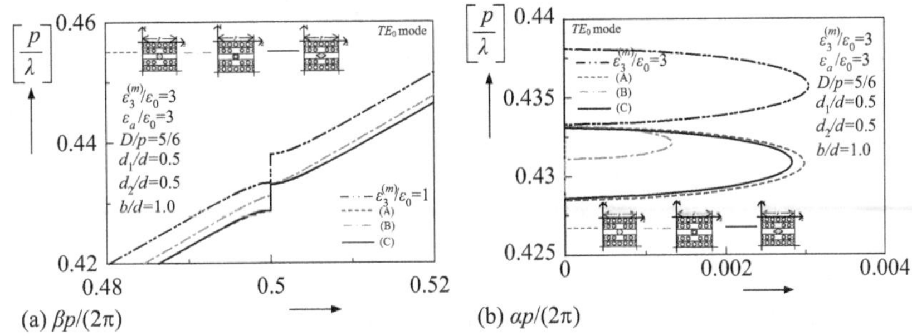

Figure 18.16 Normalized frequency p/λ versus normalized propagation constants $\gamma p/(2\pi)$: (a) normalized phase constants $\beta p/(2\pi)$ and (b) normalized attenuation constants $\alpha p/(2\pi)$

type B are shown as a red dashed-dotted line, and that of type C are shown as a black solid line. And also, the double-dotted-dashed line displays the results for the case of air ($\varepsilon_3^{(m)} = \varepsilon_0$) in the middle layer. Here, the leaky mode appears in the frequency range that is larger than normalized frequency $p/\lambda > 0.46$. From Figure 18.17, we can see the following features:

1. The energy concentration is of the same degree in the defect area for the three types.
2. In the above region of the defect area, the energy of type B becomes small compared with types A and C. We consider the effect depends on the hollow dielectric cylinder in the middle layer.

18.4.1 Conclusions

In this session, we have analyzed a new structure that combines dielectric cylinders with an air-hole cylinder array and investigated the influence of dielectric structures in the middle layer for periodic dielectric waveguides. As a result, the energy can be

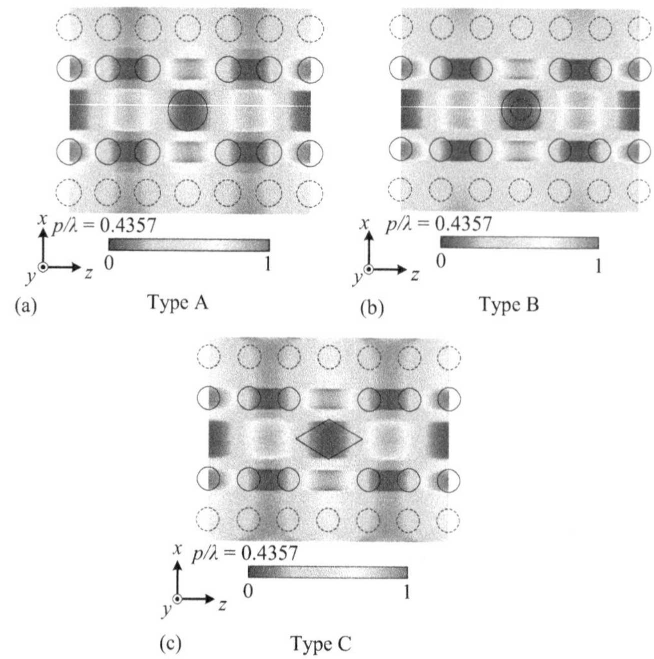

Figure 18.17 Comparison of the energy distribution $P^{(TE)}$ for dielectric waveguide structures: (a) type A, (b) type B, and (c) type C

concentrated in the defect area by using dielectric circular cylinders in the middle layer instead of hollow dielectric cylinders or rhombic dielectric cylinders with equivalent cross section. Consequently, we obtained the following conclusions for the TE_0 mode.

1. In the case of dielectric structures with equivalent cross section along a middle layer, the energy of the defect region has approximately the same values.
2. In the case of hollow dielectric structures along a middle layer, the energy outside of the defect region is affected.

18.5 Mixed positive and negative media

In this section, we propose a new method for determining the electromagnetic fields in inhomogeneous media with mixed positive and negative regions; it is a combination of the IFEM and using the extrapolation method.

One of the methods that is commonly employed in solving problems in an inhomogeneous medium is the homogeneous multilayer approximation (HMA) method. Although the method is widely known and is proved to solve the electromagnetic fields in inhomogeneous media, it cannot be applied to the positive and negative regions with oblique angle of incidences in TM wave. Yamaguchi and Hosono [44] pointed out this difficulty and applied the modified multilayer approximation (MMA) method to this problem. However, the MMA cannot be applied to the case when the permittivity profiles touch zero. In this case, HMA cannot also obtain the electromagnetic fields even if the loss term of the permittivity tends to zero.

We consider an inhomogeneous medium containing mixed positive and negative regions as shown in Figure 18.18. The structure is uniform in the y-direction and the permittivity $\varepsilon_2(z)$ has zero at $z = z_1$ and $z = z_2$ [see Figure 18.18(b)]. The permeability is assumed to be μ_0. The time dependence is $\exp(-i\omega t)$ and is suppressed throughout.

In the formulation, the TM wave (the magnetic field has only the y-component) is discussed. When the TM wave is assumed to be incident from $z < 0$ at the angle θ_0, the magnetic fields in the regions $S_1(z \leq 0)$ and $S_3(z \geq 0)$ are expressed [45] as

$$\underline{S_1(z \leq 0)} : H_y^{(1)} = e^{ik_1(x \sin\theta_0 + z\cos\theta_0)} + Re^{ik_1(x\sin\theta_0 - z\cos\theta_0)},$$

$$k_1 := \omega\sqrt{\varepsilon_1\mu_0} = k_0\sqrt{\varepsilon_1/\varepsilon_0}, \quad k_0 := 2\pi/\lambda, \tag{18.33}$$

$$\underline{S_3(z \geq d)} : H_y^{(1)} = e^{ik_1(x \sin\theta_0 + z\cos\theta_0)} + Re^{ik_1(x\sin\theta_0 - z\cos\theta_0)},$$

$$k_{3x} := \sqrt{k_3^2 - (k_1\sin\theta_0)^2}; \quad j = 1, 3, \tag{18.34}$$

where λ is the wavelength in free space, and R and T are the reflection and transmission coefficients to be determined by boundary conditions.

The inhomogeneous layer consists of periodically stratified layers which is the iteration of the permittivity $\varepsilon_d(z)[=\varepsilon_2(z)]$ in the original region ($0 < z < d$;

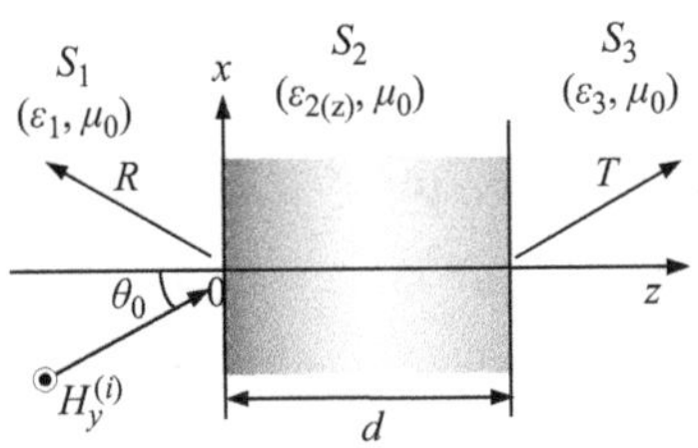

(a) *Coordinate system*

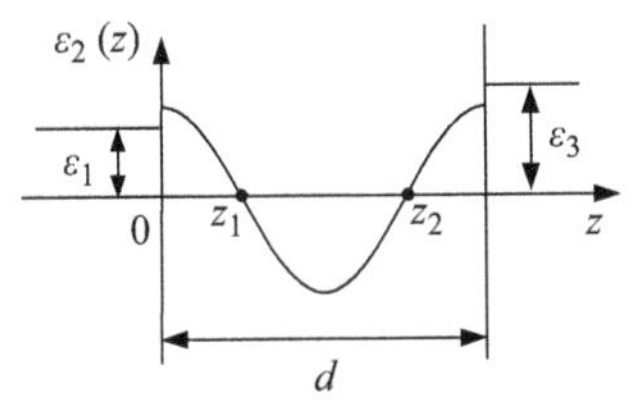

(b) *Distribution of dielectric constant*

Figure 18.18 Structure of the inhomogeneous medium mixed the positive and negative region: (a) coordinate system and (b) distribution of dielectric constant

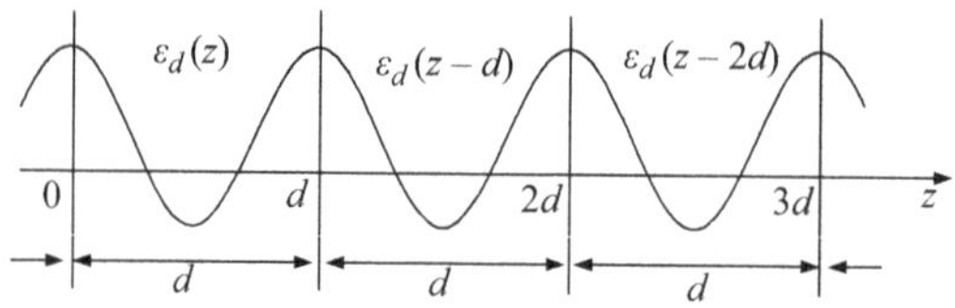

Figure 18.19 Periodically inhomogeneous layers

see Figure 18.19) [45]. The modal component of the magnetic field can be written as $H_y^{(2)} = H(z)e^{ik_1 x \sin\theta_0}$, and $H(z)$ must satisfy the following wave equation [45,46]:

$$\frac{d^2H(z)}{dz^2} - \frac{1}{\varepsilon_d(z)}\frac{d\varepsilon_d(z)}{dz}\frac{dH(z)}{dz} + \left[k_0^2\varepsilon_d(z)/\varepsilon_0 - (k_1\sin\theta_0)^2\right]H(z) = 0. \quad (18.35)$$

In the lossless case, the singularity in $\varepsilon_d(z)$ appears in the second term in (18.35). Taking into account Floquet's theorem, $H(z)$ can be approximated by the finite Fourier series as

$$H(z) = e^{ihz}\sum_{n=-N}^{N} u_n e^{i2\pi nz/d}. \quad (18.36)$$

To obtain the correct solution in the analysis of the lossless case, $\varepsilon_d(z)$ must include a loss term σ [46]:

$$\frac{\tilde{\varepsilon}_d(z)}{\varepsilon_0} := \frac{\varepsilon_d(z)}{\varepsilon_0} + i\sigma(\sigma \geq 0). \quad (18.37)$$

Substituting (18.36) and (18.37) into (18.35), multiplying both side by $\tilde{\varepsilon}_d(z)e^{-i2\pi mz/d}$, and rearranging after integrating with respect to z in the interval $0 < z < d$, we get the following equation in regard to h [45]:

$$h^2\mathbf{MU} + h\mathbf{CU} + \mathbf{KU} = 0, \quad (18.38)$$

where

$$\begin{aligned}
\mathbf{U}^{(l)} &:= [u_{-N}, \ldots, u_0, \ldots, u_N]^T \quad T: \textit{transpose}, \\
\mathbf{M} &:= [\eta_{m,n}] \quad \mathbf{C} := [\zeta_{m,n}], \mathbf{K} := [\gamma_{m,n}] \\
\zeta_{n,m} &:= \frac{2\pi}{d}\{2n + (n-m)\}\eta_{n,m}\} \\
\gamma_{n,m} &:= \left[\left(\frac{2\pi}{d}\right)^2(n^2 + n(n-m)) + (k_0\sin\theta_0)^2\right]\eta_{n,m} - \xi_{n,m} \\
m, n &= (-N, \ldots, 0, \ldots, N). \quad (18.39)
\end{aligned}$$

$$\eta_{n,m} := \frac{1}{d}\int_0^d \left\{\frac{\tilde{\varepsilon}_d(z)}{\varepsilon_0}\right\} e^{i2\pi(n-m)z/p}dz. \quad \xi_{n,m} := \frac{k_0^2}{d}\int_0^d \left\{\frac{\tilde{\varepsilon}_d(z)}{\varepsilon_0}\right\}^2 e^{i2\pi(n-m)z/p}dz.$$

Letting $\mathbf{V} := h\mathbf{U}$ and modifying (18.38), it is reduced to the following conventional eigenvalue equation:

$$\mathbf{AW} = h\mathbf{W}; \quad \mathbf{A} := \begin{bmatrix} \mathbf{0} & \mathbf{1} \\ -\mathbf{M}^{-1}\mathbf{K} & -\mathbf{M}^{-1}\mathbf{C} \end{bmatrix}, \quad \mathbf{W} := \begin{bmatrix} \mathbf{U} \\ \mathbf{V} \end{bmatrix}, \tag{18.40}$$

where $\mathbf{1}$ is the unit vector and $\mathbf{M}^{-1}$ is the inverse matrix of $\mathbf{M}$.

When an eigenvalue $h_0(=\beta + i\alpha; \alpha \geq 0)$ is obtained from (18.40) for $N \to \infty$, $-h_0(=-\beta - i\alpha)$ is also a solution, as are $(\pm h_0 \pm 2\pi n/d)$ [45]. Therefore, we select h_0 by convergence characteristics of $(\pm h_0 \pm 2\pi n/d)$ in the numerical analysis.

In the lossless case for $\sigma \to 0$, we can get the eigenvalue h_{EV} according to the following extrapolation equation ($a_0 := h_{\mathrm{EV}}$) in regard to the loss parameters $\sigma_j(j = 1 \sim 3)$ [46]:

$$h_0(\sigma_j) = a_0 + a_1 \cdot \sigma_j + a_3 \cdot (\sigma_j)^2; \quad j = 1 \sim 3. \tag{18.41}$$

In the same manner, the eigenvectors $u_n^{(EV)}$ are also obtained by the following extrapolation equation ($b_0 := u_n^{(\mathrm{EV})}$) in regard to the loss parameters $\sigma_j(j = 1 \sim 3)$ [46]:

$$u_n^{(EV)}(\sigma_j) = b_0 + b_1 \cdot \sigma_j + b_3 \cdot (\sigma_j)^2; \quad j = 1 \sim 3. \tag{18.42}$$

$\underline{S_2(0 < z < d):}$

Using h_{EV} and $-h_{EV}$, and the corresponding eigenvectors $u_n^{(EV)}$ and $u_n^{(-EV)}$, the electromagnetic fields in the inhomogeneous layer ($0 < z < d$) are expanded by a finite Fourier series [45]:

$$f_{EV}(z) := \sum_{n=-N}^{N} u_n^{(EV)} e^{i(2\pi n/d)z}, \quad f_{(-EV)}(z) := \sum_{n=-N}^{N} u_n^{(-EV)} e^{i(2\pi n/d)z}, \tag{18.43}$$

$$E_x^{(2)} = \{-i\omega\varepsilon(z)\}^{-1} \frac{\partial H_y^{(2)}}{\partial z}, \tag{18.44}$$

where $t_\nu^{(2)}$ and $r_\nu^{(2)}$ are unknown coefficients that satisfy the boundary conditions.

Using the boundary conditions $R(=R_{EV})$ and $T(=T_{EV})$ can be obtained by the following equation:

$$R_{EV}(N) = T_{EV}(N) - 1, \tag{18.45}$$

$$T_{EV}(N) = 2[f_{EV}(d)D - f_{(-EV)}(d)C]/(AD - BC), \tag{18.46}$$

where

$$A := f_{EV}(0) + \frac{\varepsilon_1 F_1}{\varepsilon_d(0)k_1 \cos\theta_0}, \quad B := f_{(-EV)}(0) + \frac{\varepsilon_1 F_2}{\varepsilon_d(0)k_1 \cos\theta_0},$$

$$C := f_{EV}(d) - \frac{\varepsilon_3 F_1}{\varepsilon_d(d)k_{3x}} e^{ih_{EV}d}, \quad D := f_{(-EV)}(d) - \frac{\varepsilon_3 F_2}{\varepsilon_d(d)k_{3x}} e^{-ih_{EV}d},$$

$$F_1 := \sum_{n=-N}^{N} \left(\frac{2\pi n}{d} + h_{EV}\right) u_n^{(EV)}, \quad F_2 := \sum_{n=-N}^{N} \left(\frac{2\pi n}{d} - h_{EV}\right) u_n^{(-EV)}. \tag{18.47}$$

From the Poynting vectors $\overline{S}, \overline{S}_x$, and $\overline{S}_z$ for various values of z/d in the inhomogeneous region our methods can be applied to the positive and negative medium [46]. The definition of Poynting vector is as follows:

$$\overline{S} = \overline{S}_x\mathbf{i} + \overline{S}_z\mathbf{j}, \quad \overline{S}_x := -\frac{\mathrm{Re}\left[E_z \times H_y^*\right]}{2}, \quad \overline{S}_z := -\frac{\mathrm{Re}\left[E_x \times H_y^*\right]}{2}, \tag{18.48}$$

where **i** and **j** are unit vectors in the direction of the x axis and z axis, respectively.

Next we consider when the distribution of permittivity $\varepsilon_2(z)$ touches $\varepsilon_2(z) = 0$. This problem is difficult to analyze by the MMA.

We consider the following sinusoidal profiles (see Figure 18.20):

$$\frac{\varepsilon_2(z)}{\varepsilon_0} = \varepsilon_A\left\{1 - \delta\cos\left(\frac{2\pi z}{d}\right)\right\} \tag{18.49}$$

$$\frac{\varepsilon_2(z)}{\varepsilon_0} = \varepsilon_A\left\{0.5 - \delta\cos\left(\frac{2\pi z}{d}\right) + 0.5\cos\left(\frac{4\pi z}{d}\right)\right\}, \tag{18.50}$$

where $\varepsilon_A := \{\varepsilon_2(\max) + \varepsilon_2(\min)\}/2, \quad \delta := \{\varepsilon_2(\max) - \varepsilon_2(\min)\}/\{\varepsilon_2(\max) + \varepsilon_2(\min)\}$.

It has three zero points at $z_1/d = 0.25$, $z_2/d = 0.5$, and $z_2/d = 0.75$. The values of the parameters ε_A and δ are chosen so $\varepsilon_1 = \varepsilon_3 = \varepsilon_0$, $\varepsilon_2(\max) = 0.1$ (profile ①), $\varepsilon_2(\max) = 0$ (profile ②) $\varepsilon_2(\min) = -0.1$ and $\sqrt{A}\lambda/d = 0.8$, respectively.

Figure 18.21(a) and (b) show the normalized magnetic fields $|H_y^{(2)}/H_y^{(i)}|$ and the normalized electric fields $|E_x^{(2)}/E_x^{(i)}|$ for various values of z/d at $\theta_0 = 30°$ for $N = 200$ using the extrapolation method. We see the following features:

1. For $|H_y^{(2)}/H_y^{(i)}|$, the influence of distribution of profiles ① and ② appears at $z/d = 0.5$;
2. The effect when the distribution of permittivity $\varepsilon_2(z)$ touches $\varepsilon_2(z) = 0$ is more significant for the electric fields $|E_x^{(2)}/E_x^{(i)}|$ at $z_2/d = 0.5$, and the limiting values

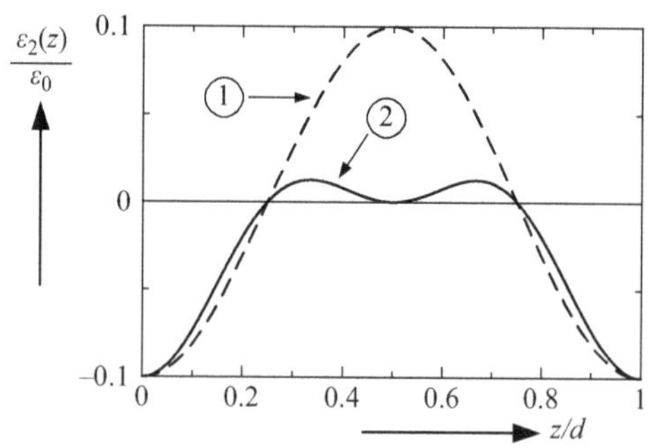

Figure 18.20 A profile of ① Eq. (18.49) and ② Eq. (18.50)

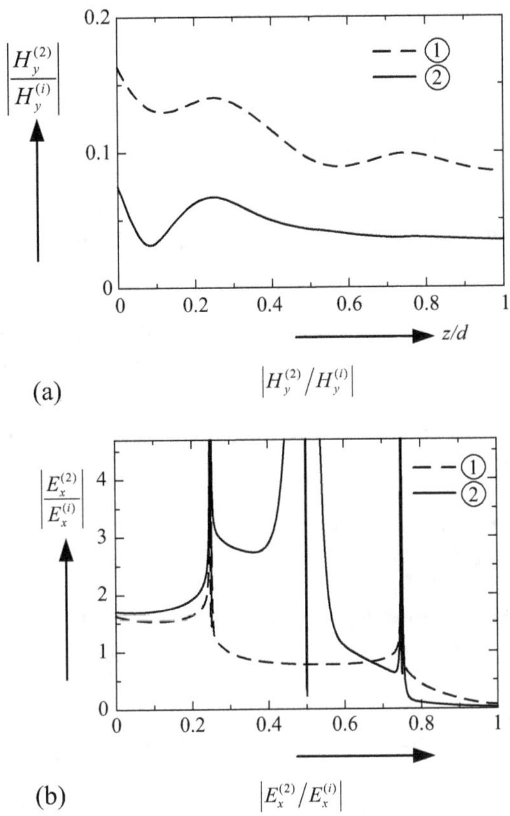

Figure 18.21 $|H_y^{(2)}/H_y^{(i)}|$ *and* $|E_x^{(2)}/E_x^{(i)}|$ *vs.* z/d

are $f_E(z_1) \cong 3.2904$, $f_E(z_2) \cong 0.35798$, and $f_E(z_3) \cong 0.65633$ for the profile ②. In the case of profile ①, the corresponding values are $f_E(z_1) \cong 1.1417$ and $f_E(z_3) \cong 0.77027$, respectively.

18.5.1 Conclusions

In this section, we have proposed a new method for determining the electromagnetic fields for inhomogeneous media with mixed positive and negative regions by the combination of the IFEM and an extrapolation method that obtains the correct value of the eigenvalue and eigenvectors for the case of TM wave.

Our results are in good agreement with the exact solution that is obtained by the MMA method for the power reflection and transmission coefficient, the energy absorption, the electromagnetic fields, and the power flow in the inhomogeneous medium, including the case when the permittivity profiles touch zero for the TM wave. This method can be also applied to metamaterials [47].

References

[1] Bertoni, H. L., Cheo, L. S. and Tamir, T., Frequency-selective reflection and transmission by a periodic dielectric layer, IEEE Trans. Antennas Propag., 1989, Vol. AP-37, No. 1, pp. 78–83.

[2] Magnusson, R., Wang, S. S., Black, T. D. and Sohn, A., Resonance properties of dielectric waveguide gratings: theory and experiments at 4-18 GHz, IEEE Trans. Antennas Propag., 1994, Vol. AP-42, No. 4, pp. 567–569.

[3] Tamir, T., ed., Integrated Optics. Springer-Verlag, Berlin, Heidelberg, 1979, pp. 110–118.

[4] Neviere, M., Vincent, P., Petit, R. and Cadilhac, M., Systematic study of resonances of holographic thin film couplers, Opt. Commun., 1973, Vol. 9, No. 1, pp. 48–53.

[5] Tremain, D. E. and Mei, K. K., Application of the unimoment method to scattering from periodic dielectric structures, J. Opt. Soc. Am., 1978, Vol. 68, No. 6, pp. 775–783.

[6] Yasuura, K. and Tomita, M., Numerical analysis of plane wave scattering from dielectric gratings, Trans. IECE Jpn., 1978, Vol. 61-B, No. 7, pp. 662–669 (in Japanese).

[7] Yamamoto, Y., Kamiya, K. and Yanai, H., Improved coupled mode analysis of corrugated waveguides and lasers, IEEE J. Quantum Electron., 1978, Vol. QE-14, No. 4, pp. 245–257.

[8] Chang, K. C., Shah, V. and Tamir, T., Scattering and guiding of waves by dielectric grating with arbitrary profiles, J. Opt. Soc. Am., 1980, Vol. 70, No. 7, pp. 804–813.

[9] Moharam, M. G. and Gaylord, T. K., Diffraction analysis of dielectric surface-relief gratings, J. Opt. Soc. Am., 1982, Vol. 72, No. 10, pp. 1385–1392.

[10] Chuang, S. L. and Kong, J. A., Wave scattering and guidance by dielectric surface-relief gratings, J. Opt. Soc. Am., 1983, Vol. 73, No. 5, pp. 669–679.

[11] Matsumoto, M., Tsutsumi, M. and Kumagai, N., Analysis of periodic waveguides using integral equation method, Tech. Rep. Electromagn. Theory, IEE Jpn., 1985, Vol. EMT-85-63, pp. 1–10 (in Japanese).

[12] Nakata, Y., Koshiba, M. and Suzuki, M., Finite-element analysis of plane wave diffraction from dielectric gratings, Trans. IECE Jpn., 1986, Vol. 69-C, No. 12, pp. 1503–1511 (in Japanese).

[13] Yasumoto, K. and Murayama, M., A rigorous analysis of a thin-film dielectric waveguide with a sinusoidal surface corrugation, Radio Sci., 1987, Vol. 22, No. 6, pp. 1013–1017.

[14] Yamakita, J., Imamura, H., Hirota, A. and Rokushima, K., Analysis of dielectric gratings using eigen-mode expansion, Trans. IEICE Jpn., 1989, Vol. J69-C-I, No. 11, pp. 740–746 (in Japanese).

[15] Raguin, D. H. and Morris, G. M., Analysis of antireflection-structured surfaces with continuous one-dimensional surface profiles, Appl. Opt., 1993, Vol. 32, No. 14, pp. 2583–2598.

[16] Lifeng, L., Multilayer modal method for diffraction gratings of arbitrary profile, depth, and permittivity, J. Opt. Soc. Am. A, 1993, Vol. 10, No. 12, pp. 2581–2591.

[17] Peng, S. T., Bertoni, H. L. and Tamir, T., Theory of periodic dielectric waveguides, IEEE Trans. Microwave Theory Tech., 1975, Vol. MTT-23, No. 1, pp. 123–133.

[18] Yamasaki, T., Hinata, T. and Hosono, T., Electromagnetic field analysis of planar gratings with periodic distribution of dielectric constant, IEICE Trans. Jpn., 1985, Vol. J69-B, No. 1, pp. 125–132 [translated Scripta Technica, INC., 1986, Vol. 69, No. 4, pp. 75–84].

[19] Yamasaki, T., Hosono, T. and Kong, J. A., Propagation characteristics of dielectric waveguides with periodic surface-relief, IEICE Trans., 1991, Vol. E74, No. 9, pp. 2839–2847.

[20] Yamasaki, T. and Tanaka, T., Scattering of electromagnetic waves by a dielectric grating with planar slanted-fringe, IEICE Trans. Jpn., 1993, Vol. E76-C, No. 10, pp. 1435–1442.

[21] Tanaka, H., Yamasaki, T. and Hosono, T., Propagation characteristics of dielectric waveguides with slanted grating structure, IEICE Trans. Jpn., 1994, Vol. E77-C, No. 11, pp. 1820–1827.

[22] Cotter, N. P. K., Preist, T. W. and Sambles, R., Scattering-matrix approach to multilayer diffraction, J. Opt. Soc. Am. A, 1995, Vol. 12, No. 5, pp. 1097–1103.

[23] Montiel, F. and Neviere, M., Differential theory of gratings: extension to deep gratings of arbitrary profile and permittivity through the R-matrix propagation algorithm, J. Opt. Soc. Am. A, 1994, Vol. 11, No. 12, pp. 3241–3250.

[24] Neviere, M., Bragg-Fresnel multilayer gratings: electromagnetic theory, J. Opt. Soc. Am. A, 1994, Vol. 11, No. 6, pp. 1835–1845.

[25] Yamasaki, T., Tanaka, T., Hinata, T. and Hosono, T., Analysis of Electromagnetic Fields in Inhomogeneous Dielectric Gratings with Periodic Surface Relief. Radio Sci., 1996, Vol. 31, No. 6, pp. 1931–1939.

[26] Yamasaki, T., Hinata, T. and Hosono, T., Scattering and Guiding of Electromagnetic Waves in Inhomogeneous Dielectric Gratings with Periodic Surface Relief, Telecommun. Radio. Eng., 2000, Vol. 5, pp. 14–27.

[27] Yamasaki, T., Hishinuma, S., Hinata, T. and Hosono, T., Propagation Characteristics of Dielectric Waveguides by Multilayer Gratings with Periodic Surface Relief, Proceedings International Symposium on Electromagnetic Theory, Vol. II, URSI, Thessaloniki, Greece, 1998, pp. 733–735.

[28] Collin, R. E. and Zucker, F. J., ed., Antenna Theory, Part 2, McGraw-Hill, New York, 1969, pp. 203–208.

[29] Conte, S. D. and Boor, C., Elementary Numerical Analysis, McGraw-Hill, New York, 1972, pp. 27–90.

[30] Chu, R. C. and Kong, J. A., Modal theory of spatially periodic media, IEEE Trans., 1977, Vol. MTT-25, No. 1, pp. 18–24.

[31] Moharam, M. G. and Gaylord, T. K., Rigorous coupled-analysis of planar-grating diffraction, J. Opt. Soc. Am., 1981, Vol. 71, No. 7, pp. 811–818.

[32] Yamasaki, T., Scattering of electromagnetic waves by inhomogeneous dielectric gratings loaded with conducting strips – matrix formulation of point matching method, IEICE Trans. Electron., 2019, Vol. E102-c, No. 1, pp. 3–11.

[33] Hinata, T. and Hosono, T., On the scattering of electromagnetic wave by plane grating placed in homogeneous medium – mathematical foundation of point matching method and numerical analysis, IECE Trans. Jpn., 1976, Vol. J59-B, No. 12, 571–578 [translated Scripta Technica, INC., 1976, Vol. 59-B, No.12, pp. 73–81].

[34] Yamasaki, T. and Hinata, T., Scattering of electromagnetic wave by plane grating with reflector, IECE Trans. Jpn., 1978, Vol. J61-B, No. 11, pp. 935–942 [translated Scripta Technica, INC., 1978, Vol. 61-B, No.11, pp. 52–60].

[35] Yamasaki, T., Hinata, T. and Hosono, T., Scattering of electromagnetic waves by multilayered dielectric gratings with elliptically layered media, Telecommun. Radio Eng., 2002, Vol. 56, No. 4–6, pp. 100–108.

[36] Yamasaki, T., Hinata, T. and Hosono, T., Scattering of electromagnetic waves by columnar dielectric gratings with elliptically layered media, Trans. IEE. Jpn., 2002, Vol. 122-A, No. 1, pp. 28–33.

[37] Jandieri, V. and Yasumoto, K., Diffraction of plane waves by multielements grating of circular cylinders, IEEJ Trans. Fundam. Mater., 2004, Vol. 24, No. 12, pp. 1148–1153.

[38] Ozaki, R., Yamasaki, T. and Hinata, T., Scattering of electromagnetic waves by multilayered inhomogeneous columnar dielectric gratings loaded rectangular dielectric constant, IEICE Trans. Jpn., 2007, Vol. E90(2)-C, pp. 295–303.

[39] Ozaki, R., Yamasaki, T. and Hinata, T., Scattering of electromagnetic waves by dielectric gratings with dielectric rectangular cylinders sandwiched between two multilayers, Trans. IEE. Jpn., 2009, Vol. 129-A, No. 10, pp. 718–724.

[40] Ozaki, R. and Yamasaki, T., Propagation characteristics of dielectric waveguides with arbitrary inhomogeneous media along the middle layer, IEICE Trans. Electron., 2012, Vol. E95-C, No. 1, pp. 53–62.

[41] Ozaki, R. and Yamasaki, T., Distribution of energy flow by dielectric waveguide with rhombic dielectric structure along a middle layer, IEICE Electron. Express, 2012, Vol. 9, No. 7, pp. 698–705.

[42] Ozaki, R. and Yamasaki, T. Distribution of energy flow by dielectric waveguide with rhombic dielectric structure along a middle layer – case of compared with deformed rhombic dielectric structure, IEICE Trans. Electron., 2013, Vol. E96-C, No. 1, pp. 68–72.

[43] Ozaki, R. and Yamasaki, T., Energy distribution of periodically dielectric waveguides by arbitrary shape of dielectric constants – the influence of dielectric structures in the middle layer, Trans. IEICE. Jpn., 2016, Vol. E99(7)-C, pp. 820–824.

[44] Yamaguchi, S. and Hosono, T., Some difficulties in Inhomogeneous multilayer approximation method and their remedy, IEICE Trans. Commun., 1981, Vol. J64-B, No. 2, pp. 1115–1122 (in Japanese) [translated Scripta Technica, INC., 1982, Vol. 65-B, No. 5, pp. 75–83].

[45] Yamasaki, T., Hinata T. and Hosono, T., Analysis of electromagnetic field in inhomogeneous media by Fourier series expansion methods, IEICE Trans. 1983, Vol. J66-B, No. 10, pp. 1239–1246 (in Japanese) [translated Scripta Technica, INC., 1984, Vol. 67-B, No. 6, pp. 61–71].

[46] Yamasaki, T., Isono, K. and Hinata, T., Analysis of electromagnetic fields in inhomogeneous media by Fourier series expansion methods – the case of a dielectric constant mixed a positive and negative regions, Trans. IEICE Jpn., 2005, Vol. E88(12)-C, pp. 2216–222.

[47] Enghta N. and Ziolkowski R. W., Metamaterials, Wiley-Interscience, IEEE Press, 2006.

Chapter 19
Methods and fast algorithms for the solution of volume singular integral equations

Alexander B. Samokhin[1]

19.1 Introduction

We will consider two families of problems of electromagnetics: (I) the medium in a finite 3D domain Q is characterized by a dielectric permittivity tensor function $\hat{\varepsilon}$ with the permittivity being constant outside Q and the permeability constant everywhere; (II) the same domain Q lies over a perfectly conducting plane. The problem is to find the electromagnetic field excited in the medium by an external field. These problems can be reduced to volume singular integral equations (VSIEs) with respect to the electric field in domain Q. Also, we study the spectrum of the operators of integral equations.

We will construct the methods and fast algorithms for the numerical solution of the initial problems by using integral equations. At the first glance, the partial differential equations are more appropriate for the numerical solution because after discretization we receive the system of linear algebraic equations (SLAE) with sparse matrix in comparison with the full matrix which we obtain in the integral equation case. But for the wave scattering problems, the solution must satisfy the radiation condition at infinity. Therefore, for the good accuracy, we need to find numerically the unknown wave field in the domain that is sufficiently greater than scattering object Q and due to 3D of the initial problems the dimension of the SLAE will be huge. Taking into account that the kernels of VSIEs depend only on the difference and sum of arguments and using regular grid for the discretization, we reduce integral equations to SLAE with matrices possessing the properties of symmetry. Then memory volume M is proportional to the dimension of SLAE. Next, using FFT techniques, we construct fast algorithms for the multiplication of the SLAE matrix by a vector. We also present iteration methods that are effective for the solution of the considered problems and demonstrate some numerical results.

[1]Department of Applied Mathematics, MIREA – Russian Technological University, Moscow, Russia

19.2 Formulation of the problems

Consider the two families of electromagnetic problems (I) and (II) introduced earlier. We look for the electromagnetic field excited in the medium by an external field $\vec{E}^0$ with the time dependence given by the factor $\exp(-i\omega t)$.

At first, consider the problem (I). The corresponding mathematical problem is stated as follows: find unknown vector functions $\vec{E}$ and $\vec{H}$ satisfying the Maxwell equations

$$\operatorname{rot}\vec{H} = -i\omega\hat{\varepsilon}\,\vec{E} + \vec{J}^0, \quad \operatorname{rot}\vec{E} = i\omega\mu_0\vec{H} \tag{19.1}$$

and the radiation condition at infinity. In (19.1), $\vec{J}^0$ is the external current generating the external field $\vec{E}^0$, $\vec{H}^0$; and $Im\,\varepsilon_0 \geq 0$, $Im\,\mu_0 \geq 0$.

Write (19.1) in the equivalent form

$$\operatorname{rot}\vec{H} = -i\omega\varepsilon_0\vec{E} + \vec{J}, \quad \operatorname{rot}\vec{E} = i\omega\mu_0\vec{H}. \tag{19.2}$$

where

$$\vec{J} = \vec{J}^0 + \vec{J}^p, \tag{19.3}$$

where

$$\vec{J}^p = -i\omega(\hat{\varepsilon} - \varepsilon_0\hat{I})\vec{E} \tag{19.4}$$

is the electric current of polarization that does not equal zero only in domain Q.

We may formally consider (19.2) as Maxwell's equations in homogeneous medium, i.e., assuming that the electromagnetic field is produced by current $\vec{J}$. Express the solution of (19.2) that satisfies the radiation condition at infinity in terms of vector potential $\vec{A}$ using the known formulas

$$\vec{A}(x) = \int \vec{J}(y)G(R)\,dy, \tag{19.5}$$

$$\vec{E} = i\omega\mu_0\vec{A} - \frac{1}{i\omega\varepsilon_0}\operatorname{grad}\operatorname{div}\vec{A}, \quad \vec{H} = \operatorname{rot}\vec{A}. \tag{19.6}$$

In (19.23), G is the Green function of the Helmholtz equation

$$G(R) = \frac{\exp(ik_0R)}{4\pi R}, \quad k_0 = \omega\sqrt{\varepsilon_0\mu_0}. \tag{19.7}$$

From (19.3) to (19.7), we obtain that unknown electromagnetic field can be presented as

$$\vec{E}(x) = \vec{E}^0(x) + k_0^2\int_Q (\hat{\varepsilon}_r - \hat{I})\vec{E}(y)G(R)\,dy$$

$$+\operatorname{grad}\operatorname{div}\int_Q (\hat{\varepsilon}_r - \hat{I})\vec{E}(y)G(R)\,dy, \; x \in E_3 \tag{19.8}$$

$$\vec{H}(x) = \vec{H}^0(x) - i\omega\varepsilon_0\operatorname{rot}\int_Q (\hat{\varepsilon}_r - \hat{I})\vec{E}(y)G(R)\,dy, \; x \in E_3, \tag{19.9}$$

where $R = |x - y|$; $x = (x_1, x_2, x_3)$; $y = (y_1, y_2, y_3)$; $\hat{\varepsilon}_r = \hat{\varepsilon}/\varepsilon_0$.

In (19.8) and (19.9), $\vec{E}^0(x)$, $\vec{H}^0(x)$ is the electromagnetic field generated by known current $\vec{J}^0$ in the homogeneous space with parameters ε_0 and μ_0. If we know the electric field, we may calculate magnetic field by using (19.9).

Because $\hat{\varepsilon}_r = \hat{I}$ outside Q we can reduce the initial problem to the volume integro-differential equation with respect to field $\vec{E}$ in the domain Q

$$\vec{E}(x) - k_0^2 \int\limits_Q (\hat{\varepsilon}_r - \hat{I})\vec{E}(y)G(R)\,dy - \operatorname{grad}\operatorname{div} \int\limits_Q (\hat{\varepsilon}_r - \hat{I})\vec{E}(y)G(R)\,dy$$
$$= \vec{E}^0(x), \quad x \in Q. \tag{19.10}$$

If we have obtained solution to (19.10) in the domain Q, then we may calculate electromagnetic field outside Q by using representations (19.8) and (19.9).

Note that we cannot apply *grad div* under the integral sign in (19.10) because in this case, one must differentiate function G twice with respect to coordinates, which yields the term $\sim 1/R^3$ in the kernel of the integral equation and the corresponding improper integrals diverge. However, the *rot* can be applied under the integral (19.9) because in this case, we will have a weak singularity $\sim 1/R^2$.

By using the theory of singular integral equations, we can reduce (19.10) to the VSIE [1,2].

$$\vec{E}(x) + \frac{1}{3}(\hat{\varepsilon}_r(x) - \hat{I})\vec{E}(x) - p.v. \int\limits_Q \Big((\hat{\varepsilon}_r(y) - \hat{I})\vec{E}(y), \operatorname{grad} \Big) \operatorname{grad} G(R)\,dy$$
$$- k_0^2 \int\limits_Q (\hat{\varepsilon}_r(y) - \hat{I})\vec{E}(y)\, G(R)\,dy = \vec{E}^0(x), \quad x \in Q. \tag{19.11}$$

Symbol $(*, *)$ denotes the inner product of the vectors and $p.v. \int$ a singular integral, i.e., an integral over a domain where an infinitesimal ball centered in a vicinity of the points $x = y$ is extracted.

Consider the second class of electromagnetic problems [3]. First, we construct Green tensor function for a current over an ideally conducting plane. Assume that, over that plane described in the Cartesian coordinate system by the equation $x_3 = 0$, in a domain Q, we have an arbitrary distribution of electric current $\vec{J}(x)$, $\vec{J} = (J_1, J_2, J_3)$, $x = (x_1, x_2, x_3)$. In the half space $x_3 \geq 0$ the medium is specified by constant values of the electric permittivity and magnetic permeability ε_0, μ_0.

The electric field of an electric dipole placed at the origin of a spherical coordinate system and oriented along the axis $\theta = 0$ is defined by the relations

$$E_r = \cos\theta f_1(r) E_\theta = \sin\theta f_2(r), \; H_\varphi = \sin\theta f_3(r), \tag{19.12}$$

where r, θ, φ are the coordinates of the spherical coordinate system. Note that the form of the function $f_1(r)$, $f_2(r)$, $f_3(r)$ makes no difference in forthcoming considerations. Consider an electric dipole (J_1) placed at the point (x_1, x_2, x_3); moreover $x_3 > 0$. We introduce the mirror dipole $(-J_1)$ placed at the point $(x_1, x_2, -x_3)$. Then, by using elementary considerations and by taking into account representations (19.12) and the equality of the distances of any point of the plane to dipoles, we find that the electromagnetic field of these dipoles in free space satisfies the radiation conditions at infinity, and on the plane $x_3 = 0$ the tangential components of the electric field is zero. Therefore, that electromagnetic field describes the field of the dipole (J_1) placed over the ideally conducting plane $x_3 = 0$. Likewise, to find the electromagnetic field of the dipole (J_2) placed over an ideally conducting plane, we introduce the mirror dipole ($-J_2$). By performing similar considerations with regard of relations (19.12), we find that the introduction of the mirror dipole (J_3) is required for the dipole (J_3).

It follows from the abovementioned considerations and formulas (19.5), (19.6) that, for the posed problem, the electromagnetic field in the half-space $x_3 \geq 0$ is described by the formulas

$$\vec{E} = i\omega\mu_0\vec{A} - \frac{1}{i\omega\varepsilon_0}\text{grad div}\,\vec{A}, \quad \vec{H} = \text{rot}\,\vec{A},$$

$$\vec{A}(x) = \int \hat{G}(x,y)\vec{J}(y)\,dy,\ y \in Q, \tag{19.13}$$

$$\hat{G}(x,y) = G\hat{I} + \hat{G}_1, \quad \hat{G}_1 = \begin{pmatrix} -G_1 & 0 & 0 \\ 0 & -G_1 & 0 \\ 0 & 0 & G_1 \end{pmatrix}, \tag{19.14}$$

where

$$G = \frac{\exp(ik_0R)}{4\pi R}, \quad G_1 = \frac{\exp(ik_0R_1)}{4\pi R_1},$$

$$R_1 = \sqrt{(x_1 - y_1)^2 + (x_2 - y_2)^2 + (x_3 + y_3)^2}. \tag{19.15}$$

where $\hat{I}$ is the unit tensor. Note that, in half-space $x_3 > 0$, the function $G_1(R_1)$ is an everywhere smooth function of coordinates. Note also that, in relations (19.14) and (19.15) the tensor $\hat{G}_1$ is represented in a Cartesian coordinate system in which the conducting plane is described by the equation, $x_3 = 0$. For other coordinate systems, the tensor $\hat{G}_1$ should be transformed in an appropriate way.

Now consider the second problem (II). The corresponding mathematical problem is stated as follows: find unknown vector functions $\vec{E}$ and $\vec{H}$ satisfying Maxwell equations (19.1), radiation condition at infinity and the condition—the tangential components of the electric field on perfectly conducting plane is zero.

By using the electric current of polarization (19.4), we can reduce the initial problem to the VSIE with respect to field $\vec{E}$ in the domain Q [3]

$$\vec{E}(x) + \frac{1}{3}(\hat{\varepsilon}_r(x) - \hat{I})\vec{E}(x) - p.v.\int_Q \Big((\hat{\varepsilon}_r(y) - \hat{I})\vec{E}(y), \text{grad}\Big)\,\text{grad}\,G(R)\,dy$$
$$- k_0^2\int_Q \hat{G}(x,y)(R)\,(\hat{\varepsilon}_r(y) - \hat{I})\vec{E}(y)\,dy$$
$$- \int_Q \text{grad}\Big(\text{grad}, \hat{G}_1(x,y)\,(\hat{\varepsilon}_r(y) - \hat{I})\vec{E}(y)\Big)\,dy = \vec{E}^0(x),\ x \in Q. \tag{19.16}$$

If we have obtained solution to (19.16) in the domain Q, then we may calculate electromagnetic field outside Q by using the following formula:

$$\vec{E}(x) = \int_Q \text{grad}\left(\text{grad}, \hat{G}(x,y)\,(\hat{\varepsilon}_r(y) - \hat{I})\vec{E}(y)\right) dy$$

$$+ k_0^2 \int_Q \hat{G}(x,y)\,(\hat{\varepsilon}_r(y) - \hat{I})\vec{E}(y)\,dy + \vec{E}^0(x),\ x \notin Q. \tag{19.17}$$

The magnetic field in the half-space is expressed via the found electric field by the integral formula:

$$\vec{H}(x) = \vec{H}^0(x) - i\omega\varepsilon_0 \int_Q [\text{grad}, \hat{G}(x,y)\,(\hat{\varepsilon}_r(y) - \hat{I})\vec{E}(y)]dy. \tag{19.18}$$

19.3 Spectrum of integral operator

The *spectrum* of the operator $\hat{A}$ on the complex plane Z is the set of points λ such that the operator $(\hat{A} - \lambda\hat{I})$ does not have an inverse defined everywhere in the Hilbert space H. The points λ such that the operator $(\hat{A} - \lambda\hat{I})$ is not Noether belong to the continuous part of the spectrum of $\hat{A}$. The points λ such that $(\hat{A} - \lambda\hat{I})$ is a Fredholm operator and there exists a nontrivial solution u, $\hat{A}u - \lambda u = 0$ belong to the discrete part of the spectrum of $\hat{A}$.

The following statement holds [1,2].

Theorem 19.1. *Assume that the Cartesian components of tensor $\hat{\varepsilon}(x)$ are Holder-continuous functions everywhere. Then the operator of the singular integral equation (19.11) is a Noether operator in Hilbert space $\vec{L}_2(Q)$ if and only if the following condition is satisfied*:

$$\sum_{n,m=1}^{3} \varepsilon_{nm}(x)\beta_n\beta_m \neq 0,\ x \in Q, \beta_1^2 + \beta_2^2 + \beta_3^2 = 1. \tag{19.19}$$

Rewrite integral equation (19.11) in the symbolic form

$$\hat{A}u \equiv u - \hat{S}\left((\hat{\varepsilon}_r - \hat{I})u\right) = f. \tag{19.20}$$

Obviously

$$\hat{A} - \lambda\hat{I} = (1 - \lambda)\left[\hat{I} - \hat{S}\left(\frac{\hat{\varepsilon}_r - \lambda I}{1 - \lambda} - I\right)\right]. \tag{19.21}$$

By comparing (19.20) and (19.21) from Theorem 19.1, we find that the continuous part of the spectrum of the operator in (19.11) contains the set σ_1 of points on the complex plane given by the formula

$$\lambda = \frac{1}{\varepsilon_0} \sum_{n,m=1}^{3} \varepsilon_{nm}(x)\beta_n\beta_m, x \in Q, \beta_1^2 + \beta_2^2 + \beta_3^2 = 1. \tag{19.22}$$

It follows that the point $\lambda = 1$ belongs to σ_1, since $\varepsilon_{nm} = \delta_{nm}\varepsilon_0$ on the boundary of the domain Q. For isotropic case ($\hat{\varepsilon}_r(x) = \varepsilon_r(x)\,\hat{I}$), we have the following formula:

$$\lambda = \varepsilon_r(x), \quad x \in Q. \tag{19.23}$$

Denote by σ minimal simply connected set on complex plane Z containing σ_1. Consider a simply connected set $\sigma^+ = Z\backslash\sigma$. Then taking into consideration the foregoing analysis, we can state that $(\hat{A} - \lambda\hat{I})$ is a Noether operator if $\lambda \in \sigma^+$. It is known [4,5] that in every connected component of the domain where an operator is of Noether type, the operators $(\hat{A} - \lambda\hat{I})$ have the same index. If $|\lambda| > \left\|\hat{A}\right\|$, then operator $(\hat{A} - \lambda\hat{I})$ has a bounded inverse defined everywhere; therefore, the index of the operator is zero. Since domain σ is bounded, there exist points $\lambda_0 \in \sigma^+$ that $|\lambda_0| > \left\|\hat{A}\right\|$; consequently, the index of operators $(\hat{A} - \lambda\hat{I})$ equals zero at $\lambda \in \sigma^+$ and they are Fredholm operators. Thus, the following statement is valid.

Theorem 19.2. *The set σ_1 on the complex plane defined by (19.22) belongs to the continuous spectrum of the operator of integral equation (19.11). In addition, $(\hat{A} - \lambda\hat{I})$ is a Fredholm operator if $\lambda \in \sigma^+ = Z\backslash\sigma$, where $\sigma$is the minima simply connected set σ_1.*

From Theorem 19.2, it follows that points $\lambda \in \sigma^+$ belong to the resolvent set of operator or its discrete spectrum.

By comparing (19.11) and (19.16), we find that their singular parts coincide. Since compact operators are neglected in the theory of singular equations, it follows that Theorems 19.1 and 19.2 are valid for integral equation (19.16).

In the general case, it is impossible to describe the localization domain of the discrete spectrum of the operator accurately. However, this can be done in a special case that is important in practice. Consider low-frequency electromagnetic wave scattering problems such that the diameter of Q is much less than the wavelength, $D \ll \lambda$, where $\lambda = 2\pi/k_0$.

19.3.1 Spectrum for low-frequency case

We will consider integral equation (19.11). Equation (19.11) can be applied when the wave number $k_0 = 0$, i.e., for the static case. Obviously, all preceding assertions remain valid in the static case. It follows from (19.11) that

$$\left(\hat{A}(k_0) - \hat{A}(0)\right)\vec{V} = -k_0^2 \int\limits_Q (\hat{\varepsilon}_r - 1)\vec{E}(y)G(R)dy$$

$$- \int\limits_Q ((\hat{\varepsilon}_r - 1)\vec{V}(y), \mathrm{grad})\,\mathrm{grad}\, G_0(R)dy,$$

$$G_0(R) = \frac{\exp(ik_0R) - 1}{4\pi R}, \quad G_1 = \frac{1}{4\pi R}, \tag{19.24}$$

where $\hat{A}(k_0)$ and $\hat{A}(0)$ are the operators in the integral equations for the stationary and static cases, respectively. The second integral operator in (19.24) does not contain a singular integral, since the kernel of this operator has no singularity at $x = y$ and is a smooth function of the coordinates. Therefore, from (19.24) we obtain

$$\lim_{k_0 \to 0} \left\| \hat{A}(k_0) - \hat{A}(0) \right\| = 0. \tag{19.25}$$

From (19.25), we have the following assertion.

Lemma 19.1. *The spectrum of the low-frequency integral operator $\hat{A}(k_0)$ tends to the spectrum of the static operator $\hat{A}(0)$ as $k_0 \to 0$.*

In the static case, the integral equation (19.10) can be represented in the form

$$\vec{E}(x) - \text{grad div} \int_Q \left((\varepsilon_r(y) - 1) \vec{E}(y) \left(\frac{1}{4\pi R} \right) \right) dy = \vec{E}^0(x). \tag{19.26}$$

The solution of the homogeneous equation (19.26) satisfies the differential equations

$$\text{rot}\, \vec{E} = 0, \quad \text{div}\, (\hat{\varepsilon}_r \vec{E}) = 0. \tag{19.27}$$

The first equation of (19.27) follows from the identity rot grad $= 0$, and the second equation follows from the identities grad div $=$ rot rot $+ \Delta$ and div rot rot $= 0$, and differential equation $\Delta \vec{A} = -\vec{J}$, which is satisfied by the volume potential $\vec{A}(x) = \int \vec{J}(y)(1/4\pi\, R)\, dy$.

From the first equation in (19.27), we have $\vec{E} = \text{grad}\, \varphi$. Then (19.27) can be reduced to a second-order differential equation for the function φ

$$\text{div}\, (\hat{\varepsilon}_r\ \text{grad}\, \varphi) = 0. \tag{19.28}$$

Let ψ be an everywhere-defined differentiable function. We have an obvious identity

$$\text{div}\, (\psi \hat{\varepsilon}_r\, \text{grad}\, \varphi) = \psi\ \text{div}\, (\hat{\varepsilon}_r\, \text{grad}\, \varphi) + (\text{grad}\, \psi,\ \hat{\varepsilon}_r\, \text{grad}\, \varphi). \tag{19.29}$$

Let $\psi = \bar{\varphi}$. Then, by integrating relation (19.29) over the space and by taking into account (19.28) and the divergence theorem, we obtain the integral relation

$$\int (\hat{\varepsilon}_r \text{grad}\, \varphi,\ \text{grad}\, \varphi)\, dv = \lim_{R \to \infty} \int_{S_R} \bar{\varphi} \frac{\partial \varphi}{\partial n}\, dS, \tag{19.30}$$

where S_R is the sphere of radius R centered at the origin and n is the normal to the sphere. Since φ is a harmonic function outside Q, it follows that $\bar{\varphi}\, \partial\varphi/\partial n$ decreases as R^{-3} at infinity. Therefore, the limit on the right-hand side of (19.30) is zero, and each solution of the homogeneous integral equation (19.26) with $\vec{E} = \text{grad}\, \varphi$ satisfies the integral relation

$$\int (\hat{\varepsilon}_r \text{grad}\, \varphi,\ \text{grad}\, \varphi)\, dv = 0. \tag{19.31}$$

It is clear that λ will be a point of the discrete spectrum of the operator of (19.26) if there exists a nonzero solution $\varphi(\lambda, x)$ of (19.28) with the permittivity function

$$\varepsilon_r^+(\lambda, x) = \frac{(\varepsilon_r(x) - \lambda)}{(1 - \lambda)}, \quad \lambda \notin \sigma_1. \tag{19.32}$$

Then, from (19.31) and (19.32) it follows that the corresponding value of λ is given by the formula

$$\lambda = \frac{\displaystyle\int \big(\hat{\varepsilon}(x) \operatorname{grad} \varphi(\lambda, x), \operatorname{grad} \varphi(\lambda, x)\big)\, dx}{\displaystyle\int |\operatorname{grad} \varphi(\lambda, x)|^2\, dx}. \tag{19.33}$$

It is impossible to find the corresponding functions $\varphi(\lambda, x)$. However, using formula (19.33), one can find the localization domain of points of the discrete spectrum on the complex plane [6].

For the isotropic case, the points of the discrete spectrum of the integral operator (19.26) can lie only inside the convex envelope of the set σ_1 given by formula (19.23). Set σ^+ lies inside convex envelope of the set σ_1. Therefore, we arrive to the statement.

Theorem 19.3. *In the case of isotropic medium, the whole spectrum of the operator* of (19.26) *and therefore of operator (19.29) at* $k_0 = 0$ *is situated only inside a convex envelope of the set* σ_1 *given by formula* (19.23).

For anisotropic case, the situation is more complicated. Write function $\hat{\varepsilon}(x)$ as

$$\hat{\varepsilon}(x) = \hat{\delta}_1(x) + i\,\hat{\delta}_2(x), \quad \hat{\delta}_1(x) = \frac{\hat{\varepsilon}(x) + \hat{\varepsilon}^*(x)}{2}, \quad \hat{\delta}_2(x) = \frac{\hat{\varepsilon}(x) - \hat{\varepsilon}^*(x)}{2i}, \tag{19.34}$$

where $\hat{\delta}_1$ and $\hat{\delta}_2$ are Hermitian tensors for all $x \in Q$ so that eigenvalues are real numbers. Denote by $a_{\min}^{(1)}(x), a_{\max}^{(1)}(x)$ and $a_{\min}^{(2)}(x),\ a_{\max}^{(2)}(x)$ the minimal and maximal eigenvalues of Hermitian tensors $\hat{\delta}_1(x)$ and $\hat{\delta}_2(x)$. The following statement is valid.

Theorem 19.4. *In the case of anisotropic medium, the whole spectrum of the operator* of (19.26) *and therefore of operator (19.29) at* $k_0 = 0$ *is situated inside a rectangle, the sides of which are parallel to the axes of the complex plane and left lower and right upper vertices have the coordinates* $(A_{\min}^{(1)}, A_{\min}^{(2)}), (A_{\max}^{(1)}, A_{\max}^{(2)})$, *where*

$$A_{\min}^{(n)} = \min a_{\min}^{(n)}(x),\ A_{\max}^{(n)} = \max a_{\max}^{(n)}(x),\ x \in Q,\ n = 1, 2.$$

Theorems 19.3 and 19.4 and Lemma 19.1 provide approximate information about the localization domain of points of the spectrum on the complex plane for the low-frequency case.

Let us illustrate theoretical finding of this study by the results of computations obtained for a representative problem: low-frequency scattering of a plane wave from homogeneous isotropic dielectric balls. There is a correspondence between the spectrum of the integral operator and its discrete (matrix) analog [5]. Of course, a matrix has only eigenvalues and has no continuous spectrum; therefore, we cannot see bold lines. However, eigenvalues will thicken along the lines of continuous spectrum.

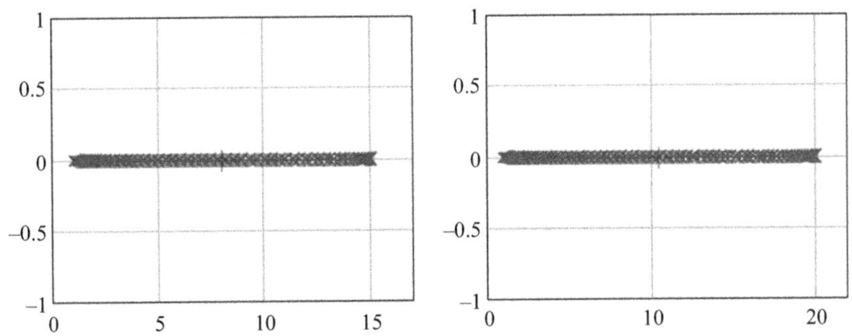

Figure 19.1 The spectra of the SLAE matrix

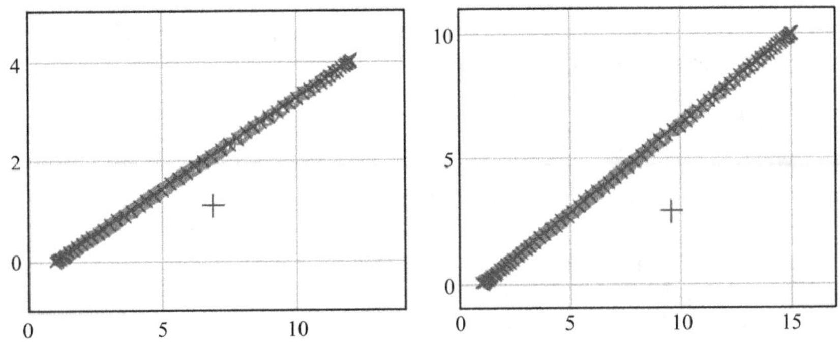

Figure 19.2 The spectra of the SLAE matrix

Discretizing integral equations (19.11), we obtain an SLAE of dimension N, where N is determined from the conditions of approximation of the integral operator.

Figure 19.1 shows the spectra of the SLAE matrix for homogeneous ball with radius $R = \lambda/30$ and dielectric permittivity $\varepsilon_r = 15$ and $\varepsilon_r = 20$.

Figure 19.2 shows the spectra of the SLAE matrix for homogeneous ball with radius $R = \lambda/30$ and dielectric permittivity $\varepsilon_r = 12 + 4i$ and $\varepsilon_r = 15 + 10i$.

It can be seen that the spectra of the operators lie on the complex plane inside the segment $[1, \varepsilon_r]$ which agree with the theory presented in this study. Note that for lossless bodies, the spectrum is situated on the real axis.

19.4 Stationary iteration methods

19.4.1 Generalized simple iteration method

In the Banach space B, we consider the linear operator equation

$$\hat{A}u = f, \tag{19.35}$$

where $\hat{A}$ is a bounded operator.

Rewrite (19.35) in the equivalent form

$$u - \hat{B}_\mu = \frac{f}{\mu}. \tag{19.36}$$

where $\hat{B}_\mu$ is the linear operator given by the formula $\hat{B}_\mu = (\mu \hat{I} - \hat{A})/\mu$ and $\mu \neq 0$ is an arbitrary complex number.

The successive approximations

$$u_{n+1} = \hat{B}_\mu u_n + \frac{f}{\mu}, \quad n = 0, 1, \ldots \tag{19.37}$$

converge to the solution of (19.36) and hence of (19.35) for any u_0, $f \in B$ provided that

$$\rho_0(\mu) = \sup |\eta(\mu)| < 1, \quad \eta(\mu) \in \sigma(\hat{B}_\mu). \tag{19.38}$$

One can readily show that there is a one-to-one correspondence between points of the spectrum $\sigma(\hat{A})$ of the operator $\hat{A}$ and points of the spectrum $\sigma(\hat{B}_\mu)$ of the operator $\hat{B}_\mu$; this correspondence is given by the formula

$$\eta = \frac{(\mu - \lambda)}{\mu}, \quad \lambda \in \sigma(\hat{A}), \quad \eta \in \sigma(\hat{B}_\mu). \tag{19.39}$$

The iterations (19.37) can be represented in the simpler form

$$u_{n+1} = u_n - \frac{1}{\mu}\left(\hat{A}u_n - f\right), \quad n = 0, 1, \ldots \tag{19.40}$$

One can prove the following statement [2].

Theorem 19.5. *A necessary and sufficient condition for the existence of complex number μ such that the iterations (19.40) converges to the solution of (19.40) for arbitrary u_0, $f \in B$ is that the origin of the complex plane lies outside a convex envelope of the spectrum of $\hat{A}$.*

Definition. The intersection of all convex sets containing a set M is called the convex envelope of M.

The convex envelope is illustrated in Figure 19.3.

The iteration converges to the solution at the rate of a geometric progression; i.e.,

$$\|u_n - u\| \leq C\,[\rho_0(\mu)]^n, \quad C = \text{const}, \tag{19.41}$$

where by (19.38) and (19.39), $\rho_0(\mu)$ is given by the formula

$$\rho_0(\mu) = \frac{\sup |\mu - \lambda|}{|\mu|}, \quad \lambda \in \sigma(\hat{A}). \tag{19.42}$$

Obviously, the best convergence of the iterations is attached at the value of μ, for which the function $\rho_0(\mu)$ takes the minimum value. By S_μ, we denote the disk on the complex plane with center μ and the least radius R that contains all points of the spectrum of $\hat{A}$. Obviously, $R = \sup |\mu - \lambda|, \ \lambda \in \sigma(\hat{A})$. From the origin, we draw the

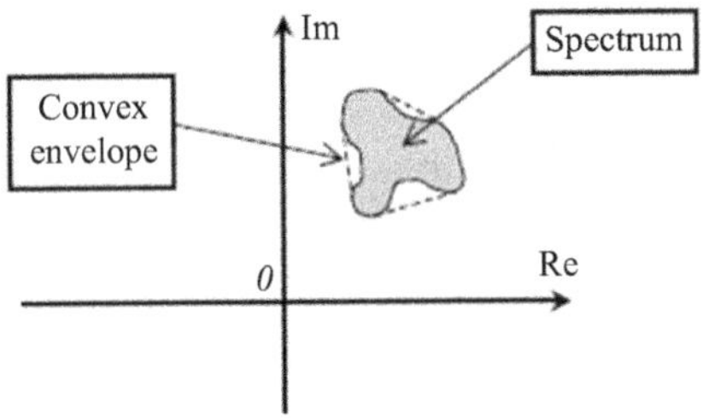

Figure 19.3 Spectrum and its convex envelope.

tangents to the disk S_μ and denote the angle between them by α. Then it follows from (19.42) that $\rho_0(\mu) = \sin(\alpha/2)$. Thus, we have proved the following assertion.

Theorem 19.6. *Let the origin of the complex plane lie outside the convex envelope of the spectrum of $\hat{A}$. Let S_0 be the disk that contains all points of the spectrum of $\hat{A}$ and is "seen" from the origin at minimal angle α_0. Then, the best convergence of the iterations (19.40) to the solution of (19.35) is attained at the complex value μ_0 that is the center of the disk S_0. The iterations converge to the solution at the rate of a geometric progression with the denominator $\rho_0 = \sin(\alpha_0/2)$.*

Now we present a finite algorithm for determination sought-for circle S_0 and corresponding iteration parameter μ_0 for a given convex envelope of the spectrum that does not contain the origin. Assume without loss of generality that this envelope is an arbitrary polygon with n vertices on the complex plane.

At first, consider segment ($n=2$) with ends z_1 and z_2 on the complex plane. Then, iteration parameter μ_0 and sought-for circle S_0 are determined by the following formulas:

$$\mu_0 = \frac{z_1 + z_2}{2} + \frac{i[\mathrm{Im}\,(z_1\,\bar{z}_2)]\,(z_2 - z_1)}{2[|z_1\,\bar{z}_2| + \mathrm{Re}\,(z_1\,\bar{z}_2)]}, \tag{19.43}$$

$$|z - \mu_0|^2 \le \frac{|z_1 - z_2|^2\,|\bar{z}_1\,z_2|}{2\,[|\bar{z}_1\,z_2| + \mathrm{Re}\,(\bar{z}_1\,z_2)]}. \tag{19.44}$$

Describe a finite algorithm of finding, for a given polygon, the sought-for circle S_0 and the best iteration parameter μ_0.

Step A. The number of all segments joining the vertices of the polygon is $n(n-1)/2$. We consider successively these segments and construct the best circle for each of them according to formulas (19.43) and (19.44). If for a certain segment this circle contains the whole polygon, then this circle will be the sought-for S_0 and the algorithm stops. If none of the circles contains the whole polygon, we go to step B.

Step B. In this case, the sought-for circle S_0 will go through three vertices of the polygon. The number of all triangles constructed on vertices of the polygon is

$n(n-1)(n-2)/6$. The sought-for S_0 will be a circle inscribed in one or several triangles. The algorithm terminates.

The convex envelope of the spectrum of an integral operator on a complex plane depending on the form of the dielectric permittivity function has been defined in Section 19.3 in the case of low-frequency electromagnetic scattering problems. It follows from Theorems 19.3 and 19.4 that the simple iteration method can be used for integral equation (19.11) for arbitrary real media; moreover, we can readily evaluate the optimal iterative parameter μ_0. Numerical experiments have shown that this method is very effective for the numerical solution of low-frequency scattering problems.

19.4.2 Generalized Chebyshev iteration method

Simple iteration method employs only one iteration parameter. However, for solving an equation with a self-adjoint operator the Chebyshev iteration turns out to be more efficient. In this method, the iteration parameters depend on the iteration number. Next, we apply this method for solving equations with nonself-adjoint operators [6].

Consider linear operator equation (19.35) in the Banach space E assuming that $\hat{A}$ is a bounded operator and there exists a bounded inverse $\hat{A}^{-1}$. We seek a solution to (19.35) using an iteration procedure

$$\begin{aligned} u^l_{n+1} &= u^l_n - \tau_{n+1}(\hat{A}u^l_n - f), \quad n = 0, \dots, m-1, \\ l &= 0, 1, \dots, u^0_0 = u_{(0)}, \quad u^{l+1}_0 = u^l_m, \quad l \geq 1, \end{aligned} \tag{19.45}$$

where $u_{(0)}$ is an initial approximation and τ_n are complex iteration parameters. From (19.45), we see that the iteration procedure constitutes a sequence of layers and each layer consists of m iterations with an identical set of parameters. Denote by h^l_n the iteration residuals defined according to the formulas

$$h^l_n = (\hat{A}u^l_n - f), \qquad n = 0, \dots, m-1, \quad l = 0, 1, \dots. \tag{19.46}$$

Now (19.45) and (19.46) yield

$$h^l_{n+1} = h^l_n - \tau_{n+1}\hat{A}h^l_n, \quad n = 0, \dots, m-1, \quad l = 0, 1, \dots. \tag{19.47}$$

From (19.47), it follows that within each layer the first and last residuals are coupled by a relationship

$$h^l_m = \left[\prod_{n=1}^{m} (\hat{I} - \tau_n\hat{A})\right] h^l_0. \tag{19.48}$$

Successively applying (19.48), we obtain that after completing k layers (19.48) residual is coupled with initial residual by the inequality

$$\|h^k_m\| \leq \left\| \left[\prod_{n=1}^{m} (\hat{I} - \tau_n\hat{A})\right]^k \right\| \|h^0_0\|. \tag{19.49}$$

Estimate (19.49) implies that the iteration parameters must minimize the norm $\left\|\left[\prod_{n=1}^{m}\left(\hat{I}-\tau_n\hat{A}\right)\right]^k\right\|$. Denote

$$\hat{B}_k(\tau_1,\ldots,\tau_m)=\left[\prod_{n=1}^{m}(\hat{I}-\tau_n\hat{A})\right]^k. \tag{19.50}$$

For any bounded operator, the following estimate [5] holds

$$\left\|\hat{B}\right\|\leq C\rho_0(\hat{B}),\quad C=const. \tag{19.51}$$

Applying (19.50) and (19.51) and the theorem of the spectrum mapping [4], we obtain

$$\left\|\hat{B}_k(\tau_1,\ldots,\tau_m)\right\|\leq C\max_{z\in\sigma(\hat{A})}\left|\prod_{n=1}^{m}(1-\tau_n z)\right|^k. \tag{19.52}$$

The following obvious inequalities are valid

$$\left\|u-u_m^k\right\|=\left\|\hat{A}\hat{A}^{-1}\left(u-u_m^k\right)\right\|\leq\left\|\hat{A}^{-1}\right\|\left\|f-\hat{A}u_m^k\right\|=\left\|\hat{A}^{-1}\right\|\left\|h_m^k\right\|, \tag{19.53}$$

where u is the sought-for to (19.35). Now from (19.49), (19.50) and (19.53), it follows

$$\left\|u-u_m^k\right\|\leq\left\|\hat{A}^{-1}\right\|C\left[\max_{z\in\sigma(\hat{A})}\left|\prod_{n=1}^{m}(1-\tau_n z)\right|\right]^k\left\|h_0^0\right\|. \tag{19.54}$$

Therefore, finding an optimal set of iteration parameters $\tau_1,\ldots,\tau_m$ reduced to the determination of a complex polynomial of degree m

$$P_m(z)=\prod_{n=1}^{m}(1-\tau_n z), \tag{19.55}$$

which has a minimal maximum of the absolute value in the domain of location of the spectrum of operator on the complex plane. Namely, we have to solve a minimax problem

$$\rho_{opt}=\min_{\{\tau_n\}}\max_{z\in\sigma(\hat{A})}\left|\prod_{n=1}^{m}(1-\tau_n z)\right|. \tag{19.56}$$

Obviously, the iteration parameters do not vanish. Therefore, performing the change of variables $\tau_n=1/\mu_n$, we can represent problem (19.56) in the form

$$\rho_{opt}=\min_{\{\mu_n\}}\frac{1}{\left|\prod_{n=1}^{m}\mu_n\right|}\max_{z\in\sigma(\hat{A})}\left|\prod_{n=1}^{m}(\mu_n-z)\right|. \tag{19.57}$$

From (19.57), we see that the case $m=1$ corresponds to simple iteration method; the determination of the corresponding iteration parameter was considered earlier.

In general, for the arbitrary spectrum localization on the complex plane, the solution to problem (19.56) is not known if $m\geq 1$.

Consider several particular cases.

If the spectrum is located on a segment $[a, b]$, $b > a > 0$, of the real axis, then we arrive at a well-known classical solution. Then the real iteration parameters are determined from the formula [7]

$$\mu_n = \frac{(b+a)}{2} + \frac{(b-a)}{2}\cos\frac{(2n-1)\pi}{2m}, \quad n = 1, \ldots, m. \tag{19.58}$$

We demonstrate expressions for parameters μ_n because for them the formulas are algebraically more convenient than for $\tau_n = 1/\mu_n$. From (19.58), it is clear that parameters μ_n belong to segment $[a, b]$.

Let the spectrum of the operator be located on a segment of a beam drawn from the origin of the complex plane. Then formula (19.57) yields $\mu_n \exp(i\varphi)$, where φ is the angle of the inclination of the beam to the real axis of the complex plane and μ_n are given by (19.58).

Assume now that the spectrum of the operator is contained in a circle S_0, and the origin of the complex plane is situated outside this domain. We have proved [6] in this case that all iteration parameters are the same for any n:

$$\tau_1 = \cdots = \tau_m = \frac{1}{\mu_0}, \tag{19.59}$$

where μ_0 is the origin of circle S_0 on the complex plane.

In the general case, when the spectrum of the operator lies in the complex plane, an algorithm of determination of iteration parameters is not known and it is necessary to apply numerical methods.

19.5 Nonstationary iteration methods

Consider the linear operator equation in the Hilbert space H

$$\hat{A}u = f \tag{19.60}$$

with a bounded operator $\hat{A}$. Define the sequence $\{u_n\}$, $u_n \in H$ by the formula

$$u_{n+1} = u_n - \tau_n(\hat{A}\,u_n - f); \quad n = 0, 1, \ldots \tag{19.61}$$

with the complex iteration parameters τ_n.

Denote

$$h_n = \hat{A}u_n - f. \tag{19.62}$$

Then multiplying both side of (19.62) by $\hat{A}$, we obtain the relationship

$$h_{n+1} = h_n - \tau_n\hat{A}h_n. \tag{19.63}$$

Therefore, we determine iteration parameters τ_n so as to provide the minimum of the functions

$$\gamma_n(\tau) = \frac{\|h_{n+1}\|}{\|h_n\|} = \frac{\left\|h_n - \tau\hat{A}\,h_n\right\|}{\|h_n\|}. \tag{19.64}$$

Represent h_n as a sum of two elements; one of them is a projection of h_n on element $\hat{A}h_n$ and the other is orthogonal to $\hat{A}h_n$. This decomposition is uniquely determined from the obvious formulas

$$h_n = \tau_n \hat{A} h_n + d_n, \qquad \tau_n = \frac{(h_n, \hat{A}\, h_n)}{(\hat{A}h_n, \hat{A}h_n)}. \tag{19.65}$$

Here (f, g) denotes the symbol of inner product for any elements f, g from Hilbert space H.

Substituting (19.65) into (19.64) and taking into account that $(\hat{A}h_n,\, d_n) = 0$, we have

$$\gamma_n^2(\tau) = \frac{|\tau_n - \tau|^2 \left\| \hat{A} h_n \right\| + \|d_n\|^2}{\|h_n\|^2}. \tag{19.66}$$

Now it is clear that (19.66) reaches the minimum at the complex value $\tau = \tau_n$. The corresponding value of $\gamma_n(\tau_n)$ is given by the expression

$$\gamma_n(\tau_n) = \sqrt{1 - \frac{\left|\left(h_n, \hat{A}h_n\right)\right|^2}{(h_n, h_n)\left(\hat{A}h_n, \hat{A}h_n\right)}}. \tag{19.67}$$

Note that $\gamma_n(\tau_n) \le 1$, equality takes place only when $(h_n,\ \hat{A}h_n) = 0$, i.e., elements h_n and $\hat{A}h_n$ are orthogonal.

The following statement is valid [2]: iterations (19.61), (19.63) and (19.65) converge to the solution (19.60) if for any $v \in H$

$$\left|\left(\hat{A}v, v\right)\right| \ge p_0\,(v, v)\,, \quad p_0 > 0. \tag{19.68}$$

By using Poynting's theorem, it can be shown that inequalities (19.68) are satisfied for (19.11) and (19.16) [2,3].

Consider a generalization of the minimal residual iteration method. The iterative parameter τ_n in (19.61) is defined from the orthogonal condition of the $(n+1)$st residual h_{n+1} to the element $\hat{A}h_n$. Consider the sequence $\{u_n\}$, $u_n \in H$, which is defined by the formulas

$$u_{n+1} = u_n - \sum_{l=1}^{m} \tau_{nl} \hat{A}^{l-1} h_n, \qquad n = 0, 1, \ldots, \qquad m = \text{const}, \qquad m \ge 1, \tag{19.69}$$

where $\{\tau_{nl}\}$ is the set of the iterative parameters used to pass from nth to $(n+1)$st layer. Because the residuals are related by $h_{n+1} = h_n - \sum_{l=1}^{m} \tau_{nl} \hat{A}^l h_n$, we define the iterative parameters from the orthogonal condition of h_{n+1} to the subspace, which is generated by the elements $\{\hat{A}h_n, \ldots, \hat{A}^m h_n\}$ (the Krylov subspace). The iterative procedure is completely defined by the formulas

$$
\begin{aligned}
& h_n^{(0)} \equiv h_n = \hat{A}u_n - f, \quad u_n^{(0)} \equiv u_n, \\
& \upsilon_l = \begin{cases} h_n^{(0)}, & l = 1, \\ h_n^{(l-1)} - \sum_{k=1}^{l-1} \beta_{lk} \upsilon_k, & l > 1, \end{cases} \\
& \beta_{lk} = \frac{\left(\hat{A}h_n^{(l-1)}, \hat{A}\upsilon_k\right)}{\left\|\hat{A}\upsilon_k\right\|^2}, \quad \tau_i = \frac{\left(h_n^{(l-1)}, \hat{A}h_n^{(l-1)}\right)}{\left\|\hat{A}\upsilon_l\right\|^2}, \\
& u_n^{(l)} = u_n^{(l-1)} - \tau_l \upsilon_l, \quad h_n^{(l)} = h_n^{(l-1)} - \tau_l \hat{A}\upsilon_l, \quad u_{n+1} \equiv u_n^{(m)}, \\
& n = 0, 1, \ldots, \quad l = 1, \ldots, m.
\end{aligned}
\tag{19.70}
$$

Iterative procedure (19.70) that may be named as multistep minimal residual method (for shot it is called GMRES method) converges to the solution of (19.60) if condition (19.68) is satisfied.

It is clear that the use of one layer in (19.70) with $m > 1$ leads to a larger decrease of the residual than that in m sequential iterations (19.61), (19.63) and (19.65). The more general assertion holds, namely, the more m in the iterative procedure the faster the convergence. Let k be the number of layers in (19.70). Then, the quantity $L = km$ defines the amount of multiplications of a matrix by a vector in the algorithm and hence L is the number of iterations in this case (see Section 19.7). However, the increase in the convergence rate is due to the increase in the amount of memory required. Therefore, when solving a particular problem, it is necessary to take into account available computational resources (e.g., speed and memory), and maybe other restrictions. Note that the use of even small $m = 2$, 3 leads to sharp increase in the convergence rate as compared with (19.61), (19.63) and (19.65).

19.6 Discretization of integral equations

The kernels of (19.11) and (19.16) depend, respectively, on the difference of Cartesian coordinates of x and y and on their difference and sum. Therefore, in the discretization, it is desirable to take account of these facts.

Introduce a grid in the rectangular Cartesian coordinate system such that the domain Q is entirely placed in the parallelepiped Π with dimensions, $N_l\, h$, $l = 1, 2, 3$, where h is the grid step along the Cartesian coordinates. The parallelepiped is divided by the grid into cells (elementary cubes). The choice of the grid is defined by the domain shape and by the requirements that the desired solution, external field and the parameters of the medium vary slightly inside the cells. The node points, at which the values of functions are given, are positioned at the cell centers (elementary cubes) $x(p)$, $p = (p_1, p_2, p_3)$ with the coordinates $x_l(p) = p_l h + \left(h/2\right)$, $l = 1, 2, 3$.

Then, integral equations are approximated by an SLAE with respect to the values of the unknown field taken at the nodes situated in domain Q.

Approximation for integral equation (19.11) has the form

$$u(p) - \sum_{y(q)\in Q} B(p-q)v(q)\,u(q) = u_0(p), \quad x(p) \in Q. \tag{19.71}$$

In (19.71), $u(p)$ and $u_0(p)$ are the vectors, $B(p-q)$ and $v(q)$ are 3×3 matrix.

Approximation for integral equation (19.16) has the following form:

$$\begin{aligned} u(p) \; - \; & \sum_{y(q)\in Q} [B_1(p-q) + B_2(p_1 - q_1, p_2 - q_2, p_3 + q_3)]v(q)\,u(q) \\ & = u_0(p), \quad x(p) \in Q. \end{aligned} \tag{19.72}$$

19.7 Fast algorithms

For numerical solution of SLAE (19.71) and (19.72), we apply an iteration method. Number T of arithmetic operations that guarantees the required accuracy of solution and memory volume M required for the implementation of the algorithm are the main efficiency criteria for any numerical algorithm. Multiplication of the SLAE matrix by a vector is the most laborious operation of the iteration method. In view of this, the number of multiplications for the implementation of a particular algorithm will be called the number of iterations. The value of T is estimated by the formula

$$T \approx L(T_A + T_0), \tag{19.73}$$

where L is the number of iterations, T_A is the number of arithmetic operations required for multiplication of a matrix by a vector and T_0 is the number of arithmetic operations required for other computations. As a rule $T_0 \ll T_A$.

Main computational expenses concerning multiplication of matrix (19.71) by a vector are connected with computing the sums

$$W(p) = \sum_{y(q)\in Q} B(p-q)\,V(q), \quad x(p) \in Q. \tag{19.74}$$

For matrix (19.72), they are connected with computing the sums

$$W(p) = \sum_{y(q)\in Q} [B_1(p-q) + B_2(p_1 - q_1, p_2 - q_2, p_3 + q_3)]V(q), x(p) \in Q. \tag{19.75}$$

Consider sums (19.74). Let us complete a definition of $V(q)$ by zero at the nodes of Π that do not belong to Q. Then we can rewrite (19.74) as

$$W(p_1, p_2, p_3) = \sum_{q_1=0}^{N_1-1} \sum_{q_2=0}^{N_2-1} \sum_{q_3=0}^{N_3-1} B(p_1 - q_1, p_2 - q_2, p_3 - q_3)\,V(q_1, q_2, q_3), \tag{19.76}$$

where the matrix function of the discrete argument $B(p)$ is defined for

$$-(N_1-1) \le p_1 \le (N_1-1); \quad -(N_2-1) \le p_2 \le (N_2-1);$$
$$-(N_3-1) \le p_3 \le (N_3-1).$$

To decrease the amount of operations, we apply the direct and inverse fast Fourier transform. Let Π_2 denote the parallelepiped with the dimensions $2N_1h$, $2N_2h$ and $2N_3h$. We continue the matrix function of the discrete argument $B(p_1,p_2,p_3)$ to all integers p_1,p_2,p_3, assuming that it is periodic with respect to each variable with the periods $2N_1, 2N_2, 2N_3$, respectively. We complete a definition of the vector function of the discrete argument $V(p_1,p_2,p_3)$ setting it zero at all nodes of Π_2 that do not belong to Π. Let us write the expression

$$W(p_1,p_2,p_3) = \sum_{q_1=0}^{2N_1-1}\sum_{q_2=0}^{2N_2-1}\sum_{q_3=0}^{2N_3-1} B(p_1-q_1,p_2-q_2,p_3-q_3)\,V(q_1,q_2,q_3). \quad (19.77)$$

It is obviously that $W(p)$ in (19.77) coincides with (19.76) for $p \in \Pi$. If we perform the discrete Fourier transform with respect to each variable in both sides of (19.77), we obtain

$$\tilde{W}(k_1,k_2,k_3) = \tilde{B}(k_1,k_2,k_3)\,\tilde{V}(k_1,k_2,k_3). \quad (19.78)$$

When calculating $W(p)$, $p \in \Pi$, from formula (19.78), the main computing expenses, without taking into the account the definition of $\tilde{B}(k)$, are connected with the fast direct and inverse Fourier transform of the vector functions of the discrete arguments. In the direct Fourier transform, the function $V(p)$ is only nonzero in Π_2 for $p \in \Pi$. On the other hand, when we apply the inverse Fourier transform, the function $W(p)$ must be defined only at the points $p \in \Pi$. Taking this into account, we find that the amount of arithmetical operations required to find $W(p)$, $p \in \Pi$ is estimated by the formula

$$T_B \sim N_1\,N_2\,N_3(LOG(N_1) + 2LOG(N_2) + 4LOG(N_3)), \quad (19.79)$$

where

$$LOG(N) = \sum_{k=1}^{m} N_k, \quad N = N_1, N_2, \ldots, N_m. \quad (19.80)$$

In (19.80), $N_1, N_2, \ldots, N_m$ are the prime numbers. If N_1, N_2, N_3 are the power of 2, we have

$$T_A \sim N_1\,N_2\,N_3(\log_2(N_1) + 2\log_2(N_2) + 4\log_2(N_3)). \quad (19.81)$$

For the sums (19.75), we have the same values (19.79) and (19.81).

19.8 Numerical results

Consider the plane wave scattering by a homogeneous dielectric cube with edge d. In Table 19.1, we present parameters of numerical calculations of these problems with

Table 19.1 Parameters of numerical calculations

D	N	Re ε	Im ε	M (Mb)	Time (s)	L	Time per iteration
1	100	0.5	0.0	1,100	25	8	3.13
2	100	0.5	0.0	1,100	47	16	2.94
5	100	0.5	0.0	1,100	68	24	2.83
1	150	0.5	0.0	3,700	97	8	12.13
2	150	0.5	0.0	3,700	192	16	12.00
3	100	1.5	0.0	1,100	47	16	2.94
4	100	2.0	0.5	1,100	69	24	2.88
4	200	2.0	0.5	8,900	771	24	32.13
6	200	4.0	2.0	8,900	1,553	56	27.73

relative error 10^{-4}. In Table 19.1, $D = \left(2\pi/\lambda\right) d$, N is the number of nodal points for each Cartesian axes and λ is the wavelength.

As we see from Table 19.1, we can solve SLAE with huge dimension (sufficiently more than one million) by using PC computer.

19.9 Conclusion

We have reduced (19.11) and (19.16) to SLAE where the type of dependence of the VSIE kernels on their arguments has been taken into account. This finding has enabled us to improve significantly the existing solution methods.

Using fast techniques, we have obtained the following estimations for the main parameters characterizing the quality of the proposed computational procedures confirming the efficiency of the proposed numerical algorithms for iterative solution of the SLAE:

Memory volume $M \approx N_1 N_2 N_3$.

Number T of arithmetic operations $T \approx LT_A$, with other parameters estimated as

$$L \approx 10 - 200,$$

$$T_A \sim N_1\, N_2\, N_3(LOG(N_1) + 2LOG(N_2) + 4LOG(N_3)),$$

where $N_1 N_2 N_3$ are the numbers of nodal points for each Cartesian axes.

The results outlined in this study are promising to elaborate new versions of iterative numerical approaches and algorithms for solving volume integral equations for the problems of electromagnetic scattering by dielectric bodies. The proposed methods and techniques can be applied for the efficient solution of large-scale problems, in particular, for SLAE of huge dimensions, especially when their matrices have certain types of symmetry connected with convolution-type integral equations.

References

[1] A.B. Samokhin, "Volume singular integral equations for problems of scattering on three-dimensional dielectric structures". Differential Equations, 2014, Vol. 50, No. 9, pp. 1201–1216.

[2] A.B. Samokhin, Integral equations and iteration methods in electromagnetic scattering. VSP, Utrecht, The Netherlands, 2001.

[3] A.B. Samokhin and A.S. Samokhina, "Volume singular integral equations for transparent structures over an ideally conducting surface". Differential Equations, 2015, Vol. 51, No. 9, pp. 1219–1224.

[4] N. Danford and J. Schwartz, Linear operators. Spectral theory. Wiley, New York, 1988.

[5] L.V. Kantorovich and G.P. Akilov. Functional analysis. Pergamon Press, Oxford, 1982.

[6] A. Samokhin, Y, Shestopalov, and K. Kobayashi, "Stationary iteration methods for solving 3D electromagnetic scattering problems". Applied Mathematics and Computations, 2013, Vol. 222, pp. 107–122.

[7] G.I. Marchuk, Methods of numerical mathematics. Springer, Berlin, 1982.

Chapter 20

Herglotz functions and applications in electromagnetics

Mitja Nedic[1], Casimir Ehrenborg[2], Yevhen Ivanenko[3], Andrei Ludvig-Osipov[4], Sven Nordebo[5], Annemarie Luger[6], Lars Jonsson[7], Daniel Sjöberg[8], and Mats Gustafsson[8]

Herglotz functions inevitably appear in pure mathematics, mathematical physics, and engineering with a wide range of applications. In particular, they are the pertinent functions to model passive systems, and thus appear in modeling of electromagnetic phenomena in circuits, antennas, materials, and scattering. In this chapter, we review the basic theory of Herglotz functions and its applications to determine sum rules and physical bounds for passive systems.

20.1 Introduction

Holomorphic mappings between certain half-planes appear in areas such as spectral theory [1,2], moment problems [3,4], passive systems [5,6], circuit synthesis [7,8], dispersion relations [9,10], and homogenization [11,12]. These functions, here referred to as Herglotz functions, are also known as Nevanlinna, Herglotz-Nevanlinna, Pick, R- [13], and positive real (PR) [7] functions. In this chapter, we review mathematical properties of Herglotz functions and discuss their applications in the modeling of passive systems, derivation of sum rules, and physical bounds.

Herglotz functions can be represented by an integral representation solely depending on scalar parameters and a positive measure. This representation is very powerful and is the starting point for many results on Herglotz functions. In particular, we use the integral representation to derive identities that relate weighted integrals of a Herglotz function with its asymptotic expansion [3]. These identities are often referred to as sum rules and have many applications in electromagnetics [14].

[1] Department of Mathematics and Statistics, University of Helsinki, Helsinki, Finland
[2] Department of Electrical Engineering, Katholieke Universiteit Leuven, Leuven, Belgium
[3] Department of Mathematics and Natural Sciences, Blekinge Institute of Technology, Karlskrona, Sweden
[4] Department of Electrical Engineering, Chalmers University of Technology, Göteborg, Sweden
[5] Department of Physics and Electrical Engineering, Linnaeus University, Växjö, Sweden
[6] Department of Mathematics, Stockholm University, Stockholm, Sweden
[7] School of Electrical Engineering and Computer Science, KTH Royal Institute of Technology, Stockholm, Sweden
[8] Department of Electrical and Information Technology, Lund University, Lund, Sweden

Passivity is instrumental in the modeling of electromagnetic phenomena. It is used to describe objects and systems that do not produce energy. Here, a system perspective is employed where passivity characterizes systems where the energy leaving the system does not exceed the energy that has entered the system for all signals and times. Linear, time-translational invariant, and passive systems have transfer functions that either are or can be transformed to Herglotz and PR functions [6]. This implies that Herglotz functions constitute a generic way to model all passive systems regardless of the complexity of the system.

Sum rules are useful in many branches of physics and engineering as they relate dynamic parameter values with their low- and high-frequency expansions [9]. Properties of the dynamic response can hence be inferred by the, in many cases much simpler, static response. These types of identities and physical bounds are of great interest in many areas of electromagnetic theory. They provide physical insight of the relation between design parameters and are useful in optimization as they provide upper bounds on the design. The approach is based on the identification of a passive system where a parameter of interest is the imaginary part of a Herglotz function. This has, e.g., been used for lossless matching networks [15], radar absorbers [16], extinction cross section [17], antennas [18], high-impedance surfaces [19,20], metamaterials [21], and array antennas [22,23].

In this chapter, we show that the integral identities for Herglotz functions constitute a unified approach to derive sum rules and illustrate the approach for the input impedance of lumped circuit networks, temporal dispersion of metamaterials, and radar absorbers.

Convex optimization can be used together or as an alternative to sum rules to determine physical bounds [24,25]. This greatly extends the class of solvable problems for the price of numerical solutions. This approach is based on the linearity and positivity of the measure in the integral representation implying that the set of Herglotz functions is a convex cone. Convex optimization can also be used to characterize rational PR functions and passive systems via the PR Lemma [5,26] and in conjunction with Nevanlinna–Pick interpolation [27].

This chapter starts with a review of the definitions, basic properties, and integral identities of Herglotz functions in Section 20.2. Passive systems in electromagnetics and their relation to Herglotz functions are discussed in Section 20.3. Sum rules and physical bounds are investigated in Section 20.4. The application of Herglotz functions in convex optimization for obtaining physical bounds is reviewed in Section 20.5.

20.2 Basics about Herglotz functions

In this section, we introduce some standard notations and review some classical results about Herglotz functions. Throughout this text, we denote the open upper and right half-planes, respectively, by

$$\mathbb{C}^+ := \{z \in \mathbb{C} \mid \mathrm{Im}[z] > 0\} \quad \text{and} \quad \mathbb{C}_+ := \{z \in \mathbb{C} \mid \mathrm{Re}[z] > 0\} \tag{20.1}$$

The open lower and left half-planes are defined analogously and are denoted by $\mathbb{C}^-$ and $\mathbb{C}_-$, respectively (see Figure 20.1). In connection with these particular subsets

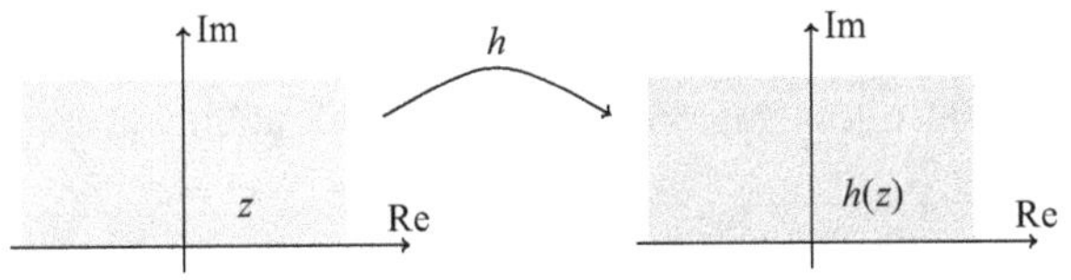

Figure 20.1 A Herglotz function maps the upper half-plane $\mathbb{C}^+$ (green) to the closed upper half-plane $\mathbb{C}^+ \cup \mathbb{R}$ (red)

of the complex plane $\mathbb{C}$, we are interested in the following classes of holomorphic (cf. [28]) functions.

Definition 20.1. A function $h\colon \mathbb{C}^+ \to \mathbb{C}$ is called a *Herglotz function* if it is holomorphic with $\mathrm{Im}[h(z)] \geq 0$ for all $z \in \mathbb{C}^+$. □

Example 20.1. Some basic examples of Herglotz functions are the following:

$$h_1(z) = 1, \quad h_2(z) = z, \quad h_3(z) = \mathrm{i}, \quad \text{and} \quad h_4(z) = -\frac{1}{z} \tag{20.2}$$

A few less trivial Herglotz functions (cf. [29]) are

$$h_5(z) = \tan(z), \quad h_6(z) = -\cot(z), \quad \text{and } h_7(z) = \frac{\Gamma'(z)}{\Gamma(z)} \tag{20.3}$$

where Γ denotes Euler's gamma function (cf. [30]) and Γ' is its derivative. ■

New Herglotz functions can be constructed using the following properties:

a. The set of all Herglotz functions is a convex cone; that is, any positive linear combination of Herglotz functions is again a Herglotz function.
b. For any two Herglotz functions h_{I} and h_{II}, such that the function h_{I} does not attain a real value, we can form the composition $z \mapsto h_{\mathrm{II}}(h_{\mathrm{I}}(z))$, which is again a Herglotz function.

For example, using only the known Herglotz functions from Example 20.1 and the two properties (a) and (b) stated previously, we can conclude that the function

$$z \mapsto 1 + z - \frac{1}{2z - \frac{3}{z} + 4\mathrm{i}} \tag{20.4}$$

is also a Herglotz function.

A common special case is to consider the composition of a Herglotz function with the function $z \mapsto \mathrm{Log}(z)$, as in the following example.

Example 20.2. First, let Log be the function defined as

$$\mathrm{Log}\colon z \mapsto \ln|z| + \mathrm{i}\,\mathrm{Arg}(z) \tag{20.5}$$

where the argument of complex number is taken from the interval $[-\pi, \pi)$. Observe that with this definition, the complex logarithm becomes a Herglotz function.

Let now h be a Herglotz function such that $h(z) \notin (-\infty, 0]$ for any $z \in \mathbb{C}^+$. The function $z \mapsto \mathrm{Log}(h(z))$ is therefore well-defined and, in fact, is also a Herglotz function, since $\mathrm{Im}[\mathrm{Log}(h(z))] = \mathrm{Arg}(h(z)) \geq 0$. ■

An important subclass of Herglotz functions is given in the following definition.

Definition 20.2. A Herglotz function h, satisfying the additional condition that

$$h(-z^*) = -h(z)^* \tag{20.6}$$

is called a *symmetric Herglotz function.* □

Herglotz functions appear, for instance, in quantum mechanics, whereas in system theory, as well as in other places, also the following class of functions is widely used.

Definition 20.3. A holomorphic function $p\colon \mathbb{C}_+ \to \mathbb{C}$ is called positive real (PR) if it has $\mathrm{Re}[p(z)] \geq 0$ for $z \in \mathbb{C}_+$ and takes real values on the positive real line (Figure 20.2). □

Any symmetric Herglotz function h gives rise to a PR function p by setting $p(z) := -\mathrm{i}\, h(\mathrm{i}\, z)$ and, conversely, any PR function p gives rise to a symmetric Herglotz function h by setting $h(z) := \mathrm{i}\, p(-\mathrm{i}\, z)$.

One of the most powerful tools in the theory of Herglotz functions is the classical integral representation theorem. The theorem was first considered over 100 years ago and has paved the way for many other results about Herglotz functions. Using modern notation, we present the theorem in the following form (see also [13]).

Theorem 20.1. A function $h\colon \mathbb{C}^+ \to \mathbb{C}$ is a Herglotz function if and only if h can be written as

$$h(z) = a + b\,z + \int_{\mathbb{R}} \left(\frac{1}{\tau - z} - \frac{\tau}{1+\tau^2} \right) \mathrm{d}\mu(\tau) \tag{20.7}$$

where μ is a positive Borel measure (cf. [32]) satisfying the growth condition

$$\int_{\mathbb{R}} \frac{1}{1+\tau^2}\, \mathrm{d}\mu(\tau) < \infty \tag{20.8}$$

and the constants $a \in \mathbb{R}$ and $b \geq 0$. □

The following properties can be derived from the above theorem [13,31]:

i. If a Herglotz function attains a real value in the open upper half-plane, then it is a real-constant function.

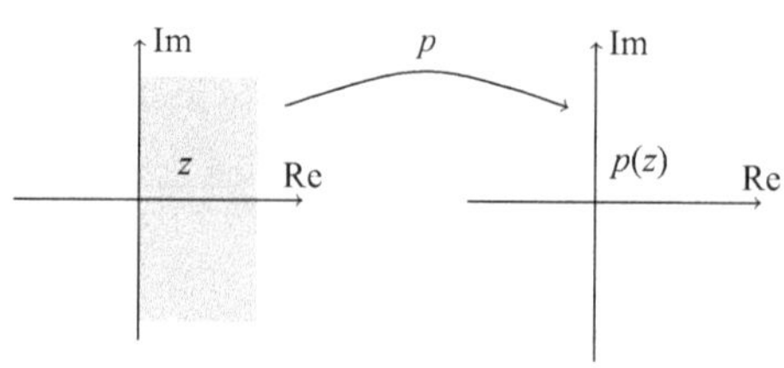

Figure 20.2 A PR function maps the right half-plane $\mathbb{C}_+$ (blue) to the closed right half-plane $\mathbb{C}_+ \cup \mathrm{i}\mathbb{R}$ (yellow) and, in particular, $\mathbb{R}^+ := (0, \infty)$ into $\mathbb{R}_0^+ := [0, \infty)$

ii. The number a from Theorem 20.1 is equal to $a = \mathrm{Re}[h(\mathrm{i})]$ while the number b is given by

$$b = \lim_{z \hat{\to} \infty} \frac{h(z)}{z} \tag{20.9}$$

Here, the symbol $\hat{\to}$ denotes a non-tangential limit (see Remark 20.1).

iii. The measure μ from Theorem 20.1 is given as the distributional boundary value of $\mathrm{Im}[h]$. More precisely, for any C^1- (i.e., continuously differentiable) function $\varphi\colon \mathbb{R} \to \mathbb{R}$ which satisfies the inequality $|\varphi(x)| \le C(1+x^2)^{-1}$ for some $C \ge 0$ and all $x \in \mathbb{R}$, we have that

$$\lim_{y \to 0^+} \frac{1}{\pi} \int_{\mathbb{R}} \varphi(x) \mathrm{Im}[h(x + \mathrm{i}\, y)]\, \mathrm{d}x = \int_{\mathbb{R}} \varphi(\tau)\, \mathrm{d}\mu(\tau) \tag{20.10}$$

iv. The measure μ has a point mass at the point $\tau_0 \in \mathbb{R}$ if and only if the limit

$$\mu(\{\tau_0\}) = \lim_{z \hat{\to} \tau_0} (\tau_0 - z) h(z) \tag{20.11}$$

is positive, [13].

v. If the measure μ is purely absolutely continuous with respect to the Lebesgue measure $\lambda_{\mathbb{R}}$ and the boundary value $\mu'(x) := \lim_{y \to 0^+} \frac{1}{\pi} \mathrm{Im}[h(x + \mathrm{i}y)]$ exists as a square integrable function, i.e., $\mu' \in \mathrm{L}^2(\mathbb{R})$, then

$$h(z) = \tilde{a} + b\, z + \int_{\mathbb{R}} \frac{\mu'(\tau)}{\tau - z} \mathrm{d}\tau, \quad \text{where } \tilde{a} := a - \int_{\mathbb{R}} \frac{\tau\, \mu'(\tau)}{1 + \tau^2} \mathrm{d}\tau \tag{20.12}$$

vi. Symmetric Herglotz functions (20.6) admit the integral representation

$$h(z) = b\, z + \lim_{R \to \infty} \int_{-R}^{R} \frac{\mathrm{d}\mu(\tau)}{\tau - z} = b\, z - \frac{c}{z} + \int_{(0,\infty)} \frac{2z}{\tau^2 - z^2} \mathrm{d}\mu(\tau) \tag{20.13}$$

where $c = \mu(\{0\})$.

Remark 20.1. The notation $z \hat{\to} \infty$ denotes the non-tangential limit $|z| \to \infty$, i.e., within some *Stoltz domain* $\{z \in \mathbb{C}^+ \,|\, \theta \le \mathrm{Arg}(z) \le \pi - \theta\}$ with parameter $\theta \in (0, \frac{\pi}{2}]$. Similarly, $z \hat{\to} \tau_0$ denotes the non-tangential limit $z \to \tau_0$ within $\{z \in \mathbb{C}^+ \mid \theta \le \mathrm{Arg}(z - \tau_0) \le \pi - \theta\}$ with $\theta \in (0, \frac{\pi}{2}]$. A Stoltz domain in the latter case is visualized in the figure below.

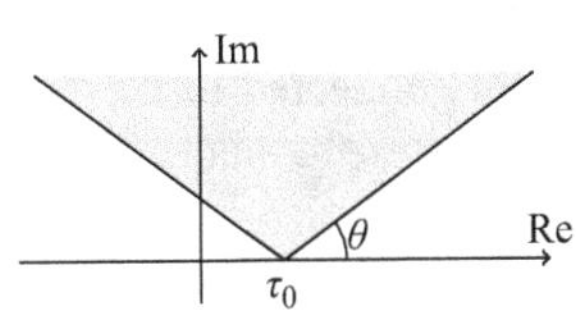

The integral representation offers also a way to synthesize Herglotz functions.

Example 20.3. The point measure δ_{x_0} at a point $x_0 \in \mathbb{R}$, along with suitably chosen constants a and b, produces the rational Herglotz function $h(z) = 1/(x_0 - z)$ with a simple pole at the point x_0; see also (20.11). Here, we note that $\mathrm{Im}[h(x)] = 0$ for

$x \in \mathbb{R} \setminus \{x_0\}$ while $h(x_0)$ is undefined. Similarly, the sum of the point masses $\delta_{(n-\frac{1}{2})\pi}$ at the points $(n-\frac{1}{2})\pi$, $n \in \mathbb{Z}$ produces the tangent function, which is thus a Herglotz function, i.e.,

$$\tan(z) = \sum_{n=-\infty}^{\infty} \left(\frac{1}{(n-\frac{1}{2})\pi - z} - \frac{(n-\frac{1}{2})\pi}{1+(n-\frac{1}{2})^2\pi^2} \right) = \sum_{n=1}^{\infty} \frac{8z}{(2n-1)^2\pi^2 - 4z^2} \tag{20.14}$$

where the first expression includes the convergence term in (20.7) and the symmetric version (20.13) is used in the second expression. ■

The class of Herglotz functions has a non-empty intersection with Hardy spaces (cf. [33]). The Hardy space H^p, where $1 \leq p \leq \infty$, is the space of holomorphic functions in $\mathbb{C}^+$ for which the norm

$$\|g\|_{\mathrm{H}^p} := \begin{cases} \sup\limits_{y>0} \left(\int_{\mathbb{R}} |g(x+\mathrm{i}y)|^p \,\mathrm{d}x \right)^{\frac{1}{p}} & 1 \leq p < \infty \\ \sup\limits_{z \in \mathbb{C}^+} |g(z)| & p = \infty \end{cases} \tag{20.15}$$

is finite. In particular, for a function $g \in \mathrm{H}^p$ for almost all $x \in \mathbb{R}$ the boundary value $\hat{g}(x) := \lim_{y \to 0^+} g(x+\mathrm{i}y)$ exist and $\hat{g} \in \mathrm{L}^p(\mathbb{R})$.

Hardy spaces are also related to the Hilbert transform [9]. The Hilbert transform of a function u is defined as

$$\mathscr{H}\{u\}(x) := \frac{1}{\pi} \,\mathrm{p.v.} \int_{\mathbb{R}} \frac{u(\tau)}{x-\tau} \,\mathrm{d}\tau \tag{20.16}$$

where p.v. means the Cauchy principal value of the integral. The operator $\mathscr{H}$ is bounded in L^p for $1 < p < \infty$; see [9] for a discussion of the L^1 and L^∞ cases.

The real and imaginary parts of the boundary value of a function $g \in \mathrm{H}^p$ are related by the Hilbert transform, i.e., for $u(x) := \mathrm{Re}[\hat{g}(x)]$ and $v(x) := \mathrm{Im}[\hat{g}(x)]$ it holds that

$$u = -\mathscr{H}\{v\} \quad \text{and} \quad v = \mathscr{H}\{u\} \tag{20.17}$$

Functions $g \in \mathrm{H}^p$ with $\mathrm{Im}[g(z)] \geq 0$ are Herglotz functions and the corresponding Hilbert transform relations are often referred to as the Kramers–Kronig relations or dispersion relations [10]. On the contrary, some Herglotz functions do not belong to any space H^p, some examples being the functions h_2, h_4, h_5, h_6, and h_7 from Example 20.1.

Another feature of Herglotz functions is that they, under certain conditions, provide integral identities closely connected to the moments of the measures with interesting applications to electromagnetic theory. Let us begin with some necessary definitions.

Definition 20.4. Let h be a Herglotz function. If for $K \geq -1$, there exist real numbers $b_1, b_0, b_{-1}, \ldots, b_{-K}$ such that h can be written as

$$h(z) = b_1 z + b_0 + \frac{b_{-1}}{z} + \cdots + \frac{b_{-K}}{z^K} + o\left(\frac{1}{z^K}\right) \qquad \text{as } z \hat{\to} \infty \tag{20.18}$$

then we say that h admits at $z = \infty$ an *asymptotic expansion of order* K. □

Remark 20.2. This means that

$$\lim_{z \hat{\to} \infty} z^K \left(h(z) - b_1 z - b_0 - \frac{b_{-1}}{z} - \cdots - \frac{b_{-K}}{z^K} \right) = 0 \tag{20.19}$$

Moreover, the coefficients b_{-j} are given by

$$b_{-j} = \lim_{z \hat{\to} \infty} z^j \left(h(z) - b_1 z - b_0 - \frac{b_{-1}}{z} - \cdots - \frac{b_{-(j-1)}}{z^{j-1}} \right) \tag{20.20}$$

Expansions at $z = 0$ are defined analogously. This can either be done explicitly, as in the following, or via the expansion at ∞ for the Herglotz function $\widetilde{h}(z) := h(-1/z)$. The above remark applies then accordingly.

Definition 20.5. Let h be a Herglotz function. If for $K \geq -1$, there exist real numbers $a_{-1}, a_0, a_1, \ldots, a_K$ such that h can be written as

$$h(z) = \frac{a_{-1}}{z} + a_0 + a_1 z + \cdots + a_K z^K + o(z^K) \qquad \text{as } z \hat{\to} 0 \tag{20.21}$$

then we say that h admits at $z = 0$ an *asymptotic expansion of order* K. □

Remark 20.3. Note that every Herglotz function has asymptotic expansions both at $z = \infty$ and at $z = 0$ of order -1. Indeed, by equalities (20.9) and (20.11), the numbers b_1 and a_1 always exist and are equal to the numbers b from Theorem 20.1 and $-\mu(\{0\})$, respectively.

The following two theorems are the central statements in this context, [14]. We start with the version with positive exponents, which are related to expansions at ∞.

Theorem 20.2. Let h be a Herglotz function. Then, for some integer $N_\infty \geq 0$, the limit

$$\lim_{\varepsilon \to 0^+} \lim_{y \to 0^+} \int_{\varepsilon < |x| < \frac{1}{\varepsilon}} x^{2N_\infty} \operatorname{Im}[h(x + \mathrm{i}y)] \, \mathrm{d}x \tag{20.22}$$

exists as a finite number if and only if the function h admits at $z = \infty$ an asymptotic expansion of order $2N_\infty + 1$. In this case,

$$\lim_{\varepsilon \to 0^+} \lim_{y \to 0^+} \frac{1}{\pi} \int_{\varepsilon < |x| < \frac{1}{\varepsilon}} x^n \operatorname{Im}[h(x + \mathrm{i}y)] \, \mathrm{d}x = \begin{cases} a_{-1} - b_{-1} & n = 0 \\ -b_{-n-1} & 0 < n \leq 2N_\infty \end{cases} \tag{20.23}$$

holds. □

The corresponding result for negative exponents, which are related to the expansions at 0, reads as follows.

Theorem 20.3. Let h be a Herglotz function. Then, for some integer $N_0 \geq 1$, the limit

$$\lim_{\varepsilon \to 0^+} \lim_{y \to 0^+} \int_{\varepsilon < |x| < \frac{1}{\varepsilon}} \frac{\operatorname{Im}[h(x + \mathrm{i}y)]}{x^{2N_0}} \, \mathrm{d}x \tag{20.24}$$

exists (as a finite number) if and only if h admits at $z = 0$ an asymptotic expansion of order $2N_0 - 1$. In this case,

$$\lim_{\varepsilon\to 0^+}\lim_{y\to 0^+}\frac{1}{\pi}\int_{\varepsilon<|x|<\frac{1}{\varepsilon}}\frac{\mathrm{Im}[h(x+\mathrm{i}y)]}{x^p}\,\mathrm{d}x = \begin{cases} a_1 - b_1 & p = 2 \\ a_{p-1} & 2 < p \le 2N_0 \end{cases} \tag{20.25}$$

holds. □

The above integral identities (20.23) and (20.25) are often called *sum rules.*

Remark 20.4. The proof of the above two theorems relies on a version of formula (20.10) for noncontinuous test functions; see, e.g., [34].

Example 20.4. The Herglotz function $h(z) = \tan(z)$ has the asymptotic expansion

$$\tan(z) = z + \frac{z^3}{3} + \frac{2z^5}{15} + \cdots \quad \text{as } z\hat{\to}0 \tag{20.26}$$

and $\tan(z) = \mathrm{i} + o(1)$ as $z\hat{\to}\infty$. We thus find that $a_1 = 1$, $a_3 = 1/3$, $a_5 = 2/15$, and $b_1 = 0$, and the following sum rules

$$\lim_{\varepsilon\to 0^+}\lim_{y\to 0^+}\frac{1}{\pi}\int_{\varepsilon\le|x|\le 1/\varepsilon}\frac{\mathrm{Im}[\tan(x+\mathrm{i}y)]}{x^p}\,\mathrm{d}x = \begin{cases} 1 & p = 2 \\ 1/3 & p = 4 \\ 2/15 & p = 6 \end{cases} \tag{20.27}$$

apply. ■

Remark 20.5. Note that the case of $p = 1$ is not included in Theorem 20.3. In order to guarantee this limit to be finite, it is required that h admits both at ∞ and at the point zero asymptotic expansions of order 1. In this case, the limit equals $a_0 - b_0$.

Remark 20.6. Note that the exponents in (20.22) and (20.24) are even. A corresponding statement for odd exponents, meaning that the existence of the limit is equivalent to the existence of the expansion, does not hold. A counterexample is given in [14, p. 9].

Example 20.5. Note that the assumption that the coefficients in expansions (20.18) and (20.21) are real is essential. Consider, e.g., the function $h(z) = \mathrm{i}$ for $z \in \mathbb{C}^+$, which admits expansions of arbitrary order if non-real coefficients are allowed. However, neither of the limits (20.22) nor (20.24) do exist. Note that this example also shows that not every Herglotz function does admit a sum rule. ■

For symmetric Herglotz functions, we note that the non-zero odd and even ordered coefficients within an asymptotic expansion of a symmetric Herglotz function (20.6) are necessarily real-valued and purely imaginary, respectively. Thus based on Theorems 20.2, 20.3 and Remark 20.5, we stop our expansion at the appearance of the first imaginary term, or the first non-existing term. If the assumptions in both theorems are satisfied, i.e., that both asymptotic series exist, are real-valued to order $2N_0 - 1$ and $2N_\infty + 1$, respectively, we can summarize Theorems 20.2, 20.3 and Remark 20.5 as

$$\frac{2}{\pi}\int_{0+}^{\infty} \frac{\mathrm{Im}[h(x)]}{x^{2n}}\,\mathrm{d}x := \lim_{\varepsilon\to 0^+}\lim_{y\to 0^+}\frac{2}{\pi}\int_{\varepsilon}^{1/\varepsilon} \frac{\mathrm{Im}[h(x+\mathrm{i}y)]}{x^{2n}}\,\mathrm{d}x = a_{2n-1} - b_{2n-1} \tag{20.28}$$

for $n = -N_\infty, \ldots, N_0$.

20.3 Passive systems

In this section, we show how symmetric Herglotz functions and the corresponding PR functions are related to passive systems.

Physical objects that cannot produce energy are usually considered as passive. However, these objects are not necessarily passive from a system point of view. The crucial point is how the input and the output of the system are defined. For the most part here, we constrain our viewpoint to one-port systems. These systems consist of one input and one output parameter, which can be measured at the *ports* of these systems (see Figure 20.3). Here, the signal enters the system and its response can be measured. A common example of such a system is an electric circuit with two nodes to which we can input a signal, e.g., a current, and measure a voltage. The one-port systems, we regard in this section are linear, continuous, and time-translationally invariant. Time-translation invariance means that the system does not explicitly depend on time; that is, if a system produces the output $v(t)$ from the input $u(t)$ at time t, then a time-shifted input $u(t+\tau)$ gives a shifted output $v(t+\tau)$. These properties characterize systems in convolution form [6].

Definition 20.6. A system with input $u(t)$ and output $v(t)$ in the time-domain is in convolution form if

$$v(t) = (w * u)(t) := \int_{\mathbb{R}} w(\tau)u(t-\tau)\,\mathrm{d}\tau \tag{20.29}$$

where $w(t)$ is the impulse response. □

In this chapter, we restrict ourselves to real-valued systems, i.e., the systems where the impulse response w is real-valued.

There are two ways to define passivity for different types of systems: admittance passivity and scattering passivity [6,35].

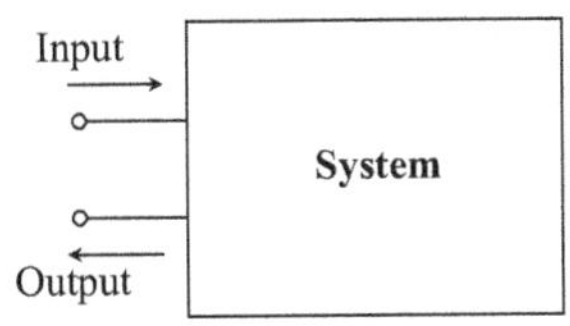

Figure 20.3 A one-port system

Definition 20.7. Consider a convolution system with input signal $u(t)$ and output signal $v(t)$, both of which in general can be complex valued. The system is called *admittance-passive* if

$$\mathscr{W}_{\mathrm{adm}}(T) := \mathrm{Re} \int_{-\infty}^{T} v(t)u(t)^* \,\mathrm{d}t \geq 0 \tag{20.30}$$

for all $T \in \mathbb{R}$ and all $u \in C_0^\infty$ (i.e., smooth functions with compact support). □

Here, $\mathscr{W}_{\mathrm{adm}}(T)$ represents all energy the system has absorbed until the time T. By requiring this quantity to be positive, we say that the system absorbs more energy than it emits, and thus, the system does not produce energy. Passivity also implies that the system is causal [6].

The crucial connection to Herglotz and PR functions is given by the following fact. It can be shown that the impulse response of a passive system has the representation [6]

$$w(t) = b\delta'(t) + \theta(t) \int_{\mathbb{R}} \cos(\xi t)\,\mathrm{d}\mu(\xi) \tag{20.31}$$

where $b \geq 0$, δ' denotes the derivative of the Dirac distribution (cf. [6]), θ is the Heaviside step function (cf. [6]), and the measure μ satisfies the growth condition (20.8). This implies that the Laplace transform of the impulse response (20.31), $W_{\mathrm{adm}}(s)$, is a PR function and that the corresponding symmetric Herglotz function (20.13) has exactly the parameters b and μ.

Let us consider a few examples of passive systems in electromagnetics.

Example 20.6. Input impedance of electrical circuit networks Consider a simple electric one-port circuit containing passive components, meaning that all lumped element components have positive values, e.g., that each resistance R, inductance L, and capacitance C are positive. The input signal to this system is the real-valued electric current $i(t)$ and its output signal is the voltage $v(t)$; see Figure 20.4(a). As an explicit example, consider the simple circuit in Figure 20.4(b). To verify that this system is passive, we evaluate the integral (20.30).

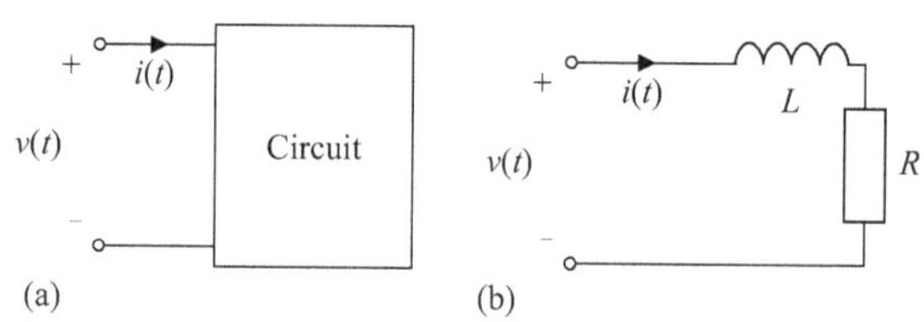

Figure 20.4 Two electrical one-port systems: (a) A general electric circuit. (b) A simple circuit example

Given the input current $i(t)$, the voltage is given by $v(t) = L\frac{\mathrm{d}i(t)}{\mathrm{d}t} + Ri(t) = (w * i)(t)$, where $w = L\delta' + R\delta$ is the impulse response, cf. (20.31). Hence, the integral (20.30) becomes

$$\mathscr{W}_{\mathrm{adm}}(T) = \int_{-\infty}^{T} \left(L\frac{\mathrm{d}i(t)}{\mathrm{d}t}i(t) + Ri(t)^2 \right) \mathrm{d}t = \frac{L}{2}i(T)^2 + R\int_{-\infty}^{T} i(t)^2\,\mathrm{d}t \geq 0 \tag{20.32}$$

showing that the system is admittance-passive. The transfer function (input impedance), which by definition is the Laplace transform of the impulse response, becomes, in this case, the PR function

$$W_{\mathrm{adm}}(s) := Z_{\mathrm{in}}(s) = sL + R \tag{20.33}$$

The input and output can be interchanged, and by using the voltage as input, the input admittance is given by $Y_{\mathrm{in}}(s) = 1/Z_{\mathrm{in}}(s)$, which is also a PR function. This simple example generalizes to circuit networks composed of arbitrary number and combinations of passive resistors, capacitances, and inductances resulting in rational PR functions [36]. Moreover, it is straightforward to include transformers and transmission lines as well as multiple input and output systems resulting in matrix-valued PR functions [37]. ■

In contrast to this example, in other situations, it is often less clear how to define a system in a suitable way so that it becomes passive. For instance, consider the constitutive relations.

Example 20.7. Constitutive relations Electromagnetic properties of isotropic materials are characterized through constitutive relations with quantities known as permittivity ε and permeability μ, which relate the frequency domain electric $\boldsymbol{E}$ and magnetic $\boldsymbol{H}$ field intensities to the electric $\boldsymbol{D} = \varepsilon\boldsymbol{E}$ and magnetic $\boldsymbol{B} = \mu\boldsymbol{H}$ flux densities, respectively. These constitutive relations are regarded as systems with field intensities and flux densities as input and output signals, respectively. The class of linear, time-translationally invariant, continuous and passive constitutive relations are, hence, related to PR and Herglotz functions. The energy relation in (20.30) is derived by the Poynting's theorem [38,39] and reveals that the pertinent passive systems are the maps from the field intensities to the temporally differentiated flux densities [14,21].

Time-domain admittance passivity implies that in the Laplace domain, the corresponding transfer function is a PR function or, equivalently, a symmetric Herglotz function after some suitable transformations; see Section 20.2. In this domain, time differentiation results in a multiplication with the angular frequency. The constitutive relation $\boldsymbol{D}(s) = \varepsilon(s)\boldsymbol{E}(s)$ is a passive system if $s\varepsilon(s)$ is a PR function. Similarly, permeabilities μ, such that $s\mu(s)$ are PR functions, can be used to model all linear, time-translationally invariant, continuous, and passive models for magnetic media. ■

Let us now consider the second definition of passivity, scattering passivity.

Definition 20.8. Consider a convolution system with input signal $u(t)$ and output signal $v(t)$. The system is called *scattering-passive* if

$$\mathscr{W}_{\mathrm{scat}}(T) := \int_{-\infty}^{T} \left(|u(t)|^2 - |v(t)|^2\right) \mathrm{d}t \geq 0 \tag{20.34}$$

for all $T \in \mathbb{R}$ and all $u \in C_0^{\infty}$. □

Requiring a system to be scattering-passive corresponds to the energy of its output signal always being less than that of its input signal. It can be shown that the transfer function $W_{\mathrm{scat}}(s)$ of a scattering-passive system satisfies the relation $|W_{\mathrm{scat}}(s)| \leq 1$ for all $s \in \mathbb{C}_+$ [6,14]. Then a suitable rational (Cayley) transformation of the transfer function, namely the function $s \mapsto (1 + W_{\mathrm{scat}}(s))/(1 - W_{\mathrm{scat}}(s))$, is a PR function.

Let us consider an example of a scattering-passive system.

Example 20.8. Reflection from an isotropic slab placed above a ground plane The reflection coefficient Γ from an isotropic slab characterized by passive permittivity ε_{r}, passive permeability μ_{r}, thickness $d > 0$, and placed above a perfect electric conducting (PEC) plane, see the figure below, can describe a scattering-passive system. Let the incident wave E_{i} be the input signal and the reflected wave E_{r} be the output signal of the system. Assume the reference plane is placed in front of the slab, i.e., $x \leq 0$.

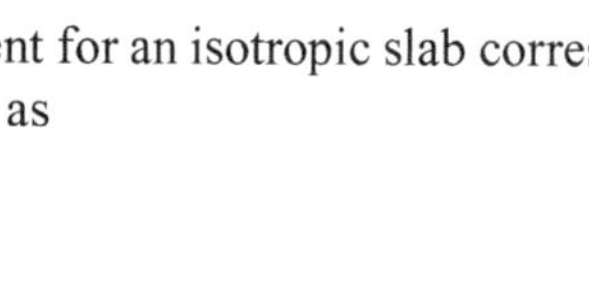

The reflection coefficient for an isotropic slab corresponds to the reflected wave $\boldsymbol{E}_{\mathrm{r}} = \Gamma \boldsymbol{E}_{\mathrm{i}}$ and is calculated as

$$\Gamma(k) = \frac{\Gamma_0 - \mathrm{e}^{2\mathrm{i}knd}}{1 - \Gamma_0 \mathrm{e}^{2\mathrm{i}knd}} \tag{20.35}$$

where $\Gamma_0 = (\eta - 1)/(\eta + 1)$ is the reflection coefficient at the air–slab interface, k is the wave number, $n = \sqrt{\varepsilon_{\mathrm{r}}}\sqrt{\mu_{\mathrm{r}}}$ is the refractive index of the slab, and $\eta = \sqrt{\mu_{\mathrm{r}}/\varepsilon_{\mathrm{r}}}$ is the relative wave impedance of the slab. This function can be shown to map the open right-half of complex plane to the closed unit disk. ■

Admittance-passive and scattering-passive systems are in close correspondence; namely given an admittance-passive system, it is possible to construct a new scattering-passive system, and conversely (see Theorem 5 in [35], as well as [37,40,41]).

20.4 Sum rules and physical bounds

The integral identities in Theorems 20.2 and 20.3 have a very useful application in the derivation of physical bounds on passive systems; see, e.g., [14]. In the engineering and physics literature, these integral identities appear in various forms and are also often referred to as *sum rules* [9,14]. Typically, the appropriate Herglotz function h is the Fourier transform of a physically realizable real-valued convolution kernel, and which, hence, possesses the symmetry $h(-z^*) = -h(z)^*$, or $p(z^*) = p(z)^*$ for a PR function. For Herglotz functions, the integral identities are given on the real axis where $z = x$ is commonly interpreted as angular frequency ω (in rad/s), wave number $k = \omega/c_0$ (in m^{-1}), or as wavelength $\lambda = 2\pi/k$ (in m). For PR functions, the imaginary axis $z = \mathrm{i}y$ is instead playing this role.

In many practical electromagnetic applications, it is reasonable to assume (or even to acquire by measurements) some partial knowledge regarding the low- and/or high-frequency asymptotic expansions of the corresponding Herglotz function, such as the static and the optical responses of a material, or a structure. In these cases, the sum rules and the corresponding integral identities can be used to obtain inequalities by constraining the integration interval to a finite bandwidth in the frequency (or wavelength) domain, and thereby yielding useful physical limitations in a variety of applications. Typical examples are with matching networks [15], radar absorbers [16], high-impedance surfaces [20], passive metamaterials [21], scattering [17,42], antennas [18,22,23,43], reflection coefficients [44], waveguides [45], periodic structures [46], etc.

Three examples are given next to illustrate typical situations where sum rules can be used to derive physical limitations on passive electromagnetic systems. In special situations, such results have previously been derived using residue calculus and Kramers–Kronig relations [16,36,38,47–49], whereas here, we consistently use the general result (20.28).

Example 20.9. The resistance-integral theorem Consider a passive circuit consisting of a parallel connection of a capacitance C and an impedance $Z_1(s)$ that does not contain a shunt capacitance (i.e., $Z_1(0)$ is finite and $Z_1(s)$ does not have a zero at $s = \infty$); see the figure below. Then the input impedance of this circuit is given by $Z(s) = 1/(sC + 1/Z_1(s))$, which is a PR function in the Laplace variable $s \in \mathbb{C}_+$, and hence the system is admittance passive.

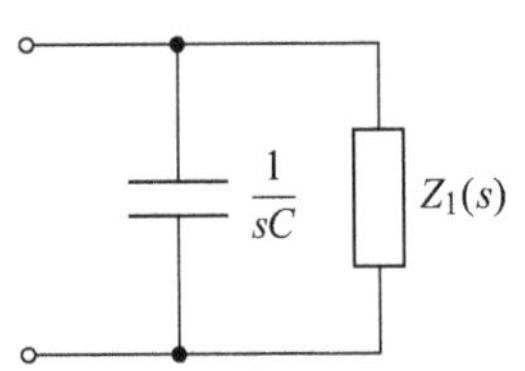

The asymptotic expansions are $Z(s) = Z_1(0) + o(s)$ as $s \hat{\to} 0$ and $Z(s) = 1/(sC) + o(s^{-1})$ as $s \hat{\to} \infty$. Here, the corresponding Herglotz function is $h(\omega) := \mathrm{i}Z(-\mathrm{i}\omega)$ where $s = -\mathrm{i}\omega$ and $\omega \in \mathbb{C}^+$. Its low- and high-frequency asymptotics are

$$h(\omega) = o(\omega^{-1}) \quad \text{as } \omega \hat{\to} 0 \quad \text{and } h(\omega) = -\frac{1}{\omega C} + o(\omega^{-1}) \quad \text{as } \omega \hat{\to} \infty \tag{20.36}$$

In terms of (20.21) and (20.18), we have $a_{-1} = 0$ and $b_{-1} = -1/C$, and thus the sum rule (20.28) with $n = 0$ gives

$$\frac{2}{\pi}\int_{0+}^{\infty} \mathrm{Re}[Z(-\mathrm{i}\omega)]\,\mathrm{d}\omega = \frac{2}{\pi}\int_{0+}^{\infty} \mathrm{Im}[h(\omega)]\,\mathrm{d}\omega = a_{-1} - b_{-1} = \frac{1}{C} \tag{20.37}$$

By integrating only over a finite frequency interval $\Omega := [\omega_1, \omega_2]$, and estimating this integral from the following, we obtain the bound

$$\Delta\omega \inf_{\omega\in\Omega} \mathrm{Re}[Z(-\mathrm{i}\omega)] \le \int_{0+}^{\infty} \mathrm{Re}[Z(-\mathrm{i}\omega)]\,\mathrm{d}\omega = \frac{\pi}{2C} \tag{20.38}$$

where $\Delta\omega := \omega_2 - \omega_1$. Consequently, inequality (20.38) limits the product between the bandwidth and the minimum resistance over the given frequency interval; see also [47]. ■

Compositions of Herglotz functions can be used to construct new Herglotz functions and, hence, also new sum rules. Here, we illustrate this for a case where the minimal temporal dispersion for metamaterials is determined. The problem is first transformed to the problem of determining the minimum amplitude of a Herglotz function over a bandwidth [14,21].

Example 20.10. Metamaterials and temporal dispersion When a dielectric medium is specified to have inductive properties (i.e., has negative permittivity) over a given bandwidth, it is regarded as a metamaterial. A given negative permittivity value at a single frequency is always possible to achieve. For instance, the plasmonic resonances in small metal particles can be readily explained by using common Drude or Lorentz models, etc. However, when a constant negative permittivity value is prescribed over a given bandwidth, the passivity of the material will imply severe bandwidth limitations; see, e.g., [21].

To derive these limitations based on the general theory of Herglotz functions, we start by considering the following general situation: Let $F(x) := -h_0(x)$ be the negative of a fixed Herglotz function h_0 that can be extended continuously to a neighborhood of the compact interval $\Omega \subset \mathbb{R}$ and has the large argument asymptotics $h_0(z) = b_1^0 z + o(z)$ as $z \hat{\to} \infty$. Let h denote an arbitrary Herglotz function with the same continuity property on the real line and satisfying the asymptotics $h(z) = b_1 z + o(z)$ as $z \hat{\to} \infty$. We aim to derive a lower bound for the error norm

$$\|h - F\|_{\mathrm{L}^\infty(\Omega)} := \sup_{x\in\Omega} |h(x) - F(x)| \tag{20.39}$$

To this end, an auxiliary Herglotz function $h_\Delta(z)$, cf. [21], is defined by

$$h_\Delta(z) := \frac{1}{\pi}\int_{-\Delta}^{\Delta}\frac{1}{\xi - z}\,\mathrm{d}\xi = \frac{1}{\pi}\mathrm{Log}\frac{z-\Delta}{z+\Delta} = \begin{cases} \mathrm{i} + o(1) & \text{as } z\hat{\to}0 \\ \dfrac{-2\Delta}{\pi z} + o(z^{-1}) & \text{as } z\hat{\to}\infty \end{cases} \tag{20.40}$$

Note that $\mathrm{Im}[h_\Delta(z)] \geq \frac{1}{2}$ for $|z| \leq \Delta$ and $\mathrm{Im}[z] \geq 0$. Next, consider the composite Herglotz function $h_1(z) := h_\Delta(h(z) + h_0(z))$, where $h(z) + h_0(z) = (b_1 + b_1^0)z + o(z)$ as $z\hat{\to}\infty$ yields the asymptotics

$$h_1(z) = o(z^{-1}) \text{ as } z\hat{\to}0 \quad \text{and} \quad h_1(z) = \frac{-2\Delta}{\pi(b_1 + b_1^0)}z^{-1} + o(z^{-1}) \text{ as } z\hat{\to}\infty \tag{20.41}$$

The sum rule (20.28) with $n = 0$ is given by

$$\frac{2}{\pi}\int_{0+}^{\infty}\mathrm{Im}[h_1(x)]\,\mathrm{d}x = a_{-1} - b_{-1} = \frac{2\Delta}{\pi(b_1 + b_1^0)} \tag{20.42}$$

Choosing $\Delta := \sup_{x\in\Omega}|h(x) + h_0(x)|$, the following integral inequalities follow

$$\frac{1}{\pi}|\Omega| \leq \frac{2}{\pi}\int_{\Omega}\underbrace{\mathrm{Im}[h_1(x)]}_{\geq\frac{1}{2}}\,\mathrm{d}x \leq \frac{2}{\pi}\int_{0+}^{\infty}\mathrm{Im}[h_1(x)]\,\mathrm{d}x = \frac{2\sup_{x\in\Omega}|h(x) + h_0(x)|}{\pi(b_1 + b_1^0)} \tag{20.43}$$

or

$$\|h + h_0\|_{\mathrm{L}^\infty(\Omega)} \geq (b_1 + b_1^0)\frac{1}{2}|\Omega|, \quad \text{where } |\Omega| = \int_{\Omega}\mathrm{d}x \tag{20.44}$$

Consider now a dielectric metamaterial with a constant, real-valued, and negative target permittivity $\varepsilon_\mathrm{t} < 0$ to be approximated over an interval Ω. In this case, $F(z) = z\varepsilon_\mathrm{t}$ and $h_0(z) = -F(z)$ with $b_1^0 = -\varepsilon_\mathrm{t}$. Let $\varepsilon(z)$ be the permittivity function of the approximating passive dielectric material, and $h(z) = z\varepsilon(z)$ the corresponding Herglotz function with $b_1 = \varepsilon_\infty$, the assumed high-frequency permittivity of the material, and the approximation interval $\Omega = \omega_0[1 - B/2, 1 + B/2]$, where ω_0 is the center frequency and B the relative bandwidth with $0 < B < 2$. The resulting physical bound obtained from (20.44) is given by

$$\|\varepsilon(\,\cdot\,) - \varepsilon_\mathrm{t}\|_{\mathrm{L}^\infty(\Omega)} \geq \frac{(\varepsilon_\infty - \varepsilon_\mathrm{t})B}{2 + B} \tag{20.45}$$

see also [21]. Note that the variable x corresponds here to angular frequency, also commonly denoted as ω (in rad/s). ■

Scattering passive systems have transfer functions that map $\mathbb{C}^+$ to the unit disk. To use (20.28), we first construct a Herglotz (or PR) function by mapping the unit disk to $\mathbb{C}^+$. This map can be made in many different ways and the particular choice depends on the asymptotic expansion and the physical interpretation of the system.

The Cayley transform, logarithm, and addition are most common in applications. Here, the reflection coefficient of a radar absorber is considered. It is desired to bound the magnitude of the reflection coefficient and, hence, the logarithm is used to construct a Herglotz function.

Example 20.11. Sum rules for passive radar absorbers The reflection coefficient $\Gamma(k)$ for an isotropic slab with permeability $\mu(k)$ and permittivity $\varepsilon(k)$ was derived in Example 20.8, where $k = \omega/c_0$ is the wave number. Assuming that $\mu(k) = \mu_s + O(k)$ and $\varepsilon(k) = \varepsilon_s + O(k)$, where μ_s and ε_s are the corresponding static values, it can be shown that the reflection coefficient has the following asymptotic expansions

$$\Gamma(k) = -1 - \mathrm{i}2kd\mu_s + O(k^2) \text{ as } k \hat{\to} 0 \quad \text{and} \quad \Gamma(k) = O(1) \text{ as } k \hat{\to} \infty \tag{20.46}$$

The function $\Gamma(k)$ is the Fourier transform of a scattering passive convolution kernel, so $\Gamma(k)$ is an analytic function with $|\Gamma(k)| \leq 1$ for $k \in \mathbb{C}^+$. Hence it has a representation of the form $\Gamma(k) = -B(k)\mathrm{e}^{\mathrm{i}h(k)}$, where $B(k)$ is a Blaschke product (cf. [50]) and $h(k)$ a Herglotz function [14]. With an appropriately chosen branch of the logarithm, h can thus be written

$$h(k) := -\mathrm{i}\,\log\left(-\Gamma(k)\prod_n \frac{1-k/k_n^*}{1-k/k_n}\right) \tag{20.47}$$

where k_n are the zeros of $\Gamma(k)$ with $\mathrm{Im}[k_n] > 0$. The Blaschke product is used to remove these zeros from the upper half plane and the negative sign is chosen here to make the Herglotz function $h(k)$ symmetric.

The asymptotic expansions of $h(k)$ are given by

$$h(k) = k\left(2d\mu_s + 2\sum_n \mathrm{Im}\left[\frac{1}{k_n}\right]\right) + o(k) \quad \text{as } k \hat{\to} 0 \tag{20.48}$$

and $h(k) = o(k)$ as $k \hat{\to} \infty$. Therefore, there is a sum rule (20.28) yielding

$$\frac{2}{\pi}\int_{0+}^{\infty} \frac{\mathrm{Im}[h(k)]}{k^2}\,\mathrm{d}k = 2d\mu_s + 2\sum_n \mathrm{Im}\left[\frac{1}{k_n}\right] \tag{20.49}$$

Since $\mathrm{Im}[1/k_n] < 0$, the following inequality is obtained:

$$\frac{2}{\pi}\int_{0+}^{\infty} \frac{-\ln|\Gamma(k)|}{k^2}\,\mathrm{d}k \leq 2d\mu_s, \quad \text{or} \quad \int_{0+}^{\infty} -\ln|\Gamma(\lambda)|\,\mathrm{d}\lambda \leq 2\pi^2 d\mu_s \tag{20.50}$$

where we used $\lambda = 2\pi/k$ to arrive at the final formulation, which was originally given in [16]. Note that the inequality in (20.50) gives also the following bound on the absorption parameter $1/|\Gamma(\lambda)|$ over the interval $[\lambda_1, \lambda_2]$

$$\Delta\lambda \inf_{\lambda\in[\lambda_1,\lambda_2]} \ln\frac{1}{|\Gamma(\lambda)|} \leq \int_{0+}^{\infty} \ln\frac{1}{|\Gamma(\lambda)|}\,\mathrm{d}\lambda \leq 2\pi^2 d\mu_s \tag{20.51}$$

where $\Delta\lambda = \lambda_2 - \lambda_1$.

The derivation above is, for simplicity, presented for homogeneous slabs similar to the multilayer case originally presented in [16]. The sum rule and physical bounds are generalized to arbitrary inhomogeneous periodic structures using the low-frequency expansion in [20] together with passivity and expansion of the reflected field in Floquet modes. This derivation shows that (20.51) is valid for the general case. An extension of the result (20.51) has applications in the evaluation of bandwidth performance of array antennas [22,23]. ■

20.5 Convex optimization and physical bounds

Convex optimization [27,51] based on the Herglotz function representation (20.7) can be used to approximate and identify passive systems. However, to facilitate the computation of a numerical solution using a software such as CVX [51], it is necessary to first impose some a-priori constraints on the class of approximating Herglotz functions. In particular, we are interested here in Herglotz functions that are known to be locally Hölder continuous on some given intervals on the real line. Hence, a passive approximation problem is considered where the target function F is an arbitrary complex-valued continuous function defined on an approximation domain $\Omega \subset \mathbb{R}$ consisting of a finite union of closed and bounded intervals of the real axis. The norms used, denoted by $\|\cdot\|_{\mathrm{L}^p(w,\Omega)}$, are weighted $\mathrm{L}^p(\Omega)$-norms [50] which are defined here by using a positive continuous weight function w on Ω, and where $1 \leq p \leq \infty$.

Here, the approximating function h is the Hölder continuous extension (to Ω) of some Herglotz function generated by an absolutely continuous measure μ having a density μ' which is Hölder continuous on the closure $\overline{U}$ of an arbitrary neighborhood $U \supset \Omega$ of the approximation domain. The function μ' is Hölder continuous with Hölder exponent α, meaning that $|\mu'(\tau) - \mu'(\varsigma)| \leq C\,|\tau - \varsigma|^{\alpha}$ for all $\tau, \varsigma \in \overline{U}$, where $0 < \alpha < 1$ is fixed, and $C > 0$ is an arbitrary constant. The corresponding Hölder space is denoted $\mathrm{C}^{0,\alpha}(\overline{U})$; see, e.g., [52, pp. 94–104]. It can be shown that the Hilbert transform integral operator $\mathscr{H}$, defined similarly as in (20.16), is a bounded operator $\mathscr{H}\colon \mathrm{C}^{0,\alpha}(\overline{U}) \to \mathrm{C}^{0,\alpha}(\Omega)$; cf., e.g., Theorem 7.6 and Corollary 7.7 on pp. 101–102 in [52] and see also [9]. By assumption, both the real and the imaginary parts of h are continuous functions on Ω. Moreover, it holds that $\mathrm{Im}[h] = \pi\mu'$ on $\overline{U}$ (cf. [13, p. 7]) and, due to the Hölder continuity of the density μ' on $\overline{U}$, the real part is given by the associated Hilbert transform [52], similar as in (20.17). Now, the impulse response of a passive system can usually be considered as real-valued. In this case, the approximating Herglotz function h can be assumed to be symmetric and its real part admits the representation

$$\mathrm{Re}[h(x)] = bx + \mathrm{p.v.}\int_{\mathbb{R}} \frac{\mu'(\tau)}{\tau - x}\,\mathrm{d}\tau \quad \text{for } x \in \Omega \tag{20.52}$$

where $\mathrm{Im}[h(x)] = \pi\mu'(x)$ for $x \in \Omega$.

The continuity of h on Ω implies that the norm $\|h\|_{L^p(w,\Omega)}$ is well-defined for $1 \leq p \leq \infty$. An approximation problem of interest can now be formulated in terms of the greatest lower bound on the approximation error, defined by

$$d := \inf_h \|h - F\|_{L^p(w,\Omega)} \tag{20.53}$$

where the infinum is taken over all Herglotz functions h generated by a measure having a Hölder continuous density on $\overline{U}$.

In general, a best approximation achieving the bound d in (20.53) does not exist. In practice, however, the problem is approached by using numerical algorithms such as CVX [51], solving finite-dimensional approximation problems using B-splines, with the number of basis functions N fixed during the optimization, cf. [24,25]. Here, a B-spline of order $m \geq 2$ is an $m - 2$ times continuously differentiable and compactly supported positive basis spline function consisting of piecewise polynomial functions of order $m - 1$, i.e., linear, quadratic, cubic, etc., and which is defined by $m + 1$ break points [53]. Let us now consider a discretization of the problem expressed in (20.53), which is based on an arbitrary, finite partition of the approximation domain Ω. Let

$$\pi \mu'(x) = \sum_{n=1}^{N} \zeta_n \left(p_n(x) + p_n(-x)\right) \tag{20.54}$$

for $x \in \mathbb{R}$ be a finite B-spline expansion of $\mathrm{Im}\,[h(x)]$, where ζ_n are optimization variables for $n = 1, \ldots, N$, and $p_n(x)$ are B-spline basis functions of fixed order m which are defined on the given partition. The real part $\mathrm{Re}[h(x)]$ for $x \in \Omega$ is then given by (20.52), and can be expressed as

$$\mathrm{Re}[h(x)] = bx - \frac{\zeta_0}{x} + \sum_{n=1}^{N} \zeta_n \left(\hat{p}_n(x) - \hat{p}_n(-x)\right), \quad x \in \Omega \tag{20.55}$$

where $\hat{p}_n(x)$ is the (negative) Hilbert transform of the B-spline function $p_n(x)$ and where a point mass at $x = 0$ with amplitude c_0 has been included; see (20.13). Any other a-priori assumed point masses can be included in a similar way. Explicit formulas for general B-splines and their Hilbert transforms are given in [24]. As an example, a piecewise linear ("roof-top") B-spline on a uniform partition is given by $p_n(x) = p(x - n\Delta)$, where

$$p(x) = \begin{cases} 1 - |x|/\Delta & |x| \leq \Delta \\ 0 & |x| > \Delta \end{cases} \tag{20.56}$$

for $x \in \mathbb{R}$ and $\Delta > 0$, and its (negative) Hilbert transform $\hat{p}_n(x) = \hat{p}(x - n\Delta)$, where

$$\hat{p}(x) = \frac{1}{\pi\Delta}\left(2x \ln|x| - (x - \Delta)\ln|x - \Delta| - (x + \Delta)\ln|x + \Delta|\right) \tag{20.57}$$

for $x \in \mathbb{R}$. Note that the logarithmic singularities in (20.57) cancel, and the function $\hat{p}(x)$ is a continuous function on $\mathbb{R}$.

Consider now the following convex optimization problem:

$$\begin{aligned}&\text{minimize} && \|h - F\|_{L^p(w,\Omega)}\\ &\text{subject to} && \zeta_n \geq 0, \ \text{for } n = 0, \ldots, N\\ &&& b \geq 0,\end{aligned} \tag{20.58}$$

where the optimization is over the variables $(\zeta_0, \zeta_1, \ldots, \zeta_N, b)$. Note that the objective function in (20.58) above is the norm of an affine form in the optimization variables. Hence, the objective function is a convex function in the variables $(\zeta_0, \zeta_1, \ldots, \zeta_N, b)$.

The uniform continuity of all functions involved implies that the solution to (20.58) can be approximated within an arbitrary accuracy by discretizing the approximation domain Ω (and the computation of the norm) using only a finite number of sample points. The corresponding numerical problem (20.58) can now be solved efficiently by using the CVX MATLAB® software for disciplined convex programming [51]. The convex optimization formulation (20.58) offers a great advantage in the flexibility in which additional or alternative convex constraints and formulations can be implemented; see also [24,25].

Example 20.12. A canonical example of convex optimization is with the passive approximation of metamaterials; see also [21,24,25]. As in Example 20.10, the variable x corresponds here to angular frequency, also commonly denoted as ω (in rad/s). A typical application is with the study of optimal plasmonic resonances in small structures (or particles) for which the absorption cross section can be approximated by

$$\sigma_{\text{abs}} \approx k\text{Im}[\gamma] \tag{20.59}$$

where $k = 2\pi/\lambda$ is the wave number of free space, λ the wavelength, and γ is the electric polarizability of the particle; see [54]. As, e.g., the polarizability of a dielectric sphere with radius a is given by $\gamma(x) = 4\pi a^3(\varepsilon(x) - 1)/(\varepsilon(x) + 2)$, where $\varepsilon(x)$ is the permittivity function of the dielectric material inside the sphere.

A surface plasmon resonance is obtained when $\varepsilon(x) \approx -2$, and, hence, we specify that the target permittivity of our metamaterial is $\varepsilon_{\text{t}} = -2$. However, a metamaterial with a negative real part cannot, in general, be implemented as a passive material over a given bandwidth, cf. [21]. Based on the theory of Herglotz functions and associated sum rules, the physical bound in (20.45) can be derived, where ε_∞ is the high-frequency permittivity of the material, $\varepsilon_{\text{t}} < \varepsilon_\infty$, $\Omega = \omega_0[1 - B/2, 1 + B/2]$, ω_0 the center frequency, and B the relative bandwidth with $0 < B < 2$, cf. [21]. The convex optimization formulation (20.58) can be used to study passive realizations (20.54) and (20.55) that satisfies the bound (20.45) as close as possible. Here, the approximating Herglotz function is $h(x) = x\varepsilon(x)$, the target function $F(x) = x\varepsilon_{\text{t}}$, ζ_0 the amplitude of a point mass at $x = 0$, $b = \varepsilon_\infty$, and a weighted norm is used defined by $\|f\|_{L^\infty(w,\Omega)} = \max_{x\in\Omega} |f(x)/x|$ assuming that $0 \notin \Omega$.

In Figure 20.5 is shown the absorption cross section $\sigma_{\text{abs}}^{\text{D}}$ of a small ($a = 5$ nm) gold nanosphere modeled by using a dielectric Drude model with static conductivity $\sigma_0 = 4.52 \cdot 10^7$ S/m and collision frequency $\nu = 4.449 \cdot 10^{12}$ Hz; see [55]. The plot also shows the maximal bandwidth (MB) for the corresponding Drude peak resonance based on the associated sum rule $\Delta\lambda = 4\pi^3 a^3/\max \sigma_{\text{abs}}^{\text{D}}$ corresponding to a relative

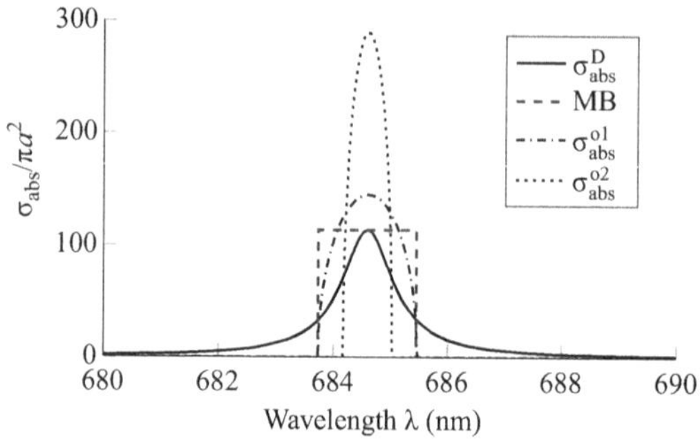

Figure 20.5 Absorption cross section σ^{D}_{abs} of a small (a = 5 nm) gold nanosphere modeled by using a dielectric Drude model. The red dashed line (MB) gives the maximal bandwidth for the corresponding peak resonance based on the associated sum rule $\Delta\lambda = 4\pi^3 a^3 / \max \sigma^{D}_{abs}$. Here, σ^{o1}_{abs} and σ^{o2}_{abs} correspond to optimized permittivity functions for relative bandwidths $B = 2.5 \cdot 10^{-3}$ and $B = 1.25 \cdot 10^{-3}$, respectively

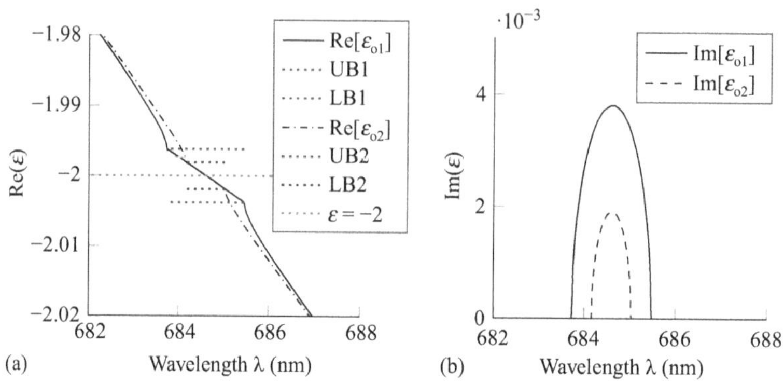

Figure 20.6 Real and imaginary parts of the optimized permittivity functions approximating a metamaterial with $\varepsilon_t = -2$. Here, ε_{o1} and ε_{o2} are optimized using (20.58) and linear B-splines, for the relative bandwidths $B = 2.5 \cdot 10^{-3}$ and $B = 1.25 \cdot 10^{-3}$, respectively. UB1-2 and LB1-2 denote the corresponding upper and lower physical bounds based on (20.45)

bandwidth of $B = 2.5 \cdot 10^{-3}$. Finally, the plot shows the absorption cross sections σ^{o1}_{abs} and σ^{o2}_{abs} in two examples, where the permittivity functions have been optimized using (20.58) and linear B-splines for the relative bandwidths $B = 2.5 \cdot 10^{-3}$ and $B = 1.25 \cdot 10^{-3}$, respectively. The resulting optimal permittivity functions are plotted in Figure 20.6 together with the fundamental bound (20.45). Note that there is also a point mass with amplitude ζ_0 at $x = 0$ which is not seen in this plot. All plots are made with respect to wavelength $\lambda = 2\pi/k$. ■

20.6 Conclusions

We have illustrated how the theory of Herglotz functions provides essential information about electromagnetic systems and engineering challenges, both by using this theory on well-known problems and recent results. Sum rules and convex optimization utilize the inherent constraints due to passivity, linearity, and time-translational invariance of the electromagnetic applications to obtain fundamental physical bounds. Furthermore, these techniques not only give physical bounds, but in several cases have been shown to be predictive as a tool in the design of electromagnetic structures.

Acknowledgments

This work was supported by the Swedish Foundation for Strategic Research (SSF) under the program Applied Mathematics and the project Complex Analysis and Convex Optimization for EM Design.

References

[1] N.I. Akhiezer and I.M. Glazman. *Theory of Linear Operators in Hilbert Space*. Vol. 1. Frederick Ungar Publishing Co., 1961.
[2] M. Reed and B. Simon. *Methods of Modern Mathematical Physics . Vol. I: Functional Analysis*. Academic Press, 1972.
[3] N.I. Akhiezer. *The Classical Moment Problem*. Oliver and Boyd, 1965.
[4] B. Simon. "The classical moment problem as a self-adjoint finite difference operator." *Advances in Mathematics* 137.1 (1998), pp. 82–203.
[5] J.C. Willems. "Dissipative dynamical systems part II: linear systems with quadratic supply rates." *Arch. Rational Mech. Anal.* 45.5 (1972), pp. 352–393.
[6] A.H. Zemanian. *Distribution Theory and Transform Analysis: An Introduction to Generalized Functions, with Applications*. McGraw-Hill, 1965.
[7] O. Brune. "Synthesis of a finite two-terminal network whose driving-point impedance is a prescribed function of frequency." *MIT J. Math. Phys.* 10 (1931), pp. 191–236.
[8] S. Darlington. "Synthesis of reactance 4-poles which produce prescribed insertion loss characteristics." *Journal of Mathematics and Physics* 18 (1939), pp. 275–353.
[9] F.W. King. *Hilbert Transforms Vol. I–II*. Cambridge University Press, 2009.
[10] H.M. Nussenzveig. *Causality and Dispersion Relations*. Academic Press, 1972.

[11] G.W. Milton, M. Cassier, O. Mattei, M. Milgrom, and A. Welters. *Extending the Theory of Composites to Other Areas of Science*. Milton-Patton Publishers, 2016.

[12] G.W. Milton. *The Theory of Composites*. Cambridge University Press, 2002.

[13] I.S. Kac and M.G. Krein. "R-functions—analytic functions mapping the upper halfplane into itself." *Amer. Math. Soc. Transl. (2)* 103 (1974), pp. 1–18.

[14] A. Bernland, A. Luger, and M. Gustafsson. "Sum rules and constraints on passive systems." *J. Phys. A: Math. Theor.* 44.14 (2011), p. 145205.

[15] R.M. Fano. "Theoretical limitations on the broadband matching of arbitrary impedances." *Journal of the Franklin Institute* 249.1,2 (1950), pp. 57–83 and 139–154.

[16] K.N. Rozanov. "Ultimate thickness to bandwidth ratio of radar absorbers." *IEEE Trans. Antennas Propag.* 48.8 (2000), pp. 1230–1234.

[17] C. Sohl, M. Gustafsson, and G. Kristensson. "Physical limitations on broadband scattering by heterogeneous obstacles." *J. Phys. A: Math. Theor.* 40 (2007), pp. 11165–11182.

[18] M. Gustafsson, C. Sohl, and G. Kristensson. "Physical limitations on antennas of arbitrary shape." *Proc. R. Soc. A* 463 (2007), pp. 2589–2607.

[19] C.R. Brewitt-Taylor. "Limitation on the bandwidth of artificial perfect magnetic conductor surfaces." *Microwaves, Antennas & Propagation, IET* 1.1 (2007), pp. 255–260.

[20] M. Gustafsson and D. Sjöberg. "Physical bounds and sum rules for high-impedance surfaces." *IEEE Trans. Antennas Propag.* 59.6 (2011), pp. 2196–2204.

[21] M. Gustafsson and D. Sjöberg. "Sum rules and physical bounds on passive metamaterials." *New Journal of Physics* 12 (2010), p. 043046.

[22] J.P. Doane, K. Sertel, and J.L. Volakis. "Matching bandwidth limits for arrays backed by a conducting ground plane." *IEEE Trans. Antennas Propag.* 61.5 (2013), pp. 2511–2518.

[23] B.L.G. Jonsson, C.I. Kolitsidas, and N. Hussain. "Array antenna limitations." *Antennas and Wireless Propagation Letters, IEEE* 12 (2013), pp. 1539–1542.

[24] Y. Ivanenko, M. Gustafsson, B.L.G. Jonsson, *et al. Passive Approximation and Optimization with B-splines*. Tech. rep. URN: urn:nbn:se:lnu:diva-63878. Linnaeus University, 2017.

[25] S. Nordebo, M. Gustafsson, B. Nilsson, and D. Sjöberg. "Optimal realizations of passive structures." *IEEE Trans. Antennas Propag.* 62.9 (2014), pp. 4686–4694.

[26] S. Boyd, L. El Ghaoui, E. Feron, and V. Balakrishnan. *Linear Matrix Inequalities in System and Control Theory*. SIAM, 1994.

[27] S.P. Boyd and L. Vandenberghe. *Convex Optimization*. Cambridge Univ. Pr., 2004.

[28] L.V. Ahlfors. *Complex Analysis*. McGraw-Hill Book Company, 1979.

[29] F. Gesztesy and E. Tsekanovskii. "On matrix-valued Herglotz functions." *Mathematische Nachrichten* 218.1 (2000), pp. 61–138.

[30] M. Abramowitz and I.A. Stegun, eds. *Handbook of Mathematical Functions*. Applied Mathematics Series No. 55. National Bureau of Standards, 1970.

[31] N.I. Akhiezer and I.M. Glazman. *Theory of Linear Operators in Hilbert Space*. Vol. 2. Frederick Ungar Publishing Co., 1963.

[32] G.E. Shilov and B.L. Gurevich. *Integral, Measure and Derivative: A Unified Approach. Dover Books on Mathematics*. Dover Publications, 2013.

[33] J.B. Garnett. *Bounded Analytic Functions*. Revised first. Springer-Verlag, 2007.

[34] P. Henrici. *Applied and Computational Complex Analysis, Volume 3: Discrete Fourier Analysis, Cauchy Integrals, Construction of Conformal Maps, Univalent Functions*. John Wiley & Sons, 1993.

[35] M. Wohlers and E. Beltrami. "Distribution theory as the basis of generalized passive-network analysis." *IEEE Transactions on Circuit Theory* 12.2 (1965), pp. 164–170.

[36] E.A. Guillemin. *Synthesis of Passive Networks*. John Wiley & Sons, 1957.

[37] D. Youla, L. Castriota, and H. Carlin. "Bounded real scattering matrices and the foundations of linear passive network theory." *IRE Transactions on Circuit Theory* 6.1 (1959), pp. 102–124.

[38] J.D. Jackson. *Classical Electrodynamics*. Third Edition. John Wiley & Sons, 1999.

[39] G. Kristensson. *Scattering of Electromagnetic Waves by Obstacles*. SciTech Publishing, an Imprint of the IET, 2016.

[40] V.S. Vladimirov. *Equations of Mathematical Physics*. Mir Publishers, 1984.

[41] A.H. Zemanian. "An *n*-port realizability theory based on the theory of distributions." *IEEE Transactions on Circuit Theory* 10.2 (1963), pp. 265–274.

[42] A. Bernland, M. Gustafsson, and S. Nordebo. "Physical limitations on the scattering of electromagnetic vector spherical waves.' *J. Phys. A: Math. Theor.* 44.14 (2011), p. 145401.

[43] C. Sohl and M. Gustafsson. "A priori estimates on the partial realized gain of ultra-wideband (UWB) antennas." *Quart. J. Mech. Appl. Math.* 61.3 (2008), pp. 415–430.

[44] M. Gustafsson. "Sum rules for lossless antennas." *IET Microwaves, Antennas & Propagation* 4.4 (2010), pp. 501–511.

[45] I. Vakili, M. Gustafsson, D. Sjoberg, R. Seviour, M. Nilsson, and S. Nordebo. "Sum rules for parallel-plate waveguides: experimental results and theory." *IEEE Trans. Microwave Theory Tech.* 62.11 (2014), pp. 2574–2582.

[46] M. Gustafsson, I. Vakili, S.E.B. Keskin, D. Sjöberg, and C. Larsson. "Optical theorem and forward scattering sum rule for periodic structures." *IEEE Trans. Antennas Propag.* 60.8 (2012), pp. 3818–3826.

[47] H.W. Bode. *Network Analysis and Feedback Amplifier Design*, 1945. Van Nostrand, 1945.

[48] L.D. Landau, E.M. Lifshitz, and L.P. Pitaevskiĭ. *Electrodynamics of Continuous Media*. Second Edition. Pergamon Press, 1984.

[49] O. Wing. *Classical Circuit Theory*. Springer, 2008.

[50] W. Rudin. *Real and Complex Analysis*. McGraw-Hill, 1987.

[51] M. Grant and S. Boyd. *CVX: A System for Disciplined Convex Programming*, Release 2.0, © 2012 CVX Research, Inc., Austin, TX.

[52] R. Kress. *Linear Integral Equations*. Second Edition. Springer-Verlag, 1999.

[53] C. De Boor. "On calculating with B-splines." *Journal of Approximation Theory* 6.1 (1972), pp. 50–62.

[54] C.F. Bohren and D.R. Huffman. *Absorption and Scattering of Light by Small Particles*. John Wiley & Sons, 1983.

[55] M.G. Blaber, M.D. Arnold, and M.J. Ford. "Search for the ideal plasmonic nanoshell: the effects of surface scattering and alternatives to gold and silver." *J. Phys. Chem. C* 113.8 (2009), pp. 3041–3045.

Chapter 21

Scattering and guidance by layered cylindrically periodic arrays of circular cylinders

Vakhtang Jandieri[1] and Kiyotoshi Yasumoto[2]

21.1 Introduction

Periodic dielectric or metallic structures are a subject of continuing interest because of their wide use for practical devices in microwaves and optical waves [1,2]. A periodic array of circular rods is typical of a discrete periodic structure. Various analytical or numerical techniques have been developed over the years to formulate the electromagnetic scattering and guidance by the periodic arrays [3–7]. However, the previous pertinent efforts have been mostly concerned with the planar arrays. An alternative of the configuration is a cylindrical array formed by circular rods periodically distributed on a circular ring. Recently, the cylindrically periodic structures have received a growing attention because of their potential applications to the designs of photonic crystal fibers [8], directive antennas or beam-switching antennas [9–12], plasmonic crystals and THz waveguides [13–15].

In this chapter, we discuss a self-contained semi-analytical approach for electromagnetic radiation, scattering and guidance by cylindrical arrays composed of circular rods periodically distributed along concentrically layered circular rings. Generalization of the problem to the eccentrically layered circular rings is straightforward. The rods can be dielectrics, perfect conductor, air holes, metals or magnetized ferrites. The method uses the transition matrix (T-matrix) of a circular rod in isolation [5,16,17], the reflection and transmission matrices of a cylindrical array based on the cylindrical harmonics expansion [18–21] and the generalized reflection and transmission matrices for a cylindrically layered structure [18–22]. The formulation is rigorous. The only approximation involved in the solutions is the truncation of the cylindrical harmonic expansion and the translation matrices related to Graf's addition theorem [23], which is necessary to solve a set of linear equations by matrix inversion. The proposed approach introduces a cylindrical layer model to the array, extracts the reflection and transmission matrices of a cylindrically periodic layer and then obtains

[1]General and Theoretical Electrical Engineering, Faculty of Engineering, University of Duisburg-Essen, and CENIDE – Center for Nanointegration Duisburg-Essen, Duisburg, Germany.
[2]Department of Computer Science and Communication Engineering, Kyushu University, Fukuoka, Japan.

the characteristics of the whole layered structure by using a recursive formula for the generalized reflection and transmission matrices. The recursive formula is based on a simple matrix multiplication [5,16,22,24], which guarantees a very short computation time for arbitrary number of the layers. The proposed semi-analytical approach is general, which takes into account all cylindrical space harmonics and their multiple interactions through the scattering process. The latter could be considered as a big advantage of the method. The formulation could be applied to various configurations of the layered cylindrical arrays with different types and locations of the excitation sources. To the best of authors' knowledge, the rigorous self-contained formulation for cylindrical structures has been proposed for the first time.

First, the proposed formulation is used to analyze the radiation characteristics of a localized source located inside the multilayered cylindrical structures—cylindrical bandgap structure [19]. Three different configurations of the cylindrical bandgap structures are studied, and the rods are taken as perfect electric conductors. We discuss in detail the relations between the transmission spectra of the cylindrical harmonic waves and the directivity of radiation by the exciting source. Radiation patterns in both principle H-plane and E-plane are analyzed. Discussions about the relation between the resonance and stopband characteristics of the transmission spectra and the radiation patterns of the localized source are given from the viewpoint of the flexible design and control of the multibeam and directive beam-forming characteristics.

Next, we investigate the light scattering by metal-coated dielectric nanocylinders periodically distributed along a cylindrical surface [25]. The problem is interesting from the viewpoint of practical applications in sensors, super-scattering and invisible cloaking [26–30]. The surface plasmons have ability to guide electromagnetic energy on length scales below the diffraction limit. This leads to developing metallic nanostructures that can control light at the nanoscale in direct analogy to traditional optical components such as lens, mirrors and waveguides. The spectral responses of the scattering cross section (SCS) and absorption cross section (ACS) and the near field distributions are analyzed. Various resonances observed in the spectral responses are discussed, and unique features of the near-field distributions related to the surface plasmon resonances are brought out for the metallic nanocylinders structure. Special attention is paid to the unique and interesting phenomena characterizing the cylindrical structure composed of the metal-coated nanocylinders such as (a) localization of the field at the outer and inner interfaces of the metal-coated nanocylinders; (b) excitement of the field in the gap region between the nanocylinders through the coupled plasmon resonance and (c) strong confinement of the field inside the cylindrical structure.

Finally, the method is applied to the modal analysis of the guided waves in a specific microstructured optical fiber [31,32]. Such kind of structure represents a new class of optical fiber that allows greater freedom in fiber design than could be achieved by the use of conventional materials. In the case of guiding problem without any initial excitation, we could assume a unique symmetric property of the mode field distribution inherent to the periodicity of the circular rods. The results of the normalized propagation constant are compared with those obtained by a variational method [33], and a very good agreement with the present rigorous method is observed.

Time dependence is assumed to be $\exp(-i\omega t)$ and is omitted in the chapter.

21.2 Formulation of the problem

21.2.1 Field expressions

Cross-sectional view of N-layered cylindrical arrays of circular rods located in a homogeneous background medium with material constants ε_s and μ_s is shown in Figure 21.1. The M circular rods of radius r_ν and the material constants ε_ν and μ_ν are symmetrically distributed on each of N-layered cylindrical surfaces with radii $R_\nu(\nu = 1, 2, 3, \ldots, N)$. The radii R_ν and r_ν must satisfy the condition $R_{\nu+1} - R_\nu > r_{\nu+1} + r_\nu$ so that consecutive layers do not interpenetrate. Rotation of the ν th layer is denoted by Θ_ν. The z dependence of all field components is given as $\exp(i\xi z)$, where ξ is the propagation constant along the z axis. The rods are infinitely long along the z axis and parallel to each other. The problem is formulated by employing electric E_z and magnetic $\hat{H}_z = (\mu_0/\varepsilon_0)^{1/2} H_z$ fields as the leading fields. Region with $R_\nu < \rho < R_{\nu+1}$ is labeled as region (ν) and the innermost region as region (0). Using the Graf addition theorem for the cylindrical functions [23], the scattered fields in the (ν)th and $(\nu - 1)$st regions coming from the νth layer are expressed in the matrix form in global coordinate system (ρ, ϕ) as follows:

$$\bar{\boldsymbol{\psi}}^{s(\nu)} = \exp(i\xi z)\bar{\boldsymbol{\Psi}}^T \cdot \sum_{j=1}^{M} \bar{\boldsymbol{\beta}}_j^\nu \cdot \bar{\boldsymbol{a}}_j^\nu \quad \text{for } \rho > R_\nu \tag{21.1}$$

$$\bar{\boldsymbol{\psi}}^{s(\nu-1)} = \exp(i\xi z)\bar{\boldsymbol{\Phi}}^T \cdot \sum_{j=1}^{M} \bar{\boldsymbol{\eta}}_j^\nu \cdot \bar{\boldsymbol{a}}_j^\nu \quad \text{for } \rho < R_\nu \tag{21.2}$$

with

$$\bar{\boldsymbol{\psi}}^{s(\nu)} = \begin{bmatrix} E_z^{s(\nu)} \\ \hat{H}_z^{s(\nu)} \end{bmatrix} \tag{21.3}$$

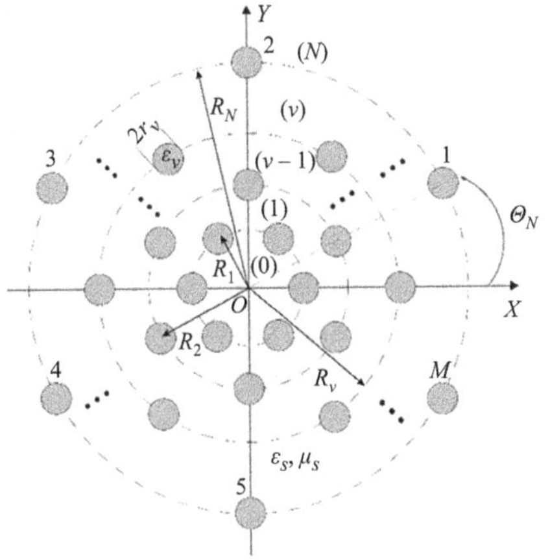

Figure 21.1 N-layered cylindrical arrays formed by M circular rods periodically distributed on each of concentric circular cylindrical surfaces

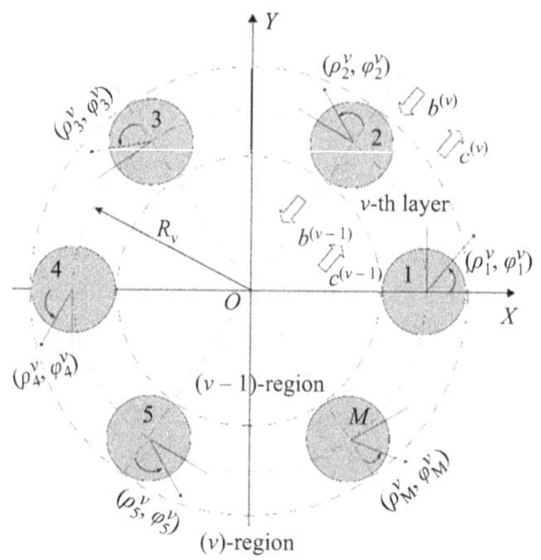

Figure 21.2 Schematic view of the scattering process by the circular rods on the νth layer and the local coordinates ρ_j^ν, φ_j^ν $(j = 1, 2, 3, \ldots M)$ fixed to each of the rods

$$\bar{\boldsymbol{a}}_j^\nu = \begin{bmatrix} \boldsymbol{a}_j^{e,\nu} \\ \boldsymbol{a}_j^{h,\nu} \end{bmatrix}, \quad \boldsymbol{a}_j^{e,\nu} = [\boldsymbol{a}_{j,m}^{e,\nu}], \; \boldsymbol{a}_j^{h,\nu} = [\boldsymbol{a}_{j,m}^{h,\nu}] \tag{21.4}$$

$$\bar{\boldsymbol{\Psi}}^T = \begin{bmatrix} \boldsymbol{\Psi}^T & \boldsymbol{0} \\ \boldsymbol{0} & \boldsymbol{\Psi}^T \end{bmatrix}, \quad \boldsymbol{\Psi} = [H_m^{(1)}(\kappa\rho)\exp(im\varphi)] \tag{21.5}$$

$$\bar{\boldsymbol{\Phi}}^T = \begin{bmatrix} \boldsymbol{\Phi}^T & \boldsymbol{0} \\ \boldsymbol{0} & \boldsymbol{\Phi}^T \end{bmatrix}, \quad \boldsymbol{\Phi} = [J_m(\kappa\rho)\exp[im\varphi]] \tag{21.6}$$

$$\bar{\boldsymbol{\beta}}_j^\nu = \begin{bmatrix} \boldsymbol{\beta}_j^\nu & \boldsymbol{0} \\ \boldsymbol{0} & \boldsymbol{\beta}_j^\nu \end{bmatrix}, \quad \boldsymbol{\beta}_j^\nu = \{J_{m-n}(\kappa R_\nu)\exp[-im(j-1)\theta_M]\} \tag{21.7}$$

$$\bar{\boldsymbol{\eta}}_j^\nu = \begin{bmatrix} \boldsymbol{\eta}_j^\nu & \boldsymbol{0} \\ \boldsymbol{0} & \boldsymbol{\eta}_j^\nu \end{bmatrix}, \quad \boldsymbol{\eta}_j^\nu = \left\{H_{m-n}^{(1)}(\kappa R_\nu)\exp[-im(j-1)\theta_M]\right\} \tag{21.8}$$

where $\kappa = \sqrt{k_s^2 - \xi^2}$, k_s is the wavenumber in a background medium, J_m and $H_m^{(1)}$ are Bessel and Hankel functions of mth order, $\theta_M = 2\pi/M$ is the separation angle between the nearest two rods along a circular ring, $\{a_{j,m}^{e,\nu}\}$ and $\{a_{j,m}^{h,\nu}\}$ denote the unknown scattering amplitudes for the electric and magnetic fields by each of M circular rods located on the νth layer, T denotes the transpose of the indicated vector. Note that m and n describe the row and column of the matrices, respectively.

21.2.2 Calculation of the scattering amplitudes

Let us consider first that the *incoming* cylindrical wave $\bar{\boldsymbol{\Phi}}^T \cdot \bar{\boldsymbol{b}}^{(\nu)}$ is incident on the ν th layer from the outer region (ν) as it is demonstrated in Figure 21.2. Scattered fields (1) and (2) must satisfy the boundary conditions on the surfaces $\rho_j^\nu = r_\nu (j = 1, 2, \ldots, M)$

of each circular rod under the presence of the incident field $\bar{\boldsymbol{\Phi}}^T \cdot \bar{\boldsymbol{b}}^{(\nu)}$. In order to apply the boundary conditions on # i rod, $\bar{\boldsymbol{\Phi}}^T \cdot \bar{\boldsymbol{b}}^{(\nu)}$ and the scattered field components expressed in terms of the local coordinates $(\rho_j^\nu,\ \varphi_j^\nu)\quad (j \neq i)$ are transformed to those expressed by the local coordinate $(\rho_i^\nu,\ \phi_i^\nu)$. This transformation is performed by using Graf's addition theorem [23]. After straightforward mathematical manipulations, a set of linear equations to determine the unknown scattering amplitude vectors $\bar{a}_j^\nu$ is obtained as follows:

$$\underbrace{\begin{bmatrix} \bar{\mathbf{I}} & \mathbf{A}_2^\nu & \cdots & \cdots & \mathbf{A}_M^\nu \\ \mathbf{A}_M^\nu & \bar{\mathbf{I}} & \ddots & & \vdots \\ \vdots & \ddots & \ddots & \ddots & \vdots \\ \vdots & & \ddots & \bar{\mathbf{I}} & \mathbf{A}_2^\nu \\ \mathbf{A}_2^\nu & \cdots & \cdots & \mathbf{A}_M^\nu & \bar{\mathbf{I}} \end{bmatrix}}_{\bar{\Gamma}^{(\nu)}} \begin{bmatrix} \bar{\boldsymbol{a}}_1^\nu \\ \bar{\boldsymbol{a}}_2^\nu \\ \vdots \\ \vdots \\ \bar{\boldsymbol{a}}_M^\nu \end{bmatrix} = \begin{bmatrix} \bar{\mathbf{T}}^\nu \bar{\boldsymbol{\alpha}}_1^\nu \cdot \bar{\boldsymbol{b}}^{(\nu)} \\ \bar{\mathbf{T}}^\nu \bar{\boldsymbol{\alpha}}_2^\nu \cdot \bar{\boldsymbol{b}}^{(\nu)} \\ \vdots \\ \vdots \\ \bar{\mathbf{T}}^\nu \bar{\boldsymbol{\alpha}}_M^\nu \cdot \bar{\boldsymbol{b}}^{(\nu)} \end{bmatrix} \tag{21.9}$$

with

$$\mathbf{A}_p^\nu = -\bar{\mathbf{T}}^\nu \bar{\mathbf{K}}_p^\nu \quad (p = 2, 3, \ldots, M) \tag{21.10}$$

$$\bar{\mathbf{K}}_p^\nu = \begin{bmatrix} \mathbf{K}_p^\nu & \mathbf{0} \\ \mathbf{0} & \mathbf{K}_p^\nu \end{bmatrix}, \quad \mathbf{K}_p^\nu = \left[H_{m-n}^{(1)}\left(\kappa d_p^\nu\right)(-1)^m e^{i(m+n)\zeta_p}\right] \tag{21.11}$$

$$\bar{\boldsymbol{\alpha}}_q^\nu = \begin{bmatrix} \boldsymbol{\alpha}_q^\nu & \mathbf{0} \\ \mathbf{0} & \boldsymbol{\alpha}_q^\nu \end{bmatrix}, \quad \boldsymbol{\alpha}_q^\nu = \left[(-1)^{m-n} J_{m-n}(\kappa R_\nu) e^{in(q-1)\theta_M}\right] \tag{21.12}$$

$$\bar{\boldsymbol{b}}^{(\nu)} = \begin{bmatrix} \boldsymbol{b}^{e(\nu)} \\ \boldsymbol{b}^{h(\nu)} \end{bmatrix}, \quad \boldsymbol{b}^{e(\nu)} = \left[\boldsymbol{b}_m^{e(\nu)}\right], \quad \boldsymbol{b}^{h(\nu)} = \left[\boldsymbol{b}_m^{h(\nu)}\right] \tag{21.13}$$

$$\zeta_p = \frac{\pi}{2} - (p-1)\frac{\theta_M}{2} \quad (p = 2, 3, \ldots, M) \tag{21.14}$$

$$d_p^\nu = 2R_\nu \sin\left[\frac{(p-1)\theta_M}{2}\right] \tag{21.15}$$

where $\bar{\mathbf{T}}^\nu$ is the T-matrix of a single circular rod located on the νth circular ring, which is obtained in closed form [2,5,16], $\boldsymbol{\alpha}_q^\nu$ $(q = 1, 2, \ldots, M)$ is the translation matrix of the regular part of cylindrical harmonics from the global coordinate $(\rho,\ \phi)$ to the #q local coordinate $(\rho_q^\nu,\ \phi_q^\nu)$, $\mathbf{K}_p^\nu$ is the translation matrix of the singular part of cylindrical harmonics from the #p local coordinate $(\rho_p^\nu,\ \phi_p^\nu)$ to the #1 local coordinate $(\rho_1^\nu,\ \phi_1^\nu)$, d_p^ν is the distance between the centers of the #p and #1 rods and $\bar{\mathbf{I}}$ is the unit matrix. Due to the periodicity of the structure, $\bar{\boldsymbol{\Gamma}}^{(\nu)}$ matrix in (21.9) is a block-circulant matrix [34] and the solutions to $\bar{a}_j^\nu$ are simply obtained in closed form using the eigenvalues and eigenvectors of the $M \times M$ circulant matrix. From a straightforward manipulation, we have

$$\bar{\boldsymbol{a}}_j^\nu = \bar{\bar{\mathbf{T}}}_{+,j}^\nu \cdot \bar{\boldsymbol{b}}^{(\nu)} \tag{21.16}$$

with

$$\bar{\bar{\mathbf{T}}}^{\nu}_{+,j} = \sum_{q=1}^{M} \boldsymbol{\sigma}^{\nu}_{q,j} \bar{\mathbf{T}}^{\nu} \bar{\boldsymbol{\alpha}}^{\nu}_{q} \quad (j = 1, 2, \ldots, M) \tag{21.17}$$

$$\boldsymbol{\sigma}^{\nu}_{q,j} = \frac{1}{M} \sum_{\ell=1}^{M} e^{-i(q-1)(\ell-1)\theta_M} (\boldsymbol{\Lambda}^{\nu}_{\ell})^{-1} e^{i(\ell-1)(j-1)\theta_M} \tag{21.18}$$

$$\boldsymbol{\Lambda}^{\nu}_{\ell} = \mathbf{I} + \sum_{p=2}^{M} e^{i(\ell-1)(p-1)\theta_M} \mathbf{A}^{\nu}_{p} \tag{21.19}$$

Next, let us consider that the *outgoing* cylindrical wave $\bar{\boldsymbol{\Psi}}^{T} \cdot \bar{c}^{(\nu-1)}$ is incident on the cylindrical array from the inner region $(\nu - 1)$ as it is shown in Figure 21.2. In this case, the local scattering amplitude vectors $\bar{a}^{\nu}_{j}$ are obtained as follows:

$$\bar{\boldsymbol{a}}^{\nu}_{j} = \bar{\bar{\mathbf{T}}}^{\nu}_{-,j} \cdot \bar{\boldsymbol{c}}^{(\nu-1)}_{j} \tag{21.20}$$

with

$$\bar{\boldsymbol{c}}^{(\nu-1)} = \begin{bmatrix} \boldsymbol{c}^{e(\nu-1)} \\ \boldsymbol{c}^{h(\nu-1)} \end{bmatrix}, \quad \boldsymbol{c}^{e(\nu-1)} = [\boldsymbol{c}^{e(\nu-1)}_{m}], \quad \boldsymbol{c}^{h(\nu-1)} = [\boldsymbol{c}^{h(\nu-1)}_{m}] \tag{21.21}$$

$$\bar{\bar{\mathbf{T}}}^{\nu}_{-,j} = \sum_{q=1}^{M} \boldsymbol{\sigma}^{\nu}_{q,j} \bar{\mathbf{T}}^{\nu} \bar{\boldsymbol{\gamma}}^{\nu}_{q} \quad (j = 1, 2, \ldots, M) \tag{21.22}$$

$$\bar{\gamma}^{\nu}_{q} = \begin{bmatrix} \gamma^{\nu}_{q} & \mathbf{0} \\ \mathbf{0} & \gamma^{\nu}_{q} \end{bmatrix}, \quad \gamma^{\nu}_{q} = \left[(-1)^{m-n} H^{(1)}_{m-n}(\kappa R_{\nu}) e^{in(q-1)\theta_M} \right] \tag{21.23}$$

where $\boldsymbol{\gamma}^{\nu}_{q}$ is the translation matrix of the singular part of cylindrical harmonics from the global coordinate (ρ, ϕ) to the #q local coordinate $(\rho^{\nu}_{q}, \phi^{\nu}_{q})$.

It is very important to note that if the periodic defects are introduced in the cylindrical arrays by removing some particular circular rods from a circular ring [35], in this case, $\bar{\boldsymbol{\Gamma}}^{(\nu)}$ matrix in (21.9) is not reduced to a block circulant matrix, and we should numerically calculate the inversion of matrix $\bar{\boldsymbol{\Gamma}}^{(\nu)}$ to obtain unknown scattering amplitudes $a^{(\nu)}_{j}$. As a result, the calculation time is increasing.

21.2.3 Reflection and transmission matrices

Substituting (21.16) into (21.1) and (21.2), the reflected $\bar{\boldsymbol{\psi}}^{r(\nu)}$ and transmitted $\bar{\boldsymbol{\psi}}^{t(\nu-1)}$ fields into the outer region (ν) and the inner region $(\nu - 1)$ for the incidence of $\bar{\boldsymbol{\Phi}}^{T} \cdot \bar{\boldsymbol{b}}^{(\nu)}$ field are obtained as follows:

$$\bar{\psi}^{r(\nu)} = \bar{\boldsymbol{\Psi}}^{T} \cdot \bar{R}_{\nu,\nu-1} \cdot \bar{\boldsymbol{b}}^{(\nu)}, \quad \bar{\psi}^{t(\nu-1)} = \bar{\boldsymbol{\Phi}}^{T} \cdot \bar{\mathbf{F}}_{\nu-1,\nu} \cdot \bar{\boldsymbol{b}}^{(\nu)} \tag{21.24}$$

with

$$\bar{\mathbf{R}}_{\nu,\nu-1} = \sum_{j=1}^{M} \bar{\boldsymbol{\beta}}^{\nu}_{j} \cdot \bar{\bar{\mathbf{T}}}^{\nu}_{+,j}, \quad \bar{\mathbf{F}}_{\nu-1,\nu} = \bar{\mathbf{I}} + \sum_{j=1}^{M} \bar{\boldsymbol{\eta}}^{\nu}_{j} \cdot \bar{\bar{\mathbf{T}}}^{\nu}_{+,j} \tag{21.25}$$

where $\bar{\mathbf{F}}_{\nu-1,\nu}$ defines the transmission matrix of the cylindrical array, which characterizes the transmission from the outer region (ν) to the inner region ($\nu - 1$) and $\bar{\mathbf{R}}_{\nu,\nu-1}$ defines the reflection matrix characterizing the reflection to the region (ν) from the region ($\nu - 1$). Note that the unit matrix $\bar{\mathbf{I}}$ contained in (21.25) indicates the contribution of the original incident field $\bar{\boldsymbol{\Phi}}^T \cdot \bar{\boldsymbol{b}}^{(\nu)}$ in the inner region.

When *outgoing* cylindrical wave $\bar{\boldsymbol{\Psi}}^T \cdot \bar{\boldsymbol{c}}^{(\nu-1)}$ is incident on the cylindrical array from the inner region ($\nu - 1$), substituting (21.20) into (21.1) and (21.2), the transmitted $\bar{\boldsymbol{\psi}}^{t(\nu)}$ and reflected $\bar{\boldsymbol{\psi}}^{r(\nu-1)}$ fields into the region (ν) and the region ($\nu - 1$) are given as

$$\bar{\boldsymbol{\psi}}^{t(\nu)} = \bar{\boldsymbol{\Psi}}^T \cdot \bar{\mathbf{F}}_{\nu,\nu-1} \cdot \bar{\boldsymbol{c}}^{(\nu-1)}, \quad \bar{\boldsymbol{\psi}}^{r(\nu-1)} = \bar{\boldsymbol{\Phi}}^T \cdot \bar{\mathbf{R}}_{\nu-1,\nu} \cdot \bar{\boldsymbol{c}}^{(\nu-1)} \tag{21.26}$$

with

$$\bar{\mathbf{F}}_{\nu,\nu-1} = \bar{\mathbf{I}} + \sum_{j=1}^{M} \bar{\boldsymbol{\beta}}_j^{\nu} \cdot \bar{\bar{\mathbf{T}}}_{-j}^{\nu}, \quad \bar{\mathbf{R}}_{\nu-1,\nu} = \sum_{j=1}^{M} \bar{\boldsymbol{\eta}}_j^{\nu} \cdot \bar{\bar{\mathbf{T}}}_{-j}^{\nu} \tag{21.27}$$

where $\bar{\mathbf{F}}_{\nu,\nu-1}$ defines the transmission matrix, which characterizes the transmission from region ($\nu - 1$) to region (ν) through the cylindrical array, and $\bar{\mathbf{R}}_{\nu-1,\nu}$ defines the reflection matrix characterizing the reflection to the region ($\nu - 1$) from region (ν). It should be noted that the reflection and transmission matrices $\bar{\mathbf{R}}_{\nu,\nu-1}$, $\bar{\mathbf{R}}_{\nu-1,\nu}$, $\bar{\mathbf{F}}_{\nu-1,\nu}$ and $\bar{\mathbf{F}}_{\nu,\nu-1}$ are expressed in terms of the cylindrical harmonic waves as the basis and hence $\bar{\mathbf{R}}_{\nu,\nu-1} \neq \bar{\mathbf{R}}_{\nu-1,\nu}$ and $\bar{\mathbf{F}}_{\nu-1,\nu} \neq \bar{\mathbf{F}}_{\nu,\nu-1}$. This situation is quite different from the reflection and transmission matrices obtained for a planar array [5,16,24] of circular rods.

If the positions of the cylinders on the νth layer are shifted counterclockwise by an angle of Θ_ν with respect to the global coordinate *XOY*, the reflection and transmission matrices are slightly modified as

$$\hat{\mathbf{R}}_{\nu,\nu-1} = \boldsymbol{\Theta}_\nu \cdot \bar{\mathbf{R}}_{\nu,\nu-1} \cdot \boldsymbol{\Theta}_\nu^{-1}, \qquad \hat{\mathbf{R}}_{\nu-1,\nu} = \boldsymbol{\Theta}_\nu \cdot \bar{\mathbf{R}}_{\nu-1,\nu} \cdot \boldsymbol{\Theta}_\nu^{-1} \tag{21.28}$$

$$\hat{\mathbf{F}}_{\nu-1,\nu} = \boldsymbol{\Theta}_\nu \cdot \bar{\mathbf{F}}_{\nu-1,\nu} \cdot \boldsymbol{\Theta}_\nu^{-1}, \qquad \hat{\mathbf{F}}_{\nu,\nu-1} = \boldsymbol{\Theta}_\nu \cdot \bar{\mathbf{F}}_{\nu,\nu-1} \cdot \boldsymbol{\Theta}_\nu^{-1} \tag{21.29}$$

where $\boldsymbol{\Theta}_\nu = [e^{-im\Theta_\nu}\delta_{mm'}]$ is a diagonal matrix. It is obvious that when the layers are not rotated, $\boldsymbol{\Theta}_\nu$ is reduced to the unit matrix and $\hat{\mathbf{R}}_{\nu,\nu-1} = \bar{\mathbf{R}}_{\nu,\nu-1}$, $\hat{\mathbf{R}}_{\nu-1,\nu} = \bar{\mathbf{R}}_{\nu-1,\nu}$, $\hat{\mathbf{F}}_{\nu-1,\nu} = \bar{\mathbf{F}}_{\nu-1,\nu}$, $\hat{\mathbf{F}}_{\nu,\nu-1} = \bar{\mathbf{F}}_{\nu,\nu-1}$.

The reflection $\hat{\mathbf{R}}_{\nu,\nu-1}$, $\hat{\mathbf{R}}_{\nu-1,\nu}$ and the transmission $\hat{\mathbf{F}}_{\nu-1,\nu}$, $\hat{\mathbf{F}}_{\nu,\nu-1}$ matrices consist of four block matrices. The diagonal block matrices $\left\{\hat{\mathbf{R}}^{EE}_{\nu,\nu-1}, \hat{\mathbf{R}}^{HH}_{\nu,\nu-1}\right\}$, $\left\{\hat{\mathbf{R}}^{EE}_{\nu-1,\nu}, \hat{\mathbf{R}}^{HH}_{\nu-1,\nu}\right\}$ and $\left\{\hat{\mathbf{F}}^{EE}_{\nu,\nu-1}, \hat{\mathbf{F}}^{HH}_{\nu,\nu-1}\right\}$, $\left\{\hat{\mathbf{F}}^{EE}_{\nu-1,\nu}, \hat{\mathbf{F}}^{HH}_{\nu-1,\nu}\right\}$ represent the reflection and transmission into the co-polarized electric and magnetic fields, while the non-diagonal block matrices $\left\{\hat{\mathbf{R}}^{EH}_{\nu,\nu-1}, \hat{\mathbf{R}}^{HE}_{\nu,\nu-1}\right\}$, $\left\{\hat{\mathbf{R}}^{EH}_{\nu-1,\nu}, \hat{\mathbf{R}}^{HE}_{\nu-1,\nu}\right\}$ and $\left\{\hat{\mathbf{F}}^{EH}_{\nu,\nu-1}, \hat{\mathbf{F}}^{HE}_{\nu,\nu-1}\right\}$, $\left\{\hat{\mathbf{F}}^{EH}_{\nu-1,\nu}, \hat{\mathbf{F}}^{HE}_{\nu-1,\nu}\right\}$ describe the reflection and transmission into the cross-polarization space-harmonic components. In the case of dielectric scatterers, both co-polarized and cross-polarized fields give the contribution in the formation of the scattered field [20]; however, in the case of perfect electric conductor scatterers, the cross-polarization scattering effect disappears

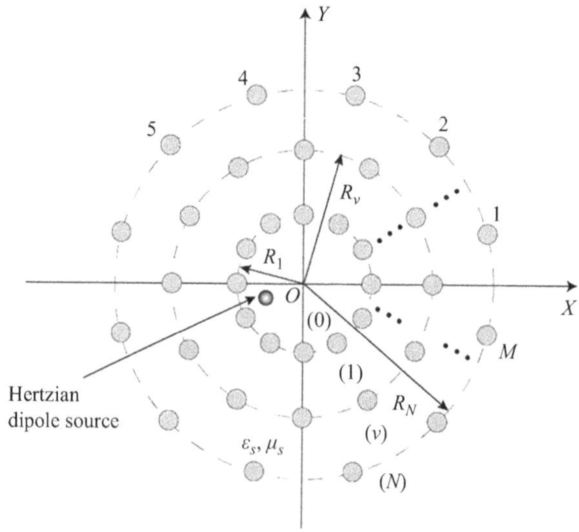

Figure 21.3 Cross-sectional view of N-layered cylindrical structure formed by M circular rods with radius r_v periodically distributed on each of N-layered arrays. Radii of the N-layered cylindrical surfaces are $R_v(v = 1, 2, 3, \ldots, N)$. Excitation by a Hertzian dipole source placed in the innermost region (0) is considered

and the problem is reduced to the consideration of TM and TE waves, separately [20]. The expressions for the reflection and transmission matrices are the key expressions in our formulation. These expressions are used in the following sections to rigorously analyze of electromagnetic radiation, scattering and guidance in cylindrical periodic and bandgap structures. Finally, we should mention that a generalization of the problem to eccentrically layered circular rings is straightforward. The readers may refer to [20].

21.2.4 Hertzian dipole source radiation in the layered cylindrical structure

Let us consider that Hertzian dipole source located in the innermost (0) region at $(d_s, \varphi_s, 0)$ is directed in z direction (Figure 21.3). Using the Fourier integral representation, the source field in the spectral domain is expressed in the global coordinate system (ρ, ϕ) as follows:

$$\tilde{E}_z^i(\rho, \phi, \xi) = -\frac{\omega \mu_0 I \ell}{4} \frac{\kappa^2}{k_s^2} \boldsymbol{\Psi}^T \cdot \mathbf{s}^e, \quad \mathbf{s}^e = [J_m(\kappa d_s) \exp{(-im\varphi_s)}], \tag{21.30}$$

where $\kappa = \sqrt{k_s^2 - \xi^2}$ and $I\ell$ is the current moment of the electric dipole. After straightforward mathematical manipulations, the transmitted fields $E_z^{t(N)}$ and $\hat{H}_z^{t(N)}$ in the far-zone are obtained as

$$\begin{bmatrix} E_z^{t(N)} \\ \hat{H}_z^{t(N)} \end{bmatrix} = -\frac{\omega\mu_0 I\ell}{4}\frac{\exp(ik_s r)}{\pi r}\sin^2\theta \sum_{m=-\infty}^{\infty} \exp(im\phi)$$

$$\times \exp\left[-i(2m+1)\frac{\pi}{4}\right]\begin{bmatrix} \boldsymbol{c}_m(\xi)^{e(N)} \\ \boldsymbol{c}_m(\xi)^{h(N)} \end{bmatrix}_{\xi=k_s\cos\theta} \tag{21.31}$$

where (r, ϕ, θ) denote the location of an observation point in the spherical coordinate system. The scattering amplitudes $\{c_m(\xi)^{e(N)}\}$ and $\{c_m(\xi)^{h(N)}\}$ in the outermost (N)th region are related to the spectral amplitude $s_m^e(\xi)$ of the source field through the generalized transmission matrix $\bar{\bar{\mathbf{F}}}_{N,0}$:

$$\begin{bmatrix} \boldsymbol{c}_m(\xi)^{e(N)} \\ \boldsymbol{c}_m(\xi)^{h(N)} \end{bmatrix} = \bar{\bar{\mathbf{F}}}_{N,0}(\xi)\begin{bmatrix} \boldsymbol{s}_m^e(\xi) \\ 0 \end{bmatrix} \tag{21.32}$$

where

$$\bar{\bar{\mathbf{F}}}_{N,0}(\xi) = \bar{\mathbf{F}}_{N,N-1}(\xi)\cdot\bar{\mathbf{\Gamma}}_{N-1,N-2}(\xi)\cdots\bar{\mathbf{\Gamma}}_{v,v-1}(\xi)\cdots\bar{\mathbf{\Gamma}}_{1,0}(\xi) \tag{21.33}$$

$$\bar{\mathbf{\Gamma}}_{v,v-1}(\xi) = \left[\bar{\mathbf{I}} - \bar{\mathbf{R}}_{v,v-1}(\xi)\cdot\bar{\bar{\mathbf{R}}}_{v,v+1}(\xi)\right]^{-1}\cdot\bar{\mathbf{F}}_{v,v-1}(\xi) \tag{21.34}$$

$$\bar{\bar{\mathbf{R}}}_{v,v+1}(\xi) = \bar{\mathbf{R}}_{v,v+1}(\xi) + \bar{\mathbf{F}}_{v,v+1}(\xi)\cdot\bar{\bar{\mathbf{R}}}_{v+1,v+2}(\xi)\cdot[\bar{\mathbf{I}} - \bar{\mathbf{R}}_{v+1,v}(\xi)$$

$$\times\bar{\bar{\mathbf{R}}}_{v+1,v+2}(\xi)]^{-1}\cdot\bar{\mathbf{F}}_{v+1,v}(\xi) \tag{21.35}$$

where $\bar{\bar{\mathbf{R}}}_{v,v+1}(\xi)$ denotes the generalized reflection matrix viewed from the th region to the outer regions and it is calculated through the N times recursion process [18–20].

21.2.5 Plane wave scattering by the layered cylindrical structure

Let us consider that a plane wave of unit amplitude with a wavevector $\mathbf{k}_s$ is obliquely incident from the outermost region (N) on the cylindrical structure, where $\mathbf{k}_s = (k_s, \theta^i, \varphi^i)$ is the spherical coordinate system with the origin O, $k_s = \omega\sqrt{\varepsilon_s\mu_s}$ (Figure 21.4). The z components of the incident field are expressed in the cylindrical coordinate system as follows:

$$\bar{\boldsymbol{\psi}}^{i(N)} = \bar{\mathbf{\Phi}}^T\cdot\bar{\boldsymbol{b}}^{(N)} \tag{21.36}$$

with

$$\bar{\boldsymbol{\psi}}^{i(N)} = \begin{bmatrix} E_z^{i(N)} \\ \hat{H}_z^{i(N)} \end{bmatrix} \tag{21.37}$$

$$\bar{\boldsymbol{b}}^{(N)} = \begin{bmatrix} \boldsymbol{a}^{e(N)} \\ \boldsymbol{a}^{h(N)} \end{bmatrix} \tag{21.38}$$

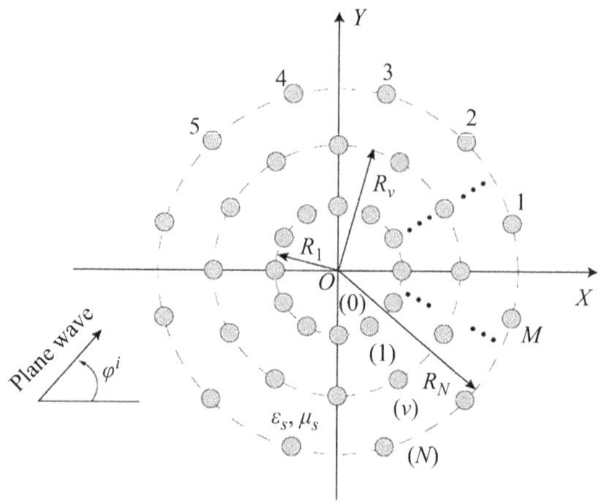

Figure 21.4 Cross-sectional view of N-layered cylindrical structure formed by M circular rods with radius r_ν periodically distributed on each of N-layered arrays. Radii of the N-layered cylindrical surfaces are $R_\nu (\nu = 1, 2, 3, \ldots, N)$. Plane wave oblique incidence from the outermost (N) region is considered

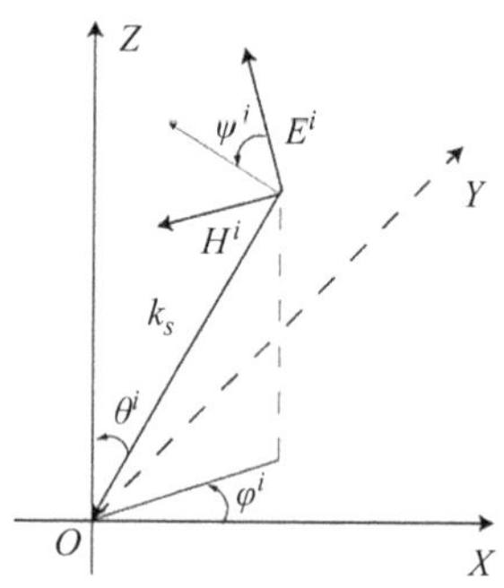

Figure 21.5 Oblique incidence of plane waves, where (θ^i, ϕ^i) denotes the angle of incidence in spherical coordinate system and ψ^i is polarization angle of the incident field

$$\boldsymbol{a}^{e(N)} = \exp(i\xi z) \sin\theta^i \cos\psi^i \, \mathbf{s} \tag{21.39}$$

$$\boldsymbol{a}^{h(N)} = \exp(i\xi z) \sin\theta^i \sin\psi^i \, \mathbf{s} \tag{21.40}$$

$$\mathbf{s} = [i^m \exp(-im\phi^i)] \tag{21.41}$$

where $\kappa = \sqrt{k_s^2 - \xi^2}$, $\xi = k_s \cos\theta^i$ and ψ^i is the polarization angle of the incident electric field as depicted in Figure 21.5. By tracing the scattering process over the layered cylindrical arrays, the generalized reflection matrices $\tilde{\bar{\mathbf{R}}}_{\nu,\nu-1}$ viewed from region (ν) to all of the inner regions could be obtained as follows:

$$\tilde{\bar{\mathbf{R}}}_{\nu,\nu-1} = \bar{\mathbf{R}}_{\nu,\nu-1} + \bar{\mathbf{F}}_{\nu,\nu-1}(\bar{\mathbf{I}} - \tilde{\bar{\mathbf{R}}}_{\nu-1,\nu-2}\bar{\mathbf{R}}_{\nu-1,\nu})^{-1}\tilde{\bar{\mathbf{R}}}_{\nu-1,\nu-2}\bar{\mathbf{F}}_{\nu-1,\nu} \tag{21.42}$$

The generalized reflection matrix could be calculated through the N times recursion process starting with $\tilde{\bar{\mathbf{R}}}_{2,1} = \bar{\mathbf{R}}_{2,1}$. Since the reflection $\bar{\mathbf{R}}_{\nu,\nu-1}$, $\bar{\mathbf{R}}_{\nu-1,\nu}$ and transmission $\bar{\mathbf{F}}_{\nu-1,\nu}$, $\bar{\mathbf{F}}_{\nu,\nu-1}$ matrices for each cylindrical array and the generalized reflection matrix $\tilde{\bar{\mathbf{R}}}_{\nu,\nu-1}$ for the N-layered cylindrical structure are derived, the total field in each region of the cylindrical structure could be defined. The total field in the outermost (N)th region is expressed as

$$\bar{\boldsymbol{\psi}}^{tot(N)} = (\bar{\boldsymbol{\Phi}}^T + \bar{\boldsymbol{\Psi}}^T \cdot \tilde{\bar{\mathbf{R}}}_{N,N-1}) \cdot \bar{\boldsymbol{b}}^{(N)} \tag{21.43}$$

In the (ν)th region inside the cylindrical structure, the total field is given by the following expression:

$$\bar{\boldsymbol{\psi}}^{tot(\nu)} = (\bar{\boldsymbol{\Phi}}^T + \bar{\boldsymbol{\Psi}}^T \cdot \tilde{\bar{\mathbf{R}}}_{\nu,\nu-1}) \cdot \tilde{\bar{\mathbf{F}}}_{\nu,N} \cdot \bar{\boldsymbol{b}}^{(N)} \tag{21.44}$$

where

$$\tilde{\bar{\mathbf{F}}}_{\nu,N} = \bar{\boldsymbol{\Lambda}}_{\nu,\nu+1} \cdot \bar{\boldsymbol{\Lambda}}_{\nu+1,\nu+2} \cdots \bar{\boldsymbol{\Lambda}}_{N-1,N} \tag{21.45}$$

$$\bar{\boldsymbol{\Lambda}}_{\nu,\nu+1} = (\bar{\mathbf{I}} - \bar{\mathbf{R}}_{\nu,\nu+1} \cdot \tilde{\bar{\mathbf{R}}}_{\nu,\nu-1})^{-1} \cdot \bar{\mathbf{F}}_{\nu,\nu+1} \tag{21.46}$$

The transmitted field in the innermost (0)th region is given as

$$\bar{\boldsymbol{\psi}}^{tot(0)} = \bar{\boldsymbol{\Phi}}^T \cdot \tilde{\bar{\mathbf{F}}}_{0,N} \cdot \bar{\boldsymbol{b}}^{(N)} \tag{21.47}$$

where $\tilde{\bar{\mathbf{F}}}_{0,N}$ is the generalized transmission matrix in the innermost (0) region from the outermost (N) region.

21.2.6 Guidance in the layered cylindrical structure

For the guiding problems, there is no source excitation and $\bar{\boldsymbol{c}}^{(0)} = \bar{\boldsymbol{b}}^{(N)} = 0$. In region (1) of the N-layered system, we have the following relation:

$$\bar{\boldsymbol{b}}^{(1)} = \bar{\bar{\mathbf{R}}}_{1,2}(\xi) \cdot \bar{\boldsymbol{c}}^{(1)}, \quad \bar{\boldsymbol{c}}^{(1)} = \hat{\mathbf{R}}_{1,0}(\xi) \cdot \bar{\boldsymbol{b}}^{(1)}. \tag{21.48}$$

Equation (48) leads to

$$[\mathbf{I} - \bar{\bar{\mathbf{R}}}_{1,2}(\xi) \cdot \hat{\mathbf{R}}_{1,0}(\xi)] \cdot \bar{\boldsymbol{b}}^{(1)} = 0 \tag{21.49}$$

where $\bar{\bar{\mathbf{R}}}_{1,2}(\xi)$ is calculated using the following recursive relation:

$$\bar{\bar{\mathbf{R}}}_{\nu,\nu+1} = \hat{\mathbf{R}}_{\nu,\nu+1} + \hat{\mathbf{F}}_{\nu,\nu+1} \cdot \bar{\bar{\mathbf{R}}}_{\nu+1,\nu+2} \cdot (\mathbf{I} - \hat{\mathbf{R}}_{\nu+1,\nu} \cdot \bar{\bar{\mathbf{R}}}_{\nu+1,\nu+2})^{-1} \cdot \hat{\mathbf{F}}_{\nu+1,\nu}, \tag{21.50}$$

starting from $\bar{\bar{\mathbf{R}}}_{N-1,N} = \hat{\mathbf{R}}_{N-1,N}$. It should be noted that the $\bar{\boldsymbol{\Psi}}$ field does not exist in the innermost region (0) and the $\bar{\boldsymbol{\Phi}}$ field does not exist in the outermost region (N).

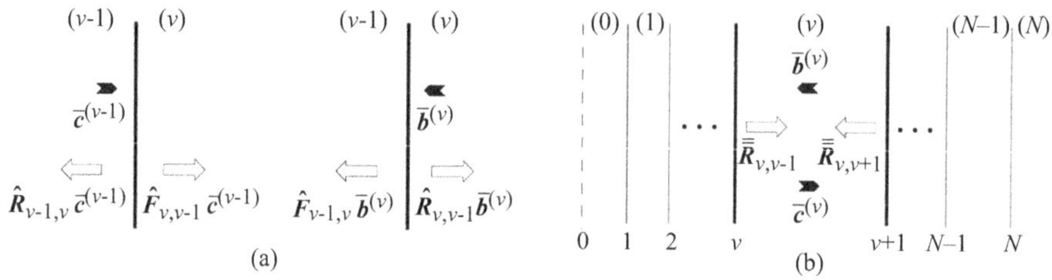

Figure 21.6 Scattering process in layered cylindrical arrays: (a) two scattering process in a single layer of array and (b) generalized reflection matrix in N-layered arrays

Equation (21.49) yields the dispersion equation for the mode propagation constant ξ as follows:

$$\det\left[\mathbf{I} - \bar{\bar{\mathbf{R}}}_{1,2}(\xi)\cdot\hat{\mathbf{R}}_{1,0}(\xi)\right] = 0. \tag{21.51}$$

The solutions to (21.51) are used in (21.48) to determine the amplitude vector $\bar{\boldsymbol{b}}^{(1)}$ and $\bar{\boldsymbol{c}}^{(1)}$. Other amplitudes vectors $\bar{b}^{(v)}$ and $\bar{\boldsymbol{c}}^{(v)}$ as shown in Figure 21.6 are obtained using the relations:

$$\bar{\boldsymbol{b}}^{(0)} = \hat{\mathbf{F}}_{01}\cdot\bar{\boldsymbol{b}}^{(1)}, \quad \bar{\boldsymbol{c}}^{(0)} = \mathbf{0} \tag{21.52}$$

$$\bar{\boldsymbol{b}}^{(\nu)} = \bar{\bar{\mathbf{R}}}_{\nu,\nu+1}\cdot\bar{\boldsymbol{c}}^{(\nu)}\ (\nu \geq 2) \tag{21.53}$$

$$\bar{\boldsymbol{c}}^{(\nu)} = \hat{\mathbf{F}}_{\nu,\nu-1}\cdot\bar{\boldsymbol{c}}^{(\nu-1)} + \hat{\mathbf{R}}_{\nu,\nu-1}\cdot\bar{\boldsymbol{b}}^{(\nu)}\ (\nu \geq 2) \tag{21.54}$$

$$\bar{\boldsymbol{b}}^{(N)} = \mathbf{0}, \quad \bar{\boldsymbol{c}}^{(N)} = \hat{\mathbf{F}}_{N,N-1}\cdot\bar{\boldsymbol{c}}^{(N-1)}. \tag{21.55}$$

21.3 Numerical results and discussions

21.3.1 Directivity of radiation of a dipole source coupled to the cylindrical EBG structure

Although a substantial number of numerical results could be generated for different configurations of cylindrical electromagnetic bandgap (EBG) structures, we consider three different configurations of three-layered ($N = 3$) cylindrical EBG structures composed of identical perfect electric conductor (PEC) circular rods with $r_1 = r_2 = r_3 = 0.05R_1$ periodically distributed along each cylindrical layer having radii R_1, $R_2 = 2R_1$, $R_3 = 3R_1$. It is very important to note that unlike the planar periodic structure, where the excitation of particular space harmonics depends on the frequency range, in the case of cylindrically periodic structure all orders of cylindrical harmonic waves are excited. However, due to the resonances and stopbands nature some of them are enhanced, whereas others are strongly suppressed (EBG region). Cylindrical

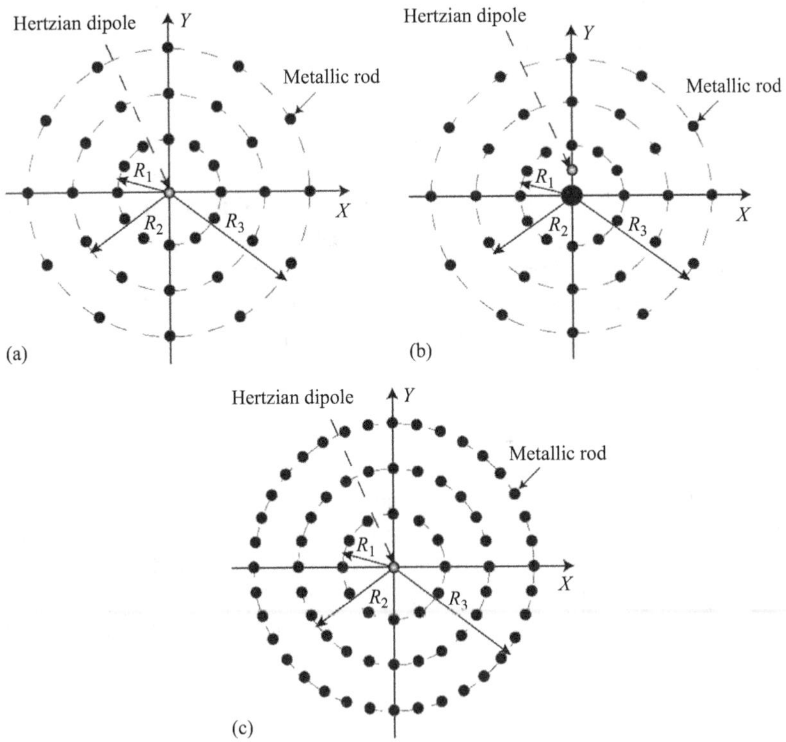

Figure 21.7 Cross-sectional view of three different configurations of three-layered cylindrical EBG structure of PEC circular rods with $r_1 = r_2 = r_3 = 0.05R_1$ periodically located on the concentric circular layers. Radii of the first, second and third circular layers are R_1, $R_2 = 2R_1$ and $R_3 = 3R_1$, respectively. (a) Twelve circular rods are symmetrically distributed along the first, second and third layers and the source is located at the global origin; (b) one additional metallic circular rod having a radius $r_{Orig} = 0.3R_1$ is additionally placed at the origin of cylindrical structure (a) and the source is located at a distance d_s from the origin; (c) 12 circular rods are symmetrically distributed along the first layer, 24 circular rods along the second layer and 36 circular rods along the third layer. The dipole source is located at the global origin. © 2017 OSA. After, with kind permission, from [19]

EBG structure shown in Figure 21.7(a) is composed of $M_1 = M_2 = M_3 = 12$ circular rods periodically distributed along each cylindrical layer, which means that the angle between the nearest circular rods along the different layers is the same. In Figure 21.7(a), the dipole source is located at the global origin $d_s = 0$. In three-layered cylindrical EBG structure shown in Figure 21.7(b) the source is located at a distance

$d_s \neq 0$ from the global origin along the positive y axis, and one PEC circular rod is additionally placed at the origin. The circular rod at the origin could serve as a metallic ground for the cylindrical EBG structure. Figure 21.7(c) shows the cylindrical EBG structure with $M_1 = 12$, $M_2 = 24$ and $M_3 = 36$ circular rods periodically distributed along the first, second and third layers, which means that the radial distance between the nearest circular rods along different cylindrical layers is the same. The source is assumed to be located at the origin $d_s = 0$. The main goal is to bring out the relation between the transmission spectra of cylindrical EBG structures and the radiation of the dipole source located inside the cylindrical EBG structures. We give a deep insight into the spectral responses of cylindrical EBG structures in terms of generalized transmission matrix $\tilde{\tilde{\mathbf{F}}}_{N,0}(\omega,\theta)$ given in (21.33)–(21.35). First, we investigate the cylindrical EBG structure shown in Figure 21.7(a). When the Hertzian dipole is located at the global origin O, the cylindrical structure is symmetric viewed from the origin and only $(0, 0)$, $(\pm 12, 0)$, elements of the generalized transmission matrix contribute in the excitation of the lowest principle cylindrical modes. In order to give a deep insight into the relation between the transmission spectra of the cylindrical space-harmonics and the radiation patterns of the Hertzian dipole source, we start our studies by investigating the one-layered cylindrical structures with 12 metallic circular rods periodically distributed. Figure 21.8 shows the dependence of the components and of the transmission matrix as a function of the normalized wavelength for different

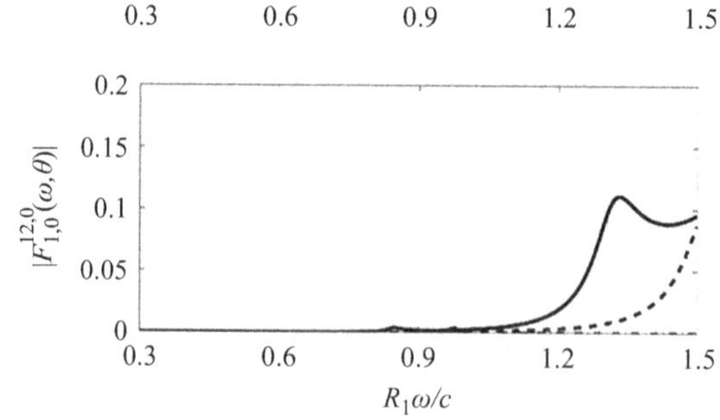

Figure 21.8 Frequency response of $\left|F_{1,0}^{m,0}(\omega,\theta)\right|$ *for one-layered cylindrical EBG structure composed of 12 PEC circular rods with* $r_1 = 0.05R_1$. *Dashed-dotted line—*$\theta = 26°$*; dashed line—*$\theta = 60°$*; solid line—*$\theta = 90°$*. © 2017 OSA. After, with kind permission, from [19]*

spherical angles . Note that $\left|\tilde{\tilde{F}}_{N,0}^{-m,0}(\omega,\theta)\right| = \left|\tilde{\tilde{F}}_{N,0}^{m,0}(\omega,\theta)\right|$. From Figure 21.8, it follows that for $\left|F_{1,0}^{0,0}(\omega,\theta = 90°)\right|$ (solid line) the main resonance occurs at $R_1\omega/c = 0.37$. When the spherical angle θ is decreasing, the resonance peak is shifting to the higher frequency regions as for $\theta = 60°$ (dashed line) and $\theta = 26°$ (dashed-dotted line). Note that $\left|F_{1,0}^{12,0}(\omega,\theta)\right|$ is a small value within the whole frequency range $0.3 < R_1\omega/c < 1.5$ and it does not make any significant contribution in the formation of the radiated field.

Radiation patterns of a dipole source located at the global origin O of the cylindrical EBG structure are presented in Figure 21.9 in both H-plane and E-plane at two resonance frequencies $R_1\omega/c = 0.37,\ 0.845$. At both frequencies, the radiation pattern in H-plane is completely omnidirectional. It could be explained by the fact that the radiation is defined exclusively by the (0,0) component of the transmission

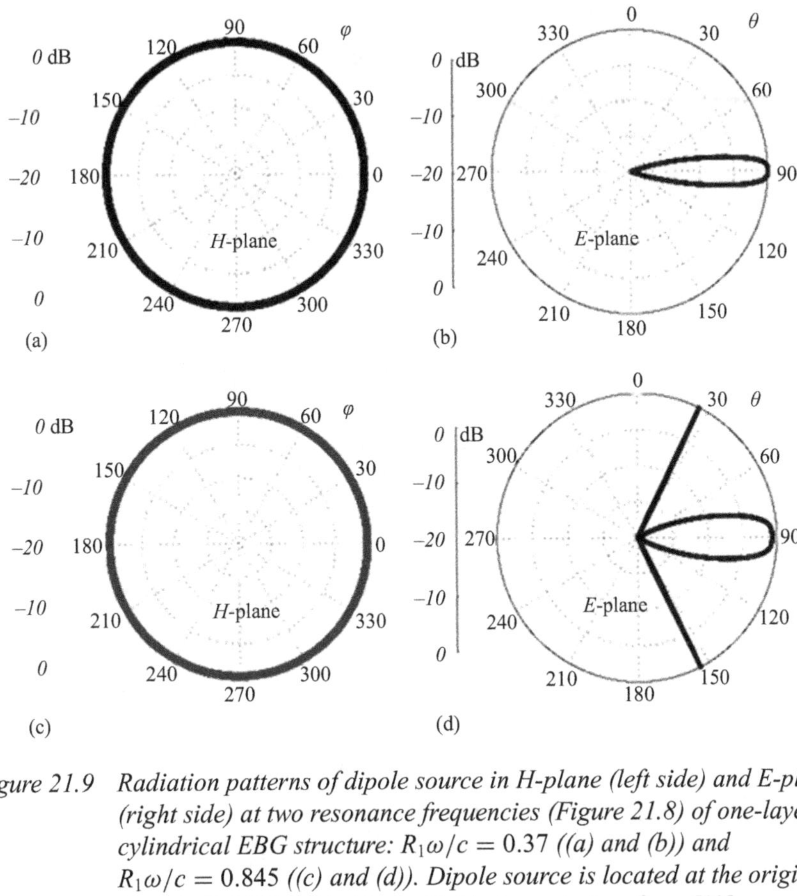

Figure 21.9 Radiation patterns of dipole source in H-plane (left side) and E-plane (right side) at two resonance frequencies (Figure 21.8) of one-layered cylindrical EBG structure: $R_1\omega/c = 0.37$ ((a) and (b)) and $R_1\omega/c = 0.845$ ((c) and (d)). Dipole source is located at the origin $d_s = 0$. © 2017 OSA. After, with kind permission, from [19]

matrix $\mathbf{F}_{1,0}(\omega, \theta)$, and the radiated field does not contain any term with ϕ dependence. The radiation pattern in the principle E-plane is strongly directed in the vertical direction at $R_1\omega/c = 0.37$; however, the strong side-lobes are observed along $\theta = 26°$ at $R_1\omega/c = 0.845$. The latter could be exclusively explained by the transmission spectra at this particular frequency $\left|F_{1,0}^{0,0}(\omega, \theta = 26°)\right| \gg \left|F_{1,0}^{0,0}(\omega, \theta = 90°)\right|$ (Figure 21.8).

Figure 21.10 illustrates the dependence of the components of the transmission matrix for the three-layered cylindrical EBG structure in Figure 21.7(a) versus $R_1\omega/c$ for different spherical angles θ. Comparison to the one-layered structure, the higher order cylindrical harmonics $\left|\tilde{\tilde{F}}_{3,0}^{12,0}(\omega, \theta)\right|$ and $\left|\tilde{\tilde{F}}_{3,0}^{24,0}(\omega, \theta)\right|$ give the contribution in the formation of the radiated field. Figure 21.10 vividly shows that there exist a series of stopband and the sharp resonance peaks for each of the excited cylindrical space-harmonics, which have not been observed in the case of one-layered structure. Transmission is extremely small in the stopband regions, whereas sharp resonance

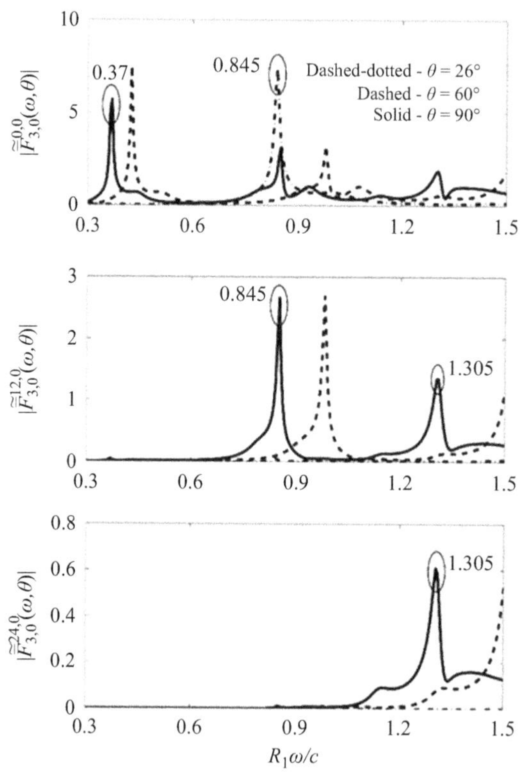

Figure 21.10 Frequency response of $\left|\tilde{\tilde{F}}_{3,0}^{m,0}(\omega, \theta)\right|$ for three-layered cylindrical EBG structure shown in Figure 21.7(a). Dashed-dotted line—$\theta = 26°$; dashed line—$\theta = 60°$; solid line—$\theta = 90°$. © 2017 OSA. After, with kind permission, from [19]

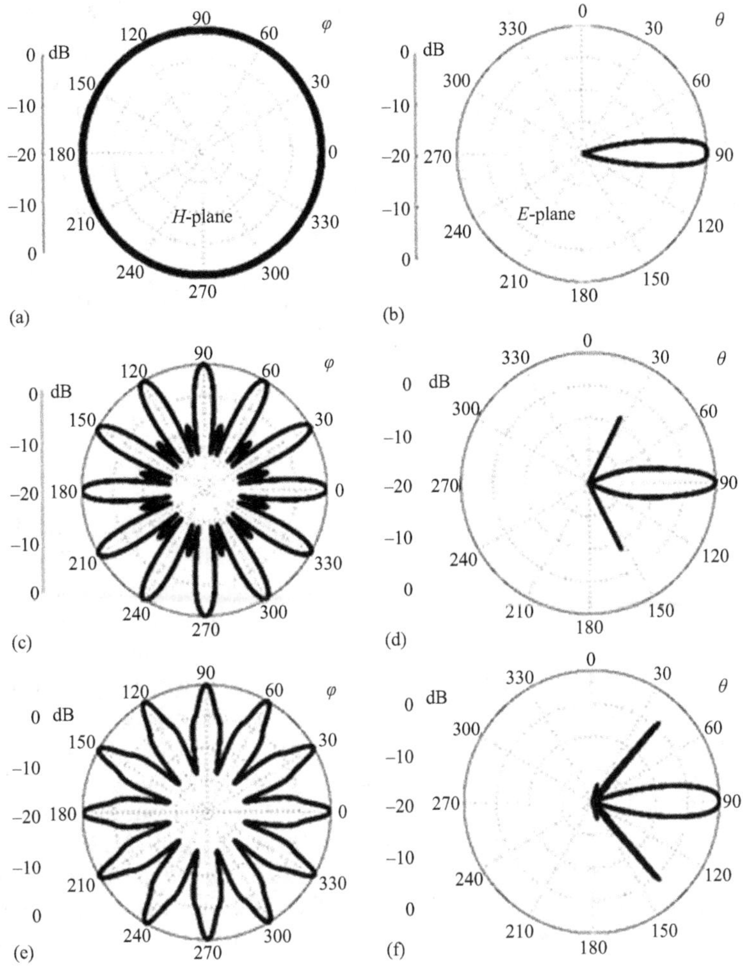

Figure 21.11 Radiation patterns of dipole source in H-plane (left side) and E-plane (right side) at three resonance frequencies of three-layered cylindrical EBG structure: $R_1\omega/c = 0.37$ *((a) and (b));* $R_1\omega/c = 0.845$ *((c) and (d)) and* $R_1\omega/c = 1.305$ *((e) and (f)). Dipole source is located at the origin* $d_s = 0$*. © 2017 OSA. After, with kind permission, from [19]*

peaks appear in the passband regions. Similarly to the one-layered cylindrical structure, two resonance frequencies for $\left|\tilde{\tilde{F}}^{0,0}_{3,0}(\omega,\theta=90^\circ)\right|$, $\left|\tilde{\tilde{F}}^{12,0}_{3,0}(\omega,\theta=90^\circ)\right|$ appear at $R_1\omega/c = 0.37, 0.845$. However, one resonance peak is additionally formed at for and $\left|\tilde{\tilde{F}}^{24,0}_{3,0}(\omega,\theta=90^\circ)\right|$. Radiation patterns of Hertzian dipole source located at the origin of the cylindrical EBG structure (Figure 21.7(a)) are presented in Figure 21.11 at these three resonance frequencies. At $R_1\omega/c = 0.37$ the radiation patterns in both H-plane and E-plane are similar to those for one-layered structure as shown in Figure 21.11(a) and (b). Interesting features for the three-layered cylindrical EBG structure appear at the higher frequency range. At $R_1\omega/c = 0.845, \ 1.305$ the radiation pattern with 12 beams is formed in H-plane, since the higher order modes $(\pm 12, 0)$, $(\pm 24, 0)$ give the substantial contribution in the formation of the radiated field. Our analysis has clearly demonstrated the principles of the formation of multibeam radiation pattern of Hertzian dipole source for cylindrical EBG structure and shown that the directivity of radiation of the source is very closely related to the resonance characteristics and stopband regions in the transmission spectra of the cylindrical bandgap structures. Another point of worth mentioning is that the strong side-lobs along $\theta = 26°$ in E-plane for one-layered structure (Figure 21.9(d)) are strongly suppressed in the case of three-layered cylindrical bandgap structure and the main beam is directed along $\theta = 90°$. It could be explained by the interference of the excited lowest three cylindrical harmonics $(0, 0)$, $(\pm 12, 0)$ at $R_1\omega/c = 0.845$ as shown in Figure 21.10 by the dashed-dotted and solid lines.

Next we analyze the case when the dipole source is located at some distance from the global origin $d_s \neq 0$ as shown in Figure 21.7(b). It is important to mention that when the excitation source is not located at the global origin O, $n - m = p$ $(p = 0, \pm 12, \pm 24, \ldots)$ components of the transmission matrix give the contribution in the formation of the radiated field. Figure 21.12 illustrates the dependence of three components of the generalized transmission matrix of the three-layered cylindrical EBG structure versus $R_1\omega/c$ for different spherical angle θ. From Figure 21.12, it follows that there exist number of stopbands and resonance peaks at different frequencies for each of the excited cylindrical space harmonics. Figure 21.13 illustrates the radiation pattern of dipole source located at a distance $d_s/R_1 = 0.35$ from the origin in the innermost region of the cylindrical EBG structure shown in Figure 21.7(b). One additional PEC circular rod having radius $r_{orig}/R_1 = 0.3$ is placed at the origin, which serves as a ground plane for the cylindrical EBG structure. Radiation patterns of the dipole source presented in Figure 21.13 are given at two resonance frequencies $R_1\omega/c = 0.37, 0.845$ (Figure 21.10). From Figure 21.13(a) and (b) it follows that the radiation patterns are directive in the vertical direction in both principle planes at $R_1\omega/c = 0.37$. It could be explained by the fact that in the lower frequency region, the contribution of the higher order cylindrical space harmonics in the radiated field is very small as shown in Figures 21.8 and 21.10 and the main contribution is given only by the (0,0) component of the transmission matrix. At $R_1\omega/c = 0.845$ the interference of the higher order cylindrical space harmonics (Figures 21.10 and 21.12) leads to the significant modification of radiation patterns in both H-plane and E-plane as presented in Figure 21.13(c) and (d), respectively. From the analysis, it follows that when the dipole source is not located at the origin, to obtain a directive beam in both principle planes, the lower excitation frequencies should be considered.

Finally, we consider three-layered cylindrical EBG structure with $M_1 = 12$, $M_2 = 24$ and $M_3 = 36$ circular rods shown in Figure 21.7(c). The Hertzian dipole source is located at the global origin O. For the cylindrical EBG structure presented

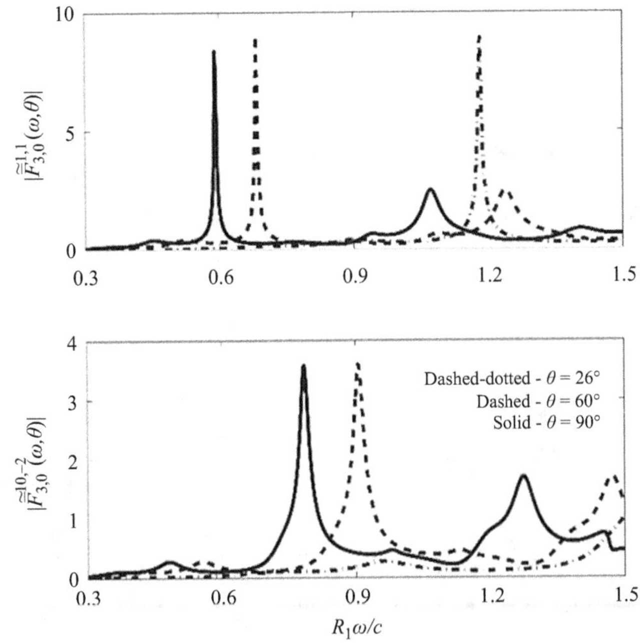

Figure 21.12 Frequency response of three components of generalized transmission matrix $\left|\tilde{\tilde{F}}_{3,0}^{m,n}(\omega,\theta)\right|$ *for three-layered cylindrical EBG structure shown in Figure 21.7(b). Dashed-dotted line—*$\theta = 26°$*; dashed line—*$\theta = 60°$*; solid line—*$\theta = 90°$*. © 2017 OSA. After, with kind permission, from [19]*

in Figure 21.7(c), we should consider the seven excited cylindrical space harmonics $\left|\tilde{\tilde{F}}_{3,0}^{0,0}(\omega,\theta)\right|$, $\left|\tilde{\tilde{F}}_{3,0}^{\pm12,0}(\omega,\theta)\right|$, $\left|\tilde{\tilde{F}}_{3,0}^{\pm24,0}(\omega,\theta)\right|$ and $\left|\tilde{\tilde{F}}_{3,0}^{\pm36,0}(\omega,\theta)\right|$. The frequency dependence of these modes versus $R_1\omega/c$ is demonstrated in Figure 21.14. One distinctive feature for the cylindrical EBG structure shown in Figure 21.7(c) is that the stopband effect is much stronger than that for the cylindrical EBG structure shown in Figure 21.7(a) and stopband regions are observed in the wider frequency range for each of the excited cylindrical space harmonic. For at $\left|\tilde{\tilde{F}}_{3,0}^{0,0}(\omega,\theta)\right|$ at $\theta = 90°$, the main resonances occur at $R_1\omega/c = 0.825$, 1.34. Radiation patterns at these two resonance frequencies are presented in Figure 21.15. At $R_1\omega/c = 0.825$ the radiation pattern in H-plane is completely omnidirectional, since only (0,0) component of the generalized transmission matrix gives the contribution in the radiated field (Figure 21.15(a) and (b)), whereas at $R_1\omega/c = 1.34$ the interference between the resonated cylindrical modes produces a very fine multibeam pattern in H-plane as it could be seen in Figure 21.15(b).

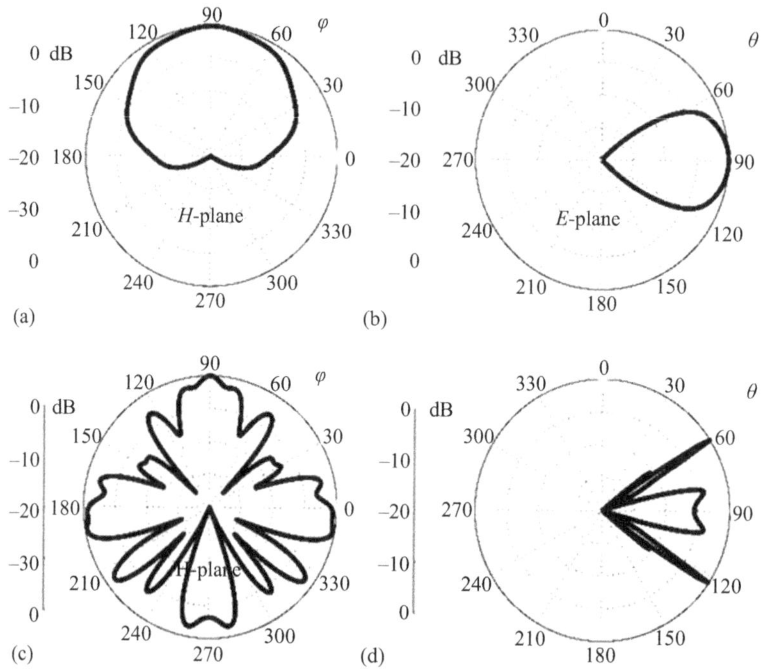

Figure 21.13 Radiation patterns of dipole source in H-plane (left side) and E-plane (right side) at two resonance frequencies of three-layered cylindrical EBG structure shown in Figure 21.7(b), where the dipole source is located at $d_s/R_1 = 0.35$ and one additional perfect electric conductor circular rod having radius $r_{orig}/R_1 = 0.3$ is placed at the origin: $R_1\omega/c = 0.37$ ((a) and (b)) and $R_1\omega/c = 0.845$ ((c) and (d)). © 2017 OSA. After, with kind permission, from [19]

In order to plot the radiation diagrams shown in Figures 21.9, 21.11, 21.13 and 21.15 we need around 4 s using the desktop 3.6 GHz Intel Core i7 with 8 GB RAM.

21.3.2 Light scattering by the metal-coated dielectric nanocylinders with angular periodicity

In the numerical examples, we assume silver (Ag) for the metal employing the Drude–Lorentz model [36]. Silver is commonly used for nanoparticle plasmonics because of the relatively easy production and small loss. An incidence of the plane waves at $\phi^i = 0°$ (from negative x axis), θ^i=90°, $\psi^i = 90°$ (see Section 21.2.5) is studied under the following structural parameters: $r_1 = 60$ nm, $r_2 = 45$ nm and $\varepsilon = 2.25\varepsilon_0$ (glass core). Geometry of the problem is illustrated in Figure 21.16. Although the

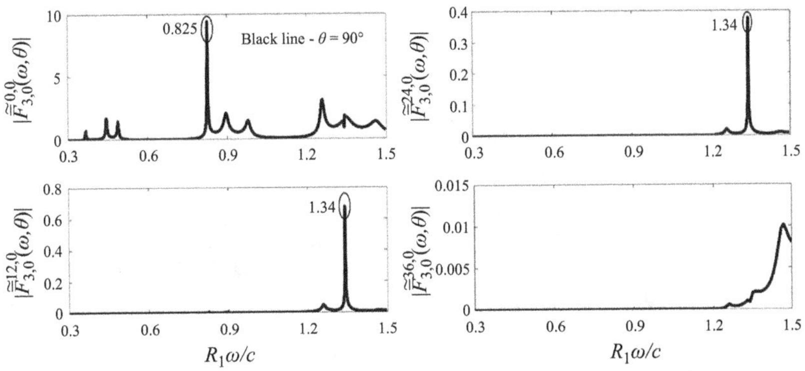

Figure 21.14 *Frequency response of the seven components of the generalized transmission matrix* $\left|\tilde{\tilde{F}}_{3,0}^{m,0}(\omega, \theta = 90^0)\right|$ *for three-layered cylindrical EBG structure shown in Figure 217(c). © 2017 OSA. After, with kind permission, from [19]*

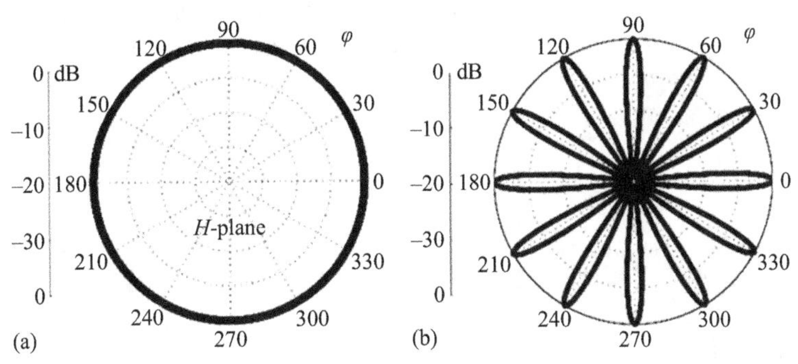

Figure 21.15 *Radiation patterns of dipole in H-plane at two resonance frequencies of three-layered cylindrical EBG structure (Figure 21.7(c)), where the Hertzian dipole source is located at the global origin O:* $R_1\omega/c = 0.825$*(a) and* $R_1\omega/c = 1.34$*(b). © 2017 OSA. After, with kind permission, from [19]*

developed formulation can be applied to an arbitrary number of nanocylinders periodically situated along a cylindrical surface, the light scattering on three (Ag)-coated nanocylinders is investigated. First, SCS and ACS [25] of three metal (Ag)-coated nanocylinders are plotted in Figure 21.17 in the wavelength range from 250 to 2,000 nm at four different values of radius R: $R = 100$ nm (nanocylinders are well separated), $R = 72$ nm and $R = 70$ nm (nanocylinders are located in a close vicinity

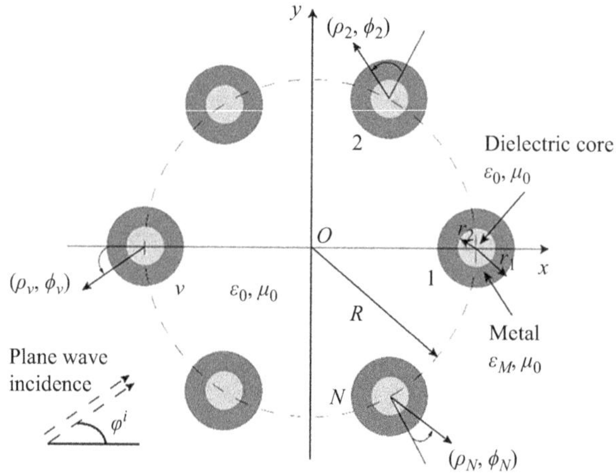

Figure 21.16 Cross-sectional view of N metal-coated dielectric nanocylinders periodically distributed on a circular ring with radius R. Material constants of the coating metal and dielectric core are denoted by $(\varepsilon_M, \mu_0, r_1)$ and $(\varepsilon, \mu_0, r_2)$, respectively, and (ρ_j, φ_j) denote the local coordinates for the jth nanocylinder. Excitation by a plane wave in XOY plane with incident angle ϕ^i is considered. A metal coating layer has a thickness $r_1 - r_2$

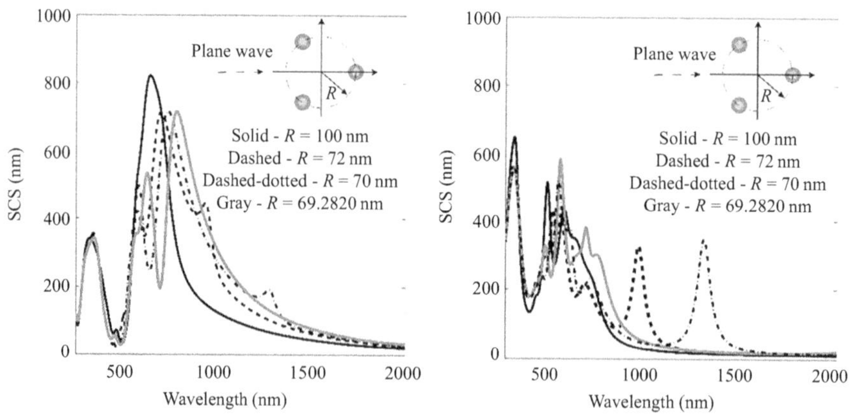

Figure 21.17 Scattering cross section (SCS) and absorption cross section (ACS) of three metal (Ag)-coated dielectric nanocylinders as functions of the wavelength at $r_1 = 60$ nm, $r_2 = 45$ nm, $\varepsilon = 2.25\varepsilon_0$ and $\phi^i = 0°$, $\theta^i = 90°$, $\psi^i = 90°$ for different values of R: R = 100 nm (solid line), R = 72 nm (dashed line), R = 70 nm (dashed-dotted line) and R = 69.2820 nm (gray line)

but not touching each other) and $R = 69.2820$ nm (nanocylinders are touching). From Figure 21.17 it follows that SCS and ACS have the resonance peaks at the wavelengths $\lambda = 330$ and $\lambda = 300$ nm for all values of radius R. Resonances in the short wavelength region characterize the plasmon resonances of the metal (Ag)-coated nanocylinders in isolation [37 (Fig. 2)] and, therefore, the resonance wavelengths do not change by changing a separation distance between the nanocylinders. It should be noted that when the nanocylinders are well separated $R = 100$ nm, SCS has one pronounced resonance peak at $\lambda = 650$ nm (black line). However, when the separation distance between the nanocylinders decreases, it splits into two peaks at $\lambda = 575$ nm, $\lambda = 705$ nm when $R = 72$ nm (dashed line), at $\lambda = 590$ nm, $\lambda = 740$ nm when $R = 70$ nm (dashed-dotted line) and at $\lambda = 630$ nm, $\lambda = 780$ nm when $R = 69.2820$ nm (gray line). Note that the first peaks are formed due to the coaxial nature of the metal-coated nanocylinders [37 (Fig. 2)], whereas the second resonance peaks in the intermediate wavelength region are due to the resonant coupling of the surface plasmon fields in the multiple nanocylinders' systems. It is important to mention that when the nanocylinders are located quite close but not touching each other at $R = 72$ nm (dashed line) and $R = 70$ nm (dashed-dotted line), SCS and ACS have the resonances in the long wavelength region at $\lambda = 950$ nm and $\lambda = 1{,}300$ nm, respectively. These resonances are characterized as the resonances intrinsic to the cylindrically periodic system. Note that there are no resonance peaks in the long wavelength region when the nanocylinders are well separated.

In order to give a physical insight into the analyses of the spectral responses, we study the near-field distributions $|H_z|$ at particular resonance wavelengths shown in Figure 21.17. The circles depicted by white lines indicate the boundary surfaces of the metal (Ag)-coated coaxial nanocylinders. From Figure 21.18(a) and (b), it follows that when the nanocylinders are well separated $R = 100$ nm, the surface plasmon supported by the interface between the metal and the free-space resonates to the incident plane wave. A strong field is excited in the illuminated side of the nanocylinders at the resonances in the short wavelength region at $\lambda = 300$ nm (resonance wavelength in ACS) and at $\lambda = 330$ nm (resonance wavelength in SCS). At the resonances in the intermediate wavelength region, an effect of the coaxial nature of the nanocylinders becomes significant. Incident waves penetrate through the metal layer and the excited field is localized along the interface between the dielectric core and the coating metal (Ag) layer at $\lambda = 472$ nm (resonance peak in the ACS) and at $\lambda = 650$ nm (resonance peak in the SCS) as shown in Figure 21.18(c) and (d), respectively. It should be noted that there is no interference between the near fields scattered from the individual nanocylinders at $\lambda = 472$ nm and the field is strongly localized only at the interfaces between the dielectric core and the coating metal (Ag) layer. On the other hand, there is a resonant coupling of the surface plasmon fields in the nanocylinders' systems at $\lambda = 650$ nm.

Figure 21.18(e)–(g) shows the near-field distributions $|H_z|$ when the nanocylinders are located very close but not touching each other $R = 72$ nm. Near-field distribution at the resonance wavelength $\lambda = 575$ nm (resonance wavelength in SCS) is shown in Figure 21.18(e). The field is strongly localized at the surface between the metal cover and the dielectric core of the nanocylinders located at the illuminating

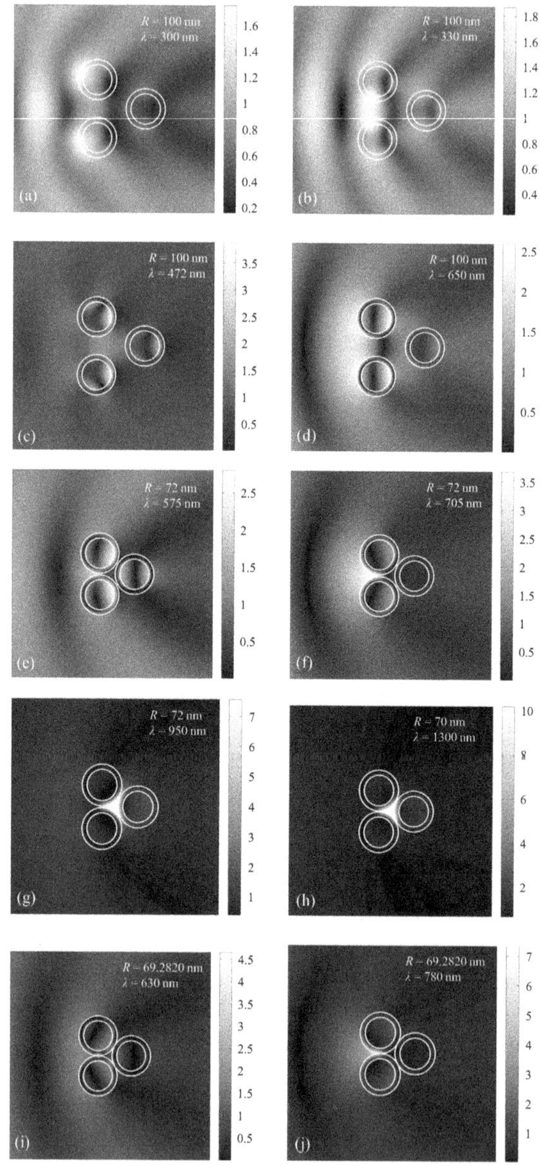

Figure 21.18 Near-field distributions $|H_z|$ of three metal (Ag)-coated nanocylinders periodically situated along the cylindrical surface at the resonance wavelengths of the spectral response (Figure 21.17) for different values of radius R: R = 100 nm (Figure 21.18(a)–(d)), R = 72 nm (Figure 21.18(e)–(g)), R = 70 nm (Figure 21.18(h)) and R = 69.2820 nm (Figure 21.18(i) and (j)). The plane waves are incident on the structure from the negative x axis

side of the incident plane waves. Multiple interactions between the nanocylinders are hardly observed. However, the incident plane wave resonates to the coupled surface plasmon modes supported in the gap between two nanocylinders at $\lambda = 705$ nm and the localized field is strongly enhanced in a small region between the nanocylinders (Figure 21.18(f)). Illumination at the right-side is blocked by two nanocylinders and, therefore, the major contribution in the field formation is governed only by these two metal scatterers. It is important to note that the field inside the region surrounded by three nanocylinders is almost zero.

A very interesting effect is observed at $\lambda = 950$ nm (resonance wavelength in SCS and ACS), when the localized field is strongly enhanced inside the region around the global origin surrounded by the three metal-coated nanocylinders. This effect is demonstrated in Figure 21.18(g). In order to study a strong field confinement inside the cylindrical periodic structure in a more detail, we further slightly decrease a distance between the nanocylinders. Near-field distribution at the resonance wavelength $\lambda = 1.300$ nm when $R = 70$ nm is illustrated in Figure 21.18(h). It could be vividly seen that the field localization inside the cylindrical structure is much stronger than that in Figure 21.18(g). It is very important to note that in order to realize a strong field confinement inside the cylindrical structure, we should have a very small gap between the nanocylinders because the surface plasmon field extends over the metal region and the free space. Effect of the strong field confinement could be considered as a unique feature for the cylindrical configuration composed of the nanocylinders. When the nanocylinders are touching each other at $R = 69.2820$ nm, there is no small gap between them and the field cannot penetrate inside the structure. The surface plasmons on the outer metal interfaces are strongly coupled and a new surface plasmon mode is formed in the touching region. This effect is demonstrated in Figure 21.18(i) and (j). All these abovementioned features could be considered as technologically important to flexibly design and fabricate the plasmonic waveguides and plasmonic resonators.

Finally, it should be noted that recently Rahmani *et al.* have numerically and experimentally analyzed the electromagnetic scattering on various configurations of the cylindrical structure composed of Au nanoparticles [38]. The authors use the conventional numerical method finite-difference time-domain (FDTD) to study the scattering and absorption characteristics, whereas our analysis is based on the rigorous formulation. Using the proposed formulation, we can easily and in a very short time analyze various models of the multilayered cylindrical structures with arbitrary number of periodically distributed nanocylinders under different geometrical parameters. When the desired characteristics are observed, we will be able to implement the geometrical parameters under which the desired characteristics are obtained, in the numerical and experimental studies of the real-world structures. It will substantially decrease time, effort and expenses needed for the detailed numerical and experimental investigations.

21.3.3 Modal analysis of specific microstructured optical fibers

Although the formulation for guiding problem in Section 21.2 was rigorously presented for the vector fields, as a numerical example, we investigate the scalar field in

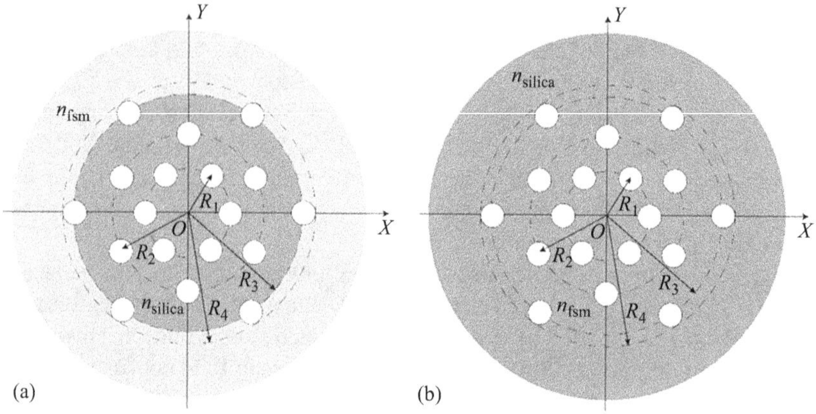

Figure 21.19 Two different configurations of the background material for three-layered cylindrical arrays with a hexagonal arrangement of circular air holes. © 2016 IEICE. After, with permission, from Reference [39], Permission No.: 17RA0051

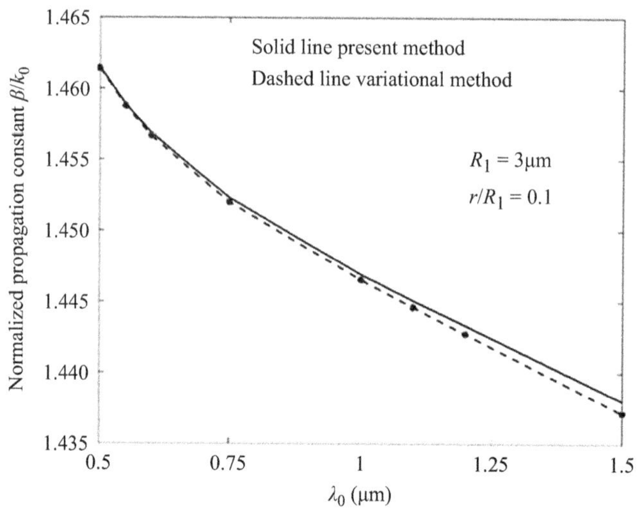

Figure 21.20 Comparison of present method (solid line) with the variational method [33] (dashed line) for the normalized propagation constant in three-layered cylindrical arrays with a hexagonal arrangement of circular air holes with $R_1 = 3\ \mu m$ and $r/R_1 = 0.1$. The outer area is n_{fsm}. © 2016 IEICE. After, with permission, from Reference [39], Permission No.: 17RA0051

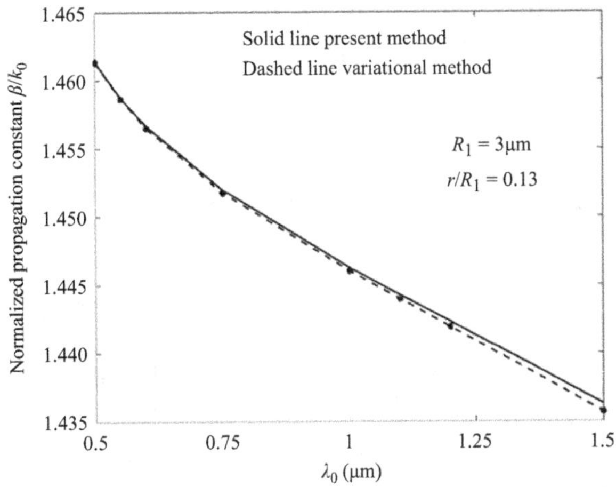

Figure 21.21 *The same as in Figure 21.20 with $R_1 = 3\ \mu m$ and $r/R_1 = 0.13$. © 2016 IEICE. After, with permission, from Reference [39], Permission No.: 17RA0051*

a microstructured fiber formed by three-layered cylindrical arrays with a hexagonal arrangement of circular air holes on the cladding of pure silica with $n_{silica}(\lambda_0)$. Geometrical parameters of the microstructured optical fiber are $R_2 = \sqrt{3}R_1$, $R_3 = 2R_1$, $r_1 = r_2 = r_3 = r$, $R_4 = R_3 + r$. The offset angles of the arrays are $\Theta_1 = 0°$, $\Theta_2 = 30°$ and $\Theta_3 = 0°$, respectively. We have considered two different situations for the background material as shown in Figure 21.19. In Figure 21.19(a), the outer area of the third layer is filled by a material of index equal to the effective index of the fundamental space-filling mode [39], whereas the background medium in Figure 21.19(b) is pure silica over the whole area. Note that the refractive indices of the background media in Figure 21.19(a) and (b) are functions of the wavelength. The calculated values of normalized propagation constant β/k_0 of fundamental mode are shown in Table 21.1 and compared with those obtained by the variational method [33]. We can see that the results of the variational method are in very close agreement with the present rigorous ones. Relative errors in the variational method are approximately 0.001%–0.01%. Comparison of the results is also graphically illustrated in Figures 21.20 and 21.21, respectively [39]. It is worth noting that there is no significant influence of the material exterior to the third layer on the propagation constant because the mode field is strongly confined within the three-layered structure.

Extension to the vector fields is straightforward. In this case, we must introduce two scalar fields representing E_z and H_z fields. The T-matrix of a circular rod, the reflection and transmission matrixes and the generalized reflection and transmission matrices are defined in terms of the cylindrical harmonics of both E_z and H_z fields as the basis. The size of matrices to be solved becomes four times larger than that of the present scalar field case.

Table 21.1 Normalized propagation constant β/k_0 of the fundamental mode calculated for three-layered cylindrical arrays with a hexagonal arrangement of circular air holes with $R_1 = 3\ \mu m$, $R_2 = \sqrt{3}R_1$, $R_3 = 2R_1$ and $r_1 = r_2 = r_3 = r$. © 2016 IEICE. After, with permission, from Reference [39], Permission No.: 17RA0051

$R_1 = 3\mu m$	λ_0 (μm)	Present method with n_{fsm}	Variational method with n_{fsm}	Present method with n_{silica}	Errors in variational method with n_{fsm}
$r/R_1 = 0.10$	0.50	1.461622	1.461443	1.461572	1.7900e-004
	0.55	1.459003	1.458789	1.458943	2.1400e-004
	0.60	1.456916	1.456664	1.456844	2.5200e-004
	0.75	1.452406	1.452019	1.452289	3.8700e-004
	1.00	1.447231	1.446564	1.447012	6.6700e-004
	1.10	1.445439	1.444645	1.445168	7.9400e-004
	1.20	1.443698	1.442771	1.443370	9.2700e-004
	1.50	1.438522	1.437179	1.437978	1.3000e-003
$r/R_1 = 0.13$	0.50	1.461457	1.461326	1.461409	1.3100e-004
	0.55	1.458803	1.458649	1.458745	1.5400e-004
	0.60	1.456677	1.456497	1.456607	1.8000e-004
	0.75	1.452027	1.451752	1.451915	2.7500e-004
	1.00	1.446532	1.446043	1.446325	4.8900e-004
	1.10	1.444580	1.443984	1.444324	5.9600e-004
	1.20	1.442659	1.441946	1.442349	7.1300e-004
	1.50	1.436837	1.435705	1.436317	1.1000e-003

21.4 Conclusion

We have proposed an accurate and self-contained semi-analytical approach for two-dimensional electromagnetic scattering, guidance and radiation by cylindrical arrays composed of circular rods periodically distributed along the layered circular rings. The rods may be dielectric, perfect electric conductor, air hole, metal or magnetized ferrite. The method uses the T-matrix of a circular rod in isolation, the reflection and transmission matrices of a cylindrical array based on the cylindrical harmonics expansion and the generalized reflection and transmission matrices for a cylindrically layered structure. The formulation is rigorous. The only approximation involved in the solutions is the truncation of the cylindrical harmonic expansion and the translation matrices related to Graf's addition theorem, which is necessary to solve a set of linear equations by matrix inversion. We have taken into account all cylindrical space-harmonics and their interactions through the scattering by each cylindrical layer. The proposed approach could be applied to various configurations of the layered cylindrical arrays with different types and locations of the excitation sources.

We have applied the method to the modal analysis of the guided waves in a specific microstructured optical fiber and demonstrated its validity. In the case of the modal analysis of the guided waves without any initial excitation, we could assume a unique symmetric property of the mode field distribution inherent to the periodicity of the circular rods. For the scattering or radiation problems with initial excitations, however, there is no such a specific symmetry in the scattered field, and a modification of the formulation is required. The proposed formulation has been used to analyze the radiation characteristics of a localized source located inside the multilayered cylindrical structures—cylindrical bandgap structure. Three different configurations of the cylindrical bandgap structures have been studied. We have discussed in detail the relations between the transmission spectra of the cylindrical harmonic waves and the directivity of radiation by the exciting source. In this chapter, we have also investigated the light scattering by metal-coated dielectric nanocylinders periodically distributed along a cylindrical surface. The spectral responses of the SCS and ACS and the near-field distributions have been analyzed. Finally, it should be mentioned that the presented rigorous formulation is accompanied with a neat implementation providing a very high computation efficiency and could be considered as an important tool for designing and optimizing cylindrical periodic and bandgap structures.

Acknowledgment

V. Jandieri acknowledges financial support from the Alexander von Humboldt Foundation.

References

[1] Petit R. *Electromagnetic Theory of Gratings* (Berlin, Heidelberg, Springer-Verlag, 1980).

[2] Yasumoto K. *Electromagnetic Theory and Applications for Photonic Crystals* (Boca Raton, FL, CRC Press, 2005).

[3] Tayeb G. and Maystre D. 'Rigorous theoretical study of finite-size two-dimensional photonic crystals doped by microcavities'. *Journal of the Optical Society of America A*. 1997; **15**(12): 3323–32.

[4] Pelosi G., Cocchi A., and Monorchio, A. 'A hybrid FEM-based procedure for the scattering from photonic crystals illuminated by a Gaussian beam'. *IEEE Transactions on Antennas and Propagation*. 2000; **48**(6): 973–80.

[5] Yasumoto K., Toyama H., and Kushta, T. 'Accurate analysis of two-dimensional electromagnetic scattering from multilayered periodic arrays of circular cylinders using lattice sums technique'. *IEEE Transactions on Antennas and Propagation*. 2004; **52**(10): 2603–11.

[6] Guida G., Maystre D., Tayeb G., and Vincent, P. 'Mean-field theory of two-dimensional metallic photonic crystals'. *Journal of the Optical Society of America B*. 1998; **15**(8): 2308–15.

[7] Watanabe K., Petit R., and Neviere, M. 'Differential theory of gratings made of anisotropic materials'. *Journal of the Optical Society of America A*. 2002; **19**(2): 325–34.

[8] Russell P. St. 'Photonic-crystal fibers'. *Journal of Lightwave Technology*. 2006; **24**(2): 4729–49.

[9] Boutayeb H. and Denidni, T. 'Metallic cylindrical EBG structures with defects: directivity analysis and design optimization'. *IEEE Transactions on Antennas and Propagation*. 2007; **55**(11): 3356–61.

[10] Boutayeb H., Tarot A.-C., and Mahdjoubi, K. 'Focusing characteristics of a metallic cylindrical electromagnetic band gap structure with defects'. *Progress in Electromagnetics Research (PIER)*. 2006; **66**: 89–103.

[11] Biancotto C. and Record, P. 'Design of a beam forming dielectric cylindrical EBG antenna'. *Progress in Electromagnetics Research (PIER)*. 2009; **18**: 327–46.

[12] Palikaras G., Feresidis A., and Vardaxoglou, J. 'Cylindrical electromagnetic bandgap structures for directive base station antennas'. *IEEE Antennas and Wireless Propagation Letters*. 2004; **3**: 87–89.

[13] Atakaramians S., Afshar S., Monro T.M., and Abbott, D. 'Terahertz dielectric waveguides'. *Advances in Optics and Photonics*. 2013; **5**(2): 169–215.

[14] Mitrofanov O., James R., Anibal Fernandez F., Mavrogordatos T.K., and Harrington, J. 'Reducing transmission losses in hollow THz waveguides'. *IEEE Transactions on Terahertz Science and Technology*. 2011; **1**(1): 124–32.

[15] Mitrofanov O. and Harrington, J. 'Dielectric-lined cylindrical metallic THz waveguides: mode structure and dispersion'. *Optics Express*. 2010; **18**(3): 1898–1903.

[16] Toyama H. and Yasumoto, K. 'Electromagnetic scattering from periodic arrays of composite circular cylinder with internal cylindrical scatterers'. *Progress in Electromagnetics Research (PIER)*. 2005; **52**: 321–33.

[17] Jandieri V., Yasumoto K., and Liu, Y. 'Corner reflector as a model of layered cylindrical structures of metallic circular rods'. *Journal of Electromagnetic Waves and Applications*. 2012; **26**: 571–82.

[18] Jandieri V. and Yasumoto, K. 'Electromagnetic scattering by layered cylindrical arrays of circular rods'. *IEEE Transactions on Antennas and Propagation*. 2011; **59**(6): 2437–41.

[19] Jandieri V., Yasumoto K., and Liu, Y. 'Directivity of radiation of a dipole source coupled to cylindrical electromagnetic bandgap structures'. *Journal of the Optical Society of America B*, 2012; **29**(9): 2622–29.

[20] Jandieri V., Yasumoto K., and Cho Y.-K. 'Rigorous analysis of electromagnetic scattering by cylindrical EBG structures'. *Progress in Electromagnetics Research (PIER)*. 2011; **121**: 317–42.

[21] Yasumoto K., Jandieri V., and Gupta, B. 'Guidance and scattering of electromagnetic waves by layered cylindrical arrays of circular rods'. *Proceedings of IEEE Applied Electromagnetics Conference (AEMC 2009)*; Kolkata, India, December 2009, pp. 1–4.

[22] Chew W.C. *Waves and Fields in Inhomogeneous Media* (New York, NY, Van Nostrand Reinhold, 1990).

[23] Abramowitz M. and Stegun, I. *Handbook of Mathematical Functions* (Dover Publications, New York, 1965).

[24] Jandieri V., Yasumoto K., and Gupta, B. 'Directivity of radiation from a localized source coupled to electromagnetic crystals'. *Journal of Infrared, Millimetre and Terahertz Waves* 2009; **30**(10): 1102–12.

[25] Jandieri V., Yasumoto K., Liu Y., and Pistora, J. 'Scattering of light by the metal-coated dielectric nanocylinders with angular periodicity'. *Progress in Electromagnetics Research M (PIER M)*. 2016; **45**: 101–11.

[26] Ruan Z. and Fan, S. 'Superscattering of light from subwavelength nanostructures'. *Physical Review Letters*. 2010; **105**: 013901.

[27] Mirzaei A., Shadrivov I.V., Miroshnichenko A.E., and Kivshar Y.S. 'Cloaking and enhanced scattering of core-shell plasmonic nanowires'. *Optics Express*. 2013; **21**(9): 10454–59.

[28] Okada N. and Cole, J. 'Cylindrical invisibility cloak based on photonic crystal layers that permits communication with the outside'. *Journal of the Optical Society of America B*. 2012; **29**(12): 3344–48.

[29] Berini, P. 'Long-range surface plasmon polaritons'. *Advances in Optics and Photonics*. 2009; **1**(3): 484–588.

[30] Hess O., Pendry J., Maier S., Oulton R., Hamm J., and Tsakmakidis, K. 'Active nanoplasmonic materials'. *Nature Materials*. 2012; **11**: 573–84.

[31] Knight J.C., Broeng J., Birks T.A., and Russell, P. St.J. 'Photonic band gap guidance in optical fibers'. *Science*. 1998; **282**: 1476–78.

[32] Birks, T.A., Knight J.C., and Russell, P. St.J. 'Endlessly single-mode photonic crystal fiber'. *Optics Letters*. 1997; **22**(13): 961–63.

[33] Sharma A. and Chauhan, H. 'Variational analysis of index-guiding microstructured fibers'. *Proceedings of the 4th Research Forum of Japan-Indo Collaboration Project on Infrastructural Communication Technologies Supporting Fully Ubiquitous Information Society*; Tokyo, Japan, July 2008, pp. 52–56.

[34] De Mazancourt T. and Gerlic, D. 'The inverse of a block-circulant matrix'. *IEEE Transactions on Antennas and Propagation*. 1982; **AP-31**(5): 808–810.

[35] Jandieri V., Yasumoto K., and Cho Y.-K. 'Analysis of radiation from line source located in cylindrical electromagnetic bandgap structures with defects'. *IEICE Transactions on Electronics* 2011; **E94-C**(8): 1245–52.

[36] Laroche T. and Girard, C. 'Near-field optical properties of single plasmonic nanowires'. *Applied Physics Letters*. 2006; **89**: 233119.

[37] Jandieri V., Meng P., Yasumoto K., and Liu, Y. 'Scattering of light by gratings of metal-coated circular nanocylinders on a dielectric substrate'. *Journal of the Optical Society of America A*. 2015; **32**(7): 1384–89.

[38] Rahmani M., Lei D., Giannini V., *et al.* 'Subgroup decomposition of plasmonic resonances in hybrid oligomers: modeling the resonance lineshape'. *Nano Letters*. 2012; **12**(4): 2101–06.

[39] Jandieri V., Yasumoto K., Sharma A., and Chauhan, H. 'Modal analysis of specific microstructured optical fibers using a model of layered cylindrical arrays of circular rods'. *IEICE Transactions on Electronics; Special Section on Recent Progress in Electromagnetic Theory and Its Application*. 2010; **E93-C**(1): 17–23.

Chapter 22

Analytical and numerical solution techniques for forward and inverse scattering problems in waveguides

E.D. Derevyanchuk[1], Yu.V. Shestopalov[2], and Yu.G. Smirnov[1]

Investigations of the wave propagation in waveguides with inhomogeneous dielectric inclusions are performed using advanced analytical and numerical methods. A particular attention is paid to the analysis of benchmark problems: forward and inverse scattering by parallel-plane dielectric diaphragms and layered structures placed in a rectangular waveguide with perfectly conducting walls. A detailed study of the properties of transmitted fields using the theory of functions of one or several complex variables leads to the development of a technique aimed at reconstructing permittivity of the inclusions from the transmission coefficient. The unique solvability of this inverse problems is proved by establishing the domains where the transmission coefficient is a one-to-one function of permittivity and other problem parameters. The obtained results are used to improve finite-difference time domain (FDTD) numerical solution to the corresponding forward and inverse scattering problems for Maxwell's equations in waveguides and create specifically developed codes. In general, the chapter contributes to the electromagnetic wave theory with several new findings and modified theoretical approaches.

22.1 Introduction

The determination of electromagnetic parameters of dielectric bodies that have complicated geometry or structure is an urgent problem arising, e.g., when nanocomposite or artificial materials and media are used as elements of various devices. However, as a rule, these parameters cannot be directly measured (because of composite character

[1]Department of Mathematics and Supercomputer Simulation, Faculty of Computer Engineering, Penza State University, Penza, Russia
[2]Faculty of Engineering and Sustainable Development, University of Gävle, Gävle, Sweden; Institute of Information Technologies, Russian Technological University MIREA, Moscow, Russia

of the material and small size of samples), which leads to the necessity of applying methods of mathematical modeling and numerical solution of the corresponding forward and inverse electromagnetic problems [1]. It is especially important to develop the solution techniques when the inverse problem for bodies of complicated shape are considered in the resonance frequency range, which is the case when permittivity of nanocomposite materials must be reconstructed [2,3].

One of the possible applications of composites is the creation of radio absorbing materials that can be used in systems that provide electromagnetic compatibility of modern electronic devices and in “Stealth”-type systems aimed at damping and decreasing reflectivity of microwave electromagnetic radiation from objects to be detected [4,5]. When calculating reflection and absorption characteristics of electromagnetic microwave radiation of radio absorbing materials, researchers use models employing the data on the material constants (permittivity, permeability, conductivity) of these materials in the microwave range. Such composites often contain carbon particles, short carbon fibers, carbon nanofibers, and multilayer carbon nanotubes as fillers for polymer dielectric matrices [4,5,6]. The use of carbon nanotubes enables one to achieve a significant (up to 10 dB) absorption of microwave electromagnetic radiation at relatively thin composite layers and low volume fractions of nanotubes and hence a small weight, in a broad frequency range (up to 5 GHz). Such characteristics are caused by both the geometrical sizes of individual nanotubes and their electrophysical properties; among the most important parameters here are permittivity and electric conductivity (which can vary over very wide ranges).

It is important to determine permittivity and conductivity not only of a composite as a solid body (as in [4,5]), but also of its components, e.g., nanotubes, whose physical characteristics can vary substantially in the process of composite formation.

This study is focused on the analysis of benchmark problems: forward and inverse scattering by parallel-plane dielectric diaphragms and layered structures placed in a rectangular waveguide with perfectly conducting walls.

The forward scattering problem for a diaphragm in a parallel-plane waveguide was considered in [7]. In papers [8,9,10,11,12,13,14,15], the inverse problem of reconstructing complex permittivity was analyzed from the measurements of the transmission coefficient; in [10,16,17], the artificial neural networks method was applied. Several techniques for the permittivity determination of homogeneous materials loaded in a waveguide are reported in [8,9,18]. The permittivity reconstruction of inhomogeneous structures are not as widely investigated and only a few studies exist for multilayered materials [19,20]. Note a recently developed advanced approach [11] that can be also applied to numerical solution of this inverse problem.

However, complete analysis of the closed-form solution to the inverse problem of permittivity determination of materials loaded in a waveguide is not available in the literature, to the best of our knowledge, even for the simplest configuration of a parallel-plane dielectric insert in a guide of rectangular cross section. This fact dictates the aim of this work: to develop a method of solution to the inverse problem of reconstructing effective permittivity of layered dielectrics in the form of diaphragms in a waveguide of rectangular cross section that would enable both obtaining solution in a closed form for benchmark problems and efficient numerical implementation.

We note that the corresponding forward problem for a one-sectional diaphragm is considered in [6,21].

In this work, we perform analysis and solution of the inverse problems of reconstructing electromagnetic parameters of multi-sectional parallel-plane dielectric diaphragms in a single-mode rectangular waveguide (when only one H_{10} wave propagates in the waveguide without attenuation and other waves (modes) decay exponentially). A particular attention is paid to the development of the methods of extraction of permittivity and permeability tensors of anisotropic multi-sectional diaphragm located in a rectangular waveguide.

The proposed method of reconstruction of the permittivity from the transmission coefficient constitutes a well-posed problem. In fact, it is proved (in Section 22.3) that the (complex-valued) transmission coefficient is a one-to-one analytical function of the permittivity considered on a certain domain in the complex plane which implies the invertibility of this function and consequently unique solvability of the equation from which the permittivity is determined.

We present a summary of mathematical techniques and their applications to the solution of inverse problems for specific multilayer structures. Detailed description of mathematical methods and proofs of the theorems can be found in [13,14,15,22,23,24,25,26,27].

22.2 Inverse problems

In this chapter, we study the inverse problems of reconstructing electromagnetic parameters of layered dielectrics placed in a single-mode rectangular waveguide.

The inverse problems under consideration are divided into three classes (Figure 22.1):

Class I: inverse problems for an isotropic one-sectional diaphragm.
Class AnI: inverse problems for an anisotropic one-sectional diaphragm.
Class M: inverse problems for both one- and multi-sectional diaphragms.

We use the following notation for the inverse problems: if the transmission coefficient F/A is given, then the inverse problem is denoted by letter P, and for the given reflection coefficient B/A, by Q. Unknown values in Figure 22.1 enter the lower index.

First, we consider the simplest case, namely, an isotropic one-sectional diaphragm when permittivity and permeability are unknown scalar values. In Class I, we consider four types of inverse problems:

(I1) reconstruction of real permittivity ($P^{\mathbb{R}}_{\varepsilon_1}$, $Q^{\mathbb{R}}_{\varepsilon_1}$) and (I2) complex permittivity ($P^{\mathbb{C}}_{\varepsilon_1}$, $Q^{\mathbb{C}}_{\varepsilon_1}$), and two kinds of inverse problems where unknown parameters are permittivity ε_1 and thickness l_1 of the diaphragm (I3) $P^{\mathbb{R}}_{\varepsilon_1, l_1}$ ($Q^{\mathbb{R}}_{\varepsilon_1, l_1}$) and (I4) $P^{\mathbb{C}}_{\varepsilon_1}(l_1 \ll 1)$ $Q^{\mathbb{C}}_{\varepsilon_1}(l_1 \ll 1)$. In the last two problems, we suppose that the thickness is sufficiently small.

In class AnI, the inverse problems are more difficult: the diaphragm is filled with anisotropic material so that the unknown permittivity and permeability are tensor

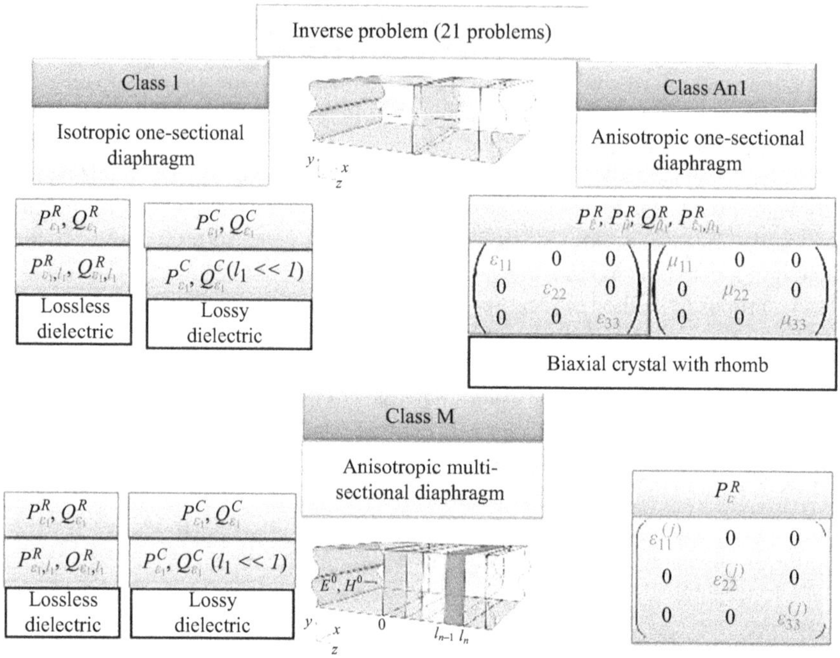

Figure 22.1 Inverse problems under consideration

quantities. Here we consider only the case of diagonal permittivity and permeability tensors.

22.2.1 General statement for inverse problems

Assume that a waveguide $PW = \{x : 0 < x_1 < a, 0 < x_2 < b, -\infty < x_3 < \infty\}$ with the perfectly conducting boundary surface ∂PW is given in the cartesian coordinate system (Figure 22.2).

A three-dimensional body $QD = \bigcup_{j=1}^{n} QD_j$, where

$$QD_j := (x, y, z) : 0 < x < a, 0 < y < b, l_{j-1} < z < l_j$$

is placed in the waveguide; the body has the form of a diaphragm adjacent to the waveguide walls; $l_j - l_{j-1}$ is the thickness of the jth section, $l_0 = 0$, $l_n = l$, and l is the total thickness of the diaphragm.

Domain $PW \backslash \overline{QD}$ is filled with an isotropic and homogeneous layered medium having constant permeability $(\mu_0 > 0)$ and constant permittivity $(\varepsilon_0 > 0)$, the first section of the diaphragm having tensor permeability and permittivity.

In the region $PW \backslash \overline{QD}$, the medium is isotropic and homogeneous with the known permittivity $\varepsilon_0 > 0$ and permeability $\mu_0 > 0$; $k_0^2 = \omega^2 \varepsilon_0 \mu_0$, so that k_0 is the

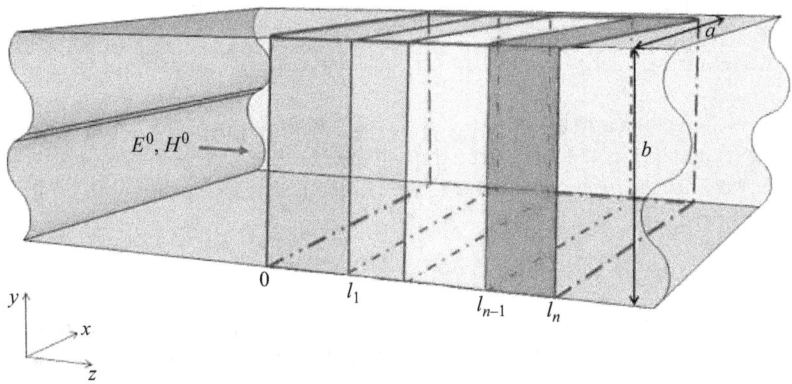

Figure 22.2 Diaphragm in the waveguide

free-space wave number and $\omega > 0$ is the circular frequency. Section QW_j is filled with anisotropic medium that has diagonal permeability and permittivity tensors

$$\widehat{\mu}^{(j)}(\omega) = \begin{pmatrix} \mu^j_{11}(\omega) & 0 & 0 \\ 0 & \mu^j_{22}(\omega) & 0 \\ 0 & 0 & \mu^j_{33}(\omega) \end{pmatrix} \tag{22.1}$$

$$\widehat{\varepsilon}^{(j)}(\omega) = \begin{pmatrix} \varepsilon^j_{11}(\omega) & 0 & 0 \\ 0 & \varepsilon^j_{22}(\omega) & 0 \\ 0 & 0 & \varepsilon^j_{33}(\omega) \end{pmatrix}. \tag{22.2}$$

The electromagnetic field $\mathbf{E}, \mathbf{H}$ inside the waveguide is governed by Maxwell's equations: outside the diaphragm,

$$\begin{cases} \nabla \times \mathbf{H} = -i\omega\varepsilon_0 \mathbf{E} \\ \nabla \times \mathbf{E} = i\omega\mu_0 \mathbf{H} \end{cases} \tag{22.3}$$

and inside the diaphragm,

$$\begin{cases} \nabla \times \mathbf{H} = -i\omega\widehat{\varepsilon}\, \mathbf{E} \\ \nabla \times \mathbf{E} = i\omega\widehat{\mu}\, \mathbf{H} \end{cases} \tag{22.4}$$

where $\mathbf{E}$ and $\mathbf{H}$ are the vectors of the electric and magnetic field intensity.

Assume that $\pi/a < k_0 < \pi/b$. In this case, only one wave H_{10} propagates in the waveguide without attenuation (we have a single-mode waveguide [20, 28]); the field structure is

$$\mathbf{E} = (0, E_y, 0), \quad \mathbf{H} = (H_x, 0, H_z) \tag{22.5}$$

The incident electric field has the form

$$\mathbf{E}^0 = A \sin\left(\frac{\pi x}{a}\right) e^{-i\gamma_0 z}\mathbf{e}_2$$

where A is the known amplitude; $\gamma_0 = \gamma_0(\omega) \neq 0$ with γ_0 the propagation constant of the wave H_{10}; and $\mathbf{e}_2$ is the basis vector along the Oy-axis. Vector $\mathbf{H}^0$ is reconstructed from the second equation of system (22.3). On the boundary surfaces, the following conditions hold:

$$[\mathbf{E}_\tau]|_L = 0, \ \ [\mathbf{H}_\tau]|_L = 0$$

where $L := \{(x, y, z) : z = 0, \ldots, z = l_j, \ldots; z = l_n\}, j = 1, \ldots, n$, $[\,\cdot\,]|_L$ denote functions' break on boundary surface L, and $\mathbf{E}_\tau$ and $\mathbf{H}_\tau$ are the tangential components of vectors $\mathbf{E}$ and $\mathbf{H}$.

We have the following recurrent formulas for the transmission coefficient F/A [25]:

$$\frac{F}{A} = \frac{2\prod\limits_{j=0}^{n}\frac{\gamma_j}{\mu_{11}^{(j)}}}{e^{-i\gamma_0 l_n}\left(\frac{\gamma_n}{\mu_{11}^{(n)}}p_{n+1}^{(+)} + \frac{\gamma_0}{\mu_0}q_{n+1}^{(+)}\right)} \tag{22.6}$$

where

$$p_{j+1}^{(+)} = \frac{\gamma_{j-1}}{\mu_{11}^{(j-1)}}p_j^{(+)}\cos\alpha_j + \frac{\gamma_j}{\mu_{11}^{(j)}}q_j^{(+)} i\sin\alpha_j, p_1^{(+)} = 1 \tag{22.7}$$

$$q_{j+1}^{(+)} = \frac{\gamma_{j-1}}{\mu_{11}^{(j-1)}}p_j^{(+)} i\sin\alpha_j + \frac{\gamma_j}{\mu_{11}^{(j)}}q_j^{(+)}\cos\alpha_j, q_1^{(+)} = 1$$

$$\gamma_j = \sqrt{\left(\omega^2\varepsilon_{22}^{(j)}\mu_{33}^{(j)} - \frac{\pi^2}{a^2}\right)\frac{\mu_{11}^{(j)}}{\mu_{33}^{(j)}}}, \ \ j = \overline{1, n} \tag{22.8}$$

$$\gamma_0 = \sqrt{k_0^2 - \frac{\pi^2}{a^2}} = \sqrt{\left(\omega^2\varepsilon_0\mu_0 - \frac{\pi^2}{a^2}\right)} \tag{22.9}$$

$$\alpha_j = \gamma_j(l_j - l_{j-1}), l_0 = 0$$

here γ_j $(j = \overline{1, n})$ are the propagation constant inside each section of the diaphragm.

Expression (22.8) follows from (22.1)–(22.2), (22.5), and Maxwell's equation (22.4); and (22.9) are obtained from (22.1)–(22.2), (22.5), and Maxwell's equation (22.3).

For solving an inverse problem where the input data is the reflection coefficient B/A, we have the following recurrent expressions:

$$\frac{B}{A} = \frac{\frac{\gamma_n}{\mu_{11}^{(n)}}p_{n+1}^{(-)} + \frac{\gamma_0}{\mu_0}q_{n+1}^{(-)}}{\frac{\gamma_n}{\mu_{11}^{(n)}}p_{n+1}^{(+)} + \frac{\gamma_0}{\mu_0}q_{n+1}^{(+)}} \tag{22.10}$$

where

$$p_1 = 1, p_2^{(\pm)} = \frac{\gamma_0}{\mu_0} p_1 \cos\alpha_1 \pm \frac{\gamma_1}{\mu_{11}^{(1)}} q_1 i \sin\alpha_1 \tag{22.11}$$

$$p_{j+1}^{(\pm)} = \frac{\gamma_{j-1}}{\mu_{11}^{(j-1)}} p_j^{(\pm)} \cos\alpha_j + \frac{\gamma_j}{\mu_{11}^{(j)}} q_j^{(\pm)} i \sin\alpha_j$$

$$q_1 = 1, q_2^{(\pm)} = \frac{\gamma_0}{\mu_0} p_1 i \sin\alpha_1 \pm \frac{\gamma_1}{\mu_{11}^{(1)}} q_1 \cos\alpha_1$$

$$q_{j+1}^{(\pm)} = \frac{\gamma_{j-1}}{\mu_{11}^{(j-1)}} q_j^{(\pm)} i \sin\alpha_j + \frac{\gamma_j}{\mu_{11}^{(j)}} q_j^{(\pm)} \cos\alpha_j,\ j = \overline{1,n}$$

here γ_j are expressed by (22.8) and (22.9).

22.3 Inverse problems. Class I (Isotropic case)

22.3.1 *Statement of the inverse problem for isotropic one-sectional diaphragm (Class I)*

Let us consider the inverse problems of Class I.

Statement of the inverse problem for isotropic one-sectional diaphragm (Class I)

$P_{\varepsilon_1}^{\mathbb{R}}$, $P_{\varepsilon_1}^{\mathbb{C}}$ ($Q_{\varepsilon_1}^{\mathbb{R}}$, $Q_{\varepsilon_1}^{\mathbb{C}}$): From transmission coefficient F/A (or reflection coefficient B/A), determine *real or complex permittivity* ε_1 of isotropic one-sectional diaphragm.

$P_{\varepsilon_1, l_1}^{\mathbb{R}}$ ($Q_{\varepsilon_1, l_1}^{\mathbb{R}}$): From transmission coefficient F/A (or reflection coefficient B/A), determine *real permittivity* ε_1 *and thickness* l_1 of isotropic one-sectional diaphragm.

$P_{\varepsilon_1}^{\mathbb{C}}(l_1 \ll 1)$ $Q_{\varepsilon_1}^{\mathbb{C}}(l_1 \ll 1)$: From transmission coefficient F/A (or reflection coefficient B/A), determine *complex permittivity* ε_1 of isotropic thin ($l_1 \ll 1$) diaphragm.

At first, we present theoretical results for inverse problems $P_{\varepsilon_1}^{\mathbb{R}}$, $P_{\varepsilon_1}^{\mathbb{C}}$ ($Q_{\varepsilon_1}^{\mathbb{R}}$, $Q_{\varepsilon_1}^{\mathbb{C}}$).

22.3.2 *Explicit solution to the inverse problem*

From (22.6)–(22.7) for a one-sectional diaphragm, we have

$$\frac{Ae^{i\gamma_0 l_1}}{F} = g(z)$$

$$g(z) = \cos z + i\left(\frac{z}{2\gamma_0 l_1} + \frac{\gamma_0 l_1}{2z}\right)\sin z \tag{22.12}$$

$$z = \gamma_1 l_1 = l_1\sqrt{k_1^2 - \frac{\pi^2}{a^2}}$$

where z is generally a complex variable. Thus, we obtain a relation for the transmission coefficient

$$F = \frac{Ae^{i\gamma_0 l_1}}{g(z)}$$

which, together with formula (22.11), gives an explicit solution to the forward problem under study.

When the inverse problem is solved, ε_1 is considered as an unknown quantity that should be determined from (22.11) in terms of F.

The most important properties of $g(z)$ follow from its explicit representation and are the following:

i. $g(z)$ is an entire function.
ii. $g(z)$ has neither real zeros nor poles. This fact is in line with physical requirements that the transmission coefficient does not vanish and is a bounded quantity at real frequencies.
iii. When $g(z)$ is considered as a function of real τ, it is not invertible locally at the origin because $g'(0) = 0$. Next, the inverse of $g(z)$ is a multi-valued function. In fact, the inverse function does not exist globally according to the statement in Remark concerning violation of uniqueness.
iv. $g(z)$ is not a fractional-linear function; therefore, $g(z)$ performs one-to-one conformal mappings only of certain regions of the complex plane onto regions of the complex plane.
v. $g'(\tau) \neq 0$ for (real) $\tau \neq 0$; hence, $g(z)$ is invertible locally at a real point $\tau \neq 0$.

Assuming that ε_1 is real, it is reasonable to introduce a real variable

$$\tau = \gamma_1 l_1 = l_1\sqrt{k_1^2 - \frac{\pi^2}{a^2}} > 0$$

which may be used for parametrization. Extracting the real and imaginary part of $g(\tau)$, denoting them by x and y, yields

$$\begin{cases} x = \cos\tau, \\ y = h(\tau)\sin\tau, \end{cases} \quad \text{where } h(\tau) = \frac{\tau}{2C} + \frac{C}{2\tau}, \; C = \gamma_0 l_1$$

Equation (22.11) is equivalent to the system

$$\begin{cases} \cos\tau = p, & \rho = \Re\left(\frac{Ae^{-i\gamma_0 l_1}}{F}\right) \\ h(\tau)\sin(\tau) = q, & \zeta = \Im\left(\frac{Ae^{-i\gamma_0 l_1}}{F}\right) \end{cases} \tag{22.13}$$

where p and q are known values. Using the results of [23,26], we prove the following theorems stating the existence and uniqueness of the solution to the inverse problem of finding the permittivity of a one-sectional diaphragm in a waveguide of rectangular cross section.

For the inverse problem $P^{\mathbb{R}}_{\varepsilon_1}$, the following theorems and statement are proved.

Theorem 1. *Define* $\rho = \Re\left(Ae^{i\gamma_0 l_1}/F\right)$ *and* $\zeta = \Im\left(Ae^{i\gamma_0 l_1}/F\right)$ *such that* $|\rho| < 1$ *and* $\rho^2 + \zeta^2 \geq 1$. *Then for any* $\mu_1 \in R$ *there exists a unique solution of inverse problem* $P^{\mathbb{R}}_{\varepsilon_1}$. *This solution can be expressed by formulas*

$$\varepsilon_1 = k_0^{-2}\left(\left(\frac{\pi}{a}\right)^2 + \left(\frac{\tau}{l_1}\right)^2\right) \tag{22.14}$$

where

$$\tau = \tau_1 = \gamma_0 l_1 \frac{|\zeta| + \sqrt{\rho^2 + \zeta^2 - 1}}{\sqrt{1-\rho^2}}$$

if $\tau_1/\gamma_0 l_1 > 1$, $\varepsilon_1 > \varepsilon_0$, $\cos\tau_1 = \rho$, *and* $sign\,(\zeta) = sign\,(\sin\tau_1)$.

If $\tau_2/\gamma_0 l_1 < 1$, $\pi^2/(a^2\omega^2\mu_0) < \varepsilon_1 < \varepsilon_0$, $\cos\tau_2 = \rho$, *and* $sign\,(\zeta) = sign\,(\sin\tau_2)$, *then inverse problem has a unique solution, expressed by*

$$\tau = \tau_2 = \gamma_0 l_1 \frac{\sqrt{1-\rho^2}}{|\zeta| + \sqrt{\rho^2 + \zeta^2 - 1}}$$

Otherwise inverse problem $P^{\mathbb{R}}_{\varepsilon_1}$ *has no solution.*

Remark 1. *If* $\rho = 1$, *then* $\zeta = 0$ *and* $\tau = 2\pi n$, $n \in \mathbb{Z}$. *If* $\rho = -1$, *then* $\zeta = 0$ *and* $\tau = \pi + 2\pi n$, $n \in \mathbb{Z}$.

In this case, the inverse problem $P^{\mathbb{R}}_{\varepsilon_1}$ *has infinitely many solutions.*

Theorem 2. *If* $F = \pm Ae^{i\gamma_0 l_1}$, *or* $F = \frac{Ae^{i\gamma_0 l_1}}{g(\tau_1^*)}$, *where*

$$\tau_1^* = \frac{(\gamma_0 l_1)^2}{\pi n + \sqrt{\pi^2 n^2 + (\gamma_0 l_1)^2}}, \quad n \geq 0$$

then solution of the inverse problem $P^{\mathbb{R}}_{\varepsilon_1}$ *is not unique.*

If both conditions are not satisfied, then inverse problem $P^{\mathbb{R}}_{\varepsilon_1}$ *has unique solution.*

Note that Theorems 1 and 2 state existence and uniqueness of solution to the inverse problem $P^{\mathbb{R}}_{\varepsilon_1}$.

In some applications, only the modulus of the transmission coefficient is known. In this case, we have no uniqueness of solution for the inverse problem.

Remark 2. *If* $|F/A|$ *(*$|F/A| < 1$*) is known and* ε_1 *is real, then the inverse problem* $P^{\mathbb{R}}_{\varepsilon_1}$ *has an infinite set of solutions.*

Theorem 3. *If*

$$\frac{Re\left(\frac{B}{A}\right)}{\left|\frac{B}{A}\right|^2} < -1, \varepsilon_1 \neq \frac{\pi^2}{\omega^2\mu_0}\left(\frac{n^2}{l_1^2} + \frac{1}{a^2}\right), \quad n \in \mathbb{Z} \tag{22.15}$$

and

$$ctg(\gamma_1 l_1) = \frac{-Im\left(\frac{B}{A}\right)}{\left|\frac{B}{A}\right|^2 \sqrt{\left(\frac{Re\left(\frac{B}{A}\right)}{\left|\frac{B}{A}\right|^2}\right)^2 - 1}}$$

then the inverse problem $Q^{\mathbb{R}}_{\varepsilon_1}$ *has a unique solution which can be expressed by form*

$$\varepsilon_1 = \frac{1}{\omega^2 \mu_0 \varepsilon_0}\left(\left(\frac{\pi}{a}\right)^2 + \gamma_0^2 \frac{\frac{\Re(B/A)}{|B/A|^2} - 1}{\frac{\Re(B/A)}{|B/A|^2} + 1}\right) \tag{22.16}$$

Otherwise, the inverse problem $Q^{\mathbb{R}}_{\varepsilon_1}$ *has no solution.*

Theorem 3 states the existence and uniqueness for the inverse problem $Q^{\mathbb{R}}_{\varepsilon_1}$.

Thus for a one-sectional diaphragm, a solution in the closed form is obtained and the uniqueness is proved.

These results make it possible to use the case of a one-sectional diaphragm in a waveguide of rectangular cross section as a benchmark test problem and perform a complete analysis of the inverse scattering problem for arbitrary n-sectional diaphragms.

Now let us consider the results of solving inverse problems $P^{\mathbb{R}}_{\varepsilon_1, l_1}(Q^{\mathbb{R}}_{\varepsilon_1, l_1})$, $P^{\mathbb{C}}_{\varepsilon_1}(l_1 \ll 1)$, and $Q^{\mathbb{C}}_{\varepsilon_1}(l_1 \ll 1)$.

For the inverse problem $Q^{\mathbb{R}}_{\varepsilon_1, l_1}$, the following statement is true.

Theorem 4. *If conditions (22.14) are satisfied, then the inverse problem* $Q^{\mathbb{R}}_{\varepsilon_1, l_1}$ *has a unique solution which is expressed by formulas (22.15) and (22.16):*

$$l_1 = \frac{1}{\gamma_1} arctg\left(\frac{2\gamma_0\gamma_1}{\gamma_1^2 - \gamma_0^2}\frac{1}{\Im\frac{A}{B}}\right) + \frac{\pi n}{\gamma_1} \tag{22.17}$$

The diaphragm's thickness l_1 *is determined uniquely if* $l_2 > l_1$ *and* $\pi/\gamma_1 < |l_2 - l_1| < 2(\pi/\gamma_1)$.

For inverse problems $P^{\mathbb{C}}_{\varepsilon_1}(l_1 \ll 1)$ and $Q^{\mathbb{C}}_{\varepsilon_1}(l_1 \ll 1)$, the following statements hold.

Theorem 5. *Any solution of the inverse problem* $P^{\mathbb{C}}_{\varepsilon_1}(l_1 \ll 1)$ *is expressed approximately as*

$$\varepsilon_1 \approx \left(\gamma_0\left(\frac{2i}{l_1}\left(1 - \frac{A}{F}e^{i\gamma_0 l_1}\right) - \gamma_0\right) + \frac{\pi^2}{a^2}\right)k_0^{-2} \tag{22.18}$$

Theorem 6. *In case of a thin one-sectional diaphragm, any solution of the inverse problem* $Q^{\mathbb{C}}_{\varepsilon_1}(l_1 \ll 1)$ *is expressed approximately as*

$$\varepsilon_1 \approx \left(i\frac{2\gamma_0}{l_1}\frac{\frac{B}{A}}{\left(\frac{B}{A} + 1\right)} - \gamma_0^2\frac{\frac{B}{A} - 1}{\frac{B}{A} + 1} + \frac{\pi^2}{a^2}\right)k_0^{-2} \tag{22.19}$$

Thus, inverse problems of Class I has explicit solutions obtained as main terms of asymptotic series w.r.t. small parameter l_1.

Generally, the inverse problems are reduced to a nonlinear equation or a nonlinear system of equations which can be solved numerically by standard methods.

22.4 Inverse problems. Class AnI (Anisotropic case)

22.4.1 Statements of inverse problems for anisotropic one-sectional diaphragm (Class AnI)

This section is devoted to the solution of inverse problems of reconstructing electromagnetic parameters of isotropic and anisotropic one-sectional diaphragms.

Statement of the inverse problems for anisotropic one-sectional diaphragm (Class AnI)

$P^{\mathbb{R}}_{\widehat{\varepsilon}_1}$: From the transmission coefficient F/A, find *the permittivity tensor* $\widehat{\varepsilon}_1$ of the anisotropic diaphragm.

$P^{\mathbb{R}}_{\widehat{\mu}_1}$ ($Q^{\mathbb{R}}_{\widehat{\mu}_1}$): From the transmission coefficient F/A (or reflection coefficient B/A), find *the tensor permeability* $\widehat{\mu}_1$ of the anisotropic diaphragm.

$P^{\mathbb{R}}_{\widehat{\varepsilon}_1,\widehat{\mu}_1}$: From the transmission coefficient F/A, find *the tensor permittivity* $\widehat{\varepsilon}_1$ *or the tensor permeability* $\widehat{\mu}_1$ of the anisotropic diaphragm.

For solving inverse problems $P^{\mathbb{R}}_{\widehat{\varepsilon}_1}$, $P^{\mathbb{R}}_{\widehat{\mu}_1}$, $Q^{\mathbb{R}}_{\widehat{\mu}_1}$, and $P^{\mathbb{R}}_{\widehat{\varepsilon}_1,\widehat{\mu}_1}$ for anisotropic one-sectional diaphragms, the so-called rotation method is developed. The main idea of the method is to find all components of the diagonal tensor by using diaphragm's rotation w.r.t. its axis from the given (measured) values of transmission, F/A, and reflection, B/A, coefficients for different positions of the diaphragm.

In the original position of the diaphragm, the main equation ((22.6) or (22.9)) contains only three components of the permittivity and permeability of diagonal tensors; therefore, we use the diaphragm's rotation. To find other components of permittivity tensors, ε_{11} and ε_{33}, we apply the tensor transformation that changes the positions of its diagonal components (in our case, ε_{11} and ε_{22} are interchanged as well as ε_{33} and ε_{22}). Note that the transformations should be realized in practice.

Introduce the notation for measurements in different positions of the diaphragm as the number in the upper index $j = 1, 2, 3$, so that $j = 1$ means the original diaphragm's position, $j = 2$ the rotation w.r.t. Oz at the angle $\varphi = \frac{\pi}{2}$ (Figure 22.3) and $j = 3$ the rotation w.r.t. Ox at the angle $\varphi = \frac{\pi}{2}$ (Figure 22.4).

22.4.2 Explicit formulas for the transmission coefficient

Solving the boundary problem for the Maxwell equations and using conditions on the boundaries, we obtain formulas for the transmission coefficient in the form

$$\frac{F}{A} = \frac{e^{i\gamma_0 l_1}}{g(\tau)}$$

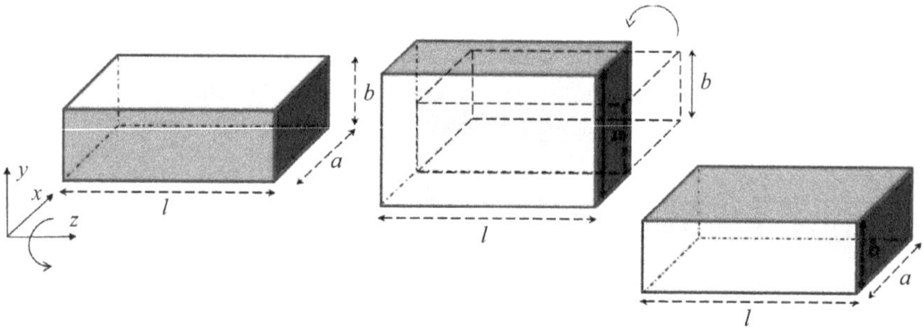

$$\gamma_1^{(1)} = \sqrt{\left(\omega^2 \varepsilon_{22}^{(1)} \mu_{33}^{(1)} - \frac{\pi^2}{a^2}\right) \frac{\mu_{11}^{(1)}}{\mu_{33}^{(1)}}} \qquad A_2 = \begin{pmatrix} 0 & -1 & 0 \\ 1 & 0 & 0 \\ 0 & 0 & 1 \end{pmatrix} \qquad \gamma_1^{(2)} = \sqrt{\left(\omega^2 \varepsilon_{11}^{(1)} \mu_{33}^{(1)} - \frac{\pi^2}{a^2}\right) \frac{\mu_{22}^{(1)}}{\mu_{33}^{(1)}}}$$

$$\hat{\varepsilon}^{(1)} = \begin{pmatrix} \varepsilon_{11} & 0 & 0 \\ 0 & \varepsilon_{22} & 0 \\ 0 & 0 & \varepsilon_{33} \end{pmatrix} \qquad \hat{\varepsilon}^{(2)} = A_2^{-1} \varepsilon^{(1)} A_2 \qquad \hat{\varepsilon}^{(2)} = \begin{pmatrix} \varepsilon_{22} & 0 & 0 \\ 0 & \varepsilon_{11} & 0 \\ 0 & 0 & \varepsilon_{33} \end{pmatrix}$$

$$\hat{\mu}^{(1)} = \begin{pmatrix} \mu_{11} & 0 & 0 \\ 0 & \mu_{22} & 0 \\ 0 & 0 & \mu_{33} \end{pmatrix} \qquad \hat{\mu}^{(2)} = A_2^{-1} \mu^{(1)} A_2 \qquad \hat{\mu}^{(2)} = \begin{pmatrix} \mu_{22} & 0 & 0 \\ 0 & \mu_{11} & 0 \\ 0 & 0 & \mu_{33} \end{pmatrix}$$

Figure 22.3 Diaphragm's rotation w.r.t. Oz

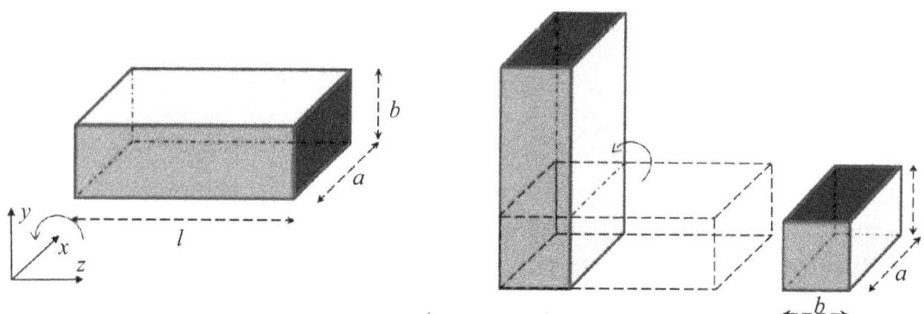

$$\gamma_1^{(1)} = \sqrt{\left(\omega^2 \varepsilon_{22}^{(1)} \mu_{33}^{(1)} - \frac{\pi^2}{a^2}\right) \frac{\mu_{11}^{(1)}}{\mu_{33}^{(1)}}} \qquad A_3 = \begin{pmatrix} 1 & 0 & 0 \\ 0 & 0 & 1 \\ 0 & -1 & 0 \end{pmatrix} \qquad \gamma_1^{(3)} = \sqrt{\left(\omega^2 \varepsilon_{33}^{(1)} \mu_{22}^{(1)} - \frac{\pi^2}{a^2}\right) \frac{\mu_{11}^{(1)}}{\mu_{22}^{(1)}}}$$

$$\hat{\varepsilon}^{(1)} = \begin{pmatrix} \varepsilon_{11} & 0 & 0 \\ 0 & \varepsilon_{22} & 0 \\ 0 & 0 & \varepsilon_{33} \end{pmatrix} \qquad \hat{\varepsilon}^{(3)} = A_3^{-1} \varepsilon^{(1)} A_3 \qquad \hat{\varepsilon}^{(3)} = \begin{pmatrix} \varepsilon_{11} & 0 & 0 \\ 0 & \varepsilon_{33} & 0 \\ 0 & 0 & \varepsilon_{22} \end{pmatrix}$$

$$\hat{\mu}^{(1)} = \begin{pmatrix} \mu_{11} & 0 & 0 \\ 0 & \mu_{22} & 0 \\ 0 & 0 & \mu_{33} \end{pmatrix} \qquad \hat{\mu}^{(3)} = A_3^{-1} \mu^{(1)} A_3 \qquad \hat{\mu}^{(3)} = \begin{pmatrix} \mu_{33} & 0 & 0 \\ 0 & \mu_{11} & 0 \\ 0 & 0 & \mu_{22} \end{pmatrix}$$

Figure 22.4 Diaphragm's rotation w.r.t. Ox

$$g(\tau) = \cos\tau + i\left(\frac{\tau}{2\gamma_0 l_1}\frac{\mu_0}{\mu_{11}^{(1)}} + \frac{\gamma_0 l_1}{2\tau}\frac{\mu_{11}^{(1)}}{\mu_0}\right)\sin\tau$$

where

$$\tau = \gamma_1 l_1 = l_1\sqrt{\omega^2\varepsilon_{22}^{(1)}\mu_{33}^{(1)} - \frac{\pi^2}{a^2}\frac{\mu_{11}^{(1)}}{\mu_{33}^{(1)}}}$$

22.4.3 The existence and uniqueness of the solution to the inverse problem

Let us present the results on existence and uniqueness of the solution to the inverse problem.

For inverse problems $P_{\widehat{\varepsilon}_1}^{\mathbb{R}}$ and $P_{\widehat{\varepsilon}_1,\widehat{\mu}_1}^{\mathbb{R}}$, the following theorems hold.

Theorem 7. *Assume that $\widehat{\mu}_1$ is known and $\rho^{(j)} = Re\left(Ae^{i\gamma_0 l_1}/F^{(j)}\right)$, $\zeta^{(j)} = Im\left(Ae^{i\gamma_0 l_1}/F^{(j)}\right)$, $\left|\rho^{(j)}\right| < 1$, $(\rho^{(j)})^2 + (\zeta^{(j)})^2 \geq 1$, $j = 1, 2, 3$. Then the solution of inverse problem $P_{\widehat{\varepsilon}_1}^{\mathbb{R}}$ exists is unique, and is expressed as*

$$\varepsilon_{11} = \frac{1}{\omega^2\mu_{22}\varepsilon_0}\left(\left(\frac{\pi}{a}\right)^2\frac{\mu_{22}}{\mu_{33}} + \left(\frac{\tau^{(2)}}{l_1}\right)^2\right)$$

$$\varepsilon_{22} = \frac{1}{\omega^2\mu_{11}\varepsilon_0}\left(\left(\frac{\pi}{a}\right)^2\frac{\mu_{11}}{\mu_{33}} + \left(\frac{\tau^{(1)}}{l_1}\right)^2\right)$$

$$\varepsilon_{33} = \frac{1}{\omega^2\mu_{11}\varepsilon_0}\left(\left(\frac{\pi}{a}\right)^2\frac{\mu_{11}}{\mu_{22}} + \left(\frac{\tau^{(3)}}{l_1}\right)^2\right)$$

where

$$\tau^{(1)} = \tau_1^{(1)} = \gamma_0 l_1\frac{\mu_{11}}{\mu_0}\widetilde{Q}^{(1)}$$

$$\tau^{(2)} = \tau_1^{(2)} = \gamma_0 l_1\frac{\mu_{22}}{\mu_0}\widetilde{Q}^{(2)} \tag{22.20}$$

$$\tau^{(3)} = \tau_1^{(3)} = \gamma_0 l_1\frac{\mu_{11}}{\mu_0}\widetilde{Q}^{(3)}$$

$$\widetilde{Q}^{(j)} = \frac{\left|\zeta^{(j)}\right| + \sqrt{(\rho^{(j)})^2 + (\zeta^{(j)})^2 - 1}}{\sqrt{1 - (\rho^{(j)})^2}},\ j = 1, 2$$

if the conditions $(\tau_1^{(j)}\mu_0)/(\gamma_0 l_1\mu_{11}) \geq 1$ $(j = 1, 3)$ and $(\tau_1^{(2)}\mu_0)/(\gamma_0 l_1\mu_{22}) \geq 1$, $\varepsilon_{11} > \varepsilon_0$, $\varepsilon_{22} > \varepsilon_0$, $\varepsilon_{33} > \varepsilon_0$, $\cos\tau_1^{(j)} = \rho^{(j)}$, $sign\left(\zeta^{(j)}\right) = sign\left(\sin\tau_1^{(j)}\right)$ $(j = 1, 2, 3)$ hold; and

$$\tau^{(1)} = \tau_2^{(1)} = \gamma_0 l_1\frac{\mu_{11}}{\mu_0}\widetilde{Q}_{(1)}^{-1},\ \tau^{(2)} = \tau_1^{(2)} = \gamma_0 l_1\frac{\mu_{22}}{\mu_0}\widetilde{Q}_{(2)}^{-1} \tag{22.21}$$

$$\tau^{(3)} = \tau_1^{(3)} = \gamma_0 l_1 \frac{\mu_{11}}{\mu_0} \widetilde{Q}_{-1}^{(3)}$$

if the conditions $(\tau_1^{(j)}\mu_0)/(\gamma_0 l_1 \mu_{11}) < 1$ $(j = 1,3)$ *and* $(\tau_1^{(2)}\mu_0)/(\gamma_0 l_1 \mu_{22}) < 1$, $\varepsilon_{11} > \varepsilon_0$, $\varepsilon_{22} > \varepsilon_0$, $\varepsilon_{33} > \varepsilon_0$, $\cos \tau_2^{(j)} = \rho^{(j)}$, *sign* $(\zeta^{(j)}) =$ *sign* $\left(\sin \tau_2^{(j)}\right)$ $(j = 1,2,3)$ *hold. Otherwise the inverse problem* $P_{\widehat{\varepsilon}_1}^{\mathbb{R}}$ *has no solution.*

In the anisotropic case, it is not possible to reconstruct all components of diagonal permittivity and permeability tensors using our method. But if we know one component, we have the following result.

Theorem 8. *Suppose* μ_{33} *is known,* $\rho^{(j)} = Re\left(Ae^{i\gamma_0 l_1}/F^{(j)}\right)$, $\zeta^{(j)} = Im\left(Ae^{i\gamma_0 l_1}/F^{(j)}\right)$ *è* $\left|\rho^{(j)}\right| < 1$, $(\rho^{(j)})^2 + (\zeta^{(j)})^2 \geq 1, j = 1,2,3$. *Then the solution of the inverse problem* $P_{\widehat{\varepsilon}_1,\widehat{\mu}_1}^{\mathbb{R}}$ *exists, is unique, and is expressed as*

$$\mu_{11} = \frac{\left(\frac{\pi}{a}\right)^2 \left(\omega_1^{-2} - \omega_2^{-2}\right)}{\left(\frac{\widetilde{Q}^{(1)}(\omega_2)}{\omega_2}\right)^2 - \left(\frac{\widetilde{Q}^{(1)}(\omega_1)}{\omega_1}\right)^2} \left(\frac{\mu_0}{\gamma_0}\right)^2 \mu_{33}^{-1} \tag{22.22}$$

$$\mu_{22} = \frac{\left(\frac{\pi}{a}\right)^2 \left(\omega_1^{-2} - \omega_2^{-2}\right)}{\left(\frac{\widetilde{Q}^{(2)}(\omega_2)}{\omega_2}\right)^2 - \left(\frac{\widetilde{Q}^{(2)}(\omega_1)}{\omega_1}\right)^2} \left(\frac{\mu_0}{\gamma_0}\right)^2 \mu_{33}^{-1} \tag{22.23}$$

where

$$\widetilde{Q}^{(j)} = \frac{\left|\zeta^{(j)}\right| + \sqrt{(\rho^{(j)})^2 + (\zeta^{(j)})^2 - 1}}{\sqrt{1 - (\rho^{(j)})^2}}, \; j = 1,2$$

the permittivity components are expressed in the form (22.20)–(22.21) if the following conditions are satisfied: $(\tau_1^{(j)}\mu_0)/(\gamma_0 l_1 \mu_{11}) \geq 1$ $(j = 1,3)$ *and* $(\tau_1^{(2)}\mu_0)/(\gamma_0 l_1 \mu_{22}) \geq 1$, $\varepsilon_{11} > \varepsilon_0$, $\varepsilon_{22} > \varepsilon_0$, $\varepsilon_{33} > \varepsilon_0$, $\cos \tau_1^{(j)} = \rho^{(j)}$, *and sign* $(\zeta^{(j)}) =$ *sign* $\left(\sin \tau_1^{(j)}\right)$ $(j = 1,2,3)$; *or as (22.20), (22.21) if* $(\tau_1^{(j)}\mu_0)/(\gamma_0 l_1 \mu_{11}) < 1$ $(j = 1,3)$ *è* $(\tau_1^{(2)}\mu_0)/(\gamma_0 l_1 \mu_{22}) < 1$, $\varepsilon_{11} > \varepsilon_0$, $\varepsilon_{22} > \varepsilon_0$, $\varepsilon_{33} > \varepsilon_0$, $\cos \tau_2^{(j)} = \rho^{(j)}$, *and sign* $(\zeta^{(j)}) =$ *sign* $\left(\sin \tau_2^{(j)}\right)$ $(j = 1,2,3)$.

Otherwise inverse problem $P_{\widehat{\varepsilon}_1,\widehat{\mu}_1}^{\mathbb{R}}$ *has no solution.*

Thus, Theorems 7 and 8 answer the questions concerning existence and uniqueness of solution to inverse problems $P_{\widehat{\varepsilon}_1}^{\mathbb{R}}$ and $P_{\widehat{\varepsilon}_1,\widehat{\mu}_1}^{\mathbb{R}}$.

22.5 Inverse problem for multi-sectional diaphragm. Class M

In this section, the inverse problems for a multi-sectional diaphragm are considered.

Consider first the statement of the inverse problem in case of an isotropic multi-sectional diaphragm.

Statement of the inverse problems for an isotropic multi-sectional diaphragm (Class M)

$P^{\mathbb{R}}_{\varepsilon_j}, P^{\mathbb{C}}_{\varepsilon_j}$ $(Q^{\mathbb{R}}_{\varepsilon_j}, Q^{\mathbb{C}}_{\varepsilon_j})$: From the transmission coefficient F/A (or the reflection coefficient B/A), find *the real or complex permittivity* ε_j $(j = 1, \ldots, n)$ of each section of an isotropic diaphragm.

$P^{\mathbb{R}}_{\varepsilon_j, l_j}(Q^{\mathbb{R}}_{\varepsilon_j, l_j})$: From the transmission coefficient F/A (or the reflection coefficient B/A), find *the real permittivity* ε_j $(j = 1, \ldots, n)$ *and the thickness* l_j $(j = 1, \ldots, n)$ of each section of an isotropic diaphragm.

$P^{\mathbb{C}}_{\varepsilon_j}(l_j \ll 1)(Q^{\mathbb{C}}_{\varepsilon_j}(l_j \ll 1))$: From the transmission coefficient F/A (or the reflection coefficient B/A) find *the complex permittivity* ε_j of each section of an isotropic thin $(l_1 \ll 1)$ multi-sectional diaphragm.

Let us formulate the statement of the inverse problem in case of *an anisotropic multi-sectional* diaphragm.

Statement of the inverse problems for an anisotropic multi-sectional diaphragm (class M)

$P^{\mathbb{R}}_{\widehat{\varepsilon}_j}$: From the transmission coefficient F/A find *the permittivity tensor* $\widehat{\varepsilon}_j$ of each section of an anisotropic multi-sectional diaphragm.

For solving the inverse problem $P^{\mathbb{C}}_{\varepsilon_j}$, we rewrite (22.24)–(22.25) in the form

$$G(h) = H, \quad H = \frac{2A\gamma_0 e^{i\gamma_0 l_n}}{F} \tag{22.24}$$

where

$$G(h) = \frac{\gamma_n p^{(+)}_{n+1} + \gamma_0 q^{(+)}_{n+1}}{\prod\limits_{j=0}^{n} \gamma_j}, \quad h := (\varepsilon_1, \ldots, \varepsilon_n) \tag{22.25}$$

Formula (22.25) defines a complex function of n complex variables. Let us formulate the inverse problem as follows.

Statement of the inverse problem $P^{\mathbb{C}}_{\varepsilon_j}$. Assume that n different frequencies $\Omega = (\omega_1, \ldots, \omega_n)$ and a vector of the function values $G_j(h) := G(h, \omega_j)$, $j = \overline{1, n}$, are given. Find the solution of the nonlinear system of n equations w.r.t. unknowns $\varepsilon_1, \ldots, \varepsilon_n$:

$$G_j(h) = H_j, \quad H_j = H(\omega_j), \quad j = \overline{1, n} \tag{22.26}$$

The following statement holds:

Theorem 9. *Function $G(h)$ is holomorphic on $\mathbb{C}^n$ as a function of n complex variables.*

We prove the existence and uniqueness of solution to the inverse problem $P_{\varepsilon_j}^{\mathbb{C}}$ by using the Hartogs' theorem.

Theorem 10. *If (22.26) is true for $h = h^*$ and if the Jacobian $\frac{\partial(G_1, G_2, \ldots, G_n)}{\partial(\varepsilon_1, \varepsilon_2, \ldots, \varepsilon_n)} \neq 0$ at a point h^*, then the function $G(h)$ is locally invertible in a vicinity of h^* and the inverse problem $\mathbf{P}_{\varepsilon_j}^{\mathbb{C}}$ has a unique solution for every h from that vicinity.*

For solving the inverse problem $Q_{\varepsilon_j}^{\mathbb{C}}$, we write the corresponding nonlinear equation system

$$G^{(B)}(h) = H, \quad H = \frac{A}{B} \tag{22.27}$$

where

$$G^{(B)}(h) = \frac{\gamma_n p_{n+1}^{(+)} + \gamma_0 q_{n+1}^{(+)}}{\gamma_n p_{n+1}^{(-)} + \gamma_0 q_{n+1}^{(-)}} \tag{22.28}$$

and $h := (\varepsilon_1, \ldots, \varepsilon_n)$.

Thus, (22.27)–(22.28) specifies a complex function of n complex variables.

Statement of the inverse problem $Q_{\varepsilon_j}^{\mathbb{C}}$. *Assume that n different frequencies $\Omega = (\omega_1, \ldots, \omega_n)$ and a vector of the function values $G_j^{(B)}(h) := G^{(B)}(h, \omega_j), j = 1, \ldots, n$ are given. Find the solution of the nonlinear system of n equations w.r.t. unknowns $\varepsilon_1, \ldots, \varepsilon_n$:*

$$G_j^{(B)}(h) = H_j, \quad H_j = H(\omega_j), \quad j = 1, \ldots, n \tag{22.29}$$

The following statements hold.

Theorem 11. *$G^{(B)}(h)$ is a meromorphic function on $\mathbb{C}^n$ as a function of n complex variables.*

Theorem 12 (existence and uniqueness of the solution to inverse problem $\mathbf{Q}_{\varepsilon_j}^{\mathbb{C}}$). *If (22.29) holds when $h = h^*$ and if the Jacobian $\frac{\partial\left(G_1^{(B)}, G_2^{(B)}, \ldots, G_n^{(B)}\right)}{\partial(\varepsilon_1, \varepsilon_2, \ldots, \varepsilon_n)} \neq 0$ at a point h^*, then the function $G^{(B)}(h)$ is locally invertible in a vicinity of h^* and the inverse problem $\mathbf{Q}_{\varepsilon_j}^{\mathbb{C}}$ has a unique solution for every h from that vicinity.*

22.6 Numerical results

In this section, we present the results obtained by numerical methods. All the inverse problems considered in this study are reduced to the nonlinear equation systems ((22.27) and (22.29)) in the real or complex domain which are solved numerically by a Levenberg–Marquardt iteration method. The choice of a good initial condition is important for the implementation of the corresponding algorithm. This choice is performed using closed-form solutions either for a thin or a one-sectional diaphragm. Giving more details, for a multi-sectional diaphragm filled with a media having real permittivity it is assumed that initially the given diaphragm has one section. Then explicit expressions are used to determine effective permittivity of the whole diaphragm and this value is applied as initial guess to solve the inverse problem for the multi-sectional diaphragm. In the next step, a nonlinear equation system of the considered inverse problem is solved by the Levenberg–Marquardt method. For a one-sectional diaphragm filled with lossy medium, the corresponding initial guess is chosen using closed-form solution for a thin diaphragm having complex permittivity. In case of anisotropic filling, the initial condition is determined in a similar manner; a diagonal tensor with the components equal to effective permittivity and permeability serves as initial guess.

22.6.1 *Example 1. Inverse problem for one-sectional anisotropic diaphragm*

Assume that the parameters of the diaphragm are $a = 2$ cm, $b = 1$ cm, and $l_1 = 2$ cm, the excitation frequencies are $f_1 = 75.02$ GHz and $f_2 = 51.02$ ($\omega = 2\pi f$) GHz, $\mu_{33} = 1.8$, and the exact permittivity $\hat{\varepsilon}$ and permeability $\hat{\mu}$ tensors are

$$\hat{\varepsilon} = \begin{pmatrix} 2 & 0 & 0 \\ 0 & 1.1 & 0 \\ 0 & 0 & 3 \end{pmatrix}, \quad \hat{\mu} = \begin{pmatrix} 1.2 & 0 & 0 \\ 0 & 1.4 & 0 \\ 0 & 0 & 1.8 \end{pmatrix} \tag{22.30}$$

The numerical results presented in Table 22.1 confirm the efficiency of the method.

22.6.2 *Example 2. Extraction of the complex permittivity of each section of three-sectional isotropic diaphragm*

In this section, we present the results of the numerical solution to the inverse problems for a three-sectional diaphragm.

Table 22.1 Inverse problem for one-sectional anisotropic diaphragm

Values of $\frac{F^{(j)}}{A}(\omega_1), j = \overline{(1,3)}$	Values of $\frac{F^{(j)}}{A}(\omega_2), j = \overline{(1,3)}$	Calculated $\hat{\varepsilon}$	Calculated $\hat{\mu}$
$0.316 - 0.945i$	$-0.035 - 0.991i$	2.03	1.16
$-0.610 + 0.709i$	$-0.596 + 0.280i$	1.07	1.42
$0.331 + 0.906i$	$-0.418 + 0.634i$	2.92	

Table 22.2 shows the test results of numerical solution to the inverse problem of reconstructing permittivities of a three-sectional diaphragm at three different frequencies. The test values of the transmission coefficient are taken from the solution to the corresponding forward problem. Parameters of the three-sectional diaphragm are $a = 2$ cm, $b = 1$ cm, $c = 2$ cm, $l_1 = 1$ cm, $l_2 = 1.5$ cm; the excitation frequencies $f_1 = 11.94$ GHz , $f_2 = 8.12$ GHz, and $f_3 = 9.55$ GHz. The first, second, and third columns of Table 22.2 shows, respectively, the values of transmission coefficient $\frac{F}{A}$, calculated values of permittivity of the section, and true values of the (real) permittivity of the section.

We see that in all examples the error of computations does not exceed 3%, which proves the high accuracy of the method.

22.6.3 Example 3. Extraction of permittivity and permeability of one-sectional anisotropic diaphragm

Parameters of the diaphragm are $a = 2$ cm, $b = 1$ cm, and $l_1 = 2$ cm, the excitation frequencies $f_1 = 75.02$ and $f_2 = 51.02$ $(\omega = 2\pi f)$ GHz, and $\mu_{33} = 1.8$. The exact permittivity and permeability tensors are the following (Table 22.3):

$$\hat{\varepsilon} = \begin{pmatrix} 2 & 0 & 0 \\ 0 & 1.1 & 0 \\ 0 & 0 & 3 \end{pmatrix}, \quad \hat{\mu} = \begin{pmatrix} 1.2 & 0 & 0 \\ 0 & 1.4 & 0 \\ 0 & 0 & 1.8 \end{pmatrix} \tag{22.31}$$

Table 22.2 Inverse problem for three-sectional isotropic diaphragm

Values of $\frac{F^{(j)}}{A}(\omega_1), j = \overline{(1,3)}$	**Calculated** $\varepsilon_{1,2,3}$	**True** $\varepsilon_{1,2,3}$
$0.012 + i \cdot 0.036$	$-1.713 + i \cdot 0.078$,	-1.7
$0.025 - i \cdot 0.012$	$1.523 - i \cdot 0.085$	1.5
$0.029 + i \cdot 0.014$	$4.01 + i \cdot 0.13$	4
$0.073 - i \cdot 0.177$	$1.702 - i \cdot 0.0004$,	1.7
$-0.269 - i \cdot 0.197$	$-1.499 + i \cdot 0.006$	-1.5
$-0.052 - i \cdot 0.22$	$3.996 + i \cdot 0.003$	4
$0.106 + i \cdot 0.061$	$1.716 - i \cdot 0.0004$,	1.7
$0.096 - i \cdot 0.023$	$1.48 - i \cdot 0.013$	1.5
$0.102 + i \cdot 0.046$	$-3.928 + i \cdot 0.037$	-4
$-0.0004 - i \cdot 0.00376$	$-1.708 - i \cdot 0.008$,	-1.7
$-0.0045 + i \cdot 0.00368$	$1.5 + i \cdot 0.0007$	1.5
$-0.0057 - i \cdot 0.00134$	$-3.982 + i \cdot 0.017$	-4
$-0.00005 - i \cdot 0.0005$	$-1.708 + i \cdot 0.03$,	-1.7
$-0.002 + i \cdot 0.001$	$-1.499 - i \cdot 0.03$	-1.5
$-0.002 - i \cdot 0.0004$	$-3.982 - i \cdot 0.044$	-4

Table 22.3 Inverse problem for anisotropic one-sectional diaphragm

Values of $\frac{F^{(j)}}{A}(\omega_1), j = \overline{(1,3)}$	Values $\frac{F^{(j)}}{A}(\omega_2), j = \overline{(1,3)}$	Calculated $\hat{\varepsilon}$	Calculated $\hat{\mu}$
$0.316 - 0.945i$	$-0.035 - 0.991i$	2.03	1.16
$-0.610 + 0.709i$	$-0.596 + 0.280i$	1.07	1.42
$0.331 + 0.906i$	$-0.418 + 0.634i$	2.92	

22.6.4 Example 4. Extraction of permittivity tensor of two-sectional anisotropic diaphragm

The waveguide parameters are $a = 2$ cm, $b = 1$ cm, the section widths are $l_1 = 1.2$ cm, $l_2 = 1.5$ cm, and $l_2 = 1.9$ cm; the excitation frequencies $f_1 = 11.94$ GHz, $f_2 = 8.12$ GHz, and $f_3 = 9.55$ GHz (or $\omega_1 = 75.0212$ GHz, $\omega_2 = 51.0195$ GHz and $\omega_2 = 60.0044$ GHz ($\omega = 2\pi f$)). The permeability tensors are

$$\hat{\mu}_1 = \begin{pmatrix} 1.7 & 0 & 0 \\ 0 & 1.9 & 0 \\ 0 & 0 & 2 \end{pmatrix}, \quad \hat{\mu}_2 = \begin{pmatrix} 3 & 0 & 0 \\ 0 & 4 & 0 \\ 0 & 0 & 4.7 \end{pmatrix}.$$

The third components of each section are $\varepsilon_{33}^{(1)} = 1.4 + i\frac{0.001}{\omega}$, $\varepsilon_{33}^{(2)} = 1.8 + i\frac{0.001}{\omega}$.

Values of $\frac{F^{(j)}}{A}(\omega_1)$, $j = (1, 2)$ in original position $(F^{(1)}/A)$ and after rotation $(F^{(2)}/A)$ at two frequencies ω_1, ω_2.

$$\frac{F^1}{A}(\omega_1) = -0.8111 - i0.5843, \frac{F^1}{A}(\omega_2) = -0.5355 - i0.8249,$$
$$\frac{F^2}{A}(\omega_1) = -0.994 + i0.0985, \frac{F^2}{A}(\omega_2) = -0.8952 - i0.4407$$

Calculated values of $\hat{\varepsilon}$ with added noise of 1%

$$\begin{pmatrix} 1.2174 - i\frac{0.00141}{\omega} & 0 & 0 \\ 0 & 1.2822 + i\frac{0.003}{\omega} & 0 \\ 0 & 0 & 1.4 + i\frac{0.001}{\omega} \end{pmatrix},$$

$$\begin{pmatrix} 1.9736 + i\frac{0.003}{\omega} & 0 & 0 \\ 0 & 1.718 + i\frac{0.01}{\omega} & 0 \\ 0 & 0 & 1.8 + i\frac{0.001}{\omega} \end{pmatrix}$$

The true permittivity tensor

$$\varepsilon_1(\omega) = \begin{pmatrix} 1.2 + i\frac{0.001}{\omega} & 0 & 0 \\ 0 & 1.3 + i\frac{0.001}{\omega} & 0 \\ 0 & 0 & 1.4 + i\frac{0.001}{\omega} \end{pmatrix},$$

Table 22.4 Inverse problem $P^{\mathbb{C}}_{\varepsilon_1}$

Frequency, GHz	Experimental $\frac{F}{A}$	Calculated ε^1_1	Calculated σ_1
8.410	$0.5357 - 0.6719i$	1.996	0.001000006
8.431	$0.5379 - 0.6723$	1.993	0.00100005
8.452	$0.5388 - 0.6751i$	1.994	0.00100005
8.473	$0.5446 - 0.6777i$	1.987	0.00100004
8.494	$0.5399 - 0.6767i$	1.994	0.00100003
8.515	$0.542 - 0.6795i$	1.994	0.00100003
8.536	$0.5402 - 0.679i$	1.996	0.00100002
8.557	$0.5432 - 0.6796i$	1.992	0.00100001
8.562	$0.5477 - 0.6868i$	1.991	0.000999997

$$\varepsilon_2(\omega) = \begin{pmatrix} 2 + i\frac{0,001}{\omega} & 0 & 0 \\ 0 & 1.7 + i\frac{0.001}{\omega} & 0 \\ 0 & 0 & 1.8 + i\frac{0.001}{\omega} \end{pmatrix}.$$

The numerical results confirm high efficiency of the developed method.

22.6.5 Example 5. Inverse problem $P^{\mathbb{C}}_{\varepsilon_1}$

For inverse problem $P^{\mathbb{C}}_{\varepsilon_1}$, we compare its test numerical solution and experimental data. The second column of Table 22.4 contains experimental values of transmission coefficient F/A obtained at different frequencies.

The waveguide parameters are $a = 2.274$ cm and $b = 1.004$ cm; the parameters of the one-sectional diaphragm are $\varepsilon_1 = \varepsilon^1_1 + i\sigma_1/\omega = 2.05 + i0.00095/\omega$ (true values are used for solving the forward diffraction problem). Table 22.4 presents in its third and forth columns the calculated values of the real part of permittivity ε and conductivity σ.

As the result of calculation, we determined the material of the diaphragm in the waveguide, which is Teflon.

One can see that relative error of the permittivity (ε) determination does not exceed 3%. It confirms the accuracy of the developed method.

We conclude that we have developed an accurate numerical–analytical method of solution to the inverse problem of reconstructing permittivities of n-sectional diaphragms in a waveguide of rectangular cross section.

22.7 Conclusion

We have developed a numerical–analytical method of solution to the inverse problems of reconstructing electromagnetic parameters of isotropic and anisotropic multilayer dielectric diaphragms in a waveguide of rectangular cross section. The method comprises explicit formulas for scalar and tensor permittivities and permeabilities as well as theoretical justification of the approach applied for obtaining these formulas. Results of numerical modeling supported by comparison with test solutions to canonical problems and experimental and measurement data confirm the efficiency of the proposed techniques.

Acknowledgments

This work was supported by the Ministry of Education and Science of the Russian Federation (Agreement No. 1.894.2017/4.6) and President's Grant (No. MK – 3604.2018.1).

References

[1] Smirnov, Yu.G., Medvedik, M.Yu., and Derevyanchuk, E.D. Collocation method of solving volume singular integral equation for diffraction by dielectric body in rectangular waveguide. Proceedings of Progress in Electromagnetics Research Symposium, Moscow, Russia (2009).

[2] Shestopalov, Yu.V., Smirnov, Yu.G., and Yakovlev, V.V. Volume singular integral equations method for determination of effective permittivity of meta- and nanomaterials. Proceedings of Progress in Electromagnetics Research Symposium, Cambridge, USA, 291–292 (2008).

[3] Shestopalov, Yu.V., Smirnov, Yu.G., and Yakovlev, V.V. Development of mathematical methods for reconstructing complex permittivity of a scatterer in a waveguide. Proceedings of 5th International Workshop on Electromagnetic Wave Scattering, Antalya, Turkey (2008).

[4] De Rosa, I.M., Dinescu, A., Sarasini, F., Sarto, M.S., and Tamburrano, A. Effect of short carbon fibers and MWCNTs on microwave absorbing properties of polyester composites containing nickel-coated carbon fibers. Composites Science and Technology **70**, 102–109 (2010).

[5] Saib, A., Bednarz, L., Daussin, R., *et al.* Carbon nanotube composites for broadband microwave absorbing materials. IEEE TMTT **54**, No. 6, 2745–2754 (2006).

[6] Kelly, J.M., Stenoien, J.O., and Isbell, D.E. Wave-guide measurements in the microwave region on metal powders suspended in paraffin wax. J. Appl. Phys. **24**, No. 3, 258–262 (1953).

[7] Lurie, K.A. and Yakovlev, V.V. Optimization of electric field in rectangular waveguide with lossy layer. IEEE Trans. on Magnetics **36**, 1094–1097 (2000).

[8] Akleman, F. Reconstruction of complex permittivity of a longitudinally inhomogeneous material loaded in a rectangular waveguide. IEEE MWCL **18**, No. 3, 158–160 (2008).

[9] Baker-Jarvis, J. and Vanzura, E.J. Improved technique for determining complex permittivity with the transmission-reflection method. IEEE TMTT **38**, No. 8, 1096–1103 (1990).

[10] Faircloth, D.L., Baginski, M.E., and Wentworth, S.M. Complex permittivity and permeability extraction for multilayered samples using S-parameter waveguide measurements. IEEE TMTT **54**, No. 3, 1201–1209 (2006).

[11] Dediu, S. and McLaughlin, J.R. Recovering inhomogeneities in a waveguide using eigensystem decomposition. Inverse Problems **22**, No. 4, 1227–1246 (2006).

[12] Lurie, K.A. and Yakovlev, V.V. Control over electric field in traveling wave applicators. Journal of Engineering Mathematics **44**, No. 2, 107–123 (2002).

[13] Smirnov, Yu.G., Shestopalov, Yu.V., and Derevyanchuk, E.D. Inverse problem method for complex permittivity reconstruction of layered media in a rectangular waveguide. Physica Status Solidi (C) Current Topics in Solid State Physics **11**, No. 5–6, 969–974 (2014).

[14] Smirnov, Yu.G., Shestopalov, Yu. V., and Derevyanchuk, E.D. Permittivity determination of thin multi-sectional diaphragms in a rectangular waveguide. Days on Diffraction 2013: Proceedings of the International Conference. St. Petersburg, 32–35 (2013).

[15] Derevyanchuk, E.D and Smirnov, Yu.G. Tensor permittivity reconstruction of two-sectional diaphragm in a rectangular waveguide. Days on Diffraction 2014 : Proceedings of the International Conference, St. Petersburg, 65–68 (2014).

[16] Grishina, E.E, Derevyanchuk, E.D., Medvedik, M.Y, and Smirnov, Y.G. Numerical and analytical solutions of electromagnetic field diffraction on the two-sectional body with different permittivity in the rectangular waveguide. Izv. Vyssh. Uchebn. Zaved. Povolzh. Region, Fiz.-Mat. Nauki No. 4, 73–81 (2010).

[17] Outifa, L., Delmotte, M., and Jullien, H. Dielectric and geometric dependence of electric field and power distribution in a waveguide heterogeneously filled with lossy dielectrics. IEEE TMTT **45**, Suppl. 1, 1154–1161 (1997).

[18] Eves, E.E., Kopyt, P., and Yakovlev, V.V. Practical aspects of complex permittivity reconstruction with neural-network-controlled FDTD modeling of a two-port fixture. J. Microwave Power & Electromagnetic Energy **41**, No. 4, 81–94 (2007).

[19] Baginski, M.E., Faircloth, D.L., and Deshpande M.D. Comparison of two optimization techniques for the estimation of complex complex permittivities of multilayered structures using waveguide measurements. IEEE TMTT **53**, No. 10, 3251–3259 (2005).

[20] Jackson, J.D. Classical Electromagnetics. John Wiley & Sons, New York (1967).

[21] Yakovlev, V.V., Murphy, E.K., and Eves, E.E. Neural networks for FDTD-backed permittivity reconstruction. COMPEL **33**, No. 1, 291–304 (2005).

[22] Shestopalov, Yu.V., Smirnov, Yu.G., and Derevyanchuk, E.D. Inverse problem method for permittivity of an n-sectional diaphragm in a rectangular waveguide. PIERS 2014 : Proceedings of Progress in Electromagnetic Research Symposium, Guangzhou, 2610–2613 (2014).

[23] Smirnov, Yu.G., Shestopalov, Yu.V., and Derevyanchuk, E.D. Permittivity reconstruction of layered dielectrics in a rectangular waveguide from the transmission coefficient at different frequencies. Inverse Problems and Large-Scale Computations, Series : Springer Proceedings in Mathematics and Statistics, Vol. 52, 169–181 (2013).

[24] Smirnov, Yu.G., Shestopalov, Yu.V., and Derevyanchuk, E.D. Permittivity determination of multi-sectional diaphragm with metamaterial layers in rectangular waveguide. PIERS 2013 : Proceedings of Progress in Electromagnetic Research Symposium, Taipei, 135–139 (2013).

[25] Smirnov, Yu.G., Shestopalov, Yu.V., and Derevyanchuk, E.D. Reconstruction of permittivity and permeability tensors of anisotropic materials in a rectangular waveguide from the reflection and transmission coefficients at different frequencies. PIERS 2013: Proceedings of Progress in Electromagnetic Research Symposium, Stockholm, 290–295 (2013).

[26] Smirnov, Yu.G., Shestopalov, Yu.V., and Derevyanchuk, E.D. Solution to the inverse problem of reconstructing permittivity of an n-sectional diaphragm in a rectangular waveguide. Algebra, Geometry and Mathematical Physics : Springer Proceedings in Mathematics and Statistics, Ser. 10533, Vol. 85, 555–567 (2014).

[27] Smirnov, Yu.G., Shestopalov, Yu.V., and Derevyanchuk, E.D. Frequency-dependent complex permittivity reconstruction of layered diaphragm in a rectangular waveguide: comparison with experiment. Proceedings of the 16th Seminar Computer Modeling in Microwave Power Engineering, Karlsruhe, 30–34 (2014).

[28] Weinschtien, L.A. Electromagnetic Waves. Moscow: Radio i Svyaz', 1988 (in Russian).

[29] Brekhovskih, L. Waves in Layered Media. Academic Press, New York (1980).

[30] Beilina, L. and Klibanov, M. Approximate Global Convergence and Adaptivity for Coefficient Inverse Problems. Springer, New York (2012).

[31] Colton, D. and Kress, R. Inverse Acoustic and Electromagnetic Scattering Theory. Springer, New York (1998).

[32] Eves, E.E., Kopyt, P., and Yakovlev, V.V. Determination of complex permittivity with neural networks and FDTD modeling. Microw. Opt. Tech. Let. **40**, No. 3, 183–188 (2004).

[33] Kiley, E.M., Yakovlev, V.V., Ishizaki, K., and Vaucher, S. Applicability study of classical and contemporary models for effective permittivity of metal powders. École nationale supérieure d'électrotechnique, d'électronique, d'informatique, d'hydraulique et des télécommunications (INP-ENSEEIHT), Toulouse, France (2011).

[34] Norgren, M. and He, S. Reconstruction of the constitutive parameters for an Ω material in a rectangular waveguide. IEEE TMTT **43**, No. 6, 1315–1321 (1995).

[35] Shabat, B. Introduction to Complex Analysis. Part II: Functions of Several Variables. American Math. Society, Town (2003).

[36] Solymar, L. and Shamonina, E. Waves in Metamaterials. Oxford University Press Inc., New York (2009).

[37] Usanov, D.A, Skripal, A.V., Abramov, A.V., and Bogolyubov, A.S. Determination of the metal nanometer layer thickness and semiconductor conductivity in metal-semiconductor structures from electromagnetic reflection and transmission spectra. Technical Physics **51**, No. 5, 644–649 (2006).

[38] Usanov, D.A, Skripal, A.V., Abramov, A.V, and Bogolyubov, A.S. Complex permittivity of composites based on dielectric matrices with carbon nanotubes. Technical Physics **56**, No. 1, 102–106 (2011).

Chapter 23

Beam-based local diffraction tomography

Ram Tuvi[1], Ehud Heyman[1], and Timor Melamed[2]

23.1 Introduction and overview

In this chapter, we present a beam-based local diffraction-tomography (DT) inversion scheme as an alternative to the conventional Green's function or plane-wave inversion schemes. The beam waves[‡] are used here as basis functions for local phase-space processing of the data and for local backpropagation and reconstruction. Beam waves may also be used for local interrogation of the medium, as, e.g., in [1], but this approach is not considered here.

There are several reasons for using beam-based schemes:

i. Beam propagators can be used to construct the wave solutions in the inhomogeneous background.
ii. The beam-domain data, obtained via the beam-based processing of the scattering data, is a-priori sparse and is structured upon the geometrical-ray skeleton and resolves the local directions and times of arrival (DoAs and ToAs, respectively).
iii. In view of the property (ii), one may focus the imaging scheme on any given sub-domains of interest (DoIs) in the image domain by retaining only those beam backpropagators that pass through that domain.
iv. Property (iii) implies that the beam algorithm not only reduces the overall complexity, but it also filters out irrelevant data and noise arriving from regions outside the DoI.

The beam-based DT algorithm that is reviewed here is the final phase in a long-term research on this subject (see [2–5] and references therein). The potential of the beam approach has already been noted and utilized in various inverse scattering applications [6–19]. These publications, however, were based on the either the GB expansion for of point source fields, or on the Gabor-series approach, reviewed in

[1]Department of Physical Electronics, Tel Aviv University, Tel Aviv, Israel
[2]School of Electrical and Computer Engineering, Ben-Gurion University of the Negev, Beersheba, Israel

[‡]Beam waves are highly collimated wave functions that propagate along ray trajectory. We use the generic name "beams" for both time-harmonic fields, where the propagators are typically Gaussian beams (GBs) or their higher-order Hermit GB counterparts, and for time-domain fields, where the propagators are the so-called iso-diffracting pulsed beams (ID-PBs).

Section 23.2.1, but none has used the ultra-wide-band beam summation (UWB-BS) theory presented in Sections 23.2.2 and 23.2.3, which provides a rigorous and self-consistent alternative to the plane-wave approaches in the time or multifrequency domains.

DT can be implemented using either an angular diversity or a frequency diversity approach. In the former, the monochromatic DT, the scattering data is measured as a function of the illumination direction, whereas in the latter, the UWB-DT, the data is measured for a single source location but over a wide frequency band (see a short review in Sections 23.3.1.3 and 23.3.1.4, respectively). Depending on the problem, one may use a combined approach [20,21]. This chapter applies the beam approach to UWB-DT. We consider two algorithms that are expressed either in the multifrequency domain or directly in the time domain (TD). The beam approach can also be formulated for the angular diversity case, but we do not consider this approach here.

Following the discussion above, the chapter is divided into three parts: Section 23.2 reviews the beam methods used, Section 23.3 reviews the DT formulations, and finally, Section 23.4 applies the beam approach to DT.

The discussion of the beam method in Section 23.2 starts with a conceptual overview of the so-called beam summation (BS) method (Section 23.2.1), followed by a more detailed discussion of the ultra-wide-band, phase-space BS (UWB-PS-BS) method that provides the mathematical framework for our data processing and inversion algorithms. Following the discussion above, this method is presented both in the multifrequency domain and in the TD (Section 23.2.2 and 23.2.3, respectively). Moving forward to the conventional DT formulations in Section 23.3, we first present the UWB-DT in the multifrequency domain (Section 23.3.1) and then directly in the TD (Section 23.3.2). We conclude in Section 23.4 with the formulation the beam-based DT in both the multifrequency domain and the TD, and by presenting some illustrative numerical examples. Initial results of this chapter were presented in [22].

23.2 The UWB-PS-BS method

23.2.1 BS methods: an overview

The BS method is a ray-based approach to construct local spectrum representations of wavefields. In this approach, the source field is expanded into a sum of beam propagators that emerge from the source in all directions. These beams are then tracked in the medium and their contributions at the observation point are summed up to form the total field there. The advantages of the BS formulations over the traditional Green's function or the plane-wave spectrum representations are given in the following (see review in [23]):

1. Unlike plane waves, the beam propagators can be tracked locally in inhomogeneous media or through interactions with interfaces, and unlike rays, they are insensitive to the geometrical optics (GO) transition zones.
2. Being a spectral summation representation, like the plane-wave integrals, BS formulations are a-priori uniform.

3. Unlike the plane-wave integrals, they are a-priori localized in the vicinity of the phase-space skeleton of GO, since only those beam propagators that pass near the observation point actually contribute there.
4. Finally, the BS approach circumnavigates the need to search for the ray connecting the source with that observer.

There are basically two classes of BS formulations, one for *localized (or point) sources* and the other for *extended (or aperture) sources*.§ For each class, there are both frequency and TD formulations. In the present context of inverse scattering, we are interested in the extended sources formulations, which are used to expand the scattering data that is measured over the measurement aperture. The relevant BS formulations are summarized in Sections 23.2.2 and 23.2.3.

In the *point-source BS formulations*, the field is expressed as a superposition of beams that emerge from the source in all directions. The method has been derived first asymptotically for monochromatic fields [26,27] but subsequently it has been set on exact footing using the complex-source method [28,29], and used in various applications [30–39]. These algorithms were subsequently extended to the TD using complex-source pulsed-beam (CSPB) propagators [29,39,40].

In the *extended-source BS formulations*, the field is expressed as a discrete superposition of shifted and tilted GBs that emanate from a lattice of sampling points and directions in the aperture-source domain. The method has been structured originally upon a Gabor-series expansion of the aperture sources [41–44]. However, the Gabor formulation suffers from the following difficulties: (i) The Gabor expansion coefficients are notoriously nonlocal and unstable. (ii) The beam lattice (the set of the beam initiation points and directions) is frequency dependent, thus making this approach inefficient for UWB applications where the beam lattice needs to be retracked for each frequency.

A major step forward has been the replacement in [45] of the Gabor series by a Gabor frame‖ (henceforth we use the term windowed Fourier transform (WFT) frame). The overcompleteness of the frame formulation renders the expansion coefficients local and stable. In addition, this overcompleteness adds a degree of freedom that has been utilized in [46,47] to construct a UWB formulation that has the following favorable properties:

a. The beam lattice is *frequency independent*, so that the beam axes do not have to be recalculated for each frequency.
b. The algorithm utilizes the so-called iso-diffracting (ID) Gaussian beam (ID-GB) propagators whose propagation parameters are frequency independent, and hence they are calculated only once and then used for all frequencies in the excitation band.
c. The use of properly tuned ID-GB windows provides the *snuggest* frame for *all* the frequencies, leading to smooth and local expansion coefficients for all frequencies.

§We do not consider here the Hermite GB summation method which is relevant only for paraxial fields [24,25].

‖A "frame" is an overcomplete set of functions used to expand functions in a given Hilbert space.

This algorithm is summarized in Section 23.2.2. Properties (a) and (b) were subsequently used to formulate a TD-BS algorithm for extended sources [48], summarized in Section 23.2.3.

The final development in this theory has been the proof in [49,50] that the UWB-PS set of beam propagators constitutes a frame not only in the aperture plane but actually everywhere in the propagation domain (see Section 23.2.2.4). This property implies that this "beam frame" (BF) can be used to expand not only aperture sources but also volume sources. As an example, the method has been used there to analyze wave propagation through a roughly fluctuating medium by using the BF for a local expansion of the volume sources induced by the interaction of the incident wave with the medium.

In the context of inverse scattering, considered in this work, the BF is used for local expansion (and reconstruction) of the medium inhomogeneities (see Section 23.4).

23.2.2 The UWB-PS-BS method: a frequency domain formulation

In this section, we briefly review the basic strategy of the UWB-PS-BS. For simplicity, it is presented in the context of radiation into the half-space $z > 0$ from a given time-harmonic aperture source distribution $\hat{u}_0(\mathbf{x};\omega)$ in the plane $z = 0$, where we use $\mathbf{r} = (x_1, x_2, z) = (\mathbf{x}, z)$. It is assumed that the source $\hat{u}_0(\mathbf{x};\omega)$ has a wide frequency band $\Omega = [\omega_{\min}, \omega_{\max}]$. We use the hat ^ to denote time-harmonic fields with a suppressed time dependence $e^{-i\omega t}$. The medium is assumed to be uniform with wave speed v_0.

The conventional representations of the radiated field are given by

$$\hat{u}(\mathbf{r}) = \int_{-\infty}^{\infty} d^2x'\, \hat{u}_0(\mathbf{x}') 2\partial_{z'} \hat{G}(\mathbf{r},\mathbf{r}') = \left(\frac{k}{2\pi}\right)^2 \int_{-\infty}^{\infty} d^2\xi\, \hat{\tilde{u}}_0(\boldsymbol{\xi}) e^{ik(\boldsymbol{\xi}\cdot\mathbf{x}+\zeta z)} \tag{23.1}$$

where $\hat{G}(\mathbf{r},\mathbf{r}') = \exp\{ik|\mathbf{r}-\mathbf{r}'|\}/4\pi|\mathbf{r}-\mathbf{r}'|$ is the 3D free-space Green's function and $k = \omega/v_0$. The second expression in (23.1) is the plane-wave integral, expressed in terms the spectral function

$$\hat{\tilde{u}}_0(\boldsymbol{\xi}) = \int_{-\infty}^{\infty} d^2x\, \hat{u}_0(\mathbf{x}) e^{-ik\boldsymbol{\xi}\cdot\mathbf{x}}, \quad \text{with} \quad \boldsymbol{\xi} = (\xi_1, \xi_2), \quad \zeta = \sqrt{1-|\boldsymbol{\xi}|^2}, \quad \operatorname{Im}\zeta \geq 0 \tag{23.2}$$

Anticipating an extension to the TD, we use the frequency-normalized spectral coordinate $\boldsymbol{\xi} = \mathbf{k}_x/k$ which relates to the plane-wave direction in the frequency independent fashion $\boldsymbol{\xi} = (\xi_1, \xi_2) = \sin\theta(\cos\phi, \sin\phi)$. The real spectral regimes $|\boldsymbol{\xi}| < 1$ and $|\boldsymbol{\xi}| > 1$ are defined, respectively, as the "propagation" and "evanescent" spectra. In the former we have $\zeta = \cos\theta$, whereas in the latter, $\zeta = i\sqrt{|\boldsymbol{\xi}|^2 - 1}$.

23.2.2.1 WFT frame

The BS representation is structured upon the WFT frame set $\{\hat{\psi}_\mu(\mathbf{x})\}$, defined by

$$\hat{\psi}_\mu(\mathbf{x}) = \hat{\psi}(\mathbf{x} - \mathbf{x_m})e^{ik\boldsymbol{\xi}_\mathbf{n}\cdot(\mathbf{x}-\mathbf{x_m})}, \quad \boldsymbol{\mu} = (\mathbf{m}, \mathbf{n}) = ((m_1, m_2), (n_1, n_2)) \in \mathbb{Z}^4 \tag{23.3}$$

where $\boldsymbol{\mu}$ is a four-index tagging the frame elements, $\hat{\psi}(\mathbf{x})$ is a localized "mother window" (typically a Gaussian) and $(\mathbf{x_m}, \boldsymbol{\xi}_\mathbf{n}) = (\mathbf{m}\bar{x}, \mathbf{n}\bar{\xi})$ define a lattice in the $(\mathbf{x}, \boldsymbol{\xi})$ phase space whereon the frame-elements $\hat{\psi}_\mu$ in (23.3) are localized, with $(\bar{x}, \bar{\xi})$ being the unit-cell dimensions (see Figure 23.1(a)). The parameter $(\bar{x}, \bar{\xi})$ chosen to provide an overcomplete phase space coverage in the sense that

$$k\bar{x}\bar{\xi} = 2\pi\nu \tag{23.4}$$

with $\nu < 1$ being the overcompleteness parameter and $\nu \uparrow 1$ defines the critically complete (Gabor) limit.

The WFT frame expansion of the aperture field $\hat{u}_0(\mathbf{x})$ is given by

$$\hat{u}_0(\mathbf{x}) = \sum_\mu \hat{a}_\mu \hat{\psi}_\mu(\mathbf{x}), \qquad \hat{a}_\mu = \langle \hat{u}_0(\mathbf{x}), \hat{\varphi}_\mu(\mathbf{x})\rangle \tag{23.5}$$

where $\langle f, g\rangle = \int d^2x f(\mathbf{x})g^*(\mathbf{x})$ is the conventional $\mathbb{L}_2$ inner product in $\mathbf{x}$. Here $\{\hat{\varphi}_\mu\}$ is the "canonical dual frame" which is a WFT frame with the same form as in (23.3) but with $\hat{\psi}$ replaced by the "dual window" $\hat{\varphi}(\mathbf{x})$.

Since $\hat{\varphi}_\mu$ has the structure in (23.3), the expansion coefficients $\hat{a}_\mu$ in (23.5) are readily recognized as the WFT of $\hat{u}_0(\mathbf{x})$ around the lattice points $(\mathbf{x_m}, \boldsymbol{\xi}_\mathbf{n})$ with respect to the window function $\hat{\varphi}(\mathbf{x})$.

In general, the dual window $\hat{\varphi}$ needs to be calculated numerically for a given $\hat{\psi}$ and lattice $(\bar{x}, \bar{\xi})$, but if the lattice is sufficiently overcomplete then $\hat{\varphi}$ can be approximated

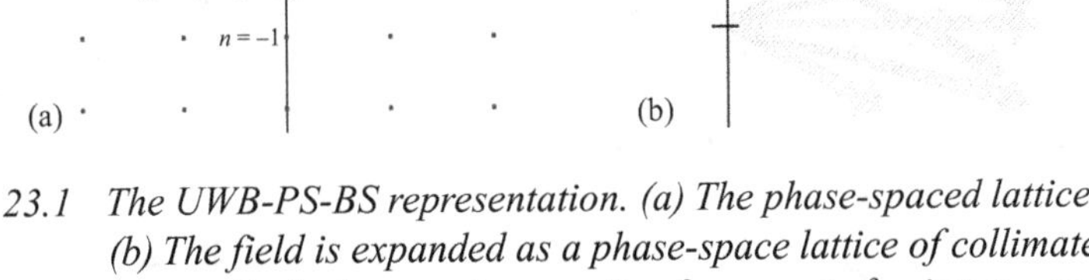

Figure 23.1 The UWB-PS-BS representation. (a) The phase-spaced lattice. (b) The field is expanded as a phase-space lattice of collimated beams (dashed arrows) emanating from a set of points $\mathbf{x}_m$ and directions $\boldsymbol{\xi}_n$. In the UWB-PS-BS, these lattices are frequency independent

by $\hat{\varphi} \approx \frac{\nu}{\|\psi\|^2}\hat{\psi}$. The reasons to prefer this overcomplete parameter regime, even though it implies more terms in the series (23.5), are the following:

1. In this regime, $\hat{\varphi}$ is proportional to $\hat{\psi}$ and therefore is localized both spatially and spectrally. As a result, $\hat{a}_\mu$ in (23.5) are local and stable, representing the WFT of $\hat{u}_0(\mathbf{x})$ around $(\mathbf{x_m}, \boldsymbol{\xi}_\mathbf{n})$.
2. Since in this parameter regime, $\hat{\varphi}$ is known analytically, it does not have to be to calculated numerically for every ω. This also enables a simple transformation of $\hat{\varphi}(\mathbf{x})$ to the TD.

As shown in [46, Fig. 3], the analytic approximation above applies for $\nu \lesssim 0.4$, whereas for smaller overcompleteness (larger ν), $\hat{\varphi}$ is less localized.

23.2.2.2 BS representation of the radiation field

The radiated field for $z > 0$ is obtained by propagating the expansion set $\hat{\psi}_\mu(\mathbf{x}, \omega)$ in (23.5), giving

$$\hat{u}(\mathbf{r}) = \sum_\mu a_\mu \hat{\Psi}_\mu(\mathbf{r}), \quad \hat{\Psi}_\mu(\mathbf{r}) = \frac{k^2}{4\pi^2}\int_{-\infty}^{\infty} d^2\xi\, \hat{\tilde{\psi}}_\mu(\boldsymbol{\xi}) e^{ik(\boldsymbol{\xi}\cdot\mathbf{x}+\zeta z)} \tag{23.6}$$

where the "beam propagators" $\hat{\Psi}_\mu$ are the fields radiating to $z > 0$ due to the fields $\hat{\psi}_\mu(\mathbf{x})$ at $z = 0$. They are expressed here in term of the plane-wave representation (23.1) with $\hat{\tilde{\psi}}_\mu(\boldsymbol{\xi}) = \hat{\tilde{\psi}}(\boldsymbol{\xi} - \boldsymbol{\xi}_\mathbf{n})e^{-ik\boldsymbol{\xi}\cdot\mathbf{x_m}}$ being the spectrum (23.2) of $\hat{\psi}_\mu$ and $\hat{\tilde{\psi}}(\boldsymbol{\xi})$ being the spectrum of $\hat{\psi}$.

If $\hat{\psi}$ is wide on a wavelength scale then the beam propagators $\hat{\Psi}_\mu$ behave like collimated beams whose propagation axes $\mathring{\boldsymbol{\kappa}}_\mu$ emerge from the points $\mathbf{x_m}$ in the plane $z = 0$ in the directions $(\theta_\mathbf{n}, \phi_\mathbf{n})$ such that $\sin\theta_\mathbf{n}(\cos\phi_\mathbf{n}, \sin\phi_\mathbf{n}) = \boldsymbol{\xi}_\mathbf{n}$ (dashed arrows in Figure 23.1(b)). Propagating beams are obtained only if $|\boldsymbol{\xi}_\mathbf{n}| \lesssim 1 - \Delta_\xi$ where Δ_ξ is the spectral width of $\hat{\psi}$ (note: $\Delta_\xi \ll 1$ for collimated beams).

In this work, we are not interested in the field near the aperture, but rather in the propagating field at ranges larger than a wavelength away from the aperture. In these ranges, the propagating field has no evanescent spectrum and the summation in (23.6) involves only the subset of "propagating beams," tagged the index subset $\boldsymbol{\mu}_{\mathscr{P}}$.

23.2.2.3 UWB considerations

For UWB applications, it is desirable to have *the same beam lattice* $(\bar{x}, \bar{\xi})$ *for all* ω (see item (a) in the discussion of Section 23.2.1). In view of (23.4), this implies that ν is a linear function of ω: $\nu(\omega) = \omega\,\frac{\nu_{\max}}{\omega_{\max}}$, where $\nu_{\max}$ is the value of ν at $\omega_{\max}$. It follows that $\nu(\omega) < \nu_{\max}$ for all $\omega \in \Omega$. Following the discussion after (23.5), we typically choose $\nu_{\max} \approx 1/3$.

Another important ingredient in this theory is the use of the "ID-GB" propagators, defined by the mother window

$$\hat{\psi}_{\text{ID}}(\mathbf{x}, \omega) = e^{-k|\mathbf{x}|^2/2b} \tag{23.7}$$

where the parameter b is the *collimation* length of the resulting GB, and is taken to be ω-independent, hence the ID designation. The ID-GB propagators have many favorable properties including the following:

i. They can be tracked analytically in inhomogeneous medium.
ii. Choosing $b = \bar{x}/\bar{\xi}$ provides the *snuggest* expansion for all $\omega \in \Omega$, with stable expansion coefficients [46, Eq. (33)].
iii. Their propagation parameters are ω-independent, hence they need to be calculated only once and then used for all ω [51].
iv. In view of item (iii), one may derive closed form expressions for the TD propagators in terms of the so-called *iso-diffracting pulsed beams (ID-PBs)* [23,51]. As a result, the entire expansion of (23.6) can be formulated directly in the TD [48] (Section 23.2.3).

A closed-form expression for the ID-GB $\hat{\Psi}_\mu(\mathbf{r})$ in (23.6) is given in [46, Sec. IV.A], along with criteria for choosing the expansion parameters b, $\nu_{\max}$ and $(\bar{x}, \bar{\xi})$ for a given physical problem. Recalling the discussion after (23.5), regarding the choice of the proper expansion-parameter set, we approximate

$$\hat{\varphi}_{\mathrm{ID}}(\mathbf{x}, \omega) \simeq \nu^2 \frac{k}{\pi b} \hat{\psi}_{\mathrm{ID}}(\mathbf{x}, \omega) = \frac{\nu_{\max}^2}{\pi b k_{\max}^2} k^3 \, \hat{\psi}_{\mathrm{ID}}(\mathbf{x}, \omega) \tag{23.8}$$

23.2.2.4 The beam frames

Next we consider the set of the *forward propagating* beams $\{\hat{\Psi}^+_\mu(\mathbf{r}, \omega)\}_{\mu \in \mathcal{P}}$ of (23.6), depicted as dashed arrows in Figure 23.2(a). Note that the definition in (23.6) applies for both positive and negative z. Likewise we define the *backward propagating* beams $\{\hat{\Psi}^-_\mu(\mathbf{r}, \omega)\}_{\mu \in \mathcal{P}}$ in Figure 23.2(b), which are given by the same spectral integrals in (23.6) but with $\zeta \to -\zeta$. As shown in Figure 23.2, both sets are structured upon the same beam lattice, converging at the $z = 0$ plane to the conventional WFT frame set which is structured upon a Cartesian phase-space lattice.

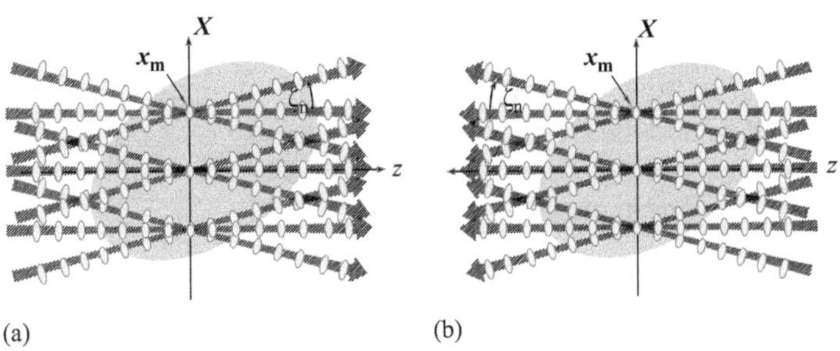

Figure 23.2 The forward and backward propagating BFs (dashed arrows) for frequency domain representations, and PBFs (ellipses along the beam axes). (a) $\Psi^+_{\mu,s}$. (b) $\Psi^-_{\mu,s}$

The following BF Theorem has been proved in [49, Theorem 1]: *The set* $\{\hat{\Psi}^{+}_{\mu}\}_{\mu_{\mathscr{P}}}$ *constitutes a frame at any* $z = const$ *plane over the Hilbert subspace* $\mathbb{L}^2_{\mathscr{P}}(\mathbf{x})$, *with* $\{\hat{\Phi}^{+}_{\mu}\}_{\mu_{\mathscr{P}}}$ *being the "canonical dual frame."* A similar statement applies also for the set $\{\hat{\Psi}^{-}_{\mu}\}_{\mu_{\mathscr{P}}}$ depicted in Figure 23.2(b).

Here $\mathbb{L}^2_{\mathscr{P}}(\mathbf{x})$ is the Hilbert subspace of functions $f(\mathbf{x})$ with *propagating spectrum only* discussed after (23.6), i.e., whose spectrum vanishes for $|\boldsymbol{\xi}| > 1$. $\hat{\Phi}^{\pm}_{\mu}$ are the beam-waves corresponding to $\hat{\varphi}_{\mu}(\mathbf{x})$ at $z = 0$. They are given by the same expression for $\hat{\Psi}^{\pm}_{\mu}$ as in (23.6), but with the mother window $\hat{\psi}(\mathbf{x})$ replaced by the canonical dual window $\hat{\varphi}(\mathbf{x})$. Choosing the expansion parameters as discussed after (23.5) implies that $\hat{\varphi}(\mathbf{x}, \omega)$ is proportional to $\hat{\psi}(\mathbf{x}, \omega)$, for all ω, hence $\hat{\Phi}^{+}_{\mu}(\mathbf{r}, \omega)$ are also proportional to $\hat{\Psi}^{+}_{\mu}(\mathbf{r}, \omega)$.

Henceforth, we refer to the sets $\{\hat{\Psi}^{\pm}_{\mu}\}$ as the BFs. Referring to Figure 23.2, they are structured upon the phase-space skeleton $(\mathbf{x_m}, \boldsymbol{\xi}_\mathbf{n})$ and extend the conventional WFT-frame from the $z = 0$ plane to any $z = const$ plane in the propagation domain, while reducing to the conventional WFT at $z = 0$.

In view of the BF Theorem, the beam expansion of the field $\hat{u}$ is expressed as in (23.6), but the expansion coefficients can now be calculated from the data at any given $z' = const$ plane, not only on the plane $z = 0$ as in (23.5). Assuming that we are given a *forward propagating* wave $u^{+}(\mathbf{r})$ on some plane $z' = const$, then it may be expressed everywhere as

$$\hat{u}^{+}(\mathbf{r}, \omega) = \sum_{\mu_{\mathscr{P}}} \hat{A}^{+}_{\mu}(\omega)\hat{\Psi}^{+}_{\mu}(\mathbf{r}, \omega), \quad \hat{A}^{+}_{\mu}(\omega) = \langle \hat{u}^{+}(\mathbf{x}, z'), \hat{\Phi}^{+}_{\mu}(\mathbf{x}, z')\rangle\big|_{z'=const} \tag{23.9}$$

It can be shown that *the coefficient set* $\{\hat{A}^{+}_{\mu}\}$ *is independent of* z' *and equals the set* $\{\hat{a}^{+}_{\mu}\}$ *of* (23.5) *calculated at* $z = 0$ (as long as the domain between these planes is source free) [49, Theorem 2].

A similar statement applies for backward propagating wave $u^{-}(\mathbf{r})$, expanded using the backward propagating BF $\{\hat{\Psi}^{-}_{\mu}(\mathbf{r})\}_{\mu_{\mathscr{P}}}$.

The main significance and advantage of the BF theory is that it can be used to decompose the field due to sources that are defined on any given plane z' using the *same* beam set $\{\hat{\Psi}^{\pm}_{\mu}\}_{\mu_{\mathscr{P}}}$. It can be used, for example, to expand the field generated by a volume-source distribution, real or induced [49, Sec. VI], as a local-spectrum alternative to the conventional plane-wave spectrum (or K-domain) approach. As such, this approach has been used in [49,50] for a beam-based solution of wave propagation through rough media.

23.2.3 *The phase-space pulsed BS method: a TD formulation*

23.2.3.1 The plan-wave spectrum in the TD

As noted in comment (iv) after (23.7), the UWB-PS-BS formulation can be expressed directly in the TD. The theory has been derived in [48] in the context of radiation from a time-dependent aperture distribution $u_0(\mathbf{x}, t)$, band limited in $\Omega = [\omega_{\min}, \omega_{\max}]$, and is extended here within the BF formulation. We start with a brief review of this theory.

We define the TD plane-wave spectrum of $u_0(\mathbf{x}, t)$ as the TD counterpart of the time harmonic spectrum $\hat{\tilde{u}}_0(\boldsymbol{\xi})$ of (23.2), viz

$$\tilde{u}_0(\boldsymbol{\xi}, \tau) = \frac{1}{2\pi} \int_{-\infty}^{\infty} d\omega\, e^{-i\omega\tau} \hat{\tilde{u}}_0(\boldsymbol{\xi}; \omega) \tag{23.10}$$

Substituting $\hat{\tilde{u}}_0(\boldsymbol{\xi})$ from (23.2) and inverting the order of integrations, we calculate $\tilde{u}_0$ directly from the TD data via

$$\tilde{u}_0(\boldsymbol{\xi}, \tau) = \int_{-\infty}^{\infty} d^2x\, u_0(\mathbf{x}; \tau + v_0^{-1}\boldsymbol{\xi}) \tag{23.11}$$

Equation (23.11) is recognized as Radon transform of u_0 in the 3D $(\mathbf{x}, t)$ data domain, referred to as the slant stack transform (SST), consisting of projections along the slanted delay surfaces $t - v_0^{-1}\boldsymbol{\xi} \cdot \mathbf{x} = \tau = const$, forming an angle $\tan\boldsymbol{\theta} = \boldsymbol{\xi}$ with respect to the $\mathbf{x}$ coordinates (see Figure 23.3(a)). Consequently it extracts from $u_0(\mathbf{x}; t)$ the signal $\tilde{u}_0(\boldsymbol{\xi}, \tau)$ which is the time-dependent plane wave that propagates in the direction $\mathring{\boldsymbol{\kappa}} = (\boldsymbol{\xi}, \zeta)$, as depicted in Figure 23.3(b). Referring to Figure 23.3(a), for $|\boldsymbol{\xi}| < 1$ (the propagating spectrum) the slant of the projection is small such that the normal to the slanted plane is confined within a $\pi/4$ cone about the $v_0 t$-axis (the light cone), whereas for $|\boldsymbol{\xi}| > 1$ the slant is larger.

Likewise, the propagating field is obtained by inverting the plane-wave integral in (23.1) into the TD. As discussed after (23.6), we are interested here only with the

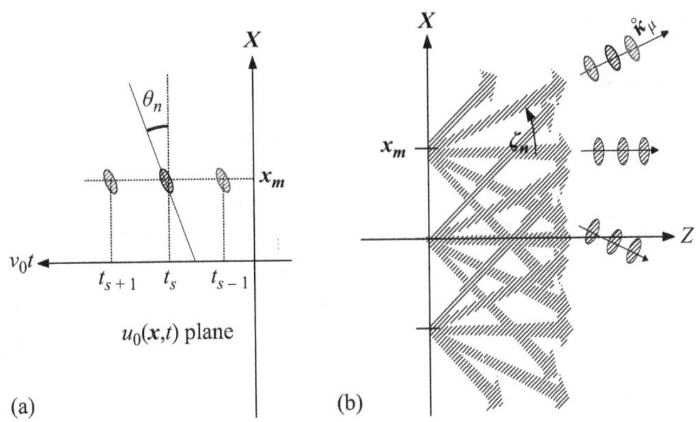

Figure 23.3 The SST and the WRT frame. (a) The data domain $u_0(\mathbf{x}, t)$: The full line represents the SST integration in (23.10), and the ellipses represent the WRT frame elements $\varphi_{\mu,s}$ in (23.14b), with spectral tilt $\boldsymbol{\xi}_n$. (b) The propagated field $u(\mathbf{r}, t)$ described as a sum of PB propagators (small ellipses) that propagate along the beam axes $\mathring{\boldsymbol{\kappa}}_\mu$ (dashed arrows), emerging at a discrete set of points and directions $(\mathbf{x}_m, \boldsymbol{\xi}_n)$ in the plane $z = 0$ at times t_s

propagating field and not the evanescent spectrum, defined by the spectral regime $|\boldsymbol{\xi}| < 1$. Thus inverting (23.1) to the TD and using (23.10)–(23.11) we obtain, noting that ζ is real here,¶

$$u_{\mathscr{P}}(\mathbf{r}, t) = -(2\pi v_0)^{-2} \int_{\mathscr{P}} d^2\xi \; \partial_t^2 \tilde{u}_0[\boldsymbol{\xi}, t - v_0^{-1}(\boldsymbol{\xi} \cdot \mathbf{x} + \zeta z)] \tag{23.12}$$

where $\mathscr{P}$ indicates that the integration spans only the propagating spectrum $|\boldsymbol{\xi}| < 1$. Equation (23.12) expresses the radiated field as a superposition of time-dependent plane waves propagating in the $\mathring{\boldsymbol{\kappa}} = (\boldsymbol{\xi}, \zeta)$ directions as schematized in Figure 23.3(a).

23.2.3.2 The WRT frame

We start with the window functions $(\hat{\psi}(\mathbf{x}, \omega), \hat{\varphi}(\mathbf{x}, \omega))$ defined for all $\omega \in \Omega$ to be a dual pair with respect to the *frequency independent* lattice $(\bar{x}, \bar{\xi})$. Outside this frequency band, they may be chosen arbitrarily, and the considerations are typically the smoothness and localization of the resulting TD windows (see examples for choosing these pulses in [48, Sec. IV-C]; see also discussion after (23.16)). The corresponding space-time "mother" and "dual" windows are defined now via

$$\left.\begin{matrix}\psi(\mathbf{x}, t) \\ \varphi(\mathbf{x}, t)\end{matrix}\right\} = \mathrm{Re}\,\frac{1}{\pi}\int_0^\infty d\omega\; e^{-i\omega t}\left\{\begin{matrix}\hat{f}(\omega)\hat{\psi}(\mathbf{x}, \omega) \\ \hat{g}(\omega)\hat{\varphi}(\mathbf{x}, \omega)\end{matrix}\right\} \tag{23.13}$$

where $\hat{f}, \hat{g}$ are introduced to add flexibility to the formulation. They are constrained by $\hat{f}(\omega)\hat{g}^*(\omega) = 1 \; \forall \omega \in \Omega$, in order to guarantee duality there. These functions are chosen such that they yield localized pulses when transformed to the TD [48, Eq. (33)]. Specific expressions for these window functions are discussed next.

Using these function, we define the function-sets $\{\psi_{\mu,s}(\mathbf{x}, t)\}$ and $\{\varphi_{\mu,s}(\mathbf{x}, t)\}$:

$$\psi_{\mu,s}(\mathbf{x}, t) = \psi[\mathbf{x} - \mathbf{x_m},\; t - t_s - v_0^{-1}\boldsymbol{\xi}_\mathbf{n} \cdot (\mathbf{x} - \mathbf{x_m})] \tag{23.14a}$$

$$\varphi_{\mu,s}(\mathbf{x}, t) = \bar{t}\varphi[\mathbf{x} - \mathbf{x_m},\; t - t_s - v_0^{-1}\boldsymbol{\xi}_n \cdot (\mathbf{x} - \mathbf{x_m})] \tag{23.14b}$$

where $(\boldsymbol{\mu}, s) = (\mathbf{m}, \mathbf{n}, s) \in \mathbb{Z}^5$ is an index, $t_s = s\bar{t}$, and $\bar{t} \leq \pi/\omega_{\max}$ is chosen based on the Nyquist sampling condition of the data. Note that $\psi_{\mu,s}$ and $\varphi_{\mu,s}$ are the inverse Fourier transform of $\hat{f}\,\hat{\psi}_\mu$ and $\hat{g}\,\hat{\varphi}_\mu$ to the time coordinate $t - t_s$.

As follows from the space and time shifts in (23.14), these functions are localized about the space-time lattice $(\mathbf{x}_m, t_s)$ with a spectral slant $\boldsymbol{\xi}_n$, as schematized in Figure 23.3(a).

The following windowed Radon transform (WRT) frame Theorem has been established in [48, Appx. B]: *The sets* $\{\psi_{\mu,s}\}$ *and* $\{\varphi_{\mu,s}\}$ *constitute dual frame sets in the* $(\mathbf{x}, t)$ *domain above the Hilbert space* $\mathbb{L}^2_\Omega$ *of band limited functions, henceforth referred to as the WRT frames.*

¶The formulation may also be extended to the evanescent spectrum using analytic signal formulation, but we do not need this extension in the present context.

The aperture distribution $u_0(\mathbf{x}, t) \in \mathbb{L}_2^{\Omega}$ can therefore be expanded as

$$u_0(\mathbf{x}, t) = \sum_{\mu,s} a_{\mu,s}\, \psi_{\mu,s}(\mathbf{x}, t), \quad a_{\mu,s} = \langle u_0(\mathbf{x}, t),\, \varphi_{\mu,s}(\mathbf{x}, t)\rangle_{(\mathbf{x},t)} \tag{23.15}$$

In view of the slanted structure of $\varphi_{\mu,s}$, the inner product in (23.15) is identified as a WRT (or a local SST) of u_0 in the 3D $(\mathbf{x}, t)$ domain, as schematized in Figure 23.3(a).

23.2.3.3 Phase-space pulsed BS representation for the field

Equation (23.15) can be propagated to the region $z > 0$ giving

$$u(\mathbf{r}, t) = \sum_{\mu,s} a_{\mu,s} \Psi_{\mu,s}(\mathbf{r}, t) \tag{23.16}$$

where the propagators $\Psi_{\mu,s}(\mathbf{r}, t)$ are the fields due to the aperture distribution $\psi_{\mu,s}(\mathbf{x}, t)$ at $z = 0$ (a general spectral integral representation for $\Psi_{\mu,s}$ is given in [48, Eq.(17)], but it is irrelevant in the present context). In view of the properties of $\psi_{\mu,s}(\mathbf{x}, t)$, discussed after (23.14), the propagators $\Psi_{\mu,s}(\mathbf{r}, t)$ follow the $(\mathbf{x_m}, \boldsymbol{\xi}_\mathbf{n})$ beam axis, and behave like collimated pulsed beams (space-time wavepackets) as schematized by the ellipses in Figure 23.3(b).

As discussed after (23.7), in practice we use the ID window because of its favorable UWB properties. Closed-form expressions for ID windows $\psi(\mathbf{x}, t)$ and $\varphi(\mathbf{x}, t)$ are derived in [48, Eqs. (25) and (28)], respectively, where f and g are the smooth pulses defined in [48, Eqs. (33)] (See also Figs. 5 and 6 there).

Closed-form expressions for the resulting ID-PB propagators $\Psi_{\mu,s}(\mathbf{r}, t)$ are given in [48, Eq. (29)]. They are *dispersion-less space-time wavepackets* that propagate along the $(\mathbf{x_m}, \boldsymbol{\xi}_\mathbf{n})$ beam axis as schematized in Figure 23.3(b). Note that these ID-PB may be tracked analytically even in an inhomogeneous medium [51].

23.2.3.4 The pulsed beam frames

In view of the UWB construction of the BFs outlined in Section 23.2.2.4, this formulation may readily be extended to the TD [22]. We start by introducing the sets of forward and backward propagating PBs $\{\Psi^{\pm}_{\mu,s}(\mathbf{r}, t)\}_{(\mu_{\mathscr{P}},s)}$ with $\mu_{\mathscr{P}}$ denoting the subset of "propagating beams," i.e., those with indices $|\boldsymbol{\xi}_\mathbf{n}| < 1$ as defined after (23.6). These PBs, schematized by the ellipses in Figure 23.2, are obtained by forward/backward propagation of the window functions $\{\psi_{\mu,s}(\mathbf{x}, t)\}_{(\mu_{\mathscr{P}},s)}$ of (23.14a) which are defined in the plane $z = 0$. Likewise we introduce the sets $\{\Phi^{\pm}_{\mu,s}(\mathbf{r}, t)\}_{(\mu_{\mathscr{P}},s)}$ obtained by forward/backward propagation of the window functions $\{\varphi_{\mu,s}(\mathbf{x}, t)\}_{(\mu_{\mathscr{P}},s)}$.

This leads to the following pulsed-beam frame (PBF) Theorem: *The sets $\{\Psi^{\pm}_{\mu,s}(\mathbf{r}, t)\}_{(\mu_{\mathscr{P}},s)}$ constitute frames at any $z = const$ plane over the Hilbert subspace $\mathbb{L}^2_{\Omega,\mathscr{P}}(\mathbf{x}, t)$ of functions band-limited in Ω with spatial spectrum band limited in $|\boldsymbol{\xi}| < 1$, with $\{\Phi^{\pm}_{\mu,s}(\mathbf{r}, t)\}_{(\mu_{\mathscr{P}},s)}$ being the canonical dual frames, respectively.*

Thus, given a forward or backward propagating wave $u^{\pm}(\mathbf{r},t)|_{z'}$ at the z'-plane, it may be described at any other z-value by (cf. (23.9))

$$u^{\pm}(\mathbf{r},t) = \sum_{\mu \mathscr{P} s} A^{\pm}_{\mu,s}\, \Psi^{\pm}_{\mu,s}(\mathbf{r},t), \quad A^{\pm}_{\mu,s} = \langle u^{\pm}(\mathbf{r},t),\, \Phi_{\mu,s}(\mathbf{r},t)\rangle_{(\mathbf{x},t)}\Big|_{z'=const} \tag{23.17}$$

As in (23.9), it can be shown that *the coefficient-set* $\{A^{\pm}_{\mu,s}\}$ *is independent of* z' *and equals the set* $\{a^{\pm}_{\mu,s}\}$ *of* (23.15) *calculated via the WRT frame at* $z=0$ (the Coefficient Independence Theorem).

23.3 UWB tomographic inverse scattering

23.3.1 *Tomographic inverse scattering: frequency domain formulation*

In the following sections, we apply the local spectrum and the BS formulations discussed above for inverse scattering. We start however with a brief review of DT with emphasis on UWB-DT and time-domain diffraction tomography (i.e., TD-DT).

In general, inverse scattering problems are nonlinear, hence exact numerical solutions typically require iterative schemes and are limited to rather small configurations. For large domains, however, practical algorithms rely on linearized weak scattering formulations using single scattering approximations such as the Born, Rytov, or physical optics approximation (see [20,21] and references therein). These approximations lead to a direct relation between the properties (shape or composition) of the scattering object and the plane-wave spectrum of the scattered field. This class of formulations is referred to as DT [20]. Here we shall consider only the first-order Born approximation, since it is more transparent for physical interpretation, but the same local algorithm may be applied for the Rytov approximation.

23.3.1.1 Problem statement

Referring to Figure 23.4(a), we consider a lossless, nondispersive medium with an inhomogeneous wave speed $v(\mathbf{r})$ in a 3D coordinate frame, which is embedded in a uniform medium v_0. The medium is characterized by the "object function"

$$O(\mathbf{r}) = n^2(\mathbf{r}) - 1, \quad n(\mathbf{r}) = v_0/v(\mathbf{r}) = \text{refractive index} \tag{23.18}$$

The unknown object is assumed to be located between two measurements planes at $z=z_1<0$ and $z=z_2>0$ planes. It is illuminated by time-dependent plane wave

$$u^i(\mathbf{r},t) = F(t - v_0^{-1}\mathring{\boldsymbol{\kappa}}^i\cdot\mathbf{r}), \quad \mathring{\boldsymbol{\kappa}}^i \stackrel{\text{def}}{=} (\boldsymbol{\xi}^i,\zeta^i) = (\sin\theta^i\cos\phi^i, \sin\theta^i\sin\phi^i, \cos\theta^i) \tag{23.19}$$

where $F(t)$ is a short pulse, whose width determines the resolution of the measurement, and the unit vector $\mathring{\boldsymbol{\kappa}}^i$ defines the plane-wave direction. The scattered field measured over these planes is denoted by $u_j(\mathbf{r},t)$, where here and henceforth we tag the fields corresponding to the measurement planes 1 and 2 by the subscript $j=1,2$, respectively

The scalar field $u(\mathbf{r},t)$ satisfies the wave equation

$$\left[\nabla^2 - n^2(\mathbf{r})v_0^{-2}\,\partial_t^2\right]u(\mathbf{r},t) = 0 \tag{23.20}$$

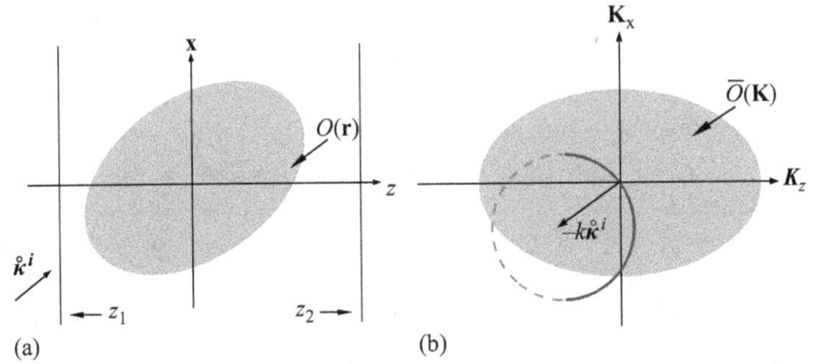

Figure 23.4 DT in the spatial and spectral domains. (a) The object is illuminated by a plane wave (23.21) propagating in the direction $\mathring{\kappa}^i$. (b) K− space representation of (23.29). The scattered field $\hat{\tilde{u}}_j(\boldsymbol{\xi})$ is related to $\bar{O}(\mathbf{K})$ on the shifted Ewald sphere $\mathbf{K} = k(\mathring{\kappa}_j - \mathring{\kappa}^i)$. The dashed and solid-line hemispheres correspond to reflection and transmission data measured on the j = 1, 2 planes, respectively

or, expressed in the frequency domain

$$\left[\nabla^2 + k^2 n^2(\mathbf{r})\right] \hat{u}(\mathbf{r}, \omega) = 0, \quad \hat{u}^i(\mathbf{r}, \omega) = e^{ik\mathring{\kappa}^i \cdot \mathbf{r}} \tag{23.21}$$

where $k = \omega / v_0$ and the hat is used to denote frequency domain constituents.

Equation (23.21) can be cast as an integral equation (Lippmann–Schwinger equation)

$$\hat{u}^s(\mathbf{r}) \stackrel{\text{def}}{=} \hat{u}(\mathbf{r}) - \hat{u}^i(\mathbf{r}) = k^2 \int_V d^3 r' \hat{u}(\mathbf{r}') O(\mathbf{r}') \hat{G}(\mathbf{r}, \mathbf{r}') \tag{23.22}$$

where $\hat{G}$, the 3D Green's function of the background medium, is defined in (23.1).

23.3.1.2 The DT identity

The DT identity expresses the direct relation between the scattered data and the target O, obtained under the weak scattering (first-order Born) approximation. In this approximation, the scattered field due to the excitation $\hat{u}^i$ is given by

$$\hat{u}^s(\mathbf{r}) = k^2 \int_V d^3 r' \, \hat{u}^i(\mathbf{r}') O(\mathbf{r}') \hat{G}(\mathbf{r}, \mathbf{r}') \tag{23.23}$$

The approximation is valid if $O_{\max} \ll 1$ and, in addition, $kLO_{\max} < 1$ where L is the support of O and $O_{\max}$ is a bound on O.

We define now the spectrum of the scattered field $\hat{u}_j(\mathbf{x})$ (cf. (23.2)) by

$$\hat{\tilde{u}}_j^s(\boldsymbol{\xi}) = e^{\pm ik\zeta z_j} \int_{-\infty}^{\infty} d^2 x \, \hat{u}_j^s(\mathbf{x}) e^{-ik\boldsymbol{\xi}\cdot x}, \quad \zeta = \sqrt{1 - |\boldsymbol{\xi}|^2}, \quad \operatorname{Im} \zeta \geq 0 \tag{23.24}$$

where $\hat{\tilde{u}}_j^s(\boldsymbol{\xi})$ describe the amplitude of the spectral plane waves propagating away from O in the directions

$$\mathring{\kappa}_j = (\boldsymbol{\xi}, \mp\zeta) = (\sin\theta_j \cos\phi_j, \sin\theta_j \sin\phi_j, \cos\theta_j) \tag{23.25}$$

where $(\theta_j(\boldsymbol{\xi}), \phi_j(\boldsymbol{\xi}))$ are the conventional spherical angles defined with respect to the $+z$-axis, and the upper/lower signs correspond to $j = 1, 2$, respectively. The extra term $e^{\pm ik\zeta z_j}$ in (23.24) is introduced conveniently in order to normalize the spectral plane waves with respect to the plane $z = 0$, instead of with respect to the measurement planes $z = z_{1,2}$ (i.e., their phase centers are at the origin). The spectral ranges $|\boldsymbol{\xi}| < 1$ and $|\boldsymbol{\xi}| > 1$ define, respectively, the propagating and the evanescent spectra, respectively. We note that DT formulations typically involve only the propagation spectrum data.

Next we insert into (23.23) the incident field $\hat{u}^i$ from (23.21), and use also the spectral representation of $\hat{G}$,

$$\hat{G}(\mathbf{r}, \mathbf{r}') = \left(\frac{k}{2\pi}\right)^2 \int_{-\infty}^{\infty} d^2\xi \frac{1}{-2ik\zeta} e^{ik(\boldsymbol{\xi}\cdot(\mathbf{x}-\mathbf{x}')+\zeta|z-z'|)} \tag{23.26}$$

Applying then the transformation in (23.24), we find that

$$\hat{\hat{u}}_j^s(\boldsymbol{\xi}) = \frac{k}{-2i\zeta} \int_V d^3r' O(\mathbf{r}') e^{-ik[(\boldsymbol{\xi}-\boldsymbol{\xi}^i)\cdot\mathbf{x}'+(\mp\zeta-\zeta^i)z']}, \quad |\boldsymbol{\xi}| < 1 \tag{23.27}$$

Defining the K-domain representation of O by

$$\bar{O}(\mathbf{K}) = \int_V d^3r\, O(\mathbf{r}) e^{-i\mathbf{K}\cdot\mathbf{r}}, \qquad \mathbf{K} = (K_1, K_2, K_z) \tag{23.28}$$

we observe that for $|\boldsymbol{\xi}| < 1$, (23.27) can be cast in the form

$$\hat{\hat{u}}_j^s(\boldsymbol{\xi}) \simeq \frac{k}{-2i\zeta} \bar{O}(\mathbf{K})\big|_{\mathbf{K}=k(\mathring{\boldsymbol{\kappa}}_j-\mathring{\boldsymbol{\kappa}}^i)} \tag{23.29}$$

where the unit vectors $\mathring{\boldsymbol{\kappa}}_j$ and $\mathring{\boldsymbol{\kappa}}^i$ are defined in (23.19) and (23.25).

Equation (23.29) is the *basic DT identity*. It states that the amplitude of the scattered plane waves in the $\mathring{\boldsymbol{\kappa}}_j$ direction is determined by the value of $\bar{O}(\mathbf{K})$ at $\mathbf{K} = k(\mathring{\boldsymbol{\kappa}}_j - \mathring{\boldsymbol{\kappa}}^i)$. These points are located on a sphere of radius k centered at $\mathbf{K} = -k\mathring{\boldsymbol{\kappa}}^i$, denoted as the *shifted Ewald sphere* (see Figure 23.4(b)). As follows from the definition of the spectral directions $\mathring{\boldsymbol{\kappa}}_j$ in (23.25), the dashed and the solid-line hemispheres correspond to data measured on the $j = 1, 2$ planes, respectively (i.e., reflection and transmission data). This relation implies that $\bar{O}(\mathbf{K})$ can be recovered by measuring the scattered field $\hat{u}_j^s$ for various values of $\mathring{\boldsymbol{\kappa}}^i$ or ω. It provides the basis for the inversion algorithms discussed next.

Note that (23.29) involves only the propagating part of the spectrum data $|\boldsymbol{\xi}| < 1$. Adding the evanescent spectrum data to improve the reconstruction suffers from several difficulties: (i) the evanescent data is exponentially weak, becoming weaker as the distance of the measurement planes $z_{1,2}$ from the target increases. (ii) Backpropagation of this data implies exponential amplification of this weak data from $z_{1,2}$ to the target domain, thereby amplifying also the noise. (iii) The evanescent spectrum data is controlled by contributions from the parts of the scatterer that are closest to $z_{1,2}$. In this work we shall not use this option.

23.3.1.3 Object reconstruction via angular diversity (monochromatic tomography)

From (23.29) and Figure 23.4(b), O may be reconstructed by measuring the scattered field for several illumination directions $\mathring{\boldsymbol{\kappa}}^i$. Each direction determines the value of $\bar{O}(\mathbf{K})$ on the corresponding shifted Ewald sphere as illustrated in Figure 23.5(a), so that full covering is obtained by sweeping over all $\mathring{\boldsymbol{\kappa}}^i$, and finally $O(\mathbf{r})$ is determined by an inverse Fourier transform of (23.28). This inversion approach is called "spectral reconstruction." Note that it is sufficient to reconstruct $\bar{O}(\mathbf{K})$ only on a half-space since it satisfies the symmetry property $\bar{O}(\mathbf{K}) = \bar{O}^*(-\mathbf{K})$. Further details and references concerning the implementation of this approach can be found in [21, Chap. 9],[20, Chap. 8].

An alternative reconstruction approach, called "filtered backpropagation," circumnavigates the spectral reconstruction of $\bar{O}(\mathbf{K})$ by performing these operations entirely in the configuration domain [52,53]. Here, the field measured for each illumination is multiplied by a spectral filter and then backpropagated to the target domain. Aggregating the backpropagated fields due to all the illuminations yields the reconstructed value of $O(\mathbf{r})$. Further details and references can be found in [20,21].

The angular diversity approach is a generalization of the conventional X-ray tomography to the finite (nonzero) wavelength case. In the X-ray case, the radius of the shifted Ewald sphere in Figure 23.5(a) becomes infinite so that the hemispheres corresponding to the transmitted data (full blue lines in Figure 23.5(a)) reduce to planar cross sectional cuts of $\bar{O}(\mathbf{K})$ through the origin of the K-space normal to the directions of incidence $\mathring{\boldsymbol{\kappa}}^i$. It follows that the results taken over all the illumination directions reconstruct $\bar{O}(\mathbf{K})$. Alternatively, in the zero-wavelength limit the transmitted field due to each illumination direction constitutes a Radon transform of the object so that $O(\mathbf{r})$ may be reconstructed via a "back-projection" algorithm. Both approaches,

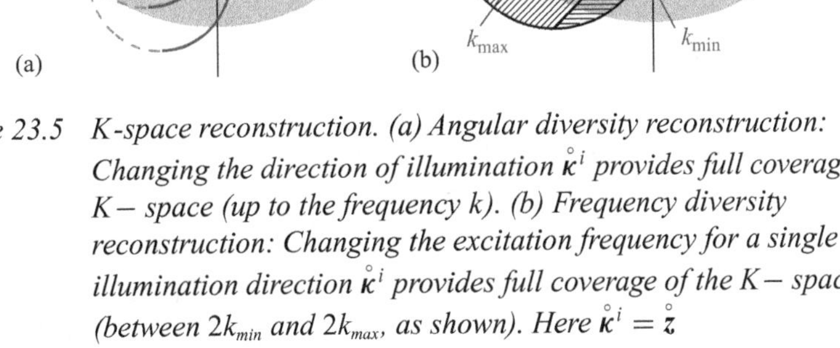

Figure 23.5 *K-space reconstruction. (a) Angular diversity reconstruction: Changing the direction of illumination $\mathring{\boldsymbol{\kappa}}^i$ provides full coverage of the $K-$ space (up to the frequency k). (b) Frequency diversity reconstruction: Changing the excitation frequency for a single illumination direction $\mathring{\boldsymbol{\kappa}}^i$ provides full coverage of the $K-$ space (between $2k_{min}$ and $2k_{max}$, as shown). Here $\mathring{\boldsymbol{\kappa}}^i = \mathring{\mathbf{z}}$*

namely the K-space representation and the Radon transform approach are related via the Fourier slice theorem [54].

23.3.1.4 Object reconstruction via frequency diversity (UWB-DT)

An alternative approach to covering the K-space is to change the excitation frequency, thereby changing the radius of the shifted Ewald sphere of (23.29). This can be done either by performing multi-monochromatic experiments, or by using an ultra-wide-band, short-pulse excitation. The data can then be analyzed in the frequency or TDs.

The frequency domain approach is illustrated in Figure 23.5(b). For simplicity, the interrogating plane waves propagates in the $+\mathring{z}$ direction and the frequency is swept over the range $\Omega = [\omega_{\min}, \omega_{\max}]$. Figure 23.5(b) depicts the K-space coverage by the shifted Ewald spheres as their radii vary between $k_{\min}$ and $k_{\max}$. One readily finds that the reflection data at z_1 (i.e., $j = 1$) covers a $\pi/4$ cone about the $-K_z$ axis that is truncated between $K = k_{\min}$ and $K = k_{\max}$ (diagonal shading in Figure 23.5(b)). The transmission data ($j = 2$) covers a smaller transversal K-space zone (horizontal shadow). Analogous results are obtained for other illumination directions $\mathring{\boldsymbol{\kappa}}^i$.

Following the discussion above, there are two alternative approaches to reconstruct the object $O(\mathbf{r})$. The first is to reconstruct its spectrum $\bar{O}(\mathbf{K})$ via (23.29) as depicted in Figure 23.5(b), and then invert the result to the $\mathbf{r}$-domain via the inverse of the 3D Fourier transform (23.28). We note, again, that it is sufficient to reconstruct $\bar{O}(\mathbf{K})$ only over a half-space since it satisfies the symmetry property $\bar{O}(\mathbf{K}) = \bar{O}^*(-\mathbf{K})$.

An alternative approach to reconstruct the object is to use filtered backpropagation of the UWB data [55]. In this section, this approach is implemented in the frequency domain, whereas in Section 23.3.2 it will be implemented directly in the TD.

The backpropagated field may be calculated via either a Green's function integration or via a plane-wave integral. Here we use conveniently the second approach. As mentioned earlier we backpropagate only the visible spectrum, obtaining

$$\hat{u}_j^b(\mathbf{r}) = \left(\frac{k}{2\pi}\right)^2 \int_{\mathscr{P}} d^2\xi \; \hat{\tilde{u}}_j^s(\boldsymbol{\xi}) e^{ik(\boldsymbol{\xi}\cdot\mathbf{x} \mp \zeta z)} \tag{23.30}$$

where the integration spans only the propagating spectrum $\mathscr{P}$ defined in (23.29).

Next we define the so-called "imaging field" or the "filtered backpropagating field" corresponding to measurements on the $j = 1, 2$ plane by

$$\hat{I}_j(\mathbf{r}, \omega) = 2v_0^{-1} k^{-2} \mathring{\boldsymbol{\kappa}}^i \cdot \nabla[e^{-ik\mathring{\boldsymbol{\kappa}}^i\cdot\mathbf{r}} \hat{u}_j^b(\mathbf{r}, \omega)] \tag{23.31}$$

The reconstructed object, $\breve{O}(\mathbf{r})$, is obtained by summing over all the frequencies, i.e.,

$$\breve{O}(\mathbf{r}) = \sum_j I_j(\mathbf{r}, t)\big|_{t=0} = \sum_j \mathrm{Re}\, \frac{1}{\pi} \int_0^\infty d\omega\, \hat{I}_j(\mathbf{r}, \omega) \tag{23.32}$$

To prove this formula, we assume that the data is described by the Born approximation (23.23) so that $\hat{\tilde{u}}_j(\boldsymbol{\xi})$ are given by (23.27). Substituting into (23.30) and then in (23.31), one ends up with (23.32).

23.3.2 Time-domain diffraction tomography

In Section 23.3.1, it was shown how the UWB data can be used for tomographic reconstruction. If the UWB data is measured in the TD, using an interrogation by pulsed plane waves (from a single or several directions), then it might be preferable to perform the tomographic reconstruction directly in the time domain via the TD-DT derived in [55]. We assume that the pulse $F(t)$ is sufficiently short to provide a proper spectral resolution (i.e., coverage of the K-domain representation as discussed after (23.28)). We shall also present simplified expressions for the case where $F(t) = \delta(t)$.

We start by calculating the TD plane-wave spectra $\tilde{u}_j^s(\boldsymbol{\xi}, \tau)$ (cf. (23.11)),

$$\tilde{u}_j^s(\boldsymbol{\xi}, \tau) = \int_{-\infty}^{\infty} d^2x \, u_j^s[\mathbf{x}, \tau + v_0^{-1}(\boldsymbol{\xi} \cdot \mathbf{x} \mp \zeta z_j)] = \frac{1}{2\pi} \int d\omega \, e^{-i\omega\tau} \hat{\tilde{u}}_j^s(\boldsymbol{\xi}; \omega) \tag{23.33}$$

where the first term expresses the calculations directly in the TD, whereas the second expression is just a restatement of the relation between $\tilde{u}_j^s(\boldsymbol{\xi}, \tau)$ and $\hat{\tilde{u}}_j^s(\boldsymbol{\xi}, \omega)$ of (23.24). The extra term $\mp\zeta z_j$ in (23.33) which does not appear in (23.11) is introduced as in (23.24) to normalize the spectra $\tilde{u}_j$ that are measured on the z_j planes so that they are referenced to the plane $z = 0$.

Within the Born approximation, the scattered transient signal, $\tilde{u}_j^s(\boldsymbol{\xi}, \tau)$ is obtained by multiplying (23.29) by the excitation spectrum $\hat{F}(\omega)$ and substituting into the second expression in (23.33), obtaining

$$\tilde{u}_j(\boldsymbol{\xi}, \tau) = \int_V d^3r \frac{1}{-2v_0\zeta} O(\mathbf{r}) \dot{F}[\tau + v_0^{-1}(\mathring{\boldsymbol{\kappa}}_j - \mathring{\boldsymbol{\kappa}}^i) \cdot \mathbf{r}] \tag{23.34}$$

where $\dot{F}(t) = \partial_t F(t)$. Denoting the angle between the $-\boldsymbol{\kappa}^i$ and $\boldsymbol{\kappa}_j$ as γ (see Figure 23.6), we define

$$\mathring{\mathbf{s}}_j = \frac{\mathring{\boldsymbol{\kappa}}_j - \mathring{\boldsymbol{\kappa}}^i}{2\cos\frac{\gamma}{2}}, \quad p_j = -\frac{v_0\tau}{2\cos\frac{\gamma}{2}} \tag{23.35}$$

$\mathring{\mathbf{s}}_j$ is readily identified as a unit vector that bisects the angle between $-\boldsymbol{\kappa}^i$ and $\boldsymbol{\kappa}_j$ (see Figure 23.6). Using these variables and assuming also $F(t) \simeq \delta(t)$, (23.34) reduces to

$$\tilde{u}_j^s(\boldsymbol{\xi}, \tau) = \frac{v_0 \partial_p}{8\zeta\cos^2\frac{\gamma}{2}} \int_V d^3r O(\mathbf{r}) \delta(p_j - \mathring{\mathbf{s}}_j \cdot \mathbf{r}) = \frac{v_0 \partial_p}{8\zeta\cos^2\frac{\gamma}{2}} \mathscr{R}\{O\}[\mathring{s}_j, p_j] \tag{23.36}$$

where $\mathscr{R}\{O\}$ is the Radon transform of O.

Relation (23.36) implies that the scattered transient signal in a given $\mathring{\boldsymbol{\kappa}}_j$ direction is related to the Radon transform of $O(\mathbf{r})$ along the axis $\mathring{\mathbf{s}}_j$ that bisects the angle γ between $\mathring{\boldsymbol{\kappa}}^i$ and $\mathring{\boldsymbol{\kappa}}_j$ (see Figures 23.6(a) and (b)). The location of the Radon

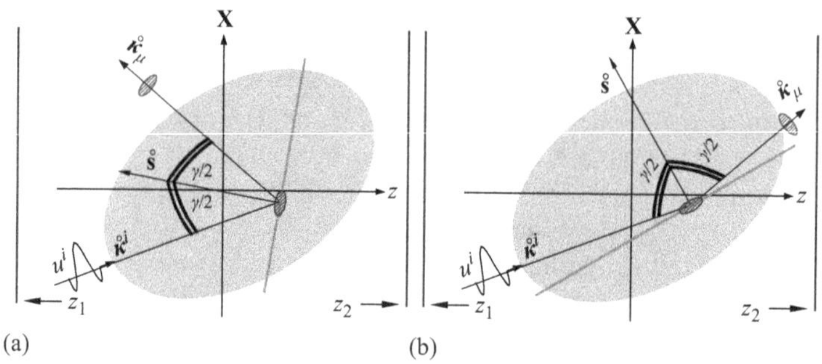

Figure 23.6 *TD-DT: The figure illustrates both the pulsed plane-wave formulation of Section 23.3.2, and the PB formulation of Section 23.4.1. (a): Reflection data at z_1. (b): Transmission data at z_2. In the TD-DT, the scattered field in the $\mathring{\boldsymbol{\kappa}}$ direction is related to the Radon transform of O(**r**) along the axis $\mathring{\mathbf{s}}$ that bisects the angle γ between $\mathring{\boldsymbol{\kappa}}^i$ and $\mathring{\boldsymbol{\kappa}}$, as schematized by the red Radon projection plane. In the PB formulation, the scattered PB along the $\boldsymbol{\mu}$ axis (the green ellipses) is related to the local Radon transform of O(**r**) (the blue ellipses), tilted at the angle γ [55]*

projection plane along this axis, marked red in these figures, is defined by the time t along the reflected signal. This equation is the basic *TD-DT* as it implies that the Radon transform of O, and thereby O, may be reconstructed from the scattering data.

This relation has a *cogent physical interpretation:* it is based on the observation that the Radon transform of O along any given direction $\mathring{\mathbf{s}}$ expresses O as a stratified medium along that axis, so that O is expressed as a collection of layered media along all $\mathring{\mathbf{s}}$ axes. Considering now a given stratification axis $\mathring{\mathbf{s}}$, then the transient field due to the incident pulsed plane wave (23.19) that impinges along the $\mathring{\boldsymbol{\kappa}}^i$ direction, is reflected at an angle that equals the angle of incidence. Note that the interpretation above applies both for data measured on the z_1 or on the z_2 planes, as illustrated in Figure 23.6(a) and (b), respectively. Thus, the TD-DT equation (23.36) may be regarded a generalized Snell's law.

In view of (23.36), the object reconstruction may be formulated as an inverse Radon transform of the $(\boldsymbol{\xi}, \tau)$ data domain. Alternatively, the object can be reconstructed using filtered backpropagation directly in the TD [55]. We first calculate the backpropagated field $u_j^b(\mathbf{r}, t)$ from the data $u_j^s(\mathbf{x}, t)$ measured on the z_j plane (cf. (23.12)) from

$$u_j^b(\mathbf{r}, t) = -(2\pi v_0)^{-2} \int_{\mathscr{P}} d^2\xi \; \partial_t^2 \tilde{u}_j^s[\boldsymbol{\xi}, t - v_0^{-1}(\boldsymbol{\xi} \cdot \mathbf{x} \mp \zeta z)] \tag{23.37}$$

where, as noted above, the integration covers only the propagating spectrum $\mathscr{P}$ defined in (23.29), and we also recall that the plane-wave spectra $\tilde{u}_j^s$ are normalized to the plane $z = 0$ (see comment following (23.33)). Next we define the so-called "imaging field" or the "filtered backpropagating field" corresponding to measurements on the $j = 1, 2$ plane by

$$-\partial_t^2 I_j(\mathbf{r}, t) = 2v_0 \mathring{\boldsymbol{\kappa}}^i \cdot \nabla u_j^b(\mathbf{r}, t + v_0^{-1} \mathring{\boldsymbol{\kappa}}^i \cdot \mathbf{r}), \quad j = 1, 2 \tag{23.38}$$

The reconstructed object $\breve{O}(\mathbf{r})$ is obtained by setting the *imaging condition* $t = 0$, obtaining

$$\breve{O}(\mathbf{r}) = I_1(\mathbf{r}, t)|_{t=0} + I_2(\mathbf{r}, t)|_{t=0} \tag{23.39}$$

One may also consider separately the properties of the partial images $\breve{O}_1$ and $\breve{O}_2$ reconstructed from data $u_{1,2}$, respectively.

23.4 Beam-based TD-DT

As noted in Section 23.2, the BF formulation provides an alternative to the conventional plane-wave spectrum approach, in the frequency or TDs. This approach has been used, for example, in [49,50] in the context of wave propagation through rough media. In this section, it is used to formulate the beam-based DT. The readers are referred, again, to references [2–19], discussed in Section 23.1, regarding other beam-based local inversion schemes.

23.4.1 The beam-domain data

We now use the BF concept introduced in Sections 23.2.2.4 and 23.2.3.4 to address the inverse problem stated in Section 23.3.1.1. We use the filtered backpropagation approach expressed either in the multifrequency or in the TD ((23.31)–(23.32) and (23.38)–(23.39), respectively). Given the scattering data $u_{1,2}^s(\mathbf{x}, t)$ on the $z_{1,2}$ planes, we find from (23.17) that the scattered fields for $z < z_1$ or for $z > z_2$ can be expanded using PB as

$$u_{1,2}^s(\mathbf{r}, t)\Big|_{\substack{z<z_1 \\ z>z_2}} = \sum_{\mu \in \mathscr{P}, s} A_{\mu,s}^{(1,2)} \Psi_{\mu,s}^{\mp}(\mathbf{r}, t), \quad A_{\mu,s}^{(1,2)} = \langle u_{1,2}^s(\mathbf{x}, t),\ \Phi_{\mu,s}^{\mp}(\mathbf{r}, t)\rangle_{(\mathbf{x},t)}\Big|_{z=z_{1,2}} \tag{23.40}$$

where the backward/forward propagators $\Psi_{\mu,s}^{\mp}$ are used for $z < z_1$ and for $z > z_2$, respectively, and the coefficients $A_{\mu,s}^{(1,2)}$ are calculated from the scattering data $u_{1,2}^s$ on the $z_{1,2}$ planes via the inner product $\langle\ \rangle_{(\mathbf{x},t)}$. These coefficients define the *beam-domain data*.

23.4.2 The beam-domain data-object relation within the Born approximation

An analytic model for the beam-domain data $A^{(1,2)}_{\mu,s}$ can be obtained by using the Born approximation for the data

$$u^s(\mathbf{r},t) = -v_0^{-2}\partial_t^2 \int_V d^3r'\, O(\mathbf{r}')\frac{u^i(\mathbf{r}',t-v_0^{-1}|\mathbf{r}-\mathbf{r}'|)}{4\pi|\mathbf{r}-\mathbf{r}'|} \tag{23.41}$$

where u^i is the pulsed plane wave in (23.19). Calculating the coefficients as discussed in Section 23.4.1, one obtains

$$A^{(1,2)}_{\mu,s} = \frac{1}{2v_0\cos\theta_{\mathbf{n}}}\langle O(\mathbf{r}), \dot{\Phi}^{\mp}_{\mu,s}(\mathbf{r},t)\rangle\big|_{t=t^i=v_0^{-1}\mathring{\boldsymbol{\kappa}}^i\cdot\mathbf{r}} \tag{23.42}$$

where $\dot{\Phi} \overset{\text{def}}{=} \partial_t\Phi$, $\cos\theta_{\mathbf{n}} = \sqrt{1-|\boldsymbol{\xi}_{\mathbf{n}}|}$ and $t^i = v_0^{-1}\mathring{\boldsymbol{\kappa}}^i\cdot\mathbf{r}$ is the wavefront time of arrival at $\mathbf{r}$. We refer to (23.42) as the "beam-domain data-object relation," as it provides, within the Born model, a direct relation between the object and the amplitudes $A^{(1,2)}_{\mu,s}$ of the scattered field. From (23.42), this operation is identified as a projection of $O(\mathbf{r})$ on the window $\dot{\Phi}^{\mp}_{\mu,s}(\mathbf{r}, t = v_0^{-1}\mathring{\boldsymbol{\kappa}}^i\cdot\mathbf{r})$ as it traverses the object. Referring to the discussion after [48, Eq. (29)], $\Phi_{\mu,s}$ is localized along the beam axis around

$$\sigma_\mu = v_0(t^i - t_s) \tag{23.43}$$

where σ_μ denotes the beam axis. To clarify further the shape of this window we refer to (23.14) and note that the propagators $\Phi^{\mp}_{\mu,s}(\mathbf{r},t)$ are localized in space around $t - t_s - v_0^{-1}(\mathring{\boldsymbol{\kappa}}_\mu\cdot\mathbf{r} - \boldsymbol{\xi}_{\mathbf{n}}\cdot\mathbf{x}_{\mathbf{m}}) = 0$. Substituting the imaging time t^i from (23.42) we obtain $v_0^{-1}(\mathring{\boldsymbol{\kappa}}^i - \mathring{\boldsymbol{\kappa}}_\mu)\cdot\mathbf{r} + v_0^{-1}\boldsymbol{\xi}_{\mathbf{n}}\cdot\mathbf{x}_{\mathbf{m}} = t_s$. This expression can be cast in the form

$$-2\cos\frac{\gamma_{\mathbf{n}}}{2}\mathring{\mathbf{s}}\cdot\mathbf{r} + \boldsymbol{\xi}_{\mathbf{n}}\cdot\mathbf{x}_{\mathbf{m}} = v_0 t_s \tag{23.44}$$

It follows that the sampling windows are localized along the beam axis, and are orthogonal to the direction $\mathring{\mathbf{s}}$ that bisects the angle $\gamma_{\mathbf{n}}$ between $\mathring{\boldsymbol{\kappa}}^i$ and the scattered PB direction $\mathring{\boldsymbol{\kappa}}_\mu$. Its location along this axis is determined by the time parameter t_s via (23.43). This operation is therefore identified as a local Radon transform of $O(\mathbf{r})$, as schematized by the red ellipses in Figure 23.6, being a localized version of the TD-DT identity discussed in (23.36). The validity of this local model for the data is verified by the results discussed in Section 23.4.4 and shown in Figure 23.9.

23.4.3 Backpropagation and local reconstruction of $O(\mathbf{r})$

Equation (23.40) expresses the scattered field u^s for $z < z_1$ and for $z > z_2$. The backpropagated fields $u^b_{1,2}$ are obtained be extending this expression to $z > z_1$ and $z < z_2$ viz

$$u^b_{1,2}(\mathbf{r},t)\Big|_{\substack{z>z_1\\ z<z_2}} = \sum_{\mu\,\mathscr{P},s} A^{(1,2)}_{\mu,s}\Psi^{\mp}_{\mu,s}(\mathbf{r},t) \tag{23.45}$$

where the coefficients $A^{(1,2)}_{\mu,s}$ (the beam-domain data) are calculated from the scattered field via (23.40). Inserting (23.45) into (23.38), the imaging fields are given as a sum of PB propagators

$$I_{1,2}(\mathbf{r},t) = -4 \sum_{\mu,\mathscr{P},s} A^{(1,2)}_{\mu,s} \cos^2 \frac{\gamma}{2} \partial_t^{-1} \Psi^{\mp}_{\mu,s}(\mathbf{r}, t + v_0^{-1} \boldsymbol{\kappa}^i \cdot \mathbf{r}) \tag{23.46}$$

Reconstruction is achieved by applying the imaging condition $t = 0$ in (23.39). With this condition, one may verify that the imaging kernels in (23.46) are localized around the sampling windows centers of (23.42). Following the discussion in (23.43)–(23.44), the local PB backpropagation can be regarded as an inverse local Radon transform of the object, thus providing local reconstruction of the object.

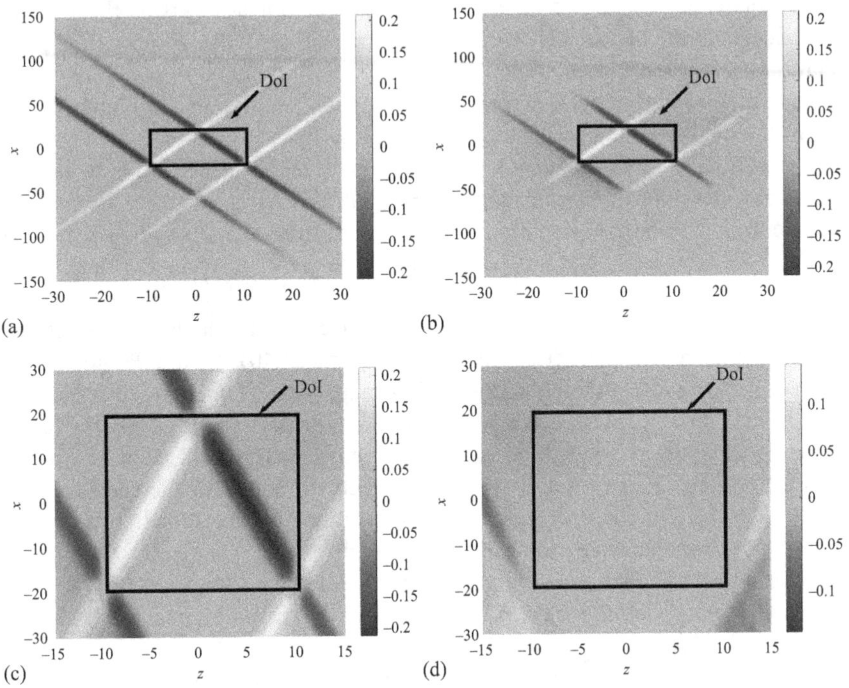

Figure 23.7 Numerical example: (a) The original and (b) reconstructed objects. (c) Zoom of Figure (b) on the reconstructed DoI. (d) Reconstruction error, i.e., difference between the original and reconstructed object. Note that the reconstruction is extremely good inside the DoI, whereas outside the DoI the reconstructed object vanishes due to the localization of the beam-based inversion

23.4.4 Numerical examples

We consider a 2*D* problem with $\mathbf{r} = (x, z)$. $O(\mathbf{r})$ consists of four 2D Gaussians, shifted and tilted at $\pm 15°$ as illustrated in Figure 23.7(a). In two of them, O is positive and in the other two, O is negative (yellow and blue in Figure 23.7(a), respectively). They are given by $\pm 0.2\, e^{-x^2/2\sigma_x^2 - z^2/2\sigma_z^2}$, with $\sigma_x = 50\sqrt{2}$ and $\sigma_z = 1$, thus describing essentially narrow layers in the range direction z, and wide in the cross-range direction x. Note that we consider here a relatively large 20% contrast.

We are interested in reconstruction of the DoI $x \in [-20, 20]$ and $z \in [-10, 10]$, marked by a black rectangle. The medium is probed by a normally incident pulsed plane wave, $u^i(\mathbf{r}, t) = F(t - v_0^{-1} z)$, where $F = 1$ inside the frequency band $\Omega \in [\omega_{\min}, \omega_{\max}] = [0.1, 1]$ and 0 outside. The scattered data is measured only on the $z_1 = -100$ plane (reflection data).

The reconstructed object $\breve{O}(\mathbf{r})$ inside the DoI is calculated via (23.39) using the PB backpropagation I_1 of (23.46), where only those PBs passing through the DoI have been retained in the $(\boldsymbol{\mu}, s)$ summation. Figure 23.7(b) depicts the reconstructed object. Clearly the part of the object inside the DoI is well reconstructed, whereas, as expected, the reconstruction gradually vanishes outside the DoI due to the localization of the PB backpropagators (see also the zoomed images in Figure 23.7(c) and (d)).

In order to get a better quantitative comparison, we explore in Figure 23.8(a) and (b) the quality of the reconstruction along the lines $x = 0$ and $x = -10$. Note that the parameter $\bar{x}$ in our beam lattice has been taken to be ~ 6.6 so that the $x = 0$ coincides with the beam axis $x_m = 0$ whereas the line $x = -10$ does not coincide with any beam axis. The resulting reconstruction $\breve{O}(\mathbf{r})$, depicted in red in Figure 23.8, agrees very well with the original $O(\mathbf{r})$ in blue, but only within the DoI.

Further insight into the inversion is gained by exploring the beam-domain data $a^{(1)}_{\mu,s}$ calculated from the scattering data $u_1(\mathbf{x}, t)$ via (23.40). Figure 23.9(a) depicts these coefficients in the three dimensional discrete (m, n, s) domain, corresponding to the spatial shift x_m, the spectral tilt ξ_n and the temporal shift t_s. Only the largest coefficients are shown, and their magnitude is tagged by colors, from the brightest to the darkest. This figure verifies the theoretical model for the coefficients in (23.42). As explained there, the coefficients constitute a local Radon transform sampling of $O(\mathbf{r})$ and therefore are localized in regions where $\theta_n \simeq \pm 30°$, representing a bistatic reflections from the direction of incidence (here $\theta^i = 0$) to θ_n. The (m, s) coordinates correspond to the (x, z) coordinates of the reflection points as explained in (23.43)–(23.44).

Finally, the localization properties of the PB formulation are demonstrated further in Figure 23.9(b), where we depict a reconstruction using only the PB backpropagators with negative n index. These indices correspond to downward scattering PBs, and the corresponding scattering coefficients $a^{(1)}_{\mu,s}$ are those with negative n in Figure 23.9(a). One readily observes that the only Gaussian that is reconstructed is the one with a negative tilt, as expected.

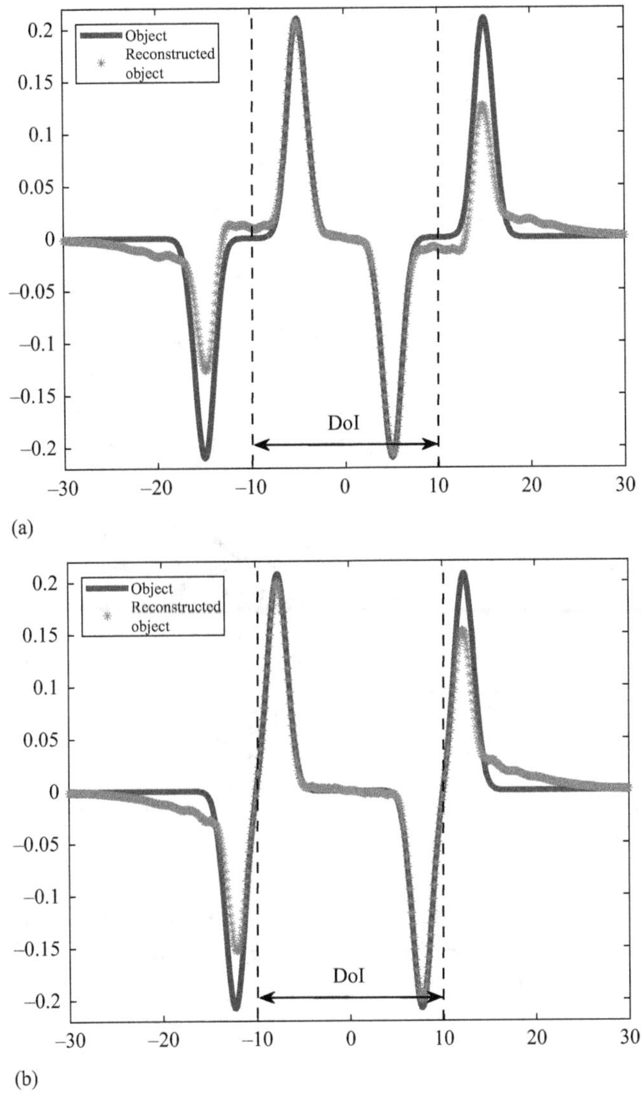

Figure 23.8 Comparison of the reconstructed object $\breve{O}$ (red) with the original (blue) along the lines (a) $x = 0$ and (b) $x = -10$. Note the excellent agreement inside the DoI

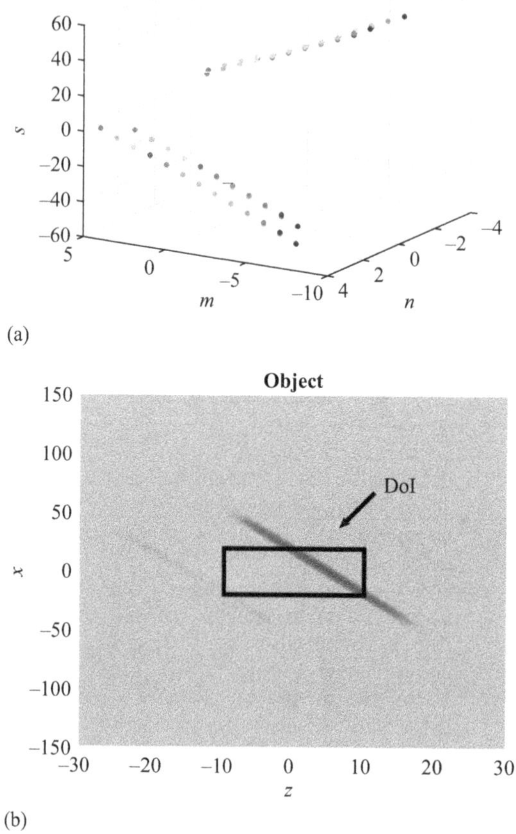

Figure 23.9 *(a) Data-object relation—the dominant $A^{(l)}_{\mu,s}$ coefficients in the discrete 3D (m, n, s) phase space. (b) Reconstruction using only beams with negative n*

23.5 Conclusions

We have considered local UWB-DT inversion of weak scatterers using beam-wave processing. The beam waves were used for both local phase-space processing of the data and for backpropagation and local reconstruction of the object. The beam formulations are based on the UWB-PS-BS method whose main features are the following:

a. The algorithm is based on an a-priori discrete phase space lattice of beams.
b. The beam lattice is *frequency independent*, so that the beam axes are calculated only once and then used for all frequencies in the excitation band.
c. The algorithm utilizes the so-called ID-GB propagators whose propagation parameters are frequency independent, and hence they are calculated only once and then used for all frequencies.
d. The use of properly tuned ID-GB windows provides the *snuggest* frame for *all* the frequencies, leading to smooth and local expansion coefficients.

e. These properties imply that the algorithm may be used either in the UWB multifrequency domain or it can be formulated directly in the TD.
f. For the multifrequency formulation, the processing is based on the WFT frame processing.

The local spectrum processing of the scattering data utilizes the BFs and the PBFs for processing in the multifrequency domain or directly in the TD. They are a generalization of the WFT and the WRT frames, respectively. Likewise, the windowed propagators are the ID-GB and the ID-PB, respectively.

Via the numerical examples in Section 23.4.4, we have demonstrated that the beam-based local processing indeed yields a sparse phase-space representation of the scattering data (see Figure 23.9(a)) that delineates the local DoA and ToA, respectively, and that by using the beam backpropagators we may focus the inversion to obtain local imaging in any desired DoI.

The theory presented her has been studied thoroughly in [56, 57], and has also applied recently for a realistic Geophysical inversion problem in [58].

Acknowledgments

This research is partially supported by the Israeli Science Foundation (ISF) under Grant Nos. 263/11 and 412/15. E.H. would like to thank Prof. George Papanicolaou from Stanford University for his kind hospitality at Stanford during the preparation of this manuscript.

References

[1] A Ishimaru, S Jaruwatanadilok and Y Kuga, "Imaging of a target through random media using a short-pulse focused beam," *IEEE Trans. Antennas Prop.*, vol. 55(6), pp. 1622–1629, 2007.
[2] T Melamed, E Heyman and LB Felsen, "Local spectral analysis of short-pulse-excited scattering from weakly inhomogenous media: Part I—forward scattering," *IEEE Trans. Antennas Propagat.*, vol. 47, pp. 1208–1217, 1999.
[3] T Melamed, E Heyman and LB Felsen, "Local spectral analysis of short-pulse-excited scattering from weakly inhomogeneous media: Part II—inverse scattering," *IEEE Trans. Antennas Propagat.*, vol. 47, pp. 1218–1227, 1999.
[4] T Heilpern, A Shlivinski and E Heyman, "Back-propagation and correlation imaging using phase-space Gaussian-beams processing," *IEEE Trans. Antennas Propagat.*, vol. 61, pp. 5676–5688, 2013.
[5] T Heilpern and E Heyman, "Music imaging using phase-space Gaussian-beams processing," *IEEE Trans. Antennas Propagat.*, vol. 62, pp. 1270–1281, 2014.
[6] NR Hill, "Gaussian beam migration," *Geophysics*, vol. 55, pp. 1416–1428, 1990.
[7] NR Hill, "Prestack Gaussian-beam depth migration," *Geophysics*, vol. 66(4), pp. 1240–1250, 2001.

[8] L Chen, RS Wu and Y Chen, "Target-oriented beamlet migration based on Gabor-Daubechies frame decomposition" *Geophysics*, vol. 71(2), pp. S37–S52, 2006.

[9] RL Nowack, S Dasgupta, GT Schuster and JM Sheng, "Correlation migration using Gaussian beams of scattered teleseismic body waves," *Bull. Seismological Soc. Am.*, vol. 96(1), pp. 1–10, 2006.

[10] RL Nowack, WP Chen and TL Tseng, "Application of Gaussian-beam migration to multiscale imaging of the lithosphere beneath the Hi-CLIMB Array in Tibet," *Bull. Seism. Soc. Am.*, vol. 100(4), pp. 1743–1754, 2010.

[11] N Bleistein and SH Gray, "Amplitude calculations for 3D Gaussian beam migration using complex-valued traveltimes," *Inverse Problems*, vol. 26, article no. 085017, 2010.

[12] K Žáček, "Gaussian packet pre-stack depth migration," *SEG Int'l Exposition and 74th Annual Meeting*, Denver, Colorado, Oct. 2004, Expanded Abstracts, vol. 23, pp. 957–960.

[13] V Bucha, "Gaussian packet prestack depth migration. Part 2: Optimized Gaussian packets," *Seismic Waves in Complex 3-D Structures*, Report 18, pp. 129–138, Dep. Geophys., Charles Univ., Prague, 2008.

[14] MM Popov, NM Semtchenok, AR Verdel and P M Popov, "Seismic migration by Gaussian beams summation," *Doklady Earth Sciences*, vol. 417(8), pp. 1236–1239, 2007.

[15] MM Popov, NM Semtchenok, PM Popov and A R Verdel, "Depth migration by the Gaussian beam summation method," *Geophysics*, vol. 75(2), pp. S81–S93, 2010.

[16] L Klimeš, "Stochastic wavefield inversion using the sensitivity Gaussian packets," *Seismic Waves in Complex 3-D Structures*, Report 18, pp. 71–85, Dep. Geophys., Charles Univ., Prague, 2008

[17] C Hu and P L Stoffa, "Slowness-driven Gaussian-beam prestack depth migration for low-fold seismic data," *Geophysics*, vol. 74(6), WCA35–WCA45, 2009.

[18] MI Protasov and VA Tcheverda, "True amplitude elastic Gaussian beam imaging of multicomponent walkaway vertical seismic profiling data," *Geophysical Prospecting*, vol. 60(6), pp. 1030–1042, 2012.

[19] B Rao and L Carin, "Beam-tracing-based inverse scattering for general aperture antennas," *J. Opt. Soc. Am. A*, vol. 16, pp. 2219–2231, 1999.

[20] AJ Devaney, *Mathematical Foundations of Imaging, Tomography and Wavefield Inversion*, Cambridge University Press, Cambridge, UK, 2012.

[21] KJ Langenberg, "Applied inverse problems for acoustic, electromagnetic and elastic wave scattering," in *Basic Methods of Tomography and Inverse Problems, Malvern Physics Series*, PC Sabatier, Ed. Adam Hilger, Bristol, England, 1987.

[22] R Tuvi, T Melamed and E Heyman, "Propagating beam frame: A novel formulation for local inverse scattering," in *32nd URSI GASS, Montreal, 19–26 August 2017.*

[23] E Heyman and LB Felsen, "Gaussian beam and pulsed-beam dynamics: complex-source and complex-spectrum formulations within and beyond paraxial asymptotics," *J. Opt. Soc. Am. A*, vol. 18, pp. 1588–1611, 2001.

[24] AE Siegman, "Hermite-Gaussian functions of complex argument as optical-beam eigenfunctions," *J. Opt. Soc. Am.*, vol. 63, pp. 1093–1094, 1973.

[25] E Heyman and I Beracha, "Complex multipole pulsed beams and Hermite pulsed beams: A novel expansion scheme for transient radiation from well collimated apertures," *J. Opt. Soc. Am. A*, vol. 9, pp. 1779–1793, 1992.

[26] MM Popov, "A new method of computation of wave fields using Gaussian beams," *Wave Motion*, vol. 4, pp. 85–97, 1982.

[27] V Ĉerveny, MM Popov and I Pŝenĉik, "Computation of wave fields in inhomogeneous media – Gaussian beam approach," *Geophys. J. Roy. Asro. Soc*, vol. 70, pp. 109–128, 1982.

[28] AN Norris, "Complex point-source representation of real sources and the Gaussian beam summation method," *J. Opt. Soc. A.*, vol. 3, pp. 205–210, 1986.

[29] E Heyman, "Complex source pulsed beam expansion of transient radiation," *Wave Motion*, vol. 11, pp. 337–349, 1989.

[30] V Ĉerveny, "Gaussian beam synthetic seismograms," *Journal of Geophysics*, vol. 58, pp. 44–72, 1985.

[31] VM Babich and MM Popov, "Gaussian beam summation method (review)," *Radio Phys. Quant. Electr.*, vol. 32, pp. 1063–1081, 1989.

[32] A Bayliss, BS White, AN Norris and R Burridge, "Some remarks on the Gaussian beam summation method," *Geophys. J.R. Astr. Soc.*, vol. 89, pp. 579–636, 1987.

[33] LB Felsen, IT Lu and YZ Ruan, "Spectral aspects of the Gaussian beam method: Reflection from a homogeneous half space," *Geophys. J.R. Astro. Soc.*, vol. 89, pp. 915–922, 1987.

[34] MB Porter and HP Bucker, "Gaussian beam tracing for computing ocean acoustic fields," *J. Acoust. Soc. Am.*, vol. 82, pp. 1349–1359, 1987.

[35] L Klimeš, "Gaussian packets in the computation of seismic wavefields," *Geophysical Journal International*, vol. 99, pp. 421–433, 1989.

[36] RL Nowack, "Calculation of synthetic seismograms with Gaussian beams," in *Seismic Motion, Lithospheric Structures, Earthquake and Volcanic Sources: The Keiiti Aki Volume*, pp. 487–507. Basel: Birkhäuser, 2003.

[37] G Ariel, B Engquist, NM Tanushev and R Tsai, "Gaussian beam decomposition of high frequency wave fields using expectation–maximization," *Journal of Computational Physics*, vol. 230, pp. 2303–2321, 2011.

[38] T Heilpern, E Heyman, and V Timchenko, "A beam summation algorithm for wave radiation and guidance in stratified media," *Journal Acoustic. Soc. Am.*, vol. 121, pp. 1856–1864, 2007.

[39] TB Hansen and AN Norris, "Exact complex source representations of transient radiation," *Wave Motion*, vol. 26, pp. 101–115, 1997.

[40] Y Gluk and E Heyman, "Pulsed beams expansion algorithms for time-dependent point-source radiation. A basic algorithm and a standard-pulsed-beams algorithm," *IEEE Trans. Antennas Propagat.*, vol. 59, pp. 1356–1371, 2011.

[41] MJ Bastiaans, "The expansion of an optical signal into a discrete set of Gaussian beams," *Optik*, vol. 57, pp. 95–102, 1980.

[42] PD Einziger, S Raz and M Shapira, "Gabor representation and aperture theory," *J. Opt. Soc. Am. A*, vol. 3, pp. 508–522, 1986.

[43] BZ Steinberg, E Heyman and LB Felsen, "Phase space beam summation for time-harmonic radiation from large apertures," *J. Opt. Soc. Am. A*, vol. 8, pp. 41–59, 1991.

[44] JJ Maciel and LB Felsen, "Gaussian beam analysis of propagation from an extended plane aperture distribution through dielectric layers, Part I – plane layer," *IEEE Trans. Antenna Propagat.*, vol. 38, pp. 1107–1617, 1990.

[45] D Lugara and C Letrou, "Alternative to Gabor's representation of plane aperture radiation," *Electronics Letters*, vol. 34, pp. 2286–2287, 1998.

[46] A Shlivinski, E Heyman, A Boag and C Letrou, "A phase-space beam summation formulation for ultra wideband radiation," *IEEE Trans. Antennas Propagat.*, vol. 52, pp. 2042–2056, 2004.

[47] A Shlivinski, E Heyman and A Boag, "A phase-space beam summation formulation for ultrawide-band radiation – Part II: A multiband scheme," *IEEE Trans. Antennas Propagat.*, vol. 53, pp. 948–957, 2005.

[48] A Shlivinski, E Heyman and A Boag, "A pulsed beam summation formulation for short pulse radiation based on windowed radon transform (WRT) frames," *IEEE Trans. Antennas Propagation*, vol. 53, pp. 3030–3048, 2005.

[49] M Leibovich and E Heyman, "Beam summation theory for waves in fluctuating media. Part I: The beam frame and the beam-domain scattering matrix," *IEEE Trans. on Antennas and Propagat.*, vol. 65, pp. 5431–5442, 2017.

[50] M Leibovich and E Heyman, "Beam summation theory for waves in fluctuating media. Part II: Stochastic field representation," *IEEE Trans. on Antennas and Propagat.*, vol. 65, pp. 5443–5452, 2017.

[51] E Heyman, "Pulsed beam propagation in an inhomogeneous medium," *IEEE Trans. on Antennas and Propagat.*, vol. 42, pp. 311–319, 1994.

[52] A J Devaney, "A filtered backpropagation algorithm for diffraction tomography," *Ultrasonic Imaging*, vol. 4, pp. 336–349, 1982.

[53] AJ Devaney, "Diffraction tomography," in *Inverse Methods in Electromagnetic Imaging, Part 2*, WM Boerner *et al.*, Ed., pp. 1107–1135. D. Reidel Publ. Comp., Dordrecht, 1985.

[54] SR Dean, *The Radon Transform and Some of Its Applications*, Wiley-Interscience, New York, 1983.

[55] T Melamed, Y Ehrlich and E Heyman, "Short-pulse inversion of inhomogeneous media: A time-domain diffraction tomography," *Inverse Problems*, vol. 12, pp. 977–993, 1996.

[56] R. Tuvi, E. Heyman, and T. Melamed, "UWB beam-based local diffraction tomography—Part I: Phase-space processing and physical interpretation," *IEEE Trans. Antennas Propag.*, vol. 68, pp. 7144–7157, 2020.

[57] R. Tuvi, E. Heyman, and T. Melamed, "UWB beam-based local diffraction tomography—Part II: The inverse problem," *IEEE Trans. Antennas Propag.*, vol. 68, pp. 7158–7169, 2020.

[58] R. Tuvi, Z. Zhao and M. K. Sen, "Multifrequency beam-based migration in inhomogeneous media using windowed Fourier transform frames," *Geophysical Journal International*, vol. 223, pp. 1086–1099, 2020.

Chapter 24

Modal expansions in dispersive material systems with application to quantum optics and topological photonics

Mário G. Silveirinha[1]

24.1 Introduction

The response of many physical systems is often described by a Hermitian operator. For such systems, the spectral theorem guarantees that an arbitrary state vector can be written in terms of the normal modes of oscillation. Thus, modal expansions are ubiquitous in many branches of physics and engineering. In particular, they enable one to construct formal solutions of many linear partial-differential equations of mathematical physics and to interpret and comprehend more easily the associated physical effects. Here, we are particularly interested in light phenomena. Modal expansions are widely used in such a context, for example, to obtain the solutions of radiation problems in waveguides, cavities or photonic crystals [1–3], to derive analytical expressions for Green's functions [1–3], in quantum optics [4], and in many analytical developments, just to name some of the most common applications.

In the literature, the use of modal expansions is typically restricted to dispersionless structures, i.e., to the case wherein the relevant materials respond instantaneously (without any time delay) to an external excitation. One of the reasons is that from the Kramers–Kronig relations, the material dispersion is typically accompanied by absorption, and for lossy structures the Maxwell equations cannot be reduced to a Hermitian operator problem. Crucially, the difficulties remain even in the absence of (or for vanishingly small) material loss. Indeed, for dispersive systems the material response is written in terms of a convolution in time, and in general it is not obvious how to reformulate the system dynamics in terms of a standard (Schrödinger-type) Hermitian operator problem.

The objective of this chapter is to highlight that for lossless material platforms formed by arbitrary inhomogeneous bianisotropic and possibly nonreciprocal materials, the natural modes of oscillation form, indeed, a complete set of expansion

[1] Instituto Superior Técnico and Instituto de Telecomunicações, University of Lisbon, Lisboa, Portugal

functions. Based on our recent work [5,6], it is proven that the Maxwell equations in dispersive systems can always be reduced to a generalized dynamical problem whose time evolution is described by a Hermitian operator. The effects of material dispersion are taken into account by introducing additional variables that may model the internal degrees of freedom of the material. With such a result, we construct formal expansions of the electromagnetic field in terms of the normal modes, and in particular it is highlighted that the modal expansion coefficients are not unique. The developed theory is used to obtain a modal expansion of the system Green function.

Furthermore, based on the equivalence between the dispersive Maxwell equations and the generalized dynamical problem, we discuss the definitions of energy and momentum in a dispersive system. In particular, it is shown that for time-harmonic solutions the energy and the total momentum can be written exclusively in terms of the electromagnetic degrees of freedom.

In addition, we illustrate the application of the proposed formalism in the context of (i) topological photonics and (ii) quantum optics. It is shown that the notions of Berry potential and Berry curvature can be generalized in a straightforward manner to a generic dispersive system, and in particular our theory enables the topological classification of a generic bianisotropic and nonreciprocal electromagnetic continuum. Moreover, we develop a simple and intuitive theory for the quantization of the electromagnetic fields in dispersive material systems and for the calculation of the quantum field correlations. Our approach yields results that agree with those obtained with other more sophisticated approaches and with the fluctuation–dissipation theorem.

24.2 Electrodynamics of dispersive media

In this chapter, we consider frequency dispersive media with vanishing (or weak) material loss. The Maxwell equations in the time domain can be written in a compact manner as

$$\hat{N} \cdot \mathbf{f} = i\left[\frac{\partial \mathbf{g}}{\partial t} + \mathbf{j}\right]. \tag{24.1}$$

We adopt six-vector notations with $\mathbf{f} = \begin{pmatrix} \mathbf{E} & \mathbf{H} \end{pmatrix}^T$, $\mathbf{g} = \begin{pmatrix} \mathbf{D} & \mathbf{B} \end{pmatrix}^T$, $\mathbf{j} = \begin{pmatrix} \mathbf{j}_e & \mathbf{j}_m \end{pmatrix}^T$, where $\mathbf{E}, \mathbf{H}$ are the electric and magnetic fields, $\mathbf{D}, \mathbf{B}$ are the electric displacement and the induction fields, $\mathbf{j}_e, \mathbf{j}_m$ are the electric and magnetic current densities. The superscript T denotes the transposition operation. The differential operator $\hat{N}$ is defined as

$$\hat{N} = \begin{pmatrix} \mathbf{0} & i\nabla \times \mathbf{1}_{3\times 3} \\ -i\nabla \times \mathbf{1}_{3\times 3} & \mathbf{0} \end{pmatrix}, \tag{24.2}$$

where $\mathbf{1}_{3\times3}$ is the identity matrix of dimension three. The relation between the **f** and **g** vector fields is determined by the material constitutive relations. For dispersive linear media, the electromagnetic response is not instantaneous, and thereby generally **f** and **g** are linked by a convolution in time. In the spectral (frequency) domain, **f** and **g** are simply related by a multiplication operator **M**:

$$\mathbf{g}(\mathbf{r},\omega) = \mathbf{M}(\mathbf{r},\omega)\cdot\mathbf{f}(\mathbf{r},\omega), \quad \text{with } \mathbf{M} = \begin{pmatrix} \varepsilon_0\bar{\varepsilon} & \frac{1}{c}\bar{\xi} \\ \frac{1}{c}\bar{\zeta} & \mu_0\bar{\mu} \end{pmatrix}. \tag{24.3}$$

The components of the 6×6 material matrix **M** are the relative permittivity $\bar{\varepsilon}$, the relative permeability $\bar{\mu}$, and the tensors $\bar{\xi}, \bar{\zeta}$ associated with magneto-electric effects (e.g., chirality or more generally bianisotropy) [7,8]. Since the electromagnetic fields are real valued in the time-domain, the material response must satisfy [9,10]

$$\mathbf{M}(\omega) = \mathbf{M}^*(-\omega^*), \tag{24.4}$$

where the superscript "*" denotes the complex conjugation operation. For a lossless medium, the material matrix must be Hermitian for ω real valued [8,11]:

$$\mathbf{M}(\omega) = \mathbf{M}^{\dagger}(\omega). \tag{24.5}$$

The superscript "†" denotes the conjugate transpose matrix. Moreover, the medium response must also guarantee that $(\partial/\partial\omega)[\omega\mathbf{M}(\omega)]$ is a positive definite matrix,

$$\frac{\partial}{\partial\omega}[\omega\mathbf{M}(\omega)] > 0, \tag{24.6}$$

so that the time-averaged stored energy density is nonnegative [8].

Due to the causality of the material response, **M** must be an analytic function of ω in the upper half frequency plane ($\mathrm{Im}\{\omega\} > 0$) [9,10]. It is further assumed that **M** is meromorphic in the complex plane so that all the singular points are poles. For a lossless system, all the poles are in the real-frequency axis. Let $\mathbf{M}_\infty = \lim_{\omega\to\infty}\mathbf{M}(\omega)$ represent the asymptotic high-frequency response of the material. Usually, $\mathbf{M}_\infty$ reduces to the material matrix of the vacuum with diagonal elements ε_0, μ_0. Using the Cauchy theorem and (24.4)–(24.6), it is straightforward to prove that for ω real-valued **M** has the following partial-fraction expansion [5,6]:

$$\mathbf{M}(\omega) = \mathbf{M}_\infty - \sum_\alpha \frac{\mathrm{sgn}(\omega_{p,\alpha})}{\omega - \omega_{p,\alpha}}\mathbf{A}_\alpha^2, \tag{24.7}$$

where $\mathrm{sgn} = \pm$ is the sign function, $\omega_{p,\alpha}$ are the (real-valued) poles of **M**, $(\mathrm{Res}\mathbf{M})_\alpha$ are the corresponding residues and $\mathbf{A}_\alpha = \left[-\mathrm{sgn}(\omega_{p,\alpha})(\mathrm{Res}\mathbf{M})_\alpha\right]^{1/2}$. It can be shown that $\mathbf{A}_\alpha$ is a positive (semi-)definite matrix ($\mathbf{A}_\alpha \geq 0$).

24.3 Hermitian formulation in the time domain

In frequency dispersive structures, it is not obvious how to obtain modal expansions of the electromagnetic fields. In the following, it is proven that in a rather general context it is possible to introduce additional variables that describe the dynamics of the medium internal degrees of freedom responsible for the non-instantaneous response. Furthermore, we show that for lossless systems the electromagnetic problem can be formulated as Schrödinger-type (eventually inhomogeneous) equation with a Hermitian time evolution operator.

As a starting point, we rewrite the Maxwell equations (24.1) in the spectral (frequency) domain:

$$\hat{N} \cdot \mathbf{f} = \omega \mathbf{M}(\mathbf{r}, \omega) \cdot \mathbf{f} + i\mathbf{j}. \tag{24.8}$$

Next, we introduce the auxiliary fields:

$$\mathbf{Q}^{(\alpha)} = \frac{|\omega_{p,\alpha}|^{1/2}}{(\omega - \omega_{p,\alpha})} \mathbf{A}_\alpha \cdot \mathbf{f}, \quad \alpha = 1, 2, \ldots, N_p \tag{24.9}$$

where $\omega_{p,\alpha}$ and $\mathbf{A}_\alpha$ are defined as in Section 24.2, and in general, depend on the spatial coordinates $\mathbf{r}$. Note that $\omega_{p,\alpha}$ and $\mathbf{A}_\alpha$ are typically discontinuous functions of $\mathbf{r}$, i.e., depend on the considered material. Furthermore, the number of poles typically depends on the material region. The number of auxiliary fields (N_p) is determined by the material with the larger number of poles. To minimize the bookkeeping, it is convenient to define $\mathbf{Q}^{(\alpha)}$ over all space. Thus, we take $\mathbf{A}_\alpha(\mathbf{r}) = 0$ for $\alpha > N_{p,i}$ and $\mathbf{r}$ in the ith material region. Here, $N_{p,i}$ is the number of poles of the ith material. From the definition of $\mathbf{Q}^{(\alpha)}$, it is clear that

$$\omega \mathbf{Q}^{(\alpha)} = \omega_{p,\alpha} \mathbf{Q}^{(\alpha)} + |\omega_{p,\alpha}|^{1/2} \mathbf{A}_\alpha \cdot \mathbf{f}. \tag{24.10}$$

On the other hand, from the partial-fraction decomposition of the material matrix (24.7) we get the following result:

$$\mathbf{g} = \mathbf{M} \cdot \mathbf{f} = \mathbf{M}_\infty \cdot \mathbf{f} - \sum_\alpha \operatorname{sgn}(\omega_{p,\alpha}) \frac{1}{|\omega_{p,\alpha}|^{1/2}} \mathbf{A}_\alpha \cdot \mathbf{Q}^{(\alpha)}. \tag{24.11}$$

Then, with the help of (24.10) it is possible to write $\omega \mathbf{g} = \omega \mathbf{M}_\infty \cdot \mathbf{f} - \sum_\alpha |\omega_{p,\alpha}|^{1/2} \mathbf{A}_\alpha \cdot \mathbf{Q}^{(\alpha)} - \sum_\alpha \operatorname{sgn}(\omega_{p,\alpha}) \mathbf{A}_\alpha^2 \cdot \mathbf{f}$. Substituting this formula into the Maxwell equations (24.8), and calculating the inverse Fourier transform in time of the resulting equation and of (24.10), one finds that the Maxwell equations

(24.1) are equivalent to a first-order in time partial differential system of the form $\hat{L}\cdot\mathbf{Q} = i(\partial/\partial t)\mathbf{M}_g\cdot\mathbf{Q} + i\mathbf{j}_g$ with:

$$\underbrace{\begin{pmatrix} \hat{N} + \sum_{\alpha} \mathrm{sgn}(\omega_{p,\alpha})\,\mathbf{A}_\alpha^2 & |\omega_{p,1}|^{1/2}\mathbf{A}_1 & |\omega_{p,2}|^{1/2}\mathbf{A}_2 & \cdots \\ |\omega_{p,1}|^{1/2}\mathbf{A}_1 & \omega_{p,1}\mathbf{1} & \mathbf{0} & \cdots \\ |\omega_{p,2}|^{1/2}\mathbf{A}_2 & \mathbf{0} & \omega_{p,2}\mathbf{1} & \cdots \\ \cdots & \cdots & \cdots & \cdots \end{pmatrix}}_{\hat{L}} \cdot \underbrace{\begin{pmatrix} \mathbf{f} \\ \mathbf{Q}^{(1)} \\ \mathbf{Q}^{(2)} \\ \cdots \end{pmatrix}}_{\mathbf{Q}}$$

$$= i\frac{\partial}{\partial t}\underbrace{\begin{pmatrix} \mathbf{M}_\infty & \mathbf{0} & \mathbf{0} & \cdots \\ \mathbf{0} & \mathbf{1} & \mathbf{0} & \cdots \\ \mathbf{0} & \mathbf{0} & \mathbf{1} & \cdots \\ \cdots & \cdots & \cdots & \cdots \end{pmatrix}}_{\mathbf{M}_g} \cdot \begin{pmatrix} \mathbf{f} \\ \mathbf{Q}^{(1)} \\ \mathbf{Q}^{(2)} \\ \cdots \end{pmatrix} + i\underbrace{\begin{pmatrix} \mathbf{j} \\ \mathbf{0} \\ \mathbf{0} \\ \cdots \end{pmatrix}}_{\mathbf{j}_g}, \tag{24.12}$$

where $\mathbf{1} \equiv \mathbf{1}_{6\times 6}$. We introduced the generalized state vector $\mathbf{Q} = \left(\mathbf{f}\ \mathbf{Q}^{(1)} \cdots \mathbf{Q}^{(\alpha)} \cdots\right)^T$ that describes the dynamics of the electromagnetic field and of the internal degrees of freedom of the material response (see also [12–16]). Note that $\hat{L}$ is a differential operator and that in an inhomogeneous material platform both $\hat{L}$ and $\mathbf{M}_g$ depend on the spatial coordinates. The operators $\hat{L}$ and $\mathbf{M}_g$ are Hermitian with respect to the canonical inner product because $\mathbf{A}_\alpha$ and $\mathbf{M}_\infty$ are Hermitian matrices. This property implies that the operator $\hat{H}_{cl}(\mathbf{r}, -i\nabla) = \mathbf{M}_g^{-1}\cdot\hat{L}$ is Hermitian with respect to the weighted inner product:

$$\langle \mathbf{Q}_B | \mathbf{Q}_A \rangle \equiv \int_V \frac{1}{2}\mathbf{Q}_B^* \cdot \mathbf{M}_g(\mathbf{r}) \cdot \mathbf{Q}_A d^3\mathbf{r}, \tag{24.13}$$

where V represents the domain of interest and it is assumed that the boundary conditions (e.g., periodic boundary conditions) ensure that $\hat{H}_{cl}$ is indeed Hermitian. Hence, in the absence of an external excitation ($\mathbf{j}_g = 0$), the generalized system (24.12) is equivalent to $\hat{H}_{cl}\cdot\mathbf{Q} = i(\partial/\partial t)\,\mathbf{Q}$, which is analogous to the Schrödinger equation with $\hbar = 1$.

In a conventional time-evolution problem, one needs to specify the value of the fields at the initial time instant, let us say $t = 0$. Clearly, in a dispersive material system the solution of the time-evolution problem requires the knowledge of the initial time electromagnetic fields $\mathbf{f}$ as well as the initial time additional variables $\mathbf{Q}^{(\alpha)}$.

24.4 Poynting theorem and stored energy

Straightforward manipulations of the dynamical equation $\hat{L}\cdot\mathbf{Q} = i\,(\partial/\partial t)\,\mathbf{M}_g\cdot\mathbf{Q} + i\mathbf{j}_g$ [Equation (24.12)] show that the state vector $\mathbf{Q}$ satisfies the following generalized

Poynting theorem:

$$\nabla \cdot \mathbf{S} + \partial_t W = -\frac{1}{2}\left(\mathbf{Q}^* \cdot \mathbf{j}_g + \mathbf{Q} \cdot \mathbf{j}_g^*\right). \tag{24.14}$$

Here $\mathbf{S}(\mathbf{r}, t) = (1/2)\,(\mathbf{E} \times \mathbf{H}^* + \mathbf{E}^* \times \mathbf{H})$ is the instantaneous Poynting vector and W is the instantaneous stored energy density:

$$W(\mathbf{r}, t) = \frac{1}{2}\mathbf{Q}^* \cdot \mathbf{M}_g \cdot \mathbf{Q}. \tag{24.15}$$

Thus, the stored energy density is determined by a weighted scalar product. Note that for completeness, we consider complex-valued solutions of the dynamical equation (24.12) in the time-domain.

Importantly, for time-harmonic solutions of (24.12) the inner product can be written simply in terms of the electromagnetic fields. To demonstrate this result, we consider two generic solutions, $\mathbf{Q}_A(\mathbf{r}, t) = \tilde{\mathbf{Q}}_A(\mathbf{r})\, e^{-i\omega_A t}$ and $\mathbf{Q}_B(\mathbf{r}, t) = \tilde{\mathbf{Q}}_B(\mathbf{r})\, e^{-i\omega_B t}$, of the generalized problem (24.12), with a time dependence of the form $e^{-i\omega t}$ and envelopes $\tilde{\mathbf{Q}}_A$ and $\tilde{\mathbf{Q}}_B$. The two oscillation frequencies ω_A and ω_B may be different. The state vectors $\mathbf{Q}_A$ and $\mathbf{Q}_B$ are not necessarily natural modes of the system and may be associated with some external current excitations, $\mathbf{j}_A$ and $\mathbf{j}_B$, respectively. However, it is required that the generalized excitation vector $\mathbf{j}_g$ in (24.12) is of the form $\mathbf{j}_g = (\mathbf{j}\ \mathbf{0}\ \cdots\ \mathbf{0}\ \cdots)^T$, i.e., the excitation can only act on the electromagnetic degrees of freedom so that (24.9) holds.

Let $\mathbf{f}_A(\mathbf{r}, t) = \mathbf{F}_A(\mathbf{r})\, e^{-i\omega_A t}$ and $\mathbf{f}_B(\mathbf{r}, t) = \mathbf{F}_B(\mathbf{r})\, e^{-i\omega_B t}$ be the corresponding solutions of the time-harmonic Maxwell equations (24.8). Using (24.9), one finds that the scalar product of $\tilde{\mathbf{Q}}_A$ and $\tilde{\mathbf{Q}}_B$ is

$$\begin{aligned} &\frac{1}{2}\tilde{\mathbf{Q}}_B^* \cdot \mathbf{M}_g(\mathbf{r}) \cdot \tilde{\mathbf{Q}}_A \\ &= \frac{1}{2}\left[\mathbf{F}_B^* \cdot \mathbf{M}_\infty \cdot \mathbf{F}_A + \mathbf{F}_B^* \cdot \sum_\alpha \mathbf{A}_\alpha \frac{\left|\omega_{p,\alpha}\right|^{1/2}}{\left(\omega_B - \omega_{p,\alpha}\right)} \cdot \frac{\left|\omega_{p,\alpha}\right|^{1/2}}{\left(\omega_A - \omega_{p,\alpha}\right)} \mathbf{A}_\alpha \cdot \mathbf{F}_A\right]. \end{aligned} \tag{24.16}$$

With the help of (24.7), the previous result can be written in a compact manner as

$$\frac{1}{2}\tilde{\mathbf{Q}}_B^* \cdot \mathbf{M}_g(\mathbf{r}) \cdot \tilde{\mathbf{Q}}_A = \frac{1}{2}\mathbf{F}_B^* \cdot \left[\frac{\omega_B \mathbf{M}(\mathbf{r}, \omega_B) - \omega_A \mathbf{M}(\mathbf{r}, \omega_A)}{\omega_B - \omega_A}\right] \cdot \mathbf{F}_A. \tag{24.17}$$

Thus, the inner product of $\tilde{\mathbf{Q}}_A$ and $\tilde{\mathbf{Q}}_B$ can be expressed simply in terms of the "electromagnetic envelopes," $\mathbf{F}_A$ and $\mathbf{F}_B$ [5,6]:

$$\left\langle \tilde{\mathbf{Q}}_B \,\middle|\, \tilde{\mathbf{Q}}_A \right\rangle = \frac{1}{2}\int_V d^3\mathbf{r}\, \mathbf{F}_B^*(\mathbf{r}) \cdot \left[\frac{\omega_B \mathbf{M}(\mathbf{r}, \omega_B) - \omega_A \mathbf{M}(\mathbf{r}, \omega_A)}{\omega_B - \omega_A}\right] \cdot \mathbf{F}_A(\mathbf{r}). \tag{24.18}$$

Clearly, when $\omega_B = \omega_A \equiv \omega$ the right-hand side of (24.17) has a singular behavior. The singularity can be formally removed by taking the limit $\omega_B \to \omega_A$ that is well defined because $\mathbf{Q}_A$ and $\mathbf{Q}_B$ are generally driven by external excitations $\mathbf{j}_A$ and $\mathbf{j}_B$

(alternatively, one may directly use (24.16) that has no singularities when $\omega_B = \omega_A \equiv \omega$). The outlined limit process yields

$$\frac{1}{2}\tilde{\mathbf{Q}}_B^* \cdot \mathbf{M}_g(\mathbf{r}) \cdot \tilde{\mathbf{Q}}_A = \frac{1}{2}\mathbf{f}_B^* \cdot \frac{\partial}{\partial \omega}[\omega \mathbf{M}(\mathbf{r}, \omega)] \cdot \mathbf{f}_A. \tag{24.19}$$

In particular, it follows that the inner product of a time-harmonic solution of (24.12) with itself satisfies [5,6]

$$\langle \mathbf{Q}|\mathbf{Q}\rangle = \frac{1}{2}\int_V d^3\mathbf{r}\, \mathbf{f}^*(\mathbf{r}) \cdot \frac{\partial}{\partial \omega}[\omega \mathbf{M}(\mathbf{r}, \omega)] \cdot \mathbf{f}(\mathbf{r}). \tag{24.20}$$

Note that the right-hand side of the previous expression represents the total energy in volume V, which due to (24.6) is always a positive quantity [2,9,10,17,18].

24.5 Canonical momentum

Let us consider again a time-harmonic solution of the generalized system (24.12), $\mathbf{Q}(\mathbf{r}, t) = \tilde{\mathbf{Q}}(\mathbf{r})\, e^{-i\omega t}$. For material structures invariant to translations along a given space direction, let us say the x-direction, we introduce a canonical (wave) momentum defined by

$$\mathcal{P}_x = \frac{1}{\omega}\langle \mathbf{Q}| - i\partial_x|\mathbf{Q}\rangle. \tag{24.21}$$

The canonical momentum of dispersive isotropic media was first discussed in [19,20]. The definition (24.21) generalizes the theory of [19,20] to arbitrary bianisotropic and nonreciprocal layered structures. Note that $-i\partial_x$ is a Hermitian operator with respect to the weighted inner product (24.13) when $\partial_x \mathbf{M}_g = 0$, i.e., when the material parameters are independent of x. Thus, $\mathcal{P}_x$ is indeed a real-valued number. As shown next, $\mathcal{P}_x$ can be written as a function of the electromagnetic component of the state vector ($\mathbf{f}(\mathbf{r}, t) = \mathbf{F}(\mathbf{r})\, e^{-i\omega t}$).

To prove this, we use (24.9) and take into account that the material parameters are independent of x to obtain (compare with (24.16))

$$\frac{1}{2}\mathbf{Q}^* \cdot \mathbf{M}_g(\mathbf{r}) \cdot (-i\partial_x)\,\mathbf{Q}$$
$$= \frac{1}{2}\left[\mathbf{f}^* \cdot \mathbf{M}_\infty \cdot (-i\partial_x)\,\mathbf{f} + \mathbf{f}^* \cdot \sum_\alpha \frac{|\omega_{p,\alpha}|}{(\omega - \omega_{p,\alpha})^2}\mathbf{A}_\alpha^2 \cdot (-i\partial_x)\,\mathbf{f}\right]. \tag{24.22}$$

But from the partial-fraction expansion (24.7), it follows that

$$\frac{1}{2}\mathbf{Q}^* \cdot \mathbf{M}_g(\mathbf{r}) \cdot (-i\partial_x)\,\mathbf{Q} = \frac{1}{2}\mathbf{f}^* \cdot \partial_\omega(\omega \mathbf{M}) \cdot (-i\partial_x)\,\mathbf{f}, \tag{24.23}$$

which implies that the canonical momentum is given by

$$\mathcal{P}_x = \frac{1}{2\omega}\int_V d^3\mathbf{r}\, \mathbf{f}^* \cdot \partial_\omega(\omega \mathbf{M}) \cdot (-i\partial_x)\,\mathbf{f}. \tag{24.24}$$

In particular, for a wave with a time-variation along x of the form $e^{ik_x x}$ (the variation along y and z may be completely arbitrary; note that the material structure can be nonuniform along y and z) it follows from (24.20) that

$$\mathcal{P}_x = \frac{k_x}{\omega}\mathcal{E}, \tag{24.25}$$

where $\mathcal{E}$ is the stored energy. This formula is consistent with the understanding that the field is formed by quanta with momenta $\hbar k_x$ because $\mathcal{E}/(\hbar\omega)$ may be understood as the number of quanta in the field [21–23].

It is interesting to connect the canonical momentum to the Minkowski momentum [22,24]. To this end, we rewrite (24.24) as

$$\mathcal{P}_x = \frac{1}{2\omega}\int_V d^3\mathbf{r}\,[\mathbf{g}^* \cdot (-i\partial_x)\,\mathbf{f} + \omega\mathbf{f}^* \cdot \partial_\omega\mathbf{M} \cdot (-i\partial_x)\,\mathbf{f}], \tag{24.26}$$

where we introduced $\mathbf{g} = \mathbf{M}(\mathbf{r},\omega)\cdot\mathbf{f}$, so that $\mathbf{g} = (\mathbf{D} \quad \mathbf{B})^T$. Furthermore, we suppose in the rest of this section that $\mathbf{f}$ is a natural mode ($\mathbf{j} = 0$). In such a case, it is possible to show by direct manipulation of the frequency domain Maxwell equations that (see [25, Ap. C])

$$\mathbf{g}^* \cdot \partial_x\mathbf{f} = \nabla\cdot\left(\mathbf{D}^*\,\mathbf{E}\cdot\hat{\mathbf{x}} + \mathbf{B}^*\,\mathbf{H}\cdot\hat{\mathbf{x}}\right) + i\omega\hat{\mathbf{x}}\cdot\left(\mathbf{D}^*\times\mathbf{B} + \mathbf{D}\times\mathbf{B}^*\right). \tag{24.27}$$

Integrating both sides of the previous equation over the volume of interest and supposing that the boundary conditions (e.g., Bloch boundary conditions) ensure that the integral of the divergence term vanishes, it is found that

$$\mathcal{P}_x = \int_V d^3\mathbf{r}\left[\mathrm{Re}\{\mathbf{D}\times\mathbf{B}^*\}\cdot\hat{\mathbf{x}} + \frac{1}{2}\mathbf{f}^*\cdot\partial_\omega\mathbf{M}\cdot(-i\partial_x)\,\mathbf{f}\right]. \tag{24.28}$$

Thus, for dispersionless media the canonical momentum is determined by the Minkowski momentum with momentum density $\mathbf{D}\times\mathbf{B}$ [22,25]. In dispersive media, the canonical momentum gains an extra term, as it was first shown in [19,20,26] for the case of isotropic dispersive dielectrics. As discussed in detail in [24], the canonical momentum must be understood as the *total* momentum of the system and has contributions from the electromagnetic field as well as from the *kinetic* degrees of freedom of matter. Indeed, the light-matter interactions typically result in optical forces that set the medium into motion. For very massive bodies, the center of mass velocity is insignificant, but the transferred kinetic momentum (i.e., the product of the total mass and of the center of mass velocity) cannot be neglected [24]. The wave momentum $\mathcal{P}_x$ is the sum of the light momentum and of the kinetic momentum of the medium [24]. It is underlined that this result applies *exclusively* to lossless media invariant to translations along x and to natural modes of the Maxwell equations. Indeed, when the medium is inhomogeneous along x, typically the kinetic momentum transferred to the medium grows linearly with time in a steady state regime (with time variation $e^{-i\omega t}$) because of the stress created at material interfaces. Finally, we point out that for real-valued fields, $\mathbf{f}(\mathbf{r},t) = \mathrm{Re}\{\mathbf{f}_\omega(\mathbf{r})\,e^{-i\omega t}\}$, the time-averaged canonical momentum is given by the right-hand side of (24.28) multiplied by an extra $1/2$ factor and with the replacement $\mathbf{f} \to \mathbf{f}_\omega$.

24.6 Modal expansions

Let us consider a generic state vector $\mathbf{Q}(\mathbf{r}) = \left(\mathbf{f}\ \mathbf{Q}^{(1)}\ \cdots\ \mathbf{Q}^{(\alpha)}\ \cdots\right)^T$ defined over the volume of interest V. For example, V may correspond to a closed cavity. Since $\hat{H}_{cl}$ is Hermitian with respect to the weighted inner product (24.13), the spectral theorem guarantees that $\mathbf{Q}(\mathbf{r})$ can be expanded into a basis of eigenmodes of $\hat{H}_{cl}$, $\mathbf{Q}_1, \mathbf{Q}_2, \ldots$, (the natural modes of oscillation), which satisfy

$$\hat{H}_{cl}\mathbf{Q}_n = \omega_n \mathbf{Q}_n, \tag{24.29}$$

where ω_n are the eigenfrequencies. Supposing that the eigenmodes are normalized such that $\langle \mathbf{Q}_n | \mathbf{Q}_m \rangle = \delta_{n,m}$ we have the modal expansion:

$$\mathbf{Q}(\mathbf{r}) = \sum_n \mathbf{Q}_n(\mathbf{r})\, c_n \quad \text{with } c_n = \langle \mathbf{Q}_n | \mathbf{Q} \rangle. \tag{24.30}$$

In particular, if we take $\mathbf{Q}(\mathbf{r}) \to \delta(\mathbf{r} - \mathbf{r}')\, \mathbf{1}_g$ with $\mathbf{1}_g$ the identity tensor (with the same dimension as $\mathbf{M}_g$), we obtain the completeness relation:

$$\delta(\mathbf{r} - \mathbf{r}')\, \mathbf{1}_g = \frac{1}{2} \sum_n \mathbf{Q}_n(\mathbf{r}) \otimes \mathbf{Q}_n^*(\mathbf{r}') \cdot \mathbf{M}_g(\mathbf{r}'), \tag{24.31}$$

where $\otimes$ represents the tensor product of two vectors.

Let $\mathbf{f}_n$ be the (envelope) electromagnetic field distribution associated with $\mathbf{Q}_n$, which thereby satisfies

$$\hat{N} \cdot \mathbf{f}_n = \omega_n \mathbf{M}(\mathbf{r}, \omega_n) \cdot \mathbf{f}_n. \tag{24.32}$$

From (24.18) and (24.20), the normalization condition $\langle \mathbf{Q}_n | \mathbf{Q}_m \rangle = \delta_{n,m}$ implies that

$$\frac{1}{2} \int_V d^3\mathbf{r}\, \mathbf{f}_n^*(\mathbf{r}) \cdot \frac{\partial}{\partial \omega} [\omega \mathbf{M}(\mathbf{r}, \omega)]_{\omega = \omega_n} \cdot \mathbf{f}_m(\mathbf{r}) = \delta_{n,m}, \quad \text{if } \omega_n = \omega_m. \tag{24.33a}$$

$$\frac{1}{2} \int_V d^3\mathbf{r}\, \mathbf{f}_n^*(\mathbf{r}) \cdot \left[\frac{\omega_n \mathbf{M}(\mathbf{r}, \omega_n) - \omega_m \mathbf{M}(\mathbf{r}, \omega_m)}{\omega_n - \omega_m} \right] \cdot \mathbf{f}_m(\mathbf{r}) = 0, \quad \text{if } \omega_n \neq \omega_m. \tag{24.33b}$$

Remarkably, the completeness relation (24.31) yields the following completeness relation for the electromagnetic modes [27]:

$$\delta(\mathbf{r} - \mathbf{r}')\, \mathbf{M}_\infty^{-1}(\mathbf{r}') = \frac{1}{2} \sum_n \mathbf{f}_n(\mathbf{r}) \otimes \mathbf{f}_n^*(\mathbf{r}'), \tag{24.34}$$

with the modes normalized as in (24.33a) and (24.33b). In particular, a generic electromagnetic field distribution $\mathbf{f}(\mathbf{r})$ has the modal expansion:

$$\mathbf{f}(\mathbf{r}) = \sum_n \mathbf{f}_n(\mathbf{r})\, c_n, \quad \text{with } c_n = \frac{1}{2}\int_V d^3\mathbf{r}'\, \mathbf{f}_n^*(\mathbf{r}') \cdot \mathbf{M}_\infty(\mathbf{r}') \cdot \mathbf{f}(\mathbf{r}'). \tag{24.35}$$

Importantly, in a dispersive medium the expansion coefficients c_n are *not* unique. In other words, an identity of the type $\mathbf{f}(\mathbf{r}) = \sum_n \mathbf{f}_n(\mathbf{r})\, c_n$ may be satisfied with different sets of coefficients c_n. For example, if we take $\mathbf{f}(\mathbf{r}) = \mathbf{f}_m(\mathbf{r})$ an obvious solution is $c_n = \delta_{n,m}$, which for a dispersive material clearly differs from the coefficients given by (24.35). The lack of uniqueness of the expansion coefficients follows from the fact that the electromagnetic component of $\mathbf{Q}$ describes only a subset of the degrees of freedom of the material. In contrast, the modal expansion (24.30) is unique.

Let us consider now that $\mathbf{f}(\mathbf{r})$ is some solution of the time-harmonic problem (24.8), and let $\mathbf{Q}$ be the corresponding solution of the generalized time-harmonic problem. Then, we may obtain a set of expansion coefficients for $\mathbf{f}(\mathbf{r})$ from (24.18) and (24.30):

$$\begin{aligned} c_n &= \frac{1}{2}\int_V d^3\mathbf{r}\, \mathbf{f}_n^*(\mathbf{r}) \cdot \left[\frac{\omega_n \mathbf{M}(\mathbf{r},\omega_n) - \omega\mathbf{M}(\mathbf{r},\omega)}{\omega_n - \omega}\right] \cdot \mathbf{f}(\mathbf{r}) \\ &= \frac{1}{2}\int_V d^3\mathbf{r}\, \frac{\left[\hat{N}\,\mathbf{f}_n\right]^* \cdot \mathbf{f} - \mathbf{f}_n^* \cdot \left(\hat{N}\,\mathbf{f} - i\mathbf{j}\right)}{\omega_n - \omega}. \end{aligned} \tag{24.36}$$

We used the fact that the material matrix is Hermitian and (24.32). But since the differential operator $\hat{N}$ is Hermitian with respect to the canonical inner product, we finally obtain

$$\mathbf{f}(\mathbf{r}) = \sum_n \mathbf{f}_n(\mathbf{r})\, c_n, \quad \text{with } c_n = \frac{1}{2}\int_V d^3\mathbf{r}\, \frac{\mathbf{f}_n^* \cdot i\mathbf{j}}{\omega_n - \omega}. \tag{24.37}$$

We emphasize that the modal expansion (24.35) is universally valid, whereas the modal expansion (24.37) only holds when $\mathbf{f}$ is a time-harmonic solution of the Maxwell equations with oscillation frequency ω.

24.7 Green's function

We define the (causal) Green function tensor in the time domain as the solution of [27]:

$$\left(\hat{N} - i\frac{\partial}{\partial t}\mathbf{M}\right) \cdot \overline{\mathbf{G}}(\mathbf{r},\mathbf{r}_0,t) = i\mathbf{1}\delta(\mathbf{r}-\mathbf{r}_0)\,\delta(t), \tag{24.38}$$

with $\mathbf{1} \equiv \mathbf{1}_{6\times 6}$. It is understood that the action of the material matrix on the time-domain Green function is given by a convolution. Evidently, in the spectral domain the Green function satisfies (with a time-harmonic variation $e^{-i\omega t}$)

$$\hat{N} \cdot \overline{\mathbf{G}}(\mathbf{r},\mathbf{r}_0,\omega) = \omega\mathbf{M}(\mathbf{r},\omega) \cdot \overline{\mathbf{G}}(\mathbf{r},\mathbf{r}_0,\omega) + i\mathbf{1}\delta(\mathbf{r}-\mathbf{r}_0). \tag{24.39}$$

It is useful to note that $\overline{\mathbf{G}}$ is a 6×6 tensor, and thus it can be decomposed into electric and magnetic terms as follows:

$$\overline{\mathbf{G}} = \begin{pmatrix} \mathbf{G}_{\text{EE}} & \mathbf{G}_{\text{EM}} \\ \mathbf{G}_{\text{ME}} & \mathbf{G}_{\text{MM}} \end{pmatrix}. \tag{24.40}$$

For material systems with a trivial magnetic response ($\bar{\mu} = \mathbf{1}_{3\times3},\ \bar{\xi} = \bar{\zeta} = 0$), the electric part of the Green function $\mathbf{G}_{\text{EE}}$ is related to the standard electric Green function $\mathcal{G}$ (i.e., the solution of $\nabla \times \nabla \times \mathcal{G} - (\omega^2/c^2)\,\bar{\varepsilon} \cdot \mathcal{G} = \mathbf{1}_{3\times3}\delta(\mathbf{r} - \mathbf{r}_0)$) as $\mathbf{G}_{\text{EE}} = i\omega\mu_0\mathcal{G}$.

By expanding $\overline{\mathbf{G}} \cdot \hat{\mathbf{u}}_i = \sum_n \mathbf{f}_n(\mathbf{r})\, c_n$ into electromagnetic modes (with $\hat{\mathbf{u}}_i$ a generic unity six-vector) and using (24.10), we find that [6,27]

$$\overline{\mathbf{G}}(\mathbf{r}, \mathbf{r}_0, \omega) = \frac{i}{2} \sum_n \frac{1}{\omega_n - \omega} \mathbf{f}_n(\mathbf{r}) \otimes \mathbf{f}_n^*(\mathbf{r}_0). \tag{24.41}$$

It is underlined that the electromagnetic modes are normalized as in (24.33a) and (24.33b), and that the modal expansion of the Green function applies only in the limit of vanishing material loss.

The summation in the right-hand side of (24.41) is over all modes, i.e., modes with positive, negative and zero frequencies. Using the completeness relation (24.34), the sum can be restricted to "transverse" modes with $\omega_n \neq 0$, so that $\overline{\mathbf{G}} = (i/2) \sum_{\omega_n \neq 0} \omega_n / \left((\omega_n - \omega)\,\omega\right) \mathbf{f}_n(\mathbf{r}) \otimes \mathbf{f}_n^*(\mathbf{r}_0) - (i/\omega)\,\delta(\mathbf{r} - \mathbf{r}_0)\,\mathbf{M}_\infty^{-1}(\mathbf{r}_0)$. Furthermore, the reality of the electromagnetic fields implies that modes with positive and negative frequencies are linked by complex conjugation. Thus, the Green function expansion can be restricted to modes with positive frequencies as follows [6,27,28]:

$$\begin{aligned} \overline{\mathbf{G}} = & \sum_{\omega_n > 0} \frac{i\omega_n}{2\omega} \left[\frac{1}{\omega_n - \omega} \mathbf{f}_n(\mathbf{r}) \otimes \mathbf{f}_n^*(\mathbf{r}_0) + \frac{1}{\omega_n + \omega} \mathbf{f}_n^*(\mathbf{r}) \otimes \mathbf{f}_n(\mathbf{r}_0) \right] \\ & - \frac{i}{\omega} \delta(\mathbf{r} - \mathbf{r}_0)\, \mathbf{M}_\infty^{-1}(\mathbf{r}_0) \end{aligned} \tag{24.42}$$

We emphasize that this result applies to arbitrary bianisotropic and possibly nonreciprocal inhomogeneous material structures. Furthermore, some of the ideas described here can be readily extended to spatially dispersive materials [5,15,29].

The following useful property can be derived by replacing $\omega \to \omega + i0^+$ and using $\left(1/\left(\omega + i0^+\right)\right) = \text{PV}(1/\omega) - i\pi\delta(\omega)$ with PV the principal value operator [30]:

$$\begin{aligned} & \left[\left(-i\omega\overline{\mathbf{G}}\right)(\mathbf{r}, \mathbf{r}_0) - \left(-i\omega\overline{\mathbf{G}}\right)^\dagger(\mathbf{r}_0, \mathbf{r}) \right]_{\omega + 0^+ i} \\ & = i\omega\pi \sum_{\omega_n > 0} \left[\delta(\omega - \omega_n)\, \mathbf{f}_n(\mathbf{r}) \otimes \mathbf{f}_n^*(\mathbf{r}_0) + \delta(\omega + \omega_n)\, \mathbf{f}_n^*(\mathbf{r}) \otimes \mathbf{f}_n(\mathbf{r}_0) \right]. \end{aligned} \tag{24.43}$$

To conclude, we note that from (24.34) and (24.41) the time-domain Green function satisfies [27]

$$\begin{aligned}\overline{\mathbf{G}}(\mathbf{r},\mathbf{r}_0,t) &= -u(t)\,\frac{1}{2}\sum_{n} e^{-i\omega_n t}\mathbf{f}_n(\mathbf{r})\otimes\mathbf{f}_n^*(\mathbf{r}_0)\\ &= -u(t)\,\frac{1}{2}\sum_{\omega_n\neq 0}\left(e^{-i\omega_n t}-1\right)\mathbf{f}_n(\mathbf{r})\otimes\mathbf{f}_n^*(\mathbf{r}_0)-\delta(\mathbf{r}-\mathbf{r}_0)\,\mathbf{M}_\infty^{-1}(\mathbf{r}_0)\,u(t).\end{aligned} \tag{24.44}$$

24.8 Positive and negative frequency components of the Green function

From the previous section, the Green function can be decomposed as [28]

$$\overline{\mathbf{G}} = \overline{\mathbf{G}}^{+} + \overline{\mathbf{G}}^{-} - \frac{i}{\omega}\delta(\mathbf{r}-\mathbf{r}_0)\,\mathbf{M}_\infty^{-1}(\mathbf{r}_0), \tag{24.45}$$

with $\overline{\mathbf{G}}^{+}, \overline{\mathbf{G}}^{-}$ the positive and negative frequency components,

$$\begin{aligned}-i\omega\overline{\mathbf{G}}^{+} &= \sum_{\omega_n>0}\frac{\omega_n}{2}\frac{1}{\omega_n-\omega}\mathbf{f}_n(\mathbf{r})\otimes\mathbf{f}_n^*(\mathbf{r}_0),\\ -i\omega\overline{\mathbf{G}}^{-} &= \sum_{\omega_n>0}\frac{\omega_n}{2}\frac{1}{\omega_n+\omega}\mathbf{f}_n^*(\mathbf{r})\otimes\mathbf{f}_n(\mathbf{r}_0),\end{aligned} \tag{24.46}$$

with poles in the $\mathrm{Re}\{\omega\} > 0$ ($\mathrm{Re}\{\omega\} < 0$) semi-spaces, respectively.

Interestingly, the components $\overline{\mathbf{G}}^{+}, \overline{\mathbf{G}}^{-}$ are relevant in some problems of quantum optics, e.g., they appear when studying the interaction of an atomic system with a quantized electromagnetic field in the framework of a Markov approximation [28]. In practice, it is convenient to evaluate $\overline{\mathbf{G}}^{+}, \overline{\mathbf{G}}^{-}$ directly in terms of the total Green function $\overline{\mathbf{G}}$. In what follows, we establish such a link and prove that for positive frequencies ($\omega > 0$) $\overline{\mathbf{G}}^{-}$ can be written in terms of an integral of $\overline{\mathbf{G}}$ over the imaginary axis.

To begin with, we note that $\overline{\mathbf{G}}^{-}$ is analytic for $\mathrm{Re}\{\omega\} > 0$, and thus from the Cauchy theorem [28]

$$\begin{aligned}\left(-i\omega\overline{\mathbf{G}}^{-}\right)\Big|_{\omega_0} &= \int_{-i\infty}^{i\infty} d\omega\,\frac{1}{2\pi i}\frac{1}{\omega_0-\omega}\left(-i\omega\overline{\mathbf{G}}^{-}\right)\\ &= \frac{1}{2\pi}\int_{-\infty}^{\infty} d\xi\,\frac{1}{\omega_0-i\xi}\left(-i\omega\overline{\mathbf{G}}^{-}\right)_{\omega=i\xi},\end{aligned} \tag{24.47}$$

where ω_0 is assumed real valued and positive. Using again the Cauchy theorem, it is clear that $0 = (1/2\pi)\int_{-\infty}^{\infty} d\xi\, (1/(\omega_0 + i\xi))\,(-i\omega\overline{\mathbf{G}}^{-})_{\omega=i\xi}$. But from (24.46), it is evident that $(-i\omega\overline{\mathbf{G}}^{-}(\mathbf{r},\mathbf{r}_0,\omega))_{\omega=i\xi} = [(-i\omega\overline{\mathbf{G}}^{+}(\mathbf{r},\mathbf{r}_0,\omega))_{\omega=i\xi}]^*$. Therefore, we conclude that

$$0 = \frac{1}{2\pi}\int_{-\infty}^{\infty} d\xi \frac{1}{\omega_0 + i\xi}\left[\left(-i\omega\overline{\mathbf{G}}^{+}(\mathbf{r},\mathbf{r}_0,\omega)\right)_{\omega=i\xi}\right]^*. \tag{24.48}$$

Conjugating both sides of the previous equation and combining it with (24.47), we obtain the desired result:

$$\left(-i\omega\overline{\mathbf{G}}^{-}(\mathbf{r},\mathbf{r}_0,\omega)\right)\Big|_{\omega=\omega_0>0} = \frac{1}{2\pi}\int_{-\infty}^{\infty} d\xi \frac{1}{\omega_0 - i\xi}\left(-i\omega\overline{\mathbf{G}}(\mathbf{r},\mathbf{r}_0,\omega) + \delta(\mathbf{r}-\mathbf{r}_0)\,\mathbf{M}_{\infty}^{-1}(\mathbf{r}_0)\right)_{\omega=i\xi}. \tag{24.49}$$

As further discussed in [28], $\overline{\mathbf{G}}^{-}$ usually is associated with nonresonant ground state interactions and consistent with that it is written in terms of an integral of the system Green function over imaginary frequencies.

24.9 Application to topological photonics

Currently, there is a great interest in the use of topological methods in problems of electromagnetic wave propagation [31,32]. Following seminal studies in the context of condensed matter theory [33–36], it has been shown that electromagnetic materials can be topologically classified by integer numbers that are insensitive to weak perturbations of the material inner structure [5,6,12,13,37–39]. Furthermore, quite remarkably it has been shown that it is possible to predict some rather fundamental wave phenomena (e.g., the propagation of scattering immune edge states) based on the topological classification of the involved materials [40–43].

The topological classification of a material is typically done relying on the eigenmodes. Both for periodic structures (e.g., photonic crystals) and electromagnetic continua (with no intrinsic periodicity), the eigenmodes, $\mathbf{f}_{n\mathbf{k}}$, are organized in bands ($n = 1, 2, \ldots$) and can be labeled by a wave vector $\mathbf{k}$. The topological numbers are found from the Berry curvature $\mathcal{F}_{\mathbf{k}}$ that is written in terms of the eigenmodes. The Berry curvature is a scalar function defined over the wave vector $\mathbf{k}$-space, and it turns out that when the $\mathbf{k}$-space is a closed manifold with no boundary the Chern number,

$$\mathcal{C} = \frac{1}{2\pi}\iint dk_x dk_y\, \mathcal{F}_{\mathbf{k}}, \tag{24.50}$$

is necessarily an integer and thereby has a topological nature [5,12]. This result is known as the Chern theorem. The integral in the previous equation is over the entire wave vector space. It is possible to introduce other topological invariants for subclasses of media with specific symmetries [35,38].

The standard definition of the Berry potential in condensed-matter systems [36] assumes that the eigenfunctions are associated with the spectrum of a Hermitian operator. Thereby, the topological classification of dispersive electromagnetic material platforms must be done based on the generalized (homogeneous) problem (24.12). Specifically, a straightforward generalization of the ideas of condensed-matter physics gives [5,12]

$$\mathcal{A}_{n\mathbf{k}} = i\langle \mathbf{Q}_{n\mathbf{k}} | \partial_{\mathbf{k}} \mathbf{Q}_{n\mathbf{k}} \rangle, \tag{24.51}$$

where $\mathbf{Q}_{n\mathbf{k}}$ represent the envelopes of a family of eigenmodes of the generalized system ($\hat{H}_{cl}(\mathbf{r}, -i\nabla + \mathbf{k}) \cdot \mathbf{Q}_{n\mathbf{k}} = \omega_{n\mathbf{k}} \mathbf{Q}_{n\mathbf{k}}$), $\partial_{\mathbf{k}} = (\partial/\partial k_x)\,\hat{\mathbf{x}} + (\partial/\partial k_y)\,\hat{\mathbf{y}}$, and it is implicit that the wave vector is restricted to the $k_z = 0$ plane. Furthermore, the eigenmode envelopes must be normalized as $\langle \mathbf{Q}_{n\mathbf{k}} | \mathbf{Q}_{n\mathbf{k}} \rangle = 1$. It should be noted that the Berry potential is gauge dependent because the gauge transformation $\mathbf{Q}_{n\mathbf{k}} \to \mathbf{Q}_{n\mathbf{k}} e^{i\theta_{n\mathbf{k}}}$ (with $e^{i\theta_{n\mathbf{k}}}$ a smooth function) transforms the Berry potential as $\mathcal{A}_{n\mathbf{k}} \to \mathcal{A}_{n\mathbf{k}} - \partial_{\mathbf{k}} \theta_{n\mathbf{k}}$. The Berry curvature is defined in terms of the Berry potential as

$$\mathcal{F}_{\mathbf{k}} = \hat{\mathbf{z}} \cdot \nabla_{\mathbf{k}} \times \mathcal{A}_{n\mathbf{k}}, \tag{24.52}$$

and is manifestly gauge independent. In general, $\mathcal{A}_{n\mathbf{k}}$ cannot be defined as a smooth vector field over the entire wave vector space. Nevertheless, it is possible to cover the $\mathbf{k}$-space with patches wherein for some gauge $\mathcal{A}_{n\mathbf{k}}$ is a smooth vector field. When $\mathcal{A}_{n\mathbf{k}}$ can be globally defined as a smooth function, it is evident from Stokes theorem that the Chern number vanishes and thereby that the material is topologically trivial. Hence, a nonzero Chern number indicates an obstruction to the application of the Stokes theorem [36]. Furthermore, it is possible to show that the Chern number always vanishes for reciprocal materials [5,12].

Interestingly, the Berry potential can be expressed in terms of the electromagnetic component ($\mathbf{f}_{n\mathbf{k}}$) of the eigenmodes. To prove this, we use (24.9) to obtain

$$\begin{aligned} &\frac{1}{2} \mathbf{Q}^*_{n\mathbf{k}} \cdot \mathbf{M}_g(\mathbf{r}) \cdot i\partial_{\mathbf{k}} \mathbf{Q}_{n\mathbf{k}} \\ &= \frac{1}{2} \left[\mathbf{f}^*_{n\mathbf{k}} \cdot \mathbf{M}_\infty \cdot i\partial_{\mathbf{k}} \mathbf{f}_{n\mathbf{k}} + \mathbf{f}^*_{n\mathbf{k}} \cdot \sum_\alpha \frac{|\omega_{p,\alpha}|}{(\omega_{n\mathbf{k}} - \omega_{p,\alpha})} \mathbf{A}^2_\alpha \cdot i\partial_{\mathbf{k}} \left(\frac{1}{\omega_{n\mathbf{k}} - \omega_{p,\alpha}} \mathbf{f}_{n\mathbf{k}} \right) \right]. \end{aligned} \tag{24.53}$$

Integrating both sides of the equation over the volume V, noting that the Berry potential is necessarily real valued, and using (24.7) it is possible to write [5,12,13]:

$$\mathcal{A}_{n\mathbf{k}} = \int_V d^3\mathbf{r} \mathrm{Re} \left\{ i\, \mathbf{f}^*_{n\mathbf{k}} \cdot \frac{1}{2} \frac{\partial}{\partial \omega} (\omega \mathbf{M})_{\omega_{n\mathbf{k}}} \cdot \partial_{\mathbf{k}} \mathbf{f}_{n\mathbf{k}} \right\}. \tag{24.54}$$

The constraint $\langle \mathbf{Q}_{n\mathbf{k}} | \mathbf{Q}_{n\mathbf{k}} \rangle = 1$ implies that the electromagnetic fields are normalized as in (24.33a) and (24.33b). The application of this result to the topological classification of dispersive electromagnetic platforms is thoroughly discussed in [5,6,42]. Furthermore, in [5,6] the result is also generalized to a subclass of spatially dispersive materials. This case is of practical importance because the application of topological concepts to electromagnetic continua typically requires the introduction of a wave vector cutoff. For more details, the reader is referred to [5,6,42].

24.10 Application to quantum optics

Modal expansions in dispersive media are of key importance in quantum optics. Indeed, the quantized electromagnetic fields in a closed cavity are expressed in terms of the electromagnetic modes. For nondispersive structures, the time evolution of the electromagnetic field is described by a Hermitian operator and therefore the quantization procedure is straightforward: each (positive frequency) natural mode of oscillation is associated with a quantum harmonic oscillator with zero-point energy $\hbar\omega_n/2$ [4,25,44]. Importantly, the homogeneous (with $\mathbf{j}_g = 0$) generalized system (24.12) is also described by a Hermitian operator ($\mathbf{M}_g^{-1} \cdot \hat{L}$) and hence the quantization of the state vector $\mathbf{Q}$ can be done by a simple generalization of the standard procedure (see [25, Ap. A]).

To show this, first we highlight that the dynamics of the homogeneous generalized system (24.12) is formally equivalent to the dynamics of an infinite set of harmonic oscillators. Following the standard procedure [25], we restrict our attention to *transverse* solutions of (24.12), i.e., solutions $\mathbf{Q}$ that can be expanded in terms of modes with $\omega_n \neq 0$. Thus, $\mathbf{Q} = \sum_{\omega_n>0} b_n\mathbf{Q}_n(\mathbf{r}) + b_{-n}\mathbf{Q}_{-n}(\mathbf{r})$, with the eigenmodes normalized as $\langle \mathbf{Q}_n|\mathbf{Q}_m\rangle = \delta_{n,m}$ and the inner product is defined as in (24.13). Here, $\mathbf{Q}_n$ is an eigenmode associated with a positive oscillation frequency ω_n, and $\mathbf{Q}_{-n}$ is the corresponding eigenmode associated with the negative frequency $-\omega_n$. Furthermore, it is supposed that the electromagnetic component of $\mathbf{Q}$ is real valued. Because of the "particle-hole" symmetry [Equation (24.4)] of the Maxwell equations, it is possible to impose that the electromagnetic component of $\mathbf{Q}_{-n}$ ($\mathbf{f}_{-n}$) is linked to that of $\mathbf{Q}_n$ ($\mathbf{f}_n$) by complex conjugation, $\mathbf{f}_{-n} = \mathbf{f}_n^*$. Then, since by hypothesis the electromagnetic component of $\mathbf{Q}$ is real valued, it follows that $b_{-n} = b_n^*$, so that

$$\mathbf{Q} = \sum_{\omega_n>0} b_n\mathbf{Q}_n(\mathbf{r}) + b_n^*\mathbf{Q}_{-n}(\mathbf{r}). \tag{24.55}$$

The system energy is evidently equal to $H = \langle \mathbf{Q}|\mathbf{Q}\rangle$. Taking into account the normalization of the modes, we find that

$$H = 2\sum_{\omega_n>0} |b_n|^2 = 2\sum_{\omega_n>0} \left(b_n'^2 + b_n''^2\right), \tag{24.56}$$

where $b_n = b_n' + i\,b_n''$. In order that $\mathbf{Q}$ satisfies the equation system (24.12) with $\mathbf{j}_g = 0$, it is necessary that $b_n(t) = b_n(0)\,e^{-i\omega_n t}$, so that the real and imaginary parts of b_n must satisfy $\dot{b}'_n = \omega_n\, b_n''$ and $\dot{b}''_n = -\omega_n\, b_n'$ (the dot represents derivation in time).

Let us now introduce the variables q_n and p_n such that $b'_n = (\omega_n/2)\sqrt{m}\,q_n$ and $b''_n = \left(1/2\sqrt{m}\right)p_n$, where m has dimensions of mass (and can be chosen arbitrarily). Then, from the previous results it is clear that the system energy can be written as

$$H = \sum_{\omega_n>0}\left(\frac{1}{2}m\omega_n^2 q_n^2 + \frac{1}{2m}p_n^2\right) \tag{24.57}$$

with $\dot{q}_n = (1/m)\,p_n = \partial H/\partial p_n$ and $\dot{p}_n = -m\omega_n^2\,q_n = -\partial H/\partial q_n$. Therefore, the dynamics of the classical problem is indeed equivalent to that of an infinite set of decoupled harmonic oscillators with classical Hamiltonian given by (24.57).

The generalized state vector $\mathbf{Q}$ can now be quantized simply by quantizing the associated harmonic oscillators. To this end, q_n and p_n are promoted to operators that satisfy canonical commutation relations $\left[\hat{q}_n, \hat{p}_n\right] = i\hbar$. Furthermore, introducing the annihilation operator $\hat{a}_n = \sqrt{m\omega_n/2\hbar}\,\hat{q}_n + i\sqrt{1/2\hbar m\omega_n}\,\hat{p}_n$, it is simple to verify that the Hamiltonian of the quantized system is

$$\hat{H} = \sum_{\omega_n>0}\hbar\omega_n\left(\hat{a}_n^\dagger\hat{a}_n + \frac{1}{2}\right), \tag{24.58}$$

with the quantized state vector given by (the Schrödinger picture is implicit):

$$\hat{\mathbf{Q}}(\mathbf{r}) = \sum_{\omega_n>0}\sqrt{\frac{\hbar\omega_n}{2}}\left(\hat{a}_n\mathbf{Q}_n(\mathbf{r}) + \hat{a}_n^\dagger\mathbf{Q}_{-n}(\mathbf{r})\right). \tag{24.59}$$

The "hat" indicates that a given quantity must be understood as a quantum operator. As expected, $\hat{a}_n, \hat{a}_n^\dagger$ are standard creation and annihilation operators that satisfy the usual bosonic commutation relations:

$$\left[\hat{a}_n, \hat{a}_m\right] = \left[\hat{a}_n^\dagger, \hat{a}_m^\dagger\right] = 0, \quad \left[\hat{a}_n, \hat{a}_m^\dagger\right] = \delta_{n,m}. \tag{24.60}$$

Since the electromagnetic fields are determined by the first element of $\hat{\mathbf{Q}}$, it is clear that (24.59) and $\mathbf{f}_{-n} = \mathbf{f}_n^*$ give [27,30]

$$\hat{\mathbf{f}} = \begin{pmatrix}\hat{\mathbf{E}}\\ \hat{\mathbf{H}}\end{pmatrix} = \sum_{\omega_n>0}\sqrt{\frac{\hbar\omega_n}{2}}\left(\hat{a}_n\mathbf{f}_n(\mathbf{r}) + \hat{a}_n^\dagger\mathbf{f}_n^*(\mathbf{r})\right). \tag{24.61}$$

The condition $\langle\mathbf{Q}_n|\mathbf{Q}_m\rangle = \delta_{n,m}$ implies that the electromagnetic field natural modes are normalized as in (24.33a) and (24.33b). The derived formula for the quantized electromagnetic field is consistent with the results obtained with other more sophisticated approaches [45–47].

Evidently, the quantized electric displacement vector and magnetic induction are given by

$$\hat{\mathbf{g}} = \begin{pmatrix}\hat{\mathbf{D}}\\ \hat{\mathbf{B}}\end{pmatrix} = \sum_{\omega_n>0}\sqrt{\frac{\hbar\omega_n}{2}}\left(\hat{a}_n\mathbf{g}_n(\mathbf{r}) + \hat{a}_n^\dagger\mathbf{g}_n^*(\mathbf{r})\right), \tag{24.62}$$

with $\mathbf{g}_n = \mathbf{M}(\mathbf{r}, \omega_n) \cdot \mathbf{f}_n$. Using the commutation relations (24.60) and (24.32), it is simple to check that the equal-time commutator of the fields satisfies

$$\begin{aligned}\left[\hat{\mathbf{g}}(\mathbf{r}), \hat{\mathbf{f}}(\mathbf{r}')\right] &= \sum_{\omega_n>0} \frac{\hbar\omega_n}{2} \{\mathbf{g}_n(\mathbf{r}) \otimes \mathbf{f}_n^*(\mathbf{r}') - \mathbf{g}_n^*(\mathbf{r}) \otimes \mathbf{f}_n(\mathbf{r}')\} \\ &= \hat{N} \cdot \frac{\hbar}{2} \sum_{\omega_n>0} \{\mathbf{f}_n(\mathbf{r}) \otimes \mathbf{f}_n^*(\mathbf{r}') + \mathbf{f}_n^*(\mathbf{r}) \otimes \mathbf{f}_n(\mathbf{r}')\} \\ &= \hat{N} \cdot \frac{\hbar}{2} \sum_n \mathbf{f}_n(\mathbf{r}) \otimes \mathbf{f}_n^*(\mathbf{r}')\end{aligned} \tag{24.63}$$

where $[\hat{\mathbf{g}}(\mathbf{r}), \hat{\mathbf{f}}(\mathbf{r}')]$ represents a tensor with elements $[\hat{\mathbf{g}}_i(\mathbf{r}), \hat{\mathbf{f}}_j(\mathbf{r}')]$, $i,j = 1, \ldots, 6$. Using the completeness relation (24.7), we find that

$$\left[\hat{\mathbf{g}}(\mathbf{r}), \hat{\mathbf{f}}(\mathbf{r}')\right] = \hbar\hat{N} \cdot \{\delta(\mathbf{r} - \mathbf{r}')\, \mathbf{M}_\infty^{-1}(\mathbf{r}')\}. \tag{24.64}$$

This commutation relation applies to general bianisotropic possibly nonreciprocal and inhomogeneous material structures. Since $\mathbf{M}_\infty$ is real valued and symmetric, the commutation relation may be rewritten as $[\hat{\mathbf{g}}(\mathbf{r}), \mathbf{M}_\infty(\mathbf{r}') \cdot \hat{\mathbf{f}}(\mathbf{r}')] = \hbar\hat{N} \cdot \{\delta(\mathbf{r} - \mathbf{r}')\, \mathbf{1}_{6\times6}\}$. In particular, when the source point $\mathbf{r}'$ lies in a nondispersive material region, we find that $[\hat{\mathbf{g}}(\mathbf{r}), \hat{\mathbf{g}}(\mathbf{r}')] = \hbar\hat{N} \cdot \{\delta(\mathbf{r} - \mathbf{r}')\, \mathbf{1}_{6\times6}\}$, which are the standard field commutation relations in nondispersive bianisotropic media [25,44]. Note that the commutation relations for the macroscopic fields are not coincident with the commutation relations in vacuum for the microscopic fields [25].

Using the developed theory, it is also straightforward to obtain the spectral density of the quantized field correlations in terms of a modal expansion. To this end, in the rest of this section we adopt the Heisenberg picture so that from (24.61) one finds that $\hat{\mathbf{f}}(\mathbf{r}, t) = \sum_{\omega_n>0} \sqrt{\hbar\omega_n/2}\left(\hat{a}_n \mathbf{f}_n(\mathbf{r})\, e^{-i\omega_n t} + \hat{a}_n^\dagger \mathbf{f}_n^*(\mathbf{r})\, e^{+i\omega_n t}\right)$. Thus, the Fourier transform of the quantized fields satisfies

$$\hat{\mathbf{f}}(\mathbf{r}, \omega) = 2\pi \sum_{\omega_n>0} \sqrt{\frac{\hbar\omega_n}{2}} \left(\hat{a}_n \mathbf{f}_n(\mathbf{r})\, \delta(\omega - \omega_n) + \hat{a}_n^\dagger \mathbf{f}_n^*(\mathbf{r})\, \delta(\omega + \omega_n)\right). \tag{24.65}$$

For two generic scalar operators, we introduce the symmetrized product as $\{\hat{A}\hat{B}\} = 1/2(\hat{A}\hat{B} + \hat{B}\hat{A})$. Some algebra shows that the quantum vacuum expectation ($\langle\rangle_0$) of the tensor operator $\{\hat{\mathbf{f}}(\mathbf{r}, \omega)\, \hat{\mathbf{f}}^\dagger(\mathbf{r}', \omega')\}$ is [30]

$$\begin{aligned}&\frac{1}{(2\pi)^2}\left\langle\left\{\hat{\mathbf{f}}(\mathbf{r}, \omega)\, \hat{\mathbf{f}}^\dagger(\mathbf{r}', \omega')\right\}\right\rangle_0 \\ &= \delta(\omega - \omega')\varepsilon_{0,\omega} \sum_{\omega_n>0} \frac{1}{2}\left[\delta(\omega - \omega_n)\, \mathbf{f}_n(\mathbf{r}) \otimes \mathbf{f}_n^*(\mathbf{r}') + \delta(\omega + \omega_n)\, \mathbf{f}_n^*(\mathbf{r}) \otimes \mathbf{f}_n(\mathbf{r}')\right],\end{aligned} \tag{24.66}$$

where $\varepsilon_{0,\omega} = \hbar|\omega|/2$ is the zero-point energy of a harmonic oscillator. With the help of (24.16), it is possible to write the field correlations in terms of the system Green function [30] (note that the Green function defined in [30] differs from the Green function used here):

$$\frac{1}{(2\pi)^2}\left\langle\left\{\hat{\mathbf{f}}(\mathbf{r},\omega)\,\hat{\mathbf{f}}^{\dagger}\left(\mathbf{r}',\omega'\right)\right\}\right\rangle_0 = \delta\left(\omega-\omega'\right)\varepsilon_{0,\omega}\frac{-1}{2\pi}\left[\overline{\mathbf{G}}\left(\mathbf{r},\mathbf{r}',\omega\right)+\overline{\mathbf{G}}^{\dagger}\left(\mathbf{r}',\mathbf{r},\omega\right)\right]_{\omega+0^+i}. \tag{24.67}$$

This result corresponds to the fluctuation–dissipation theorem in the limit of a zero temperature [48]. The formula can be readily extended to the case of thermally induced fluctuations simply by replacing $\varepsilon_{0,\omega}$ by the energy of a harmonic oscillator at temperature T, i.e., $\varepsilon_{T,\omega} = (\hbar\omega/2)\coth\left(\hbar\omega/2k_BT\right)$ [48].

To conclude, we note that by calculating the inverse Fourier transforms of the two operators in (24.39), it is found that in the time domain [30]

$$\left\langle\left\{\hat{\mathbf{f}}(\mathbf{r},t)\,\hat{\mathbf{f}}^{\dagger}\left(\mathbf{r}',t\right)\right\}\right\rangle_T = \int_0^{+\infty} d\omega\;\varepsilon_{T,\omega}\sum_{\omega_n>0}\delta\left(\omega-\omega_n\right)\frac{1}{2}\left[\mathbf{f}_n(\mathbf{r})\otimes\mathbf{f}_n^*\left(\mathbf{r}'\right)+\mathbf{f}_n^*(\mathbf{r})\otimes\mathbf{f}_n\left(\mathbf{r}'\right)\right]. \tag{24.68}$$

The subscript T indicates that the expectation is taken at the temperature T, and therefore $\varepsilon_{0,\omega}$ was replaced by $\varepsilon_{T,\omega}$.

24.11 Summary

It was proven that in the lossless case, the electrodynamics of a generic inhomogeneous possibly bianisotropic and possibly nonreciprocal system may be described by an augmented state vector, with the time evolution of the system determined by a Hermitian operator [5,14]. It was shown that the electromagnetic natural modes form a complete set of expansion functions, and we derived different modal expansions for a generic field distribution. The modal expansions in dispersive systems are not unique because the electromagnetic degrees of freedom span a subspace of the entire Hilbert space wherein the relevant Hermitian operator is defined. Moreover, it was shown that the natural modes satisfy generalized orthogonality relations [Equations (24.33a) and (24.33b)]. The developed theory gives the system Green function in terms of the natural modes and has been used to find the solution of standard radiation problems [28,29] and to characterize the light emission by moving sources [15,27,49].

The stored energy and canonical momentum in dispersive media were studied. We recovered a well-known textbook formula for the stored energy [Equation (24.20)] directly from the generalized Schrödinger-type formulation of Maxwell's equations. We also obtained an explicit formula for the canonical momentum in translation-invariant dispersive material platforms, which generalizes the findings of previous works [19,20,23–26].

Furthermore, it was highlighted that the Hermitian-type formulation of the dispersive Maxwell equations enables one to extend the powerful ideas of topological photonics to a wide range of electromagnetic systems, and to introduce concepts such as the Berry potential, Berry curvature and Chern numbers that are essential to characterize the topological phases of photonic platforms [5,12,13]. In addition, we illustrated how the developed formalism can be applied to quantum optics. In particular, we presented a simple procedure to quantize the electromagnetic field in a generic bianisotropic and nonreciprocal closed cavity and derived the quantum correlations of the electromagnetic fields. The developed methods are useful in problems of radiative heat transport [30,43,50], in the study of the interactions of atomic systems with the quantized electromagnetic field [27,28,51], and to characterize the quantum Casimir–Polder forces [27,28,51].

We hope that the described methods and ideas may stimulate further studies, developments and applications of modal expansions in photonics, plasmonics and metamaterials.

Acknowledgments

This work is supported in part by the IET under the A F Harvey Prize, by the Simons Foundation under the award 733700 (Simons Collaboration in Mathematics and Physics, "Harnessing Universal Symmetry Concepts for Extreme Wave Phenomena"), and by Instituto de Telecomunicações (IT) under project UIDB/50008/2020

References

[1] Collin R. E., *Field Theory of Guided Waves*, 2nd Ed., IEEE Press, New York, NY, 1991.

[2] Jackson J. D., *Classical Electrodynamics*, Wiley, New York, NY, 1998.

[3] Sakoda K., *Optical Properties of Photonic Crystals*, Springer, Berlin, Heidelberg, New York, NY, 20024.

[4] Milonni P. W., *The Quantum Vacuum: An Introduction to Quantum Electrodynamics*, Academic Press, San Diego, CA, 1994.

[5] Silveirinha M. G., "Chern invariants for continuous media", *Phys. Rev. B*, 2015, vol. 92, pp. 125153.

[6] Silveirinha M. G., "Topological classification of Chern-type insulators by means of the photonic Green function", *Phys. Rev. B*, 2018, vol. 97, pp. 115146.

[7] Lindell I. V., Sihvola A. H., Tretyakov S. A., and Viitanen A. J., *Electromagnetic Waves in Chiral and Bi-Isotropic Media*, Artech House, Boston, MA, 1994.

[8] Serdyukov A., Semchenko I., Tretyakov S., and Sihvola A., *Electromagnetics of Bianisotropic Materials Theory and Applications*, 1st Ed., Gordon and Breach Science Publishers, Amsterdam, 2001.

[9] Landau L. D. and Lifshitz E. M., *Electrodynamics of Continuous Media*, Course of Theoretical Physics, vol. 8, Elsevier Butterworth-Heinemann, Burlington MA, 2004.

[10] Agranovich V. and Ginzburg V., *Spatial Dispersion in Crystal Optics and the Theory of Excitons*, Wiley-Interscience, New York, NY, 1966.

[11] Silveirinha M. G. and Maslovski S. I., Comment on "Repulsive Casimir force in chiral metamaterials", *Phys. Rev. Lett.*, 2010, vol. 105, pp. 189301.

[12] Haldane F. D. M. and Raghu S., "Possible realization of directional optical waveguides in photonic crystals with broken time-reversal symmetry", *Phys. Rev. Lett.*, 2008, vol. 100, pp. 013904.

[13] Raghu S. and Haldane F. D. M., "Analogs of quantum-Hall-effect edge states in photonic crystals", *Phys. Rev. A*, 2008, vol. 78, pp. 033834.

[14] Raman A. and Fan S., "Photonic band structure of dispersive metamaterials formulated as a Hermitian eigenvalue problem", *Phys. Rev. Lett.*, 2010, vol. 104, pp. 087401.

[15] Morgado T. A., Fernandes D. E., and Silveirinha M. G., "Analytical solution for the stopping power of the Cherenkov radiation in a uniaxial nanowire material", *Photonics*, 2015, vol. 2, pp. 702.

[16] Dodin I. Y., "Geometric view on noneikonal waves", *Phys. Lett. A*, 2014, vol. 378, pp. 1598–1621.

[17] Silveirinha M. G., "Poynting vector, heating rate, and stored energy in structured materials: A first-principles derivation", *Phys. Rev. B*, 2009, vol. 80, pp. 235120.

[18] Costa J. T., Silveirinha M. G., and Alù A., "Poynting vector in negative-index metamaterials", *Phys. Rev. B*, 2011, vol. 83, pp. 165120.

[19] Bliokh K. Y., Bekshaev A. Y., and Nori F., "Optical momentum and angular momentum in dispersive media: From the Abraham-Minkowski debate to unusual properties of surface plasmon-polaritons", *New J. Phys.*, 2017, vol. 19, pp. 123014.

[20] Bliokh K. Y., Bekshaev A. Y., and Nori F., "Optical momentum, spin, and angular momentum in dispersive media", *Phys. Rev. Lett.*, 2017, vol. 119, pp. 073901.

[21] Barnett S. M. and Loudon R., "The enigma of optical momentum in a medium", *Phil. Trans. R. Soc. A*, 2010, vol. 368, pp. 927.

[22] Barnett S. M., "Resolution of the Abraham-Minkowski Dilemma", *Phys. Rev. Lett.*, 2010, vol. 104, pp. 070401.

[23] Dodin I. Y. and Fisch N. J., "Axiomatic geometrical optics, Abraham-Minkowski controversy, and photon properties derived classically", *Phys. Rev. A*, 2012, vol. 86. pp. 053834.

[24] Silveirinha M. G., "Reexamination of the Abraham-Minkowski Dilemma", *Phys. Rev. A*, 2017, vol. 96, pp. 033831.

[25] Silveirinha M. G. and Maslovski S. I., "Exchange of momentum between moving matter induced by the zero-point fluctuations of the electromagnetic field", *Phys. Rev. A*, 2012, vol. 86, pp. 042118.

[26] Philbin T. G., "Electromagnetic energy-momentum in dispersive media", *Phys. Rev. A*, 2011, vol. 83, pp. 013823; Erratum: *Phys. Rev. A*, 2012, vol. 85, pp. 059902(E).

[27] Lannebère S. and Silveirinha M. G., "Negative spontaneous emission by a moving two-level atom", *J. Opt.*, 2017, vol. 19, pp. 014004.

[28] Silveirinha M. G., Gangaraj S. A. H., Hanson G. W., and Antezza M., "Fluctuation-induced forces on an atom near a photonic topological material", *Phys. Rev. A*, 2018, vol. 97, pp. 022509.

[29] Morgado T. A. and Silveirinha M. G., "Super collimation of the radiation by a point source in a uniaxial wire medium", *EPJ Appl. Metamater.*, 2015, vol. 2, pp. 14.

[30] Silveirinha M. G., "Topological angular momentum and radiative heat transport in closed orbits", *Phys. Rev. B*, 2017, vol. 95, pp. 115103.

[31] Lu L., Joannopoulos J. D., and Soljaèiæ M., "Topological photonics", *Nat. Photonics*, 2014, vol. 8, pp. 821.

[32] Lu L., Joannopoulos J. D., and Soljaèiæ M., "Topological states in photonic systems", *Nat. Phys.*, 2016, vol. 12, pp. 626.

[33] Thouless D. J., Kohmoto M., Nightingale M. P., and den Nijs M., "Quantized Hall conductance in a two-dimensional periodic potential", *Phys. Rev. Lett.*, 1982, vol. 49, pp. 405.

[34] Haldane F. D. M., "Model for a quantum Hall effect without landau levels: condensed-matter realization of the parity anomaly", *Phys. Rev. Lett.* 1988, vol. 61, pp. 2015.

[35] Kane C. L. and Mele E. J., "Z_2 topological order and the quantum spin hall effect", *Phys. Rev. Lett.* 2005, vol. 95, pp. 146802.

[36] Bernervig B. A. and Hughes T., *Topological Insulators and Topological Superconductors*, Princeton University Press, Princeton, NJ, 2013.

[37] Khanikaev A. B., Mousavi S. H., Tse W.-K., Kargarian M., MacDonald A. H., and Shvets G., "Photonic topological insulators", *Nat. Mater.*, 2013, vol. 12, pp. 233.

[38] Silveirinha M. G., "Z_2 topological index for continuous photonic materials", *Phys. Rev. B*, 2016, vol. 93, pp. 075110.

[39] Silveirinha M. G., "Topological theory of non-Hermitian photonic systems", *Phys. Rev. B*, 2019, vol. 99, pp. 125155

[40] Wang Z., Chong Y., Joannopoulos J. D., and Soljacic M., "Observation of unidirectional backscattering immune topological electromagnetic states", *Nature*, 2009, vol. 461, pp. 772.

[41] Silveirinha M. G., "PTD symmetry protected scattering anomaly in optics", *Phys. Rev. B*, 2017, vol. 95, pp. 035153.

[42] Silveirinha M. G., "Bulk edge correspondence for topological photonic continua", *Phys. Rev. B*, 2016, vol. 94, pp. 205105.

[43] Silveirinha M. G., "Proof of the bulk-edge correspondence through a link between topological photonics and fluctuation-electrodynamics", *Phys. Rev. X*, 2019, vol. 9, pp. 011037.

[44] Silveirinha M. G., "Quantization of the electromagnetic field in non-dispersive polarizable moving media above the Cherenkov threshold", *Phys. Rev. A*, 2013, vol. 88, pp. 043846.

[45] Knoll L., Vogel W., and Welsch D. G., "Action of passive, lossless optical systems in quantum optics", *Phys. Rev. A*, 1987, vol. 36, pp. 3803.

[46] Glauber R. J. and Lewenstein M., "Quantum optics of dielectric media", *Phys. Rev. A*, 1991, vol. 43, pp. 467.

[47] Bhat N. A. R. and Sipe J. E., "Hamiltonian treatment of the electromagnetic field in dispersive and absorptive structured media", *Phys. Rev. A*, 2006, vol. 73, pp. 063808.

[48] Landau L. D. and Lifshitz E. M., *Statistical Physics Part 2*, vol. 9 of Course on Theoretical Physics, Pergamon Press, Oxford, 1981 (Chapter VIII, p. 320).

[49] Prudêncio F. R. and Silveirinha M. G., "Asymmetric Cherenkov emission in a topological plasmonic waveguide", *Phys. Rev. B*, 2018, vol. 98, pp. 115136.

[50] Silveirinha M. G., "Quantized angular momentum in topological optical systems", *Nat. Commun.*, 2019, vol. 10, pp. 349.

[51] Gangaraj S. A. H., Hanson G. W., Antezza M., and Silveirinha M. G., "Spontaneous lateral atomic recoil force close to a photonic topological material", *Phys. Rev. B*, 2018, vol. 97, pp. 201108(R).

Chapter 25

Multiple scattering by a collection of randomly located obstacles distributed in a dielectric slab

Gerhard Kristensson[1] and Niklas Wellander[2]

25.1 Introduction

Multiple scattering of electromagnetic waves by a discrete collection of scatterers is a well-studied subject, and many excellent treatments are found in the literature, see *e.g.*, [1,2] and references therein. The journal literature is vast, and too extensive to list here, and we refer to the list of references in [3] for further consultations. In some of previous treatments, the Null-field approach (Waterman's method) has been employed to solve the complex deterministic electromagnetic scattering problem, which serves as the starting point for the stochastic analysis of the problem. The Null-field approach is well documented [4,5], and a comprehensive database of the method has been collected in [6] and in the previously published updates listed of this reference. A simple introduction to the Null-field approach is found in [7].

The deterministic analysis of the scattering problem in this chapter is an extension of the problems treated in [8,9]. Moreover, the present analysis generalizes the established results in two previous papers [3,10] to a geometry with a more general background material, which is practical for a controlled experimental verification of the final result. The transmitted and reflected intensities are conveniently represented as a sum of two terms—the coherent and the incoherent contribution. In this chapter, we focus on the analysis of the coherent term.

The chapter is organized as follows. In Section 25.2, the geometry of the multiple electromagnetic scattering problem is given, and in Section 25.3, the main tool to solve the problem—the integral representation—is introduced. The integral representations are exploited in the various homogeneous regions of the problem in Section 25.4, and the appropriate expansions of the surface fields are introduced in Section 25.5. The final goal of the chapter is to calculate the transmitted and reflected coherent fields of the problem. This is done in Section 25.6. The stochastic description of the many-body problem in this chapter is made in Section 25.7, and two natural and

[1] Department of Electrical and Information Technology, Lund University, Lund, Sweden
[2] Swedish Defence Research Agency, Linköping, Sweden

important approximations—the tenuous media and low-frequency approximations—are developed in Section 25.8. The chapter ends with a short conclusion in Section 25.9 and several useful appendices.

25.2 The geometry

We study a collection of N different scatterers, where each scatterer is centered at the location $\boldsymbol{r}_p$, which defines the position of the local origin O_p, $p = 1, 2, \ldots, N$, relative the global origin O, see Figure 25.1. The volume of the individual scatterer is denoted as V_{sp}, bounded by the surface $S_{sp}, p = 1, 2, \ldots, N$. The radii of the maximum inscribed and minimum circumscribed spheres, both centered at the local origin, of each scatterer are denoted as a_p and A_p, $p = 1, 2, \ldots, N$, respectively. We assume that no minimum circumscribed spheres of the scatterers intersect. Each scatterer has its own material properties, which do not have to be the same for all scatterers.

The two half spaces are denoted as $V_1 = \{\boldsymbol{r} : z < z_1\}$ and $V_3 = \{\boldsymbol{r} : z > z_2\}$, and they are characterized by the lossless, homogeneous, isotropic (relative) material parameters ε_i and μ_i, $i = 1, 3$. All scatterers are located in a homogeneous, isotropic slab, volume V_2—bounded by the planes $z = z_1$ and $z = z_2$—with relative material parameters ε_2 and μ_2, which can be lossless or lossy. No minimum circumscribed sphere of the scatterers is assumed to intersect with the planes $z = z_1$ and $z = z_2$. Notice that the volume V_2 is confined by the two planes, and excludes the finite scatterers V_{sp}, $p = 1, 2, \ldots, N$, *i.e.*, $V_2 = \{\boldsymbol{r} \in \mathbb{R}^3 : z_1 < z < z_2, \boldsymbol{r} \notin \overline{V_{sp}}, p = 1, 2, \ldots, N\}$.

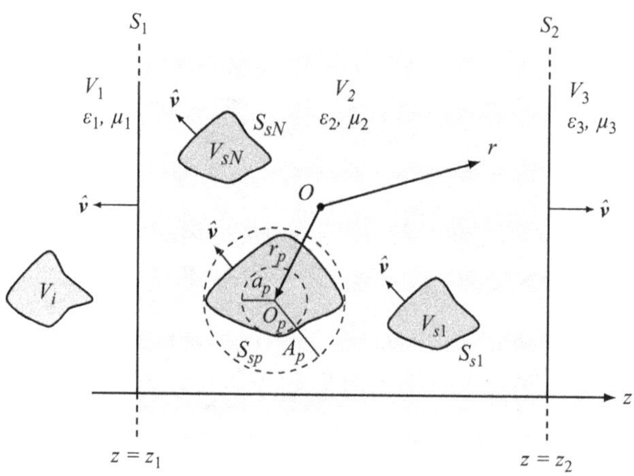

Figure 25.1 The geometry of a collection of the N scatterers and the region of prescribed sources V_i. The figure also shows the orientation of the unit normals of the surfaces.

The prescribed sources of the configuration are located in the region V_i, which is located in V_1. These sources generate an electric field $\boldsymbol{E}_\mathrm{i}(\boldsymbol{r})$ and a magnetic field $\boldsymbol{H}_\mathrm{i}(\boldsymbol{r})$ everywhere outside V_i. Generalizations to other locations of the sources, such as in V_2, are possible, but not pursued in this chapter.

With no loss of generality, we start with a collection of perfectly conducting obstacles in V_2 and then generalize to more complex scatterers later. The reason for this assumption is to make the derivation of the results more simple. The final result holds for a collection of scatterers with arbitrary materials and geometry, as long as the transition matrices of the scatterers are known.

In each domain, V_i, $i = 1, 2, 3$, the Maxwell equations are satisfied (the time convention $\mathrm{e}^{-\mathrm{i}\omega t}$ is used, where ω is the angular frequency of the problem)

$$\begin{cases} \nabla \times \boldsymbol{E}(\boldsymbol{r}) = \mathrm{i}\omega\mu_0\mu_i\boldsymbol{H}(\boldsymbol{r}) = \mathrm{i}k_i\eta_0\eta_i\boldsymbol{H}(\boldsymbol{r}) \\ \eta_0\eta_i\nabla \times \boldsymbol{H}(\boldsymbol{r}) = -\mathrm{i}k_i\boldsymbol{E}(\boldsymbol{r}) \end{cases} \quad \boldsymbol{r} \in V_i, \quad i = 1, 2, 3$$

where the wave number $k_i = \omega(\varepsilon_0\mu_0\varepsilon_i\mu_i)^{1/2}$, and the wave impedances are given by $\eta_0 = (\mu_0/\varepsilon_0)^{1/2}$ and $\eta_i = (\mu_i/\varepsilon_i)^{1/2}$, $i = 1, 2, 3$, and ε_0 and μ_0 denote the permittivity and the permeability of vacuum, respectively.

Each scatterer is characterized by the boundary condition $\hat{\boldsymbol{\nu}} \times \boldsymbol{E} = \boldsymbol{0}$ on all $S_{\mathrm{s}p}$, $p = 1, 2, \ldots, N$. The boundary condition on the interfaces $z = z_1$ and $z = z_2$ are $\hat{\boldsymbol{\nu}} \times \boldsymbol{E}_- = \hat{\boldsymbol{\nu}} \times \boldsymbol{E}_+$ and $\hat{\boldsymbol{\nu}} \times \boldsymbol{H}_- = \hat{\boldsymbol{\nu}} \times \boldsymbol{H}_+$. The subscript $\pm$ denotes the limit values of the fields on the surface w.r.t. to the direction of the unit normal vector, see Figure 25.1.

25.3 Integral representation

The deterministic scattering problem, which is the first problem we address, is solved by the systematic employment of two main tools—the integral representation of the solution in a homogeneous domain and the decomposition of the Green dyadic for the electric field in free space [7].

Our starting point is the three integral representations of the electric field in the domains V_i, $i = 1, 2, 3$. The appropriate integral representation in V_1 is

$$-\frac{1}{\mathrm{i}k_1}\nabla \times \left\{\nabla \times \iint_{S_1} \mathbf{G}_\mathrm{e}(k_1, |\boldsymbol{r} - \boldsymbol{r}'|) \cdot \left(\hat{\boldsymbol{\nu}} \times \eta_0\eta_1\boldsymbol{H}(\boldsymbol{r}')\right)_+ \mathrm{d}S'\right\}$$
$$+\nabla \times \iint_{S_1} \mathbf{G}_\mathrm{e}(k_1, |\boldsymbol{r} - \boldsymbol{r}'|) \cdot \left(\hat{\boldsymbol{\nu}} \times \boldsymbol{E}(\boldsymbol{r}')\right)_+ \mathrm{d}S' = \begin{cases} \boldsymbol{E}_\mathrm{r}(\boldsymbol{r}), & \boldsymbol{r} \text{ inside } V_1 \\ -\boldsymbol{E}_\mathrm{i}(\boldsymbol{r}), & \boldsymbol{r} \text{ outside } V_1 \end{cases} \tag{25.1}$$

where the boundary values are taken as limits from the positive side of the surface S_1 w.r.t. the unit normal vector. The total fields on the left-hand side of the integral representation are the sum of the incident and reflected fields, *i.e.*, $\boldsymbol{E} = \boldsymbol{E}_\mathrm{i} + \boldsymbol{E}_\mathrm{r}$. To close the surface, we have added a half sphere in the left half space of V_1. This surface integral, due to radiation conditions, gives zero contribution [7]. The Green dyadic for the electric field in free space $\mathbf{G}_e(k, |\boldsymbol{r} - \boldsymbol{r}'|)$ is [7]

$$\mathbf{G}_e(k, |\boldsymbol{r} - \boldsymbol{r}'|) = \left(\mathbf{I}_3 + \frac{1}{k^2}\nabla\nabla\right)\frac{e^{ik|\boldsymbol{r}-\boldsymbol{r}'|}}{4\pi|\boldsymbol{r} - \boldsymbol{r}'|}$$

The domain V_2 is bounded by the planes $z = z_1$ and $z = z_2$ and excludes all scatterers $V_{sp}, p = 1, 2, \ldots, N$. The pertinent integral representation is

$$\begin{aligned}
&-\frac{1}{ik_2}\nabla\times\left\{\nabla\times\sum_{p=1}^{N}\iint_{S_{sp}}\mathbf{G}_e(k_2, |\boldsymbol{r} - \boldsymbol{r}'|)\cdot\left(\hat{\boldsymbol{\nu}}\times\eta_0\eta_2\boldsymbol{H}(\boldsymbol{r}')\right)_+\,dS'\right\}\\
&+\frac{1}{ik_2}\nabla\times\left\{\nabla\times\sum_{i=1}^{2}\iint_{S_i}\mathbf{G}_e(k_2, |\boldsymbol{r} - \boldsymbol{r}'|)\cdot\left(\hat{\boldsymbol{\nu}}\times\eta_0\eta_2\boldsymbol{H}(\boldsymbol{r}')\right)_-\,dS'\right\}\\
&-\nabla\times\sum_{i=1}^{2}\iint_{S_i}\mathbf{G}_e(k_2, |\boldsymbol{r} - \boldsymbol{r}'|)\cdot\left(\hat{\boldsymbol{\nu}}\times\boldsymbol{E}(\boldsymbol{r}')\right)_-\,dS' = \begin{cases}\boldsymbol{E}(\boldsymbol{r}), & \boldsymbol{r}\text{ inside }V_2\\ \mathbf{0}, & \boldsymbol{r}\text{ outside }V_2\end{cases}
\end{aligned} \tag{25.2}$$

where the boundary condition $\hat{\boldsymbol{\nu}} \times \boldsymbol{E} = \boldsymbol{0}$ on each $S_{sp}, p = 1, 2, \ldots, N$, has been used. The boundary values on the planar surfaces are taken as limits from negative side of the surfaces w.r.t. to their unit normal vectors. In particular, the lower row in (25.2) holds for all points inside a particular scatterer bounded by S_{sp}.

The final integral representation of the domain $V_3 = \{\boldsymbol{r} : z > z_2\}$ is

$$\begin{aligned}
&-\frac{1}{ik_3}\nabla\times\left\{\nabla\times\iint_{S_2}\mathbf{G}_e(k_3, |\boldsymbol{r} - \boldsymbol{r}'|)\cdot\left(\hat{\boldsymbol{\nu}}\times\eta_0\eta_3\boldsymbol{H}(\boldsymbol{r}')\right)_+\,dS'\right\}\\
&+\nabla\times\iint_{S_2}\mathbf{G}_e(k_3, |\boldsymbol{r} - \boldsymbol{r}'|)\cdot\left(\hat{\boldsymbol{\nu}}\times\boldsymbol{E}(\boldsymbol{r}')\right)_+\,dS' = \begin{cases}\boldsymbol{E}(\boldsymbol{r}), & \boldsymbol{r}\text{ inside }V_3\\ \mathbf{0}, & \boldsymbol{r}\text{ outside }V_3\end{cases}
\end{aligned} \tag{25.3}$$

where the boundary values are taken as limits from the right-hand side of the surface S_2, and the total field in V_3 is denoted as $\boldsymbol{E}_t(\boldsymbol{r}) = \boldsymbol{E}(\boldsymbol{r})$ (the transmitted field).

A simple count of the number of unknown surface fields gives four (two surface fields on S_1 and two surface fields on S_2) plus N unknown surface fields on the N scatterers in the slab. In this count, the boundary conditions on each interface have been used. The number of extinction parts (the number of different, distinct locations of the position vector $\boldsymbol{r}$ in the lower rows) in the integral representations is one in (25.1), two plus N in (25.2), and one in (25.3). Consequently, the number of unknowns and equations matches. The top rows in the integral representations then give the fields in each specific volume. In particular, the reflected and transmitted fields are the main fields of interest in this chapter.

25.4 Exploiting the integral representations

The position vector, $\boldsymbol{r}$, can take several principal positions. We exploit each of these cases to connect the fields in the three different regions to each other.

Let the position vector $\boldsymbol{r}$ be located to the right of the surface $z = z_1$, *i.e.*, $z > z_1$. The incident electric field has the form

$$\boldsymbol{E}_{\mathrm{i}}(\boldsymbol{r}) = \sum_{j=1,2} \iint_{\mathbb{R}^2} a_j(\boldsymbol{k}_{\mathrm{t}}) \boldsymbol{\varphi}_j^{+}(\boldsymbol{k}_{\mathrm{t}}; k_1, \boldsymbol{r})\, \mathrm{d}k_{\mathrm{x}}\, \mathrm{d}k_{\mathrm{y}}, \quad z > z_1 \tag{25.4}$$

where the expansion coefficients $a_j(\boldsymbol{k}_{\mathrm{t}})$, for a given incident field, are obtained by the use of (C.1) in Appendix C. These coefficients are the excitation of the scattering problem in this chapter. The lower part in integral representation (25.1) and the decomposition of the Green dyadic (D.2) imply that the surface fields satisfy

$$\begin{aligned} a_j(\boldsymbol{k}_{\mathrm{t}}) &= -\frac{2\mathrm{i}k_1}{k_{1\mathrm{z}}} \iint_{S_1} \boldsymbol{\varphi}_j^{-\dagger}(\boldsymbol{k}_{\mathrm{t}}; k_1, \boldsymbol{r}') \cdot \left(\hat{\boldsymbol{\nu}} \times \mathrm{i}\eta_0\eta_1 \boldsymbol{H}(\boldsymbol{r}')\right)_+ \mathrm{d}S' \\ &\quad - \frac{2\mathrm{i}k_1}{k_{1\mathrm{z}}} \iint_{S_1} \boldsymbol{\varphi}_{\bar{j}}^{-\dagger}(\boldsymbol{k}_{\mathrm{t}}; k_1, \boldsymbol{r}') \cdot \left(\hat{\boldsymbol{\nu}} \times \boldsymbol{E}(\boldsymbol{r}')\right)_+ \mathrm{d}S', \quad \boldsymbol{k}_{\mathrm{t}} \in \mathbb{R}^2, j = 1, 2 \end{aligned} \tag{25.5}$$

where $k_{1\mathrm{z}} = \left(k_1^2 - k_{\mathrm{t}}^2\right)^{1/2}$ and where $\bar{j}$ is the dual index of j, *i.e.*, $\bar{1} = 2$ and $\bar{2} = 1$. Equation (25.4) is a representation of the incident field in the region $z > z_1$. The null-field equation (25.5) is an equation for the unknown surface fields. The expansion functions of the incident field on the left-hand side drive the equation and the whole system, since the null-field equations are coupled via the boundary conditions on the surface S_1.

Similarly, by expanding the reflected wave in planar vector waves for a position vector to the left of $z = z_1$, $z < z_1$, we have (using (25.1) and (D.2) in Appendix D and interchanging the order of integration)

$$\boldsymbol{E}_{\mathrm{r}}(\boldsymbol{r}) = \sum_{j=1,2} \iint_{\mathbb{R}^2} r_j(\boldsymbol{k}_{\mathrm{t}}) \boldsymbol{\varphi}_j^{-}(\boldsymbol{k}_{\mathrm{t}}; k_1, \boldsymbol{r})\, \mathrm{d}k_{\mathrm{x}}\, \mathrm{d}k_{\mathrm{y}}, \quad z < z_1 \tag{25.6}$$

where the reflection coefficients are given by

$$\begin{aligned} r_j(\boldsymbol{k}_{\mathrm{t}}) &= \frac{2\mathrm{i}k_1}{k_{1\mathrm{z}}} \iint_{S_1} \boldsymbol{\varphi}_j^{+\dagger}(\boldsymbol{k}_{\mathrm{t}}; k_1, \boldsymbol{r}') \cdot \left(\hat{\boldsymbol{\nu}} \times \mathrm{i}\eta_0\eta_1 \boldsymbol{H}(\boldsymbol{r}')\right)_+ \mathrm{d}S' \\ &\quad + \frac{2\mathrm{i}k_1}{k_{1\mathrm{z}}} \iint_{S_1} \boldsymbol{\varphi}_{\bar{j}}^{+\dagger}(\boldsymbol{k}_{\mathrm{t}}; k_1, \boldsymbol{r}') \cdot \left(\hat{\boldsymbol{\nu}} \times \boldsymbol{E}(\boldsymbol{r}')\right)_+ \mathrm{d}S', \quad \boldsymbol{k}_{\mathrm{t}} \in \mathbb{R}^2, j = 1, 2 \end{aligned} \tag{25.7}$$

once the surface fields on S_1 are known.

We proceed with the integral representation in V_2. The extinction part of the integral representation (25.2) has three principal regions, corresponding to the position vector $\boldsymbol{r}$ not in V_2, (1) inside any of the scatterers, (2) to the left of $z = z_1$, and (3) to the right of $z = z_2$. We exploit the first case in detail.

To prepare the analysis later, we decompose the Green dyadic for the electric field in free space, see (D.1) in Appendix D. For an observation point inside the inscribed sphere of a particular scatterer, say the pth scatterer, we have $|\boldsymbol{r} - \boldsymbol{r}_p| < a_p$, where the vector $\boldsymbol{r}_p$ denotes the local origin depicted in Figure 25.1. The Green dyadic is then decomposed as, see (D.1),

$$\mathbf{G}_{\mathrm{e}}(k_2, |\boldsymbol{r} - \boldsymbol{r}'|) = \mathrm{i}k_2 \sum_n \boldsymbol{v}_n(k_2(\boldsymbol{r} - \boldsymbol{r}_p))\boldsymbol{u}_n(k_2(\boldsymbol{r}' - \boldsymbol{r}_p)), \quad \boldsymbol{r}' \in S_{\mathrm{s}p}$$

for this particular scatterer. On all other scatterers, we have $|\boldsymbol{r} - \boldsymbol{r}_q| > |\boldsymbol{r}' - \boldsymbol{r}_q|$, where $\boldsymbol{r}$ is inside the inscribed sphere of $S_{\mathrm{s}p}$ and $\boldsymbol{r}' \in S_{\mathrm{s}q}$, $q = 1, 2, \ldots, N$, $q \neq p$. The Green dyadic is then decomposed as

$$\mathbf{G}_{\mathrm{e}}(k_2, |\boldsymbol{r} - \boldsymbol{r}'|) = \mathrm{i}k_2 \sum_n \boldsymbol{u}_n(k_2(\boldsymbol{r} - \boldsymbol{r}_q))\boldsymbol{v}_n(k_2(\boldsymbol{r}' - \boldsymbol{r}_q)) \tag{25.8}$$

where $\boldsymbol{r}' \in S_{\mathrm{s}q}$, $q = 1, 2, \ldots, N$, and $q \neq p$.

For this choice of position of $\boldsymbol{r}$, the integral representation in (25.2) implies with the two decompositions of the Green dyadic, and the plane wave decomposition of the Green dyadic (D.2) ($p = 1, 2, \ldots, N$)

$$\begin{aligned}\boldsymbol{E}_{\mathrm{exc}p}(\boldsymbol{r}) \stackrel{\mathrm{def}}{=} \sum_n \alpha_n^p \boldsymbol{v}_n(k_2(\boldsymbol{r} - \boldsymbol{r}_p)) &= \sum_{\substack{q=1\\ q\neq p}}^{N} \sum_n f_n^q \boldsymbol{u}_n(k_2(\boldsymbol{r} - \boldsymbol{r}_q)) \\ &+ \sum_{j=1,2} \iint_{\mathbb{R}^2} \varphi_j^+(\boldsymbol{k}_{\mathrm{t}}; k_2, \boldsymbol{r})\alpha_j^+(\boldsymbol{k}_{\mathrm{t}})\,\mathrm{d}k_{\mathrm{x}}\,\mathrm{d}k_{\mathrm{y}} \\ &+ \sum_{j=1,2} \iint_{\mathbb{R}^2} \varphi_j^-(\boldsymbol{k}_{\mathrm{t}}; k_2, \boldsymbol{r})\alpha_j^-(\boldsymbol{k}_{\mathrm{t}})\,\mathrm{d}k_{\mathrm{x}}\,\mathrm{d}k_{\mathrm{y}}, \quad |\boldsymbol{r} - \boldsymbol{r}_p| < a_p\end{aligned} \tag{25.9}$$

where the expansion coefficients, α_n^p and f_n^p, are defined as

$$\alpha_n^p = k_2^2 \iint_{S_{\mathrm{s}p}} \boldsymbol{u}_n(k_2(\boldsymbol{r}' - \boldsymbol{r}_p)) \cdot \left(\hat{\boldsymbol{\nu}}(\boldsymbol{r}') \times \eta_0\eta_2\boldsymbol{H}(\boldsymbol{r}')\right)_+ \,\mathrm{d}S', \quad p = 1, 2, \ldots, N \tag{25.10}$$

and

$$f_n^p = -k_2^2 \iint_{S_{\mathrm{s}p}} \boldsymbol{v}_n(k_2(\boldsymbol{r}' - \boldsymbol{r}_p)) \cdot \left(\hat{\boldsymbol{\nu}}(\boldsymbol{r}') \times \eta_0\eta_2\boldsymbol{H}(\boldsymbol{r}')\right)_+ \,\mathrm{d}S', \quad p = 1, 2, \ldots, N \tag{25.11}$$

and where the expansion functions $\alpha_j^{\pm}(\boldsymbol{k}_t)$ are ($k_{2z} = \left(k_2^2 - k_t^2\right)^{1/2}$)

$$\begin{aligned}\alpha_j^{+}(\boldsymbol{k}_t) &= -\frac{2ik_2}{k_{2z}}\iint_{S_1} \varphi_j^{-\dagger}(\boldsymbol{k}_t;k_2,\boldsymbol{r}')\cdot\left(\hat{\boldsymbol{\nu}}\times i\eta_0\eta_2\boldsymbol{H}(\boldsymbol{r}')\right)_-\,dS' \\ &\quad -\frac{2ik_2}{k_{2z}}\iint_{S_1} \varphi_{\bar{j}}^{-\dagger}(\boldsymbol{k}_t;k_2,\boldsymbol{r}')\cdot\left(\hat{\boldsymbol{\nu}}\times \boldsymbol{E}(\boldsymbol{r}')\right)_-\,dS', \quad \boldsymbol{k}_t\in\mathbb{R}^2,\ j=1,2\end{aligned} \tag{25.12}$$

and

$$\begin{aligned}\alpha_j^{-}(\boldsymbol{k}_t) &= -\frac{2ik_2}{k_{2z}}\iint_{S_2} \varphi_j^{+\dagger}(\boldsymbol{k}_t;k_2,\boldsymbol{r}')\cdot\left(\hat{\boldsymbol{\nu}}\times i\eta_0\eta_2\boldsymbol{H}(\boldsymbol{r}')\right)_-\,dS' \\ &\quad -\frac{2ik_2}{k_{2z}}\iint_{S_2} \varphi_{\bar{j}}^{+\dagger}(\boldsymbol{k}_t;k_2,\boldsymbol{r}')\cdot\left(\hat{\boldsymbol{\nu}}\times \boldsymbol{E}(\boldsymbol{r}')\right)_-\,dS', \quad \boldsymbol{k}_t\in\mathbb{R}^2,\ j=1,2\end{aligned} \tag{25.13}$$

The exciting field $\boldsymbol{E}_{\mathrm{exc}p}(\boldsymbol{r})$ in (25.9) is the exciting field at the position of the scatterer located at $\boldsymbol{r}_p$. This field consists of the contributions from the two planar surfaces S_1 and S_2 plus the scattered fields from all scatterers except the contribution from the pth scatterer itself.

The transition matrix of the pth scatterer, $T^p_{nn'}$, connects the expansion coefficients α^p_n in (25.10) and f^p_n in (25.11) to each other, *viz.*

$$f^p_n = \sum_{n'} T^p_{nn'}\alpha^p_{n'}, \quad p = 1,2,\ldots,N \tag{25.14}$$

The transition matrix $T^p_{nn'}$ is the linear relation between the expansion coefficients of the scattered field, f^p_n, in terms of outgoing spherical vector waves, $\boldsymbol{u}_n(k_2(\boldsymbol{r}-\boldsymbol{r}_p))$, at the local origin $\boldsymbol{r}_p$, and the expansion coefficients, α^p_n, of the local excitation in terms of the regular spherical vector waves, $\boldsymbol{v}_n(k_2(\boldsymbol{r}-\boldsymbol{r}_p))$, at the local origin $\boldsymbol{r}_p$, see (25.9). Some known properties of the transition matrix can be found in [7,11].

To proceed, expand the unknown surface field on the pth scatterer, $i\hat{\boldsymbol{\nu}}\times\eta_0\eta_2\boldsymbol{H}$, in the transverse regular spherical vector waves. This is a complete set of expansion functions in the space of square integrable tangential functions [12]. We have

$$i\eta_0\eta_2\,\hat{\boldsymbol{\nu}}(\boldsymbol{r})\times\boldsymbol{H}(\boldsymbol{r})\big|_+ = \sum_n \beta_n\boldsymbol{\nu}(\boldsymbol{r})\times\boldsymbol{v}_{\bar{n}}(k_2(\boldsymbol{r}-\boldsymbol{r}_p)), \quad \boldsymbol{r}\in S_p,\ p=1,2,\ldots,N$$

where $\bar{n}$ denotes the index set $\{\bar{\tau},\sigma,m,l\}$, and where $\bar{\tau}$ is the dual index of τ, *i.e.*, $\bar{1}=2$ and $\bar{2}=1$. Introduce the two matrices

$$R^p_{nn'} = -ik_2^2\iint_{S_{sp}} \boldsymbol{v}_n(k_2(\boldsymbol{r}-\boldsymbol{r}_p))\cdot\left(\hat{\boldsymbol{\nu}}(\boldsymbol{r})\times\boldsymbol{v}_{\overline{n'}}(k_2(\boldsymbol{r}-\boldsymbol{r}_p))\right)\,dS$$

and

$$Q^p_{nn'} = -\mathrm{i}k_2^2 \iint_{S_{\mathrm{s}p}} \boldsymbol{u}_n(k_2(\boldsymbol{r}-\boldsymbol{r}_p)) \cdot \left(\hat{\boldsymbol{\nu}}(\boldsymbol{r}) \times \boldsymbol{v}_{\overline{n'}}(k_2(\boldsymbol{r}-\boldsymbol{r}_p))\right) \, \mathrm{d}S$$

Then, from (25.10) and (25.11)

$$f^p_n = -\sum_{n'} R^p_{nn'} \beta^p_{n'}, \quad \alpha^p_n = \sum_{n'} Q^p_{nn'} \beta^p_{n'}, \quad p = 1, 2, \ldots, N$$

and formal elimination of the expansion coefficients β^p_n gives

$$T^p_{nn'} = -\sum_{n''} R^p_{nn''} (Q^p)^{-1}_{n''n'}$$

This is the classic construction of the transition matrix by means of the Null-field approach (Waterman's method). The mathematical justification of the Null-field approach has been the subject of intensive research. Some of the efforts are reported in the literature [11,13]. Next, rewrite (25.9) by translating the planar vector waves to the local origin O_p, *i.e.*,

$$\begin{aligned}\sum_n \alpha^p_n \boldsymbol{v}_n(k_2(\boldsymbol{r}-\boldsymbol{r}_p)) &= \sum_{\substack{q=1\\ q\neq p}}^{N} \sum_n f^q_n \boldsymbol{u}_n(k_2(\boldsymbol{r}-\boldsymbol{r}_q)) \\ &+ \sum_{j=1,2} \iint_{\mathbb{R}^2} \boldsymbol{\varphi}^+_j(\boldsymbol{k}_\mathrm{t}; k_2, \boldsymbol{r}-\boldsymbol{r}_p) \mathrm{e}^{\mathrm{i}\boldsymbol{k}_2^+ \cdot \boldsymbol{r}_p} \alpha^+_j(\boldsymbol{k}_\mathrm{t}) \, \mathrm{d}k_\mathrm{x} \, \mathrm{d}k_\mathrm{y} \\ &+ \sum_{j=1,2} \iint_{\mathbb{R}^2} \boldsymbol{\varphi}^-_j(\boldsymbol{k}_\mathrm{t}; k_2, \boldsymbol{r}-\boldsymbol{r}_p) \mathrm{e}^{\mathrm{i}\boldsymbol{k}_2^- \cdot \boldsymbol{r}_p} \alpha^-_j(\boldsymbol{k}_\mathrm{t}) \, \mathrm{d}k_\mathrm{x} \, \mathrm{d}k_\mathrm{y}, \quad |\boldsymbol{r}-\boldsymbol{r}_p| < a_p\end{aligned}$$

where $\boldsymbol{k}_2^\pm = \boldsymbol{k}_\mathrm{t} \pm k_{2z}\hat{\boldsymbol{z}}$. We now use the translation properties of the spherical vector waves, see Appendix B and [7]

$$\boldsymbol{u}_n(k_2(\boldsymbol{r}-\boldsymbol{r}_q)) = \sum_{n'} \mathcal{P}_{nn'}(k_2(\boldsymbol{r}_p-\boldsymbol{r}_q)) \boldsymbol{v}_{n'}(k_2(\boldsymbol{r}-\boldsymbol{r}_p)), \quad |\boldsymbol{r}-\boldsymbol{r}_p| < |\boldsymbol{r}_p-\boldsymbol{r}_q|$$

and the expansion of planar waves in regular spherical vector waves, see Appendix D

$$\boldsymbol{\varphi}^\pm_j(\boldsymbol{k}_\mathrm{t}; k_2, \boldsymbol{r}-\boldsymbol{r}_p) = \sum_n B^{\pm\dagger}_{nj}(\boldsymbol{k}_\mathrm{t}) \boldsymbol{v}_n(k_2(\boldsymbol{r}-\boldsymbol{r}_p))$$

We get

$$\sum_n \alpha_n^p \boldsymbol{v}_n(k_2(\boldsymbol{r}-\boldsymbol{r}_p)) = \sum_{\substack{q=1\\q\neq p}}^N \sum_{nn'} f_n^q \mathcal{P}_{nn'}(k_2(\boldsymbol{r}_p-\boldsymbol{r}_q))\boldsymbol{v}_{n'}(k_2(\boldsymbol{r}-\boldsymbol{r}_p))$$
$$+\sum_n \sum_{j=1,2} \iint_{\mathbb{R}^2} B_{nj}^{+\dagger}(\boldsymbol{k}_\mathrm{t})\boldsymbol{v}_n(k_2(\boldsymbol{r}-\boldsymbol{r}_p))\mathrm{e}^{\mathrm{i}\boldsymbol{k}_2^+\cdot\boldsymbol{r}_p}\alpha_j^+(\boldsymbol{k}_\mathrm{t})\,\mathrm{d}k_\mathrm{x}\,\mathrm{d}k_\mathrm{y}$$
$$+\sum_n \sum_{j=1,2} \iint_{\mathbb{R}^2} B_{nj}^{-\dagger}(\boldsymbol{k}_\mathrm{t})\boldsymbol{v}_n(k_2(\boldsymbol{r}-\boldsymbol{r}_p))\mathrm{e}^{\mathrm{i}\boldsymbol{k}_2^-\cdot\boldsymbol{r}_p}\alpha_j^-(\boldsymbol{k}_\mathrm{t})\,\mathrm{d}k_\mathrm{x}\,\mathrm{d}k_\mathrm{y}, \quad |\boldsymbol{r}-\boldsymbol{r}_p| < a_p$$

The orthogonality of the regular spherical vector waves on a spherical surface implies

$$\alpha_n^p = \sum_{\substack{q=1\\q\neq p}}^N \sum_{n'} f_{n'}^q \mathcal{P}_{n'n}(k_2(\boldsymbol{r}_p-\boldsymbol{r}_q)) + \sum_{j=1,2} \iint_{\mathbb{R}^2} B_{nj}^{+\dagger}(\boldsymbol{k}_\mathrm{t})\mathrm{e}^{\mathrm{i}\boldsymbol{k}_2^+\cdot\boldsymbol{r}_p}\alpha_j^+(\boldsymbol{k}_\mathrm{t})\,\mathrm{d}k_\mathrm{x}\,\mathrm{d}k_\mathrm{y}$$
$$+\sum_{j=1,2} \iint_{\mathbb{R}^2} B_{nj}^{-\dagger}(\boldsymbol{k}_\mathrm{t})\mathrm{e}^{\mathrm{i}\boldsymbol{k}_2^-\cdot\boldsymbol{r}_p}\alpha_j^-(\boldsymbol{k}_\mathrm{t})\,\mathrm{d}k_\mathrm{x}\,\mathrm{d}k_\mathrm{y}, \quad p=1,2,\ldots,N \tag{25.15}$$

Equation (25.14) is used in (25.15), when we in Section 25.5 derive the final system of equations for the deterministic system for a finite number of scatterers in the slab.

Now, we let the position vector $\boldsymbol{r}$ be located in V_2. Moreover, assume that the observation point lies outside all minimum circumscribed spheres of the scatterers, *i.e.*, the position vector $\boldsymbol{r}$ satisfies $|\boldsymbol{r}-\boldsymbol{r}_p| > A_p$ for all $p = 1,2,\ldots,N$, and the Green dyadic for the electric field in free space is decomposed as in (25.8).

Outside all minimum circumscribed spheres, we get, using (25.8) in the upper row in (25.2) and using (D.2) in the integrals over S_1 and S_2 ($\boldsymbol{r} \in V_2$ and $|\boldsymbol{r}-\boldsymbol{r}_p| > A_p$ for all $p = 1,2,\ldots,N$)

$$\boldsymbol{E}(\boldsymbol{r}) = \sum_n \sum_{p=1}^N f_n^p \boldsymbol{u}_n(k_2(\boldsymbol{r}-\boldsymbol{r}_p)) + \sum_{j=1,2} \iint_{\mathbb{R}^2} \boldsymbol{\varphi}_j^+(\boldsymbol{k}_\mathrm{t};k_2,\boldsymbol{r})\alpha_j^+(\boldsymbol{k}_\mathrm{t})\,\mathrm{d}k_\mathrm{x}\,\mathrm{d}k_\mathrm{y}$$
$$+\sum_{j=1,2} \iint_{\mathbb{R}^2} \boldsymbol{\varphi}_j^-(\boldsymbol{k}_\mathrm{t};k_2,\boldsymbol{r})\alpha_j^-(\boldsymbol{k}_\mathrm{t})\,\mathrm{d}k_\mathrm{x}\,\mathrm{d}k_\mathrm{y} \tag{25.16}$$

where f_n^p and $a_j^\pm(\boldsymbol{k}_\mathrm{t})$ are given in (25.11), (25.12), and (25.13), respectively. This expression gives the field inside the slab provided the unknowns, f_n^p and $\alpha_j^\pm(\boldsymbol{k}_\mathrm{t})$, can be found.

Finally, we exploit the integral representation in V_3. Let the position vector $\boldsymbol{r}$ be located to the left of $z = z_2$, $z < z_2$. The integral representation (25.3) and the decomposition of the Green dyadic (D.2) imply ($k_{3z} = \left(k_3^2 - k_t^2\right)^{1/2}$)

$$0 = \frac{2\mathrm{i}k_3}{k_{3z}} \iint_{S_2} \varphi_j^{+\dagger}(\boldsymbol{k}_t; k_3, \boldsymbol{r}') \cdot \left(\hat{\boldsymbol{\nu}} \times \mathrm{i}\eta_0\eta_3 \boldsymbol{H}(\boldsymbol{r}')\right)_+ \, \mathrm{d}S' + \frac{2\mathrm{i}k_3}{k_{3z}} \iint_{S_2} \varphi_{\bar{j}}^{+\dagger}(\boldsymbol{k}_t; k_3, \boldsymbol{r}') \cdot \left(\hat{\boldsymbol{\nu}} \times \boldsymbol{E}(\boldsymbol{r}')\right)_+ \, \mathrm{d}S', \quad \boldsymbol{k}_t \in \mathbb{R}^2, \ j = 1, 2 \tag{25.17}$$

Similarly, for a position vector to the right of the surface $z = z_2$, $z > z_2$, we get a representation of the transmitted fields in planar vector waves. Analogously to the previous analysis, we obtain

$$\boldsymbol{E}_t(\boldsymbol{r}) = \sum_{j=1,2} \iint_{\mathbb{R}^2} t_j(\boldsymbol{k}_t) \varphi_j^{+}(\boldsymbol{k}_t; k_3, \boldsymbol{r}) \, \mathrm{d}k_x \, \mathrm{d}k_y, \quad z > z_2 \tag{25.18}$$

where the transmission coefficients are given by

$$t_j(\boldsymbol{k}_t) = \frac{2\mathrm{i}k_3}{k_{3z}} \iint_{S_2} \varphi_j^{-\dagger}(\boldsymbol{k}_t; k_3, \boldsymbol{r}') \cdot \left(\hat{\boldsymbol{\nu}} \times \mathrm{i}\eta_0\eta_3 \boldsymbol{H}(\boldsymbol{r}')\right)_+ \, \mathrm{d}S' + \frac{2\mathrm{i}k_3}{k_{3z}} \iint_{S_2} \varphi_{\bar{j}}^{-\dagger}(\boldsymbol{k}_t; k_3, \boldsymbol{r}') \cdot \left(\hat{\boldsymbol{\nu}} \times \boldsymbol{E}(\boldsymbol{r}')\right)_+ \, \mathrm{d}S', \quad \boldsymbol{k}_t \in \mathbb{R}^2, \ j = 1, 2 \tag{25.19}$$

25.5 Expansions of surface fields

To solve the reflection and transmission problem, we need to eliminate the unknown surface fields $\hat{\boldsymbol{\nu}} \times \boldsymbol{E}(\boldsymbol{r})\big|_-$ and the corresponding tangential magnetic fields on S_1 and S_2. We also have to eliminate the tangential magnetic fields on the finite scatterers S_{s_p}, which in (25.15) are contained in the unknown coefficients α_n^p and f_n^p for the regular and radiating spherical vector waves that are used in the expression for the exciting field (25.9) at each scatterer. One of these sets of coefficients, α_n^p, is eliminated by the use of (25.14).

In fact, if the scatterers are not perfectly conducting conductors, as assumed above, the previous results still hold. The main reason for the assumption of perfectly conducting scatterers was to simplify the theoretical work. If a more general scatterer is present, replace the transition matrix of the scatterer with the appropriate one. Therefore, the previous results hold for any set of scatterers—single or multiple, transparent or not, homogeneous or not—only the individual transition matrices of the scatterers (nonintersecting minimum circumscribed spheres) are known.

The electric field close to the surface $z = z_1$ has an expansion given by (25.16). This expansion is assumed valid in the domain $z \geq z_1$ and $z < \min_p\{\hat{\boldsymbol{z}} \cdot \boldsymbol{r}_p - A_p\}$, *i.e.*, we assume no minimum circumscribed spheres of the scatterers intersect the surface S_1. Similarly, the electric field close to the surface $z = z_2$ is also assumed to be given by (25.16) in the domain $z \leq z_2$ and $z > \max_p\{\hat{z} \cdot \boldsymbol{r}_p + A_p\}$.

Representation (25.16) gives us an expression of the traces (limit values) of the tangential electric fields on $z = z_1$ and $z = z_2$, *i.e.*,

$$\begin{aligned}
\hat{\boldsymbol{\nu}} \times \boldsymbol{E}(\boldsymbol{r})\big|_- &= \sum_n \sum_{p=1}^N f_n^p \hat{\boldsymbol{\nu}} \times \boldsymbol{u}_n(k_2(\boldsymbol{r} - \boldsymbol{r}_p)) \\
&\quad + \sum_{j=1,2} \iint_{\mathbb{R}^2} \alpha_j^+(\boldsymbol{k}_t) \hat{\boldsymbol{\nu}} \times \boldsymbol{\varphi}_j^+(\boldsymbol{k}_t; k_2, \boldsymbol{r}) \, \mathrm{d}k_x \, \mathrm{d}k_y \\
&\quad + \sum_{j=1,2} \iint_{\mathbb{R}^2} \alpha_j^-(\boldsymbol{k}_t) \hat{\boldsymbol{\nu}} \times \boldsymbol{\varphi}_j^-(\boldsymbol{k}_t; k_2, \boldsymbol{r}) \, \mathrm{d}k_x \, \mathrm{d}k_y, \quad z = z_1, z_2
\end{aligned} \tag{25.20}$$

It is convenient to introduce a new notation that contains the unknown f_n^p. We define

$$F_j^\pm(\boldsymbol{k}_t) = \frac{2}{k_2 k_{2z}} \sum_n \sum_{p=1}^N f_n^p B_{nj}^\pm(\boldsymbol{k}_t) \mathrm{e}^{-\mathrm{i}\boldsymbol{k}_2^\pm \cdot \boldsymbol{r}_p} \tag{25.21}$$

With the use of (D.3) and this notation, we rewrite (25.20)

$$\begin{aligned}
\hat{\boldsymbol{\nu}} \times \boldsymbol{E}(\boldsymbol{r})\big|_- &= \sum_{j=1,2} \iint_{\mathbb{R}^2} \alpha_j^\pm(\boldsymbol{k}_t) \hat{\boldsymbol{\nu}} \times \boldsymbol{\varphi}_j^\pm(\boldsymbol{k}_t; k_2, \boldsymbol{r}) \, \mathrm{d}k_x \, \mathrm{d}k_y \\
&\quad + \sum_{j=1,2} \iint_{\mathbb{R}^2} \left(F_j^\mp(\boldsymbol{k}_t) + \alpha_j^\mp(\boldsymbol{k}_t)\right) \hat{\boldsymbol{\nu}} \times \boldsymbol{\varphi}_j^\mp(\boldsymbol{k}_t; k_2, \boldsymbol{r}) \, \mathrm{d}k_x \, \mathrm{d}k_y
\end{aligned} \tag{25.22}$$

where the upper (lower) sign holds on $z = z_1$ ($z = z_2$).

We also obtain the corresponding tangential magnetic fields. The use of the Maxwell equations and the representation (25.16) imply

$$\begin{aligned}
\mathrm{i}\eta_0\eta_2 \, \hat{\boldsymbol{\nu}} \times \boldsymbol{H}(\boldsymbol{r})\big|_- &= \sum_{j=1,2} \iint_{\mathbb{R}^2} \alpha_j^\pm(\boldsymbol{k}_t) \hat{\boldsymbol{\nu}} \times \boldsymbol{\varphi}_{\bar{j}}^\pm(\boldsymbol{k}_t; k_2, \boldsymbol{r}) \, \mathrm{d}k_x \, \mathrm{d}k_y \\
&\quad + \sum_{j=1,2} \iint_{\mathbb{R}^2} \left(F_j^\mp(\boldsymbol{k}_t) + \alpha_j^\mp(\boldsymbol{k}_t)\right) \hat{\boldsymbol{\nu}} \times \boldsymbol{\varphi}_{\bar{j}}^\mp(\boldsymbol{k}_t; k_2, \boldsymbol{r}) \, \mathrm{d}k_x \, \mathrm{d}k_y
\end{aligned} \tag{25.23}$$

where the upper (lower) sign holds on $z = z_1$ ($z = z_2$).

The unknowns in the scattering problem are $\alpha_j^\pm(\boldsymbol{k}_t)$, f_n^p and α_n^p. The aim is to eliminate the expansion coefficients $\alpha_j^\pm(\boldsymbol{k}_t)$, f_n^p and α_n^p and express them in the known coefficients $a_j(\boldsymbol{k}_t)$. The coefficients $r_j(\boldsymbol{k}_t)$ and $t_j(\boldsymbol{k}_t)$ in (25.7) and (25.19) can then be expressed in $a_j(\boldsymbol{k}_t)$, which solves the deterministic reflection and transmission problems. This elimination is accomplished by the use of (25.5), (25.14), (25.15), and (25.17).

We start the elimination of the unknown quantities with (25.5), the boundary conditions $\hat{\boldsymbol{\nu}} \times \boldsymbol{E}\big|_- = \hat{\boldsymbol{\nu}} \times \boldsymbol{E}\big|_+$ and $\hat{\boldsymbol{\nu}} \times \boldsymbol{H}\big|_- = \hat{\boldsymbol{\nu}} \times \boldsymbol{H}\big|_+$ on S_1, and insert the surface expansions (25.22) and (25.23). Changing the order of integration and the use of the orthogonality of the planar vector waves imply (details are given in [14])

$$T_j^1(\boldsymbol{k}_t)a_j(\boldsymbol{k}_t) = \alpha_j^+(\boldsymbol{k}_t) - \mathcal{R}_j^1(\boldsymbol{k}_t)\left(F_j^-(\boldsymbol{k}_t) + \alpha_j^-(\boldsymbol{k}_t)\right), \quad \boldsymbol{k}_t \in \mathbb{R}^2,\ j = 1,2 \tag{25.24}$$

where we used the Fresnel reflection and transmission coefficients of the surface $z = z_1$ (reflection from the left side of S_1 and transmission from left to right) [7, Sec. 10.6.1.2]

$$\begin{cases} R_j^1(\boldsymbol{k}_t) = (-1)^{\bar{j}} \dfrac{\eta_2 k_{\bar{j}} k_{j_z} - \eta_1 k_j k_{\bar{j}_z}}{\eta_2 k_{\bar{j}} k_{j_z} + \eta_1 k_j k_{\bar{j}_z}} \mathrm{e}^{2\mathrm{i}k_{1z}z_1} \\[2ex] T_j^1(\boldsymbol{k}_t) = \dfrac{2\eta_2 k_2 k_{1z}}{\eta_2 k_{\bar{j}} k_{j_z} + \eta_1 k_j k_{\bar{j}_z}} \mathrm{e}^{\mathrm{i}k_{1z}z_1} \mathrm{e}^{-\mathrm{i}k_{2z}z_1} \end{cases} \quad \boldsymbol{k}_t \in \mathbb{R}^2,\ j = 1,2,$$

respectively, where the exponential factors are due to phase corrections, since the surface S_1 is not located at $z = 0$. Similarly, the reflection coefficient from the right side of S_1 and the transmission coefficient from right to left are

$$\begin{cases} \mathcal{R}_j^1(\boldsymbol{k}_t) = (-1)^j \dfrac{\eta_2 k_{\bar{j}} k_{j_z} - \eta_1 k_j k_{\bar{j}_z}}{\eta_2 k_{\bar{j}} k_{j_z} + \eta_1 k_j k_{\bar{j}_z}} \mathrm{e}^{-2\mathrm{i}k_{2z}z_1} \\[2ex] \mathcal{T}_j^1(\boldsymbol{k}_t) = \dfrac{2\eta_1 k_1 k_{2z}}{\eta_2 k_{\bar{j}} k_{j_z} + \eta_1 k_j k_{\bar{j}_z}} \mathrm{e}^{-\mathrm{i}k_{2z}z_1} \mathrm{e}^{\mathrm{i}k_{1z}z_1} \end{cases} \quad \boldsymbol{k}_t \in \mathbb{R}^2,\ j = 1,2$$

We proceed with (25.17). Using the boundary conditions, $\hat{\boldsymbol{\nu}} \times \boldsymbol{E}\big|_- = \hat{\boldsymbol{\nu}} \times \boldsymbol{E}\big|_+$ and $\hat{\boldsymbol{\nu}} \times \boldsymbol{H}\big|_- = \hat{\boldsymbol{\nu}} \times \boldsymbol{H}\big|_+$ on S_2, and inserting the surface expansions (25.22) and (25.23), we get (again, details are given in [14])

$$\alpha_j^-(\boldsymbol{k}_t) = R_j^2(\boldsymbol{k}_t)\left(F_j^+(\boldsymbol{k}_t) + \alpha_j^+(\boldsymbol{k}_t)\right), \quad \boldsymbol{k}_t \in \mathbb{R}^2,\ j = 1,2 \tag{25.25}$$

where the Fresnel reflection and transmission coefficients of the surface $z = z_2$ (reflection from the left side of S_2 and transmission from left to right) are*

$$\begin{cases} R_j^2(\boldsymbol{k}_t) = (-1)^{\bar{j}} \dfrac{\eta_3 k_{\overline{j+1}} k_{j+1_z} - \eta_2 k_{j+1} k_{\overline{j+1}_z}}{\eta_3 k_{\overline{j+1}} k_{j+1_z} + \eta_2 k_{j+1} k_{\overline{j+1}_z}} \mathrm{e}^{2\mathrm{i}k_{2z}z_2} \\[2ex] T_j^2(\boldsymbol{k}_t) = \dfrac{2\eta_3 k_3 k_{2z}}{\eta_3 k_{\overline{j+1}} k_{j+1_z} + \eta_2 k_{j+1} k_{\overline{j+1}_z}} \mathrm{e}^{\mathrm{i}k_{2z}z_2} \mathrm{e}^{-\mathrm{i}k_{3z}z_2} \end{cases} \quad \boldsymbol{k}_t \in \mathbb{R}^2,\ j = 1,2$$

*Do not misinterpret the super index as the square of the reflection or transmission coefficient. It denotes the surface number—in this case S_2.

Combine (25.24) and (25.25) and solve for the unknown $\alpha_j^+(\boldsymbol{k}_t)$ and $\alpha_j^-(\boldsymbol{k}_t)$. The result after some algebra is ($\boldsymbol{k}_t \in \mathbb{R}^2$, $j = 1, 2$)

$$\alpha_j^+(\boldsymbol{k}_t) = \frac{T_j^1(\boldsymbol{k}_t)a_j(\boldsymbol{k}_t)}{1 - \mathcal{R}_j^1(\boldsymbol{k}_t)R_j^2(\boldsymbol{k}_t)} + \mathcal{R}_j^1(\boldsymbol{k}_t)\frac{F_j^-(\boldsymbol{k}_t) + R_j^2(\boldsymbol{k}_t)F_j^+(\boldsymbol{k}_t)}{1 - \mathcal{R}_j^1(\boldsymbol{k}_t)R_j^2(\boldsymbol{k}_t)} \tag{25.26}$$

and

$$\alpha_j^-(\boldsymbol{k}_t) = \frac{R_j^2(\boldsymbol{k}_t)T_j^1(\boldsymbol{k}_t)a_j(\boldsymbol{k}_t)}{1 - \mathcal{R}_j^1(\boldsymbol{k}_t)R_j^2(\boldsymbol{k}_t)} + R_j^2(\boldsymbol{k}_t)\frac{\mathcal{R}_j^1(\boldsymbol{k}_t)F_j^-(\boldsymbol{k}_t) + F_j^+(\boldsymbol{k}_t)}{1 - \mathcal{R}_j^1(\boldsymbol{k}_t)R_j^2(\boldsymbol{k}_t)} \tag{25.27}$$

The coefficients $\alpha_j^\pm(\boldsymbol{k}_t)$ are now expressed in the reflection and transmission properties of the surfaces S_1 and S_2, the given coefficient $a_j(\boldsymbol{k}_t)$, and the factors $F_j^\pm(\boldsymbol{k}_t)$. The latter quantity contains the unknown coefficients f_n^p, which contain all interaction contributions from the N finite scatterers. The elimination of the coefficients f_n^p is done by solving a matrix equation.

We multiply (25.15) with $T_{nn'}^p$ and sum over the free index (notice change in index p and q in the translation matrix and the use of $\mathcal{P}^t(k\boldsymbol{d}) = \mathcal{P}(-k\boldsymbol{d})$). The use of (25.14) leads to the result ($p = 1, 2, \ldots, N$)

$$f_n^p = \sum_{n'} T_{nn'}^p \left\{ \sum_{\substack{q=1 \\ q \neq p}}^N \sum_{n''} \mathcal{P}_{n'n''}(k_2(\boldsymbol{r}_q - \boldsymbol{r}_p)) f_{n''}^q \right. \\ \left. + \sum_{j=1,2} \iint_{\mathbb{R}^2} \left\{ B_{n'j}^{+\,\dagger}(\boldsymbol{k}_t) e^{\mathrm{i}\boldsymbol{k}_2^+ \cdot \boldsymbol{r}_p} \alpha_j^+(\boldsymbol{k}_t) + B_{n'j}^{-\,\dagger}(\boldsymbol{k}_t) e^{\mathrm{i}\boldsymbol{k}_2^- \cdot \boldsymbol{r}_p} \alpha_j^-(\boldsymbol{k}_t) \right\} \mathrm{d}k_x \, \mathrm{d}k_y \right\}$$

Insert the expression of $\alpha_j^+(\boldsymbol{k}_t)$ and $\alpha_j^-(\boldsymbol{k}_t)$ given in (25.26) and (25.27) and use (25.21). The result has the form

$$f_n^p - \sum_{n'n''} \sum_{q=1}^N T_{nn''}^p A_{n''n'}^{pq} f_{n'}^q = d_n^p, \quad p = 1, 2, \ldots, N \tag{25.28}$$

where

$$A_{n''n'}^{pq} = \mathcal{P}_{n''n'}(k_2(\boldsymbol{r}_q - \boldsymbol{r}_p))(1 - \delta_{pq}) + \sum_{j=1,2} \iint_{\mathbb{R}^2} \frac{2 \, \mathrm{d}k_x \, \mathrm{d}k_y}{k_2 k_{2z}} \\ \times \left\{ B_{n''j}^{+\,\dagger}(\boldsymbol{k}_t) e^{\mathrm{i}\boldsymbol{k}_2^+ \cdot \boldsymbol{r}_p} \mathcal{R}_j^1(\boldsymbol{k}_t) \frac{B_{n'j}^-(\boldsymbol{k}_t) e^{-\mathrm{i}\boldsymbol{k}_2^- \cdot \boldsymbol{r}_q} + R_j^2(\boldsymbol{k}_t) B_{n'j}^+(\boldsymbol{k}_t) e^{-\mathrm{i}\boldsymbol{k}_2^+ \cdot \boldsymbol{r}_q}}{1 - \mathcal{R}_j^1(\boldsymbol{k}_t) R_j^2(\boldsymbol{k}_t)} \right. \\ \left. + B_{n''j}^{-\,\dagger}(\boldsymbol{k}_t) e^{\mathrm{i}\boldsymbol{k}_2^- \cdot \boldsymbol{r}_p} R_j^2(\boldsymbol{k}_t) \frac{B_{n'j}^+(\boldsymbol{k}_t) e^{-\mathrm{i}\boldsymbol{k}_2^+ \cdot \boldsymbol{r}_q} + \mathcal{R}_j^1(\boldsymbol{k}_t) B_{n'j}^-(\boldsymbol{k}_t) e^{-\mathrm{i}\boldsymbol{k}_2^- \cdot \boldsymbol{r}_q}}{1 - \mathcal{R}_j^1(\boldsymbol{k}_t) R_j^2(\boldsymbol{k}_t)} \right\} \tag{25.29}$$

which contains only known geometrical material properties for a given scattering problem, and the source that drives the system is given by

$$d_n^p = \sum_{n'} T_{nn'}^p \sum_{j=1,2} \iint_{\mathbb{R}^2} \left\{ \frac{B_{n'j}^{+\dagger}(\boldsymbol{k}_t)\mathrm{e}^{\mathrm{i}\boldsymbol{k}_2^+\cdot\boldsymbol{r}_p}}{1-\mathcal{R}_j^1(\boldsymbol{k}_t)R_j^2(\boldsymbol{k}_t)} + \frac{R_j^2(\boldsymbol{k}_t)B_{n'j}^{-\dagger}(\boldsymbol{k}_t)\mathrm{e}^{\mathrm{i}\boldsymbol{k}_2^-\cdot\boldsymbol{r}_p}}{1-\mathcal{R}_j^1(\boldsymbol{k}_t)R_j^2(\boldsymbol{k}_t)} \right\} T_j^1(\boldsymbol{k}_t)a_j(\boldsymbol{k}_t)\,\mathrm{d}k_x\,\mathrm{d}k_y$$

We are now in a position to summarize the solution procedure of the deterministic problem presented in this chapter. The complete solution of the problem, for a given scattering configuration and incident field, *i.e.*, given $a_j(\boldsymbol{k}_t)$, and consequently known d_n^p, is found by solving (25.28) for f_n^p. Then, (25.21) is used to get $F_j^\pm(\boldsymbol{k}_t)$ and (25.26) and (25.27) to get $\alpha_j^\pm(\boldsymbol{k}_t)$. The surface fields on S_1 and S_2 are now known and can be obtained by using (25.22) and (25.23).

The coefficients for the expansions of the reflected and transmitted fields, $r_j(\boldsymbol{k}_t)$ and $t_j(\boldsymbol{k}_t)$, are given by (25.7) and (25.19), respectively. The reflected and transmitted fields are then determined by (25.6) and (25.18), respectively. The details of this analysis are given in the next section.

25.6 The transmitted and reflected fields

We are now in a position of calculating the transmitted and reflected fields from the entire scattering configuration.

The transmitted field is given in (25.18), and the transmission coefficients are (use (25.19), the boundary conditions on S_2, (25.22) and (25.23), orthogonality of the planar vector waves, and insert (25.26) and (25.27))

$$t_j(\boldsymbol{k}_t) = T_j^2(\boldsymbol{k}_t)\frac{\mathcal{R}_j^1(\boldsymbol{k}_t)F_j^-(\boldsymbol{k}_t)+F_j^+(\boldsymbol{k}_t)}{1-\mathcal{R}_j^1(\boldsymbol{k}_t)R_j^2(\boldsymbol{k}_t)} + T_j(\boldsymbol{k}_t)a_j(\boldsymbol{k}_t), \quad \boldsymbol{k}_t \in \mathbb{R}^2,\ j=1,2 \tag{25.30}$$

where the total transmission coefficient of the slab reads [7, Sec. 10.6.1.1]

$$T_j(\boldsymbol{k}_t) = \frac{T_j^1(\boldsymbol{k}_t)T_j^2(\boldsymbol{k}_t)}{1-\mathcal{R}_j^1(\boldsymbol{k}_t)R_j^2(\boldsymbol{k}_t)} \tag{25.31}$$

The amplitude of the transmitted field, $t_j(\boldsymbol{k}_t)$, consists of two terms—the last term $T_j(\boldsymbol{k}_t)a_j(\boldsymbol{k}_t)$ is the direct transmitted contribution of the slab itself, and the first term is the additional contribution to the transmitted field from the scatterers inside the slab.

Similarly, the reflected field is given in (25.6), where the reflection coefficients are given by (25.7). The result is

$$r_j(\boldsymbol{k}_t) = \mathcal{T}_j^1(\boldsymbol{k}_t)\frac{F_j^-(\boldsymbol{k}_t) + R_j^2(\boldsymbol{k}_t)F_j^+(\boldsymbol{k}_t)}{1 - \mathcal{R}_j^1(\boldsymbol{k}_t)R_j^2(\boldsymbol{k}_t)} + R_j(\boldsymbol{k}_t)a_j(\boldsymbol{k}_t), \quad \boldsymbol{k}_t \in \mathbb{R}^2,\ j = 1, 2 \tag{25.32}$$

where the total reflection coefficient of the slab reads [7, Sec. 10.6.1.1]

$$R_j(\boldsymbol{k}_t) = R_j^1(\boldsymbol{k}_t) + T_j^1(\boldsymbol{k}_t)\frac{R_j^2(\boldsymbol{k}_t)}{1 - \mathcal{R}_j^1(\boldsymbol{k}_t)R_j^2(\boldsymbol{k}_t)}\mathcal{T}_j^1(\boldsymbol{k}_t)$$

The amplitude of the reflected field, $r_j(\boldsymbol{k}_t)$, consists of two terms—the last term $R_j(\boldsymbol{k}_t)a_j(\boldsymbol{k}_t)$ is the direct reflected contribution of the slab itself, and the first term is the additional contribution to the reflected field from the scatterers inside the slab.

25.7 Statistical problem—ensemble average

The solution of the set of equations in (25.28) for the unknowns, f_n^p, is an unsurmountable task, if the number of scatterers is large. Fortunately, there are statistical methods that apply if the location and state (material properties, size, shape, orientation, etc.) of the scatterers are randomly distributed. Moreover, with a large number of scatterers, we rarely have complete information about the position and the state of each scatterer, and we are often not interested in the physical quantities of a particular configuration, but ensemble averages suffice. In particular, the average of the electric field is an appropriate quantity in several radio and radar applications, but it is of limited values as an optical observable. The presentation in this section follows to some extent the one presented in [15–17] but deviates in the method of solving the problem.

A statistical evaluation of the relevant physical quantities involves ensemble averages, which we denote by the symbol $\langle\cdot\rangle$. The relation between the ensemble average and the time average and use of the ergodic hypothesis are found in the comprehensive review article [18] and in the excellent textbook [19].

The scatterer locations $\boldsymbol{r}_p, p = 1, 2, \ldots, N$, are now random variables. Moreover, the properties of each scatterer (geometry and material) are also random variables that we collect in a state variable $\xi_p, p = 1, 2, \ldots, N$. The common N-particle probability density function (PDF) is denoted as $P(\boldsymbol{r}_1, \ldots, \boldsymbol{r}_N; \xi_1, \ldots, \xi_N)$. More details on this PDF are collected in Appendix E.

The explicit assumptions made about the scatterers in this section are as follows:

1. The number of scatterers N is large, so that statistical methods are appropriate to apply.
2. The N scatterers are characterized by a common PDF.
3. The scatterers are indistinguishable insofar as the numbering of the scatterers is arbitrary.

4. The state variables of the scatterers, ξ_p, are independent between different scatterers and of the position variables, $\boldsymbol{r}_p, p = 1, 2, \ldots, N$.
5. No minimum circumscribed spheres of the individual scatterers intersect.

The position variables cannot be statistically independent variables for finite size scatterers, due to the assumption of nonintersecting circumscribed spheres. The assumption of independence in Item 4. makes the averaging of scatterer position separate from the averaging over their states in the transition matrix entries. The random variables $\boldsymbol{r}_p$ and ξ_p are now dummy variables in the integration over all possible positions and states. Consequently, the index p is finally dropped in the analysis below.

The ensemble average and the use of the conditional PDF, see (E.1) in Appendix E, imply the following expressions of the reflected and transmitted fields from (25.18) and (25.6)

$$\begin{cases} \langle \boldsymbol{E}_\mathrm{t}(\boldsymbol{r})\rangle = \sum\limits_{j=1,2} \iint\limits_{\mathbb{R}^2} \langle t_j(\boldsymbol{k}_\mathrm{t})\rangle \, \boldsymbol{\varphi}_j^+(\boldsymbol{k}_\mathrm{t}; k_3, \boldsymbol{r}) \, \mathrm{d}k_\mathrm{x} \, \mathrm{d}k_\mathrm{y}, \quad z > z_2 \\ \langle \boldsymbol{E}_\mathrm{r}(\boldsymbol{r})\rangle = \sum\limits_{j=1,2} \iint\limits_{\mathbb{R}^2} \langle r_j(\boldsymbol{k}_\mathrm{t})\rangle \, \boldsymbol{\varphi}_j^-(\boldsymbol{k}_\mathrm{t}; k_1, \boldsymbol{r}) \, \mathrm{d}k_\mathrm{x} \, \mathrm{d}k_\mathrm{y}, \quad z < z_1 \end{cases} \tag{25.33}$$

where the ensemble average is denoted by $\langle\cdot\rangle$, see (E.2). These fields are the average or coherent contribution of the electric field outside the slab, and the computations of these fields are the main purpose of this chapter.

To evaluate the averaged transmitted and reflected fields, we need to obtain $\langle t_j(\boldsymbol{k}_\mathrm{t})\rangle$ and $\langle r_j(\boldsymbol{k}_\mathrm{t})\rangle$, which both contain the factors $\langle F_j^\pm(\boldsymbol{k}_\mathrm{t})\rangle$, see (25.30) and (25.32). All other quantities in $\langle t_j(\boldsymbol{k}_\mathrm{t})\rangle$ and $\langle r_j(\boldsymbol{k}_\mathrm{t})\rangle$ are deterministic. We have ($\boldsymbol{k}_\mathrm{t} \in \mathbb{R}^2$, $j = 1, 2$)

$$\langle t_j(\boldsymbol{k}_\mathrm{t})\rangle = T_j^2(\boldsymbol{k}_\mathrm{t}) \frac{\mathcal{R}_j^1(\boldsymbol{k}_\mathrm{t})\langle F_j^-(\boldsymbol{k}_\mathrm{t})\rangle + \langle F_j^+(\boldsymbol{k}_\mathrm{t})\rangle}{1 - \mathcal{R}_j^1(\boldsymbol{k}_\mathrm{t})R_j^2(\boldsymbol{k}_\mathrm{t})} + T_j(\boldsymbol{k}_\mathrm{t})a_j(\boldsymbol{k}_\mathrm{t}) \tag{25.34}$$

and

$$\langle r_j(\boldsymbol{k}_\mathrm{t})\rangle = \mathcal{T}_j^1(\boldsymbol{k}_\mathrm{t}) \frac{\langle F_j^-(\boldsymbol{k}_\mathrm{t})\rangle + R_j^2(\boldsymbol{k}_\mathrm{t})\langle F_j^+(\boldsymbol{k}_\mathrm{t})\rangle}{1 - \mathcal{R}_j^1(\boldsymbol{k}_\mathrm{t})R_j^2(\boldsymbol{k}_\mathrm{t})} + R_j(\boldsymbol{k}_\mathrm{t})a_j(\boldsymbol{k}_\mathrm{t}) \tag{25.35}$$

The average of the functions $F_j^\pm(\boldsymbol{k}_\mathrm{t})$ in (25.21) has to be computed. The conditional PDF implies, see (E.1)

$$\langle F_j^\pm(\boldsymbol{k}_\mathrm{t})\rangle = \frac{2N}{k_2 k_{2z}} \sum_n B_{nj}^\pm(\boldsymbol{k}_\mathrm{t}) \iiint\limits_{V_\mathrm{s}} P(\boldsymbol{r}) \langle f_n\rangle (\boldsymbol{r}) \mathrm{e}^{-\mathrm{i}\boldsymbol{k}_2^\pm \cdot \boldsymbol{r}} \, \mathrm{d}v \tag{25.36}$$

since the variable $\boldsymbol{r}_p$ is a dummy variable and all integrals in the expression on the right-hand side are identical. We can also skip the superscript p on f_n. The volume

V_s is the volume of possible locations of the local origins $\boldsymbol{r}_p$, $p = 1, 2, \ldots, N$. The notation $\langle f_n \rangle (\boldsymbol{r})$ is the ensemble average of the coefficient f_n where one spatial variable is held fixed, *cf.* (E.3). If the number of scatterers $N \to \infty$ and the scatterers are randomly filled in the entire slab (this requires that an appropriate limit procedure is employed), the volume V_s is $\{\boldsymbol{r} : z_1 + a \leq z \leq z_2 - a\}$, where $a = \max_p A_p$. This limit process implies that $NP(\boldsymbol{r}) \to n_0$ as $N \to \infty$, where n_0 is the number density of the scatterers (number of scatterers per unit volume), see Appendix E.

To obtain $\langle f_n \rangle (\boldsymbol{r})$, we need to take the conditional average of the set of matrix equations in (25.28). Using the conditional PDF, we obtain

$$\begin{aligned}
&\left\langle f_n^p \right\rangle (\boldsymbol{r}_p; \xi_p) - \sum_{n'n''} \sum_{q=1}^{N} \iiint_{V_s} P(\boldsymbol{r}_q | \boldsymbol{r}_p) T^p_{nn''}(\xi_p) A^{pq}_{n''n'} \left\langle f_{n'}^q \right\rangle (\boldsymbol{r}_p, \boldsymbol{r}_q; \xi_p) \,\mathrm{d}v_q \\
&\quad = \langle d_n \rangle (\boldsymbol{r}_p; \xi_p) = \sum_{n'} T^p_{nn'}(\xi_p) \sum_{j=1,2} \iint_{\mathbb{R}^2} \left\{ \frac{B^{+\,\dagger}_{n'j}(\boldsymbol{k}_t) \mathrm{e}^{\mathrm{i}\boldsymbol{k}_2^+ \cdot \boldsymbol{r}_p}}{1 - \mathcal{R}^1_j(\boldsymbol{k}_t) R^2_j(\boldsymbol{k}_t)} \right. \\
&\quad + \left. \frac{R^2_j(\boldsymbol{k}_t) B^{-\,\dagger}_{n'j}(\boldsymbol{k}_t) \mathrm{e}^{\mathrm{i}\boldsymbol{k}_2^- \cdot \boldsymbol{r}_p}}{1 - \mathcal{R}^1_j(\boldsymbol{k}_t) R^2_j(\boldsymbol{k}_t)} \right\} T^1_j(\boldsymbol{k}_t) a_j(\boldsymbol{k}_t) \,\mathrm{d}k_x \,\mathrm{d}k_y, \quad p = 1, 2, \ldots, N \qquad (25.37)
\end{aligned}$$

where $A^{pq}_{n''n'}$ is defined in (25.29).

In practice, higher order density functions are harder to obtain. To break the hierarchy in (25.37), we introduce the quasi-crystalline approximation (QCA), which states that the conditional average with two positions held fixed, $\left\langle f_{n'}^q \right\rangle (\boldsymbol{r}_p, \boldsymbol{r}_q; \xi_p)$, is replaced with the conditional average with one position held fixed, $\left\langle f_{n'}^q \right\rangle (\boldsymbol{r}_q; \xi_p)$. We have [20]

$$\left\langle f_{n'}^q \right\rangle (\boldsymbol{r}_p, \boldsymbol{r}_q; \xi_p) \approx \left\langle f_{n'}^q \right\rangle (\boldsymbol{r}_q; \xi_p).$$

This approximation has been successfully applied in a range of concentrations from tenuous to dense media, and from the low frequency to intermediate frequency range [21].

The QCA in (25.37) leads to a set of integral equations in the unknowns, $\langle f_n \rangle (\boldsymbol{r}'; \xi)$, *viz.* (the indices p and q are now superfluous)

$$\langle f_n \rangle (\boldsymbol{r}; \xi) - k_2^3 \sum_{n'} \iiint_{V_s} \mathcal{K}_{nn'}(\boldsymbol{r}, \boldsymbol{r}'; \xi) \langle f_{n'} \rangle (\boldsymbol{r}'; \xi) \,\mathrm{d}v' = \langle d_n \rangle (\boldsymbol{r}; \xi), \quad \boldsymbol{r} \in V_s \qquad (25.38)$$

where the general form of the kernel $\mathcal{K}_{nn'}(\boldsymbol{r}, \boldsymbol{r}'; \xi)$ can be found from (25.37). Equation (25.38) is a system of integral equations for the unknown $\langle f_n \rangle (\boldsymbol{r}; \xi)$, which can be simplified under the assumption made in this chapter. We solve the system of integral equations for each value of the state variable ξ and take the ensemble average.

In this chapter, we apply the result to a plane wave incidence, *i.e.*,

$$\boldsymbol{E}_i(\boldsymbol{r}) = \boldsymbol{E}_0 \mathrm{e}^{\mathrm{i}k_1 \hat{\boldsymbol{k}}_i \cdot \boldsymbol{r}}$$

where the complex-valued vector $\boldsymbol{E}_0$ satisfies $\boldsymbol{E}_0 \cdot \hat{\boldsymbol{k}}_{\rm i} = 0$ and the real-valued incident direction $\hat{\boldsymbol{k}}_{\rm i}$ satisfies $\hat{\boldsymbol{k}}_{\rm i} \cdot \hat{\boldsymbol{z}} > 0$. The explicit form of the expansion function $a_j(\boldsymbol{k}_{\rm t})$ for a plane wave incident field is

$$a_j(\boldsymbol{k}_{\rm t}) = \delta(\boldsymbol{k}_{\rm t} - \boldsymbol{k}_{\rm it})A_j$$

The factor A_j is a short-hand notation for

$$A_j = 4\pi \boldsymbol{E}_0 \cdot \begin{Bmatrix} -\dfrac{\hat{\boldsymbol{z}} \times \boldsymbol{k}_{\rm it}}{{\rm i}k_{\rm it}}, & j = 1 \\ \dfrac{\hat{\boldsymbol{z}}k_{\rm it}^2 - \boldsymbol{k}_{\rm it}k_{\rm 1iz}}{k_1 k_{\rm it}}, & j = 2 \end{Bmatrix}$$

where $\boldsymbol{k}_{\rm it} = k_1 \left(\hat{\boldsymbol{x}}\hat{\boldsymbol{x}} + \hat{\boldsymbol{y}}\hat{\boldsymbol{y}}\right) \cdot \hat{\boldsymbol{k}}_{\rm i} = k_1 \mathbf{I}_2 \cdot \hat{\boldsymbol{k}}_{\rm i}$, $k_{\rm 1iz} = \left(k_1^2 - k_{\rm it}^2\right)^{1/2}$ and $k_{\rm it} = |\boldsymbol{k}_{\rm it}|$. The delta function implies that $\boldsymbol{k}_{\rm t} = \boldsymbol{k}_{\rm it}$ everywhere. This is usually called Snell's law.

For a plane wave incidence, the right-hand sides of (25.37) and (25.38) simplify to

$$\begin{aligned} \langle d_n \rangle (\boldsymbol{r}; \xi) &= d_n(z; \xi){\rm e}^{{\rm i}\boldsymbol{k}_{\rm it} \cdot \boldsymbol{r}_{\rm c}} \\ &= {\rm e}^{{\rm i}\boldsymbol{k}_{\rm it} \cdot \boldsymbol{r}_{\rm c}} \sum_{n'} T_{nn'}(\xi) \sum_{j=1,2} T_j^1(\boldsymbol{k}_{\rm it})A_j \frac{B_{n'j}^{+\,\dagger}(\boldsymbol{k}_{\rm it}){\rm e}^{{\rm i}k_{\rm 2iz}z} + R_j^2(\boldsymbol{k}_{\rm it})B_{n'j}^{-\,\dagger}(\boldsymbol{k}_{\rm it}){\rm e}^{-{\rm i}k_{\rm 2iz}z}}{1 - \mathcal{R}_j^1(\boldsymbol{k}_{\rm it})R_j^2(\boldsymbol{k}_{\rm it})} \end{aligned}$$

where $\boldsymbol{r}_{\rm c} = x\hat{\boldsymbol{x}} + y\hat{\boldsymbol{y}}$ and $k_{\rm 2iz} = \left(k_2^2 - k_{\rm it}^2\right)^{1/2}$.

If the scatterers are randomly distributed within the volume between the planes, the solution of the problem shows lateral invariance, which implies that the coefficients have the form $\langle f_n \rangle (\boldsymbol{r}; \xi) = f_n(z; \xi){\rm e}^{{\rm i}\boldsymbol{k}_{\rm it} \cdot \boldsymbol{r}_{\rm c}}$, where $f_n(z; \xi)$ only depends on the depth z of the slab.† The integral equation (25.38) then has the form (for simplicity, we suppress the state variable ξ which acts as a parameter)

$$f_n(z) - k_2 \sum_{n'} \int_{z_1+a}^{z_2-a} K_{nn'}(z, z')f_n(z')\,{\rm d}z' = d_n(z), \quad z \in [z_1 + a, z_2 - a] \tag{25.39}$$

where $a = \max_p A_p$, and where

$$K_{nn'}(z, z') = k_2^2 \iint_{\mathbb{R}^2} \mathcal{K}_{nn'}(\boldsymbol{r}, \boldsymbol{r}'){\rm e}^{{\rm i}\boldsymbol{k}_{\rm it} \cdot (\boldsymbol{r}_{\rm c}' - \boldsymbol{r}_{\rm c})}\,{\rm d}x'\,{\rm d}y'$$

†We require the integrand to vanish in an appropriate way as the lateral variables, x and y, approach infinity. One way to accomplish this is to assume an infinitely small positive imaginary part of the wave number k.

Note that, due to invariance under translation in the lateral variables, this expression of the kernel does not depend on $\boldsymbol{r}_c$, see the explicit expression later. The explicit expression of $K_{nn'}(z,z')$ is ($N \to \infty$ is assumed)

$$K_{nn'}(z,z') = \frac{n_0}{k_2^3} \sum_{n''} T_{nn''} \Bigg\{ C_{n''n'}(\boldsymbol{k}_{\mathrm{it}}; z - z') + \sum_{j=1,2} \iint_{\mathbb{R}^2} \frac{2\, \mathrm{d}k_x\, \mathrm{d}k_y}{k_2 k_{2z}}$$

$$\times\, G(\boldsymbol{k}_{\mathrm{t}} - \boldsymbol{k}_{\mathrm{it}}; z - z') \Bigg\{ B^{+\,\dagger}_{n''j}(\boldsymbol{k}_{\mathrm{t}}) \mathcal{R}^1_j(\boldsymbol{k}_{\mathrm{t}}) \mathrm{e}^{\mathrm{i}k_{2z}z}$$

$$\times \frac{B^-_{n'j}(\boldsymbol{k}_{\mathrm{t}}) \mathrm{e}^{\mathrm{i}k_{2z}z'} + R^2_j(\boldsymbol{k}_{\mathrm{t}}) B^+_{n'j}(\boldsymbol{k}_{\mathrm{t}}) \mathrm{e}^{-\mathrm{i}k_{2z}z'}}{1 - \mathcal{R}^1_j(\boldsymbol{k}_{\mathrm{t}}) R^2_j(\boldsymbol{k}_{\mathrm{t}})}$$

$$+ B^{-\,\dagger}_{n''j}(\boldsymbol{k}_{\mathrm{t}}) R^2_j(\boldsymbol{k}_{\mathrm{t}}) \mathrm{e}^{-\mathrm{i}k_{2z}z} \frac{B^+_{n'j}(\boldsymbol{k}_{\mathrm{t}}) \mathrm{e}^{-\mathrm{i}k_{2z}z'} + \mathcal{R}^1_j(\boldsymbol{k}_{\mathrm{t}}) B^-_{n'j}(\boldsymbol{k}_{\mathrm{t}}) \mathrm{e}^{\mathrm{i}k_{2z}z'}}{1 - \mathcal{R}^1_j(\boldsymbol{k}_{\mathrm{t}}) R^2_j(\boldsymbol{k}_{\mathrm{t}})} \Bigg\} \Bigg\}$$

where we simplified the notation by introducing the following dimensionless quantity:

$$C_{nn'}(\boldsymbol{k}_{\mathrm{t}}; z) = k_2^2 \iint_{\mathbb{R}^2} g(|\boldsymbol{r}_{\mathrm{c}} - \hat{\boldsymbol{z}}z|) \mathcal{P}_{nn'}(k_2(\boldsymbol{r}_{\mathrm{c}} - \hat{\boldsymbol{z}}z)) \mathrm{e}^{\mathrm{i}\boldsymbol{k}_{\mathrm{t}}\cdot\boldsymbol{r}_{\mathrm{c}}}\, \mathrm{d}x\, \mathrm{d}y, \quad |z| \le z_2 - z_1 - 2a \tag{25.40}$$

and where we expressed the conditional probability density $P(\boldsymbol{r}'|\boldsymbol{r})$ in terms of the pair distribution function $g(\boldsymbol{r},\boldsymbol{r}')$, *i.e.*, $(N-1)P(\boldsymbol{r}'|\boldsymbol{r}) = n_0 g(\boldsymbol{r},\boldsymbol{r}')$, see (E.4). Moreover, for symmetric scatterers, the pair distribution function depends only on the distance between two scatterers $|\boldsymbol{r}-\boldsymbol{r}'|$, *i.e.*, $g(\boldsymbol{r},\boldsymbol{r}') = g(|\boldsymbol{r}-\boldsymbol{r}|)$, and the spatial integral is invariant w.r.t. lateral translations. The assumption $N \to \infty$ simplifies the factor $N/(N-1) \to 1$ that occurs in the second term. The lateral Fourier transform of the pair correlation function is

$$G(\boldsymbol{k}_{\mathrm{t}}; z) = k_2^2 \iint_{\mathbb{R}^2} g(|\boldsymbol{r}_{\mathrm{c}} - \hat{\boldsymbol{z}}z|) \mathrm{e}^{-\mathrm{i}\boldsymbol{k}_{\mathrm{t}}\cdot\boldsymbol{r}_{\mathrm{c}}}\, \mathrm{d}x\, \mathrm{d}y, \quad |z| \le z_2 - z_1 - 2a \tag{25.41}$$

The simplest model for the pair correlation function $g(r)$ is the hole correction function, *i.e.*, $g(r) = \mathrm{H}(r-2a)$, where $\mathrm{H}(x)$ denotes the Heaviside function. The function $C_{nn'}(\boldsymbol{0}; z)$ in (25.40) is investigated in [3]:

$$C_{nn'}(\boldsymbol{0}; z) = \sum_{\lambda=|l-l'|+|\tau-\tau'|}^{l+l'} I_\lambda(-z) \mathcal{A}_{nn'\lambda}, \quad |z| \le z_2 - z_1 - 2a$$

where the azimuthal average $\mathcal{A}_{nn'\lambda}$ is explicitly evaluated in Appendix B, and where the important integral $I_\lambda(z)$ is [22]

$$I_l(z) = k_2^2 \int_{d(z)}^{\infty} g(|\boldsymbol{r}_{\mathrm{c}} + z\hat{\boldsymbol{z}}|) \mathrm{h}_l^{(1)}\left(k_2\sqrt{r_{\mathrm{c}}^2 + z^2}\right) \mathrm{P}_l\left(\frac{z}{\sqrt{r_{\mathrm{c}}^2 + z^2}}\right) r_{\mathrm{c}}\, \mathrm{d}r_{\mathrm{c}}, \quad z \in \mathbb{R}$$

where $d(z) = \mathrm{H}(2a - |z|)\sqrt{4a^2 - z^2}$. Some effective iteration schemes to compute the integrals $I_\lambda(z)$ are presented in [22].

The lateral Fourier transform of the pair correlation function, see (25.41), for the hole correction becomes

$$\begin{aligned} G(\boldsymbol{k}_\mathrm{t}; z) &= k_2^2 \iint\limits_{\mathbb{R}^2} \mathrm{H}(|\boldsymbol{r}_\mathrm{c} - \hat{z}z| - 2a)\mathrm{e}^{-\mathrm{i}\boldsymbol{k}_\mathrm{t}\cdot\boldsymbol{r}_\mathrm{c}}\, \mathrm{d}x\, \mathrm{d}y \\ &= 4\pi^2 k_2^2 \delta(\boldsymbol{k}_\mathrm{t}) - 2\pi(k_2 d(z))^2 \frac{\mathrm{J}_1(k_\mathrm{t} d(z))}{k_\mathrm{t} d(z)}, \quad |z| \le z_2 - z_1 - 2a \end{aligned}$$

The plane wave incidence also simplifies the evaluation of $\langle F_j^\pm(\boldsymbol{k}_\mathrm{t})\rangle$ in (25.36):

$$\langle F_j^\pm(\boldsymbol{k}_\mathrm{t})\rangle = \frac{8\pi^2 n_0 \delta(\boldsymbol{k}_\mathrm{t} - \boldsymbol{k}_\mathrm{it})}{k_2 k_{2\mathrm{iz}}} \sum_n B_{nj}^\pm(\boldsymbol{k}_\mathrm{it}) \int_{z_1+a}^{z_2-a} f_n(z)\mathrm{e}^{\mp\mathrm{i}k_{2\mathrm{iz}}z}\, \mathrm{d}z$$

and from (25.34) and (25.35), we obtain the transmitted and the reflected fields in (25.33)

$$\begin{aligned} \langle \boldsymbol{E}_\mathrm{t}(\boldsymbol{r})\rangle &= \sum_{j=1,2} T_j(\boldsymbol{k}_\mathrm{it}) A_j \varphi_j^+(\boldsymbol{k}_\mathrm{it}; k_3, \boldsymbol{r}) \\ &\quad + \frac{8\pi^2 n_0}{k_2 k_{2\mathrm{iz}}} \sum_{j=1,2} T_j^2(\boldsymbol{k}_\mathrm{it}) \frac{\mathcal{R}_j^1(\boldsymbol{k}_\mathrm{it}) C_j^-(\boldsymbol{k}_\mathrm{it}) + C_j^+(\boldsymbol{k}_\mathrm{it})}{1 - \mathcal{R}_j^1(\boldsymbol{k}_\mathrm{it}) R_j^2(\boldsymbol{k}_\mathrm{it})} \varphi_j^+(\boldsymbol{k}_\mathrm{it}; k_3, \boldsymbol{r}), \quad z > z_2 \end{aligned} \tag{25.42}$$

and

$$\begin{aligned} \langle \boldsymbol{E}_\mathrm{r}(\boldsymbol{r})\rangle &= \frac{8\pi^2 n_0}{k_2 k_{2\mathrm{iz}}} \sum_{j=1,2} T_j^1(\boldsymbol{k}_\mathrm{it}) \left\{ \frac{C_j^-(\boldsymbol{k}_\mathrm{it}) + R_j^2(\boldsymbol{k}_\mathrm{it}) C_j^+(\boldsymbol{k}_\mathrm{it})}{1 - \mathcal{R}_j^1(\boldsymbol{k}_\mathrm{it}) R_j^2(\boldsymbol{k}_\mathrm{it})} \right\} \varphi_j^-(\boldsymbol{k}_\mathrm{it}; k_1, \boldsymbol{r}) \\ &\quad + \sum_{j=1,2} R_j(\boldsymbol{k}_\mathrm{it}) A_j \varphi_j^-(\boldsymbol{k}_\mathrm{it}; k_1, \boldsymbol{r}), \quad z < z_1 \end{aligned}$$

where

$$C_j^\pm(\boldsymbol{k}_\mathrm{it}) = \sum_n B_{nj}^\pm(\boldsymbol{k}_\mathrm{it}) \int_{z_1+a}^{z_2-a} f_n(z')\mathrm{e}^{\mp\mathrm{i}k_{2\mathrm{iz}}z'}\, \mathrm{d}z'$$

Both expressions of the transmitted and reflected fields contain one deterministic contribution of the slab itself without scatterers, and one stochastic contribution from the scatterers and their interactions with the slab boundaries and all other scatterers. The latter effects are contained in $C_j^\pm(\boldsymbol{k}_\mathrm{it})$.

25.8 Approximations

Two different approximations of the final expression of the transmission and reflection coefficients are of interest—tenuous (sparse) media and the low-frequency approximations. We focus on the coherent transmitted field, since this field occurs in most applications. The coherent reflected field follows analogously.

If the number density of the scatterers is small, we can approximate the solution to the system of integral equations in (25.39) by the first iteration (first order in the number density n_0)

$$f_n(z) \approx d_n(z), \quad z \in [z_1 + a, z_2 - a]$$

which, after some tedious but straightforward algebra, implies an expression of the coherent (averaged) transmitted field $\langle \boldsymbol{E}_{\rm t}(\boldsymbol{r}) \rangle$ with a plane wave impinging normally to the slab, *i.e.*, $\boldsymbol{k}_{\rm it} = \boldsymbol{0}$, see (25.42). We get

$$\langle \boldsymbol{E}_{\rm t}(\boldsymbol{r}) \rangle = t\boldsymbol{E}_0 {\rm e}^{{\rm i}k_3(z-z_2)}, \quad z > z_2$$

where the transmission coefficient t is (details are given in [14])

$$t = t_{\rm slab}{\rm e}^{{\rm i}k_1 z_1} + \frac{2\pi {\rm i} n_0}{k_2^2} \frac{t_{\rm slab}{\rm e}^{{\rm i}k_1 z_1}}{1 + r_{z_1} r_{z_2} {\rm e}^{2{\rm i}k_2(z_2 - z_1)}} \{ -r_{z_1} \mathbf{S}(-\hat{z}, \hat{z}) {\rm e}^{-2{\rm i}k_2 z_1} I_+ \\ - r_{z_1} \mathbf{S}(-\hat{z}, -\hat{z}) r_{z_2} {\rm e}^{2{\rm i}k_2(z_2 - z_1)} k_2 D + \mathbf{S}(\hat{z}, \hat{z}) k_2 D + \mathbf{S}(\hat{z}, -\hat{z}) r_{z_2} {\rm e}^{2{\rm i}k_2 z_2} I_- \} \cdot \boldsymbol{E}_0$$

where the transmission coefficient for the homogeneous slab is

$$t_{\rm slab} = \frac{t_{z_1} t_{z_2} {\rm e}^{{\rm i}k_2(z_2 - z_1)}}{1 + r_{z_1} r_{z_2} {\rm e}^{2{\rm i}k_2(z_2 - z_1)}}$$

and

$$\begin{cases} r_{z_1} = \dfrac{\eta_2 - \eta_1}{\eta_2 + \eta_1} \\ r_{z_2} = \dfrac{\eta_3 - \eta_2}{\eta_3 + \eta_2} \end{cases} \qquad \begin{cases} t_{z_1} = \dfrac{2\eta_2}{\eta_2 + \eta_1} = 1 + r_{z_1} \\ t_{z_2} = \dfrac{2\eta_3}{\eta_3 + \eta_2} = 1 + r_{z_2} \end{cases}$$

and where

$$I_\pm = {\rm e}^{\pm {\rm i}k_2(z_2 + z_1)} \sin(k_2 D)$$

and $D = z_2 - z_1 - 2a$. This is an explicit expression of the coherent transmitted field in the approximation of tenuous scatterers expressed in the scattering dyadic of a single scatterer [7, Sec. 7.8.1]

$$\mathbf{S}(\hat{\boldsymbol{r}}, \hat{\boldsymbol{k}}_{\rm i}) = \frac{4\pi}{{\rm i}k} \sum_{n,n'} {\rm i}^{(l'-l)-(\tau'-\tau)} \boldsymbol{A}_n(\hat{\boldsymbol{r}}) T_{nn'} \boldsymbol{A}_{n'}(\hat{\boldsymbol{k}}_{\rm i})$$

The first term in the expression of the transmission coefficient is the direct transmission through the slab, which is obvious. The second term, however, is more complex, and not easy to derive from physical arguments. In this term, the transition matrix appears to the first power, which indicates that single scattering approximation has been adopted. However, all multiple reflections between the two interfaces are included in this expression of the transmission coefficient. The denominator in the common factor in the second term quantifies these multiple reflections. Moreover, the second and third terms in the parenthesis contain forward scattering terms, and these two terms lack the extra phase correction $I_\pm$ present in the first and last terms.

If we restrict to spherical, dielectric obstacles of radius a and material parameters ε and μ so that $T_{nn'}$ is diagonal in its indices (we adopt the notion $t_{\tau l}$ for the transition matrix entries [7, Chap. 8]), the expressions simplify [7, Ex. 7.3]. We get

$$\mathbf{S}(\hat{z},\hat{z}) = \mathbf{S}(-\hat{z},-\hat{z}) = \frac{\mathbf{I}_2}{2ik_2}\sum_{l=1}^{\infty}(2l+1)\,(t_{1l}+t_{2l})$$

and

$$\mathbf{S}(\hat{z},-\hat{z}) = \mathbf{S}(-\hat{z},\hat{z}) = \frac{\mathbf{I}_2}{2ik_2}\sum_{l=1}^{\infty}(2l+1)(-1)^l\,(t_{1l}-t_{2l})$$

The transmission coefficient t becomes

$$\begin{aligned} t = {} & \frac{3f}{4(k_2a)^3}\frac{t_{\text{slab}}\mathrm{e}^{\mathrm{i}k_1z_1}}{1+r_{z_1}r_{z_2}\mathrm{e}^{2\mathrm{i}k_2(z_2-z_1)}}\left\{\left(1-r_{z_1}r_{z_2}\mathrm{e}^{2\mathrm{i}k_2(z_2-z_1)}\right)k_2D\sum_{l=1}^{\infty}(2l+1)\,(t_{1l}+t_{2l})\right. \\ & \left. +\left(r_{z_2}-r_{z_1}\right)\mathrm{e}^{\mathrm{i}k_2(z_2-z_1)}\sin(k_2D)\sum_{l=1}^{\infty}(-1)^l(2l+1)\,(t_{1l}-t_{2l})\right\} + t_{\text{slab}}\mathrm{e}^{\mathrm{i}k_1z_1} \end{aligned}$$

where we used the dimensionless volume fraction, $f = n_0 4\pi a^3/3$. A numerical implementation of this result, simulating air bubbles ($\varepsilon = \varepsilon_1 = \varepsilon_3$ and $\mu = \mu_1 = \mu_3$) in a dielectric slab of resin, is illustrated in Figure 25.2. The data of the slab is $(z_2 - z_1)/a = 10, f = 0.01, \varepsilon_2/\varepsilon_1 = 2.4(1 + 0.001\mathrm{i})$ and $\mu_2/\mu_1 = 1$.

We now face the task of solving the integral equation (25.39) for low frequencies for spherical particles of radius a, or, more precisely, small k_2a. To simplify the analysis, we assume $\varepsilon_1 = \varepsilon_3$ and $\mu_1 = \mu_3$ (the same materials on both sides of the slab) and we assume the plane wave impinges normally to the slab, *i.e.*, $\boldsymbol{k}_{\text{it}} = \mathbf{0}$.

If the spherical particles of radius a are nonmagnetic, $\mu = 1$, and with a permittivity ε, the transition matrix entries of the scatterers to leading order in powers of k_2a are [7] (lossless materials assumed)

$$t_{21} = T_{2\sigma 11,2\sigma 11} = \frac{2\mathrm{i}(k_2a)^3}{3}y, \quad \sigma = \mathrm{e},\mathrm{o}$$

where

$$y = \frac{\varepsilon - \varepsilon_2}{\varepsilon + 2\varepsilon_2}$$

All other entries of the transition matrix contribute with higher order terms in k_2a.

Keeping only the dominant contribution in k_2a leads after some cumbersome calculations to a transmission coefficient t of the following form (details are given in [14]):

$$t = t_{\text{slab}}\mathrm{e}^{\mathrm{i}k_1 z_1}\left(1 + \frac{3\mathrm{i}}{2}\frac{fy}{1-fy}\left\{\frac{-2r_{z_1}\mathrm{e}^{\mathrm{i}k_2 d}\sin(k_2 d)}{1-r_{z_1}^2\mathrm{e}^{2\mathrm{i}k_2 d}} + \frac{\left(1+r_{z_1}^2\mathrm{e}^{2\mathrm{i}k_2 d}\right)k_2 d}{1-r_{z_1}^2\mathrm{e}^{2\mathrm{i}k_2 d}}\right\}\right)$$

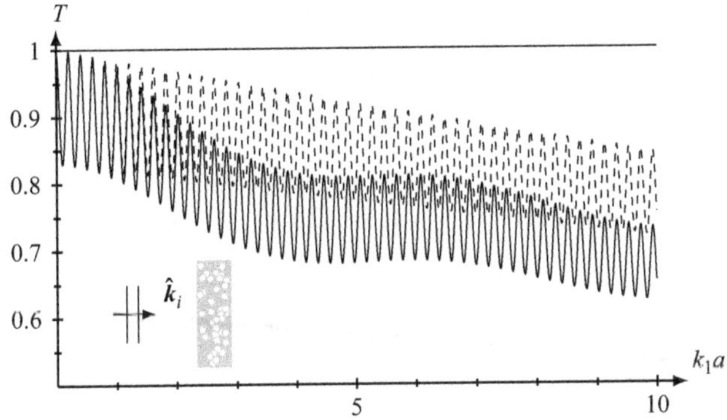

Figure 25.2 The transmissivity $T = |t|^2$ as a function of the electrical size k_1a. The tenuous approximation of the transmissivity with a slab background is depicted as a solid curve. The dashed curve corresponds to the transmissivity without scatterers, and the curve $T = 1$ is also shown.

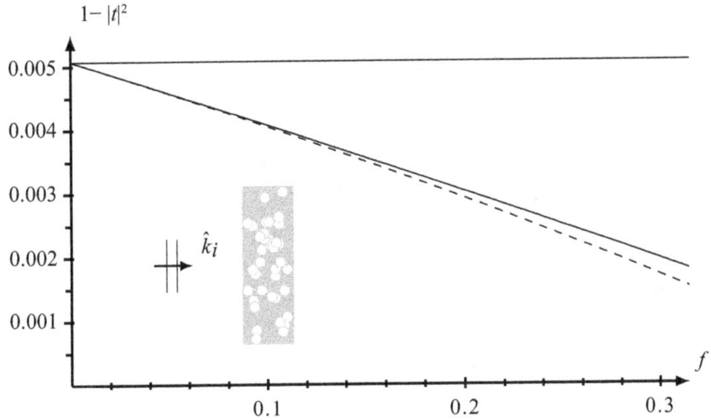

Figure 25.3 The transmissivity $|t|^2$ as a function of the volume fraction f. The low-frequency approximation of the transmissivity is displayed (solid curve) and the tenuous approximation is also given (dashed curve). The transmissivity in a slab without scatterers is also shown.

This expression shows many similarities with the corresponding result with no slab present [3]. The result in this section is illustrated in Figure 25.3 for a slab of thickness $(z_2 - z_1)/a = 10$ and constant electric size $k_1 a = 0.01$, simulating air bubbles in a dielectric slab of resin. Data of the slab is identical to the ones in Figure 25.2.

The dominant contribution in powers of the electric thickness $k_1(z_2 - z_1)$ is also of interest. The transmission coefficient t can be compared with the transmission field of a nonmagnetic, dielectric slab of thickness $d = z_2 - z_1$ with permittivity ε' in a background material ε_1, *i.e.*,

$$\langle \boldsymbol{E}_{\mathrm{t}}(\boldsymbol{r})\rangle = \left(1 + \mathrm{i}\frac{\varepsilon' - \varepsilon_1}{2\varepsilon_1} k_1 d + O\left((k_1 d)^2\right)\right) \boldsymbol{E}_0 \mathrm{e}^{\mathrm{i}k_1 z} = t\boldsymbol{E}_0 \mathrm{e}^{\mathrm{i}k_1(z-z_2)} \tag{25.43}$$

Equating the transmission coefficients t above (approximated to leading order in the thickness $k_1(z_2 - z_1)$ and frequency $k_2 a$) and (25.43) gives

$$\epsilon' = \epsilon_2 + \frac{3fy\epsilon_2}{1 - fy} = \epsilon_2 \frac{1 + 2fy}{1 - fy}$$

which is the Clausius–Mossotti relation [23].

25.9 Conclusions

The results presented in this chapter generalize the analysis reported in [3], where the background material was identical everywhere. The generalization reported in this chapter extends the results such that defects in a slab can be handled, *e.g.*, air bubbles. The analysis solves the boundary value with an arbitrary number of general scatterers inside a slab with different material parameters. Moreover, the results explicitly identify the scattering contribution from the particles themselves. In particular, a potentially lossy background material hosting the particles is covered. A lossy background material has been causing controversial arguments over time regarding extinction [24–26]. The method developed in this chapter avoids this controversy.

The complex scattering problem of randomly located obstacles in a slab with different materials is solved employing a systematic use of two main tools: (1) the integral representation of the solution of the Maxwell equations, (2) decomposition of the Green dyadic for the electric field in free space in spherical and planar vector waves. In addition, we also employ transformations between planar and spherical vector waves. The statistical treatment is general, but the details are only analyzed in full detail for the uniformly distributed obstacles and for the hole correction. Several generalizations of the results presented in this chapter are possible as well as a numerical implementation of the theory.

Appendix A Spherical vector waves

The vector spherical harmonics are defined as [7]

$$\begin{cases} \boldsymbol{A}_{1n}(\hat{\boldsymbol{r}}) = \dfrac{1}{\sqrt{l(l+1)}}\nabla\times\left(\boldsymbol{r}\mathrm{Y}_n(\hat{\boldsymbol{r}})\right) = \dfrac{1}{\sqrt{l(l+1)}}\nabla\mathrm{Y}_n(\hat{\boldsymbol{r}})\times\boldsymbol{r} \\ \boldsymbol{A}_{2n}(\hat{\boldsymbol{r}}) = \dfrac{1}{\sqrt{l(l+1)}}r\nabla\mathrm{Y}_n(\hat{\boldsymbol{r}}) \\ \boldsymbol{A}_{3n}(\hat{\boldsymbol{r}}) = \hat{\boldsymbol{r}}\mathrm{Y}_n(\hat{\boldsymbol{r}}) \end{cases}$$

where the spherical harmonics are denoted by $\mathrm{Y}_n(\hat{\boldsymbol{r}})$ defined as

$$\mathrm{Y}_n(\hat{\boldsymbol{r}}) = \mathrm{Y}_n(\theta,\phi) = \sqrt{\frac{2-\delta_{m0}}{2\pi}}\sqrt{\frac{2l+1}{2}\frac{(l-m)!}{(l+m)!}}\mathrm{P}_l^m(\cos\theta)\begin{Bmatrix}\cos m\phi\\ \sin m\phi\end{Bmatrix}$$

where $\mathrm{P}_l^m(x)$ are the associated Legendre functions. The index n is a multi-index for the integer indices $l = 1,2,3,\ldots$, $m = 0,1,\ldots,l$, and $\sigma = \mathrm{e,o}$ (even and odd in the azimuthal angle).‡

The radiating solutions to the Maxwell equations in vacuum are defined as (outgoing spherical vector waves) [7]

$$\begin{cases} \boldsymbol{u}_{1n}(k\boldsymbol{r}) = \dfrac{\xi_l(kr)}{kr}\boldsymbol{A}_{1n}(\hat{\boldsymbol{r}}) \\ \boldsymbol{u}_{2n}(k\boldsymbol{r}) = \dfrac{\xi_l'(kr)}{kr}\boldsymbol{A}_{2n}(\hat{\boldsymbol{r}}) + \sqrt{l(l+1)}\dfrac{\xi_l(kr)}{(kr)^2}\boldsymbol{A}_{3n}(\hat{\boldsymbol{r}}) \end{cases}$$

where the Riccati–Bessel functions $\xi_l(x) = x\mathrm{h}_l^{(1)}(x)$, where $\mathrm{h}_l^{(1)}(x)$ is the spherical Hankel function of the first kind. The regular spherical vector waves $\boldsymbol{v}_{\tau n}(k\boldsymbol{r})$ are identical to the outgoing ones, but with $\xi_l(x)$ replaced with $\psi_l(x) = x\mathrm{j}_l(x)$, where $\mathrm{j}_l(x)$ is the spherical Bessel function.

Appendix B The translation matrices

The translation properties of the spherical vector waves are instrumental for the formulation and the solution of the scattering problem of many individual scatterers. Let $\boldsymbol{r}' = \boldsymbol{r} + \boldsymbol{d}$. These translation properties are well known, and we refer to, *e.g.*, [7, p. 675] for details. We have

$$\boldsymbol{u}_n(k\boldsymbol{r}') = \sum_{n'}\mathcal{P}_{nn'}(k\boldsymbol{d})\boldsymbol{v}_{n'}(k\boldsymbol{r}), \quad r < d$$

‡The index set at several places in this chapter also denotes a four index set and includes the τ index. That is, the index n can denote $n = \{\sigma, l, m\}$ or $n = \{\tau, \sigma, l, m\}$.

Translation in the opposite direction is identical to the transpose of the translation matrices, *i.e.*,

$$\mathcal{P}^t(k\boldsymbol{d}) = \mathcal{P}(-k\boldsymbol{d})$$

The translation matrices have the form

$$\int_0^{2\pi} \mathcal{P}_{nn'}(k\boldsymbol{d})\,\mathrm{d}\psi = \sum_{\lambda=|l-l'|+|\tau-\tau'|}^{l+l'} \mathcal{A}_{nn'\lambda}\mathrm{h}_\lambda^{(1)}(kd)\mathrm{P}_\lambda(\cos\eta)$$

where $\mathrm{P}_l^m(x)$ are the associated Legendre functions, and where the spherical coordinates of $\boldsymbol{r}$, $\boldsymbol{r}'$, and $\boldsymbol{d}$ are denoted by (r,θ,ϕ), (r',θ',ϕ'), and (d,η,ψ), respectively.

The integral of the translational matrix w.r.t. the azimuthal variable ψ is relevant. Explicitly, we have [3]

$$\mathcal{A}_{\tau\sigma ml,\tau'\sigma'm'l'\lambda} = 2\pi(-1)^m\delta_{mm'} \begin{array}{c} \\ 1\mathrm{e} \\ 2\mathrm{o} \\ 1\mathrm{o} \\ 2\mathrm{e} \end{array} \begin{array}{c} \begin{array}{cccc} 1\mathrm{e} & 2\mathrm{o} & 1\mathrm{o} & 2\mathrm{e} \end{array} \\ \begin{pmatrix} C & D & 0 & 0 \\ -D & C & 0 & 0 \\ 0 & 0 & C & -D \\ 0 & 0 & D & C \end{pmatrix} \end{array}, \quad m = 1,2,3,\ldots$$

$$\mathcal{A}_{\tau\sigma ml,\tau'\sigma'm'l'\lambda} = 2\pi\delta_{mm'} \begin{array}{c} \\ 1\mathrm{e} \\ 2\mathrm{o} \\ 1\mathrm{o} \\ 2\mathrm{e} \end{array} \begin{array}{c} \begin{array}{cccc} 1\mathrm{e} & 2\mathrm{o} & 1\mathrm{o} & 2\mathrm{e} \end{array} \\ \begin{pmatrix} C & 0 & 0 & 0 \\ 0 & 0 & 0 & 0 \\ 0 & 0 & 0 & 0 \\ 0 & 0 & 0 & C \end{pmatrix} \end{array}, \quad m = 0$$

where

$$C = \frac{1}{2}\mathrm{i}^{l'-l+\lambda}(2\lambda+1)\sqrt{\frac{(2l+1)(2l'+1)}{l(l+1)l'(l'+1)}} \times \begin{pmatrix} l & l' & \lambda \\ 0 & 0 & 0 \end{pmatrix} \begin{pmatrix} l & l' & \lambda \\ m & -m & 0 \end{pmatrix} \left[l(l+1) + l'(l'+1) - \lambda(\lambda+1)\right]$$

$$D = \frac{1}{2}\mathrm{i}^{l'-l+\lambda+1}(2\lambda+1)\sqrt{\frac{(2l+1)(2l'+1)}{l(l+1)l'(l'+1)}} \times \begin{pmatrix} l & l' & \lambda-1 \\ 0 & 0 & 0 \end{pmatrix} \begin{pmatrix} l & l' & \lambda \\ m & -m & 0 \end{pmatrix} \sqrt{\lambda^2-(l-l')^2}\sqrt{(l+l'+1)^2-\lambda^2}$$

and where $\begin{pmatrix} \cdot & \cdot & \cdot \\ \cdot & \cdot & \cdot \end{pmatrix}$ denotes Wigner's 3j symbol [27]. Note that the factors $\mathrm{i}^{l'-l+\lambda}$ in C and $\mathrm{i}^{l'-l+\lambda+1}$ in D are always real numbers, due to the conditions on the Wigner's 3j symbol.

Appendix C Planar vector waves

Planar vector waves are commonly used in this chapter. The planar vector waves $\varphi_j^{\pm}(\boldsymbol{k}_t; k, \boldsymbol{r}), j = 1, 2$, are defined as [7]

$$\varphi_1^{\pm}(\boldsymbol{k}_t; k, \boldsymbol{r}) = \frac{\hat{\boldsymbol{z}} \times \boldsymbol{k}_t}{4\pi \mathrm{i} k_t} \mathrm{e}^{\mathrm{i}\boldsymbol{k}_t \cdot \boldsymbol{r}_c \pm \mathrm{i}k_z z}, \quad \varphi_2^{\pm}(\boldsymbol{k}_t; k, \boldsymbol{r}) = \frac{\mp \boldsymbol{k}_t k_z + k_t^2 \hat{\boldsymbol{z}}}{4\pi k k_t} \mathrm{e}^{\mathrm{i}\boldsymbol{k}_t \cdot \boldsymbol{r}_c \pm \mathrm{i}k_z z}$$

where the lateral distance $\boldsymbol{r}_c = x\hat{\boldsymbol{x}} + y\hat{\boldsymbol{y}}$, the lateral wave vector $\boldsymbol{k}_t = k_x\hat{\boldsymbol{x}} + k_y\hat{\boldsymbol{y}}$, the lateral wave number, $k_t = |\boldsymbol{k}_t|$, and the longitudinal wave number k_z is defined by

$$k_z = \left(k^2 - k_t^2\right)^{1/2} = \begin{cases} \sqrt{k^2 - k_t^2} & \text{for } k_t < k \\ \mathrm{i}\sqrt{k_t^2 - k^2} & \text{for } k_t > k \end{cases}$$

From these definitions, we identify $j = 1$ with a TE mode or $\perp$ electric polarization, and $j = 2$ with a TM or $\parallel$ electric polarization.

The planar vector waves satisfy $\nabla \times \varphi_j^{\pm}(\boldsymbol{k}_t; k, \boldsymbol{r}) = k\varphi_{\bar{j}}^{\pm}(\boldsymbol{k}_t; k, \boldsymbol{r}), j = 1, 2$, where the dual index $\bar{j}$ of j is $\bar{1} = 2$ and $\bar{2} = 1$.

There is an alternative representation of the expansion functions $a_j(\boldsymbol{k}_t)$ for the incident fields $\boldsymbol{E}_i$ and $\boldsymbol{H}_i$ in (25.5), see [7]

$$\begin{aligned} a_j(\boldsymbol{k}_t) = &-\frac{2\mathrm{i}k_1}{k_{1z}} \iint_S \varphi_j^{-\dagger}(\boldsymbol{k}_t; k_1, \boldsymbol{r}) \cdot \left(\hat{\boldsymbol{\nu}} \times \mathrm{i}\eta_0\eta_1 \boldsymbol{H}_i(\boldsymbol{r})\right) \mathrm{d}S \\ &-\frac{2\mathrm{i}k_1}{k_{1z}} \iint_S \varphi_{\bar{j}}^{-\dagger}(\boldsymbol{k}_t; k_1, \boldsymbol{r}) \cdot \left(\hat{\boldsymbol{\nu}} \times \boldsymbol{E}_i(\boldsymbol{r})\right) \mathrm{d}S, \quad \boldsymbol{k}_t \in \mathbb{R}^2, \ j = 1, 2 \end{aligned} \tag{C.1}$$

where the surface S is any plane surface to the left of S_1, but to the right of the volume V_i, which contains the sources of the incident field. For a given incident field, this expression gives an explicit formula for the computation of the expansion functions $a_j(\boldsymbol{k}_t)$.

Appendix D The Green dyadic

An important tool is the decomposition of the Green dyadic for the electric field in free space. We decompose the Green dyadic in spherical vector waves, see [7].

$$\mathbf{G}_e(k, |\boldsymbol{r} - \boldsymbol{r}'|) = \mathrm{i}k \sum_n \boldsymbol{v}_n(k\boldsymbol{r}_<)\boldsymbol{u}_n(k\boldsymbol{r}_>) \tag{D.1}$$

where the index n is a multi-index, and where $\boldsymbol{r}_<$ ($\boldsymbol{r}_>$) is the position vector with the smallest (largest) distance to the origin, *i.e.*, if $r < r'$ then $\boldsymbol{r}_< = \boldsymbol{r}$ and $\boldsymbol{r}_> = \boldsymbol{r}'$. This expansion is uniformly convergent in finite domains, provided $r \neq r'$ in the domain [28]. The regular spherical vector waves $\boldsymbol{v}_n(k\boldsymbol{r})$ and the radiating (out-going) spherical vector waves $\boldsymbol{u}_n(k\boldsymbol{r})$ are defined in Appendix A.

The Green dyadic can also be decomposed in planar vector waves [7].

$$\mathbf{G}_{\mathrm{e}}(k,|\boldsymbol{r}-\boldsymbol{r}'|)=2\mathrm{i}k\sum_{j=1,2}\iint_{\mathbb{R}^2}\varphi_j^{\pm}(\boldsymbol{k}_{\mathrm{t}};k,\boldsymbol{r})\varphi_j^{\mp\dagger}(\boldsymbol{k}_{\mathrm{t}};k,\boldsymbol{r}')\frac{k}{k_{\mathrm{z}}}\frac{\mathrm{d}k_{\mathrm{x}}\,\mathrm{d}k_{\mathrm{y}}}{k^2} \tag{D.2}$$

where the upper (lower) is used if $z>z'$ $(z<z')$, and where $\varphi_j^{\pm\dagger}(\boldsymbol{k}_{\mathrm{t}};k,\boldsymbol{r})=\varphi_j^{\pm}(-\boldsymbol{k}_{\mathrm{t}};k,\boldsymbol{r})$.

The outgoing spherical vector waves, $\boldsymbol{u}_n(k\boldsymbol{r})$, can be expressed in the planar vector waves, $\varphi_j^{\pm}(\boldsymbol{k}_{\mathrm{t}};\boldsymbol{r}), j=1,2$. This transformation reads, see [7, p. 508]

$$\boldsymbol{u}_n(k\boldsymbol{r})=2\sum_{j=1,2}\iint_{\mathbb{R}^2}B_{nj}^{\pm}(\boldsymbol{k}_{\mathrm{t}})\varphi_j^{\pm}(\boldsymbol{k}_{\mathrm{t}};k,\boldsymbol{r})\frac{k}{k_{\mathrm{z}}}\frac{\mathrm{d}k_{\mathrm{x}}\,\mathrm{d}k_{\mathrm{y}}}{k^2},\qquad z\gtrless 0 \tag{D.3}$$

where

$$B_{nj}^{+}(\boldsymbol{k}_{\mathrm{t}})=\mathrm{i}^{-l}C_{lm}\left\{-\mathrm{i}\delta_{\tau j}\Delta_l^m\left(\frac{k_{\mathrm{z}}}{k}\right)\begin{Bmatrix}\cos m\beta\\ \sin m\beta\end{Bmatrix}-\delta_{\tau\bar{j}}\pi_l^m\left(\frac{k_{\mathrm{z}}}{k}\right)\begin{Bmatrix}-\sin m\beta\\ \cos m\beta\end{Bmatrix}\right\}$$

and

$$B_{nj}^{-}(\boldsymbol{k}_{\mathrm{t}})=(-1)^{l+m+\tau+j+1}B_{nj}^{+}(\boldsymbol{k}_{\mathrm{t}})$$

where again the index $\bar{j}$ is the dual index of j, defined by $\bar{1}=2$ and $\bar{2}=1$, and where $\boldsymbol{k}_{\mathrm{t}}=k_{\mathrm{t}}(\hat{\boldsymbol{x}}\cos\beta+\hat{\boldsymbol{y}}\sin\beta)$, and

$$C_{lm}=\sqrt{\frac{2-\delta_{m0}}{2\pi}}\sqrt{\frac{2l+1}{2}\frac{(l-m)!}{(l+m)!}}$$

and

$$\Delta_l^m(t)=-\frac{\left(1-t^2\right)^{1/2}}{\sqrt{l(l+1)}}\mathrm{P}_l^{m\prime}(t),\quad \pi_l^m(t)=\frac{m}{\sqrt{l(l+1)}\left(1-t^2\right)^{1/2}}\mathrm{P}_l^m(t)$$

where $\mathrm{P}_l^m(t)$ are the associated Legendre functions. Notice that the argument t can take complex values.

Moreover, we make use of [7, p. 509]

$$\varphi_j^{\pm}(\boldsymbol{k}_{\mathrm{t}};k,\boldsymbol{r})=\sum_n B_{nj}^{\pm\dagger}(\boldsymbol{k}_{\mathrm{t}})\boldsymbol{v}_n(k\boldsymbol{r})$$

where $B_{nj}^{\pm\dagger}(\boldsymbol{k}_{\mathrm{t}})=(-1)^{l+\tau+j+1}B_{nj}^{\pm}(\boldsymbol{k}_{\mathrm{t}})$.

Appendix E Probability density functions

The statistical distribution of the N scatterers, positioned at $\boldsymbol{r}_p$, $p=1,2,\dots,N$, and the state, which is collected in one symbol ξ_p, $p=1,2,\dots,N$, is described by the N-particle PDF $P(\boldsymbol{r}_1,\dots,\boldsymbol{r}_N;\xi_1,\dots,\xi_N)=P(\boldsymbol{r}_1,\dots,\boldsymbol{r}_N)\prod_{p=1}^N P_{\mathrm{s}}(\xi_p)$, where we assumed the state variables are independent stochastic variables.

The PDF$P(\boldsymbol{r}_1, \ldots, \boldsymbol{r}_N)$ is expressed in terms of conditional probability densities according to the definition of conditional PDFs[§][‖]

$$P(\boldsymbol{r}_1, \ldots, \boldsymbol{r}_N) = P(\boldsymbol{r}_1)P(\boldsymbol{r}_2, \ldots, \boldsymbol{r}_N|\boldsymbol{r}_1)$$
$$= P(\boldsymbol{r}_1)P(\boldsymbol{r}_2|\boldsymbol{r}_1)P(\boldsymbol{r}_3, \ldots, \boldsymbol{r}_N|\boldsymbol{r}_1, \boldsymbol{r}_2) \tag{E.1}$$

The average of a function $f(\boldsymbol{r}_1, \ldots, \boldsymbol{r}_N; \xi_1, \ldots, \xi_N)$ over all position variables is denoted by

$$\langle f \rangle = \int_{\Omega_s^N} \int_{V_s^N} P(\boldsymbol{r}_1, \ldots, \boldsymbol{r}_N) \prod_{p=1}^{N} P_s(\xi_p) f(\boldsymbol{r}_1, \ldots, \boldsymbol{r}_N; \xi_1, \ldots, \xi_N) \prod_{p=1}^{N} \mathrm{d}v_p \, \mathrm{d}\xi_p \tag{E.2}$$

where the integration is taken over V_s, which is the volume that the local origins, $\boldsymbol{r}_p$, can occupy, and the state variable is assumed to take values in the abstract space Ω_s.

If the position and state of the pth scatterer are held fixed and all other scatterers are averaged over, we use the notion $\langle f \rangle (\boldsymbol{r}_p; \xi_p)$, where all other variables have been integrated over V_s^{N-1} and Ω_s^{N-1}, *i.e.*,

$$\langle f \rangle (\boldsymbol{r}_1; \xi_1) = \int_{\Omega_s^{N-1}} \int_{V_s^{N-1}} P(\boldsymbol{r}_2, \ldots, \boldsymbol{r}_N|\boldsymbol{r}_1) \prod_{p=2}^{N} P_s(\xi_p)$$
$$\times f(\boldsymbol{r}_1, \ldots, \boldsymbol{r}_N; \xi_1, \ldots, \xi_N) \prod_{p=2}^{N} \mathrm{d}v_p \, \mathrm{d}\xi_p \tag{E.3}$$

Similarly, $\langle f \rangle (\boldsymbol{r}_p)$ and $\langle f \rangle (\boldsymbol{r}_p, \boldsymbol{r}_q; \xi_p, \xi_q)$ denote the averages when the pth scatterer is fixed and the position and state of the pth and qth variables are fixed, respectively.

If the medium is statistically homogeneous, the single-particle PDF is simple (denote the volume of V_s by $|V_s|$)

$$P(\boldsymbol{r}) = \begin{cases} \dfrac{1}{|V_s|} = \dfrac{n_0}{N}, & \boldsymbol{r} \in V_s \\ 0, & \boldsymbol{r} \notin V_s \end{cases}$$

where the number density is $n_0 = N/|V_s|$.

We can express the conditional PDF $P(\boldsymbol{r}|\boldsymbol{r}')$ in terms of the pair distribution function $g(\boldsymbol{r}, \boldsymbol{r}')$ as

$$P(\boldsymbol{r}'|\boldsymbol{r}) = \frac{NP(\boldsymbol{r}')g(\boldsymbol{r}, \boldsymbol{r}')}{N-1} = \frac{n_0 g(\boldsymbol{r}, \boldsymbol{r}')}{N-1} \tag{E.4}$$

[§]The conditional PDF is defined as [29, Sec. 7.2]

$$P(\boldsymbol{r}_{k+1}, \ldots, \boldsymbol{r}_N|\boldsymbol{r}_1, \ldots, \boldsymbol{r}_k) = \frac{P(\boldsymbol{r}_1, \ldots, \boldsymbol{r}_N)}{P(\boldsymbol{r}_1, \ldots, \boldsymbol{r}_k)}$$

[‖]Since the order of numbering is arbitrary, we specialize to scatterers 1 and 2. Any other combination of scatterers follows with a similar notation.

References

[1] Tsang L and Kong JA. Scattering of Electromagnetic Waves: Advanced Topics. New York, NY: John Wiley & Sons; 2001.

[2] Mishchenko MI, Travis LD, and Lacis AA. Multiple Scattering of Light by Particles: Radiative Transfer and Coherent Backscattering. Cambridge: Cambridge University Press; 2006.

[3] Kristensson G. Coherent scattering by a collection of randomly located obstacles—an alternative integral equation formulation. J Quant Spectrosc Radiat Transfer. 2015;164:97–108.

[4] Waterman PC. Matrix formulation of electromagnetic scattering. Proc IEEE. 1965;53(8):805–812.

[5] Waterman PC. Symmetry, unitarity, and geometry in electromagnetic scattering. Phys Rev D. 1971;3(4):825–839.

[6] Mishchenko MI, Zakharova NT, Khlebtsov NG, *et al.* Comprehensive thematic T-matrix reference database: a 2015–2017 update. J Quant Spectrosc Radiat Transfer. 2017;202:240–246.

[7] Kristensson G. Scattering of Electromagnetic Waves by Obstacles. Edison, NJ: SciTech Publishing, an imprint of the IET; 2016.

[8] Kristensson G. Electromagnetic scattering from buried inhomogeneities—a general three-dimensional formalism. J Appl Phys. 1980;51(7):3486–3500.

[9] Karlsson A and Kristensson G. Electromagnetic scattering from subterranean obstacles in a stratified ground. Radio Sci. 1983;18(3):345–356.

[10] Gustavsson M, Kristensson G, and Wellander N. Multiple scattering by a collection of randomly located obstacles—numerical implementation of the coherent fields. J Quant Spectrosc Radiat Transfer. 2016;185:95–100.

[11] Doicu A, Eremin Y, and Wriedt T. Acoustic and Electromagnetic Scattering Analysis Using Discrete Sources. London: Academic Press; 2000.

[12] Aydin K and Hizal A. On the completeness of the spherical vector wave functions. J Math Anal Appl. 1986;117(2):428–440.

[13] Kristensson G, Ramm AG, and Ström S. Convergence of the T-matrix approach in scattering theory. II. J Math Phys. 1983;24(11):2619–2631.

[14] Kristensson G and Wellander N. Multiple Scattering by a Collection of Randomly Located Obstacles. Part III: Theory—Slab Geometry. Lund, Sweden: Department of Electrical and Information Technology, Lund University; 2017. LUTEDX/(TEAT-7252)/1–67/(2017). http://www.eit.lth.se.

[15] Varadan VV and Varadan VK. Multiple scattering of electromagnetic waves by randomly distributed and oriented dielectric scatterers. Phys Rev D. 1980;21(2):388–394.

[16] Tishkovets VP, Petrova EV, and Mishchenko MI. Scattering of electromagnetic waves by ensembles of particles and discrete random media. J Quant Spectrosc Radiat Transfer. 2011;112:2095–2127.

[17] Mathur NC and Yeh KC. Multiple scattering of electromagnetic waves by random scatterers of finite size. J Math Phys. 1964;5:1619–1628.

[18] Mishchenko MI, Dlugach JM, Yurkin MA, *et al.* First-principles modeling of electromagnetic scattering by discrete and discretely heterogeneous random media. Phys Rep. 2016;632:1–75.

[19] Mishchenko MI. Electromagnetic Scattering by Particles and Particle Groups. An Introduction. New York, NY: Cambridge University Press; 2014.

[20] Lax M. Multiple scattering of waves. II. The effective field in dense systems. Phys Rev. 1952;85:621–629.

[21] West R, Gibbs D, Tsang L, *et al.* Comparison of optical scattering experiments and the quasi-crystalline approximation for dense media. JOSA A. 1994;11(6):1854–1858.

[22] Kristensson G. Evaluation of some integrals relevant to multiple scattering by randomly distributed obstacles. J Math Anal Appl. 2015;432(1):324–337.

[23] Jackson JD. Classical Electrodynamics. 3rd ed. New York, NY: John Wiley & Sons; 1999.

[24] Bohren CF and Gilra DP. Extinction by a spherical particle in an absorbing medium. J Colloid Interface Sci. 1979;72(2):215–221.

[25] Mishchenko MI. Multiple scattering by particles embedded in an absorbing medium. 1. Foldy-Lax equations, order-of-scattering expansion, and coherent field. Opt Express. 2008;16(3):2288–2301.

[26] Mishchenko MI. Multiple scattering by particles embedded in an absorbing medium. 2. Radiative transfer equation. J Quant Spectrosc Radiat Transfer. 2008;109(14):2386–2390.

[27] Edmonds AR. Angular Momentum in Quantum Mechanics. 3rd ed. Princeton University Press, Princeton, NJ; 1974.

[28] Colton D and Kress R. Inverse Acoustic and Electromagnetic Scattering Theory. 3rd ed. Berlin: Springer-Verlag; 2013.

[29] Papoulis A and Pillai SU. Probability, Random Variables, and Stochastic Processes. 4th ed. Boston, MA: McGraw-Hill; 2002.

Chapter 26
Electromagnetics of complex environments applied to geophysical and biological media

Akira Ishimaru[1]*, Yasuo Kuga*[1]*, and Max Bright*[2]

One of the important areas of research is the study of electromagnetic and related wave theories, which have a wide range of practical applications in complex environments, such as microwave remote sensing of the Earth, object detection and imaging in clutter, medical optics and ultrasound imaging, characterization of metamaterials and composite and porous media, and communication through complex clutter environments. This chapter gives a review of wave theories applied to imaging in geophysical and biological media, including imaging through air turbulence and particulate matter, imaging near-ocean rough surfaces and communication and signal processing in clutter, coherence in multiple scattering and super resolution, time-reversal (TR) imaging, radiative transfer, waves in porous media, seismic CODA waves, and the memory effect.

26.1 Introduction

There have been great advances made in understanding electromagnetic phenomena in complex environments, in particular geophysical and biological media. This chapter gives a review of recent progress in imaging, super-resolution, communication through complex media, radiative transfer, bio EM, optics and ultrasound, coherence in multiple scattering, metamaterials, transformation EM, porous medium, and seismic codas. We discuss statistical multiple scattering theories, as applied to turbulence, particulate matter, rough surfaces, and remote sensing and imaging of objects in random clutter environments, and integration of signal processing, including seismic coda.

26.2 Stochastic wave theories

Many problems in electromagnetics can be classified as either "deterministic" or "random." For deterministic problems, we deal with media and objects that are

[1]Department of Electrical and Computer Engineering, University of Washington, Seattle, WA, USA
[2]Department of Electrical and Computer Engineering, University of Michigan, Ann Arbor, MI, USA

well defined in shape, position, and material characteristics. There are many cases where the media characteristics, positions, and slopes vary randomly; these are called "random media," and the waves propagating and scattering in random media vary randomly in space and time. It should be stated that though the waves may vary randomly, there are well-defined theories underlying the random phenomena. Statistical wave theories are therefore aimed at discerning these well-defined theories concerning these randomly varying phenomena [1–4].

Multiple-scattering theories for random media have been developed over many years. The theories have been developed in astrophysics, and in scattering and disordered media, Dyson's equation for the coherent field, Bethe–Salpeter equation for the second moment, and the Feynman diagram are the key theoretical basis for scattering in random media. These theories were developed through Foldy, Lax, and others. It should also be noted that radiative transfer theories have been developed which have been shown to conserve the power and have been used extensively in many practical problems in geophysical and biomedical applications. Radiative transfer theories are related to neutron transport theories and applied to tissue optics, optical diffusion, imaging, remote sensing, and communications and signal processing. It is, however, noted that radiative transfer is an approximate theory and there are several cases where radiative transfer is not applicable. Examples are coherence in multiple scattering, memory effects, and enhanced backscattering.

Multiple scattering theories have been applied to scattering by rough surfaces, ocean surfaces, terrain, and low-grazing angle scattering. Small perturbation theory, Kirchhoff approximation, and small and high slope rough surfaces have been studied [5].

Keys to the study of waves in random media are the "mutual coherence function" (MCF), the second moments, the fourth moment, the scintillation index, and the stochastic Green function.

Several important topics need to be noted. Geophysical media include terrain, atmosphere, ocean, and the planetary atmosphere. Applications of the statistical theories are many and include climate, hydrological, water, agriculture and ocean, thermal emissions, soils, clouds, snow, ice, vegetation, wind, and soil moisture issues.

26.3 Time-reversal imaging

The phase-conjugate mirror and retro reflectance are early examples of "TR" imaging. When a wave from a source is incident on a "TR mirror," it is reflected back to the original source. It is equivalent to producing the reversed field, which propagates back to focus on the original source even if the medium is complex and multiple-scattering. TR imaging is related to DORT (decomposition of time-reversal operator), singular value decomposition, and TR-MUSIC (multiple signal classification) [6–8].

Imaging through random multiple scattering media requires a combination of wave scattering and signal processing. Consider an antenna array, random medium, and point target (Figure 26.1). First we consider the "multistatic matrix" T and TR matrix $T = KK'$. K' is the conjugate-transpose of K. This is a symmetric $N \times N$

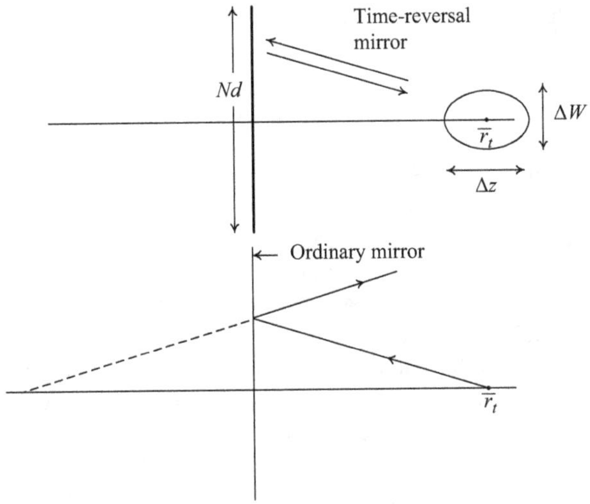

Figure 26.1 Time-reversal (conjugate) mirror and ordinary mirror (from [5], Fig. 22.2)

matrix, and its element K_{ij} is the signal received at the ith antenna located at r'' when the signal is emitted from the jth antenna and reflected by targets. Consider the wave v ($N \times 1$ matrix) applied to the N elements. The wave propagates and is reflected by targets and received by the N receivers.

The received signal is then given by the $N \times 1$ matrix Kv where K is the $N \times N$ multistatic data matrix. This received signal is then time reversed at each terminal giving the next transmitted signal $(Kv)^*$.

This process can continue and if this sequence converges, then except for a constant σ, we should get the transmitted signal $(Kv)^*$ equal to the original transmitted signal v. The constant σ represents the power lost by scattering and diffraction.

$$(Kv)^* = \sigma v \tag{26.1}$$

Rewriting this we have

$$Kv = \sigma^* v^*$$

This can be converted to an eigenvalue equation by multiplying both sides by K^*

$$K^* K v = \lambda v$$

$$\lambda = |\sigma^2| = \text{eigenvalue}$$

The $N \times N$ matrix $T = K^* K$ is called the TR matrix and is a Hermitian matrix

$$T' = T$$

where T' is the transpose conjugate of T and K is symmetric, $K' = K^*$.

The eigenvalue problem can be written as

$$Tv = \lambda v \tag{26.2}$$

Once we obtain K (multistatic data matrix) and the eigenvectors v and eigenvalues λ, we can obtain the imaging function.

In order to obtain the image, we use the signal σv and focus the wave to the search point $\bar{r}_s$ by using the steering vector g_s. If the steering vector is the Green function for the imaging problem then the wave will be focused on the target. In general, however, we may not know the Green function and therefore we use an approximate g_s such as Green's function in a homogeneous background. The steering function g_s is the same as the focusing matched filter function used in SAR (synthetic-aperture radar). See Figure 26.1 for the image formed by the TR mirror and an ordinary mirror.

DORT is the French acronym for "décomposition de l'opérateur de retournement temporel" (decomposition of time-reversal operator) and was proposed by Prada and Fink [6,9]. It provides a theoretical study of the iterative TR process and shows that in a multi-target media, the brightest target is associated with the eigenvector of greatest eigenvalue.

26.4 Imaging through random multiple scattering clutter

There are several imaging techniques in random media making use of the multistatic matrix, TR matrix, eigenvalues, steering vector, and MCF. Imaging functions can be developed using several techniques such as TR imaging, TR-MUSIC, modified beamformer, and SAR techniques. Imaging through random multiple scattering media requires a combination of wave scattering and signal scattering [10].

Imaging functions for a point object using different techniques are shown as follows (details are shown in [10]):

$$\begin{aligned}
&TR\ imaging = \int d\omega_1 d\omega_2 U_1^2 U_2^2 \lambda_1^2 \sum_i \sum_j \langle G_i G_j^* \rangle \left(G_{si}^* G_{sj} \right) \\
&TR\ MUSIC = \psi_{MUSIC} = \frac{1}{\sum_{p=M+2}^{N} d\omega U^2 \left| [g_s]' v_p \right|} \\
&Modified\ beamformer = \int d\omega_1 d\omega_2 U_1^2 U_2^2 \sum_i \sum_j \sum_p \sum_q \langle G_i G_j G_p^* G_q^* \rangle \\
&\qquad \left(G_{si} G_{sj} G_{sp}^* G_{sq}^* \right) \\
&SAR = \int d\omega_1 d\omega_2 \left| U_1 \right|^2 \left| U_2 \right|^2 \sum_i \sum_j \langle G_i^2(\omega_1)\, G_j^{*2}(\omega_2) \rangle \left(G_{si}^{*2}(\omega_1)\, G_{sj}^2(\omega_2) \right)
\end{aligned} \tag{26.3}$$

26.5 Geophysical remote sensing and imaging, and super resolution

Satellite or aircraft observations of geological media, the atmosphere, terrain, and the ocean give useful information on an environment. Active and passive radars and sensors can measure the wave signals, amplitude, phase, intensity, space–time correlations, and polarizations that are used to determine the medium characterization [11].

Microwave remote sensing is one of the most important areas of research related to environmental issues of the Earth. Microwaves are most effective in probing and sensing of snow, ice, soil moisture, vegetation cover, forestry, clouds, rainfall, terrain, and ocean. The fundamental theories of microwave remote sensing have been developed by Tsang and Ulaby [3,12], including vector radiative transfer, Dyson and Bethe–Salpeter equation, Feynman's diagram, and the dense media theory and pair-correlations. The spot size W is reduced in a multiple scattering medium because the coherence length is reduced, which gives the increase of effective aperture.

It is first necessary to obtain the coherent propagation constant through a random medium consisting of isotropic or spherical as well as nonspherical particles.

The coherent component $\langle E_1, E_2\rangle$ satisfies the following differential equation:

$$\frac{d}{ds}\begin{bmatrix}\langle E_1\rangle \\ \langle E_2\rangle\end{bmatrix} = [M]\begin{bmatrix}\langle E_1\rangle \\ \langle E_2\rangle\end{bmatrix} \tag{26.4}$$

where

$$[M] = [M_0] + [M^1]$$

$$[M_0] = ik\begin{bmatrix}1 & 0 \\ 0 & 1\end{bmatrix}$$

$$[M^1] = i\frac{2\pi}{k}\begin{bmatrix}\rho f_{11} & \rho f_{12} \\ \rho f_{21} & \rho f_{22}\end{bmatrix} = \begin{bmatrix}M_{11} & M_{12} \\ M_{21} & M_{22}\end{bmatrix}$$

f_{ij} are the scattering amplitudes in the forward direction and ρf is the integration over the size distribution

$$\rho f_{ij} = \int_0^\infty f_{11} n(D) dD.$$

The complete vector radiative transfer equation is then given by

$$\frac{d}{ds}[I] = -[T][I] + \int [S(\hat{s}, \hat{s}')][I(\hat{s}')]d\Omega'$$

The space–time vector radiative transfer equation by two-frequency specific intensity is given by

$$\frac{d}{ds}I = i[K(\omega_1) - K^*(\omega_2)]I + \int S(\omega_1, \omega_2)\, I' d\Omega' \tag{26.5}$$

$K(\omega_1) - K^*(\omega_2)$ is approximately given by

$$K(\omega_1) - K^*(\omega_2) = \omega_d \left.\frac{\partial K}{\partial \omega}\right|_{\omega_c} + i\rho\sigma_t.$$

It is expected that in multiple scattering the images will be degraded and resolution becomes poor. So it is surprising that the resolution is actually improved in multiple scattering. This is called "super resolution" This has been numerically proved. It can be explained by considering that multiple scattering creates the increase in effective aperture as shown in Figure 26.2.

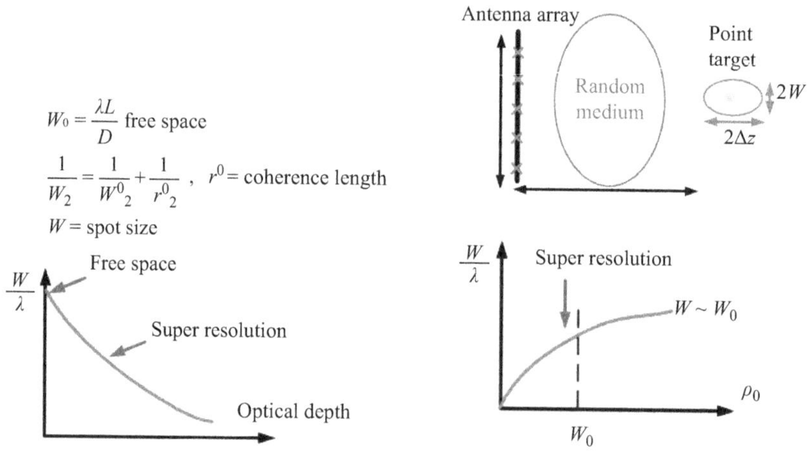

Figure 26.2 Super resolution

26.6 Wigner distribution function and specific intensity

It should be noted that the formulations given in the last section can be viewed from a more fundamental formulation of Wigner distribution.

Let us now start with the CW MCF given by

$$\begin{aligned}\Gamma(\bar{r}_1,\bar{r}_2) &= U(\bar{r}_1)U^*(\bar{r}_2)\\ &= \Gamma(\bar{r},\bar{r}_d)\end{aligned}$$

where $\bar{r} = 1/2(\bar{r}_1 + \bar{r}_2)$ and $\bar{r}_d = (\bar{r}_1 - \bar{r}_2)$.

The Wigner distribution function is given by

$$\begin{aligned}W(\bar{r},\bar{K}) &= \int \Gamma(\bar{r},\bar{r}_d)e^{-i\bar{K}\cdot\bar{r}_d}dV_d\\ &= \int U\left(\bar{r}+\frac{\bar{r}_d}{2}\right)U^*\left(\bar{r}-\frac{\bar{r}_d}{2}\right)e^{-i\bar{K}\cdot\bar{r}_d}dV_d\end{aligned} \tag{26.6}$$

The Wigner function W has certain properties as shown here. First, the integral of W with respect to the wave vector $\bar{K}$ is the "energy density"

$$\begin{aligned}\frac{1}{(2\pi)^3}\int W(\bar{r},\bar{K})d\bar{K} &= \Gamma(\bar{r},\bar{r}_d = 0)\\ &= \langle|U(\bar{r})|^2\rangle\end{aligned} \tag{26.7}$$

Note that

$$\int e^{-i\bar{K}\cdot\bar{r}_d}d\bar{K} = (2\pi)^3\delta(\bar{r}_d)$$

Also note that W is real.

$$W^*(\bar{r}, \bar{K}) = \int \left\langle U^*\left(\bar{r} + \frac{\bar{r}_d}{2}\right) U\left(\bar{r} - \frac{\bar{r}_d}{2}\right)\right\rangle e^{+i\bar{K}\cdot\bar{r}_d} dV_d$$

We let $\bar{r}_d = -\bar{r}'_d$ and get

$$\begin{aligned} W^*(\bar{r}, \bar{K}) &= \int \left\langle U\left(\bar{r} + \frac{\bar{r}_d}{2}\right) U^*\left(\bar{r} - \frac{\bar{r}_d}{2}\right)\right\rangle e^{+i\bar{K}\cdot\bar{r}_d} dV_d \\ &= W(\bar{r}, \bar{K}) \end{aligned}$$

showing that $W^*(\bar{r}, \bar{K})$ is real.

The Wigner distribution function W is real and is closely related to the specific intensity $I(\bar{r}, \hat{s})$. The difference is that the specific intensity I is real and positive as it represents the real power flow in the direction of $\hat{s}$. $W(\bar{r}, \bar{K})$ is identical to $I(\bar{r}, \hat{s})$ if $\bar{K}$ is limited to $\bar{K} = k\hat{s}$ where $\hat{s}$ is the unit vector with $|\hat{s}| = 1$. While I is real and positive, W is real but need not be positive.

The temporal Wigner function can be obtained similarly. We have

$$W(t, \omega) = \int_{-\infty}^{\infty} \langle U(t_1) U^*(t_2)\rangle e^{+i\omega t_d} dt_d \tag{26.8}$$

where $t_1 = t + (t_d/2), t_2 = t - (t_d/2)$.

We can express this using the Fourier transform

$$\begin{aligned} U(t) &= \frac{1}{2\pi}\int \bar{U}(\omega) e^{-i\omega t} d\omega \\ W(t, \omega) &= \frac{1}{(2\pi)^2}\iint \langle \bar{U}(\omega_1)\bar{U}^*(\omega_2)\rangle e^{-i\omega_1 t_1 + i\omega_2 t_2 + iwt_d} d\omega_1 d\omega_2 dt_d \end{aligned}$$

where $\langle \bar{U}(\omega_1)\bar{U}^*(\omega_2)\rangle$ is the two-frequency MCF.

Noting that $\omega_1 t_1 - \omega_2 t_2 = \omega_c t_d + \omega_d t_d$, we get

$$W(t, \omega) = \frac{1}{(2\pi)}\int \left\langle U\left(\omega + \frac{\omega_d}{2}\right) U^*\left(\omega - \frac{\omega_d}{2}\right)\right\rangle e^{-i\omega_d t} d\omega_d \tag{26.9}$$

The Wigner distribution function $W(t, \omega)$ is related to the "ambiguity function" $\chi(\omega_d, t_d)$

$$\chi(\omega_d, t_d) = \frac{1}{(2\pi)}\int U(t_1)\, U^*(t_2)\, e^{i\omega_d t} dt \tag{26.10}$$

From this, we can show that

$$W(t, \omega) = \int \chi(\omega_d, t_d)\, e^{i(\omega_d t - \omega t_d)} d\omega_d dt_d$$

26.7 Biomedical electromagnetism and optics

Imaging and detection of tissue characteristics are important in health care. In biomedical electromagnetism, extensive work has been done on the effects of the specific absorption rate of cellphones, and heat diffusion in tissues. For bio-optics, optical scattering and imaging in tissue, optical diffusion, optical coherence tomography (OCT), and the photon density wave have been studied extensively. Ultrasound imaging of tissues and blood have also been studied [13–15]

Interactions of electromagnetic fields with biological systems date back to the late 1800s, when d'Arsonval applied the high frequency (10 kHz) current on himself and found that it produced warming of muscles [13]. Interest in wireless communication and cellular phones, in particular, when electromagnetic radiation is close to the body, is important in health and medical issues. In order to study the biological effects, vivo (within the living) and vitro (outside the living) experiments are used with static electric and magnetic fields, radio frequency (RF) fields dosimetry, energy absorption, specific absorption rate, and ultrawideband pulses. The typical pulses range from milliseconds with 10 kV/m to microseconds with 100 kV/m. Also noted are biological effects and medical application of magnetic fields. Extensive recent discussions include the effects of mobile phone and RF electromagnetic fields on brain function, the human body and safety as detailed in "bioelectromagnetics" [16].

26.8 Heat diffusion in tissues

As tissues are illuminated by EM waves, EM power is dissipated in the lossy tissue medium and converted into heat energy. The heat energy is then diffused in the tissue and the temperature varies.

The heat flow per unit area is proportional to the temperature gradient

$$\bar{F} = -k\nabla T \tag{26.11}$$

where k is the thermal conductivity and $\bar{F}$ is the heat flow per unit area.

The heat generated per unit volume creates the temperature rise and gives the diffusion equation

$$\frac{\partial T}{\partial t} = a^2\nabla^2 T + \frac{q}{\rho C_s} \tag{26.12}$$

where a is the diffusion constant and C_s is the specific heat. q is the energy introduced per unit volume (W/m^3).

It is then possible to solve the diffusion equation with boundary conditions to obtain the space–time distribution of the temperature [5] The optical diffusion in tissues can be calculated by using the diffusion approximation of radiative transfer.

The same theory can then be used to determine the temperature rise in objects close to an optical focused beam through atmospheric turbulence near the ground and ocean surface [17].

26.9 Ultrasound in tissues and blood

In the normal frequency range of a few thousand kilohertz to 10 MHz of ultrasound, the tissues are absorbing and the scattering is small. Therefore, the single scattering approximation can be used. This is in contrast to the optical propagation in tissue where the optical beam is strongly scattered and the diffusion approximation is appropriate. Ultrasound imaging has been extensively studied [18] (Figure 26.3 shows the focused beam spot size and depth of focus of the typical imaging geometry).

Consider a volume dv of tissues with density ρ_e and compressibility κ_e which are different from the surrounding average density ρ and compressibility κ. Under the assumption that the medium ρ_e and κ_e are only slightly different from ρ and κ, we can use the Born approximation to obtain the following well-known formula for the scattering amplitude:

$$f(\hat{o},\hat{i}) = \frac{k^2}{4\pi}\int_{\delta v} (\gamma_k + \gamma_p \cos\theta) e^{i\hat{k}_s\cdot\hat{r}'}\, dv', \tag{26.13}$$

where

$$\gamma_k = \frac{\kappa_e - \kappa}{\kappa} = \text{compressibility fluctuation,}$$

$$\gamma_p = \frac{\rho_e - \rho}{\rho} = \text{density fluctuation.}$$

We then obtain the differential scattering coefficient σ_d or the differential cross section per unit volume of the tissue.

$$\sigma_d(\hat{o},\hat{i}) = \frac{\langle ff^*\rangle}{\delta v} = \left(\frac{k^2}{4\pi}\right)^2 \frac{1}{\delta v}\iint \langle \gamma(\bar{r}_1)\gamma(\bar{r}_2)\rangle e^{i\bar{k}_s\cdot(\bar{r}_1-\bar{r}_2)} dv_1 dv_2 \tag{26.14}$$

where

$$\gamma(\bar{r}) = \gamma_k(\bar{r}) + \gamma_p(\bar{r})\cos\theta,$$

and

$$\bar{k}_s = k(\hat{i},\hat{o}).$$

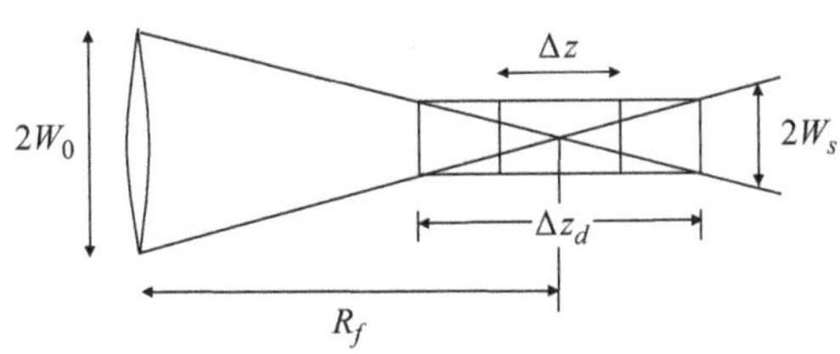

Figure 26.3 Focused beam spot size W_0, axial spot size Δz, depth of focus Δz_d (from [5] Fig. 20.7)

If the medium is assumed to be statistically homogeneous and isotropic, then the covariance $\langle\gamma(\bar{r}_1)\gamma(\bar{r}_2)\rangle$ is a function of the magnitude of the difference $|\bar{r}_1 - \bar{r}_2|$. The double integral can then be expressed as a Fourier transform of the covariance function $\langle\gamma(\bar{r}_1)\gamma(\bar{r}_2)\rangle$ that is called the spectral density as noted by the Wiener–Khinchin theorem.

We can express (26.14) using the spectral densities

$$S_\gamma(\bar{k}_s) = \frac{1}{(2\pi)^3}\int B_\gamma(\bar{r}_d)e^{i\bar{k}_s\cdot\bar{r}_d}dv_d$$

and $B_\gamma(\bar{r}_d)$ is the correlation function given by

$$B_\gamma(\bar{r}_d) = \gamma(\bar{r}_1)\gamma(\bar{r}_2) = B_k(\bar{r}_d) + B_\rho(\bar{r}_d)\cos^2\theta + 2B_{\kappa\rho}(\bar{r}_d)\cos\theta.$$

We therefore have an expression for σ_d

$$\sigma_d(\hat{o},\hat{i}) = \left(\frac{\pi}{2}\right)k^4[S_k(k_S) + S_\rho(k_S)\cos^2\theta + 2S_{\kappa\rho}\cos\theta]$$

The unit commonly used for σ_d (differential cross section per unit volume) of the tissue is $\mathrm{cm}^{-2}/(\mathrm{cm}^3\ \mathrm{sr}) = \mathrm{cm}^{-1}\ \mathrm{sr}^{-1}$ where sr = steradian (unit solid angle).

Extensive studies have been made on ultrasonic properties of blood. For the normal frequency range of a few hundred kilohertz to 10 MHz used in biological media, the wavelength is much greater than the size of red blood cells. Therefore the Rayleigh formula for a sphere with the same volume as that of a red blood cell should give a good approximation of the absorption and scattering characteristics. The scattering amplitude $f(\hat{0},\hat{i})$ is given by

$$f(\hat{0},\hat{i}) = \frac{k^2a^3}{3}\left(\frac{\kappa_e - \kappa}{\kappa} + \frac{3\rho_e - 3\rho}{2\rho_e + \rho}\cos\theta\right) \tag{26.15}$$

where $k = 2\pi/\lambda$, λ is the wavelength of the surrounding medium (plasma), a the radius of the equivalent sphere, κ_e and ρ_e are the adiabatic compressibility and density of the red blood cell, respectively, κ and ρ are those of plasma, and θ is the angle between $\hat{0}$ and $\hat{i}$. For a normal blood cell, the volume is 87 $\mu\mathrm{m}^3$, and the equivalent radius $a = 2.75$ μm. The compressibility and density of a red blood cell are $\kappa_e = 34.1\times10^{-12}$ cm^2/dyne and $\rho_e = 1.092$ g/cm^3 and those of plasma are $\kappa = 40.9\times10^{-12}$ cm^2/dyne and $\rho_e = 1.021$ g/cm^3.

The scattering cross section σ_s is given by

$$\frac{\sigma_s}{\pi a^2} = \frac{4\,(ka)^2}{9}\left|\left|\frac{\kappa_e - \kappa}{\kappa}\right|^2 + \frac{1}{3}\left|\frac{3\rho_e - 3\rho}{2\rho_e + \rho}\right|^2\right| \tag{26.16}$$

Using the values of $\kappa_e, \rho_e, \kappa, \rho$, and a, we get

$$\sigma_s = 0.47\times10^{16}f^4\ \mathrm{cm}^2,$$

where f is the frequency measured in megahertz. The backscattering cross section is approximately given by $\sigma_b \cong 1.86\sigma_s$.

The absorption cross section σ_a is proportional to the frequency and is given by

$$\frac{\sigma_a}{\pi a^2} = \frac{4ka}{3}\mathrm{Im}\left(\frac{\kappa_e - \kappa}{\kappa} + \frac{3\rho_e - 3\rho}{2\rho_e + \rho}\right) \tag{26.17}$$

where Im designates the "imaginary part of." The imaginary parts of $\kappa_e, \rho_e, \kappa,$ and ρ are not known. However, it is known that the absorption cross section σ_a is much greater than the scattering cross section σ_s in the frequency range 0.1–10 MHz, and therefore the attenuation through blood is mainly due to the absorption, and not the scattering. It is also known that the attenuation of a plane wave in random scatterers is given by $\rho\sigma_a$ nepers per unit distance (Np/cm). The attenuation constant is proportional to the hematocrit H and the frequency f (in megahertz) and is approximately given by

$$\alpha = (5 - 7) \times 10^{-2} Hf \ \mathrm{Np/cm} = 0.3Hf \ \mathrm{dB/cm}.$$

Since $\alpha = \rho\sigma_a = (H/V_e)\sigma_a$, we obtain the absorption cross section of a single erythrocyte

$$\sigma_a = 6 \times 10^{-12} f \ \mathrm{cm}^2$$

where f is measured in megahertz. The effect of viscosity of the plasma and the erythrocyte on the scattering and absorption characteristics may be significant [4,18].

The differential scattering cross section per unit volume of the blood is therefore

$$\sigma(\hat{o}, \hat{i}) = \frac{H f_P(H)}{V_e} |f(\hat{o}, \hat{i})|^2,$$

where H is the hematocrit (0.4 for human), V_e is the volume of the single cell $\left(4\pi a^3/3\right)$, and $f_P(H)$ is the packing factor. The Percus–Yevick packing factor for hard spheres is often used as an approximation

$$f_P(H) = \frac{(1 - H)^4}{(1 + 2H)^2} \tag{26.18}$$

26.10 Low coherence interferometry and optical coherence tomography (OCT)

Low coherence interferometry is equivalent to short pulses but it uses continuous waves without the need for ultrashort pulses. It has been effectively used in OCT and passive coherent scatter radar. Here we discuss OCT that is now commercially available in ophthalmology.

OCT was introduced to obtain noninvasive images of biological tissues with a resolution of 1–15 μm [5,14]. This is made possible by using low coherence interferometry.

The images are generated by measuring the echo time delay and the light intensity backscattered from tissues. The echo time delay is measured by using low coherence light. The backscattered light from the sample is interfered with low coherence light.

The interference occurs only when the two path lengths match somewhere within the coherence length. By varying the path length of the low coherence light, the echo time delay can be varied. This is equivalent to short pulses but it uses continuous-wave light without the need for ultrashort pulses. A sketch of OCT is shown in Figure 26.4.

Let us next consider the transverse resolution. If the aperture with the diameter $2W_0$ transmits a beam wave focused at R_f, the transverse spot size and the focal plane $(z = R_f)$ is given by

$$W_s = \frac{\lambda R_f}{\pi W_0}. \tag{26.19}$$

The axial spot size (resolution) is given by the coherence length l_c

$$\Delta z = l_c = 0.664 \left(\frac{\lambda_0^2}{\Delta \lambda_0} \right). \tag{26.20}$$

In addition, the depth of focus Δz_d is given by

$$\Delta z_d = \frac{\lambda R_f^2}{\pi W_0^2}. \tag{26.21}$$

The incident field U_0 with coherence time and coherence length l_c is now incident on the Michelson interferometer shown in Figure 26.4. The incident wave $U_0(t)$ propagates and splits into U_1 and U_2 in Figure 26.4. U_1 travels to the reference mirror and is reflected back to become $K_1 U_0$, and U_2 is reflected back by the sample and becomes $K_2 U_0$. K_1 and K_2 include the effects of the beam splitter and reflections.

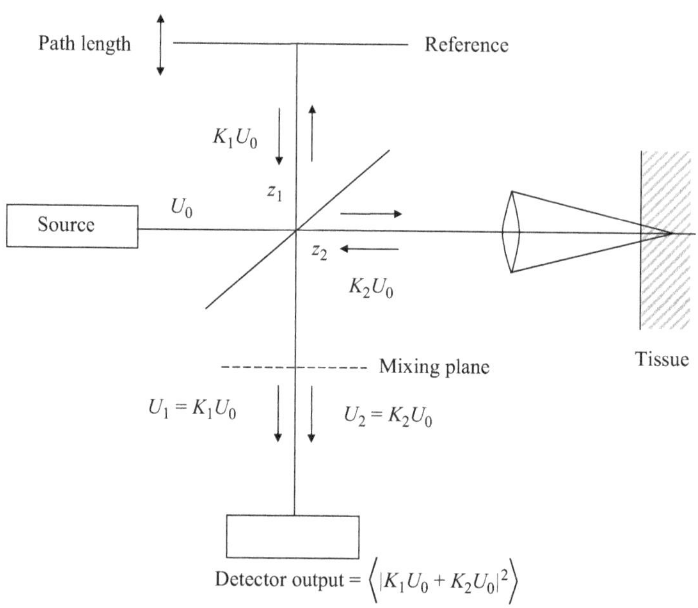

Figure 26.4 OCT and low coherence interferometry (from [5] Fig. 20.5)

The total $U_1 + U_2$ is incident on the detector. The output from the detector I_d is then given by

$$I_d = \langle |K_1 U_0 + K_2 U_0|^2 \rangle = |U_0|^2 \left(K_1^2 + K_2^2 + 2\mathrm{Re}(K_1 K_2)\,\Gamma_{12} \right) \tag{26.22}$$

where Γ_{12} is the MCF.

26.11 Waves in metamaterials and electromagnetic and acoustic Brewster's angle

There has been extensive work reported on metamaterials. "Meta" means "beyond" indicating the material is beyond those found in nature [19–23] The waves in metamaterials have been studied extensively [5] Generalized constitutive relations, space–time wave packet, backward lateral waves, negative Goos–Hanchen shift, perfect lens and Brewster's angle and acoustic Brewster's angle are among those studied extensively. Transformation EM, invisible cloak and surface flattening coordinate transform are studied.

Waves in metamaterials have been studied and there are many books on the subject. Here in this section, we will briefly discuss "acoustic Brewster's angle" It has been well known that Brewster's angle θ_b for p polarization is given by

$$\sin\theta_{bp} = \left(\frac{\varepsilon^2 - \mu\varepsilon}{\varepsilon^2 - 1} \right)^{1/2} \tag{26.23}$$

while Brewster's angle for s-polarization is given by

$$\sin\theta_{bs} = \left(\frac{\mu^2 - \mu\varepsilon}{\mu^2 - 1} \right)^{1/2} \tag{26.24}$$

Note that for $\mu = 1$, the Brewster angle does not exist for s-polarization

For acoustic waves, we have

$$\sin\theta_{ba} = \left(\frac{(\rho/\rho_0)^2 - (c_0/c)}{(\rho/\rho_0)^2 - 1} \right)^{1/2} \tag{26.25}$$

where ρ_0 and c_0 are the density and the velocity of the medium where the wave is incident and ρ and c are those of the second medium.

The details of these characteristics are shown in [5].

26.12 Coherence in multiple scattering

As a wave propagates through random media such as turbulence, rain, fog or biological media, the wave experiences random variation in amplitude and phase and the wave becomes progressively incoherent. In particular in the back direction, the wave is expected to be incoherent, and this is consistent with radiative transfer theory.

However, an optical experiment by Kuga and Ishimaru [24] showed a sharp peak in the back direction indicating coherent back scattering. This was one of the optical experiments and was shown to be an optical equivalent of weak Anderson [25,26] localization of electron diffusion in a disordered medium [27].

Enhanced backscattering occurs in many areas of engineering including scattering from particles, rough surfaces, and turbulence. It is related to "retro reflectance" or "opposition" effects and observed enhanced backscattering by soils and vegetation. In this chapter, we present an introduction to "backscattering enhancement" Some historical and additional accounts are given by Ishimaru [27].

As an example, we show an enhanced radar cross section (RCS) in turbulence. It may seem surprising that the RCS increases through turbulence, as we have the conventional view that RCS may decrease in turbulence. However, this increase has been theoretically studied by noting that the backscattered intensity in turbulence is proportional to the fourth-order moment and approximately twice the multiple-scattered intensity. This has also been verified experimentally.

The apparent RCS of an object is

$$\sigma_{ap} = \sigma_b |e|^4 \tag{26.26}$$

where e is the product of uplink and downlink and $|e|^4$ is the average power that is proportional to the fourth-order moment.

We note that for the Rayleigh distribution, we have

$$\langle I^N \rangle = (N!)\, \langle I \rangle^N$$

Therefore we get

$$|e|^4 = 2|e|^2.$$

The apparent RCS is then given by

$$\sigma_{ap} = 2\sigma_b. \tag{26.27}$$

This shows that the RCS is twice the actual cross section. The scintillation index S_4^2 is given by

$$S_4^2 = \frac{\langle (I^2 - \langle I^2 \rangle)^2 \rangle}{\langle I^2 \rangle^2} = \frac{\langle I^4 \rangle - \langle I^2 \rangle^2}{\langle I^2 \rangle^2}. \tag{26.28}$$

Now for the Rayleigh distribution, noting (26.28), we get

$$S_4^2 = \frac{4! - 2^2}{2^2} = 5$$

Summarizing, the apparent cross section is twice the actual cross section and the scintillation index is 5 assuming the Rayleigh distribution. Experimental evidence for the enhanced RCS and the scintillation index has been given [27].

Coherence in multiple scattering occurs in "enhanced backscattering" from rough surfaces, from particles, photon localization, and memory effects. The memory effects are not limited to rough-surface scattering and apply to all multiple scattering from random media. More general memory effects proposed by Fend can be stated as follows:

A wave with the wave number vector $\bar{K}_i$ is incident on a random medium at frequency $\omega(k = \omega/c)$ and the scattered wave in the direction of wave number vector $\bar{K}$ is observed at ω. Another wave with wave number vector $\bar{K}_i'$ at frequency $\omega'(k' = \omega'/c)$ is incident upon the same random medium, and the scattered wave in the direction of the wave number $\bar{K}'$ is observed at ω'. Then there is a strong correlation between these two scattered waves at $\bar{K}$ and $\bar{K}'$ if the incident and scattered wave number vectors satisfy

$$(\bar{K} - \bar{K}_i)_t = (\bar{K}' - \bar{K}_i')_t. \tag{26.29}$$

The subscript (vector)$_t$ indicates the component of the vector along the surface.

26.13 Porous media

We are familiar with the Maxwell–Gannett mixing formula, Rayleigh mixing formula and Bruggeman–Polder van Santen mixing formula. Equivalent mixing formulae could be studied for "porous media" Normally, porous media are described by a "matrix" and "porosity ϕ" The conductivity and permittivity of porous medium (shale) are extensively studied and the empirical "Archie law" has been used for conductivity of shale.

For the mixing formula, we use the fractional volume of the ith species as f_i with

$$\sum_i f_i = 1 \tag{26.30}$$

For a porous medium, we use the porosity ϕ that is related to the fractional volume of the matrix f_m:

$$\phi = 1 - f_m. \tag{26.31}$$

The porosity ϕ is therefore the fractional volume of the pore (or void). One of the important examples of a porous medium is oil shale.

The ground penetrating radar (GPR) technique is used for nearsurface geophysical subsurface imaging. Critical parameters for GPR include the dielectric constant, porosity, and water saturation, and frequencies are usually 25–1500 MHz. There have been extensive studies in modeling and experimental data [28]. It is important to study mixing models of geophysical materials, and the effective medium models such as Bruggeman are useful for studies of the bulk dielectric constant, water saturation, and porosity. Carcione and Seriani [29] discuss the detection of hydrocarbons in the subsoil. At radar frequencies (50 MHz to 1 GHz), the hydrocarbons have a relative permittivity ranging from 2 to 30 while the permittivity of water is 80. The conductivity of hydrocarbons ranges from 0 to 10 mS/m while the conductivity of saltwater is 200 mS/m or more. (The unit for conductivity is (A/V m) or (S/m), where S is siemens.) Therefore there is sufficient contrast between hydrocarbons and water for detection and mapping of hydrocarbons.

Determination of the dielectric constant of the porous medium in terms of the composition of the mixtures is of great importance for characterization, monitoring, and evaluation. In Figure 26.5 a simplified sketch of a porous medium is shown, where the fractional volume of rock (matrix) is f_m. The pore or void is partly filled with hydrocarbon with a fractional volume of $f_h = (1 - S_w)\phi$ and partly filled with water with a volume fraction of $f_w = S_w\phi$, where S_w is the fraction of pore space saturated with water. This is a mixture of solid f_m and two liquids f_h and f_w. So, we have

$$f_m = \text{matrix} = 1 - \phi : \varepsilon_m$$

$$f_h = (1 - S_w)\,\phi = \text{hydrocarbon} : \varepsilon_h$$

$$f_w = S_w\phi = \text{water} : \varepsilon_w$$

$$S_w = \text{fraction of pore space saturated with water}$$

$$0 < S_w < 1$$

$$f_h + f_w = \phi$$

$$f_m + f_h + f_w = 1$$

Additional complications include complex shapes, other minerals and materials. In this section, we give a simplified theory of the mixture based on the Bruggeman model (Polder-van Santen).

$$f_m \frac{\varepsilon_m - \varepsilon_l}{\varepsilon_m + 2\varepsilon_l} + f_h \frac{\varepsilon_h - \varepsilon_l}{\varepsilon_h + 2\varepsilon_l} + f_w \frac{\varepsilon_w - \varepsilon_l}{\varepsilon_w + 2\varepsilon_l} = 0, \tag{26.32}$$

where ε_m is the dielectric constant of the matrix, ε_h is the dielectric constant of hydrocarbon, and ε_w is the dielectric constant of water. ε_l is the effective dielectric constant of the medium.

Note that

$$\varepsilon_m = \varepsilon_m' + i\varepsilon_m''$$

$$\varepsilon_h = \varepsilon_h' + i\varepsilon_h''$$

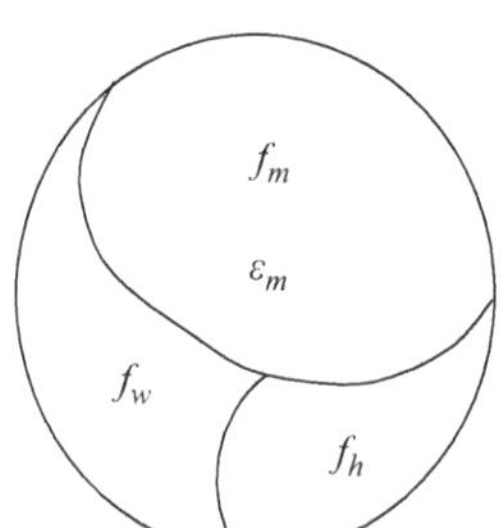

Figure 26.5 Simplified sketch of solid (matrix) and two solids (from [5] Fig. 26.2)

$$\varepsilon_w = \varepsilon_w' + i\left(\varepsilon_m'' + \frac{\sigma_w}{\omega\varepsilon_0}\right)$$

$$\varepsilon_l = \varepsilon_l' + i\left(\varepsilon_l'' + \frac{\sigma_l}{\omega\varepsilon_0}\right) = \text{effective dielectric constant}$$

σ_w = conductivity of water

σ_l = conductivity of effective medium.

It is then possible to calculate ε_l for a given ε_m, ε_h, and ε_w (or S_w and ϕ). As a special case, we consider a limiting case when $\omega \to 0$. Then the conductivity of the effective dielectric constant ε_l dominated by $(i\sigma_l/\omega\varepsilon_0)$ and the dielectric constant of water ε_w is dominated by $(i\sigma_w/\omega\varepsilon_0)$. As a limiting case $(\omega \to 0)$, we get

$$\sigma_e \cong \sigma_w\left(\frac{3}{2}\phi S_w - \frac{1}{2}\right). \tag{26.33}$$

This is valid only if $\phi S_w > 1/3$.

The previous simplified results do not include different processes of forming the porous medium such as adding oil and rock gains separately or simultaneously resulting in different results for (σ/σ_w) [29–31].

A more realistic model of effective conductivity σ_e was discovered by Archie [32]. It is an empirical law relating porosity, electrical conductivity, and brine saturation of rocks

$$\sigma = \frac{1}{a}\sigma_w\phi^m S_w^n \tag{26.34}$$

where σ and σ_w are the conductivities of the medium and the formation saline water, m is the porosity/cementation exponent with common values of $1.8 < m < 2.0$, n is the saturation exponent, usually close to 2, and a is the tortuosity factor that corrects for variations in compaction, pore structure, and grain size, ranging from 0.5 to 1.5. The bulk dielectric permittivity is affected by factors such as the air–solid–water interface [33]. A constitutive law for the electric response of saturated and unsaturated porous media is presented [34].

The study of porous media is closely related to the "percolation theory" [35] and the "fractal" [36,37].

26.14 Seismic coda

The upper 100 km of the Earth is called the lithosphere and the acoustic (seismic) wave has been investigated using the layered media model. However, the crust is heterogeneous with scales of a few to tens of kilometers, which excite wave trains called "coda" The seismic wave consists of P wave (pressure), S wave (shear wave) and the Rayleigh surface wave [38]

Acoustic waves in homogeneous earth have been studied extensively. However, acoustic waves in heterogeneous earth have been studied only recently. The study requires the space–time Fourier transform and the MCF. The wave can be expressed

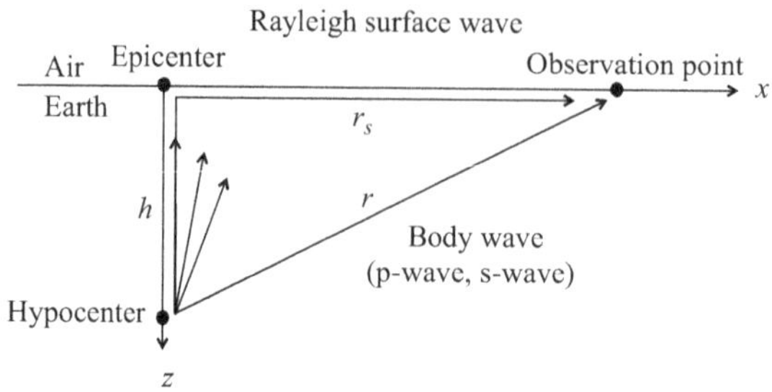

Figure 26.6 Body wave and Rayleigh surface wave (from [5] Fig. 26.3)

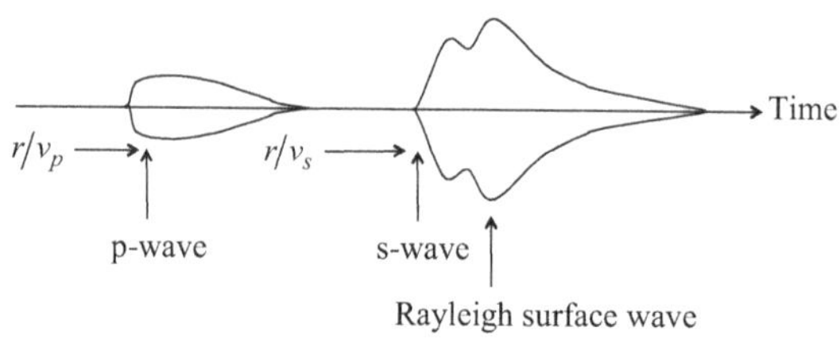

Figure 26.7 P Wave arrives first, S wave second, Rayleigh surface wave arrives slightly later (from [5] Fig. 26.4)

by scalar and vector potential and the boundary conditions. The P wave (pressure) and S wave (shear) need to be considered and result in the pole in the complex plane, which gives Rayleigh surface waves. Figure 26.6 shows these three waves and Figure 26.7 shows a sketch of these three waves with coda (train). The coda is important to determine the total energy of a point source. Further studies are needed to examine the details, which has not yet been done completely.

26.15 Conclusion

This chapter gives a review of some of the recent advances in electromagnetics in complex media with emphasis on geophysical and biological media. Included are stochastic wave theories, TR imaging, imaging through multiple scattering media, remote sensing, the Wigner transform, heat diffusion in tissues, ultrasound in tissues and blood, OCT, metamaterials, coherence in multiple scattering, porous media, and seismic coda. References are listed for most of the topics.

Acknowledgments

This chapter is in part supported by NSF and ONR. Editorial assistance by John Ishimaru is gratefully appreciated. This work was presented at AFOSR/AOARD in Tokyo, 2016.

References

[1] Tatarskii, V. I., A. Ishimaru, and V. U. Zavorotny. (1993). Wave Propagation in Random Media. Bellingham, WA: SPIE.

[2] Tsang, L., and J. A. Kong. (2001). Scattering of Electromagnetic Waves. New York, NY: John Wiley & Sons.

[3] Tsang, L., J. A. Kong, and R. T. Shin. (1985). Theory of Microwave Remote Sensing. New York, NY: Wiley-Interscience.

[4] Ishimaru, A. (1997). Wave Propagation and Scattering in Random Media. Piscataway, NJ: Wiley-IEEE Press.

[5] Ishimaru, A. (2017). Electromagnetic Wave Propagation, Radiation, and Scattering. Second Edition, Piscataway, NJ: Wiley-IEEE Press.

[6] Prada, C., and M. Fink. (1994). Eigenmodes of the time reversal operator: a solution to selective focusing in multiple-target media. Wave Motion, 20(2), 151–163.

[7] Devaney, A. J. (2012). Mathematical Foundations of Imaging, Tomography and Wavefield Inversion. Cambridge: Cambridge University Press.

[8] Yavuz, M. E., and F. L. Teixeira. (2008). Space-frequency ultrawideband time-reversal imaging. IEEE Trans. Geosci. Remote Sens., 46(4), 1115–1124.

[9] Prada, C., S. Manneville, D. Spoliansky, and M. Fink. (1996). Decomposition of the time reversal operator: detection and selective focusing on two scatterers. J. Acoust. Soc. Am., 99(4), 2067–2076.

[10] Ishimaru, A., S. Jaruwatanadilok, and Y. Kuga. (2012). Imaging through random multiple scattering media using integration of propagation and array signal processing. Waves Random Complex Media, 22(2), 24–39.

[11] van Zyl, J. J., H. A. Zebker, and C. Elachi. (1987). Imaging radar polarization signatures: theory and observation. Radio Sci., 22(4), 529–543.

[12] Ulaby, F. T., R. K. Moore, and A. K. Fung. (1981). Microwave Remote Sensing, Vols. 1, 2, and 3. London: Addison-Wesley.

[13] Lin, J. C. (2012). Electromagnetic Fields in Biological Systems. New York, NY: CRC Press.

[14] Fujimoto, J. G. (2001). Optical Coherence Tomography. C. R. Acad. Sci. Paris, t.2, Series IV, 1099–1111.

[15] Shung, K. K., and G. A. Thieme. (1993). Ultrasonic Scattering in Biological Tissues. Boca Raton, FL: CRC Press.

[16] Lin, J. C., ed. (2017). Electromagnetics. Hoboken, NJ, Wiley and Sons.

[17] Stoneback, M., A. Ishimaru, C. Reinhardt, and Y. Kuga. (2013). Temperature rise in objects due to optical focused beam through atmospheric turbulence near ground and ocean surface. Opt. Eng., 52(3), 36001–36008.

[18] Ishimaru, A. (2001). Acoustical and optical scattering and imaging of tissues: an overview. Proc. SPIE 4325, Medical Imaging. DOI: 10.1117/12.428184.

[19] Veselago, V. G. (1968). The electrodynamics of substances with simultaneously negative values of ε and μ. Sov. Phys. Usp., 10, 509–514.

[20] Eleftheriades, G. V., and K. G. Balmain. (2005). Negative Refraction Metamaterials: Fundamental Principles and Applications. New York, NY: John Wiley & Sons.

[21] Sihvola, A. (2007). Metamaterials in electromagnetics. Metamaterials, 1, 2–11.

[22] Engheta, N., and R. W. Ziolkowski, eds. (2006). Metamaterials Physics and Engineering Explanations. New York, NY: John Wiley & Sons.

[23] Caloz, C., and T. Itoh. (2006). Electromagnetic Metamaterials: Transmission Line Theory and Microwave Applications. New York, NY: Wiley-Interscience.

[24] Kuga, Y., and A. Ishimaru. (1984). Retroreflectance from a dense distribution of spherical particles. J. Opt. Soc. Am. A., 1(8), 831–835.

[25] Anderson, P. W. (1958). Absence of diffusion in certain random lattices. Phys. Rev., 109, 1492–1505.

[26] Anderson, P. W. (1985). The questions of classical localization, a theory of white paint? Philos. Mag. B, 52(3), 505–509.

[27] Ishimaru, A. (1991). Backscattering enhancement: from radar cross sections to electron and light localizations to rough surface scattering. IEEE Trans. Antennas Propag., 33(5), 7–11.

[28] Martinez, A., and A. P. Byrnes. (2001). Modeling dielectric-constant values of geologic materials: an aid to ground-penetrating radar data collection and interpretation. Curr. Res. Earth Sci. Bull., 247(Part 1).

[29] Carcione, J. M., and G. Seriani. (2000). An electromagnetic modeling tool for the detection of hydrocarbons in the subsoil. Geophys. Prospect., 48(2), 231–256.

[30] Feng, S., and P. N. Sen. (1985). Geometrical model of conductive and dielectric properties of partially saturated rocks. J. Appl. Phys., 58(8), 3236–3243.

[31] Sen, P. N., C. Scala, and M. H. Cohen. (1981). A self-similar model for sedimentary rocks with application to the dielectric constant of fused glass beads. Geophysics, 46(5), 781–795.

[32] Gao, G., A. Abubakar, and T. M. Habashy. (2012). Joint petrophysical inversion of electromagnetic and full-waveform seismic data. Geophysics, 77(3), WA 3–WA 18.

[33] Chen, Y., and D. Or. (2006). Geometrical factors and interfacial processes affecting complex dielectric permittivity of partially saturated porous media. Water Resour. Res., 42(6), WO6423. DOI: 1029 2005 WR004714.

[34] Brovelli, A., and G. Cassiani. (2010). A combination of the Hashin–Shtrikman bounds aimed at modeling electrical conductivity and permittivity of variably saturated porous media. Geophys. J. Int., 180(1), 225–237.

[35] Stauffer, D., and A. Aharony. (1991). Introduction to Percolation Theory. Boca Raton, FL: CRC Press.

[36] Mandelbrot, B. B. (1977). The Fractal Geometry of Nature. New York, NY: W. H. Freeman and Company.

[37] Falconer, K. (1990). Fractal Geometry. Mathematical Foundations and Applications. Hoboken, NJ: John Wiley & Sons.

[38] Sato, H., M. C. Fehler, and T. Maeda. (2012). Seismic Wave Propagation and Scattering in the Heterogeneous Earth. Second Edition, Berlin/Heidelberg: Springer-Verlag.

Chapter 27
Innovative tools for SI units in solving various problems of electrodynamics

Oleg A. Tretyakov[1,2], Oleksandr Butrym[2], and Fatih Erden[3]

27.1 Introduction

The goal of this chapter is rather ambitious: it proposes to "reboot" the standard system of Maxwell's equations in SI units in a new, much simpler, format, but executed within the SI metric system, as well. Ever since the SI metric system has been introduced in general usage, the electric and magnetic field vectors, $\mathcal{E}$ and $\mathcal{H}$, carry units of *volt per meter* $\lfloor \mathrm{V/m} \rfloor$ and *ampere per meter* $\lfloor \mathrm{A/m} \rfloor$, respectively. That is why the free-space constants, ε_0 and μ_0, specified as

$$\varepsilon_0 = \frac{1}{\mathrm{c}_0^2\, 4\pi \times 10^{-7}} \left\lfloor \frac{\mathrm{F}}{\mathrm{m}} \equiv \frac{\mathrm{A^2 s^2}}{\mathrm{Nm^2}} \right\rfloor \text{ and } \mu_0 = 4\pi \times 10^{-7} \left\lfloor \frac{\mathrm{H}}{\mathrm{m}} \equiv \frac{\mathrm{N}}{\mathrm{A^2}} \right\rfloor, \tag{27.1}$$

have *empirically* been installed in Maxwell's equations. They carry the dimensions of *farad per meter* $\lfloor \mathrm{F/m} \rfloor$ and *henry per meter* $\lfloor \mathrm{H/m} \rfloor$, respectively. Herein, $\mathrm{N} = \mathrm{kg\,m/s^2}$ is the force unit, *newton,* and $\mathrm{c}_0 = 2.9979\,2458 \times 10^8$ is the *quantity symbol* for the speed of light in vacuum, c, see [1,2]. That is to say, $c = 1/\sqrt{\varepsilon_0 \mu_0} = \mathrm{c}_0 \lfloor \mathrm{m/s} \rfloor$ as it follows from (27.1).

One can verify that the combination of N (what implies a force of 1 (one) *newton*) and the dimension of μ_0, taken as $\sqrt{\mathrm{N}/(\mathrm{H/m})}$, results in *ampere* $\lfloor \mathrm{A} \rfloor$, the basic SI unit. The combination of the same N and the dimension of ε_0 yields $\sqrt{\mathrm{N}/(\mathrm{F/m})} = \lfloor \mathrm{Nm}/(\mathrm{As}) \rfloor \equiv \lfloor \mathrm{V} \rfloor$ that is *volt.*

Proposition: *The latter observations suggest to introduce two new constants, $\varepsilon_0^{\mathrm{V}}$ and μ_0^{A}, and define them as follows:*

$$\varepsilon_0^{\mathrm{V}} \overset{def}{=} \sqrt{\frac{\mathrm{N}}{\varepsilon_0}} \cong 3.3607 \times 10^5 \left\lfloor \frac{\mathrm{Nm}}{\mathrm{As}} \equiv \mathrm{V} \right\rfloor \text{ and } \mu_0^{\mathrm{A}} \overset{def}{=} \sqrt{\frac{\mathrm{N}}{\mu_0}} \cong 8.9206 \times 10^2 \lfloor \mathrm{A} \rfloor. \tag{27.2}$$

[1]Department of Electronics Engineering, Gebze Technical University, Gebze, Turkey
[2]Department of Theoretical Radiophysics, V.N. Karazin Kharkiv National University, Kharkiv, Ukraine
[3]Department of Electronics Engineering, National Defense University, Naval Academy, Istanbul, Turkey

Scaling the original field vectors, $\mathcal{E}$ and $\mathcal{H}$, with these new constants enables the separation of their dimensions as

$$\boxed{\underbrace{\mathcal{E}(\mathbf{r},t)}_{\mathrm{V/m}} = \underbrace{\varepsilon_0^{\mathrm{V}}}_{\mathrm{V}}\underbrace{\mathbb{E}(\mathbf{r},t)}_{1/\mathrm{m}} \text{ and } \underbrace{\mathcal{H}(\mathbf{r},t)}_{\mathrm{A/m}} = \underbrace{\mu_0^{\mathrm{A}}}_{\mathrm{A}}\underbrace{\mathbb{H}(\mathbf{r},t)}_{1/\mathrm{m}}.} \tag{27.3}$$

Thus, the dimensions of volt, V, *and ampere,* A, *are shifted to the scaling coefficients,* $\varepsilon_0^{\mathrm{V}}$ and μ_0^{A}, *respectively, whereas the new field vectors* $\mathbb{E}$ *and* $\mathbb{H}$ *both have the common dimension of inverse meter,* $1/\mathrm{m}$. *Substitution of* (27.3) *to original Maxwell's equations* (27.4) *(see later) changes thoroughly their format and also* (27.9) *as given later. Indeed,* (27.9) *involves the only fundamental constant, c, and all the field vectors* ($\mathbb{E}$, $\mathbb{H}$, *and* $\mathbb{P}$), *each of which has a dimension of inverse meter,* $1/\mathrm{m}$.

27.2 Novel format of Maxwell's equations in SI units: Energetic and mechanical field characteristics

27.2.1 Novel format of Maxwell's equations in SI units

The standard system of Maxwell's equations is overloaded by various physical dimensions of the field quantities and the coefficients. Particularly, Maxwell's equations, which govern by electromagnetic fields in a lossy dielectric, look as follows:

$$\begin{aligned} \nabla \times \underbrace{\mathcal{E}(\mathbf{r},t)}_{\mathrm{V/m}} &= -\underbrace{\mu_0}_{\mathrm{H/m}}\,\partial_t\underbrace{\mathcal{H}(\mathbf{r},t)}_{\mathrm{A/m}} \quad \text{and} \\ \nabla \times \underbrace{\mathcal{H}(\mathbf{r},t)}_{\mathrm{A/m}} &= \partial_t[\underbrace{\varepsilon_0}_{\mathrm{F/m}}\underbrace{\mathcal{E}(\mathbf{r},t)}_{\mathrm{V/m}} + \underbrace{\mathcal{P}(\mathbf{r},t)}_{\mathrm{C/m^2}}] + \underbrace{\sigma}_{\mathrm{S/m}}\underbrace{\mathcal{E}(\mathbf{r},t)}_{\mathrm{V/m}} \end{aligned} \tag{27.4}$$

where $\mathcal{P}$ is the polarization vector with its dimension of *coulomb per meter*^{2} $\lfloor \mathrm{C/m^2} = \mathrm{As/m^2} \rfloor$. The last term in (27.4), $\sigma\mathcal{E}$, where σ has a dimension of *siemens per meter* $\lfloor \mathrm{S/m} = \mathrm{A^2 s/(Nm^2)} \rfloor$, is installed heuristically for modeling possible ohmic losses in the dielectric. A relationship between the polarization vector, $\mathcal{P}$, and the applied field, $\mathcal{E}$, is called in electrodynamics a constitutive equation. In the time domain, the latter has a *dynamic* form of Newton's equation of motion as

$$\frac{d^2}{dt^2}\mathcal{P}(\mathbf{r},t) + 2\Gamma\frac{d}{dt}\mathcal{P}(\mathbf{r},t) + \left(\frac{\omega_r^2 - \omega_p^2}{3}\right)\mathcal{P}(\mathbf{r},t) = N_e\frac{q_e^2}{m_e}\mathcal{E}(\mathbf{r},t) \tag{27.5}$$

where Γ and ω_r^2 are the intrinsic parameters of a dielectric medium, q_e and m_e are charge and mass of electron, N_e is the volumetric density of the polarizable electrons of the material, and $\omega_p^2 = N_e(q_e^2/m_e\varepsilon_0)$; ω_p is so-called plasma frequency. Equation (27.5) is valid for Lorentz temporally dispersive lossy media with electronic mechanism of polarization [3]. Maxwell's equations (27.4) and the *dynamic* constitutive relation (27.5) originate simultaneously a closed (i.e., solvable) system of differential equations.

Substitutions of (27.3) to standard Maxwell's equations (27.4) give

$$\begin{aligned}\sqrt{\tfrac{\mathrm{N}}{\varepsilon_0}}\nabla\times\mathbb{E}(\mathbf{r},t) &= -\sqrt{\mathrm{N}\mu_0}\partial_t\mathbb{H}(\mathbf{r},t) \quad \text{and} \\ \sqrt{\tfrac{\mathrm{N}}{\mu_0}}\nabla\times\mathbb{H}(\mathbf{r},t) &= \sqrt{\mathrm{N}\varepsilon_0}\partial_t\left[\mathbb{E}(\mathbf{r},t)+\tfrac{1}{\sqrt{\mathrm{N}\varepsilon_0}}\mathcal{P}(\mathbf{r},t)\right]+\sigma\sqrt{\tfrac{\mathrm{N}}{\varepsilon_0}}\mathbb{E}(\mathbf{r},t)\,.\end{aligned} \tag{27.6}$$

Multiplying the first equation in (27.6) by $\sqrt{\varepsilon_0/\mathrm{N}}$ and the second one by $\sqrt{\mu_0/\mathrm{N}}$ results in

$$\begin{aligned}\nabla\times\mathbb{E}(\mathbf{r},t) &= -\tfrac{1}{c}\partial_t\mathbb{H}(\mathbf{r},t) \quad \text{and} \\ \nabla\times\mathbb{H}(\mathbf{r},t) &= \tfrac{1}{c}\partial_t\left[\mathbb{E}(\mathbf{r},t)+\tfrac{1}{\sqrt{\mathrm{N}\varepsilon_0}}\mathcal{P}(\mathbf{r},t)\right]+\sigma\sqrt{\tfrac{\mu_0}{\varepsilon_0}}\mathbb{E}\,(\mathbf{r},t)\end{aligned} \tag{27.7}$$

where $\partial_t=\partial/\partial t$, $1/c=\sqrt{\varepsilon_0\mu_0}$. The two items in (27.7) placed in the square brackets are *summable.* So, the vector $\mathcal{P}$ can be scaled as

$$\underbrace{\mathcal{P}\,(\mathbf{r},t)}_{\mathrm{C/m^2}} = \underbrace{\sqrt{\mathrm{N}\varepsilon_0}}_{\mathrm{C/m}}\underbrace{\mathbb{P}\,(\mathbf{r},t)}_{1/\mathrm{m}} \cong \underbrace{2.9756\times10^{-6}}_{\mathrm{C/m}}\underbrace{\mathbb{P}\,(\mathbf{r},t)}_{1/\mathrm{m}}. \tag{27.8}$$

In the long run, substitution of (27.8) to (27.7) results in the novel simple format of Maxwell's equations in SI units as

$$\boxed{\nabla\times\mathbb{E}(\mathbf{r},t) = -\tfrac{1}{c}\partial_t\mathbb{H}(\mathbf{r},t) \text{ and } \nabla\times\mathbb{H}(\mathbf{r},t) = \tfrac{1}{c}\partial_t[\mathbb{E}(\mathbf{r},t)+\mathbb{P}(\mathbf{r},t)]+2\gamma\mathbb{E}(\mathbf{r},t)} \tag{27.9}$$

where $2\gamma=376.73\,\sigma_0$ has a dimension of *inverse meter* provided that σ_0 is the *quantity symbol* for the parameter σ given in (27.4) . It is worthwhile to notice again that all the vectors (including $\mathbb{P}$) in (27.9) have their *common* dimension of *inverse meter.*

Finally, the substitution of formulas (27.8) to the motion equation (27.5) simplifies its form as

$$\boxed{\tfrac{d^2}{dt^2}\mathbb{P}(\mathbf{r},t)+2\Gamma\tfrac{d}{dt}\mathbb{P}(\mathbf{r},t)+\left(\tfrac{\omega_r^2-\omega_p^2}{3}\right)\mathbb{P}(\mathbf{r},t)=\omega_p^2\mathbb{E}(\mathbf{r},t)\,.} \tag{27.10}$$

Thus, a researcher has to solve the simplified equations (27.9) , (27.10) and then one can recover the standard field vectors (when it is needed) as $\mathcal{E}=\varepsilon_0^{\mathrm{V}}\mathbb{E}$, $\mathcal{H}=\mu_0^{\mathrm{A}}\mathbb{H}$, $\mathcal{P}=\sqrt{\mathrm{N}\varepsilon_0}\mathbb{P}$ by formulas (27.3) and (27.8) .

Plasma frequency, $\omega_p=\sqrt{N_e q_e^2/\,(m_e\varepsilon_0)}$, involves the free parameter, N_e. The value $N_e=0$ corresponds to the free space. Substitution of $\omega_p=0$ to (27.10) gives the solution $\mathbb{P}=0$ for the free space. The coefficient γ in (27.9) is proportional to the conductivity σ, which is a free parameter, as well. The value $\sigma=0$ corresponds to the free space. Substitution of $\mathbb{P}=0$ and $\gamma=0$ to (27.9) yields

$$\nabla\times\mathbb{E}(\mathbf{r},t) = -\tfrac{1}{c}\partial_t\mathbb{H}(\mathbf{r},t) \text{ and } \nabla\times\mathbb{H}(\mathbf{r},t) = \tfrac{1}{c}\partial_t\mathbb{E}(\mathbf{r},t)\,, \tag{27.11}$$

which is a new format of Maxwell's equations in SI units for free space. Mathematically, the Heaviside–Lorentz equations have the same form, which were derived for free space within the frameworks of their version of CGS metric system [1].

The standard definition of Lorentz force, $\mathcal{F}_L$, is specified via the field vectors $\mathcal{E}$ and $\mathcal{H}$. Rewrite $\mathcal{F}_L$ in terms of the new fields, $\mathbb{E}$ and $\mathbb{H}$, and leave implicit their argument $(\mathbf{r},t)$ for brevity. After substituting (27.3), the Lorentz force law becomes

$$\begin{aligned} \mathcal{F}_L(\mathbf{r},t) &= q(\mathcal{E} + \mathbf{v} \times \mu_0 \mathcal{H}) = q\left(\varepsilon_0^{\mathrm{V}} \mathbb{E} + \mu_0 \mu_0^{\mathrm{A}} \mathbf{v} \times \mathbb{H}\right) \quad \text{yields} \\ \mathbb{F}_L(\mathbf{r},t) &= 3.3607 \times 10^5 \left(\tfrac{q}{\mathrm{C}}\right) \left[\left(\mathbb{E} + \tfrac{\mathbf{v}}{c} \times \mathbb{H}\right)\mathrm{m}\right] \lfloor \mathrm{N} \rfloor \end{aligned} \tag{27.12}$$

where positive charge q of a particle is measurable in *coulomb,* C = As, and $\mathbf{v}$ is a velocity of the charged particle in SI units. It is evident that the factor placed in square brackets for $\mathbb{F}_L$ is dimension-free, and hence, the Lorentz force, $\mathbb{F}_L$, is measurable in *newton,* N. It is worthwhile to notice that the charge of unsigned electron, q_e, is equal to $1.6021\,7662 \times 10^{-19} \lfloor \mathrm{C} \rfloor$.

27.2.2 *Energetic characteristics of the electromagnetic field*

Hereinafter, we assume that all the fields are real-valued.[§] Rearrangement of the standard electromagnetic field energy density, $\mathcal{U}$, results in a new quantity, U, expressed via $\mathbb{E}$ and $\mathbb{H}$, as

$$\begin{aligned} \mathcal{U}(\mathbf{r},t) &= \tfrac{1}{2}(\varepsilon_0 \mathcal{E} \cdot \mathcal{E} + \mu_0 \mathcal{H} \cdot \mathcal{H}) = \tfrac{1}{2}\left(\varepsilon_0 \varepsilon_0^{\mathrm{V}} \mathbb{E} \cdot \varepsilon_0^{\mathrm{V}} \mathbb{E} + \mu_0 \mu_0^{\mathrm{A}} \mathbb{H} \cdot \mu_0^{\mathrm{A}} \mathbb{H}\right) \quad \text{yields} \\ U(\mathbf{r},t) &= \tfrac{1}{2}\left[(\mathbb{E} \cdot \mathbb{E} + \mathbb{H} \cdot \mathbb{H})\, \mathrm{m}^2\right] \left\lfloor \tfrac{\mathrm{N}}{\mathrm{m}^2} \equiv \tfrac{\mathrm{J}}{\mathrm{m}^3} \right\rfloor \end{aligned} \tag{27.13}$$

where the quantity placed in square brackets in the formula for U is dimension-free, and J = Nm is joule.

The power flow of electromagnetic energy density is specified by the Poynting vector, $\mathcal{S}$, which is transformed to $\mathbb{S}$ as follows:

$$\begin{aligned} \mathcal{S}(\mathbf{r},t) &= [\mathcal{E} \times \mathcal{H}] = \left[\varepsilon_0^{\mathrm{V}} \mathbb{E} \times \mu_0^{\mathrm{A}} \mathbb{H}\right] \quad \text{yields} \\ \mathbb{S}(\mathbf{r},t) &= \mathrm{c}_0\left[(\mathbb{E} \times \mathbb{H})\, \mathrm{m}^2\right] \left\lfloor \tfrac{\mathrm{N}}{\mathrm{m}^2} \tfrac{\mathrm{m}}{\mathrm{s}} = \tfrac{\mathrm{J/s}}{\mathrm{m}^2} = \tfrac{\mathrm{W}}{\mathrm{m}^2} \right\rfloor \end{aligned} \tag{27.14}$$

where $\mathrm{c}_0 = 2.9979\,2458 \times 10^8$ is the *quantity symbol* for the speed of light in vacuum, c (i.e., $c = \mathrm{c}_0 \lfloor \mathrm{m/s} \rfloor$), the factor $[(\mathbb{E} \times \mathbb{H})\mathrm{m}^2]$ in the formula for $\mathbb{S}$ is dimension-free, and J/s = W is watt.

Umov defined the velocity of energy transportation, $\mathcal{V}$, in any wave process as the ratio where its power flow stands in numerator and its energy stands in denominator [4]. Poynting proved that for the electromagnetic process $\mathcal{V} = \mathcal{S}/\mathcal{U}$ [5]. So,

$$\mathcal{V}(\mathbf{r},t) = \frac{\mathcal{S}(\mathbf{r},t)}{\mathcal{U}(\mathbf{r},t)} = \frac{2[\mathcal{E} \times \mathcal{H}]}{\varepsilon_0 \mathcal{E} \cdot \mathcal{E} + \mu_0 \mathcal{H} \cdot \mathcal{H}} \text{ yields } \mathbb{V}(\mathbf{r},t) = c\, \frac{2[\mathbb{E}(\mathbf{r},t) \times \mathbb{H}(\mathbf{r},t)]}{\mathbb{E}(\mathbf{r},t) \cdot \mathbb{E}(\mathbf{r},t) + \mathbb{H}(\mathbf{r},t) \cdot \mathbb{H}(\mathbf{r},t)} \le c. \tag{27.15}$$

[§]In the time domain, operations with real-valued fields are preferable in studies of the energetic and mechanical field characteristics as functions of time.

27.2.3 *Mechanical equivalents of the energetic field characteristics*

The quantities $U = 1/2(\mathbb{E}\cdot\mathbb{E}+\mathbb{H}\cdot\mathbb{H})$ and $\bar{\mathbb{S}} = \mathbb{S}/c = \mathbb{E}\times\mathbb{H}$ have the *common* dimension of *joule per meter*³ $\lfloor \mathrm{J/m^3} \rfloor$. Following Kaiser's technique (see [6]), a pair of new *scalars* composed of the scalar $\mathbb{U}$ and the vector $\bar{\mathbb{S}}$ is

$$\begin{aligned}\mathfrak{U}(\mathbf{r},t) &= \sqrt{\tfrac{1}{4}\left(\mathbb{E}^2+\mathbb{H}^2\right)^2} = \sqrt{\tfrac{1}{4}\left(\mathbb{E}^4+2\mathbb{E}^2\mathbb{H}^2+\mathbb{H}^4\right)} \quad \text{and} \\ \mathfrak{I}(\mathbf{r},t) &= \sqrt{(\mathbb{E}\times\mathbb{H})\cdot(\mathbb{E}\times\mathbb{H})} = \sqrt{\mathbb{E}^2\mathbb{H}^2-(\mathbb{E}\cdot\mathbb{H})^2}\end{aligned} \tag{27.16}$$

where the dot product, $[\mathbb{E}\times\mathbb{H}]\cdot[\mathbb{E}\times\mathbb{H}]$, is calculated via applying identity from [3, eq. (B.8)]. The combination of $\mathfrak{U}$ and $\mathfrak{I}$ as

$$R(\mathbf{r},t) = \sqrt{\mathfrak{U}^2-\mathfrak{I}^2} = \frac{1}{2}\sqrt{\left(\mathbb{E}^2-\mathbb{H}^2\right)^2+4(\mathbb{E}\cdot\mathbb{H})^2}\ \lfloor \mathrm{J/m^3} \rfloor \tag{27.17}$$

has been defined by Kaiser as the *reactive (rest) energy density* in [6].

To derive a mechanical equivalent of R, first recall that the general relationship from relativistic mechanics between energy, E, and momentum, $\mathfrak{p}$, of a particle with its mass, $\mathfrak{m}$, in the state of rest, is

$$E = c\sqrt{\mathfrak{p}^2+\mathfrak{m}^2c^2} \quad \text{and} \quad \mathfrak{p} = \mathfrak{m}\,|\mathbf{v}|\,/\sqrt{1-\mathbf{v}^2/c^2} \tag{27.18}$$

where $\mathbf{v}$ is the velocity of the particle in the chosen inertial reference frame. The energy expressed in terms of the momentum is called the Hamiltonian function [7]. Let us start with the reference frame of rest, where $\mathbf{v} = \mathbf{0}$, and substitute that value of the velocity to (27.18). This results in Einstein's famous formula $E = \mathfrak{m}c^2$. Replacing E by R herein yields

$$\mathfrak{m}(\mathbf{r},t) = \frac{R(\mathbf{r},t)}{c^2} = \frac{1}{2c_0^2}\sqrt{\left(\mathbb{E}^2-\mathbb{H}^2\right)^2+4(\mathbb{E}\cdot\mathbb{H})^2}\ \lfloor \mathrm{kg/m^3} \rfloor \tag{27.19}$$

where $\mathfrak{m}$ has the sense of measure of electromagnetic inertia. As long as $\mathbb{E}^2-\mathbb{B}^2$ and $\mathbb{E}\cdot\mathbb{B}$ are the pair of Lorentz invariants, the formula (27.19) is valid for *any* inertial reference frame. Specifically, the plane waves propagating in the free space have their invariants as $\mathbb{E}^2-\mathbb{B}^2=0$ and $\mathbb{E}\cdot\mathbb{B}=0$. Hence, electromagnetic mass of the plane waves is zero in free space.

Consider now the case when energy of the particle, E, is large compared to its rest energy, $\mathfrak{m}c^2$. Then the formula for Hamiltonian (27.18) can be read as $\mathfrak{p} = E/c$. The replacement herein of E by $\mathfrak{I}$ (see (27.16)) yields

$$\mathfrak{p}(\mathbf{r},t) = \tfrac{\mathfrak{I}(\mathbf{r},t)}{c} \text{ yields } \mathfrak{p}(\mathbf{r},t) = \tfrac{1}{c_0}\sqrt{\mathbb{E}^2\mathbb{H}^2-(\mathbb{E}\cdot\mathbb{H})^2}\ \lfloor \mathrm{kg\tfrac{m}{s}/m^3} \rfloor. \tag{27.20}$$

This is the mechanical momentum distribution of the electromagnetic field. Notice that the plane electromagnetic waves in free space have their mechanical momentum $\mathfrak{p} = |\mathbb{E}|\,|\mathbb{H}|/c_0 \neq 0$ whereas their electromagnetic mass, $\mathfrak{m}$, is equal to zero.

Incidentally, the modal waveguide waves have their electric and magnetic field components orthogonal. That is to say, $\mathbb{E}(\mathbf{r},t)\cdot\mathbb{H}(\mathbf{r},t)=0$ *locally.* Meanwhile, $\mathbb{E}^2-\mathbb{H}^2\neq 0$. Thereby, (27.19) and (27.20) result in

$$\mathfrak{m}(\mathbf{r},t)=\tfrac{1}{2c_0^2}\left|\mathbb{E}^2-\mathbb{H}^2\right|\left\lfloor \mathrm{kg/m^3}\right\rfloor \text{ and } \mathfrak{p}(\mathbf{r},t)=\tfrac{1}{c_0}|\mathbb{E}||\mathbb{H}|\left\lfloor \mathrm{kg\tfrac{m}{s}/m^3}\right\rfloor. \quad (27.21)$$

The definitions given in (27.21) are valid for any closed cylindrical waveguide of *arbitrary* cross section, which is geometrically regular along its axis, and the waveguide surface has the properties of the perfect electric conductor.

27.3 Exact solutions for polarization of Lorentz media associated with a signal of finite duration

27.3.1 Rearrangement of the motion equation (27.10) *to its equivalent matrix format and solving a vector Cauchy problem*

For the sake of simplicity, take the applied field, $\mathbb{E}(\mathbf{r},t)$, which stands on the right-hand side of (27.10), in a factorized form as $\mathbb{E}(\mathbf{r},t)=E(t)\vec{\mathbb{E}}(\mathbf{r})$. The scalar factor, $E(t)$, is the field amplitude, which should be specified as a given function of time. The other factor, $\vec{\mathbb{E}}(\mathbf{r})$, should be known, as well, as a vector function dependent on coordinates only. As long as (27.10) is linear, its solution is presentable via a similar factorized form as $\mathbb{P}(\mathbf{r},t)=p(t)\vec{\mathbb{E}}(\mathbf{r})$ where the amplitude, $p(t)$, is the subject of our study in what follows. Suppose that the applied field has a beginning in time, say, $E(t)=0$ when $-\infty<t\leq 0$. Hence, the differential equation (27.10) of second order should be supplemented with a pair of appropriate initial conditions, i.e., $p(0)=0$ and $(d/dt)p(t)_{t=0}=0$. Under these assumptions, we arrive at the Cauchy problem for the amplitude, $p(t)$ as

$$\begin{aligned}&\tfrac{d^2}{dt^2}p(t)+2\Gamma\tfrac{d}{dt}p(t)+\omega_0^2p(t)=\omega_p^2E(t)\\&\text{provided that}\quad p(0)=0\quad\text{and}\quad \tfrac{d}{dt}p(t)_{t=0}=0\end{aligned}\qquad(27.22)$$

where Γ is a constant parameter, $\omega_0^2=\omega_r^2-\omega_p^2/3$. It is appropriate to divide (27.22) by ω_0^2 and then introduce a "dimension-free" time as $\tau\doteq\omega_0 t$ and the other dimension-free coefficients are as follows: $\alpha\doteq\Gamma/\omega_0$, $\varpi\doteq\omega_p^2/\omega_0^2$. Ultimately, it results in

$$\begin{aligned}&\tfrac{d^2}{d\tau^2}p(\tau)+2\alpha\tfrac{d}{d\tau}p(\tau)+p(\tau)=\varpi E(\tau)\\&\text{provided that}\quad p(0)=0\quad\text{and}\quad \tfrac{d}{d\tau}p(\tau)_{\tau=0}=0.\end{aligned}\qquad(27.23)$$

Denote $(d/d\tau)p(\tau)$ in (27.23) as $q(\tau)$ and rewrite differential equation (27.23) as a pair of the first-order ones. That yields

$$\begin{cases}\tfrac{d}{d\tau}p(\tau)-q(\tau)=0\\ \tfrac{d}{d\tau}q(\tau)+p(\tau)+2\alpha q(\tau)=\varpi E(\tau)\end{cases}\quad\text{and}\quad\begin{matrix}p(0)=0\\ q(0)=0\end{matrix}\qquad(27.24)$$

Introduce a two-component column vector $X(\tau)$ composed of $p(\tau)$ and $q(\tau)$ as $X(\tau) = \mathrm{col}(\mathrm{p}(\tau), \mathrm{q}(\tau))$ where "col" symbolizes "column." Evidently, equations (27.24) are equivalent to the vector Cauchy problem for $X(\tau)$ as

$$\left.\begin{array}{c} \underbrace{\frac{d}{d\tau}\begin{pmatrix} p(\tau) \\ q(\tau) \end{pmatrix}}_{X(\tau)} + \underbrace{\begin{pmatrix} 0 & -1 \\ 1 & 2\alpha \end{pmatrix}}_{Q}\underbrace{\begin{pmatrix} p(\tau) \\ q(\tau) \end{pmatrix}}_{X(\tau)} = \underbrace{\begin{pmatrix} 0 \\ \varpi E(\tau) \end{pmatrix}}_{f(\tau)} \text{ and } \underbrace{X(\tau)|_{\tau=0}}_{X(0)} = \begin{pmatrix} 0 \\ 0 \end{pmatrix} \\ \boxed{\frac{d}{d\tau}X(\tau) + QX(\tau) = f(\tau) \text{ and } X(0) = 0} \end{array}\right\} \tag{27.25}$$

where Q is a constant non-singular matrix. To solve the vector equation (27.25), apply a substitution for the vector $X(\tau)$ as $X(\tau) = e^{-\tau Q}Y(\tau)$ where $Y(\tau) = \mathrm{col}(\mathrm{u}(\tau), \mathrm{v}(\tau))$ is a new vector with its components, $u(\tau)$ and $v(\tau)$, to be sought (instead of $p(\tau)$ and $q(\tau)$, originally). Applying this substitution to the vector equation (27.25) results in

$$e^{-\tau Q}\frac{d}{d\tau}Y(\tau) = f(\tau) \text{ yielding } Y(\tau) = \int_0^\tau \left(e^{-xQ}\right)^{-1} f(x)\,dx \tag{27.26}$$

where $e^{xQ} = (e^{-\tau Q})^{-1}|_{(-\tau)\to x}$ is the matrix inverse to e^{-tQ} with a replacement of $(-\tau)$ by x. Finally,

$$\boxed{X(\tau) = e^{-\tau Q}\int_0^\tau \left(e^{-xQ}\right)^{-1} f(x)\,dx.} \tag{27.27}$$

This is a formal solution to the Cauchy problem (27.25) with an arbitrary *integrable* function $f(x)$ and initial condition $X(0) = 0$.

27.3.1.1 Formulas for calculation of $e^{-\tau Q}$

First separate the matrix Q onto two parts, A and B, as

$$Q = \begin{pmatrix} 0 & -1 \\ 1 & 2\alpha \end{pmatrix} = A + B \text{ where } A = \begin{pmatrix} \alpha & 0 \\ 0 & \alpha \end{pmatrix},\ B = \begin{pmatrix} -\alpha & -1 \\ 1 & \alpha \end{pmatrix}. \tag{27.28}$$

If and only if the matrices A and B commute (i.e., $AB = BA$), the matrix exponential, $e^{-\tau Q}$, is presentable as $e^{-\tau(A+B)} = e^{-\tau A}e^{-\tau B}$. The first exponential, $e^{-\tau A}$, with diagonal matrix A can be easily defined by the convergent power series as

$$e^{-\tau A} = \sum_{k=0}^{\infty}\frac{(-\tau A)^k}{k!} = \mathbf{I} + (-\tau A) + \frac{(-\tau A)^2}{2} + \frac{(-\tau A)^3}{6} + \cdots \tag{27.29}$$

where $(-\tau)^0 = 1$, $0! = 1$, and $A^0 = \mathbf{I}$ is the identity matrix of the same order as the matrix A. This is the generalization of the Taylor series expansion for the standard exponential matrix function. The series (27.29) converges for *any* constant matrix of a finite order and uniformly in τ, see [8]. Calculations of $e^{-\tau A}$ by formula (27.29) result in

$$e^{-\tau A} = \begin{pmatrix} e^{-\tau\alpha} & 0 \\ 0 & e^{-\tau\alpha} \end{pmatrix} = e^{-\tau\alpha}\,\mathbf{I} \quad \text{where } \mathbf{I} = \begin{pmatrix} 1 & 0 \\ 0 & 1 \end{pmatrix}. \tag{27.30}$$

Let us consider now the eigenvalues of the matrix B via solving its characteristic equation for the eigenvalues, λ, that is, $\det(\lambda\,\mathbf{I}-B)=\lambda^2-\alpha^2+1=0$. That yields a pair of *distinct* eigenvalues as $\lambda_1=\sqrt{\alpha^2-1}$ and $\lambda_2=-\sqrt{\alpha^2-1}$ provided $\alpha\neq 1$. If so, we can apply *Lagrange interpolation* (see [9,10]) of the matrix exponential $e^{-\tau B}$ which yields

$$e^{-\tau B}=e^{-\tau\lambda_1}\mathcal{K}(\lambda_1)+e^{-\tau\lambda_2}\mathcal{K}(\lambda_2)\ \text{ where } \mathcal{K}(\lambda_1)=\tfrac{\lambda_2\mathbf{I}-B}{\lambda_2-\lambda_1},\ \mathcal{K}(\lambda_2)=\tfrac{\lambda_1\mathbf{I}-B}{\lambda_1-\lambda_2}. \tag{27.31}$$

The matrix coefficients have remarkable properties, which are useful for the control of calculations, namely,

$$\begin{aligned}\mathcal{K}(\lambda_1)+\mathcal{K}(\lambda_2)&=\mathbf{I},\mathcal{K}(\lambda_1)\,\mathcal{K}(\lambda_1)=\mathcal{K}(\lambda_1)\,,\mathcal{K}(\lambda_2)\,\mathcal{K}(\lambda_2)=\mathcal{K}(\lambda_2)\,,\\ \mathcal{K}(\lambda_1)\,\mathcal{K}(\lambda_2)&=\mathcal{O},\end{aligned} \tag{27.32}$$

where $\mathcal{O}$ symbolizes 2×2 matrix with zero-value in all its elements.

27.3.1.2 Matrix exponential $e^{-\tau Q}$ for three different cases

Case 1. Consider first the case when the eigenvalues of the matrix B are *distinct* and complex-conjugated as $\lambda_{1,2}=\pm i\sqrt{1-\alpha^2}$. In other words, the lossy parameter, $\alpha=\Gamma/\omega_0$ (see (27.22)), is enough that inequality $\alpha<1$ holds. Calculation of the matrix exponential by formulas (27.30) and (27.31) results in

$$\begin{aligned}e^{-\tau Q}&=\frac{e^{-\tau\alpha}}{\cos(\theta)}\begin{pmatrix}\cos(\tau\beta-\theta) & \sin(\tau\beta)\\ -\sin(\tau\beta) & \cos(\tau\beta+\theta)\end{pmatrix},\\ \left(e^{-xQ}\right)^{-1}&=\frac{e^{x\alpha}}{\cos\theta}\begin{pmatrix}\cos(x\beta+\theta) & -\sin x\beta\\ \sin x\beta & \cos(x\beta-\theta)\end{pmatrix}\end{aligned} \tag{27.33}$$

where $\beta=\sqrt{1-\alpha^2}<1$ and $\theta=\cos^{-1}(\beta)=\sin^{-1}(\alpha)$. Notice that $e^{-\tau Q}|_{\tau=0}=\mathbf{I}$ and $\left(e^{-xQ}\right)^{-1}|_{x=0}=\mathbf{I}$, i.e., the identity matrix (see (27.30)). Mathematicians call any matrix with that property at its initial instant a *matrizant.*

Case 2. Consider now the case when the eigenvalues are *distinct* and real. Calculations by the same algorithm yield

$$\begin{aligned}e^{-\tau Q}&=\frac{e^{-\alpha\tau}}{\sinh(\vartheta)}\begin{pmatrix}\sinh(\vartheta+\gamma\tau) & \sinh(\gamma\tau)\\ -\sinh(\gamma\tau) & \sinh(\vartheta-\gamma\tau)\end{pmatrix},\\ \left(e^{-xQ}\right)^{-1}&=\frac{e^{\alpha x}}{\sinh\vartheta}\begin{pmatrix}\sinh(\vartheta-\gamma x) & -\sinh\gamma x\\ \sinh\gamma x & \sinh(\vartheta+\gamma x)\end{pmatrix}\end{aligned} \tag{27.34}$$

where $\alpha=\Gamma/\omega_0>1$ is supposed, $\gamma=\sqrt{\alpha^2-1}>0$ and $\vartheta=\cosh^{-1}(\alpha)=\sinh^{-1}(\gamma)$. One can verify that the matrix exponentials in (27.34), $e^{-\tau Q}$ and $(e^{-xQ})^{-1}$, are the matrizants.

Case 3. Lastly, if $\alpha=\Gamma/\omega_0=1$, then the eigenvalues of B coincide as $\lambda_1=\lambda_2=0$. Matrix B becomes singular: i.e., $\det(B)=0$. Calculation of $\mathrm{e}^{-\tau B}$ by (27.31) is impossible (since $\lambda_1=\lambda_2$), but the series expansion (27.29) for B is available. Hence,

$$e^{-\tau Q}=e^{-\tau\alpha}\begin{pmatrix}1+\tau & \tau\\ -\tau & 1-\tau\end{pmatrix},\quad \left(e^{-xQ}\right)^{-1}=e^{x\alpha}\begin{pmatrix}1-x & -x\\ x & 1+x\end{pmatrix}. \tag{27.35}$$

It is evident that the matrix exponentials, $e^{-\tau Q}$ and $\left(e^{-xQ}\right)^{-1}$ in (27.35) are the matrizants, as well.

27.3.2 Exact explicit solutions for the amplitudes of the polarization vector

The formal solution (27.27) to the problem (27.25) involves the matrizants, which are already calculated earlier. Besides, the given vector force function, $f(x) = f(\tau)|_{\tau=x}$, appeared in the integrand in (27.27). The vector $f(\tau)$ is specified in (27.25) via the scalar $E(\tau)$ which is an amplitude of a given applied signal (see (27.22)), physically. Suppose that the applied signal has a finite duration: say as $0 \le t \le T$ in real time, which is equivalent to the dimensionless time, $\tau = \omega_0 t$, as $0 \le \tau \le \Upsilon$ where $\Upsilon = \omega_0 T$.

In the analysis that follows, we suppose implicitly that $E(\tau)$ is any *integrable* function‖ specified by

$$E(\tau) = 0 \text{ if } \tau \le 0,\ E(\tau) \ne 0 \text{ is given if } 0 \le \tau \le \Upsilon,\ E(\tau) = 0 \text{ if } \tau > \Upsilon. \tag{27.36}$$

First consider solution (27.27), which corresponds to the **Case 1** of the matrizant (27.33). That has the form

$$\begin{aligned} p_1'(\tau) &= M_1'(\tau)\cos(\tau\beta - \theta) + N_1'(\tau)\sin(\tau\beta) \quad \text{and} \\ q_1'(\tau) &= N_1'(\tau)\cos(\tau\beta + \theta) - M_1'(\tau)\sin(\tau\beta) \end{aligned} \tag{27.37}$$

where subscript ($_1$) symbolizes that this solution corresponds to the **Case 1**; the prime, ($'$), implies that (27.37) is related to the temporal interval $0 \le \tau \le \Upsilon$. The coefficients, $M_1'(\tau)$ and $N_1'(\tau)$, are specified by the external signal, $E(x)$, as follows:

$$\begin{aligned} M_1'(\tau) &= -\frac{e^{-\tau\alpha}}{\cos^2(\theta)} \int_0^\tau e^{x\alpha} \varpi E(x)\sin(x\beta)\,dx \quad \text{and} \\ N_1'(\tau) &= \frac{e^{-\tau\alpha}}{\cos^2(\theta)} \int_0^\tau e^{x\alpha} \varpi E(x)\cos(x\beta - \theta)\,dx \end{aligned} \tag{27.38}$$

where $\varpi = \omega_p^2/\omega_0^2$. Physically, these coefficients play the role of the modulators for the harmonic functions $\cos(\tau\beta \mp \theta)$ and $\sin(\tau\beta)$ in (27.37). Formula (27.37) is valid until the end of the interval, $\tau = \Upsilon$, which yields a vector $X'(\Upsilon) = \mathrm{col}(p_1'(\Upsilon), q_1'(\Upsilon))$. When $\tau > \Upsilon$, the external signal, $E(x)$, vanishes, but we can specify the solution $X''(\tau) = \mathrm{col}(p_1''(\tau), q_1''(\tau))$ for $\tau > \Upsilon$ by

$$\begin{aligned} X''(\tau) = \begin{pmatrix} p_1''(\tau) \\ q_1''(\tau) \end{pmatrix} &= H(\tau - \Upsilon)\,\frac{e^{-(\tau-\Upsilon)\alpha}}{\cos(\theta)} \\ &\times \begin{pmatrix} \cos((\tau-\Upsilon)\beta - \theta) & \sin((\tau-\Upsilon)\beta) \\ -\sin((\tau-\Upsilon)\beta) & \cos((\tau-\Upsilon)\beta + \theta) \end{pmatrix} \begin{pmatrix} p_1'(\Upsilon) \\ q_1'(\Upsilon) \end{pmatrix} \end{aligned} \tag{27.39}$$

where $H(\tau - \Upsilon)$ is the Heaviside unit step function and the vector $X'(\Upsilon)$ plays the role of the initial condition for the solution $X''(\tau)$. The double prime symbolizes

‖For example, $E(t) = \sin(\Omega t)$ is for a harmonic signal of finite duration; $E(t) = 1$, where $0 \le t < T$, corresponds to a fragment of a Walsh digital signal.

that solution (27.39) is valid just for $\tau > \Upsilon$. Simple manipulations with (27.39) result in

$$p_1''(\tau) = H(\tau - \Upsilon)\frac{e^{-(\tau-\Upsilon)\alpha}}{\cos(\theta)} \times \left[p_1'(\Upsilon)\cos((\tau-\Upsilon)\beta - \theta) + q_1'(\Upsilon)\sin((\tau-\Upsilon)\beta)\right] \tag{27.40a}$$

$$q_1''(\tau) = H(\tau - \Upsilon)\frac{e^{-(\tau-\Upsilon)\alpha}}{\cos(\theta)}\left[q_1'(\Upsilon)\cos((\tau-\Upsilon)\beta + \theta)\right] - p_1'(\Upsilon)\sin((\tau-\Upsilon)\beta). \tag{27.40b}$$

In the long run, the solution (27.27) , which is associated with the matrizant (27.33) , is as follows:

$$p_1(\tau) = H(\Upsilon - \tau)\left[M_1'(\tau)\cos(\tau\beta - \theta) + N_1'(\tau)\sin(\tau\beta)\right] + p_1''(\tau) \tag{27.41a}$$

$$\frac{d}{dt}p_1(\tau) = H(\Upsilon - \tau)\left[N_1'(\tau)\cos(\tau\beta + \theta) - M_1'(\tau)\sin(\tau\beta)\right] + q_1''(\tau). \tag{27.41b}$$

Now derive the solution (27.27) by applying the matrizants given in (27.34) , which corresponds to the **Case 2.** The term $\int_0^\tau (e^{-xQ})^{-1}f(x)\,dx$ in (27.27) yields the modulating factors, $M_2'(\tau)$ and $N_2'(\tau)$, as

$$\begin{aligned} M_2'(\tau) &= -\frac{\varpi}{\sinh^2\vartheta}\int_0^\tau e^{\alpha x}E(x)\sinh\gamma x\,dx \quad \text{and} \\ N_2'(\tau) &= \frac{\varpi}{\sinh^2\vartheta}\int_0^\tau e^{\alpha x}E(x)\sinh(\vartheta + \gamma x)\,dx. \end{aligned} \tag{27.42}$$

So, the solution available in the time interval, $0 \le \tau \le \Upsilon$, where the applied signal acts, $E(x)$, is as follows:

$$\begin{aligned} p_2'(\tau) &= e^{-\alpha\tau}\left[M_2'(\tau)\sinh(\vartheta + \gamma\tau) + N_2'(\tau)\sinh(\gamma\tau)\right], \\ q_2'(\tau) &= e^{-\alpha\tau}\left[N_2'(\tau)\sinh(\vartheta - \gamma\tau) - M_2'(\tau)\sinh(\gamma\tau)\right]. \end{aligned} \tag{27.43}$$

At the instant, when the signal is turned off, the values of $p_2'(\Upsilon)$ and $q_2'(\Upsilon)$ can be calculated by formulas (27.43). They can be used as the initial conditions for calculations of the solution when $\tau > \Upsilon$. Ultimately, this solution for the **Case 2** is

$$p_2''(\tau) = H(\tau - \Upsilon)e^{-\alpha(\tau-\Upsilon)} \times \left[p_2'(\Upsilon)\sinh(\vartheta + \gamma(\tau-\Upsilon)) + q_2'(\Upsilon)\sinh(\gamma(\tau-\Upsilon))\right] \tag{27.44a}$$

$$q_2''(\tau) = H(\tau - \Upsilon)e^{-\alpha(\tau-\Upsilon)} \times \left[q_2'(\Upsilon)\sinh(\vartheta - \gamma(\tau-\Upsilon)) - p_2'(\Upsilon)\sinh(\gamma(\tau-\Upsilon))\right]. \tag{27.44b}$$

Finally, the complete solution for the **Case 2** in the domain $\tau \geq 0$ is as follows:

$$p_2(\tau) = H(\Upsilon - \tau)\, e^{-\alpha\tau}\big[M_2'(\tau)\sinh(\vartheta + \gamma\tau) + N_2'(\tau)\sinh(\gamma\tau)\big] + p_2''(\tau) \tag{27.45a}$$

$$\frac{d}{dt}p_2(\tau) = H(\Upsilon - \tau)\, e^{-\alpha\tau}\big[N_2'(\tau)\sinh(\vartheta - \gamma\tau) - M_2'(\tau)\sinh(\gamma\tau)\big] + q_2''(\tau). \tag{27.45b}$$

Finally, we exhibit the solution (27.27) where the matrizants (27.35) that correspond to the **Case 3** will be used. Calculate first the modulators, $M_3'(\tau)$ and $N_3'(\tau)$, via operations with the term $\int_0^\tau (e^{-x\mathcal{Q}})^{-1} f(x)\, dx$ of (27.27). This yields

$$M_3'(\tau) = -\varpi \int_0^\tau e^{x\alpha} x E(x)\, dx \text{ and } N_3'(\tau) = \varpi \int_0^\tau e^{x\alpha}(1 + x)\, E(x)\, dx. \tag{27.46}$$

In turn, the solutions, $p_3'(\tau)$ and $q_3'(\tau)$, within the finite time interval, $0 \leq \tau \leq \Upsilon$, are

$$\begin{aligned} p_3'(\tau) &= e^{-\tau\alpha}\big[(1+\tau)\, M_3'(\tau) + \tau N_3'(\tau)\big] \quad \text{and} \\ q_3'(\tau) &= e^{-\tau\alpha}\big[(1-\tau)\, N_3'(\tau) - \tau M_3'(\tau)\big]. \end{aligned} \tag{27.47}$$

At the end of this interval, $\tau = \Upsilon$, one can calculate the values of $p_3'(\Upsilon)$ and $q_3'(\Upsilon)$. If $\tau > \Upsilon$, then the solution is

$$p_3''(\tau) = H(\tau - \Upsilon)\, e^{-\alpha(\tau-\Upsilon)}\,\big[(1 + (\tau - \Upsilon))\, p_3'(\Upsilon) + (\tau - \Upsilon)\, q_3'(\Upsilon)\big] \tag{27.48a}$$

$$q_3''(\tau) = H(\tau - \Upsilon)\, e^{-\alpha(\tau-\Upsilon)}\,\big[(1 - (\tau - \Upsilon))\, q_3'(\Upsilon) - (\tau - \Upsilon)\, p_3'(\Upsilon)\big]. \tag{27.48b}$$

The combination of (27.47) and (27.48a) and (27.48b) yields the final complete solution as

$$p_3(\tau) = H(\Upsilon - \tau)\, e^{-\alpha\tau}\big[(1+\tau)\, M_3'(\tau) + \tau N_3'(\tau)\big] + p_3''(\tau) \tag{27.49a}$$

$$\frac{d}{dt}p_3(\tau) = H(\Upsilon - \tau)\, e^{-\alpha\tau}\big[(1-\tau)\, N_3'(\tau) - \tau M_3'(\tau)\big] + q_3''(\tau). \tag{27.49b}$$

In closing, it is worthwhile to notice that the term "*matrizant*" has an equivalent name of the "*evolutionary matrix.*" Indeed, formula (27.39) shows how the initial vector $X'(\Upsilon)$, which is obtained at the instant $\tau = \Upsilon$ as $X'(\Upsilon) = \mathrm{col}(p_1'(\Upsilon), q_1'(\Upsilon))$, evolves under action of the matrizant and results in (27.40a)–(27.40b) while $\tau > \Upsilon$. In the same way, the formulas (27.44a)–(27.44b) and (27.48a)–(27.48b) were obtained.

27.4 Upgrading the evolutionary approach to electrodynamics (*EAE*)

27.4.1 Comparison of two alternative approaches to the electromagnetic field theory

Let us start with the statement of the problem under study. Consider a hollow cavity of volume V bounded by a closed simply connected surface S, which has the properties of a perfect electric conductor. Suppose that the surface is smooth enough in the sense that none of its possible *inner* angles (i.e., measured *within* V) exceeds π. As long as the cavity is hollow (i.e., dielectric-free), eliminate the polarization vector, $\mathbb{P}$, from Maxwell's equations (27.9) and supplement them with the boundary conditions over the surface S and the initial conditions. This yields the statement of the problem as

$$\left.\begin{array}{ll} \nabla\times\mathbb{E}(\mathbf{r},t) = -\frac{1}{c}\partial_t\mathbb{H}(\mathbf{r},t)\,, & \nabla\times\mathbb{H}(\mathbf{r},t) = \frac{1}{c}\partial_t\mathbb{E}(\mathbf{r},t) + 2\gamma\,\mathbb{E}(\mathbf{r},t) \\ \underline{\mathbf{n}\times\mathbb{E}|_S = \mathbf{0},\ \mathbf{n}\cdot\mathbb{H}|_S = 0\,;} & \underline{\mathbb{E}|_{t=0} = \mathbb{E}_0(\mathbf{r})\,,\ \mathbb{H}|_{t=0} = \mathbb{H}_0(\mathbf{r})} \end{array}\right\} \quad (27.50)$$

where $\mathbf{n}$ is outward normal to the surface, $\mathbf{r}$ is the position vector of a point of observation within V, t is an observation time. Maxwell's equations with ∂_t in (27.50) belong to the class of partial differential equations (PDE) of the *hyperbolic* type. Hence, the initial conditions at a fixed instant of time (say, at $t = 0$) are added with necessity to (27.50) where the initial fields, $\mathbb{E}_0(\mathbf{r})$ and $\mathbb{H}_0(\mathbf{r})$, should be given. The values as $\mathbb{E}_0(\mathbf{r}) = \mathbf{0}$ and $\mathbb{H}_0(\mathbf{r}) = \mathbf{0}$ are admissible.

Theoretical studies in electromagnetics were pioneered by Lord Rayleigh (John William Strutt) in 1897 [11]. He proved that waveguides are capable of supporting propagation of electromagnetic waves varying harmonically in time. Later on, that fact has been interpreted as a postulate that motivated presentation of the field vectors in the form

$$\mathbb{E}(\mathbf{r},t) = \mathrm{Re}\{\mathrm{e}^{\mathrm{i}\omega t}\mathbf{E}(\mathbf{r},\omega)\} \text{ and } \mathbb{H}(\mathbf{r},\mathrm{t}) = \mathrm{Re}\{\mathrm{e}^{\mathrm{i}\omega t}\mathbf{H}(\mathbf{r},\omega)\} \quad (27.51)$$

where $i = \sqrt{-1}$ is the imaginary unit, and ω is a frequency parameter, $-\infty < \omega < +\infty$. The new field vectors, $\mathbf{E}$ and $\mathbf{H}$ (so-called phasor), become the subject for studies. After the substitution of the fields (27.51) in Maxwell's equations (27.50) they lose their time derivative, indeed ∂_t is replaced by the constant $i\omega$. Furthermore, Maxwell's equations for the phasors, $\mathbf{E}$ and $\mathbf{H}$, transit to the class of the *elliptic* PDE, which do not need initial conditions. Solutions of these elliptic Maxwell's equations have the physical sense of cavity modes. In the 1940s, it was proved that the set of modes is *complete* in an energetic space of solutions and originate a *basis,* which has been employed for various problems. Due the well-developed basis concept, the time-harmonic field method (THM) has become as a powerful tool of electromagnetics in the last century. Sometimes, the area of applicability of the THM is called electromagnetics in the frequency domain (FD). However, all the time-harmonic solutions have a common defect, namely, they do not satisfy the causality principle. As follows from the format (27.51), time-harmonic fields have neither a beginning nor an end in time while $-\infty < t < +\infty$.

An evolutionary approach to electrodynamics (EAE) was proposed at the end of the 1980s to develop an adequate time-domain theory of the transient processes in cavities and of propagation in waveguides of the nonharmonic (e.g., digital) signals [12–15]. Development of the EAE was stimulated by several profitable ideas, which the first author discovered in early publications on electronics, electrodynamics, and functional analysis, and borrowed from them as a background for his approach [15–19].

Outlines of the EAE can be shortly described as follows. We start with extraction from the system of Maxwell's equations of a self-adjoint operator, $\mathfrak{R}$, which acts on the space variables, $\mathbf{r}$, only. The operator is composed as a consolidation of both the differential operators $\nabla\times$, which are present in the pair of vectorial Maxwell's equations, and the boundary conditions over the perfectly conducting cavity surface. The operator acts on the six-component column vector, $\mathbb{X}$, composed of the three-component field vectors, $\mathbb{E}$ and $\mathbb{H}$, as $\mathbb{X} = \mathrm{col}(\mathbb{E}, \mathbb{H})$ where col symbolizes "column." The set of the eigenvectors of the operator $\mathfrak{R}$ is complete and originates an *orthonormal basis* in a functional space with an energetic $\mathfrak{L}_2$ metric. Completeness of the basis was proven by applying the Weyl theorem [20]. The elements of the basis have the physical sense of cavity modes. As long as the self-adjoint operator is *diagonal,* it can be inverted analytically in the form of an eigenvector series. The series have a physical sense of the modal field expansion where the time-dependent scalar modal amplitudes are unknown as yet. A problem for the amplitudes can be obtained via the substitution of the modal field expansions in Maxwell's equations with time derivative and applying the orthonormal conditions. This yields a system of ordinary differential equations (ODE) for the modal amplitudes. The electromagnetic fields from the given initial conditions (27.50) should also be presented in the form of modal field expansions. This results in the initial conditions for the ODEs, which we derived earlier for the modal amplitudes. The ODE supplemented with the appropriate initial conditions jointly define a Cauchy problem well studied in mathematics. Any solution to the Cauchy problem exhibits how a process *progresses in time* (i.e., *evolves,* shortly) starting from its initial instant and up to the time of observation. That is why we call our method of time-domain studies of Maxwell's equations with ∂_t the EAE.

One can find in [14] the both versions of the EAE, which were developed individually for studies of the time-domain processes in cavities and in waveguides as well. In [14], the standard Maxwell equations in SI units were used and the solutions were obtained in the class of complex-valued functions. These were made for facilitating comparison of the solutions obtained in the time domain and known in the FD. In the present chapter, the EAE will be upgraded in the following two aspects. (1) We shall implement the scheme of the evolutionary approach starting with Maxwell's equations in SI units in the *new format* (27.50) . As long as the new format of the equations remains much simpler than the standard one, the upgraded version of the EAE will be more convenient for its practical applications. (2) We shall find the solutions in the class of the *real-valued* functions. Hence, the energetic and mechanical field characteristics can be obtained as the *functions of time.* So, this opens a way to studying the *energetic* wave processes that should inevitably accompany that for oscillations in cavities and waves in waveguides. Meanwhile, the complex-valued solutions result in the energetic field quantities *averaged* (only) over a period of the field variations in time.

27.4.2 Separation of a self-adjoint operator from the vectorial Maxwell's equations (27.50)

The problem (27.50) involves two differential vector equations, each of which is a three-component one. In order to extract the operator, it is convenient to present this pair as single, but six-component vectorial equation as

$$\begin{pmatrix} \nabla\times\mathbb{H}(\mathbf{r},t) \\ \nabla\times\mathbb{E}(\mathbf{r},t) \end{pmatrix} = \begin{pmatrix} \mathbb{U}(\mathbf{r},t) \\ \mathbb{V}(\mathbf{r},t) \end{pmatrix} \text{ where } \left.\begin{matrix} \mathbb{U}(\mathbf{r},t) = \frac{1}{c}\partial_t\mathbb{E}(\mathbf{r},t) + 2\gamma\,\mathbb{E}(\mathbf{r},t) \\ \mathbb{V}(\mathbf{r},t) = -\frac{1}{c}\partial_t\mathbb{H}(\mathbf{r},t) \end{matrix}\right\} \tag{27.52}$$

The left-hand side of (27.52) can be rewritten as

$$\underbrace{\begin{pmatrix} \nabla\times\mathbb{H}(\mathbf{r},t) \\ \nabla\times\mathbb{E}(\mathbf{r},t) \end{pmatrix}}_{\mathfrak{R}'\mathbb{X}(\mathbf{r},t)} = \underbrace{\begin{pmatrix} \mathcal{O} & \nabla\times \\ \nabla\times & \mathcal{O} \end{pmatrix}}_{\mathfrak{R}'}\underbrace{\begin{pmatrix} \mathbb{E}(\mathbf{r},t) \\ \mathbb{H}(\mathbf{r},t) \end{pmatrix}}_{\mathbb{X}(\mathbf{r},t)} \tag{27.53}$$

where $\mathbb{X}$ a six-component "electromagnetic" vector composed of the three-component field vectors, $\mathbb{E}$ and $\mathbb{H}$, and $\mathfrak{R}'$ is a 6×6 matrix differential operation, which acts on the space variables (coordinates) only. In turn, the operation $\mathfrak{R}'$ consists of

$$\mathcal{O} = \begin{pmatrix} 0 & 0 & 0 \\ 0 & 0 & 0 \\ 0 & 0 & 0 \end{pmatrix} \text{ and } \nabla\times \text{ is } \begin{pmatrix} 0 & -\partial_z & \partial_y \\ \partial_z & 0 & -\partial_x \\ -\partial_y & \partial_x & 0 \end{pmatrix} \tag{27.54}$$

where $\mathcal{O}$ is 3×3 zero-valued matrix and $\nabla\times$ is composed of the partial derivatives by Cartesian coordinates.

The observation of (27.53) and (27.54) shows that the time variable, t, in the argument of $\mathbb{X}(\mathbf{r},t)$ plays the role of a parameter with respect to the action of $\mathfrak{R}'$ on $\mathbb{X}(\mathbf{r},t)$. This fact suggests to introduce an *operator*, $\mathfrak{R}$, via a unification of the differential procedure, $\mathfrak{R}'$, and the boundary conditions over the surface S given in (27.50). Within the volume V (except for its boundary S), this operator acts on the variables $(\mathbf{r})$ via the differential operation $\mathfrak{R}'$. Over S, the operator $\mathfrak{R}$ acts via the boundary conditions. Mathematically, it can be written as

$$\boxed{\mathfrak{R}\mathbb{X}(\mathbf{r},t) = \begin{cases} \mathfrak{R}'\mathbb{X}(\mathbf{r},t) & \text{where } \mathbf{r}\in V,\ \mathbf{r}\notin S \\ \mathbf{n}\times\mathbb{E}(\mathbf{r},t) = \mathbf{0},\ \mathbf{n}\cdot\mathbb{H}(\mathbf{r},t) = 0 & \text{where } \mathbf{r}\in S. \end{cases}} \tag{27.55}$$

Define a domain of the operator $\mathfrak{R}$ as a manifold of the *real-valued* six-component vectors $\mathfrak{X}(\mathbf{r})$ dependent on coordinates, *only*, and specify that domain via introducing the inner product, $\langle *, *\rangle$, of a pair of such vectors, $\mathfrak{X}_1$ and $\mathfrak{X}_2$, as follows:

$$\begin{aligned} \langle\mathfrak{X}_1,\mathfrak{X}_2\rangle &= \frac{1}{V}\int_V (\mathbf{E}_1\cdot\mathbf{E}_2 + \mathbf{H}_1\cdot\mathbf{H}_2)\,dV \quad \text{where } \mathfrak{X}_1(\mathbf{r}) = \begin{pmatrix} \mathbf{E}_1(\mathbf{r}) \\ \mathbf{H}_1(\mathbf{r}) \end{pmatrix}, \\ \mathfrak{X}_2(\mathbf{r}) &= \begin{pmatrix} \mathbf{E}_2(\mathbf{r}) \\ \mathbf{H}_2(\mathbf{r}) \end{pmatrix}. \end{aligned} \tag{27.56}$$

It is evident that the domain of $\mathfrak{R}$ is the Hilbert functional space, $\mathcal{L}_2$, with an energetic metric. Every element in $\mathcal{L}_2$ is a six-component vector, $\mathfrak{X}(\mathbf{r})$, composed of the three-component vectors, $\mathbf{E}(\mathbf{r})$ and $\mathbf{H}(\mathbf{r})$, each of which is a twice differentiable function in the open domain $\mathbf{r} \in V$, but satisfies the same boundary conditions as in (27.55) over the boundary S.

In order to prove that $\mathfrak{R}$ is a self-adjoint operator, let us take two pairs of the vectors composed as follows:

$$\mathfrak{X}_1 = \begin{pmatrix} \mathbf{E}_1 \\ \mathbf{H}_1 \end{pmatrix}, \ \mathfrak{R}\mathfrak{X}_1 = \begin{pmatrix} \nabla \times \mathbf{H}_1 \\ \nabla \times \mathbf{E}_1 \end{pmatrix} \ \text{and} \ \mathfrak{X}_2 = \begin{pmatrix} \mathbf{E}_2 \\ \mathbf{H}_2 \end{pmatrix}, \ \mathfrak{R}\mathfrak{X}_2 = \begin{pmatrix} \nabla \times \mathbf{H}_2 \\ \nabla \times \mathbf{E}_2 \end{pmatrix}, \tag{27.57}$$

and calculate the inner products $\langle \mathfrak{X}_1, \mathfrak{R}\mathfrak{X}_2 \rangle$ and $\langle \mathfrak{R}\mathfrak{X}_1, \mathfrak{X}_2 \rangle$ which yields

$$\begin{aligned} \langle \mathfrak{X}_1, \mathfrak{R}\mathfrak{X}_2 \rangle &= \tfrac{1}{V} \int_V (\mathbf{E}_1 \cdot \nabla \times \mathbf{H}_2 + \mathbf{H}_1 \cdot \nabla \times \mathbf{E}_2)\, dV \quad \text{and} \\ \langle \mathfrak{R}\mathfrak{X}_1, \mathfrak{X}_2 \rangle &= \tfrac{1}{V} \int_V (\mathbf{E}_2 \cdot \nabla \times \mathbf{H}_1 + \mathbf{H}_2 \cdot \nabla \times \mathbf{E}_1)\, dV. \end{aligned} \tag{27.58}$$

Calculation of the difference, $\langle \mathfrak{X}_1, \mathfrak{R}\mathfrak{X}_2 \rangle - \langle \mathfrak{R}\mathfrak{X}_1, \mathfrak{X}_2 \rangle$, taking into account Gauss's theorem and the boundary conditions, yields

$$\begin{aligned} \langle \mathfrak{X}_1, \mathfrak{R}\mathfrak{X}_2 \rangle - \langle \mathfrak{R}\mathfrak{X}_1, \mathfrak{X}_2 \rangle &= \frac{1}{V} \int_V (\mathbf{E}_1 \cdot \nabla \times \mathbf{H}_2 - \mathbf{H}_2 \cdot \nabla \times \mathbf{E}_1)\, dV \\ &+ \frac{1}{V} \int_V (\mathbf{H}_1 \cdot \nabla \times \mathbf{E}_2 - \mathbf{E}_2 \cdot \nabla \times \mathbf{H}_1)\, dV = 0. \end{aligned} \tag{27.59}$$

The fact that $\langle \mathfrak{X}_1, \mathfrak{R}\mathfrak{X}_2 \rangle - \langle \mathfrak{R}\mathfrak{X}_1, \mathfrak{X}_2 \rangle = 0$ means that operator $\mathfrak{R}$ is symmetric and the operator eigenvalue equation can be formulated as

$$\mathfrak{R}\mathfrak{X}_n(\mathbf{r}) = k_n \mathfrak{X}_n(\mathbf{r}) \quad \text{which is equivalent to} \quad \begin{cases} \nabla \times \mathbf{H}_n(\mathbf{r}) = k_n \mathbf{E}_n(\mathbf{r}), & \mathbf{n} \cdot \mathbf{H}_n|_S = 0 \\ \nabla \times \mathbf{E}_n(\mathbf{r}) = k_n \mathbf{H}_n(\mathbf{r}), & \mathbf{n} \times \mathbf{E}_n|_S = \mathbf{0} \end{cases} \tag{27.60}$$

where k_n is an eigenvalue of the operator $\mathfrak{R}$ and $\mathfrak{X}_n = \mathrm{col}(\mathbf{E}_\mathrm{n}, \mathbf{H}_\mathrm{n})$ is an eigenvector, which corresponds to this eigenvalue.

The spectrum of the operator $\mathfrak{R}$ is real¶ and discrete of form $\{k_n\}_{n=-\infty}^{n=+\infty}$, where $n = 0, \pm 1, \pm 2, \ldots$, because the domain $\mathbf{r} \in V$ is finite. Furthermore, the discrete eigenvalues are distributed on the real axis symmetrically with respect to the point $k_0 = 0$ as a center of their "mirror symmetry." To prove this fact, let us write down

¶Any self-adjoint operator has all its eigenvalues real-valued.

the boundary eigenvalue problem (27.60) (provided that $k_n \neq 0$) in two formats, (a) and (b) , as follows:

$$(a)\begin{cases}\nabla\times\mathbf{H}_{+n}=k_{+n}\mathbf{E}_{+n}, & \mathbf{n}\cdot\mathbf{H}_{+n}|_S=0;\\ \nabla\times\mathbf{E}_{+n}=k_{+n}\mathbf{H}_{+n}, & \mathbf{n}\times\mathbf{E}_{+n}|_S=\mathbf{0};\end{cases}\quad\text{and}$$
$$(b)\begin{cases}\nabla\times(-\mathbf{H}_{+n})=(-k_{+n})\,\mathbf{E}_{+n}, & \mathbf{n}\cdot(-\mathbf{H}_{+n})\,|_S=0\\ \nabla\times\mathbf{E}_{+n}=(-k_{+n})\,(-\mathbf{H}_{+n}), & \mathbf{n}\times\mathbf{E}_{+n}|_S=\mathbf{0}\end{cases}\tag{27.61}$$

One can interpret the version (a) in (27.61) as an equivalent to the operator eigenvalue equation, $\mathfrak{R}\mathfrak{X}_{+n}=k_{+n}\mathfrak{X}_{+n}$, where the eigenvalue is positive ($k_{+n}>0$) and the appropriate eigenvector is composed as $\mathfrak{X}_{+n}=\mathrm{col}(\mathbf{E}_{+\mathrm{n}},\mathbf{H}_{+\mathrm{n}})$. On the other hand, the version (b) can be interpreted as an *another* operator equation, $\mathfrak{R}\mathfrak{X}_{-n}=k_{-n}\mathfrak{X}_{-n}$, where the eigenvalue, $k_{-n}=-k_{+n}$, is located symmetrically to k_{+n} on the real axis, and the appropriate eigenvector, $\mathfrak{X}_{-n}$, is composed of the *same vectors,* $\mathbf{E}_{+n}$ and $\mathbf{H}_{+n}$, as follows: $\mathfrak{X}_{-n}=\mathrm{col}(\mathbf{E}_{+\mathrm{n}},(-\mathbf{H}_{+\mathrm{n}}))$. It is evident, that these eigenvectors, $\mathfrak{X}_{+n}$ and $\mathfrak{X}_{-n}$, are *linearly independent.*

The symmetrical eigenvectors, $\mathfrak{X}_{+n}$ and $\mathfrak{X}_{-n}$, can be presented as sum and difference, respectively, of two six-component vectors, $\mathfrak{X}_n^{\mathrm{E}}$ and $\mathfrak{X}_n^{\mathrm{H}}$, where $\mathfrak{X}_n^{\mathrm{E}}$ is a pure electric vector and the other, $\mathfrak{X}_n^{\mathrm{H}}$, is a pure magnetic vector. They are specified as

$$k_{\pm n}=\pm k_{+n}\gtrless 0:\ \mathfrak{X}_{\pm n}=\mathfrak{X}_n^{\mathrm{E}}\pm\mathfrak{X}_n^{\mathrm{H}}\ \text{ where }\ \mathfrak{X}_n^{\mathrm{E}}=\begin{pmatrix}\mathbf{E}_{+n}\\ \mathbf{0}\end{pmatrix},\ \mathfrak{X}_n^{\mathrm{H}}=\begin{pmatrix}\mathbf{0}\\ \mathbf{H}_{+n}\end{pmatrix}.\tag{27.62}$$

Herein, $\mathbf{0}$ symbolizes the three-component zero-valued vector. The operator eigenvalue equations, $\mathfrak{R}\mathfrak{X}_{\pm n}=k_{\pm n}\mathfrak{X}_{\pm n}$, being rewritten in terms of the vectors $\mathfrak{X}_n^{\mathrm{E}}$ and $\mathfrak{X}_n^{\mathrm{H}}$, appear as follows:

$$\mathfrak{R}\mathfrak{X}_{+n}=k_{+n}\mathfrak{X}_{+n}\ \text{ is equivalent to }\ \mathfrak{R}(\mathfrak{X}_n^{\mathrm{E}}+\mathfrak{X}_n^{\mathrm{H}})=k_{+n}(\mathfrak{X}_n^{\mathrm{E}}+\mathfrak{X}_n^{\mathrm{H}})\tag{27.63a}$$
$$\mathfrak{R}\mathfrak{X}_{-n}=k_{-n}\mathfrak{X}_{-n}\ \text{ is equivalent to }\ \mathfrak{R}(\mathfrak{X}_n^{\mathrm{E}}-\mathfrak{X}_n^{\mathrm{H}})=-k_{+n}(\mathfrak{X}_n^{\mathrm{E}}-\mathfrak{X}_n^{\mathrm{H}}).\tag{27.63b}$$

The sum and difference of (27.64a) and (27.64b) result in, respectively,

$$\mathfrak{R}\mathfrak{X}_n^{\mathrm{E}}=k_{+n}\mathfrak{X}_n^{\mathrm{H}}\ \text{ and }\ \mathfrak{R}\mathfrak{X}_n^{\mathrm{H}}=k_{+n}\mathfrak{X}_n^{\mathrm{E}}.\tag{27.64}$$

Thus, the six-component vectors, $\mathfrak{X}_n^{\mathrm{E}}$ and $\mathfrak{X}_n^{\mathrm{H}}$, individually, are *not* the eigenvectors of the operator $\mathfrak{R}$ if $k_{+n}\neq 0$. Only their combinations, being taken as $\mathfrak{X}_n^{\mathrm{E}}+\mathfrak{X}_n^{\mathrm{H}}$ and $\mathfrak{X}_n^{\mathrm{E}}-\mathfrak{X}_n^{\mathrm{H}}$, create the symmetric linearly independent eigenvectors of $\mathfrak{R}$.

In the definition (27.55) of the operator $\mathfrak{R}$ and in the equivalent boundary eigenvalue problem (27.60), two boundary conditions participate as $\mathbf{n}\times\mathbf{E}_n|_S=\mathbf{0}$ and $\mathbf{n}\cdot\mathbf{H}_n|_S=0$. Let us prove now that the second boundary condition, $\mathbf{n}\cdot\mathbf{H}_n|_S=0$, is a consequence of the first one, and vice versa. To this end, introduce an auxiliary surface S', which is located within the volume V infinitesimally close to the real surface S, but does not coincide with that one as yet. Notice that *differential* equation $\nabla\times\mathbf{E}_n=k_n\mathbf{H}_n$ (see (27.60)) holds everywhere over S'. Let us select on the surface S' a part, s', bounded by a closed contour ℓ'. The surface s' should be small enough to consider it to be a plane. Let the unit vector $\mathbf{n}'$ be an outward normal to the plane s'. Consider the integral relationship $\int_{s'} \mathbf{n}' \cdot \nabla \times \mathbf{E}_n \, ds' = k_n \int_{s'} \mathbf{n}' \cdot \mathbf{H}_n \, ds'$ from applying Stokes' theorem. This yields an *algebraic* relationship $\oint_{\ell'} \vec{\ell}' \cdot \mathbf{E}_n \, d\ell' = k_n \int_{s'} \mathbf{n}' \cdot \mathbf{H}_n \, ds'$ where $\vec{\ell}'$ is a unit vector *tangential* to the contour ℓ'. As long as the new relationship is algebraic, one can take the limit when the surface s' together with its boundary ℓ' coincide with appropriate parts of the cavity surface S, which yields $\oint_{\ell} \vec{\ell} \cdot \mathbf{E}_n \, d\ell = k_n \int_{\underline{s}} \mathbf{n} \cdot \mathbf{H}_n \, ds$. So, if $\vec{\ell} \cdot \mathbf{E}_n|_\ell = 0$ holds then $\mathbf{n} \cdot \mathbf{H}_n|_s = 0$ holds, and vice versa. Notice that $\vec{\ell} \cdot \mathbf{E}_n|_\ell = 0$ is equivalent to $\mathbf{n} \times \mathbf{E}_n|_s = \mathbf{0}$. The position of the small surface s on the cavity surface S can be *arbitrary*. Hence, if $\mathbf{n} \times \mathbf{E}_n|_S = \mathbf{0}$ holds everywhere on S, then $\mathbf{n} \cdot \mathbf{H}_n|_S = 0$ holds automatically, and vice versa.

In the case of a *cylindrical* cavity, the boundary value problem in (27.60) should be interpreted as two independent problems. One of them can be considered as the statement of problem for the transverse-electric (TE) (with respect to an axis of the cylinder) cavity modes as

$$(\text{TE}) \begin{cases} \nabla \times \mathbf{H}'_n(\mathbf{r}) = k'_n \mathbf{E}'_n(\mathbf{r}), & k'_n > 0 \\ \nabla \times \mathbf{E}'_n(\mathbf{r}) = k'_n \mathbf{H}'_n(\mathbf{r}), & n = 1, 2, \ldots \\ \nabla \cdot \mathbf{E}'_n(\mathbf{r}) = \nabla \cdot \mathbf{H}'_n(\mathbf{r}) = 0, & \underline{\mathbf{n} \times \mathbf{E}'_n|_S = \mathbf{0}} \end{cases} \tag{27.65}$$

where $k'_n > 0$ are the real-valued eigenvalues and $\mathbf{E}'_n$ and $\mathbf{H}'_n$ are the real-valued solutions. The boundary condition $\mathbf{n} \cdot \mathbf{H}'_n|_S = 0$ holds automatically. The other one is the statement of problem for the transverse-magnetic (TM) cavity modes as

$$(\text{TM}) \begin{cases} \nabla \times \mathbf{H}''_n(\mathbf{r}) = k''_n \mathbf{E}''_n(\mathbf{r}), & k''_n > 0 \\ \nabla \times \mathbf{E}''_n(\mathbf{r}) = k''_n \mathbf{H}''_n(\mathbf{r}), & n = 1, 2, \ldots \\ \nabla \cdot \mathbf{E}''_n(\mathbf{r}) = \nabla \cdot \mathbf{H}''_n(\mathbf{r}) = 0, & \underline{\mathbf{n} \cdot \mathbf{H}''_n|_S = 0} \end{cases} \tag{27.66}$$

where $k''_n > 0$, $\mathbf{E}''_n$, and $\mathbf{H}''_n$ are the real-valued quantities. The boundary condition $\mathbf{n} \times \mathbf{E}''_n|_S = \mathbf{0}$ holds automatically. Standard manipulations within (27.65) and (27.66) result in familiar statements of the problems for the *solenoidal* cavity modes as

$$\boxed{(\text{TE}) \begin{cases} \nabla^2 \mathbf{E}'_n(\mathbf{r}) + k'^2_n \mathbf{E}'_n(\mathbf{r}) = 0, \; \nabla \cdot \mathbf{E}'_n(\mathbf{r}) = 0, & \underline{\mathbf{n} \times \mathbf{E}'_n|_S = \mathbf{0}} \\ \mathbf{H}'_n(\mathbf{r}) = \nabla \times \mathbf{E}'_n(\mathbf{r}) / k'_n, \; k'_n > 0, & n = 1, 2, \ldots \end{cases}} \tag{27.67a}$$

$$\boxed{(\text{TM}) \begin{cases} \nabla^2 \mathbf{H}''_n(\mathbf{r}) + k''^2_n \mathbf{H}''_n(\mathbf{r}) = 0, \; \nabla \cdot \mathbf{H}''_n(\mathbf{r}) = 0, & \underline{\mathbf{n} \cdot \mathbf{H}''_n|_S = 0} \\ \mathbf{E}''_n(\mathbf{r}) = \nabla \times \mathbf{H}''_n(\mathbf{r}) / k''_n, \; k''_n > 0, & n = 1, 2, \ldots \end{cases}} \tag{27.67b}$$

Let us now appeal to the boundary eigenvalue problem (27.60) again and put therein $n = 0$ and $k_0 = 0$. This yields

$$\boxed{\nabla \times \mathbf{E}_0(\mathbf{r}) = \mathbf{0}, \; \mathbf{n} \times \mathbf{E}_0|_S = \mathbf{0}} \text{ and } \boxed{\nabla \times \mathbf{H}_0(\mathbf{r}) = \mathbf{0}, \; \mathbf{n} \cdot \mathbf{H}_0|_S = 0.} \tag{27.68}$$

It is evident that these differential equations have solutions distinct from zero of form $\mathbf{E}_0(\mathbf{r}) = \nabla\Phi(\mathbf{r})$ and $\mathbf{H}_0(\mathbf{r}) = \nabla\Psi(\mathbf{r})$ where the potentials, $\Phi(\mathbf{r})$ and $\Psi(\mathbf{r})$, may be *arbitrary* twice differentiable functions. The fields, $\mathbf{E}_0$ and $\mathbf{H}_0$, are the *irrotational* vectors.** Keeping in mind the Sturm–Liouville theorems from mathematical physics [21], we can define the *six-component* eigenvectors of $\mathfrak{R}$, each of which corresponds to the same eigenvalue, $k_0 = 0$, as follows:

$$\boxed{\mathring{\mathfrak{X}}^{\mathrm{E}}_m(\mathbf{r}) = \mathrm{col}(\nabla\Phi_{\mathrm{m}}(\mathbf{r}), \mathbf{0}) : \ \nabla^2\Phi_m(\mathbf{r}) + \kappa_m^2\Phi_m(\mathbf{r}) = 0, \ \Phi_m|_S = 0, \ \kappa_m^2 > 0, \ m = 1, 2, \ldots} \tag{27.69a}$$

$$\boxed{\mathring{\mathfrak{X}}^{\mathrm{H}}_m(\mathbf{r}) = \mathrm{col}(\mathbf{0}, \nabla\Psi_{\mathrm{m}}(\mathbf{r})) : \ \nabla^2\Psi_m(\mathbf{r}) + \nu_m^2\Psi_m(\mathbf{r}) = 0, \ \mathbf{n}\cdot\nabla\Psi_m|_S = 0, \ \nu_m^2 > 0, \ m = 1, 2, \ldots} \tag{27.69b}$$

where $\mathbf{0}$ is the three-component zero-valued vector, $\mathbf{n}$ is the outward unit normal to the simply connected cavity surface S, κ_m^2 and ν_m^2 are the eigenvalues in Dirichlet and Neumann boundary-eigenvalue problems for the Laplacian, ∇^2, respectively.

Lastly, consider the *irrotational* eigenvectors of $\mathfrak{R}$, which correspond to the eigenvalues $\kappa_0^2 = 0$ and $\nu_0^2 = 0$ in the problems (27.69a) and (27.69b). They are specified via the solutions Φ_0 and Ψ_0 to the Laplace equation as follows:

$$\begin{aligned} \mathring{\mathfrak{X}}^{\mathrm{E}}_0(\mathbf{r}) &= \mathrm{col}(\nabla\Phi_0(\mathbf{r}), \mathbf{0}) \colon \nabla^2\Phi_0(\mathbf{r}) = 0 \quad \text{and} \\ \mathring{\mathfrak{X}}^{\mathrm{H}}_0(\mathbf{r}) &= \mathrm{col}(\mathbf{0}, \nabla\Psi_0(\mathbf{r})) \colon \nabla^2\Psi_0(\mathbf{r}) = 0. \end{aligned} \tag{27.70}$$

So, the potentials Φ_0 and Ψ_0 are *harmonic* functions. It is worthwhile to recall that the cavity surface S is a closed *simply connected* domain. In accordance with the minimum–maximum theorem for harmonic functions, the potentials $\Phi_0(\mathbf{r})$ and $\Psi_0(\mathbf{r})$ are constants†† in the *closed* domain $\{\mathbf{r} \in V, \mathbf{r} \in S\}$. The gradient of those constants is zero in (27.70) and hence,

$$\mathring{\mathfrak{X}}^{\mathrm{E}}_0(\mathbf{r}) = \mathrm{col}(\mathbf{0}, \mathbf{0}) \quad \text{and} \quad \mathring{\mathfrak{X}}^{\mathrm{H}}_0(\mathbf{r}) = \mathrm{col}(\mathbf{0}, \mathbf{0}) \ \text{if} \ \mathbf{r} \in V, \ \mathbf{r} \in S. \tag{27.71}$$

Observation of the problems (27.69a) and (27.69b) suggests the following conclusions: (1) the eigenvalue k_0 of the self-adjoint operator $\mathfrak{R}$ is multiple, and furthermore, k_0 has an infinite degree of degeneration; (2) the sets of solutions to the Dirichlet and Neumann problems, $\{\Phi_m\}$ and $\{\Psi_m\}$, are complete: see [21]; (3) any potential, Φ and Ψ, which participates in the solutions to (27.68), can be represented via the eigenfunction sets, $\{\Phi_m\}$ and $\{\Psi_m\}$, respectively; (4) topologically, there are two mutually orthogonal *subspaces* of the irrotational eigenvectors, $\mathring{\mathfrak{X}}^{\mathrm{E}}_m$ and $\mathring{\mathfrak{X}}^{\mathrm{H}}_m$, in the space of solutions, $\mathcal{L}_2$; (5) all the irrotational and solenoidal eigenvectors are mutually orthogonal as well, as long as they correspond to *distinct* eigenvalues of the self-adjoint operator, namely, $k_0 = 0$ and $k_n \neq 0$, respectively.

**Equation $\nabla \times \mathbf{E}_0 = \mathbf{0}$ was defined originally as $\mathrm{curl}\mathbf{E}_0 = \mathbf{0}$ where "curl" is a synonym of "rotation." So, the fields like $\mathbf{E}_0$ were named *ir*rotational.

††It is true if and only if the surface S is a simply connected domain. If S is multiply connected, the Laplace equation has a finite number of linearly independent solutions varying in V. They generate a subspace, which involves a finite number of the harmonic eigenvectors in $\mathcal{L}_2$.

27.4.3 Normalization of the eigenvectors of operator $\mathfrak{R}$

Let us begin with the normalization of the irrotational eigenvector $\mathring{\mathfrak{X}}_m^{\mathrm{E}}$, which is specified via problem (27.69a). Substitution of $\mathfrak{X}_1 = \mathfrak{X}_2 = \mathring{\mathfrak{X}}_m^{\mathrm{E}}$ in the definition of the inner product (27.56) results in

$$\left\langle \mathring{\mathfrak{X}}_m^{\mathrm{E}}, \mathring{\mathfrak{X}}_m^{\mathrm{E}} \right\rangle = \frac{1}{V} \int_V \nabla\Phi_m \cdot \nabla\Phi_m \, dV = 1. \tag{27.72}$$

One can simplify the integrand in (27.72) by applying the vector identity $\nabla(\varphi\mathbf{A}) = \nabla\varphi{\cdot}\mathbf{A} + \varphi\nabla{\cdot}\mathbf{A}$ where φ is a scalar and $\mathbf{A}$ is a vector. To this end, let us take φ as $\Phi_m(\mathbf{r})$ and $\mathbf{A}$ as $\nabla\Phi_m(\mathbf{r})$ which yields

$$\nabla\Phi_m \cdot \nabla\Phi_m = \Phi_m\left(-\nabla^2\Phi_m\right) + \nabla\cdot(\Phi_m\nabla\Phi_m). \tag{27.73}$$

One can utilize the Helmholtz equation in problem (27.69a) as a direct formula to define $(-\nabla^2\Phi_m)$ as $\kappa_m^2\Phi_m$. Notice that the eigenvector $\mathring{\mathfrak{X}}_m^{\mathrm{E}}$ has the required physical dimension of *inverse meter* provided that the potential Φ_m is a dimensionless quantity. This observation suggests to introduce a normalization of the potential Φ_m as follows:

$$\Phi_m(\mathbf{r}) = \left(\frac{\mathring{N}_m^{\mathrm{E}}}{\kappa_m}\right)\bar{\Phi}_m(\mathbf{r}) \tag{27.74}$$

where $\mathring{N}_m^{\mathrm{E}}$ is the required normalization constant. Substitution of (27.73) , (27.74) in (27.72) and utilizing Gauss's theorem results in

$$\left\langle \mathring{\mathfrak{X}}_m^{\mathrm{E}}, \mathring{\mathfrak{X}}_m^{\mathrm{E}} \right\rangle = (\mathring{N}_m^{\mathrm{E}})^2 \tfrac{1}{V} \int_V \bar{\Phi}_m^2(\mathbf{r})\, dV = 1 \text{ yields } \mathring{N}_m^{\mathrm{E}} = \frac{1}{\sqrt{\frac{1}{V}\int_V \bar{\Phi}_m^2(\mathbf{r})\, dV}}. \tag{27.75}$$

Normalization of the *irrotational* eigenvector $\mathring{\mathfrak{X}}_m^{\mathrm{H}}$ can be performed in the same three steps, (1)–(3), namely,

$$\begin{aligned}&(1)\left\langle \mathring{\mathfrak{X}}_m^{\mathrm{H}}, \mathring{\mathfrak{X}}_m^{\mathrm{H}} \right\rangle = \tfrac{1}{V}\int_V \nabla\Psi_m{\cdot}\nabla\Psi_m\, dV = 1,\ (2)\ \Psi_m(\mathbf{r}) = \tfrac{\mathring{N}_m^{\mathrm{H}}}{\nu_m}\bar{\Psi}_m(\mathbf{r}),\\ &(3)\ \mathring{N}_m^{\mathrm{H}} = \frac{1}{\sqrt{\frac{1}{V}\int_V \bar{\Psi}_m^2(\mathbf{r})\, dV}}.\end{aligned} \tag{27.76}$$

It is convenient to normalize the *solenoidal* eigenvectors, $\mathfrak{X}_{\pm n}(\mathbf{r})$, taking into account their symmetry. A pair of the eigenvectors, which correspond to the symmetrical eigenvalues, $k_{+n} > 0$ and $k_{-n} = -k_{+n} < 0$, was specified in (27.62) as

$$\begin{aligned}&k_{+n}{>}0{:}\ \mathfrak{X}_{+n} = \mathfrak{X}_n^{\mathrm{E}} + \mathfrak{X}_n^{\mathrm{H}},\ k_{-n} = -k_{+n}{:}\ \mathfrak{X}_{-n} = \mathfrak{X}_n^{\mathrm{E}} - \mathfrak{X}_n^{\mathrm{H}}\\ &\text{where } \mathfrak{X}_n^{\mathrm{E}} = \begin{pmatrix}\mathbf{E}_n\\ \mathbf{0}\end{pmatrix},\ \mathfrak{X}_n^{\mathrm{H}} = \begin{pmatrix}\mathbf{0}\\ \mathbf{H}_n\end{pmatrix}.\end{aligned} \tag{27.77}$$

Herein, $\mathbf{E}_n \equiv \mathbf{E}_{+n}$ and $\mathbf{H}_n \equiv \mathbf{H}_{+n}$, see (27.62) . Let the eigenvector $\mathfrak{X}_{+n}$ be normalized so that $\langle \mathfrak{X}_{+n}, \mathfrak{X}_{+n} \rangle = 2$. The inner product $\langle \mathfrak{X}_{+n}, \mathfrak{X}_{-n} \rangle$ is zero as long as the eigenvectors $\mathfrak{X}_{+n}$ and $\mathfrak{X}_{-n}$ correspond to the eigenvalues of $\mathfrak{R}$, k_{+n} and k_{-n}, which are *distinct* a fortiori. Substitution of the definitions (27.77) for $\mathfrak{X}_{+n}$ and $\mathfrak{X}_{-n}$ via $\mathfrak{X}_n^{\mathrm{E}}$ and $\mathfrak{X}_n^{\mathrm{H}}$ in the inner products yields

$$\langle \mathfrak{X}_{+n}, \mathfrak{X}_{+n} \rangle = \frac{1}{V} \int_V \mathbf{E}_n \cdot \mathbf{E}_n \, dV + \frac{1}{V} \int_V \mathbf{H}_n \cdot \mathbf{H}_n \, dV = 2 \tag{27.78a}$$

$$\langle \mathfrak{X}_{+n}, \mathfrak{X}_{-n} \rangle = \frac{1}{V} \int_V \mathbf{E}_n \cdot \mathbf{E}_n \, dV - \frac{1}{V} \int_V \mathbf{H}_n \cdot \mathbf{H}_n \, dV = 0. \tag{27.78b}$$

The combination of (27.78a) and (27.78b) yields individual normalization of the three-component "electric" and "magnetic" parts of the six-component solenoidal eigenvectors denoted as $\mathfrak{X}_n = \mathrm{col}\,(\mathbf{E}_\mathrm{n}, \mathbf{H}_\mathrm{n})$ in the form of

$$\frac{1}{V} \int_V \mathbf{E}_n \cdot \mathbf{E}_n \, dV = 1 \quad \text{and} \quad \frac{1}{V} \int_V \mathbf{H}_n \cdot \mathbf{H}_n \, dV = 1. \tag{27.79}$$

Normalization by formulas (27.79) of the three-component vectors, which are specified in (27.67a) and (27.67b), should provide their physical dimension of *inverse meter* at the endpoint.

27.4.4 Configurational orthonormal modal basis in the space of solutions $\mathcal{L}_2$

Weyl has studied the Hilbert space $\mathcal{L}_2$ in connection with the vectorial boundary eigenvalue problems for the Laplacian in [20]. In the aspect of functional analysis, his main result is known at present as Weyl's theorem about the topological splitting of $\mathcal{L}_2$ into three mutually orthogonal subspaces, which can be shortly written as

$$\boxed{\mathcal{L}_2 = \mathcal{J} \oplus \mathcal{G} \oplus \mathcal{U}} \tag{27.80}$$

where $\oplus$ symbolizes *direct summation* of the subspaces $\mathcal{J}$, $\mathcal{G}$, and $\mathcal{U}$. The subspace $\mathcal{J}$ is the closure of the linear space of the *three-component* solenoidal vectors (say, $\mathbf{A}$) varying within a finite volume V, which is bounded by a closed surface S over which either tangential or normal components of the vectors vanish:

$$\mathcal{J}\colon \{\mathbf{A}\colon (\nabla^2 + k^2)\mathbf{A} = \mathbf{0},\ \nabla \cdot \mathbf{A} = \mathbf{0},\ \text{either}\ [\mathbf{n} \times \mathbf{A}]\,|_S = 0\ \text{or}\ (\mathbf{n} \cdot \mathbf{A})\,|_S = 0\} \tag{27.81}$$

where $\mathbf{n}$ is a normal to S. The subspace $\mathcal{G}$ is the closure of the linear space of gradients of such smooth functions (say, φ) that either the functions themselves or their normal derivatives vanish over the boundary S:

$$\mathcal{G}\colon \{\nabla\varphi\colon (\nabla^2 + \varkappa^2)\,\varphi = 0,\ \text{either}\ \varphi|_S = 0,\ \text{or}\ \mathbf{n} \cdot \nabla\varphi|_S = 0\}. \tag{27.82}$$

The subspace $\mathcal{U}$ is the closure of the linear space of gradients of harmonic functions continuously differentiable in the domain V:

$$\mathcal{U}\colon \{\nabla u\colon\ \nabla^2 u = 0\}. \tag{27.83}$$

Weyl's theorem (27.80) , which was proved for the *three*-component vectors, can be extended on the *equationally definable class* (i.e., a manifold) of the *six*-component eigenvectors (27.67a)–(27.70) of the self-adjoint operator $\mathfrak{R}$. The solenoidal eigenvectors, which are equationally defined in (27.67a) and (27.67b), are mutually orthogonal as long as they correspond to distinct eigenvalues, k'_n and k''_n, of the self-adjoint operator $\mathfrak{R}$. So, a manifold of the six-component eigenvectors, $\mathcal{J}$, is a direct sum

$$\mathcal{J} = \mathcal{J}' \oplus \mathcal{J}'' \text{ where } \mathcal{J}' = \begin{pmatrix} \{\mathbf{E}'_n\} \\ \{\mathbf{H}'_n\} \end{pmatrix} \text{ and } \mathcal{J}'' = \begin{pmatrix} \{\mathbf{E}''_n\} \\ \{\mathbf{H}''_n\} \end{pmatrix} \tag{27.84}$$

where notation $\{*\}$ symbolizes a manifold. Herein, the elements of the sets $\mathcal{J}'$ and $\mathcal{J}''$ are specified by the problems (27.67a) and (27.67b), respectively. There is a one-to-one correspondence in $\mathcal{J}'$ of the three-component vectors $\mathbf{H}'_n$ to the vectors $\mathbf{E}'_n$, which are specified via the boundary eigenvalue problem for the Laplacian in (27.67a). In turn, there is a one-to-one correspondence in $\mathcal{J}''$ of the three-component vectors $\mathbf{E}''_n$ to the vectors $\mathbf{H}''_n$, which are specified via *another* boundary eigenvalue problem for the Laplacian in (27.67b).

The *equationally definable class* of the *six*-component *irrotational* eigenvectors of the self-adjoint operator $\mathfrak{R}$ is presented in (27.69a) and (27.69b). Every eigenvector of the electric kind, $\mathring{\mathfrak{X}}^{\mathrm{E}}_m$, is orthogonal to the eigenvector $\mathring{\mathfrak{X}}^{\mathrm{H}}_m$ in accordance with their definitions. Hence, the subspace $\mathcal{G}$ of the six-component irrotational eigenvectors of $\mathfrak{R}$ is representable as a direct sum as follows:

$$\mathcal{G} = \mathcal{G}_{\mathrm{E}} \oplus \mathcal{G}_{\mathrm{H}} \text{ where } \mathcal{G}_{\mathrm{E}} = \begin{pmatrix} \{\nabla\Phi_m\} \\ \{\emptyset\} \end{pmatrix} \text{ and } \mathcal{G}_{\mathrm{H}} = \begin{pmatrix} \{\emptyset\} \\ \{\nabla\Psi_m\} \end{pmatrix}. \tag{27.85}$$

Herein, the set, which is symbolized as $\{\emptyset\}$, involves only the *single three-*component zero-valued vector. Besides, the subspace of the six-component harmonic eigenvectors of $\mathfrak{R}$ involves only the *single six*-component zero-valued vector: see (27.71).

Let us denote the multiplicity of eigenvectors of the operator $\mathfrak{R}$ as $\mathcal{M}^{(6)}$ where superscript $(^{(6)})$ symbolizes that all the elements of $\mathcal{M}^{(6)}$ are *six-component* vectors. The previous analysis shows that $\mathcal{M}^{(6)} = \mathcal{J} \oplus \mathcal{G}$. Direct summations of the subspaces given in (27.84) and (27.85) exhibit that $\mathcal{M}^{(6)}$ is $\mathcal{L}_2^{(6)}$. Indeed,

$$\boxed{\mathcal{M}^{(6)} = \begin{pmatrix} \{\mathbf{E}'_n\} \oplus \{\mathbf{E}''_n\} \oplus \{\nabla\Phi_m\} \\ \{\mathbf{H}'_n\} \oplus \{\mathbf{H}''_n\} \oplus \{\nabla\Psi_m\} \end{pmatrix} = \begin{pmatrix} \mathcal{L}_2 \\ \mathcal{L}_2 \end{pmatrix} = \mathcal{L}_2^{(6)}} \tag{27.86}$$

where $\mathcal{L}_2^{(6)}$ is the Hilbert space $\mathcal{L}_2$, the elements of which are six-component vectors, whereas $\mathcal{L}_2$ in (27.80) has the three-component vectors as its elements. From a physical point of view, the basis elements of $\mathcal{L}_2^{(6)}$ are specified completely by the shape and geometrical parameters of the cavity surface S: see (27.67a)–(27.70). So it is natural to name the set of normalized eigenvectors of the self-adjoint operator $\mathfrak{R}$ as a *configurational* basis.

27.4.5 Projecting the field vectors and Maxwell's equations onto the modal basis

The self-adjoint operator, $\mathfrak{R}$, is *diagonal* and hence, $\mathfrak{R}$ is invertible *explicitly* in the form of the eigenvector series. The series have the physical sense of modal field expansions. Every element in the series is a product of two factors. One of them is a vectorial element of the modal basis and the other one is a scalar coefficient. All elements of the orthonormal basis are already specified in (27.67a)–(27.67b) and (27.69a)–(27.69b) as functions of the coordinates, whereas the scalar coefficients in that product have physical sense of a modal amplitude dependent on time. In order to obtain a problem for the modal amplitudes, Maxwell's equations (with time derivative!) should be projected onto the same basis elements.

27.4.5.1 Projecting the field vectors onto the modal basis

The composition of the six-component electromagnetic field vector as $\mathbb{X}(\mathbf{r},t) = \mathrm{col}(\mathbb{E}(\mathbf{r},t), \mathbb{H}(\mathbf{r},t))$ is specified in (27.53). The projection of $\mathbb{X}(\mathbf{r},t)$ onto the basis can be formally written as

$$\mathbb{X}(\mathbf{r},t) = \begin{pmatrix} \mathbb{E}(\mathbf{r},t) \\ \mathbb{H}(\mathbf{r},t) \end{pmatrix} = \sum_{n=-\infty}^{\infty} c'_n(t)\, \mathfrak{X}'_n(\mathbf{r}) + \sum_{n=-\infty}^{\infty} c''_n(t)\, \mathfrak{X}''_n(\mathbf{r}) + \sum_{m=1}^{\infty} e_m(t)\, \mathring{\mathfrak{X}}^{\mathrm{E}}_m(\mathbf{r}) + \sum_{m=1}^{\infty} h_m(t)\, \mathring{\mathfrak{X}}^{\mathrm{H}}_m(\mathbf{r}) \tag{27.87}$$

where the time-dependent scalar coefficients are the modal amplitudes. The first series in (27.87) can be simplified as follows:

$$\begin{aligned}
\sum_{n=-\infty}^{\infty} c'_n(t)\, \mathfrak{X}'_n(\mathbf{r}) &= \sum_{n=1}^{\infty} [c'_n \mathfrak{X}'_n + c'_{-n} \mathfrak{X}'_{-n}] \\
&= \sum_{n=1}^{\infty} [c'_n(\mathfrak{X}'^{\mathrm{E}}_n + \mathfrak{X}'^{\mathrm{H}}_n) + c'_{-n}(\mathfrak{X}'^{\mathrm{E}}_n - \mathfrak{X}'^{\mathrm{H}}_n)] \\
&= \sum_{n=1}^{\infty} [(c'_n + c'_{-n})\, \mathfrak{X}'^{\mathrm{E}}_n + (c'_n - c'_{-n})\, \mathfrak{X}'^{\mathrm{H}}_n] \\
&= \sum_{n=1}^{\infty} \left[(c'_n + c'_{-n}) \begin{pmatrix} \mathbf{E}'_n \\ \mathbf{0} \end{pmatrix} + (c'_n - c'_{-n}) \begin{pmatrix} \mathbf{0} \\ \mathbf{H}'_n \end{pmatrix}\right] \\
&= \begin{pmatrix} \sum_{n=1}^{\infty} e'_n(t)\, \mathbf{E}'_n(\mathbf{r}) \\ \sum_{n=1}^{\infty} h'_n(t)\, \mathbf{H}'_n(\mathbf{r}) \end{pmatrix}
\end{aligned} \tag{27.88}$$

where $e'_n(t) = c'_n + c'_{-n}$, $h'_n(t) = c'_n - c'_{-n}$, and the formulas (27.62) are used. Simplification of the other series in (27.87) yields

$$\begin{aligned}
\sum_{n=-\infty}^{\infty} c''_n \mathfrak{X}''_n &= \begin{pmatrix} \sum_{n=1}^{\infty} e''_n \mathbf{E}''_n \\ \sum_{n=1}^{\infty} h''_n \mathbf{H}''_n \end{pmatrix} \quad \text{and} \\
\sum_{m=1}^{\infty} e_m \mathring{\mathfrak{X}}^{\mathrm{E}}_m + \sum_{m=1}^{\infty} h_m \mathring{\mathfrak{X}}^{\mathrm{H}}_m &= \begin{pmatrix} \sum_{m=1}^{\infty} e_m(t)\, \nabla\Phi_m(\mathbf{r}) \\ \sum_{m=1}^{\infty} h_m(t)\, \nabla\Psi_m(\mathbf{r}) \end{pmatrix}.
\end{aligned} \tag{27.89}$$

Finally, the three-component vectors $\mathbb{E}(\mathbf{r},t)$ and $\mathbb{H}(\mathbf{r},t)$ have the modal field expansions in the form of

$$\mathbb{E}(\mathbf{r},t) = \sum_{n=1}^{\infty} e'_n(t)\, \mathbf{E}'_n(\mathbf{r}) + \sum_{n=1}^{\infty} e''_n(t)\, \mathbf{E}''_n(\mathbf{r}) + \sum_{m=1}^{\infty} e_m(t)\, \nabla\Phi_m(\mathbf{r}) \tag{27.90a}$$

$$\mathbb{H}(\mathbf{r},t) = \sum_{n=1}^{\infty} h'_n(t)\, \mathbf{H}'_n(\mathbf{r}) + \sum_{n=1}^{\infty} h''_n(t)\, \mathbf{H}''_n(\mathbf{r}) + \sum_{m=1}^{\infty} h_m(t)\, \nabla\Psi_m(\mathbf{r}). \tag{27.90b}$$

Notice in passing that the modal field expansions (27.90a) and (27.90b) correlate with Weyl's theorem (27.86).

27.4.5.2 Projecting Maxwell's equations onto the same basis elements

A formal procedure of projecting Maxwell's equations is based on the identity (27.59), that is,

$$\langle \mathfrak{R}\mathfrak{X}_1, \mathfrak{X}_2\rangle - \langle \mathfrak{X}_1, \mathfrak{R}\mathfrak{X}_2\rangle = 0 \tag{27.91}$$

where both the electromagnetic vectors, $\mathfrak{X}_1$ and $\mathfrak{X}_2$, should belong to the space of solutions. So, we can take $\mathbb{X}(\mathbf{r},t)$ as $\mathfrak{X}_1$ in (27.91) and the *right-hand side* of Maxwell's equations $\mathfrak{R}\mathbb{X} = \mathrm{col}(\mathbb{U}, \mathbb{V})$ from (27.52) as the vector $\mathfrak{R}\mathfrak{X}_1$ (see (27.55)). So,

$$\mathfrak{X}_1 \equiv \mathbb{X}(\mathbf{r},t) = \begin{pmatrix} \mathbb{E}(\mathbf{r},t) \\ \mathbb{H}(\mathbf{r},t) \end{pmatrix} \text{ and } \mathfrak{R}\mathfrak{X}_1 \equiv \mathfrak{R}\mathbb{X}(\mathbf{r},t) = \begin{pmatrix} \mathbb{U}(\mathbf{r},t) + \dddot{\gamma}\, \dddot{\mathbb{E}}(\mathbf{r},t) \\ \mathbb{V}(\mathbf{r},t) \end{pmatrix} \tag{27.92}$$

where the term $\dddot{\gamma}\,\dddot{\mathbb{E}}$ is added *heuristically* as a given external force function,[‡‡] which is capable of exciting forced oscillations in the cavity. By analogy with (27.90a), it can be written in the form of a modal field expansion as

$$\dddot{\gamma}\,\dddot{\mathbb{E}} = \dddot{\gamma}\left(\sum_{n=1}^{\infty} \dddot{e}'_n(t)\, \mathbf{E}'_n(\mathbf{r}) + \sum_{n=1}^{\infty} \dddot{e}''_n(t)\, \mathbf{E}''_n(\mathbf{r}) + \sum_{m=1}^{\infty} \dddot{e}_m(t)\, \nabla\Phi_m(\mathbf{r}) \right) \tag{27.93}$$

where the constant $\dddot{\gamma}$ and the time-dependent modal amplitudes should be given. The vectors $\mathbb{U}$ and $\mathbb{V}$ were specified earlier via the field vectors $\mathbb{E}$ and $\mathbb{H}$ in (27.52) as follows:

$$\mathbb{U}(\mathbf{r},t) = \tfrac{1}{c}\partial_t \mathbb{E}(\mathbf{r},t) + 2\gamma\, \mathbb{E}(\mathbf{r},t) \text{ and } \mathbb{V}(\mathbf{r},t) = -\tfrac{1}{c}\partial_t \mathbb{H}(\mathbf{r},t). \tag{27.94}$$

In order to facilitate the calculations that follow, we represent the six-component vectors $\mathbb{X}$ and $\mathfrak{R}\mathbb{X}$ as

$$\mathbb{X}(\mathbf{r},t) = \mathbb{X}^{\mathrm{E}} + \mathbb{X}^{\mathrm{H}} \text{ where } \mathbb{X}^{\mathrm{E}} = \begin{pmatrix} \mathbb{E}(\mathbf{r},t) \\ \mathbf{0} \end{pmatrix} \text{ and } \mathbb{X}^{\mathrm{H}} = \begin{pmatrix} \mathbf{0} \\ \mathbb{H}(\mathbf{r},t) \end{pmatrix} \tag{27.95a}$$

$$\begin{aligned} \mathfrak{R}\mathbb{X}(\mathbf{r},t) &= \mathfrak{A}^{\mathrm{E}} + \mathfrak{B}^{\mathrm{H}} \text{ where } \mathfrak{A}^{\mathrm{E}} = \begin{pmatrix} \mathbb{U}(\mathbf{r},t) + \dddot{\gamma}\,\dddot{\mathbb{E}}(\mathbf{r},t) \\ \mathbf{0} \end{pmatrix} \text{ and} \\ \mathfrak{B}^{\mathrm{H}} &= \begin{pmatrix} \mathbf{0} \\ -\tfrac{1}{c}\partial_t \mathbb{H}(\mathbf{r},t) \end{pmatrix} \end{aligned} \tag{27.95b}$$

[‡‡]The physical dimension of $\dddot{\gamma}$ and $\dddot{\mathbb{E}}$ should each be of inverse meter, $\lfloor 1/\mathrm{m} \rfloor$.

where $\mathbf{0}$ is the three-component zero-valued vector. Thus, the identity (27.91) takes the form

$$\langle(\mathfrak{A}^{\mathrm{E}}+\mathfrak{B}^{\mathrm{H}}),\mathfrak{X}_2\rangle-\langle(\mathbb{X}^{\mathrm{E}}+\mathbb{X}^{\mathrm{H}}),\mathfrak{R}\mathfrak{X}_2\rangle=0. \tag{27.96}$$

As for the vector $\mathfrak{X}_2$ in (27.96), one should treat sequentially all the elements of the modal basis, that is, $\mathfrak{X}'_n$, $\mathfrak{X}''_n$, where $n=1,2,\dots$, and $\mathring{\mathfrak{X}}^{\mathrm{E}}_m$, $\mathring{\mathfrak{X}}^{\mathrm{H}}_m$, where $m=1,2,\dots$. First, let $\mathfrak{X}_2\equiv\mathfrak{X}'_n=\mathrm{col}(\mathbf{E}'_n,\mathbf{H}'_n)$. As for the vector $\mathfrak{R}\mathfrak{X}_2$ in (27.96), one can use the operator eigenvalue equation, $\mathfrak{R}\mathfrak{X}'_{\pm n}=k'_{\pm n}\mathfrak{X}'_{\pm n}$, as a direct formula. In accordance with the mirror symmetry of the eigenvalues, $k_{\pm n}\neq 0$, and the appropriate eigenvectors, $\mathfrak{X}_{\pm n}$ (see (27.61) and (27.62)), one can write down

$$k'_{+n}\equiv k'_n>0:\ \mathfrak{X}'_{+n}=\mathfrak{X}'^{\mathrm{E}}_n+\mathfrak{X}'^{\mathrm{H}}_n$$

$$\text{where } \mathfrak{X}'^{\mathrm{E}}_n=\begin{pmatrix}\mathbf{E}'_n(\mathbf{r})\\ \mathbf{0}\end{pmatrix},\ \mathfrak{X}'^{\mathrm{H}}_n=\begin{pmatrix}\mathbf{0}\\ \mathbf{H}'_n(\mathbf{r})\end{pmatrix},\ \mathfrak{R}\mathfrak{X}'_{+n}=k'_n\left(\mathfrak{X}'^{\mathrm{E}}_n+\mathfrak{X}'^{\mathrm{H}}_n\right) \tag{27.97a}$$

$$k'_{-n}=-k'_n:\ \mathfrak{X}'_{-n}=\mathfrak{X}'^{\mathrm{E}}_n-\mathfrak{X}'^{\mathrm{H}}_n$$

$$\text{where } \mathfrak{X}'^{\mathrm{E}}_n=\begin{pmatrix}\mathbf{E}'_n(\mathbf{r})\\ \mathbf{0}\end{pmatrix},\ \mathfrak{X}'^{\mathrm{H}}_n=\begin{pmatrix}\mathbf{0}\\ \mathbf{H}'_n(\mathbf{r})\end{pmatrix},\ \mathfrak{R}\mathfrak{X}'_{-n}=k'_n\left(\mathfrak{X}'^{\mathrm{H}}_n-\mathfrak{X}'^{\mathrm{E}}_n\right) \tag{27.97b}$$

where $k'_n\neq 0$ corresponds to the case of the *solenoidal* modes.

Substitution of (27.97a)–(27.97b) in (27.96) and simple algebraic manipulations result in

$$\langle\mathfrak{A}^{\mathrm{E}},\mathfrak{X}'^{\mathrm{E}}_n\rangle-k'_n\langle\mathbb{X}^{\mathrm{H}},\mathfrak{X}'^{\mathrm{H}}_n\rangle=0 \text{ and } \langle\mathfrak{B}^{\mathrm{H}},\mathfrak{X}'^{\mathrm{H}}_n\rangle-k'_n\langle\mathbb{X}^{\mathrm{E}},\mathfrak{X}'^{\mathrm{E}}_n\rangle=0. \tag{27.98}$$

Substitution of the functions $\mathfrak{A}^{\mathrm{E}}$ and $\mathfrak{B}^{\mathrm{H}}$ in equations (27.98), in which the modal field expansions (27.90a)–(27.90b) and (27.93) should be used, yields a pair of the evolutionary equations for the sought modal amplitudes, e'_n and h'_n,

$$\tfrac{1}{c}\tfrac{d}{dt}e'_n(t)+2\gamma\, e'_n(t)-k'_nh'_n(t)=-\dddot{\gamma}\,\ddot{e}'_n(t) \text{ and } \tfrac{1}{c}\tfrac{d}{dt}h'_n(t)+k'_ne'_n(t)=0. \tag{27.99}$$

If one now takes $\mathfrak{X}_2$ in (27.96) as $\mathfrak{X}_2\equiv\mathfrak{X}''_n=\mathrm{col}(\mathbf{E}''_n,\mathbf{H}''_n)$ and $\mathfrak{R}\mathfrak{X}_2$ as $\mathfrak{R}\mathfrak{X}''_{\pm n}=k''_{\pm n}\mathfrak{X}''_{\pm n}$, then similar manipulations result in another pair of evolutionary equations for the sought modal amplitudes, $e''_n(t)$ and $h''_n(t)$,

$$\tfrac{1}{c}\tfrac{d}{dt}e''_n(t)+2\gamma\, e''_n(t)-k''_nh''_n(t)=-\dddot{\gamma}_n\,\ddot{e}''_n(t) \text{ and } \tfrac{1}{c}\tfrac{d}{dt}h''_n(t)+k''_ne''_n(t)=0. \tag{27.100}$$

In the modal field expansion (27.87) (and its equivalent (27.90a) and (27.90b)), the basis elements $\mathring{\mathfrak{X}}^{\mathrm{E}}_m$ and $\mathring{\mathfrak{X}}^{\mathrm{H}}_m$ participate. They are specified in (27.69a) and (27.69b) and correspond to the *irrotational* modes. In these cases, the operator eigenvalue equation takes the following form: $\mathfrak{R}\mathring{\mathfrak{X}}^{\mathrm{E}}_m=\mathbf{0}$ and $\mathfrak{R}\mathring{\mathfrak{X}}^{\mathrm{H}}_m=\mathbf{0}$. Substitution of the latter (both) in (27.96) as $\mathfrak{R}\mathfrak{X}_2$ yields, respectively,

$$\langle\mathfrak{A}^{\mathrm{E}},\mathring{\mathfrak{X}}^{\mathrm{E}}_m\rangle=0 \text{ and } \langle\mathfrak{B}^{\mathrm{H}},\mathring{\mathfrak{X}}^{\mathrm{H}}_m\rangle=0. \tag{27.101}$$

Calculations of the inner products in (27.101) yield the evolutionary equations for the modal amplitudes, e_m and h_m, which participate in the modal field expansions (27.90a) and (27.90b),

$$\frac{1}{c}\frac{d}{dt}e_m(t) + 2\gamma\, e_m(t) = -\dddot{\gamma}_m\, \dddot{e}_m(t) \text{ and } \frac{1}{c}\frac{d}{dt}h_m(t) = 0 \tag{27.102}$$

where $m = 1, 2, \ldots$ and $\dddot{\gamma}_m, \dddot{e}_m$ should be known in the modal expansion (27.93) for a given external applied field.

27.4.5.3 Exact explicit solutions for the modal amplitudes

The solution to the second equation in (27.102) is evident, that is, $h_m(t) = C$ where C is a constant of integration. If at time $t = 0$ the initial condition is given as $h_m(0) = 0$, then $C = 0$ and hence, $h_m(t) = 0$ for $t \geq 0$. The solution to the first equation in (27.102) has the form

$$e_m(t) = -\dddot{\gamma}_m c \int_0^t e^{2\gamma c(\tau - t)}\dddot{e}_m(\tau)\, d\tau \quad \text{for } t \geq 0 \tag{27.103}$$

provided that the initial condition at $t = 0$ is chosen as $e_m(0) = 0$.

The pair of simultaneous equations (27.99) should be supplemented with initial conditions (say, at time $t = 0$), which we choose as $e'_n(0) = 0$ and $h'_n(0) = 0$. In total, this yields the Cauchy problem, which can be solved by the method of the matrix exponential. The final solution to the Cauchy problem for $t \geq 0$ has the form

$$\begin{aligned} e'_n(t) &= \mathcal{U}_n(t)\cos(\theta + \varpi_n t) + \mathcal{V}_n(t)\sin(\varpi_n t) \quad \text{and} \\ h'_n(t) &= \mathcal{V}_n(t)\cos(\theta - \varpi_n t) - \mathcal{U}_n(t)\sin(\varpi_n t) \end{aligned} \tag{27.104}$$

where $\varpi_n = k'_n c\sqrt{1 - (\gamma/k'_n)^2}$, $\theta = \sin^{-1}(\gamma/k'_n) = \cos^{-1}(\sqrt{1 - (\gamma/k'_n)^2})$, $k'_n > 0$, $\gamma \geq 0$, and also

$$\begin{aligned} \mathcal{U}_n(t) &= -\frac{\dddot{\gamma}\, c}{\cos^2\theta}\int_0^t e^{\gamma c(\tau - t)}\dddot{e}'_n(\tau)\cos(\theta - \varpi_n \tau)\, d\tau \quad \text{and} \\ \mathcal{V}_n(t) &= -\frac{\dddot{\gamma}\, c}{\cos^2\theta}\int_0^t e^{\gamma c(\tau - t)}\dddot{e}'_n(\tau)\sin(\varpi_n \tau)\, d\tau. \end{aligned} \tag{27.105}$$

The system of evolutionary equations (27.100) (being supplemented with the same homogeneous initial conditions) yields a similar solution to (27.99)–(27.105) with due changes to notation.

27.5 Present state of art and recent advances

Now we provide a concise overview of different branches that grew from the idea of mode basis expansions used directly in time domain for solving transient problems.

One can find implementation of the method available for studying a wide class of time-variant hollow cavity systems in [22]. Explicit solutions are obtained for the cavity oscillations excited by a set of the digital signals [23]. Excitation of the electromagnetic fields by a wide-band current surge in a hollow cavity is studied in [24], and in a cavity filled with Debye and Lorentz kind dispersive medium in [25], and is presented in [26–31]. Herein, the Debye and Lorentz equations play the role of the dynamic constitutive relation between the polarization vector and the electric field. In [32], the instantaneous and dynamic component parts of polarization for Debye medium are presented separately. Recently, the interaction of the cavity fields with plasma is studied in the time domain, making use of the motion equation for plasma in [33–36].

The cavity analysis proposed here was generalized to a homogeneous filling with dispersive medium, and the derivations are compared with the FDTD calculations [37]. A cavity with double negative medium was considered in [38] as an example. It was shown that due to the interaction of a spatial resonance in the cavity (due to waves bouncing between the walls) with local medium resonances (due to molecular oscillations), there are several possible frequency resonances for a single spatial mode distribution. In [39] some problems with transient homogeneous medium in a cavity were considered.

As was already mentioned, following the pioneering work by Kysun'ko [17] the method was further pushed into the waveguide realm in papers [12,13]. For a regular hollow waveguide this formulation leads to a mode expansion with modes being the same as are commonly used in FD, while the mode amplitudes being governed by the Klein–Gordon equation (KGE) (κ_n is the eigenvalue of the nth mode):

$$\left(c^{-2}\partial_t^2 - \partial_z^2 + \kappa_n^2\right)f_n(z,t) = 0. \tag{27.106}$$

The excitation and propagation problem of digital signals in a hollow waveguide is considered in [40] making use of the method. Solving the KGE explicitly, a special case has been considered in [41] where the modal amplitudes of the waveguide fields are expressible explicitly via the Airy functions from mathematical physics. The complete set of *TE*—and *TM*—time-domain modal waves is established and studied in detail in [42,43]. As a result of dealing with real-valued functions in the time domain, power flow and energy densities of the modal waveforms are obtained explicitly in [44,45]. The velocity of transportation of the modal field energy, defined by the power flow and energy densities, is derived as a function of time and axial coordinate and is presented in [46–49]. The electromagnetic inertia, $\mathfrak{m}(\mathbf{r},t)$, and the mechanical momentum, $\mathfrak{p}(\mathbf{r},t)$, of the electromagnetic fields for any closed cylindrical waveguide are obtained and presented in [50–56].

Within this approach, the problems of short pulse propagation and diffraction from a permittivity step in a waveguide were solved analytically in [57] with a detailed time-domain energy flow analysis. Later, in [58], the same problem was considered for a conductive medium boundary with emphasis on the effect of transient charge displacement by a pulsed E-wave.

Next the method was applied to free space considered as a regular structure in a cylindrical coordinate system. Modes for this case had already been found in the FD. In [59,60] this approach was shifted to the time domain for analysis of transient beam radiation and propagation using an expansion over Bessel modes with mode amplitudes being governed by the Klein–Gordon equation (KGE) formulated earlier. The main peculiarity in this case is that since the transverse domain is now unbounded, the mode spectrum is continuous so that leads to integrals instead of sums in the mode expansion. In [61] the approach was applied to a physical analysis of transient waveform curvature transformation in the near-field zone of a pulsed wave beam. Finally, in [62], the diffraction operator approach of [59,60] was united with the beam description method for the beam diffraction problem.

The next main improvement of the method for waveguides was introduced in [63], where in a inhomogeneous waveguide the permittivity and permeability functions of the filling are factorized as $\varepsilon(\mathbf{r},t) = \varepsilon_{\perp}(x,y)\,\varepsilon_{\parallel}(z,t)$, i.e., as some function of transverse coordinates (which goes into the BVP for the modes) and a function of the longitudinal coordinate and time. The latter goes into the evolutionary waveguide equations as coefficients that govern the modes' amplitude transformation with propagation. In this case, the mode coupling is unavoidable, and a very brief explanation of it is given in [64]. A detailed description of the approach is available in [65]. In this case the governing equations can be considered as the KGE with matrix coefficients [65]:

$$\left(\mathbf{K}\partial_{\tau}^{2} - \mathbf{L}\partial_{z}^{2} + \mathbf{Q}^{2}\right)\mathbf{e} = 0. \tag{27.107}$$

Here $\mathbf{K}$, $\mathbf{L}$ are the matrices of mode coupling, which degenerate into unit matrices if the inhomogeneity is eliminated in a continuous way, $\mathbf{Q}$ is a diagonal matrix of eigenvalues, and $\mathbf{e}(z,t) = \mathrm{col}(\mathrm{e}_1\,(\mathrm{z},\mathrm{t}), \mathrm{e}_2\,(\mathrm{z},\mathrm{t}), \ldots)$ is a column-vector of mode amplitudes. Thus, the chosen basis diagonalizes one of three matrices in this equation leading to minimum coupling, because for a band limited signal only a finite number of coupled modes should be taken into account, and adding more modes results in exponentially small corrections [65–67]. Thus, the sought fields are presented as a series over "modes," which are eigenfunctions of some BVP in transverse coordinates [67]:

$$\mathbb{E}(\mathbf{r},t) = \sum_{n} \upsilon_n(z,t)\,\mathbf{E}_n(x,y) + \mathbf{z}_0 \sum_{n} e_n(z,t) f_n(x,y) \tag{27.108}$$

where $\mathbf{E}_n$ are the vector mode distributions of the TE field, and f_n are the scalar mode distributions of the electric field longitudinal component. It should be noted that, despite the waveguide being inhomogeneous, the modes are frequency independent since no conception of frequency is introduced yet. In [63,68] it was proposed to use these frequency-independent "modes" to present the classical modes known in FD. To this end, we seek the solution to the matrix KGE above in the form $e(z,t) \sim \exp(-ikr + i\beta z)$; $\tau = ct$; $k = \omega/c$. Then the KGE transforms from a PDE to a homogeneous infinite system of linear algebraic equations; equating its determinant to zero, we obtain a dispersion equation for the waveguide in FD in the following form:

$$\det\left(\mathbf{K}k^2 - \mathbf{L}\beta^2 - \mathbf{Q}^2\right) = 0. \tag{27.109}$$

When these matrices are truncated to account for a finite number N of modes, the determinant is a polynomial in k^2, β^2, and its roots determine the approximation for the first N FD modes. In contrast to the transcendental equations typically obtained in FD, many analytical instruments are available for the analysis of these polynomial form dispersion equations.

The next step in developing the mode basis method has been made in the direction of spherical geometry. Expansion in spherical harmonics was among the first problems solved analytically in FD [69]. In the TD, propagation of such modes is governed by the Klein–Gordon–Fock equation (KGFE):

$$\left(\partial_\tau^2 - \partial_r^2 + b_n r^{-2}\right) e_n(r,t) = 0. \qquad (27.110)$$

Comparing this equation to the KGE for a cylindrical waveguide one can see that now the parameter that determines the cutoff condition reduces with the radial coordinate. It means that near the origin most of the spectrum is below cutoff and only outside some radius do the waves become propagating. In [70], a generalized approach for considering mode decompositions in the time domain for regular conical structures with angular and radial inhomogeneous filling was proposed. It is built similarly with the same technique in cylindrical coordinates, but now the radial coordinate is chosen as the longitudinal (propagation) coordinate, while the cross-section is determined in the angular domain. The filling is supposed to be in the factorized form as $\varepsilon(\mathbf{r},t) = \varepsilon_\perp(\theta,\varphi)\,\varepsilon_\parallel(r,t)$. The resulting governing equations are the KGFE with matrix coefficients describing mode coupling. In the case of radial inhomogeneity, there is no coupling, and a single mode transient analysis allows one to study problems such as radial layers in a biconical line [71,72] or radiation of a dipole in a dielectric sphere [73]. The latter problem is closely related to that of the diffraction from a permittivity step in a waveguide mentioned earlier [57]. It should be noted that for free space the separation parameter has specific values, $b_n = n(n+1)$, that result in closed-form presentations for incoming/outgoing spherical multipole waves; in contrast for an arbitrary conical line, this parameter is determined from a BVP that can yield any real numbers. The study of this equation conducted in [74] revealed classes of transient solutions to this equation with non-separated coordinates.

Next, we should mention the more complex diffraction problem of junctions of waveguides with different cross-sections. In FD, this kind of problems can be treated with the mode matching technique [75]. A similar technique was proposed in the TD [76,77] for solving not only waveguide diffraction problems but also the problem of transient wave diffraction at the junction of a biconical line with free space that actually is the antenna radiation problem [72,78]. The technique was also used for modeling diffraction at the end of a dielectric rod antenna in cylindrical coordinates [79].

The final work to mention is the most complex in formulation—that of open dielectric waveguides [80]. This case combines the complexity of mode coupling due to the inhomogeneous dielectric with a continuous spectrum due to the open boundary, thus leading to coupled integral–differential equations. In some cases, the BVP in such structures can have both a discrete and a continuous spectrum [81]. The analytical method for solving the BVP in this case is described in [82]. This approach can be combined with the above-described technique for building dispersion relations in FD for fast calculation and analysis of FD modes in fibers. When analyzing transient wave propagation in a circular dielectric waveguide [83] with this technique, a new effect of fast pulse guided wave propagation has been revealed [84,85].

27.6 Ongoing and future research

This approach can be applied to solve the Maxwell equations in format (27.9) simultaneously with the motion equation (27.10), where the plasma frequency, ω_p^2, is time dependent. In an electromagnetic sense, this means that the relative permittivity of the dielectric, $\varepsilon \equiv \varepsilon(t)$, is a function of time. Such materials are important for development of nonreciprocal components with modulated capacitors.

Among unsolved problems and ongoing research within the framework of the mode basis method, we should mention works by Dumin in considering nonlinear phenomena for wave beams in open space [86,87]. Next, the physical analysis of non-separable transient waves revealed in [74] yet to be done. Mode matching in the time-domain technique [77] was implemented in its simplest form thus far, and its advancement based on recent progress in this field [88] as well as using new mathematical tools such as frames (a generalization of basis in vector space) is also on the way. The problems of transient radiation from small dipole antennas depicted in the conclusion section of [73] still require completion. The next big step that should be done is in the direction of non-regular waveguides; the very first steps are designated in [89]. Studying transient excitation of non-regular conical waveguides is very promising from the point of view of designing short pulse antennas based on flared horns, tapered slot antennas, Vivaldi antennas, Volcano antennas, and similar TEM-conical line-based structures. One further subject requiring deeper study is the analysis (and synthesis) of dispersion relations for an inhomogeneous waveguide, based on expansions over frequency-independent modes [68]. Finally, we should mention our ongoing research on the application of this technique to the analysis of periodic structures (Floquet channels), with the aim of modeling metamaterials, frequency selective surfaces, etc.

References

[1] Carron, N.J., "Babel of Units. The evolution of units systems in classical electromagnetism," 2015. http://arxiv.org/abs/1506.01951v1

[2] Newell, D.B. and Tiesinga, E., "The International System of Units (SI)," Gaithersburg, MD: US National Institute of Standards and Technology (NIST) Special Publication 330, 2019.

[3] Rothwell, E.J. and Cloud, M.J., "Electromagnetics," New York, NY: CRC Press, 2001.

[4] Umov, N.A., "Ein Theorem über die Wechselwirkung in endlichen Entfernungen," Zeitschrift für Mathematik und Physik, vol. 19, pp. 97–114, 1874.

[5] Poynting, J.H., "On the transfer of energy in the electromagnetic field," Philosophical Transactions of the Royal Society of London, vol. 175, pp. 343–361, 1884.

[6] Kaiser, G., "Electromagnetic inertia, reactive energy and energy flow velocity," Journal of Physics A: Mathematical and Theoretical, vol. 44, 345206, 15 pp., 2011.

[7] Landau, L.D. and Lifshitz, E.M., "The classical theory of fields," the 3rd edition, Elmsford, NY: Pergamon Press Ltd., 1971.
[8] Lappo-Danilevsky, I.A., "Theory of functions on matrices and systems of linear differential equations," (in Russian), Leningrad: ONTI State Techn.-Theor. Press, 1934.
[9] Hayashi, S., "Surges on transmission systems," Kyoto: Denki-Shoin Inc., 1955.
[10] Moler, C. and Van Loan, C., "Nineteen dubious ways to compute the exponential of a matrix, twenty-five years later," SIAM Review, 45, no. 1, pp. 3–49, 2003.
[11] Strutt, J.W., "On the passage of electric waves through tubes, or the vibrations of dielectric cylinders," Philosophical Magazine, vol. 43, pp. 125–132, 1897.
[12] Tretyakov, O.A., "Evolutionary waveguide equations," Radiotekhnika i Electronika, (in Russian), vol. 34, no. 5, pp. 917–926, 1989.
[13] Tretyakov, O.A., "Evolutionary waveguide equations," Proceedings of the IEEE AP-S Int. Symp., Seattle, WA, USA, pp. 2465–2471, 1994.
[14] Tretyakov, O.A., "Essentials of nonstationary and nonlinear electromagnetic field theory," Chapter 3 in Analytical and Numerical Methods in Electromagnetic Wave Theory, M. Hashimoto, M. Idemen, and O.A. Tretyakov, Eds., Tokyo: Science House Co. Ltd., 1993.
[15] Tretyakov, O.A. and Erden, F., "Evolutionary approach to electromagnetics as an alternative to the time-harmonic field method," IEEE AP-S/USNC URSI Symposium, Chicago, IL, USA, 2012. http://dx.doi.org/10.13140/2.1.2283.4242.
[16] Slater, J.C., "Microwave electronics," Reviews of Modern Physics, vol. 18, no 1, pp. 441–512, 1946.
[17] Kysun'ko, G.V., "Electrodynamics of hollow systems," (in Russian), Leningrad: VKAS, 426 p., 1949.
[18] Nazyrov, Z.F. and Shmatko, A.A., "Method of mode basis in the problems of electrodynamics with piecewise-homogeneous nonstationary media," Radiotekhnika i Elektronika (in Russian), 1988, vol. 33, no. 5, pp. 104–1049, 1988.
[19] Nazirov, Z.F. and Shmatko, A.A. "Separation of nonsinusoidal oscillations in resonant cavities with nonstationary medium," Radiotekhnika i Elektronika (in Russian), 1988, vol. 33, no. 5, pp. 107–1081, 1988.
[20] Weyl, H., "The method of orthogonal projection in potential theory," Duke Mathematical Journal, vol. 7, pp. 411–444, 1940.
[21] Morse, P.M. and Feshbach, H., "Methods of theoretical physics," Parts 1 and 2, New York, NY: McGraw-Hill, 1953.
[22] Aksoy, S. and Tretyakov, O.A., "Study of a time variant cavity system," Journal of Electromagnetic Waves and Applications, vol. 16, pp. 1535–1553, 2002.
[23] Aksoy, S. and Tretyakov, O.A., "The evolution equations in study of the cavity oscillations excited by a digital signal," IEEE Transactions on Antennas and Propagation, vol. 52, pp. 263–270, 2004.

[24] Erden, F. and Tretyakov, O.A., "Excitation by a transient signal of the real-valued electromagnetic fields in a cavity," Physical Review E, vol. 77, pp. 056605, 2008.

[25] Tretyakov, O.A. and Erden, F., "Temporal cavity oscillations caused by a wide-band waveform," Progress in Electromagnetics Research B, vol. 6, pp. 183–204, 2008.

[26] Tretyakov, O.A. and Erden, F., "Temporal cavity oscillations caused by a wide-band double-exponential waveform," 4th International Workshop on Electromagnetic Wave Scattering, Gebze, Turkey, 2006.

[27] Erden, F. and Tretyakov, O.A., "Temporal evolution of the irrotational and solenoidal cavity modes," XXXth URSI General Assembly and Scientific Symposium, Istanbul, Turkey, 2011.

[28] Erden, F. and Tretyakov, O.A., "Analytical approach for studying a time-domain cavity problem," IEEE Antennas and Propagation Society International Symposium (APSURSI), Memphis, TN, USA, 2014.

[29] Erden, F. and Tretyakov, O.A., "Studying oscillations caused by a current surge via Evolutionary Approach to Electromagnetics", International Conference on Software, Telecommunications and Computer Networks, Split, Croatia, 2014. http://doi.org/10.13140/2.1.3548.6722.

[30] Erden, F., Tretyakov, O.A. and Bicer, U., "Evolution of the solenoidal and irrotational modal fields in a cavity excited by a signal", (in Turkish), Signal Processing and Communication Application Conference (SIU), Zonguldak, Turkey, 2016.

[31] Erden, F. and Tretyakov, O.A., "Causal dynamic constitutive relations for lossy dispersive dielectrics," URSI Commission B International Symposium on Electromagnetic Theory, San Diego, USA, 2019.

[32] Tretyakov, O.A. and Erden, F., "Separation of the instantaneous and dynamic polarizations in studies of dispersive dielectrics," The Sixth International Kharkov Symposium on Physics and Engineering of Microwaves, Millimeter and Submillimeter Waves, Kharkov, Ukraine, vol. 1, pp. 42–48, 2007.

[33] Erden, F., "Evolutionary approach to solve a novel time-domain cavity problem," IEEE Transactions on Antennas and Propagation, vol. 65, no. 11, pp. 5918–5931, 2017.

[34] Erden, F., "Study of the surge signals in a plasma-filled rectangular cavity," Physics of Wave Phenomena, vol. 26, no. 2, pp. 139–149, 2018.

[35] Erden, F., "Evolution of a double-exponential pulse signal in a rectangular cavity filled with a plasma," International Conference on Software, Telecommunications and Computer Networks (SoftCOM), Split, Croatia, 2015.

[36] Erden, F. and Bicer, U., "Evolution of the fields excited by a pulse signal in a cavity filled with a plasma," (in Turkish), Dokuz Eylul University Engineering Faculty Science and Engineering Journal, vol. 18, no. 2:53, pp. 164–177, 2016. http://doi.org/10.21205/deufmd.20165318378

[37] Aksoy, S., Antyufeyeva, M., Basaran, E., Ergin, A.A. and Tretyakov, O.A., "Time-domain cavity oscillations supported by a temporally dispersive dielectric," IEEE Transactions on Microwave Theory and Techniques, vol. 53, no. 8, pp. 2465–2471, 2005.

[38] Antyufeyeva, M.S., Butrym, A.Y. and Tretyakov, O.A., "Transient electromagnetic fields in a cavity with dispersive double negative medium," Progress in Electromagnetics Research M, vol. 8, pp. 51–65, 2009.

[39] Antyufeyeva, M.S. and Tretyakov, O.A., "Electromagnetic fields in a cavity filled with some nonstationary media," Progress in Electromagnetics Research B, vol. 19, pp. 177–203, 2010.

[40] Aksoy, S. and Tretyakov, O.A., "Evolution equations for analytical study of digital signals in waveguides," Journal of Electromagnetic Waves and Applications, vol. 17, pp. 1665–1682, 2003.

[41] Tretyakov, O.A. and Akgun, O., "Derivation of Klein-Gordon equation from Maxwell's equations and study of relativistic time-domain waveguide modes," PIER, vol. 105, pp. 171–191, 2010.

[42] Tretyakov, O.A. and Kaya, M., "The real-valued time-domain TE-modes in lossy waveguides," PIER, vol. 127, pp. 405–426, 2012.

[43] Tretyakov, O.A. and Kaya, M., "Time-domain real-valued TM-modal waves in lossy waveguides," PIER, vol. 138, pp. 675–696, 2013.

[44] Akgun, O. and Tretyakov, O.A., "Solution to the Klein–Gordon equation for the study of time-domain waveguide fields and accompanying energetic processes," IET Microwaves, Antennas & Propagation, vol. 9, pp. 1337–1344, 2015.

[45] Erden, F., "Study of the energetic field characteristics of the TE-modal waves in waveguides," Turkish Journal of Physics, vol. 41, pp. 47–54, 2017.

[46] Erden, F., Cosan, A.A. and Tretyakov, O.A., "Properties of the time-domain waveguide modes," IEEE Antennas and Propagation Society International Symposium (APSURSI), Fajardo, Puerto Rico, 2016.

[47] Tretyakov, O.A. and Erden, F., "Evolution of the quadratic functions of the time-domain waveguide fields," URSI Commission B International Symposium on Electromagnetic Theory, Espoo, Finland, 2016.

[48] Erden, F. and Tretyakov, O.A., "Analytical study of the TE–waveguide modes in time domain," IEEE International Symposium on Antennas and Propagation & USNC/URSI National Radio Science Meeting, San Diego, CA, USA, 2017.

[49] Cosan, A.A., Erden, F. and Tretyakov, O.A.,"Derivation of energetic properties of the waveguide fields in time-domain," Journal of the Faculty of Engineering and Architecture of Gazi University (in Turkish), vol. 34, no.1, pp. 327–338, 2019. http://dx.doi.org/10.17341/gazimmfd.416495.

[50] Tretyakov, O.A., "Factorizing physical dimensions of the quantities ingressed in Maxwell's equations in SI units," PIERS-2017, St Petersburg, Russia, May 2017. http://dx.doi.org/10.13140/RG.2.2.30148.73604.

[51] Erden, F. and Tretyakov, O.A., "Mechanical properties of the waveguide modal fields in the time domain," PIERS-2017, St Petersburg, Russia, May 2017. http://dx.doi.org/10.13140/RG.2.2.20636.18569.

[52] Erden, F., Tretyakov, O.A. and Cosan, A.A.,"Inertial properties of the TE waveguide fields," Progress in Electromagnetics Research M (PIER M), vol. 68, pp. 11–19, 2018.

[53] Erden, F, Cosan, A.A. and Tretyakov, O.A., "Time-domain energetic properties of the TM-modes in a lossy waveguide", (in Turkish), IEEE Signal Processing and Communications Applications Conference, Antalya, Turkey, 2017.

[54] Erden, F. and Tretyakov, O.A., "Electromagnetic inertia of the waveguide modes," URSI Atlantic Radio Science Conference, Gran Canaria, Spain, 2018. http://doi.org/10.13140/RG.2.2.30508.31367

[55] Erden, F., Cosan, A.A. and Tretyakov, O.A., "Massic properties of the TE-modes in a lossless waveguide", (in Turkish), IEEE Signal Processing and Communications Applications Conference, Izmir, Turkey, 2018.

[56] Tretyakov, O.A. and Erden, F., "A new simple format of Maxwell's equations in SI units", IEEE International Kharkov Symposium on Physics and Engineering of Microwaves, Millimeter and Submillimeter Waves, Kharkiv, Ukraine, 2020. http://doi.org/10.13140/RG.2.2.27799.80806

[57] Butrym, A.Y., Zheng, Y. and Tretyakov, O.A., "Transient diffraction on a permittivity step in a waveguide: closed-form solution in time domain," Journal of Electromagnetic Waves and Applications, vol: 18, no. 7, pp. 861–876, 2004.

[58] Butrym, A.Y. and Legenkiy, M.N., "Charge transport by a pulse E-wave in a waveguide with conductive medium," Progress In Electromagnetics Research B, vol. 15, pp. 325–346, 2009.

[59] Tretyakov, O.A. and Dumin, A.N., "Emission of nonstationary electromagnetic fields by a plane radiator," Telecommunications and Radio Engineering, vol. 54, no. 1, 2000.

[60] Dumin, O.M., Dumina, O.O., Zheng, Y., Shishkova, A.V. and Katrich, V.O., "Transient radiation from the open end of circular waveguide," IEEE 2007 6th International Conference on Antenna Theory and Techniques, Sevastopol, Ukraine, pp. 181–183, 2007.

[61] Butrym, A. and Zheng, Y., "Time and amplitude centre of an impulse wavebeam," IEEE The Third International Conference on Ultrawideband and Ultrashort Impulse Signals, Sevastopol, Ukraine, pp. 350–352, 2006.

[62] Butrym, A.Y., Zheng, Y., Dumin, A.N. and Tretyakov O.A., "Transient wave beam diffraction by lossy dielectric half space," IEEE 10th International Conference on Mathematical Methods in Electromagnetic Theory, Dnepropetrovsk, Ukraine, pp. 321–323, 2004.

[63] Butrym, A.Y. and Tretyakov, O.A., "Electromagnetic signals in a waveguide filled with an inhomogeneous time-variant medium," IEEE International Conference on Mathematical Methods in Electromagnetic Theory, Kiev, Ukraine, pp. 213–215, 2002.

[64] Butrym, A.Y., "Mode expansions in time domain," IEEE 6th International Conference on Ultrawideband and Ultrashort Impulse Signals (UWBUSIS), Sevastopol, Ukraine, pp. 47–50, 2012.

[65] Butrym, A.Y. and Kochetov, B.A., "Time domain mode basis method for a waveguide with transverse inhomogeneous multi-connected cross-section, 1. The general theory of the method," Radio Physics and Radio Astronomy, vol. 14, no. 2, p. 162, 2012.

[66] Kochetov, B.A. and Butrym, A.Y., "Rigorous calculation of ultra short pulse propagation in a shielded microstrip line using coupled mode expansion in time domain," 2012 6th International Conference on Ultrawideband and Ultrashort Impulse Signals (UWBUSIS), Sevastopol, Ukraine, pp. 284–287, 2012.

[67] Butrym, A.Y. and Kochetov, B.A., "Time domain mode basis method for a waveguide with transverse inhomogeneous multi-connected cross-section. 2. Example of numerical implementation of the method," Radio Physics and Radio Astronomy, vol. 14, no. 3, p. 266, 2012.

[68] Butrym, A.Y. and Tretyakov, O.A., "Application of evolutionary waveguide equations method to analysis of inhomogeneous waveguides in frequency domain," Bulletin of Karazin Kharkiv National University, vol. 570, no. 2, pp. 284–286, 2002.

[69] Hansen, W.W., "A new type of expansion in radiation problems," Physical Review, vol. 47, no. 2, p. 139, 1935.

[70] Butrym, A.Y. and Kochetov, B.A., "Mode expansion in time domain for conical lines with angular medium inhomogeneity," Progress in Electromagnetics Research B, vol. 19, pp. 151–176, 2010.

[71] Kochetov, B.A. and Butrym, A.Y., "Axially symmetric transient electromagnetic fields in a radially inhomogeneous biconical transmission line," Progress in Electromagnetics Research B, vol. 48, pp. 375–394, 2013.

[72] Legenkiy, M.M., "Features of biconical antenna pulse radiation," Radio Physics and Radio Astronomy, vol. 20, no. 2, pp. 133, 2015.

[73] Butrym, A.Y. and Legenkiy, M.N., "Time domain analysis of transient radiation of dipole in dielectric sphere," IEEE International Conference on Antenna Theory and Techniques (ICATT), Kharkiv, Ukraine, pp. 1–3, 2015.

[74] Kochetov, B.A., "Lie group symmetries and Riemann function of Klein–Gordon–Fock equation with central symmetry," Communications in Nonlinear Science and Numerical Simulation, vol. 19, no. 6, pp. 1723–1728, 2014.

[75] Mittra, R. and Lee, S.-W., "Analytical techniques in the theory of guided waves," New York, NY: The Macmillan Company, 1971.

[76] Legenkiy, M.N., Zheng, Y. and Butrym, A.Y., "Mode-matching in time domain," IEEE 4th International Conference on Ultrawideband and Ultrashort Impulse Signals, Sevastopol, Ukraine, pp. 225–227, 2008.

[77] Legenkiy, M.N. and Butrym, A.Y., "Method of mode matching in time domain," Progress in Electromagnetics Research B, vol. 22, pp. 257–283, 2010.

[78] Butrym, A.Y., Kochetov, B.A. and Legenkiy, M.N., "Numerical analysis of simple TEM conical-like antennas using mode matching in time domain," IEEE 3rd European Conference on Antennas and Propagation, Berlin, Germany, pp. 3471–3475, 2009.

[79] Legenkiy, M. and Butrym, A., "Modeling pulse radiation of rod antenna by method of mode matching in time domain," IEEE International Conference on Mathematical Methods in Electromagnetic Theory, Kyiv, Ukraine, pp. 1–4, 2010.

[80] Legenkiy, M. and Butrym, A., "Mode basis method for open dielectric waveguides," IEEE International Conference on Mathematical Methods in Electromagnetic Theory, Odesa, Ukraine, pp. 392–394, 2008.

[81] Legenkiy, M.N. and Butrym, A.Y., "Physical features of mode basis in open dielectric structures with discrete and continuous spectrum," Telecommunications and Radio Engineering, vol. 73, no. 10, pp. 863–880, 2014.

[82] Legenkiy, M.N., "Mode basis derivation by using integral equation technique for a circular dielectric waveguide," Radio Physics and Radio Astronomy, vol. 2, no. 2, pp. 171–180, 2011.

[83] Legenkiy, M.N. and Butrym, A.Y., "Impulse signal propagation in open dielectric circular waveguide," IEEE 5th International Conference on Ultrawideband and Ultrashort Impulse Signals, Sevastopol, Ukraine, pp. 119–121, 2010.

[84] Legenkiy, M.N. and Butrym, A.Y., "Excitation and propagation of a fast pulse guided wave in a circular dielectric waveguide," Radio Physics and Radio Astronomy, vol. 18, no. 2, pp. 147–151, 2013.

[85] Legenkiy, M.N. and Butrym, A.Y., "Fast pulse guided wave in an optical waveguide," IEEE International Kharkov Symposium on Physics and Engineering of Microwaves, Millimeter and Submillimeter Waves, Kharkov, Ukraine, pp. 207–209, 2013.

[86] Dumin, O.M., Tretyakov, O.A., Akhmedov, R.D. and Dumina, O.O., "Transient electromagnetic field propagation through nonlinear medium in time domain," IEEE International Conference on Antenna Theory and Techniques, Kharkiv, Ukraine, pp. 1–3, 2015.

[87] Dumin, O.M., Akhmedov, R.D., Katrich, V.A. and Dumina, O.O., "Propagation of transient field radiated from plane disk in nonlinear medium," IEEE 8th International Conference on Ultrawideband and Ultrashort Impulse Signals, Odessa, Ukraine, pp. 77–80, 2016.

[88] Petrusenko, I.V. and Sirenko, Y.K., "Generalized mode-matching technique in the theory of guided wave diffraction. Part 1: Fresnel formulas for scattering operators," Telecommunications and Radio Engineering, vol. 72, no. 5, pp. 369–384, 2013.

[89] Butrym, A.Y., "Mode expansion in time domain technique for short pulse propagation in a regular waveguide with inhomogeneous smooth varying filling," IEEE 8th International Conference on Ultrawideband and Ultrashort Impulse Signals, Odessa, Ukraine, pp. 117–119, 2016.

Index

Abel's integral transform 332
absorption cross section (ACS) 516
ad-hoc quadrature techniques 371
adjacency matrices, Green's theorem for 99
  incidence matrices 101
  matrix of adjacency 1 100
  vertex Laplacian matrix 100
  weighted adjacency matrix 100
admittance-passive system 500–2
analytical regularization method (ARM) 9–10, 303, 308–9
  applications 324–5
  considerations for implementation 323–4
arbitrary bounded sources above passive/active impedance plane 32
  expression of potentials for bounded sources 34
  formulation of the problem 32–3
  impedance plane 34–6
arbitrary wedge angle 30
  curved half plane to discontinuity of curvature 30–1
  discontinuity of curvature in impedance surface 31
asymptotic currents on cylinder with ogival cross-section 161
  asymptotic currents outside grazing incidence 162
    edge excited currents on shadowed edge 163
    illuminated edge 162–3
  geometry and analysis of the problem 161–2
  grazing incidence 163–4
    asymptotic current for grazing incidence 166
    fringe current contribution to diffracted field 164
    radiation of the PO current 164–6
asymptotic currents on elliptic cylinder with truncated strongly elongated cross-section 147
  analysis of interactions 147–8
  asymptotic currents 160–1
  asymptotic field in boundary layer
    due to incident plane wave 157–60
    due to magnetic line current 148–52
  radiated field and total field 152–3
  spectral decomposition of field in boundary layer 154–6
  truncated elliptic cylinder 156–7
asymptotic expressions in region with creeping waves terms 27
  plane wave illumination and observation
    at finite distance 27–9
    at infinity 27
  point source illumination and observation at finite distance 29–30
asymptotic representation in a region without creeping waves 25
  first term of f influenced by curvatures 26
  Maliuzhinets type representation 25–6
augmented electric and magnetic field equations 361

beam-based local diffraction tomography 571
 beam-based TD-DT 589
  backpropagation and local reconstruction of O (r) 590–1
  beam-domain data 589
  beam-domain data-object relation within the Born approximation 590
  numerical examples 591–4
 ultra-wide-band, phase-space BS (UWB-PS-BS) method
  beam frames 577–8
  BS methods 572–4
  phase-space pulsed BS method 578–82
  radiation field, BS representation of 576
  UWB considerations 576–7
  windowed Fourier transform (WFT) frame 575–6
 UWB tomographic inverse scattering
  DT identity 583–4
  object reconstruction via angular diversity 585–6
  object reconstruction via frequency diversity (UWB-DT) 586
  problem statement 582–3
  time-domain diffraction tomography 587–9
beam frame (BF) 574
beam summation (BS) method 572
Berry potential 612
Bessel–Hankel Fourier series 75, 81–6
biomedical electromagnetism and optics 660
Blaschke product 506
boundary element method (BEM) 365, 381
boundary value problem (BVP)
 Dirichlet BVP: *see* Dirichlet BVP
 formulation of 242–4
 Neumann BVP: *see* Neumann BVP
Brewster angles 69, 665
Brillouin diagrams 448–9
Bruggeman model 668

Calderón-based techniques 383
canonical problems and function theoretic methods 5
Cartesian coordinate system 473–4
Casimir–Polder forces 617
Cauchy formula 291–2, 294
Cauchy problem 697
Cauchy–Schwarz inequality 238
Chebyshev GM 418
Chebyshev iteration method 482–4
Chern number 611
classical potential theory 331
Clausius–Mossotti relation 644
combined-field integral equation 361, 363
combined source integral equation 361
complex eigenvalues 329
 computation, for various open cavities 336
  circular cylinder with longitudinal slit 336–7
  elliptic cavity with moveable longitudinal slit 337–43
  open rectangular cavity with finite flanges 343–7
 scheme for finding 333–5
complex-source beam (CSB) 189, 198
 converging and diverging CSB 198–200
 uniform CSB 200–1
complex-valued eigen-wavenumbers, numerical calculation of 322–3
converging CSB 200
convex envelope 480
convex optimization and physical bounds 507–11
curl-conforming (CC) bases 366
cut-off frequencies 329
CVX MATLAB software 509

DC instabilities 382, 398–401
Debye approximation 149

Debye potentials 36
defect layers, energy distribution of 458–60
diakoptic principle 117
dielectric gratings 440
dielectric permittivity tensor function 471
dielectric-wedge Fourier series 73
 Bessel–Hankel Fourier series 81–6
 diffraction problem 75
  Hilbert-space problem 75–6
  singular-field problem 76
 incident plane waves 87–8
 integral equations 77–9
  solution of 79–81
 numerical results 88–91
diffraction by semi-transparent obstacle 316
 BVP description 316–17
 integral representation for $u^{s(+)}$ and $\partial nu^{s(+)}$ 317–18
 reduction of the BVP to system of integral equations 318–19
 reduction of the system of integral equations to infinite system of linear algebraic equations 319–20
diffraction by slit on thick conducting screen 177
 background 177
 diffraction by thick and loaded slit 182–5
 diffraction by thin slit 182
 diffraction by trough 185–6
 formulation 177–8
  edge-diffracted rays 178–9
  modal excitation 179
  modal reradiation and reflection coupling 179–80
  total diffracted field 180–2
diffraction of waves with complex frequencies and spectral theory of open cavities 320
 description of the BVP 320–1
 Dirichlet BVP for complex-valued wave numbers 322
 numerical calculation of complex-valued eigen-wavenumbers and eigenmodes 322–3
 qualitative features of Dirichlet BVP 322
Dirichlet and Neumann eigenvalues 195, 202
Dirichlet boundary condition 222
Dirichlet BVP 313–15
 for complex-valued wave numbers 322
 qualitative features of 322
discrete Maxwell equations 110–11
discretization 365
dispersive material systems, modal expansions in 599
 canonical momentum 605–7
 electrodynamics of dispersive media 600–1
 Green's function 608–10
  positive and negative frequency components of 610–11
 Hermitian formulation in the time domain 602–3
 modal expansions 607–8
 Poynting theorem and stored energy 603–5
 quantum optics, application to 613–16
 topological photonics, application to 611–13
dispersive wedge-shaped region, causal time domain representation of a field above 23–4
divergence bases 366
diverging CSB 199
double extremum 350
double Laplace transform of the electric field
 singular integral equation for 244–6
dual IEs for resistive disk MSA 426
dual integral equations (DIE) 418

dual surface integral equations 361
dyadic Green's function 203–4

edge-diffracted rays 178–9
edge effect 348
effective numerical methods 10
eigenfrequencies 320
eigenmodes, numerical calculation of 322–3
electric-field integral equation (EFIE) 3, 330, 332, 381
  definition 383–4
  discretization strategy for 387–8
  low-frequency breakdown 382
  for perfectly conducting bodies 359–61
electrodynamics, tools for SI units in solving various problems of 673
  evolutionary approach to electrodynamics (EAE), upgrading
    configurational orthonormal modal basis in the space of solutions 692–3
    electromagnetic field theory, comparison of two alternative approaches to 684–5
    normalization of the eigenvectors of operator 691–2
    projecting the field vectors and Maxwell's equations onto the modal basis 693–7
    vectorial Maxwell's equations, separation of self-adjoint operator from 686–90
  Lorentz media, exact solutions for polarization of 678–83
  novel format of Maxwell's equations in SI units 674–6
    energetic characteristics of electromagnetic field 676
    energetic field characteristics, mechanical equivalents of 677–8
  ongoing and future research 701
  present state of art and recent advances 697–700
electrodynamics of dispersive media 600–1
electromagnetic diffraction analysis 177, 241
electromagnetic fields as differential forms 111–12
electromagnetic surface waves
  on curved surface with varying surface impedance 52–4
  excitation of, by dipolc located near plane impedance surface 56–9
  on right circular conical surface 54–6
  scattering of skew incident surface wave by edge on impedance wedge 59–61
    edge diffraction, beyond the critical angle of 65–7
    far-field expansion 62–3
    incident surface wave at the edge of impedance wedge 63–5
    integral equations for the spectra 61–2
  supported by planar impedance surfaces 50–2
electromagnetic theory 1, 6
electromagnetic wave propagation 6
elliptically layered and columnar media 454–5
elliptic cylinder with strongly elongated cross-section 145
  asymptotic currents on cylinder with ogival cross-section 161
    asymptotic currents outside grazing incidence 162–3
    geometry and analysis of the problem 161–2
    grazing incidence 163–6
  asymptotic currents on elliptic cylinder with truncated strongly elongated cross-section 147
    analysis of interactions 147–8
    asymptotic currents 160–1

asymptotic field in boundary layer due to incident plane wave 157–60
asymptotic field in boundary layer due to magnetic line current 148–52
diffraction by the edge of truncated elliptic cylinder 156–7
radiated field and total field 152–3
spectral decomposition of the field in boundary layer 154–6
energy density 658
E polarization
asymptotic solution of certain integral equation 263–6
factorization of the Kernel functions 259–61
formal solution of Wiener–Hopf equation 261–3
formulation of the problem 256–9
high-frequency asymptotic solution 266–8
scattered far field 268–9
E polarization case 183
E-polarized line source 241
E-polarized plane waves, resonance scattering of 329
computation of complex eigenvalues for various open cavities 336
circular cylinder with longitudinal slit 336–7
elliptic cavity with moveable longitudinal slit 337–43
open rectangular cavity with finite flanges 343–7
development of a systematic approach 330
mathematical background
schematic description of the MAR 331–2
scheme for finding the complex eigenvalues 333–5
resonance response of slotted cavities 347
far-field calculations 350–5
surface current calculations 348–50
Euler–Poincaré formula 386
Euler's gamma function 493
evolutionary approach to electrodynamics (EAE) 16
configurational orthonormal modal basis in the space of solutions 692–3
electromagnetic field theory, comparison of two alternative approaches to 684–5
exact explicit solutions for the modal amplitudes 697
normalization of eigenvectors of operator 691–2
projecting Maxwell's equations onto the same basis elements 695–7
projecting the field vectors onto the modal basis 694–5
vectorial Maxwell's equations, separation of self-adjoint operator from 686–90
evolutionary matrix 683

far-field calculations 350–5
far-field near-field mapping 84–6
far-field pattern's changes induced rounding corners of scatterer
analytic bounds for far-field difference 222
approximate integral equation 223–36
far-field difference 236–8
integral equations for difference in surface quantities 222–3
numerical results and discussion 221–2
problem formulation 218–21
fast Fourier transform (FFT) 332
FDTD 417–18
filtered backpropagating field 586
finite element method (FEM) 365, 381
first kind, equation of 306
first kind Fredholm equation 331

Floquet modes 507
Floquet's theorem 442
formulations that remediate fictitious internal resonances for dielectric targets 363–4
forward propagating beams 577
Fourier–Chebyshev transforms 324
Fourier method 74
Fourier series expansion method 439
Fourier transform 662
of quantized fields 615
Fredholm alternative 332
Fredholm equations 283–92
Fredholm factorization 279, 283–4, 292
Fredholm integral equations 65
Fredholm operator 475–6
Fredholm second-kind equations, theory of 418
Fredholm second-kind matrix equation 418
frequency domain (FD) 684
definition of EFIE 383–4
definition of LS matrices 384–7
definition of RWG elements 384
discretization strategy for EFIE 387–8
frequency-domain (FD) EFIE (FD-EFIE) 381–2

Gabor expansion coefficients 573
Galerkin method (GM) 417, 435
Galerkin's technique 330
Gauss's theorem 691
generalization of the WHT (GWHT) 279, 301
generalized boundary conditions (GBC) 417
generalized functions 119
application 138
extension of inverse initial value problem to range 139–40
solution of extended problem 140–3
basic properties of δ distribution 124–8
Green's functions associated with wave equation 128
ingoing Green's function 131–3
outgoing Green's function 130–3
generalized functions 120
generalized WH equations (GWHEs) 282
geometrical boundary conditions 191
geometrical optical (GO) analysis 282
Geometrical Optics laws 47
geometrical theory of diffraction (GTD) 4, 145, 172, 190
geophysical and biological media, complex environments' electromagnetics applied to 653
biomedical electromagnetism and optics 660
coherence in multiple scattering 665–7
geophysical remote sensing and imaging, and super resolution 656–8
heat diffusion in tissues 660
low coherence interferometry and optical coherence tomography (OCT) 663–5
porous media 667–9
random multiple scattering clutter, imaging through 656
seismic coda 669–70
stochastic wave theories 653–4
time-reversal imaging 654–6
ultrasound in tissues and blood 661–3
waves in metamaterials and electromagnetic and acoustic Brewster's angle 665
Wigner distribution function and specific intensity 658–9
Graf's addition theorem 515, 517, 519, 542
Green dyadic 623–6, 629–30, 647–8

Green function of Helmholtz equation 472
Green's formulae technique 313
Green's function 586, 599, 608–10, 615
  associated with wave equation
    ingoing Green's function 131–3
    outgoing Green's function 130–3
  positive and negative frequency components of 610–11
  for r-forms 109
  of semi-infinite elliptic cone 202
    dyadic Green's function 203–4
    scalar Green's function 202–3
Green's theorem 97, 242, 244
  for adjacency matrices 99
    incidence matrices 101
    matrix of adjacency 1 100
    vertex Laplacian matrix 100
    weighted adjacency matrix 100
  difference forms and discrete exterior calculus 104
    contextual algebraic notation of forms 106–7
    dual forms 105–6
    essentials of cell decomposition 108
    hypercube decomposition 106
    manifolds, graphs and lattices 107–8
    simplicial decomposition 104–5
  discrete Green's theorem and Green's functions in computational field theory 116
    diakoptics 117
    exterior–interior connection 116
  discrete time 114
  dynamical systems on topological vector spaces 109
    discrete Maxwell equations 110–11
    electromagnetic fields as differential forms 111–12
  Kirchhoff's theorem for r-forms 109
  for r-forms 108
  time-domain Green's functions for dynamical systems 112–13
  on topological vector space 101–4
ground penetrating radar (GPR) technique 667
Gudermann function 62

Hamilton–Jacobi equation 54
Hankel function kernel 244
heat diffusion in tissues 660
Heaviside function 639
Heaviside–Lorentz equations 675
Helmholtz decomposition 365, 367, 405, 418
Helmholtz equation 39, 219, 691
  Green function of 472
  in sphero-conal coordinates 192–6
Herglotz functions 12, 491
  basics about 492–9
  convex optimization and physical bounds 507–11
  passive systems 499–502
  sum rules and physical bounds 503–7
Herglotz-Nevanlinna function: *see* Herglotz functions
Hermitian formulation in the time domain 602–3
Hermitian tensors 478
Hermitian time evolution operator 602, 613
Hertzian dipole 47, 211
Hertzian dipole source radiation in the layered cylindrical structure 522–3
hierarchical vector basis functions 367–9
high-frequency hybrid ray–mode techniques 169
  diffraction by slit on thick conducting screen 177
    background 177
    diffraction by thick and loaded slit 182–5
    diffraction by thin slit 182

diffraction by trough 185–6
formulation 177–82
modal excitation at aperture 171
formulation 171–4
numerical results 174–7
ray–mode conversion technique 170–1
high-frequency techniques 6
and function theoretic methods 145
high-refinement breakdown 383
Hilbert–Schmidt matrix 311
Hilbert space 475, 484, 692–3
Hilbert space integral equation 74
Hilbert-space problem 75–6
Hilbert transform 496
holomorphic function 494
homogeneous dielectric bodies, integral equations for 361–3
homogeneous multilayer approximation (HMA) method 460–1
H polarization 269–72
H-polarization diffraction problem 73
Huygens' principle in discrete electromagnetic 98, 107, 115
hybrid ray–mode techniques 7
hyper-singular IE for VED-excited resistive disk 421–2

IFEM (improved Fourier series expansion method) 439–40, 449
imaging field 586
incidence matrices 101
incident angle, characteristic of 446–7
incident plane waves 87–8
infinite algebraic systems
instability in the numerical solution of 304–8
inhomogeneous media 11
integral equations 77–9, 359
discretization of 486–7
for electromagnetics, numerical solutions of 359
alternative formulations to remediate fictitious internal resonances 361
EFIE and MFIE for perfectly conducting bodies 359–61
formulations that remediate fictitious internal resonances for dielectric targets 363–4
integral equations for homogeneous dielectric bodies 361–3
interpolatory and hierarchical vector basis functions 367–9
low-frequency breakdown of integral equations 365
numerical solution of integral equations 365–6
single-source integral equations for dielectric bodies 364–5
singular vector basis functions 369–71
vector basis functions 366–7
low-frequency breakdown of 365
solution of 79–81
integral operator, spectrum of 475
low-frequency case, spectrum for 476–9
integral representation theory 17
arbitrary bounded sources above passive or active impedance plane 32
expression of potentials for bounded sources 34
formulation of the problem 32–3
impedance plane, expression of the potentials for 34–6
arbitrary wedge angle 30
curved half plane to discontinuity of curvature 30–1
discontinuity of curvature in impedance surface 31
asymptotic expressions in region with creeping waves terms 27
plane wave illumination and observation at finite distance 27–9
plane wave illumination and observation at infinity 27

point source illumination and observation at finite distance 29–30
asymptotic representation in a region without creeping waves 25
first term of f influenced by the curvatures 26–7
Maliuzhinets type representation 25–6
spectral function in frequency and time domain 18
causal time domain representation of a field above dispersive wedge-shaped region 23–4
Sommerfeld–Maliuzhinets representation and properties 18–19
spectral function from far field radiation of one face with arbitrary shape 22–3
spectral functions 20–1
spectral representation of field for 3D conical scatterers 36
formulation 36–8
Kontorovich–Lebedev integrals 38–41
potentials and properties for incident plane wave 41–4
integro-differential equation: *see* integral equations
interface boundary condition 317
interpolatory vector basis functions 367–9
inverse problems 549
anisotropic case (Class AnI)
anisotropic one-sectional diaphragm 557
existence and uniqueness of solution to inverse problem 559–60
transmission coefficient, explicit formulas for 557–9
extraction of complex permittivity of each section of three-sectional isotropic diaphragm (example) 563–4
extraction of permittivity and permeability of one-sectional anisotropic diaphragm (example) 564–5
extraction of permittivity tensor of two-sectional anisotropic diaphragm (example) 565–6
general statement for 550–3
inverse problem for one-sectional anisotropic diaphragm (example) 563
isotropic case (Class I)
explicit solution to the inverse problem 553–7
statement of the inverse problem for isotropic one-sectional diaphragm (Class I) 553
multi-sectional diaphragm, problem for (Class M) 560–2
inverse scattering 13
inverse scattering problems 5
iso-diffracting (ID) Gaussian beam (ID-GB) propagators 573
iso-diffracting pulsed beams (ID-PBs) 577

Kernel functions, factorization of 259–61
kernel singularities 290
Kirchhoff's theorem for r-forms 109
Klein–Gordon equation (KGE) 698–9
Klein–Gordon–Fock equation (KGFE) 700
Kobayashi Potential (KP) method 174
Kontorovich–Lebedev (KL) integrals 17, 38
equality and compatibility conditions 39–41
integral and nonintegral terms 38–9
Kontorovich–Lebedev transform 73
Kramers–Kronig relations 599

Lagrange interpolation 680
large time step breakdown 382
  solution to 405–11
  in time domain 398
layered cylindrically periodic arrays of circular cylinders, scattering and guidance by 515
  directivity of radiation of dipole source coupled to cylindrical EBG structure 526–34
  field expressions 517–18
  guidance in layered cylindrical structure 525–6
  Hertzian dipole source radiation in layered cylindrical structure 522–3
  light scattering by metal-coated dielectric nanocylinders with angular periodicity 534–9
  plane wave scattering by layered cylindrical structure 523–5
  reflection and transmission matrices 520–2
  scattering amplitudes, calculation of 518–20
  specific microstructured optical fibers, modal analysis of 539–42
layered cylindrical structure, plane wave scattering by 523–5
Levenberg–Marquardt iteration method 563
Liouville's theorem 262
"loop-flower" bases 365
Loop–Star (LS) decomposition 382, 396–8
Loop–Star (LS) matrices 384–7
"loop-star" basis functions 365
"loop-tree" representations 365
Lorentz force 676
Lorentz media, exact solutions for polarization of 678–83
low coherence interferometry 663–5
low-frequency breakdown 382–3, 392
  effective solution to the low-frequency breakdown for EFIE 402–5
  low-frequency ill conditioning 395–6
  numerical instability 393–5
low-frequency integral operator 477

magnetic field integral equation (MFIE) 3
  for perfectly conducting bodies 359–61
magnetic surface wave 51
Maliuzhinets inversion theorem 26
Malyuzhinets function 301
Mathematica 282, 290
matrix equation for resistive disk MSA 426–8
matrix of adjacency 1 100
Matrizant 683
maximal bandwidth (MB) 509
Maxwell's equations 1, 109, 472, 547, 552, 600, 694
  discrete Maxwell equations 110–11
  electromagnetic fields as differential forms 111–12
  novel format in SI units 674–6
    energetic characteristics of electromagnetic field 676
    energetic field characteristics, mechanical equivalents of 677–8
  projecting Maxwell's equations onto the same basis elements 695–7
  in sphero-conal coordinates 191
    Helmholtz equation in sphero-conal coordinates 192–6
    sphero-conal coordinates 191–2
    vector spherical-multipole expansion of the electromagnetic field 196–7
  vectorial Maxwell's equations, separation of self-adjoint operator from 686–90
metasurfaces 122

method of analytical regularization (MAR) 330–2, 417
method of moments (MM) 330, 365, 417–18
microstrip antennas (MSA) 417
Minkowski momentum 606
mixed positive and negative media 460–5
MKS system, Maxwell's equations in 242
modal excitation 179
modal expansions 14
modal reradiation and reflection coupling 179–80
modelling at arbitrarily low frequency via qH projectors 381
 DC instabilities 398–401
 effective solution to the low-frequency breakdown for EFIE
  implementation details 404–5
  leveraging the qH projectors 402–4
 frequency domain
  definition of EFIE 383–4
  definition of LS matrices 384–7
  definition of RWG elements 384
  discretization strategy for EFIE 387–8
 large time step breakdown in TD 398
 low-frequency breakdown in FD 391
  analysis of 392–6
  illustration of the problem 391–2
  traditional LS decomposition 396–8
 qH projectors 401–2
 solution to the large time step breakdown and DC instability for TD-EFIE 405
  numerical results 409–11
  preconditioning 406–7
  time discretization 407–9
 time domain 388
  definition of TD-EFIE equation 388–9
  spatial discretization of TD-EFIE 389
  temporal discretization of TD-EFIE 389–91
modified geometrical optics (MGO) scattered field 87
modified multilayer approximation (MMA) method 461
modulated refractive index type grating 440
Moore–Penrose pseudo-inverse 401
Müller system of equations 364
multilayered dielectric grating with elliptically layered media 454
multilayer method (MLM) 439
multiple scattering, coherence in 665–7
multiple scattering by a collection of randomly located obstacles distributed in a dielectric slab 621
 approximations 641–4
 geometry 622–3
 integral representation 623–4
  exploiting 625–30
 statistical problem—ensemble average 635–41
 surface fields, expansions of 630–4
 transmitted and reflected fields 634–5
multistep minimal residual method 486
mutual coherence function (MCF) 654

Neumann BVP 315–16
Neumann series GM 418
Nevanlinna function: *see* Herglotz functions
Nevanlinna–Pick interpolation 492
Nikolskii–Atkinson criterion 307
Noether operator 475–6
nonstationary iteration methods 484–6
normalized frequency, characteristic of 447–8
null-field approach 621, 628
numerical catastrophe 305
numerical quadrature rules 388

Nystrom-type algorithms 419
Nyström method 216

open-ended parallel-plane waveguide 171
optical coherence tomography (OCT) 663–5

parallel-plane waveguide modes 175
paraxial approximation 198
passive radar absorbers, sum rules for 506
Percus–Yevick packing factor 663
perfectly electrically conducting (PEC) surfaces 331, 381, 418, 502
periodic structures 13, 324
Petrov–Galerkin method 4
phase-space pulsed BS method
  phase-space pulsed BS representation for the field 581
  plan-wave spectrum in the TD 578–80
  pulsed beam frames 581–2
  windowed Radon transform (WRT) frame 580–1
physical optics (PO) 4, 145
physical theory of diffraction (PTD) 4, 145
Pick function: *see* Herglotz functions
piece-wise homogeneous bounded domain 129
planar vector waves 647
plane wave scattering by layered cylindrical structure 523–5
PMCHWT system of equations 364
Poisson summation formula 169
positive real (PR) function: *see* Herglotz functions
potentials and their pseudodifferential representations 309–13
Poynting theorem and stored energy 603–5
Poynting vector 458
probability density function (PDF) 635, 648–9
propagation characteristics 448–9
pulse basis—delta testing 421

qH projectors 401–2
  leveraging 402–4
quasi-crystalline approximation (QCA) 637
quasi-Helmholtz (qH) projectors 11, 382

radar cross section (RCS) 666
  for E polarization 272–6
  RCS studies 330
radiated far field 249–50
radiation losses 329
Radon transform 587, 590
random multiple scattering clutter, imaging through 656
Rao–Wilton–Glisson (RWG) elements 384
Rayleigh distribution 666
ray–mode conversion technique 170–1
rectangular media 455–7
reflection and transmission matrices 520–2
refractive index 440
resistance-integral theorem 503
resistive and thin dielectric disk antennas 417
  formulation and GBC 419–20
  numerical results 429
    radiation characteristics of resistive MSA 430–2
    radiation characteristics of thin disk DA 432–5
  resistive disk MSA excited by VED
    dual IEs for resistive disk MSA 426
    matrix equation for resistive disk MSA 426–8
  singular IEs and solution by MAR
    eigenfunctions of the IE operator
      static limit for VED-excited PEC and resistive disks 422–3

hyper-singular IE for VED-excited resistive disk 421–2
log-singular IE for VMD-excited resistive disk 424–6
matrix equation and DIE for VED-excited disk 423–4
thin disk DA excited by VED
coupled set of DIEs for thin disk DA 428
matrix equation for thin disk DA 428–9
resistive disk MSA
dual IEs for 426
matrix equation for 426–8
resistive MSA
radiation characteristics of 430–2
resonance doublet 350–2, 354
"resonance" frequencies 329
R-function: *see* Herglotz functions
rhombic media with strips 450–3
Riemann–Hilbert problem 5

SAR (synthetic-aperture radar) 656
scalar and electromagnetic waves, scattering and diffraction of 189
complex-source beams (CSB) 198
converging and diverging CSB 198–200
uniform CSB 200–1
Green's function of semi-infinite elliptic cone 202
dyadic Green's function 203–4
scalar Green's function 202–3
Maxwell's equations in sphero-conal coordinates 191
Helmholtz equation in sphero-conal coordinates 192–6
sphero-conal coordinates 191–2
vector spherical-multipole expansion of the electromagnetic field 196–7
numerical evaluation 204
for acoustically soft or hard semi-infinite elliptic cone 206–8
convergence analysis 204–6
for a perfectly conducting semi-infinite elliptic cone 208–12
scalar Green's function 202–3
scattered far field 268–9
scattering and diffraction 7
scattering and guiding problems of electromagnetic waves 439
elliptically layered and columnar media 454–5
energy distribution of defect layers 458–60
formulation 440
guiding problem 444–5
scattering problem 441–4
mixed positive and negative media 460–5
numerical results 445
characteristic of incident angle 446–7
characteristic of normalized frequency 447–8
propagation characteristics 448–9
rectangular media 455–7
rhombic media with strips 450–3
slanted layer 449–50
scattering cross section (SCS) 516
scattering-passive system 502
second kind, equation of 306
semi-inversion procedure (SIP) 10, 303
semi-transparent boundary conditions 317
shifted Ewald sphere 584
single-source combined field integral equation 365
single-source integral equations for dielectric bodies 364–5
singular-field problem 76
singular IEs and solution by MAR 421–6
singular integral equation (SIE) 417, 473, 475, 477
approximate solution of 246–8

for double Laplace transform of the electric field 244–6
singular vector basis functions 369–71
skew incident surface wave scattering by edge on impedance wedge 59–61
edge diffraction, beyond critical angle of 65–7
far-field expansion 62–3
incident surface wave at the edge of impedance wedge 63–5
integral equations for the spectra 61–2
slanted layer 449–50
slant stack transform (SST) 579
slotted cavities
plane wave excitation of 347
resonance response of 347–55
solenoidal eigenvectors 691
Sommerfeld contours 5
Sommerfeld integrals 42
Sommerfeld–Malyuzhinets technique 5, 18, 42, 49
basic integral representation 18–19
basic properties of total field and its spectral function 19
specific absorption rate (SAR) 660
spectral functions 20
attached to the Sommerfeld–Maliuzhinets representation 20
from far field radiation of one face with arbitrary shape 22
deformation and simple exact expression of the spectral function 23
simple exact expression of the spectral function 22–3
radiation of single face of wedge-shaped region 20
single-face spectral function, simple exact expression of 20–1
spectral reconstruction 585
spectral representation of field for 3D conical scatterers 36
formulation 36–8
Kontorovich–Lebedev integrals 38
equality and compatibility conditions 39–41
integral and nonintegral terms 38–9
potentials and properties for incident plane wave 41
complex properties of $g^{i}_{e,h}$ 43–4
efficient expression 41–3
spectrum of integral operator 475
low-frequency case, spectrum for 476–9
spherical vector waves 645
sphero-conal coordinates 191–2
stationary iteration methods
generalized Chebyshev iteration method 482–4
generalized simple iteration method 479–82
stochastic wave theories 653–4
Stokes theorem 612
Stoltz domain 495
sum rules 492
and physical bounds 503–7
super resolution 657
surface current calculations 348–50
surface plasmon resonance 509
surface-relief grating 440
symmetric Herglotz function 494
system of linear algebraic equations (SLAE) 471, 479, 489

tangential-field EFIE 361
thin disk DA
coupled set of DIEs for 428
matrix equation for a thin disk DA 428–9
radiation characteristics of 432–5
Tikhonov regularization 307
time domain 388–91
time-domain EFIE (TD-EFIE) 381–2
definition of TD-EFIE equation 388–9

solution to the large time step breakdown and DC instability for 405
numerical results 409–11
preconditioning 406–7
time discretization 407–9
spatial discretization of 389
temporal discretization of 389–91
time-domain admittance passivity 501
time-domain perspective 16
time-harmonic field method (THM) 684
time-reversal imaging 654–6
TM-polarized eigenmodes 331
TM wave 461
topological vector space, Green's theorem on 101–4
total diffracted field 180–2
transformed wave equation 258
transition matrix (*T*-matrix) 515
translation matrices 645–6
two-dimensional (2D) hollow cylinders 329

ultrasound in tissues and blood 661–3
ultra-wide-band, phase-space BS (UWB-PS-BS) method
beam frames 577–8
BS methods 572–4
phase-space pulsed BS method
phase-space pulsed BS representation for the field 581
plan-wave spectrum in the TD 578–80
pulsed beam frames 581–2
windowed Radon transform (WRT) frame 580–1
radiation field, BS representation of 576
UWB considerations 576–7
windowed Fourier transform (WFT) frame 575–6
uniform complex-source beams (CSBs) 7
uniform theory of diffraction (UTD) 190

vector basis functions 366–7
interpolatory and hierarchical vector basis functions 367–9
singular vector basis functions 369–71
vectorial Maxwell's equations, separation of self-adjoint operator from 686–90
vector spherical-multipole expansion of the electromagnetic field 196–7
vertical electric dipole (VED)
VED-excited disk, matrix equation and DIE for 423–4
VED-excited PEC and resistive disks, eigenfunctions of the IE operator static limit for 422–3
vertex Laplacian matrix 100
vertical magnetic dipole (VMD)
log-singular IE for VMD-excited resistive disk 424–6
volume singular integral equations (VSIEs) 12, 471
discretization of integral equations 486–7
fast algorithms 487–8
formulation of the problems 472–5
nonstationary iteration methods 484–6
numerical results 488–9
spectrum of integral operator 475
low-frequency case, spectrum for 476–9
stationary iteration methods
generalized Chebyshev iteration method 482–4
generalized simple iteration method 479–82

Waterman's method 621, 628
waves in complex media 15

waves in metamaterials and
    electromagnetic and acoustic
        Brewster's angle 665
Weber–Schafheitlin discontinuous
    integrals 177
Weber–Schafheitlin formulas 428
weighted adjacency matrix 100
weighted residual method 365
Weyl's theorem 692
WHFT 280
Whittaker functions 152
    incident field in terms of 150–2,
        158–9
WH technique 9
Wiener–Hopf (WH) method 5, 73, 241
    E polarization
        asymptotic solution of certain
            integral equation 263–6
        factorization of the Kernel
            functions 259–61
        formal solution of Wiener–Hopf
            equation 261–3
        formulation of the problem 256–9
        high-frequency asymptotic
            solution 266–8
        scattered far field 268–9
    H polarization 269–72
    numerical results and discussion
        272–6
Wiener–Hopf Fredholm factorization
    279
    analytical continuation of the
        numerical solution
        294–8
    novel test case 298–300
    reduction of WH equations to FIEs
        283
        Fredholm equation of the angular
            region (a) 288–92
        Fredholm equation of the region
            (c) 284–6
        Fredholm equations of the region
            (b) 286–8
    solution of the FIE 292–4
    WH equations of the problem
        281–3
Wiener–Khinchin theorem 662
Wigner distribution function 659
    and specific intensity 658–9
windowed Fourier transform (WFT)
        frame 575–6,
        580–1

www.ingramcontent.com/pod-product-compliance
Lightning Source LLC
LaVergne TN
LVHW011853200926
848484LV00002B/449